GALAXY FORMATION AND EVOLUTION

The rapidly expanding field of galaxy formation lies at the interfaces of astronomy, particle physics, and cosmology. Covering diverse topics from these disciplines, all of which are needed to understand how galaxies form and evolve, this book is ideal for researchers entering the field.

Individual chapters explore the evolution of the Universe as a whole and its particle and radiation content; linear and nonlinear growth of cosmic structures; processes affecting the gaseous and dark matter components of galaxies and their stellar populations; the formation of spiral and elliptical galaxies; central supermassive black holes and the activity associated with them; galaxy interactions; and the intergalactic medium.

Emphasizing both observational and theoretical aspects, this book provides a coherent introduction for astronomers, cosmologists, and astroparticle physicists to the broad range of science underlying the formation and evolution of galaxies.

HOUJUN MO is Professor of Astrophysics at the University of Massachusetts. He is known for his work on the formation and clustering of galaxies and their dark matter halos.

FRANK VAN DEN BOSCH is Assistant Professor at Yale University, and is known for his studies of the formation, dynamics, and clustering of galaxies.

SIMON WHITE is Director at the Max Planck Institute for Astrophysics in Garching. He is one of the originators of the modern theory of galaxy formation and has received numerous international prizes and honors.

Jointly and separately the authors have published almost 500 papers in the refereed professional literature, most of them on topics related to the subject of this book.

GALAXY FORMATION AND EVOLUTION

HOUJUN MO
University of Massachusetts

FRANK VAN DEN BOSCH
Yale University

SIMON WHITE
Max Planch Institute for Astrophysics

Shaftesbury Road, Cambridge CB2 8EA, United Kingdom

One Liberty Plaza, 20th Floor, New York, NY 10006, USA

477 Williamstown Road, Port Melbourne, VIC 3207, Australia

314–321, 3rd Floor, Plot 3, Splendor Forum, Jasola District Centre, New Delhi – 110025, India

103 Penang Road, #05–06/07, Visioncrest Commercial, Singapore 238467

Cambridge University Press is part of Cambridge University Press & Assessment, a department of the University of Cambridge.

We share the University's mission to contribute to society through the pursuit of education, learning and research at the highest international levels of excellence.

www.cambridge.org
Information on this title: www.cambridge.org/9780521857932

© H. Mo, F. van den Bosch and S. White 2010

This publication is in copyright. Subject to statutory exception and to the provisions of relevant collective licensing agreements, no reproduction of any part may take place without the written permission of Cambridge University Press & Assessment.

First published 2010 (version 10, March 2024)

Printed in Great Britain by CPI Group (UK) Ltd, Croydon CR0 4YY, March 2024

A catalogue record for this publication is available from the British Library

Library of Congress Cataloging-in-Publication data
Mo, Houjun.
Galaxy formation and evolution / Houjun Mo, Frank van den Bosch, Simon White.
p. cm.
Includes bibliographical references and index.
ISBN 978-0-521-85793-2
1. Galaxies – Formation. 2. Galaxies – Evolution. I. Van den Bosch, Frank, 1969– II. White, S. (Simon D. M.) III. Title.
QB857.M63   2010
523.1´12–dc22
2009054020

ISBN   978-0-521-85793-2   Hardback

Cambridge University Press & Assessment has no responsibility for the persistence or accuracy of URLs for external or third-party internet websites referred to in this publication and does not guarantee that any content on such websites is, or will remain, accurate or appropriate.

Contents

Preface		*page* xvii
1	**Introduction**	1
1.1	The Diversity of the Galaxy Population	2
1.2	Basic Elements of Galaxy Formation	5
	1.2.1 The Standard Model of Cosmology	6
	1.2.2 Initial Conditions	6
	1.2.3 Gravitational Instability and Structure Formation	7
	1.2.4 Gas Cooling	8
	1.2.5 Star Formation	8
	1.2.6 Feedback Processes	9
	1.2.7 Mergers	10
	1.2.8 Dynamical Evolution	12
	1.2.9 Chemical Evolution	12
	1.2.10 Stellar Population Synthesis	13
	1.2.11 The Intergalactic Medium	13
1.3	Time Scales	14
1.4	A Brief History of Galaxy Formation	15
	1.4.1 Galaxies as Extragalactic Objects	15
	1.4.2 Cosmology	16
	1.4.3 Structure Formation	18
	1.4.4 The Emergence of the Cold Dark Matter Paradigm	20
	1.4.5 Galaxy Formation	22
2	**Observational Facts**	25
2.1	Astronomical Observations	25
	2.1.1 Fluxes and Magnitudes	26
	2.1.2 Spectroscopy	29
	2.1.3 Distance Measurements	32
2.2	Stars	34
2.3	Galaxies	37
	2.3.1 The Classification of Galaxies	38
	2.3.2 Elliptical Galaxies	41
	2.3.3 Disk Galaxies	49

	2.3.4	The Milky Way	55
	2.3.5	Dwarf Galaxies	57
	2.3.6	Nuclear Star Clusters	59
	2.3.7	Starbursts	60
	2.3.8	Active Galactic Nuclei	60
2.4	Statistical Properties of the Galaxy Population		61
	2.4.1	Luminosity Function	62
	2.4.2	Size Distribution	63
	2.4.3	Color Distribution	64
	2.4.4	The Mass–Metallicity Relation	65
	2.4.5	Environment Dependence	65
2.5	Clusters and Groups of Galaxies		67
	2.5.1	Clusters of Galaxies	67
	2.5.2	Groups of Galaxies	71
2.6	Galaxies at High Redshifts		72
	2.6.1	Galaxy Counts	73
	2.6.2	Photometric Redshifts	75
	2.6.3	Galaxy Redshift Surveys at $z \sim 1$	75
	2.6.4	Lyman-Break Galaxies	77
	2.6.5	Lyα Emitters	78
	2.6.6	Submillimeter Sources	78
	2.6.7	Extremely Red Objects and Distant Red Galaxies	79
	2.6.8	The Cosmic Star-Formation History	80
2.7	Large-Scale Structure		81
	2.7.1	Two-Point Correlation Functions	82
	2.7.2	Probing the Matter Field via Weak Lensing	84
2.8	The Intergalactic Medium		85
	2.8.1	The Gunn–Peterson Test	85
	2.8.2	Quasar Absorption Line Systems	86
2.9	The Cosmic Microwave Background		89
2.10	The Homogeneous and Isotropic Universe		92
	2.10.1	The Determination of Cosmological Parameters	94
	2.10.2	The Mass and Energy Content of the Universe	95
3	**Cosmological Background**		**100**
3.1	The Cosmological Principle and the Robertson–Walker Metric		102
	3.1.1	The Cosmological Principle and its Consequences	102
	3.1.2	Robertson–Walker Metric	104
	3.1.3	Redshift	106
	3.1.4	Peculiar Velocities	107
	3.1.5	Thermodynamics and the Equation of State	108
	3.1.6	Angular-Diameter and Luminosity Distances	110
3.2	Relativistic Cosmology		112
	3.2.1	Friedmann Equation	113
	3.2.2	The Densities at the Present Time	114

	3.2.3	Explicit Solutions of the Friedmann Equation	115
	3.2.4	Horizons	119
	3.2.5	The Age of the Universe	119
	3.2.6	Cosmological Distances and Volumes	121
3.3	The Production and Survival of Particles		124
	3.3.1	The Chronology of the Hot Big Bang	125
	3.3.2	Particles in Thermal Equilibrium	127
	3.3.3	Entropy	129
	3.3.4	Distribution Functions of Decoupled Particle Species	132
	3.3.5	The Freeze-Out of Stable Particles	133
	3.3.6	Decaying Particles	137
3.4	Primordial Nucleosynthesis		139
	3.4.1	Initial Conditions	139
	3.4.2	Nuclear Reactions	140
	3.4.3	Model Predictions	142
	3.4.4	Observational Results	144
3.5	Recombination and Decoupling		146
	3.5.1	Recombination	146
	3.5.2	Decoupling and the Origin of the CMB	148
	3.5.3	Compton Scattering	150
	3.5.4	Energy Thermalization	151
3.6	Inflation		152
	3.6.1	The Problems of the Standard Model	152
	3.6.2	The Concept of Inflation	154
	3.6.3	Realization of Inflation	156
	3.6.4	Models of Inflation	158
4	**Cosmological Perturbations**		**162**
4.1	Newtonian Theory of Small Perturbations		162
	4.1.1	Ideal Fluid	162
	4.1.2	Isentropic and Isocurvature Initial Conditions	166
	4.1.3	Gravitational Instability	166
	4.1.4	Collisionless Gas	168
	4.1.5	Free-Streaming Damping	171
	4.1.6	Specific Solutions	172
	4.1.7	Higher-Order Perturbation Theory	176
	4.1.8	The Zel'dovich Approximation	177
4.2	Relativistic Theory of Small Perturbations		178
	4.2.1	Gauge Freedom	179
	4.2.2	Classification of Perturbations	181
	4.2.3	Specific Examples of Gauge Choices	183
	4.2.4	Basic Equations	185
	4.2.5	Coupling between Baryons and Radiation	189
	4.2.6	Perturbation Evolution	191
4.3	Linear Transfer Functions		196
	4.3.1	Adiabatic Baryon Models	198

	4.3.2	Adiabatic Cold Dark Matter Models	200
	4.3.3	Adiabatic Hot Dark Matter Models	201
	4.3.4	Isocurvature Cold Dark Matter Models	202
4.4	Statistical Properties		202
	4.4.1	General Discussion	202
	4.4.2	Gaussian Random Fields	204
	4.4.3	Simple Non-Gaussian Models	205
	4.4.4	Linear Perturbation Spectrum	206
4.5	The Origin of Cosmological Perturbations		209
	4.5.1	Perturbations from Inflation	209
	4.5.2	Perturbations from Topological Defects	213

5 Gravitational Collapse and Collisionless Dynamics — 215

5.1	Spherical Collapse Models		215
	5.1.1	Spherical Collapse in a $\Lambda = 0$ Universe	215
	5.1.2	Spherical Collapse in a Flat Universe with $\Lambda > 0$	218
	5.1.3	Spherical Collapse with Shell Crossing	219
5.2	Similarity Solutions for Spherical Collapse		220
	5.2.1	Models with Radial Orbits	220
	5.2.2	Models Including Non-Radial Orbits	224
5.3	Collapse of Homogeneous Ellipsoids		226
5.4	Collisionless Dynamics		230
	5.4.1	Time Scales for Collisions	230
	5.4.2	Basic Dynamics	232
	5.4.3	The Jeans Equations	233
	5.4.4	The Virial Theorem	234
	5.4.5	Orbit Theory	236
	5.4.6	The Jeans Theorem	240
	5.4.7	Spherical Equilibrium Models	240
	5.4.8	Axisymmetric Equilibrium Models	244
	5.4.9	Triaxial Equilibrium Models	247
5.5	Collisionless Relaxation		248
	5.5.1	Phase Mixing	249
	5.5.2	Chaotic Mixing	250
	5.5.3	Violent Relaxation	251
	5.5.4	Landau Damping	253
	5.5.5	The End State of Relaxation	254
5.6	Gravitational Collapse of the Cosmic Density Field		257
	5.6.1	Hierarchical Clustering	257
	5.6.2	Results from Numerical Simulations	258

6 Probing the Cosmic Density Field — 262

6.1	Large-Scale Mass Distribution		262
	6.1.1	Correlation Functions	262
	6.1.2	Particle Sampling and Bias	264
	6.1.3	Mass Moments	266

6.2	Large-Scale Velocity Field		270
	6.2.1 Bulk Motions and Velocity Correlation Functions		270
	6.2.2 Mass Density Reconstruction from the Velocity Field		271
6.3	Clustering in Real Space and Redshift Space		273
	6.3.1 Redshift Distortions		273
	6.3.2 Real-Space Correlation Functions		276
6.4	Clustering Evolution		278
	6.4.1 Dynamics of Statistics		278
	6.4.2 Self-Similar Gravitational Clustering		280
	6.4.3 Development of Non-Gaussian Features		282
6.5	Galaxy Clustering		283
	6.5.1 Correlation Analyses		284
	6.5.2 Power Spectrum Analysis		288
	6.5.3 Angular Correlation Function and Power Spectrum		290
6.6	Gravitational Lensing		292
	6.6.1 Basic Equations		292
	6.6.2 Lensing by a Point Mass		295
	6.6.3 Lensing by an Extended Object		297
	6.6.4 Cosmic Shear		300
6.7	Fluctuations in the Cosmic Microwave Background		302
	6.7.1 Observational Quantities		302
	6.7.2 Theoretical Expectations of Temperature Anisotropy		304
	6.7.3 Thomson Scattering and Polarization of the Microwave Background		311
	6.7.4 Interaction between CMB Photons and Matter		314
	6.7.5 Constraints on Cosmological Parameters		316
7	**Formation and Structure of Dark Matter Halos**		**319**
7.1	Density Peaks		321
	7.1.1 Peak Number Density		321
	7.1.2 Spatial Modulation of the Peak Number Density		323
	7.1.3 Correlation Function		324
	7.1.4 Shapes of Density Peaks		325
7.2	Halo Mass Function		326
	7.2.1 Press–Schechter Formalism		327
	7.2.2 Excursion Set Derivation of the Press–Schechter Formula		328
	7.2.3 Spherical versus Ellipsoidal Dynamics		331
	7.2.4 Tests of the Press–Schechter Formalism		333
	7.2.5 Number Density of Galaxy Clusters		334
7.3	Progenitor Distributions and Merger Trees		336
	7.3.1 Progenitors of Dark Matter Halos		336
	7.3.2 Halo Merger Trees		336
	7.3.3 Main Progenitor Histories		339
	7.3.4 Halo Assembly and Formation Times		340
	7.3.5 Halo Merger Rates		342
	7.3.6 Halo Survival Times		343

7.4	Spatial Clustering and Bias	345
	7.4.1 Linear Bias and Correlation Function	345
	7.4.2 Assembly Bias	348
	7.4.3 Nonlinear and Stochastic Bias	348
7.5	Internal Structure of Dark Matter Halos	351
	7.5.1 Halo Density Profiles	351
	7.5.2 Halo Shapes	354
	7.5.3 Halo Substructure	355
	7.5.4 Angular Momentum	358
7.6	The Halo Model of Dark Matter Clustering	362
8	**Formation and Evolution of Gaseous Halos**	366
8.1	Basic Fluid Dynamics and Radiative Processes	366
	8.1.1 Basic Equations	366
	8.1.2 Compton Cooling	367
	8.1.3 Radiative Cooling	367
	8.1.4 Photoionization Heating	369
8.2	Hydrostatic Equilibrium	371
	8.2.1 Gas Density Profile	371
	8.2.2 Convective Instability	373
	8.2.3 Virial Theorem Applied to a Gaseous Halo	374
8.3	The Formation of Hot Gaseous Halos	376
	8.3.1 Accretion Shocks	376
	8.3.2 Self-Similar Collapse of Collisional Gas	379
	8.3.3 The Impact of a Collisionless Component	383
	8.3.4 More General Models of Spherical Collapse	384
8.4	Radiative Cooling in Gaseous Halos	385
	8.4.1 Radiative Cooling Time Scales for Uniform Clouds	385
	8.4.2 Evolution of the Cooling Radius	387
	8.4.3 Self-Similar Cooling Waves	388
	8.4.4 Spherical Collapse with Cooling	390
8.5	Thermal and Hydrodynamical Instabilities of Cooling Gas	393
	8.5.1 Thermal Instability	393
	8.5.2 Hydrodynamical Instabilities	396
	8.5.3 Heat Conduction	397
8.6	Evolution of Gaseous Halos with Energy Sources	398
	8.6.1 Blast Waves	399
	8.6.2 Winds and Wind-Driven Bubbles	404
	8.6.3 Supernova Feedback and Galaxy Formation	406
8.7	Results from Numerical Simulations	408
	8.7.1 Three-Dimensional Collapse without Radiative Cooling	408
	8.7.2 Three-Dimensional Collapse with Radiative Cooling	409

8.8	Observational Tests	410
	8.8.1 X-ray Clusters and Groups	410
	8.8.2 Gaseous Halos around Elliptical Galaxies	414
	8.8.3 Gaseous Halos around Spiral Galaxies	416
9	**Star Formation in Galaxies**	**417**
9.1	Giant Molecular Clouds: The Sites of Star Formation	418
	9.1.1 Observed Properties	418
	9.1.2 Dynamical State	419
9.2	The Formation of Giant Molecular Clouds	421
	9.2.1 The Formation of Molecular Hydrogen	421
	9.2.2 Cloud Formation	422
9.3	What Controls the Star-Formation Efficiency	425
	9.3.1 Magnetic Fields	425
	9.3.2 Supersonic Turbulence	426
	9.3.3 Self-Regulation	428
9.4	The Formation of Individual Stars	429
	9.4.1 The Formation of Low-Mass Stars	429
	9.4.2 The Formation of Massive Stars	432
9.5	Empirical Star-Formation Laws	433
	9.5.1 The Kennicutt–Schmidt Law	434
	9.5.2 Local Star-Formation Laws	436
	9.5.3 Star-Formation Thresholds	438
9.6	The Initial Mass Function	440
	9.6.1 Observational Constraints	441
	9.6.2 Theoretical Models	443
9.7	The Formation of Population III Stars	446
10	**Stellar Populations and Chemical Evolution**	**449**
10.1	The Basic Concepts of Stellar Evolution	449
	10.1.1 Basic Equations of Stellar Structure	450
	10.1.2 Stellar Evolution	453
	10.1.3 Equation of State, Opacity, and Energy Production	453
	10.1.4 Scaling Relations	460
	10.1.5 Main-Sequence Lifetimes	462
10.2	Stellar Evolutionary Tracks	463
	10.2.1 Pre-Main-Sequence Evolution	463
	10.2.2 Post-Main-Sequence Evolution	464
	10.2.3 Supernova Progenitors and Rates	468
10.3	Stellar Population Synthesis	470
	10.3.1 Stellar Spectra	470
	10.3.2 Spectral Synthesis	471
	10.3.3 Passive Evolution	472
	10.3.4 Spectral Features	474
	10.3.5 Age–Metallicity Degeneracy	475

	10.3.6 K and E Corrections	475
	10.3.7 Emission and Absorption by the Interstellar Medium	476
	10.3.8 Star-Formation Diagnostics	482
	10.3.9 Estimating Stellar Masses and Star-Formation Histories of Galaxies	484
10.4	Chemical Evolution of Galaxies	486
	10.4.1 Stellar Chemical Production	486
	10.4.2 The Closed-Box Model	488
	10.4.3 Models with Inflow and Outflow	490
	10.4.4 Abundance Ratios	491
10.5	Stellar Energetic Feedback	492
	10.5.1 Mass-Loaded Kinetic Energy from Stars	492
	10.5.2 Gas Dynamics Including Stellar Feedback	493

11 Disk Galaxies — 495

11.1	Mass Components and Angular Momentum	495
	11.1.1 Disk Models	496
	11.1.2 Rotation Curves	498
	11.1.3 Adiabatic Contraction	501
	11.1.4 Disk Angular Momentum	502
	11.1.5 Orbits in Disk Galaxies	503
11.2	The Formation of Disk Galaxies	505
	11.2.1 General Discussion	505
	11.2.2 Non-Self-Gravitating Disks in Isothermal Spheres	505
	11.2.3 Self-Gravitating Disks in Halos with Realistic Profiles	507
	11.2.4 Including a Bulge Component	509
	11.2.5 Disk Assembly	509
	11.2.6 Numerical Simulations of Disk Formation	511
11.3	The Origin of Disk Galaxy Scaling Relations	512
11.4	The Origin of Exponential Disks	515
	11.4.1 Disks from Relic Angular Momentum Distribution	515
	11.4.2 Viscous Disks	517
	11.4.3 The Vertical Structure of Disk Galaxies	518
11.5	Disk Instabilities	521
	11.5.1 Basic Equations	521
	11.5.2 Local Instability	523
	11.5.3 Global Instability	525
	11.5.4 Secular Evolution	528
11.6	The Formation of Spiral Arms	531
11.7	Stellar Population Properties	534
	11.7.1 Global Trends	535
	11.7.2 Color Gradients	537
11.8	Chemical Evolution of Disk Galaxies	538
	11.8.1 The Solar Neighborhood	538
	11.8.2 Global Relations	540

12	**Galaxy Interactions and Transformations**		544
12.1	High-Speed Encounters		545
12.2	Tidal Stripping		548
	12.2.1	Tidal Radius	548
	12.2.2	Tidal Streams and Tails	549
12.3	Dynamical Friction		553
	12.3.1	Orbital Decay	556
	12.3.2	The Validity of Chandrasekhar's Formula	559
12.4	Galaxy Merging		561
	12.4.1	Criterion for Mergers	561
	12.4.2	Merger Demographics	563
	12.4.3	The Connection between Mergers, Starbursts and AGN	564
	12.4.4	Minor Mergers and Disk Heating	565
12.5	Transformation of Galaxies in Clusters		568
	12.5.1	Galaxy Harassment	569
	12.5.2	Galactic Cannibalism	570
	12.5.3	Ram-Pressure Stripping	571
	12.5.4	Strangulation	572
13	**Elliptical Galaxies**		574
13.1	Structure and Dynamics		574
	13.1.1	Observables	575
	13.1.2	Photometric Properties	576
	13.1.3	Kinematic Properties	577
	13.1.4	Dynamical Modeling	579
	13.1.5	Evidence for Dark Halos	581
	13.1.6	Evidence for Supermassive Black Holes	582
	13.1.7	Shapes	584
13.2	The Formation of Elliptical Galaxies		587
	13.2.1	The Monolithic Collapse Scenario	588
	13.2.2	The Merger Scenario	590
	13.2.3	Hierarchical Merging and the Elliptical Population	593
13.3	Observational Tests and Constraints		594
	13.3.1	Evolution of the Number Density of Ellipticals	594
	13.3.2	The Sizes of Elliptical Galaxies	595
	13.3.3	Phase-Space Density Constraints	598
	13.3.4	The Specific Frequency of Globular Clusters	599
	13.3.5	Merging Signatures	600
	13.3.6	Merger Rates	601
13.4	The Fundamental Plane of Elliptical Galaxies		602
	13.4.1	The Fundamental Plane in the Merger Scenario	604
	13.4.2	Projections and Rotations of the Fundamental Plane	604
13.5	Stellar Population Properties		606
	13.5.1	Archaeological Records	606
	13.5.2	Evolutionary Probes	609

	13.5.3 Color and Metallicity Gradients	610
	13.5.4 Implications for the Formation of Elliptical Galaxies	610
13.6	Bulges, Dwarf Ellipticals and Dwarf Spheroidals	613
	13.6.1 The Formation of Galactic Bulges	614
	13.6.2 The Formation of Dwarf Ellipticals	616

14 Active Galaxies — 618

14.1	The Population of Active Galactic Nuclei	619
14.2	The Supermassive Black Hole Paradigm	623
	14.2.1 The Central Engine	623
	14.2.2 Accretion Disks	624
	14.2.3 Continuum Emission	626
	14.2.4 Emission Lines	631
	14.2.5 Jets, Superluminal Motion and Beaming	633
	14.2.6 Emission-Line Regions and Obscuring Torus	637
	14.2.7 The Idea of Unification	638
	14.2.8 Observational Tests for Supermassive Black Holes	639
14.3	The Formation and Evolution of AGN	640
	14.3.1 The Growth of Supermassive Black Holes and the Fueling of AGN	640
	14.3.2 AGN Demographics	644
	14.3.3 Outstanding Questions	647
14.4	AGN and Galaxy Formation	648
	14.4.1 Radiative Feedback	649
	14.4.2 Mechanical Feedback	650

15 Statistical Properties of the Galaxy Population — 652

15.1	Preamble	652
15.2	Galaxy Luminosities and Stellar Masses	654
	15.2.1 Galaxy Luminosity Functions	654
	15.2.2 Galaxy Counts	658
	15.2.3 Extragalactic Background Light	660
15.3	Linking Halo Mass to Galaxy Luminosity	663
	15.3.1 Simple Considerations	663
	15.3.2 The Luminosity Function of Central Galaxies	665
	15.3.3 The Luminosity Function of Satellite Galaxies	666
	15.3.4 Satellite Fractions	668
	15.3.5 Discussion	669
15.4	Linking Halo Mass to Star-Formation History	670
	15.4.1 The Color Distribution of Galaxies	670
	15.4.2 Origin of the Cosmic Star-Formation History	673
15.5	Environmental Dependence	674
	15.5.1 Effects within Dark Matter Halos	675
	15.5.2 Effects on Large Scales	677
15.6	Spatial Clustering and Galaxy Bias	679
	15.6.1 Application to High-Redshift Galaxies	683

15.7	Putting it All Together	684
	15.7.1 Semi-Analytical Models	684
	15.7.2 Hydrodynamical Simulations	686
16	**The Intergalactic Medium**	**689**
16.1	The Ionization State of the Intergalactic Medium	690
	16.1.1 Physical Conditions after Recombination	690
	16.1.2 The Mean Optical Depth of the IGM	690
	16.1.3 The Gunn–Peterson Test	692
	16.1.4 Constraints from the Cosmic Microwave Background	694
16.2	Ionizing Sources	695
	16.2.1 Photoionization versus Collisional Ionization	695
	16.2.2 Emissivity from Quasars and Young Galaxies	697
	16.2.3 Attenuation by Intervening Absorbers	699
	16.2.4 Observational Constraints on the UV Background	701
16.3	The Evolution of the Intergalactic Medium	702
	16.3.1 Thermal Evolution	702
	16.3.2 Ionization Evolution	704
	16.3.3 The Epoch of Re-ionization	705
	16.3.4 Probing Re-ionization with 21-cm Emission and Absorption	707
16.4	General Properties of Absorption Lines	709
	16.4.1 Distribution Function	709
	16.4.2 Thermal Broadening	710
	16.4.3 Natural Broadening and Voigt Profiles	711
	16.4.4 Equivalent Width and Column Density	712
	16.4.5 Common QSO Absorption Line Systems	714
	16.4.6 Photoionization Models	714
16.5	The Lyman α Forest	714
	16.5.1 Redshift Evolution	715
	16.5.2 Column Density Distribution	716
	16.5.3 Doppler Parameter	717
	16.5.4 Sizes of Absorbers	718
	16.5.5 Metallicity	719
	16.5.6 Clustering	720
	16.5.7 Lyman α Forests at Low Redshift	721
	16.5.8 The Helium Lyman α Forest	722
16.6	Models of the Lyman α Forest	723
	16.6.1 Early Models	723
	16.6.2 Lyman α Forest in Hierarchical Models	724
	16.6.3 Lyman α Forest in Hydrodynamical Simulations	731
16.7	Lyman-Limit Systems	732
16.8	Damped Lyman α Systems	733
	16.8.1 Column Density Distribution	734
	16.8.2 Redshift Evolution	734
	16.8.3 Metallicities	736
	16.8.4 Kinematics	738

16.9	Metal Absorption Line Systems	738
	16.9.1 MgII Systems	739
	16.9.2 CIV and OVI Systems	740

A Basics of General Relativity — 741

A1.1 Space-time Geometry — 741

A1.2 The Equivalence Principle — 743

A1.3 Geodesic Equations — 744

A1.4 Energy–Momentum Tensor — 746

A1.5 Newtonian Limit — 747

A1.6 Einstein's Field Equation — 747

B Gas and Radiative Processes — 748

B1.1 Ideal Gas — 748

B1.2 Basic Equations — 749

B1.3 Radiative Processes — 751
- B1.3.1 Einstein Coefficients and Milne Relation — 752
- B1.3.2 Photoionization and Photo-excitation — 755
- B1.3.3 Recombination — 756
- B1.3.4 Collisional Ionization and Collisional Excitation — 757
- B1.3.5 Bremsstrahlung — 758
- B1.3.6 Compton Scattering — 759

B1.4 Radiative Cooling — 760

C Numerical Simulations — 764

C1.1 N-Body Simulations — 764
- C1.1.1 Force Calculations — 766
- C1.1.2 Issues Related to Numerical Accuracy — 767
- C1.1.3 Boundary Conditions — 769
- C1.1.4 Initial Conditions — 769

C1.2 Hydrodynamical Simulations — 770
- C1.2.1 Smoothed-Particle Hydrodynamics (SPH) — 770
- C1.2.2 Grid-Based Algorithms — 772

D Frequently Used Abbreviations — 775

E Useful Numbers — 776

References — 777

Index — 806

Preface

The vast ocean of space is full of starry islands called galaxies. These objects, extraordinarily beautiful and diverse in their own right, not only are the localities within which stars form and evolve, but also act as the lighthouses that allow us to explore our Universe over cosmological scales. Understanding the majesty and variety of galaxies in a cosmological context is therefore an important, yet daunting task. Particularly mind-boggling is the fact that, in the current paradigm, galaxies only represent the tip of the iceberg in a Universe dominated by some unknown 'dark matter' and an even more elusive form of 'dark energy'.

How do galaxies come into existence in this dark Universe, and how do they evolve? What is the relation of galaxies to the dark components? What shapes the properties of different galaxies? How are different properties of galaxies correlated with each other and what physics underlies these correlations? How do stars form and evolve in different galaxies? The quest for the answers to these questions, among others, constitutes an important part of modern cosmology, the study of the structure and evolution of the Universe as a whole, and drives the active and rapidly evolving research field of extragalactic astronomy and astrophysics.

The aim of this book is to provide a self-contained description of the physical processes and the astronomical observations which underlie our present understanding of the formation and evolution of galaxies in a Universe dominated by dark matter and dark energy. Any book on this subject must take into account that this is a rapidly developing field; there is a danger that material may rapidly become outdated. We hope that this can be avoided if the book is appropriately structured. Our premises are the following. In the first place, although observational data are continually updated, forcing revision of the theoretical models used to interpret them, the general principles involved in building such models do not change as rapidly. It is these principles, rather than the details of specific observations or models, that are the main focus of this book. Secondly, galaxies are complex systems, and the study of their formation and evolution is an applied and synthetic science. The interest of the subject is precisely that there are so many unsolved problems, and that the study of these problems requires techniques from many branches of physics and astrophysics – the formation of stars, the origin and dispersal of the elements, the link between galaxies and their central black holes, the nature of dark matter and dark energy, the origin and evolution of cosmic structure, and the size and age of our Universe. A firm grasp of the basic principles and the main outstanding issues across this full breadth of topics is needed by anyone preparing to carry out her/his own research, and this we hope to provide.

These considerations dictated both our selection of material and our style of presentation. Throughout the book, we emphasize the principles and the important issues rather than the details of observational results and theoretical models. In particular, special attention is paid to bringing out the physical connections between different parts of the problem, so that the reader will not lose the big picture while working on details. To this end, we start in each chapter with an introduction describing the material to be presented and its position in the overall scenario. In a field as broad as galaxy formation and evolution, it is clearly impossible to include all relevant material. The selection of the material presented in this book is therefore unavoidably biased by our prejudice, taste, and limited knowledge of the literature, and we apologize to anyone whose important work is not properly covered.

This book can be divided into several parts according to the material contained. Chapter 1 is an introduction, which sketches our current ideas about galaxies and their formation processes. Chapter 2 is an overview of the observational facts related to galaxy formation and evolution. Chapter 3 describes the cosmological framework within which galaxy formation and evolution must be studied. Chapters 4–8 contain material about the nature and evolution of the cosmological density field, both in collisionless dark matter and in collisional gas. Chapters 9 and 10 deal with topics related to star formation and stellar evolution in galaxies. Chapters 11–15 are concerned with the structure, formation, and evolution of individual galaxies and with the statistical properties of the galaxy population, and Chapter 16 gives an overview of the intergalactic medium. In addition, we provide appendixes to describe the general concepts of general relativity (Appendix A), basic hydrodynamic and radiative processes (Appendix B), and some commonly used techniques of N-body and hydrodynamical simulations (Appendix C).

The different parts are largely self-contained, and can be used separately for courses or seminars on specific topics. Chapters 1 and 2 are particularly geared towards novices to the field of extragalactic astronomy. Chapter 3, combined with parts of Chapters 4 and 5, could make up a course on cosmology, while a more advanced course on structure formation might be constructed around the material presented in Chapters 4–8. Chapter 2 and Chapters 11–15 contain material suited for a course on galaxy formation. Chapters 9, 10 and 16 contain special topics related to the formation and evolution of galaxies, and could be combined with Chapters 11–15 to form an extended course on galaxy formation and evolution. They could also be used independently for short courses on star formation and stellar evolution (Chapters 9 and 10), and on the intergalactic medium (Chapter 16).

Throughout the book, we have adopted a number of abbreviations that are commonly used by galaxy-formation practitioners. In order to avoid confusion, these abbreviations are listed in Appendix D along with their definitions. Some important physical constants and units are listed in Appendix E.

References are provided at the end of the book. Although long, the reference list is by no means complete, and we apologize once more to anyone whose relevant papers are overlooked. The number of references citing our own work clearly overrates our own contribution to the field. This is again a consequence of our limited knowledge of the existing literature, which is expanding at such a dramatic pace that it is impossible to cite all the relevant papers. The references given are mainly intended to serve as a starting point for readers interested in a more detailed literature study. We hope, by looking for the papers cited by our listed references, one can find relevant papers published in the past, and by looking for the papers citing the listed references, one can find relevant papers published later. Nowadays this is relatively easy to do with the use of the search engines provided by *The SAO/NASA Astrophysics Data System*[1] and the *arXiv e-print server*.[2]

We would not have been able to write this book without the help of many people. We benefitted greatly from discussions with and comments by many of our colleagues, including E. Bell, A. Berlind, G. Börner, A. Coil, J. Dalcanton, A. Dekel, M. Hähnelt, M. Heyer, W. Hu, Y. Jing, N. Katz, R. Larson, M. Longair, M. Mac Low, C.-P. Ma, S. Mao, E. Neistein, A. Pasquali, J. Peacock, M. Rees, H.-W. Rix, J. Sellwood, E. Sheldon, R. Sheth, R. Somerville, V. Springel, R. Sunyaev, A. van der Wel, R. Wechsler, M. Weinberg, and X. Yang. We are also deeply indebted to our many students and collaborators who made it possible for us to continue to publish

---

[1] *http://adsabs.harvard.edu/abstract_service.html*
[2] *http://arxiv.org/*

scientific papers while working on the book, and who gave us many new ideas and insights, some of which are presented in this book.

Many thanks to the following people who provided us with figures and data used in the book: M. Bartelmann, F. Bigiel, M. Boylan-Kolchin, S. Charlot, S. Courteau, J. Dalcanton, A. Dutton, K. Gebhardt, A. Graham, P. Hewett, G. Kauffmann, Y. Lu, L. McArthur, A. Pasquali, R. Saglia, S. Shen, Y. Wang, and X. Yang.

We thankfully acknowledge the (almost) inexhaustible amount of patience of the people at Cambridge University Press, in particular our editor, Vince Higgs.

We also thank the following institutions for providing support and hospitality to us during the writing of this book: the University of Massachusetts, Amherst; the Max-Planck Institute for Astronomy, Heidelberg; the Max-Planck Institute for Astrophysics, Garching; the Swiss Federal Institute of Technology, Zürich; the University of Utah; Shanghai Observatory; the Aspen Center for Physics; and the Kavli Institute of Theoretical Physics, Santa Barbara.

Last but not least we wish to thank our loved ones, whose continuous support has been absolutely essential for the completion of this book. HM would like to thank his wife, Ling, and son, Ye, for their support and understanding during the years when the book was drafted. FB gratefully acknowledges the love and support of Anna and Daka, and apologizes for the times they felt neglected because of 'the book'.

May 2009
Houjun Mo
Frank van den Bosch
Simon White

1
Introduction

This book is concerned with the physical processes related to the formation and evolution of galaxies. Simply put, a galaxy is a dynamically bound system that consists of many stars. A typical bright galaxy, such as our own Milky Way, contains a few times 10^{10} stars and has a diameter ($\sim 20\,\mathrm{kpc}$) that is several hundred times smaller than the mean separation between bright galaxies. Since most of the visible stars in the Universe belong to a galaxy, the number density of stars within a galaxy is about 10^7 times higher than the mean number density of stars in the Universe as a whole. In this sense, galaxies are well-defined, astronomical identities. They are also extraordinarily beautiful and diverse objects whose nature, structure and origin have intrigued astronomers ever since the first galaxy images were taken in the mid-nineteenth century.

The goal of this book is to show how physical principles can be used to understand the formation and evolution of galaxies. Viewed as a physical process, galaxy formation and evolution involve two different aspects: (i) initial and boundary conditions; and (ii) physical processes which drive evolution. Thus, in very broad terms, our study will consist of the following parts:

- Cosmology: Since we are dealing with events on cosmological time and length scales, we need to understand the space-time structure on large scales. One can think of the cosmological framework as the stage on which galaxy formation and evolution take place.
- Initial conditions: These were set by physical processes in the early Universe which are beyond our direct view, and which took place under conditions far different from those we can reproduce in Earth-bound laboratories.
- Physical processes: As we will show in this book, the basic physics required to study galaxy formation and evolution includes general relativity, hydrodynamics, dynamics of collisionless systems, plasma physics, thermodynamics, electrodynamics, atomic, nuclear and particle physics, and the theory of radiation processes.

In a sense, galaxy formation and evolution can therefore be thought of as an application of (relatively) well-known physics with cosmological initial and boundary conditions. As in many other branches of applied physics, the phenomena to be studied are diverse and interact in many different ways. Furthermore, the physical processes involved in galaxy formation cover some 23 orders of magnitude in physical size, from the scale of the Universe itself down to the scale of individual stars, and about four orders of magnitude in time scales, from the age of the Universe to that of the lifetime of individual, massive stars. Put together, it makes the formation and evolution of galaxies a subject of great complexity.

From an empirical point of view, the study of galaxy formation and evolution is very different from most other areas of experimental physics. This is due mainly to the fact that even the shortest time scales involved are much longer than that of a human being. Consequently, we cannot witness the actual evolution of individual galaxies. However, because the speed of light is finite, looking at galaxies at larger distances from us is equivalent to looking at galaxies when the Universe was younger. Therefore, we may hope to infer how galaxies form and evolve by comparing their properties, in a statistical sense, at different epochs. In addition, at each epoch we can try to identify regularities and correspondences among the galaxy population. Although galaxies span a wide range in masses, sizes, and morphologies, to the extent that no two galaxies are alike, the structural parameters of galaxies also obey various scaling relations, some of which are remarkably tight. These relations must hold important information regarding the physical processes that underlie them, and any successful theory of galaxy formation has to be able to explain their origin.

Galaxies are not only interesting in their own right, they also play a pivotal role in our study of the structure and evolution of the Universe. They are bright, long-lived and abundant, and so can be observed in large numbers over cosmological distances and time scales. This makes them unique tracers of the evolution of the Universe as a whole, and detailed studies of their large scale distribution can provide important constraints on cosmological parameters. In this book we therefore also describe the large scale distribution of galaxies, and discuss how it can be used to test cosmological models.

In Chapter 2 we start by describing the observational properties of stars, galaxies and the large scale structure of the Universe as a whole. Chapters 3 through 10 describe the various physical ingredients needed for a self-consistent model of galaxy formation, ranging from the cosmological framework to the formation and evolution of individual stars. Finally, in Chapters 11–16 we combine these physical ingredients to examine how galaxies form and evolve in a cosmological context, using the observational data as constraints.

The purpose of this introductory chapter is to sketch our current ideas about galaxies and their formation process, without going into any detail. After a brief overview of some observed properties of galaxies, we list the various physical processes that play a role in galaxy formation and outline how they are connected. We also give a brief historical overview of how our current views of galaxy formation have been shaped.

1.1 The Diversity of the Galaxy Population

Galaxies are a diverse class of objects. This means that a large number of parameters is required in order to characterize any given galaxy. One of the main goals of any theory of galaxy formation is to explain the full probability distribution function of all these parameters. In particular, as we will see in Chapter 2, many of these parameters are correlated with each other, a fact which any successful theory of galaxy formation should also be able to reproduce.

Here we list briefly the most salient parameters that characterize a galaxy. This overview is necessarily brief and certainly not complete. However, it serves to stress the diversity of the galaxy population, and to highlight some of the most important observational aspects that galaxy formation theories need to address. A more thorough description of the observational properties of galaxies is given in Chapter 2.

(a) Morphology One of the most noticeable properties of the galaxy population is the existence of two basic galaxy types: spirals and ellipticals. Elliptical galaxies are mildly flattened, ellipsoidal systems that are mainly supported by the random motions of their stars. Spiral galaxies, on the other hand, have highly flattened disks that are mainly supported by rotation. Consequently, they are also often referred to as disk galaxies. The name 'spiral' comes from the fact that the gas and stars in the disk often reveal a clear spiral pattern. Finally, for historical reasons, ellipticals and spirals are also called early- and late-type galaxies, respectively.

Most galaxies, however, are neither a perfect ellipsoid nor a perfect disk, but rather a combination of both. When the disk is the dominant component, its ellipsoidal component is generally called the bulge. In the opposite case, of a large ellipsoidal system with a small disk, one typically talks about a disky elliptical. One of the earliest classification schemes for galaxies, which is still heavily used, is the Hubble sequence. Roughly speaking, the Hubble sequence is a sequence in the admixture of the disk and ellipsoidal components in a galaxy, which ranges from early-type ellipticals that are pure ellipsoids to late-type spirals that are pure disks. As we will see in Chapter 2, the important aspect of the Hubble sequence is that many intrinsic properties of galaxies, such as luminosity, color, and gas content, change systematically along this sequence. In addition, disks and ellipsoids most likely have very different formation mechanisms. Therefore, the morphology of a galaxy, or its location along the Hubble sequence, is directly related to its formation history.

For completeness, we stress that not all galaxies fall in this spiral vs. elliptical classification. The faintest galaxies, called dwarf galaxies, typically do not fall on the Hubble sequence. Dwarf galaxies with significant amounts of gas and ongoing star formation typically have a very irregular structure, and are consequently called (dwarf) irregulars. Dwarf galaxies without gas and young stars are often very diffuse, and are called dwarf spheroidals. In addition to these dwarf galaxies, there is also a class of brighter galaxies whose morphology neither resembles a disk nor a smooth ellipsoid. These are called peculiar galaxies and include, among others, galaxies with double or multiple subcomponents linked by filamentary structure and highly distorted galaxies with extended tails. As we will see, they are usually associated with recent mergers or tidal interactions. Although peculiar galaxies only constitute a small fraction of the entire galaxy population, their existence conveys important information about how galaxies may have changed their morphologies during their evolutionary history.

(b) Luminosity and Stellar Mass Galaxies span a wide range in luminosity. The brightest galaxies have luminosities of $\sim 10^{12} L_\odot$, where $L_\odot$ indicates the luminosity of the Sun. The exact lower limit of the luminosity distribution is less well defined, and is subject to regular changes, as fainter and fainter galaxies are constantly being discovered. In 2007 the faintest galaxy known was a newly discovered dwarf spheroidal Willman I, with a total luminosity somewhat below $1000 L_\odot$.

Obviously, the total luminosity of a galaxy is related to its total number of stars, and thus to its total stellar mass. However, the relation between luminosity and stellar mass reveals a significant amount of scatter, because different galaxies have different stellar populations. As we will see in Chapter 10, galaxies with a younger stellar population have a higher luminosity per unit stellar mass than galaxies with an older stellar population.

An important statistic of the galaxy population is its luminosity probability distribution function, also known as the luminosity function. As we will see in Chapter 2, there are many more faint galaxies than bright galaxies, so that the faint ones clearly dominate the number density. However, in terms of the contribution to the total luminosity density, neither the faintest nor the brightest galaxies dominate. Instead, it is the galaxies with a characteristic luminosity similar to that of our Milky Way that contribute most to the total luminosity density in the present-day Universe. This indicates that there is a characteristic scale in galaxy formation, which is accentuated by the fact that most galaxies that are brighter than this characteristic scale are ellipticals, while those that are fainter are mainly spirals (at the very faint end dwarf irregulars and dwarf spheroidals dominate). Understanding the physical origin of this characteristic scale has turned out to be one of the most challenging problems in contemporary galaxy formation modeling.

(c) Size and Surface Brightness As we will see in Chapter 2, galaxies do not have well-defined boundaries. Consequently, several different definitions for the size of a galaxy can be found in the literature. One measure often used is the radius enclosing a certain fraction (e.g. half) of the total luminosity. In general, as one might expect, brighter galaxies are bigger. However, even for a fixed luminosity, there is a considerable scatter in sizes, or in surface brightness, defined as the luminosity per unit area.

The size of a galaxy has an important physical meaning. In disk galaxies, which are rotation supported, the sizes are a measure of their specific angular momenta (see Chapter 11). In the case of elliptical galaxies, which are supported by random motions, the sizes are a measure of the amount of dissipation during their formation (see Chapter 13). Therefore, the observed distribution of galaxy sizes is an important constraint for galaxy formation models.

(d) Gas Mass Fraction Another useful parameter to describe galaxies is their cold gas mass fraction, defined as $f_{\rm gas} = M_{\rm cold}/[M_{\rm cold} + M_\star]$, with $M_{\rm cold}$ and $M_\star$ the masses of cold gas and stars, respectively. This ratio expresses the efficiency with which cold gas has been turned into stars. Typically, the gas mass fractions of ellipticals are negligibly small, while those of disk galaxies increase systematically with decreasing surface brightness. Indeed, the lowest surface brightness disk galaxies can have gas mass fractions in excess of 90 percent, in contrast to our Milky Way which has $f_{\rm gas} \sim 0.1$.

(e) Color Galaxies also come in different colors. The color of a galaxy reflects the ratio of its luminosity in two photometric passbands. A galaxy is said to be red if its luminosity in the redder passband is relatively high compared to that in the bluer passband. Ellipticals and dwarf spheroidals generally have redder colors than spirals and dwarf irregulars. As we will see in Chapter 10, the color of a galaxy is related to the characteristic age and metallicity of its stellar population. In general, redder galaxies are either older or more metal rich (or both). Therefore, the color of a galaxy holds important information regarding its stellar population. However, extinction by dust, either in the galaxy itself, or along the line-of-sight between the source and the observer, also tends to make a galaxy appear red. As we will see, separating age, metallicity and dust effects is one of the most daunting tasks in observational astronomy.

(f) Environment As we will see in §§2.5–2.7, galaxies are not randomly distributed throughout space, but show a variety of structures. Some galaxies are located in high-density clusters containing several hundreds of galaxies, some in smaller groups containing a few to tens of galaxies, while yet others are distributed in low-density filamentary or sheet-like structures. Many of these structures are gravitationally bound, and may have played an important role in the formation and evolution of the galaxies. This is evident from the fact that elliptical galaxies seem to prefer cluster environments, whereas spiral galaxies are mainly found in relative isolation (sometimes called the field). As briefly discussed in §1.2.8 below, it is believed that this morphology–density relation reflects enhanced dynamical interaction in denser environments, although we still lack a detailed understanding of its origin.

(g) Nuclear Activity For the majority of galaxies, the observed light is consistent with what we expect from a collection of stars and gas. However, a small fraction of all galaxies, called active galaxies, show an additional non-stellar component in their spectral energy distribution. As we will see in Chapter 14, this emission originates from a small region in the centers of these galaxies, called the active galactic nucleus (AGN), and is associated with matter accretion onto a supermassive black hole. According to the relative importance of such non-stellar emission, one can separate active galaxies from normal (or non-active) galaxies.

(h) Redshift Because of the expansion of the Universe, an object that is farther away will have a larger receding velocity, and thus a larger redshift. Since the light from high-redshift galaxies was emitted when the Universe was younger, we can study galaxy evolution by observing the galaxy population at different redshifts. In fact, in a statistical sense the high-redshift galaxies are the progenitors of present-day galaxies, and any changes in the number density or intrinsic properties of galaxies with redshift give us a direct window on the formation and evolution of the galaxy population. With modern, large telescopes we can now observe galaxies out to redshifts beyond six, making it possible for us to probe the galaxy population back to a time when the Universe was only about 10 percent of its current age.

1.2 Basic Elements of Galaxy Formation

Before diving into details, it is useful to have an overview of the basic theoretical framework within which our current ideas about galaxy formation and evolution have been developed. In this section we give a brief overview of the various physical processes that play a role during the formation and evolution of galaxies. The goal is to provide the reader with a picture of the relationships among the various aspects of galaxy formation to be addressed in greater detail in the chapters to come. To guide the reader, Fig. 1.1 shows a flow chart of galaxy formation, which illustrates how the various processes to be discussed below are intertwined. It is important to stress, though, that this particular flow chart reflects our current, undoubtedly incomplete view of galaxy formation. Future improvements in our understanding of galaxy formation and evolution may add new links to the flow chart, or may render some of the links shown obsolete.

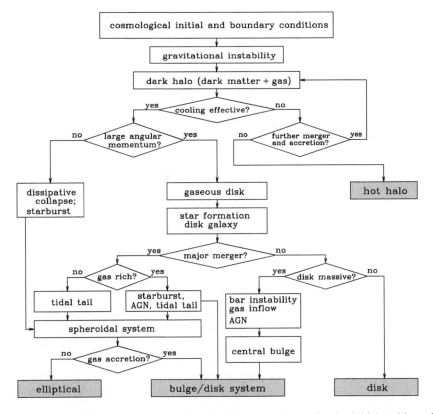

Fig. 1.1. A logic flow chart for galaxy formation. In the standard scenario, the initial and boundary conditions for galaxy formation are set by the cosmological framework. The paths leading to the formation of various galaxies are shown along with the relevant physical processes. Note, however, that processes do not separate as neatly as this figure suggests. For example, cold gas may not have the time to settle into a gaseous disk before a major merger takes place.

1.2.1 The Standard Model of Cosmology

Since galaxies are observed over cosmological length and time scales, the description of their formation and evolution must involve cosmology, the study of the properties of space-time on large scales. Modern cosmology is based upon the cosmological principle, the hypothesis that the Universe is spatially homogeneous and isotropic, and Einstein's theory of general relativity, according to which the structure of space-time is determined by the mass distribution in the Universe. As we will see in Chapter 3, these two assumptions together lead to a cosmology (the standard model) that is completely specified by the curvature of the Universe, K, and the scale factor, $a(t)$, describing the change of the length scale of the Universe with time. One of the basic tasks in cosmology is to determine the value of K and the form of $a(t)$ (hence the space-time geometry of the Universe on large scales), and to show how observables are related to physical quantities in such a universe.

Modern cosmology not only specifies the large-scale geometry of the Universe, but also has the potential to predict its thermal history and matter content. Because the Universe is expanding and filled with microwave photons at the present time, it must have been smaller, denser and hotter at earlier times. The hot and dense medium in the early Universe provides conditions under which various reactions among elementary particles, nuclei and atoms occur. Therefore, the application of particle, nuclear and atomic physics to the thermal history of the Universe in principle allows us to predict the abundances of all species of elementary particles, nuclei and atoms at different epochs. Clearly, this is an important part of the problem to be addressed in this book, because the formation of galaxies depends crucially on the matter/energy content of the Universe.

In currently popular cosmologies we usually consider a universe consisting of three main components. In addition to the 'baryonic' matter, the protons, neutrons and electrons[1] that make up the *visible* Universe, astronomers have found various indications for the presence of dark matter and dark energy (see Chapter 2 for a detailed discussion of the observational evidence). Although the nature of both dark matter and dark energy is still unknown, we believe that they are responsible for more than 95 percent of the energy density of the Universe. Different cosmological models differ mainly in (i) the relative contributions of baryonic matter, dark matter, and dark energy, and (ii) the nature of dark matter and dark energy. At the time of writing, the most popular model is the so-called ΛCDM model, a flat universe in which ~ 75 percent of the energy density is due to a cosmological constant, ~ 21 percent is due to 'cold' dark matter (CDM), and the remaining 4 percent is due to the baryonic matter out of which stars and galaxies are made. Chapter 3 gives a detailed description of these various components, and describes how they influence the expansion history of the Universe.

1.2.2 Initial Conditions

If the cosmological principle held perfectly and the distribution of matter in the Universe were perfectly uniform and isotropic, there would be no structure formation. In order to explain the presence of structure, in particular galaxies, we clearly need some deviations from perfect uniformity. Unfortunately, the standard cosmology does not in itself provide us with an explanation for the origin of these perturbations. We have to go beyond it to search for an answer.

A classical, general relativistic description of cosmology is expected to break down at very early times when the Universe is so dense that quantum effects are expected to be important. As we will see in §3.6, the standard cosmology has a number of conceptual problems when applied to the early Universe, and the solutions to these problems require an extension of the standard

---

[1] Although an electron is a lepton, and not a baryon, in cosmology it is standard practice to include electrons when talking of baryonic matter

cosmology to incorporate quantum processes. One generic consequence of such an extension is the generation of density perturbations by quantum fluctuations at early times. It is believed that these perturbations are responsible for the formation of the structures observed in today's Universe.

As we will see in §3.6, one particularly successful extension of the standard cosmology is the inflationary theory, in which the Universe is assumed to have gone through a phase of rapid, exponential expansion (called inflation) driven by the vacuum energy of one or more quantum fields. In many, but not all, inflationary models, quantum fluctuations in this vacuum energy can produce density perturbations with properties consistent with the observed large scale structure. Inflation thus offers a promising explanation for the physical origin of the initial perturbations. Unfortunately, our understanding of the very early Universe is still far from complete, and we are currently unable to predict the initial conditions for structure formation entirely from first principles. Consequently, even this part of galaxy formation theory is still partly phenomenological: typically initial conditions are specified by a set of parameters that are constrained by observational data, such as the pattern of fluctuations in the microwave background or the present-day abundance of galaxy clusters.

1.2.3 Gravitational Instability and Structure Formation

Having specified the initial conditions and the cosmological framework, one can compute how small perturbations in the density field evolve. As we will see in Chapter 4, in an expanding universe dominated by non-relativistic matter, perturbations grow with time. This is easy to understand. A region whose initial density is slightly higher than the mean will attract its surroundings slightly more strongly than average. Consequently, over-dense regions pull matter towards them and become even more over-dense. On the other hand, under-dense regions become even more rarefied as matter flows away from them. This amplification of density perturbations is referred to as gravitational instability and plays an important role in modern theories of structure formation. In a static universe, the amplification is a run-away process, and the density contrast $\delta\rho/\rho$ grows exponentially with time. In an expanding universe, however, the cosmic expansion damps accretion flows, and the growth rate is usually a power law of time, $\delta\rho/\rho \propto t^\alpha$, with $\alpha > 0$. As we will see in Chapter 4, the exact rate at which the perturbations grow depends on the cosmological model.

At early times, when the perturbations are still in what we call the linear regime ($\delta\rho/\rho \ll 1$), the physical size of an over-dense region increases with time due to the overall expansion of the universe. Once the perturbation reaches over-density $\delta\rho/\rho \sim 1$, it breaks away from the expansion and starts to collapse. This moment of 'turn-around', when the physical size of the perturbation is at its maximum, signals the transition from the mildly nonlinear regime to the strongly nonlinear regime.

The outcome of the subsequent nonlinear, gravitational collapse depends on the matter content of the perturbation. If the perturbation consists of ordinary baryonic gas, the collapse creates strong shocks that raise the entropy of the material. If radiative cooling is inefficient, the system relaxes to hydrostatic equilibrium, with its self-gravity balanced by pressure gradients. If the perturbation consists of collisionless matter (e.g. cold dark matter), no shocks develop, but the system still relaxes to a quasi-equilibrium state with a more-or-less universal structure. This process is called violent relaxation and will be discussed in Chapter 5. Nonlinear, quasi-equilibrium dark matter objects are called dark matter halos. Their predicted structure has been thoroughly explored using numerical simulations, and they play a pivotal role in modern theories of galaxy formation. Chapter 7 therefore presents a detailed discussion of the structure and formation of dark matter halos. As we shall see, halo density profiles, shapes, spins and internal substructure all depend very weakly on mass and on cosmology, but the abundance and characteristic density of halos depend sensitively on both of these.

In cosmologies with both dark matter and baryonic matter, such as the currently favored CDM models, each initial perturbation contains baryonic gas and collisionless dark matter in roughly their universal proportions. When an object collapses, the dark matter relaxes violently to form a dark matter halo, while the gas shocks to the virial temperature, $T_{\rm vir}$ (see §8.2.3 for a definition) and may settle into hydrostatic equilibrium in the potential well of the dark matter halo if cooling is slow.

1.2.4 Gas Cooling

Cooling is a crucial ingredient of galaxy formation. Depending on temperature and density, a variety of cooling processes can affect gas. In massive halos, where the virial temperature $T_{\rm vir} \gtrsim 10^7$ K, gas is fully collisionally ionized and cools mainly through bremsstrahlung emission from free electrons. In the temperature range $10^4\,{\rm K} < T_{\rm vir} < 10^6\,{\rm K}$, a number of excitation and de-excitation mechanisms can play a role. Electrons can recombine with ions, emitting a photon, or atoms (neutral or partially ionized) can be excited by a collision with another particle, thereafter decaying radiatively to the ground state. Since different atomic species have different excitation energies, the cooling rates depend strongly on the chemical composition of the gas. In halos with $T_{\rm vir} < 10^4\,{\rm K}$, gas is predicted to be almost completely neutral. This strongly suppresses the cooling processes mentioned above. However, if heavy elements and/or molecules are present, cooling is still possible through the collisional excitation/de-excitation of fine and hyperfine structure lines (for heavy elements) or rotational and/or vibrational lines (for molecules). Finally, at high redshifts ($z \gtrsim 6$), inverse Compton scattering of cosmic microwave background photons by electrons in hot halo gas can also be an effective cooling channel. Chapter 8 will discuss these cooling processes in more detail.

Except for inverse Compton scattering, all these cooling mechanisms involve two particles. Consequently, cooling is generally more effective in higher density regions. After nonlinear gravitational collapse, the shocked gas in virialized halos may be dense enough for cooling to be effective. If cooling times are short, the gas never comes to hydrostatic equilibrium, but rather accretes directly onto the central protogalaxy. Even if cooling is slow enough for a hydrostatic atmosphere to develop, it may still cause the denser inner regions of the atmosphere to lose pressure support and to flow onto the central object. The net effect of cooling is thus that the baryonic material segregates from the dark matter, and accumulates as dense, cold gas in a protogalaxy at the center of the dark matter halo.

As we will see in Chapter 7, dark matter halos, as well as the baryonic material associated with them, typically have a small amount of angular momentum. If this angular momentum is conserved during cooling, the gas will spin up as it flows inwards, settling in a cold disk in centrifugal equilibrium at the center of the halo. This is the standard paradigm for the formation of disk galaxies, which we will discuss in detail in Chapter 11.

1.2.5 Star Formation

As the gas in a dark matter halo cools and flows inwards, its self-gravity will eventually dominate over the gravity of the dark matter. Thereafter it collapses under its own gravity, and in the presence of effective cooling, this collapse becomes catastrophic. Collapse increases the density and temperature of the gas, which generally reduces the cooling time more rapidly than it reduces the collapse time. During such runaway collapse the gas cloud may fragment into small, high-density cores that may eventually form stars (see Chapter 9), thus giving rise to a visible galaxy.

Unfortunately, many details of these processes are still unclear. In particular, we are still unable to predict the mass fraction of, and the time scale for, a self-gravitating cloud to be transformed into stars. Another important and yet poorly understood issue is concerned with the mass distribution with which stars are formed, i.e. the initial mass function (IMF). As we will see in Chapter 10, the evolution of a star, in particular its luminosity as function of time and its eventual fate, is largely determined by its mass at birth. Predictions of observable quantities for model galaxies thus require not only the birth rate of stars as a function of time, but also their IMF. In principle, it should be possible to derive the IMF from first principles, but the theory of star formation has not yet matured to this level. At present one has to assume an IMF ad hoc and check its validity by comparing model predictions to observations.

Based on observations, we will often distinguish two modes of star formation: quiescent star formation in rotationally supported gas disks, and starbursts. The latter are characterized by much higher star-formation rates, and are typically confined to relatively small regions (often the nucleus) of galaxies. Starbursts require the accumulation of large amounts of gas in a small volume, and appear to be triggered by strong dynamical interactions or instabilities. These processes will be discussed in more detail in §1.2.8 below and in Chapter 12. At the moment, there are still many open questions related to these different modes of star formation. What fraction of stars formed in the quiescent mode? Do both modes produce stellar populations with the same IMF? How does the relative importance of starbursts scale with time? As we will see, these and related questions play an important role in contemporary models of galaxy formation.

1.2.6 Feedback Processes

When astronomers began to develop the first dynamical models for galaxy formation in a CDM dominated universe, it immediately became clear that most baryonic material is predicted to cool and form stars. This is because in these 'hierarchical' structure formation models, small dense halos form at high redshift and cooling within them is predicted to be very efficient. This disagrees badly with observations, which show that only a relatively small fraction of all baryons are in cold gas or stars (see Chapter 2). Apparently, some physical process must either prevent the gas from cooling, or reheat it after it has become cold.

Even the very first models suggested that the solution to this problem might lie in feedback from supernovae, a class of exploding stars that can produce enormous amounts of energy (see §10.5). The radiation and the blast waves from these supernovae may heat (or reheat) surrounding gas, blowing it out of the galaxy in what is called a galactic wind. These processes are described in more detail in §§8.6 and 10.5.

Another important feedback source for galaxy formation is provided by active galactic nuclei (AGN), the active accretion phase of supermassive black holes (SMBH) lurking at the centers of almost all massive galaxies (see Chapter 14). This process releases vast amounts of energy – this is why AGN are bright and can be seen out to large distances, which can be tapped by surrounding gas. Although only a relatively small fraction of present-day galaxies contain an AGN, observations indicate that virtually all massive spheroids contain a nuclear SMBH (see Chapter 2). Therefore, it is believed that virtually all galaxies with a significant spheroidal component have gone through one or more AGN phases during their life.

Although it has become clear over the years that feedback processes play an important role in galaxy formation, we are still far from understanding which processes dominate, and when and how exactly they operate. Furthermore, to make accurate predictions for their effects, one also needs to know how often they occur. For supernovae this requires a prior understanding of the star-formation rates and the IMF. For AGN it requires understanding how, when and where supermassive black holes form, and how they accrete mass.

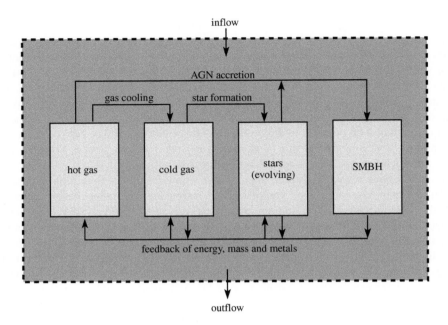

Fig. 1.2. A flow chart of the evolution of an individual galaxy. The galaxy is represented by the dashed box which contains hot gas, cold gas, stars and a supermassive black hole (SMBH). Gas cooling converts hot gas into cold gas, star formation converts cold gas into stars, and dying stars inject energy, metals and gas into the gas components. In addition, the SMBH can accrete gas (both hot and cold) as well as stars, producing AGN activity which can release vast amounts of energy which affect primarily the gaseous components of the galaxy. Note that in general the box will not be closed: gas can be added to the system through accretion from the intergalactic medium and can escape the galaxy through outflows driven by feedback from the stars and/or the SMBH. Finally, a galaxy may merge or interact with another galaxy, causing a significant boost or suppression of all these processes.

It should be clear from the above discussion that galaxy formation is a subject of great complexity, involving many strongly intertwined processes. This is illustrated in Fig. 1.2, which shows the relations between the four main baryonic components of a galaxy: hot gas, cold gas, stars, and a supermassive black hole. Cooling, star formation, AGN accretion, and feedback processes can all shift baryons from one of these components to another, thereby altering the efficiency of all the processes. For example, increased cooling of hot gas will produce more cold gas. This in turn will increases the star-formation rate, hence the supernova rate. The additional energy injection from supernovae can reheat cold gas, thereby suppressing further star formation (negative feedback). On the other hand, supernova blast waves may also compress the surrounding cold gas, so as to boost the star-formation rate (positive feedback). Understanding these various feedback loops is one of the most important and intractable issues in contemporary models for the formation and evolution of galaxies.

1.2.7 Mergers

So far we have considered what happens to a single, isolated system of dark matter, gas and stars. However, galaxies and dark matter halos are not isolated. For example, as illustrated in Fig. 1.2, systems can accrete new material (both dark and baryonic matter) from the intergalactic medium, and can lose material through outflows driven by feedback from stars and/or AGN. In addition, two (or more) systems may merge to form a new system with very different properties from its progenitors. In the currently popular CDM cosmologies, the initial density fluctuations

1.2 Basic Elements of Galaxy Formation

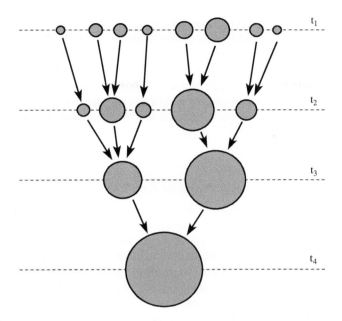

Fig. 1.3. A schematic merger tree, illustrating the merger history of a dark matter halo. It shows, at three different epochs, the progenitor halos that at time t_4 have merged to form a single halo. The size of each circle represents the mass of the halo. Merger histories of dark matter halos play an important role in hierarchical theories of galaxy formation.

have larger amplitudes on smaller scales. Consequently, dark matter halos grow hierarchically, in the sense that larger halos are formed by the coalescence (merging) of smaller progenitors. Such a formation process is usually called a hierarchical or 'bottom-up' scenario.

The formation history of a dark matter halo can be described by a 'merger tree' that traces all its progenitors, as illustrated in Fig. 1.3. Such merger trees play an important role in modern galaxy formation theory. Note, however, that illustrations such as Fig. 1.3 can be misleading. In CDM models part of the growth of a massive halo is due to merging with a large number of much smaller halos, and to a good approximation, such mergers can be thought of as smooth accretion. When two similar mass dark matter halos merge, violent relaxation rapidly transforms the orbital energy of the progenitors into the internal binding energy of the quasi-equilibrium remnant. Any hot gas associated with the progenitors is shock-heated during the merger and settles back into hydrostatic equilibrium in the new halo. If the progenitor halos contained central galaxies, the galaxies also merge as part of the violent relaxation process, producing a new central galaxy in the final system. Such a merger may be accompanied by strong star formation or AGN activity if the merging galaxies contained significant amounts of cold gas. If two merging halos have very different mass, the dynamical processes are less violent. The smaller system orbits within the main halo for an extended period of time during which two processes compete to determine its eventual fate. Dynamical friction transfers energy from its orbit to the main halo, causing it to spiral inwards, while tidal effects remove mass from its outer regions and may eventually dissolve it completely (see Chapter 12). Dynamical friction is more effective for more massive satellites, but if the mass ratio of the initial halos is large enough, the smaller object (and any galaxy associated with it) can maintain its identity for a long time. This is the process for the build-up of clusters of galaxies: a cluster may be considered as a massive dark matter halo hosting a relatively massive galaxy near its center and many satellites that have not yet dissolved or merged with the central galaxy.

As we will see in Chapters 12 and 13, numerical simulations show that the merger of two galaxies of roughly equal mass produces an object reminiscent of an elliptical galaxy, and the result is largely independent of whether the progenitors are spirals or ellipticals. Indeed, current hierarchical models of galaxy formation assume that most, if not all, elliptical galaxies are merger remnants. If gas cools onto this merger remnant with significant angular momentum, a new disk may form, producing a disk–bulge system like that in an early-type spiral galaxy.

It should be obvious from the above discussion that mergers play a crucial role in galaxy formation. Detailed descriptions of halo mergers and galaxy mergers are presented in Chapter 7 and Chapter 12, respectively.

1.2.8 Dynamical Evolution

When satellite galaxies orbit within dark matter halos, they experience tidal forces due to the central galaxy, due to other satellite galaxies, and due to the potential of the halo itself. These tidal interactions can remove dark matter, gas and stars from the galaxy, a process called tidal stripping (see §12.2), and may also perturb its structure. In addition, if the halo contains a hot gas component, any gas associated with the satellite galaxy will experience a drag force due to the relative motion of the two fluids. If the drag force exceeds the restoring force due to the satellite's own gravity, its gas will be ablated, a process called ram-pressure stripping. These dynamical processes are thought to play an important role in driving galaxy evolution within clusters and groups of galaxies. In particular, they are thought to be partially responsible for the observed environmental dependence of galaxy morphology (see Chapter 15).

Internal dynamical effects can also reshape galaxies. For example, a galaxy may form in a configuration which becomes unstable at some later time. Large scale instabilities may then redistribute mass and angular momentum within the galaxy, thereby changing its morphology. A well-known and important example is the bar-instability within disk galaxies. As we shall see in §11.5, a thin disk with too high a surface density is susceptible to a non-axisymmetric instability, which produces a bar-like structure similar to that seen in barred spiral galaxies. These bars may then buckle out of the disk to produce a central ellipsoidal component, a so-called 'pseudo-bulge'. Instabilities may also be triggered in otherwise stable galaxies by interactions. Thus, an important question is whether the sizes and morphologies of galaxies were set at formation, or are the result of later dynamical process ('secular evolution', as it is termed). Bulges are particularly interesting in this context. They may be a remnant of the first stage of galaxy formation, or, as mentioned in §1.2.7, may reflect an early merger which has grown a new disk, or may result from buckling of a bar. It is likely that all these processes are important for at least some bulges.

1.2.9 Chemical Evolution

In astronomy, all chemical elements heavier than helium are collectively termed 'metals'. The mass fraction of a baryonic component (e.g. hot gas, cold gas, stars) in metals is then referred to as its metallicity. As we will see in §3.4, the nuclear reactions during the first three minutes of the Universe (the epoch of primordial nucleosynthesis) produced primarily hydrogen ($\sim 75\%$) and helium ($\sim 25\%$), with a very small admixture of metals dominated by lithium. All other metals in the Universe were formed at later times as a consequence of nuclear reactions in stars. When stars expel mass in stellar winds, or in supernova explosions, they enrich the interstellar medium (ISM) with newly synthesized metals.

Evolution of the chemical composition of the gas and stars in galaxies is important for several reasons. First of all, the luminosity and color of a stellar population depend not only on its age and IMF, but also on the metallicity of the stars (see Chapter 10). Secondly, the cooling efficiency of gas depends strongly on its metallicity, in the sense that more metal-enriched gas cools faster (see §8.1). Thirdly, small particles of heavy elements known as dust grains, which are mixed with the interstellar gas in galaxies, can absorb significant amounts of the starlight and reradiate it in infrared wavelengths. Depending on the amount of the dust in the ISM, which scales roughly linearly with its metallicity (see §10.3.7), this interstellar extinction can significantly reduce the brightness of a galaxy.

As we will see in Chapter 10, the mass and detailed chemical composition of the material ejected by a stellar population as it evolves depend both on the IMF and on its initial metallicity. In principle, observations of the metallicity and abundance ratios of a galaxy can therefore be used to constrain its star-formation history and IMF. In practice, however, the interpretation of the observations is complicated by the fact that galaxies can accrete new material of different metallicity, that feedback processes can blow out gas, perhaps preferentially metals, and that mergers can mix the chemical compositions of different systems.

1.2.10 Stellar Population Synthesis

The light we receive from a given galaxy is emitted by a large number of stars that may have different masses, ages, and metallicities. In order to interpret the observed spectral energy distribution, we need to predict how each of these stars contributes to the total spectrum. Unlike many of the ingredients in galaxy formation, the theory of stellar evolution, to be discussed in Chapter 10, is reasonably well understood. This allows us to compute not only the evolution of the luminosity, color and spectrum of a star of given initial mass and chemical composition, but also the rates at which it ejects mass, energy and metals into the interstellar medium. If we know the star-formation history (i.e. the star-formation rate as a function of time) and IMF of a galaxy, we can then synthesize its spectrum at any given time by adding together the spectra of all the stars, after evolving each to the time under consideration. In addition, this also yields the rates at which mass, energy and metals are ejected into the interstellar medium, providing important ingredients for modeling the chemical evolution of galaxies.

Most of the energy of a stellar population is emitted in the optical, or, if the stellar population is very young ($\lesssim 10\,\mathrm{Myr}$), in the ultraviolet (see §10.3). However, if the galaxy contains a lot of dust, a significant fraction of this optical and UV light may get absorbed and re-emitted in the infrared. Unfortunately, predicting the final emergent spectrum is extremely complicated. Not only does it depend on the amount of the radiation absorbed, it also depends strongly on the properties of the dust, such as its geometry, its chemical composition, and (the distribution of) the sizes of the dust grains (see §10.3.7).

Finally, to complete the spectral energy distribution emitted by a galaxy, we also need to add the contribution from a possible AGN. Chapter 14 discusses various emission mechanisms associated with accreting SMBHs. Unfortunately, as we will see, we are still far from being able to predict the detailed spectra for AGN.

1.2.11 The Intergalactic Medium

The intergalactic medium (IGM) is the baryonic material lying between galaxies. This is and has always been the dominant baryonic component of the Universe and it is the material from which galaxies form. Detailed studies of the IGM can therefore give insight into the properties of the pregalactic matter before it condensed into galaxies. As illustrated in Fig. 1.2, galaxies do not evolve as closed boxes, but can affect the properties of the IGM through exchanges of mass, energy and heavy elements. The study of the IGM is thus an integral part of understanding how galaxies form and evolve. As we will see in Chapter 16, the properties of the IGM can be probed most effectively through the absorption it produces in the spectra of distant quasars (a certain class of active galaxies; see Chapter 14). Since quasars are now observed out to redshifts beyond 6, their absorption line spectra can be used to study the properties of the IGM back to a time when the Universe was only a few percent of its present age.

1.3 Time Scales

As discussed above, and as illustrated in Fig. 1.1, the formation of an individual galaxy in the standard, hierarchical formation scenario involves the following processes: the collapse and virialization of dark matter halos, the cooling and condensation of gas within the halo, and the conversion of cold gas into stars and a central supermassive black hole. Evolving stars and AGN eject energy, mass and heavy elements into the interstellar medium, thereby determining its structure and chemical composition and perhaps driving winds into the intergalactic medium. Finally, galaxies can merge and interact, reshaping their morphology and triggering further starbursts and AGN activity. In general, the properties of galaxies are determined by the competition among all these processes, and a simple way to characterize the relative importance of these processes is to use the time scales associated with them. Here we give a brief summary of the most important time scales in this context.

- **Hubble time:** This is an estimate of the time scale on which the Universe as a whole evolves. It is defined as the inverse of the Hubble constant (see §3.2), which specifies the current cosmic expansion rate. It would be equal to the time since the Big Bang if the Universe had always expanded at its current rate. Roughly speaking, this is the time scale on which substantial evolution of the galaxy population is expected.
- **Dynamical time:** This is the time required to orbit across an equilibrium dynamical system. For a system with mass M and radius R, we define it as $t_{\rm dyn} = \sqrt{3\pi/16G\bar{\rho}}$, where $\bar{\rho} = 3M/4\pi R^3$. This is related to the free-fall time, defined as the time required for a uniform, pressure-free sphere to collapse to a point, as $t_{\rm ff} = t_{\rm dyn}/\sqrt{2}$.
- **Cooling time:** This time scale is the ratio between the thermal energy content and the energy loss rate (through radiative or conductive cooling) for a gas component.
- **Star-formation time:** This time scale is the ratio of the cold gas content of a galaxy to its star-formation rate. It is thus an indication of how long it would take for the galaxy to run out of gas if the fuel for star formation is not replenished.
- **Chemical enrichment time:** This is a measure for the time scale on which the gas is enriched in heavy elements. This enrichment time is generally different for different elements, depending on the lifetimes of the stars responsible for the bulk of the production of each element (see §10.1).
- **Merging time:** This is the typical time that a halo or galaxy must wait before experiencing a merger with an object of similar mass, and is directly related to the major merger frequency.
- **Dynamical friction time:** This is the time scale on which a satellite object in a large halo loses its orbital energy and spirals to the center. As we will see in §12.3, this time scale is proportional to $M_{\rm main}/M_{\rm sat}$, where $M_{\rm sat}$ is the mass of the satellite object and $M_{\rm main}$ is that of the main halo. Thus, more massive galaxies will merge with the central galaxy in a halo more quickly than smaller ones.

These time scales can provide guidelines for incorporating the underlying physical processes in models of galaxy formation and evolution, as we describe in later chapters. In particular, comparing time scales can give useful insights. As an illustration, consider the following examples:

- Processes whose time scale is longer than the Hubble time can usually be ignored. For example, satellite galaxies with mass less than a few percent of their parent halo normally have

dynamical friction times exceeding the Hubble time (see §12.3). Consequently, their orbits do not decay significantly. This explains why clusters of galaxies have so many 'satellite' galaxies – the main halos are so much more massive than a typical galaxy that dynamical friction is ineffective.
- If the cooling time is longer than the dynamical time, hot gas will typically be in hydrostatic equilibrium. In the opposite case, however, the gas cools rapidly, losing pressure support, and collapsing to the halo center on a free-fall time without establishing any hydrostatic equilibrium.
- If the star formation time is comparable to the dynamical time, gas will turn into stars during its initial collapse, a situation which may lead to the formation of something resembling an elliptical galaxy. On the other hand, if the star formation time is much longer than the cooling and dynamical times, the gas will settle into a centrifugally supported disk before forming stars, thus producing a disk galaxy (see §1.4.5).
- If the relevant chemical evolution time is longer than the star-formation time, little metal enrichment will occur during star formation and all stars will end up with the same, initial metallicity. In the opposite case, the star-forming gas is continuously enriched, so that stars formed at different times will have different metallicities and abundance patterns (see §10.4).

So far we have avoided one obvious question, namely, what is the time scale for galaxy formation itself? Unfortunately, there is no single useful definition for such a time scale. Galaxy formation is a process, not an event, and as we have seen, this process is an amalgam of many different elements, each with its own time scale. If, for example, we are concerned with its stellar population, we might define the formation time of a galaxy as the epoch when a fixed fraction (e.g. 1% or 50%) of its stars had formed. If, on the other hand, we are concerned with its structure, we might want to define the galaxy's formation time as the epoch when a fixed fraction (e.g. 50% or 90%) of its mass was first assembled into a single object. These two 'formation' times can differ greatly for a given galaxy, and even their ordering can change from one galaxy to another. Thus it is important to be precise about definition when talking about the formation times of galaxies.

1.4 A Brief History of Galaxy Formation

The picture of galaxy formation sketched above is largely based on the hierarchical cold dark matter model for structure formation, which has been the standard paradigm since the beginning of the 1980s. In the following, we give an historical overview of the development of ideas and concepts about galaxy formation up to the present time. This is not intended as a complete historical account, but rather as a summary for young researchers of how our current ideas about galaxy formation were developed. Readers interested in a more extensive historical review can find some relevant material in the book *The Cosmic Century: A History of Astrophysics and Cosmology* by Malcolm Longair (2006).

1.4.1 Galaxies as Extragalactic Objects

By the end of the nineteenth century, astronomers had discovered a large number of astronomical objects that differ from stars in that they are fuzzy rather than point-like. These objects were collectively referred to as 'nebulae'. During the period 1771 to 1784 the French astronomer Charles Messier cataloged more than 100 of these objects in order to avoid confusing them with the comets he was searching for. Today the Messier numbers are still used to designate a number of bright galaxies. For example, the Andromeda Galaxy is also known as M31, because it is the 31st nebula in Messier's catalog. A more systematic search for nebulae was carried out by the Herschels, and in 1864 John Herschel published his *General Catalogue of Galaxies* which contains 5079 nebular objects. In 1888, Dreyer published an expanded version as his *New General Catalogue of Nebulae and Clusters of Stars*. Together with its two supplementary *Index Catalogues*, Dreyer's catalogue contained about 15,000 objects. Today, NGC and IC numbers are still widely used to refer to galaxies.

For many years after their discovery, the nature of the nebular objects was controversial. There were two competing ideas: one assumed that all nebulae are objects within our Milky Way, the other that some might be extragalactic objects, individual 'island universes' like the Milky Way. In 1920 the National Academy of Sciences in Washington invited two leading astronomers, Harlow Shapley and Heber Curtis, to debate this issue, an event which has passed into astronomical folklore as 'The Great Debate'. The controversy remained unresolved until 1925, when Edwin Hubble used distances estimated from Cepheid variables to demonstrate conclusively that some nebulae are extragalactic, individual galaxies comparable to our Milky Way in size and luminosity. Hubble's discovery marked the beginning of extragalactic astronomy. During the 1930s, high-quality photographic images of galaxies enabled him to classify galaxies into a broad sequence according to their morphology. Today Hubble's sequence is still widely adopted to classify galaxies.

Since Hubble's time, astronomers have made tremendous progress in systematically searching the skies for galaxies. At present deep CCD imaging and high-quality spectroscopy are available for about a million galaxies.

1.4.2 Cosmology

Only four years after his discovery that galaxies truly are extragalactic, Hubble made his second fundamental breakthrough: he showed that the recession velocities of galaxies are linearly related to their distances (Hubble, 1929; see also Hubble & Humason 1931), thus demonstrating that our Universe is expanding. This is undoubtedly the greatest single discovery in the history of cosmology. It revolutionized our picture of the Universe we live in.

The construction of mathematical models for the Universe actually started somewhat earlier. As soon as Albert Einstein completed his theory of general relativity in 1916, it was realized that this theory allowed, for the first time, the construction of self-consistent models for the Universe as a whole. Einstein himself was among the first to explore such solutions of his field equations. To his dismay, he found that all solutions require the Universe either to expand or to contract, in contrast with his belief at that time that the Universe should be static. In order to obtain a static solution, he introduced a cosmological constant into his field equations. This additional constant of gravity can oppose the standard gravitational attraction and so make possible a static (though unstable) solution. In 1922 Alexander Friedmann published two papers exploring both static and expanding solutions. These models are today known as Friedmann models, although this work drew little attention until Georges Lemaitre independently rediscovered the same solutions in 1927.

An expanding universe is a natural consequence of general relativity, so it is not surprising that Einstein considered his introduction of a cosmological constant as 'the biggest blunder of my life' once he learned of Hubble's discovery. History has many ironies, however. As we will see later, the cosmological constant is now back with us. In 1998 two teams independently used the distance–redshift relation of Type Ia supernovae to show that the expansion of the Universe is accelerating at the present time. Within general relativity this requires an additional mass/energy component with properties very similar to those of Einstein's cosmological constant. Rather than just counterbalancing the attractive effects of 'normal' gravity, the cosmological constant today overwhelms them to drive an ever more rapid expansion.

1.4 A Brief History of Galaxy Formation

Since the Universe is expanding, it must have been denser and perhaps also hotter at earlier times. In the late 1940s this prompted George Gamow to suggest that the chemical elements may have been created by thermonuclear reactions in the early Universe, a process known as primordial nucleosynthesis. Gamow's model was not considered a success, because it was unable to explain the existence of elements heavier than lithium due to the lack of stable elements with atomic mass numbers 5 and 8. We now know that this was not a failure at all; all heavier elements are a result of nucleosynthesis within stars, as first shown convincingly by Fred Hoyle and collaborators in the 1950s. For Gamow's model to be correct, the Universe would have to be hot as well as dense at early times, and Gamow realized that the residual heat should still be visible in today's Universe as a background of thermal radiation with a temperature of a few degrees kelvin, thus with a peak at microwave wavelengths. This was a remarkable prediction of the cosmic microwave background radiation (CMB), which was finally discovered in 1965. The thermal history suggested by Gamow, in which the Universe expands from a dense and hot initial state, was derisively referred to as the Hot Big Bang by Fred Hoyle, who preferred an unchanging steady state cosmology. Hoyle's cosmological theory was wrong, but his name for the correct model has stuck.

The Hot Big Bang model developed gradually during the 1950s and 1960s. By 1964, it had been noticed that the abundance of helium by mass is everywhere about one third that of hydrogen, a result which is difficult to explain by nucleosynthesis in stars. In 1964, Hoyle and Tayler published calculations that demonstrated how the observed helium abundance could emerge from the Hot Big Bang. Three years later, Wagoner et al. (1967) made detailed calculations of a complete network of nuclear reactions, confirming the earlier result and suggesting that the abundances of other light isotopes, such as helium-3, deuterium and lithium, could also be explained by primordial nucleosynthesis. This success provided strong support for the Hot Big Bang. The 1965 discovery of the cosmic microwave background showed it to be isotropic and to have a temperature (2.7K) exactly in the range expected in the Hot Big Bang model (Penzias & Wilson, 1965; Dicke et al., 1965). This firmly established the Hot Big Bang as the standard model of cosmology, a status which it has kept up to the present day. Although there have been changes over the years, these have affected only the exact matter/energy content of the model and the exact values of its characteristic parameters.

Despite its success, during the 1960s and 1970s it was realized that the standard cosmology had several serious shortcomings. Its structure implies that the different parts of the Universe we see today were never in causal contact at early times (e.g. Misner, 1968). How then can these regions have contrived to be so similar, as required by the isotropy of the CMB? A second shortcoming is connected with the spatial flatness of the Universe (e.g. Dicke & Peebles, 1979). It was known by the 1960s that the matter density in the Universe is not very different from the critical density for closure, i.e. the density for which the spatial geometry of the Universe is flat. However, in the standard model any tiny deviation from flatness in the early Universe is amplified enormously by later evolution. Thus, extreme fine tuning of the initial curvature is required to explain why so little curvature is observed today. A closely related formulation is to ask how our Universe has managed to survive and to evolve for billions of years, when the time scales of all physical processes in its earliest phases were measured in tiny fractions of a nanosecond. The standard cosmology provides no explanations for these puzzles.

A conceptual breakthrough came in 1981 when Alan Guth proposed that the Universe may have gone through an early period of exponential expansion (inflation) driven by the vacuum energy of some quantum field. His original model had some problems and was revised in 1982 by Linde and by Albrecht & Steinhardt. In this scenario, the different parts of the Universe we see today were indeed in causal contact *before* inflation took place, thereby allowing physical processes to establish homogeneity and isotropy. Inflation also solves the flatness/time-scale problem, because the Universe expanded so much during inflation that its curvature radius grew to be much larger than the presently observable Universe. Thus, a generic prediction of the inflation scenario is that today's Universe should appear flat.

1.4.3 Structure Formation

(a) Gravitational Instability In the standard model of cosmology, structures form from small initial perturbations in an otherwise homogeneous and isotropic universe. The idea that structures can form via gravitational instability in this way originates from Jeans (1902), who showed that the stability of a perturbation depends on the competition between gravity and pressure. Density perturbations grow only if they are larger (heavier) than a characteristic length (mass) scale [now referred to as the Jeans' length (mass)] beyond which gravity is able to overcome the pressure gradients. The application of this Jeans criterion to an expanding background was worked out by, among others, Gamow & Teller (1939) and Lifshitz (1946), with the result that perturbation growth is power-law in time, rather than exponential as for a static background.

(b) Initial Perturbations Most of the early models of structure formation assumed the Universe to contain two energy components, ordinary baryonic matter and radiation (CMB photons and relativistic neutrinos). In the absence of any theory for the origin of perturbations, two distinct models were considered, usually referred to as adiabatic and isothermal initial conditions. In adiabatic initial conditions all matter and radiation fields are perturbed in the same way, so that the total density (or local curvature) varies, but the ratio of photons to baryons, for example, is spatially invariant. Isothermal initial conditions, on the other hand, correspond to initial perturbations in the ratio of components, but with no associated spatial variation in the total density or curvature.[2]

In the adiabatic case, the perturbations can be considered as applying to a single fluid with a constant specific entropy as long as the radiation and matter remain tightly coupled. At such times, the Jeans' mass is very large and small-scale perturbations execute acoustic oscillations driven by the pressure gradients associated with the density fluctuations. Silk (1968) showed that towards the end of recombination, as radiation decouples from matter, small-scale oscillations are damped by photon diffusion, a process now called Silk damping. Depending on the matter density and the expansion rate of the Universe, the characteristic scale of Silk damping falls in the range of 10^{12}–$10^{14} M_\odot$. After radiation/matter decoupling the Jeans' mass drops precipitously to $\simeq 10^6 M_\odot$ and perturbations above this mass scale can start to grow,[3] but there are no perturbations left on the scale of galaxies at this time. Consequently, galaxies must form 'top-down', via the collapse and fragmentation of perturbations larger than the damping scale, an idea championed by Zel'dovich and colleagues.

In the case of isothermal initial conditions, the spatial variation in the ratio of baryons to photons remains fixed before recombination because of the tight coupling between the two fluids. The pressure is spatially uniform, so that there is no acoustic oscillation, and perturbations are not influenced by Silk damping. If the initial perturbations include small-scale structure, this survives until after the recombination epoch, when baryon fluctuations are no longer supported by photon pressure and so can collapse. Structure can then form 'bottom-up' through hierarchical clustering. This scenario of structure formation was originally proposed by Peebles (1965).

By the beginning of the 1970s, the linear evolution of both adiabatic and isothermal perturbations had been worked out in great detail (e.g. Lifshitz, 1946; Silk, 1968; Peebles & Yu, 1970; Sato, 1971; S. Weinberg, 1971). At that time, it was generally accepted that observed structures must have formed from finite amplitude perturbations which were somehow part of the

---

[2] Note that the nomenclature 'isothermal', which is largely historical, is somewhat confusing; the term 'isocurvature' would be more appropriate.

[3] Actually, as we will see in Chapter 4, depending on the gauge adopted, perturbations can also grow before they enter the horizon.

initial conditions set up at the Big Bang. Harrison (1970) and Zel'dovich (1972) independently argued that only one scaling of the amplitude of initial fluctuations with their wavelength could be consistent with the formation of galaxies from fluctuations imposed at very early times. Their suggestion, now known as the Harrison–Zel'dovich initial fluctuation spectrum, has the property that structure on every scale has the same dimensionless amplitude, corresponding to fluctuations in the equivalent Newtonian gravitational potential, $\delta\Phi/c^2 \sim 10^{-4}$.

In the early 1980s, immediately after the inflationary scenario was proposed, a number of authors realized almost simultaneously that quantum fluctuations of the scalar field (called the inflaton) that drives inflation can generate density perturbations with a spectrum that is close to the Harrison–Zel'dovich form (Hawking, 1982; Guth & Pi, 1982; Starobinsky, 1982; Bardeen et al., 1983). In the simplest models, inflation also predicts that the perturbations are adiabatic and that the initial density field is Gaussian. When parameters take their natural values, however, these models generically predict fluctuation amplitudes that are much too large, of order unity. This apparent fine-tuning problem is still unresolved.

In 1992 anisotropy in the cosmic microwave background was detected convincingly for the first time by the Cosmic Background Explorer (COBE) (Smoot et al., 1992). These anisotropies provide an image of the structure present at the time of radiation/matter decoupling, $\sim$400,000 years after the Big Bang. The resolved structures are all of very low amplitude and so can be used to probe the properties of the initial density perturbations. In agreement with the inflationary paradigm, the COBE maps were consistent with Gaussian initial perturbations with the Harrison–Zel'dovich spectrum. The fluctuation amplitudes are comparable to those inferred by Harrison and Zel'dovich. The COBE results have since been confirmed and dramatically refined by subsequent observations, most notably by the Wilkinson Microwave Anisotropy Probe (WMAP) (Bennett et al., 2003; Hinshaw et al., 2007). The agreement with simple inflationary predictions remains excellent.

(c) Nonlinear Evolution In order to connect the initial perturbations to the nonlinear structures we see today, one has to understand the outcome of nonlinear evolution. In 1970 Zel'dovich published an analytical approximation (now referred to as the Zel'dovich approximation) which describes the initial nonlinear collapse of a coherent perturbation of the cosmic density field. This model shows that the collapse generically occurs first along one direction, producing a sheet-like structure, often referred to as a 'pancake'. Zel'dovich imagined further evolution to take place via fragmentation of such pancakes. At about the same time, Gunn & Gott (1972) developed a simple spherically symmetric model to describe the growth, turn-around (from the general expansion), collapse and virialization of a perturbation. In particular, they showed that dissipationless collapse results in a quasi-equilibrium system with a characteristic radius that is about half the radius at turn-around. Although the nonlinear collapse described by the Zel'dovich approximation is more realistic, since it does not assume any symmetry, the spherical collapse model of Gunn & Gott has the virtue that it links the initial perturbation directly to the final quasi-equilibrium state. By applying this model to a Gaussian initial density field, Press & Schechter (1974) developed a very useful formalism (now referred to as Press–Schechter theory) that allows one to estimate the mass function of collapsed objects (i.e. their abundance as a function of mass) produced by hierarchical clustering.

Hoyle (1949) was the first to suggest that perturbations (and the associated protogalaxies) might gain angular momentum through the tidal torques from their neighbors. A linear perturbation analysis of this process was first carried out correctly and in full generality by Doroshkevich (1970), and was later tested with the help of numerical simulations (Peebles, 1971; Efstathiou & Jones, 1979). The study of Efstathiou and Jones showed that clumps formed through gravitational collapse in a cosmological context typically acquire about 15% of the angular momentum needed for full rotational support. Better simulations in more recent years have shown that the correct value is closer to 10%. In the case of 'top-down' models, it was suggested that objects could acquire angular momentum not only through gravitational torques as pancakes fragment, but also via oblique shocks generated by their collapse (Doroshkevich, 1973).

1.4.4 The Emergence of the Cold Dark Matter Paradigm

The first evidence that the Universe may contain dark matter (undetected through electromagnetic emission or absorption) can be traced back to 1933, when Zwicky studied the velocities of galaxies in the Coma Cluster and concluded that the total mass required to hold the Cluster together is about 400 times larger than the luminous mass in stars. In 1937 he reinforced this analysis and noted that galaxies associated with such large amounts of mass should be detectable as gravitational lenses producing multiple images of background galaxies. These conclusions were substantially correct, but remarkably it took more than 40 years for the existence of dark matter to be generally accepted. The tide turned in the mid-1970s with papers by Ostriker et al. (1974) and Einasto et al. (1974) extending Zwicky's analysis and noting that massive halos are required around our Milky Way and other nearby galaxies in order to explain the motions of their satellites. These arguments were supported by continually improving 21 cm and optical measurements of spiral galaxy rotation curves which showed no sign of the fall-off at large radius expected if the visible stars and gas were the only mass in the system (Roberts & Rots, 1973; Rubin et al., 1978, 1980). During the same period, numerous suggestions were made regarding the possible nature of this dark matter component, ranging from baryonic objects such as brown dwarfs, white dwarfs and black holes (e.g. White & Rees, 1978; Carr et al., 1984), to more exotic, elementary particles such as massive neutrinos (Gershtein & Zel'dovich, 1966; Cowsik & McClelland, 1972).

The suggestion that neutrinos might be the unseen mass was partly motivated by particle physics. In the 1960s and 1970s, it was noticed that grand unified theories (GUTs) permit the existence of massive neutrinos, and various attempts to measure neutrino masses in laboratory experiments were initiated. In the late 1970s, Lyubimov et al. (1980) and Reines et al. (1980) announced the detection of a mass for the electron neutrino at a level of cosmological interest (about 30 eV). Although the results were not conclusive, they caused a surge in studies investigating neutrinos as dark matter candidates (e.g. Bond et al., 1980; Sato & Takahara, 1980; Schramm & Steigman, 1981; Klinkhamer & Norman, 1981), and structure formation in a neutrino-dominated universe was soon worked out in detail. Since neutrinos decouple from other matter and radiation fields while still relativistic, their abundance is very similar to that of CMB photons. Thus, they must have become non-relativistic at the time the Universe became matter-dominated, implying thermal motions sufficient to smooth out all structure on scales smaller than a few tens of Mpc. The first nonlinear structures are then Zel'dovich pancakes of this scale, which must fragment to make smaller structures such as galaxies. Such a picture conflicts directly with observation, however. An argument by Tremaine & Gunn (1979), based on the Pauli exclusion principle, showed that individual galaxy halos could not be made of neutrinos with masses as small as 30 eV, and simulations of structure formation in neutrino-dominated universes by White et al. (1984) demonstrated that they could not produce galaxies without at the same time producing much stronger galaxy clustering than is observed. Together with the failure to confirm the claimed neutrino mass measurements, these problems caused a precipitous decline in interest in neutrino dark matter by the end of the 1980s.

In the early 1980s, alternative models were suggested, in which dark matter is a different kind of weakly interacting massive particle. There were several motivations for this. The amount of baryonic matter allowed by cosmic nucleosynthesis calculations is far too little to provide the flat universe preferred by inflationary models, suggesting that non-baryonic dark matter may be present. In addition, strengthening upper limits on temperature anisotropies in the CMB made it increasingly difficult to construct self-consistent, purely baryonic models for structure formation; there is simply not enough time between the recombination epoch and the present day to grow the structures we see in the nearby Universe from those present in the high-redshift photon–baryon fluid. Finally, by the early 1980s, particle physics models based on the idea of supersymmetry had provided a plethora of dark matter candidates, such as neutralinos, photinos and gravitinos, that could dominate the mass density of the Universe. Because of their much larger mass, such particles would initially have much smaller velocities than a 30 eV neutrino, and so they were generically referred to as warm or cold dark matter (WDM or CDM, the former corresponding to a particle mass of order 1 keV, the latter to much more massive particles) in contrast to neutrino-like hot dark matter (HDM). The shortcomings of HDM motivated consideration of a variety of such scenarios (e.g. Peebles, 1982; Blumenthal et al., 1982; Bond et al., 1982; Bond & Szalay, 1983).

Lower thermal velocities result in the survival of fluctuations of galactic scale (for WDM and CDM) or below (for CDM). The particles decouple from the radiation field long before recombination, so perturbations in their density can grow at early times to be substantially larger than the fluctuations visible in the CMB. After the baryons decouple from the radiation, they quickly fall in these dark matter potential wells, causing structure formation to occur sufficiently fast to be consistent with observed structure in today's Universe. M. Davis et al. (1985) used simulations of the CDM model to show that it could provide a good match to the observed clustering of galaxies provided either the mass density of dark matter is well below the critical value, or (their preferred model) that galaxies are biased tracers of the CDM density field, as expected if they form at the centers of the deepest dark matter potential wells (e.g. Kaiser, 1984). By the mid-1980s, the 'standard' CDM model, in which dark matter provides the critical density, Hubble's constant has a value $\sim 50 \,\mathrm{km\,s^{-1}\,Mpc^{-1}}$, and the initial density field was Gaussian with a Harrison–Zel'dovich spectrum, had established itself as the 'best bet' model for structure formation.

In the early 1990s, measurements of galaxy clustering, notably from the APM galaxy survey (Maddox et al., 1990a; Efstathiou et al., 1990), showed that the standard CDM model predicts less clustering on large scales than is observed. Several alternatives were proposed to remedy this. One was a mixed dark matter (MDM) model, in which the universe is flat, with $\sim 30\%$ of the cosmic mass density in HDM and $\sim 70\%$ in CDM and baryons. Another flat model assumed all dark matter to be CDM, but adopted an enhanced radiation background in relativistic neutrinos (τCDM). A third possibility was an open model, in which today's Universe is dominated by CDM and baryons, but has only about 30% of the critical density (OCDM). A final model assumed the same amounts of CDM and baryons as OCDM but added a cosmological constant in order to make the universe flat (ΛCDM).

Although all these models match observed galaxy clustering on large scales, it was soon realized that galaxy formation occurs too late in the MDM and τCDM models, and that the open model has problems in matching the perturbation amplitudes measured by COBE. ΛCDM then became the default 'concordance' model, although it was not generally accepted until Garnavich et al. (1998) and Perlmutter et al. (1999) used the distance–redshift relation of Type Ia supernovae to show that the cosmic expansion is accelerating, and measurements of small-scale CMB fluctuations showed that our Universe is flat (de Bernardis et al., 2000). It seems that the present-day Universe is dominated by a dark energy component with properties very similar to those of Einstein's cosmological constant.

At the beginning of this century, a number of ground-based and balloon-borne experiments measured CMB anisotropies, notably Boomerang (de Bernardis et al., 2000), MAXIMA (Hanany et al., 2000), DASI (Halverson et al., 2002) and CBI (Sievers et al., 2003). They successfully detected features, known as acoustic peaks, in the CMB power spectrum, and showed their wavelengths and amplitudes to be in perfect agreement with expectations for a ΛCDM cosmology. In 2003, the first year data from WMAP not only confirmed these results, but also allowed much more precise determinations of cosmological parameters. The values obtained were in remarkably good agreement with independent measurements; the baryon density matched that estimated from cosmic nucleosynthesis, the Hubble constant matched that found by direct measurement, the dark-energy density matched that inferred from Type Ia supernovae, and the implied large-scale clustering in today's Universe matched that measured using large galaxy surveys and weak gravitational lensing (see Spergel et al., 2003, and references therein). Consequently, the ΛCDM model has now established itself firmly as the standard paradigm for structure formation. With further data from WMAP and from other sources, the parameters of this new paradigm are now well constrained (Spergel et al., 2007; Komatsu et al., 2009).

1.4.5 Galaxy Formation

(a) Monolithic Collapse and Merging Although it was well established in the 1930s that there are two basic types of galaxies, ellipticals and spirals, it would take some 30 years before detailed models for their formation were proposed. In 1962, Eggen, Lynden-Bell & Sandage considered a model in which galaxies form from the collapse of gas clouds, and suggested that the difference between ellipticals and spirals reflects the rapidity of star formation during the collapse. If most of the gas turns into stars as it falls in, the collapse is effectively dissipationless and infall motions are converted into the random motion of stars, resulting in a system which might resemble an elliptical galaxy. If, on the other hand, the cloud remains gaseous during collapse, the gravitational energy can be effectively dissipated via shocks and radiative cooling. In this case, the cloud will shrink until it is supported by angular momentum, leading to the formation of a rotationally supported disk. Gott & Thuan (1976) took this picture one step further and suggested that the amount of dissipation during collapse depends on the amplitude of the initial perturbation. Based on the empirical fact that star-formation efficiency appears to scale as ρ^2 (Schmidt, 1959), they argued that protogalaxies associated with the highest initial density perturbations would complete star formation more rapidly as they collapse, and so might produce an elliptical. On the other hand, protogalaxies associated with lower initial density perturbations would form stars more slowly and so might make spirals.

Larson (1974a,b, 1975, 1976) carried out the first numerical simulations of galaxy formation, showing how these ideas might work in detail. Starting from near-spherical rotating gas clouds, he found that it is indeed the ratio of the star-formation time to the dissipation/cooling time which determines whether the system turns into an elliptical or a spiral. He also noted the importance of feedback effects during galaxy formation, arguing that in low-mass galaxies, supernovae would drive winds that could remove most of the gas and heavy elements from a system before they could turn into stars. He argued that this mechanism might explain the low surface brightnesses and low metallicities of dwarf galaxies. However, he was unable to obtain the high observed surface brightnesses of bright elliptical galaxies without requiring his gas clouds to be much more slowly rotating than predicted by the tidal torque theory; otherwise they would spin up and make a disk long before they became as compact as the observed galaxies. The absence of highly flattened ellipticals and the fact that many bright ellipticals show little or no rotation (Bertola & Capaccioli, 1975; Illingworth, 1977) therefore posed a serious problem for this scenario. As we now know, its main defect was that it left out the effects of the dark matter.

In a famous 1972 paper, Toomre & Toomre used simple numerical simulations to demonstrate convincingly that some of the extraordinary structures seen in peculiar galaxies, such as long tails, could be produced by tidal interactions between two normal spirals. Based on the observed frequency of galaxies with such signatures of interactions, and on their estimate of the time scale over which tidal tails might be visible, Toomre & Toomre (1972) argued that most elliptical galaxies could be merger remnants. In an extreme version of this picture, all galaxies initially form as disks, while all ellipticals are produced by mergers between pre-existing galaxies. A virtue of this idea was that almost all known star formation occurs in disk gas. Early simulations showed that the merging of two spheroids produces remnants with density profiles that agree with observed ellipticals (e.g. White, 1978). The more relevant (but also the more difficult) simulations of mergers between disk galaxies were not carried out until the early 1980s (Gerhard, 1981; Farouki & Shapiro, 1982; Negroponte & White, 1983; Barnes, 1988). These again showed merger remnants to have properties similar to those of observed ellipticals.

Although the merging scenario fits nicely into a hierarchical formation scheme, where larger structures grow by mergers of smaller ones, the extreme picture outlined above has some problems. Ostriker (1980) pointed out that observed giant ellipticals, which are dense and can have velocity dispersions as high as $\sim 300\,\mathrm{km\,s^{-1}}$, could not be formed by mergers of present-day spirals, which are more diffuse and almost never have rotation velocities higher than $300\,\mathrm{km\,s^{-1}}$. As we will see below, this problem may be resolved by considering the dark halos of the galaxies, and by recognizing that the high-redshift progenitors of ellipticals were more compact than present-day spirals. The merging scenario remains a popular scenario for the formation of (bright) elliptical galaxies.

(b) The Role of Radiative Cooling An important question for galaxy formation theory is why galaxies with stellar masses larger than $\sim 10^{12}\,\mathrm{M_\odot}$ are absent or extremely rare. In the adiabatic model, this mass scale is close to the Silk damping scale and could plausibly set a *lower* limit to galaxy masses. However, in the presence of dark matter Silk damping leaves no imprint on the properties of galaxies, simply because the dark matter perturbations are not damped. Press & Schechter (1974) showed that there is a characteristic mass also in the hierarchical model, corresponding to the mass scale of the typical nonlinear object at the present time. However, this mass scale is relatively large, and many objects with mass above $10^{12}\,\mathrm{M_\odot}$ are predicted, and indeed are observed as virialized groups and clusters of galaxies. Apparently, the mass scale of galaxies is not set by gravitational physics alone.

In the late 1970s, Silk (1977), Rees & Ostriker (1977) and Binney (1977) suggested that radiative cooling might play an important role in limiting the mass of galaxies. They argued that galaxies can form effectively only in systems where the cooling time is comparable to or shorter than the collapse time, which leads to a characteristic scale of $\sim 10^{12}\,\mathrm{M_\odot}$, similar to the mass scale of massive galaxies. They did not explain why a typical galaxy should form with a mass near this limit, nor did they explicitly consider the effects of dark matter. Although radiative cooling plays an important role in all current galaxy formation theories, it is still unclear if it alone can explain the characteristic mass scale of galaxies, or whether various feedback processes must also be invoked.

(c) Galaxy Formation in Dark Matter Halos By the end of the 1970s, several lines of argument had led to the conclusion that dark matter must play an important role in galaxy formation. In particular, observations of rotation curves of spiral galaxies indicated that these galaxies are embedded in dark halos which are much more extended than the galaxies themselves. This motivated White & Rees (1978) to propose a two-stage theory for galaxy formation: dark halos form first through hierarchical clustering; the luminous content of galaxies then results from cooling and condensation of gas within the potential wells provided by these dark halos. The mass function of galaxies was calculated by applying these ideas within the Press & Schechter model for the growth of nonlinear structure. The model of White and Rees contains many of the basic ideas of the modern theory of galaxy formation. They noticed that feedback is required to explain the low overall efficiency of galaxy formation, and invoked Larson's (1974a) model for supernova feedback in dwarf galaxies to explain this. They also noted, but did not emphasize, that even with strong feedback, their hierarchical model predicts a galaxy luminosity function with far too many faint galaxies. This problem is alleviated but not solved by adopting CDM initial conditions rather than the simple power-law initial conditions they adopted. In 1980, Fall & Efstathiou developed a model of disk formation in dark matter halos, incorporating the angular momentum expected from tidal torques, and showed that many properties of observed disk galaxies can be understood in this way.

Many of the basic elements of galaxy formation in the CDM scenario were already in place in the early 1980s, and were summarized nicely by Efstathiou & Silk (1983) and in Blumenthal et al. (1984). Blumenthal et al. invoked the idea of biased galaxy formation, suggesting that disk galaxies may be associated with density peaks of typical heights in the CDM density field, while giant ellipticals may be associated with higher density peaks. Efstathiou & Silk (1983) discussed in some detail how the two-stage theory of White & Rees (1978) can solve some of the problems in earlier models based on the collapse of gas clouds. In particular, they argued that, within an extended halo, cooled gas can settle into a rotation-supported disk of the observed scale in a fraction of the Hubble time, whereas without a dark matter halo it would take too long for a perturbation to turn around and shrink to form a disk (see Chapter 11 for details). They also argued that extended dark matter halos around galaxies make mergers of galaxies more likely, a precondition for Toomre & Toomre's (1972) merger scenario of elliptical galaxy formation to be viable.

Since the early 1990s many studies have investigated the properties of CDM halos using both analytical and N-body methods. Properties studied include the progenitor mass distributions (Bond et al., 1991), merger histories (Lacey & Cole, 1993), spatial clustering (Mo & White, 1996), density profiles (Navarro et al., 1997), halo shapes (e.g. Jing & Suto, 2002), substructure (e.g. Moore et al., 1998a; Klypin et al., 1999), and angular-momentum distributions (e.g. Warren et al., 1992; Bullock et al., 2001a). These results have paved the way for more detailed models for galaxy formation within the CDM paradigm. In particular, two complementary approaches have been developed: semi-analytical models and hydrodynamical simulations. The semi-analytical approach, originally developed by White & Frenk (1991) and subsequently refined in a number of studies (e.g. Kauffmann et al., 1993; Cole et al., 1994; Dalcanton et al., 1997; Mo et al., 1998; Somerville & Primack, 1999), uses knowledge about the structure and assembly history of CDM halos to model the gravitational potential wells within which galaxies form and evolve, treating all the relevant physical processes (cooling, star formation, feedback, dynamical friction, etc.) in a semi-analytical fashion. The first three-dimensional, hydrodynamical simulations of galaxy formation including dark matter were carried out by Katz in the beginning of the 1990s (Katz & Gunn, 1991; Katz, 1992) and focused on the collapse of a homogeneous, uniformly rotating sphere. The first simulation of galaxy formation by hierarchical clustering from proper cosmological initial conditions was that of Navarro & Benz (1991), while the first simulation of galaxy formation from CDM initial conditions was that of Navarro & White (1994). Since then, numerical simulations of galaxy formation with increasing numerical resolution have been carried out by many authors.

It is clear that the CDM scenario has become the preferred scenario for galaxy formation, and we have made a great deal of progress in our quest towards understanding the structure and formation of galaxies within it. However, as we will see later in this book, there are still many important unsolved problems. It is precisely the existence of these outstanding problems that makes galaxy formation such an interesting subject. It is our hope that this book will help you to equip yourself for your own explorations in this area.

2
Observational Facts

Observational astronomy has developed at an extremely rapid pace. Until the end of the 1940s observational astronomy was limited to optical wavebands. Today we can observe the Universe at virtually all wavelengths covering the electromagnetic spectrum, either from the ground or from space. Together with the revolutionary growth in computer technology and with a dramatic increase in the number of professional astronomers, this has led to a flood of new data. Clearly it is impossible to provide a complete overview of all this information in a single chapter (or even in a single book). Here we focus on a number of selected topics relevant to our forthcoming discussion, and limit ourselves to a simple description of some of the available data. Discussion regarding the interpretation and/or implication of the data is postponed to Chapters 11–16, where we use the physical ingredients described in Chapters 3–10 to interpret the observational results presented here. After a brief introduction of observational techniques, we present an overview of some of the observational properties of stars, galaxies, clusters and groups, large scale structure, the intergalactic medium, and the cosmic microwave background. We end with a brief discussion of cosmological parameters and the matter/energy content of the Universe.

2.1 Astronomical Observations

Almost all information we can obtain about an astronomical object is derived from the radiation we receive from it, or by the absorption it causes in the light of a background object. The radiation from a source may be characterized by its spectral energy distribution (SED), $f_\lambda \, d\lambda$, which is the total energy of emitted photons with wavelengths in the range λ to $\lambda + d\lambda$. Technology is now available to detect electromagnetic radiation over an enormous energy range, from low frequency radio waves to high energy gamma rays. However, from the Earth's surface our ability to detect celestial objects is seriously limited by the transparency of our atmosphere. Fig. 2.1 shows the optical depth for photon transmission through the Earth's atmosphere as a function of photon wavelength, along with the wavelength ranges of some commonly used wavebands. Only a few relatively clear windows exist in the optical, near-infrared and radio bands. In other parts of the spectrum, in particular the far-infrared, ultraviolet, X-ray and gamma-ray regions, observations can only be carried out by satellites or balloon-borne detectors.

Although only a very restricted range of frequencies penetrate our atmosphere, celestial objects actually emit over the full range accessible to our instruments. This is illustrated in Fig. 2.2, a schematic representation of the average brightness of the sky as a function of wavelength as seen from a vantage point well outside our own galaxy. With the very important exception of the cosmic microwave background (CMB), which dominates the overall photon energy content of the Universe, the dominant sources of radiation at all energies below the hard gamma-ray regime are related to galaxies, their evolution, their clustering and their nuclei. At radio, far-UV, X-ray and soft gamma-ray wavelengths the emission comes primarily from active galactic

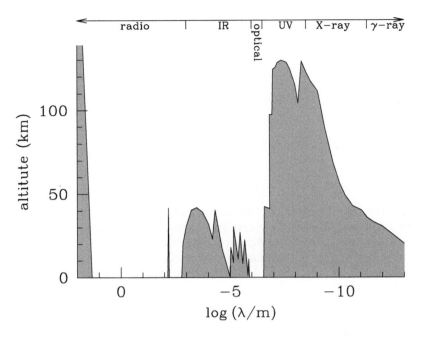

Fig. 2.1. The altitude above sea level at which a typical photon is absorbed as a function of the photon's wavelength. Only radio waves, optical light, the hardest γ-ray, and infrared radiation in a few wavelength windows can penetrate the atmosphere to reach sea level. Observations at all other wavebands have to be carried out above the atmosphere.

nuclei. Galactic starlight dominates in the near-UV, optical and near-infrared, while dust emission from star-forming galaxies is responsible for most of the far-infrared emission. The hot gas in galaxy clusters emits a significant but non-dominant fraction of the total X-ray background and is the only major source of emission from scales larger than an individual galaxy. Such large structures can, however, be seen in absorption, for example in the light of distant quasars.

2.1.1 Fluxes and Magnitudes

The image of an astronomical object reflects its surface brightness distribution. The surface brightness is defined as the photon energy received by a unit area at the observer per unit time from a unit solid angle in a specific direction. Thus if we denote the surface brightness by I, its units are $[I] = \mathrm{erg\,s^{-1}\,cm^{-2}\,sr^{-1}}$. If we integrate the surface brightness over the entire image, we obtain the flux of the object, f, which has units $[f] = \mathrm{erg\,s^{-1}\,cm^{-2}}$. Integrating the flux over a sphere centered on the object and with radius equal to the distance r from the object to the observer, we obtain the bolometric luminosity of the object:

$$L = 4\pi r^2 f, \quad (2.1)$$

with $[L] = \mathrm{erg\,s^{-1}}$. For the Sun, $L = 3.846 \times 10^{33}\,\mathrm{erg\,s^{-1}}$.

The image size of an extended astronomical object is usually defined on the basis of its isophotal contours (curves of constant surface brightness), and the characteristic radius of an isophotal contour at some chosen surface brightness level is usually referred to as an isophotal radius of the object. A well-known example is the Holmberg radius defined as the length of the semimajor axis of the isophote corresponding to a surface brightness of 26.5 mag arcsec^{-2} in the B-band. Two other commonly used size measures in optical astronomy are the core radius, defined as the

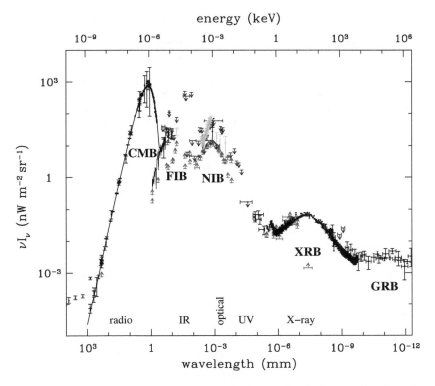

Fig. 2.2. The energy density spectrum of cosmological background radiation as a function of wavelength. The value of νI_ν measures the radiation power per decade of wavelength. This makes it clear that the cosmic microwave background (CMB) contributes most to the overall background radiation, followed by the far- (FIB) and near-infrared (NIB) backgrounds, the X-ray background (XRB) and the gamma-ray background (GRB). [Courtesy of D. Scott; see Scott (2000)]

radius where the surface brightness is half of the central surface brightness, and the half-light radius (also called the effective radius), defined as the characteristic radius that encloses half of the total observed flux. For an object at a distance r, its physical size, D, is related to its angular size, θ, by

$$D = r\theta. \qquad (2.2)$$

Note, though, that relations (2.1) and (2.2) are only valid for relatively small distances. As we will see in Chapter 3, for objects at cosmological distances, r in Eqs. (2.1) and (2.2) has to be replaced by the luminosity distance and angular diameter distance, respectively.

(a) Wavebands and Bandwidths Photometric observations are generally carried out in some chosen waveband. Thus, the observed flux from an object is related to its SED, f_λ, by

$$f_X = \int f_\lambda F_X(\lambda) R(\lambda) T(\lambda) \, d\lambda. \qquad (2.3)$$

Here $F_X(\lambda)$ is the transmission of the filter that defines the waveband (denoted by X), $T(\lambda)$ represents the atmospheric transmission, and $R(\lambda)$ represents the efficiency with which the telescope plus instrument detects photons. In the following we will assume that f_X has been corrected for atmospheric absorption and telescope efficiency (the correction is normally done by calibrating the data using standard objects with known f_λ). In this case, the observed flux depends only on the spectral energy distribution and the chosen filter. Astronomers have constructed a variety of

Table 2.1. Filter characteristics of the UBVRI photometric system.

Band:	U	B	V	R	I	J	H	K	L	M
λ_{eff} (nm):	365	445	551	658	806	1220	1630	2190	3450	4750
FWHM (nm):	66	94	88	138	149	213	307	390	472	460
$\mathcal{M}_\odot$:		5.61	5.48	4.83	4.42	4.08	3.64	3.32	3.28	3.25	–
$L_\odot (10^{32}\,\text{erg/s})$:	1.86	4.67	4.64	6.94	4.71	2.49	1.81	0.82	0.17	–

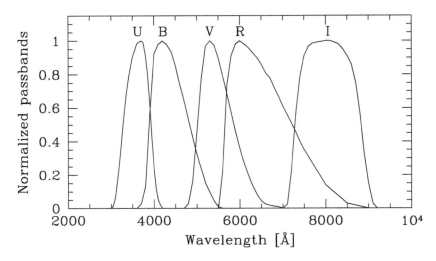

Fig. 2.3. The transmission characteristics of Johnson *UBV* and Kron Cousins *RI* filter systems. [Based on data published in Bessell (1990)]

photometric systems. A well-known example is the standard *UBV* system originally introduced by Johnson. The filter functions for this system are shown in Fig. 2.3. In general, a filter function can be characterized by an effective wavelength, λ_{eff}, and a characteristic bandwidth, usually quoted as a full width at half maximum (FWHM). The FWHM is defined as $|\lambda_1 - \lambda_2|$, with $F_X(\lambda_1) = F_X(\lambda_2) = $ half the peak value of $F_X(\lambda)$. Table 2.1 lists λ_{eff} and the FWHM for the filters of the standard *UBVRI* photometric system. In this system, the FWHM are all of order 10% or larger of the corresponding λ_{eff}. Such 'broad-band photometry' can be used to characterize the overall shape of the spectral energy distribution of an object with high efficiency. Alternatively, one can use 'narrow-band photometry' with much narrower filters to image objects in a particular emission line or to study its detailed SED properties.

(b) Magnitude and Color For historical reasons, the flux of an astronomical object in the optical band (and also in the near-infrared and near-ultraviolet bands) is usually quoted in terms of apparent magnitude:

$$m_X = -2.5 \log(f_X/f_{X,0}), \tag{2.4}$$

where the flux zero-point $f_{X,0}$ has traditionally been taken as the flux in the X band of the bright star Vega. In recent years it has become more common to use 'AB-magnitudes', for which

$$f_{X,0} = 3.6308 \times 10^{-20}\,\text{erg s}^{-1}\text{cm}^{-2}\text{Hz}^{-1} \int F_X(c/\nu) d\nu. \tag{2.5}$$

Here ν is the frequency and c is the speed of light. Similarly, the luminosities of objects (in waveband X) are often quoted as an absolute magnitude: $\mathcal{M}_X = -2.5 \log(L_X) + C_X$, where C_X

is a zero-point. It is usually convenient to write L_X in units of the solar luminosity in the same band, $L_{\odot X}$. The values of $L_{\odot X}$ in the standard *UBVRI* photometric system are listed in Table 2.1. It then follows that

$$\mathcal{M}_X = -2.5 \log \left(\frac{L_X}{L_{\odot X}} \right) + \mathcal{M}_{\odot X}, \tag{2.6}$$

where $\mathcal{M}_{\odot X}$ is the absolute magnitude of the Sun in the waveband in consideration. Using Eq. (2.1), we have

$$m_X - \mathcal{M}_X = 5 \log(r/r_0), \tag{2.7}$$

where r_0 is a fiducial distance at which m_X and $\mathcal{M}_X$ are defined to have the same value. Conventionally, r_0 is chosen to be 10 pc (1 pc = 1 parsec = 3.0856×10^{18} cm; see §2.1.3 for a definition). According to this convention, the Vega absolute magnitudes of the Sun in the *UBVRI* photometric system have the values listed in Table 2.1.

The quantity $(m_X - \mathcal{M}_X)$ for an astronomical object is called its distance modulus. If we know both m_X and $\mathcal{M}_X$ for an object, then Eq. (2.7) can be used to obtain its distance. Conversely, if we know the distance to an object, a measurement of its apparent magnitude (or flux) can be used to obtain its absolute magnitude (or luminosity).

Optical astronomers usually express surface brightness in terms of magnitudes per square arcsecond. In such 'units', the surface brightness in a band X is denoted by μ_X, and is related to the surface brightness in physical units, I_X, according to

$$\mu_X = -2.5 \log \left(\frac{I_X}{L_\odot \, \mathrm{pc}^{-2}} \right) + 21.572 + \mathcal{M}_{\odot X}. \tag{2.8}$$

Note that it is the flux, not the magnitude, that is additive. Thus in order to obtain the total (apparent) magnitude from an image, one must first convert magnitude per unit area into flux per unit area, integrate the flux over the entire image, and then convert the total flux back to a total magnitude.

If observations are made for an object in more than one waveband, then the difference between the magnitudes in any two different bands defines a color index (which corresponds to the slope of the SED between the two wavebands). For example,

$$(B-V) \equiv m_B - m_V = \mathcal{M}_B - \mathcal{M}_V \tag{2.9}$$

is called the $(B-V)$ color of the object.

2.1.2 Spectroscopy

From spectroscopic observations one obtains spectra for objects, i.e. their SEDs f_λ or f_ν defined so that $f_\lambda \, d\lambda$ and $f_\nu \, d\nu$ are the fluxes received in the elemental wavelength and frequency ranges $d\lambda$ at λ and $d\nu$ at ν. From the relation between wavelength and frequency, $\lambda = c/\nu$, we then have that

$$f_\nu = \lambda^2 f_\lambda / c \quad \text{and} \quad f_\lambda = \nu^2 f_\nu / c. \tag{2.10}$$

At optical wavelengths, spectroscopy is typically performed by guiding the light from an object to a spectrograph where it is dispersed according to wavelength. For example, in multi-object fiber spectroscopy, individual objects are imaged onto the ends of optical fibers which take the light to prism or optical grating where it is dispersed. The resulting spectra for each individual fiber are then imaged on a detector. Such spectroscopy loses all information about the distribution of each object's light within the circular aperture represented by the end of the fiber. In long-slit spectroscopy, on the other hand, the object of interest is imaged directly onto the spectrograph slit, resulting in a separate spectrum from each point of the object falling on the slit. Finally, in an integral field unit (or IFU) the light from each point within the image of an extended object is led to a different point on the slit (for example, by optical fibers) resulting in a three-dimensional data cube with two spatial dimensions and one dimension for the wavelength.

At other wavelengths quite different techniques can be used to obtain spectral information. For example, at infrared and radio wavelengths the incoming signal from a source may be Fourier analyzed in time in order to obtain the power at each frequency, while at X-ray wavelengths the energy of each incoming photon can be recorded and the energies of different photons can be binned to obtain the spectrum.

Spectroscopic observations can give us a lot of information which photometric observations cannot. A galaxy spectrum usually contains a slowly varying component called the continuum, with localized features produced by emission and absorption lines (see Fig. 2.12 below for some examples). It is a superposition of the spectra of all the individual stars in the galaxy, modified by emission and absorption from the gas and dust lying between the stars. From the ultraviolet through the near-infrared the continuum is due primarily to bound–free transitions in the photospheres of the stars, in the mid- and far-infrared it is dominated by thermal emission from dust grains, in the radio it is produced by diffuse relativistic and thermal electrons within the galaxy, and in the X-ray it comes mainly from accretion of gas onto compact stellar remnants or a central black hole. Emission and absorption lines are produced by bound–bound transitions within atoms, ions and molecules, both in the outer photospheres of stars and in the interstellar gas. By analyzing a spectrum, we may infer the relative importance of these various processes, thereby understanding the physical properties of the galaxy. For example, the strength of a particular emission line depends on the abundance of the excited state that produces it, which in turn depends not only on the abundance of the corresponding element but also on the temperature and ionization state of the gas. Thus emission line strengths can be used to measure the temperature, density and chemical composition of interstellar gas. Absorption lines, on the other hand, mainly arise in the atmospheres of stars, and their relative strengths contain useful information regarding the age and metallicity of the galaxy's stellar population. Finally, interstellar dust gives rise to continuum absorption with broad characteristic features. In addition, since dust extinction is typically more efficient at shorter wavelengths, it also causes reddening, a change of the overall slope of the continuum emission.

Spectroscopic observations have another important application. The intrinsic frequency of photons produced by electron transitions between two energy levels E_1 and E_2 is $\nu_{12} = (E_2 - E_1)/h_P$, where h_P is Planck's constant, and we have assumed $E_2 > E_1$. Now suppose that these photons are produced by atoms moving with velocity $\mathbf{v}$ relative to the observer. Because of the Doppler effect, the observed photon frequency will be (assuming $v \ll c$),

$$\nu_{\rm obs} = \left(1 - \frac{\mathbf{v} \cdot \hat{\mathbf{r}}}{c}\right) \nu_{12}, \tag{2.11}$$

where $\hat{\mathbf{r}}$ is the unit vector of the emitting source relative to the observer. Thus, if the source is receding from the observer, the observed frequency is redshifted, $\nu_{\rm obs} < \nu_{12}$; conversely, if the source is approaching the observer, the observed frequency is blueshifted, $\nu_{\rm obs} > \nu_{12}$. It is convenient to define a redshift parameter to characterize the change in frequency,

$$z \equiv \frac{\nu_{12}}{\nu_{\rm obs}} - 1. \tag{2.12}$$

For the Doppler effect considered here, we have $z = \mathbf{v} \cdot \hat{\mathbf{r}}/c$. Clearly, by studying the properties of spectral lines from an object, one may infer the kinematics of the emitting (or absorbing) material.

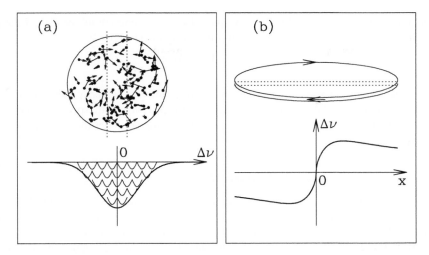

Fig. 2.4. (a) An illustration of the broadening of a spectral line by the velocity dispersion of stars in a stellar system. A telescope collects light from all stars within a cylinder through the stellar system. Each star contributes a narrow spectral line with rest frequency ν_{12}, which is Doppler shifted to a different frequency $\nu = \nu_{12} + \Delta\nu$ due to its motion along the line-of-sight. The superposition of many such line profiles produces a broadened line, with the profile given by the convolution of the original stellar spectral line and the velocity distribution of the stars in the cylinder. (b) An illustration of long-slit spectroscopy of a thin rotating disk along the major axis of the image. In the plot, the rotation speed is assumed to depend on the distance from the center as $V_{\rm rot}(x) \propto \sqrt{x/(1+x^2)}$.

As an example, suppose that the emitting gas atoms in an object have random motions along the line-of-sight drawn from a velocity distribution $f(v)\,dv$. The observed photons will then have the following frequency distribution:

$$F(\nu_{\rm obs})\,d\nu_{\rm obs} = f(v)(c/\nu_{12})\,d\nu_{\rm obs}, \qquad (2.13)$$

where v is related to $\nu_{\rm obs}$ by $v = c(1 - \nu_{\rm obs}/\nu_{12})$, and we have neglected the natural width of atomic spectral lines. Thus, by observing $F(\nu_{\rm obs})$ (the emission line profile in frequency space), we can infer $f(v)$. If the random motion is caused by thermal effects, we can infer the temperature of the gas from the observed line profile. For a stellar system (e.g. an elliptical galaxy) the observed spectral line is the convolution of the original stellar line profile $S(\nu)$ (which is a luminosity weighted sum of the spectra of all different stellar types that contribute to the flux) with the line-of-sight velocity distribution of all the stars in the observational aperture,

$$F(\nu_{\rm obs}) = \int S[\nu_{\rm obs}(1+v/c)] f(v)\,dv. \qquad (2.14)$$

Thus, each narrow, stellar spectral line is broadened by the line-of-sight velocity dispersion of the stars that contribute to that line (see Fig. 2.4a). If we know the type of stars that dominate the spectral lines in consideration, we can estimate $S(\nu)$ and use the above relation to infer the properties of $f(v)$, such as the mean velocity, $\overline{v} = \int v f(v)\,dv$, and the velocity dispersion, $\sigma = [\int (v-\overline{v})^2 f(v)\,dv]^{1/2}$.

Similarly, long-slit and IFU spectroscopy of extended objects can be used not only to study random motions along each line-of-sight through the source, but also to study large-scale flows in the source. An important example here is the rotation of galaxy disks. Suppose that the rotation of a disk around its axis is specified by a rotation curve, $V_{\rm rot}(R)$, which gives the rotation velocity as a function of distance to the disk center. Suppose further that the inclination angle between the rotation axis and the line-of-sight is i. If we put a long slit along the major axis of the image of the disk, it is easy to show that the frequency shift along the slit is

$$\nu_{\rm obs}(R) - \nu_{12} = \pm \frac{V_{\rm rot}(R)\sin i}{c}\nu_{12}, \qquad (2.15)$$

where the $+$ and $-$ signs correspond to points on opposite sides of the disk center (see Fig. 2.4b). Thus the rotation curve of the disk can be measured from its long-slit spectrum and from its apparent shape (which allows the inclination angle to be estimated under the assumption that the disk is intrinsically round).

2.1.3 Distance Measurements

A fundamental task in astronomy is the determination of the distances to astronomical objects. As we have seen above, the direct observables from an astronomical object are its angular size on the sky and its energy flux at the position of the observer. Distance is therefore required in order to convert these observables into physical quantities. In this subsection we describe the principles behind some of the most important methods for estimating astronomical distances.

(a) Trigonometric Parallax The principle on which this distance measure is based is very simple. We are all familiar with the following: when walking along one direction, nearby and distant objects appear to change their orientation with respect to each other. If the walked distance b is much smaller than the distance to an object d (assumed to be perpendicular to the direction of motion), then the change of the orientation of the object relative to an object at infinity is $\theta = b/d$. Thus, by measuring b and θ we can obtain the distance d. This is called the trigonometric parallax method, and can be used to measure distances to some relatively nearby stars. In principle, this can be done by measuring the change of the position of a star relative to one or more background objects (assumed to be at infinity) at two different locations. Unfortunately, the baseline provided by the Earth's diameter is so short that even the closest stars do not have a measurable trigonometric parallax. Therefore, real measurements of stellar trigonometric parallax have to make use of the baseline provided by the diameter of the Earth's orbit around the Sun. By measuring the trigonometric parallax, π_t, which is half of the angular change in the position of a star relative to the background as measured over a six month interval, we can obtain the distance to the star as

$$d = \frac{A}{\tan(\pi_t)}, \qquad (2.16)$$

where $A = 1\,{\rm AU} = 1.49597870 \times 10^{13}$ cm is the length of the semimajor axis of the Earth's orbit around the Sun. The distance corresponding to a trigonometric parallax of 1 arcsec is defined as 1 parsec (or 1 pc). From the Earth the accuracy with which π_t can be measured is restricted by atmospheric seeing, which causes a blurring of the images. This problem is circumvented when using satellites. With the Hipparcos satellite reliable distances have been measured for nearby stars with $\pi_t \gtrsim 10^{-3}$ arcsec, or with distances $d \lesssim 1$ kpc. The GAIA satellite, which is currently scheduled for launch in 2012, will be able to measure parallaxes for stars with an accuracy of $\sim 2 \times 10^{-4}$ arcsec, which will allow distance measurements to 10% accuracy for $\sim 2 \times 10^8$ stars.

(b) Motion-Based Methods The principle of this distance measurement is also very simple. We all know that the angle subtended by an object of diameter l at a distance d is $\theta = l/d$ (assuming $l \ll d$). If we measure the angular diameters of the same object from two distances, d_1 and d_2, then the difference between them is $\Delta\theta = l\Delta d/d^2 = \theta\Delta d/d$, where $\Delta d = |d_1 - d_2|$ is assumed to be much smaller than both d_1 and d_2, and $d = (d_1 d_2)^{1/2}$ can be considered the distance to the object. Thus, we can estimate d by measuring $\Delta\theta$ and Δd. For a star cluster consisting of many stars, the change of its distance over a time interval Δt is given by $\Delta d = v_r \Delta t$, where v_r is the mean radial velocity of the cluster and can be measured from the shift of its spectrum. If we can measure the change of the angular size of the cluster during the same time interval, $\Delta\theta$, then the distance to the cluster can be estimated from $d = \theta v_r \Delta t / \Delta\theta$. This is called the moving-cluster method.

Another distance measure is based on the angular motion of cluster stars caused by their velocity with respect to the Sun. If all stars in a star cluster had the same velocity, the extensions of their proper motion vectors would converge to a single point on the celestial sphere (just like the two parallel rails of a railway track appear to converge to a point at large distance). By measuring the proper motions of the stars in a star cluster, this convergent point can be determined. Because of the geometry, the line-of-sight from the observer to the convergent point is parallel to the velocity vector of the star cluster. Hence, the angle, ϕ, between the star cluster and its convergent point, which can be measured, is the same as that between the proper motion vector and its component along the line-of-sight between the observer and the star cluster. By measuring the cluster's radial velocity v_r, one can thus obtain the transverse velocity $v_t = v_r \tan\phi$. Comparing v_t to the proper motion of the star cluster then yields its distance. This is called the convergent-point method and can be used to estimate accurate distances of star clusters up to a few hundred parsec.

(c) Standard Candles and Standard Rulers As shown by Eqs. (2.1) and (2.2), the luminosity and physical size of an object are related through the distance to its flux and angular size, respectively. Since the flux and angular size are directly observable, we can estimate the distance to an object if its luminosity or its physical size can be obtained in a distance-independent way. Objects whose luminosities and physical sizes can be obtained in such a way are called standard candles and standard rulers, respectively. These objects play an important role in astronomy, not only because their distances can be determined, but more importantly, because they can serve as distance indicators to calibrate the relation between distance and redshift, allowing the distances to other objects to be determined from their redshifts, as we will see below.

One important class of objects in cosmic distance measurements is the Cepheid variable stars (or Cepheids for short). These objects are observed to change their apparent magnitudes regularly, with periods ranging from 2 to 150 days. The period is tightly correlated with the star's luminosity, such that

$$\mathcal{M} = -a - b\log P, \tag{2.17}$$

where P is the period of light variation in days, and a and b are two constants which can be determined using nearby Cepheids whose distances have been measured using another method. For example, using the trigonometric parallaxes of Cepheids measured with the Hipparcos satellite, Feast & Catchpole (1997) obtained the following relation between P and the absolute magnitude in the V band: $\mathcal{M}_V = -1.43 - 2.81\log P$, with a standard error in the zero-point of about 0.10 magnitudes (see Madore & Freedman, 1991, for more examples of such calibrations). Once the luminosity–period relation is calibrated, and if it is universally valid, it can be applied to distant Cepheids (whose distances cannot be obtained from trigonometric parallax or proper motion) to obtain their distances from measurements of their variation periods. Since Cepheids are relatively bright, with absolute magnitudes $\mathcal{M}_V \sim -3$, telescopes with sufficiently high spatial resolution, such as the Hubble Space Telescope (HST), allow Cepheid distances to be determined for objects out to $\sim 10\,\mathrm{Mpc}$.

Another important class of objects for distance measurements are Type Ia supernovae (SNIa), which are exploding stars with well-calibrated light profiles. Since these objects can reach peak luminosities up to $\sim 10^{10}\,\mathrm{L}_\odot$ (so that they can outshine an entire galaxy), they can be observed out to cosmological distances of several thousand megaparsecs. Empirically it has been found that the peak luminosities of SNIa are remarkably similar (e.g. Branch & Tammann, 1992). In fact, there is a small dispersion in peak luminosities, but this has been found to be correlated with the rate at which the luminosity decays and so can be corrected (e.g. Phillips et al., 1999). Thus, one can obtain the *relative* distances to Type Ia supernovae by measuring their light curves. The absolute distances can then be obtained once the absolute values of the light curves of some nearby Type Ia supernovae are calibrated using other (e.g. Cepheid) distances. As we will see in §2.10.1, SNIa play an important role in constraining the large scale geometry of the Universe.

(d) Redshifts as Distances One of the most important discoveries in modern science was Hubble's (1929) observation that almost all galaxies appear to move away from us, and that their recession velocities increase in direct proportion to their distances from us, $v_r \propto r$. This relation, called the Hubble law, is explained most naturally if the Universe as a whole is assumed to be expanding. If the expansion is homogeneous and isotropic, then the distance between any two objects comoving with the expanding background can be written as $r(t) = a(t)r(t')/a(t')$, where $a(t)$ is a time-dependent scale factor of the Universe, describing the expansion. It then follows that the relative separation velocity of the objects is

$$v_r = \dot{r} = H(t)r, \quad \text{where} \quad H(t) \equiv \dot{a}(t)/a(t). \tag{2.18}$$

This relation applied at the present time gives $v_r = H_0 r$, as observed by Hubble. Since the recession velocity of an object can be measured from its redshift z, the distance to the object simply follows from $r = cz/H_0$ (assuming $v_r \ll c$). In practice, the object under consideration may move relative to the background with some (gravitationally induced) peculiar velocity, v_{pec}, so that its observed velocity is the sum of this peculiar velocity along the line-of-sight, $v_{\text{pec},r}$, and the velocity due to the Hubble expansion:

$$v_r = H_0 r + v_{\text{pec},r}. \tag{2.19}$$

In this case, the redshift is no longer a precise measurement of the distance, unless $v_{\text{pec},r} \ll H_0 r$. Since for galaxies the typical value for v_{pec} is a few hundred kilometers per second, redshifts can be used to approximate distances for $cz \gg 1000\,\text{km}\,\text{s}^{-1}$.

In order to convert redshifts into distances, we need a value for the Hubble constant, H_0. This can be obtained if the distances to some sufficiently distant objects can be measured independently of their redshifts. As mentioned above, such objects are called distance indicators. For many years, the value of the Hubble constant was very uncertain, with estimates ranging from $\sim 50\,\text{km}\,\text{s}^{-1}\,\text{Mpc}^{-1}$ to $\sim 100\,\text{km}\,\text{s}^{-1}\,\text{Mpc}^{-1}$ (current constraints on H_0 are discussed in §2.10.1). To parameterize this uncertainty in H_0 it has become customary to write

$$H_0 = 100 h\,\text{km}\,\text{s}^{-1}\,\text{Mpc}^{-1}, \tag{2.20}$$

and to express all quantities that depend on redshift-based distances in terms of the reduced Hubble constant h. For example, distance determinations based on redshifts often contain a factor of h^{-1}, while luminosities based on these distances contain a factor h^{-2}, etc. If these factors are not present, it means that a specific value for the Hubble constant has been assumed, or that the distances were not based on measured redshifts.

2.2 Stars

As we will see in §2.3, the primary visible constituent of most galaxies is the combined light from their stellar population. Clearly, in order to understand galaxy formation and evolution it is important to know the main properties of stars. In Table 2.1 we list some of the photometric properties of the Sun. These, as well as the Sun's mass and radius, $M_\odot = 2 \times 10^{33}\,\text{g}$ and

Table 2.2. Solar abundances in number relative to hydrogen.

Element:	H	He	C	N	O	Ne	Mg	Si	Fe
$(N/N_H) \times 10^5$:	10^5	9800	36.3	11.2	85.1	12.3	3.80	3.55	4.68

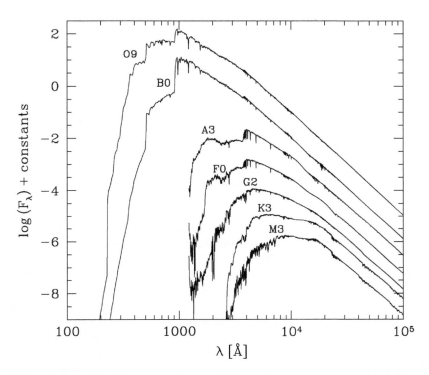

Fig. 2.5. Spectra for stars of different spectral types. f_λ is the flux per angstrom, and an arbitrary constant is added to each spectrum to avoid confusion. [Based on data kindly provided by S. Charlot]

$R_\odot = 7 \times 10^{10}$ cm, are usually used as fiducial values when describing other stars. The abundance by number of some of the chemical elements in the solar system is given in Table 2.2. The fraction in mass of elements heavier than helium is referred to as the metallicity and is denoted by Z, and our Sun has $Z_\odot \approx 0.02$. The relative abundances in a star are usually specified relative to those in the Sun:

$$[A/B] \equiv \log \left[\frac{(n_A/n_B)_\star}{(n_A/n_B)_\odot} \right], \tag{2.21}$$

where $(n_A/n_B)_\star$ is the number density ratio between element A and element B in the star, and $(n_A/n_B)_\odot$ is the corresponding ratio for the Sun.

Since all stars, except a few nearby ones, are unresolved (i.e. they appear as point sources), the only intrinsic properties that are directly observable are their luminosities, colors and spectra. These vary widely (some examples of stellar spectra are shown in Fig. 2.5) and form the basis for their classification. The most often used classification scheme is the Morgan–Keenan (MK) system, summarized in Tables 2.3 and 2.4. These spectral classes are further divided into decimal subclasses [e.g. from B0 (early) to B9 (late)], while luminosity classes are divided into subclasses such as Ia, Ib, etc. The importance of this classification is that, although entirely based

Table 2.3. MK spectral classes.

Class	Temperature	Spectral characteristics
O	28,000–50,000 K	Hot stars with He II absorption; strong UV continuum
B	10,000–28,000 K	He I absorption; H developing in later classes
A	7,500–10,000 K	Strong H lines for A0, decreasing thereafter; Ca II increasing
F	6,000–7,500 K	Ca II stronger; H lines weaker; metal lines developing
G	5,000–6,000 K	Ca II strong; metal lines strong; H lines weaker
K	3,500–5,000 K	Strong metal lines; CH and CN developing; weak blue continuum
M	2,500–3,500 K	Very red; TiO bands developing strongly

Table 2.4. MK luminosity classes.

I	Supergiants
II	Bright giants
III	Normal giants
IV	Subgiants
V	Dwarfs (main-sequence stars)

on observable properties, it is closely related to the basic physical properties of stars. For example, the luminosity classes are related to surface gravities, while the spectral classes are related to surface temperatures (see e.g. Cox, 2000).

Fig. 2.6 shows the color–magnitude relation of a large number of stars for which accurate distances are available (so that their absolute magnitudes can be determined). Such a diagram is called a Hertzsprung–Russell diagram (abbreviated as H-R diagram), and features predominantly in studies of stellar astrophysics. The MK spectral and luminosity classes are also indicated. Clearly, stars are not uniformly distributed in the color–magnitude space, but lie in several well-defined sequences. Most of the stars lie in the 'main sequence' (MS) which runs from the lower-right to the upper-left. Such stars are called main-sequence stars and have MK luminosity class V. The position of a star in this sequence is mainly determined by its mass. Above the main sequence one finds the much rarer but brighter giants, making up the MK luminosity classes I to IV, while the lower-left part of the H-R diagram is occupied by white dwarfs. The Sun, whose MK type is G2V, lies in the main sequence with V-band absolute magnitude 4.8 and (atmospheric) temperature 5780K.

As a star ages it moves off the MS and starts to traverse the H-R diagram. The location of a star in the H-R diagram as function of time is called its evolutionary track which, again, is determined mainly by its mass. An important property of a stellar population is therefore its initial mass function (IMF), which specifies the abundance of stars as function of their initial mass (i.e. the mass they have at the time when reach the MS shortly after their formation). For a given IMF, and a given star-formation history, one can use the evolutionary tracks to predict the abundance of stars in the H-R diagram. Since the spectrum of a star is directly related to its position in the H-R diagram, this can be used to predict the spectrum of an entire galaxy, a procedure which is called spectral synthesis modeling. Detailed calculations of stellar evolution models (see Chapter 10) show that a star like our Sun has a MS lifetime of about 10 Gyr, and that the MS lifetime scales with mass roughly as M^{-3}, i.e. more massive (brighter) stars spend less time on the MS. This strong dependence of MS lifetime on mass has important observational consequences, because it implies that the spectrum of a stellar system (a galaxy) depends on its star-formation history. For a system where the current star-formation rate is high, so that many young massive stars are still on the main sequence, the stellar spectrum is expected to have a strong blue continuum produced by O and B stars. On the other hand, for a system where star

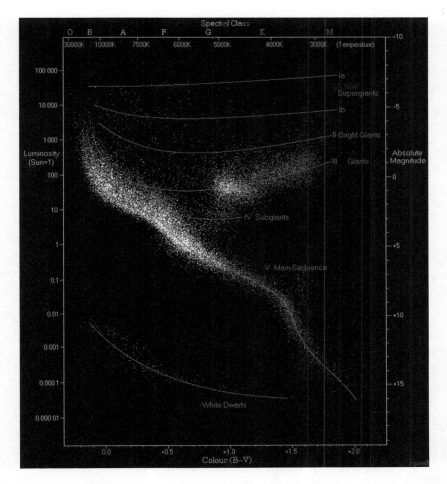

Fig. 2.6. The color–magnitude diagram (i.e. the H-R diagram) of 22,000 stars from the Hipparcos Catalogue together with 1,000 low-luminosity stars (red and white dwarfs) from the Gliese Catalogue of Nearby Stars. The MK spectral and luminosity classes are also indicated, as are the luminosities in solar units. [Diagram from R. Powell, taken from Wikipedia]

formation has been terminated a long time ago, so that all massive stars have already evolved off the MS, the spectrum (now dominated by red giants and the low-mass MS stars) is expected to be red.

2.3 Galaxies

Galaxies, whose formation and evolution is the main topic of this book, are the building blocks of the Universe. They not only are the cradles for the formation of stars and metals, but also serve as beacons that allow us to probe the geometry of space-time. Yet it is easy to forget that it was not until the 1920s, with Hubble's identification of Cepheid variable stars in the Andromeda Nebula, that most astronomers became convinced that the many 'nebulous' objects cataloged by John Dreyer in his 1888 *New General Catalogue of Nebulae and Clusters of Stars* and the two supplementary *Index Catalogues* are indeed galaxies. Hence, extragalactic astronomy is a relatively new science. Nevertheless, as we will see, we have made tremendous progress: we

Fig. 2.7. Examples of different types of galaxies. From left to right and top to bottom, NGC 4278 (E1), NGC 3377 (E6), NGC 5866 (S0), NGC 175 (SBa), NGC 6814 (Sb), NGC 4565 (Sb, edge on), NGC 5364 (Sc), Ho II (Irr I), NGC 520 (Irr II). [All images are obtained from the NASA/IPAC Extragalactic Database (NED) which is operated by the Jet Propulsion Laboratory, California Institute of Technology, under contract with the National Aeronautics and Space Administration]

have surveyed the local population of galaxies in exquisite detail covering the entire range of wavelengths, we have constructed redshift surveys with hundreds of thousands of galaxies to probe the large scale structure of the Universe, and we have started to unveil the population of galaxies at high redshifts, when the Universe was only a small fraction of its current age.

2.3.1 The Classification of Galaxies

Fig. 2.7 shows a collage of images of different kinds of galaxies. Upon inspection, one finds that some galaxies have smooth light profiles with elliptical isophotes, others have spiral arms together with an elliptical-like central bulge, and still others have irregular or peculiar morphologies. Based on such features, Hubble ordered galaxies in a morphological sequence, which is

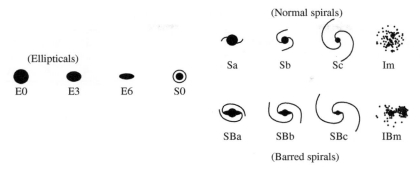

Fig. 2.8. A schematic representation of the Hubble sequence of galaxy morphologies. [Courtesy of R. Abraham; see Abraham (1998)]

now referred to as the Hubble sequence or Hubble tuning-fork diagram (see Fig. 2.8). Hubble's scheme classifies galaxies into four broad classes:

(i) Elliptical galaxies: These have smooth, almost elliptical isophotes and are divided into subtypes E0, E1, ..., E7, where the integer is the one closest to $10(1-b/a)$, with a and b the lengths of the semimajor and semiminor axes.
(ii) Spiral galaxies: These have thin disks with spiral arm structures. They are divided into two branches, barred spirals and normal spirals, according to whether or not a recognizable bar-like structure is present in the central part of the galaxy. On each branch, galaxies are further divided into three classes, a, b and c, according to the following three criteria:
- the fraction of the light in the central bulge;
- the tightness with which the spiral arms are wound;
- the degree to which the spiral arms are resolved into stars, HII regions and ordered dust lanes.

These three criteria are correlated: spirals with a pronounced bulge component usually also have tightly wound spiral arms with relatively faint HII regions, and are classified as Sa. On the other hand, spirals with weak or absent bulges usually have open arms and bright HII regions and are classified as Sc. When the three criteria give conflicting indications, Hubble put most emphasis on the openness of the spiral arms.
(iii) Lenticular or S0 galaxies: This class is intermediate between ellipticals and spirals. Like ellipticals, lenticulars have a smooth light distribution with no spiral arms or HII regions. Like spirals they have a thin disk and a bulge, but the bulge is more dominant than that in a spiral galaxy. They may also have a central bar, in which case they are classified as SB0.
(iv) Irregular galaxies: These objects have neither a dominating bulge nor a rotationally symmetric disk and lack any obvious symmetry. Rather, their appearance is generally patchy, dominated by a few HII regions. Hubble did not include this class in his original sequence because he was uncertain whether it should be considered an extension of any of the other classes. Nowadays irregulars are usually included as an extension to the spiral galaxies.

Ellipticals and lenticulars together are often referred to as early-type galaxies, while the spirals and irregulars make up the class of late-type galaxies. Indeed, traversing the Hubble sequence from the left to the right the morphologies are said to change from early- to late-type. Although somewhat confusing, one often uses the terms 'early-type spirals' and 'late-type spirals' to refer to galaxies at the left or right of the spiral sequence. We caution, though, that this historical nomenclature has no direct physical basis: the reference to 'early' or 'late' should not be interpreted as reflecting a property of the galaxy's evolutionary state. Another largely historical

Table 2.5. Galaxy morphological types.

Hubble	E	E-SO	SO	SO-Sa	Sa	Sa-b	Sb	Sb-c	Sc	Sc-Irr	Irr
deV	E	SO^-	SO^0	SO^+	Sa	Sab	Sb	Sbc	Scd	Sdm	Im
T	-5	-3	-2	0	1	2	3	4	6	8	10

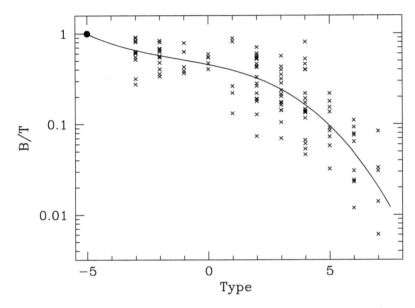

Fig. 2.9. Fractional luminosity of the spheroidal bulge component in a galaxy as a function of morphological type (based on the classification of de Vaucouleurs). Data points correspond to individual galaxies, and the curve is a fit to the mean. Elliptical galaxies (Type $= -5$) are considered to be pure bulges. [Based on data presented in Simien & de Vaucouleurs (1986)]

nomenclature, which can be confusing at times, is to refer to faint galaxies with $\mathcal{M}_B \gtrsim -18$ as 'dwarf galaxies'. In particular, early-type dwarfs are often split into dwarf ellipticals (dE) and dwarf spheroidals (dSph), although there is no clear distinction between these types – often the term dwarf spheroidals is simply used to refer to early-type galaxies with $\mathcal{M}_B \gtrsim -14$.

Since Hubble, a variety of other classification schemes have been introduced. A commonly used one is due to de Vaucouleurs (1974). He put spirals in the Hubble sequence into a finer gradation by adding new types such as SOa, Sab, Sbc (and the corresponding barred types). After finding that many of Hubble's irregular galaxies in fact had weak spiral arms, de Vaucouleurs also extended the spiral sequence to irregulars, adding types Scd, Sd, Sdm, Sm, Im and I0, in order of decreasing regularity. (The m stands for 'Magellanic' since the Magellanic Clouds are the prototypes of this kind of irregulars.) Furthermore, de Vaucouleurs used numbers between -6 and 10 to represent morphological types (the de Vaucouleurs' T types). Table 2.5 shows the correspondence between de Vaucouleurs' notations and Hubble's notations – note that the numerical T types do not distinguish between barred and unbarred galaxies. As shown in Fig. 2.9, the morphology sequence according to de Vaucouleurs' classification is primarily a sequence in the importance of the bulge.

The Hubble classification and its revisions encompass the morphologies of the majority of the observed galaxies in the local Universe. However, there are also galaxies with strange appearances which defy Hubble's classification. From their morphologies, these 'peculiar'

2.3 Galaxies

Fig. 2.10. The peculiar galaxy known as the Antennae, a system exhibiting prominent tidal tails (the left inlet), a signature of a recent merger of two spiral galaxies. The close-up of the center reveals the presence of large amounts of dust and many clusters of newly formed stars. [Courtesy of B. Whitmore, NASA, and Space Telescope Science Institute]

galaxies all appear to have been strongly perturbed in the recent past and to be far from dynamical equilibrium, indicating that they are undergoing a transformation. A good example is the Antennae (Fig. 2.10) where the tails are produced by the interaction of the two spiral galaxies, NGC 4038 and NGC 4039, in the process of merging.

The classifications discussed so far are based only on morphology. Galaxies can also be classified according to other properties. For instance, they can be classified into *bright* and *faint* according to luminosity, into *high* and *low surface brightness* according to surface brightness, into *red* and *blue* according to color, into *gas-rich* and *gas-poor* according to gas content, into *quiescent* and *starburst* according to their current level of star formation, and into *normal* and *active* according to the presence of an active nucleus. All these properties can be measured observationally, although often with some difficulty. An important aspect of the Hubble sequence (and its modifications) is that many of these properties change systematically along the sequence (see Figs. 2.11 and 2.12), indicating that it reflects a sequence in the basic physical properties of galaxies. However, we stress that the classification of galaxies is far less clear cut than that of stars, whose classification has a sound basis in terms of the H-R diagram and the evolutionary tracks.

2.3.2 Elliptical Galaxies

Elliptical galaxies are characterized by smooth, elliptical surface brightness distributions, contain little cold gas or dust, and have red photometric colors, characteristic of an old stellar population. In this section we briefly discuss some of the main, salient observational properties. A more in-depth discussion, including an interpretation within the physical framework of galaxy formation, is presented in Chapter 13.

(a) Surface Brightness Profiles The one-dimensional surface brightness profile, $I(R)$, of an elliptical galaxy is usually defined as the surface brightness as a function of the isophotal

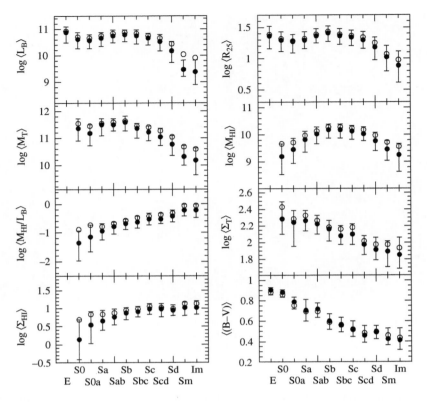

Fig. 2.11. Galaxy properties along the Hubble morphological sequence based on the RC3-UGC sample. Filled circles are medians, open ones are mean values. The bars bracket the 25 and 75 percentiles. Properties plotted are L_B (blue luminosity in $L_\odot$), R_{25} (the radius in kpc of the 25 mag arcsec^{-2} isophote in the B-band), M_T (total mass in solar units within a radius $R_{25}/2$), M_{HI} (HI mass in solar units), M_{HI}/L_B, Σ_T (total mass surface density), Σ_{HI} (HI mass surface density), and the $B-V$ color. [Based on data presented in Roberts & Haynes (1994)]

semimajor axis length R. If the position angle of the semimajor axis changes with radius, a phenomenon called isophote twisting, then $I(R)$ traces the surface brightness along a curve that connects the intersections of each isophote with its own major axis.

The surface brightness profile of spheroidal galaxies is generally well fit by the Sérsic profile (Sérsic, 1968), or $R^{1/n}$ profile,[1]

$$I(R) = I_0 \exp\left[-\beta_n \left(\frac{R}{R_e}\right)^{1/n}\right] = I_e \exp\left[-\beta_n \left\{\left(\frac{R}{R_e}\right)^{1/n} - 1\right\}\right], \quad (2.22)$$

where I_0 is the central surface brightness, n is the so-called Sérsic index which sets the concentration of the profile, R_e is the effective radius that encloses half of the total light, and $I_e = I(R_e)$. Surface brightness profiles are often expressed in terms of $\mu \propto -2.5 \log(I)$ (which has the units of mag arcsec^{-2}), for which the Sérsic profile takes the form

$$\mu(R) = \mu_e + 1.086 \beta_n \left[\left(\frac{R}{R_e}\right)^{1/n} - 1\right]. \quad (2.23)$$

---

[1] A similar formula, but with R denoting 3-D rather than projected radius, was used by Einasto (1965) to describe the stellar halo of the Milky Way.

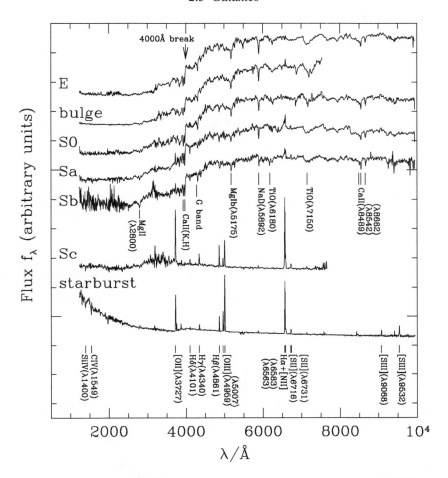

Fig. 2.12. Spectra of different types of galaxies from the ultraviolet to the near-infrared. From ellipticals to late-type spirals, the blue continuum and emission lines become systematically stronger. For early-type galaxies, which lack hot, young stars, most of the light emerges at the longest wavelengths, where one sees absorption lines characteristic of cool K stars. In the blue, the spectrum of early-type galaxies show strong H and K absorption lines of calcium and the G band, characteristic of solar type stars. Such galaxies emit little light at wavelengths shorter than 4000 Å and have no emission lines. In contrast, late-type galaxies and starbursts emit most of their light in the blue and near-ultraviolet. This light is produced by hot young stars, which also heat and ionize the interstellar medium giving rise to strong emission lines. [Based on data kindly provided by S. Charlot]

The value for β_n follows from the definition of R_e and is well approximated by $\beta_n = 2n - 0.324$ (but only for $n \gtrsim 1$). Note that Eq. (2.22) reduces to a simple exponential profile for $n = 1$. The total luminosity of a spherical system with a Sérsic profile is

$$L = 2\pi \int_0^\infty I(R)\, R\, dR = \frac{2\pi n\, \Gamma(2n)}{(\beta_n)^{2n}} I_0 R_e^2, \qquad (2.24)$$

with $\Gamma(x)$ the gamma function. Early photometry of the surface brightness profiles of normal giant elliptical galaxies was well fit by a de Vaucouleurs profile, which is a Sérsic profile with $n = 4$ (and $\beta_n = 7.67$) and is therefore also called a $R^{1/4}$-profile. With higher accuracy photometry and with measurements of higher and lower luminosity galaxies, it became clear that ellipticals as a class are better fit by the more general Sérsic profile. In fact, the best-fit values for n have

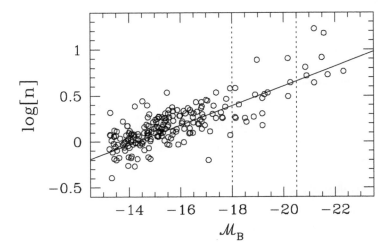

Fig. 2.13. Correlation between the Sérsic index, n, and the absolute magnitude in the B-band for a sample of elliptical galaxies. The vertical dotted lines correspond to $\mathcal{M}_B = -18$ and $\mathcal{M}_B = -20.5$ and are shown to facilitate a comparison with Fig. 2.14. [Data compiled and kindly made available by A. Graham (see Graham & Guzmán, 2003)]

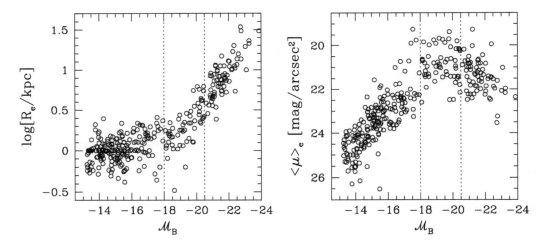

Fig. 2.14. The effective radius (left panel) and the average surface brightness within the effective radius (right panel) of elliptical galaxies plotted against their absolute magnitude in the B-band. The vertical dotted lines correspond to $\mathcal{M}_B = -18$ and $\mathcal{M}_B = -20.5$. [Data compiled and kindly made available by A. Graham (see Graham & Guzmán, 2003), combined with data taken from Bender et al. (1992)]

been found to be correlated with the luminosity and size of the galaxy: while at the faint end dwarf ellipticals have best-fit values as low as $n \sim 0.5$, the brightest ellipticals can have Sérsic indices $n \gtrsim 10$ (see Fig. 2.13).

Instead of I_0 or I_e, one often characterizes the surface brightness of an elliptical galaxy via the average surface brightness within the effective radius, $\langle I \rangle_e = L/(2\pi R_e^2)$, or, in magnitudes, $\langle \mu \rangle_e$. Fig. 2.14 shows how R_e and $\langle \mu \rangle_e$ are correlated with luminosity. At the bright end ($\mathcal{M}_B \lesssim -18$), the sizes of elliptical galaxies increase strongly with luminosity. Consequently, the average surface brightness actually decreases with increasing luminosity. At the faint end ($\mathcal{M}_B \gtrsim -18$), however, all ellipticals have roughly the same effective radius ($R_e \sim 1\,\mathrm{kpc}$), so that the average surface brightness *increases* with increasing luminosity. Because of this apparent change-over in properties, ellipticals with $\mathcal{M}_B \gtrsim -18$ are typically called 'dwarf' ellipticals, in order to distinguish them from the 'normal' ellipticals (see §2.3.5). However, this alleged 'dichotomy' between dwarf and normal ellipticals has recently been challenged. A number of studies have argued that there is actually a smooth and continuous sequence of increasing surface brightness with increasing luminosity, except for the very bright end ($\mathcal{M}_B \lesssim -20.5$) where this trend is reversed (e.g. Jerjen & Binggeli, 1997; Graham & Guzmán, 2003).

The fact that the photometric properties of elliptical galaxies undergo a transition around $\mathcal{M}_B \sim -20.5$ is also evident from their central properties (in the inner few hundred parsec). High spatial resolution imaging with the HST has revealed that the central surface brightness profiles of elliptical galaxies are typically not well described by an inward extrapolation of the Sérsic profiles fit to their outer regions. Bright ellipticals with $\mathcal{M}_B \lesssim -20.5$ typically have a deficit in $I(R)$ with respect to the best-fit Sérsic profile, while fainter ellipticals reveal excess surface brightness. Based on the value of the central cusp slope $\gamma \equiv \mathrm{d}\log I/\mathrm{d}\log r$ the population of ellipticals has been split into 'core' ($\gamma < 0.3$) and 'power-law' ($\gamma \geq 0.3$) systems. The majority of bright galaxies with $\mathcal{M}_B \lesssim -20.5$ have cores, while power-law galaxies typically have $\mathcal{M}_B > -20.5$ (Ferrarese et al., 1994; Lauer et al., 1995). Early results, based on relatively small samples, suggested a bimodal distribution in γ, with virtually no galaxies in the range $0.3 < \gamma < 0.5$. However, subsequent studies have significantly weakened the evidence for a clear dichotomy, finding a population of galaxies with intermediate properties (Rest et al., 2001; Ravindranath et al., 2001). In fact, recent studies, using significantly larger samples, have argued for a smooth transition in nuclear properties, with no evidence for any dichotomy (Ferrarese et al., 2006b; Côté et al., 2007; see also §13.1.2).

(b) Isophotal Shapes The isophotes of elliptical galaxies are commonly fitted by ellipses and characterized by their minor-to-major axis ratios b/a (or, equivalently, by their ellipticities $\varepsilon = 1 - b/a$) and by their position angles. In general, the ellipticity may change across the system, in which case the overall shape of an elliptical is usually defined by some characteristic ellipticity (e.g. that of the isophote which encloses half the total light). In most cases, however, the variation of ε with radius is not large, so that the exact definition is of little consequence. For normal elliptical galaxies the axis ratio lies in the range $0.3 \lesssim b/a \leq 1$, corresponding to types E0 to E7. In addition to the ellipticity, the position angle of the isophotes may also change with radius, a phenomenon called isophote twisting.

Detailed modeling of the surface brightness of elliptical galaxies shows that their isophotes are generally not exactly elliptical. The deviations from perfect ellipses are conveniently quantified by the Fourier coefficients of the function

$$\Delta(\phi) \equiv R_{\mathrm{iso}}(\phi) - R_{\mathrm{ell}}(\phi) = a_0 + \sum_{n=1}^{\infty} (a_n \cos n\phi + b_n \sin n\phi), \tag{2.25}$$

where $R_{\mathrm{iso}}(\phi)$ is the radius of the isophote at angle ϕ and $R_{\mathrm{ell}}(\phi)$ is the radius of an ellipse at the same angle (see Fig. 2.15). Typically one considers the ellipse that best fits the isophote in question, so that a_0, a_1, a_2, b_1 and b_2 are all consistent with zero within the errors. The deviations from this best-fit isophote are then expressed by the higher-order Fourier coefficients a_n and b_n with $n \geq 3$. Of particular importance are the values of the a_4 coefficients, which indicate whether the isophotes are 'disky' ($a_4 > 0$) or 'boxy' ($a_4 < 0$), as illustrated in Fig. 2.15. The *diskiness* of an isophote is defined as the dimensionless quantity, a_4/a, where a is the length of the semimajor axis of the isophote's best-fit ellipse. We caution that some authors use an alternative method to specify the deviations of isophotes from pure ellipses. Instead of using

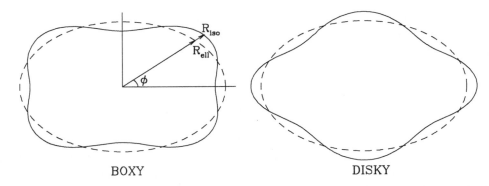

Fig. 2.15. An illustration of boxy and disky isophotes (solid curves). The dashed curves are the corresponding best-fit ellipses.

isophote deviation from an ellipse, they quantify how the *intensity* fluctuates along the best-fit ellipse:

$$I(\phi) = I_0 + \sum_{n=1}^{\infty} (A_n \cos n\phi + B_n \sin n\phi), \qquad (2.26)$$

with I_0 the intensity of the best-fit ellipse. The coefficients A_n and B_n are (approximately) related to a_n and b_n according to

$$A_n = a_n \left| \frac{dI}{dR} \right|, \qquad B_n = b_n \left| \frac{dI}{dR} \right|, \qquad (2.27)$$

where $R = a\sqrt{1-\varepsilon}$, with ε the ellipticity of the best-fit ellipse.

The importance of the disky/boxy classification is that boxy and disky ellipticals turn out to have systematically different properties. Boxy ellipticals are usually bright, rotate slowly, and show stronger than average radio and X-ray emission, while disky ellipticals are fainter, have significant rotation and show little or no radio and X-ray emission (e.g. Bender et al., 1989; Pasquali et al., 2007). In addition, the diskiness is correlated with the nuclear properties as well; disky ellipticals typically have steep cusps, while boxy ellipticals mainly harbor central cores (e.g. Jaffe et al., 1994; Faber et al., 1997).

(c) Colors Elliptical galaxies in general have red colors, indicating that their stellar contents are dominated by old, metal-rich stars (see §10.3). In addition, the colors are tightly correlated with the luminosity such that brighter ellipticals are redder (Sandage & Visvanathan, 1978). As we will see in §13.5, the slope and (small) scatter of this color–magnitude relation puts tight constraints on the star-formation histories of elliptical galaxies. Ellipticals also display color gradients. In general, the outskirts have bluer colors than the central regions. Peletier et al. (1990) obtained a mean logarithmic gradient of $\Delta(U-R)/\Delta \log r = -0.20 \pm 0.02$ mag in $U-R$, and of $\Delta(B-R)/\Delta \log r = -0.09 \pm 0.02$ mag in $B-R$, in good agreement with the results obtained by Franx et al. (1989b).

(d) Kinematic Properties Giant ellipticals generally have low rotation velocities. Observationally, this may be characterized by the ratio of maximum line-of-sight streaming motion v_m (relative to the mean velocity of the galaxy) to $\overline{\sigma}$, the average value of the line-of-sight velocity dispersion interior to $\sim R_e/2$. This ratio provides a measure of the relative importance of ordered and random motions within the galaxy. For isotropic, oblate galaxies flattened by the centrifugal force generated by rotation, $v_m/\overline{\sigma} \approx \sqrt{\varepsilon/(1-\varepsilon)}$, with ε the ellipticity of the spheroid (see §13.1.7). As shown in Fig. 2.16a, for bright ellipticals, $v_m/\overline{\sigma}$ lies well below this prediction,

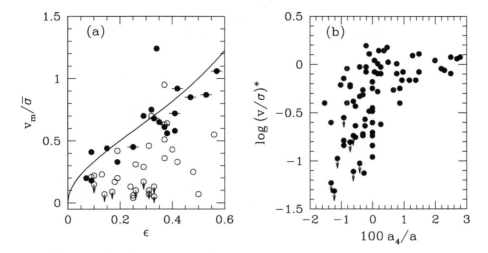

Fig. 2.16. (a) The ratio $v_\mathrm{m}/\overline{\sigma}$ for ellipticals and bulges (with bulges marked by horizontal bars) versus ellipticity. Open circles are for bright galaxies with $\mathcal{M}_B \leq -20.5$, with upper limits marked by downward arrows; solid circles are for early types with $-20.5 < \mathcal{M}_B < -18$. The solid curve is the relation expected for an oblate galaxy flattened by rotation. [Based on data published in Davies et al. (1983)] (b) The rotation parameter $(v/\sigma)^*$ (defined as the ratio of $v_\mathrm{m}/\overline{\sigma}$ to the value expected for an isotropic oblate spheroid flattened purely by rotation) versus the average diskiness of the galaxy. [Based on data published in Kormendy & Bender (1996)]

indicating that their flattening must be due to velocity anisotropy, rather than rotation. In contrast, ellipticals of intermediate luminosities (with absolute magnitude $-20.5 \lesssim \mathcal{M}_B \lesssim -18.0$) and spiral bulges have $v_\mathrm{m}/\overline{\sigma}$ values consistent with rotational flattening. Fig. 2.16b shows, as noted above, that disky and boxy ellipticals have systematically different kinematics: while disky ellipticals are consistent with rotational flattening, rotation in boxy ellipticals is dynamically unimportant.

When the kinematic structure of elliptical galaxies is examined in more detail a wide range of behavior is found. In most galaxies the line-of-sight velocity dispersion depends only weakly on position and is constant or falls at large radii. Towards the center the dispersion may drop weakly, remain flat, or rise quite sharply. The behavior of the mean line-of-sight streaming velocity is even more diverse. While most galaxies show maximal streaming along the major axis, a substantial minority show more complex behavior. Some have non-zero streaming velocities along the minor axis, and so it is impossible for them to be an oblate body rotating about its symmetry axis. Others have mean motions which change suddenly in size, in axis, or in sign in the inner regions, the so-called kinematically decoupled cores. Such variations point to a variety of formation histories for apparently similar galaxies.

At the very center of most nearby ellipticals (and also spiral and S0 bulges) the velocity dispersion is observed to rise more strongly than can be understood as a result of the gravitational effects of the observed stellar populations alone. It is now generally accepted that this rise signals the presence of a central supermassive black hole. Such a black hole appears to be present in virtually every galaxy with a significant spheroidal component, and to have a mass which is roughly 0.1% of the total stellar mass of the spheroid (Fig. 2.17). A more detailed discussion of supermassive black holes is presented in §13.1.4.

(e) Scaling Relations The kinematic and photometric properties of elliptical galaxies are correlated. In particular, ellipticals with a larger (central) velocity dispersion are both brighter, known as the Faber–Jackson relation, and larger, known as the D_n-σ relation (D_n is the isophotal

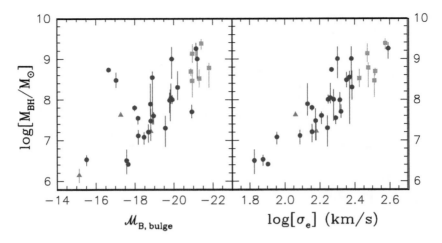

Fig. 2.17. The masses of central black holes in ellipticals and spiral bulges plotted against the absolute magnitude (left) and velocity dispersion (right) of their host spheroids. [Adapted from Kormendy (2001)]

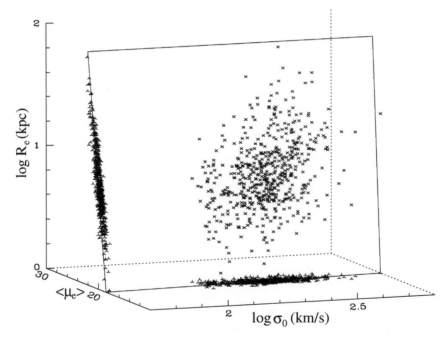

Fig. 2.18. The fundamental plane of elliptical galaxies in the $\log R_e$-$\log \sigma_0$-$\langle \mu \rangle_e$ space (σ_0 is the central velocity dispersion, and $\langle \mu \rangle_e$ is the mean surface brightness within R_e expressed in magnitudes per square arcsecond). [Plot kindly provided by R. Saglia, based on data published in Saglia et al. (1997) and Wegner et al. (1999)]

diameter within which the average, enclosed surface brightness is equal to a fixed value). Furthermore, when plotted in the three-dimensional space spanned by $\log \sigma_0$, $\log R_e$ and $\log \langle I \rangle_e$, elliptical galaxies are concentrated in a plane (see Fig. 2.18) known as the fundamental plane. In mathematical form, this plane can be written as

$$\log R_e = a \log \sigma_0 + b \log \langle I \rangle_e + \text{constant}, \tag{2.28}$$

where $\langle I \rangle_e$ is the mean surface brightness within R_e (not to be confused with I_e, which is the surface brightness at R_e). The values of a and b have been estimated in various photometric bands. For example, Jørgensen et al. (1996) obtained $a = 1.24 \pm 0.07$, $b = -0.82 \pm 0.02$ in the optical, while Pahre et al. (1998) obtained $a = 1.53 \pm 0.08$, $b = -0.79 \pm 0.03$ in the near-infrared. More recently, using 9,000 galaxies from the Sloan Digital Sky Survey (SDSS), Bernardi et al. (2003b) found the best fitting plane to have $a = 1.49 \pm 0.05$ and $b = -0.75 \pm 0.01$ in the SDSS r-band with a *rms* of only 0.05. The Faber–Jackson and D_n-σ relations are both two-dimensional projections of this fundamental plane. While the D_n-σ projection is close to edge-on and so has relatively little scatter, the Faber–Jackson projection is significantly tilted resulting in somewhat larger scatter. These relations can not only be used to determine the distances to elliptical galaxies, but are also important for constraining theories for their formation (see §13.4).

(f) Gas Content Although it was once believed that elliptical galaxies contain neither gas nor dust, it has become clear over the years that they actually contain a significant amount of interstellar medium which is quite different in character from that in spiral galaxies (e.g. Roberts et al., 1991; Buson et al., 1993). Hot ($\sim 10^7$ K) X-ray emitting gas usually dominates the interstellar medium (ISM) in luminous ellipticals, where it can contribute up to $\sim 10^{10} \, \mathrm{M_\odot}$ to the total mass of the system. This hot gas is distributed in extended X-ray emitting atmospheres (Fabbiano, 1989; Mathews & Brighenti, 2003), and serves as an ideal tracer of the gravitational potential in which the galaxy resides (see §8.2).

In addition, many ellipticals also contain small amounts of warm ionized (10^4 K) gas as well as cold (< 100 K) gas and dust. Typical masses are 10^2–$10^4 \, \mathrm{M_\odot}$ in ionized gas and 10^6–$10^8 \, \mathrm{M_\odot}$ in the cold component. Contrary to the case for spirals, the amounts of dust and of atomic and molecular gas are not correlated with the luminosity of the elliptical. In many cases, the dust and/or ionized gas is located in the center of the galaxy in a small disk component, while other ellipticals reveal more complex, filamentary or patchy dust morphologies (e.g. van Dokkum & Franx, 1995; Tran et al., 2001). This gas and dust either results from accumulated mass loss from stars within the galaxy or has been accreted from external systems. The latter is supported by the fact that the dust and gas disks are often found to have kinematics decoupled from that of the stellar body (e.g. Bertola et al., 1992).

2.3.3 Disk Galaxies

Disk galaxies have a far more complex morphology than ellipticals. They typically consist of a thin, rotationally supported disk with spiral arms and often a bar, plus a central bulge component. The latter can dominate the light of the galaxy in the earliest types and may be completely absent in the latest types. The spiral structure is best seen in face-on systems and is defined primarily by young stars, HII regions, molecular gas and dust absorption. Edge-on systems, on the other hand, give a better handle on the vertical structure of the disk, which often reveals two separate components: a thin disk and a thick disk. In addition, there are indications that disk galaxies also contain a spheroidal, stellar halo, extending out to large radii. In this subsection we briefly summarize the most important observational characteristics of disk galaxies. A more in-depth discussion, including models for their formation, is presented in Chapter 11.

(a) Surface Brightness Profiles Fig. 2.19 shows the surface brightness profiles of three disk galaxies, as measured along their projected, major axes. A characteristic of these profiles is that they typically reveal a range over which $\mu(R)$ can be accurately fitted by a straight line. This corresponds to an exponential surface brightness profile

$$I(R) = I_0 \exp(-R/R_d), \quad I_0 = \frac{L}{2\pi R_d^2}, \tag{2.29}$$

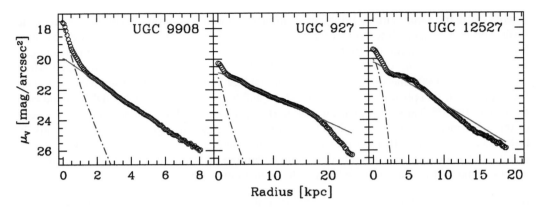

Fig. 2.19. The surface brightness profiles of three disk galaxies plus their decomposition in an exponential disk (solid line) and a Sérsic bulge (dot-dashed line). [Based on data published in MacArthur et al. (2003) and kindly made available by L. MacArthur]

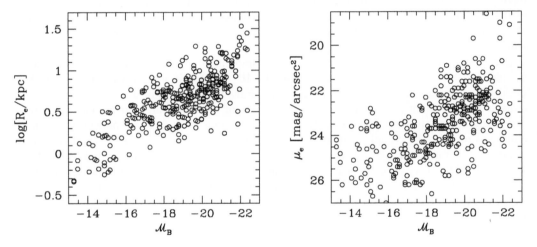

Fig. 2.20. The effective radius (left panel) and the surface brightness at the effective radius (right panel) of disk dominated galaxies plotted against their absolute magnitude in the B-band. [Based on data published in Impey et al. (1996b)]

(i.e. a Sérsic profile with $n = 1$). Here R is the cylindrical radius, R_d is the exponential scale-length, I_0 is the central luminosity surface density, and L is the total luminosity. The effective radius enclosing half of the total luminosity is $R_e \simeq 1.67 R_d$. Following Freeman (1970) it has become customary to associate this exponential surface brightness profile with the actual disk component. The central regions of the majority of disk galaxies show an excess surface brightness with respect to a simple inward extrapolation of this exponential profile. This is interpreted as a contribution from the bulge component, and such interpretation is supported by images of edge-on disk galaxies, which typically reveal a central, roughly spheroidal, component clearly thicker than the disk itself (see e.g. NGC 4565 in Fig. 2.7). At large radii, the surface brightness profiles often break to a much steeper (roughly exponential) profile (an example is UGC 927, shown in Fig. 2.19). These breaks occur at radii $R_b = \alpha R_d$ with α in the range 2.5 to 4.5 (e.g. Pohlen et al., 2000; de Grijs et al., 2001).

Fig. 2.20 shows R_e and μ_e as functions of the absolute magnitude for a large sample of disk dominated galaxies (i.e. with a small or negligible bulge component). Clearly, as expected, more luminous galaxies tend to be larger, although there is large scatter, indicating that galaxies of a given luminosity span a wide range in surface brightnesses. Note that, similar to ellipticals with $\mathcal{M}_B \gtrsim -20.5$, more luminous disk galaxies on average have a higher surface brightness (see Fig. 2.14).

When decomposing the surface brightness profiles of disk galaxies into the contributions of disk and bulge, one typically fits $\mu(R)$ with the sum of an exponential profile for the disk and a Sérsic profile for the bulge. We caution, however, that these bulge–disk decompositions are far from straightforward. Often the surface brightness profiles show clear deviations from a simple sum of an exponential plus Sérsic profile (e.g. UGC 12527 in Fig. 2.19). In addition, seeing tends to blur the central surface brightness distribution, which has to be corrected for, dust can cause significant extinction, and bars and spiral arms represent clear deviations from perfect axisymmetry. In addition, disks are often lop-sided (the centers of different isophotes are offset from each other in one particular direction) and can even be warped (the disk is not planar, but different disk radii are tilted with respect to each other). These difficulties can be partly overcome by using the full two-dimensional information in the image, by using color information to correct for dust, and by using kinematic information. Such studies require much detailed work and even then ambiguities remain.

Despite these uncertainties, bulge–disk decompositions have been presented for large samples of disk galaxies (e.g. de Jong, 1996a; Graham, 2001; MacArthur et al., 2003). These studies have shown that more luminous bulges have a larger best-fit Sérsic index, similar to the relation found for elliptical galaxies (Fig. 2.13): while the relatively massive bulges of early-type spirals have surface brightness profiles with a best-fit Sérsic index $n \sim 4$, the surface brightness profiles of bulges in late-type spirals are better fit with $n \lesssim 1$. In addition, the ratio between the effective radius of the bulge and the disk scale length is found to be roughly independent of Hubble type, with an average of $\langle r_{e,b}/R_d \rangle = 0.22 \pm 0.09$. The fact that the bulge-to-disk ratio increases from late-type to early-type therefore indicates that brighter bulges have a higher surface brightness.

Although the majority of bulges have isophotes that are close to elliptical, a non-negligible fraction of predominantly faint bulges in edge-on, late-type disk galaxies have isophotes that are extremely boxy, or sometimes even have the shape of a peanut. As we will see in §11.5.4, these peanut-shaped bulges are actually bars that have been thickened out of the disk plane.

(b) Colors In general, disk galaxies are bluer than elliptical galaxies of the same luminosity. As discussed in §11.7, this is mainly owing to the fact that disk galaxies are still actively forming stars (young stellar populations are blue). Similar to elliptical galaxies, more luminous disks are redder, although the scatter in this color–magnitude relation is much larger than that for elliptical galaxies. Part of this scatter is simply due to inclination effects, with more inclined disks being more extincted and hence redder, although the intrinsic scatter (corrected for dust extinction) is still significantly larger than for ellipticals. In general, disk galaxies also reveal color gradients, with the outer regions being bluer than the inner regions (e.g. de Jong, 1996b).

Although it is often considered standard lore that disks are blue and bulges are red, this is not supported by actual data. Rather, the colors of bulges are in general very similar to, or at least strongly correlated with, the central colors of their associated disks (e.g. de Jong, 1996a; Peletier & Balcells, 1996; MacArthur et al., 2004). Consequently, bulges also span a wide range in colors.

(c) Disk Vertical Structure Galaxy disks are not infinitesimally thin. Observations suggest that the surface brightness distribution in the 'vertical' (z-) direction is largely independent of the distance R from the disk center. The three-dimensional luminosity density of the disk is therefore typically written in separable form as

$$\nu(R,z) = \nu_0 \exp(-R/R_d) f(z). \tag{2.30}$$

A general fitting function commonly used to describe the luminosity density of disks in the z-direction is

$$f_n(z) = \text{sech}^{2/n}\left(\frac{n|z|}{2z_\text{d}}\right), \qquad (2.31)$$

where n is a parameter controlling the shape of the profile near $z = 0$ and z_d is called the scale height of the disk. Note that all these profiles project to face-on surface brightness profiles given by Eq. (2.29) with $I_0 = a_n \nu_0 z_\text{d}$, with a_n a constant. Three values of n have been used extensively in the literature:

$$f_n(z) = \begin{cases} \text{sech}^2(z/2z_\text{d}) & a_n = 4 \quad n = 1 \\ \text{sech}(z/z_\text{d}) & a_n = \pi \quad n = 2 \\ \exp(-|z|/z_\text{d}) & a_n = 2 \quad n = \infty. \end{cases} \qquad (2.32)$$

The sech^2-form for $n = 1$ corresponds to a self-gravitating isothermal sheet. Although this model has been used extensively in dynamical modeling of disk galaxies (see §11.1), it is generally recognized that the models with $n = 2$ and $n = \infty$ provide better fits to the observed surface brightness profiles. Note that all $f_n(z)$ decline exponentially at large $|z|$; they only differ near the mid-plane, where larger values of n result in steeper profiles. Unfortunately, since dust is usually concentrated near the mid-plane, it is difficult to accurately constrain n. The typical value of the ratio between the vertical and radial scale lengths is $z_\text{d}/R_\text{d} \sim 0.1$, albeit with considerable scatter.

Finally, it is found that most (if not all) disks have excess surface brightness, at large distances from the mid-plane, that cannot be described by Eq. (2.31). This excess light is generally ascribed to a separate 'thick disk' component, whose scale height is typically a factor of 3 larger than for the 'thin disk'. The radial scale lengths of thick disks, however, are remarkably similar to those of their corresponding thin disks, with typical ratios of $R_\text{d,thick}/R_\text{d,thin}$ in the range 1.0–1.5, while the stellar mass ratios $M_\text{d,thick}/M_\text{d,thin}$ decrease from ~ 1 for low mass disks with $V_\text{rot} \lesssim 75\,\text{km}\,\text{s}^{-1}$ to ~ 0.2 for massive disks with $V_\text{rot} \gtrsim 150\,\text{km}\,\text{s}^{-1}$ (Yoachim & Dalcanton, 2006).

(d) Stellar Halos The Milky Way contains a halo of old, metal-poor stars with a density distribution that falls off as a power law, $\rho \propto r^{-\alpha}$ ($\alpha \sim 3$). In recent years, however, it has become clear that the stellar halo reveals a large amount of substructure in the form of stellar streams (e.g. Helmi et al., 1999; Yanny et al., 2003; Bell et al., 2008). These streams are associated with material that has been tidally stripped from satellite galaxies and globular clusters (see §12.2), and in some cases they can be unambiguously associated with their original stellar structure (e.g. Ibata et al., 1994; Odenkirchen et al., 2002). Similar streams have also been detected in our neighbor galaxy, M31 (Ferguson et al., 2002).

However, the detection of stellar halos in more distant galaxies, where the individual stars cannot be resolved, has proven extremely difficult due to the extremely low surface brightnesses involved (typically much lower than that of the sky). Nevertheless, using extremely deep imaging, Sackett et al. (1994) detected a stellar halo around the edge-on spiral galaxy NGC 5907. Later and deeper observations of this galaxy suggest that this extraplanar emission is once again associated with a ring-like stream of stars (Zheng et al., 1999). By stacking the images of hundreds of edge-on disk galaxies, Zibetti et al. (2004) were able to obtain statistical evidence for stellar halos around these systems, suggesting that they are in fact rather common. On the other hand, recent observations of the nearby late-type spiral M33 seem to exclude the presence of a significant stellar halo in this galaxy (Ferguson et al., 2007). Currently the jury is still out as to what fraction of (disk) galaxies contain a stellar halo, and as to what fraction of the halo stars are associated with streams versus a smooth, spheroidal component.

(e) Bars and Spiral Arms More than half of all spirals show bar-like structures in their inner regions. This fraction does not seem to depend significantly on the spiral type, and indeed S0 galaxies are also often barred. Bars generally have isophotes which are more squarish than ellipses and can be fit by the 'generalized ellipse' formula, $(|x|/a)^c + (|y|/b)^c = 1$, where a, b and c are constants and c is substantially larger than 2. Bars are, in general, quite elongated, with axis ratios in their equatorial planes ranging from about 2.5 to 5. Since it is difficult to observe bars in edge-on galaxies, their thickness is not well determined. However, since bars are so common, some limits may be obtained from the apparent thickness of the central regions of edge-on spirals. Such limits suggest that most bars are very flat, probably as flat as the disks themselves, but the bulges complicate this line of argument and it is possible that some bulges (for example, the peanut-shaped bulges) are directly related to bars (see §11.5.4).

Galaxy disks show a variety of spiral structure. 'Grand-design' systems have arms (most frequently two) which can be traced over a wide range of radii and in many, but far from all, cases are clearly related to a strong bar or to an interacting neighbor. 'Flocculent' systems, on the other hand, contain many arm segments and have no obvious large-scale pattern. Spiral arms are classified as leading or trailing according to the sense in which the spiral winds (moving from center to edge) relative to the rotation sense of the disk. Almost all spirals for which an unambiguous determination can be made are trailing.

Spiral structure is less pronounced (though still present) in red light than in blue light. The spiral structure is also clearly present in density maps of atomic and molecular gas and in maps of dust obscuration. Since the blue light is dominated by massive and short-lived stars born in dense molecular clouds, while the red light is dominated by older stars which make up the bulk of the stellar mass of the disk, this suggests that spiral structure is not related to the star-formation process alone, but affects the structure of all components of disks, a conclusion which is more secure for grand-design than for flocculent spirals (see §11.6 for details).

(f) Gas Content Unlike elliptical galaxies which contain gas predominantly in a hot and highly ionized state, the gas component in spiral galaxies is mainly in neutral hydrogen (HI) and molecular hydrogen (H_2). Observations in the 21-cm lines of HI and in the mm-lines of CO have produced maps of the distribution of these components in many nearby spirals (e.g. Young & Scoville, 1991). The gas mass fraction increases from about 5% in massive, early-type spirals (Sa/SBa) to as much as 80% in low mass, low surface brightness disk galaxies (McGaugh & de Blok, 1997). In general, while the distribution of molecular gas typically traces that of the stars, the distribution of HI is much more extended and can often be traced to several Holmberg radii. Analysis of emission from HII regions in spirals provides the primary means for determining their metal abundance (in this case the abundance of interstellar gas rather than of stars). Metallicity is found to decrease with radius. As a rule of thumb, the metal abundance decreases by an order of magnitude for a hundred-fold decrease in surface density. The mean metallicity also correlates with luminosity (or stellar mass), with the metal abundance increasing roughly as the square root of stellar mass (see §2.4.4).

(g) Kinematics The stars and cold gas in galaxy disks move in the disk plane on roughly circular orbits. Therefore, the kinematics of a disk are largely specified by its rotation curve $V_{\rm rot}(R)$, which expresses the rotation velocity as a function of galactocentric distance. Disk rotation curves can be measured using a variety of techniques, most commonly optical long-slit or IFU spectroscopy of HII region emission lines, or radio or millimeter interferometry of line emission from the cold gas. Since the HI gas is usually more extended than the ionized gas associated with HII regions, rotation curves can be probed out to larger galactocentric radii using spatially resolved 21-cm observations than using optical emission lines. Fig. 2.21 shows two examples of disk rotation curves. For massive galaxies these typically rise rapidly at small radii and then are almost constant over most of the disk. In dwarf and lower surface brightness systems a slower central rise is common. There is considerable variation from system to system, and features in rotation curves are often associated with disk structures such as bars or spiral arms.

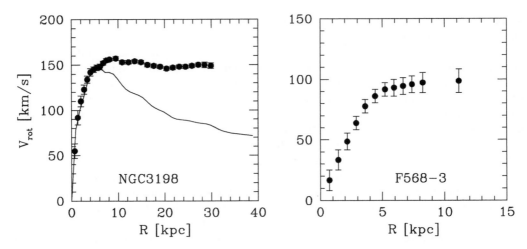

Fig. 2.21. The rotation curves of the Sc galaxy NGC 3198 (left) and the low surface brightness galaxy F568-3 (right). The curve in the left panel shows the contribution from the disk mass assuming a mass-to-light ratio of $3.8\,{\rm M}_\odot/{\rm L}_\odot$. [Based on data published in Begeman (1989) and Swaters et al. (2000)]

The rotation curve is a direct measure of the gravitational force within a disk. Assuming, for simplicity, spherical symmetry, the total enclosed mass within radius r can be estimated from

$$M(r) = rV_{\rm rot}^2(r)/G. \qquad (2.33)$$

In the outer region, where $V_{\rm rot}(r)$ is roughly a constant, this implies that $M(r) \propto r$, so that the enclosed mass of the galaxy (unlike its enclosed luminosity) does not appear to be converging. For the rotation curve of NGC 3198 shown in Fig. 2.21, the last measured point corresponds to an enclosed mass of $1.5 \times 10^{11}\,{\rm M}_\odot$, about four times larger than the stellar mass. Clearly, the asymptotic total mass could even be much larger than this. The fact that the observed rotation curves of spiral galaxies are flat at the outskirts of their disks is evidence that they possess massive halos of unseen, dark matter. This is confirmed by studies of the kinematics of satellite galaxies and of gravitational lensing, both suggesting that the enclosed mass continues to increase roughly with radius out to at least 10 times the Holmberg radius.

The kinematics of bulges are difficult to measure, mainly because of contamination by disk light. Nevertheless, the existing data suggests that the majority are rotating rapidly (consistent with their flattened shapes being due to the centrifugal forces), and in the same sense as their disk components.

(h) Tully–Fisher Relation Although spiral galaxies show great diversity in luminosity, size, rotation velocity and rotation-curve shape, they obey a well-defined scaling relation between luminosity L and rotation velocity (usually taken as the maximum of the rotation curve well away from the center, $V_{\rm max}$). This is known as the Tully–Fisher relation, an example of which is shown in Fig. 2.22. The observed Tully–Fisher relation is usually expressed in the form $L = AV_{\rm max}^\alpha$, where A is the zero-point and α is the slope. The observed value of α is between 2.5 and 4, and is larger in redder bands (e.g. Pierce & Tully, 1992). For a fixed $V_{\rm max}$, the scatter in luminosity is typically 20%. This tight relation can be used to estimate the distances to spiral galaxies, using the principle described in §2.1.3(c). However, as we show in Chapter 11, the Tully–Fisher relation is also important for our understanding of galaxy formation and evolution, as it defines a relation between dynamical mass (due to stars, gas, and dark matter) and luminosity.

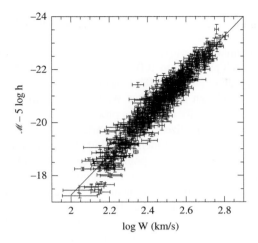

Fig. 2.22. The Tully–Fisher relation in the I-band. Here W is the linewidth of the HI 21-cm line which is roughly equal to twice the maximum rotation velocity, $V_{\max}$. [Adapted from Giovanelli et al. (1997) by permission of AAS]

2.3.4 The Milky Way

We know much more about our own Galaxy, the Milky Way, than about most other galaxies, simply because our position within it allows its stellar and gas content to be studied in considerable detail. This 'internal perspective' also brings disadvantages, however. For example, it was not demonstrated until the 1920s and 30s that the relatively uniform brightness of the Milky Way observed around the sky does not imply that we are close to the center of the system, but rather is a consequence of obscuration of distant stars by dust. This complication, combined with the problem of measuring distances, is the main reason why many of the Milky Way's large scale properties (e.g. its total luminosity, its radial structure, its rotation curve) are still substantially more uncertain than those of some external galaxies.

Nevertheless, we believe that the Milky Way is a relatively normal spiral galaxy. Its main baryonic component is the thin stellar disk, with a mass of $\sim 5 \times 10^{10} \, M_\odot$, a radial scale length of $\sim 3.5 \, \text{kpc}$, a vertical scale height of $\sim 0.3 \, \text{kpc}$, and an overall diameter of $\sim 30 \, \text{kpc}$. The Sun lies close to the mid-plane of the disk, about 8 kpc from the Galactic center, and rotates around the center of the Milky Way with a rotation velocity of $\sim 220 \, \text{km s}^{-1}$. In addition to this thin disk component, the Milky Way also contains a thick disk whose mass is 10–20% of that of the thin disk. The vertical scale height of the thick disk is $\sim 1 \, \text{kpc}$, but its radial scale length is remarkably similar to that of the thin disk. The thick disk rotates slower than the thin disk, with a rotation velocity at the solar radius of $\sim 175 \, \text{km s}^{-1}$.

In addition to the thin and thick disks, the Milky Way also contains a bulge component with a total mass of $\sim 10^{10} \, M_\odot$ and a half-light radius of $\sim 1 \, \text{kpc}$, as well as a stellar halo, whose mass is only about 3% of that of the bulge despite its much larger radial extent. The stellar halo has a radial number density distribution $n(r) \propto r^{-\alpha}$, with $2 \lesssim \alpha \lesssim 4$, reaches out to at least 40 kpc, and shows no sign of rotation (i.e. its structure is supported against gravity by random rather than ordered motion). The structure and kinematics of the bulge are more complicated. The near-infrared image of the Milky Way, obtained with the COBE satellite, shows a modest, somewhat boxy bulge. As discussed in §11.5.4, it is believed that these boxy bulges are actually bars. This bar-like nature of the Milky Way bulge is supported by the kinematics of atomic and molecular gas in the inner few kiloparsecs (Binney et al., 1991), by microlensing measurements of the bulge (Zhao et al., 1995), and by asymmetries in the number densities of various types of stars (Whitelock & Catchpole, 1992; Stanek et al., 1994; Sevenster, 1996). The very center of the Milky Way is also known to harbor a supermassive black hole with a mass approximately $2 \times 10^6 M_\odot$. Its presence is unambiguously inferred from the radial velocities, proper motions and accelerations of stars which pass within 100 astronomical units (1.5×10^{15} cm) of the central object (Genzel et al., 2000; Schödel et al., 2003; Ghez et al., 2005).

During World War II the German astronomer W. Baade was interned at Mount Wilson in California, where he used the unusually dark skies produced by the blackout to study the stellar populations of the Milky Way. He realized that the various components are differentiated not only by their spatial distributions and their kinematics, but also by their age distributions and their chemical compositions. He noted that the disk population (which he called Population I) contains stars of all ages and with heavy element abundances ranging from about 0.2 to 1 times solar. The spheroidal component (bulge plus halo), which he called Population II, contains predominantly old stars and near the Sun its heavy element abundances are much lower than in the disk. More recent work has shown that younger disk stars are more concentrated to the mid-plane than older disk stars, that disk stars tend to be more metal-rich near the Galactic center than at large radii, and that young disk stars tend to be somewhat more metal-rich than older ones. In addition, it has become clear that the spheroidal component contains stars with a very wide range of metal abundances. Although the majority are within a factor of 2 or 3 of the solar value, almost the entire metal-rich part of the distribution lies in the bulge. At larger radii the stellar halo is predominantly metal-poor with a metallicity distribution reaching down to very low values: the current record holder has an iron content that is about 200,000 times smaller than that of the Sun! Finally, the relative abundances of specific heavy elements (for example, Mg and Fe) differ systematically between disk and spheroid. As we will see in Chapter 10, all these differences indicate that the various components of the Milky Way have experienced very different star-formation histories (see also §11.8).

The Milky Way also contains about $5 \times 10^9 M_\odot$ of cold gas, almost all of which is moving on circular orbits close to the plane of the disk. The majority of this gas ($\sim 80\%$) is neutral, atomic hydrogen (HI), which emits radio emission at 21 cm. The remaining $\sim 20\%$ the gas is in molecular form and is most easily traced using millimeter-wave line emission from carbon monoxide (CO). The HI has a scale height of ~ 150 pc and a velocity dispersion of ~ 9 km s^{-1}. Between 4 and 17 kpc its surface density is roughly constant, declining rapidly at both smaller and larger radii. The molecular gas is more centrally concentrated than the atomic gas, and mainly resides in a ring-like distribution at ~ 4.5 kpc from the center, and with a FWHM of ~ 2 kpc. Its scale height is only ~ 50 pc, while its velocity dispersion is ~ 7 km s^{-1}, somewhat smaller than that of the atomic gas. The molecular gas is arranged in molecular cloud complexes with typical masses in the range 10^5–$10^7 M_\odot$ and typical densities of order 100 atoms per cm^3. New stars are born in clusters and associations embedded in the dense, dust-enshrouded cores of these molecular clouds (see Chapter 9). If a star-forming region contains O and B stars, their UV radiation soon creates an ionized bubble, an 'HII region', in the surrounding gas. Such regions produce strong optical line emission which makes them easy to identify and to observe. Because of the (ongoing) star formation, the ISM is enriched with heavy elements. In the solar neighborhood, the metallicity of the ISM is close to that of the Sun, but it decreases by a factor of a few from the center of the disk to its outer edge.

Three other diffuse components of the Milky Way are observed at levels which suggest that they may significantly influence its evolution. Most of the volume of the Galaxy near the Sun is occupied by hot gas at temperatures of about 10^6 K and densities around 10^{-4} atoms per cm^3. This gas is thought to be heated by stellar winds and supernovae and contains much of the energy density of the ISM. A similar energy density resides in relativistic protons and electrons (cosmic rays) which are thought to have been accelerated primarily in supernova shocks. The third component is the Galactic magnetic field which has a strength of a few μG, is ordered on large scales, and is thought to play a significant role in regulating star formation in molecular clouds.

The final and dominant component of the Milky Way appears to be its dark halo. Although the 'dark matter' out of which this halo is made has not been observed directly (except perhaps for a small fraction in the form of compact objects, see §2.10.2), its presence is inferred from the outer rotation curve of the Galaxy, from the high velocities of the most extreme local Population II stars, from the kinematics of globular star clusters and dwarf galaxies in the stellar halo, and from the infall speed of our giant neighbor, the Andromeda Nebula. The estimated total mass of this unseen distribution of dark matter is about $10^{12} M_\odot$ and it is thought to extend well beyond 100 kpc from the Galactic center.

2.3.5 Dwarf Galaxies

For historical reasons, galaxies with $\mathcal{M}_B \gtrsim -18$ are often called dwarf galaxies (Sandage & Binggeli, 1984). These galaxies span roughly six orders of magnitude in luminosity, although the faint end is subject to regular changes as fainter and fainter galaxies are constantly being discovered. The current record holder is Willman I, a dwarf spheroidal galaxy in the local group with an estimated magnitude of $\mathcal{M}_V \simeq -2.6$ (Willman et al., 2005; Martin et al., 2007).

By number, dwarfs are the most abundant galaxies in the Universe, but they contain a relatively small fraction of all stars. Their structure is quite diverse, and they do not fit easily into the Hubble sequence. The clearest separation is between gas-rich systems with ongoing star formation – the dwarf irregulars (dIrr) – and gas-poor systems with no young stars – the dwarf ellipticals (dE) and dwarf spheroidals (dSph). Two examples of them are shown in Fig. 2.23.

Fig. 2.24 sketches the regions in the parameter space of effective radius and absolute magnitude that are occupied by different types of galaxies. Spirals and dwarf irregulars cover roughly four orders of magnitude in luminosity, almost two orders of magnitude in size, and about three orders of magnitude in surface brightness. As their name suggests, dwarf irregulars have highly irregular structures, often being dominated by one or a few bright HII regions. Their gas content increases with decreasing mass and in extreme objects, such as blue compact dwarfs, the so-called 'extragalactic HII regions', the HI extent can be many times larger than the visible galaxy. The larger systems seem to approximate rotationally supported disks, but the smallest systems show quite chaotic kinematics. The systems with regular rotation curves

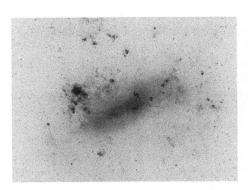
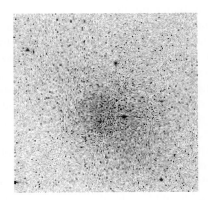

Fig. 2.23. Images of two dwarf galaxies: the Large Magellanic Cloud (LMC, left panel), which is a proto typical dwarf irregular, and the dwarf spheroidal Fornax (right panel). [Courtesy of NASA/IPAC Extragalactic Database]

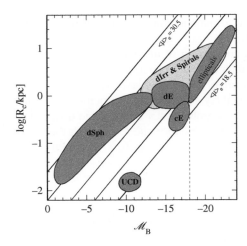

Fig. 2.24. A sketch of the regions in the parameter space of effective radius and absolute magnitude (both in the B-band) occupied by different types of galaxies. The spheroidal systems are split into ellipticals, dwarf ellipticals (dE), compact ellipticals (cE), dwarf spheroidals (dSph), and ultra-compact dwarfs (UCD). The dashed, vertical line corresponds to $\mathscr{M}_B = -18$, and reflects the magnitude limit below which galaxies are often classified as dwarfs. The diagonal lines are lines of constant surface brightness; galaxies roughly span five orders of magnitude in surface brightness, from $\langle \mu_B \rangle_e \sim -18.5$ to $\langle \mu_B \rangle_e \sim -30.5$.

often appear to require substantial amounts of dark matter even within the visible regions of the galaxy.

Dwarf ellipticals are gas-poor systems found primarily in groups and clusters of galaxies. Their structure is regular, with luminosity profiles closer to exponential than to the de Vaucouleurs law (see Fig. 2.13). In addition, they have lower metallicities than normal ellipticals, although they seem to follow the same relation between metallicity and luminosity.

Dwarf spheroidals (dSphs) are faint objects of very low surface brightness, which have so far only been identified unambiguously within the Local Group (see §2.5.2). Their structure is relatively regular and they appear to contain no gas and no, or very few, young stars with ages less than about 1 Gyr. However, several dSphs show unambiguous evidence for several distinct bursts of star formation. Their typical sizes range from a few tens to several hundreds of parsec, while their luminosities span almost five orders of magnitude. Their kinematics indicate dynamical mass-to-light ratios that can be as large as several hundred times that of the Sun, which is usually interpreted as implying a large dark matter content (Mateo, 1998; Gilmore et al., 2007). One of the most luminous dSphs, the Sagittarius dwarf, currently lies only about 20 kpc from the center of the Milky Way and is being torn apart by the Milky Way's tidal forces.

The distinction between 'dwarf' and 'regular' galaxies had its origin in the observation that ellipticals with $\mathscr{M}_B \gtrsim -18$ are not well described by the de Vaucouleurs $R^{1/4}$ law. Instead, their surface brightness profiles were found to be closer to exponential (e.g. Faber & Lin, 1983; Binggeli et al., 1984). This distinction was further strengthened by the work of Kormendy (1985) who found that bright ellipticals have their surface brightness decrease with increasing luminosity, while dEs have increasing surface brightness with increasing luminosity (see Fig. 2.14). This gave rise to the concept of a clear dichotomy between dwarf and regular ellipticals. More recently, however, it has been argued that this 'dichotomy', with a characteristic scale at $\mathscr{M}_B \simeq -18$, is an artefact of sample selection and of the fact that the surface brightness profiles were fit with either an $R^{1/4}$ profile or an exponential. Fitting with the more general Sérsic profiles instead indicates clearly that there is a smooth trend between the best-fit Sérsic index and absolute magnitude (see Fig. 2.13) and an equally smooth trend between absolute magnitude and *central* surface brightness (see Graham & Guzmán, 2003, and references therein). Hence, there seems to be no clear distinction between dEs and 'regular' ellipticals. Neither is there a clear distinction between dEs and dSphs; the latter simply make up the low luminosity extreme of the dEs, typically with $\mathcal{M}_B \gtrsim -14$. Although we will adhere to the 'historical' nomenclature throughout this book, we caution that there is no clear physical motivation for discriminating between dSphs, dEs, and 'regular' ellipticals (but see §13.6).

Fig. 2.24 also sketches the location in size–luminosity space occupied by a special class of (dwarf) galaxies known as compact ellipticals (cEs). These are characterized by unusually high surface brightness for their luminosity, although they do seem to form a smooth continuation of the size–luminosity relation of 'regular' ellipticals. The prototypical example is M32, a companion of the Andromeda Galaxy, M31. Compact ellipticals are very rare, and only a handful of these systems are known. Some authors have argued that the bulges of (early-type) disk galaxies occupy the same region in parameter space as the cEs, suggesting that these two types of objects are somehow related (e.g. Bender et al., 1992). Finally, Drinkwater et al. (2003) have recently identified a new class of (potential) galaxies, called ultra-compact dwarfs (UCDs). They typically have $\mathcal{M}_B \sim -11$ and effective radii of 10–20 pc, giving them an average surface brightness comparable to that of cEs. Their nature is still very uncertain. In particular, it is still unclear whether they should be classified as galaxies, or whether they merely reflect the bright end of the population of globular clusters. Alternatively, they may also be the remnant nuclei of disrupted low surface brightness galaxies (see below).

2.3.6 Nuclear Star Clusters

In their landmark study of the Virgo Cluster, Binggeli et al. (1987) found that $\sim 25\%$ of the dEs contain a massive star cluster at their centers (called the nucleus), which clearly stands out against the low surface brightness of its host galaxy. Following this study it has become customary to split the population of dEs into 'nucleated' and 'non-nucleated'. Binggeli et al. (1987) did not detect any nuclei in the more luminous ellipticals, although they cautioned that these might have been missed in their photographic survey due to the high surface brightness of the underlying galaxy. Indeed, more recent studies, capitalizing on the high spatial resolution afforded by the HST, have found that as much as $\sim 80\%$ of all early-type galaxies with $\mathcal{M}_B \lesssim -15$ are nucleated (e.g. Grant et al., 2005; Côté et al., 2006). In addition, HST imaging of late-type galaxies has revealed that 50–70% of these systems also have compact stellar clusters near their photometric centers (e.g. Phillips et al., 1996; Böker et al., 2002). These show a remarkable similarity in luminosity and size to those detected in early-type galaxies. However, the nuclear star clusters in late-type galaxies seem to have younger stellar ages than their counterparts in early-type galaxies (e.g. Walcher et al., 2005; Côté et al., 2006). Thus a large fraction of all galaxies, independent of their morphology, environment or gas content, contain a nuclear star cluster at their photometric center. The only exception seem to be the brightest ellipticals, with $\mathcal{M}_B \lesssim -20.5$, which seem to be devoid of nuclear star clusters. Note that this magnitude corresponds to the transition from disky, power-law ellipticals to boxy, core ellipticals (see §2.3.2), supporting the notion of a fundamental transition at this luminosity scale.

On average, nuclear star clusters are an order of magnitude more luminous than the peak of the globular cluster luminosity function of their host galaxies, have stellar masses in the range $\sim 10^6$–$10^8 \, \mathrm{M_\odot}$, and typical radii of $\sim 5 \, \mathrm{pc}$. This makes nuclear star clusters the densest stellar systems known (e.g. Geha et al., 2002; Walcher et al., 2005). In fact, they are not that dissimilar to the ultra-compact dwarfs, suggesting a possible relation (e.g. Bekki et al., 2001).

As discussed in §2.3.2 (see also §13.1.4), the majority of bright spheroids (ellipticals and bulges) seem to contain a supermassive black hole (SMBH) at their nucleus. The majority of spheroids with secure SMBH detections have magnitudes in the range $-22 \lesssim \mathcal{M}_B \lesssim -18$.

Although it is unclear whether (the majority of) fainter spheroids also harbor SMBHs, current data seems to support a view in which bright galaxies ($\mathcal{M}_B \lesssim -20$) often, and perhaps always, contain SMBHs but not stellar nuclei, while at the faint end ($\mathcal{M}_B \gtrsim -18$) stellar nuclei become the dominant feature. Intriguingly, Ferrarese et al. (2006a) have shown that stellar nuclei and SMBHs obey a common scaling relation between their mass and that of their host galaxy, with $M_{\rm CMO}/M_{\rm gal} = 0.018^{+0.034}_{-0.012}$ (where CMO stands for central massive object), suggesting that SMBHs and nuclear clusters share a common origin. This is somewhat clouded, though, by the fact that nuclear star clusters and SMBHs are not mutually exclusive. The two best known cases in which SMBHs and stellar nuclei coexist are M32 (Verolme et al., 2002) and the Milky Way (Ghez et al., 2003; Schödel et al., 2003).

2.3.7 Starbursts

In normal galaxies like the Milky Way, the specific star-formation rates are typically of order $0.1\,{\rm Gyr}^{-1}$, which implies star-formation time scales (defined as the ratio between the total stellar mass and the current star-formation rate) that are comparable to the age of the Universe. There are, however, systems in which the (specific) star-formation rates are 10 or even 100 times higher, with implied star-formation time scales as short as 10^8 years. These galaxies are referred to as starbursts. The star-formation activity in such systems (at least in the most massive ones) is often concentrated in small regions, with sizes typically about 1 kpc, much smaller than the disk sizes in normal spiral galaxies.

Because of the large current star-formation rate, a starburst contains a large number of young stars. Indeed, for blue starbursts where the star-formation regions are not obscured by dust, their spectra generally have strong blue continuum produced by massive stars, and show strong emission lines from HII regions produced by the UV photons of O and B stars (see Fig. 2.12). Since the formation of stars is, in general, associated with the production of large amounts of dust,[2] most of the strong starbursts are not observed directly via their strong UV emission. Rather, the UV photons produced by the young stars are absorbed by dust and re-emitted in the far-infrared. In extreme cases these starbursting galaxies emit the great majority of their light in the infrared, giving rise to the population of infrared luminous galaxies (LIRGs) discovered in the 1980s with the Infrared Astronomical Satellite (IRAS). A LIRG is defined as a galaxy with a far-infrared luminosity exceeding $10^{11}\,L_\odot$ (Soifer et al., 1984). If its far-infrared luminosity exceeds $10^{12}\,L_\odot$ it is called an ultraluminous infrared galaxy (ULIRG).

The fact that starbursts are typically confined to a small region (usually the nucleus) of the starbursting galaxy, combined with their high star-formation rates, requires a large amount of cold gas to be accumulated in a small region in a short time. The most efficient way of achieving this is through mergers of gas-rich galaxies, where the interstellar media of the merging systems can be strongly compressed and concentrated by tidal interactions (see §12.4.3). This scenario is supported by the observation that massive starbursts (in particular ULIRGs) are almost exclusively found in strongly interacting systems with peculiar morphologies.

2.3.8 Active Galactic Nuclei

The centers of many galaxies contain small, dense and luminous components known as active galactic nuclei (AGN). An AGN can be so bright that it outshines its entire host galaxy, and differs from a normal stellar system in its emission properties. While normal stars emit radiation primarily in a relatively narrow wavelength range between the near-infrared and the near-UV, AGN are powerful emitters of non-thermal radiation covering the entire electromagnetic spectrum from the

---

[2] It is believed that dust is formed in the atmospheres of evolved stars and in supernova explosions.

radio to the gamma-ray regime. Furthermore, the spectra of many AGN contain strong emission lines and so contrast with normal stellar spectra which are typically dominated by absorption lines (except for galaxies with high specific star-formation rates). According to their emission properties, AGN are divided into a variety of subclasses, including radio sources, Seyferts, liners, blazars and quasars (see Chapter 14 for definitions).

Most of the emission from an AGN comes from a very small, typically unresolved region; high-resolution observations of relatively nearby objects with HST or with radio interferometry demonstrate the presence of compact emitting regions with sizes smaller than a few parsecs. These small sizes are consistent with the fact that some AGN reveal strong variability on time scales of only a few days, indicating that the emission must emanate from a region not much larger than a few light-days across. The emission from these nuclei typically reveals a relatively featureless power-law continuum at radio, optical and X-ray wavelengths, as well as broad emission lines in the optical and X-ray bands. On somewhat larger scales, AGN often manifest themselves in radio, optical and even X-ray jets, and in strong but narrow optical emission lines from hot gas. The most natural explanation for the energetics of AGN, combined with their small sizes, is that AGN are powered by the accretion of matter onto a supermassive black hole (SMBH) with a mass of 10^6–$10^9 \, M_\odot$. Such systems can be extremely efficient in converting gravitational energy into radiation. As mentioned in §2.3.2, virtually all spheroidal galaxy components (i.e. ellipticals and bulges) harbor a SMBH whose mass is tightly correlated with that of the spheroid, suggesting that the formation of SMBHs is tightly coupled to that of their host galaxies. Indeed, the enormous energy output of AGN may have an important feedback effect on the formation and evolution of galaxies. Given their importance for galaxy formation, Chapter 14 is entirely devoted to AGN, including a more detailed overview of their observational properties.

2.4 Statistical Properties of the Galaxy Population

So far our description has focused on the properties of separate classes of galaxies. We now turn our attention to statistics that describe the galaxy population as a whole, i.e. that describe how galaxies are distributed with respect to these properties. As we will see in §§2.5 and 2.7, the galaxy distribution is strongly clustered on scales up to ~ 10 Mpc, which implies that one needs to probe a large volume in order to obtain a sample that is representative of the entire population. Therefore, the statistical properties of the galaxy population are best addressed using large galaxy redshift surveys. Currently the largest redshift surveys available are the two-degree Field Galaxy Redshift Survey (2dFGRS; Colless et al., 2001a) and the Sloan Digital Sky Survey (SDSS; York et al., 2000), both of which probe the galaxy distribution at a median redshift $z \sim 0.1$. The 2dFGRS has measured redshifts for $\sim 220,000$ galaxies over ~ 2000 square degrees down to a limiting magnitude of $b_j \sim 19.45$. The source catalogue for the survey is the APM galaxy catalogue, which is based on Automated Plate Measuring machine (APM) scans of photographic plates (Maddox et al., 1990b). The SDSS consists of a photometrically and astrometrically calibrated imaging survey covering more than a quarter of the sky in five broad optical bands (u, g, r, i, z) that were specially designed for the survey (Fukugita et al., 1996), plus a spectroscopic survey of $\sim 10^6$ galaxies ($r < 17.77$) and $\sim 10^5$ quasars detected in the imaging survey.

The selection function of these and other surveys plays an important role in the observed sample properties. For example, most surveys select galaxies above a given flux limit (i.e. the survey is complete down to a given apparent magnitude). Since intrinsically brighter galaxies will reach the flux limit at larger distances, a flux limited survey is biased towards brighter galaxies. This is called the Malmquist bias and needs to be corrected for when trying to infer the intrinsic probability distribution of galaxies. There are two ways to do this. One is to construct a volume limited sample, by only selecting galaxies brighter than a given absolute magnitude limit, M_{lim},

Table 2.6. Relative number densities of galaxies in the local Universe.

Type of object	Number density
Spirals	1
Lenticulars	0.1
Ellipticals	0.2
Irregulars	0.05
Dwarf galaxies	10
Peculiar galaxies	0.05
Starbursts	0.1
Seyferts	10^{-2}
Radio galaxies	10^{-4}
QSOs	10^{-5}
Quasars	10^{-7}

and below a given redshift, z_{lim}, where z_{lim} is the redshift at which a galaxy with absolute magnitude M_{lim} has an apparent magnitude equal to the survey limit. Alternatively, one can weight each galaxy by the inverse of V_{max}, defined as the survey volume out to which the specific galaxy in question could have been detected given the flux limit of the survey. The advantage of this method over the construction of volume-limited samples is that one does not have to discard any data. However, the disadvantage is that intrinsically faint galaxies can only be seen over a relatively small volume (i.e. V_{max} is small), so that they get very large weights. This tends to make the measurements extremely noisy.

As a first example of a statistical description of the galaxy population, Table 2.4 lists the number densities of the various classes of galaxies described in the previous section, relative to that of spiral galaxies. Note, however, that these numbers are only intended as a rough description of the galaxy population in the nearby Universe. The real galaxy population is extremely diverse, and an accurate description of the galaxy number density is only possible for a well-defined sample of galaxies.

2.4.1 Luminosity Function

Arguably one of the most fundamental properties of a galaxy is its luminosity (in some waveband). An important statistic of the galaxy distribution is therefore the luminosity function, $\phi(L)dL$, which describes the number density of galaxies with luminosities in the range $L \pm dL/2$. Fig. 2.25 shows the luminosity function in the photometric b_j-band obtained from the 2dFGRS. At the faint end $\phi(L)$ seems to follow a power-law which truncates at the bright end, where the number density falls roughly exponentially. A similar behavior is also seen in other wavebands, so that the galaxy luminosity function is commonly fitted by a Schechter function (Schechter, 1976) of the form

$$\phi(L)dL = \phi^* \left(\frac{L}{L^*}\right)^\alpha \exp\left(-\frac{L}{L^*}\right) \frac{dL}{L^*}. \qquad (2.34)$$

Here L^* is a characteristic luminosity, α is the faint-end slope, and ϕ^* is an overall normalization. As shown in Fig. 2.25, this function fits the observed luminosity function over a wide range. From the Schechter function, we can write the mean number density, n_g, and the mean luminosity density, $\mathscr{L}$, of galaxies in the Universe as

$$n_g \equiv \int_0^\infty \phi(L)\,dL = \phi^* \Gamma(\alpha+1), \qquad (2.35)$$

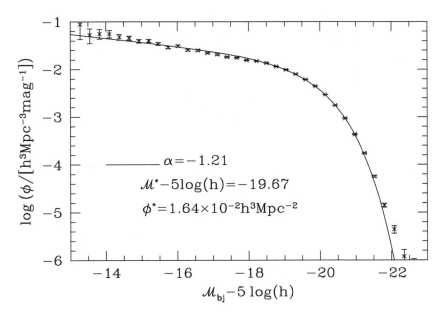

Fig. 2.25. The luminosity function of galaxies in the b_j-band as obtained from the 2-degree Field Galaxy Redshift Survey. [Based on data published in Norberg et al. (2002b)]

and

$$\mathscr{L} \equiv \int_0^\infty \phi(L) L \, \mathrm{d}L = \phi^* L^* \Gamma(\alpha + 2), \qquad (2.36)$$

where $\Gamma(x)$ is the gamma function. Note that n_g diverges for $\alpha \leq -1$, while $\mathscr{L}$ diverges for $\alpha \leq -2$. Observations from the near-UV to the near-infrared show that $-2 < \alpha < -1$, indicating that the number density is dominated by faint galaxies while the luminosity density is dominated by bright ones.

As we will see in Chapter 15, the luminosity function of galaxies depends not only on the waveband, but also on the morphological type, the color, the redshift, and the environment of the galaxy. One of the most challenging problems in galaxy formation is to explain the general shape of the luminosity function and the dependence on other galaxy properties.

2.4.2 Size Distribution

Size is another fundamental property of a galaxy. As shown in Figs. 2.14 and 2.20, galaxies of a given luminosity may have very different sizes (and therefore surface brightnesses). Based on a large sample of galaxies in the SDSS, Shen et al. (2003) found that the size distribution for galaxies of a given luminosity L can roughly be described by a log-normal function,

$$P(R|L) \, \mathrm{d}R = \frac{1}{\sqrt{2\pi}\sigma_{\ln R}} \exp\left[-\frac{\ln^2(R/\overline{R})}{2\sigma_{\ln R}^2}\right] \frac{\mathrm{d}R}{R}, \qquad (2.37)$$

where $\overline{R}$ is the median and $\sigma_{\ln R}$ the dispersion. Fig. 2.26 shows that $\overline{R}$ increases with galaxy luminosity roughly as a power law for both early-type and late-type galaxies, and that the dependence is stronger for early types. The dispersion $\sigma_{\ln R}$, on the other hand, is similar for both early-type

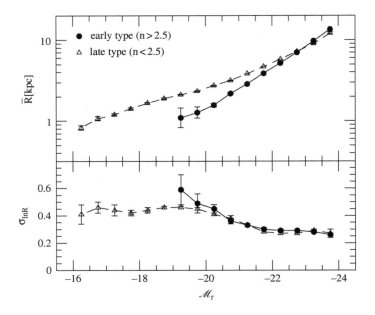

Fig. 2.26. The median (upper panel) and dispersion (lower panel) of the size distribution of galaxies in the SDSS as function of their r-band magnitude. Results are shown separately for early-type (solid dots) and late-type (open triangles) galaxies defined according to the Sérsic index n. [Kindly provided to us by S. Shen, based on data published in Shen et al. (2003)]

and late-type galaxies, decreasing from ~ 0.5 for galaxies with $M_r \gtrsim -20.5$ to ~ 0.25 for brighter galaxies.

2.4.3 Color Distribution

As shown in Fig. 2.5, massive stars emit a larger fraction of their total light at short wavelengths than low-mass stars. Since more massive stars are in general shorter-lived, the color of a galaxy carries important information about its star-formation history. However, the color of a star also depends on its metallicity, in the sense that stars with higher metallicities are redder. In addition, dust extinction is more efficient at bluer wavelengths, so that the color of a galaxy also contains information regarding its chemical composition and dust content.

The left panel of Fig. 2.27 shows the distribution of the $^{0.1}(g-r)$ colors of galaxies in the SDSS, where the superscript indicates that the magnitudes have been converted to the same rest-frame wavebands at $z = 0.1$. The most salient characteristic of this distribution is that it is clearly bimodal, revealing a relatively narrow peak at the red end of the distribution plus a significantly broader distribution at the blue end. To first order, this simply reflects that galaxies come in two different classes: early-type galaxies, which have relatively old stellar populations and are therefore red, and late-type galaxies, which have ongoing star formation in their disks and are therefore blue. However, it is important to realize that this color–morphology relation is not perfect: a disk galaxy may be red due to extensive dust extinction, while an elliptical may be blue if it had a small amount of star formation in the recent past.

The bimodality of the galaxy population is also evident from the color–magnitude relation, plotted in the right-hand panel of Fig. 2.27. This shows that the galaxy population is divided into a red sequence and a blue sequence (also sometimes called the blue cloud). Two trends are noteworthy. First of all, at the bright end the red sequence dominates, while at the faint end the majority of the galaxies are blue. As we will see in Chapter 15, this is consistent with the fact

2.4 Statistical Properties of the Galaxy Population

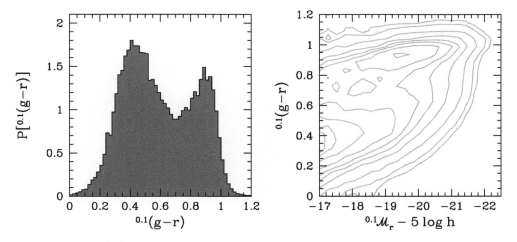

Fig. 2.27. The probability density of galaxy colors (left) and the color–magnitude relation (right) of ∼ 365,000 galaxies in the SDSS. Each galaxy has been weighted by $1/V_{\max}$ to correct for Malmquist bias. Note the pronounced bimodality in the color distribution, and the presence of both a red sequence and a blue sequence in the color–magnitude relation.

that the bright (faint) end of the galaxy luminosity function is dominated by early-type (late-type) galaxies. Secondly, within each sequence brighter galaxies appear to be redder. As we will see in Chapters 11 and 13 this most likely reflects that the stellar populations in brighter galaxies are both older and more metal rich, although it is still unclear which of these two effects dominates, and to what extent dust plays a role.

2.4.4 The Mass–Metallicity Relation

Another important parameter to characterize a galaxy is its average metallicity, which reflects the amount of gas that has been reprocessed by stars and exchanged with its surroundings. One can distinguish two different metallicities for a given galaxy: the average metallicity of the stars and that of the gas. Depending on the star-formation history and the amount of inflow and outflow, these metallicities can be significantly different. Gas-phase metallicities can be measured from the emission lines in a galaxy spectrum, while the metallicity of the stars can be obtained from the absorption lines which originate in the atmospheres of the stars.

Fig. 2.28 shows the relation between the gas-phase oxygen abundance and the stellar mass of SDSS galaxies. The oxygen abundance is expressed as $12 + \log[(\mathrm{O/H})]$, where O/H is the abundance by number of oxygen relative to hydrogen. Since the measurement of gas-phase abundances requires the presence of emission lines in the spectra, all these galaxies are still forming stars, and the sample is therefore strongly biased towards late-type galaxies. Over about three orders of magnitude in stellar mass the average gas-phase metallicity increases by an order of magnitude. The relation is remarkably tight and reveals a clear flattening above a few times $10^{10} \, \mathrm{M_\odot}$. The average stellar metallicity follows a similar trend with stellar mass but with much larger scatter at the low-mass end (Gallazzi et al., 2005). An interpretation of these results in terms of the chemical evolution of galaxies is presented in Chapter 10.

2.4.5 Environment Dependence

As early as the 1930s it was realized that the morphological mix of galaxies depends on environment, with denser environments (e.g. clusters, see §2.5.1) hosting larger fractions of early-type

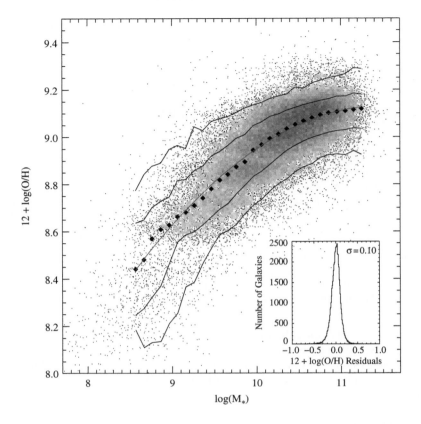

Fig. 2.28. The relation between stellar mass, in units of solar masses, and the gas-phase oxygen abundance for ∼53,400 star-forming galaxies in the SDSS. For comparison, the Sun has $12 + \log[(O/H)] = 8.69$. The large black points represent the median in bins of 0.1 dex in mass. The solid lines are the contours which enclose 68% and 95% of the data. The gray line shows a polynomial fit to the data. The inset shows the residuals of the fit. [Adapted from Tremonti et al. (2004) by permission of AAS]

galaxies (Hubble & Humason, 1931). This morphology–density relation was quantified more accurately in a paper by Dressler (1980b), who studied the morphologies of galaxies in 55 clusters and found that the fraction of spiral galaxies decreases from ∼ 60% in the lowest density regions to less than 10% in the highest density regions, while the elliptical fraction basically reveals the opposite behavior (see Fig. 2.29). Note that the fraction of S0 galaxies is significantly higher in clusters than in the general field, although there is no strong trend of S0 fraction with density within clusters.

More recently, the availability of large galaxy redshift surveys has paved the way for far more detailed studies into the environment dependence of galaxy properties. It is found that in addition to a larger fraction of early-type morphologies, denser environments host galaxies that are on average more massive, redder, more concentrated, less gas-rich, and have lower specific star-formation rates (e.g. Kauffmann et al., 2004; Baldry et al., 2006; Weinmann et al., 2006b). Interpreting these findings in terms of galaxy formation processes, however, is complicated by the fact that various galaxy properties are strongly correlated even at a fixed environment. An important outstanding question, therefore, is which relationship with environment is truly causal, and which are just reflections of other correlations that are actually independent of environment (see §15.5 for a more detailed discussion).

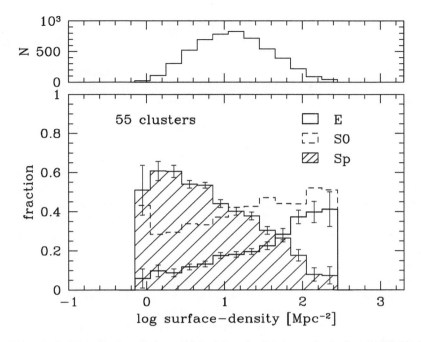

Fig. 2.29. The morphology–density relation, which shows the fractions of galaxies of individual morphological types as functions of galaxy surface number density. The lower panel shows such relations for 55 clusters, while the upper panel shows the number of galaxies in each density bin. [After Dressler (1980a)]

2.5 Clusters and Groups of Galaxies

A significant fraction of the galaxies in the present-day Universe is collected into groups and clusters in which the number density of galaxies is a few tens to a few hundred times higher than the average. The densest and most populous of these aggregations are called galaxy clusters, which typically contain more than 50 relatively bright galaxies in a volume only a few megaparsecs across. The smaller, less populous aggregations are called 'groups', although there is no well-defined distinction. Groups and clusters are the most massive, virialized objects in the Universe, and they are important laboratories to study the evolution of the galaxy population. Because of their high surface densities and large number of very luminous member galaxies, they can be identified out to very large distances, making them also useful as cosmological probes. In this section we summarize some of their most important properties, focusing in particular on their populations of galaxies.

2.5.1 Clusters of Galaxies

In order to select clusters (or groups) of galaxies from the observed galaxy distribution, one needs to adopt some selection criteria. In order for the selected clusters to be dynamically significant, two selection criteria are usually set. One is that the selected system must have high enough density, and the other is that the system must contain a sufficiently large number of galaxies.

According to these criteria, Abell (1958) selected 1,682 galaxy clusters from the Palomar Sky Survey, which are now referred to as the Abell clusters. The two selection criteria set by Abell are

(i) Richness criterion: each cluster must have at least 50 member galaxies with apparent magnitudes $m < m_3 + 2$, where m_3 is the apparent magnitude of the third brightest member. The richness of a cluster is defined to be the number of member galaxies with apparent magnitudes between m_3 and $m_3 + 2$. Rich Abell clusters are those with richness greater than 50, although Abell also listed poor clusters with richness in the range from 30 to 50.

(ii) Compactness criterion: only galaxies with distances to the cluster center smaller than $1.5 h^{-1}$Mpc (the Abell radius) are selected as members. Given the richness criterion, the compactness criterion is equivalent to a density criterion.

Abell also classified a cluster as regular if its galaxy distribution is more or less circularly symmetric and concentrated, otherwise as irregular. The two most well-studied clusters, because of their proximity, are the Virgo Cluster and the Coma Cluster. The Virgo Cluster, which is the rich cluster nearest to our Galaxy, is a very representative example. It lacks clear symmetry, and reveals significant substructure, indicating that the dynamical relaxation on the largest scales is not yet complete. The Coma Cluster, on the other hand, is a fairly rare species. It is extremely massive, and is richer than 95% of all clusters catalogued by Abell. Furthermore, it appears remarkably relaxed, with a highly concentrated and symmetric galaxy distribution with no sign of significant subclustering.

The Abell catalogue was constructed using visual inspections of photographic sky plates. Since its publication, this has been improved upon using special purpose scanning machines (such as the APM at Cambridge and COSMOS at Edinburgh), which resulted in digitized versions of the photographic plates allowing for a more objective identification of clusters (e.g. Lumsden et al., 1992; Dalton et al., 1997). More recently, several cluster catalogues have been constructed from large galaxy redshift surveys such as the 2dFGRS and the SDSS (e.g. Bahcall et al., 2003; Miller et al., 2005; Koester et al., 2007). Based on all these catalogues it is now well established that the number density of rich clusters is of the order of $10^{-5} h^3 \text{Mpc}^{-3}$, about 1000 times smaller than that of L^* galaxies.

(a) Galaxy Populations As we have seen in §2.4.5, clusters are in general rich in early-type galaxies. The fraction of E+S0 galaxies is about 80% in regular clusters, and about 50% in irregular clusters, compared to about 30% in the general field. This is generally interpreted as evidence that galaxies undergo morphological transformations in dense (cluster) environments, and various mechanisms have been suggested for such transformations (see §12.5).

The radial number density distribution of galaxies in clusters is well described by $n(r) \propto 1/[r^\gamma (r+r_s)^{3-\gamma}]$, where r_s is a scale radius and γ is the logarithmic slope of the inner profile. The value of γ is typically ~ 1 and the scale radius is typically $\sim 20\%$ of the radius of the cluster (e.g. van der Marel et al., 2000; Lin et al., 2004b). As we will see in Chapter 7 this is very similar to the density distribution of dark matter halos, suggesting that within clusters galaxies are a reasonably fair tracer of the mass distribution. There is, however, evidence for some segregation by mass and morphology/color, with more massive, red, early-type galaxies following a more concentrated number density distribution than less massive, blue, late-type galaxies (e.g. Quintana, 1979; Carlberg et al., 1997; Adami et al., 1998; Yang et al., 2005b; van den Bosch et al., 2008b).

Often the brightest cluster galaxy (BCG) has an extraordinarily diffuse and extended outer envelope, in which case it is called a cD galaxy (where the 'D' stands for diffuse). They typically have best-fit Sérsic indices that are much larger than 4, and are often located at or near the center of the cluster (because of this, it is a useful mnemonic to think of 'cD' as meaning 'centrally dominant'). cD galaxies are the most massive galaxies known, with stellar masses often exceeding $10^{12} \text{M}_\odot$, and their light can make up as much as $\sim 30\%$ of the entire visible light of a rich cluster of galaxies. However, it is unclear whether the galaxy's diffuse envelope should be considered part of the galaxy or as 'intracluster light' (ICL), stars associated with the cluster itself rather than with any particular galaxy. In a few cD galaxies the velocity dispersion appears to rise strongly in the extended envelope, approaching a value similar to that of the cluster in which the galaxy is embedded. This supports the idea that these stars are more closely associated with the cluster than with the galaxy (i.e. they are the cluster equivalent of the stellar halo in the Milky Way). cD galaxies are believed to have grown through the accretion of multiple galaxies in the cluster, a process called galactic cannibalism (see §12.5.2). Consistent with this, nearby cDs frequently appear to have multiple nuclei (e.g. Schneider et al., 1983)

(b) The Butcher–Oemler Effect When studying the galaxy populations of clusters at intermediate redshifts ($0.3 \lesssim z \lesssim 0.5$), Butcher & Oemler (1978) found a dramatic increase in the fraction of blue galaxies compared to present-day clusters, which has become known as the Butcher–Oemler effect. Although originally greeted with some skepticism (see Dressler, 1984, for a review), this effect has been confirmed by numerous studies. In addition, morphological studies, especially those with the HST, have shown that the Butcher–Oemler effect is associated with an increase of the spiral fraction with increasing redshift, and that many of these spirals show disturbed morphologies (e.g. Couch et al., 1994; Wirth et al., 1994).

In addition, spectroscopic data has revealed that a relatively large fraction of galaxies in clusters at intermediate redshifts have strong Balmer lines in absorption and no emission lines (Dressler & Gunn, 1983). This indicates that these galaxies were actively forming stars in the past, but had their star formation quenched in the last 1–2 Gyr. Although they were originally named 'E+A' galaxies, currently they are more often referred to as 'k+a' galaxies or as post-starburst galaxies (since their spectra suggest that they must have experienced an elevated amount of star formation prior to the quenching). Dressler et al. (1999) have shown that the fraction of k+a galaxies in clusters at $z \sim 0.5$ is significantly larger than in the field at similar redshifts, and that they have mostly spiral morphologies.

All these data clearly indicate that the population of galaxies in clusters is rapidly evolving with redshift, most likely due to specific processes that operate in dense environments (see §12.5).

(c) Mass Estimates Galaxies are moving fast in clusters. For rich clusters, the typical line-of-sight velocity dispersion, $\sigma_{\rm los}$, of cluster member galaxies is of the order of $1{,}000 \, {\rm km \, s}^{-1}$. If the cluster has been relaxed to a static dynamical state, which is roughly true for regular clusters, one can infer a dynamical mass estimate from the virial theorem (see §5.4.4) as

$$M = A \frac{\sigma_{\rm los}^2 R_{\rm cl}}{G}, \tag{2.38}$$

where A is a pre-factor (of order unity) that depends on the density profile and on the exact definition of the cluster radius $R_{\rm cl}$. Using this technique one obtains a characteristic mass of $\sim 10^{15} h^{-1} \, {\rm M}_\odot$ for rich clusters of galaxies. Together with the typical value of the total luminosity in a cluster, this implies a typical mass-to-light ratio for clusters,

$$(M/L_B)_{\rm cl} \sim 350 h ({\rm M}_\odot/{\rm L}_\odot)_B. \tag{2.39}$$

Hence, only a small fraction of the total gravitational mass of a cluster is associated with galaxies.

Ever since the first detection by the UHURU satellite in the 1970s, it has become clear that clusters are bright X-ray sources, with characteristic luminosities ranging from $L_X \sim 10^{43}$ to $\sim 10^{45} \, {\rm erg \, s}^{-1}$. This X-ray emission is spatially extended, with detected sizes of $\sim 1 \, {\rm Mpc}$, and so it cannot originate from the individual member galaxies. Rather, the spectral energy distribution of the X-ray emission suggests that the emission mechanism is thermal bremsstrahlung (see §B1.3) from a hot plasma. The inferred temperatures of this intracluster medium (ICM) are in the range 10^7–$10^8 \, {\rm K}$, corresponding to a typical photon energy of 1–10 keV, so that the gas is expected to be fully ionized.

Fig. 2.30. Hubble Space Telescope image of the cluster Abell 2218. The arcs and arclets around the center of the cluster are images of background galaxies that are strongly distorted due to gravitational lensing. [Courtesy of W. Couch, R. Ellis, NASA, and Space Telescope Science Institute]

For a fully ionized gas, the thermal bremsstrahlung emissivity, i.e. the emission power per unit frequency per unit volume, is related to its density and temperature roughly as

$$\varepsilon_{\rm ff}(\nu) \propto n^2 T^{-1/2} \exp\left(-\frac{h_{\rm P}\nu}{k_{\rm B}T}\right). \tag{2.40}$$

The quantity we observe from a cluster is the X-ray surface brightness, which is the integration of the emissivity along the line-of-sight:[3]

$$S_\nu(x,y) \propto \int \varepsilon_{\rm ff}(\nu;x,y,z)\,{\rm d}z. \tag{2.41}$$

If S_ν is measured as a function of ν (i.e. photon energy), the temperature at a given projected position (x,y) can be estimated from the shape of the spectrum. Note that this temperature is an emissivity-weighted mean along the line-of-sight, if the temperature varies with z. Once the temperature is known, the amplitude of the surface brightness can be used to estimate $\int n^2\,{\rm d}z$ which, together with a density model, can be used to obtain the gas density distribution. Thus, X-ray observations of clusters can be used to estimate the corresponding masses in hot gas. These are found to fall in the range $(10^{13}$–$10^{14})h^{-5/2}\,{\rm M}_\odot$, about 10 times as large as the total stellar mass in member galaxies. Furthermore, as we will see in §8.2.1, if the X-ray gas is in hydrostatic equilibrium with the cluster potential, so that the local pressure gradient is balanced by the gravitational force, the observed temperature and density distribution of the gas can also be used to estimate the *total* mass of the cluster.

Another method to measure the total mass of a cluster of galaxies is through gravitational lensing. According to general relativity, the light from a background source is deflected when it passes a mass concentration in the foreground, an effect called gravitational lensing. As discussed in more detail in §6.6, gravitational lensing can have a number of effects: it can create multiple images on the sky of the same background source, it can magnify the flux of the source, and it can distort the shape of the background source. In particular, the image of a circular source is distorted into an ellipse if the source is not close to the line-of-sight to the lens so that the lensing effect is weak (weak lensing). Otherwise, if the source is close to the line-of-sight to the lens, the image is stretched into an arc or an arclet (strong lensing).

---

[3] Here we ignore redshifting and surface brightness dimming due to the expansion of the Universe; see §3.1.

Both strong and weak lensing can be used to estimate the total gravitational mass of a cluster. In the case of strong lensing, one uses giant arcs and arclets, which are the images of background galaxies lensed by the gravitational field of the cluster (see Fig. 2.30). The location of an arc in a cluster provides a simple way to estimate the projected mass of the cluster within the circle traced by the arc. Such analyses have been carried out for a number of clusters, and the total masses thus obtained are in general consistent with those based on the internal kinematics, the X-ray emission, or weak lensing. Typically the total cluster masses are found to be an order of magnitude larger than the combined masses of stars and hot gas, indicating that clusters are dominated by dark matter, as first pointed out by Fritz Zwicky in the 1930s.

2.5.2 Groups of Galaxies

By definition, groups are systems of galaxies with richness less than that of clusters, although the dividing line between groups and clusters is quite arbitrary. Groups are selected by applying certain richness and compactness criteria to galaxy surveys, similar to what Abell used for selecting clusters. Typically, groups selected from redshift surveys include systems with at least three galaxies and with a number density enhancement of the order of 20 (e.g. Geller & Huchra, 1983; Nolthenius & White, 1987; Eke et al., 2004; Yang et al., 2005b; Berlind et al., 2006; Yang et al., 2007). Groups so selected typically contain $3-30\,L^*$ galaxies, have a total B-band luminosity in the range $10^{10.5}-10^{12}h^{-2}\,L_\odot$, have radii in the range $(0.1-1)\,h^{-1}\mathrm{Mpc}$, and have typical (line-of-sight) velocity dispersion of the order of $300\,\mathrm{km\,s^{-1}}$. As for clusters, the total dynamical mass of a group can be estimated from its size and velocity dispersion using the virial theorem (2.38), and masses thus obtained roughly cover the range $10^{12.5}-10^{14}h^{-1}\,\mathrm{M_\odot}$. Therefore, the typical mass-to-light ratio of galaxy groups is $(M/L_B) \sim 100h(\mathrm{M_\odot}/\mathrm{L_\odot})_B$, significantly lower than that for clusters.

(a) **Compact Groups** A special class of groups are the so-called compact groups. Each of these systems consists of only a few galaxies but with an extremely high density enhancement. A catalogue of about 100 compact groups was constructed by Hickson (1982) from an analysis of photographic plates. These Hickson Compact Groups (HCGs) typically consist of only four or five galaxies and have a projected radius of only 50–100 kpc. A large fraction ($\sim 40\%$) of the galaxies in HCGs show evidence for interactions, and based on dynamical arguments, it is expected that the HCGs are each in the process of merging to perhaps form a single bright galaxy.

(b) **The Local Group** The galaxy group that has been studied in most detail is the Local Group, of which the Milky Way and M31 are the two largest members. The Local Group is a loose association of galaxies which fills an irregular region just over 1 Mpc across. Because we are in it, we can probe the members of the Local Group down to much fainter magnitudes than is possible in any other group. Table 2.7 lists the 30 brightest members of the Local Group, while Fig. 2.31 shows their spatial distribution. Except for a few of the more distant objects, the majority of the Local Group members can be assigned as satellites of either the Milky Way or M31. The largest satellite of the Milky Way is the Large Magellanic Cloud (LMC). Its luminosity is about one tenth of that of its host and it is currently actively forming stars. Together with its smaller companion, the Small Magellanic Cloud (SMC), it follows a high angular momentum orbit almost perpendicular to the Milky Way's disk and currently lies about 50 kpc from the Galactic center. Both Magellanic Clouds have metallicities significantly lower than that of the Milky Way. All the other satellites of our Galaxy are low mass, gas-free and metal-poor dwarf spheroidals. The most massive of these are the Fornax and Sagittarius systems. The latter lies only about 20 kpc from the Galactic center and is in the process of being disrupted by the tidal

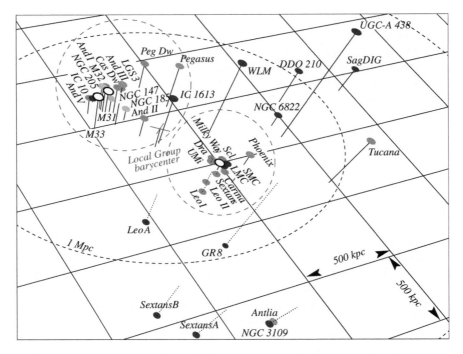

Fig. 2.31. Schematic distribution of galaxies in the local group. [Courtesy of E. Grebel, see Grebel (1999)]

effects of its host. Several of the dwarf spheroidals contain stellar populations with a range of ages, some being 10 times younger than typical Population II stars.

The Andromeda Nebula itself is similar to but more massive than the Milky Way, with a more prominent bulge population and somewhat less active current star formation. Its largest satellite is the bulge-less dwarf spiral M33, which is only slightly brighter than the LMC and is actively forming stars. M31 also has two close dwarf elliptical companions, M32 and NGC 205, and two similar satellites, NGC 147 and NGC 185, at somewhat larger distances. These galaxies are denser and more luminous than dwarf spheroidals, but are also devoid of gas and young stars. (NGC 205 actually has a small star-forming region in its nucleus.) Finally M31 has its own retinue of dwarf spheroidal satellites.

The more distant members of the Local Group are primarily dwarf irregular galaxies with active star formation, similar to but less luminous than the Magellanic Clouds. Throughout the Local Group there is a very marked tendency for galaxies with a smaller stellar mass to have a lower metallicity, with the smallest dwarfs having metallicities about one-tenth of the solar value (Mateo, 1998).

2.6 Galaxies at High Redshifts

Since galaxies at higher redshifts are younger, a comparison of the (statistical) properties of galaxies at different redshifts provides a direct window on their formation and evolution. However, a galaxy of given luminosity and size is both fainter and of lower surface brightness when located at higher redshifts (see §3.1.6). Thus, if high-redshift galaxies have similar luminosities and sizes as present-day galaxies, they would be extremely faint and of very low surface brightness, making them very difficult to detect. Indeed, until the mid-1990s, the known high-redshift galaxies with $z \gtrsim 1$ were almost exclusively active galaxies, such as quasars, QSOs and radio galaxies, simply because these were the only galaxies sufficiently bright to be observable with

2.6 Galaxies at High Redshifts

Table 2.7. Local Group members.

Name	Type	$\mathcal{M}_V$	l,b	Distance (kpc)
Milky Way (Galaxy)	Sbc	-20.6	0,0	8
LMC	Irr	-18.1	280,-33	49
SMC	Irr	-16.2	303,-44	58
Sagittarius	dSph/E7	-14.0	6,-14	24
Fornax	dSph/E3	-13.0	237,-65	131
Leo I (DDO 74)	dSph/E3	-12.0	226,49	270
Sculptor	dSph/E3	-10.7	286,-84	78
Leo II (DDO 93)	dSph/E0	-10.2	220,67	230
Sextans	dSph/E4	-10.0	243,42	90
Carina	dSph/E4	-9.2	260,-22	87
Ursa Minor (DDO 199)	dSph/E5	-8.9	105,45	69
Draco (DDO 208)	dSph/E3	-8.6	86,35	76
M 31 (NGC 224)	Sb	-21.1	121,-22	725
M 33 (NGC 598)	Sc	-18.9	134,-31	795
IC 10	Irr	-17.6	119,-03	1250
NGC 6822 (DDO 209)	Irr	-16.4	25,-18	540
M 32 (NGC 221)	dE2	-16.4	121,-22	725
NGC 205	dE5	-16.3	121,-21	725
NGC 185	dE3	-15.3	121,-14	620
IC 1613 (DDO 8)	Irr	-14.9	130,-60	765
NGC 147 (DDO 3)	dE4	-14.8	120,-14	589
WLM (DDO 221)	Irr	-14.0	76,-74	940
Pegasus (DDO 216)	Irr	-12.7	94,-43	759
Leo A	Irr	-11.7	196, 52	692
And I	dSph/E0	-11.7	122,-25	790
And II	dSph/E3	-11.7	129,-29	587
And III	dSph/E6	-10.2	119,-26	790
Phoenix	Irr	-9.9	272,-68	390
LGC 3	Irr	-9.7	126,-41	760
Tucana	dSph/E5	-9.6	323,-48	900

the facilities available then. Thanks to a number of technological advancements in both telescopes and detectors, we have made enormous progress, and today the galaxy population can be probed out to $z \gtrsim 6$.

The search for high-redshift galaxies usually starts with a photometric survey of galaxies in multiple photometric bands down to very faint magnitude limits. Ideally, one would like to have redshifts for all these galaxies and study the entire galaxy population at all different redshifts. In reality, however, it is extremely time-consuming to obtain spectra of faint galaxies even with the 10-meter class telescopes available today. In order to make progress, different techniques have been used, which basically fall in three categories: (i) forsake the use of spectra and only use photometry either to analyze the number counts of galaxies down to very faint magnitudes or to derive photometric redshifts; (ii) use broad-band color selection to identify target galaxies likely to be at high redshift for follow-up spectroscopy; and (iii) use narrow-band photometry to find objects with a strong emission line in a narrow redshift range. Here we give a brief overview of these different techniques.

2.6.1 Galaxy Counts

In the absence of redshifts, some information about the evolution of the galaxy population can be obtained from galaxy counts, $\mathcal{N}(m)$, defined as the number of galaxies per unit apparent magnitude (in a given waveband) per unit solid angle:

$$\mathrm{d}^2 N(m) = \mathcal{N}(m)\,\mathrm{d}m\,\mathrm{d}\omega. \tag{2.42}$$

Although the measurement of $\mathcal{N}(m)$ is relatively straightforward from any galaxy catalogue with uniform photometry, interpreting the counts in terms of galaxy number density as a function of redshift is far from trivial. First of all, the waveband in which the apparent magnitudes are measured corresponds to different rest-frame wavebands at different redshifts. To be able to test for evolution in the galaxy population with redshift, this shift in waveband needs to be corrected for. But such correction is not trivial to make, and can lead to large uncertainties (see §10.3.6). Furthermore, both cosmology and evolution can affect $\mathcal{N}(m)$. In order to break this degeneracy, and to properly test for evolution, accurate constraints on cosmological parameters are required.

Despite these difficulties, detailed analyses of galaxy counts have resulted in a clear detection of evolution in the galaxy population. Fig. 2.32 shows the galaxy counts in four wavebands obtained from a variety of surveys. The solid dots are obtained from the Hubble Deep Fields (Ferguson et al., 2000) imaged to very faint magnitudes with the HST. The solid lines in Fig. 2.32 show the predictions for a realistic cosmology in which it is assumed that the galaxy population does not evolve with redshift. A comparison with the observed counts shows that this model

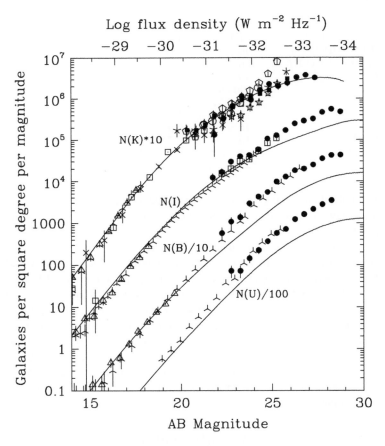

Fig. 2.32. Galaxy counts in the U, B, I and K bands obtained from the Hubble deep fields (solid symbols) and a number of other ground-based surveys (other symbols). The solid lines show the predictions for a realistic cosmology in which it is assumed that the galaxy population does not evolve with redshift. [Adapted from Ferguson et al. (2000) by permission of ARAA]

severely underpredicts the galaxy counts of faint galaxies, especially in the bluer wavebands. The nature of this excess of faint blue galaxies will be discussed in §15.2.2.

2.6.2 Photometric Redshifts

Since spectroscopy relies on dispersing the light from an object according to wavelength, accurate redshifts, which require sufficient signal-to-noise in individual emission and/or absorption lines, can only be obtained for relatively bright objects. An alternative, although less reliable, technique to measure redshifts relies on broad-band photometry. By measuring the flux of an object in a relatively small number of wavebands, one obtains a very crude sampling of the object's SED. As we have seen, the SEDs of galaxies reveal a number of broad spectral features (see Fig. 2.12). An important example is the 4000Å break, which is due to a sudden change in the opacity at this wavelength in the atmospheres of low mass stars, and therefore features predominantly in galaxies with stellar population ages $\gtrsim 10^8$ yr. Because of this 4000Å break and other broad spectral features, the colors of a population of galaxies at a given redshift only occupy a relatively small region of the full multi-dimensional color space. Since this region changes as function of redshift, the broad-band colors of a galaxy can be used to estimate its redshift.

In practice one proceeds as follows. For a given template spectrum, either from an observed galaxy or computed using population synthesis models, one can determine the relative fluxes expected in different wavebands for a given redshift. By comparing these expected fluxes with the observed fluxes one can determine the best-fit redshift and the best-fit template spectrum (which basically reflects the spectral type of the galaxy). The great advantage of this method is that photometric redshifts can be measured much faster than their spectroscopic counterparts, and that it can be extended to much fainter magnitudes. The obvious downside is that photometric redshifts are far less reliable. While a spectroscopic redshift can easily be measured to a relative error of less than 0.1%, photometric errors are typically of the order of 3–10%, depending on which and how many wavebands are used. Furthermore, the error is strongly correlated with the spectral type of the galaxy. It is typically much larger for star-forming galaxies, which lack a pronounced 4000Å break, than for galaxies with an old stellar population.

A prime example of a photometric redshift survey, illustrating the strength of this technique, is the COMBO-17 survey (Wolf et al., 2003), which comprises a sample of $\sim 25,000$ galaxies with photometric redshifts obtained from photometry in 17 relatively narrow optical wavebands. Because of the use of a relatively large number of filters, this survey was able to reach an average redshift accuracy of $\sim 3\%$, sufficient to study various statistical properties of the galaxy population as a function of redshift.

2.6.3 Galaxy Redshift Surveys at $z \sim 1$

In order to investigate the nature of the excess of faint blue galaxies detected with galaxy counts, a number of redshift surveys out to $z \sim 1$ were carried out in the mid-1990s using 4-m class telescopes, including the Canada–France Redshift Survey (CFRS; Lilly et al., 1995) and the Autofib-LDSS survey (Ellis et al., 1996). These surveys, containing the order of 1,000 galaxies, allowed a determination of galaxy luminosity functions (LFs) covering the entire redshift range $0 < z \lesssim 1$. The results, although limited by small-number statistics, confirmed that the galaxy population is evolving with redshift, in agreement with the results obtained from the galaxy counts.

With the completion of a new class of 10-meter telescopes, such as the KECK and the VLT, it became possible to construct much larger redshift samples at intermediate to high redshifts. Currently the largest redshift survey at $z \sim 1$ is the DEEP2 Redshift Survey (Davis et al., 2003),

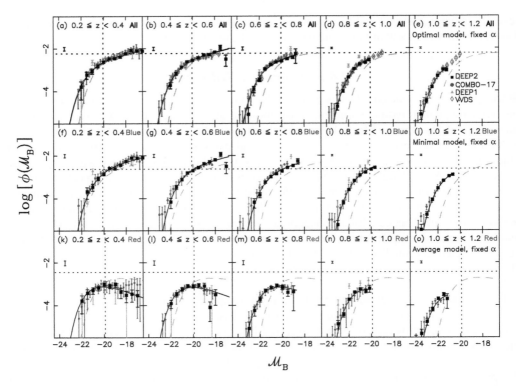

Fig. 2.33. Luminosity functions measured in different redshift bins for 'All' galaxies (top row), 'Blue' galaxies (middle row), and 'Red' galaxies (bottom row). Different symbols correspond to results obtained from different redshift surveys (DEEP1, DEEP2, COMBO-17 and VVDS, as indicated). The solid black lines indicate Schechter functions fitted to the DEEP2 results. For comparison, the dashed gray lines show the Schechter functions for local samples obtained from the SDSS. Overall the agreement between the different surveys is very good. [Adapted from Faber et al. (2007) by permission of AAS]

which contains about 50,000 galaxies brighter than $R_{AB} \approx 24.1$ in a total of ~ 3 square degrees in the sky. The adopted color criteria ensure that the bulk of the galaxies selected for spectroscopy have redshifts in the range $0.7 \lesssim z \lesssim 1.4$. Results from DEEP2 show, among others, that the color bimodality observed in the local Universe (see §2.4.3) is already present at $z \sim 1$ (Bell et al., 2004; Willmer et al., 2006; Cooper et al., 2007). Together with COMBO-17, the DEEP2 survey has provided accurate measurements of the galaxy luminosity function, split according to color, out to $z \sim 1.2$. As shown in Fig. 2.33, the different surveys yield results in excellent mutual agreement. In particular, they show that the characteristic luminosity, L^*, of the galaxy population in the rest-frame B-band becomes fainter by ~ 1.3 mag from $z = 1$ to $z = 0$ for both the red and blue populations. However, the number density of L^* galaxies, ϕ^*, behaves very differently for red and blue galaxies: while ϕ^* of blue galaxies has roughly remained constant since $z = 1$, that of red galaxies has nearly quadrupled (Bell et al., 2004; Brown et al., 2007; Faber et al., 2007). As we will see in §13.2, this puts important constraints on the formation history of elliptical galaxies.

Another large redshift survey, which is being conducted at the time of writing, is the VIRMOS VLT Deep Survey (VVDS; Le Fèvre et al., 2005) which will ultimately acquire $\sim 150,000$ redshifts over ~ 4 square degrees in the sky. Contrary to DEEP2, the VVDS does not apply any color selection; rather, spectroscopic candidates are purely selected on the basis of their apparent magnitude in the I_{AB} band. Consequently the redshift distribution of VVDS galaxies is very broad: it peaks at $z \sim 0.7$, but has a long high-redshift tail extending all the way out to $z \sim 5$.

The luminosity functions obtained from ∼ 8,000 galaxies in the first data of the VVDS are in excellent agreement with those obtained from DEEP2 and COMBO-17 (see Fig. 2.33).

2.6.4 Lyman-Break Galaxies

As discussed above, broad features in the SEDs of galaxies allow for the determination of photometric redshifts, and for a very successful pre-selection of candidate galaxies at $z \sim 1$ for follow-up spectroscopy. The same principle can also be used to select a special subset of galaxies at much higher redshifts. A star-forming galaxy has a SED roughly flat down to the Lyman limit at $\lambda \sim 912\,\text{Å}$, beyond which there is a prominent break due to the spectra of the stellar population (see the spectra of the O9 and B0 stars in Fig. 2.5) and to intervening absorption. Physically this reflects the large ionization cross-section of neutral hydrogen. A galaxy revealing a pronounced break at the Lyman limit is called a Lyman-break galaxy (LBG), and is characterized by a relatively high star-formation rate.

For a LBG at $z \sim 3$, the Lyman break falls in between the U and B bands (see Fig. 2.34). Therefore, by selecting those galaxies in a deep multi-color survey that are undetected (or extremely faint) in the U band, but detected in the B and redder bands, one can select candidate star-forming galaxies in the redshift range $z = 2.5$–3.5 (Steidel et al., 1996). Galaxies selected this way are called UV drop-outs. Follow-up spectroscopy of large samples of UV drop-out candidates has confirmed that this Lyman-break technique is very effective, with the vast majority of the candidates being indeed star-forming galaxies at $z \sim 3$.

To date more than 1,000 LBGs with $2.5 \lesssim z \lesssim 3.5$ have been spectroscopically confirmed. The comoving number density of bright LBGs is estimated to be comparable to that of present-day bright galaxies. However, contrary to typical bright galaxies at $z \sim 0$, which are mainly early-type galaxies, LBGs are actively forming stars (note that they are effectively selected in the B band,

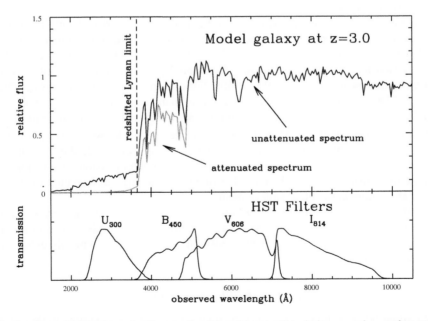

Fig. 2.34. An illustration of how the 'Lyman-break' or 'drop-out' technique can be used to select star-forming galaxies at redshifts $z \sim 3$. The spectrum of a typical star-forming galaxy has a break at the Lyman limit (912 Å), which is redshifted to a wavelength $\lambda \sim 4000\,\text{Å}$ if the galaxy is at $z \sim 3$. As a result, the galaxy appears very faint (or may even be undetectable) in the U band, but bright in the redder bands. [Courtesy of M. Dickinson; see Dickinson (1998)]

corresponding to rest-frame UV at $z \sim 3$) with inferred star-formation rates in the range of a few times $10 M_\odot \, yr^{-1}$ up to $\sim 100 M_\odot \, yr^{-1}$, depending on the uncertain amount of dust extinction (Adelberger & Steidel, 2000).

The Lyman break (or drop-out) technique has also been applied to deep imaging surveys in redder bands to select galaxies that drop out of the B band, V band and even the I band. If these are indeed LBGs, their redshifts correspond to $z \sim 4$, $z \sim 5$, and $z \sim 6$, respectively. Deep imaging surveys with the HST and ground-based telescopes have already produced large samples of these drop-out galaxies. Unfortunately, most of these galaxies are too faint to follow-up spectroscopically, so that it is unclear to what extent these samples are contaminated by low redshift objects. With this caveat in mind, the data have been used to probe the evolution of the galaxy luminosity function (LF) in the rest-frame UV all the way from $z \sim 0$ (using data from the GALEX satellite) to $z \sim 6$. Over the redshift range $4 \lesssim z \lesssim 6$ this LF is found to have an extremely steep faint-end slope, while the characteristic luminosity L^*_{UV} is found to brighten significantly from $z = 6$ to $z = 4$ (Bouwens et al., 2007).

2.6.5 Lyα Emitters

In addition to the broad-band selection techniques mentioned above, one can also search for high-redshift galaxies using narrow-band photometry. This technique has been used extensively to search for Lyα emitters (LAEs) at redshifts $z \gtrsim 3$ for which the Lyα emission line ($\lambda = 1216 \text{Å}$) appears in the optical.

Objects with strong Lyα are either QSOs or galaxies actively forming stars. However, since the Lyα flux is easily quenched by dust extinction, not all star-forming galaxies feature Lyα emission. In fact, a large fraction of LBGs, although actively forming stars, lack an obvious Lyα emission line. Therefore, by selecting LAEs one is biased towards star-forming galaxies with relatively little dust, or in which the dust has a special geometry so that part of the Lyα flux can leave the galaxy unextincted.

One can search for LAEs at a particular redshift, z_{LAE}, using a narrow-band filter centered on a wavelength $\lambda = 1216 \text{Å} \times (1 + z_{LAE})$ plus another, much broader filter centered on the same λ. The objects in question then show up as being particularly bright in the narrow-band filter in comparison to the broad-band image. A potential problem is that one might also select emission-line galaxies at very different redshifts. For example, a galaxy with strong [OII] emission ($\lambda = 3727 \text{Å}$) would shift into the same narrow-band filter if the galaxy is at a redshift $z_{[OII]} = 0.33 z_{LAE} - 0.67$. To minimize this kind of contamination one generally only selects systems with a large equivalent width[4] in the emission line ($\gtrsim 150 \text{Å}$), which excludes all but the rarest [OII] emitters. Another method to check whether the object is indeed a LAE at z_{LAE} is to use follow-up spectroscopy to see whether (i) there are any other emission lines visible that help to determine the redshift, and (ii) the emission line is asymmetric, as expected for Lyα due to preferential absorption in the blue wing of the line.

This technique can be used to search for high-redshift galaxies in several narrow redshift bins ranging from $z \sim 3$ to $z \sim 6.5$, and at the time of writing ~ 100 LAEs covering this redshift range have been spectroscopically confirmed. Since these systems are typically extremely faint, the nature of these objects is still unclear.

2.6.6 Submillimeter Sources

Since the Lyman-break technique and Lyα imaging select galaxies according to their rest-frame UV light, they may miss dust-enshrouded star-forming galaxies, the high-redshift counterparts of

---

[4] The equivalent width of an emission line, a measure for its strength, is defined as the width of the wavelength range over which the continuum needs to be integrated to have the same flux as measured in the line (see §16.4.4).

local starbursts. Most of the UV photons from young stars in such galaxies are absorbed by dust and re-emitted in the far-infrared. Such galaxies can therefore be detected in the submillimeter (sub-mm) band, which corresponds to rest-frame far-infrared at $z \sim 3$. Deep surveys in the sub-mm bands only became possible in the mid-1990s with the commissioning of the Submillimeter Common-User Bolometer Array (SCUBA; see Holland et al., 1999), operating at 450 μm and 850 μm, on the James Clerk Maxwell Telescope (JCMT). This led to the discovery of an unexpectedly large population of faint sub-mm sources (Smail et al., 1997). An extensive and difficult observational campaign to identify the optical counterparts and measure their redshifts has shown that the majority of these sources are indeed starburst galaxies at a median redshift of $z \sim 2.5$. Some of the strong sub-mm sources with measured redshifts have inferred star-formation rates as high as several $100 \, M_\odot \, \mathrm{yr}^{-1}$, similar to those of ULIRGS at $z \simeq 0$. Given the large number density of SCUBA sources, and their inferred star-formation rates, the total number of stars formed in these systems may well be larger than that formed in the Lyman-break galaxies at the same redshift (Blain et al., 1999).

2.6.7 Extremely Red Objects and Distant Red Galaxies

Another important step forward in the exploration of the galaxy population at high redshift came with the development of large format near-infrared (NIR) detectors. Deep, wide-field surveys in the K band led to the discovery of a class of faint galaxies with extremely red optical-to-NIR colors ($R - K > 5$). Follow-up spectroscopy has shown that these extremely red objects (EROs) typically have redshifts in the range $0.7 \lesssim z \lesssim 1.5$. There are two possible explanations for their red colors: either they are galaxies dominated by old stellar populations with a pronounced 4000Å break that has been shifted red-wards of the R-band filter, or they are starbursts (or AGN) strongly reddened due to dust extinction. Spectroscopy of a sample of ~ 50 EROs suggests that they are a roughly equal mix of both (Cimatti et al., 2002a).

Deep imaging in the NIR can also be used to search for the equivalent of 'normal' galaxies at $z \gtrsim 2$. As described above, the selections of LBGs, LAEs and sub-mm sources are strongly biased towards systems with relatively high star-formation rates. Consequently, the population of high-redshift galaxies picked out by these selections is very different from the typical, present-day galaxies whose light is dominated by evolved stars. In order to select high-redshift galaxies in a way similar to how 'normal' galaxies are selected at low redshift, one has to go to the rest-frame optical, which corresponds to the NIR at $z \sim 2$–3. Using the InfraRed ExtraGalactic Survey (FIRES; Labbé et al., 2003), Franx et al. (2003) identified a population of galaxies on the basis of their red NIR color, $J_s - K_s > 2.3$, where the K_s and J_s filters are similar to the classical J and K filters, but centered on somewhat shorter wavelengths. The galaxies so selected are now referred to as distant red galaxies (DRGs). The color criterion efficiently isolates galaxies with prominent Balmer or 4000Å breaks at $z \gtrsim 2$, and can therefore be used to select galaxies with the oldest stellar populations at these redshifts. However, the NIR color criterion alone also selects galaxies with significant current star formation, even dusty starbursts. The brightest DRGs ($K_s < 20$) are among the most massive galaxies at $z \gtrsim 2$, with stellar masses $\gtrsim 10^{11} \, M_\odot$, likely representing the progenitors of present-day massive ellipticals. As EROs, DRGs are largely missed in UV-selected (e.g. LBG) samples. Yet, as shown by van Dokkum et al. (2006), among the most massive population of galaxies in the redshift range $2 \lesssim z \lesssim 3$, DRGs dominate over LBGs both in number density and in stellar mass density.

Using photometry in the B, z, and K bands, Daddi et al. (2004) introduced a selection criterion which allows one to recover the bulk of the galaxy population in the redshift range $1.4 \lesssim z \lesssim 2.5$, including both active star-forming galaxies as well as passively evolving galaxies, and to distinguish between the two classes. In particular, the color criterion $BzK \equiv (z - K)_{AB} - (B - z)_{AB} > -0.2$ is very efficient in selecting star-forming galaxies with $1.4 \lesssim z \lesssim 2.5$, independently of their dust reddening, while the criteria $BzK < -0.2$ and $(z-K)_{AB} > 2.5$ predominantly select passively evolving galaxies in the same redshift interval. At $z \sim 2$ the BzK-selected star-forming galaxies typically have higher reddening and higher star-formation rates than UV-selected galaxies. A comparison of BzK galaxies with DRGs in the same redshift range shows that many of the DRGs are reddened starbursts rather than passively evolving galaxies.

2.6.8 The Cosmic Star-Formation History

The data on star-forming galaxies at different redshifts can in principle be used to map out the production rate of stars in the Universe as a function of redshift. If we do not care where stars form, the star-formation history of the Universe can be characterized by a global quantity, $\dot{\rho}_\star(z)$, which is the total gas mass that is turned into stars per unit time per unit volume at redshift z.

In order to estimate $\dot{\rho}_\star(z)$ from observation, one requires estimates of the number density of galaxies as a function of redshift and their (average) star-formation rates. In practice, one observes the number density of galaxies as a function of luminosity in some waveband, and estimates $\dot{\rho}_\star(z)$ from

$$\dot{\rho}_\star(z) = \int d\dot{M}_\star \dot{M}_\star \int P(\dot{M}_\star|L,z)\phi(L,z)\,dL = \int \langle \dot{M}_\star \rangle (L,z)\,\phi(L,z)\,dL, \qquad (2.43)$$

where $P(\dot{M}_\star|L,z)\,d\dot{M}_\star$ is the probability for a galaxy with luminosity L (in a given band) at redshift z to have a star-formation rate in the range $(\dot{M}_\star, \dot{M}_\star + d\dot{M}_\star)$, and $\langle \dot{M}_\star \rangle (L,z)$ is the mean star-formation rate for galaxies with luminosity L at redshift z. The luminosity function $\phi(L,z)$ can be obtained from deep redshift surveys of galaxies, as summarized above. The transformation from luminosity to star-formation rate depends on the rest-frame waveband used to measure

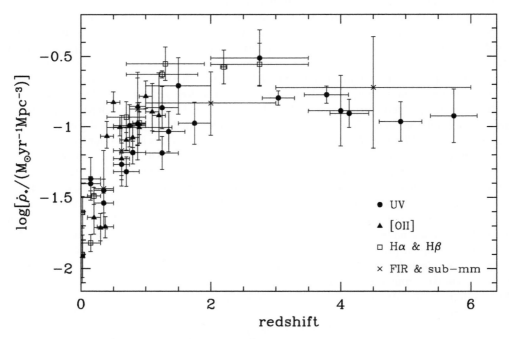

Fig. 2.35. The global star-formation rate (in $M_\odot\,yr^{-1}\,Mpc^{-3}$) as a function of redshift. Different symbols correspond to different rest-frame wavelength ranges used to infer the star-formation rates, as indicated. [Based on the data compilation of Hopkins (2004)]

the luminosity function, and typically involves many uncertainties (see §10.3.8 for a detailed discussion).

Fig. 2.35 shows a compilation of various measurements of the global star-formation rate at different redshifts, obtained using different techniques. Although there is still considerable scatter, and the data may be plagued by systematic errors due to uncertain extinction corrections, it is now well established that the cosmic star-formation rate has dropped by roughly an order of magnitude from $z \sim 2$ to the present. Integrating this cosmic star-formation history over time, one can show that the star-forming populations observed to date are already sufficient to account for the majority of stars observed at $z \sim 0$ (e.g. Dickinson et al., 2003).

2.7 Large-Scale Structure

An important property of the galaxy population is its overall spatial distribution. Since each galaxy is associated with a large amount of mass, one might naively expect that the galaxy distribution reflects the large-scale mass distribution in the Universe. On the other hand, if the process of galaxy formation is highly stochastic, or galaxies only form in special, preferred environments, the relation between the galaxy distribution and the matter distribution may be far from straightforward. Therefore, detailed studies of the spatial distribution of galaxies in principle can convey information regarding both the overall matter distribution, which is strongly cosmology dependent, and the physics of galaxy formation.

Fig. 2.36 shows the distribution of more than 80,000 galaxies in the 2dFGRS, where the distances of the galaxies have been estimated from their redshifts. Clearly the distribution of galaxies in space is not random, but shows a variety of structures. As we have already seen in §2.5, some galaxies are located in high-density clusters containing several hundreds of galaxies, or in smaller groups containing a few to tens of galaxies. The majority of all galaxies, however, are distributed in low-density filamentary or sheet-like structures. These sheets and filaments surround large

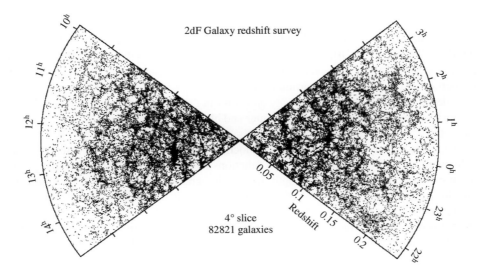

Fig. 2.36. The spatial distribution of $\sim 80,000$ galaxies in the 2dFGRS in a $4°$ slice projected onto the redshift/right-ascension plane. Clearly galaxies are not distributed randomly, but are clumped together in groups and clusters connected by large filaments that enclose regions largely devoid of galaxies. [Adapted from Peacock (2002)]

voids, which are regions with diameters up to $\sim 100\,\mathrm{Mpc}$ that contain very few, or no, galaxies. One of the challenges in studying the spatial distribution of galaxies is to properly quantify the complexity of this 'cosmic web' of filaments, sheets and voids. In this section we consider the galaxy distribution as a point set in space and study the spatial correlations among these points in a statistical sense.

2.7.1 Two-Point Correlation Functions

One of the most important statistics used to characterize the spatial distribution of galaxies is the two-point correlation function, defined as the excess number of galaxy pairs of a given separation, r, relative to that expected for a random distribution:

$$\xi(r) = \frac{DD(r)\Delta r}{RR(r)\Delta r} - 1. \tag{2.44}$$

Here $DD(r)\Delta r$ is the number of galaxy pairs with separations in the range $r \pm \Delta r/2$, and $RR(r)\Delta r$ is the number that would be expected if galaxies were randomly distributed in space. Galaxies are said to be positively correlated on scale r if $\xi(r) > 0$, to be anticorrelated if $\xi(r) < 0$, and to be uncorrelated if $\xi(r) = 0$. Since it is relatively straightforward to measure, the two-point correlation function of galaxies has been estimated from various samples. In many cases, redshifts are used as distances and the corresponding correlation function is called the correlation function in redshift space. Because of peculiar velocities, this redshift-space correlation is different from that in real space. The latter can be estimated from the projected two-point correlation function, in which galaxy pairs are defined by their separations projected onto the plane perpendicular to the line-of-sight so that it is not affected by using redshift as distance (see Chapter 6 for details). Fig. 2.37 shows an example of the redshift-space correlation function and the corresponding real-space correlation function. On scales smaller than about $10\,h^{-1}\mathrm{Mpc}$ the real-space correlation function can well be described by a power law,[5]

$$\xi(r) = (r/r_0)^{-\gamma}, \tag{2.45}$$

with $\gamma \sim 1.8$ and with a correlation length $r_0 \approx 5\,h^{-1}\mathrm{Mpc}$. This shows that galaxies are strongly clustered on scales $\lesssim 5\,h^{-1}\mathrm{Mpc}$, and the clustering strength becomes weak on scales much larger than $\sim 10\,h^{-1}\mathrm{Mpc}$. The exact values of γ and r_0 are found to depend significantly on the properties of the galaxies. In particular the correlation length, r_0, defined by $\xi(r_0) = 1$, is found to depend on both galaxy luminosity and color in the sense that brighter and redder galaxies are more strongly clustered than their fainter and bluer counterparts (e.g. Norberg et al., 2001, 2002a; Zehavi et al., 2005; Wang et al., 2008b).

One can apply exactly the same correlation function analysis to groups and clusters of galaxies. This shows that their two-point correlation functions has a logarithmic slope, γ, that is similar to that of galaxies, but a correlation length, r_0, which increases strongly with the richness of the systems in question, from about $5\,h^{-1}\mathrm{Mpc}$ for poor groups to about $20\,h^{-1}\mathrm{Mpc}$ for rich clusters (e.g. Yang et al., 2005c).

Another way to describe the clustering strength of a certain population of objects is to calculate the variance of the number counts within randomly placed spheres of given radius r:

$$\sigma^2(r) \equiv \frac{1}{(\bar{n}V)^2}\frac{1}{M}\sum_{i=1}^{M}(N_i - \bar{n}V)^2, \tag{2.46}$$

---

[5] Note that, because of the definition of the two-point correlation function, $\xi(r)$ has to become negative on large scales. Therefore, a power law can only fit the data up to a finite scale.

where $\bar{n}$ is the mean number density of objects, $V = 4\pi r^3/3$, and N_i ($i = 1,\ldots,M$) are the number counts of objects in M randomly placed spheres. For optically selected galaxies with a luminosity of the order of L^* one finds that $\sigma \sim 1$ on a scale of $r = 8\,h^{-1}$Mpc and decreases to $\sigma \sim 0.1$ on a scale of $r = 30\,h^{-1}$Mpc. This confirms that the galaxy distribution is strongly inhomogeneous on scales of $\lesssim 8\,h^{-1}$Mpc, but starts to approach homogeneity on significantly larger scales.

Since galaxies, groups and clusters all contain large amounts of matter, we expect their spatial distribution to be related to the mass distribution in the Universe to some degree. However, the fact that different objects have different clustering strengths makes one wonder if any of them are actually fair tracers of the matter distribution. The spatial distribution of luminous objects, such as galaxies, groups and clusters, depends not only on the matter distribution in the Universe, but also on how they form in the matter density field. Therefore, without a detailed understanding of galaxy formation, it is unclear which, if any, population of galaxies accurately traces the matter distribution. It is therefore very important to have independent means to probe the matter density field.

One such probe is the velocity field of galaxies. The peculiar velocities of galaxies are generated by the gravitational field, and therefore contain useful information regarding the matter distribution in the Universe. In the past, two different methods have been used to extract this information from observations. One is to estimate the peculiar velocities of many galaxies by measuring both their receding velocities (i.e. redshifts) and their distances. The peculiar velocities then follow from Eq. (2.19), which can then be used to trace out the matter distribution. Such analyses not only yield constraints on the mean matter density in the Universe, but also on how galaxies trace the mass distribution. Unfortunately, although galaxy redshifts are easy to measure, accurate distance measurements for a large sample of galaxies are very difficult to obtain, severely impeding the applicability of this method. Another method, which is more statistical in nature, extracts information about the peculiar velocities of galaxies from a comparison of the real-space and redshift-space two-point correlation functions. This method is based on the fact that an isotropic distribution in real space will appear anisotropic in redshift space due to the presence of peculiar velocities. Such redshift-space distortions are the primary reason why the redshift-space correlation function has a shape different from that of the real-space correlation function (see Fig. 2.37). As described in detail in §6.3, by carefully modeling the redshift space distortions one can obtain useful constraints on the matter distribution in the Universe.

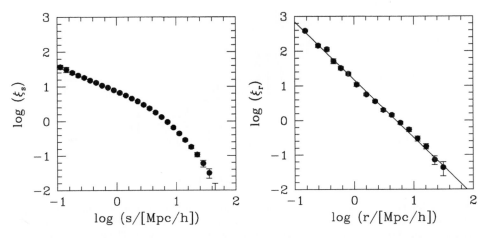

Fig. 2.37. The two-point correlation function of galaxies in redshift space (left) and real space (right). The straight line is a power law, $\xi(r) = (r/r_0)^{-\gamma}$, with $r_0 = 5.05\,h^{-1}$Mpc and $\gamma = 1.67$. [Based on data published in Hawkins et al. (2003)]

2.7.2 Probing the Matter Field via Weak Lensing

A very promising way to probe the mass distribution in the Universe is through weak gravitational lensing. Any light beam we observe from a distant source has been deflected and distorted due to the gravitational tidal field along the line-of-sight. This cumulative gravitational lensing effect due to the inhomogeneous mass distribution between source and observer is called cosmic shear, and holds useful information about the statistical properties of the matter field. The great advantage of this technique over the clustering analysis discussed above is that it does not have to make assumptions about the relation between galaxies and matter.

Unless the beam passes very close to a particular overdensity (i.e. a galaxy or cluster), in which case we are in the strong lensing regime, these distortions are extremely weak. Typical values for the expected shear are of the order of one percent on angular scales of a few arcminutes, which means that the distorted image of an intrinsically circular source has an ellipticity of 0.01. Even if one could accurately measure such a small ellipticity, the observed ellipticity holds no information without prior knowledge of the intrinsic ellipticity of the source, which is generally unknown. Rather, one detects cosmic shear via the spatial correlations of image ellipticities. The light beams from two distant sources that are close to each other on the sky have roughly encountered the same large-scale structure along their lines-of-sight, and their distortions (i.e. image ellipticities) are therefore expected to be correlated (both in magnitude and in orientation). Such correlations have been observed (see Fig. 2.38), and detailed modeling of these results

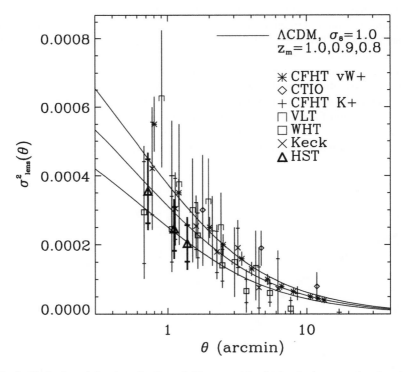

Fig. 2.38. In the limit of weak lensing, the shear field at a position in the sky is proportional to the ellipticity of the image of a circular source at that position. This plot shows the mean square of the shear field averaged within circular regions of given radius, θ, obtained from various observations. The non-zero values of this 'cosmic shear' are due to gravitational lensing induced by the line-of-sight projected mass distribution in the Universe. The solid curves are theoretical predictions (see §6.6) and are in good agreement with the data. [Adapted from Refregier et al. (2002) by permission of AAS]

shows that the variance of the matter density field on scales of $8h^{-1}$Mpc is about 0.7–0.9 (e.g. Van Waerbeke et al., 2001), slightly lower than that of the distribution of bright galaxies.

Since the matter distribution around a given galaxy or cluster will cause a distortion of its background galaxies, weak lensing can also be used to probe the matter distributions around galaxies and clusters. In the case of clusters, one can often detect a sufficient number of background galaxies to reliably measure the shear induced by its gravitational potential. Weak lensing therefore offers a means of measuring the total gravitational mass of an individual (massive) cluster. In the case of individual galaxies, however, one typically has only a few background galaxies available. Consequently, the weak lensing signal is far too weak to detect around individual galaxies. However, by stacking the images of many foreground galaxies (for example, according to their luminosity), one obtains sufficient signal-to-noise to measure the shear, which reflects the *average* mass distribution around the stacked galaxies. This technique is called galaxy–galaxy lensing, and has been used to demonstrate that galaxies are surrounded by extended dark matter halos with masses 10–100 times more massive than the galaxies themselves (e.g. Mandelbaum et al., 2006b).

2.8 The Intergalactic Medium

The intergalactic medium (IGM) is the medium that permeates the space in between galaxies. In the framework laid out in Chapter 1, galaxies form by the gravitational aggregation of gas in a medium which was originally quite homogeneous. In this scenario, the study of the IGM is an inseparable part of galaxy formation, because it provides us with the properties of the gas from which galaxies form.

The properties of the IGM can be probed observationally by its emission and by its absorption of the light from background sources. If the medium is sufficiently dense and hot, it can be observed in X-ray emission, as is the case for the intracluster medium described in §2.5.1. However, in general the density of the IGM is too low to produce detectable emission, and its properties have to be determined from absorption studies.

2.8.1 The Gunn–Peterson Test

Much information about the IGM has been obtained through its absorption of light from distant quasars. Quasars are not only bright, so that they can be observed out to large distances, but also have well-behaved continua, against which absorption can be analyzed relatively easily. One of the most important tests of the presence of intergalactic neutral hydrogen was proposed by Gunn & Peterson (1965). The Gunn–Peterson test makes use of the fact that the Lyα absorption of neutral hydrogen at $\lambda_\alpha = 1216$ Å has a very large cross-section. When the ultraviolet continuum of a distant quasar (assumed to have redshift z_Q) is shifted to 1216 Å at some redshift $z < z_Q$, the radiation would be absorbed at this redshift if there were even a small amount of neutral hydrogen. Thus, if the Universe were filled with a diffuse distribution of neutral hydrogen, photons bluer than Lyα would be significantly absorbed, causing a significant decrement of flux in the observed quasar spectrum at wavelengths shorter than $(1+z_Q)\lambda_\alpha$. Using the hydrogen Lyα cross-section and the definition of optical depth (see Chapter 16 for details), one obtains that the proper number density of HI atoms obeys

$$n_{\rm HI}(z) \sim 2.42 \times 10^{-11} \tau(z) h H(z)/H_0 \, {\rm cm}^{-3}, \qquad (2.47)$$

where $H(z)$ is Hubble's constant at redshift z, and $\tau(z)$ is the absorption optical depth out to z that can be determined from the flux decrements in quasar spectra. Observations show that the Lyα absorption optical depth is much smaller than unity out to $z \lesssim 6$. The implied density of neutral hydrogen in the diffuse IGM is thus much lower than the mean gas density in the Universe (which is about $10^{-7}\,\mathrm{cm}^{-3}$). This suggests that the IGM must be highly ionized at redshifts $z \lesssim 6$.

As we will show in Chapter 3, the IGM is expected to be highly neutral after recombination, which occurs at a redshift $z \sim 1000$. Therefore, the fact that the IGM is highly ionized at $z \sim 6$ indicates that the Universe must have undergone some phase transition, from being largely neutral to being highly ionized, a process called re-ionization. It is generally believed that photo-ionization due to energetic photons (with energies above the Lyman limit) are responsible for the re-ionization. This requires the presence of effective emitters of UV photons at high redshifts. Possible candidates include quasars, star-forming galaxies and the first generation of stars. But to this date the actual ionizing sources have not yet been identified, nor is it clear at what redshift re-ionization occurred. The highest redshift quasars discovered to date, which are close to $z = 6.5$, show almost no detectable flux at wavelengths shorter than $(1+z)\lambda_\alpha$ (Fan et al., 2006). Although this seems to suggest that the mass density of neutral hydrogen increases rapidly at around this redshift, it is not straightforward to convert such flux decrements into an absorption optical depth or a neutral hydrogen fraction, mainly because any $\tau \gg 1$ can result in an almost complete absorption of the flux. Therefore it is currently still unclear whether the Universe became (re-)ionized at a redshift just above 6 or at a significantly higher redshift. At the time of writing, several facilities are being constructed that will attempt to detect 21cm line emission from neutral hydrogen at high redshifts. It is anticipated that these experiments will shed important light on the detailed re-ionization history of the Universe, as we discuss in some detail in §16.3.4.

2.8.2 Quasar Absorption Line Systems

Although the flux blueward of $(1+z_Q)\lambda_\alpha$ is not entirely absorbed, quasar spectra typically reveal a large number of absorption lines in this wavelength range (see Fig. 2.39). These absorption lines are believed to be produced by intergalactic clouds that happen to lie along the line-of-sight from the observer to the quasar, and can be used to probe the properties of the IGM. Quasar absorption line systems are grouped into several categories:

- Lyα forest: These are narrow lines produced by HI Lyα absorption. They are numerous and appear as a 'forest' of lines blueward of the Lyα emission line of a quasar.
- Lyman-limit systems (LLS): These are systems with HI column densities $N_\mathrm{HI} \gtrsim 10^{17}\,\mathrm{cm}^{-2}$, at which the absorbing clouds are optically thick to the Lyman-limit photons (912Å). These systems appear as continuum breaks in quasar spectra at the redshifted wavelength $(1+z_a) \times$ 912Å, where z_a is the redshift of the absorber.
- Damped Lyα systems (DLAs): These systems are produced by HI Lyα absorption of gas clouds with HI column densities, $N_\mathrm{HI} \gtrsim 2 \times 10^{20}\,\mathrm{cm}^{-2}$. Because the Ly$\alpha$ absorption optical depth at such column densities is so large, the quasar continuum photons are completely absorbed near the line center and the line profile is dominated by the damping wing due to the natural (Lorentz) broadening of the absorption line. DLAs with column densities in the range $10^{19}\,\mathrm{cm}^{-2} < N_\mathrm{HI} < 2 \times 10^{20}\,\mathrm{cm}^{-2}$ also exhibit damping wings, and are sometimes called sub-DLAs (Péroux et al., 2002). They differ from the largely neutral DLAs in that they are still significantly ionized.
- Metal absorption line systems: In addition to the hydrogen absorption line systems listed above, QSO spectra also frequently show absorption lines due to metals. The best-known examples are MgII systems and CIV systems, which are caused by the strong resonance-line doublets MgII$\lambda\lambda$2796,2800 and CIV$\lambda\lambda$1548,1550, respectively. Note that both doublets have rest-frame wavelengths longer than $\lambda_\mathrm{Ly\alpha} = 1216$Å. Consequently, they can appear on the

2.8 The Intergalactic Medium

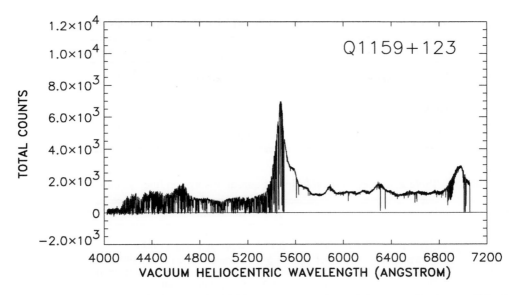

Fig. 2.39. The spectrum of a QSO that reveals a large number of absorption lines due to the IGM. The strongest peak at 5473 Å is the emission line due to Lyα at a rest-frame wavelength of 1216 Å. The numerous absorption lines at $\lambda < 5473$ Å make up the Lyα forest which is due to Lyα absorption of neutral hydrogen clouds between the QSO and the Earth. The break at 4150 Å is due to a Lyman-limit cloud which is optically thick at the hydrogen Lyman edge (rest-frame wavelength of 912 Å). The relatively sparse lines to the right of the Lyα emission line are due to absorption by metal atoms associated with the absorbing clouds. [Adapted from Songaila (1998) by permission of AAS]

red side of the Lyα emission line of the QSO, which makes them easily identifiable because of the absence of confusion from the Lyα forest.

Note that a single absorber may be detected as more than one absorption system. For example, an absorber at z_a may be detected as a HI Lyα line at $\lambda = (1+z_a) \times 1216$ Å, as a CIV system at $\lambda = (1+z_a) \times 1548$ Å, if it has a sufficiently large abundance of CIV ions, and as a Lyman-limit system at $\lambda = (1+z_a) \times 912$ Å, if its HI column density is larger than $\sim 10^{17}$ cm^{-2}.

In addition to the most common absorption systems listed above, other line systems are also frequently identified in quasar spectra. These include low ionization lines of heavy elements, such as CII, MgI, FeII, etc., and the more highly ionized lines, such as SiIV and NV. Highly ionized lines such as OVI and OVII are also detected in the UV and/or X-ray spectra of quasars. Since the ionization state of an absorbing cloud depends on its temperature, highly ionized lines, such as OVI and OVII, in general signify the existence of hot ($\sim 10^6$ K) gas, while low-ionization lines, such as HI, CII and MgII, are more likely associated with relatively cold ($\sim 10^4$ K) gas.

For a given quasar spectrum, absorption line systems are identified by decomposing the spectrum into individual lines with some assumed profiles (e.g. the Voigt profile, see §16.4.3). By modeling each system in detail, one can in principle obtain its column density, b parameter (defined as $b = \sqrt{2}\sigma$, where σ is the velocity dispersion of the absorbing gas), ionization state, and temperature. If both hydrogen and metal systems are detected, one may also estimate the metallicity of the absorbing gas. Table 2.8 lists the typical values of these quantities for the most commonly detected absorption systems mentioned above.

The evolution of the number of absorption systems is described by the number of systems per unit redshift, $d\mathcal{N}/dz$, as a function of z. This relation is usually fitted by a power law $d\mathcal{N}/dz \propto (1+z)^\gamma$, and the values of γ for different systems are listed in Table 2.9. The distribution of absorption line systems with respect to the HI column density is shown in Fig. 2.40. Over the

Table 2.8. Properties of common absorption lines in quasar spectra.

System	$\log(N_{\rm HI}/{\rm cm}^{-2})$	$b/({\rm km\,s^{-1}})$	$Z/Z_\odot$	$\log(N_{\rm HI}/N_{\rm H})$
Lyα forest	12.5 – 17	15 – 40	< 0.01	< -3
Lyman limit	> 17	~ 100	~ 0.1	> -2
sub-DLA	19 – 20.3	~ 100	~ 0.1	> -1
DLA	> 20.3	~ 100	~ 0.1	~ 0
CIV	> 15.5	~ 100	~ 0.1	> -3
MgII	> 17	~ 100	~ 0.1	> -2

Table 2.9. Redshift evolution of quasar absorption line systems.

System	z range	γ	Reference
Lyα forest	2.0 – 4.0	~ 2.5	Kim et al. (1997)
Lyα forest	0.0 – 1.5	~ 0.15	Weymann et al. (1998)
Lyman limit	0.3 – 4.1	~ 1.5	Stengler-Larrea et al. (1995)
Damped Lyα	0.1 – 4.7	~ 1.3	Storrie-Lombardi et al. (1996a)
CIV	1.3 – 3.4	~ -1.2	Sargent et al. (1988)
MgII	0.2 – 2.2	~ 0.8	Steidel & Sargent (1992)

whole observed range, this distribution follows roughly a power law, $d\mathcal{N}/dN_{\rm HI} \propto N_{\rm HI}^{-\beta}$, with $\beta \sim 1.5$.

From the observed column density distribution, one can estimate the mean mass density of neutral hydrogen that is locked up in quasar absorption line systems:

$$\rho_{\rm HI}(z) = \left(\frac{dl}{dz}\right)^{-1} m_{\rm H} \int N_{\rm HI} \frac{d^2 \mathcal{N}}{dN_{\rm HI}\, dz} dN_{\rm HI}, \qquad (2.48)$$

where dl/dz is the physical length per unit redshift at z (see §3.2.6). Given that $d\mathcal{N}/dN_{\rm HI}$ is a power law with index ~ -1.5, $\rho_{\rm HI}$ is dominated by systems with the highest $N_{\rm HI}$, i.e. by damped Lyα systems. Using the observed HI column density distribution, one infers that about 5% of the baryonic material in the Universe is in the form of HI gas at $z \sim 3$ (e.g. Storrie-Lombardi et al., 1996b). In order to estimate the total hydrogen mass density associated with quasar absorption line systems, however, one must know the neutral fraction, $N_{\rm HI}/N_{\rm H}$, as a function of $N_{\rm HI}$. This fraction depends on the ionization state of the IGM. Detailed modeling shows that the Lyα forest systems are highly ionized, and that the main contribution to the total (neutral plus ionized) gas density comes from absorption systems with $N_{\rm HI} \sim 10^{14}\,{\rm cm}^{-2}$. The total gas mass density at $z \sim 3$ thus inferred is comparable to the total baryon density in the Universe (e.g. Rauch et al., 1997; Weinberg et al., 1997b).

Quasar absorption line systems with the highest HI column densities are expected to be gas clouds in regions of high gas densities where galaxies and stars may form. It is therefore not surprising that these systems contain metals. Observations of damped Lyα systems show that they have typical metallicities about 1/10 of that of the Sun (e.g. Pettini et al., 1990; Kulkarni et al., 2005), lower than that of the ISM in the Milky Way. This suggests that these systems may be associated with the outer parts of galaxies, or with galaxies in which only a small fraction of the gas has formed stars. More surprising is the finding that most, if not all, of the Lyα forest lines also contain metals, although the metallicities are generally low, typically about 1/1000 to 1/100 of that of the Sun (e.g. Simcoe et al., 2004). There is some indication that the metallicity increases with HI column density, but the trend is not strong. Since star formation requires relatively high column densities of neutral hydrogen (see Chapter 9), the metals observed in absorption line

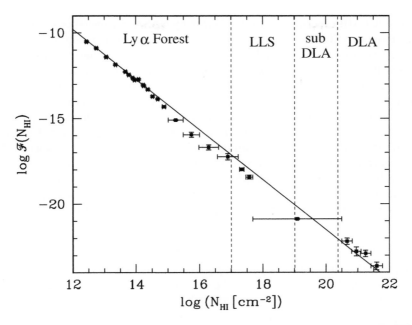

Fig. 2.40. The HI column density distribution of QSO absorption line systems. Here $\mathscr{F}(N_{HI})$ is defined as the number of absorption lines per unit column density, per unit X (which is a quantity that is related to redshift according to Eq. [16.92]). The solid line corresponds to $\mathscr{F}(N_{HI}) \propto N_{HI}^{-1.46}$, which fits the data reasonably well over the full 10 orders of magnitude in column density. [Based on data published in Petitjean et al. (1993) and E. M. Hu et al. (1995)]

systems with low HI column densities most likely originate from, and have been expelled by, galaxies at relatively large distances.

2.9 The Cosmic Microwave Background

The cosmic microwave background (CMB) was discovered by Penzias and Wilson in 1965 when they were commissioning a sensitive receiver at centimeter wavelengths in Bell Telephone Laboratories. It was quickly found that this radiation background was highly isotropic on the sky and has a spectrum close to that of a blackbody with a temperature of about 3 K. The existence of such a radiation background was predicted by Gamow, based on his model of a Hot Big Bang cosmology (see §1.4.2), and it therefore did not take long before the cosmological significance of this discovery was realized (e.g. Dicke et al., 1965).

The observed properties of the CMB are most naturally explained in the standard model of cosmology. Since the early Universe was dense, hot and highly ionized, photons were absorbed and re-emitted many times by electrons and ions and so a blackbody spectrum could be established in the early Universe. As the Universe expanded and cooled and the density of ionized material dropped, photons were scattered less and less often and eventually could propagate freely to the observer from a last-scattering surface, inheriting the blackbody spectrum.

Because the CMB is so important for our understanding of the structure and evolution of the Universe, there have been many attempts in the 1970s and 1980s to obtain more accurate measurements of its spectrum. Since the atmospheric emission is quite close to the peak wavelength of a 3 K blackbody spectrum, most of these measurements were carried out using high-altitude balloon experiments (for a discussion of early CMB experiments, see Partridge, 1995).

A milestone in CMB experiments was the launch by NASA in November 1989 of the Cosmic Background Explorer (COBE), a satellite devoted to accurate measurements of the CMB over the entire sky. Observations with the Far InfraRed Absolute Spectrophotometer (FIRAS) on board COBE showed that the CMB has a spectrum that is perfectly consistent with a blackbody spectrum, to exquisite accuracy, with a temperature $T = 2.728 \pm 0.002\,\mathrm{K}$. As we will see in §3.5.4 the lack of any detected distortions from a pure blackbody spectrum puts strong constraints on any processes that may change the CMB spectrum after it was established in the early Universe.

Another important observational result from COBE is the detection, for the first time, of anisotropy in the CMB. Observations with the Differential Microwave Radiometers (DMR) on board COBE have shown that the CMB temperature distribution is highly isotropic over the sky, confirming earlier observational results, but also revealed small temperature fluctuations (see Fig. 2.41). The observed temperature map contains a component of anisotropy on very large angular scales, which is well described by a dipole distribution over the sky,

$$T(\alpha) = T_0 \left(1 + \frac{v}{c}\cos\alpha\right), \qquad (2.49)$$

where α is the angle of the line-of-sight relative to a specific direction. This component can be explained as the Doppler effect caused by the motion of the Earth with a velocity $v = 369 \pm 3\,\mathrm{km\,s^{-1}}$ towards the direction $(l,b) = (264.31° \pm 0.20°, 48.05° \pm 0.10°)$ in Galactic coordinates

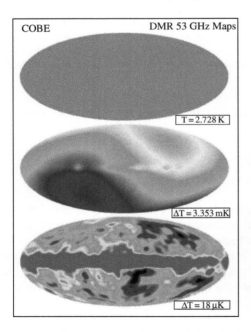
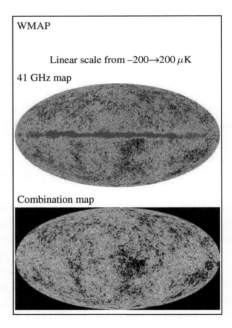

Fig. 2.41. Temperature maps of the CMB in galactic coordinates. The three panels on the left show the temperature maps obtained by the DMR on board the COBE satellite [Courtesy of NASA Goddard Space Flight Center]. The upper panel shows the near-uniformity of the CMB brightness; the middle panel is the map after subtraction of the mean brightness, showing the dipole component due to our motion with respect to the background; and the bottom panel shows the temperature fluctuations after subtraction of the dipole component. Emission from the Milky Way is evident in the bottom image. The two right panels show the temperature maps observed by WMAP from the first year of data [Courtesy of WMAP Science Team]; one is from the 41 GHz channel and the other is a linear combination of five channels. Note that the large-scale temperature fluctuations in the COBE map at the bottom are clearly seen in the WMAP maps, and that the WMAP angular resolution (about 0.5°) is much higher than that of COBE (about 7°).

(Lineweaver et al., 1996). Once this dipole component is subtracted, the map of the temperature fluctuations looks like that shown in the lower left panel of Fig. 2.41. In addition to emission from the Milky Way, it reveals fluctuations in the CMB temperature with an amplitude of the order of $\Delta T/T \sim 2 \times 10^{-5}$.

Since the angular resolution of the DMR is about $7°$, COBE observations cannot reveal anisotropy in the CMB on smaller angular scales. Following the detection by COBE, there have been a large number of experiments to measure small-scale CMB anisotropies, and many important results have come out in recent years. These include the results from balloon-borne experiments such as Boomerang (de Bernardis et al., 2000) and Maxima (Hanany et al., 2000), from ground-based interferometers such as the Degree Angular Scale Interferometer (DASI; Halverson et al., 2002) and the Cosmic Background Imager (CBI; Mason et al., 2002), and from an all-sky satellite experiment called the Wilkinson Microwave Anisotropy Probe (WMAP; Bennett et al., 2003; Hinshaw et al., 2007). These experiments have provided us with extremely detailed and accurate maps of the anisotropies in the CMB, such as that obtained by WMAP shown in the right panels of Fig. 2.41.

In order to quantify the observed temperature fluctuations, a common practice is to expand the map in spherical harmonics,

$$\frac{\Delta T}{T}(\vartheta,\varphi) \equiv \frac{T(\vartheta,\varphi) - \overline{T}}{\overline{T}} = \sum_{\ell,m} a_{\ell m} Y_{\ell,m}(\vartheta,\varphi). \quad (2.50)$$

The angular power spectrum, defined as $C_\ell \equiv \langle |a_{\ell m}|^2 \rangle^{1/2}$ (where $\langle \ldots \rangle$ denotes averaging over m), can be used to represent the amplitudes of temperature fluctuations on different angular scales. Fig. 2.42 shows the temperature power spectrum obtained by the WMAP satellite. As one can

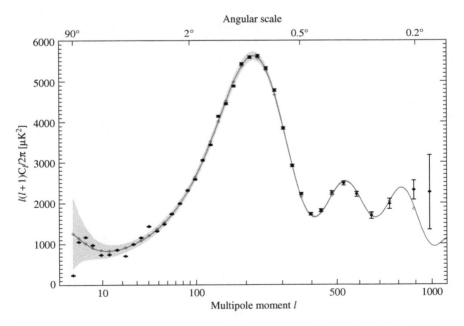

Fig. 2.42. The angular power spectrum, C_ℓ, of the CMB temperature fluctuations in the WMAP full-sky map. This shows the relative brightness of the 'spots' in the CMB temperature map vs. the size of the spots. The shape of this curve contains a wealth of information about the geometry and matter content of the Universe. The curve is the model prediction for the best-fit ΛCDM cosmology. [Adapted from Hinshaw et al. (2007) by permission of AAS]

see, the observed C_ℓ as a function of ℓ shows complex features. These observational results are extremely important for our understanding of the structure formation in the Universe. First of all, the observed high degree of isotropy in the CMB gives strong support for the assumption of the standard cosmology that the Universe is highly homogeneous and isotropic on large scales. Second, the small temperature fluctuations observed in the CMB are believed to be caused by the density perturbations at the time when the Universe became transparent to CMB photons. These same density perturbations are thought to be responsible for the formation of structures in the Universe. So the temperature fluctuations in the CMB may be used to infer the properties of the initial conditions for the formation of galaxies and other structures in the Universe. Furthermore, the observations of CMB temperature fluctuations can also be used to constrain cosmological parameters. As we will discuss in detail in Chapter 6, the peaks and valleys in the angular power spectrum are caused by acoustic waves present at the last scattering surface of the CMB photons. The heights (depths) and positions of these peaks (valleys) depend not only on the density of baryonic matter, but also on the total mean density of the Universe, Hubble's constant and other cosmological parameters. Modeling the angular power spectrum of the CMB temperature fluctuations can therefore provide constraints on all of these cosmological parameters.

2.10 The Homogeneous and Isotropic Universe

As we will see in Chapter 3, the standard cosmological model is based on the 'cosmological principle' according to which the Universe is homogeneous and isotropic on large scales. As we have seen, observations of the CMB and of the large-scale spatial distribution of galaxies offer strong support for this cosmological principle. Since according to Einstein's general relativity the space-time geometry of the Universe is determined by the matter distribution in the Universe, this large-scale distribution of matter has important implications for the large-scale geometry of space-time.

For a homogeneous and isotropic universe, its global properties (such as density and pressure) at any time must be the same as those in any small volume. This allows one to study the global properties of the Universe by examining the properties of a small volume within which Newtonian physics is valid. Consider a (small) spherical region of fixed mass M. Since the Universe is homogeneous and isotropic, the radius R of the sphere should satisfy the Newtonian equation[6]

$$\ddot{R} = -\frac{GM}{R^2}. \qquad (2.51)$$

Note that, because of the homogeneity, there is no force due to pressure gradients and that only the mass within the sphere is relevant for the motion of R. This follows directly from Birkhoff's theorem, according to which the gravitational acceleration at any radius in a spherically symmetric system depends only on the mass within that radius. For a given M, the above equation can be integrated once to give

$$\frac{1}{2}\dot{R}^2 - \frac{GM}{R} = E, \qquad (2.52)$$

[6] As we will see in Chapter 3, in general relativity it is the combination of energy density ρ and pressure P, $\rho + 3P/c^2$, instead of ρ, that acts as the source of gravitational acceleration. Therefore, Eq. (2.51) is not formally valid, even though Eq. (2.53), which derives from it, happens to be correct.

where E is a constant, equal to the specific energy of the spherical shell. For simplicity, we write $R = a(t)R_0$, where R_0 is independent of t. It then follows that

$$\frac{\dot{a}^2}{a^2} - \frac{8\pi G \overline{\rho}}{3} = -\frac{Kc^2}{a^2}, \qquad (2.53)$$

where $\overline{\rho}$ is the mean density of the Universe and $K = -2E/(cR_0)^2$. Unless $E = 0$, which corresponds to $K = 0$, we can always choose the value of R_0 so that $|K| = 1$. So defined, K is called the curvature signature, and takes the value $+1$, 0, or -1. With this normalization, the equation for a is independent of M. As we will see in Chapter 3, Eq. (2.53) is identical to the Friedmann equation based on general relativity. For a universe dominated by a non-relativistic fluid, this is not surprising, as it follows directly from the assumption of homogeneity and isotropy. However, as we will see in Chapter 3, it turns out that Eq. (2.53) also holds even if relativistic matter and/or the energy density associated with the cosmological constant are included.

The quantity $a(t)$ introduced above is called the scale factor, and describes the change of the distance between any two points fixed in the cosmological background. If the distance between a pair of points is l_1 at time t_1, then their distance at some later time t_2 is related to l_1 through $l_2 = l_1 a(t_2)/a(t_1)$. It then follows that at any time t the velocity between any two (comoving) points can be written as

$$\dot{l} = [\dot{a}(t)/a(t)]l, \qquad (2.54)$$

where l is the distance between the two points at time t. Thus, $\dot{a} > 0$ corresponds to an expanding universe, while $\dot{a} < 0$ corresponds to a shrinking universe; the universe is static only when $\dot{a} = 0$. The ratio $\dot{a}/a$ evaluated at the present time, t_0, is called the Hubble constant,

$$H_0 \equiv \dot{a}_0/a_0, \qquad (2.55)$$

where $a_0 \equiv a(t_0)$, and the relation between velocity and distance, $\dot{l} = H_0 l$, is known as Hubble's expansion law. Another quantity that characterizes the expansion of the Universe is the deceleration parameter, defined as

$$q_0 \equiv -\frac{\ddot{a}_0 a_0}{\dot{a}_0^2}. \qquad (2.56)$$

This quantity describes whether the expansion rate of the Universe is accelerating ($q_0 < 0$) or decelerating ($q_0 > 0$) at the present time.

Because of the expansion of the Universe, waves propagating in the Universe are stretched. Thus, photons with a wavelength λ emitted at an earlier time t will be observed at the present time t_0 with a wavelength $\lambda_{\rm obs} = \lambda a_0/a(t)$. Since $a_0 > a(t)$ in an expanding universe, $\lambda_{\rm obs} > \lambda$ and so the wavelength of the photons is redshifted. The amount of redshift z between time t and t_0 is given by

$$z \equiv \frac{\lambda_{\rm obs}}{\lambda} - 1 = \frac{a_0}{a(t)} - 1. \qquad (2.57)$$

Note that $a(t)$ is a monotonically increasing function of t in an expanding universe, and so redshift is uniquely related to time through the above equation. If an object has redshift z, i.e. its observed spectrum is shifted to the red relative to its rest-frame (intrinsic) spectrum by $\Delta \lambda = \lambda_{\rm obs} - \lambda = z\lambda$, then the photons we observe today from the object were actually emitted at a time t that is related to its redshift z by Eq. (2.57). Because of the constancy of the speed of light, an object's redshift can also be used to infer its distance.

From Eq. (2.53) one can see that the value of K is determined by the mean density $\bar{\rho}_0$ at the present time t_0 and the value of Hubble's constant. Indeed, if we define a critical density

$$\rho_{\text{crit},0} \equiv \frac{3H_0^2}{8\pi G}, \tag{2.58}$$

and write the mean density in terms of the density parameter,

$$\Omega_0 \equiv \bar{\rho}_0/\rho_{\text{crit},0}, \tag{2.59}$$

then $K = H_0^2 a_0^2 (\Omega_0 - 1)$. So $K = -1, 0$ and $+1$ corresponds to $\Omega_0 < 1, = 1$ and > 1, respectively. Before discussing the matter content of the Universe, it is illustrative to write the mean density as a sum of several possible components:

(i) non-relativistic matter whose (rest-mass) energy density changes as $\rho_m \propto a^{-3}$;
(ii) relativistic matter (such as photons) whose energy density changes as $\rho_r \propto a^{-4}$ (the number density changes as a^{-3} while energy is redshifted according to a^{-1});
(iii) vacuum energy, or the cosmological constant Λ, whose density $\rho_\Lambda = c^2 \Lambda/8\pi G$ is a constant.

Thus,

$$\Omega_0 = \Omega_{m,0} + \Omega_{r,0} + \Omega_{\Lambda,0}, \tag{2.60}$$

and Eq. (2.53) can be written as

$$\left(\frac{\dot{a}}{a}\right)^2 = H_0^2 E^2(z), \tag{2.61}$$

where

$$E(z) = \left[\Omega_{\Lambda,0} + (1-\Omega_0)(1+z)^2 + \Omega_{m,0}(1+z)^3 + \Omega_{r,0}(1+z)^4\right]^{1/2} \tag{2.62}$$

with z related to $a(t)$ by Eq. (2.57). In order to solve for $a(t)$, we must know the value of H_0 and the energy (mass) content $(\Omega_{m,0}, \Omega_{r,0}, \Omega_{\Lambda,0})$ at the present time. The deceleration parameter defined in Eq. (2.56) is related to these parameters by

$$q_0 = \frac{\Omega_{m,0}}{2} + \Omega_{r,0} - \Omega_{\Lambda,0}. \tag{2.63}$$

A particularly simple case is the Einstein–de Sitter model in which $\Omega_{m,0} = 1$, $\Omega_{r,0} = \Omega_{\Lambda,0} = 0$ (and so $q_0 = 1/2$). It is then easy to show that $a(t) \propto t^{2/3}$. Another interesting case is a flat model in which $\Omega_{m,0} + \Omega_{\Lambda,0} = 1$ and $\Omega_{r,0} = 0$. In this case, $q_0 = 3\Omega_{m,0}/2 - 1$, so that $q_0 < 0$ (i.e. the expansion is accelerating at the present time) if $\Omega_{m,0} < 2/3$.

2.10.1 The Determination of Cosmological Parameters

As shown above, the geometry of the Universe in the standard model is specified by a set of cosmological parameters. The values of these cosmological parameters can therefore be estimated by measuring the geometrical properties of the Universe. The starting point is to find two observables that are related to each other only through the geometrical properties of the Universe. The most important example here is the redshift–distance relation. As we will see in Chapter 3, two types of distances can be defined through observational quantities. One is the luminosity distance, d_L, which relates the luminosity of an object, L, to its flux, f, according to $L = 4\pi d_L^2 f$. The other is the angular-diameter distance, d_A, which relates the physical size of an object, D, to its angular size, θ, via $D = d_A \theta$. In general, the redshift–distance relation can formally be written as

$$d(z) = \frac{cz}{H_0}[1 + \mathscr{F}_d(z; \Omega_{m,0}, \Omega_{\Lambda,0}, \ldots)], \tag{2.64}$$

where d stands for either d_L or d_A, and by definition $\mathscr{F}_d \ll 1$ for $z \ll 1$. For redshifts much smaller than 1, the redshift–distance relation reduces to the Hubble expansion law $cz = H_0 d$, and so the Hubble constant H_0 can be obtained by measuring the redshift and distance of an object (ignoring, for the moment, that objects can have peculiar velocities). Redshifts are relatively easy to obtain from the spectra of objects, and in §2.1.3 we have seen how to measure the distances of a few classes of astronomical objects. The best estimate of the Hubble constant at the present comes from Cepheids observed by the HST, and the result is

$$H_0 = 100 h \, \text{km s}^{-1} \, \text{Mpc}^{-1}, \quad \text{with} \quad h = 0.72 \pm 0.08 \tag{2.65}$$

(Freedman et al., 2001).

In order to measure other cosmological parameters, one has to determine the nonlinear terms in the redshift–distance relation, which typically requires objects at $z \gtrsim 1$. For example, measuring the light curves of Type Ia supernovae out to $z \sim 1$ has yielded the following constraints:

$$0.8\Omega_{m,0} - 0.6\Omega_{\Lambda,0} \sim -0.2 \pm 0.1 \tag{2.66}$$

(e.g. Perlmutter et al., 1999). Using Eq. (2.63) and neglecting $\Omega_{r,0}$ because it is small, the above relation gives $q_0 \sim -0.33 - 0.83\Omega_{m,0}$. Since $\Omega_{m,0} > 0$, we have $q_0 < 0$, i.e. the expansion of the Universe is speeding up at the present time.

Important constraints on cosmological parameters can also be obtained from the angular spectrum of the CMB temperature fluctuations. As shown in Fig. 2.42, the observed angular spectrum C_ℓ contains peaks and valleys, which are believed to be produced by acoustic waves in the baryon–photon fluid at the time of photon–matter decoupling. As we will see in §6.7, the heights/depths and positions of these peaks/valleys depend not only on the density of baryonic matter in the Universe, but also on the total mean density, Hubble's constant and other cosmological parameters. In particular, the position of the first peak is sensitive to the total density parameter Ω_0 (or the curvature K). Based on the observational results shown in Fig. 2.42, one obtains

$$\begin{aligned} \Omega_0 &= 1.02 \pm 0.02; \quad \Omega_{m,0} h^2 = 0.14 \pm 0.02; \\ h &= 0.72 \pm 0.05; \quad \Omega_{b,0} h^2 = 0.024 \pm 0.001, \end{aligned} \tag{2.67}$$

where $\Omega_{m,0}$ and $\Omega_{b,0}$ are the density parameters of total matter and of baryonic matter, respectively (Spergel et al., 2007). Note that this implies that the Universe has an almost flat geometry, that matter accounts for only about a quarter of its total energy density, and that baryons account for only $\sim 17\%$ of the matter.

2.10.2 The Mass and Energy Content of the Universe

There is a fundamental difficulty in directly observing the mass (or energy) densities in different mass components: all that is gold does not glitter. There may well exist matter components with significant mass density which give off no detectable radiation. The only interaction which all components are guaranteed to exhibit is gravity, and thus gravitational effects must be studied if the census is to be complete. The global gravitational effect is the curvature of space-time which we discussed above. Independent information on the amount of gravitating mass can only be derived from the study of the inhomogeneities in the Universe, even though such studies may never lead to an unambiguous determination of the total matter content. After all, one can imagine adding a smooth and invisible component to any amount of inhomogeneously distributed mass, which would produce no detectable effect on the inhomogeneities.

The most intriguing result of such dynamical studies has been the demonstration that the total mass in large-scale structures greatly exceeds the amount of material from which emission can be detected. This unidentified 'dark matter' (or 'invisible matter') is almost certainly the dominant contribution to the total mass density $\Omega_{m,0}$. Its nature and origin remain one of the greatest mysteries of contemporary astronomy.

(a) Relativistic Components One of the best observed relativistic components of the Universe is the CMB radiation. From its blackbody spectrum and temperature, $T_{CMB} = 2.73\,\mathrm{K}$, it is easy to estimate its energy density at the present time:

$$\rho_{\gamma,0} \approx 4.7 \times 10^{-34}\,\mathrm{g\,cm^{-3}}, \quad \text{or} \quad \Omega_{\gamma,0} = 2.5 \times 10^{-5} h^{-2}. \tag{2.68}$$

As we have seen in Fig. 2.2, the energy density of all other known photon backgrounds is much smaller. The only other relativistic component which is almost certainly present, although not yet directly detected, is a background of neutrinos. As we will see in Chapter 3, the energy density in this component can be calculated directly from the standard model, and it is expected to be 0.68 times that of the CMB radiation. Since the total energy density of the Universe at the present time is not much smaller than the critical density (see the last subsection), the contribution from these relativistic components can safely be ignored at low redshift.

(b) Baryonic Components Stars are made up of baryonic matter, and so a lower limit on the mass density of baryonic matter can be obtained by estimating the mass density of stars in galaxies. The mean luminosity density of stars in galaxies can be obtained from the galaxy luminosity function (see §2.4.1). In the B band, the best-fit Schechter function parameters are $\alpha \approx -1.2$, $\phi^* \approx 1.2 \times 10^{-2} h^3\,\mathrm{Mpc^{-3}}$ and $\mathscr{M}^* \approx -20.05 + 5\log h$ (corresponding to $L^* = 1.24 \times 10^{10} h^{-2} L_\odot$), so that

$$\mathscr{L}_B \approx 2 \times 10^8 h L_\odot\,\mathrm{Mpc^{-3}}. \tag{2.69}$$

Dividing this into the critical density leads to a value for the mass per unit observed luminosity of galaxies required for the Universe to have the critical density. This critical mass-to-light ratio is

$$\left(\frac{M}{L}\right)_{B,\mathrm{crit}} = \frac{\rho_{\mathrm{crit}}}{\mathscr{L}_B} \approx 1500 h \left(\frac{M_\odot}{L_\odot}\right)_B. \tag{2.70}$$

Mass-to-light ratios for the visible parts of galaxies can be estimated by fitting their spectra with appropriate models of stellar populations. The resulting mass-to-light ratios tend to be in the range of 2 to $10(M_\odot/L_\odot)$. Adopting $M/L = 5(M_\odot/L_\odot)$ as a reasonable mean value, the global density contribution of stars is

$$\Omega_{*,0} \sim 0.003 h^{-1}. \tag{2.71}$$

Thus, the visible parts of galaxies provide less than 1% of the critical density. In fact, combined with the WMAP constraints on $\Omega_{b,0}$ and the Hubble constant, we find that stars only account for less than 10% of all baryons.

So where are the other 90% of the baryons? At low redshifts, the baryonic mass locked up in cold gas (either atomic or molecular), and detected via either emission or absorption, only accounts for a small fraction, $\Omega_{\mathrm{cold}} \sim 0.0005 h^{-1}$ (Fukugita et al., 1998). A larger contribution is due to the hot intracluster gas observed in rich galaxy clusters through their bremsstrahlung emission at X-ray wavelengths (§2.5.1). From the number density of X-ray clusters and their typical gas mass, one can estimate that the total amount of hot gas in clusters is about $(\Omega_{HII})_{cl} \sim 0.0016 h^{-3/2}$ (Fukugita et al., 1998). The total gas mass in groups of galaxies is uncertain. Based on X-ray data, Fukugita et al. obtained $(\Omega_{HII})_{\mathrm{group}} \sim 0.003 h^{-3/2}$. However, the plasma in groups is expected to be colder than that in clusters, which makes it more difficult to detect in X-ray radiation. Therefore, the low X-ray emissivity from groups may also be due to low temperatures rather than due to small amounts of plasma. Indeed, if we assume that the gas/total mass ratio in groups is comparable to that in clusters, then the total gas mass in groups could be larger by a factor of two to three. Even then, the total baryonic mass detected in stars, cold gas and hot gas only accounts for less than 50% of the total baryonic mass inferred from the CMB.

The situation is very different at higher redshifts. As discussed in §2.8, the average density of hydrogen inferred from quasar absorption systems at $z \sim 3$ is roughly equal to the total baryon density as inferred from the CMB data. Hence, although we seem to have detected the majority of all baryons at $z \sim 3$, at low redshifts roughly half of the expected baryonic mass is unaccounted for observationally. One possibility is that the gas has been heated to temperatures in the range 10^5–10^6 K at which it is very difficult to detect. Indeed, recent observations of OVI absorption line systems seem to support the idea that a significant fraction of the IGM at low redshift is part of such a warm-hot intergalactic medium (WHIM), whose origin may be associated with the formation of large-scale sheets and filaments in the matter distribution (see Chapter 16).

An alternative explanation for the 'missing baryons' is that a large fraction of the gas detected at $z \sim 3$ has turned into 'invisible' compact objects, such as brown dwarfs or black holes. The problem, though, is that most of these objects are stellar remnants, and their formation requires a star-formation rate between $z = 3$ and $z = 0$ that is significantly higher than normally assumed. Not only is this inconsistent with the observation of the global star-formation history of the Universe (see §2.6.8), but it would also result in an over-production of metals. This scenario thus seems unlikely. Nevertheless, some observational evidence, albeit controversial, does exist for the presence of a population of compact objects in the dark halo of our Milky Way. In 1986 Bohdan Paczyński proposed to test for the presence of massive compact halo objects (MACHOs) using gravitational lensing. Whenever a MACHO in our Milky Way halo moves across the line-of-sight to a background star (for example, a star in the LMC), it will magnify the flux of the background star, an effect called microlensing. Because of the relative motion of source, lens and observer, this magnification is time-dependent, giving rise to a characteristic light curve of the background source. In the early 1990s two collaborations (MACHO and EROS) started campaigns to monitor millions of stars in the LMC for a period of several years. This has resulted in the detection of about 20 events in total. The analysis by the MACHO collaboration suggests that about 20% of the mass of the halo of the Milky Way could consist of MACHOs with a characteristic mass of $\sim 0.5 \, \mathrm{M_\odot}$ (Alcock et al., 2000). The nature of these objects, however, is still unclear. Furthermore, these results are inconsistent with those obtained by the EROS collaboration, which obtained an upper limit for the halo mass fraction in MACHOs of 8%, and rule out MACHOs in the mass range $0.6 \times 10^{-7} \, \mathrm{M_\odot} < M < 15 \, \mathrm{M_\odot}$ as the primary occupants of the Milky Way halo (Tisserand et al., 2007).

(c) Non-Baryonic Dark Matter As is evident from the CMB constraints given by Eq. (2.67) on $\Omega_{m,0}$ and $\Omega_{b,0}$, baryons can only account for ~ 15–20% of the total matter content in the Universe, and this is supported by a wide range of observations. As we will see in the following chapters, constraints from a number of other measurements, such as cosmic shear, the abundance of massive clusters, large-scale structure, and the peculiar velocity field of galaxies, all agree that $\Omega_{m,0}$ is of the order of 0.3. At the same time, the total baryonic matter density inferred from CMB observations is in excellent agreement with independent constraints from nucleosynthesis and the observed abundances of primordial elements. The inference is that the majority of the matter in the Universe (75–80%) must be in some non-baryonic form.

One of the most challenging tasks for modern cosmology is to determine the nature and origin of this dark matter component. Particle physics in principle allows for a variety of candidate particles, but without a direct detection it is and will be difficult to discriminate between the various candidates. One thing that is clear from observations is that the distribution of dark matter is typically more extended than that of the luminous matter. As we have seen above, the mass-to-light ratios increase from $M/L \sim 30h(M/L)_\odot$ at a radius of about $30h^{-1}$ kpc as inferred from the extended rotation curves of spiral galaxies, to $M/L \sim 100h(M/L)_\odot$ at the scale of a few hundred kpc, as inferred from the kinematics of galaxies in groups, to $M/L \sim 350h(M/L)_\odot$ in galaxy clusters, probing scales of the order of 1 Mpc. This latter value is comparable to that of the Universe as a whole, which follows from multiplying the critical mass-to-light ratio given by Eq. (2.70) with $\Omega_{m,0}$, and suggests that the content of clusters, which are the largest virialized structures known, is representative of that of the entire Universe.

All these observations support the idea that galaxies reside in extended halos of dark matter. This in turn puts some constraints on the nature of the dark matter, namely that it has to be relatively cold (i.e. it needs to have initial peculiar velocities that are much smaller than the typical velocity dispersion within an individual galaxy). This coldness is required because otherwise the dark matter would not be able to cluster on galactic scales to form the dark halos around galaxies. Without a better understanding of the nature of the dark matter, we have to live with the vague term, cold dark matter (or CDM), when talking about the main mass component of the Universe.

(d) Dark Energy As we have seen above, the observed temperature fluctuations in the CMB show that the Universe is nearly flat, implying that the mean energy density of the Universe must be close to the critical density, $\rho_{\rm crit}$. However, studies of the kinematics of galaxies and of large-scale structure in the Universe give a mean mass density that is only about 1/4 to 1/3 of the critical density, in good agreement with the constraints on $\Omega_{m,0}$ from the CMB itself. This suggests that the dominant component of the mass/energy content of the Universe must have a homogenous distribution so that it affects the geometry of the Universe but does not follow the structure in the baryonic and dark matter. An important clue about this dominant component is provided by the observed redshift–distance relation of high-redshift Type Ia supernovae. As shown in §2.10.1, this relation implies that the expansion of the Universe is speeding up at the present time. Since all matter, both baryonic and non-baryonic, decelerates the expansion of the Universe, the dominant component must be an energy component. It must also be extremely dark, because otherwise it would have been observed.

The nature of this dark energy component is a complete mystery at the present time. As far as its effect on the expansion of the Universe is concerned, it is similar to the cosmological constant introduced by Einstein in his theory of general relativity to achieve a stationary Universe (Einstein, 1917). The cosmological constant can be considered as an energy component whose density does not change with time. As the Universe expands, it appears as if more and more energy is created to fill the space. This strange property is due to its peculiar equation of state that relates its pressure, P, to its energy density, ρ. In general, we may write $P = w\rho c^2$, and so $w = 0$ for a pressureless fluid and $w = 1/3$ for a radiation field (see §3.1.5). For a dark energy component with constant energy density, $w = -1$, which means that the fluid actually gains internal energy as it expands, and acts as a gravitational source with a negative effective mass density ($\rho + 3P/c^2 = -2\rho < 0$), causing the expansion of the Universe to accelerate. In addition to the cosmological constant, dark energy may also be related to a scalar field (with $-1 < w < -1/3$). Such a form of dark energy is called quintessence, which differs from a cosmological constant in that it is dynamic, meaning that its density and equation of state can vary through both space and time. It has also been proposed that dark energy has an equation of state parameter $w < -1$, in which case it is called phantom energy. Clearly, a measurement of the value of w will allow us to discriminate between these different models. Currently, the value of w is constrained by a number of observations to be within a relatively narrow range around -1 (e.g. Spergel et al., 2007), consistent with a cosmological constant, but also with both quintessence and phantom energy. The next generation of galaxy redshift surveys and Type Ia supernova searches aim to constrain the value of w to a few percent, in the hope of learning more about the nature of this mysterious and dominant energy component of our Universe.

3
Cosmological Background

Cosmology, the branch of science dealing with the origin, evolution and structure of the Universe on large scales, is closely related to the study of galaxy formation and evolution. Cosmology provides not only the space-time frame within which galaxy formation and evolution ought to be described, but also the initial conditions for the formation of galaxies. Modern cosmology is founded upon Einstein's theory of general relativity (GR), according to which the space-time structure of the Universe is determined by the matter distribution within it. This perspective on space-time is very different from that in classical physics, where space-time is considered eternal and absolute, independent of the existence of matter.

A complete description of GR is beyond the scope of this book. As a remedy, we provide a brief summary of the basics of GR in Appendix A and we refer the reader to the references cited there for details. It should be emphasized, however, that modern cosmology is a very simple application of GR, so simple that even a reader with little knowledge of GR can still learn it. This simplicity is owing to the simple form of the matter distribution in the Universe, which, as we have seen in the last chapter, is observed to be approximately homogeneous and isotropic on large scales. We do not yet have sufficient evidence to rule out inhomogeneity or anisotropy on very large scales, but the assumption of homogeneity and isotropy is no doubt a good basis for studying the observable Universe. If indeed the matter distribution in the Universe is completely homogeneous and isotropic, as is the ansatz on which modern cosmology is based,[1] GR would imply that space itself must also be homogeneous and isotropic. Such a space is the simplest among all possibilities. To see this more clearly, let us consider a two-dimensional space, i.e. a surface. We all know that the properties of a general two-dimensional surface can be very complicated. But if the surface is homogeneous and isotropic, we are immediately reminded of an infinite plane and a sphere. These two surfaces differ in their overall curvature. The plane is flat, while the sphere is said to have a positive curvature. In both cases the distance between any two infinitesimally close points on the surface can be written as

$$\mathrm{d}l^2 = a^2 \left(\frac{\mathrm{d}r^2}{1 - Kr^2} + r^2 \mathrm{d}\vartheta^2 \right), \qquad (3.1)$$

where $K = 0$ for a plane and $K = 1$ for a sphere. In the case of a plane, (r, ϑ) are just the polar coordinates and a is a length scale (scale factor) relating the coordinate radius r to distance. To see that $K = 1$ corresponds to a sphere, we make the coordinate transformation $r = \sin \chi$. In terms of (χ, ϑ), the distance measure becomes $\mathrm{d}l^2 = a^2 (\mathrm{d}\chi^2 + \sin^2 \chi \, \mathrm{d}\vartheta^2)$, which is clearly that of a sphere in terms of the spherical coordinates, with a being the radius of the sphere. In this case, r is a spherical coordinate in the three-dimensional space in which the two-dimensional surface is embedded; r is *not* a distance measure on the surface, but

---

[1] Although on relatively small scales the present-day Universe deviates strongly from homogeneity and isotropy, we will see in Chapter 4 that these structures arise from small perturbations of an otherwise homogeneous and isotropic matter distribution.

rather a coordinate used to label positions on the surface. Actual distances have to be computed from the metric (3.1). Only in the case with $K = 0$ is r both coordinate and distance measure.

Mathematically it can be shown that there is another two-dimensional homogeneous and isotropic surface for which $K = -1$. Changing r to $\sinh\chi$, we can write the distance measure on such a surface as $dl^2 = a^2(d\chi^2 + \sinh^2\chi\, d\vartheta^2)$, where the factor a is again a length scale relating coordinates to distance. This negatively curved, hyperbolic surface, which is locally similar to the surface near a saddle point, is not very familiar to us because it cannot be embedded in a three-dimensional Euclidean space. The existence of a low-dimensional 'space' which cannot be embedded in a space of higher dimensionality is, however, not as strange as it might seem; for example it is easy to envision that it is impossible to embed a hairspring (an intrinsically one-dimensional object) into a plane.

These examples show that the description of a homogeneous and isotropic two-dimensional surface is extremely simple. What we need to do is just to determine the value of K (1, 0, or -1), which specifies the global geometry of the surface, and the scale factor a, which relates coordinates to distances. In general, the scale factor a can change with time without violating the requirement of homogeneity and isotropy, corresponding to a surface that is uniformly expanding or contracting.

The above discussion can be extended to three-dimensional spaces. As we will see in §3.1, a homogeneous and isotropic space is also completely determined by the curvature signature K (again equal to 1 or 0 or -1), which determines the global geometry of the space, and the scale factor $a(t)$ as a function of time. Thus, as far as the space-time geometry is concerned, the task of modern cosmology is simply to determine the value of K and the functional form of $a(t)$ from the matter content of the Universe (see §3.2).

According to GR, the relationships among cosmological events are assumed to be governed by the physical laws that we are familiar with, while the effects of gravity are included in the properties of the space-time (i.e. in the transformations of reference frames). This equivalence principle (that a local gravitational field can be transformed away by choosing an appropriate frame of reference) allows one to derive physical equations in GR from their ordinary forms by general coordinate transformations (see Appendix A). Hence, once the value of K and the functional form of $a(t)$ are known, the relationships among cosmological events can be described in terms of physical laws. Similarly, if we believe that physical laws are applicable on cosmological scales, the predictions for these relationships will depend only on the space-time geometry, and so observations of such relationships can be used to test cosmological models.

One of the most important observations in cosmology is that the Universe is expanding [i.e. $a(t)$ increases with time], which implies that it must have been smaller in the past. Together with the observational fact that our Universe is filled with microwave photons, this time evolution of the scale factor determines the thermal history of the Universe. Because the Universe was denser in the past, it must also have been hotter. Since high density and temperature imply high probabilities for particles to collide with each other with high energy, the early Universe is an ideal place for the creation and transmutation of matter. As we will see in §3.3–§3.5, the applications of particle, nuclear and atomic physics to the thermal history of the early Universe lead to important predictions for the current matter content of the Universe. Although many of these predictions are still uncertain, they provide the basis for calculating relations between the dominant mass components of the Universe. Finally, in §3.6, we discuss some of the most fundamental problems of the standard model and show how the 'inflationary hypothesis' may help to solve them. Although this chapter gives a fairly detailed description of modern cosmology, readers interested in more details are referred to the textbooks by Kolb & Turner (1990), Peebles (1993), Peacock (1999), Coles & Lucchin (2002), Padmanabhan (2002), Börner (2003) and Weinberg (2008).

3.1 The Cosmological Principle and the Robertson–Walker Metric

3.1.1 The Cosmological Principle and its Consequences

The cosmological principle is the hypothesis that, on sufficiently large scales, the Universe can be considered spatially homogeneous and isotropic. While this may appear a reasonable extrapolation from current observations (see Chapter 2), it was originally proposed for quite different reasons. As stated by Milne (1935), this hypothesis follows from the belief that 'Not only the laws of Nature, but also the events occurring in Nature, the world itself, must appear the same to all observers.' In this sense, the cosmological principle can be thought of as a generalized Copernican principle: our location in the Universe should be typical, and should not be distinguished in any fundamental way from any other. The cosmological principle is, however, stronger than this simple statement implies, since it also eliminates the possibility of a self-similar, fractal structure on the largest scales. All points of such a structure are equivalent, but there are no scales on which it approaches homogeneity. Milne's statement is also incomplete, since it is possible to have a universe which appears the same from each point but is anisotropic, as in Gödel's model (Gödel, 1949).

An even stronger hypothesis is the perfect cosmological principle of Bondi & Gold (1948) and Hoyle (1948). This requires invariance not only under rotations and displacements in space, but also under displacements in time. The Universe looks the same in all directions, from all locations, and at all times. This hypothesis led to the steady state cosmology which requires a continuous creation of matter to keep the mean matter density constant with time. However, the discovery of the cosmic microwave background radiation, and in particular the demonstration that it has a perfect blackbody spectrum, has proven an unsurmountable problem for this cosmology. Additional evidence against the steady state cosmology comes from numerous detections of evolution in the galaxy population. We therefore will not discuss this theory further in this book.

What are the consequences of the cosmological principle for the geometric structure of the Universe? To answer this question, we put the cosmological principle in a slightly different form. The cosmological principle can also be stated as the existence of a *fundamental observer* at each location, to whom the Universe appears isotropic. The concept of a fundamental observer is required because two observers at the same point, but in relative motion, cannot both see the surrounding Universe as isotropic. The fundamental observer thus defines a cosmological 'rest frame' at each location in space. To better understand the meaning of a fundamental observer, let us define the fundamental observer, or the cosmological rest frame, in our neighborhood. As discussed in Chapter 2, galaxies in the Universe are strongly clustered on scales $\lesssim 10 h^{-1}$Mpc, and have random motions of the order of 100 to $1{,}000 \,\mathrm{km\,s^{-1}}$ with respect to each other. It is thus unlikely that our own Galaxy defines a cosmological rest frame. On the other hand, we expect the mean motion of galaxies within a radius much larger than $10 h^{-1}$Mpc around us to be small with respect to the cosmological rest frame. In particular the cosmic microwave background (CMB) should appear isotropic to such a frame. As shown in Chapter 2, the CMB map given by the COBE satellite appears very isotropic around us, when the dipole component is subtracted. The dipole in the CMB map is best explained by the motion of the Local Group of galaxies relative to the CMB with a velocity $(627 \pm 22)\,\mathrm{km\,s^{-1}}$ (Lineweaver et al., 1996). Thus, an observer in our neighborhood, traveling at the same speed relative to the Local Group but in the opposite direction, should be close to a fundamental observer. If the cosmological principle is correct, then the rest frame defined by the mean motion of galaxies within a large radius around us should converge to the one defined by the CMB. There are indeed indications of such convergence in present observational data (e.g. Schmoldt et al., 1999).

Since the Universe is isotropic to a fundamental observer, the velocity field in her neighborhood cannot have any preferred direction. The only allowed motion is therefore pure expansion (or pure contraction),

$$\delta \mathbf{v} = H \, \delta \mathbf{x}, \tag{3.2}$$

where $\delta \mathbf{x}$ and $\delta \mathbf{v}$ are the position and velocity of a particle relative to the fundamental observer, and H is a constant. Once some definition of distance is adopted, we can consider the set of all observers, O', which are equidistant from a given observer O at some given local time of O. Because of the isotropy, all the observers O' must measure the same local values of density, temperature, expansion rate, and other physical quantities. Furthermore, they must remain equidistant from O at any later time recorded by the clock of O. Thus they can in principle synchronize their clocks using a light signal from O, and once synchronized, the clocks must remain so. Since the original fundamental observer O is arbitrary, this argument shows that there exists a three-dimensional hypersurface in space-time, on which density, temperature, expansion rate, and all other locally defined properties are uniform and evolve according to a universally agreed time. Such a time is called the *cosmic time*. Since quantities such as the temperature of the CMB and the mean density of the Universe are monotonic functions of cosmic expansion, the value of these quantities can be used to label the cosmic time, as we will see below.

The isotropic and homogeneous three-dimensional hypersurfaces discussed above are maximally symmetric. As a result their metric can be written as

$$dl^2 = a^2(t) \left[\frac{dr^2}{1 - Kr^2} + r^2(d\vartheta^2 + \sin^2\vartheta \, d\varphi^2) \right]. \tag{3.3}$$

A proof of this can be found in Weinberg (1972). In this formula $a(t)$ is a time-dependent scale factor which relates the coordinate labels (r, ϑ, φ) of the fundamental observers to true physical distances, and K is a constant which can take the values $+1, 0$, and -1. The radial coordinate r is dimensionless in Eq. (3.3). When physical distances are required, a length scale can be assigned to the scale factor.

To understand better the geometric meanings of $a(t)$ and K, consider an expanding or contracting three-sphere (the three-dimensional analog of the two-dimensional surface of an expanding or shrinking spherical balloon) whose radius is $R(t) = a(t)R_0$ at time t. The scale factor $a(t)$ therefore simply relates the radius of the three-sphere at time t to its *comoving* radius, R_0, whose value does not change as the sphere expands or contracts. (Thus the comoving radius is just the true radius measured in units of the scale factor.) In Cartesian coordinates (x, y, z, w), this three-surface is defined by

$$x^2 + y^2 + z^2 + w^2 = a^2(t) R_0^2. \tag{3.4}$$

With the change of coordinates from (x, y, z, w) to the polar coordinates (r, ϑ, φ):

$$\begin{cases} x = a(t)\, r \sin\vartheta \cos\varphi \\ y = a(t)\, r \sin\vartheta \sin\varphi \\ z = a(t)\, r \cos\vartheta \\ w = a(t) \left(R_0^2 - r^2\right)^{1/2}, \end{cases} \tag{3.5}$$

the line element in the four-dimensional Euclidean space is

$$\begin{aligned} dl^2 &= dx^2 + dy^2 + dz^2 + dw^2 \\ &= a^2(t) \left[\frac{dr^2}{1 - r^2/R_0^2} + r^2(d\vartheta^2 + \sin^2\vartheta \, d\varphi^2) \right]. \end{aligned} \tag{3.6}$$

The curvature scalar of such a three-sphere is

$$\mathcal{R} = \frac{6}{R_0^2 a^2(t)} \tag{3.7}$$

(see Appendix A). Comparing Eqs. (3.3) and (3.6) we immediately see that Eq. (3.3) with $K = +1$ is the metric of a three-sphere with comoving radius $R_0 = 1$, and with the true radius at time t given by the value of $a(t)$. This three-sphere has a finite volume $V = 2\pi^2 a^3(t)$, and the dimensionless radial coordinate $r \in [0, 1]$.

For $K = 0$, metric (3.3) is the same as that given by Eq. (3.6) with $R_0 \to \infty$, and so it describes a Euclidean flat space with infinite volume. In this case the scale factor $a(t)$ describes the change of the length scale due to the uniform expansion (or contraction) of the space.

Metric (3.3) with $K = -1$ can be obtained by the replacement $R_0 \to i$ in Eq. (3.6). The same replacement in Eq. (3.7) shows that such a metric describes a negatively curved three-surface with curvature radius set by $a(t)$. Such a three-surface cannot be embedded in a four-dimensional Euclidean space, but can be embedded in a four-dimensional Minkowski space with line element $dl^2 = dx^2 + dy^2 + dz^2 - dw^2$. In this space, the negatively curved three-surface with curvature radius $a(t)$ can be written as $w^2 - x^2 - y^2 - z^2 = a^2(t)$. Thus, the metric (3.3) with $K = -1$ describes a hyperbolic three-surface, with unit comoving curvature radius, embedded in a four-dimensional Minkowski space. Such a three-surface has no boundaries and has infinite volume.

3.1.2 Robertson–Walker Metric

Since the isotropic and homogeneous three-dimensional surfaces described above are the space-like hypersurfaces corresponding to a constant cosmic time t, the four-metric of the space-time can be written as

$$\begin{aligned}ds^2 &= c^2 dt^2 - dl^2 \\ &= c^2 dt^2 - a^2(t)\left[\frac{dr^2}{1 - Kr^2} + r^2(d\vartheta^2 + \sin^2\vartheta\, d\varphi^2)\right],\end{aligned} \tag{3.8}$$

with c the speed of light. This is the Robertson–Walker metric. As in special relativity, the space-time interval, ds, is real for two events with a time-like separation, is zero for two events on the same light path (null geodesic), and is imaginary for two events with a space-like separation. As before, the coordinates (r, ϑ, φ), which label fundamental observers, are called comoving coordinates, and the function $a(t)$ is the cosmic scale factor. If we define the proper time of an observer as the one recorded by the clock at rest with the observer, then the cosmic time t is the proper time of all fundamental observers. A proper distance l can be defined for any two fundamental observers at any given cosmic time t: $l = \int dl$. Without losing generality we can assume one of the observers to be at the origin $r = 0$ and the other at $(r_1, \vartheta, \varphi)$. The proper distance can then be written as

$$l = a(t) \int_0^{r_1} \frac{dr}{\sqrt{1 - Kr^2}} = a(t)\chi(r_1), \tag{3.9}$$

where

$$\chi(r) = \begin{cases} \sin^{-1} r & (K = +1) \\ r & (K = 0) \\ \sinh^{-1} r & (K = -1). \end{cases} \tag{3.10}$$

3.1 The Cosmological Principle and the Robertson–Walker Metric

The χ in the above equations is called the comoving distance between the two fundamental observers; it is the proper distance l measured in units of the scale factor. It is often useful to change the time variable from proper time t to a conformal time,

$$\tau(t) = \int_0^t \frac{c\,dt'}{a(t')}. \tag{3.11}$$

In terms of χ and τ the Robertson–Walker metric can be written in another useful form:

$$ds^2 = a^2(\tau)\left[d\tau^2 - d\chi^2 - f_K^2(\chi)(d\vartheta^2 + \sin^2\vartheta\,d\varphi^2)\right], \tag{3.12}$$

where

$$f_K(\chi) = r = \begin{cases} \sin\chi & (K = +1) \\ \chi & (K = 0) \\ \sinh\chi & (K = -1). \end{cases} \tag{3.13}$$

This form of the metric is especially useful to gain insight into the causal properties of space-time.

It is instructive to look at the metric on a hypersurface with constant φ. In the $K = +1$ case the spatial part of the metric is $dl^2 = a^2(\tau)(d\chi^2 + \sin^2\chi\,d\vartheta^2)$, which is just the metric of a two-dimensional sphere in terms of the 'polar angle' χ and the 'azimuthal angle' ϑ (see Fig. 3.1). We see that χ is the (comoving) geodesic distance, because it measures the length of the shortest path (arc) connecting two points on the hypersurface, while the radial coordinate r is *not* a distance measure on the surface. This conclusion is also true for the case of $K = -1$. Only for a flat space ($K = 0$) where $r = \chi$, is the radial coordinate r also a geodesic distance.

The Hubble parameter, $H(t)$, at a cosmic time t is defined to be the rate of change of the proper distance l between any two fundamental observers at time t in units of l: $dl/dt \equiv H(t)l$. It then follows from Eq. (3.9) that

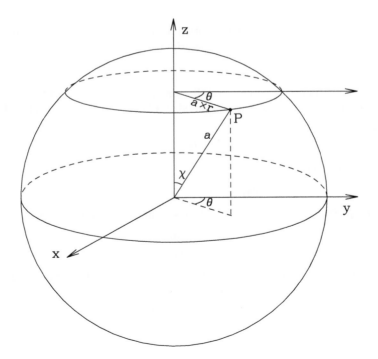

Fig. 3.1. The φ = constant section of a Robertson–Walker metric with $K = 1$, showing the geometric meanings of various coordinates.

$$H(t) = \frac{\dot{a}(t)}{a(t)}, \tag{3.14}$$

where an over-dot denotes the derivative with respect to t. The Hubble parameter at the present time is called the Hubble constant, and is denoted by H_0. Quantities that depend on the value of H_0 are often expressed in terms of

$$h \equiv \frac{H_0}{100\,\mathrm{km\,s^{-1}\,Mpc^{-1}}}. \tag{3.15}$$

The time dependence of the scale factor $a(t)$ is determined by general relativity and the equation of state appropriate for the matter content of the Universe. This will be discussed in §3.2. However, some kinematic properties of an isotropic and homogeneous universe can already be inferred from the form of the metric [either Eq. (3.8) or Eq. (3.12)] without specifying the form of $a(t)$. Such discussion is useful, because it is based only on the cosmological principle, and is valid even if general relativity fails on cosmological scales or if our knowledge about the matter content of the Universe is incomplete. In the following four subsections, we examine these 'kinematic' properties of the Robertson–Walker metric.

3.1.3 Redshift

Almost all observations about astronomical objects are made through light signals. It is therefore important to understand how photons propagate in a homogeneous and isotropic universe. Without losing generality, consider a light signal propagating to the origin along a radial direction ($\mathrm{d}\vartheta = \mathrm{d}\varphi = 0$). Since photons travel along null geodesics on which $\mathrm{d}s = 0$, their trajectories can be written as

$$\mathrm{d}\tau = \mathrm{d}\chi \tag{3.16}$$

[see Eq. (3.12)]. Thus, if a wave crest is emitted at the time t_e from a fundamental observer $(r_e, \vartheta_e, \varphi_e)$, then the time t_0 when it reaches the origin is given by

$$\tau(t_0) - \tau(t_e) = \chi(r_e) - \chi(0) = \chi(r_e). \tag{3.17}$$

Since the comoving distance $\chi(r_e)$ between the fundamental observer and the origin does not change with time, a successive wave crest emitted at a later time $t_e + \delta t_e$ reaches the origin at a time $t_0 + \delta t_0$ given by

$$\tau(t_0 + \delta t_0) - \tau(t_e + \delta t_e) = \chi(r_e). \tag{3.18}$$

Combining Eqs. (3.17) and (3.18) gives

$$\tau(t_0 + \delta t_0) - \tau(t_0) = \tau(t_e + \delta t_e) - \tau(t_e). \tag{3.19}$$

In real applications $\delta t_e \ll t_e$ and $\delta t_0 \ll t_0$, and so we can use the definition of τ to obtain

$$\frac{\delta t_0}{a(t_0)} = \frac{\delta t_e}{a(t_e)}. \tag{3.20}$$

Thus the period of the wave, and hence its wavelength, increases (or its frequency decreases) in proportion to the scale factor:

$$\frac{\lambda_0}{\lambda_e} = \frac{\nu_e}{\nu_0} = \frac{\delta t_0}{\delta t_e} = \frac{a(t_0)}{a(t_e)}. \tag{3.21}$$

Defining the relative change of wavelength by a redshift parameter, $z \equiv (\lambda_0 - \lambda_e)/\lambda_e$, we have

$$1 + z \equiv \frac{\lambda_0}{\lambda_e} = \frac{a(t_0)}{a(t_e)}. \tag{3.22}$$

If the light wave is emitted from the transitions of a given kind of atoms between two energy levels E_1 and E_2, and if these atoms are at rest with respect to the fundamental observer $(r_e, \vartheta_e, \varphi_e)$ at time t_e, then $\nu_e = |E_1 - E_2|/h_P$ (where h_P is Planck's constant). Eq. (3.21) then describes the relation between the observed wavelength and the rest-frame wavelength which can be determined by the observer in his local laboratory. In an expanding universe $a(t_0) > a(t_e)$ so that $z > 0$ and spectral features are shifted redwards (redshift). On the other hand, in a contracting universe $a(t_0) < a(t_e)$, so that $z < 0$ and spectral features are shifted bluewards (blueshift). As we have seen in Chapter 2, distant galaxies in the Universe are all observed to show redshifted spectra, indicating that the Universe is expanding.

3.1.4 Peculiar Velocities

As we will see later in Chapter 4, small perturbations in the background energy density distribution cause the growth of structures, which in turn induce velocities that deviate from pure expansion. These velocities with respect to the cosmological rest frame of fundamental observers are called peculiar velocities.

The proper velocity of a particle with respect to a fundamental observer at the origin is defined as $v = dl/dt$, with $l(t)$ the proper distance between the particle and observer. Using Eq. (3.9) we can write this as

$$v(t) = \dot{a}(t)\chi(t) + a(t)\dot{\chi}(t) = v_{\exp} + v_{\text{pec}}, \tag{3.23}$$

where $v_{\exp} = H(t)l(t)$ is the velocity component due to the universal expansion, and v_{pec} is the peculiar velocity.

Let $\mathcal{O}_1$ be a fundamental observer at the same location as a particle $\mathcal{P}$ which has a peculiar velocity v_{pec} with respect to $\mathcal{O}_1$. Since locally the geometry at $\mathcal{O}_1$ is that of a Minkowski space, $\mathcal{O}_1$ will observe the light from $\mathcal{P}$ with a Doppler redshift

$$1 + z_{\text{pec}} = \sqrt{\frac{1 + v_{\text{pec}}/c}{1 - v_{\text{pec}}/c}}. \tag{3.24}$$

But what is the redshift of $\mathcal{P}$ observed by a fundamental observer $\mathcal{O}_2$, located at a proper distance δl_{12} from $\mathcal{O}_1$? For simplicity we assume that the peculiar velocity of $\mathcal{P}$ is along the geodesic connecting $\mathcal{O}_1$ and $\mathcal{O}_2$. Using the definition of redshift in Eq. (3.22) we can write for the observed redshift

$$1 + z_{\text{obs}} = \frac{\lambda_2}{\lambda_P} = \frac{\lambda_1}{\lambda_P} \frac{\lambda_2}{\lambda_1}, \tag{3.25}$$

where λ_P is the wavelength emitted by $\mathcal{P}$, and λ_1 and λ_2 are the wavelengths observed by $\mathcal{O}_1$ and $\mathcal{O}_2$, respectively. The physical correspondence of the second equality is a simple relay station at $\mathcal{O}_1$ that passes the information from $\mathcal{P}$ on to $\mathcal{O}_2$. The first factor on the right-hand side of Eq. (3.25) is simply the Doppler redshift of Eq. (3.24), while the second factor corresponds to the cosmological redshift $z_{\cos}$ of $\mathcal{O}_1$, and thus also of $\mathcal{P}$. Therefore

$$1 + z_{\text{obs}} = (1 + z_{\text{pec}})(1 + z_{\cos}), \tag{3.26}$$

which shows that the observed redshift of any object consists of a contribution due to the universal expansion and one due to its peculiar velocity along the line-of-sight. In the non-relativistic case we can approximate Eq. (3.24) with $z_{\text{pec}} = v_{\text{pec}}/c$, so that Eq. (3.26) reduces to

$$z_{\text{obs}} = z_{\cos} + \frac{v_{\text{pec}}}{c}(1 + z_{\cos}). \tag{3.27}$$

Thus, for a cluster of galaxies at redshift z, the (peculiar) velocity dispersion of galaxies, σ_v, is related to the observed dispersion in redshifts, σ_z, as

$$\sigma_v = \sigma_z \frac{c}{1+z}. \tag{3.28}$$

Next let us consider the motion of a non-relativistic particle $\mathscr{P}$ in a homogeneous and isotropic universe. Consider once again the fundamental observers $\mathscr{O}_1$ and $\mathscr{O}_2$, and let $\mathscr{P}$ pass $\mathscr{O}_1$ at time t_1 with a peculiar velocity v_1 in the direction of $\mathscr{O}_2$. If $\mathscr{P}$ moves freely to $\mathscr{O}_2$, what is the peculiar velocity of $\mathscr{P}$ when it passes $\mathscr{O}_2$ at time t_2? To answer this question, focus first on the velocity of $\mathscr{P}$ at $t = t_2$ with respect to $\mathscr{O}_1$. This velocity consists of two components: a peculiar velocity v_2 as well as a velocity $v_{\exp} = H(t_2)\delta l_{12}$ due to the universal expansion. Since $\mathscr{P}$ has not been accelerated with respect to $\mathscr{O}_1$, the sum of these two velocities has to be equal to v_1, such that from the perspective of $\mathscr{O}_1$ the line-of-sight velocity of $\mathscr{P}$ has not changed. Therefore

$$\delta v \equiv v_2 - v_1 = -\frac{\dot{a}(t_2)}{a(t_2)} \delta l_{12}. \tag{3.29}$$

Using Taylor expansion we can write, to first order in $\delta t = t_2 - t_1$, the proper distance between $\mathscr{O}_1$ and $\mathscr{O}_2$ as $\delta l_{12} = v_1 \delta t$. Substitution in Eq. (3.29) and integration then yields

$$v_2 = v_1 \frac{a(t_1)}{a(t_2)}. \tag{3.30}$$

Therefore, the peculiar velocity of a free, non-relativistic particle decreases as the inverse of the scale factor:

$$v_{\text{pec}}(t) \propto a^{-1}(t). \tag{3.31}$$

Since the momentum p of a non-relativistic particle is proportional to its peculiar velocity, Eq. (3.31) also implies that $p(t) \propto a^{-1}(t)$. Note that for a photon with zero rest mass $pc = E = h_P \nu$. As is evident from Eq. (3.21) $\nu \propto a^{-1}$, so that the decay law $p \propto a^{-1}$ holds for photons as well as for massive particles.

3.1.5 Thermodynamics and the Equation of State

The homogeneous and isotropic properties of the expanding Universe also allow an analysis of its thermodynamic properties. Let us consider a uniform, perfect gas contained in a (small) comoving volume $V \propto a^3(t)$ which expands with the Universe. Since the Universe is homogeneous and isotropic, there should not be any net heat flow across the boundaries of V. This implies that we can consider V as an adiabatic system, and since V can be chosen arbitrarily small, no GR is required to describe its thermodynamic properties.

According to the first law of thermodynamics, the increase in internal energy, dU, is equal to the heat, dQ, transferred into the system plus the work, dW, done *on* the system: $dU = dQ + dW$. The second law of thermodynamics is related to the entropy S, and states that $dS = dQ/T$, with T the temperature. For our adiabatically expanding volume V we therefore have

$$dU + PdV = 0; \quad dS = 0, \tag{3.32}$$

with P the pressure. This shows that the entropy per unit comoving volume is conserved, and that the expansion of the Universe causes a decrease or increase of its internal energy depending on whether $P > 0$ or $P < 0$.

In order to be able to apply the first law to both relativistic and non-relativistic fluids, we write the internal energy, U, in terms of the energy density ρc^2. In principle there may be many different sources contributing to the energy density of the Universe: matter (both non-relativistic and relativistic), radiation, vacuum energy, scalar fields, etc. As we shall see later in this chapter, the Universe transited from a radiation dominated phase early on to a matter dominated phase at later stages. In addition, the Universe may have become dominated by vacuum energy in the recent past. In what follows we therefore focus on these three energy components only, so that the total energy density may be written as

$$\rho c^2 = \rho_m c^2 + \rho_m \varepsilon + \frac{4\sigma_{SB}}{c} T^4 + \rho_{vac} c^2. \tag{3.33}$$

Here ρ_m is the matter density, and ε the internal energy per unity mass ($\varepsilon = \frac{3}{2} k_B T/m$ for a monatomic ideal gas, with k_B Boltzmann's constant). The first two terms of Eq. (3.33) therefore express the energy density due to non-relativistic matter, split in a contribution of rest-mass energy and internal energy. The third term indicates the energy density of the radiation, with σ_{SB} the Stefan–Boltzmann constant.[2] Finally, $\rho_{vac} c^2$ is the energy density of the vacuum.

In terms of the energy density, the first law of thermodynamics for our adiabatically expanding volume can now be written as

$$V\, d\rho + (\rho + P/c^2) dV = 0. \tag{3.34}$$

Using that $V \propto a^3$, and differentiating with respect to a we obtain

$$\frac{d\rho}{da} + 3\left(\frac{\rho + P/c^2}{a}\right) = 0. \tag{3.35}$$

For a given equation of state, $P(\rho)$, this equation gives the density and pressure as functions of a. It is common practice to introduce the equation of state parameter w and to write

$$P = w\rho c^2. \tag{3.36}$$

If w is time-independent, then substitution of Eq. (3.36) into Eq. (3.35) gives

$$\rho \propto a^{-3(1+w)}. \tag{3.37}$$

To describe the evolution of ρ, P, and T during the matter dominated phase, we approximate the Universe as an ideal gas, for which $PV = N k_B T$, with N the number of atoms of the gas. For a monatomic gas consisting of particles of mass m we have $\rho_m = mN/V$, so that

$$P_m = \frac{k_B T}{mc^2} \rho_m c^2. \tag{3.38}$$

Note that since $\rho_m \neq \rho$, this does *not* imply that $w = k_B T/mc^2$. To determine the true equation of state parameter, it is useful to write the equation of state as function of the adiabatic index γ (for a monatomic gas $\gamma = 5/3$):

$$P_m = (\gamma - 1)(\rho - \rho_m) c^2. \tag{3.39}$$

Note that Eq. (3.39) makes it explicit that, in the non-relativistic limit, the rest-mass energy does not contribute to the pressure of the gas. Combining Eqs. (3.38)–(3.39) we can write the pressure in the form of Eq. (3.36) with

$$w = w(T) = \frac{k_B T}{mc^2} \left(1 + \frac{1}{\gamma - 1} \frac{k_B T}{mc^2}\right)^{-1}. \tag{3.40}$$

Since $k_B T \ll mc^2$ we immediately see that $w(T) \ll 1$. A non-relativistic gas is thus well approximated by a fluid of zero pressure ($w = 0$), often referred to as a dust fluid. Since $\rho \propto a^{-3(1+w)}$ a

---

[2] Since the Universe is homogeneous and isotropic, the radiation fluid is in thermal equilibrium, and its energy density follows from integrating the Planck function corresponding to a blackbody of temperature T.

Table 3.1. Thermodynamics of a homogeneous and isotropic universe.

Dominant component	w	ρ	P	T
matter	0	a^{-3}	a^{-5}	a^{-2}
radiation	1/3	a^{-4}	a^{-4}	a^{-1}
vacuum energy	-1	a^0	a^0	

dust fluid has $\rho_m \propto a^{-3}$, as expected. To obtain the relation between T and a we use kinetic theory which relates the gas temperature to the peculiar motions of the gas particles: $k_B T_m \propto m \langle v^2 \rangle$. Since $v \propto a^{-1}$ [see Eq. (3.31)], we have that $T_m \propto a^{-2}$. Finally, using Eq. (3.38) we find that $P_m \propto a^{-5}$. This rapid decrease of pressure with the scale factor indicates that the Universe quickly approaches a dust fluid once it becomes matter dominated.

At early times the Universe is radiation dominated. To investigate how ρ, P and T scale with a during this period, we approximate the fluid as an ultra-relativistic radiation fluid for which $w = 1/3$. This implies that $\rho_r \propto a^{-4}$, which is consistent with the fact that the number density of photons scales as a^{-3}, while the energy per photon, $E = h_P \nu$, scales as a^{-1} [see Eq. (3.21)]. From the equation of state we obtain that $P_r \propto a^{-4}$, while the scaling relation for the temperature, $T_r \propto a^{-1}$, follows from the fact that for radiation $\rho \propto T^4$ [see Eq. (3.33)]. As a result, a blackbody radiation field remains blackbody with a temperature decreasing as a^{-1}. This is an important result which explains how the cosmic microwave background radiation maintains its blackbody form as the Universe expands.

Finally, if the energy density is dominated by vacuum energy, it only depends on the energy difference between the true and false vacua and so is independent of a. It then follows from Eq. (3.35) that

$$P_{\rm vac} = -\rho_{\rm vac} c^2, \tag{3.41}$$

i.e. $w = -1$. This equation of state can be understood as follows: in order to keep a constant energy density $\rho_{\rm vac}$ as the Universe expands, the pressure $P_{\rm vac}$ must be negative so that the PdV work in Eq. (3.32) is a positive contribution to the total internal energy in a given comoving volume as it expands.

Although the above relations are derived from the application of thermodynamics to a small volume in the Universe, they are applicable to the Universe as a whole, because the Universe is assumed to be homogeneous and isotropic. These relations are important, because they allow us to obtain the mean density, temperature and pressure of the Universe at any redshift from their values at the present time. Table 3.1 summarizes how energy density, pressure, and temperature evolve with the scale factor a for different dominating components of the energy density. Before we continue, it is important to emphasize that these scaling relations only hold while the equation of state remains constant. In the early Universe, however, the adiabatic cooling due to the expansion of the Universe may cause various particle species to change from relativistic to non-relativistic. During these transitions, the *true* scaling relations follow from an application of the entropy conservation law (see § 3.3).

3.1.6 Angular-Diameter and Luminosity Distances

The comoving distance χ and the proper distance $a(t)\chi$ from a source are not directly observable, because the light from a distant source observed at the present time was emitted at an earlier time. In this subsection we consider two other distances that can be measured directly from astronomical observations. Consider an object of size D and intrinsic luminosity L at some distance d. The observable properties of such an object are the angular size ϑ subtended by the object, and the flux F. These allow us to define the angular-diameter distance, d_A, and the luminosity distance, d_L, according to

$$\vartheta = \frac{D}{d_A}, \tag{3.42}$$

and

$$F = \frac{L}{4\pi d_L^2}. \tag{3.43}$$

In a static space, $d_A = d_L = d$, consistent with our everyday experience. However, when cosmic distances are concerned in an expanding Universe, d_A, d_L, and d may all have different values, as we will see in the following.

To obtain an expression for d_A in a Robertson–Walker metric, we recall that the proper size D can be considered as the proper distance between two light signals, sent from two points with the same radial coordinate r_e at a given cosmic time t_e, and reaching the origin at the time t_0. Thus, the value of D is just the integral of dl in Eq. (3.8) over the transverse direction:

$$D = a_e r_e \int d\vartheta = \frac{a_0 r_e}{1+z} \vartheta, \tag{3.44}$$

where $a_0 = a(t_0)$ and $a_e = a(t_e)$. It then follows from Eq. (3.42) that

$$d_A = \frac{a_0 r_e}{1+z_e} = a_e r_e. \tag{3.45}$$

To get an expression for d_L, we consider a proper area, $\mathscr{A}$, which is at the origin (the position of the observer) and subtends a solid angle, ω, at the object. By definition of the angular-diameter distance d_A, such a solid angle at the origin corresponds to a proper area ωd_A^2 at the position of the object. If the universe were static, this area would, by symmetry arguments, be equal to $\mathscr{A}$. Because of expansion, however, the proper area at the origin subtended by a fixed solid angle at a given object is stretched by a factor in proportion to the square of the scale factor, and so

$$\mathscr{A} = \omega d_A^2 (a_0/a_e)^2 = (a_0 r_e)^2 \omega. \tag{3.46}$$

Without losing generality, we can assume that the object emits monochromatic radiation with rest-frame frequency ν_e. The number of photons emitted from the object into the solid angle ω within a time interval δt_e is $L \delta t_e \omega / (4\pi h_P \nu_e)$. If the same number of photons pass through the area $\mathscr{A}$ in a time interval δt_0, we have

$$\frac{L \delta t_e \omega}{4\pi h_P \nu_e} = \frac{F \delta t_0 \mathscr{A}}{h_P \nu_0}, \tag{3.47}$$

where ν_0 is the observed frequency of the photons at the origin. It then follows from Eqs. (3.21) and (3.46) that

$$F = \frac{\omega}{4\pi} \frac{L}{\mathscr{A}} \left[\frac{a_e}{a_0}\right]^2 = \frac{L}{4\pi [a_0 r_e (1+z)]^2}. \tag{3.48}$$

The luminosity distance defined in Eq. (3.43) can thus be written as

$$d_L = a_0 r_e (1+z). \tag{3.49}$$

Since we observe the object using photons, the quantity $a_0 r_e$ in the expressions of d_L and d_A is related to the redshift z by Eqs. (3.17) and (3.22).[3] This relation can be obtained once the dynamical equations have been solved to specify $a(t)$. Although we will address the dynamical

---

[3] In the case of a flat universe ($K = 0$), $a_0 r_e$ is equal to the proper distance between object and observer at the time of observation.

behavior of $a(t)$ in detail in §3.2, we can make a simple approximation by using the first few terms of its Taylor expansion:

$$a(t) = a_0 \left[1 + H_0(t-t_0) - \frac{1}{2} q_0 H_0^2 (t-t_0)^2 + \ldots \right], \quad (3.50)$$

where

$$q_0 \equiv -\frac{\ddot{a}_0 a_0}{\dot{a}_0^2} \quad (3.51)$$

is known as the deceleration parameter. Using Eq. (3.17), the power series can be manipulated to give

$$a_0 r_e \approx \frac{c}{H_0} \left[z - \frac{1}{2} z^2 (1+q_0) + \ldots \right]. \quad (3.52)$$

Inserting this into Eqs. (3.44) and (3.48), we can obtain D as a function of ϑ and z, and L as a function of F and z, respectively. Thus, for given values of H_0 and q_0, the proper size D (or the intrinsic luminosity L) of an object can be obtained by measuring its redshift z and its angular size ϑ (or its flux F). Similarly, if the proper sizes (or intrinsic luminosities) of a set of objects are known, one can estimate the values of H_0 and q_0 by measuring ϑ (or F) as a function of redshift. Although this way of using Eq. (3.52) to interpret observational data is common practice, it is valid only for $z \ll 1$. It is therefore preferable to use the exact equations for $a_0 r$ as function of z (derived in the next section) rather than this small-z approximation. Nevertheless, the present values of H_0 and q_0 are often used to characterize cosmological models.

Finally, Eqs. (3.44) and (3.48) can be combined to give the apparent surface brightness of an object,

$$S \equiv \frac{F}{\frac{1}{4}\pi \vartheta^2} = \frac{L}{\pi^2 D^2} (1+z)^{-4}. \quad (3.53)$$

Unlike d_A and d_L, the apparent surface brightness S is independent of the relationship between $a_0 r_e$ and z_e, and so is independent of the dynamical evolution of $a(t)$. This arises because Eq. (3.53) depends only on the local thermodynamics of the radiation field, and follows, in fact, directly from $S \propto T^4$. For given L and D, the apparent surface brightness decreases with redshift as $(1+z)^{-4}$, which is usually referred to as cosmological surface brightness dimming.

3.2 Relativistic Cosmology

In general relativity, the geometric properties of space-time are determined by the distribution of matter/energy. The standard model of cosmology arises from the application of general relativity to the very special class of matter/energy distributions implied by the cosmological principle, i.e. homogeneous and isotropic distributions. As we have seen above, the geometric properties of a homogeneous and isotropic universe are described by the Robertson–Walker metric which, in turn, is specified by the scale factor $a(t)$ and the curvature signature K. The task of this section is to obtain an expression for $a(t)$ and the value of K for any given homogeneous and isotropic matter/energy content.

3.2.1 Friedmann Equation

In the standard model of cosmology, the geometry of space-time is determined by the matter/energy content of the Universe through the Einstein field equation (see Appendix A):

$$R_{\mu\nu} - \frac{1}{2}g_{\mu\nu}R - g_{\mu\nu}\Lambda = \frac{8\pi G}{c^4}T_{\mu\nu}. \tag{3.54}$$

Here $R_{\mu\nu}$ is the Ricci tensor, describing the local curvature of space-time, R is the curvature scalar, $g_{\mu\nu}$ is the metric, $T^{\mu\nu}$ is the energy–momentum tensor of the matter content of the Universe, and Λ is the cosmological constant, which was introduced by Einstein to obtain a static universe. Contracting Eq. (3.54) with $g^{\mu\nu}$ yields the trace of the field equation,

$$R + 4\Lambda = -\frac{8\pi G}{c^4}T, \tag{3.55}$$

where $T = T^\lambda{}_\lambda$. This allows the field equation to be written in the form

$$R_{\mu\nu} + g_{\mu\nu}\Lambda = \frac{8\pi G}{c^4}\left(T_{\mu\nu} - \frac{1}{2}g_{\mu\nu}T\right). \tag{3.56}$$

For a uniform ideal fluid,

$$T^{\mu\nu} = (\rho + P/c^2)U^\mu U^\nu - g^{\mu\nu}P, \tag{3.57}$$

with ρc^2 the energy density, P the pressure, and $U^\mu = c dx^\mu/ds$ the four velocity of the fluid. In a homogeneous and isotropic universe, the density and pressure depend only on the cosmic time, and the four-velocity is $U^\mu = (c,0,0,0)$ (i.e. no peculiar motion is allowed). This implies that $T^\mu{}_\nu = \mathrm{diag}(\rho c^2, -P, -P, -P)$ and $T = \rho c^2 - 3P$.

For a homogeneous and isotropic universe, $g_{\mu\nu}$ is given by the Robertson–Walker metric, which allows the Ricci tensor $R_{\mu\nu}$ and curvature scalar R to be expressed in terms of the scale factor $a(t)$ and the curvature signature K (see Appendix A). Inserting the results into Eq. (3.56), and using the energy–momentum tensor of a perfect fluid given in Eq. (3.57), one obtains

$$\frac{\ddot{a}}{a} = -\frac{4\pi G}{3}\left(\rho + 3\frac{P}{c^2}\right) + \frac{\Lambda c^2}{3} \tag{3.58}$$

for the time-time component, and

$$\frac{\ddot{a}}{a} + 2\frac{\dot{a}^2}{a^2} + 2\frac{Kc^2}{a^2} = 4\pi G\left(\rho - \frac{P}{c^2}\right) + \Lambda c^2 \tag{3.59}$$

for the space-space components. It then follows from substituting Eq. (3.58) into Eq. (3.59) that

$$\left(\frac{\dot{a}}{a}\right)^2 = \frac{8\pi G}{3}\rho - \frac{Kc^2}{a^2} + \frac{\Lambda c^2}{3}. \tag{3.60}$$

As one sees from Eqs. (3.58)–(3.60), the cosmological constant can be considered as an energy component with 'mass' density $\rho_\Lambda = \Lambda c^2/8\pi G$ and pressure $P_\Lambda = -\rho_\Lambda c^2$. Indeed, the term of Einstein's cosmological constant in Eq. (3.54) can be included as an energy–momentum tensor, $T_{\mu\nu} = (c^4\Lambda/8\pi G)g_{\mu\nu}$, on the right-hand side of the field equation.

Eq. (3.60) is the Friedmann equation, and a cosmology that obeys it is called a Friedmann–Robertson–Walker (FRW) cosmology. Together with Eq. (3.35), an equation of state, and an initial condition, it determines the time dependence of a, ρ, P, and other properties of the Universe.

It is interesting to note that one can derive the Friedmann equation (without the cosmological constant term) for a matter dominated universe purely from Newtonian gravity (see §2.10). This follows from the assumption that the Universe is homogeneous and isotropic so that the global properties of the Universe can be represented by those in a small region where Newtonian physics applies. The Newtonian derivation, however, does not contain the pressure term, $3P/c^2$, in the equation for the acceleration, which can be considered a relativistic correction. As is evident from Eq. (3.58), in general relativity this pressure term acts as a source of gravity.

The density which appears in Eq. (3.60) can be made up of various components. At the moment we distinguish a non-relativistic matter component, a radiation component, and a possible vacuum energy (cosmological constant) component. We denote their energy densities (written in terms of mass densities) at the present time t_0 by $\rho_{m,0}$, $\rho_{r,0}$ and $\rho_{\Lambda,0}$, respectively. As the Universe expands, these quantities scale with a in different ways, as described in §3.1.5. We can then write the Friedmann equation as

$$\left(\frac{\dot{a}}{a}\right)^2 = H^2(t) = \frac{8\pi G}{3}\left[\rho_{m,0}\left(\frac{a_0}{a}\right)^3 + \rho_{r,0}\left(\frac{a_0}{a}\right)^4 + \rho_{\Lambda,0}\right] - \frac{Kc^2}{a^2}, \qquad (3.61)$$

where $a_0 = a(t_0)$.[4] Using the fact that the Universe is in its expanding phase at the present time (i.e. $H_0 = \dot{a}_0/a_0 > 0$), we can examine the behavior of $a(t)$ in various cases, even without solving the Friedmann equation explicitly.

If $\Lambda \geq 0$ and if $K = 0$ or $K = -1$, the right-hand side of Eq. (3.61) is always larger than zero, and $a(t)$ always increases with t. If $K = +1$ and $\Lambda = 0$, the right-hand side of Eq. (3.61) becomes zero in the future as the scale factor increases until the curvature term, K/a^2, is as large as the sum of the matter and radiation terms. Thereafter $a(t)$ decreases with t, and the Universe contracts until $a = 0$. If $K = +1$ and $\Lambda > 0$, the situation is similar to that with $K = +1$ and $\Lambda = 0$, provided that the Λ term in Eq. (3.61) is smaller than the matter plus radiation terms at the present time. If the Λ term is sufficiently large at the present time, there may have been a minimum value of a at some previous epoch. This corresponds to a time when the right-hand side of Eq. (3.61) is equal to zero, and an initially contracting universe 'bounced' on its vacuum energy density and started to re-expand. As one can see from Eq. (3.61), this re-expansion will continue forever. For positive Λ a static (but unstable) solution is also possible – Einstein's original static model – as are solutions which asymptotically approach this model in the infinite future or infinite past. Finally, if $\Lambda < 0$, the expansion will eventually halt and be followed by recollapse, giving a history qualitatively similar to that of a $K = +1$, $\Lambda = 0$ universe.

3.2.2 The Densities at the Present Time

To solve Eq. (3.61), we need to know K and the various densities at the present time, $\rho_{m,0}$, $\rho_{r,0}$ and $\rho_{\Lambda,0}$. Here we summarize constraints on these quantities based on observational and theoretical considerations.

The total rest mass density of non-relativistic matter in the Universe is conventionally expressed as

$$\rho_{m,0} = \Omega_{m,0}\rho_{\mathrm{crit},0} \approx 1.88 \times 10^{-29}\Omega_{m,0}h^2\,\mathrm{g\,cm^{-3}}, \qquad (3.62)$$

where, for reasons that will soon become clear, the density

$$\rho_{\mathrm{crit}}(t) \equiv \frac{3H^2(t)}{8\pi G} \qquad (3.63)$$

is known as the *critical* density at time t. The subscript '0' denotes the values at the present time. The dimensionless quantity, $\Omega_{m,0}$, is the present cosmic density parameter for non-relativistic

---

[4] Note, however, that Eq. (3.61) only applies if there is no transformation from one density component to another. If such transformation occurs, the time dependence of the equation of state must be taken into account.

matter, and h is defined in Eq. (3.15). As discussed in §2.10, current observational constraints suggest

$$\Omega_{m,0} = 0.27 \pm 0.05; \quad h = 0.72 \pm 0.05. \tag{3.64}$$

The current density in the relativistic component appears to be dominated by the cosmic microwave background which is, to high accuracy, a blackbody at temperature $T_\gamma = 2.73\,\text{K}$. Thus, using $\rho_\gamma = 4\sigma_{SB} T^4/c^3$ with σ_{SB} the Stefan–Boltzmann constant, we have

$$\rho_{\gamma,0} \approx 4.7 \times 10^{-34}\,\text{g cm}^{-3} \quad \text{or} \quad \Omega_{\gamma,0} \equiv \rho_{\gamma,0}/\rho_{\text{crit},0} \approx 2.5 \times 10^{-5} h^{-2}. \tag{3.65}$$

In addition, if the three species of neutrinos and their antiparticles are all massless (or relativistic at the present time), they will have a temperature $T_\nu = (4/11)^{1/3} T_\gamma$ (see §3.3). Because each neutrino has only one spin state (while a photon has two) and because neutrinos are fermions (and so for a given temperature the statistical weight of each degree of freedom is only 7/8 of that for photons; see §3.3 for details), the energy density in neutrinos at the present time is $3 \times (7/8) \times (4/11)^{4/3}$ times that of the CMB photons. This brings the total energy density in the relativistic component to

$$\rho_{r,0} \approx 7.8 \times 10^{-34}\,\text{g cm}^{-3} \quad \text{or} \quad \Omega_{r,0} \approx 4.2 \times 10^{-5} h^{-2}. \tag{3.66}$$

Combining Eqs. (3.62) and (3.66) shows that the ratio of the energy densities in the non-relativistic and relativistic components varies with redshift as

$$\frac{\rho_m}{\rho_r} \approx 2.4 \times 10^4 \Omega_{m,0} h^2 (1+z)^{-1}, \tag{3.67}$$

where we have used that $\rho_m \propto a^{-3}$ and $\rho_r \propto a^{-4}$ (see Table 3.1). Thus, provided the Universe did not bounce in the recent past due to a large cosmological constant, it has been matter dominated and effectively pressure-free since the epoch of matter/radiation equality defined by $\rho_r = \rho_m$, i.e. since the redshift given by

$$1 + z_{\text{eq}} \approx 2.4 \times 10^4 \Omega_{m,0} h^2. \tag{3.68}$$

To constrain the present day energy density provided by the cosmological constant, we use the Friedmann equation (3.61), which we rewrite as

$$\frac{8\pi G}{3} \rho_{\Lambda,0} = H_0^2 [1 - \Omega_{m,0} - \Omega_{r,0}] + \frac{Kc^2}{a_0^2}. \tag{3.69}$$

As discussed in §2.9, observations of the microwave background show that our Universe is almost flat and that the current density in non-relativistic matter is significant [see Eq. (3.64)]. This excludes the possibility of a bounce in the recent past due to a large cosmological constant. Such an expansion history is also excluded by the observation of objects out to redshifts beyond 6, so we will not consider such cosmological models any further. Setting $K = 0$ in Eq. (3.69) we obtain

$$\rho_{\Lambda,0} = \rho_{\text{crit},0}(1 - \Omega_{m,0} - \Omega_{r,0}) \quad \text{i.e.} \quad \Omega_{\Lambda,0} = 1 - \Omega_{m,0} - \Omega_{r,0}. \tag{3.70}$$

Data from WMAP combined with other observations give $\Omega_{\Lambda,0} \sim 0.75 \pm 0.02$ (Spergel et al., 2007).

3.2.3 Explicit Solutions of the Friedmann Equation

(a) The Evolution of Cosmological Quantities Taking $t = t_0$, the Friedmann equation can be rewritten as

$$\Omega_{K,0} \equiv -\frac{Kc^2}{H_0^2 a_0^2} = 1 - \Omega_0, \tag{3.71}$$

where
$$\Omega_0 = \Omega_{m,0} + \Omega_{\Lambda,0} + \Omega_{r,0} \tag{3.72}$$

is the total density parameter at the present time. As is immediately evident from Eq. (3.71), the curvature of space-time depends on the matter density of the Universe. In particular, Ω_0 is less than 1 for a negatively curved, open universe, is equal to 1 for a flat universe, and is bigger than 1 for a positively curved, closed universe. The terminology 'open' and 'closed' only has a logical meaning for a $\Lambda = 0$ universe; open (and flat) universes expand forever, while closed universes recollapse in the future. For non-zero Λ, however, open and flat universes can recollapse and closed universes can expand forever, depending on the values of the various density parameters (see discussion at the end of §3.2.1). Since Ω_0 is just the total energy density of the Universe in units of $\rho_{\mathrm{crit},0}$, it follows that $\rho_{\mathrm{crit},0}$ defines a critical density for closure. Note that Eq. (3.71) defines the scale factor a_0 at the present time:

$$a_0 = \frac{c}{H_0} \sqrt{\frac{K}{\Omega_0 - 1}}, \tag{3.73}$$

which goes to infinity as Ω_0 approaches 1 from either side. This follows from our definition of the coordinate r in Eqs. (3.3) and (3.8). Since a_0 is only a scale factor, its value does not have physical meaning and so can be set to any positive value. A choice for the value of a_0 corresponds to a choice in the definition of the coordinate r. In fact, physical distances are all related to a_0 through the combination $a_0 r$, which is well behaved near $\Omega_0 = 1$ and independent of the choice of a_0. It is common practice to adopt $a_0 = 1$.

Substituting Eq. (3.71) into Eq. (3.61) gives

$$H(z) \equiv \left(\frac{\dot{a}}{a}\right)(z) = H_0 E(z), \tag{3.74}$$

where
$$E(z) = \left[\Omega_{\Lambda,0} + (1-\Omega_0)(1+z)^2 + \Omega_{m,0}(1+z)^3 + \Omega_{r,0}(1+z)^4\right]^{1/2}. \tag{3.75}$$

Defining the cosmic density parameters at cosmic time t as

$$\Omega(t) \equiv \frac{\rho(t)}{\rho_{\mathrm{crit}}(t)}, \tag{3.76}$$

we have
$$\Omega_\Lambda(z) = \frac{\Omega_{\Lambda,0}}{E^2(z)}; \quad \Omega_m(z) = \frac{\Omega_{m,0}(1+z)^3}{E^2(z)}; \quad \Omega_r(z) = \frac{\Omega_{r,0}(1+z)^4}{E^2(z)}. \tag{3.77}$$

Thus, once H, Ω_Λ, Ω_m and Ω_r are known at the present time, Eqs. (3.74)–(3.77) can be used to obtain their values at any given redshift. It is also clear from Eqs. (3.61) and (3.71) that the geometry of a FRW universe is completely determined by the values of H_0, $\Omega_{\Lambda,0}$, $\Omega_{m,0}$ and $\Omega_{r,0}$. Since $\Omega_{r,0} \ll \Omega_{m,0}$ (see §3.2.2), the deceleration parameter, q_0, defined in Eq. (3.51) can be written as

$$q_0 = \Omega_{m,0}/2 - \Omega_{\Lambda,0}, \tag{3.78}$$

where we have used Eq. (3.58) with $P = 0$, as appropriate for a matter dominated universe.

Finally, using Eq. (3.71) and the definition of $E(z)$, we can write down the redshift evolution of the total density parameter

$$\Omega(z) - 1 = (\Omega_0 - 1)\frac{(1+z)^2}{E^2(z)}. \tag{3.79}$$

As long as $\Omega_{m,0}$ or $\Omega_{r,0}$ are non-zero, $\Omega(z)$ always approaches unity at high redshifts, independent of the present day values of H_0, $\Omega_{\Lambda,0}$, $\Omega_{m,0}$ and $\Omega_{r,0}$. Therefore, every FRW universe with non-zero matter or radiation content must have started out with a total density parameter very close to unity. As we will see in §3.6, this results in the so-called flatness problem.

(b) Radiation Dominated Epoch In the absence of a contracting phase in the past, the right-hand side of Eq. (3.61) is dominated by the radiation term at $z \gg z_{eq}$. In this case, integration of Eq. (3.61) yields

$$\frac{a}{a_0} = \left(\frac{32\pi G \rho_{r,0}}{3}\right)^{1/4} t^{1/2}. \tag{3.80}$$

Using that $\rho_r \propto a^{-4}$, $\rho_m \propto a^{-3}$ and $T_r \propto a^{-1}$ (see Table 3.1), this gives the following rough scalings for the early Universe:

$$\frac{T}{10^{10}\,\mathrm{K}} \sim \frac{k_B T}{1\,\mathrm{MeV}} \sim \left[\frac{\rho}{10^7\,\mathrm{g\,cm}^{-3}}\right]^{1/4} \sim \left[\frac{\rho_m}{1\,\mathrm{g\,cm}^{-3}}\right]^{1/3} \sim \frac{1+z}{10^{10}} \sim \left[\frac{t}{1\,\mathrm{s}}\right]^{-1/2}. \tag{3.81}$$

These relations are approximately correct for $0 < t < 10^{10}$ s, or $z > 10^5$. The arbitrarily high temperatures and densities which are achieved at sufficiently early times have given this standard cosmological model its generic name, the Hot Big Bang.

(c) Matter Dominated Epoch and $\Omega_{\Lambda,0} = 0$ At redshift $z \ll z_{eq}$, the radiation content of the Universe has little effect on its global dynamics, and assuming $\Lambda = 0$, Eq. (3.61) reduces to

$$\left(\frac{\dot{a}}{a}\right)^2 = H_0^2 \left[\Omega_{m,0}\left(\frac{a_0}{a}\right)^3 - \frac{Kc^2}{H_0^2 a_0^2}\left(\frac{a_0}{a}\right)^2\right]. \tag{3.82}$$

For $K = 0$ the solution is particularly simple:

$$\frac{a}{a_0} = \left(\frac{3}{2}H_0 t\right)^{2/3}. \tag{3.83}$$

This is the solution for an Einstein–de Sitter (EdS) universe. For $K = -1$, the solution can be expressed in parametric form:

$$\frac{a}{a_0} = \frac{1}{2}\frac{\Omega_{m,0}}{(1-\Omega_{m,0})}(\cosh\vartheta - 1); \quad H_0 t = \frac{1}{2}\frac{\Omega_{m,0}}{(1-\Omega_{m,0})^{3/2}}(\sinh\vartheta - \vartheta), \tag{3.84}$$

where ϑ goes from 0 to ∞. At early epochs, $a \propto t^{2/3}$, which follows directly from the fact that the curvature term in Eq. (3.82) can be neglected when a is sufficiently small. At later epochs when $\vartheta \gg 1$ and $\sinh\vartheta = \cosh\vartheta$ so that $a \propto t$, the universe enters a phase of free expansion, unretarded by gravity.

The corresponding parametric solution for a $K = +1$ universe is

$$\frac{a}{a_0} = \frac{1}{2}\frac{\Omega_{m,0}}{(\Omega_{m,0}-1)}(1-\cos\vartheta); \quad H_0 t = \frac{1}{2}\frac{\Omega_{m,0}}{(\Omega_{m,0}-1)^{3/2}}(\vartheta - \sin\vartheta), \tag{3.85}$$

where $0 \leq \vartheta \leq 2\pi$. Such models reach a maximum size, a_{max}, at a time, t_{max}, given by

$$\frac{a_{max}}{a_0} = \frac{\Omega_{m,0}}{\Omega_{m,0}-1}; \quad H_0 t_{max} = \frac{\pi}{2}\frac{\Omega_{m,0}}{(\Omega_{m,0}-1)^{3/2}}. \tag{3.86}$$

This maximum expansion is followed by recollapse to a singularity. At early epochs, $a \propto t^{2/3}$, for the same reason as that for the $K = -1$ case.

Note that $H_0 t_0$ depends only on $\Omega_{m,0}$ in these models. Since the normalization time, t_0, can be chosen arbitrarily, it is easy to see that $H(t)t$ depends only on $\Omega_m(t)$.

(d) Flat ($\Omega_{m,0}+\Omega_{\Lambda,0}=1$) Models at $z \ll z_{eq}$ In this case Eq. (3.61) can be written as

$$\left(\frac{\dot{a}}{a}\right)^2 = H_0^2 \left[\Omega_{m,0}\left(\frac{a_0}{a}\right)^3 + \Omega_{\Lambda,0}\right]. \tag{3.87}$$

When the matter term is negligible, the model is called a de Sitter universe for which the solution of Eq. (3.87) is particularly simple:

$$\frac{a}{a_0} = \exp[H_0(t-t_0)], \tag{3.88}$$

and the universe expands exponentially without an initial singularity. For $0 < \Omega_{m,0} < 1$, using the fact that $H_0 \equiv \dot{a}/a > 0$, Eq. (3.87) can be easily solved to give

$$\frac{a}{a_0} = \left(\frac{\Omega_{m,0}}{\Omega_{\Lambda,0}}\right)^{1/3} \left[\sinh\left(\frac{3}{2}\Omega_{\Lambda,0}^{1/2}H_0 t\right)\right]^{2/3}. \tag{3.89}$$

At early epochs, $a \propto t^{2/3}$ as in an Einstein–de Sitter universe; when t is large, $a \propto \exp(\Omega_{\Lambda,0}^{1/2}H_0 t)$ so that the universe approximates the de Sitter model.

(e) Open and Closed Models with $\Omega_{\Lambda,0} \neq 0$ at $z \ll z_{eq}$ The Friedmann equation in this case is

$$\left(\frac{\dot{a}}{a}\right)^2 = H_0^2 \left[\Omega_{m,0}\left(\frac{a_0}{a}\right)^3 - \frac{Kc^2}{H_0^2 a_0^2}\left(\frac{a_0}{a}\right)^2 + \Omega_{\Lambda,0}\right]. \tag{3.90}$$

This equation can be cast into a dimensionless form:

$$\frac{1}{2}\left(\frac{dx}{d\eta}\right)^2 = \frac{1}{x} - \kappa + \lambda x^2, \tag{3.91}$$

where $x = a/a_0$, $\eta = \sqrt{\Omega_{m,0}/2}\,H_0 t$, $\lambda = \Omega_{\Lambda,0}/\Omega_{m,0}$, and $\kappa = Kc^2/(H_0^2 a_0^2 \Omega_{m,0})$. The evolution of x can thus be discussed in terms of the Newtonian motion of a particle with total energy $\varepsilon = -\kappa$ in a potential $\phi(x) = -1/x - \lambda x^2$.

When $\lambda < 0$, the potential $\phi(x)$ monotonically increases from 0 to ∞ so that x is confined, and all solutions evolve from an initial singularity into a final singularity.

When $\lambda > 0$, the potential $\phi(x)$ is always negative, and x can go to infinity if $\varepsilon > 0$ or $K = -1$. Hence an open universe with $\Omega_{\Lambda,0} > 0$ expands from an initial singularity forever. If $\lambda > 0$ and $K = +1$, the potential $\phi(x)$ has a maximum, $\phi_{max} = -(27\lambda/4)^{1/3}$, at $x = x_{max} = 1/(2\lambda)^{1/3}$. In this case, if the total energy $\varepsilon > \phi_{max}$, i.e.

$$\lambda > \lambda_c \equiv \frac{4}{27}\left[\frac{c^2}{H_0^2 a_0^2 \Omega_{m,0}}\right]^3, \tag{3.92}$$

the universe still expands forever, starting from an initial singularity. If, however, $\varepsilon < \phi_{max}$ or $\lambda < \lambda_c$, then there is the possibility that the universe contracts from large radii to a minimum radius, a_{min}, given by $\phi(a_{min}/a_0) = \varepsilon$, and expands thereafter to infinity. This happens if the universe starts with a radius $a > a_0 x_{max}$. If the universe starts with an initial singularity, then it will evolve into a final singularity, giving a situation similar to that of a closed universe without cosmological constant.

If $\lambda > 0$ and $K = +1$, a special situation occurs when $\varepsilon = \phi_{max}$ or $\lambda = \lambda_c$. In this case, there is a static solution with a constant radius $a_E = a_0/(2\lambda_c)^{1/3}$. Such a model is called the 'Einstein universe'. If the universe expands from an initial singularity, or contracts from a large radius, it will coast asymptotically towards the radius a_E. If the universe expands from an initial radius larger than a_E, it will do so forever.

3.2.4 Horizons

A light ray emitted by an event (r_e, t_e) reaches an observer at the origin at time, t_0, given by

$$\chi(r_e) = \int_{t_e}^{t_0} \frac{c\,dt}{a(t)} = \int_{a_e}^{a_0} \frac{da}{a} \left[\frac{8\pi G \rho(a) a^2}{3c^2} - K \right]^{-1/2}, \qquad (3.93)$$

with $\rho = \rho_m + \rho_r + \rho_\Lambda$. The second equality follows from substituting $dt = da/\dot{a}$ and using the Friedmann equation (3.60). If the t (or a) integral converges, as $t_e \to 0$, to a value $\chi_h = \chi(r_h)$, then there may exist particles (fundamental observers) for which $\chi(r) > \chi_h$ and from which no communication can have reached the origin by time t_0. Such particles (or values of r) are said to lie beyond the particle horizon of the origin at time t_0. From the form of the last integral in Eq. (3.93) it is clear that convergence requires $\rho a^2 \to \infty$ as $a \to 0$. Thus particle horizons exist in a universe which is matter or radiation dominated at the earliest times, but do not exist in a universe which was initially dominated by vacuum energy density. As t_0 increases, χ_h becomes bigger, all particle horizons expand, and signals can be received from more and more distant particles.

If the t integral in Eq. (3.93) converges as $t_0 \to \infty$ (or as t_0 approaches the recollapse time for a universe with a finite lifetime), there may exist events which the observer at the origin will never see, and which therefore can never influence him/her by any physical means. Such events are said to lie beyond the event horizon of this observer. Event horizons exist in closed models and in models that are vacuum dominated at late times, but do not exist in flat or open universes with zero cosmological constant. In the latter case, therefore, any event will eventually be able to influence every fundamental observer in the Universe.

The existence of particle horizons in the Big Bang model has important implications, because it means that many parts of the presently observable Universe may not have been in causal contact at early times. This gives rise to certain difficulties, as we will see in §3.6.

3.2.5 The Age of the Universe

In currently viable models the Universe has been expanding since the Big Bang, so that $\dot{a} > 0$ holds over its entire history. The age of the Universe at redshift z can then be obtained from Eqs. (3.22) and (3.74):

$$t(z) \equiv \int_0^{a(z)} \frac{da}{\dot{a}} = \frac{1}{H_0} \int_z^{\infty} \frac{dz}{(1+z)E(z)}, \qquad (3.94)$$

where $E(z)$ is given by Eq. (3.75). With this, the lookback time at redshift z, defined as $t_0 - t(z)$, can also be obtained. For a given set of cosmological parameters, $t(z)$ can be calculated easily from Eq. (3.94) by numerical integration. In some special cases, the integration can even be carried out analytically.

In the radiation dominated epoch (i.e. at $z \gg z_{eq}$), the solution of the Friedmann equation is given by Eq. (3.80), and the age of the Universe is

$$t(z) \approx \left(\frac{1+z}{10^{10}} \right)^{-2} \text{ s}. \qquad (3.95)$$

In the matter dominated epoch ($z \ll z_{eq}$), we can neglect the radiation term in $E(z)$. It can then be shown that for an EdS universe (i.e. for $\Omega_{m,0} = 1$ and $\Omega_{\Lambda,0} = 0$),

$$t(z) = \frac{1}{H_0} \frac{2}{3} (1+z)^{-3/2} \approx \frac{2}{3} (1+z)^{-3/2} \times 10^{10} h^{-1} \text{ yr}. \qquad (3.96)$$

For an open universe with $\Omega_{\Lambda,0} = 0$ and $\Omega_0 = \Omega_{m,0} < 1$,

$$t(z) = \frac{1}{H_0} \frac{\Omega_0}{2(1-\Omega_0)^{3/2}} \left[\frac{2\sqrt{(1-\Omega_0)(\Omega_0 z+1)}}{\Omega_0(1+z)} - \cosh^{-1}\left(\frac{\Omega_0 z - \Omega_0 + 2}{\Omega_0 z + \Omega_0}\right) \right]. \quad (3.97)$$

For a closed universe with $\Omega_{\Lambda,0} = 0$ and $\Omega_0 = \Omega_{m,0} > 1$,

$$t(z) = \frac{1}{H_0} \frac{\Omega_0}{2(\Omega_0-1)^{3/2}} \left[-\frac{2\sqrt{(\Omega_0-1)(\Omega_0 z+1)}}{\Omega_0(1+z)} + \cos^{-1}\left(\frac{\Omega_0 z - \Omega_0 + 2}{\Omega_0 z + \Omega_0}\right) \right]. \quad (3.98)$$

Finally, for a flat universe with $\Omega_{m,0} + \Omega_{\Lambda,0} = 1$,

$$t(z) = \frac{1}{H_0} \frac{2}{3\sqrt{\Omega_{\Lambda,0}}} \ln\left[\frac{\sqrt{\Omega_{\Lambda,0}(1+z)^{-3}} + \sqrt{\Omega_{\Lambda,0}(1+z)^{-3} + \Omega_{m,0}}}{\sqrt{\Omega_{m,0}}} \right]. \quad (3.99)$$

In all these cases, the behavior at $z \gg 1$ is

$$t(z) \approx \frac{2}{3H_0} \Omega_{m,0}^{-1/2} (1+z)^{-3/2}. \quad (3.100)$$

Fig. 3.2 shows the product of the Hubble parameter, h, defined by Eq. (3.15), and the lookback time, $t_0 - t(z)$, as a function of $(1+z)$ for models with $\Omega_{\Lambda,0} = 0$, and for flat models with a

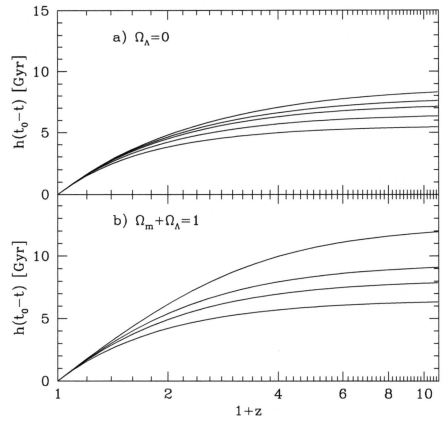

Fig. 3.2. The lookback time as a function of redshift for (a) models with $\Omega_{\Lambda,0} = 0$ and $\Omega_{m,0} = 0.1, 0.3, 0.5, 1, 2$ (from top down); and (b) flat models ($\Omega_{m,0} + \Omega_{\Lambda,0} = 1$) with $\Omega_{m,0} = 0.1, 0.3, 0.5, 1$ (from top down).

cosmological constant ($\Omega_{m,0} + \Omega_{\Lambda,0} = 1$). It is clear that for given h and $\Omega_{m,0}$, the age of the Universe is larger in models with a cosmological constant. By definition, the age of the Universe at the present time should be larger than that of the oldest objects it contains. The oldest objects whose ages can be determined reliably are a class of star clusters called globular clusters, which have ages ranging up to 13 Gyr (e.g. Carretta et al., 2000). This requires $h \lesssim 0.5$ for an EdS universe, and $h \lesssim 0.7$ for a flat universe with $\Omega_{m,0} = 0.3$ and $\Omega_{\Lambda,0} = 0.7$.

3.2.6 Cosmological Distances and Volumes

As defined in §3.1.6, the luminosity distance, d_L, and the angular-diameter distance, d_A, are related to the redshift, z, and the comoving coordinate, r, by

$$d_L = \left(\frac{L}{4\pi F}\right)^{1/2} = a_0 r (1+z); \quad d_A = \frac{D}{\vartheta} = \frac{a_0 r}{1+z}. \tag{3.101}$$

In order to write d_L and d_A in terms of observable quantities, we need to express the unobservable coordinate r as a function of z. To do this, recall that $r(t)$ is the comoving coordinate of a light signal (an event) that originates at cosmic time t and reaches us at the origin at the present time t_0. It then follows from Eq. (3.17) that the comoving distance corresponding to r is

$$\chi(r) = \tau(t_0) - \tau(t) = c\int_{a(t)}^{a_0} \frac{da}{a\dot{a}}, \tag{3.102}$$

where we have used the definition of the conformal time in Eq. (3.11) and the fact that $dt = da/\dot{a}$. Using Eq. (3.74) and the fact that $a(z) = a_0/(1+z)$ this can be rewritten as

$$\chi(r) = \frac{c}{H_0 a_0}\int_0^z \frac{dz}{E(z)}, \tag{3.103}$$

where $E(z)$ is given by Eq. (3.75). Using Eqs. (3.10) and (3.13), this gives

$$r = f_K\left[\frac{c}{H_0 a_0}\int_0^z \frac{dz}{E(z)}\right]. \tag{3.104}$$

Note that r is the angular-diameter distance in comoving units. In general Eq. (3.103) can be integrated numerically for a given set of cosmological parameters. When $z \ll z_{eq}$ and $\Omega_{\Lambda,0} = 0$, a closed expression can be derived for all three values of K,

$$a_0 r = \frac{2c}{H_0}\frac{\Omega_0 z + (2-\Omega_0)\left[1-(\Omega_0 z+1)^{1/2}\right]}{\Omega_0^2(1+z)}, \tag{3.105}$$

which is known as Mattig's formula (Mattig, 1958). For a flat ($\Omega_{m,0} + \Omega_{\Lambda,0} = 1$) universe $r = \chi$, so that for $z \ll z_{eq}$

$$a_0 r = \frac{c}{H_0}\int_0^z \frac{dz}{[\Omega_{\Lambda,0} + \Omega_{m,0}(1+z)^3]^{1/2}}. \tag{3.106}$$

Luminosity (or angular-diameter) distances can be measured directly for objects of known intrinsic luminosity (or proper size). Such objects are known as 'standard candles' (or 'standard rulers'). Since the relation of redshift to these distances depends on cosmological parameters, in particular on H_0, $\Omega_{m,0}$ and $\Omega_{\Lambda,0}$, measuring the redshift of properly calibrated standard candles (or standard rulers) can provide estimates of these parameters.

One of the most reliable and historically most important standard candles is a class of pulsating stars known as Cepheids, for which the pulsation period is tightly correlated with their mean intrinsic luminosity (see §2.1.3). Using the HST, Cepheids have been measured out to distances of about 10 Mpc. At such distances, the d_L-z relation is still linear, $d_L \approx cz/H_0$, so interesting

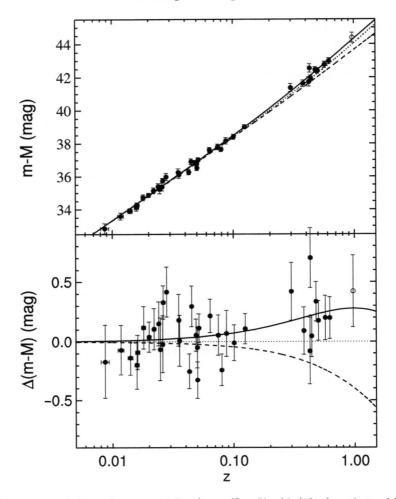

Fig. 3.3. The upper panel shows distance modulus, $(m - \mathcal{M}) = 5\log(d_L/10\,\mathrm{pc})$, against redshift for Type Ia supernovae for which the light curve shape has been used to estimate their absolute magnitudes (data points). The predicted relations for three cosmological models are indicated by dashed ($\Omega_{m,0} = 1$, $\Omega_{\Lambda,0} = 0$), dotted ($\Omega_{m,0} = 0.2$, $\Omega_{\Lambda,0} = 0$) and solid ($\Omega_{m,0} = 0.28$, $\Omega_{\Lambda,0} = 0.72$) curves. The lower panel shows the difference between the distance modulus and the prediction for the ($\Omega_{m,0} = 0.2$, $\Omega_{\Lambda,0} = 0$) model. [Adapted from Riess et al. (1998) by permission of AAS]

constraints can be obtained only for the Hubble constant. The current best estimate is $H_0 = (72 \pm 8)\,\mathrm{km\,s^{-1}\,Mpc^{-1}}$ (e.g. Freedman et al., 2001).

In order to measure other cosmological parameters we must go to sufficiently large distances so that nonlinear terms in the distance–redshift relation are important, i.e. to $z \gtrsim 1$. In Chapter 2 we have seen that Type Ia supernovae can be used as standard candles and that they have now been observed out to $z \sim 1$. In Fig. 3.3 the observed luminosity distance–redshift relation for Type Ia supernovae is compared with the predictions of a number of cosmological models. Detailed analyses of these data give the following constraint:

$$0.8\Omega_{m,0} - 0.6\Omega_{\Lambda,0} \simeq -0.2 \pm 0.1 \tag{3.107}$$

(Perlmutter et al., 1999).

The proper-distance element at time t is $dl = a(t)\,d\chi$. Using Eq. (3.103) we have

$$\frac{dl}{dz} = \frac{c}{H_0} \frac{1}{(1+z)} \frac{1}{E(z)}. \qquad (3.108)$$

This gives the proper distance per unit redshift at redshift z. Suppose that there is a population of objects with proper number density $n(z) = n_0(z)(1+z)^3$ (so that n_0 is a constant if the number of the objects is conserved) and with average proper cross-section $\sigma(z)$. The number of intersections between such objects and a sightline in a unit redshift interval around z is

$$\frac{d\mathcal{N}}{dz} = n_0(z)(1+z)^3 \sigma(z) \frac{dl}{dz} = n_0(z) \sigma(z) \frac{c}{H_0} \frac{(1+z)^2}{E(z)}. \qquad (3.109)$$

The 'optical depth' for the intersection of objects up to redshift z is therefore

$$\tau(z) = \int_0^z d\mathcal{N}(z) = \frac{c}{H_0} \int_0^z n_0(z) \sigma(z) \frac{(1+z)^2}{E(z)} dz. \qquad (3.110)$$

These quantities are relevant for the discussion of QSO absorption line systems (see Chapter 16). In this case $n_0(z)$ is the comoving number density, $\sigma(z)$ is the average absorption cross-section of absorbers, and $d\mathcal{N}/dz$ is just the expected number of absorption systems per unit redshift. Another application of Eqs. (3.109) and (3.110) concerns the interpretation of the observed number of gravitational lensing events caused by foreground objects. In this case, $n_0(z)$ is the comoving number density of lenses, and $\sigma(z)$ is the average lensing cross-section. A third application is to the scattering of the microwave background by ionized intergalactic gas. Here, $\sigma(z)$ is the Thomson cross-section and $n_0(z)$ is the comoving number density of free electrons.

Consider next the proper volume element at a redshift z. The proper-length element in the radial direction is again $a(t)\,d\chi$, and the proper distance subtended by an angle element $d\vartheta$ is $a(t)r\,d\vartheta$. The proper-volume element at redshift z corresponding to a solid angle $d\omega = d\vartheta^2$ and a depth dz is thus

$$d^2 V_p = a^3(t) r^2 \, d\chi\, d\omega = \frac{c}{H_0} \frac{dz}{(1+z)E(z)} \frac{[a_0 r(z)]^2 \, d\omega}{(1+z)^2}, \qquad (3.111)$$

where $r(z)$ is related to z by Eq. (3.104). Using Eq. (3.111), the total, proper volume out to redshift z is

$$V_p(z) = 4\pi a^3(t) \int_0^{r(z)} \frac{r'^2 \, dr'}{\sqrt{1 - Kr'^2}}$$

$$= \begin{cases} 2\pi a^3(t) \left(\sin^{-1} r - r\sqrt{1-r^2} \right) & (K = +1) \\ \frac{4\pi}{3} a^3(t) r^3 & (K = 0) \\ 2\pi a^3(t) \left(r\sqrt{1+r^2} - \sinh^{-1} r \right) & (K = -1). \end{cases} \qquad (3.112)$$

We can also use Eq. (3.111) to compute the total number of objects per unit volume. Assuming the proper number density of objects at redshift z to be $n(z) = n_0(z)(1+z)^3$, the predicted count of objects per unit redshift and per unit solid angle is

$$\frac{d^2 N}{dz\,d\omega} = n(z) \frac{d^2 V_p}{dz\,d\omega} = n_0(z) \frac{c}{H_0} \frac{[a_0 r(z)]^2}{E(z)}. \qquad (3.113)$$

Thus, if the z dependence of n_0 is known, one can use Eq. (3.113) to put constraints on cosmological parameters by simply counting objects (e.g. galaxies) as a function of z (see Loh & Spillar (1986) for a discussion).

Another important quantity in cosmology is the comoving distance between any two observed objects in the Robertson–Walker metric. Suppose that these two objects (labeled $\mathcal{O}_1$ and $\mathcal{O}_2$) are at redshifts z_1 and z_2, and are separated by an angle α on the sky. Their comoving distances from an observer at the origin are given by Eq. (3.102), and we denote them by χ_1 and χ_2, respectively. As shown in §3.1.2, for $K = +1$ the comoving distance, χ_{12}, between $\mathcal{O}_1$ and $\mathcal{O}_2$ is equal to the distance on the unit sphere between two points with polar angles χ_1 and χ_2 and with azimuthal angles differing by α. Thus

$$\cos \chi_{12} = \cos \chi_1 \cos \chi_2 + \sin \chi_1 \sin \chi_2 \cos \alpha. \tag{3.114}$$

The corresponding equation for $K = -1$ is

$$\cosh \chi_{12} = \cosh \chi_1 \cosh \chi_2 - \sinh \chi_1 \sinh \chi_2 \cos \alpha, \tag{3.115}$$

and for $K = 0$ is

$$\chi_{12}^2 = \chi_1^2 + \chi_2^2 - 2\chi_1 \chi_2 \cos \alpha. \tag{3.116}$$

Finally, consider the case in which α is zero (or very small). In this case, the angular-diameter distance from $\mathcal{O}_1$ to $\mathcal{O}_2$ can be written as

$$d_{A,12} = \frac{a_0 r_{12}}{1 + z_2}, \tag{3.117}$$

where

$$r_{12} \equiv f_K(\chi_{12}) = f_K(\chi_2 - \chi_1) = r(z_2)\sqrt{1 - Kr^2(z_1)} - r(z_1)\sqrt{1 - Kr^2(z_2)}. \tag{3.118}$$

For the $\Omega_{\Lambda,0} = 0$ case this gives

$$d_{A,12} = \frac{2c}{H_0} \frac{\sqrt{1 + \Omega_0 z_1}(2 - \Omega_0 + \Omega_0 z_2) - \sqrt{1 + \Omega_0 z_2}(2 - \Omega_0 + \Omega_0 z_1)}{\Omega_0^2 (1 + z_2)^2 (1 + z_1)} \tag{3.119}$$

(Refsdal, 1966). Note that $|d_{A,12}| \neq |d_{A,21}|$, and that, as required, Eq. (3.119) reduces to Mattig's formula for $z_1 = 0$. Eq. (3.119) plays an important role in gravitational lensing, where z_1 and z_2 are the redshifts of the lens and the source, respectively (see §6.6).

3.3 The Production and Survival of Particles

An important feature of the standard cosmology is that the temperature of the Universe was arbitrarily high at the beginning of the Big Bang [see Eq. (3.81)] and has decreased continuously as the Universe expanded to its present state. As we have seen in §3.1.5, the thermal history of the Universe follows from a simple application of thermodynamics to a small patch of the homogeneous and isotropic Universe. In this section we show that this thermal history, together with particle, nuclear and atomic physics, allows a detailed prediction of the matter content of the Universe at each epoch. The reason for this is simple: when the temperature of the Universe was higher than the rest mass of a kind of charged particles, the photon energy is high enough to create these particles and their antiparticles. This, in turn, could give rise to other kinds of particles. For example, when the temperature of the Universe was higher than the rest mass of an electron, i.e. $k_B T > m_e c^2 \approx 0.511\,\mathrm{MeV}$ (corresponding to $T \sim 5.8 \times 10^9\,\mathrm{K}$), electrons and positrons could be generated via pair production, $\gamma + \gamma \leftrightarrow e + \bar{e}$, and electronic neutrinos could be produced via neutral current reactions, such as $e + \bar{e} \leftrightarrow \nu_e + \bar{\nu}_e$. When the density of the Universe was sufficiently high, the creation and annihilation of $(e, \bar{e})$ pairs, and the Compton scattering between $(e, \bar{e})$ and photons, could establish a thermal equilibrium among these particles, while the neutrinos established such an equilibrium via their neutral current coupling to the electrons. Consequently, the Universe was filled with a hot plasma that included γ, e, $\bar{e}$, ν_e and $\bar{\nu}_e$, all in thermal equilibrium at the same temperature.

In order to maintain thermodynamic equilibrium the frequency of interactions among the various particle species involved needs to be sufficiently high. The interaction rate is $\Gamma \equiv n \langle v\sigma \rangle$, where n is the number density of particles, v is their relative velocity, and σ is the interaction cross-section (which usually depends on v). As the Universe expands and the temperature drops, this rate in general decreases. When it becomes smaller than the expansion rate of the Universe, given by the Hubble parameter, $H(t)$, the particles 'decouple' from the photon fluid, and, as long as the particles are stable, their comoving number density 'freezes-out' at its current value. Except for possible particle species created out of thermal equilibrium (e.g. axions), and for particles that have been created more recently in high-energy processes, all elementary particles in the present-day Universe are thermal relics that have decoupled from the photon fluid at some time in the past.

In what follows we first present a brief outline of the chronology of the early Universe, and then discuss the production and survival of particles during a number of important epochs. Since the early Universe was dominated by relativistic particles, Eq. (3.81) can be used to relate temperature T (or energy $k_B T$) to the cosmic time. As this section is concerned with high energy physics, we will use the natural unit system in which the speed of light c, Boltzmann's constant k_B, and Planck's constant $\hbar = h_P/2\pi$ are all set to 1. In cgs units $[c] = \mathrm{cm\,s^{-1}}$, $[\hbar] = \mathrm{g\,cm^2\,s^{-1}}$, and $[k_B] = \mathrm{g\,cm^2\,s^{-2}\,K^{-1}}$. Therefore, making these constants dimensionless implies that

$$[\text{energy}] = [\text{mass}] = [\text{temperature}] = [\text{time}]^{-1} = [\text{length}]^{-1}, \tag{3.120}$$

and all physical quantities can be expressed in one unit, usually mass or energy. However, they can also be expressed in one of the other units using the following conversion factors: $1\,\mathrm{MeV} = 1.602 \times 10^{-6}\,\mathrm{erg} = 1.161 \times 10^{10}\,\mathrm{K} = 1.783 \times 10^{-27}\,\mathrm{g} = 5.068 \times 10^{10}\,\mathrm{cm^{-1}} = 1.519 \times 10^{21}\,\mathrm{s^{-1}}$. Whenever needed, the 'missing' powers of c, k_B, and $\hbar$ in equations can be reinserted straightforwardly from a simple dimensional analysis.

3.3.1 The Chronology of the Hot Big Bang

Since our understanding of particle physics is only robust below energies of $\sim 1\,\mathrm{GeV}$ ($\sim 10^{13}\,\mathrm{K}$), the physics of the very early Universe ($t \lesssim 10^{-6}\,\mathrm{s}$) is still very uncertain. In popular, although speculative, extensions of the standard model for particle physics, this era is characterized by a number of symmetry-breaking phase transitions. Particle physicists have developed a number of models which suggest the existence of many exotic particles as a result of these symmetry breakings, and it is a popular idea that the elusive dark matter consists of one or more of such particle species. However, it should be kept in mind that the theories predicting the existence of these exotic particles are not well established and that there is not yet any convincing, direct experimental evidence for their existence.

For the purpose of the discussion here, the two most important events that (probably) took place during this early period after the Big Bang are inflation and baryogenesis. Inflation is a period of exponential expansion that resulted from a phase transition associated with some unknown scalar field. Inflation is invoked to solve several important problems for the standard Hot Big Bang cosmology, and is described in detail in §3.6. Baryogenesis is a mechanism that is needed to explain the observed asymmetry between baryons and antibaryons: one does not observe a significant abundance of antibaryons. If they were there, their continuous annihilation with baryons would produce a much greater gamma-ray background than observed, unless they are spatially segregated from the baryons, which is extremely contrived. Apparently, the Universe has a non-zero baryon number. If baryon number is conserved, this asymmetry between baryons and antibaryons must have originated at very early times through a process called baryogenesis. The details of this process are still poorly understood, and will not be discussed in this book (see Kolb & Turner, 1990).

In what follows we give a brief overview of some of the most important events that took place in the early Universe after it had cooled down to a temperature of $\sim 10^{13}$ K. At this point in time, the temperature of the Universe was still higher than the binding energy of hadrons (baryons and mesons). Quarks were not yet bound into hadronic states. Instead, the matter in the Universe was in a form referred to as quark soup, which consists of quarks, leptons and photons.

- At $T \sim 3 \times 10^{12}$ K ($t \sim 10^{-5}$ s), corresponding to an energy of 200–300 MeV, the quark–hadron phase transition occurs, confining quarks into hadrons. If the phase transition was strongly first order, it may have induced significant inhomogeneities in the baryon-to-photon ratio, and affected the later formation of elements, a topic discussed further in §3.4. Once the transition was complete, the Universe was filled with a hot plasma consisting of three types of (relativistic) pions (π^+, π^-, π^0), (non-relativistic) nucleons (protons, p, and neutrons, n), charged leptons (e, $\bar{e}$, μ, $\bar{\mu}$; the τ and $\bar{\tau}$ have already annihilated), their associated neutrinos (ν_e, $\bar{\nu}_e$, ν_μ, $\bar{\nu}_\mu$), and photons, all in thermal equilibrium. In addition, the Universe comprises several decoupled species, such as the tau-neutrinos (ν_τ and $\bar{\nu}_\tau$) – their coupling has to be through their reactions with τ and $\bar{\tau}$, and possible exotic particles that make up the (non-baryonic) dark matter.
- At $T \sim 10^{12}$ K ($t \sim 10^{-4}$ s) the (π^+, π^-) pairs annihilate while the neutral pions π^0 decay into photons. From this point on the nucleons (and a small abundance of their antiparticles which escaped annihilation) are the only hadronic species left. At around the same time, the muons start to annihilate, and their number density becomes negligibly small as T drops to about 10^{11} K. At this time, ν_μ and $\bar{\nu}_\mu$ also decouple from the hot plasma, and expand freely with the Universe.
- When T drops below 10^{11} K, the number of neutrons becomes smaller than that of protons by a factor of about $\exp(-\Delta m/T)$, where $\Delta m \approx 1.3$ MeV is the mass difference between a neutron and a proton. This asymmetry in the numbers of n and p continues to grow until the reaction rate between neutrons and protons becomes negligible.
- At $T \sim 5 \times 10^9$ K ($t \sim 4$ s), the annihilations of (e, $\bar{e}$) pairs begins. As the number density of (e, $\bar{e}$) pairs drops, ν_e and $\bar{\nu}_e$ decouple from the hot plasma. Since the (e, $\bar{e}$) annihilations heat the photons but not the decoupled neutrinos, the neutrinos expand freely with a temperature that is lower than that of the photons. Because of the reduction in the number of (e, $\bar{e}$) pairs and the cooling of ν_e and $\bar{\nu}_e$, reactions such as $n + \nu_e \leftrightarrow p + e$ and $n + \bar{e} \leftrightarrow p + \bar{\nu}_e$ are no longer effective. Consequently, the n/p ratio freezes out at a value of about $\exp(-\Delta m/T) \sim 1/10$. Note that this ratio does not change much due to beta decay of the neutrons, because the half-time of the decay (about 10 minutes) is much longer than the age of the Universe at this time.
- At $T \sim 10^9$ K ($t \sim$ few minutes), nucleosynthesis starts, synthesizing protons and neutrons to produce D, He and a few other elements. Since the temperature is still too high for the formation of neutral atoms, all these elements are highly ionized. Consequently, the Universe is now filled with freely expanding neutrinos (and possibly exotic particles) and a plasma of electrons and highly ionized atoms (mainly protons and He^{++}). However, as the temperature continues to decrease, electrons start to combine with the ions to produce neutral atoms.
- At $T \sim 4000$ K ($t \sim 2 \times 10^5$ yr) roughly 50% of the baryonic matter is in the form of neutral atoms. This point in time is often called the time of recombination. Because of the resulting drop in the number density of free electrons, the Universe suddenly becomes transparent to photons. These photons are observed today as the cosmic microwave background. From this point on, photons, neutrinos, H, He and other atoms all expand freely with the Universe. At around the same time, the energy density in relativistic particles has become smaller than that in the rest mass of non-relativistic matter, and the Universe enters the matter dominated epoch.

Once the processes involved are known from particle, nuclear and atomic physics, it is in principle straightforward to calculate the matter content at different epochs summarized above. A detailed treatment of such calculations is beyond the scope of this book, and can be found in Börner (2003) and Kolb & Turner (1990), for example. In what follows, we present a brief discussion about the basic principles involved and their applications to some important examples.

3.3.2 Particles in Thermal Equilibrium

As discussed above, at any given epoch, some particles are in thermal equilibrium with the hot plasma, some in free expansion with the Universe, and others are in transition between the two states. The number density n, energy density ρ, and pressure P of a given particle species can be written in terms of its distribution function $f(\mathbf{x},\mathbf{p},t)$. Since the Universe is homogeneous and isotropic $f(\mathbf{x},\mathbf{p},t) = f(p,t)$, with $p = |\mathbf{p}|$, so that

$$n(t) = 4\pi \int f(p,t)\, p^2\, dp, \tag{3.121}$$

$$\rho(t) = 4\pi \int E(p) f(p,t)\, p^2\, dp, \tag{3.122}$$

$$P(t) = 4\pi \int \frac{p^2}{3E(p)} f(p,t)\, p^2\, dp, \tag{3.123}$$

where the energy E is related to the momentum p as $E(p) = (p^2 + m^2)^{1/2}$. Eq. (3.123) follows from kinetic theory, according to which the pressure is related to momentum and velocity as $P = \frac{1}{3} n \langle pv \rangle$. Using the components of the four-momentum, we have $v = pc^2/E$, so that $P = n \langle p^2 c^2 / 3E \rangle$.

For a particle species in thermal equilibrium

$$f(\mathbf{p},t)\,d^3\mathbf{p} = \frac{g}{(2\pi)^3} \left\{ \exp\left[\frac{E(p) - \mu}{T(t)}\right] \pm 1 \right\}^{-1} d^3\mathbf{p}, \tag{3.124}$$

where μ is the chemical potential of the species, and $T(t)$ is its temperature at time t. The signature, $\pm$, takes the positive sign for Fermi–Dirac species and the negative sign for Bose–Einstein species. The factor $1/(2\pi)^3$ is due to Heisenberg's uncertainty principle, which states that no particle can be localized in a phase-space volume smaller than the fundamental element $(2\pi\hbar)^3$ (recall that we use $\hbar = c = k_B = 1$), and g is the spin-degeneracy factor (neutrinos have $g = 1$, photons and charged leptons have $g = 2$, and quarks have $g = 6$).

Substituting Eq. (3.124) in Eqs. (3.121)–(3.123) yields

$$n_{\rm eq} = \frac{g}{2\pi^2} \int_m^\infty \frac{(E^2 - m^2)^{1/2} E\, dE}{\exp[(E-\mu)/T] \pm 1}; \tag{3.125}$$

$$\rho_{\rm eq} = \frac{g}{2\pi^2} \int_m^\infty \frac{(E^2 - m^2)^{1/2} E^2\, dE}{\exp[(E-\mu)/T] \pm 1}; \tag{3.126}$$

$$P_{\rm eq} = \frac{g}{6\pi^2} \int_m^\infty \frac{(E^2 - m^2)^{3/2}\, dE}{\exp[(E-\mu)/T] \pm 1}. \tag{3.127}$$

Let us consider two special cases. In the non-relativistic limit, i.e. when $T \ll m$, the number density, the energy density and pressure are the same for both Bose–Einstein and Fermi–Dirac species, and can be written in the following analytic forms:

$$n_{\rm eq} = g \left(\frac{mT}{2\pi}\right)^{3/2} e^{(\mu-m)/T}, \tag{3.128}$$

$$\rho_{\rm eq} = nm, \quad P_{\rm eq} = nT. \tag{3.129}$$

For a relativistic ($T \gg m$ and $E = p$), non-degenerate ($\mu \ll T$) gas, the corresponding analytical expressions are

$$n_{\rm eq} = \begin{cases} [\zeta(3)/\pi^2] gT^3 & \text{(Bose–Einstein)} \\ (3/4)[\zeta(3)/\pi^2] gT^3 & \text{(Fermi–Dirac),} \end{cases} \tag{3.130}$$

$$\rho_{\rm eq} = \begin{cases} (\pi^2/30) gT^4 & \text{(Bose–Einstein)} \\ (7/8)(\pi^2/30) gT^4 & \text{(Fermi–Dirac).} \end{cases} \tag{3.131}$$

$$P_{\rm eq} = \rho_{\rm eq}/3, \tag{3.132}$$

where $\zeta(3) \approx 1.2021...$ is the Riemann zeta function of 3.

In general, in order to use Eqs. (3.125)–(3.127) to calculate the density and pressure, one needs to know the chemical potential μ. The principle for determining the chemical potential of a species is that chemical potential is an additive quantity which is conserved during a 'chemical' reaction (e.g. Landau & Lifshitz, 1959). Thus, if species 'i' takes part in a reaction like $i + j \leftrightarrow k + l$, then $\mu_i + \mu_j = \mu_k + \mu_l$. The values of the chemical potentials therefore depend on the various conservation laws under which the various reactions take place. For example, since the number of photons is not a conserved quantity for a thermodynamic system, the chemical potential of photons must be zero. This is consistent with the fact that photons at thermal equilibrium have the Planck distribution. It then follows that the chemical potential for a particle is the negative of that for its antiparticle (because particle–antiparticle pairs can be annihilated to photons). Put differently, the difference in the number density of particles and antiparticles depends only on the chemical potential. Similar to electric charge, particle reactions are thought to generally conserve baryon number (which explains the long lifetime of the proton, of $> 10^{34}$ years) and lepton number. Since the number densities of baryons and leptons are found to be (or, in the case of leptons, believed to be) much smaller than the number density of photons, the chemical potential of all species may be set to zero to good approximation in computing the mean energy density and pressure in the early Universe.

There is one caveat, however. Since the chemical potential of a particle is the negative of that of its antiparticle, it follows from Eq. (3.121) that, for fermions, their difference in number densities is given by

$$n - \bar{n} = \frac{gT^3}{6\pi^2}\left[\pi^2\left(\frac{\mu}{T}\right) + \left(\frac{\mu}{T}\right)^3\right]. \tag{3.133}$$

When the Universe cools to temperatures below the rest mass of the particles, all particle–antiparticle pairs will be annihilated[5] leaving only this small excess, which is zero when $\mu = 0$. Therefore, the fact that we do have non-zero baryon and lepton number densities in the Universe today implies that μ cannot have been strictly zero at all times. In the early Universe, some physics must have occurred that did not conserve baryon number or lepton number, and that resulted in the present-day number densities of protons and electrons. The actual physics of this

---

[5] In principle, because of the expansion of the Universe, tiny fractions of particles and antiparticles may survive, but their number densities are negligibly small.

baryon- and lepton-genesis are poorly understood, and will not be discussed further in this book. Detailed descriptions can be found in Kolb & Turner (1990).

With all chemical potentials set to zero, it is evident from Eqs. (3.125) and (3.130) that the number density of non-relativistic particles is suppressed exponentially with respect to that of relativistic species. This reflects the coupling to the photon fluid. When $T \gg m$ the photons have sufficient energy to create a thermal background number density of particle–antiparticle pairs. However, when $T \ll m$ only an exponential tail of the photon distribution function has sufficient energy for pair creation, causing a similar suppression of their number density. Consequently, particles in thermal equilibrium with the photon gas can only contribute significantly to the energy density and pressure when they are relativistic. Thus, to good accuracy, we can write the total energy density, number density and pressure of the Universe, in the radiation dominated era, as

$$\rho(T) = \frac{\pi^2}{30} g_* T^4, \quad n(T) = \frac{\zeta(3)}{\pi^2} g_{*,n} T^3, \quad P(T) = \rho(T)/3, \qquad (3.134)$$

with

$$g_* = \sum_{i \in \text{Boson}} g_i \left(\frac{T_i}{T}\right)^4 + \frac{7}{8} \sum_{i \in \text{Fermion}} g_i \left(\frac{T_i}{T}\right)^4, \qquad (3.135)$$

$$g_{*,n} = \sum_{i \in \text{Boson}} g_i \left(\frac{T_i}{T}\right)^3 + \frac{3}{4} \sum_{i \in \text{Fermion}} g_i \left(\frac{T_i}{T}\right)^3. \qquad (3.136)$$

Note that we have included the possibility that the temperature of a species T_i may be different from that of the radiation background T. The values of g_* and $g_{*,n}$ at a given time can be calculated once the existing relativistic species are identified. For example, at $T \ll 1\,\text{MeV}$, the only relativistic species are photons at temperature T and three species of neutrinos and their antiparticles (all assumed to be massless) at temperature $T_\nu = (4/11)^{1/3} T$ (as we will see in §3.3.3). Therefore $g_* = g_\gamma + (7/8)(3 \times 2 \times g_\nu)(T_\nu/T)^4 \approx 3.36$. At higher T (earlier times) more species are relativistic, so that the degeneracy factors are larger. Fig. 3.4 shows g_* as a function of T obtained from the standard model of particle physics. It increases from 3.36 at the present-day temperature of $2.73\,\text{K}$ to 106.75 at $T \gtrsim 300\,\text{GeV}$.

3.3.3 Entropy

An important thermodynamic quantity for describing the early Universe is the entropy $S = S(V,T)$. If we continue to ignore the chemical potential, the second law of thermodynamics, as applied to a comoving volume $V \propto a^3(t)$, states that

$$dS(V,T) = \frac{1}{T} \{ d[\rho(T)V] + P(T)\,dV \}, \qquad (3.137)$$

where ρ is the equilibrium energy density of the gas.

Alternatively, we can write the differential of S in terms of its general form

$$dS(V,T) = \frac{\partial S}{\partial V} dV + \frac{\partial S}{\partial T} dT. \qquad (3.138)$$

Using Eq. (3.137) to identify the two partial derivatives, the integrability condition,

$$\frac{\partial^2 S}{\partial T \partial V} = \frac{\partial^2 S}{\partial V \partial T}, \qquad (3.139)$$

yields

$$\frac{dP}{dT} = \frac{\rho(T) + P(T)}{T}. \qquad (3.140)$$

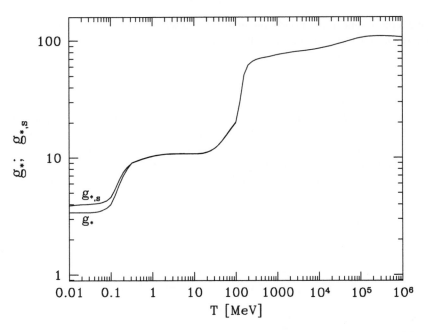

Fig. 3.4. The statistical weights g_* and $g_{*,s}$ as functions of temperature, T, in the standard $SU(3) \times SU(2) \times U(1)$ model of particle physics.

Inserting this in Eq. (3.137) we obtain

$$dS(V,T) = \frac{1}{T} d\{[\rho(T)+P(T)]V\} - \frac{V}{T^2}[\rho(T)+P(T)]dT, \qquad (3.141)$$

which may be integrated to show that, up to an additive constant, the entropy density, $s(T) \equiv S(V,T)/V$, is given by

$$s(T) = \frac{\rho(T)+P(T)}{T}. \qquad (3.142)$$

It is easy to show with the use of Eqs. (3.35) and (3.140) that

$$\frac{dS}{da} \propto \frac{d(sa^3)}{da} = 0, \qquad (3.143)$$

which is the 'entropy conservation law', owing to the adiabaticity of the universal expansion (see §3.1.5).

Using Eqs. (3.131) and (3.132) the entropy density for non-degenerate, relativistic particles in thermal equilibrium is

$$s_{eq}(T) = \frac{2\pi^2}{45} g T^3. \qquad (3.144)$$

The entropy density of non-relativistic particles in thermal equilibrium with the photon fluid can be expressed in terms of the entropy density of photons, $s_\gamma(T)$, as

$$\frac{s_{eq}(T)}{s_\gamma(T)} = \frac{3}{4} \frac{\rho(T)}{\rho_\gamma(T)} \left(1+\frac{P}{\rho}\right). \qquad (3.145)$$

Since $\rho \ll \rho_\gamma$,[6] the contribution of non-relativistic particles to the total entropy density is negligible. To good accuracy, therefore, the total entropy density of the Universe is obtained by summing over all relativistic species:

$$s(T) = \frac{2\pi^2}{45} g_{*,s} T^3 \qquad (3.146)$$

with

$$g_{*,s} = \sum_{i \in \text{Boson}} g_i \left(\frac{T_i}{T}\right)^3 + \frac{7}{8} \sum_{i \in \text{Fermion}} g_i \left(\frac{T_i}{T}\right)^3. \qquad (3.147)$$

Combining Eq. (3.146) with the entropy conservation law we see that $g_{*,s} T^3 a^3$ is a conserved quantity, so that

$$g_{*,s}^{1/3}(T) T \propto a^{-1}. \qquad (3.148)$$

Therefore, as long as $g_{*,s}$ remains constant, $T \propto a^{-1}$, consistent with the thermodynamic derivation in §3.1.5. However, as the Universe cools, every now and then particle species become non-relativistic and stop contributing (significantly) to the entropy density of the Universe. Their entropy is transferred to the remaining relativistic particle species, causing T to decrease somewhat slower. An interesting application of this is the decoupling of light neutrinos. Although neutrinos do not couple directly to the photons, they can maintain thermal equilibrium via weak reactions such as $e + \bar{e} \leftrightarrow \nu_e + \bar{\nu}_e$, etc. At a freeze-out temperature of $T_f \sim 1\,\text{MeV}$ the interaction rate for these reactions drops below the expansion rate of the Universe, and the neutrinos decouple from the photon fluid. From this point on, their temperature will decrease strictly as $T_\nu \propto a^{-1}$, while the photon temperature, T_γ, obeys Eq. (3.148). Since the neutrinos are relativistic both before and after decoupling, their freeze-out leaves $g_{*,s}$ invariant. Consequently, despite being decoupled, the temperature of the neutrinos remains exactly the same as that of the photons. This changes a little time later, when the temperature has dropped to $T \sim 0.51\,\text{MeV}$ and electrons start to annihilate and freeze-out from the photon fluid. The entropy released in this process is given to the photons, but not to the decoupled neutrinos (who conserve their entropy density separately). Consequently, after electron annihilation, $T_\gamma > T_\nu$. Their ratio follows from the entropy conservation law, according to

$$\frac{T_{\gamma,\text{after}}}{T_{\nu,\text{after}}} = \frac{T_{\gamma,\text{after}}}{T_{\gamma,\text{before}}} = \left[\frac{g_{*,s}(T_\text{before})}{g_{*,s}(T_\text{after})}\right]^{1/3}, \qquad (3.149)$$

where we have used that $T_{\nu,\text{after}} = T_{\nu,\text{before}} = T_{\gamma,\text{before}}$. Before electron annihilation, the relativistic species in the Universe are photons, electrons, positrons, and three flavors of neutrinos with their antiparticles, all at the same temperature. Therefore, $g_{*,s}(T_\text{before}) = g_\gamma + (7/8)(g_e + g_{\bar{e}} + 3g_\nu + 3g_{\bar{\nu}}) = 2 + (7/8)(2 + 2 + 3 + 3) = 43/4$. After electron annihilation, $g_{*,s}(T_\text{after}) = g_\gamma + (7/8)(3g_\nu + 3g_{\bar{\nu}})(T_{\nu,\text{after}}/T_{\gamma,\text{after}})^3$. Substitution of these degeneracy parameters into Eq. (3.149) yields

$$T_{\nu,\text{after}} = \left(\frac{4}{11}\right)^{1/3} T_{\gamma,\text{after}}. \qquad (3.150)$$

It is thus expected that the present-day Universe contains a relic neutrino background with a temperature of $T_{\nu,0} \simeq 0.71 \times 2.73\,\text{K} = 1.95\,\text{K}$. This difference in the temperature of the two relativistic species (neutrinos and photons) is also apparent from Fig. 3.4. At $T \gtrsim 0.5\,\text{MeV}$, $g_{*,s}(T)$ is identical to g_*, indicating that all relativistic particle species have a common temperature. At

---

[6] The rest-mass density of particles should not be included as part of the equilibrium energy density of the gas, because there is no creation or annihilation of particles.

lower temperatures, however, electron annihilation has increased T_γ with respect to T_ν, causing an offset of $g_{*,s}$ with respect to g_*.

3.3.4 Distribution Functions of Decoupled Particle Species

In §3.3.2 we discussed the distribution functions of particles in thermal equilibrium. We now turn our attention to species that have dropped out of thermal equilibrium, and have decoupled from the hot plasma. If particle species i decoupled at a time t_f, where the subscript 'f' stands for 'freeze-out', its temperature is approximately equal to the photon temperature at that time, i.e. $T \approx T_f \equiv T_\gamma(t_f)$. After decoupling, the mean interaction rate of the particle drops below the expansion rate, and the particle basically moves on a geodesic. As we have seen in §3.1.4, the momentum of the particle then scales as $p \propto a^{-1}$, which is valid for both relativistic and non-relativistic species. Since the *relative* momenta are conserved, the actual distribution function at $t > t_f$ can be written as

$$f(\mathbf{p},t) = f\left(\mathbf{p}\frac{a(t)}{a(t_f)}, t_f\right). \tag{3.151}$$

In other words, the form of the distribution function is 'frozen-in' the moment the particles decouple from the hot plasma.

If a species is still relativistic after decoupling, we have $E = p$, so that

$$f(\mathbf{p},t)d^3\mathbf{p} = \frac{g}{(2\pi)^3}\left\{\exp\left[\frac{pa(t)}{T_f a(t_f)}\right] \pm 1\right\}^{-1} d^3\mathbf{p}. \tag{3.152}$$

Thus, the distribution function of a decoupled, relativistic species is self-similar to that of a relativistic species in thermal equilibrium, but with a temperature

$$T = T_f \frac{a(t_f)}{a(t)}. \tag{3.153}$$

Note that this differs from the temperature scaling of species still in thermal equilibrium, which is instead given by Eq. (3.148). As we discussed in §3.3.3 this explains why the present-day temperature of the neutrino background is lower than that of the CMB.

If the species is already non-relativistic when it decouples, its energy is given by $E = m + p^2/2m$. Since for non-relativistic species we can ignore the ± 1 term, the distribution function is given by

$$f(\mathbf{p},t)d^3\mathbf{p} = \frac{g}{(2\pi)^3}\exp\left[-\frac{m}{T_f}\right]\exp\left[-\frac{p^2}{2mT}\right] d^3\mathbf{p} \tag{3.154}$$

with

$$T = T_f\left[\frac{a(t_f)}{a(t)}\right]^2. \tag{3.155}$$

Note that Eq. (3.154) is a Maxwell–Boltzmann distribution, and that the temperature scales as expected from kinetic theory (see §3.1.5).

As is immediately evident from substituting Eq. (3.151) in Eq. (3.121), the number density of decoupled particles (both relativistic and non-relativistic) is given by

$$n(t) = \left[\frac{a(t_f)}{a(t)}\right]^3 n_{eq}(t_f), \tag{3.156}$$

so that $n \propto a^{-3}$, as expected. For relativistic species, we can contrast this number density against that of the photons:

$$\frac{n(t)}{n_\gamma(t)} = \frac{g_{\text{eff}}}{2} \left(\frac{T_{\text{f}}}{T}\right)^3 \left[\frac{a(t_{\text{f}})}{a(t)}\right]^3 = \frac{g_{\text{eff}}}{2} \frac{g_{*,s}(T)}{g_{*,s}(T_{\text{f}})} \qquad (3.157)$$

with $g_{\text{eff}} = g$ for bosons and $g_{\text{eff}} = (3/4)g$ for fermions, where we have used that the photon temperature, T, scales as in Eq. (3.148). This illustrates that the number density of any relic background of relativistic particles is comparable to the number density of photons. Note that Eq. (3.156) remains valid even if the particles become non-relativistic some time after decoupling.

3.3.5 The Freeze-Out of Stable Particles

Having discussed the distribution functions of particles before and after decoupling, we now turn to discuss the actual process by which a species decouples ('freezes out') from the hot plasma. We first consider cases where the particles involved are stable (i.e. their half-time of decay is much longer than the age of the Universe), and derive their relic abundances. We distinguish between 'hot' relics, which correspond to species that decouple in the relativistic regime, and 'cold' relics, whose decoupling takes place when the particles have already become non-relativistic.

The evolution of the particle number density is governed by the Boltzmann equation, which, for a given species 'i', can be written as

$$\frac{\mathrm{d}f_i}{\mathrm{d}t} = C_i[f], \qquad (3.158)$$

where $C_i[f]$ (called the collisional term) describes the change of the distribution function of species 'i' due to the interactions with other species. Since the Universe is homogeneous and isotropic, f_i depends only on the cosmic time, t, and the value of the momentum, $p \propto a^{-1}(t)$. It then follows from Eq. (3.158) that

$$\frac{\partial f_i}{\partial t} - H(t) p \frac{\partial f_i}{\partial p} = C_i[f], \qquad (3.159)$$

where $H = \dot{a}/a$ is the Hubble parameter. Integrating both sides of Eq. (3.159) over momentum space, and using the definition of n_i, we obtain

$$\frac{\mathrm{d}n_i}{\mathrm{d}t} + 3H(t) n_i = \int C_i[f] \, \mathrm{d}^3\mathbf{p}. \qquad (3.160)$$

Here the second term on the left-hand side (often called the Hubble drag term) describes the dilution of the number density due to the expansion of the Universe, while the right-hand side describes the change in number density due to interactions. Note that in the limit $C_i[f] \to 0$ the number density scales as $n_i \propto a^{-3}$, as expected.

In general, the collisional term $C_i[f]$ depends on f_i and on the distribution functions of all other species that interact with 'i'. If the cross-sections of all these interactions are known (from relevant physics), we can obtain the functional form of $C_i[f]$. Species that do not have any channel to interact with 'i' collisionally can still affect the distribution function of 'i' via their contributions to the general expansion of the Universe. Thus, the evolution of the matter content of the Universe is described by a coupled set of Boltzmann equations for all important species in the Universe, which can in principle be solved once the initial conditions are given.

For illustration, consider a case in which species 'i' takes part *only* in the following two-body interactions:

$$i + j \leftrightarrow a + b. \qquad (3.161)$$

If the production and destruction rates of 'i' due to this reaction are $\alpha(T)$ and $\beta(T)$, respectively, then Eq. (3.160) can be written as

$$\frac{dn_i}{dt} + 3H(t)n_i = \alpha(T)n_a n_b - \beta(T)n_i n_j. \tag{3.162}$$

The meaning of this equation is clear: particles of species 'i' are destroyed due to their reactions with species 'j', and are created due to the reactions between species 'a' and 'b'. A similar equation can be written for 'j'. Subtracting these two equations gives $(n_i - n_j)a^3 = $ constant. Now suppose that 'a' and 'b' are in thermal equilibrium with the general hot plasma, so that their distribution functions are given by Eq. (3.124) with $T_a = T_b = T$, while 'i' and 'j' are coupled to the hot plasma through their reactions with 'a' and 'b'. We define an equilibrium density for 'i', $n_{i,\mathrm{eq}}$, and an equilibrium density for 'j', $n_{j,\mathrm{eq}}$, so that

$$\beta(T)n_{i,\mathrm{eq}}n_{j,\mathrm{eq}} = \alpha(T)n_a n_b. \tag{3.163}$$

Thus defined, $n_{i,\mathrm{eq}}$ and $n_{j,\mathrm{eq}}$ are just the number densities of 'i' and 'j' under the assumption that they are in thermal equilibrium with the hot plasma. Consider the case in which 'j' and 'b' are the antiparticles of 'i' and 'a', respectively. As long as the chemical potential of 'i' is small, the number densities of 'i' and 'j' will be virtually identical [see Eq. (3.133)]. In what follows we therefore set $n_i = n_j$, but note that the discussion is easily extended to cases where $n_i \neq n_j$ by using that $(n_i - n_j)a^3 = $ constant. With these definitions, we can write the rate equation (3.162) as

$$\frac{dn_i}{dt} + 3H(t)n_i = \beta(T)(n_{i,\mathrm{eq}}^2 - n_i^2). \tag{3.164}$$

Since the entropy density s is proportional to a^{-3} (see §3.3.3), it is convenient to define both n_i and $n_{i,\mathrm{eq}}$ in units of s:

$$Y_i \equiv \frac{n_i}{s}, \quad Y_{i,\mathrm{eq}} \equiv \frac{n_{i,\mathrm{eq}}}{s}. \tag{3.165}$$

Using $ds/dt = -3Hs$, Eq. (3.164) becomes

$$\frac{dY_i}{dt} = \beta(T)s(T)(Y_{i,\mathrm{eq}}^2 - Y_i^2). \tag{3.166}$$

If we now introduce the dimensionless variable, $x \equiv m_i/T$, and use the fact that, in the radiation dominated era, $t \propto a^2 \propto T^{-2}$ (or $t = t_\mathrm{m} x^2$, where t_m is the cosmic time when $x = 1$), the rate equation can be written in the following form:

$$\frac{x}{Y_{i,\mathrm{eq}}}\frac{dY_i}{dx} = -\frac{\Gamma(x)}{H(x)}\left[\left(\frac{Y_i}{Y_{i,\mathrm{eq}}}\right)^2 - 1\right], \tag{3.167}$$

where $\Gamma(x) \equiv n_{i,\mathrm{eq}}(x)\beta(x)$ and $H = (2t)^{-1} = (2t_\mathrm{m} x^2)^{-1}$ (which follows from $a \propto t^{1/2}$).

Given a particle species' rest mass m_i and its interaction cross-section $\beta(T) = \langle \sigma v \rangle(T)$, thermally averaged over all reactions in which 'i' partakes, the rate equation (3.167) can be solved for $Y_i(x)$ numerically. The initial conditions follow from the fact that for $x \ll 1$ the solution is given by $Y_i = Y_{i,\mathrm{eq}}$. Fig. 3.5 shows the solutions of Y_i thus obtained for different values of β [here assumed to be constant, $\beta(T) = \beta_0$]. A larger interaction cross-section (larger β_0) implies that the species can maintain thermal equilibrium for a longer time. As long as β_0 is such that decoupling occurs in the relativistic regime ($x \ll 1$), the final freeze-out abundance will be comparable to that of the photons [see Eq. (3.157)], and depend very little on the exact value of β_0. For sufficiently large β_0, the particles remain in thermal equilibrium well into the non-relativistic regime ($x \gg 1$), causing an exponential suppression of their final freeze-out abundance. In this regime the relic abundances are extremely sensitive to β, and thus to the exact epoch of decoupling.

3.3 The Production and Survival of Particles

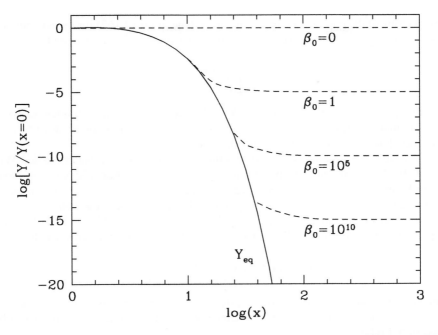

Fig. 3.5. The solution of Eq. (3.167) assuming a constant annihilation cross-section; $\beta = \beta_0$ (dashed curves). The solid curve shows the equilibrium abundance.

In what follows we present a simple, but relatively accurate, estimate of the relic abundances of various particle species. Rather than solving Eq. (3.167), which needs to be done numerically or by other approximate methods, we make the assumption that freeze-out occurs at a temperature T_f, corresponding to x_f, when $\Gamma/H = 1$, and that the relic abundance is simply given by $Y_i(x \to \infty) = Y_{i,eq}(x_f)$. Using Eqs. (3.128) and (3.130) for $n_{i,eq}$, and Eq. (3.146) for s, we have

$$Y_{i,eq}(x) = \begin{cases} (45\zeta(3)/2\pi^4)[g_{i,\text{eff}}/g_{*,s}(x)] & (x \ll 1) \\ (90/(2\pi)^{7/2})[g_i/g_{*,s}(x)]x^{3/2}e^{-x} & (x \gg 1), \end{cases} \quad (3.168)$$

where $g_{i,\text{eff}} = g_i$ for bosons and $g_{i,\text{eff}} = (3/4)g_i$ for fermions. The freeze-out temperature follows from $\Gamma(x_f) = n_{i,eq}(x_f)\beta(x_f) = H(x_f)$. From Eq. (3.61) we have that in the radiation dominated era $H^2(t) = (8\pi G/3)\rho_r(t)$. Substitution of Eq. (3.134) then gives

$$H(x) = \left(\frac{m_i m_{\text{Pl}}}{x}\right)^2 \sqrt{\frac{4\pi^3 g_*(x)}{45}}, \quad (3.169)$$

where $m_{\text{Pl}} = G^{-1/2}$ is the Planck mass in the natural units used here. Our definition of freeze-out then yields

$$x_f = \sqrt{\frac{45}{\pi^7}} \frac{\zeta(3)}{2} \frac{g_{i,\text{eff}}}{\sqrt{g_{*,s}(x_f)}} m_{\text{Pl}} m_i \beta(x_f) \quad (x_f \ll 1);$$

$$x_f^{-1/2} e^{x_f} = \sqrt{\frac{45}{32\pi^6}} \frac{g_i}{\sqrt{g_{*,s}(x_f)}} m_{\text{Pl}} m_i \beta(x_f) \quad (x_f \gg 1). \quad (3.170)$$

Note that since x_f appears on both sides of these equations, they typically need to be solved numerically.

Let us first consider the case of hot relics that have remained relativistic to the present day, i.e. their rest mass $m_i \ll T_0 = 2.4 \times 10^{-4}$ eV. Its energy density follows from Eq. (3.131), which can be written in terms of the photon energy density, as is done in Eq. (3.157) for the number density. Expressing this energy density in terms of the critical density for closure, we obtain

$$\Omega_{i,0}h^2 = \frac{g_{i,\text{eff}}}{2}\left[\frac{g_{*,s}(x)}{g_{*,s}(x_f)}\right]^{4/3}\Omega_{\gamma,0}h^2. \quad (3.171)$$

Since $g_{*,s}(x) \leq g_{*,s}(x_f)$, and since $\Omega_{\gamma,0}h^2 = 2.5 \times 10^{-5}$ [see Eq. (3.65)], we immediately see that a relic particle that is still relativistic today (e.g. zero mass neutrinos) contributes negligibly to the total energy density of the Universe at the present time.

Next we consider the case of weakly interacting massive particles, usually called WIMPs. Examples of WIMPs are massive neutrinos and stable, light supersymmetric particles. Note that WIMPs can be either hot or cold, depending on whether $x_f \ll 1$ or $x_f \gg 1$. The present-day mass density of massive relics is $\rho_{i,0} = m_i Y_{i,\text{eq}}(x_f) s_0$, with s_0 the present-day value of the entropy density. After electron annihilation, $g_{*,s} = 2 + (7/8) \times 3 \times 2 \times 1 \times (4/11) = 3.91$. Substituting this in Eq. (3.146) and using $T_0 = 2.73\,\text{K}$ gives $s_0 = 2{,}906\,\text{cm}^{-3}$. For hot relics, we then obtain

$$\Omega_{i,0}h^2 \approx 7.64 \times 10^{-2}\left[\frac{g_{i,\text{eff}}}{g_{*,s}(x_f)}\right]\left(\frac{m_i}{\text{eV}}\right). \quad (3.172)$$

This abundance depends only very weakly on the exact moment of freeze-out, x_f, reflecting the fact that $Y_i(x)$ is virtually constant for $x \ll 1$. Since $\Omega_0 h^2 \lesssim 1$, we obtain a cosmological bound to the mass of hot relics,

$$m_i \lesssim 13.1\,\text{eV}\left[\frac{g_{*,s}(x_f)}{g_{i,\text{eff}}}\right]. \quad (3.173)$$

For massive neutrinos, $g_{*,s}(x_f) = 43/4$ and $g_{i,\text{eff}} = 6/4$ (assuming $g_i = 2$ to account for antiparticles), the limit is $m_i \lesssim 93.8\,\text{eV}$.

Finally we examine cold WIMPs, which are considered to be candidates for the cold dark matter. Solving Eq. (3.170) for e^{-x_f}, and substituting the result in Eq. (3.168) gives

$$Y_{i,\text{eq}}(x) = \sqrt{\frac{45}{\pi}}\frac{x_f}{\sqrt{g_{*,s}(x_f)}}[m_{\text{pl}} m_i \beta(x_f)]^{-1}. \quad (3.174)$$

Using the present-day entropy density s_0 we obtain a density parameter for cold relics:

$$\Omega_{i,0}h^2 \approx 0.86\frac{x_f}{\sqrt{g_{*,s}(x_f)}}\left[\frac{\beta(x_f)}{10^{10}\,\text{GeV}^{-2}}\right]^{-1}. \quad (3.175)$$

Contrary to the case of hot relics, $\Omega_{i,0}h^2$ now depends strongly on the interaction cross-section, owing to the exponential decrease of $Y_{i,\text{eq}}(x)$ in the non-relativistic regime. As an example, consider a (hypothetical) stable neutrino species with $m_i \gg 1\,\text{MeV}$ but less than $m_Z \sim 100\,\text{GeV}$ (the mass of the Z boson). Because of its large mass, $x_f \gg 1$ and its relic abundance follows from Eq. (3.175). For neutrinos, the annihilation rate can be approximately written as

$$\beta(x) \approx \frac{c_2}{2\pi}G_F^2 m_i^2 x^{-b}, \quad (3.176)$$

with G_F the Fermi coupling constant, and c_2 a constant depending on the type of neutrinos ('Dirac' or 'Majorana'). The value of b is determined by the details of the annihilation processes involved, but is typically of the order unity. Substituting Eq. (3.176) in Eq. (3.175) yields

$$\Omega_{i,0}h^2 \approx \frac{3.95}{c_2}\frac{x_f^{b+1}}{\sqrt{g_{*,s}(x_f)}}\left[\frac{m_i}{\text{GeV}}\right]^{-2}. \quad (3.177)$$

For Dirac-type neutrinos $c_2 \sim 5$ and $b = 0$ (Kolb & Turner, 1990). Taking $g_i = 2$ (to also account for the antiparticles) and $g_{*,s} \sim 60$ at around the time of freeze-out, and solving Eq. (3.170) for x_f gives

$$x_f \approx 17.8 + 3\ln(m_i/\text{GeV}), \qquad (3.178)$$

so that

$$\Omega_{i,0}h^2 \approx 1.82 \left(\frac{m_i}{\text{GeV}}\right)^{-2} \left[1 + 0.17\ln\left(\frac{m_i}{\text{GeV}}\right)\right]. \qquad (3.179)$$

The cosmological bound, $\Omega_0 h^2 \lesssim 1$, to the mass of massive neutrinos is thus

$$m_i \gtrsim 1.4\,\text{GeV}. \qquad (3.180)$$

Note that $\Omega_{i,0}$ *decreases* with increasing particle mass. This reflects the fact that the annihilation cross-section in Eq. (3.176) increases as m_i^2, so that more massive species can stay in thermal equilibrium longer, resulting in a lower freeze-out abundance. The cross-section will not continue to grow as m_i^2 indefinitely, however. For particles with $m_i \gg m_Z \simeq 100\,\text{GeV}$ the cross-section actually *decreases* with particle mass as m_i^{-2}. Using the same argument as above and inserting the appropriate numbers, we find

$$\Omega_{i,0}h^2 \approx \left(\frac{m_i}{3\,\text{TeV}}\right)^2. \qquad (3.181)$$

Therefore, the cosmological bound to the mass of such species is

$$m_i \lesssim 3\,\text{TeV}. \qquad (3.182)$$

Fig. 3.6 summarizes the relation between the WIMP mass (assumed to interact as a Dirac-type neutrino) and its relic contribution to the cosmological density parameter. At $m_\text{wimp} \lesssim \text{MeV}$ the WIMPs produce 'hot' relics for which $\Omega_\text{wimp} h^2 \propto m_\text{wimp}$.[7] At particle masses above $\sim 1\,\text{MeV}$, decoupling occurs in the non-relativistic regime, resulting in 'cold' relics for which $\Omega_\text{wimp} h^2 \propto m_\text{wimp}^{-2}$. Finally, for particle masses above that of the Z boson ($m_\text{wimp} \gtrsim 100\,\text{GeV}$) the scaling changes to $\Omega_\text{wimp} h^2 \propto m_\text{wimp}^2$. Combining these results with observational constraints on the cosmological density parameter ($0.1 \lesssim \Omega_0 h^2 \lesssim 1.0$), we find that there are only three narrow mass ranges of WIMPs allowed, at $\sim 30\,\text{eV}$, $\sim 2\,\text{GeV}$ and $\sim 2\,\text{TeV}$ (see Fig. 3.6). Note, however, that these constraints are only valid under the assumption that the WIMPs have the same interaction cross-sections as neutrinos. Since the nature of the dark matter particles is still unknown, there are large uncertainties regarding the possible interaction cross-sections. Consequently, the observational constraints on $\Omega_0 h^2$ currently only constrain the combination of interaction cross-section and WIMP mass, and large ranges of WIMP masses are still allowed.

3.3.6 Decaying Particles

So far we have discussed the freeze-out of stable particles (those with a lifetime much larger than the age of the Universe) and their cosmological consequences. For unstable particles, the situation is different. In particular, if massive particles decay into photons and other relativistic particles, they will release energy into the Universe, and depending on how effectively this energy is thermalized, the decay may produce a radiation background, increasing the entropy of the Universe. Consider a heavy particle, 'h', with mass m_h and with a mean lifetime τ_h, which decays

---

[7] When their mass is this low, one normally would not speak of WIMPs, but of weakly interacting particles instead. For brevity, we also refer to these particles as WIMPs in Fig. 3.6.

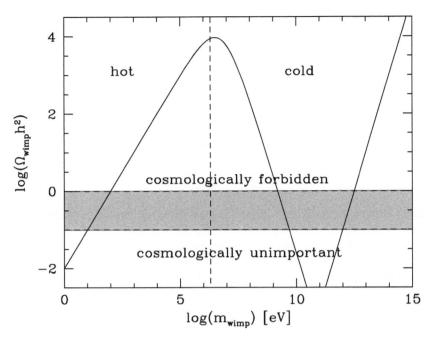

Fig. 3.6. Cosmological constraints on the mass of weakly interactive dark matter particles under the assumption that they interact as a Dirac-type neutrino. The solid curve shows the predicted cosmological density parameter of the WIMPs as a function of WIMP mass, while the shaded area roughly brackets the observed range of the cosmological density parameter. The mass ranges in which the particles make up 'hot' and 'cold' dark matter are indicated.

into light particles while it is non-relativistic. The number of decay events per proper volume at any time t is $n_h(t)/\tau_h$, with $n_h(t)$ given by

$$\frac{\mathrm{d}n_h}{\mathrm{d}t} + 3H(t)n_h = \alpha(T)n_a n_b - \beta(T)n_h n_j - n_h/\tau_h, \qquad (3.183)$$

where, as an example, we assume that 'h' takes part in the reaction $h + j \leftrightarrow a + b$ in addition to the decay. Without implicitly solving Eq. (3.183), we can directly infer the evolution of $n_h(t)$ at two extremes. At early time when the reaction rate ($\sim \beta n_j$) is higher than both the decay rate ($1/\tau_h$) and the expansion rate (H), the species 'h' has the equilibrium abundance, and basically behaves as stable particles. At later times, when the right-hand side of Eq. (3.183) is dominated by decay, it is easy to show that

$$n_h(t) = n_h(t_{\mathrm{D}}) \left[\frac{a(t)}{a(t_{\mathrm{D}})}\right]^{-3} \exp(-t/\tau_h), \qquad (3.184)$$

where t_{D} is the time when the decay becomes more important than other reactions. If the rest mass of the decaying particles is thermalized, then the entropy density per unit comoving volume (see §3.1.5) increases with time as

$$\mathrm{d}S = -\frac{\mathrm{d}\left(n_h m_h a^3\right)}{T} = \frac{\rho_h a^3}{T}\frac{\mathrm{d}t}{\tau_h}, \qquad (3.185)$$

where $\rho_h \equiv m_h n_h$. Using Eqs. (3.134) and (3.146), we have

$$\frac{\mathrm{d}S}{S} = \frac{3}{4}\frac{g_*}{g_{*,s}}\frac{\rho_h}{\rho_{\mathrm{r}}}\frac{\mathrm{d}t}{\tau_h}, \qquad (3.186)$$

where ρ_r is the total energy density in relativistic particles. The entropy of the Universe can therefore be increased significantly if $\rho_h(\tau_h) \gtrsim \rho_r(\tau_h)$, i.e. if the Universe is dominated by species 'h' at the time of decay. Since $\rho_h \propto a^{-3}$ while $\rho_r \propto a^{-4}$, we can define a time of equality for species 'h' by $\rho_r(t_{\text{eq},h}) = \rho_h(t_{\text{eq},h})$, and express the relative increase in ρ_r due to the decay of 'h' in terms of the ratio of $t_{\text{eq},h}$ to the decay time τ_h:

$$\frac{\Delta \rho_r}{\rho_r} = \frac{\rho_h(\tau_h)}{\rho_r(\tau_h)} = \frac{a(\tau_h)}{a(t_{\text{eq},h})} = \left(\frac{\tau_h}{t_{\text{eq},h}}\right)^{2/3}. \tag{3.187}$$

Therefore, any species with a decay time $t_{\text{eq},h} \lesssim \tau_h \lesssim t_0$ can have caused a significant increase of ρ_r. Such an increase can have profound impacts on the evolution of the Universe. If it occurs before radiation–matter equality it may cause a delay in the time t_{eq} when the Universe eventually becomes dominated by matter. Since perturbations cannot grow before the Universe becomes matter dominated, as we will see in the next chapter, such a particle decay can have a significant impact on the development of large-scale structure. An increase in ρ_r also causes the Universe to expand faster in the period $\tau_h \lesssim t \lesssim t_{\text{eq}}$, affecting the production of other particle species during that era. For example, as we will see in the next section, the abundance of helium can be significantly affected if the decay occurs before primordial nucleosynthesis.

If the decay product contains photons, there are additional stringent limits on the mass and lifetime of the decaying particle. If the lifetime were comparable to the present age of the Universe, we would observe a strong radiation background in X-ray and gamma-ray produced by the decay. The lack of such background requires that either $\tau_h \gg t_0$ (i.e. the particle is almost stable) or that the decay occurs at a time when the Universe is still opaque to high-energy photons (so that they can be down-graded by scattering with matter). Another stringent constraint comes from the fact that the observed CMB has a blackbody spectrum to a very high degree of accuracy. This requires the decay occur at a time when high-energy photons can be effectively thermalized (see §3.5).

3.4 Primordial Nucleosynthesis

We all know that the Universe contains not only hydrogen (whose nuclei are single protons) but also heavier elements like helium, lithium, etc. An important question is therefore how these heavier elements were synthesized. Since nuclear reactions are known to be taking place in stars – for example, the luminosity of the Sun is powered mainly by the burning of hydrogen into helium – one possibility is that all heavier elements are synthesized in stars. However, the observed mass fraction of helium is roughly a constant everywhere in the Universe, suggesting that most of the helium is in fact primordial. In this section we examine how nucleosynthesis proceeds in the early Universe.

3.4.1 Initial Conditions

All nuclei are built up of protons and neutrons. Before we explore the nuclear reactions that synthesize deuterium, helium, lithium, etc., we therefore examine the abundances of their building blocks. Protons and neutrons have a very comparable rest mass of $\sim 940\,\text{MeV}$, which implies that they become non-relativistic at very early times ($t \simeq 10^{-6}\,\text{s}$, $T \simeq 10^{13}\,\text{K}$). Down to a temperature of $\sim 0.8\,\text{MeV}$ they maintain thermal equilibrium through weak interactions like

$$\text{p} + \text{e} \leftrightarrow \text{n} + \nu_e, \quad \text{n} + \bar{\text{e}} \leftrightarrow \text{p} + \bar{\nu}_e. \tag{3.188}$$

In thermal equilibrium, their number densities follow from Eq. (3.128):

$$n_{n,p} = 2 \left(\frac{m_{n,p} T}{2\pi}\right)^{3/2} \exp\left[-\frac{m_{n,p} - \mu_{n,p}}{T}\right], \tag{3.189}$$

where we have used that both protons and neutrons have two helicity states ($g_n = g_p = 2$). Writing the mass difference as $Q \equiv m_n - m_p = 1.294\,\text{MeV}$, and using that $m_n/m_p \simeq 1$, we obtain the ratio between the number densities of protons and neutrons in thermal equilibrium:

$$\frac{n_n}{n_p} = \exp\left(-\frac{Q}{T} + \frac{\mu_n - \mu_p}{T}\right) \approx \exp\left(-\frac{Q}{T}\right), \tag{3.190}$$

where $\mu_n - \mu_p = \mu_e - \mu_\nu \approx 0$ (see §3.3). When $T \gg 10^{10}\,\text{K}$ the reactions (3.188) go equally fast in both directions and there are as many protons as neutrons. When the temperature decreases towards $\sim 1\,\text{MeV}$, however, the number density of neutrons starts to drop with respect to that of protons, because neutron is slightly more massive. If thermal equilibrium were to be maintained, the ratio would continue to decrease to very small values. However, as we have seen in §3.3.3, at about the same temperature of $\sim 1\,\text{MeV}$, neutrinos start to decouple. Therefore, the rate of the weak reactions (3.188) is no longer fast enough to establish thermal equilibrium against the expansion rate of the Universe, and the ratio n_n/n_p will eventually 'freeze out' at a value of $\sim \exp(-1.294/0.8) \sim 0.2$. However, neutrons are unstable to beta decay,

$$n \to p + e + \overline{\nu}_e, \tag{3.191}$$

so that even after freeze-out the neutron-to-proton ratio continuous to decrease. If we define the neutron abundance as

$$X_n \equiv \frac{n_n}{n_n + n_p}, \tag{3.192}$$

then it evolves due to the neutron decay as

$$X_n \propto \exp\left[-\frac{t}{\tau_n}\right], \tag{3.193}$$

where $\tau_n = (887 \pm 2)\,\text{s}$ is the mean lifetime of neutrons. The main reason that the present-day Universe contains a large abundance of neutrons is that, shortly before the Universe reaches an age $t = \tau_n$, most neutrons have already ended up in helium nuclei (which stabilizes them against beta decay due to Pauli's exclusion principle) through the process of nucleosynthesis to be described below.

3.4.2 Nuclear Reactions

Nuclei can form in abundant amounts as soon as the temperature of the Universe has cooled down to temperatures corresponding to their binding energy, and the number densities of protons and neutrons are sufficiently high. For a (non-relativistic) species with mass number A and charge number Z [such a species will be called A(Z), and contains Z protons and $A - Z$ neutrons], the equilibrium number density can be obtained from Eq. (3.128):

$$n_A = g_A \left(\frac{m_A T}{2\pi}\right)^{3/2} \exp\left(-\frac{m_A - \mu_A}{T}\right). \tag{3.194}$$

The chemical potential μ_A is related to those of protons and neutrons as

$$\mu_A = Z\mu_p + (A - Z)\mu_n, \tag{3.195}$$

which allows us to rewrite Eq. (3.194) as

$$n_A = g_A \left(\frac{m_A T}{2\pi}\right)^{3/2} \exp\left(-\frac{m}{T}\right) \left[\exp\left(\frac{\mu_p}{T}\right)\right]^Z \left[\exp\left(\frac{\mu_n}{T}\right)\right]^{(A-Z)}. \quad (3.196)$$

Writing $\exp(\mu_p/T)$ and $\exp(\mu_n/T)$ in terms of the proton and neutron mass densities given by Eq. (3.189), respectively, and defining the nucleon mass $m_N \equiv m_A/A \approx m_n \approx m_p$, we obtain

$$n_A = \frac{g_A A^{3/2}}{2^A} n_p^Z n_n^{A-Z} \left(\frac{m_N T}{2\pi}\right)^{3(1-A)/2} \exp\left(\frac{B_A}{T}\right), \quad (3.197)$$

where

$$B_A \equiv Z m_p + (A-Z) m_n - m_A \quad (3.198)$$

is the binding energy of the species A(Z). Next we define the 'mass fraction' or 'abundance' of nucleus A as

$$X_A \equiv \frac{A n_A}{n_b}. \quad (3.199)$$

Here $n_b \equiv n_n + n_p + \sum_i A_i n_{A,i}$ is the number density of baryons in the Universe, with the summation over all nuclear species so that $\sum_i X_{A,i} = 1$. Substituting Eq. (3.197) in Eq. (3.199) we obtain

$$X_A = \frac{g_A}{2} A^{5/2} \left[\frac{4\zeta(3)}{\sqrt{2\pi}}\right]^{A-1} X_p^Z X_n^{A-Z} \eta^{A-1} \left(\frac{m_N}{T}\right)^{3(1-A)/2} \exp\left(\frac{B_A}{T}\right), \quad (3.200)$$

where $\eta \equiv n_b/n_\gamma$ is the present-day baryon-to-photon ratio. Since $n_\gamma = [2\zeta(3)/\pi^2] T^3$ [see Eq. (3.130)], and $T_0 = 2.73$ K, we have

$$\eta \equiv n_b/n_\gamma \approx 2.72 \times 10^{-8} \Omega_{b,0} h^2, \quad (3.201)$$

where $\Omega_{b,0}$ is the present-day baryon density in terms of the critical density for closure. Eq. (3.200) reveals that species A(Z), with $A > 1$, can only be produced in appreciable amounts once the temperature has dropped to a value T_A given by

$$T_A \sim \frac{|B_A|}{(A-1)\left[|\ln \eta| + \frac{3}{2}\ln(m_N/T)\right]}. \quad (3.202)$$

The binding energies of the lightest nuclei, such as deuterium and helium, are all of the order of a few MeV, corresponding to a temperatures of a few $\times 10^{10}$ K. However, because of the small number of baryons per photon ($10^{-10} \lesssim \eta \lesssim 10^{-9}$), or, in other words, the high entropy per baryon, their synthesis has to wait until the Universe has cooled down to temperatures of the order of $(1 \to 3) \times 10^9$ K.

At such low temperatures, however, the number densities of protons and neutrons are already much too low to form heavy elements by direct many-body reactions, such as $2n + 2p \to {}^4\text{He}$. Therefore, nucleosynthesis must proceed through a chain of two-body reactions. The dominant reactions in this chain are:

$$p(n,\gamma)D \quad (3.203)$$

$$D(n,\gamma)^3H, \quad D(D,p)^3H \quad (3.204)$$

$$D(p,\gamma)^3\text{He}, \quad D(D,n)^3\text{He}, \quad {}^3H(p,n)^3\text{He}, \quad {}^3H(\,,e\bar{\nu}_e)^3\text{He} \quad (3.205)$$

$$^3H(p,\gamma)^4\text{He}, \quad {}^3H(D,n)^4\text{He}, \quad {}^3\text{He}(n,\gamma)^4\text{He}, \quad {}^3\text{He}(D,p)^4\text{He}, \quad (3.206)$$

$$2^3\text{He}(\,,2p)^4\text{He}, \quad {}^7\text{Li}(p,\,)2^4\text{He} \quad (3.207)$$

$$^4\text{He}(^3H,\gamma)^7\text{Li}, \quad {}^4\text{He}(^3\text{He},\gamma)^7\text{Be}, \quad {}^7\text{Be}(e,\nu_e)^7\text{Li}. \quad (3.208)$$

Here the notation $X(a,b)Y$ indicates a reaction of the form $X + a \to Y + b$. Since the cross-sections for almost all these reactions are accurately known, the reaction network can be integrated numerically to compute the final abundances of all elements.

Note that the reaction network does not produce any elements heavier than lithium. This is a consequence of the fact that there are no stable nuclei with atomic weight 5 or 8. Since direct many-body reactions at earlier epoch are very inefficient in producing heavy elements, we can conclude that elements heavier than lithium are not produced by primordial nucleosynthesis. Indeed, as we will see in Chapter 10, heavy elements can be synthesized in stars where the density of helium is so high that a short-lived ^{8}Be, formed through ^{4}He–^{4}He collisions, can quickly capture another ^{4}He to form a stable carbon nucleus (^{12}C), thus allowing further nuclear reactions to proceed.

Inspection of the reaction network of primordial nucleosynthesis given above reveals that it can only proceed if the first step, the production of deuterium, is sufficiently efficient. Since deuterium has the lowest binding energy of all nuclei in the network, its production serves as a 'bottleneck' to get nucleosynthesis started. The production of deuterium through p(n, γ)D has a rate per free neutron given by

$$\Gamma = (4.55 \times 10^{-20}\,\mathrm{cm^3\,s^{-1}}) n_p$$
$$\approx 2.9 \times 10^4 X_p \Omega_{b,0} h^2 \left(\frac{T}{10^{10}\,\mathrm{K}}\right)^3 \mathrm{s^{-1}}, \quad (3.209)$$

which is much larger than the expansion rate $H \sim (T/10^{10}\,\mathrm{K})^2\,\mathrm{s^{-1}}$. Therefore, for temperatures $T \gtrsim 5 \times 10^8$ K, deuterium nuclei are always produced with the equilibrium abundance:

$$X_D \approx 16.4 X_n X_p \eta \left(\frac{m_N}{T}\right)^{-3/2} \exp\left(\frac{2.22\,\mathrm{MeV}}{T}\right). \quad (3.210)$$

From this we see that large amounts of deuterium are only produced once the temperature drops to $T_D \sim 10^9$ K [see also Eq. (3.202)]. This occurs when the Universe is about 100 seconds old, and signals the onset of primordial nucleosynthesis. The subsequent reaction chain proceeds very quickly, because at $T \simeq T_D$ all nuclei heavier than deuterium can possess high equilibrium abundances. However, nuclei heavier than helium are still rare because of the instability of nuclei with $A = 5$ and $A = 8$, and because the temperature is already too low to effectively overcome the large Coulomb barrier in reactions like ^{4}He(^{3}H, γ)^{7}Li and ^{4}He(^{3}He, γ)^{7}Be. As a result, almost all free neutrons existing at the onset of nucleosynthesis will be bound into ^{4}He, the most tightly bound species with $A < 5$. The mass fraction of ^{4}He can therefore be approximately written as

$$Y \equiv X_{^4\mathrm{He}} \approx \frac{4(n_n/2)}{n_n + n_p} = \frac{2(n_n/n_p)_D}{1 + (n_n/n_p)_D}, \quad (3.211)$$

where $(n_n/n_p)_D$ is the neutron-to-proton ratio at $T = T_D$.

3.4.3 Model Predictions

Once the relevant reactions are specified and their cross-sections are given, the nucleosynthesis reaction network can be integrated forwards from the initial conditions at early times to make detailed predictions for the abundances of all species. This was first done with a complete network by Wagoner et al. (1967), and subsequent work using updated cross-sections and modernized computer codes (e.g. Wagoner, 1973; Walker et al., 1991; Cyburt et al., 2008) has modified their conclusions rather little. Detailed calculations show that the bulk of nucleosynthesis occurs at $t \approx 300$ s ($T \approx 0.8 \times 10^9$ K $= 0.07$ MeV), in agreement with the qualitative

3.4 Primordial Nucleosynthesis

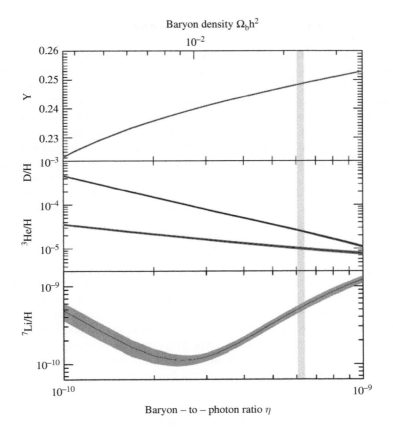

Fig. 3.7. Primordial abundances of light elements as a function of the baryon-to-photon ratio, η. The line thicknesses in each panel reflect the remaining theoretical uncertainties, while the vertical shaded band shows the range of η consistent with the WMAP measurements of fluctuations in the microwave background. [Courtesy of R. Cyburt; see Cyburt et al. (2008)]

arguments given above. At this point in time, the neutron-to-proton ratio n_n/n_p is about $1/7$. Using Eq. (3.211) this implies a final abundance of primordial ^{4}He of

$$Y_p \equiv X_{^4\text{He}} \approx 1/4. \tag{3.212}$$

Observations of the mass fraction of helium everywhere and always give values of about 24%, which would be very difficult to understand if such an abundance were not primordial. This prediction (3.212) is therefore considered a great success of the standard Big Bang model.

The primordial abundances predicted by an updated version of the code of Wagoner et al. (1967) are shown in Fig. 3.7. Note that the abundances of deuterium and ^{3}He are about three orders of magnitude below that of ^{4}He, while that of ^{7}Li is nine orders of magnitude smaller; all other nuclei are expected to be much less abundant. The predicted abundances of light elements depend on three parameters: the baryon-to-photon ratio, η, the mean lifetime of the neutron, τ_n, and $g_*(T \sim 10^{10}\,\text{K})$, which measures the number of degrees of freedom of effectively massless particles at the relevant temperature $T \sim 10^{10}\,\text{K}$ [see Eq. (3.135)]. Given the discussion earlier in this section, we can understand the sensitivity of the abundances to all these parameters.

As η increases ($|\ln \eta|$ decreases), nucleosynthesis of D, ^{3}He and ^{3}H starts slightly earlier [see Eq. (3.202)]. As a result, the synthesis of ^{4}He commences at an earlier epoch when the depletion of neutrons by beta decay is less significant, and so more neutrons are bound into

^{4}He. This explains why the ^{4}He abundance increases with η. Since the age of the Universe at temperature T_D is smaller than τ_n, the neutron-to-proton ratio decreases only slowly at the time when nucleosynthesis begins. Therefore, the η-dependence of the ^{4}He abundance is weak. However, since the burning rates of D and ^{3}He are proportional to their equilibrium abundances, which increase with η as $X_A \propto \eta^{A-1}$ [see Eq. (3.200)], a larger baryon-to-photon ratio results in a smaller abundance of these two nuclei. The more complex behavior of the ^{7}Li abundance is a result of competition between the formation and destruction reactions in the network. Direct formation dominates at small η, and formation via ^{7}Be dominates at large η.

The neutron mean lifetime affects the predicted helium abundance by influencing the number density of neutrons at the onset of nucleosynthesis. An increase of τ_n leads to an increase in the number of neutrons and so to an increase in Y_p. In the relevant range of η, $\Delta Y_p \sim 2 \times 10^{-4} (\Delta \tau_n / 1 \, \text{s})$, implying that the uncertainty in Y_p arising from that of τ_n is quite small.

Finally, substituting Eq. (3.134) in Eq. (3.80), and using that $\rho_r \propto a^{-4}$, one finds

$$t = \left(\frac{45}{16\pi^3 G g_*} \right)^{1/2} T^{-2}. \qquad (3.213)$$

Since $H = (2t)^{-1}$, the expansion rate $H \propto \sqrt{g_*} T^2$. Consequently, an increase of g_* leads to a faster expansion rate for given T. This raises the temperature at which the reaction rates equal the expansion rate, thus increasing the neutron-to-photon ratio at 'freeze-out'. Consequently, the predicted ^{4}He abundance increases with increasing g_*. In the relevant range of η, $\Delta Y_p \sim 0.01 \Delta g_*$. Therefore, the abundance of primordial helium provides a stringent constraint on the number of relativistic species at $T \gtrsim 10^9$ K. The standard model of primordial nucleosynthesis assumes these species to be photons and three species of massless neutrinos.

3.4.4 Observational Results

The predictions of primordial nucleosynthesis are of vital importance in the standard cosmology, and therefore much effort has been devoted to the observational determination of the primordial abundances of the light elements. Such determination can be used not only to constrain the number of relativistic species at the time of nucleosynthesis, but also to constrain η and so the number density of baryons in the Universe through Eq. (3.201). Unfortunately, precise determination of the primordial abundances is far from trivial. They usually rely on the emission or absorption of gas clouds due to the ions of the element in consideration. Turning this into an abundance often requires careful modeling of the properties of the observed cloud. An even greater problem comes from the fact that the material we observe today may have been processed through stars, so that (often uncertain) corrections have to be applied in order to derive a 'primordial' abundance. In the following we give a brief summary of the present observational situation.

- **Helium-4:** Because the abundance of helium is large, it is relatively easy to determine. Most measurements are made from HII clouds where the gas is highly ionized, and the abundance of both helium and hydrogen can be inferred from the strengths of their recombination lines. Since ^{4}He is also synthesized in stars, some of the observed ^{4}He may not be primordial. In order to reduce this contamination, it is desirable to use metal-poor clouds, as stars which produce the ^{4}He contamination also produce metals. Observations have been made for clouds with different metalicities, and an extrapolation to zero metallicity gives $Y_p = 0.24 \pm 0.01$ (e.g. Fields & Olive, 1998). From Fig. 3.7 we see that this observational result requires $\eta = (1.2 \rightarrow 8) \times 10^{-10}$. Since the predicted Y_p depends only weakly on η, extremely precise measurements are needed to give a more stringent constraint.

- **Deuterium:** Because of its strong dependence on η, the measurement of the primordial deuterium abundance is crucial in determining $\Omega_{b,0}$. Accurate determinations of the deuterium abundance have been obtained from UV absorption measurements in the local interstellar medium (ISM). The deuterium-to-hydrogen ratio (in mass) is found to be $[D/H]_{ISM} \approx 1.6 \times 10^{-5}$ (e.g. Linsky et al., 1995). Since deuterium is weakly bound, it is easy to destroy but hard to produce in stars. Therefore, this observed ISM value represents a lower limit on the primordial abundance. An alternative estimate of the deuterium abundance can be obtained from the absorption strength in Lyman-α clouds along the line-of-sight to quasars at high redshift. Since these high-redshift clouds are metal poor and perhaps not yet severely contaminated by stars, the deuterium abundance thus derived may actually be close to the primordial one. The observational data are still relatively sparse. The values of $[D/H]$ obtained originally ranged from $\sim 2.4 \times 10^{-5}$ (Tytler et al., 1996) to $\sim 2 \times 10^{-4}$ (Webb et al., 1997) but now seem to have settled at $2.82 \pm 0.53 \times 10^{-5}$ (Pettini et al., 2008). This agrees well with the value of η inferred from WMAP data on microwave background fluctuations.
- **Helium-3:** The abundance of ^{3}He has been measured both in the solar system (using meteorites and the solar wind) and in HII regions (based on the strength of the ^{3}He$^+$ hyperfine line, the equivalent of the 21 cm hyperfine line of neutral hydrogen). The abundance inferred from HII regions is $[^3\text{He}/\text{H}] = (1.3 \rightarrow 3.0) \times 10^{-5}$ (e.g. Gloeckler & Geiss, 1996). A similar abundance, $[^3\text{He}/\text{H}] = (1.4 \pm 0.4) \times 10^{-5}$, is obtained from the oldest meteorites, the carbonaceous chondrites. Since these meteorites are believed to have formed at about the same time as the solar system, the observed abundance may be representative of pre-solar material. The abundance of ^{3}He in the solar wind has been determined by analyzing gas-rich meteorites and lunar soil. Because D is burned to ^{3}He during the Sun's approach towards the main sequence, the observed ^{3}He in the solar system may be a good measure of the pre-solar sum (D + ^{3}He). All the measurements are consistent with $[(D + ^3\text{He})/\text{H}] \approx (4.1 \pm 0.6) \times 10^{-5}$. Although ^{3}He can be reduced by stellar burning, it is much more difficult to destroy than deuterium and the reduction factor is no more than a factor of 2. The measurements in the solar system therefore give an upper limit on the primordial abundance of $[(D + ^3\text{He})/\text{H}]_p \lesssim 10^{-4}$, corresponding to a lower limit of $\eta \gtrsim 3 \times 10^{-10}$.
- **Lithium-7:** Estimates of the ^{7}Li abundance come from stellar atmospheres. Since ^{7}Li is quite fragile, it can be depleted by circulation through the centers of stars. The observational estimates therefore vary from one stellar population to another. Since mass circulation (convection) does not go as deep in metal-poor stars as in metal-rich ones, it is desirable to use metal-poor stars where the depletion of ^{7}Li from the atmosphere is expected to be smaller. There have been attempts to observe ^{7}Li lines in the atmospheres of old stars with very low metallicity (e.g. Spite & Spite, 1982), from which the primordial ^{7}Li abundance was originally inferred to be $[^7\text{Li}/\text{H}]_p \approx (1.1 \pm 0.4) \times 10^{-10}$. More recent attempts paying close attention to systematics give values in the range 1.0 to 1.5×10^{-10} (Asplund et al., 2006). From Fig. 3.7 we see that this abundance is inconsistent by a factor of about 4 with the value $[^7\text{Li}/\text{H}]_p = 5.24 \pm 0.7 \times 10^{-10}$ inferred from the five-year WMAP data on fluctuations in the microwave background.

At the present time, Big Bang nucleosynthesis is essentially a parameter-free theory. Improvements in experimental determinations of the neutron lifetime have shrunk the uncertainties so that they are no longer significant for this problem; the standard model of particle physics is now sufficiently constrained by accelerator experiments that the number of light particle species present at nucleosynthesis cannot differ significantly from the standard value; and WMAP measurements of the power spectrum of the cosmic microwave background lead to a photon-to-baryon ratio estimate, $\eta = 6.23 \pm 0.17 \times 10^{-10}$ (see §2.10.1). With these parameters the theory gives quite precise predictions for all the light element abundances. These agree with observational estimates of the observed abundance of ^{4}He and D, where the first is only weakly constraining because of its logarithmic dependence on parameters, but the second can be considered a major success. The situation with ^{3}He is too complex for a meaningful comparison to be possible, and the results for ^{7}Li appear to disagree with observation. While this discrepancy may still reflect observational difficulties in inferring the primordial abundance of ^{7}Li, it may also be an indicator of unexpected physics in the early Universe. Notice that independent of inferences from microwave background observations, the baryon density required for successful primordial nucleosynthesis is much too small to be consistent with the large amounts of dark matter required to bind groups and clusters of galaxies, thus providing an independent argument in favor of non-baryonic dark matter (see §2.5).

3.5 Recombination and Decoupling

Immediately after primordial nucleosynthesis (when $T \sim 0.1\,\mathrm{MeV} \sim 10^9\,\mathrm{K}$) the Universe consists mainly of the following particles: hydrogen nuclei (i.e. protons), ^{4}He nuclei, electrons, photons, and decoupled neutrinos. Since the temperature is already lower than $m_e = 0.51\,\mathrm{MeV}$, baryons and electrons can all be considered non-relativistic. All the particles (except the decoupled neutrinos) interact through electromagnetic processes, such as free–free interactions among charged particles, Compton scattering between charged particles and photons, and the recombinations[8] of ions with electrons to form atoms. In this section, we examine these processes in connection to several important cosmological events at $T < 10^9\,\mathrm{K}$.

3.5.1 Recombination

As soon as the temperature of the Universe drops below $\sim 13.6\,\mathrm{eV}$, electrons and protons start to combine to form hydrogen atoms. Here we examine how this 'recombination' process proceeds. In addition, we compute the fractions of electrons and protons that remain unbound after recombination, namely the 'freeze-out' abundances of free electrons and protons. For simplicity, we ignore all elements heavier than hydrogen.

Let us start from an early enough time when recombination and ionization can maintain equilibrium among the reacting particles. The number densities of electrons, protons, and hydrogen atoms are then all given by Eq. (3.128) with $i = e$, p or H. As we will see below, the temperatures of all three species are identical to that of the photons, so that $T_i = T$. Since the chemical potentials are related by $\mu_\mathrm{H} = \mu_\mathrm{p} + \mu_\mathrm{e}$, we can write the equilibrium density of H as the Saha equation:

$$n_{\mathrm{H,eq}} = \left(\frac{g_\mathrm{H}}{g_\mathrm{p} g_\mathrm{e}}\right) n_{\mathrm{p,eq}} n_{\mathrm{e,eq}} \left(\frac{m_\mathrm{e} T}{2\pi}\right)^{-3/2} \exp\left(\frac{B_\mathrm{H}}{T}\right), \qquad (3.214)$$

where $B_\mathrm{H} = m_\mathrm{p} + m_\mathrm{e} - m_\mathrm{H} = 13.6\,\mathrm{eV}$ is the binding energy of a hydrogen atom, and we have used $(m_\mathrm{H}/m_\mathrm{p})^{3/2} \approx 1$ in the prefactor. Expressing the particle number densities in terms of the baryon number density, $n_\mathrm{b} = n_\mathrm{p} + n_\mathrm{H}$, and the ionization fraction, $X_\mathrm{e} \equiv n_\mathrm{e}/n_\mathrm{b} = n_\mathrm{p}/n_\mathrm{b}$, then yields

$$\frac{1 - X_{\mathrm{e,eq}}}{X_{\mathrm{e,eq}}^2} = \sqrt{\frac{32}{\pi}}\,\zeta(3)\,\eta \left(\frac{m_\mathrm{e}}{T}\right)^{-3/2} \exp\left(\frac{B_\mathrm{H}}{T}\right), \qquad (3.215)$$

---

[8] Note that the term 'recombination' is somewhat unfortunate, as this will be the *first* time in the history of the Universe that the electrons combine with nuclei to form atoms.

3.5 Recombination and Decoupling

where we have used that $g_e = g_p = 2$, $g_H = 4$, and $n_b = \eta n_\gamma$. This is the Saha equation for the ionization fraction in thermal equilibrium, which holds as long as the reaction rate $p + e \leftrightarrow H$ is larger than the expansion rate.

Assuming for the moment that thermal equilibrium holds, we can use Eq. (3.215) to compute the temperature, $T_{\rm rec}$, and redshift, $z_{\rm rec}$, of recombination. For example, if we define recombination as the epoch at which $X_e = 0.1$, we obtain that

$$\theta_{\rm rec}^{3/2} \exp(13.6/\theta_{\rm rec}) = 3.2 \times 10^{17} \left(\Omega_{b,0} h^2\right)^{-1}, \quad (3.216)$$

where

$$\theta \equiv (T/1\,{\rm eV}) \approx (1+z)/4250. \quad (3.217)$$

Taking logarithms and iterating once we get an approximate solution for $\theta_{\rm rec}$:

$$\theta_{\rm rec}^{-1} \approx 3.084 - 0.0735 \ln\left(\Omega_{b,0} h^2\right), \quad (3.218)$$

which corresponds to a redshift given by

$$(1+z_{\rm rec}) \approx 1367 \left[1 - 0.024 \ln\left(\Omega_{b,0} h^2\right)\right]^{-1}. \quad (3.219)$$

Assuming $\Omega_{b,0} h^2 = 0.02$, we get $T_{\rm rec} \approx 0.3\,{\rm eV}$ and $z_{\rm rec} \approx 1,300$. Note that $T_{\rm rec} \ll B_H$, which is a reflection of the high entropy per baryon (i.e. the small value of η); since there are many times more photons than baryons, there can still be sufficient photons with $h_P \nu > 13.6\,{\rm eV}$ in the Wien tail of the blackbody spectrum to keep the majority of the hydrogen atoms ionized, even when the temperature has dropped below the ionization value.

As the Universe expands and the number densities of electrons and protons decrease, the rate at which recombination and ionization can proceed may become smaller than the expansion rate. The assumption of equilibrium will then no longer be valid. In order to examine in detail how recombination proceeds, we need to understand the main reactions involved. In a normal cloud of ionized hydrogen (HII cloud), recombination occurs mainly via two processes: (i) direct recombination to the ground state, and (ii) the capture of an electron to an excited state which then cascades to the ground level. In the first case, a Lyman continuum photon (with energy larger than 13.6 eV) is produced, while in the second case one of the recombination photons must have an energy higher than or equal to that of Lyα. If the cloud is optically thin, all recombination photons can escape and do not contribute to further ionization. In the case of cosmological recombination, however, recombination photons will be absorbed again because they cannot escape from the Universe. In fact, the direct capture of electrons to the ground state does not contribute to the net recombination, because the resulting photon is energetic enough to ionize another hydrogen atom from its ground state. The normal cascade process is also ineffective, because the Lyman series photons produced can excite hydrogen atoms from their ground states, so that multiple absorptions lead to re-ionization. Therefore, recombination in the early Universe must have proceeded by different means.

There are two main channels by which cosmological recombination can proceed. One is the two-photon decay from the metastable 2S level to the ground state (1S). In this process two photons must be emitted in order to conserve both energy and angular momentum, and it is possible that the energies of the emitted photons fall below the ionization threshold. This process is forbidden to first order and so it has a slow rate: $\Gamma_{2\gamma} \approx 8.23\,{\rm s}^{-1}$. The second process is the elimination of Lyα photons by cosmological redshift. Once redshifted to a lower energy, the Lyα photons produced in the cascade will no longer be able to excite hydrogen atoms from their ground state. The details of these recombination processes have been worked out by several authors (Peebles, 1968; Zel'dovich et al., 1968; Peebles, 1993). They show that, of the two processes discussed,

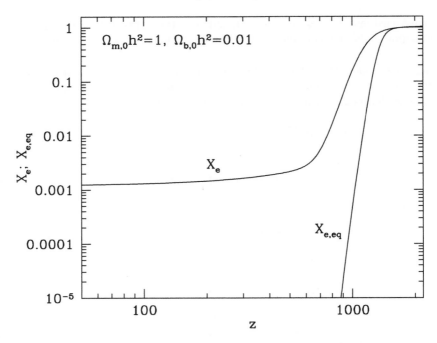

Fig. 3.8. The ionization fraction as a function of redshift, z. The curve marked $X_{e,eq}$ shows the redshift evolution of the equilibrium ionization fraction, while the one marked X_e shows the actual ionization fraction for a cosmology with $\Omega_{m,0}h^2 = 1$ and $\Omega_{b,0}h^2 = 0.01$.

the two-photon emission dominates, and that the ionization fraction drops from approximately unity at $z \gtrsim 2000$, to a 'freeze-out' value of

$$X_e \approx 1.2 \times 10^{-5} \left(\frac{\sqrt{\Omega_{m,0}}}{\Omega_{b,0}h} \right) \quad (3.220)$$

at $z \lesssim 200$. An example of the evolution of X_e with redshift is shown in Fig. 3.8.

3.5.2 Decoupling and the Origin of the CMB

Charged particles and photons interact with each other via Thomson scattering. The rate at which a photon collides with an electron is $\Gamma_T = n_e \sigma_T c$, where

$$\sigma_T = \frac{8\pi}{3} \left(\frac{q_e^2}{m_e c^2} \right)^2 \approx 6.65 \times 10^{-25} \, \text{cm}^2 \quad (3.221)$$

is the Thomson cross-section, with q_e the charge of an electron. In what follows we only consider this scattering between electrons and photons, since the interaction rate with ions is much lower. Substituting the electron number density with $n_e = X_e \eta n_\gamma$, and using the Saha equation to compute X_e in the limit $X_e \ll 1$, we obtain

$$\Gamma_T = 1.01 \left(\Omega_{b,0} h^2 \right)^{1/2} \theta^{9/4} \exp(-6.8/\theta) \, \text{s}^{-1}. \quad (3.222)$$

In order to estimate at what redshift the photons decouple from the matter, we compare this interaction rate with the expansion rate. At $z \gg 1$, we can use Eq. (3.74) to write

$$H(T) = \begin{cases} 3.8 \times 10^{-13} \theta^2 \, \text{s}^{-1} & \text{(for } z > z_{\text{eq}}) \\ 9.0 \times 10^{-13} \left(\Omega_{m,0} h^2\right)^{1/2} \theta^{3/2} \, \text{s}^{-1} & \text{(for } z < z_{\text{eq}}), \end{cases} \quad (3.223)$$

where z_{eq} is the redshift at which the Universe becomes matter dominated, and we have used $g_* = 3.36$ in calculating the energy density of relativistic species. Equating Eqs. (3.222) and (3.223) with the assumption that decoupling occurs as $z < z_{\text{eq}}$, we obtain the decoupling temperature:

$$\theta_{\text{dec}}^{-1} \approx 3.927 + 0.074 \ln\left(\Omega_{b,0}/\Omega_{m,0}\right). \quad (3.224)$$

Taking $\Omega_{b,0}/\Omega_{m,0} = 0.1$ we get

$$T_{\text{dec}} \approx 0.26 \, \text{eV}; \qquad (1+z_{\text{dec}}) \approx 1,100. \quad (3.225)$$

As expected, the decoupling of matter and radiation occurs shortly after the number density of free electrons has suddenly decreased due to recombination.

A somewhat more accurate derivation of the redshift of decoupling can be obtained by defining an optical depth of Thomson scattering from an observer at $z = 0$ to a surface at a redshift z:

$$\tau(z) = \int_0^z n_e \sigma_T \frac{dt}{dz} \, dz. \quad (3.226)$$

Using the solution of $X_e(z)$ shown in Fig. 3.8, rather than the equilibrium ionization fraction used in the previous estimate, one finds to good approximation

$$\tau(z) = 0.37 \, (z/1,000)^{14.25}. \quad (3.227)$$

The probability that a photon was last scattered in the redshift interval $z \pm dz/2$ can be approximated as

$$P(z) = e^{-\tau} \frac{d\tau}{dz} \approx 5.26 \times 10^{-3} \left(\frac{z}{1,000}\right)^{13.25} \exp\left[-0.37\left(\frac{z}{1,000}\right)^{14.25}\right]. \quad (3.228)$$

This distribution peaks sharply at $z \approx 1,067$ and has a width $\Delta z \approx 80$ (e.g. Jones & Wyse, 1985). This represents the last scattering surface of photons, which is the surface probed by the cosmic microwave background (CMB) radiation. Similar to the photosphere of the Sun, it acts as a kind of photon barrier. No information carried by photons originating from $z \gtrsim 1100$ can reach the Earth, as the photons involved will be scattered many times.

As discussed in §2.9, one of the most important properties of the observed CMB is that its spectrum is very close to that of a blackbody. This implies that the emission must have originated when the Universe was highly opaque. In the standard cosmology, such an epoch is expected because photons and other particles are tightly coupled at $z > 10^6$. However, the CMB photons have been scattered many times by electrons and ions between their redshift of origin and the last scattering surface. An important question therefore is whether the background radiation can retain a blackbody spectrum during this process. The answer is yes and the reason is, as we show below, that the high entropy content of the Universe can keep the gas particles at the same temperature as that of the photons. In this case, there is no net energy transfer between the photons and electrons, ensuring that the radiation field remains blackbody. Furthermore, although the finite thickness of the last scattering surface ($\Delta z \approx 80$) implies a spread in photon temperatures at their last scattering event, this does not lead to observable distortions in the CMB temperature spectrum. The reason is that the higher initial temperature of a photon that decoupled somewhat earlier is exactly compensated by the larger redshift it experiences before reaching the observer. Thus, the blackbody nature of the CMB is naturally explained in the standard cosmology. In what follows we examine more closely the temperature evolution of matter and radiation from the epoch of electron–positron annihilation to that of decoupling.

3.5.3 Compton Scattering

By far the most dominant electromagnetic interaction during the era of decoupling is the Coulomb interaction, which is sufficiently strong to maintain thermal equilibrium among various matter components. In the absence of any interactions between matter and radiation, the temperature of the former would decrease as $T_m \propto a^{-2}$, while the photon temperature $T_\gamma \propto a^{-1}$ (see Table 3.1). However, as we have seen above, photons and electrons interact with each other via Compton scattering. As long as $T_e > T_\gamma$ there will be a net energy transfer from the electrons to the photons, and vice versa. The mean free path is $l_\gamma = 1/(n_e \sigma_T)$ for the photons, and $l_e = 1/(n_\gamma \sigma_T)$ for electrons. Their ratio can be expressed in terms of the ionization fraction $X_e = n_e/n_b$ as

$$\frac{l_e}{l_\gamma} = X_e \eta = 2.72 \times 10^{-8} (X_e \Omega_{b,0} h^2). \tag{3.229}$$

Since $X_e \leq 1$, we have $l_e \ll l_\gamma$. This shows that it is much easier for photons to change the energy distribution of the electrons than the other way around. This is, once again, a consequence of the high entropy per baryon, or, put differently, of the fact that the heat capacity of the radiation is many orders of magnitude larger than that of the electrons. Therefore, as long as the Compton interaction rate is sufficiently large compared to the expansion rate, the matter temperature will follow that of the photons.

To compute the redshift at which the matter temperature will finally decouple from that of the radiation, we proceed as follows. The average energy transfer per Compton collision is

$$\Delta E = \frac{4}{3} \left(\frac{v}{c}\right)^2 h_P \bar{\nu} = 4 \left(\frac{k_B T_e}{m_e c^2}\right) \frac{\varepsilon_\gamma}{n_\gamma}, \tag{3.230}$$

where we have used that the average electron energy is $\frac{1}{2} m_e v^2 = \frac{3}{2} k_B T_e$ and that the mean energy of the photons is $h_P \bar{\nu} = \varepsilon_\gamma / n_\gamma$ with ε_γ the photon energy density (see §B1.3.6). The rate at which the energy density of the matter, ε_m, changes due to Compton interactions with the radiation field is

$$\frac{d\varepsilon_m}{dt} = n_e n_\gamma \sigma_T c \Delta E = 4 n_e \sigma_T \varepsilon_\gamma \left(\frac{k_B T_e}{m_e c}\right). \tag{3.231}$$

This allows us to define the Compton rate at which electrons can adjust their energy density to that of the photons as

$$\Gamma_{\gamma \to e} \equiv \frac{1}{\varepsilon_m} \frac{d\varepsilon_m}{dt} = 8.9 \times 10^{-6} \left(\frac{X_e}{X_e + 1}\right) \theta^4 \, \text{s}^{-1}, \tag{3.232}$$

where we have used that $\varepsilon_m = \frac{3}{2} n k_B T_e$, with $n = n_e + n_b$, and $\varepsilon_\gamma = (4\sigma_{SB}/c) T_\gamma^4$. Comparing this to the expansion rate given by Eq. (3.223), we find that decoupling of matter from radiation occurs at a redshift

$$1 + z = 6.8 \left(\frac{X_e}{X_e + 1}\right)^{-2/5} (\Omega_{m,0} h^2)^{1/5}. \tag{3.233}$$

As we have seen above, before the onset of the first ionizing sources, the residual ionization fraction at $z \lesssim 200$ is $X_e \sim 10^{-5} \Omega_{m,0}^{1/2}/(\Omega_{b,0} h)$ (see Fig. 3.8). Substituting this in Eq. (3.233), and adopting $\Omega_{b,0} h^2 = 0.02$, yields a redshift, $z \simeq 150$, at which the temperatures of matter and radiation decouple. This is a much lower redshift than the redshift of decoupling defined by an

3.5.4 Energy Thermalization

In addition to $\Gamma_{\gamma \to e}$ defined above, we can also define the rate at which Compton scattering can adjust the photon energy density to that of the electrons:

$$\Gamma_{e \to \gamma} \equiv \frac{1}{\varepsilon_\gamma} \frac{d\varepsilon_\gamma}{dt} = 1.3 \times 10^{-13} \left(X_e \Omega_{b,0} h^2\right) \theta^4 \, \text{s}^{-1}, \quad (3.234)$$

where we have used that in thermal equilibrium $|d\varepsilon_\gamma/dt| = |d\varepsilon_m/dt|$. This is equal to the expansion rate in Eq. (3.223) at a redshift

$$1 + z = 7.2 \times 10^3 \left(X_e \Omega_{b,0} h^2\right)^{-1/2}. \quad (3.235)$$

At $z \gtrsim 2{,}000$, $X_e = 1$ to good approximation. Using $\Omega_{b,0} h^2 = 0.02$ we thus find that Compton scattering can significantly modify the energy distribution of the photon fluid at $z \gtrsim 5 \times 10^4$. Since Compton scattering ($e + \gamma \to e + \gamma$) does not change the number of photons, this process alone cannot lead to a Planck distribution. However, since the photon fluid starts out in thermal equilibrium with the matter, it will remain properly thermalized (i.e. the Compton scattering does not lead to any net energy transfer between matter and radiation). On the other hand, one might envision scenarios in which physical processes (e.g. turbulence, black hole evaporation, decay of heavy unstable leptons) heat the electrons to a temperature above that of the photons. If this occurs at $z \gtrsim 5 \times 10^4$, Compton scattering is sufficiently efficient that photons experience multiple scattering events, which bring them into thermal equilibrium with the electrons. Since there is no change in the photon number, such scattering results in a modification of the photon energy distribution from a Planck distribution to a Bose–Einstein distribution with a negative chemical potential ($\mu < 0$). Such a distortion is usually referred to as a μ-distortion. In the absence of any photon-producing processes, an increase of the electron temperature therefore leads to a Comptonization of the CMB, which is observable as a μ-distortion of its spectrum.

Two examples of photon producing processes, which may thermalize the injected energy and bring the photon energy distribution back to that of a blackbody, are bremsstrahlung (also called free–free emission) and the double-photon Compton process ($e + \gamma \to e + 2\gamma$). In a medium with relatively high photon density, such as that in the radiation dominated era, double Compton emission is the dominant photon producing process and its rate is higher than the expansion rate of the Universe at

$$z \gtrsim 2.0 \times 10^6 \left(\frac{\Omega_{b,0} h^2}{0.02}\right)^{-2/5} \left(1 - \frac{Y_p}{2}\right)^{-2/5}, \quad (3.236)$$

where Y_p is the helium abundance in mass (e.g. Danese & de Zotti, 1982). Thus, any energy input into the radiation field at $z > 2 \times 10^6$ can be effectively thermalized into a blackbody distribution. If the energy ejection occurs at $z < 5 \times 10^4$ the Compton rate is insufficient to establish a new thermal equilibrium. Therefore, only energy injection in the redshift range $5 \times 10^4 \lesssim z \lesssim 2 \times 10^6$ can lead to a μ-distortion in the CMB. Detailed observations with the COBE satellite have established that the CMB has a blackbody spectrum to very high accuracy; the corresponding limit on the chemical potential is $|\mu| \leq 9 \times 10^{-5}$ (Fixsen et al., 1996). Apparently, there have not been any major energy ejections into the baryonic gas in the above mentioned redshift interval.

Because *multiple* Compton scattering becomes rare at $z \lesssim 5 \times 10^4$, any energy input into the electron distribution no longer drives the photon field towards a Bose–Einstein distribution to produce a μ-distortion. However, *single* Compton scattering of low-energy photons in the Rayleigh–Jeans tail of the CMB can still cause those photons to gain energy. Although this does not bring the photons in thermal equilibrium with the electrons, it does result in a distortion of the photon energy distribution. This kind of distortion is called y-distortion, because it is proportional to the Compton y-parameter defined in §B1.3.6. Such distortions can be produced by the hot intracluster medium, which is called the Sunyaev–Zel'dovich (SZ) effect, and is discussed in detail in §6.7.4.

3.6 Inflation

So far we have seen that the standard relativistic cosmology provides a very successful framework for interpreting observations. There are, however, a number of problems that cannot be solved within the standard framework. Here we summarize some of these problems and show how an 'inflationary hypothesis' can help to solve them.

3.6.1 The Problems of the Standard Model

(a) The Horizon Problem As shown in §3.2.4, the comoving radius of the particle horizon for a fundamental observer, $\mathscr{O}$, at the origin at cosmic time t is

$$\chi_h = \int_0^t \frac{c \, dt'}{a(t')}. \tag{3.237}$$

For a universe which did not have a contracting phase in its history, radiation was the dominant component of the cosmic energy density at $z > z_{\rm eq}$, and the scale factor $a(t) \propto t^{1/2}$. In this case χ_h has a finite value, so that there must be fundamental observers (denoted by $\mathscr{O}'$) whose comoving distances to $\mathscr{O}$ are larger than χ_h. No physical processes at any $\mathscr{O}'$ could have influenced $\mathscr{O}$ by time t. To get a rough idea of the size of the particle horizon at the time of decoupling, assume for simplicity an Einstein–de Sitter universe ($\Omega_{\rm m,0} = 1$), and ignore for the moment that the Universe was radiation dominated at $z > z_{\rm eq}$. It then follows from Eqs. (3.74)–(3.75) that

$$\chi_h(z) = 6{,}000 \, h^{-1} {\rm Mpc} \, (1+z)^{-1/2}. \tag{3.238}$$

At the time of decoupling, which occurs at a redshift of $\sim 1{,}100$ (see §3.5), the comoving radius of the particle horizon is $\sim 180 h^{-1}{\rm Mpc}$. The comoving distance from us to the last scattering surface is $\sim 5{,}820 h^{-1}{\rm Mpc}$, so that the particle horizon at decoupling subtends an angle of about 1.8 degrees on the sky. This implies that many regions that we observe on the CMB sky have not been in causal contact. Yet, as discussed in §2.9, once measurements are corrected to the frame of the fundamental observer at the position of the Sun, the temperature of the CMB radiation is the same in all directions to an accuracy of better than one part in 10^5. The problem is how all these causally disconnected regions can have extremely similar temperatures. This problem is known as the horizon problem of the standard model.

(b) The Flatness Problem This problem concerns the processes which determine the density, age, and size of the Universe at the present time. In the standard model, these properties are assumed to 'arise' as initial conditions at the Planck time, when the Universe emerged from the quantum gravity epoch. The problem arises if $\Omega = \Omega_{\rm m} + \Omega_\Lambda + \Omega_{\rm r}$ differs mildly from unity at the present time, because such a universe requires extreme 'fine-tuning' of Ω at the Planck time. A simple way to illustrate the situation is to focus on the quantity, $\Omega^{-1} - 1$, which measures the fractional deviation of the total density from the critical density. Using the Friedmann equation, we can write that

$$\Omega(a)^{-1} - 1 = -\frac{3Kc^2}{8\pi G\rho(a)a^2}, \qquad (3.239)$$

which is proportional to a^2 at $z > z_{\rm eq}$ and to a at $z < z_{\rm eq}$. Therefore, in the standard model,

$$\frac{\Omega_{\rm Pl}^{-1} - 1}{\Omega_0^{-1} - 1} \sim \frac{T_0}{T_{\rm eq}} \left(\frac{T_{\rm eq}}{T_{\rm Pl}}\right)^2 \sim 10^{-60}, \qquad (3.240)$$

where subscripts 'eq' and 'Pl' denote the values at the time of radiation/matter equality, $t_{\rm eq} \sim 10^4$ yr (corresponding to a temperature $T_{\rm eq} \sim 10^4$ K), and the Planck time, $t_{\rm Pl} \equiv (\hbar G/c^5)^{1/2} \sim 10^{-43}$ s (corresponding to a temperature $T_{\rm Pl} \sim 10^{32}$ K), respectively. This demonstrates that $\Omega_{\rm Pl}$ is about 60 orders of magnitude closer to unity than Ω_0. For example, if $\Omega = 0.1$ today, it must have been $1 - 10^{-59}$ at the Planck time, which clearly constitutes a fine-tuning problem. A 'trivial' way out of this problem is to postulate that Ω_0 is exactly equal to unity, in which case it has been exactly unity throughout the history of the Universe. However, this cannot be considered a proper solution unless it has a proper physical explanation. This problem is known as the flatness problem.

(c) Monopole Problem In the early stages of the Hot Big Bang, particle energies are well above the threshold at which grand unification (GUT) is expected to occur ($T_{\rm GUT} \sim 10^{14}$–10^{15} GeV). As the temperature drops through this threshold, a phase transition associated with spontaneous symmetry breaking (SSB) can occur. One speaks of SSB when the fundamental equations of a system possesses a symmetry which the ground state does not have. For example, one may have a situation in which the Lagrangian density is invariant under a gauge transformation, while the vacuum state, the state of the least energy, does not possess this symmetry. SSB plays a crucial role in quantum field theory, where it provides a mechanism for assigning masses to the gauge bosons without destroying the gauge invariance. As we will see below, SSB also plays an important role in inflation.

Depending on the properties of the symmetry breaking, the phase transition can produce topological defects, such as magnetic monopoles, strings, domain walls or textures (see Vilenkin & Shellard (1994) for a detailed description). In the case of the GUT phase transition, one expects the formation of magnetic monopoles with a density of about one per horizon volume at that epoch. The mass of each monopole is expected to be of the order of the energy scale in consideration, i.e. $m \sim T_{\rm GUT}$. This predicts a present-day energy density in magnetic monopoles of

$$\rho_{\rm mono,0} \sim \frac{T_{\rm GUT}}{t_{\rm GUT}^3} \left(\frac{T_0}{T_{\rm GUT}}\right)^3 \sim \left(\frac{T_{\rm GUT}}{10^{11}\,{\rm GeV}}\right)^4 \rho_{\gamma,0}, \qquad (3.241)$$

where T_0 and $\rho_{\gamma,0}$ are the temperature and energy density of the cosmic microwave background at the present time, and we have used Eq. (3.81) to relate $t_{\rm GUT}$ to $T_{\rm GUT}$. With $\Omega_{\gamma,0} \approx 2.5 \times 10^{-5} h^{-2}$ and $T_{\rm GUT} \sim 10^{15}$ GeV, we see that monopoles are expected to completely dominate the present matter density with $\Omega_0 \sim 5 \times 10^{11}$, in fatal conflict with observations. Since monopoles are expected to arise in almost any GUT, there is a monopole problem in the standard cosmology.

(d) Structure Formation Problem This problem concerns the origin of the large-scale structure in the Universe. The observed structures such as the clusters of galaxies have an amplitude which may be characterized by their dimensionless binding energy per unit mass, $\mathscr{E}/c^2 \sim 10^{-5}$. Such structures are coherent over a mass of about $10^{15}\,{\rm M}_\odot$ (corresponding to $\sim 10\,{\rm Mpc}$ in comoving size) and are presumed to have grown via gravitational instability from small initial perturbations. Since both the mass and binding energy of a perturbation are approximately conserved during gravitational evolution, the perturbation must have been generated while its entire mass was within the particle horizon (i.e. when $\chi_h > 10\,\mathrm{Mpc}$), in order to explain its coherence. This requires that the perturbations associated with present-day clusters be generated at $z \lesssim 10^6$. Since the standard scenario of structure formation via gravitational instability does not include any processes which could produce the binding energy of clusters at such low redshift, the origin of large, coherent density perturbations constitutes another problem for the standard cosmology.

It should be pointed out, however, that this particular problem is not fully generic for the standard cosmology. In particular, the problem may be avoided if we abandon the assumption that structures form via gravitational instability. For example, density perturbations with large amplitudes may be generated in the early Universe within patches of the horizon size at the time of generation. If these perturbations collapse and form objects which can eject energy to large distances, structures of much larger scales may form out of the perturbations created by these ejecta. Such non-gravitational models for the formation of large-scale structure have, for example, been considered by Ostriker & Cowie (1981). However, as we will see later in the book, the large-scale structure observed in the Universe is best explained by gravitational instability, implying that the structure formation problem must be considered seriously.

(e) Initial Condition Problem It should be pointed out that the problems mentioned above do not falsify the standard cosmology in any way. All of these problems can be incorporated into the standard cosmology as initial conditions, even though the standard cosmology does not explain them. In this sense, standard cosmology only provides a consistent theory to explain the state of the observable Universe with some assumed initial conditions, but does not explain their origin.

For many years it was believed that the initial conditions for standard cosmology would arise from quantum cosmology (a quantum treatment of space-time) at very early times when the Universe was so small that classical cosmology is no longer valid. Unfortunately, such theory is still highly incomplete and no reliable predictions can be presented. However, the situation changed dramatically in the early 1980s when it was realized that a new concept, called inflation, can solve all the aforementioned problems within the classical theory of space-time. Inflation basically provides an explanation for the initial conditions, and it operates at an energy scale that is much lower than the Planck scale, so that gravity can be treated classically. In what follows we present a brief overview of cosmological inflation, and illustrate how it solves the problems mentioned here. A more detailed treatment of this topic can be found in Kolb & Turner (1990) and Liddle & Lyth (2000).

3.6.2 The Concept of Inflation

As discussed above, the horizon problem arises because the comoving radius of the particle horizon of a fundamental observer (at time t),

$$\chi_h = \int_0^t \frac{\mathrm{d}ct'}{a(t')} = \int_0^a \frac{\mathrm{d}a'}{a'} \left[\frac{8\pi G \rho(a')a'^2}{3c^2} - K\right]^{-1/2}, \qquad (3.242)$$

is finite in the standard model, where $\rho(a) \propto a^{-4}$ as $a \to 0$. To get rid of this problem, χ_h must diverge, making the radius of the particle horizon infinite. From Eq. (3.242) one sees that this requires $\rho(a) \propto a^{-\beta}$ with $\beta < 2$ as $a \to 0$. Inserting this a-dependence of ρ into the first law of thermodynamics, Eq. (3.35), one obtains

$$\rho + 3P/c^2 < 0, \qquad (3.243)$$

which, in Eq. (3.58), gives $\ddot{a} > 0$. Such a phase of accelerated expansion is called inflation, and arises when the Universe is dominated by an energy component whose equation of state satisfies Eq. (3.243).

3.6 Inflation

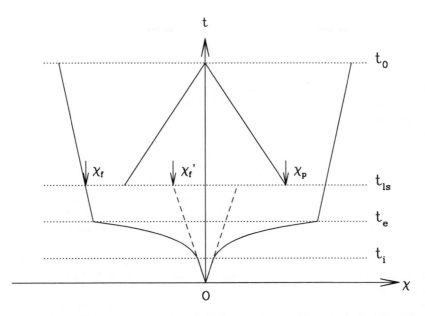

Fig. 3.9. A sketch of the light-cone structure in an inflationary universe. The cosmic time flows from bottom up (with the Big Bang labeled by O) and the horizontal axis marks the comoving radius, χ, of the light cone. In the absence of inflation, the forward light cone (the dashed lines) would be smaller than our past light cone, χ_p, at the last scattering surface (corresponding to $t = t_{ls}$), resulting in the causality problem discussed in §3.6.1. With a period of inflation (from t_i to t_e), however, the forward light cone can be (much) larger than the past light cone at t_{ls} (i.e. $\chi_f > \chi_p$).

An example of such an energy component is vacuum energy, whose equation of state is $P = -\rho_{vac}c^2$. In this case, the solution of the Friedmann equation corresponds to an exponentially expanding universe,

$$a \propto e^{Ht} \quad \text{where} \quad H = \sqrt{8\pi G \rho_{vac}/3} \quad (3.244)$$

(see §3.2.3). Fig. 3.9 illustrates how such an inflationary period can solve the horizon problem. Suppose that inflation begins at some very early time t_i and ends at some later time t_e. The period of inflation is therefore $\Delta t = t_e - t_i$. During inflation, the forward light cone expands exponentially, whereas the past light cone of an observer at the present time t_0 is not affected by the exponential expansion for $t > t_e$. Therefore, if Δt is sufficiently long, the size of the forward light cone on the last scattering surface of the CMB photons, $\chi_f(t_{ls})$, can be larger than the size of the past light cone, $\chi_p(t_{ls})$. Since $t_{ls} \ll t_0$, the size of the past light cone is $\chi_p(t_{ls}) = \int_{t_{ls}}^{t_0} dt/a(t) \approx 3t_0$. The size of the forward light cone at t_{ls} is $\chi_f(t_{ls}) \sim \int_{t_i}^{t_{ls}} dt/a(t) = (1/H)[e^{H\Delta t} - 1]a^{-1}(t_e)$. The condition that $\chi_f(t_{ls}) > \chi_p(t_{ls})$ therefore requires

$$e^{H\Delta t} > 3Ha(t_e)t_0 \sim a(t_e)\frac{t_0}{t_e} \sim \frac{1}{\sqrt{1+z_{eq}}}\frac{T_e}{T_0} \sim 10^{25}, \quad (3.245)$$

where the final value is for $T_e \sim 10^{14}$ GeV (roughly the GUT energy scale) and $T_0 \sim 10^{-13}$ GeV (the temperature of the CMB). Thus, in order to solve the horizon problem, an inflationary period of

$$\Delta t \gtrsim 60 H^{-1} \quad (3.246)$$

is required, corresponding to 60 e-foldings in the scale factor.

Inflation can also solve the flatness problem. To see this we use Eq. (3.79) to write

$$\frac{\Omega^{-1}(t_e) - 1}{\Omega^{-1}(t_i) - 1} = \left[\frac{a(t_i)}{a(t_e)}\right]^2 \lesssim 10^{-52}, \qquad (3.247)$$

where we have inserted the number of e-foldings implied by Eq. (3.245). Therefore, even when $\Omega(t_i)$ deviates substantially from unity, at the end of inflation $\Omega(t_e) \simeq 1$ to very high accuracy. If we assume that inflation ends at about the GUT time (i.e. $T_e \sim 10^{15}$ GeV), then the present-day value of Ω is related to that at the beginning of inflation according to

$$\frac{\Omega^{-1}(t_0) - 1}{\Omega^{-1}(t_i) - 1} = \frac{\Omega^{-1}(t_0) - 1}{\Omega^{-1}(t_e) - 1} \times \frac{\Omega^{-1}(t_e) - 1}{\Omega^{-1}(t_i) - 1} \lesssim 10^{-52} \left(\frac{T_{eq}}{T_0}\right)\left(\frac{T_e}{T_{eq}}\right)^2 \sim 1. \qquad (3.248)$$

Thus the same number of e-foldings needed to solve the horizon problem also solves the flatness problem. Since the value of Ω at the present time depends very sensitively on the number of e-foldings, unless it is exactly unity, extreme fine-tuning is required to give $\Omega \neq 1$. In this sense, inflation predicts that the Universe is spatially flat, with $\Omega_{m,0} + \Omega_{\Lambda,0} = 1$. This can be understood as follows. Because of inflation the curvature radius (measured in physical scale) increases exponentially, and the observed piece of space in the past light cone looks essentially flat after inflation even if it had a large curvature before.

If monopoles are produced before inflation, their number density will be diluted exponentially during inflation. At the end of inflation, the number density would be reduced by a factor $\sim \left(e^{H\Delta t}\right)^3 \sim 10^{78}$, making the contribution of monopoles to the cosmic density completely negligible. Thus, inflation also solves the monopole problem discussed in §3.6.1.

Finally, inflation also provides a mechanism to explain why structures like clusters can form in a causal way. Because of inflation, small-scale structures present before and during inflation can be blown up exponentially. Thus, the different parts of a perturbation responsible for a cluster and a larger-scale structure, although not in causal contact after inflation, were actually in causal contact before or during inflation. It is therefore possible to have causality if the perturbations responsible for the formation of clusters were generated before or during inflation. In fact, inflation not only allows such perturbations to exist, but also provides a mechanism to generate them, as we will discuss in detail in §4.5.1.

3.6.3 Realization of Inflation

The above discussion shows that inflation can solve many problems (or, more appropriately, puzzles) regarding the initial conditions of the Big Bang cosmology, as long as it operates for a sufficiently long time. All that is needed is a dominant energy component with an equation of state obeying Eq. (3.243). As already mentioned, the cosmological constant is an example of such a component. However, it cannot serve to describe inflation for the simple reason that it will never stop. By definition, Λ is a constant, and once it dominates the energy density of the Universe it will continue to do so eternally. A successful inflation model, however, needs to stop after some time, and it needs to end in a particular way. After all, at the end of inflation the matter and radiation density of the Universe will be virtually zero, so will be its temperature: the Universe is basically a vacuum. We thus need a mechanism, called reheating, which at the end of inflation creates matter and radiation. In other words, at the end of inflation the Universe needs to undergo a cosmological phase transition. The temperature at the end of this phase transition has to be sufficiently low so that during the subsequent evolution no new phase transition can create large quantities of magnetic monopoles. Otherwise we are back to where we started. In addition, the temperature needs to be sufficiently high so that the process of baryogenesis can still operate.

It was Guth (1981) who first realized that all these requirements can be realized in a natural way with scalar fields. These quantum fields, which describe scalar (spin-0) particles, play an

3.6 Inflation

important role in quantum field theory. Their dominant role is to cause spontaneous symmetry breaking (SSB) via the Higgs mechanism. These so-called Higgs fields have a non-zero vacuum expectation value. As a result, the interactions of the fermion and boson fields with the Higgs field give a finite potential energy to the fermions and bosons, which is expressed as an effective mass. Before the symmetry is broken, the Higgs field has a zero expectation value, and the fermions and bosons are massless. In what follows we show that under certain conditions, a similar scalar field can also cause inflation. The key point here is that the zero-point energy (vacuum energy) of such fields can mimic a cosmological constant. A scalar field that causes inflation is generally called an inflaton.

The Lagrangian density of a scalar field $\varphi(\mathbf{x},t)$ is

$$\mathscr{L} = \frac{1}{2}\partial_\mu \varphi \partial^\mu \varphi - V(\varphi), \tag{3.249}$$

where $V(\varphi)$ is the potential of the field. Different inflationary models, i.e. different inflatons, correspond to different choices for $V(\varphi)$. The energy–momentum tensor of the inflaton is

$$T^{\mu\nu} = \partial^\mu \varphi \partial^\nu \varphi - g^{\mu\nu}\mathscr{L}. \tag{3.250}$$

If the inhomogeneity in φ is small, this $T^{\mu\nu}$ has the form of a perfect fluid, Eq. (3.57), with energy density and pressure given by

$$\rho = \frac{\dot\varphi^2}{2} + \frac{(\nabla\varphi)^2}{2a^2} + V(\varphi) \quad \text{and} \quad P = \frac{\dot\varphi^2}{2} - \frac{(\nabla\varphi)^2}{6a^2} - V(\varphi), \tag{3.251}$$

where $\dot\varphi \equiv (\partial\varphi/\partial t)$, and ∇ is the derivative with respect to the comoving coordinates $\mathbf{x}$. We therefore have

$$\rho + 3P = 2\left[\dot\varphi^2 - V(\varphi)\right], \tag{3.252}$$

and the condition for inflation becomes

$$\dot\varphi^2 \ll V(\varphi), \tag{3.253}$$

which is called the slow-roll approximation. Note that in this case $\rho = V(\varphi)$, and for inflation to happen $V(\varphi)$ thus needs to be sufficiently large to dominate the total energy density. As soon as inflation operates it drives the curvature to zero so that the Friedmann equation (3.60) becomes

$$H = \sqrt{8\pi G V(\varphi)/3} = \frac{1}{m_{\text{Pl}}}\sqrt{\frac{8\pi V}{3}}, \tag{3.254}$$

where $m_{\text{Pl}} \equiv (\hbar c/G)^{1/2}$ is the Planck mass and we have used $\hbar = c = 1$.

Since the scale factor a increases exponentially during inflation, the spatial derivative term $\nabla\varphi/a$ in ρ and P rapidly becomes negligible, provided V is large enough for inflation to start in the first place. Therefore any spatial inhomogeneities can be neglected and, under the slow-roll approximation, one obtains $P = -\rho$, an equation of state similar to that for the cosmological constant.

To translate the slow-roll requirement into a requirement for the shape of the potential, $V(\varphi)$, which is the 'free parameter' in the construction of inflation models, we need to look at the dynamics of a scalar field. The classical equation of motion is obtained by writing down the action

$$S = \int \mathscr{L}\sqrt{-g}\, d^4x, \tag{3.255}$$

where g is the determinant of the metric tensor $g_{\mu\nu}$. The Euler–Lagrange equation, which follows from the least-action principle, $\delta S = 0$, then yields

$$\ddot\varphi + 3H\dot\varphi + dV/d\varphi = 0, \tag{3.256}$$

where we have ignored any spatial inhomogeneity ($\nabla \varphi = 0$). Equivalently, Eq. (3.256) follows from conservation of the energy–momentum tensor ($T^{\mu\nu}{}_{;\nu} = 0$), or from substituting Eq. (3.251) in the continuity equation for a FRW cosmology, $\dot{\rho} = -3H(\rho + P)$ [see Eq. (3.35)]. Note that Eq. (3.256) is similar to the equation of motion of a ball moving under the influence of a potential V in the presence of friction (the Hubble drag) proportional to $3H$. Using that $\dot{\varphi} \sim \varphi/t$, so that $\ddot{\varphi} \sim \varphi/t^2 \sim \dot{\varphi}^2/\varphi$, the slow-roll approximation implies that $\ddot{\varphi} \ll V(\varphi)/\varphi \sim dV/d\varphi$. Therefore, the first term in Eq. (3.256) is negligible, and

$$3H\dot{\varphi} + dV/d\varphi = 0. \tag{3.257}$$

This equation expresses that the acceleration due to the gradient in the potential is balanced by the Hubble drag due to the expansion. This together with Eq. (3.253) leads to the following slow-roll condition:

$$\varepsilon \equiv \frac{m_{\rm Pl}^2}{16\pi} \left(\frac{dV/d\varphi}{V}\right)^2 = \frac{m_{\rm Pl}^2}{16\pi} \frac{(3H\dot{\varphi})^2}{V^2} \ll m_{\rm Pl}^2 \frac{(3H)^2}{V} \sim 1, \tag{3.258}$$

where we have used the Friedmann equation (3.254). Similarly, since $(d^2V/d\varphi^2)/V \sim (dV/d\varphi)/(\varphi V) \sim (dV/d\varphi)^2/V^2$, we have

$$\eta \equiv \frac{m_{\rm Pl}^2}{8\pi} \frac{1}{V} \frac{d^2V}{d\varphi^2} \ll 1. \tag{3.259}$$

Conditions (3.258) and (3.259) indicate the intuitively obvious, namely that for the slow-roll condition to be satisfied the potential needs to be very flat. Any scalar field that obeys these two constraints will cause an inflationary phase, whose duration increases with the flatness of $V(\varphi)$.

As emphasized above, inflation is only successful if it can also stop and reheat the Universe. Below we illustrate how this comes about with scalar fields using three specific examples. In each of these the end of inflation and the reheating mechanism are somewhat different.

3.6.4 Models of Inflation

(a) Old Inflation The 'old inflation' model, proposed by Guth (1981), is based on a scalar field which initially gets trapped in a false vacuum at $\varphi = 0$ and which at some point undergoes spontaneous symmetry breaking to its true vacuum state via a first order phase transition. The prototype of such a potential has the form

$$V(\varphi) = \frac{1}{4}\varphi^4 - \frac{1}{3}(\alpha+\beta)|\varphi|^3 + \frac{1}{2}\alpha\beta\varphi^2 + V_0, \tag{3.260}$$

where $V_0 = \alpha^3(\alpha - 2\beta)/12 > 0$ and $\alpha > 2\beta > 0$. The field is assumed to be in thermal equilibrium with a radiation field at temperature T, and so the effective potential of the field is

$$V_{\rm eff}(\varphi) = V(\varphi) + \frac{1}{2}\tilde{\lambda} T^2 \varphi^2 \tag{3.261}$$

according to finite-temperature field theory (e.g. Brandenberger, 1995), where $\tilde{\lambda}$ is a coupling constant. Fig. 3.10a shows $V_{\rm eff}(\varphi)$ at different temperatures. When the temperature is high, the effective potential has a single minimum at $\varphi = 0$. As the temperature decreases, two other minima develop. This occurs at a critical temperature $T_{\rm c} = (\alpha - \beta)/(2\tilde{\lambda}^{1/2})$. For $T \ll T_{\rm c}$, the three minima are at $\varphi = 0, \pm\alpha$, and the values of the potential at these points are $V_{\rm eff}(0) = V_0$ and $V_{\rm eff}(\pm\alpha) = 0$. Thus, the vacua at $\varphi = \pm\alpha$ represent two true vacua of the field, while the one at $\varphi = 0$ is called a false vacuum state. When $T \gg T_{\rm c}$ the expectation value of the inflaton is $\varphi = 0$. At this stage no inflation occurs simply because the energy density of the radiation still exceeds that of the inflaton. When the temperature drops below $T_{\rm c}$, the field gets trapped in the

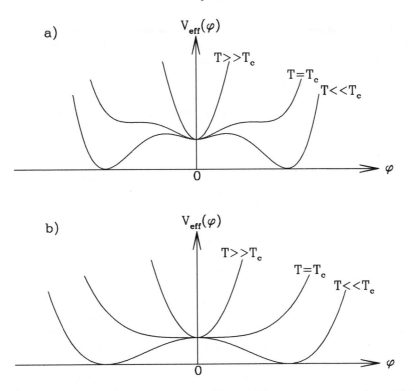

Fig. 3.10. Two examples of effective scalar potentials, at three different temperatures, that can lead to inflation. In example (a) φ experiences a first-order phase transition, characteristic of the old inflation models, while in (b) the phase transition is of second order.

false vacuum at $\varphi = 0$, and the system is said to undergo supercooling. At this point, the slow-roll condition is satisfied, and $\rho \sim V(\varphi = 0)$ is dominated by the energy density of the inflaton. Consequently, $\varphi(\mathbf{x})$ acts like a cosmological constant, the Universe enters a de Sitter phase with a Friedmann equation of the form (3.254), and the Universe expands exponentially. The epoch of inflation only ends when thermal fluctuations or quantum tunneling moves φ across the barrier so that it can proceed towards its true vacuum. This transition is a spontaneous symmetry breaking, and since the field value changes discontinuously, it is of first order. During the transition the energy $V(\varphi = 0)$ associated with the inflaton field, the so-called latent heat, is rapidly liberated and can be used for reheating. If the system stays in the false vacuum sufficiently long, the Universe can be inflated by a sufficiently large number of e-foldings. It thus appears that this model fulfills all requirements.

However, it was realized that this model has a 'graceful exit' problem (Guth, 1981; Guth & Weinberg, 1981). Because the transition is of first order, it proceeds through the nucleation of bubbles of the true vacuum in a surrounding sea of false vacuum. These bubbles must grow in a causal way, and so their sizes at the end of inflation cannot be larger than the horizon size at that time, which is much smaller than our past light cone. In addition, the latent heat needed for reheating is stored in the kinetic energy of the nucleated bubbles, and reheating only occurs when this kinetic energy is thermalized via bubble collisions. Thus, unless bubbles can collide and homogenize in the Hubble radius, the model will predict too large inhomogeneities to match the observed isotropy of the CMB and too little reheating. However, since the space between the bubbles is filled with exponentially expanding false vacuum, while the volume of a bubble expands only with a low power of time, percolation and homogenization of bubbles can never happen. Instead, inflation continues indefinitely, and the bubbles of true vacuum have only a small volume filling factor at any time. The volume filling factor can be increased by increasing the nucleation rate of true-vacuum bubbles, but this would require a high tunneling rate, making the inflation period too short.

(b) New Inflation Because of the 'graceful exit' problem, a modified scenario has been proposed by Linde (1982) and Albrecht & Steinhardt (1982). The prototype potential in this scenario has the form

$$V(\varphi) = \frac{1}{4}\lambda \left(\varphi^2 - \sigma^2\right)^2, \qquad (3.262)$$

and the effective potential $V_{\text{eff}}(\varphi)$ is plotted in Fig. 3.10 b for different temperatures. At high temperature, the effective potential has a single minimum at $\varphi = 0$, but when the temperature drops below a critical value, $T_c = \sigma(\lambda/\tilde{\lambda})^{1/2}$, the minimum at $\varphi = 0$ disappears (and becomes a local maxima) while two new minima develop. As in old inflation, the scalar field is confined to the neighborhood of $\varphi(\mathbf{x}) = 0$ by the thermal force at $T \gg T_c$, when the Universe is dominated by radiation. As the temperature drops to $T \sim T_c$ (when the vacuum energy of φ begins to dominate over radiation), the field configuration at $\varphi(\mathbf{x}) = 0$ becomes unstable and it evolves towards $\varphi = \pm\sigma$ as the temperature decreases. The change from $\varphi = 0$ to $\varphi = \pm\sigma$ is smooth everywhere, and so the spontaneous symmetry breaking occurs via a second-order phase transition. As long as the evolution obeys the slow-roll requirements derived above, inflation will occur. When φ approaches σ (or $-\sigma$), the field rolls rapidly towards the minimum (because of the large potential gradient). Since this violates the slow-roll requirement, it signals the end of inflation. The inflaton φ subsequently oscillates around the minimum with a frequency ω given by $\omega^2 = (\mathrm{d}^2 V/\mathrm{d}\varphi^2)_\sigma = \lambda\sigma^2$. If the field is coupled to the radiation field, these oscillations will be damped by the decay of φ into photons and other particles, and the Universe is reheated to a temperature $T \sim \omega \sim T_i$, with T_i the temperature at the onset of inflation. The Universe then enters the radiation dominated era of the ordinary FRW cosmology.

The spatial fluctuations in $\varphi(\mathbf{x})$ are expected to be correlated over some microphysical scale and, as a result, the field is homogeneous within domains with sizes typically of the correlation length. Since the correlated domains are established before the onset of inflation, any domain boundaries are inflated outside the present Hubble radius and the inflation in a domain stops when $|\varphi| \sim \sigma$. Since our Universe is thus contained within a single domain, there is no 'graceful exit' problem in this model. Hence the new inflation model is an improvement over the old one. Unfortunately it also has problems. In order to obtain inflation, we must have $\mathrm{d}^2 V/\mathrm{d}\varphi^2 \ll V/m_{\text{Pl}}^2$ [see Eq. (3.259)] which, for $V = (\varphi^2 - \sigma^2)^2/4$, requires $\sigma \gg m_{\text{Pl}}$. This is obviously an unnatural condition, since m_{Pl} is the highest energy scale expected in particle physics. There is also a more general problem. In order to ensure a large-enough number of e-foldings, the initial value of φ must satisfy $|\varphi_i| \ll \sigma$. However, since the thermal fluctuations of φ at the initial time (when $T = T_i$) are expected to be of the order $\lambda^{-1/4} T_i \sim [V(0)/\lambda]^{1/4} \sim \sigma$, fine-tuning is needed to get the required initial condition, $|\varphi_i| \ll \sigma$.

(c) Chaotic Inflation Chaotic inflation was proposed to give a more natural explanation for the initial conditions leading to inflation (Linde, 1986). Unlike in the old and new inflation models, no phase transition is involved here so that no initial thermal bath is required. In this model, one starts with a simple potential, e.g. $V(\varphi) = m\varphi^2/2$, and inflation simply arises because of the slow motion of φ from some initial value, φ_i, towards the potential minimum. At any given point $\mathbf{x}$, the initial field configuration is assumed to be set by some chaotic processes. The values of φ are expected to be the same within regions with a size set by the correlation length. Inflation only occurs in those regions where the conditions needed for inflation are attained; the other regions simply never inflate. Chaotic inflation therefore predicts that the Universe is locally homogeneous, but globally inhomogeneous. In a region where inflation persists for a sufficiently long period, the boundary of this region can be blown out of the current particle horizon, leaving a universe in which the initial inhomogeneities generated by the chaotic processes have no observable consequences. In this scenario our Universe is assumed to have emerged from one of such regions.

In order to solve the horizon and flatness problems, the number of e-foldings must be $N \gtrsim 60$. Using the slow-roll approximation, we can write the number of e-foldings between t_i and t_e (the times when inflation starts and terminates) as

$$N = \int_{t_i}^{t_e} H \, dt \sim -\frac{1}{m_{Pl}^2} \int_{\varphi_i}^{\varphi_e} \frac{V}{|dV/d\varphi|} d\varphi. \tag{3.263}$$

For a smooth potential such as $V(\varphi) = m^2 \varphi^2/2$, $|dV/d\varphi| \sim V/\varphi$ and so $N \sim (\varphi_i/m_{Pl})^2$ (assuming that $\varphi_e \ll \varphi_i$). It then follows that $\varphi_i \gg m_{Pl}$ is needed to have successful inflation. If inflation starts near the Planck time, the fluctuations in V are about m_{Pl}^4, and for the potential in consideration $m \ll m_{Pl}$ is required. It is unclear if such a small mass scale can be achieved in a Planck time, because the most natural mass scale at this time is m_{Pl}. Indeed, if inflation happened at the Planck time, it may not be really possible to construct a realistic inflation model without a proper understanding of quantum gravity. In this sense, our initial hope that inflation models would solve some of the problems in the standard model within the classical space-time framework is not realized.

The schemes and problems discussed above are typical of many other inflation models suggested. At the present time, it is fair to say that, although the concept of inflation can help to solve several outstanding problems in standard cosmology, a truly successful model is still lacking.

4
Cosmological Perturbations

In the standard model of cosmology described in the previous chapter, the Universe is assumed to be highly homogeneous at early times. The structures observed today, such as galaxies and the clusters of galaxies, are assumed to have grown from small initial density perturbations due to the action of gravity. In this scenario, structure formation in the Universe involves the following two aspects: (i) the properties and origin of the initial density perturbations, and (ii) the time evolution of the cosmological perturbations in an expanding Universe.

In this chapter we examine the origin of cosmological perturbations and their evolution in the regime where the amplitudes of the perturbations are small. We begin in §4.1 with a description of Newtonian perturbation theory in the linear regime. This applies to structures with sizes much smaller than the horizon size, so that causality can be considered instantaneous, and with a density contrast relative to the background much smaller than unity. The relativistic theory of small perturbations is dealt with in §4.2. This is an extension of the Newtonian perturbation theory, and is required when considering perturbations larger than the horizon size or when the matter content of the perturbations cannot be treated as a non-relativistic fluid. For a Universe with a given matter content, the theories presented in these two sections allow one to trace the time evolution of the cosmological perturbations in the linear regime. The nonlinear evolution will be discussed in Chapter 5.

If we decompose the cosmological perturbations into Fourier modes, we will find that some modes are amplified during the linear evolution while others are damped. The evolution therefore acts as a filter of the primordial density perturbations generated at some time in the early Universe. In §4.3 we show that the results of the evolution at a later time are most conveniently represented by a linear transfer function which describes the change in the perturbation amplitude as a function of the Fourier mode. The importance of the linear transfer function is that, once the perturbation spectrum (i.e. the perturbation amplitude as a function of Fourier mode) is set at some time in the early Universe, it allows us to calculate the linear perturbation power spectrum at any later times. In §4.4 we describe how to characterize the statistical properties of the cosmological perturbations. As we will see, if the density perturbations have a Gaussian distribution, then the density field is completely specified by the power spectrum. Finally, in §4.5, we briefly discuss the origin of cosmological perturbations.

4.1 Newtonian Theory of Small Perturbations

4.1.1 Ideal Fluid

Consider the Newtonian theory for the evolution of the density ρ and velocity $\mathbf{u}$ of a non-relativistic fluid under the influence of a gravitational field with potential ϕ. The fluid description is valid as long as the mean free path of the particles in consideration is much smaller than the

4.1 Newtonian Theory of Small Perturbations

scales of interest (see §B1.2). This applies to a baryonic gas, in which the frequent collisions among the particles can establish local thermal equilibrium. The fluid description is also valid for a pressureless dust (i.e. for collisionless dark matter), as long as the local velocity dispersion of the dark matter particles is sufficiently small that particle diffusion can be neglected on the scales of interest (see §4.1.4).

The time evolution of a fluid is given by the equation of continuity (which describes mass conservation), the Euler equations (the equations of motion) and the Poisson equation (describing the gravitational field):

$$\frac{D\rho}{Dt} + \rho \nabla_{\mathbf{r}} \cdot \mathbf{u} = 0 \quad \text{(continuity)}, \tag{4.1}$$

$$\frac{D\mathbf{u}}{Dt} = -\frac{\nabla_{\mathbf{r}} P}{\rho} - \nabla_{\mathbf{r}} \phi \quad \text{(Euler)}, \tag{4.2}$$

$$\nabla_{\mathbf{r}}^2 \phi = 4\pi G \rho \quad \text{(Poisson)}, \tag{4.3}$$

where $\mathbf{r}$ is the *proper* coordinate, $\partial/\partial t$ is the partial derivative for fixed $\mathbf{r}$, and

$$\frac{D}{Dt} \equiv \frac{\partial}{\partial t} + \mathbf{u} \cdot \nabla_{\mathbf{r}} \tag{4.4}$$

is the convective time derivative that describes the time derivative as a quantity moves with the fluid. Note that Eqs. (4.1)–(4.3) describe five relations for six unknowns (ρ, u_x, u_y, u_z, P, and ϕ). Therefore, in order to close the set of equations, they need to be supplemented by the equation of state to specify the fluid pressure P (see below).

To discuss the time evolution of the perturbations in an expanding Friedmann–Robertson–Walker (FRW) universe, it is best to use the comoving coordinates $\mathbf{x}$ defined as

$$\mathbf{r} = a(t)\mathbf{x}. \tag{4.5}$$

The proper velocity, $\mathbf{u} = \dot{\mathbf{r}}$, at a point $\mathbf{x}$ can then be written as

$$\mathbf{u} = \dot{a}(t)\mathbf{x} + \mathbf{v}, \quad \mathbf{v} \equiv a\dot{\mathbf{x}}, \tag{4.6}$$

where an overdot denotes derivative with respect to t, and $\mathbf{v}$ is the peculiar velocity describing the motion of the fluid element relative to the fundamental observer (the one comoving with the background) at $\mathbf{x}$. With $(\mathbf{x},t)$ replacing $(\mathbf{r},t)$ as the space-time coordinates, the time and spatial derivatives transform as

$$\nabla_{\mathbf{r}} \to \frac{1}{a}\nabla_{\mathbf{x}}; \quad \frac{\partial}{\partial t} \to \frac{\partial}{\partial t} - \frac{\dot{a}}{a}\mathbf{x} \cdot \nabla_{\mathbf{x}}. \tag{4.7}$$

Expressing the density ρ in terms of the density perturbation contrast against the background,

$$\rho(\mathbf{x},t) = \overline{\rho}(t)[1 + \delta(\mathbf{x},t)], \tag{4.8}$$

and using the fact that $\overline{\rho} \propto a^{-3}$, one can write Eqs. (4.1)–(4.3) in comoving coordinates as

$$\frac{\partial \delta}{\partial t} + \frac{1}{a}\nabla \cdot [(1+\delta)\mathbf{v}] = 0, \tag{4.9}$$

$$\frac{\partial \mathbf{v}}{\partial t} + \frac{\dot{a}}{a}\mathbf{v} + \frac{1}{a}(\mathbf{v} \cdot \nabla)\mathbf{v} = -\frac{\nabla \Phi}{a} - \frac{\nabla P}{a\overline{\rho}(1+\delta)}, \tag{4.10}$$

$$\nabla^2 \Phi = 4\pi G \overline{\rho} a^2 \delta, \quad \Phi \equiv \phi + a\ddot{a}x^2/2, \tag{4.11}$$

where $\nabla \equiv \nabla_{\mathbf{x}}$, and $\partial/\partial t$ is now for fixed $\mathbf{x}$. For a given cosmology [which specifies $a(t)$], and a given equation of state, the above set of equations can in principle be solved.

The above description, based on the assumption that the matter content of the Universe is a non-relativistic fluid, can be extended to cases where the Universe contains a smooth background of relativistic particles (photons and neutrinos) or vacuum energy (e.g. the cosmological constant). In such a case, both the continuity and Euler equations for a Newtonian fluid have the same forms as Eqs. (4.1) and (4.2), but the gravitational potential ϕ should include the contributions from these additional energy components. This can be done by replacing the Poisson equation (4.3) by

$$\nabla_r^2 \phi = 4\pi G (\rho + \tilde{\rho}_r + \tilde{\rho}_v), \tag{4.12}$$

where $\tilde{\rho}_r$ and $\tilde{\rho}_v$ are the effective gravitational mass densities of the relativistic background and the vacuum, respectively. As we have seen in §3.2.1, both the energy density and pressure of a relativistic fluid act as sources of the gravitational acceleration, and in general the source term is $\tilde{\rho} = \rho + 3P/c^2$. Thus, $\tilde{\rho}_r = \rho_r + 3P_r/c^2 = 2\rho_r$, and $\tilde{\rho}_v = \rho_v + 3P_v/c^2 = -2\rho_v$. These terms are exactly the same as those in the equation for $\ddot{a}$ [see Eq. (3.58)], and are therefore subtracted from the potential perturbation, Φ, defined in Eq. (4.11). Thus, the density perturbation δ (defined against the mean density of the non-relativistic fluid, rather than the total mean density), the peculiar velocity $\mathbf{v}$, and the potential perturbation Φ still obey Eqs. (4.9)–(4.11). The effect of adding a smooth component is only to change the general expansion of the background, i.e. to change the form of $a(t)$.

The equation of state of the fluid is determined by the thermodynamic process that the fluid undergoes. In special cases where the pressure of the fluid depends only on the density ρ, the set of fluid equations is completed with the equation of state, $P = P(\rho)$. More generally, however, we can write the equation of state as

$$P = P(\rho, S), \tag{4.13}$$

where S is the specific entropy. Since this introduces a new quantity, we need an extra equation in order to complete the set of equations. From the definition, $dS = dQ/T$ (where dQ is the amount of heat added to a fluid element with unit mass), we get

$$T \frac{dS}{dt} = \frac{\mathcal{H} - \mathcal{C}}{\rho}, \tag{4.14}$$

where $\mathcal{H}$ and $\mathcal{C}$ are the heating and cooling rates per unit volume, respectively. As described in Chapter 8, the forms of $\mathcal{H}$ and $\mathcal{C}$ are determined by physical processes such as radiative emission and absorption, and can be obtained from physical principles. If the evolution is adiabatic, then $dS/dt = 0$.

For an ideal non-relativistic monatomic gas, the first law of thermodynamics applied to a unit mass is

$$T \, dS = d\left(\frac{3}{2}\frac{P}{\rho}\right) + P \, d\left(\frac{1}{\rho}\right). \tag{4.15}$$

Using $P = (\rho/\mu m_p) k_B T$, with μ the mean molecular weight in units of the proton mass m_p, to substitute the temperature yields

$$d\ln P = \frac{5}{3} d\ln\rho + \frac{2}{3} \frac{\mu m_p}{k_B} S \, d\ln S, \tag{4.16}$$

and thus

$$P \propto \rho^{5/3} \exp\left(\frac{2}{3} \frac{\mu m_p}{k_B} S\right). \tag{4.17}$$

From this we have

$$\frac{\nabla P}{\bar{\rho}} = \frac{1}{\bar{\rho}}\left[\left(\frac{\partial P}{\partial \rho}\right)_S \nabla \rho + \left(\frac{\partial P}{\partial S}\right)_\rho \nabla S\right]$$
$$= c_s^2 \nabla \delta + \frac{2}{3}(1+\delta)T\nabla S, \qquad (4.18)$$

where

$$c_s = \left(\frac{\partial P}{\partial \rho}\right)_S^{1/2} \qquad (4.19)$$

is the adiabatic sound speed. Thus, the Euler equation (4.10) can be written as

$$\frac{\partial \mathbf{v}}{\partial t} + \frac{\dot{a}}{a}\mathbf{v} + \frac{1}{a}(\mathbf{v}\cdot\nabla)\mathbf{v} = -\frac{\nabla \Phi}{a} - \frac{c_s^2}{a}\frac{\nabla \delta}{(1+\delta)} - \frac{2T}{3a}\nabla S. \qquad (4.20)$$

In special cases where both δ and $\mathbf{v}$ are small so that the nonlinear terms in Eqs. (4.9) and (4.10) can be neglected, we get the following set of linear differential equations:

$$\frac{\partial \delta}{\partial t} + \frac{1}{a}\nabla\cdot\mathbf{v} = 0, \qquad (4.21)$$

$$\frac{\partial \mathbf{v}}{\partial t} + \frac{\dot{a}}{a}\mathbf{v} = -\frac{\nabla \Phi}{a} - \frac{c_s^2}{a}\nabla \delta - \frac{2\bar{T}}{3a}\nabla S, \qquad (4.22)$$

where $\bar{T}$ is the temperature of the background, and c_s is the sound speed evaluated using the background quantities. Operating $\nabla \times$ on both sides of Eq. (4.22) gives

$$\nabla \times \mathbf{v} \propto a^{-1}. \qquad (4.23)$$

Thus, in the linear regime, the curl of the peculiar velocity field dies off with the expansion and can be neglected at late times. This basically expresses the conservation of angular momentum in an expanding universe. Note that since the source terms in Eq. (4.22) are all gradients of scalars, which are curl-free, there is no source for vorticity, at least not in the linear, Newtonian regime considered here.

Differentiating Eq. (4.21) once with respect to t and using Eqs. (4.11) and (4.22) we obtain

$$\frac{\partial^2 \delta}{\partial t^2} + 2\frac{\dot{a}}{a}\frac{\partial \delta}{\partial t} = 4\pi G \bar{\rho}\delta + \frac{c_s^2}{a^2}\nabla^2 \delta + \frac{2}{3}\frac{\bar{T}}{a^2}\nabla^2 S. \qquad (4.24)$$

The second term on the left-hand side is the Hubble drag term, which tends to suppress perturbation growth due to the expansion of the Universe. The first term on the right-hand side is the gravitational term, which causes perturbations to grow via gravitational instability. The last two terms on the right-hand side are both pressure terms: the $\nabla^2 \delta$ term is due to the spatial variations in density, while the $\nabla^2 S$ term is caused by the spatial variations of specific entropy.

In the linear regime, the equations governing the evolution of the perturbations are all linear in perturbation quantities. It is then useful to expand the perturbation fields in some suitably chosen mode functions. If the curvature of the Universe can be neglected, as is the case when the Universe is flat or when the scales of interest are much smaller than the horizon size, the mode functions can be chosen to be plane waves and the perturbation fields can be represented by their Fourier transforms. For example, for $\delta(\mathbf{x},t)$ we can write

$$\delta(\mathbf{x},t) = \sum_{\mathbf{k}} \delta_{\mathbf{k}}(t)\exp(i\mathbf{k}\cdot\mathbf{x}); \quad \delta_{\mathbf{k}}(t) = \frac{1}{V_u}\int \delta(\mathbf{x},t)\exp(-i\mathbf{k}\cdot\mathbf{x})\,d^3\mathbf{x}, \qquad (4.25)$$

where V_u is the volume of a large box on which the perturbations are assumed to be periodic. This convention of Fourier transformation is used throughout this book. Since we will always write the perturbation fields in terms of the *comoving* coordinates **x**, the wavevectors **k** are also in comoving units.

We can obtain the evolution equation for each of the individual modes, $\delta_\mathbf{k}(t)$ and $S_\mathbf{k}(t)$, corresponding to $\delta(\mathbf{x},t)$ and $S(\mathbf{x},t)$, by Fourier transforming Eq. (4.24). Using the fact that ∇ can be replaced by $i\mathbf{k}$ and ∇^2 by $-k^2$ in the Fourier transformation, we obtain

$$\frac{d^2\delta_\mathbf{k}}{dt^2} + 2\frac{\dot{a}}{a}\frac{d\delta_\mathbf{k}}{dt} = \left[4\pi G\overline{\rho} - \frac{k^2 c_s^2}{a^2}\right]\delta_\mathbf{k} - \frac{2}{3}\frac{\overline{T}}{a^2}k^2 S_\mathbf{k}. \quad (4.26)$$

In addition, the Fourier transformed Poisson equation is

$$-k^2\Phi_\mathbf{k} = 4\pi G\overline{\rho}a^2\delta_\mathbf{k}, \quad (4.27)$$

where $\Phi_\mathbf{k}$ is the Fourier transform of the perturbed gravitational potential field. Finally, since the velocity field can be considered curl-free, we can write **v** as the gradient of a velocity potential: $\mathbf{v} = \nabla\mathscr{V}$, and so $\mathbf{v_k} = i\mathbf{k}\mathscr{V}_\mathbf{k}$. It then follows from the Fourier transformation of Eq. (4.21) that

$$\mathbf{v_k} = \frac{ia\mathbf{k}}{k^2}\frac{d\delta_\mathbf{k}}{dt}. \quad (4.28)$$

4.1.2 Isentropic and Isocurvature Initial Conditions

As one can see from Eq. (4.24), both the density perturbation, $\delta_\mathbf{k}$, and the entropy perturbation, $S_\mathbf{k}$, act as sources for the evolution of the density fluctuations. Entropy perturbations correspond to spatial variations in pressure and can generate density fluctuations through adiabatic expansion and compression. Therefore, there are two distinct initial perturbations that can seed density fluctuations: isentropic perturbations, for which the initial perturbation is in the density ($\delta_i \neq 0$) but not in the specific entropy ($\delta S_i = 0$), and isocurvature perturbations, for which $\delta_i = 0$ but $\delta S_i \neq 0$. In general, both isentropic and isocurvature perturbations may be present in the initial conditions. Since perturbations in the space-time metric are associated with perturbations in the energy density, isentropic perturbations correspond to perturbations in space-time curvature, while isocurvature perturbations do not. Because of this, isentropic perturbations are also called curvature perturbations, in contrast to isocurvature perturbations.

Note that 'isentropic' and 'isocurvature' are nomenclature used to indicate the nature of initial perturbations rather than the properties of the evolution. Even if the initial perturbation is isentropic, entropy fluctuations can still be generated through non-adiabatic processes. In cosmology, one sometimes uses 'adiabatic perturbations' to refer to isentropic initial conditions combined with adiabatic evolution. Strictly speaking, however, the term 'adiabatic' should be used to specify the evolutionary process, while 'isentropic' should be used to indicate the nature of the spatial fluctuations.

As we will see later in this chapter, isentropic initial perturbations are naturally predicted by inflationary models, whereas isocurvature initial conditions may be generated in the early Universe as spatial variations in the abundance ratios (e.g. the photon/baryon ratio) while keeping the total energy density uniform in space.

4.1.3 Gravitational Instability

For isentropic initial perturbations with adiabatic evolution, one can set $k^2 S_\mathbf{k} = 0$. If we ignore for the moment the expansion of the Universe, then Eq. (4.26) becomes

$$\frac{d^2\delta_\mathbf{k}}{dt^2} = -\omega^2\delta_\mathbf{k} \quad \text{with} \quad \omega^2 = \frac{k^2 c_s^2}{a^2} - 4\pi G\overline{\rho}. \quad (4.29)$$

This defines a characteristic *proper* length, the Jeans length,

$$\lambda_{\rm J} \equiv \frac{2\pi a}{k_{\rm J}} = c_{\rm s}\sqrt{\frac{\pi}{G\rho}}, \qquad (4.30)$$

which expresses the distance a sound wave can travel in a gravitational free-fall time $t_{\rm ff} \simeq (G\rho)^{-1/2}$. For $\lambda < \lambda_{\rm J}$ ($k > k_{\rm J}$) we have $\omega^2 > 0$ and the solution of Eq. (4.29), given by $\delta_{\bf k}(t) = \exp(\pm i\omega t)$, corresponds to a sound wave (also called acoustic wave), that propagates with the sound speed. If $\lambda > \lambda_{\rm J}$ ($k < k_{\rm J}$), then $\omega^2 < 0$. In this case, the pressure can no longer support the gravity and the solution of Eq. (4.29) represents a non-propagating, stationary wave with an amplitude that either increases (growing mode) or decreases (decaying mode) with time exponentially. This growing mode reflects the gravitational instability, or Jeans instability. If the expansion is taken into account, the Hubble drag modifies these solutions: it causes a slow damping of the acoustic waves when $\lambda < \lambda_{\rm J}$, while the growth of the unstable modes in the case with $\lambda > \lambda_{\rm J}$ is slowed down (see §4.1.6).

It is clear from the above that only perturbations with $k < k_{\rm J}$ can grow. After recombination, when baryonic matter decouples from radiation (i.e. at $z \lesssim z_{\rm dec} \approx 1{,}100$), the relevant sound speed can be approximated as that of a non-relativistic monatomic gas:

$$c_{\rm s} = \left(\frac{5k_{\rm B}T}{3m_{\rm p}}\right)^{1/2}, \qquad (4.31)$$

where $m_{\rm p}$ is the proton mass. The *comoving* Jeans length is then

$$\lambda_{\rm J} \approx 0.01(\Omega_{\rm b,0}h^2)^{-1/2}\,{\rm Mpc}. \qquad (4.32)$$

If we define the mass corresponding to a wavelength λ as the mass within a sphere of radius $\lambda/2$, the corresponding Jeans mass is

$$M_{\rm J} \equiv \frac{\pi}{6}\overline{\rho}_{\rm m,0}\lambda_{\rm J}^3 \approx 1.5 \times 10^5 (\Omega_{\rm b,0}h^2)^{-1/2}\,{\rm M}_\odot. \qquad (4.33)$$

Thus, shortly after recombination the Jeans mass is comparable to that of a globular cluster.

Before recombination, however, electrons and photons are tightly coupled via Compton scattering. In this case, baryons and photons act like a single fluid with $\rho = \rho_{\rm b} + \rho_{\rm r}$ and $P = P_{\rm r} = \frac{1}{3}\rho_{\rm r}c^2$ (see §3.1.5). For an adiabatic process, the energy densities $\rho_{\rm b}$ and $\rho_{\rm r}$ change due to adiabatic compression or contraction of volume elements as $\rho_{\rm b} \propto V^{-1}$ and $\rho_{\rm r} \propto V^{-4/3}$, and so the adiabatic sound speed is

$$c_{\rm s} = \frac{c}{\sqrt{3}}\left[\frac{3}{4}\frac{\overline{\rho}_{\rm b}(z)}{\overline{\rho}_{\gamma}(z)} + 1\right]^{-1/2}. \qquad (4.34)$$

Unfortunately, since the content of the fluid is not non-relativistic, the Newtonian treatment considered above is no longer a good approximation. Nevertheless, for an order-of-magnitude analysis, we may still use Eq. (4.30) to define a Jeans length. At the time when matter and radiation have equal energy density, which is just prior to recombination, the Jeans mass is

$$M_{\rm J} \approx 1.2 \times 10^{16}(\Omega_{\rm b,0}h^2)^{-2}\,{\rm M}_\odot. \qquad (4.35)$$

Thus, isentropic baryonic perturbations with scale sizes smaller than that of a supercluster cannot grow before recombination. At recombination, however, the Jeans mass rapidly decreases by about 10 orders of magnitude, to the scale of globular clusters, and all perturbations of intermediate scales can start to grow after recombination.

To see how isocurvature perturbations seed gravitational instability, let us consider a case where the evolution is adiabatic so that $S_{\bf k}$ is independent of time. If we denote the two general solutions for the homogeneous equation corresponding to Eq. (4.26) (i.e. with the $S_\mathbf{k}$ term set to zero) by δ_+ (the growing mode) and δ_- (the decaying mode), it can be shown that

$$\dot{\delta}_+ \delta_- - \dot{\delta}_- \delta_+ = Ca^{-2}, \qquad (4.36)$$

where C is a constant. One can then show that the special solution to Eq. (4.26) is

$$\delta_\mathbf{k}^{\mathrm{iso}}(t) = -\frac{2}{3}k^2 S_\mathbf{k}$$
$$\times \left[\delta_+(\mathbf{k},t) \int_{t_i}^t T(t') \delta_-(\mathbf{k},t') \,\mathrm{d}t' - \delta_-(\mathbf{k},t) \int_{t_i}^t T(t') \delta_+(\mathbf{k},t') \,\mathrm{d}t' \right], \qquad (4.37)$$

where the amplitudes of δ_+ and δ_- are chosen so that $C = 1$. Hence, the general solution to Eq. (4.26) can be written as $\delta_\mathbf{k}(t) = \delta_\mathbf{k}^{\mathrm{iso}}(t) + A_+ \delta_+(\mathbf{k},t) + A_- \delta_-(\mathbf{k},t)$, where A_+ and A_- are constants to be determined by initial conditions. For isocurvature initial conditions, $A_+ = A_- = 0$, and so $\delta_\mathbf{k}(t) = \delta_\mathbf{k}^{\mathrm{iso}}(t)$. Thus, isocurvature initial conditions induce both growing and decaying modes of density perturbations. At late times, when the decaying mode can be neglected, these density perturbations evolve in exactly the same way as in the isentropic case.

4.1.4 Collisionless Gas

The fluid approximation discussed above is valid only when the mean free path of particles is much smaller than the spatial scale of the perturbations in consideration. In other cases where the collisions between particles are rare and the mean free path is large, the evolution of perturbations is specified by the particle distribution function $f(\mathbf{x},\mathbf{p},t)$, which gives the number of particles per unit volume in phase space:

$$\mathrm{d}N = f(\mathbf{x},\mathbf{p},t) \, \mathrm{d}^3\mathbf{x} \, \mathrm{d}^3\mathbf{p}. \qquad (4.38)$$

Here $p_i = \partial \mathscr{L}/\partial \dot{x}^i$ is the *canonical* momentum conjugate to the comoving coordinates x^i. To obtain $\mathbf{p}$, we use the Lagrangian of a particle with mass m in an expanding universe:

$$\mathscr{L} = \frac{1}{2}m(a\dot{\mathbf{x}} + \dot{a}\mathbf{x})^2 - m\phi(\mathbf{x},t), \qquad (4.39)$$

which can be transformed into

$$\mathscr{L} = \frac{1}{2}ma^2\dot{\mathbf{x}}^2 - m\Phi \qquad (4.40)$$

by a canonical transformation $\mathscr{L} \to \mathscr{L} - \mathrm{d}X/\mathrm{d}t$ with $X = ma\dot{a}x^2/2$. It then follows that the canonical momentum and the equation of motion are

$$\mathbf{p} = \frac{\partial \mathscr{L}}{\partial \dot{\mathbf{x}}} = ma^2\dot{\mathbf{x}} = ma\mathbf{v} \quad \text{and} \quad \frac{\mathrm{d}\mathbf{p}}{\mathrm{d}t} = -m\nabla\Phi. \qquad (4.41)$$

According to Liouville's theorem, the phase-space density f of a collisionless gas is a constant along each particle trajectory and therefore obeys the Vlasov equation:

$$\frac{\partial f}{\partial t} + \frac{1}{ma^2}\mathbf{p}\cdot\nabla f - m\nabla\Phi\cdot\frac{\partial f}{\partial \mathbf{p}} = 0. \qquad (4.42)$$

This equation is just a result of conservation of particle number: the rate of change in particle number in a unit phase-space volume is equal to the net flux of particles across its surface.

Although the full Vlasov equation is difficult to solve, one can obtain useful insights by considering the $\mathbf{p}$ moments (or the velocity moments) of f. Generally, the average value of a quantity Q in the neighborhood of $\mathbf{x}$ is

$$\langle Q \rangle = \langle Q \rangle(\mathbf{x}) \equiv \frac{1}{n}\int Q f \, \mathrm{d}^3\mathbf{p}, \qquad (4.43)$$

where

$$n = n(\mathbf{x}) \equiv \int f \, d^3\mathbf{p} = \overline{\rho} a^3 \left[1 + \delta(\mathbf{x})\right]/m \qquad (4.44)$$

is the *comoving* number density of particles at $\mathbf{x}$, and for simplicity we assume the density of the Universe to be dominated by the species of collisionless particles in question. Multiplying Eq. (4.42) by Q and integrating over $\mathbf{p}$ we get

$$\frac{\partial}{\partial t}\left[n\langle Q\rangle\right] + \frac{1}{ma^2}\nabla \cdot \left[n\langle Q\mathbf{p}\rangle\right] + mn\nabla\Phi \cdot \left\langle \frac{\partial Q}{\partial \mathbf{p}}\right\rangle = 0, \qquad (4.45)$$

where we have assumed $f = 0$ for $\mathbf{p} \to \infty$ so that the surface terms equal zero. Setting $Q = m$ in Eq. (4.45) and using Eq. (4.44) gives

$$\frac{\partial \delta}{\partial t} + \frac{1}{a}\sum_j \frac{\partial}{\partial x_j}\left[(1+\delta)\langle v_j\rangle\right] = 0, \qquad (4.46)$$

which is just a result of mass conservation. This is equivalent to the continuity equation for a non-relativistic fluid given by Eq. (4.9), but with $\mathbf{v}$ replaced by the mean streaming motion $\langle \mathbf{v}\rangle$. Note that $\langle \mathbf{v}\rangle$ is the average velocity of particles in the neighborhood of $\mathbf{x}$ and can be much smaller than the typical velocity of individual particles.

The equation of motion can be obtained by setting $Q = v_i$:

$$\frac{\partial}{\partial t}\left[(1+\delta)\langle v_i\rangle\right] + \frac{\dot{a}}{a}(1+\delta)\langle v_i\rangle = -\frac{1+\delta}{a}\frac{\partial \Phi}{\partial x_i} - \frac{1}{a}\sum_j \frac{\partial}{\partial x_j}\left[(1+\delta)\langle v_i v_j\rangle\right]. \qquad (4.47)$$

Multiplying the continuity equation (4.46) with $\langle v_i\rangle$ and subtracting the result from Eq. (4.47) yields the collisionless equivalent of the Euler equations in an expanding universe:

$$\frac{\partial \langle v_i\rangle}{\partial t} + \frac{\dot{a}}{a}\langle v_i\rangle + \frac{1}{a}\sum_j \langle v_j\rangle \frac{\partial \langle v_i\rangle}{\partial x_j} = -\frac{1}{a}\frac{\partial \Phi}{\partial x_i} - \frac{1}{a(1+\delta)}\sum_j \frac{\partial}{\partial x_j}\left[(1+\delta)\sigma_{ij}^2\right], \qquad (4.48)$$

where we have defined

$$\sigma_{ij}^2 \equiv \langle v_i v_j\rangle - \langle v_i\rangle\langle v_j\rangle. \qquad (4.49)$$

A comparison with Eq. (4.10) shows that $\rho\sigma_{ij}^2$, called the stress tensor, plays the role of pressure.

In principle, one can set $Q = v_i v_j$ and obtain the dynamical equation for $\langle v_i v_j\rangle$ which, in turn, will depend on the third velocity moment. As a result, the complete dynamics is given by an infinite number of equations of velocity moments. In practice, we can truncate the hierarchy by making some assumptions. If the velocity stress σ_{ij}^2 is small, so that the right-hand side of Eq. (4.48) is dominated by the gravitational term, then to first order in δ (note that $\langle v\rangle \sim \delta$ in the linear regime), Eqs. (4.46) and (4.48) can be combined to give

$$\frac{\partial^2 \delta}{\partial t^2} + 2\frac{\dot{a}}{a}\frac{\partial \delta}{\partial t} = 4\pi G\overline{\rho}\delta. \qquad (4.50)$$

This equation is the same as Eq. (4.24) with $c_s = 0$. Thus, on scales where the velocity stress is negligible, a collisionless gas can be treated as an ideal fluid with zero pressure. In general, however, the velocity dispersion of a collisionless gas is not negligible, and the particles can stream away from one place to another (see §4.1.5 below), making the fluid treatment invalid.

In the general case, therefore, one needs to consider the evolution of the full distribution function (or, equivalently, the full infinite hierarchy of moment equations). Here we will solve the full Vlasov equation using perturbation theory. Without losing generality we can write

$$f = f_0 + f_1, \qquad (4.51)$$

where f_0 is the unperturbed distribution function and f_1 is the perturbation. Note that f_0 is independent of $\mathbf{x}$ in a homogeneous and isotropic background. The *comoving* number density of particles at $\mathbf{x}$ is $n \equiv \int f \, \mathrm{d}^3 \mathbf{p}$, and so the mass density at $\mathbf{x}$ is

$$\rho(\mathbf{x},t) = \frac{m}{a^3} \int f(\mathbf{x},\mathbf{p},t) \, \mathrm{d}^3 \mathbf{p} = \overline{\rho}(t) \left[\frac{m n_0}{\overline{\rho} a^3} + \delta(\mathbf{x},t) \right], \tag{4.52}$$

where $n_0 \equiv \int f_0 \, \mathrm{d}^3 \mathbf{p}$ is the mean number density of particles (in comoving units), and

$$\delta(\mathbf{x},t) = \frac{m}{\overline{\rho} a^3} \int f_1 \, \mathrm{d}^3 \mathbf{p} \tag{4.53}$$

is the density contrast with respect to the background. The gravitational potential Φ due to the density perturbation is given by the Poisson equation (4.11). For a homogeneous and isotropic background, f_0 depends only on the magnitude of $\mathbf{p}$ and, since Φ is a first-order perturbation, the unperturbed distribution function f_0 obeys

$$\left(\frac{\partial f_0}{\partial t} \right)_q = 0, \tag{4.54}$$

where $q = (\sum_i p_i^2)^{1/2}$ is the magnitude of $\mathbf{p}$. [Note that p is reserved to denote $(-g^{ij} p_i p_j)^{1/2}$, which is equal to q/a in a flat universe.] As we have seen in §3.3.2, the unperturbed particle distribution function has the form $f_0 \propto \left[e^{E/k_B T(a)} \pm 1 \right]^{-1}$, where $E = p^2/2m \propto q^2/a^2 \propto q^2 T(a)$ for non-relativistic particles, and $E = p \propto q/a \propto q T(a)$ for relativistic particles. We thus see that f_0 is independent of a for fixed q (or fixed $|\mathbf{p}|$), instead of for fixed p.

To first order in the perturbation quantities, the equation for f_1 is

$$\frac{\partial f_1}{\partial t} + \frac{1}{ma^2} \mathbf{p} \cdot \nabla f_1 - m \nabla \Phi \cdot \frac{\partial f_0}{\partial \mathbf{p}} = 0 \tag{4.55}$$

or, in terms of Fourier transforms,

$$\frac{\partial f_\mathbf{k}}{\partial \xi} + \frac{i \mathbf{k} \cdot \mathbf{p}}{m} f_\mathbf{k}(\mathbf{p},\xi) = ma^2 \left(i \mathbf{k} \cdot \frac{\partial f_0}{\partial \mathbf{p}} \right) \Phi_\mathbf{k}(t), \tag{4.56}$$

where $\mathrm{d}\xi = \mathrm{d}t/a^2$. This equation can be written in the form

$$\frac{\partial}{\partial \xi} \left[f_\mathbf{k} \exp\left(\frac{i \mathbf{k} \cdot \mathbf{p}}{m} \xi \right) \right] = ma^2 \left(i \mathbf{k} \cdot \frac{\partial f_0}{\partial \mathbf{p}} \right) \Phi_\mathbf{k} \exp\left(\frac{i \mathbf{k} \cdot \mathbf{p}}{m} \xi \right). \tag{4.57}$$

Integrating both sides from some initial time ξ_i to ξ, we get

$$f_\mathbf{k}(\mathbf{p},\xi) = f_\mathbf{k}(\mathbf{p},\xi_i) \exp\left[-\frac{i\mathbf{k} \cdot \mathbf{p}}{m}(\xi - \xi_i) \right]$$
$$+ mi\mathbf{k} \cdot \left(\frac{\partial f_0}{\partial \mathbf{p}} \right) \int_{\xi_i}^{\xi} \mathrm{d}\xi' a^2(\xi') \Phi_\mathbf{k}(\xi') \exp\left[-\frac{i\mathbf{k} \cdot \mathbf{p}}{m}(\xi - \xi') \right]. \tag{4.58}$$

Since the gravitational potential Φ depends on f_1 through the Poisson equation, Eq. (4.58) is an integral equation for $f_\mathbf{k}$ and can be solved iteratively. The first term on the right-hand side of Eq. (4.58) is a kinematic term due to the propagation of the initial condition, while the second term describes the dynamical evolution of the perturbation due to gravitational interaction.

Inserting Eq. (4.58) into Eq. (4.53), it is straightforward to show that the dynamical part of $\delta_\mathbf{k}$ is given by

$$\delta_\mathbf{k}(\xi) = -\frac{mk^2}{\overline{\rho} a^3} \int_{\xi_i}^{\xi} \mathrm{d}\xi'(\xi - \xi') a^2(\xi') \Phi_\mathbf{k}(\xi') \mathscr{G}\left[k(\xi - \xi')/m \right], \tag{4.59}$$

where $\mathscr{G}$ is the Fourier transform of f_0:

$$\mathscr{G}(\mathbf{s}) = \int d^3\mathbf{p} f_0(\mathbf{p}) e^{-i\mathbf{p}\cdot\mathbf{s}}. \qquad (4.60)$$

Since f_0 depends only on q (the amplitude of $\mathbf{p}$), the angular part of the integration over $\mathbf{p}$ can be carried out, giving

$$\delta_{\mathbf{k}}(\xi) = -\frac{4\pi k m^2}{\overline{\rho}a^3} \int_{\xi_i}^{\xi} d\xi' a^2(\xi')\Phi_{\mathbf{k}}(\xi')I_{\mathbf{k}}(\xi - \xi'), \qquad (4.61)$$

where

$$I_{\mathbf{k}}(\xi - \xi') \equiv \int_0^{\infty} dq\, q f_0(q) \sin\left[\frac{kq(\xi - \xi')}{m}\right]. \qquad (4.62)$$

4.1.5 Free-Streaming Damping

If $(\mathbf{k}\cdot\mathbf{p}/m) \gg \xi^{-1}$, i.e. $(a/k) \ll vt$, the phase of the dynamical part in Eq. (4.58) changes so rapidly with ξ' that the integration over ξ' makes little contribution, implying that the perturbation cannot grow with time. For the same reason, the contribution of the kinematic term to the density perturbation, which is proportional to the integration of this term over $\mathbf{p}$, is close to zero because the integrand oscillates rapidly with $\mathbf{p}$. The initial perturbation is therefore damped. What happens physically is that, because of their large random velocities and collisionless nature, particles originally in the crests can move to the troughs and vice versa, leading to damping of the density perturbations. This effect, owing to the random motions of the collisionless particles, is called free-streaming damping and is similar to Landau damping in a plasma (see §5.5.4). The proper length scale below which free-streaming damping becomes important is of the order vt, where t is the age of the Universe, and v is the typical particle velocity at t. More precisely, we define the free-streaming length as the comoving distance traveled by a particle before time t, which can be written as

$$\lambda_{\mathrm{fs}} = \int_0^t \frac{v(t')}{a(t')} dt'. \qquad (4.63)$$

If the particle becomes non-relativistic at t_{nr}, then the peculiar velocity is $v \sim c$ at $t < t_{\mathrm{nr}}$, and $v \propto a^{-1}$ at $t > t_{\mathrm{nr}}$. We will assume the Universe to be radiation dominated before t_{nr}, i.e. $t_{\mathrm{nr}} < t_{\mathrm{eq}}$, as is almost always true in real applications. Using $a(t) \propto t^{1/2}$ at $t < t_{\mathrm{eq}}$ (see §3.2.3), one finds that, at matter–radiation equality,

$$\lambda_{\mathrm{fs}} = \frac{2ct_{\mathrm{nr}}}{a_{\mathrm{nr}}}\left[1 + \ln(a_{\mathrm{eq}}/a_{\mathrm{nr}})\right]. \qquad (4.64)$$

For light neutrinos, the time at which the particles become non-relativistic is given by $3k_B T_\nu(t_{\mathrm{nr}}) \simeq m_\nu c^2$. With $T_\nu = (4/11)^{1/3} T_\gamma$ (see §3.3.3) this implies that $(1 + z_{\mathrm{nr}}) \simeq 6 \times 10^4 (m_\nu/30\,\mathrm{eV})$. Using Eqs. (3.68) and (3.80) we then obtain

$$\lambda_{\mathrm{fs}} \simeq 31 \left(\frac{m_\nu}{30\,\mathrm{eV}}\right)^{-1} \mathrm{Mpc}, \qquad (4.65)$$

which corresponds to a free-streaming mass

$$M_{\mathrm{fs}} \equiv \frac{\pi}{6}\overline{\rho}_{\mathrm{m},0}\lambda_{\mathrm{fs}}^3 \simeq 1.3 \times 10^{15} \left(\frac{m_\nu}{30\,\mathrm{eV}}\right)^{-2} \mathrm{M}_\odot, \qquad (4.66)$$

where we have used that $\Omega_\nu h^2 = 0.32(m_\nu/30\,\mathrm{eV})$ (see §3.3.5). Thus, if the Universe is dominated by light neutrinos, all perturbations with masses smaller than that of a massive cluster are damped out in the linear regime, and the first objects to form are superclusters.

4.1.6 Specific Solutions

(a) Pressureless Fluid For isentropic perturbations in a pressureless fluid (or when $k \ll k_J$), Eq. (4.26) reduces to

$$\frac{d^2\delta_k}{dt^2} + 2\frac{\dot{a}}{a}\frac{d\delta_k}{dt} = 4\pi G\overline{\rho}_m \delta_k, \qquad (4.67)$$

where $\overline{\rho}_m$ is the mean density of the fluid. It can be easily shown that if $\delta_1(t)$ and $\delta_2(t)$ are two solutions then

$$\delta_2\dot{\delta}_1 - \delta_1\dot{\delta}_2 \propto a^{-2}. \qquad (4.68)$$

This is even true if the pressure term is included. Thus, if one solution of Eq. (4.67) is known, the other one can be obtained by solving this first-order differential equation. To solve Eq. (4.67), we recall that the Hubble constant, $H(t) \equiv \dot{a}/a$, obeys

$$\frac{dH}{dt} + H^2 = -\frac{4\pi G}{3}(\overline{\rho}_m - 2\overline{\rho}_v) \qquad (4.69)$$

(see §3.2.1). Since $\overline{\rho}_m \propto a^{-3}$ and $\overline{\rho}_v = $ constant, differentiating the above equation once with respect to t gives

$$\frac{d^2H}{dt^2} + 2\frac{\dot{a}}{a}\frac{dH}{dt} = 4\pi G\overline{\rho}_m H. \qquad (4.70)$$

Thus, both $\delta_k(t)$ and $H(t)$ obey the same equation. Since $H(t)$ decreases with t [see Eq. (3.74)],

$$\delta_- \propto H(t) \qquad (4.71)$$

represents the decaying mode of $\delta(t)$. By directly substituting the solution into Eq. (4.68), one finds that the growing mode can be written as

$$\delta_+ \propto H(t) \int_0^t \frac{dt'}{a^2(t')H^2(t')} \propto H(z) \int_z^\infty \frac{(1+z')}{E^3(z')}dz', \qquad (4.72)$$

where $E(z)$ is given by Eq. (3.75).

For an Einstein–de Sitter universe,

$$\delta_+ \propto t^{2/3}; \quad \delta_- \propto H(t) \propto t^{-1}, \qquad (4.73)$$

which can be obtained directly by solving Eq. (4.67) with $a(t) \propto t^{2/3}$. For an open universe with zero cosmological constant ($\Omega_{m,0} < 1$, $\Omega_{\Lambda,0} = 0$), the growing mode can be written in the following closed form:

$$\delta_+ \propto 1 + \frac{3}{x} + \frac{3(1+x)^{1/2}}{x^{3/2}} \ln\left[(1+x)^{1/2} - x^{1/2}\right], \qquad (4.74)$$

where $x \equiv (\Omega_{m,0}^{-1} - 1)/(1+z)$. This growing mode has the asymptotic behavior that $\delta_+ \propto (1+z)^{-1}$ as $x \to 0$ and $\delta_+ \to 1$ as $x \to \infty$. In general, the growing mode can be obtained from Eq. (4.72) numerically. A good approximation has been found by Carroll et al. (1992):

$$\delta_+ \propto D(z) \propto g(z)/(1+z), \qquad (4.75)$$

where

$$g(z) \approx \frac{5}{2}\Omega_m(z)\left\{\Omega_m^{4/7}(z) - \Omega_\Lambda(z) + [1+\Omega_m(z)/2][1+\Omega_\Lambda(z)/70]\right\}^{-1}, \qquad (4.76)$$

with $\Omega_m(z)$ and $\Omega_\Lambda(z)$ given by Eq. (3.77). Fig. 4.1 shows the linear growth rate $D(z)$. Note that linear perturbations grow faster in an Einstein–de Sitter (EdS) universe than in a universe with $\Omega_{m,0} < 1$. Physically this is due to the fact that in open universes, or in universes with a non-zero

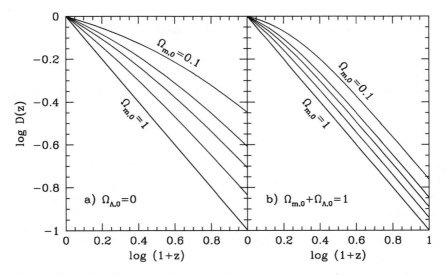

Fig. 4.1. The linear growth rates in cosmological models with $\Omega_{m,0} = 0.1, 0.2, 0.3, 0.5$ and 1. The left panel assumes $\Omega_{\Lambda,0} = 0$ while the right panel assumes $\Omega_{m,0} + \Omega_{\Lambda,0} = 1$. The rates are all normalized so that $D(z=0) = 1$.

cosmological constant, the expansion rate is larger than in an EdS universe, and the perturbation growth is reduced because of the enhanced Hubble drag.

It is also useful to look at the linear evolution of the *potential* perturbations during the matter dominated regime. From the Poisson equation (4.27) we have that $\Phi_\mathbf{k} \propto a^2 D(a) \overline{\rho}_m(a) \propto D(a)/a$. In an EdS universe where $D(a) \propto a$, the potentials do not evolve (they are frozen in). In an open universe, or in a flat universe with a cosmological constant, however, the linear growth rate is suppressed and the potentials decay as the universe expands.

Inserting the growing mode of δ in Eq. (4.28), we have

$$\mathbf{v_k} = \frac{i\mathbf{k}}{k^2} H a \delta_\mathbf{k} f(\Omega_m), \tag{4.77}$$

where

$$f(\Omega_m) \equiv -\frac{d\ln D(z)}{d\ln(1+z)}. \tag{4.78}$$

To a good approximation $f(\Omega_m) \approx \Omega_m^{0.6}$ (Peebles, 1980). Thus, for a fixed matter power spectrum, a cosmology with a larger mean density produces larger peculiar velocities. Note also that larger scale perturbations (smaller k) cause larger peculiar velocities than smaller perturbations of the same magnitude. Thus, if we could somehow measure both the large scale (linear) density field and the peculiar velocity field (i.e. the deviations from the smooth Hubble flow), then Eq. (4.77) would allow us to measure the mean density of the Universe. We will return to this in Chapter 6.

(b) Perturbations in Two Non-Relativistic Components Next we consider isentropic perturbations in a fluid consisting of two non-relativistic components, one with and one without pressure. An interesting example is the evolution of density perturbations in baryons induced by a pressureless (cold) dark matter component. For simplicity, we assume the mean density of the Universe to be dominated by the dark matter. In this case, the density perturbations in baryons, δ_b, obeys the following equation (in Fourier space):

$$\frac{d^2\delta_b}{dt^2} + 2\frac{\dot{a}}{a}\frac{d\delta_b}{dt} + \frac{k^2 c_s^2 a}{a^3}\delta_b = 4\pi G \frac{\overline{\rho}_0 a_0^3}{a^3}\delta_{dm}, \quad (4.79)$$

where δ_{dm}, the density perturbations in the dark matter, obey Eq. (4.67), and $\overline{\rho}_0 \simeq \overline{\rho}_{dm,0}$ is the mean matter density at the present time. In the special case where $c_s^2 a =$ constant (namely for a polytropic fluid with $P \propto \rho^{4/3}$) and $a(t) \propto t^{2/3}$ (i.e. for an EdS universe), a special solution of this equation is

$$\delta_b(\mathbf{k},t) = \frac{\delta_{dm}(\mathbf{k},t)}{1 + k^2/k_J^2}, \quad \text{with} \quad k_J^2 = \frac{3a^2 H^2}{2c_s^2}. \quad (4.80)$$

On large scales ($k \ll k_J$) the baryonic perturbations closely follow the dark matter perturbations ($\delta_b \to \delta_{dm}$); this simply reflects that baryons are like a collisionless fluid when its pressure can be neglected. On small scales ($k \gg k_J$), however, the baryonic pressure cannot be neglected, causing baryon oscillations (i.e. acoustic waves, see below) that slowly damp due to the expansion of the Universe ($\delta_b \to 0$). Although Eq. (4.80) is only a solution of Eq. (4.79) when the baryon component is a polytropic fluid with adiabatic index $\gamma = 4/3$, it can be shown that the general behavior is very similar for other values of γ.

The general solution of the homogeneous equation corresponding to Eq. (4.79), for which the right-hand side is equal to zero, is $\delta_b \propto t^{-(1\pm\varepsilon)/6}$ with $\varepsilon = [1 - 12(k/k_J)^2]^{1/2}$. Since $1 - 12(k/k_J)^2 < 1$, these two solutions correspond to modes which are either decaying or oscillating with time. Thus, in the absence of perturbations in the dark matter ($\delta_{dm} = 0$), there is no growing mode for δ_b, simply because there is no source term (remember that we are assuming that $\rho_{dm} \gg \rho_b$). The only source term for δ_b in this case is δ_{dm}, and the corresponding growing mode is given by Eq. (4.80).

(c) Acoustic Waves Consider once again the case of isentropic perturbations in a fluid consisting of dark matter (with zero pressure) and baryons. As above, we assume that the mean density of the Universe is dominated by the dark matter. If the time scale of interest is much shorter than the Hubble time we can neglect the expansion and Eq. (4.26) reduces to

$$\frac{d^2\delta_b}{dt^2} + \frac{k^2 c_s^2}{a^2}\delta_b = -\frac{k^2}{a^2}\Phi_{dm}. \quad (4.81)$$

Note that this differs from the case discussed in §4.1.3 in that the perturbations in the baryons are not self-gravitating; the gravitational source term is due to the perturbations in the dark matter.

Eq. (4.81) is the equation of motion for a forced oscillator. If the potential $\Phi_{dm}(\mathbf{k},t)$ is constant over the time of interest, the solution can be written as

$$\delta_b(\eta) = \left[\delta_b(0) + \frac{\Phi_{dm}}{c_s^2}\right]\cos(kc_s\eta) + \frac{1}{kc_s}\frac{d\delta_b}{d\eta}(0)\sin(kc_s\eta) - \frac{\Phi_{dm}}{c_s^2}, \quad (4.82)$$

where we have introduced a new time variable $\eta = t/a$. Thus, $\delta_b(\mathbf{k},t)$ oscillates around a zero-point, $-\Phi_{dm}(\mathbf{k},t)/c_s^2$, with a frequency $\omega = kc_s$, and with amplitude and phase set by the initial conditions $\delta_b(0) = \delta_b(\mathbf{k},0)$ and $d\delta_b/d\eta(0) = d\delta_b/d\eta(\mathbf{k},0)$. Using the continuity equation (4.28), we can also obtain the time evolution of the corresponding velocity perturbations:

$$\mathbf{v}_b(t) = \frac{i\mathbf{k}}{k^2}\frac{d\delta_b}{d\eta} = -\frac{ic_s\mathbf{k}}{k}\left[\delta_b(0) + \frac{\Phi_{dm}}{c_s^2}\right]\sin(kc_s\eta) + \frac{i\mathbf{k}}{k^2}\frac{d\delta_b}{d\eta}(0)\cos(kc_s\eta). \quad (4.83)$$

Note that there is a phase difference of $\pi/2$ between the velocity and density perturbations, characteristic of a longitudinal (acoustic) wave.

Acoustic waves also play a role during the pre-recombination era. As shown in §3.5, photons and baryons are tightly coupled before recombination and can be considered as a single fluid with a sound speed given by Eq. (4.34). In this case the acoustic waves are driven by the photon pressure, and for a given mode, the oscillation frequency, amplitude and zero-point all depend on the density ratio between baryons and photons through the sound speed c_s. As we will show in Chapter 6, the acoustic waves in the photon–baryon fluid at the epoch of decoupling give rise to oscillations in the power spectrum of the cosmic microwave background. The amplitudes and separations between peaks (or valleys) of these oscillations can thus be used to constrain the baryon density of the Universe.

(d) Collisional Damping Although photons and baryons are tightly coupled to each other by Compton scattering before recombination, the coupling is imperfect in the sense that the photon mean-free path, $\lambda = (\sigma_T n_e)^{-1}$, is not zero. Consequently, photons can diffuse from high-density to low-density regions, thereby damping the perturbations in the photon distribution. Since the acoustic waves in the pre-recombination era are driven by photon pressure, this photon diffusion also leads to damping of the acoustic oscillations in the baryonic fluid. This damping mechanism is known as Silk damping.

The scale on which Silk damping is important depends on the typical distance a photon can diffuse in a Hubble time. To estimate this distance (which we denote by λ_d), consider the motion of a photon to be a random walk with a mean 'step length' λ. The mean number of 'steps' a photon takes over a time t is therefore $N = ct/\lambda$. It then follows from kinetic theory that

$$\lambda_d = (N/3)^{1/2}\lambda = (ct/3\sigma_T n_e)^{1/2}. \qquad (4.84)$$

At recombination ($z \sim z_{rec} \sim 1{,}100$), this defines a Silk damping mass scale

$$M_d \equiv \frac{\pi}{6}\rho_{m,0}\lambda_d^3 \sim 2.8 \times 10^{12}(\Omega_{b,0}/\Omega_{m,0})^{-3/2}(\Omega_{m,0}h^2)^{-5/4}\,\mathrm{M}_\odot, \qquad (4.85)$$

where we have used an ionization fraction of $X_e = 0.1$ (see §3.5) and we have assumed that the Universe is dominated by matter at $z = z_{rec}$. Note that this is only a rough estimate; more accurate estimates yield a damping mass that is about an order of magnitude larger (see §4.2.5). The perturbations in the baryon–photon fluid with masses below M_d are expected to be damped exponentially during the pre-recombination era. This may have an important impact on galaxy formation. Consider for example a baryonic Einstein–de Sitter cosmology ($\Omega_{b,0} = \Omega_{m,0} = 1$). Then Silk damping will erase all perturbations with masses smaller than $\sim 10^{13}\,\mathrm{M}_\odot$. The only way in which galaxies (which have baryonic masses $\lesssim 10^{11}\,\mathrm{M}_\odot$) can form in such a cosmology is through the fragmentation of non-damped structures with masses larger than $\sim 10^{13}\,\mathrm{M}_\odot$. As we will see later, this requires initial perturbations in the baryon component that are too large to match the observed temperature fluctuations in the cosmic microwave background.

This problem can be circumvented if the matter component of the Universe is dominated by non-baryonic, cold dark matter. Since cold dark matter is pressureless, it does not experience Silk damping. Furthermore, because it decouples when already non-relativistic, cold dark matter does not experience any significant amount of free-streaming either. Thus, at the end of recombination, when the small-scale perturbations in the baryons have been damped away, the perturbations in the dark matter component are still present. Since the baryons basically have become a pressureless fluid after recombination, they rapidly fall into the potential wells associated with these dark matter perturbations. Consequently, in the presence of a significant amount of (cold) dark matter, Silk damping is basically irrelevant for galaxy formation. The damped baryonic perturbations can be 'recreated' at the end of recombination, as is expressed by Eq. (4.80).

(e) Perturbations on a Relativistic Background In the presence of a uniform relativistic background, the perturbations in the non-relativistic component (assumed to be pressureless) obey Eq. (4.67), while the scale factor a is given by

$$\left(\frac{\dot{a}}{a}\right)^2 = \frac{8\pi G}{3}(\overline{\rho}_m + \overline{\rho}_r), \qquad (4.86)$$

where $\bar{\rho}_m \propto a^{-3}$ is the mean density in the non-relativistic component, $\bar{\rho}_r \propto a^{-4}$ is that in the relativistic background, and for simplicity we assume an Einstein–de Sitter universe. Defining a new time variable, $\zeta \equiv \bar{\rho}_m/\bar{\rho}_r = a/a_{eq}$, the equation of motion for $\delta_{\mathbf{k}}$ can be written as

$$\frac{d^2\delta_{\mathbf{k}}}{d\zeta^2} + \frac{(2+3\zeta)}{2\zeta(1+\zeta)}\frac{d\delta_{\mathbf{k}}}{d\zeta} = \frac{3}{2}\frac{\delta_{\mathbf{k}}}{\zeta(1+\zeta)}. \tag{4.87}$$

The two solutions of this equation are

$$\delta_+ \propto 1 + \frac{3}{2}\zeta; \tag{4.88}$$

$$\delta_- \propto \left(1+\frac{3}{2}\zeta\right)\ln\left[\frac{(1+\zeta)^{1/2}+1}{(1+\zeta)^{1/2}-1}\right] - 3(1+\zeta)^{1/2}. \tag{4.89}$$

This shows that perturbations in the non-relativistic component cannot grow ($\delta_+ =$ constant) if the relativistic component dominates the energy density. As long as $\zeta \ll 1$, the Hubble drag therefore causes a stagnation of perturbation growth, known as the Mészáros effect (Mészáros, 1974). As is evident from Eq. (4.88), once the Universe becomes dominated by non-relativistic matter, $\delta_+ \propto a$, in agreement with Eq. (4.73).

4.1.7 Higher-Order Perturbation Theory

When deriving Eq. (4.24) we only used the linearized versions of the continuity equation (4.21) and the Euler equation (4.22). In what followed we have therefore only considered perturbations to linear order. However, some important new insights can be obtained by using a higher order description.

Multiplying Eq. (4.9) by $\mathbf{v}$ and Eq. (4.10) by $(1+\delta)$, adding and then taking the divergence gives

$$\frac{\partial^2\delta}{\partial t^2} + 2\frac{\dot{a}}{a}\frac{\partial\delta}{\partial t} = \frac{\nabla^2 P}{\bar{\rho}a^2} + \frac{1}{a^2}\nabla\cdot(1+\delta)\nabla\Phi + \frac{1}{a^2}\sum_{i,j}\frac{\partial^2}{\partial x^i \partial x^j}\left[(1+\delta)v^i v^j\right], \tag{4.90}$$

which is equivalent to Eq. (4.24), but now without linearization. For simplicity, we consider only pressureless gas. In this case, the above equation in Fourier space can be written as

$$\frac{d^2\delta_{\mathbf{k}}}{dt^2} + 2\frac{\dot{a}}{a}\frac{d\delta_{\mathbf{k}}}{dt} = 4\pi G\bar{\rho}\delta_{\mathbf{k}} + \mathscr{A} - \mathscr{C}, \tag{4.91}$$

where

$$\mathscr{A} = 2\pi G\bar{\rho}\sum_{\mathbf{k}'\neq(0,\mathbf{k})}\left[\frac{\mathbf{k}\cdot\mathbf{k}'}{k'^2}+\frac{\mathbf{k}\cdot(\mathbf{k}-\mathbf{k}')}{|\mathbf{k}-\mathbf{k}'|^2}\right]\delta_{\mathbf{k}'}\delta_{\mathbf{k}-\mathbf{k}'}, \tag{4.92}$$

$$\mathscr{C} = \frac{1}{V_u}\int (1+\delta)\cdot\left(\frac{\mathbf{k}\cdot\mathbf{v}}{a}\right)^2 e^{-i\mathbf{k}\cdot\mathbf{x}}d^3\mathbf{x}, \tag{4.93}$$

and the Euler equation (3.8) (with $P=0$) can be written as

$$\frac{d\mathbf{v_k}}{dt} + \frac{\dot{a}}{a}\mathbf{v_k} = 4\pi G\bar{\rho}a\delta_{\mathbf{k}}\frac{i\mathbf{k}}{k^2} + \mathbf{E}, \tag{4.94}$$

with

$$\mathbf{E} = -\frac{i}{a}\sum_{\mathbf{k}'}\left(\mathbf{k}'\cdot\mathbf{v_{k-k'}}\right)\mathbf{v_{k'}}. \tag{4.95}$$

It is clear from these equations that different Fourier modes are coupled, something that is absent in the first-order perturbation theory. However, it is also evident from these equations that these higher-order terms are negligibly small in the limit $\delta \ll 1$. Mode coupling therefore only becomes important when $\delta \sim 1$.

In general, the nonlinear equations (4.91)–(4.95) are difficult to solve. However, in the quasi-linear regime, where $\delta \lesssim 1$, one can use high-order perturbation theory. One therefore expands the density contrast (and the peculiar velocity) in a perturbation series like

$$\delta(\mathbf{x},t) = \delta^{(1)}(\mathbf{x},t) + \delta^{(2)}(\mathbf{x},t) + \delta^{(3)}(\mathbf{x},t) + \cdots, \quad (4.96)$$

where $\delta^{(1)}$ is the linear solution and $\delta^{(\ell)} = \mathcal{O}[(\delta^{(1)})^\ell]$ is the ℓth-order term obtained by inserting the lower-order solutions into the nonlinear terms, $\mathscr{A}$, $\mathscr{C}$ and $\mathbf{E}$. In principle, such a procedure is straightforward, although the calculations can be rather tedious. We refer the reader to Peebles (1980) for a discussion on the second-order term, and to Bernardeau (1994) for some higher-order solutions.

4.1.8 The Zel'dovich Approximation

Given that all fluctuations were small at early times, it is reasonable to assume that at more recent epochs only the growing mode has a significant amplitude. Under this assumption, the linear evolution of density perturbations reduces to the very simple form

$$\delta(\mathbf{x},a) = D(a)\delta_i(\mathbf{x}), \quad (4.97)$$

where $\delta_i(\mathbf{x})$ is the density perturbation at some initial time t_i, and $D(a)$ is normalized such that $D(a_i) = 1$. Thus the density field grows self-similarly with time. The same is also true both for the gravitational acceleration and the peculiar velocity. This is easily seen by substituting Eq. (4.97) into the Poisson equation (4.11). The scaling of the result with the expansion factor then implies that

$$\Phi(\mathbf{x},a) = \frac{D(a)}{a}\Phi_i(\mathbf{x}) \quad \text{where} \quad \nabla^2 \Phi_i = 4\pi G \overline{\rho}_m a^3 \delta_i(\mathbf{x}). \quad (4.98)$$

In an Einstein–de Sitter universe, where $D \propto a$, this equation implies that Φ is independent of a. Thus, the linearized Euler equation, $\dot{\mathbf{v}} + (\dot{a}/a)\mathbf{v} = -\nabla\Phi/a$, can be integrated for fixed $\mathbf{x}$ to give

$$\mathbf{v} = -\frac{\nabla \Phi_i}{a}\int \frac{D}{a}\,\mathrm{d}t. \quad (4.99)$$

Because, by definition, $D(a)$ satisfies the fluctuation growth equation, $\ddot{\delta} + (2\dot{a}/a)\dot{\delta} = 4\pi G \overline{\rho}_m \delta$, so that $\int (D/a)\,\mathrm{d}t = \dot{D}/4\pi G \overline{\rho}_m a$, Eq. (4.99) can be written as

$$\mathbf{v} = -\frac{\dot{D}}{4\pi G \overline{\rho}_m a^2}\nabla \Phi_i(\mathbf{x}) = -\frac{1}{4\pi G \overline{\rho}_m a}\frac{\dot{D}}{D}\nabla \Phi, \quad (4.100)$$

which shows that the peculiar velocity is proportional to the current gravitational acceleration. Since $\mathbf{v} = a\dot{\mathbf{x}}$, integrating the above equation once again and to the first order of the perturbations, so that $\nabla \Phi_i(\mathbf{x})$ can be replaced by $\nabla \Phi_i(\mathbf{x}_i)$ (with $\mathbf{x}_i$ the initial position of the mass element at $\mathbf{x}$), we obtain

$$\mathbf{x} = \mathbf{x}_i - \frac{D(a)}{4\pi G \overline{\rho}_m a^3}\nabla \Phi_i(\mathbf{x}_i). \quad (4.101)$$

This formulation of linear perturbation theory, which is applicable to a pressureless fluid, is due to Zel'dovich (1970). It is a Lagrangian description in that it specifies the growth of structure by giving the displacement $\mathbf{x} - \mathbf{x}_i$ and the peculiar velocity $\mathbf{v}$ of each mass element in terms of the initial position $\mathbf{x}_i$. Zel'dovich suggested that this formulation could be used to extrapolate the evolution of structures into the regime where the displacements are no longer small (i.e. up to $\delta \sim 1$). This procedure is known as the Zel'dovich approximation. Eqs. (4.100) and (4.101) show that it is a *kinematic* approximation; particle trajectories are straight lines, with the distance traveled proportional to D. The corresponding density field is, by mass conservation, simply the Jacobian of the mapping $\mathbf{x}_i \to \mathbf{x}$. Thus,

$$1+\delta = \left|\frac{\partial \mathbf{x}}{\partial \mathbf{x}_i}\right|^{-1} = \frac{1}{(1-\lambda_1 D)(1-\lambda_2 D)(1-\lambda_3 D)}, \qquad (4.102)$$

where $\lambda_1 \geq \lambda_2 \geq \lambda_3$ are the three eigenvalues of the deformation tensor $\partial_j \partial_k (\Phi_i/4\pi G \overline{\rho}_m a^3)$. In the linear case, where $\lambda_1 D \ll 1$, $\delta(\mathbf{x}) = D(a)(\lambda_1 + \lambda_2 + \lambda_3) = D(a)\delta_i(\mathbf{x})$, as expected. Zel'dovich proposed that Eq. (4.102) applies even for $\lambda_1 D(a) \sim 1$. In this case, the density will become infinite at a time when $\lambda_1 D(a) = 1$. The first nonlinear structures to form will then be two-dimensional sheets, often called 'pancakes'.

Clearly, the Zel'dovich approximation will not be valid after the formation of pancakes, when shell crossing will occur. In reality, particles falling into pancakes will oscillate in the gravitational potential, rather than move out along the directions of their initial velocities as predicted by the approximation. This difficulty can be overcome to some degree in an improved scheme where particles are assumed to stick together at shell crossing. This adhesion model is described in considerable detail by Shandarin & Zel'dovich (1989).

4.2 Relativistic Theory of Small Perturbations

The Newtonian perturbation theory of small perturbations described above applies if (i) the length scales of the perturbations are much smaller than the horizon size at the time in question (so that causality can be considered instantaneous), and (ii) the matter content can be treated as a non-relativistic fluid. If perturbations were created at an early time t_i, then all perturbations with scale sizes $\lambda > ct_i$ would, at some time in the past, have failed to fulfill the first condition. The second condition is not fulfilled for perturbations in relativistic particles, such as photons and (massless) neutrinos.

In principle, a relativistic treatment of the evolution of small perturbations is straightforward. The perturbations in the space-time metric are related to the density perturbation field through Einstein's field equation, $[\delta G]_{\mu\nu} = (8\pi G/c^4)[\delta T]_{\mu\nu}$, while the evolution of the perturbations in the matter content follows from the conservation of energy–momentum: $T^{\mu\nu}{}_{;\nu} = 0$. For a perfect, non-relativistic fluid the energy–momentum conservation law translates into a set of simple fluid equations. In general, however, the fluid description is not appropriate and one has to resort to the Boltzmann equation to describe the evolution of the full distribution functions in phase space. The energy–momentum tensor can be related to this distribution function via a set of moment equations. Thus, the Einstein field equations, combined with the energy–momentum conservation law, basically are all that is needed to evolve cosmological perturbations in general relativity. However, as we discuss below, the interpretation of fluctuations on scales larger than the horizon size can sometimes cause confusion, because of the freedom in choosing coordinate systems in general relativity.

Since the treatment here is quite tedious, the reader may want to skip this section on a first reading and directly proceed to the next section where the results are summarized in terms of linear transfer functions. Although some derivations there may depend on the material presented in this section, the final forms of the linear transfer functions are sufficient for many of our later discussions. Readers seeking an even more detailed description than what is given here are referred to the references in the text and to the textbooks by Liddle & Lyth (2000) and Dodelson (2003).

Throughout this section, we use the symbol δ to indicate the density perturbation field, and $[\delta A]$ to refer to $A(\mathbf{x}) - \overline{A}$. We use the conformal time $\tau = \int c\,dt/a$ as our time coordinate, and we distinguish between $A' \equiv dA/d\tau$ and $\dot{A} \equiv dA/dt$. The Kronecker symbol is written in three different forms, $\delta^i{}_j$, δ_{ij} and δ^{ij}. The summation rule is implied whenever there is a pair of repeated indices in a term. Finally, we set $c = 1$ unless otherwise stated.

4.2.1 Gauge Freedom

To understand the meaning of gauge freedom, let us consider perturbations on a flat background space-time. If we choose a coordinate system so that the metric of the background space-time has the form

$$ds^2 = \overline{g}_{\mu\nu} dx^\mu dx^\nu = a^2(\tau)\left(d\tau^2 - \delta_{ij} dx^i dx^j\right), \tag{4.103}$$

then the coordinates have explicit physical meaning: $a(\tau)(\delta_{ij}dx^i dx^j)^{1/2}$ is the proper distance, and $a(\tau)d\tau$ is the cosmic time. The importance of a background space-time in perturbation theory is that all perturbations are defined with respect to the background. For example, the density perturbation of a fluid is defined as $\delta(\mathbf{x},\tau) \equiv [\rho(\mathbf{x},\tau) - \overline{\rho}(\tau)]/\overline{\rho}(\tau)$, where the density $\rho(\mathbf{x},\tau)$ is the mass per unit *proper* volume as measured by an observer comoving with the fluid element at $\mathbf{x}$ at time τ, while $\overline{\rho}(\tau)$ is the mean density at this time. Thus, perturbations are defined by comparing quantities in the physical space-time with those in a fictitious background, and a (one-to-one) correspondence between points in these two space-times has to be chosen in order to define perturbation quantities. Unfortunately, the choice of such correspondence (gauge choice) is not unique, because general coordinate transformations on the perturbed space-time can change the correspondence even for a given background space-time. As a result, different gauge choices will typically yield different values for the same perturbation quantities (such as δ). In addition, a perturbation quantity may be constant in one gauge, but increase or decrease with time in another.

Because of curvature perturbations, the metric of the space-time becomes

$$g_{\mu\nu} = \overline{g}_{\mu\nu} + [\delta g]_{\mu\nu}, \tag{4.104}$$

where $[\delta g]_{\mu\nu}$ is the metric perturbation relative to the background. Since one is allowed to change coordinate systems in general relativity, one can use another set of coordinates to label the same space-time. Suppose that the new coordinates $\tilde{x}^\mu$ are related to x^μ through an infinitesimal transformation:

$$x^\mu \to \tilde{x}^\mu = x^\mu + \xi^\mu(x), \tag{4.105}$$

where ξ^μ is small. The transformation of a quantity Q of the kind

$$Q(x) \to \tilde{Q}(x) \tag{4.106}$$

under the coordinate transformation (4.105) is called a gauge transformation. Note that a gauge transformation is different from a general coordinate transformation $Q(x) \to \tilde{Q}(\tilde{x})$. In a gauge transformation, $\tilde{Q}$ and Q are both calculated at the same coordinate value corresponding to two different space-time points, while in a general coordinate transformation both Q and x are transformed so that we are dealing with the values of a quantity at the *same* space-time point observed in the two systems.

From Eq. (4.105) we have

$$\frac{\partial \tilde{x}^\mu}{\partial x^\nu} = \delta^\mu{}_\nu + \frac{\partial \xi^\mu}{\partial x^\nu}(x) \quad \text{and} \quad \frac{\partial x^\mu}{\partial \tilde{x}^\nu} = \delta^\mu{}_\nu - \frac{\partial \xi^\mu}{\partial x^\nu}(x) + \mathcal{O}(\xi^2), \tag{4.107}$$

and so to first order in ξ the gauge transformations for scalars (S), vectors (V) and rank-2 tensors (T) are

$$S(x) \to \tilde{S}(x) = S(x) - \xi^\mu S_{,\mu};$$
$$V_\mu(x) \to \tilde{V}_\mu(x) = V_\mu(x) - V_{\mu,\alpha}\xi^\alpha - V_\alpha \xi^\alpha{}_{,\mu}; \quad (4.108)$$
$$T_{\mu\nu}(x) \to \tilde{T}_{\mu\nu}(x) = T_{\mu\nu}(x) - T_{\mu\nu,\alpha}\xi^\alpha - T_{\mu\alpha}\xi^\alpha{}_{,\nu} - T_{\alpha\nu}\xi^\alpha{}_{,\mu},$$

where $Q_{,\mu} \equiv \partial Q/\partial x^\mu$. The partial derivatives can all be replaced by covariant differentiations to the first order of ξ, and so $\tilde{S}$, $\tilde{V}_\mu$ and $\tilde{T}_{\mu\nu}$ are still scalars, vectors and tensors, respectively. Applying these equations to the density (which is a scalar by definition), the four-velocity $U^\mu = dx^\mu/ds$ ($U_\mu U^\mu = \tilde{U}_\mu \tilde{U}^\mu = 1$), and the metric tensor, we have

$$\tilde{\rho}(x) = \overline{\rho}[1+\delta(x)] - \xi^\mu \frac{\partial [\overline{\rho}(1+\delta)]}{\partial x^\mu}, \text{ and so } \tilde{\delta}(x) = \delta(x) - \frac{\overline{\rho}'}{\overline{\rho}}\xi^0; \quad (4.109)$$

$$\tilde{U}^0(x) = U^0(x) + \frac{1}{a}\partial_\tau \xi^0 + \frac{a'}{a^2}\xi^0, \quad \tilde{U}^i(x) = U^i(x) + \frac{1}{a}\partial_\tau \xi^i; \quad (4.110)$$

$$\tilde{g}_{00}(x) = g_{00}(x) - 2aa'\xi^0 - 2a^2 \partial_\tau \xi^0,$$
$$\tilde{g}_{0i}(x) = g_{0i}(x) + a^2 \partial_\tau \xi^i - a^2 \partial_i \xi^0, \quad (4.111)$$
$$\tilde{g}_{ij}(x) = g_{ij}(x) + 2aa' \delta_{ij}\xi^0 + a^2(\partial_i \xi^j + \partial_j \xi^i),$$

where a prime denotes derivative with respect to the conformal time τ. The transformed metric is still a valid solution of the Einstein equation for a new energy–momentum tensor $\tilde{T}_{\mu\nu}$. Thus in the new coordinate system, it looks as if extra perturbations (the terms containing ξ^μ) have been introduced. We know that these extra 'deviations' are not physical, because they are just a result of using inhomogeneous coordinates. For example, the extra term in the density perturbation in Eq. (4.109) arises because we transform to a time coordinate $\tau + \xi^0(x)$ but define the density contrast $\tilde{\delta}$ relative to the mean background density at the time τ. The problem is how to get rid of these spurious perturbations.

Since the spurious perturbations arise from the freedom in gauge transformations (i.e. the gauge freedom), one may impose some gauge conditions on the coordinate system (i.e. on the form of the metric) to try to eliminate the freedom. Of course, one should impose as many gauge conditions as there are degrees of gauge freedom. Since the gauge transformation, $x^\mu \to \tilde{x}^\mu = x^\mu + \xi^\mu(x)$, contains four arbitrary fields, $\xi^\mu(x)$, in general one should impose four independent gauge conditions on the metric. In other words, of the 10 fields associated with the metric perturbation $[\delta g]_{\mu\nu}$ (which is symmetric), only six are physical.

There are different ways of imposing the gauge conditions (i.e. in choosing the space-time coordinate systems). This freedom can sometimes be exploited to simplify the particular problem at hand (just as one sometimes prefers to use spherical rather than Cartesian coordinates). There are, however, two further problems related to a gauge choice. The first is the existence of residual gauge freedom. As we will see below, even when four gauge conditions are imposed, not all gauge freedom is necessarily eliminated. For example, even if we choose the metric to have the form Eq. (4.103) (which is, of course, only valid for a flat space-time), there is still the freedom in choosing the origin and length scale of the coordinates. In this simple case, the residual gauge freedom is trivial, but in general it may appear in more complicated forms and must be taken into account. Obviously, we prefer a gauge in which this residual gauge freedom is easy to deal with. In general, we can single out the gauge part (gauge mode) in a perturbation quantity by studying its response to the residual gauge freedom; the part that can be transformed away is the gauge mode and should be discarded. The second problem is related to the interpretation of the perturbation quantities. Unless it is gauge-invariant, a perturbation quantity can reveal different behavior in different gauges. This is not a problem of principle, because each observable is defined only relative to a coordinate system specified by the corresponding measurement. For example, the density and pressure of a fluid at a given point are defined as those measured by an observer comoving with the fluid element at that point, while quantities like the temperature anisotropy in the CMB may be defined to be that measured by an observer to whom the expansion of the Universe appears isotropic. Once defined, these observables should be independent of gauge. However, an observable may not correspond to a single perturbation quantity in a given gauge; it may be a combination of several quantities.

The above discussion suggests the following approach for dealing with the gauge freedom: (i) choose a specific gauge and evolve all perturbation quantities in this gauge; (ii) eliminate all residual gauge modes; and (iii) interpret the perturbation quantities in terms of physical observables. This approach was taken by Lifshitz (1946) in his classical paper on cosmological perturbations. Alternatively, one can form gauge-invariant variables from linear combinations of the perturbation quantities and study the evolution of these variables in any convenient gauge. Being gauge-invariant, the results of the evolution can be interpreted in any given coordinate system. Such a gauge-invariant approach is discussed in considerable detail in Bardeen (1980).

In what follows, we first describe some specific examples of gauge choices. These are the synchronous gauge, first used by Lifshitz (1946) and subsequently adopted by many others (e.g. Landau & Lifshitz, 1975; Peebles & Yu, 1970; Peebles, 1980), the Poisson gauge, advocated by Bombelli et al. (1991) and Bertschinger (1996), and which has the advantage that the residual gauge freedom is simple, and finally the conformal Newtonian gauge (Mukhanov et al., 1992), which is a special case of the Poisson gauge with the advantage that it significantly simplifies the equations and the corresponding interpretation.

4.2.2 Classification of Perturbations

Before describing specific examples of gauge choices, we first discuss some mathematical properties of perturbations. In its most general form, the perturbed metric can be written as

$$\mathrm{d}s^2 = a^2(\tau)\left\{(1+2\Psi)\,\mathrm{d}\tau^2 - 2w_i\,\mathrm{d}\tau\,\mathrm{d}x^i - [(1-2\Phi)\delta_{ij} + H_{ij}]\,\mathrm{d}x^i\,\mathrm{d}x^j\right\}. \tag{4.112}$$

Note that for simplicity we have assumed the background to be flat; in general δ_{ij} in Eq. (4.112) should be replaced by the spatial part of a Robertson–Walker metric. The perturbation quantities w_i are the components of a vector in the three-dimensional space, so that we can write $\mathbf{w} = w_i \mathbf{e}^i$, with $\mathbf{e}^i$ a unit directional vector. Similarly, H_{ij} are the components of a traceless tensor (i.e. $\delta^{ij}H_{ij} = 0$) in the three-dimensional space, $\mathrm{H} = H_{ij}\mathbf{e}^i \otimes \mathbf{e}^j$.

It is convenient to decompose $\mathbf{w}$ and H into longitudinal and transverse parts that are, respectively, parallel and perpendicular to the gradient (or to the wavevector $\mathbf{k}$ in Fourier space). For example, we can write

$$\mathbf{w} = \mathbf{w}^{\|} + \mathbf{w}^{\perp}, \tag{4.113}$$

so that

$$\nabla \times \mathbf{w}^{\|} = 0, \quad \nabla \cdot \mathbf{w}^{\perp} = 0. \tag{4.114}$$

Being curl-free, $\mathbf{w}^{\|}$ can be written as the gradient of a scalar potential:

$$\mathbf{w}^{\|} \equiv \nabla \mathscr{W}. \tag{4.115}$$

Similarly, we can write

$$\mathrm{H} = \mathrm{H}^{\|} + \mathrm{H}^{\perp} + \mathrm{H}^{\mathrm{T}}, \tag{4.116}$$

where $H^{\|}$ has both components parallel to the gradient, H^T has both components perpendicular to the gradient, and $H^\perp$ has one component parallel and one component perpendicular to the gradient. It is straightforward to show that

$$\varepsilon_{ijk}\partial_j\partial_l H^{\|}_{lk} = 0, \quad \partial_i\partial_j H^\perp_{ij} = 0, \quad \partial_i H^T_{ij} = 0, \tag{4.117}$$

where ε_{ijk} is the three-dimensional Levi-Civita tensor. With these relations, we can write $H^{\|}_{ij}$ in terms of the derivatives of a scalar field $\mathscr{H}$, and $H^\perp_{ij}$ in terms of the derivatives of a vector field $\mathscr{H}_i$:

$$H^{\|}_{ij} = \left(\partial_i\partial_j - \frac{1}{3}\delta_{ij}\nabla^2\right)\mathscr{H}; \quad H^\perp_{ij} = \partial_i\mathscr{H}_j - \partial_j\mathscr{H}_i \quad (\text{with } \partial_i\mathscr{H}_i = 0). \tag{4.118}$$

Thus, the metric perturbations are decomposed into parts of three different types: the scalar modes, Ψ and Φ (both are scalars), and $w^{\|}_i$ and $H^{\|}_{ij}$ (both are spatial derivatives of scalar fields); the vector modes, $w^\perp_i$ (the transverse vector) and $H^\perp_{ij}$ (which is the gradient of a vector field); and the tensor modes, H^T_{ij}. The importance of decomposing metric perturbations into these different modes is that they represent distinct physical phenomena. The scalar mode is connected to the gravitational potential, the vector mode to gravito-magnetism, and the tensor mode to gravitational radiation. In turn they describe density perturbations, vorticity perturbations, and gravitational waves, respectively. Furthermore, in the linear regime, the scalar, vector, and tensor modes evolve independently.

If we decompose the spatial part of the infinitesimal coordinate transformation (4.105) into longitudinal and transverse parts and rewrite the transformation as

$$\tau \to \tilde{\tau} = \tau + \alpha(x); \quad x^i \to \tilde{x}^i = x^i + \partial_i\beta(x) + \varepsilon^i(x) \quad (\text{where } \partial_i\varepsilon^i = 0), \tag{4.119}$$

then the gauge transformations of the perturbation quantities are

$$\tilde{\Psi} = \Psi - \partial_\tau\alpha - \frac{a'}{a}\alpha, \quad \tilde{\Phi} = \Phi + \frac{1}{3}\nabla^2\beta + \frac{a'}{a}\alpha, \tag{4.120}$$

$$\tilde{w}^{\|}_i = w^{\|}_i + \partial_i(\alpha - \partial_\tau\beta), \quad \tilde{w}^\perp_i = w^\perp_i - \partial_\tau\varepsilon_i, \tag{4.121}$$

$$\tilde{H}^{\|}_{ij} = H^{\|}_{ij} - 2\left(\partial_i\partial_j - \frac{\delta_{ij}}{3}\nabla^2\right)\beta, \quad \tilde{H}^\perp_{ij} = H^\perp_{ij} - (\partial_i\varepsilon_j + \partial_j\varepsilon_i), \quad \tilde{H}^T_{ij} = H^T_{ij}. \tag{4.122}$$

As one can see, the scalar part of the transformation (α,β) only affects the scalar modes, the vector part (ε) only affects the vector modes, while the tensor mode is gauge-invariant. For the density and velocity perturbations, the transformations are

$$\tilde{\delta} = \delta - \frac{\overline{\rho}'}{\overline{\rho}}\alpha, \quad \tilde{\theta} = \theta + \partial_\tau\nabla^2\beta, \quad (\theta \equiv a\partial_j U^j = \partial_j v^j). \tag{4.123}$$

From these transformations, it can be shown that the following quantities (written in terms of Fourier transforms) are gauge-invariant:

$$\Phi_A = \Psi + \mathscr{W}' + \frac{a'}{a}\mathscr{W} - \frac{1}{2}\mathscr{H}'' - \frac{1}{2}\frac{a'}{a}\mathscr{H}', \tag{4.124}$$

$$\Phi_H = -\Phi + \frac{a'}{a}\mathscr{W} + \frac{k^2}{6}\mathscr{H} - \frac{1}{2}\frac{a'}{a}\mathscr{H}', \tag{4.125}$$

$$\varepsilon_g = \delta + \frac{\overline{\rho}'}{\overline{\rho}}\left(\mathscr{W} - \frac{1}{2}\mathscr{H}'\right), \quad \theta_g = \theta - \frac{k^2}{2}\mathscr{H}'. \tag{4.126}$$

In fact, these are the gauge-invariant potential, density and velocity perturbations defined by Bardeen (1980), and these gauge-invariant quantities are the perturbations defined from the point of view of the conformal Newtonian gauge, as we will see below. Another useful gauge-invariant combination is

$$\varepsilon = \varepsilon_g - \frac{\bar{\rho}'}{\bar{\rho}k^2}\theta_g = \delta + \frac{\bar{\rho}'}{\bar{\rho}}\left(\mathscr{W} - \theta/k^2\right), \tag{4.127}$$

which, as we will see below, is the density perturbation in the synchronous gauge.

4.2.3 Specific Examples of Gauge Choices

(a) Synchronous Gauge The synchronous gauge, first used by Lifshitz (1946), assumes that the perturbed metric has the form

$$ds^2 = g_{\mu\nu}\,dx^\mu\,dx^\nu = a^2(\tau)\left[d\tau^2 - (\delta_{ij} + h_{ij})\,dx^i dx^j\right]. \tag{4.128}$$

The synchronous gauge therefore imposes the following four restrictions on the coordinate system:

$$[\delta g]_{00} = [\delta g]_{0j} = 0, \quad \text{or equivalently,} \quad \Psi = w_i = 0. \tag{4.129}$$

The construction of such a coordinate system for a given space-time is discussed in Landau & Lifshitz (1975, §97) and Peebles (1980, §81). In such a construction, one starts with an initial spacelike hypersurface on which each point is assigned a clock and a set of spatial coordinates. The clocks are synchronized on the initial spacelike hypersurface and move in free fall. The space-time coordinates assigned to an event are the three spatial coordinates initially assigned to the clock that happens to be at the location of the event and the time recorded by this clock.

Mathematically, we can transform an arbitrary metric (4.112) into the synchronous gauge (which will be taken to be the $\tilde{x}$-system). The condition $\tilde{\Psi} = 0$ leads to

$$\partial_\tau \alpha + \frac{a'}{a}\alpha = \Psi, \tag{4.130}$$

which can be solved to give

$$a\alpha = \mathscr{A}(\mathbf{x}) + \int a\Psi\,d\tau, \tag{4.131}$$

with $\mathscr{A}(\mathbf{x})$ an arbitrary function of $\mathbf{x}$. The condition $\tilde{w}_i = 0$ gives

$$w_i + \partial_i(\alpha - \partial_\tau \beta) - \partial_\tau \varepsilon_i = 0. \tag{4.132}$$

Separated into longitudinal and transverse parts, this equation gives

$$\beta = \int (\alpha + \mathscr{W})\,d\tau + \mathscr{B}(\mathbf{x}), \qquad \varepsilon_i = \int w_i^\perp\,d\tau + \mathscr{E}_i(\mathbf{x}), \tag{4.133}$$

where $\mathscr{B}(\mathbf{x})$ (an arbitrary scalar) and $\mathscr{E}_i(\mathbf{x})$ (an arbitrary transverse vector) depend only on $\mathbf{x}$. Notice that adding a purely time-dependent function to β is irrelevant, since the transformation depends only on $\partial_i \beta$. Clearly, the synchronous gauge does not fix the coordinate system completely: we still have the freedom of choosing the forms of $\mathscr{A}(\mathbf{x})$, $\mathscr{B}(\mathbf{x})$ and $\mathscr{E}_i(\mathbf{x})$. Using the gauge transformations given in Eqs. (4.120)–(4.123) it is easy to show that these residual gauge modes appear in perturbation quantities (in Fourier space) as the terms containing $\mathscr{A}$ and $\mathscr{B}$ in the following expressions:

$$\delta - \frac{\bar{\rho}'}{\bar{\rho}}\frac{\mathscr{A}}{a}, \qquad \theta - k^2\mathscr{A}/a, \tag{4.134}$$

$$h - \left(\frac{6a'}{a^2} - 2k^2\int\frac{d\tau}{a}\right)\mathscr{A} + 2k^2\mathscr{B}, \qquad \eta + \frac{a'\mathscr{A}}{a^2}, \tag{4.135}$$

where we have separated the scalar mode of the metric perturbation, h_{ij}, into a trace part and a traceless part,

$$h_{ij} = \frac{k_i k_j}{k^2} h + \left(\frac{k_i k_j}{k^2} - \frac{1}{3}\delta_{ij} \right) 6\eta, \tag{4.136}$$

which defines the quantities h and η. These gauge modes arise from the freedom in defining the initial spacelike hypersurface of simultaneity and from assigning the initial spatial coordinates on this hypersurface.

(b) Poisson Gauge In the Poisson gauge (with coordinates denoted by $\tilde{x}^\mu$), the following four conditions are imposed on the metric:

$$\partial_i \tilde{w}_i = 0, \quad \partial_i \tilde{H}_{ij} = 0. \tag{4.137}$$

Thus, the transformation from an arbitrary coordinate system (x^μ) to the Poisson gauge must satisfy the conditions

$$\partial_i w_i + \nabla^2(\alpha - \partial_\tau \beta) = 0, \quad \partial_i H_{ij} - \frac{4}{3}\nabla^2 \partial_j \beta - \nabla^2 \varepsilon_j = 0. \tag{4.138}$$

In terms of the decompositions of w_i and H_{ij}, these conditions imply

$$\alpha = -\mathscr{W} + \frac{\partial_\tau \mathscr{H}}{2} + \mathscr{A}(\tau), \quad \beta = \frac{\mathscr{H}}{2} + \mathscr{B}(\tau), \quad \varepsilon_i = \mathscr{H}_i + \mathscr{E}_i(\tau), \tag{4.139}$$

where $\mathscr{A}(\tau)$, $\mathscr{B}(\tau)$ and $\mathscr{E}_i(\tau)$ are arbitrary functions of the conformal time τ. Therefore, the Poisson gauge also contains residual gauge freedom (arising from the arbitrariness in $\mathscr{A}$ and $\mathscr{E}_i$; $\mathscr{B}$ is irrelevant since the gauge transformation involves only $\partial_i \beta$). However, this residual freedom is trivial: a uniform change of α in space is equivalent to the change of time and length units, while a uniform change of ε_i is equivalent to a shift of the coordinate system.

(c) Conformal Newtonian Gauge A special case of the Poisson gauge is that with $w_i = 0$ and $H_{ij} = 0$. In this case the metric reduces to

$$ds^2 = a^2(\tau) \left[(1 + 2\Psi) d\tau^2 - (1 - 2\Phi) \delta_{ij} dx^i dx^j \right]. \tag{4.140}$$

This metric has a form similar to that in the Newtonian limit of gravity (see Appendix A1.5) and is called the conformal Newtonian gauge or longitudinal gauge (Mukhanov et al., 1992). One advantage of this gauge is that the metric tensor $g_{\mu\nu}$ is diagonal, which significantly simplifies the calculations. However, since it imposes more conditions than there are gauge freedoms, the conformal Newtonian gauge is a restrictive gauge which eliminates physical perturbations. In fact, this gauge only permits scalar perturbations, and is invalid when vector and tensor perturbations are concerned. However, because of its very simple form, the conformal Newtonian gauge is extremely useful for dealing with scalar perturbations. As one can see from Eq. (4.140), the perturbation Ψ causes time dilation, while the perturbation Φ corresponds to an isotropic stretching of space. With w_i and H_{ij} all set to zero in Eqs. (4.124)–(4.126), we see that the perturbations in this gauge are directly related to the gauge-invariant variables: $\Psi = \Phi_A$, $\Phi = -\Phi_H$, $\delta = \varepsilon_g$ and $\theta = \theta_g$. Finally, in terms of the perturbation quantities defined in the synchronous gauge, the perturbations in the conformal Newtonian gauge can be written (in Fourier space) as:

$$\Psi = \varpi' + \frac{a'}{a}\varpi, \quad \Phi = \eta - \frac{a'}{a}\varpi, \tag{4.141}$$

$$\delta_{\text{con}} = \delta_{\text{syn}} + \varpi \frac{\overline{\rho}'}{\overline{\rho}}, \quad [\delta P]_{\text{con}} = [\delta P]_{\text{syn}} + \varpi \overline{P}', \quad \theta_{\text{con}} = \theta_{\text{syn}} + k^2 \varpi, \tag{4.142}$$

where $\varpi = (h' + 6\eta')/(2k^2)$.

4.2.4 Basic Equations

Given that different gauge choices are physically equivalent, we can choose any convenient gauge to describe the evolution of perturbation quantities. In what follows, we derive and solve the dynamical equations in the conformal Newtonian gauge. This gauge has the advantage that the metric is very simple, but is valid only for scalar perturbations. However, it can be easily generalized to include vector and tensor perturbations (e.g. Bertschinger, 1996).

(a) Perturbed Einstein Equation As discussed in Chapter 3, the properties of space-time are determined by the energy content through the Einstein field equation:

$$G_{\mu\nu} = 8\pi G T_{\mu\nu}, \tag{4.143}$$

where the energy–momentum tensor satisfies the conservation law,

$$T^{\mu\nu}{}_{;\nu} = 0 \tag{4.144}$$

(see Appendix A). For the metric given by Eq. (4.140), it is not difficult to work out the linearized Einstein field equation (see §A1.1). The time-time, longitudinal time-space, trace space-space, and longitudinal traceless space-space parts of the Einstein equations give the following four linear equations in k-space:

$$k^2\Phi + \frac{3a'}{a}\left(\Phi' + \frac{a'}{a}\Psi\right) = -4\pi G a^2 [\delta T]^0{}_0, \tag{4.145}$$

$$k^2\left(\Phi' + \frac{a'}{a}\Psi\right) = 4\pi G a^2 (\bar{\rho} + \bar{P})\theta, \tag{4.146}$$

$$\Phi'' + \frac{a'}{a}(\Psi' + 2\Phi') + \left(\frac{2a''}{a} - \frac{a'^2}{a^2}\right)\Psi + \frac{k^2}{3}(\Phi - \Psi) = -\frac{4\pi}{3}G a^2 [\delta T]^i{}_i, \tag{4.147}$$

$$k^2(\Phi - \Psi) = 8\pi G a^2 \bar{P}\Pi, \tag{4.148}$$

where

$$(\bar{\rho} + \bar{P})\theta \equiv -ik_j[\delta T]^0{}_j, \quad \bar{P}\Pi \equiv \frac{3}{2}\left[\frac{k_i k_j}{k^2} - \frac{1}{3}\delta_{ij}\right]\left[T^i{}_j - \frac{1}{3}\delta^i{}_j T^l{}_l\right]. \tag{4.149}$$

Note that there are more equations than variables, which reflects the fact that the Einstein field equation has local conservation laws already built in.

(b) Fluid Equations For a single ideal fluid with density ρ and pressure P, the energy–momentum tensor is

$$T^{\mu\nu} = (\rho + P)U^\mu U^\nu - g^{\mu\nu} P, \tag{4.150}$$

where $U^\mu = dx^\mu/ds$ is the four-velocity. In this case, the perturbations of $T^\mu{}_\nu$ are related to the perturbations in ρ, P and $v^j \equiv aU^j$ by:

$$[\delta T]^0{}_0 = \bar{\rho}\delta, \quad [\delta T]^0{}_j = -\bar{\rho}(1+w)v^j, \quad [\delta T]^i{}_j = -[\delta P]\delta^i{}_j, \tag{4.151}$$

where

$$w \equiv \bar{P}/\bar{\rho}. \tag{4.152}$$

In this case, Eqs. (4.145) and (4.146) can be combined to give

$$k^2\Phi = -4\pi G a^2 \bar{\rho}\varepsilon, \tag{4.153}$$

where
$$\varepsilon = \delta - \frac{3a'}{a}(1+w)\theta/k^2 \qquad (4.154)$$
is the gauge-invariant density perturbation defined in Eq. (4.127). Note that Eq. (4.153) has the same form as the Poisson equation (4.11).

To linear order, the time and spatial components of the conservation law (4.144) give, respectively, the following two equations for the density and velocity perturbations of the fluid:

$$\delta' + (1+w)\left(\theta - 3\Phi'\right) = \frac{3a'}{a}\left(w - c_s^2\right)\delta, \qquad (4.155)$$

$$\theta' + \frac{a'}{a}(1-3w)\theta + \frac{w'}{1+w}\theta + \frac{\nabla^2[\delta P]}{\bar{\rho}(1+w)} - k^2\Psi = 0. \qquad (4.156)$$

Once the equation of state, w, is specified, Eqs. (4.155) and (4.156) can be solved together with the perturbed Einstein equations.

(c) Boltzmann Equation The fluid approximation described above is valid for an ideal fluid, but is inadequate for some important applications. For example, to describe the evolution of perturbations in photons and neutrinos, or to describe the interactions between photons and baryons, we need to specify the evolution of the full distribution function, $f(\mathbf{x},\mathbf{p},\tau)$, which gives the number density of particles in phase-space:

$$dN = f(\mathbf{x},\mathbf{p},\tau)d^3\mathbf{x}d^3\mathbf{p}. \qquad (4.157)$$

As in classical statistical mechanics, the phase space is described by the three positions x^i and their conjugate momenta p_i. Note that f is a scalar and is invariant under canonical transformation. Note also that the geodesic equation of a free particle can be derived from the action principle through a Lagrangian $\mathscr{L} = m(g_{\mu\nu}\dot{x}^\mu\dot{x}^\nu)^{1/2}$ where m is the mass of the particle (e.g. Peebles, 1980). It is then easy to prove that the conjugate momentum $p_i = \partial\mathscr{L}/\partial\dot{x}^i$ is just the spatial part of the four-momentum with lower indices (i.e. $p_i = mU_i$, where $U_i = dx_i/ds$), and so it obeys the geodesic equation

$$p^0\frac{dp_\mu}{d\tau} = \frac{1}{2}\frac{\partial g_{\alpha\beta}}{\partial x^\mu}p^\alpha p^\beta, \qquad (4.158)$$

where $p^0 = mU^0$ (see Appendix A). The energy–momentum tensor corresponding to the distribution function f is

$$T^{\mu\nu} = \int \frac{d^4 p}{\sqrt{-g}} 2\delta^{(D)}\left(g^{\alpha\beta}p_\alpha p_\beta - m^2\right)p^\mu p^\nu f, \qquad (4.159)$$

with $\delta^{(D)}(x)$ the Dirac delta-function.

Although the distribution function is defined by x^i and the conjugate momenta p_i, it is useful to write the distribution function in terms of the momenta (P^i) and energy (E) defined in a local Minkowski space. Note that the index of P^i is lowered by δ_{ij} and so $P_i = P^i$. Since by definition

$$E^2 = P^2 + m^2, \quad \text{where} \quad P \equiv \left(\delta_{ij}P^iP^j\right)^{1/2}, \qquad (4.160)$$

and since $g_{\mu\nu}p^\mu p^\nu = m^2$, we see that P_i and E are related to p^μ by

$$P_i = P^i = a(1-\Phi)p^i, \quad E = a(1+\Psi)p^0 = (1-\Psi)p_0/a. \qquad (4.161)$$

In practice, it is more convenient to work with another set of energy–momentum variables:

$$q_i = aP_i \quad \text{and} \quad E_q = aE. \qquad (4.162)$$

4.2 Relativistic Theory of Small Perturbations

It then follows that

$$E_q^2 = q^2 + a^2 m^2, \quad \text{where} \quad q^2 = \delta^{ij} q_i q_j. \tag{4.163}$$

Because $P \propto a^{-1}$, q_i so defined has the property that q^2 is independent of a for uniform expansion. Moreover, we shall write q_j in terms of q and its direction cosines:

$$q_j = q\gamma_j, \quad \text{where} \quad \delta^{ij} \gamma_i \gamma_j = 1. \tag{4.164}$$

Thus, we will label phase space by (x^i, q, γ_j, τ) instead of (x^i, p_j, τ). Note that this is not a canonical transformation (since q_j is not the momentum conjugate to x^j) and that we do not transform f. Hence $d^3\mathbf{x} d^3\mathbf{q}$ is not the phase-space volume element and $f d^3\mathbf{x} d^3\mathbf{q}$ is not the particle number.

For particles that are collisionless the phase-space density is conserved, i.e. $df/d\tau = 0$. However, in general, a collisional term (which describes the change of the distribution function due to collisions) should be included in the conservation law: $df/d\tau = (\partial f/\partial \tau)_c$. In terms of the variables (x^i, q, γ_j, τ), this conservation law can be written as

$$\frac{\partial f}{\partial \tau} + \frac{dx^i}{d\tau} \frac{\partial f}{\partial x^i} + \frac{dq}{d\tau} \frac{\partial f}{\partial q} + \frac{d\gamma_i}{d\tau} \frac{\partial f}{\partial \gamma_i} = \left(\frac{\partial f}{\partial \tau}\right)_c. \tag{4.165}$$

This is just the Boltzmann equation, but in non-canonical variables. Since f is independent of γ_i, and γ_i is independent of τ, for a uniform and isotropic background, the last term on the left-hand side of this equation is of second order in the perturbed quantities and can be neglected in a first-order treatment. To complete the derivation we need to express $dx^i/d\tau$ and $dq/d\tau$ in terms of (x^i, q, γ_j, τ) to first order in the perturbation quantities. By definition $p^i = mU^i = p^0 dx^i/d\tau$. It then follows that, to first order in Φ and Ψ,

$$\frac{dx^i}{d\tau} = \frac{p^i}{p^0} = \frac{q}{E_q}(1 + \Psi + \Phi)\gamma^i, \tag{4.166}$$

where $\gamma^i = \delta^{ij} \gamma_j$. Using the geodesic equation for p_0, Eq. (4.158), it is straightforward to show that, to the first order in Φ and Ψ,

$$\frac{dq}{d\tau} = q \partial_\tau \Phi - E_q \gamma^i \partial_i \Psi. \tag{4.167}$$

To study the perturbation of the distribution function, we write

$$f = f_0 + f_1, \tag{4.168}$$

where f_0 is the unperturbed distribution function and f_1 is the perturbation. The unperturbed distribution function is the Fermi–Dirac distribution for fermions (+ sign) and the Bose–Einstein distribution for bosons (− sign):

$$f_0(q) = \frac{1}{\exp(q/k_B T_0) \pm 1}, \tag{4.169}$$

where $T_0 = aT$ is the temperature of the particles at the present time. Note that f_0 is independent of a for fixed q. To first order in the perturbed quantities, the Boltzmann equation (4.165), with $dx^i/d\tau$ and $dq/d\tau$ given above, yields the following equation for f_1 (in Fourier space):

$$f_1' + ik\mu \frac{q}{E_q} f_1 - \frac{q}{4} \frac{\partial f_0}{\partial q} \Psi_q = \left(\frac{\partial f_1}{\partial \tau}\right)_c, \tag{4.170}$$

where

$$\Psi_q \equiv -4[\Phi' - ik(E_q/q)\mu\Psi], \quad \text{and} \quad \mu \equiv \mathbf{k} \cdot \mathbf{q}/|\mathbf{q}||\mathbf{k}|. \tag{4.171}$$

In terms of f_1, the contributions of the species to the source terms in the Einstein equations can be written as

$$T^0{}_0 = \frac{1}{a^4} \int q^2 \, dq \, d\omega \, E_q (f_0 + f_1),$$
$$T^0{}_i = -\frac{1}{a^4} \int q^2 \, dq \, d\omega \, q \gamma_i f_1, \qquad (4.172)$$
$$T^i{}_j = -\frac{1}{a^4} \int q^2 \, dq \, d\omega \, \frac{q^2 \gamma_i \gamma_j}{E_q} (f_0 + f_1),$$

where ω is the solid angle of $\mathbf{q}$.

Multiplying both sides of Eq. (4.170) by $\exp(ik\mu\tau q/E_q)$ and integrating it from an initial time τ_i to some later time τ, one finds

$$f_1(\chi) = f_1(\chi_i) e^{-ik\mu(\chi - \chi_i)} + \frac{E_q}{q} \int_{\chi_i}^{\chi} \left[\frac{q}{4} \frac{\partial f_0}{\partial q} \Psi_q + \left(\frac{\partial f_1}{\partial \tau}\right)_c \right]_{\chi'} e^{-ik\mu(\chi - \chi')} d\chi', \qquad (4.173)$$

where $\chi = q\tau/E_q$. It is clear from this expression that perturbations in the distribution function of a species can be produced (with a retarded time given by the exponential terms) by both gravitational and collisional interactions. Since Ψ_q depends on f_1 through the Einstein equations, and $(\partial f_1/\partial \tau)_c$ may also depend on f_1, Eq. (4.173) is an integral equation for f_1 and can be solved iteratively for given q, k and μ. Note that this solution reduces to Eq. (4.58) in the Newtonian limit.

For photons ($E_q = q$), it is convenient to consider the brightness perturbation,

$$\Delta(k, q, \mu, t) = -f_1 \left(\frac{q}{4} \frac{\partial f_0}{\partial q}\right)^{-1} = f_1 \left(\frac{T_0}{4} \frac{\partial f_0}{\partial T_0}\right)^{-1}, \qquad (4.174)$$

instead of the distribution function. Note that, for a Planck distribution, the brightness perturbation is related to the temperature perturbation as

$$\Delta = 4\Theta, \quad \text{where} \quad \Theta \equiv \frac{\Delta T}{T}. \qquad (4.175)$$

It then follows from Eq. (4.170) that

$$\Delta' + ik\mu\Delta + \Psi_q = \left(\frac{\partial \Delta}{\partial \tau}\right)_c. \qquad (4.176)$$

As we have done for f_1 in Eq. (4.173), we can write

$$\Delta(\tau) = \Delta(\tau_i) e^{-ik\mu(\tau - \tau_i)} + \int_{\tau_i}^{\tau} \left[\Psi_q - \left(\frac{\partial \Delta}{\partial \tau}\right)_c\right]_{\tau'} e^{-ik\mu(\tau - \tau')} d\tau'. \qquad (4.177)$$

This equation can be solved iteratively together with the perturbed Einstein equations, once the form of the collisional term is known.

Alternatively, one can expand the μ-dependence of $\Delta(k, q, \mu, t)$ in Legendre polynomials:

$$\Delta(k, q, \mu, t) = \sum_{\ell=0}^{\infty} (-i)^\ell (2\ell + 1) \Delta_\ell(k, q, t) \mathscr{P}_\ell(\mu). \qquad (4.178)$$

Substituting this into Eq. (4.176), using the orthonormality and recursion relation of $\mathscr{P}_\ell$, and with the collisional term omitted for the sake of brevity, we obtain

$$\delta'_\gamma = -\frac{4}{3} \theta_\gamma + 4\Phi', \qquad (4.179)$$

$$\theta'_\gamma = \frac{k^2}{4} (\delta_\gamma - 2\Delta_2) + k^2 \Psi, \qquad (4.180)$$

$$\Delta_2' = \frac{8}{15}\theta_\gamma - \frac{3k}{5}\Delta_3, \tag{4.181}$$

$$\Delta_\ell' = \frac{k}{(2\ell+1)}\left[\ell\Delta_{\ell-1} - (\ell+1)\Delta_{\ell+1}\right] \quad (\ell \geq 3), \tag{4.182}$$

where we have used the relations

$$\Delta_0 = \frac{1}{2}\int_{-1}^{1}\Delta\,\mathrm{d}\mu = \delta_\gamma; \quad \Delta_1 = \frac{1}{2}\int_{-1}^{1} i\Delta\mu\,\mathrm{d}\mu = \frac{4}{3k}\theta_\gamma. \tag{4.183}$$

With this expansion, the Boltzmann equation (4.176) is transformed into an infinite hierarchy of coupled equations. However, since Δ_ℓ decreases rapidly with ℓ for $\ell \gtrsim k\tau$, sufficiently accurate solutions can be obtained for each k by truncating the hierarchy at $\ell_{\max} \gg k\tau$.

4.2.5 Coupling between Baryons and Radiation

Because dark matter particles are expected to have become non-interacting at very early times, non-gravitational interactions are expected to be important only for baryonic matter and photons. An important example is the coupling between baryons and photons during the pre-recombination era.

As discussed in §3.5, before recombination photons and baryons are tightly coupled via Compton scattering. The linearized collisional term for Compton scattering is

$$\left(\frac{\partial\Delta}{\partial\tau}\right)_c = \sigma_T n_e a \left(\delta_\gamma + 4\mathbf{v}_e \cdot \hat{\mathbf{q}} - \Delta\right), \tag{4.184}$$

where σ_T is the Thomson cross-section, n_e is the electron density, $\mathbf{v}_e$ is the peculiar velocity of electrons in proper units, $\hat{\mathbf{q}} \equiv \mathbf{q}/|\mathbf{q}|$, and for simplicity we have neglected the polarization dependence of the scattering (see §6.7.3). Note that $\mathbf{v}_e$ is parallel to $\mathbf{k}$ for a curl-free velocity field, and so $\mathbf{v}_e \cdot \hat{\mathbf{q}} = -(i\theta_b/k)\mu$. Therefore, the Boltzmann equation for photons including Compton scattering can be written as

$$\Delta' + ik\mu\Delta + \Psi_q = \sigma_T n_e a \left(\delta_\gamma - 4i\mu\theta_b/k - \Delta\right). \tag{4.185}$$

The equation of motion for the baryonic matter (assumed to have negligible pressure) is

$$\theta_b' + \frac{a'}{a}\theta_b = k^2\Psi + \frac{4}{3}\frac{\overline{\rho}_\gamma}{\overline{\rho}_b}a\sigma_T n_e \left(\theta_\gamma - \theta_b\right), \tag{4.186}$$

where the right-hand side describes the momentum transfer from the photons to the baryons. The density perturbation in the baryonic component obeys

$$\delta_b' + \theta_b = 3\Phi'. \tag{4.187}$$

To complete the description, we need to supplement these equations with the equations for the metric perturbations, Eqs. (4.145)–(4.148). In order to integrate the above set of perturbation equations through recombination, one also needs to know how n_e changes with time. This can be done by solving the set of ionization equations described in §3.5. In general, the full set of equations has to be solved through numerical calculations (e.g. Bond & Efstathiou, 1984; Ma & Bertschinger, 1995). In what follows, we examine the behavior of the system in some limiting cases.

(a) Tight-Coupling Limit If baryons and photons are tightly coupled, i.e. the mean free time between collisions, $t_c \equiv 1/(\sigma_T n_e) \simeq a\tau_c$, is much smaller than the age of the Universe t, we can solve the perturbed quantities such as Δ in powers of τ_c/τ. For example, to solve for Δ we write Eq. (4.185) as

$$\Delta = X - \tau_c\left(\Delta' + ik\mu\Delta + \Psi_q\right), \quad \text{where} \quad X \equiv \delta_\gamma - 4i\mu\theta_b/k. \tag{4.188}$$

Iterating once we get

$$\Delta \approx X - \tau_c \left(X' + ik\mu X + \Psi_q + \ldots \right). \tag{4.189}$$

Integrating both sides over μ, and using Eqs. (4.183) and (4.187), one obtains

$$\delta_\gamma' = \frac{4}{3} \left(3\Phi' - \theta_b \right) = \frac{4}{3} \delta_b', \tag{4.190}$$

which has the solution $\delta_\gamma = 4\delta_b/3 + \text{constant}$. Thus, the entropy fluctuation

$$\delta_S \equiv \frac{3}{4} \delta_\gamma - \delta_b, \tag{4.191}$$

which is gauge-invariant according to Eq. (4.123), is independent of time. That is, the evolution is adiabatic in the tight-coupling limit. If the initial perturbation is isentropic, so that $\delta_S = 0$ (i.e. $\delta_b = 3\delta_\gamma/4$), it will remain so during the subsequent evolution in this limit.

(b) Damping of Small-Scale Perturbations during Recombination Before recombination, baryons and photons are tightly coupled and they act like a single fluid. In this case, perturbations with scale sizes smaller than the Jeans length oscillate like acoustic waves with amplitudes that are declining slightly due to the Hubble expansion (see §4.1.3). However, because τ_c is not exactly zero this coupling is not perfect. This imperfect coupling becomes more and more significant as the Universe approaches the recombination epoch, when the number density of free electrons starts to drop rapidly with time. Consequently, the photons can diffuse from high to low density regions, leading to Silk damping of small-scale perturbations.

To show this, consider a plane-wave perturbation with wavelength $\lambda = 2\pi a/k \ll ct$ during the pre-recombination era. Since the wavelength is much smaller than the Jeans length $\lambda_J \sim ct$, we can neglect the effect of gravity. Furthermore, since baryons and photons are still tightly coupled during the pre-recombination era, we have that $\tau_c \ll \tau$. For simplicity, we will also neglect the expansion of the Universe and set $a' = 0$, which is a valid approximation as long as the damping time scale is shorter than the expansion time scale. With all these assumptions and iterating Eq. (4.188) twice we obtain

$$\Delta = X - \tau_c \left(X' + ik\mu X \right) + \tau_c^2 \left(X'' + 2ik\mu X' - k^2 \mu^2 X \right). \tag{4.192}$$

Integrating both sides over μ gives

$$\delta_\gamma' \approx -\frac{4}{3} \theta_b + \frac{\tau_c}{3} \left(4\theta_b' - k^2 \delta_\gamma \right). \tag{4.193}$$

Multiplying both sides of Eq. (4.192) by μ and integrating over μ we obtain

$$\theta_b' \approx \frac{\overline{\rho}_\gamma}{\overline{\rho}_b} \left[-\frac{4}{3} \theta_b' + \frac{k^2}{3} \delta_\gamma + \tau_c \left(\frac{4}{3} \theta_b'' - \frac{2k^2}{3} \delta_\gamma' - \frac{4k^2}{5} \theta_b \right) \right], \tag{4.194}$$

where we have used Eq. (4.186) with $a' = 0$. Eqs. (4.193) and (4.194) can be solved to give

$$\delta_\gamma \propto \theta_b \propto \exp(-\mathscr{T} \tau), \tag{4.195}$$

where

$$\mathscr{T} = \pm \frac{ik}{\sqrt{3R}} + \frac{k^2 \tau_c}{6} \left(1 - \frac{6}{5R} + \frac{1}{R^2} \right), \quad R \equiv 1 + \frac{3}{4} \frac{\overline{\rho}_b}{\overline{\rho}_\gamma}. \tag{4.196}$$

The first term in $\mathscr{T}$ describes an acoustic oscillation, while the second term, which vanishes in the perfect coupling limit ($\tau_c \to 0$), describes the effect of Silk damping.

The damping becomes significant for $\text{Re}(\mathscr{T}\tau) \gtrsim 1$. This defines a characteristic wavenumber $k_d = \sqrt{6/(\tau \tau_c)}$ so that all modes with $k > k_d$ are expected to suffer significant damping. Note that for these modes the damping time scale is shorter than the expansion time scale, and so our assumption that $a' = 0$ is justified. The comoving damping scale corresponding to k_d at the redshift of decoupling is

$$\lambda_d \sim \frac{1}{k_d} \simeq 5.7 \left(\Omega_{m,0} h^2\right)^{-3/4} \left(\frac{\Omega_{b,0}}{\Omega_{m,0}}\right)^{-1/2} \left(\frac{X_e}{0.1}\right)^{-1/2} \left(\frac{1+z_{dec}}{1100}\right)^{-5/4} \text{Mpc}, \quad (4.197)$$

and the corresponding damping mass is

$$M_d \simeq 2.7 \times 10^{13} \left(\Omega_{m,0} h^2\right)^{-5/4} \left(\frac{\Omega_{b,0}}{\Omega_{m,0}}\right)^{-3/2} \left(\frac{X_e}{0.1}\right)^{-3/2} \left(\frac{1+z_{dec}}{1100}\right)^{-15/4} M_\odot, \quad (4.198)$$

where X_e is the ionization fraction at $z = z_{dec}$. Note that this damping mass is about an order of magnitude larger than that obtained in §4.1.6(d) from a simple argument based on photon diffusion. The implications of Silk damping for structure formation have been discussed in §4.1.6(d).

4.2.6 Perturbation Evolution

Once the initial conditions are specified, the equations described in §4.2.4 can be used to compute the evolution of perturbations in the linear regime. In general, the problem is complicated and numerical computations are required to find accurate solutions. In what follows, we describe several analytical solutions based on various approximations.

Although any realistic analysis of cosmological perturbations must take into account that the Universe contains several mass (energy) components, useful insight can be gained by considering a model in which the Universe is assumed to contain only radiation and dark matter. We assume that the dark matter is a collisionless fluid that has decoupled from the hot plasma when already non-relativistic (i.e. we assume the dark matter to be 'cold'; see §3.3.5). We also assume that the radiation is an ideal fluid, which is a valid treatment before recombination when the photons are tightly coupled to the baryonic component. The baryonic matter density, however, is here assumed to be negligible compared to that of the dark matter. Finally we assume the Universe to be flat, which is a valid assumption at sufficiently early times when the curvature and cosmological constant terms in the Friedmann equation can be neglected.

With all these assumptions, the fluid equations (4.155)–(4.156) describing the perturbations are

$$\delta'_{dm} + \theta_{dm} = 3\Phi', \quad \theta'_{dm} + \frac{a'}{a}\theta_{dm} = k^2 \Phi; \quad (4.199)$$

$$\delta'_\gamma + \frac{4}{3}\theta_\gamma = 4\Phi', \quad \theta'_\gamma - \frac{1}{4}k^2 \delta_\gamma = k^2 \Phi. \quad (4.200)$$

Note that for an ideal fluid the stress tensor $\overline{P}\Pi = 0$ and therefore, using Eq. (4.148), we have that $\Psi = \Phi$. The metric perturbations are related to δ_{dm}, δ_γ, θ_{dm} and θ_γ through the three perturbed Einstein equations (4.145)–(4.147). The first and third equations can be written as

$$k^2 \Phi + \frac{3a'}{a}\left(\Phi' + \frac{a'}{a}\Phi\right) = -4\pi G a^2 \left(\overline{\rho}_{dm}\delta_{dm} + \overline{\rho}_\gamma \delta_\gamma\right), \quad (4.201)$$

$$\Phi'' + \frac{3a'}{a}\Phi' + \left(\frac{2a''}{a} - \frac{a'^2}{a^2}\right)\Phi = \frac{4\pi G}{3} a^2 \overline{\rho}_\gamma \delta_\gamma, \quad (4.202)$$

and the combination of the first and second equations gives

$$k^2 \Phi = -4\pi G a^2 \left(\overline{\rho}_{dm}\varepsilon_{dm} + \overline{\rho}_\gamma \varepsilon_\gamma\right), \quad (4.203)$$

where ε_{dm} and ε_γ are defined according to Eq. (4.154). Note that the metric perturbations are described by a single function, Φ, and so only one of these equations is needed to solve the fluid equations. However, as we will see below, it is sometimes more convenient to use one equation than another.

As all the equations are linear in the perturbation quantities, their evolution can be analyzed in terms of any linear combination. Since we distinguish isentropic and isocurvature perturbations, two quantities that are particularly important for our later discussion are the metric perturbation Φ and the entropy perturbation

$$\delta_S \equiv \frac{\delta n_\gamma}{\overline{n}_\gamma} - \frac{\delta n_{dm}}{\overline{n}_{dm}} = \frac{3}{4}\delta_\gamma - \delta_{dm}, \qquad (4.204)$$

where the second equation follows from $n_{dm} \propto \rho_{dm}$ and $n_\gamma \propto T^3 \propto \rho_\gamma^{3/4}$. So defined, δ_S is zero if $n_\gamma/n_{dm} = $ constant. Using the equations given above, one can show that these two quantities obey the following two equations:

$$\frac{1}{3c_s^2}\Phi'' + \left(1 + \frac{1}{c_s^2}\right)\frac{a'}{a}\Phi' + \left(\frac{k^2}{3} + \frac{1}{4\zeta \tau_e^2}\right)\Phi = \frac{\delta_S}{2\zeta \tau_e^2}; \qquad (4.205)$$

$$\frac{1}{3c_s^2}\delta_S'' + \frac{a'}{a}\delta_S' + \frac{k^2\zeta}{4}\delta_S = \frac{1}{6}\zeta^2 k^4 \tau_e^2 \Phi, \qquad (4.206)$$

where τ_e is a characteristic, conformal time at around the time of matter–radiation equality, t_{eq},

$$\tau_e \equiv \frac{1}{H_0\sqrt{\Omega_{dm,0}(1+z_{eq})}} \sim \int_0^{t_{eq}} \frac{dt}{a} = \tau_{eq}, \qquad (4.207)$$

and we have used Eq. (3.100). The sound speed is given by

$$c_s \equiv \left(\frac{\overline{P}'}{\overline{\rho}'}\right)^{1/2} = \frac{1}{\sqrt{3}}\left(1 + \frac{3\zeta}{4}\right)^{-1/2}, \qquad (4.208)$$

and

$$\zeta \equiv \frac{\overline{\rho}_{dm}}{\overline{\rho}_\gamma} = \frac{a}{a_{eq}} = \frac{\tau}{\tau_e} + \left(\frac{\tau}{2\tau_e}\right)^2. \qquad (4.209)$$

Eq. (4.205) can be derived by replacing δ_{dm} in Eq. (4.201) with $(3/4)\delta_\gamma - \delta_S$, eliminating the term $\overline{\rho}_\gamma \delta_\gamma$ with the use of Eq. (4.202), and using the equations for the scale factor a. One way to derive Eq. (4.206) is to use $\delta_S' = (3/4)\delta_\gamma' - \delta_{dm}' = \theta_{dm} - \theta_\gamma$ and $\delta_S'' = \theta_{dm}' - \theta_\gamma' = -(a'/a)\theta_{dm} - (1/4)k^2\delta_\gamma$. These two relations, together with the definition of δ_S, give $\varepsilon_\gamma = -4[\delta_S'' + (a'/a)\delta_S']/k^2$ and $\varepsilon_{dm} = -3(\delta_S'' + k^2\delta_S/3)/k^2$. Eq. (4.205) then follows from inserting these expressions for ε_γ and ε_{dm} into Eq. (4.203).

In addition to τ_e, there is another important time scale in the evolution of Φ and δ_S, namely the time when the perturbations enter the sound horizon. The corresponding conformal time is $\tau_h \equiv \pi/(kc_s)$. In what follows we discuss the solutions of the evolution equations in various time regimes defined by τ_e and τ_h.

(a) Initial Conditions In order to connect the perturbation evolution to be described below to the initial conditions of perturbations, let us first examine the solutions of Eqs. (4.205) and (4.206) in the early radiation dominated era when $\zeta = \tau/\tau_e \ll 1$, $a \propto \tau$ and $c_s^2 = 1/3$. In this case, we can neglect the terms $\Phi/(4\zeta \tau_e^2)$ and $\delta_S/(2\zeta \tau_e^2)$ because $1/(\zeta \tau_e^2) \ll 1/\tau^2$, and so Eq. (4.205) reduces to

$$\Phi'' + \frac{4}{\tau}\Phi' + \frac{k^2}{3}\Phi = 0. \qquad (4.210)$$

Defining $u = \Phi\tau$, we can convert the above equation into $u'' + (2/\tau)u' + (k^2/3 - 2/\tau^2)u = 0$, which is the spherical Bessel equation of order 1 with solutions $j_1(k\tau/\sqrt{3})$, the spherical Bessel function, and $n_1(k\tau/\sqrt{3})$, the spherical Neumann function. Since the latter solution diverges for $\tau \to 0$, it should be discarded on the basis of initial conditions. Thus, the relevant solution of Eq. (4.210) is

$$\Phi(\mathbf{k}, \tau) = \frac{3}{(\omega\tau)^3}(\sin\omega\tau - \omega\tau\cos\omega\tau)A(\mathbf{k}), \tag{4.211}$$

where $\omega \equiv k/\sqrt{3}$, and $A(\mathbf{k})$ is the integration constant chosen so that $\Phi(\mathbf{k}, \tau \to 0) = A(\mathbf{k})$.

In the early radiation dominated era, Eq. (4.206) reduces to

$$\delta_S'' + \delta_S'/\tau = 0, \tag{4.212}$$

which has as solutions $\delta_S = $ constant and $\delta_S = \ln\tau$. The second solution should be discarded because it diverges as $\tau \to 0$, and so the relevant solution is

$$\delta_S(\mathbf{k}, \tau) = I(\mathbf{k}), \tag{4.213}$$

where $I(\mathbf{k})$ is time-independent.

Thus, the initial conditions for the metric and entropy perturbations are described by the two functions, $A(\mathbf{k})$ and $I(\mathbf{k})$, with $I(\mathbf{k}) = 0$ specifying isentropic initial conditions, and $A(\mathbf{k}) = 0$ specifying isocurvature initial conditions. In general both isentropic and isocurvature modes may exist in the initial perturbations. The relative importance of the two modes depends on how the initial perturbations are generated.

In order to see how isocurvature initial conditions generate curvature perturbations, we need to solve Eq. (4.205) including the source term $\delta_S/(2\zeta\tau_e^2)$, with the initial condition $A(\mathbf{k}) = 0$. In the radiation dominated era considered here, we can replace δ_S in the source term by $I(\mathbf{k})$, because the source term is already a higher order term in τ/τ_e compared to the Φ'' and Φ' terms. Similarly, we can replace Φ in $\Phi/(4\zeta\tau_e^2)$ by $A(\mathbf{k}) = 0$. Thus the equation to be solved is Eq. (4.210) with a source term, $I(\mathbf{k})/(2\zeta\tau_e^2)$, on the right-hand side. The particular solution of this equation is

$$\Phi(\mathbf{k}, \tau) = \frac{\tau}{\tau_e}\frac{1}{(\omega\tau)^4}\left[1 + \frac{(\omega\tau)^2}{2} - (\cos\omega\tau + \omega\tau\sin\omega\tau)\right]I(\mathbf{k}), \tag{4.214}$$

which can be obtained straightforwardly using Green's function method. For $\tau \to 0$, this solution gives $\Phi(\mathbf{k}, \tau) \sim (1/8)(\tau/\tau_e)I(\mathbf{k})$, which shows that isocurvature initial conditions can give rise to significant metric perturbations at the time when the Universe becomes matter dominated.

Similarly, the generation of entropy perturbations by isentropic initial conditions can be analyzed by studying the particular solution of Eq. (4.212) with a source term, $(1/6)\zeta^2 k^4 \tau_e A(\mathbf{k})$. The particular solution of this equation, again obtained through Green's function method, is

$$\delta_S(\mathbf{k}, \tau) = 9\left[\ln\omega\tau + \mathscr{C} - \text{Ci}(\omega\tau) + \frac{1}{2}(\cos\omega\tau - 1)\right]A(\mathbf{k}), \tag{4.215}$$

where $\mathscr{C} = 0.5772\ldots$ is the Euler constant, and

$$\text{Ci}(x) = \mathscr{C} + \ln x + \int_0^x \frac{\cos t - 1}{t}\,dt \tag{4.216}$$

is the cosine integral. This solution scales as $(k\tau)^4 A(\mathbf{k})$ for $k\tau \to 0$, which shows that the isentropic initial conditions can give rise to a significant mode of entropy perturbation at a time when the horizon size τ becomes comparable to the size of the mode, $1/k$. As we will see below, once a perturbation in the radiation component approaches the horizon, the pressure gradient in the perturbation causes it to decay. Since the same perturbation mode in the matter component remains roughly constant in the radiation dominated regime due to the Mészáros effect, entropy fluctuations are produced.

(b) Super-Horizon Evolution of Isentropic Perturbations At sufficiently early times, all modes of perturbations are super-horizon in the sense that their wavelengths are much larger than the horizon size so that $k\tau \to 0$. In such limit, δ_S remains constant according to Eq. (4.206). Thus, for isentropic initial conditions, δ_S remains zero throughout the super-horizon evolution. Setting $\delta_S = 0$ and neglecting the term containing k^2 in Eq. (4.205), and using ζ defined in Eq. (4.209) to replace τ as the time variable, we can cast Eq. (4.205) into the following form:

$$\frac{d^2\Phi}{d\zeta^2} + \frac{21\zeta^2 + 54\zeta + 32}{2\zeta(\zeta+1)(3\zeta+4)} \frac{d\Phi}{d\zeta} + \frac{\Phi}{\zeta(\zeta+1)(3\zeta+4)} = 0. \tag{4.217}$$

An analytical solution of this equation was found by Kodama & Sasaki (1984). In terms of the new variable, $u = \Phi\zeta^3/\sqrt{\zeta+1}$, the above equation can be converted into

$$\frac{d^2u}{d\zeta^2} - \left[\frac{2}{\zeta} - \frac{3/2}{1+\zeta} + \frac{3}{3\zeta+4}\right]\frac{du}{d\zeta} = 0, \tag{4.218}$$

which can be integrated to give $du/d\zeta = C_1\zeta^2(3\zeta+4)/(1+\zeta)^{3/2}$ with C_1 a constant. The general solution for Φ is therefore

$$\Phi = C_1 \frac{\sqrt{1+\zeta}}{\zeta^3} \int_0^\zeta \frac{y^2(3y+4)}{(1+y)^{3/2}} dy + C_2 \frac{\sqrt{1+\zeta}}{\zeta^3}, \tag{4.219}$$

where C_2 is another constant. It can then be shown that the solution for Φ with the initial condition $\Phi(\mathbf{k}, \zeta \to 0) = A(\mathbf{k})$ is

$$\Phi(\mathbf{k}, \tau) = \frac{A(\mathbf{k})}{10} \frac{1}{\zeta^3} \left[16\sqrt{1+\zeta} + 9\zeta^3 + 2\zeta^2 - 8\zeta - 16\right]. \tag{4.220}$$

In the radiation dominated era when $\zeta \ll 1$, this solution reduces to $\Phi = A(\mathbf{k})$, the same as that given by Eq. (4.211) in the same limit. Once the Universe becomes matter dominated so that $\zeta \gg 1$, the above solution gives $\Phi \to (9/10)A(\mathbf{k})$. Hence, for perturbations that enter the horizon after the epoch of matter–radiation equality, i.e. for $k\tau_e < 1$, their amplitudes are reduced by a factor of $1/10$ through the epoch of radiation–matter equality.

(c) Sub-Horizon Evolution of Isentropic Perturbations Consider the case in which a mode enters the horizon at a time when the Universe is dominated by radiation so that $\zeta = \tau/\tau_e \ll 1$ and $\overline{\rho}_\gamma \delta_\gamma \gg \overline{\rho}_{\rm dm} \delta_{\rm dm}$. In this limit, the evolution of isentropic perturbations is given by Eq. (4.211). In the limit $k\tau \gg 1$ (i.e. for perturbations well inside the horizon), the solution given by Eq. (4.211) reduces to

$$\Phi(\mathbf{k}, \tau) = -3\frac{\cos\omega\tau}{(\omega\tau)^2} A(\mathbf{k}). \tag{4.221}$$

In the same limit, Eq. (4.201) reduces to the Poisson equation $k^2\Phi = -4\pi G a^2 \overline{\rho}_\gamma \delta_\gamma$. Inserting Eq. (4.221) into this Poisson equation gives

$$\delta_\gamma \approx -\frac{k^2\Phi}{4\pi G a^2 \overline{\rho}_\gamma} \approx 6A(\mathbf{k})\cos\omega\tau. \tag{4.222}$$

This represents acoustic waves in the photon–baryon plasma. In this limit, Newtonian perturbation theory applies, and it is thus not surprising that the results are similar to those obtained in §4.1.6(c). Because of the radiation pressure, δ_γ oscillates with roughly constant amplitude, causing the potential to oscillate and to decay as τ^{-2}. Using the above solution for δ_γ and the solution (4.215) for δ_S, we have

$$\delta_{\rm dm} = \frac{3}{4}\delta_\gamma - \delta_S \approx -9A(\mathbf{k})\left[\ln(\omega\tau) + \mathscr{C} - 1/2\right], \tag{4.223}$$

where we have used the fact that $\text{Ci}(\omega\tau) \approx 0$ in the limit $k\tau \gg 1$.

The solutions discussed above are obtained with the assumption that $\overline{\rho}_\gamma \delta_\gamma \gg \overline{\rho}_{dm} \delta_{dm}$. Since δ_{dm} increases with time while δ_γ oscillates with a constant amplitude, and since $\overline{\rho}_\gamma$ decreases with time faster than $\overline{\rho}_{dm}$, the above assumption may not be valid at the later stages of the radiation dominated era. In order to study the evolution of δ_{dm} in this regime, we can combine the two equations in Eq. (4.199) to obtain

$$\delta_{dm}'' + \frac{a'}{a}\delta_{dm}' = 3\Phi'' + \frac{3a'}{a}\Phi' - k^2\Phi \approx -k^2\Phi, \qquad (4.224)$$

where the last step follows from $k\tau \gg 1$. In the limit $k\tau \gg 1$, Eq. (4.201) reduces to the Poisson equation $k^2\Phi = -4\pi G a^2 (\overline{\rho}_\gamma \delta_\gamma + \overline{\rho}_{dm}\delta_{dm})$. Substituting this in the above equation leads to an equation for δ_{dm} that is exactly the same as that in Newtonian perturbation theory. Thus, if the mean density of the Universe is dominated by radiation but $\overline{\rho}_\gamma \delta_\gamma \ll \overline{\rho}_{dm}\delta_{dm}$, the growing mode solution is that given by Eq. (4.88), and the growth of perturbations in the matter component is stagnated due to the Mészáros effect.

Once the Universe becomes matter dominated, sub-horizon perturbations evolve with time according to the Newtonian theory described in §4.1: $\delta_{dm} \propto D(a)$ and $\Phi \propto D(a)/a$. Note that in an Einstein–de Sitter universe, where $D(a) = a$, the potentials are frozen in at their values at around the time of matter–radiation equality. If the expansion of the Universe at late times becomes dominated by the curvature term, or by a cosmological constant, the growth rate slows down causing the potential to decay.

(d) Super-Horizon Evolution of Isocurvature Perturbations Since δ_S remains constant during super-horizon evolution according to Eq. (4.206), we can set δ_S to be its initial value, $I(\mathbf{k})$, throughout this era of evolution. Replacing δ_S by $I(\mathbf{k})$ in Eq. (4.205) and neglecting the term containing k^2 (because $k\tau \ll 1$), we obtain

$$\Phi'' + 3(1 + c_s^2)\frac{a'}{a}\Phi' + \frac{3c_s^2}{4\zeta \tau_e^2}\Phi = \frac{3c_s^2}{2\zeta \tau_e^2}I(\mathbf{k}). \qquad (4.225)$$

It is easy to see that this equation has the particular solution $\Phi = 2I(\mathbf{k})$, but this solution does not satisfy the isocurvature initial condition, $\Phi(\mathbf{k}, \tau \to 0) = 0$. The solution that satisfies this initial condition can be obtained by a linear combination of the particular solution with the general solution of the homogeneous equation. The latter is given by Eq. (4.219). It is then straightforward to show that the solution we are seeking is

$$\Phi = \left(\frac{x}{5}\right) \frac{x^2 + 6x + 10}{(x+2)^3} I(\mathbf{k}), \qquad (4.226)$$

where $x \equiv \tau/(2\tau_e)$.

During the early radiation dominated era when $\tau/\tau_e \ll 1$, the above solution reduces to

$$\Phi(\mathbf{k}, \tau) = \frac{1}{8}I(\mathbf{k})\frac{\tau}{\tau_e}\left[1 - \frac{(\omega\tau)^2}{18}\right], \qquad (4.227)$$

where we have kept the first non-zero term in $\omega\tau$. Inserting this and $\delta_S = \frac{3}{4}\delta_\gamma - \delta_{dm} = I(\mathbf{k})$ into Eq. (4.201) gives

$$\delta_\gamma \approx \frac{1}{2}I(\mathbf{k})\left[1 - \frac{7}{18}(\omega\tau)^2\right]\zeta, \quad \delta_{dm} \approx -I(\mathbf{k})\left(1 - \frac{3}{8}\zeta\right) - \frac{7\zeta}{48}I(\mathbf{k})(\omega\tau)^2. \qquad (4.228)$$

Using the first equation in Eq. (4.200) we see that the $(\omega\tau)^2$ terms in δ_γ and δ_{dm} are due to θ_γ, i.e. to the pressure gradient in the radiation component. Initially these terms are negligibly small. Since $\zeta \propto a \propto \tau$ the above solutions imply that the density perturbations in the dark matter component decrease, while those in the radiation field increase with time. However, once a mode starts to approach the horizon size ($k\tau \lesssim 1$) the $(\omega\tau)^2$ terms can no longer be neglected. In fact, the pressure gradient reduces the growth rate of δ_γ, adding a growing term to $\delta_{\rm dm}$. This effect becomes important when $\omega\tau \sim 1$, and ultimately causes a reversal in the decay of $\delta_{\rm dm}$.

In the matter dominated era ($\tau/\tau_e \gg 1$), solution (4.226) gives

$$\Phi = \frac{1}{5}I(\mathbf{k}). \tag{4.229}$$

Inserting this solution and $\delta_S = (3/4)\delta_\gamma - \delta_{\rm dm} = I(\mathbf{k})$ into Eq. (4.201), and using the fact that the k^2 term can be neglected for super-horizon evolution and that $\overline{\rho}_\gamma \ll \overline{\rho}_{\rm dm}$ in the matter dominated era, we obtain

$$\delta_{\rm dm} = -2\Phi = -\frac{2}{5}I(\mathbf{k}), \quad \delta_\gamma = \frac{4}{5}I(\mathbf{k}). \tag{4.230}$$

(e) Sub-Horizon Evolution of Isocurvature Perturbations In the early radiation dominated era when $\tau/\tau_e \ll 1$, the evolution of Φ with isocurvature initial conditions is given by Eq. (4.214). In the limit $k\tau \gg 1$, i.e. for perturbations well inside the horizon, we have

$$\Phi \approx \frac{I(\mathbf{k})}{(\omega\tau)^3}\left(\frac{\omega\tau}{2} - \sin\omega\tau\right)\zeta. \tag{4.231}$$

In the same limit, Eq. (4.201) reduces to the Poisson equation $k^2\Phi = -4\pi G a^2(\overline{\rho}_\gamma\delta_\gamma + \overline{\rho}_{\rm dm}\delta_{\rm dm})$. Since $\delta_S = I(\mathbf{k})$ according to Eq. (4.213), the above solution for Φ gives

$$\delta_\gamma \approx \zeta I(\mathbf{k}) - 2(\omega\tau)^2\Phi \sim \frac{2\sin\omega\tau}{\omega\tau}I(\mathbf{k})\zeta, \quad \delta_{\rm dm} \approx \left[\frac{3}{2}\frac{\sin\omega\tau}{\omega\tau}\zeta - 1\right]I(\mathbf{k}). \tag{4.232}$$

Thus, the potential, which builds up during horizon crossing, subsequently decays again. This is similar to the isentropic case, and is due to the fact that perturbation growth is inhibited during the radiation dominated era. The solution of δ_γ represents an acoustic wave with an amplitude of $2I(\mathbf{k})/(\omega\tau_e) \ll I(\mathbf{k})$. Note that the isocurvature modes correspond to the sine part of the acoustic solutions (4.82); they thus have a phase difference of $\pi/2$ with respect to the acoustic waves associated with the isentropic modes (4.222). The dark matter perturbations $\delta_{\rm dm} \sim -I(\mathbf{k})$, as long as $\zeta < 1$, because of the Mészáros effect described in §4.1.6(e). Note that, unlike in the isentropic case, there is no logarithmic growth before the onset of the Mészáros effect in the isocurvature case, because here $\overline{\rho}_{\rm dm}\delta_{\rm dm}$ is always larger than $\overline{\rho}_\gamma\delta_\gamma$ in the radiation dominated era. Once the Universe becomes matter dominated, sub-horizon perturbations evolve with time according to the Newtonian theory described in §4.1.

4.3 Linear Transfer Functions

In the last two sections we have seen how perturbations in the metric and the density field evolve with time. We now address the relation between the initial conditions and the density perturbations that we observe in the post-recombination Universe. A convenient way to describe this relation is through a linear transfer function, $T(k)$, which relates the amplitudes of sub-horizon Fourier modes in the post-recombination era to the initial conditions. Different definitions have been used for the transfer function in the literature. We define the linear transfer function for isentropic and isocurvature perturbations as

$$\Phi(k,t) = \mathcal{K}\beta(k)T(k,t_{\rm m})\frac{D(t)}{a(t)}\frac{a(t_{\rm m})}{D(t_{\rm m})}, \tag{4.233}$$

where $D(t)$ is the linear growth factor in the post-recombination era (see §4.1.6), and $t_{\rm m}$ is a time when the Universe is already matter dominated but the cosmological constant and curvature are still negligible, i.e. the Universe is in the Einstein–de Sitter (EdS) phase. The function $\beta(k)$ specifies the initial conditions, and is equal to $A(k)$ and $I(k)$ for isentropic and isocurvature initial conditions, respectively. A constant $\mathscr{K}$ is included in the definition to normalize the transfer function. As shown in §4.2.6, for isentropic perturbations that enter the horizon in the matter dominated era, the amplitudes of the metric perturbations are $\Phi = (9/10)A(k)$ at horizon-crossing and remain so in the Einstein–de Sitter phase. We thus choose $\mathscr{K} = 9/10$ for isentropic initial conditions so that $T(k)$ is normalized for long-wavelength models. Similarly, since $\Phi = (1/5)I(k)$ for isocurvature perturbations that enter the horizon in the matter-dominated era, we set $\mathscr{K} = 1/5$ for isocurvature perturbations.

Note that for an Einstein–de Sitter universe $D(t) = a(t)$, and so $a(t_m)/D(t_m) = 1$ in Eq. (4.233) and the transfer function is independent of t_m. For all models of interest, the redshift corresponding to t_m, z_m, can be chosen to be $\lesssim 10$, so that almost all modes of interest have already entered the horizon at $z > z_m$ and Eq. (4.233) conveniently separates the evolution into an early EdS phase and a later phase represented by the k-independent linear growth factor. In a flat universe where Λ dominates at late time, Eq. (4.233) is well defined also for modes that enter the horizon after z_m, because $\Phi(k,t)$ evolves with time as $D(t)/a(t)$ even for super-horizon modes. However, for an open or a closed universe, Eq. (4.233) is not valid for perturbation modes that have scales comparable or larger than the curvature radius of the universe, because the plane waves are no longer the normal modes of such space (e.g. Lyth & Woszczyna, 1995).

The post-recombination density perturbations are related to the metric perturbation through the Poisson equation:

$$\delta(k,t) = -\frac{k^2 \Phi(k,t)}{4\pi G a^2 \overline{\rho}} \tag{4.234}$$

$$= -\frac{2}{3} \frac{\mathscr{K}}{H_0^2 \Omega_{m,0}} k^2 \beta(k) T(k) D(t). \tag{4.235}$$

Thus, the power spectrum in the density perturbations can be written as

$$P(k,t) = \langle |\delta(k,t)|^2 \rangle = P_i(k) T^2(k) D^2(t), \tag{4.236}$$

where

$$P_i(k) \equiv \frac{4}{9} \frac{\mathscr{K}^2}{H_0^4 \Omega_{m,0}^2} k^4 \langle |\beta(k)|^2 \rangle \tag{4.237}$$

may be considered as the initial power spectrum of density perturbations. Thus, once the transfer function is known, one can calculate the post-recombination power spectrum from the initial conditions.

The linear transfer function can be calculated using the definition

$$T(k) = \frac{\Phi(k, t_m)}{\mathscr{K} \beta(k)}. \tag{4.238}$$

There are basically two kinds of effects that can affect $T(k)$ during linear evolution. The first is due to the damping processes, such as Silk damping for the baryons and free-streaming damping for (collisionless) dark matter, which reduce the small-scale perturbations relative to large-scale ones. The second is due to the fact that sub-horizon perturbations grow differently during the radiation and matter dominated eras. Before considering more realistic models, we first show how this second effect introduces a characteristic scale in the linear transfer function. For simplicity, we assume that the Universe contains only radiation and a dark matter component for which free streaming can be neglected. We start by defining the characteristic wavenumber

$$k_{eq} \equiv \frac{2\pi}{\tau_{eq}} \propto \Omega_{m,0} h^2, \tag{4.239}$$

where $\tau_{eq} \approx ct_{eq}/a_{eq}$ is the conformal time at matter–radiation equality. The length scale corresponding to k_{eq} characterizes the horizon size (in comoving units) at the time of matter–radiation equality.

Based on the results obtained in the previous section, the linear evolution of Φ is characterized by the following properties: (i) long wavelength modes ($k \ll k_{eq}$) that enter the horizon in the matter dominated era have constant amplitudes, $\Phi \sim (9/10)A(\mathbf{k})$ at $t < t_m$; (ii) since short-wavelength density perturbations with $k \gg k_{eq}$ grow logarithmically with time during the radiation dominated era according to Eq. (4.223), and since the metric perturbations remain constant over the time interval $t_{eq} < t < t_m$, we have that

$$\Phi(k,t) \sim \Phi(k,t_{eq}) \sim (27/2)H_0^2\Omega_{m,0}k^{-2}A(k)[\ln(k\tau_{eq}) - 0.47]/a(t_{eq});$$

(iii) at $t > t_m$ sub-horizon density perturbations grow with time as $\delta_{dm} \propto D(t)$ and so $\Phi \propto D(t)/a(t)$. Note that the growth in (iii) is independent of k and so does not affect the transfer function. It then follows that the post-recombination transfer function for isentropic perturbations has the properties

$$T(\mathbf{k}) = \begin{cases} 1 & \text{for } (k/k_{eq}) \ll 1 \\ C_A(k/k_{eq})^{-2}\ln(k/k_{eq}) & \text{for } (k/k_{eq}) \gg 1, \end{cases} \quad (4.240)$$

where C_A is a constant. Thus, the transition of the Universe from being radiation dominated to being matter dominated introduces a characteristic scale in the linear transfer function for isentropic perturbations.

For the isocurvature modes, the evolution of Φ has the following properties: (i) long wavelength modes ($k \ll k_{eq}$) entering the horizon in the matter dominated era have constant amplitudes $\Phi \sim \frac{1}{5}I(\mathbf{k})$ at $t < t_m$; (ii) short-wavelength density perturbations with $k \gg k_{eq}$ remain constant, $\delta_{dm} \sim I(k)$, in the radiation dominated era, so $\Phi(k,t) \sim \Phi(k,t_{eq}) \sim \frac{3}{2}H_0^2\Omega_{m,0}k^{-2}I(k)/a(t_{eq})$ at $t_{eq} < t < t_m$; (iii) sub-horizon perturbations evolve as $\Phi \propto D(t)/a(t)$ at $t > t_m$. The situation is thus similar to that in the isentropic case except that the dark matter perturbations do not grow (not even logarithmically) during radiation domination. Therefore the post-recombination transfer function for isocurvature perturbations has the properties

$$T(\mathbf{k}) = \begin{cases} 1 & \text{for } (k/k_{eq}) \ll 1 \\ C_I(k/k_{eq})^{-2} & \text{for } (k/k_{eq}) \gg 1, \end{cases} \quad (4.241)$$

where C_I is a constant. Detailed calculations show that C_I is smaller than $C_A \ln(k/k_{eq})$ for modes with $k/k_{eq} \gg 1$.

The above discussion illustrates how the horizon scale at matter–radiation equality introduces a characteristic scale in the transfer function, and thus in the power spectrum at $t > t_{eq}$. We now turn to more realistic models. After discussing a pure baryonic model (without dark matter), we turn to various dark matter models. In particular, we distinguish between hot dark matter (HDM) and cold dark matter (CDM) models that differ in the extent to which free streaming operates.

4.3.1 Adiabatic Baryon Models

Consider a universe consisting of baryons, photons and relativistic (effectively massless) neutrinos. In addition to the horizon effect described above, two additional processes play a role here. Before recombination the Jeans length is

$$\lambda_J = c_s\sqrt{\pi/G\overline{\rho}} \simeq 6ct, \quad (4.242)$$

where we have used that $c_s \sim c/\sqrt{3}$ [see Eq. (4.34)], $\overline{\rho} \simeq \rho_{crit}$ and $H(t) = \dot{a}/a = (2t)^{-1}$. Comparing this to the *proper* size of the particle horizon,

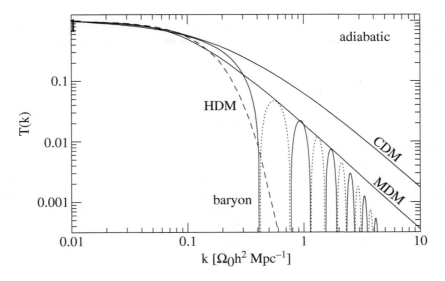

Fig. 4.2. The transfer functions for adiabatic perturbations calculated with the CMBFAST code (Seljak & Zaldarriaga, 1996): Results are shown for a purely baryonic model [the dotted parts of the curve indicate negative values of $T(k)$], for a CDM model, a HDM model, and a so-called mixed dark matter (MDM) model, consisting of 30% HDM and 70% CDM.

$$\lambda_{\rm H} = a \int_0^t \frac{c {\rm d}t}{a} = a\tau = 2ct \qquad (4.243)$$

(see §3.2.4), we see that all sub-horizon perturbations are smaller than the Jeans length. Therefore, as soon as an adiabatic baryon perturbation enters the horizon, it starts to oscillate due to the large pressure of the photon–baryon fluid. These oscillations continue until recombination, after which the perturbations start to grow via gravitational instability. However, this only applies for fluctuations with sizes larger than the Silk damping scale; fluctuations on smaller scales will have damped out before recombination.

Detailed calculations of the transfer function for adiabatic baryon models (i.e. models with isentropic initial perturbations in the baryons and photons that evolve adiabatically) have been carried out by Peebles (1981). An example of the post-recombination transfer function is shown in Fig. 4.2, along with several other transfer functions to be discussed below. On scales $k > k_{\rm eq}$ the transfer function drops rapidly due to the horizon effect and due to Silk damping. The oscillations on these scales in the post-recombination transfer function reflect the phases at recombination of the perturbations that have not been entirely damped. The deep troughs (between the solid and dotted peaks) reflect the scales on which this phase happens to be such that $\delta_{\rm b} = 0$, and are separated by $\Delta k \sim \pi a(t_{\rm rec})/(c_{\rm s} t_{\rm rec}) \sim 0.3 (\Omega_{\rm b,0} h^2) {\rm Mpc}^{-1}$.

Because of Silk damping, structure formation models based on isentropic, baryonic perturbations require large initial fluctuations in order to be able to form structures with masses comparable to the damping scale ($M \sim M_{\rm d} \sim 10^{14} \, {\rm M}_\odot$). As we will see in Chapter 5, nonlinear structures form when their corresponding perturbations have grown to $\delta_{\rm m} \sim 1$. Since $\delta_{\rm m}$ grows with a rate $\propto (1+z)$ in an EdS universe (or slower if $\Omega_0 < 1$) during the matter dominated era, this implies amplitudes of the order of $\delta_{\rm m} \gtrsim 10^{-3}$ at $z \sim z_{\rm rec}$. In the case of isentropic perturbations, the temperature fluctuations in the photon field are related to the density fluctuations as

$$\frac{\delta T}{T} = \frac{1}{4} \delta_\gamma = \frac{1}{3} \delta_{\rm b}. \qquad (4.244)$$

Thus, if clusters and superclusters (with masses $\gtrsim 10^{14} M_\odot$) formed out of isentropic baryonic perturbations, the expected temperature fluctuations in the cosmic microwave background (CMB) on angular scales of a few arcminutes would be of the order of $\delta T/T \gtrsim 10^{-3}$. This is much larger than what has been observed (see §2.9), providing strong evidence against this class of models.

4.3.2 Adiabatic Cold Dark Matter Models

It is possible that the Universe is dominated by weakly interacting massive particles (WIMPs) with masses in the range $1\,\text{GeV} \lesssim m \lesssim 3\,\text{Tev}$, or by particles which are produced without thermal velocities (see §3.3.5). Models in which the dark matter is made up of these kind of particles are called cold dark matter (CDM) models. Since the velocity stress tensor of the corresponding particles is negligible, free streaming does not play a role in these models, at least not on the scales of interest for galaxy formation. The behavior of the linear transfer function of isentropic (or 'adiabatic') CDM models is therefore given by Eq. (4.240). Detailed calculations of the linear transfer function for adiabatic CDM models have been carried out by many authors (e.g. Peebles, 1982; Bond & Efstathiou, 1984; Bardeen et al., 1986). One example is shown in Fig. 4.2. In the limit that $\Omega_{b,0} \ll \Omega_{m,0}$, the linear transfer function is well fitted by

$$T(k) = \frac{\ln(1+2.34q)}{2.34q} \left[1 + 3.89q + (16.1q)^2 + (5.46q)^3 + (6.71q)^4\right]^{-1/4}, \qquad (4.245)$$

where

$$q \equiv \frac{1}{\Gamma}\left(\frac{k}{h\text{Mpc}^{-1}}\right) \quad \text{and} \quad \Gamma = \Omega_{m,0} h \qquad (4.246)$$

(Bardeen et al., 1986). Note that Γ is a shape parameter characterizing the horizon scale at t_{eq}.

A realistic CDM model also needs to include baryons, the substance out of which galaxies are made. The presence of a non-negligible baryonic matter component influences the transfer function. Increasing the baryonic mass fraction largely leaves the shape of $T(k)$ intact, but it causes a reduction of the shape parameter, which is well approximated by

$$\Gamma = \Omega_{m,0} h \exp\left[-\Omega_{b,0}(1+\sqrt{2h}/\Omega_{m,0})\right] \qquad (4.247)$$

(Sugiyama, 1995). However, if the baryonic mass fraction becomes sufficiently large, the transfer function starts to develop oscillations (due to baryon acoustic fluctuations) and the fitting functions (4.245) and (4.247) are no longer appropriate. In this case one has to resort to more sophisticated fitting functions such as those of Eisenstein & Hu (1998). The baryon acoustic oscillations (BAO) in the transfer function produces oscillatory features in the matter power spectrum. The typical scale of the oscillations is that of the sound horizon at decoupling, which is about $150(\Omega_{b,0} h^2/0.02)^{-1}$Mpc in comoving units. Such features have indeed been observed in the galaxy distribution on large scales (e.g. Percival et al., 2007).

As we have seen, the baryonic perturbations cannot grow until after decoupling, and experience Silk damping. In the absence of dark matter, this results in problems as it predicts CMB temperature fluctuations of the order 10^{-3} or larger in order to form galaxies. The addition of cold dark matter helps in two respects. First of all, the dark matter perturbations can already start to grow after radiation–matter equality. This means that by the time of decoupling the dark matter density perturbations have already grown by a factor $\sim 20(\Omega_{m,0} h^2)$ with respect to the perturbations in the photon–baryon fluid (or even a little bit more if one takes into account that isentropic dark matter perturbations can grow logarithmically during the radiation dominated era). Secondly, after decoupling the baryons simply collapse gravitationally into the CDM perturbations, so that Silk damping no longer plays a vital role.

Adiabatic CDM models with a scale-invariant initial power spectrum, $P_i \propto k$, have proven remarkably successful in explaining the large-scale structure of the Universe (e.g. Blumenthal et al., 1984; Davis et al., 1985). They also provide a framework for galaxy formation that seems to be largely consistent with the data. Detailed analyses show that $\Gamma \sim 0.2$ is required in order for the predicted mass distribution to match the observed two-point correlation function of galaxies on large scales. Since current observational constraints suggest that $h \simeq 0.7$ (see §2.10.1), the implied value of Ω_m is ~ 0.3. If the Universe is flat, as suggested by inflationary models, the rest of the energy density, about 70% of the critical density, has to be in another energy component, such as the cosmological constant Λ. Currently, many studies are investigating the so-called ΛCDM models, in which $\Omega_{b,0} \sim 0.04$, $\Omega_{\mathrm{CDM},0} \sim 0.26$ and $\Omega_{\Lambda,0} = 1 - \Omega_{b,0} - \Omega_{\mathrm{CDM},0} \sim 0.7$.

4.3.3 Adiabatic Hot Dark Matter Models

If neutrinos have masses in the range $10 \lesssim m_\nu \lesssim 100\,\mathrm{eV}$ they can dominate the matter density of the Universe (see §3.3). In particular, if only a single species is massive then $\Omega_{\nu,0}h^2 = 0.32(m_\nu/30\,\mathrm{eV})$. Since massive neutrinos decouple from the hot plasma in the relativistic regime (see §3.3.5), models in which the dark matter consists of massive neutrinos with masses in the aforementioned range are called hot dark matter (HDM) models. The evolution of neutrino perturbations in such a HDM cosmogony is given by the Boltzmann equation (4.170), which can be solved by iterating the formal solution (4.173). Fig. 4.3 shows the evolution of adiabatic neutrino perturbations on different scales. The massive neutrinos become non-relativistic at a time when $3k_B T_\nu = m_\nu c^2$, i.e. at a redshift $z_{\mathrm{nr}} = 57300(m_\nu/30\,\mathrm{eV})$. At $z \gg z_{\mathrm{nr}}$, they move with a speed $v \sim c$, while at $z \ll z_{\mathrm{nr}}$ they are slowed down by the expansion as $v \propto a^{-1}$ (see §4.1.6). Thus, neutrino perturbations which enter the horizon at $z > z_{\mathrm{nr}}$ are significantly damped by free-streaming, while large-scale perturbations which enter the horizon at $z < z_{\mathrm{nr}}$ are not. As a result,

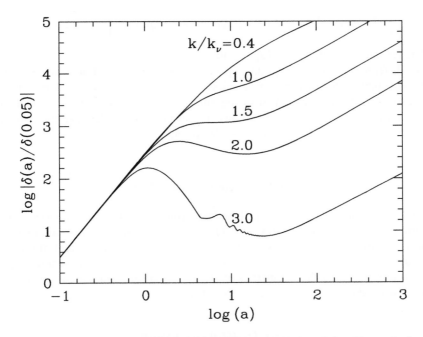

Fig. 4.3. The evolution for adiabatic neutrino perturbations with various k/k_ν. The scale factor is normalized to be unity at a redshift $z = z_{\mathrm{nr}} = 57{,}300(m_\nu/30\,\mathrm{eV})$. [Adapted from Bond & Szalay (1983) by permission of AAS]

the present-day transfer function shows a sharp decline at the high-k end (see Fig. 4.2). To good approximation, the transfer function can be written as

$$T(k) = \exp(-3.9q - 2.1q^2), \quad \text{where} \quad q \equiv k/k_\nu, \tag{4.248}$$

and $2\pi/k_\nu = 31(m_\nu/30\,\text{eV})^{-1}$ Mpc is the characteristic scale of free-streaming damping.

The free-streaming scale corresponds to a mass $M_{\text{fs}} \sim 1.3 \times 10^{14}(\Omega_{\nu,0}h^2)^{-2}$, and perturbations with smaller masses are damped. As a result, the first objects to form in HDM models are pancakes (§4.1.8) with masses $M \sim M_{\text{fs}}$, and smaller objects can only form through the subsequent fragmentation of these pancakes. In order to allow for sufficient time for this fragmentation process to produce galaxies, a large fraction of mass must have already collapsed into pancakes at an earlier epoch (say $z \gtrsim 1$). Numerical simulations by White et al. (1983), however, show that this implies a large-scale clustering strength which is much larger than observed.

In order to allow density perturbations on scales smaller than the free-streaming scale, a mixed dark matter (MDM) model, in which $\Omega_{\text{HDM},0} \sim 0.3$ and $\Omega_{\text{CDM},0} \sim 0.7$, has been considered (e.g. Ma, 1996). The linear transfer function of such a model is shown in Fig. 4.2.

4.3.4 Isocurvature Cold Dark Matter Models

The evolution of isocurvature perturbations in CDM models has been discussed in some detail in §4.2.6, and the properties of the post-recombination transfer function of the CDM component are summarized in Eq. (4.241). Detailed computations of the linear transfer function for isocurvature CDM models are given by Efstathiou & Bond (1986), for example. In the limit that $\Omega_{b,0} \ll \Omega_0$ and for three species of massless neutrinos, the linear transfer function is found to be well approximated by

$$T(k) = \left\{ 1 + \left[15.0q + (0.9q)^{3/2} + (5.6q)^2 \right]^{1.24} \right\}^{-1/1.24}, \tag{4.249}$$

where q has the same definition as in Eq. (4.246).

The main difference of this model with respect to the adiabatic CDM models described earlier is that the dark matter fluctuations here cannot grow during the radiation dominated era due to the Mészáros effect, whereas they grow logarithmically in the adiabatic case. Consequently, isocurvature CDM models have somewhat less power on small scales than the corresponding adiabatic CDM models with the same large-scale power.

4.4 Statistical Properties

In the preceding sections we have examined the time evolution of individual Fourier modes of the density perturbation field $\delta(\mathbf{x},t) \equiv \rho(\mathbf{x},t)/\overline{\rho}(t) - 1$. In the linear regime these different modes evolve independently and the amplitudes of a given Fourier mode at different times are simply related by the linear transfer function and the linear growth rate. In this section we describe how to characterize the statistical properties of the cosmological perturbations. Such a statistical description is needed in order to be able to relate theory to observation.

4.4.1 General Discussion

How can one specify a cosmic density field? In principle, one can do this by specifying the density perturbation $\delta(\mathbf{x})$ at every point in space (or, equivalently, to specify $\delta_\mathbf{k}$ for all $\mathbf{k}$). However,

4.4 Statistical Properties

this is impractical since there is an infinite number of field values (or Fourier modes) to be specified. This is also unnecessary, because we consider the mass density field in the Universe as one realization of a random process and seek a description of the cosmic density field only in a statistical sense. In this case, one aims to specify the random process that generates the cosmic density field, rather than the specific realization of the density field itself. The situation is quite similar to that in statistical mechanics: to describe the properties of a gas we do not seek to trace the positions and velocities of individual molecules but, instead, we are interested in the statistical properties given by some distribution functions.

In analogy, the statistical properties of a random perturbation field (at some given time) are specified if the probability for any particular realization of $\delta(\mathbf{x})$ is known. To see this more clearly, let us divide the Universe into n infinitesimal cells which are centered at $\mathbf{x}_1, \mathbf{x}_2, \ldots, \mathbf{x}_n$. The random perturbation field $\delta(\mathbf{x})$ is then characterized by the probability distribution function,

$$\mathscr{P}_x(\delta_1, \delta_2, \ldots, \delta_n) \, \mathrm{d}\delta_1 \, \mathrm{d}\delta_2 \ldots \mathrm{d}\delta_n, \tag{4.250}$$

which gives the probability that the field δ has values in the range δ_i to $\delta_i + \mathrm{d}\delta_i$ at positions $\mathbf{x}_i$ ($i = 1, 2, \ldots, n$). This distribution function is completely determined if we know all of its moments

$$\left\langle \delta_1^{\ell_1} \delta_2^{\ell_2} \ldots \delta_n^{\ell_n} \right\rangle \equiv \int \delta_1^{\ell_1} \delta_2^{\ell_2} \ldots \delta_n^{\ell_n} \, \mathscr{P}_x(\delta_1, \delta_2, \ldots, \delta_n) \, \mathrm{d}\delta_1 \, \mathrm{d}\delta_2 \ldots \mathrm{d}\delta_n, \tag{4.251}$$

where ℓ_i are non-negative integers. Since the cosmological principle requires all positions and directions in the Universe to be equivalent, the cosmological density field must be statistically homogeneous and isotropic. This implies that all the moments are invariant under spatial translation and rotation. The first moment $\langle \delta(\mathbf{x}) \rangle = 0$, which follows directly from the definition of the perturbation field. The variance of the perturbation field is $\sigma^2 = \langle \delta^2(\mathbf{x}) \rangle$, which, because of ergodicity, is independent of $\mathbf{x}$. Another important moment is

$$\xi(x) = \langle \delta_1 \delta_2 \rangle, \quad \text{with} \quad x \equiv |\mathbf{x}_1 - \mathbf{x}_2|, \tag{4.252}$$

which is called the two-point correlation function. Note that $\xi(0) = \sigma^2$ and that $\xi(x)$ only depends on the *distance* between $\mathbf{x}_1$ and $\mathbf{x}_2$.

A density perturbation field $\delta(\mathbf{x})$ can also be represented by its Fourier transform:

$$\delta_\mathbf{k} = \frac{1}{V_\mathrm{u}} \int \delta(\mathbf{x}) \exp(-i \mathbf{k} \cdot \mathbf{x}) \, \mathrm{d}^3 \mathbf{x}, \tag{4.253}$$

where $V_\mathrm{u} = L_\mathrm{u}^3$ is the volume of a large box on which the perturbation field is assumed periodic, and $\mathbf{k} = (2\pi/L_\mathrm{u})(i_x, i_y, i_z)$ (where i_x, i_y, i_z are integers). Note that $\delta_\mathbf{k}$ are complex quantities, and we therefore write

$$\delta_\mathbf{k} = A_\mathbf{k} + i B_\mathbf{k} = |\delta_\mathbf{k}| \exp(i\varphi_\mathbf{k}). \tag{4.254}$$

Thus, the statistical properties of $\delta_\mathbf{k}$ [and hence of $\delta(\mathbf{x})$] can also be obtained from the distribution function,

$$\mathscr{P}_k(\delta_{\mathbf{k}_1}, \delta_{\mathbf{k}_2}, \ldots, \delta_{\mathbf{k}_n}) \, \mathrm{d}|\delta_{\mathbf{k}_1}| \, \mathrm{d}|\delta_{\mathbf{k}_2}| \ldots \mathrm{d}|\delta_{\mathbf{k}_n}| \, \mathrm{d}\varphi_{\mathbf{k}_1} \, \mathrm{d}\varphi_{\mathbf{k}_2} \ldots \mathrm{d}\varphi_{\mathbf{k}_n}, \tag{4.255}$$

which gives the probability that the modes $\delta_{\mathbf{k}_i}$ have amplitudes in the range $|\delta_{\mathbf{k}_i}|$ to $|\delta_{\mathbf{k}_i}| + \mathrm{d}|\delta_{\mathbf{k}_i}|$ and phases in the range $\varphi_{\mathbf{k}_i}$ to $\varphi_{\mathbf{k}_i} + \mathrm{d}\varphi_{\mathbf{k}_i}$.

Similar to $\mathscr{P}_x$, the distribution function $\mathscr{P}_k$ is determined if all of its moments are known. In particular, the second moment,

$$P(k) \equiv V_\mathrm{u} \langle |\delta_\mathbf{k}|^2 \rangle \equiv V_\mathrm{u} \langle \delta_\mathbf{k} \delta_{-\mathbf{k}} \rangle, \tag{4.256}$$

is the power spectrum of the perturbation field. Inserting Eq. (4.253) into the above equation and using the definition of $\xi(\mathbf{r})$, we have

$$P(k) = \int \xi(\mathbf{x}) e^{-i\mathbf{k}\cdot\mathbf{x}} d^3\mathbf{x} = 4\pi \int_0^\infty \xi(r) \frac{\sin kr}{kr} r^2 \, dr. \quad (4.257)$$

The inverse relation is

$$\xi(r) = \frac{1}{(2\pi)^3} \int P(k) e^{i\mathbf{k}\cdot\mathbf{r}} d^3\mathbf{k} = \frac{1}{2\pi^2} \int_0^\infty P(k) \frac{\sin kr}{kr} k^2 \, dk. \quad (4.258)$$

We thus see that the two-point correlation function is the Fourier transform of the power spectrum, and vice versa.

4.4.2 Gaussian Random Fields

It is clear from the above discussion that it is quite difficult to specify a general random field, because it involves the determination of an infinite number of moments. Fortunately, the initial density field in the Universe is found to be well approximated by a homogeneous and isotropic Gaussian random field which, as we show below, is completely determined, in a statistical sense, by its power spectrum or its two-point correlation function.

A random field $\delta(\mathbf{x})$ is said to be Gaussian if the distribution of the field values, $(\delta_1, \delta_2, \ldots, \delta_n)$, at an arbitrary set of n points is an n-variate Gaussian:

$$\mathscr{P}_x(\delta_1, \delta_2, \ldots, \delta_n) = \frac{\exp(-\mathscr{Q})}{[(2\pi)^n \det(\mathscr{M})]^{1/2}}; \quad \mathscr{Q} \equiv \frac{1}{2} \sum_{i,j} \delta_i \left(\mathscr{M}^{-1}\right)_{ij} \delta_j, \quad (4.259)$$

where $\mathscr{M}_{ij} \equiv \langle \delta_i \delta_j \rangle$ is the covariance matrix. For a homogeneous and isotropic field, all the multivariate Gaussian distribution functions are invariant under spatial translation and rotation, and so are completely determined by the two-point correlation function $\xi(x)$. In particular, the one-point distribution function of the field itself is

$$\mathscr{P}_x(\delta) \, d\delta = \frac{1}{(2\pi\sigma^2)^{1/2}} \exp\left(-\frac{\delta^2}{2\sigma^2}\right) d\delta, \quad (4.260)$$

where $\sigma^2 = \xi(0)$ is the variance of the density perturbation field.

Any linear combination of Gaussian variates also has a Gaussian distribution. This allows us to obtain the distribution function for the Fourier transforms, $\delta_\mathbf{k} = A_\mathbf{k} + iB_\mathbf{k}$, which, after all, are linear combinations of $\delta(\mathbf{x})$. Since $\delta(\mathbf{x})$ is real, we have that $\delta_\mathbf{k}^* = \delta_{-\mathbf{k}}$ and thus $A_\mathbf{k} = A_{-\mathbf{k}}$ and $B_\mathbf{k} = -B_{-\mathbf{k}}$. This implies that we only need Fourier modes with $\mathbf{k}$ in the upper half-space to fully specify $\delta(\mathbf{x})$. It is then straightforward to prove that, for $\mathbf{k}$ in the upper half-space,

$$\langle A_\mathbf{k} A_{\mathbf{k}'} \rangle = \langle B_\mathbf{k} B_{\mathbf{k}'} \rangle = \frac{1}{2} V_u^{-1} P(k) \delta_{\mathbf{k}\mathbf{k}'}^{(\mathrm{D})}; \quad \langle A_\mathbf{k} B_{\mathbf{k}'} \rangle = 0, \quad (4.261)$$

where

$$\delta_{\mathbf{k}\mathbf{k}'}^{(\mathrm{D})} = \frac{1}{V_u} \int e^{i(\mathbf{k}-\mathbf{k}')\cdot\mathbf{x}} d^3\mathbf{x} \quad (4.262)$$

is the Kronecker delta function. As a result, the multivariate distribution functions of $A_\mathbf{k}$ and $B_\mathbf{k}$ are factorized according to $\mathbf{k}$, each factor being a Gaussian:

$$\mathscr{P}_k(\alpha_\mathbf{k}) \, d\alpha_\mathbf{k} = \frac{1}{[\pi V_u^{-1} P(k)]^{1/2}} \exp\left[-\frac{\alpha_\mathbf{k}^2}{V_u^{-1} P(k)}\right] d\alpha_\mathbf{k}, \quad (4.263)$$

where $\alpha_\mathbf{k}$ stands for either $A_\mathbf{k}$ or $B_\mathbf{k}$. Thus, for a Gaussian random field, different Fourier modes are mutually independent, and so are the real and imaginary parts of individual modes. This, in turn, implies that the phases $\varphi_\mathbf{k}$ of different modes are mutually independent, and have a random distribution over the interval between 0 and 2π. In fact, in terms of $|\delta_\mathbf{k}|$ and $\varphi_\mathbf{k}$, the distribution function for each mode can be written as

$$\mathscr{P}_k(|\delta_\mathbf{k}|,\varphi_\mathbf{k})\,\mathrm{d}|\delta_\mathbf{k}|\,\mathrm{d}\varphi_\mathbf{k} = \exp\left[-\frac{|\delta_\mathbf{k}|^2}{2V_\mathrm{u}^{-1}P(k)}\right]\frac{|\delta_\mathbf{k}|\,\mathrm{d}|\delta_\mathbf{k}|}{V_\mathrm{u}^{-1}P(k)}\frac{\mathrm{d}\varphi_\mathbf{k}}{2\pi}. \qquad (4.264)$$

Note that the power spectrum $P(k)$, which is related to the two-point correlation function $\xi(r)$ by Eq. (4.257), is the only function needed to completely specify a Gaussian random field (in a statistical sense). Note also that, during linear evolution, different Fourier modes evolve independently and so Eq. (4.261) always holds. It is then easy to see that a Gaussian perturbation field remains Gaussian in the linear regime.

Gaussian random fields are thus particularly easy to handle. The important question, of course, is whether the initial density field is Gaussian or not. At the moment, there are at least three reasons to prefer a Gaussian field to a non-Gaussian one. First of all, as discussed in §4.5, a Gaussian perturbation field arises naturally from quantum fluctuations during inflation. Since a Gaussian field remains Gaussian during linear evolution, the generic prediction of inflationary models is thus that $\delta(\mathbf{x})$ in the linear regime follows Gaussian statistics. Secondly, according to the central limit theorem, the distribution of the sum of a large number of independent variates approaches a Gaussian distribution without regard to the distribution functions of the individual variates. The initial density perturbation field $\delta(\mathbf{x})$ is a sum over a large number of Fourier modes, and so the central limit theorem guarantees a Gaussian distribution, as long as the phases of the Fourier modes are independent of each other. And thirdly, there is currently no convincing observational evidence to suggest that the linear density field is non-Gaussian.

4.4.3 Simple Non-Gaussian Models

Although there are clear theoretical and practical reasons to prefer Gaussian random fields, it is important to keep an open mind and not to exclude the possibility that the initial, linear density field has non-Gaussian statistics. For example, it is possible that some topological defects might be produced during some phase transitions in the early Universe. These defects are regions of trapped energy and could therefore act as seeds for structure formation. The density perturbation fields in defect models are generally non-Gaussian, as we briefly discuss in §4.5.

As mentioned above, a non-Gaussian density field is generally difficult to describe. However, if a non-Gaussian field is a simple transformation of an underlying Gaussian field, it is still easy to handle because in this case we only need to specify the transformation of the underlying Gaussian field. Along this line, a few simple non-Gaussian models have been proposed. One example is the χ^2 model, for which $\delta(\mathbf{x})$ is defined via a Gaussian field $\delta_\mathrm{G}(\mathbf{x})$ as

$$\delta_\chi(\mathbf{x}) = \delta_\mathrm{G}^2(\mathbf{x}) - \sigma_\mathrm{G}^2, \qquad (4.265)$$

where $\sigma_\mathrm{G}^2 = \langle\delta_\mathrm{G}^2\rangle$, and the subtraction of σ_G^2 ensures that $\langle\delta_\chi\rangle = 0$. Another simple example is the log-normal model, defined as

$$\delta_\mathrm{LN}(\mathbf{x}) = \exp\left[\delta_\mathrm{G}(\mathbf{x}) - \sigma_\mathrm{G}^2/2\right] - 1 \qquad (4.266)$$

(Coles & Jones, 1991). In both cases, all the moments of $\delta(\mathbf{x})$ can be written in terms of the moments of $\delta_\mathrm{G}(\mathbf{x})$, and so $\delta(\mathbf{x})$ is completely specified by the power spectrum of δ_G (or the two-point correlation function ξ_G). For example, the two-point correlation functions corresponding to Eqs. (4.265) and (4.266) are $\xi_\chi(r) = 2\xi_\mathrm{G}^2(r)$ and $\xi_\mathrm{LN} = \exp(\xi_\mathrm{G}) - 1$, respectively.

4.4.4 Linear Perturbation Spectrum

As discussed above, the power spectrum $P(k)$ is an important quantity characterizing a random field. In fact it is the only quantity required to specify a homogeneous and isotropic Gaussian random field. As we have seen in §4.3, because different Fourier modes evolve independently of each other in the linear regime, the linear power spectrum at any given time can be simply related to the initial power spectrum via the linear transfer function. We now take a more detailed look at the initial power spectrum.

(a) The Initial Power Spectrum In the absence of a complete theory for the origin of the density perturbations, the initial (untransferred) perturbation spectrum is commonly assumed to be a power law,

$$P_\mathrm{i}(k) \propto k^n, \tag{4.267}$$

where n is usually called the spectral index. As we will show in §4.5, the power spectra predicted by inflation models generally have this form.

It is often useful to define the dimensionless quantity,

$$\Delta^2(k) \equiv \frac{1}{2\pi^2} k^3 P(k), \tag{4.268}$$

which expresses the contribution to the variance by the power in a unit logarithmic interval of k. In terms of $\Delta^2(k)$ we have that

$$\Delta^2(k) \propto k^{3+n}. \tag{4.269}$$

The corresponding quantity for the gravitational potential is

$$\Delta_\Phi^2(k) \equiv \frac{1}{2\pi^2} k^3 P_\Phi(k) \propto k^{-4} \Delta^2(k) \propto k^{n-1}, \tag{4.270}$$

which is independent of k for $n = 1$. Thus, for the special case of $n = 1$, which is called the Harrison–Zel'dovich spectrum or scale-invariant spectrum, the gravitational potential is finite on both large and small scales. This is clearly desirable, because divergence of the gravitational potential on small or large scales would lead to perturbations on these scales that are too large.

As we will see in §4.5, in inflation models the metric (potential) perturbations are generated by quantum fluctuations during inflation. At the end of inflation, all perturbations become super-horizon because of the huge amount of expansion caused by inflation. Since metric perturbations remain roughly constant during super-horizon evolution, the amplitude of a metric perturbation at the time when it re-enters the horizon should be approximately the same as the initial amplitude. Thus, the amplitude of $\Delta_\Phi^2(k)$ evaluated at the time of horizon re-entry is proportional to k^{n-1}, which is independent of k for a scale-invariant spectrum.

(b) The Amplitude of the Linear Power Spectrum So far we have only discussed the shape of the linear power spectrum. To completely specify $P(k)$, we also need to fix its overall amplitude. Because we do not yet have a refined theory for the origin of the cosmological perturbations, the amplitude of $P(k)$ is not predicted a priori but rather has to be fixed by observations. Even for inflation models, where we can make detailed predictions for the shape of the initial power spectrum, the current theory has virtually no predictive power regarding the amplitude (see next section).

For a power spectrum with a given shape, the amplitude is fixed if we know the value of $P(k)$ at any k, or the value of any statistic that depends only on $P(k)$. Not surprisingly, many observational results can be used to normalize $P(k)$. Different observations may probe the power spectrum at different scales, providing additional constraints on the shape of $P(k)$. In fact, trying to determine the shape and amplitude of the linear power spectrum is one of the most important tasks of observational cosmology (see Chapter 6).

One historical prescription for normalizing a theoretical power spectrum involves the variance of the galaxy distribution when sampled with randomly placed spheres of radii R. The predicted variance of the density field is related to the power spectrum by

$$\sigma^2(R) = \frac{1}{2\pi^2} \int P(k)\widehat{W}_R^2(k)k^2 dk, \tag{4.271}$$

where

$$\widehat{W}_R(k) = \frac{3}{(kR)^2}[\sin(kR) - kR\cos(kR)] \tag{4.272}$$

is the Fourier transform of the spherical top-hat window function

$$W_R(r) = \begin{cases} 3/(4\pi R^3) & \text{if } r \leq R \\ 0 & \text{otherwise.} \end{cases} \tag{4.273}$$

The value of $\sigma(R)$ derived from the distribution of galaxies is about unity for $R = 8h^{-1}\text{Mpc}$ (§2.7.1). Thus, one could in principle normalize the theoretical power spectrum by requiring $\sigma(R) = 1$ at $R = 8h^{-1}\text{Mpc}$. However, there are several problems with this approach. First of all, since $\sigma(R) \sim 1$ we are not accurately probing the linear regime for which $\delta \ll 1$. Secondly, this normalization is based on the assumption that galaxies are accurate tracers of the fluctuations in the mass distribution. This may not be true if, for example, galaxies formed preferentially in high density regions. Indeed, if we adopt the less restrictive assumption that the fluctuations in the galaxy distribution are proportional (but not necessarily equal) to the fluctuations in the mass distribution, then

$$\delta_{\text{gal}} = b\,\delta_{\text{m}}, \tag{4.274}$$

where b = constant is a *bias parameter* whose value depends on how galaxies have formed in the mass density field. In this case

$$\sigma_{\text{m}}(8h^{-1}\text{Mpc}) = \frac{\sigma_{\text{gal}}(8h^{-1}\text{Mpc})}{b} \approx \frac{1}{b}. \tag{4.275}$$

Since an accurate theory for galaxy formation is still lacking at the present time, the value of b is still uncertain. In fact, as we will see in Chapter 15, b is found to be a function of various galaxy properties, such as luminosity and color.

To accurately normalize the linear power spectrum thus requires a method that is not affected by nonlinear evolution and that does not depend on the assumption of galaxies tracing the mass distribution. In Chapter 6 we will describe various statistical measures that can be used to probe the power spectrum. Some of these methods are much better suited for normalizing the linear power spectrum than $\sigma_{\text{gal}}(8h^{-1}\text{Mpc})$ is. However, as a convention, and largely for historical reasons, the amplitude of a power spectrum is usually represented by the value

$$\sigma_8 \equiv \sigma_{\text{m}}(8h^{-1}\text{Mpc}). \tag{4.276}$$

It is important to realize that σ_8 is evaluated from the initial power spectrum evolved to the present time according to linear theory. Since perturbations on scales of $\sim 8h^{-1}\text{Mpc}$ may well have gone nonlinear by the present time, this is not necessarily the same as the variance of the actual, present-day mass distribution.

Table 4.1 lists a number of adiabatic models and their power spectrum normalization as obtained from the temperature fluctuations in the CMB obtained by COBE on large scales (see §6.7 for details) and from the abundance of rich clusters of galaxies (see §7.2.5 for details). The

Table 4.1. The values of σ_8 in different cosmogonies.

Model	$\Omega_{m,0}$	$\Omega_{\nu,0}$	$\Omega_{\Lambda,0}$	h	n	σ_8(COBE)	σ_8(cluster)
SCDM	1	0	0	0.5	$n=1$	1.3	0.6
HDM	1	1	0	0.5	$n=1$	1.3	–
MDM	1	0.3	0	0.5	$n=1$	0.5	0.5
OCDM	0.3	0	0	0.7	$n=1$	0.5	0.9
ΛCDM	0.3	0	0.7	0.7	$n=1$	1.0	1.0

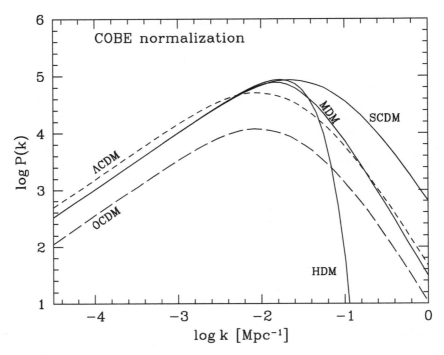

Fig. 4.4. The power spectra, $P(k)$, as a function of k for various cosmogonic models. The initial power spectrum is assumed to be scale invariant (i.e. $n = 1$), and the spectra are normalized to reproduce the COBE observations of CMB anisotropies.

COBE-normalized, linear, dark matter power spectra are shown in Fig. 4.4. Not all models match at small k, despite the fact that they are all normalized at these scales. This arises because the observed temperature fluctuations are in angular scales and the conversion from angles to distances is cosmology-dependent. Note that the COBE and cluster-abundance normalizations are only in agreement with each other for the ΛCDM model and the MDM model. Since they have been obtained from measurements of the power spectrum amplitude at vastly different scales, a discrepancy in the inferred values of σ_8 signals that the *shape* of the model power spectrum is inconsistent with the data. Indeed, as we shall see in Chapter 6, stringent constraints on the shape and amplitude of the linear power spectrum can now be obtained from a variety of observations, and the model that is currently favored is the ΛCDM model with parameters similar, but not identical, to those listed in Table 4.1.

4.5 The Origin of Cosmological Perturbations

Having discussed the (linear) evolution of cosmological perturbations we now turn to their origin. Broadly speaking, two different mechanisms have been proposed, namely inflation and phase transitions. In inflation models the perturbations arise from quantum fluctuations of the inflaton scalar field. As we will see, these models typically predict isentropic, Gaussian perturbations with a close to scale-invariant power-spectrum (although deviations from this typical prediction are certainly possible). In the alternative model, perturbations arise from cosmological defects that originate from phase transitions in the early Universe. Contrary to inflation models, these models typically predict non-Gaussian perturbations.

4.5.1 Perturbations from Inflation

As discussed in §3.6, the concept of inflation is introduced to solve a number of nagging problems related to the initial conditions of the standard cosmology. In inflation models, the Universe experiences an early period of exponential expansion, driven by the false vacuum state of a scalar field, called the inflaton. Because of quantum fluctuations the energy density of the inflaton is expected to be inhomogeneous. These inhomogeneities are initially inflated to super-horizon scales by the exponential expansion, but re-enter the horizon after the inflation is over to seed structure formation.

(a) Heuristic Arguments Without going into details, we may use some simple arguments to understand some of the most important properties of the density perturbations generated during inflation. The arguments consist of the following four important points. First of all, since the Universe is assumed to have gone through a phase of very fast expansion (inflation), all structures that we observe in the present-day Universe had sizes that were much smaller than the horizon size before inflation started. Therefore, inflation provides a mechanism for generating the initial perturbations in a causal way. Second, if the scalar field driving inflation has negligible self-coupling (as is assumed in most models), then different modes in the quantum fluctuations of the field should be independent of each other. Consequently, the density perturbations are expected to follow Gaussian statistics. Third, the perturbations produced during the inflation phase are perturbations in the energy density of the scalar field. At the end of inflation, reheating converts this energy density into photons and other particles. Since we do not expect any segregation between different particle species, the resulting perturbations are expected to be isentropic. Finally, since during inflation space is invariant under time translation (i.e. it is a de Sitter space), the perturbations generated by inflation are expected to be scale-invariant.

This final point is not very straightforward, and requires some more discussion. As shown in §3.6, in order for inflation to solve the horizon problem, the period of inflation must last long enough. This translates into a slow-roll condition, which states that the potential of the inflaton must be sufficiently flat. As a result, the Hubble constant H, the inflaton expectation value $\langle \varphi \rangle$, and the potential energy $V(\varphi)$, are all roughly time-independent during inflation. Therefore, if some physical process can generate perturbations in the inflaton, the properties of these perturbations should be independent of their time of generation. Assume, for example, that some physical process generates perturbations at all times (during inflation) on a fixed physical scale λ_i. These perturbations will then all be generated with approximately the same amplitude. Because of the (exponential) increase of the scale factor, $a(t)$, perturbations generated at different times are inflated into different scales. For example, for two perturbations generated at times t_1 and t_2, the ratio between their scales at any later time during the exponential expansion is

$$\frac{\lambda_1}{\lambda_2} = e^{H(t_2 - t_1)}. \tag{4.277}$$

The time t_H when a perturbation of comoving scale λ reaches the horizon is given by $\lambda \approx c/a[t_H(\lambda)]H[t_H(\lambda)]$. Since H is approximately constant, we can write

$$\frac{\lambda_1}{\lambda_2} \approx \frac{a[t_H(\lambda_2)]}{a[t_H(\lambda_1)]} = e^{H[t_H(\lambda_2)-t_H(\lambda_1)]}. \tag{4.278}$$

Comparing Eqs. (4.277) and (4.278), we see that

$$t_H(\lambda_2) - t_2 = t_H(\lambda_1) - t_1. \tag{4.279}$$

Thus, the time between generation and horizon-exit is the same for all perturbations. Since the properties of space change little during inflation, the amplitudes of all perturbations should be approximately the same at horizon-exit, without depending on their scale at that time. Note that the above discussion makes no reference to the value of λ_i, and so the conclusion is true even if perturbations with a range of physical scales are generated at each time. Thus, the expectation values of the perturbation amplitudes are scale-independent at horizon-exit. When such a perturbation re-enters the horizon some time well after inflation has ended, its amplitude will be about the same as it was at horizon-exit, since causal physics cannot act on super-horizon perturbations to produce observable consequences.[1] The perturbations are therefore scale-independent at horizon-reentry. Thus, inflation generically predicts scale-invariant perturbations.

(b) Some Detailed Considerations To make detailed predictions for the density perturbations in an inflation model, one needs to consider a specific model for the scalar field, which we denote by $\varphi(\mathbf{x},t)$. We write

$$\varphi(\mathbf{x},t) = \varphi_0(t) + \psi(\mathbf{x},t), \tag{4.280}$$

where $\varphi_0(t)$ is the background field and $\psi(\mathbf{x},t)$ (with $|\psi| \ll |\varphi_0|$) is the perturbation. The dynamical properties of the scalar field are described in §3.6.3. Under the slow-roll condition, which is required for inflation to occur, the evolution of φ is given by

$$\ddot{\varphi} + 3H\dot{\varphi} - a^{-2}\nabla^2\varphi = 0. \tag{4.281}$$

Thus the equation of motion for the Fourier modes of the perturbation field, $\psi_\mathbf{k}$, can be written as

$$\psi_k'' + \frac{2a'}{a}\psi_k' + k^2\psi_k = 0, \tag{4.282}$$

where again a prime denotes derivative with respect to the conformal time τ. If the expansion of the Universe is neglected (i.e. $a'/a = 0$), the above equation is that for a harmonic oscillator, and the solutions are $e^{\pm ik\tau}/\sqrt{2k}$. Upon quantization, ψ_k becomes an operator:

$$\hat{\psi}_k = Q(k,\tau)\hat{a} + Q^*(k,\tau)\hat{a}^\dagger, \tag{4.283}$$

where $\hat{a}$ and $\hat{a}^\dagger$ are the operators annihilating and creating a particle, respectively, and $Q = e^{-ik\tau}/\sqrt{2k}$. The expected quantum fluctuation of ψ_k in the ground state, $|0\rangle$, is then characterized by the dispersion $\langle 0|\hat{\psi}_k^\dagger \hat{\psi}_k|0\rangle$. Using the properties of $\hat{a}$ and $\hat{a}^\dagger$, it can be shown that

$$\langle|\psi_\mathbf{k}|^2\rangle \equiv \langle 0|\hat{\psi}_k^\dagger \hat{\psi}_k|0\rangle = |Q(k,\tau)|^2 = 1/(2k). \tag{4.284}$$

When the expansion term is included, Eq. (4.282) can be converted into the form

$$\tilde{\psi}_k'' + (k^2 - a''/a)\tilde{\psi}_k = 0, \tag{4.285}$$

---

[1] Note, however, that the amplitudes of super-horizon perturbations may appear to be evolving with time in some gauges (see §4.2.1). Such evolution, however, is completely geometrical and should not have any observable consequence, such as changing the amplitude of a physical perturbation at horizon entry.

where $\tilde{\psi}_k = a\psi_k$. Since during inflation $H = a'/a^2$ is roughly constant, we have that $\tau = \int_{a_e}^a da/(Ha^2) \approx -1/Ha$, where a_e is the expansion factor at the end of the inflation and we have set $\tau = 0$ at this time. Using the fact that $a \ll a_e$ during inflation, we have that $a''/a \approx 2H^2a^2 \approx 2/\tau^2$. Inserting this into Eq. (4.285) one finds that its two solutions are Q and Q^* with $Q = e^{-ik\tau}(1 - i/k\tau)$. Thus,

$$\langle |\tilde{\psi}_\mathbf{k}|^2 \rangle = |Q(k,\tau)|^2 = \frac{1}{2k}\left[1 + \frac{1}{(k\tau)^2}\right]. \tag{4.286}$$

After inflation, when $k\tau \ll 1$, this gives

$$\langle |\psi_\mathbf{k}|^2 \rangle \approx \frac{1}{2k^3(a\tau)^2} \approx \frac{H^2}{2k^3}. \tag{4.287}$$

The typical amplitude of ψ_k due to quantum fluctuations in the ground state is then $H/(\sqrt{2}k^{3/2})$. The question is then how such fluctuations generate the metric perturbations responsible for structure formation in the Universe.

A convenient way to make connections between $\langle |\psi_\mathbf{k}|^2 \rangle$ and the metric perturbations $\Phi(k,t)$ is to consider the following quantity:

$$\xi \equiv \frac{Ha\theta}{k^2} + \Phi = -\frac{ik^j Ha[\delta T]^0{}_j}{k^2(\bar{\rho} + \bar{P})} + \Phi = \frac{2}{3}\frac{a\Phi'/a' + \Phi}{1+w} + \Phi, \tag{4.288}$$

where the second relation follows from the definition of θ in Eq. (4.149) and the third relation follows from Eq. (4.146) with $\Psi = \Phi$. For the scalar field considered here, the energy–momentum tensor defined in §3.6.3 leads to $\theta = -k^2\psi_k/\varphi_0'$, and so ξ defined above is a combination of the metric perturbation Φ and the perturbation in the inflaton. One important property of ξ is that it is conserved, i.e. $\xi' = 0$, during super-horizon evolution. This can be proven using the definition of ξ together with Eqs. (4.146) and (4.156) (again with $\Psi = \Phi$) in the limit $k\tau \ll 1$. Since after inflation all perturbations are super-horizon and the Universe becomes radiation dominated (so that $w = 1/3$), we obtain $\xi = 3\Phi/2$ from the last relation in Eq. (4.288). Because the amplitudes of the metric perturbations remain roughly constant during super-horizon evolution, the amplitude of ξ at the time when the perturbation re-enters the horizon in the post-inflation era should be equal to its value at the time when the perturbation exits the horizon after it is generated during inflation. At the time of horizon exit, metric perturbations are still negligible, and so $\xi = -aH\psi/\varphi_0'$. It then follows that the post-inflation metric perturbations are

$$\Phi = \left[\frac{2}{3}H\frac{\psi_k}{\dot{\varphi}_0}\right]_{\text{horizon-exit}}, \tag{4.289}$$

where the subscript 'horizon-exit' indicates that the quantity is evaluated at the time when the mode exits the horizon, i.e. at $t \sim a/k$. The post-inflation power spectrum is therefore

$$P_\Phi(k) = \left[\frac{4}{9}H^2\frac{\psi_k^2}{\dot{\varphi}_0^2}\right]_{\text{horizon-exit}} = \frac{2}{9}\frac{1}{k^3}\left[\frac{H^4}{\dot{\varphi}_0^2}\right]_{\text{horizon-exit}}, \tag{4.290}$$

where the second equation follows from replacing ψ_k^2 with its quantum expectation value in Eq. (4.287). Thus, if $H^4/\dot{\varphi}_0^2$ is time-independent, the resulting post-inflation power spectrum is scale-invariant, with $n = 1$ [see Eq. (4.270)].

A crucial requirement for obtaining the scale-invariant power spectrum is that the Universe expands exponentially during the period of inflation. Deviations from a purely exponential expansion will cause deviations from a pure scale-invariant power spectrum. To see this, consider the quantity

$$\Delta_\Phi^2 \equiv \frac{1}{2\pi^2} k^3 P_\Phi(k) = \frac{1}{9\pi^2} \frac{H^4}{\dot{\varphi}_0^2} \approx \frac{1}{27\pi^2} \left(\frac{8\pi}{m_{\rm Pl}^2}\right)^3 \frac{V^3}{(dV/d\varphi)^2}, \qquad (4.291)$$

where we have used Eqs. (3.253) and (3.254). We define the tilt of the power spectrum as

$$1 - n \equiv -\frac{d \ln \Delta_\Phi^2}{d \ln k}, \qquad (4.292)$$

so that it is zero for a scale-invariant spectrum. At the time when a perturbation exits the horizon, its wavelength is equal to the Hubble radius, and so $a/k = H^{-1}$. Since H is nearly constant during inflation, we can write

$$\frac{d}{d \ln k} \approx \frac{d}{d \ln a} = \frac{\dot{\varphi}}{H} \frac{d}{d\varphi} = -\frac{m_{\rm Pl}^2}{8\pi} \frac{dV/d\varphi}{V} \frac{d}{d\varphi}. \qquad (4.293)$$

The tilt can then be written as

$$1 - n = 6\varepsilon - 2\eta, \qquad (4.294)$$

where ε and η are defined in Eqs. (3.258) and (3.259). Unless the potential V is perfectly flat, the power spectrum is expected to be tilted slightly with respect to the scale-invariant form. However, since the slow-roll condition demands that both ε and η are small, the tilt is expected to be small as well.

The above analysis of quantum fluctuations in a scalar field is valid for all free quantum fields with dynamics similar to that of a free oscillator. The expected amplitude of the quantum fluctuations given by Eq. (4.287) is independent of the details of the field in consideration and so all such fields are expected to have similar quantum fluctuations during inflation. In particular, under the weak-field assumption, the gravitational Lagrangian is $\mathscr{L} = R/16\pi G \approx \partial_\alpha h^{\mu\nu} \partial^\alpha h_{\mu\nu}/32\pi G$, where R is the perturbation in the curvature scalar and $h_{\mu\nu}$ is the perturbation in the metric tensor. This suggests that $h^{\mu\nu}/\sqrt{16\pi G}$ can be effectively treated as a scalar field and so the amplitude of the fluctuation in $h_{\mu\nu}$ at horizon crossing is given by

$$\Delta_h^2 \equiv \frac{k^3}{2\pi^2} \langle |h|^2 \rangle \sim \frac{4}{\pi} \frac{H^2}{m_{\rm Pl}^2}. \qquad (4.295)$$

The propagation of the tensor mode of the metric perturbations produces gravitational waves, and so inflation models generically predict the existence of a background of gravitational waves. Note that this mode of perturbations does not generate density perturbations (which correspond to scalar mode), because these two modes evolve independently (see §4.2.2). At the time of horizon crossing, the energy density of the gravitational waves is

$$\rho_{\rm GW} \sim \frac{\dot{h}_{\mu\nu}^2}{16\pi G} \sim m_{\rm Pl}^2 \Delta_h^2 H^2. \qquad (4.296)$$

Once the mode re-enters the horizon, $\rho_{\rm GW}$ evolves as a^{-4}, like radiation. For modes that re-enter the horizon while the Universe is dominated by radiation, $\rho_{\rm GW}/\rho_{\rm r}$ is a constant, and so

$$\Omega_{\rm GW} \sim \Omega_{\rm r}(H/m_{\rm Pl})^2 \sim 10^{-4} \left(V/m_{\rm Pl}^4\right). \qquad (4.297)$$

This relation follows from the fact that $\Omega_{\rm r} = 8\pi \overline{\rho}_{\rm r}/(H m_{\rm Pl})^2 \sim 1$ at $t < t_{\rm eq}$, which gives $\overline{\rho}_{\rm r} \sim (H m_{\rm Pl})^2$ and $(\rho_{\rm GW}/\rho_{\rm r}) \sim (H/m_{\rm Pl})^2$.

The tensor metric perturbations generated during inflation can affect the cosmic microwave background (CMB) anisotropies. For modes that re-enter the horizon after decoupling, the tensor modes produce a CMB spectrum with the same scale dependence as the scalar modes. The importance of the tensor (T) modes relative to the scalar (S) modes is represented by the ratio

$$\frac{\Delta_T^2}{\Delta_S^2} \sim \frac{\Delta_h^2}{\Delta_\Phi^2} \sim 10\varepsilon, \tag{4.298}$$

where the numerical coefficient is based on the detailed calculations of Starobinsky (1985). Thus, the contribution of the gravitational waves to the CMB anisotropy can become significant in an inflation model that has a relatively large gradient in the inflaton potential. Since such a gradient also causes a tilt in the power spectrum, the ratio Δ_T^2/Δ_S^2 is related to the tilt $1-n$. Detailed calculations show that $\Delta_T^2/\Delta_S^2 \approx 7(1-n)$ (e.g. Davis et al., 1992).

So far we have seen that inflation is quite successful in predicting the initial density perturbations for structure formation. There is, however, a severe problem concerning the overall amplitude of the perturbations predicted by such a scheme. As an illustration, consider a slowly varying inflaton potential, $V(\varphi) = V_0 - (\kappa/4)\varphi^4$, where κ is a coupling constant describing the flatness of the potential. Using the slow-roll equation (3.253) the number of e-foldings of inflation is

$$N = \int_{t_s}^{t_e} H\,\mathrm{d}t = \int_{t_s}^{t_e} \frac{H}{\dot{\varphi}_0}\,\mathrm{d}\varphi_0 = -\int_{t_s}^{t_e} \frac{3H^2}{\mathrm{d}V/\mathrm{d}\varphi_0}\,\mathrm{d}\varphi_0 \sim \frac{H^2}{\kappa \varphi_0^2(t_s)}, \tag{4.299}$$

where t_s and t_e are the times at which inflation starts and ends, respectively, and we have used the fact that $|\varphi_0(t_s)| \ll |\varphi_0(t_e)|$. Thus, the amplitudes of the metric perturbations predicted by this model can be expressed as

$$\Delta_\Phi^2 \sim \frac{H^6}{(\mathrm{d}V/\mathrm{d}\varphi_0)^2} \sim \kappa N^3. \tag{4.300}$$

This model therefore predicts an initial power spectrum that is scale invariant, with an amplitude that depends on the coupling constant κ. Since $N \gtrsim 50$ for a successful inflation (see §3.6.2), the observed amplitude $\Delta_\Phi \sim 10^{-5}$ requires that $\kappa \lesssim 10^{-15}$ according to Eq. (4.300). Such a small coupling constant does not come naturally from current particle physics. Thus, although inflation provides an attractive scheme to explain the origin of the cosmological density field, a truly viable model has yet to be found.

4.5.2 Perturbations from Topological Defects

An alternative class of mechanisms for generating cosmological density perturbations is provided by topological defects which can be produced during some phase transitions in the early Universe. According to the current view of particle physics, matter in the very early Universe is described in terms of fields, and the theory governing the motions of these fields is symmetric under certain transformations. As the Universe expands and the temperature drops, spontaneous breaking of these internal symmetries can occur. During a symmetry breaking, the field makes a phase transition from its original configuration to some final configuration with lower energy. In fact, as discussed in §3.6, the inflaton is an example of such a field undergoing spontaneous symmetry breaking. If there are more than one topologically distinct vacuum states for the final configuration, different regions in the Universe can end up in different vacuum states after the phase transition. These regions are separated by topological defects, which are still on the original field configuration. The energy trapped in such defects can then produce density perturbations.

Depending on the symmetry of the field, the vacuum manifold can be in one of the following forms: monopoles, domain walls, cosmic strings, and textures. Monopoles and domain walls can be ruled out immediately as the seeds for cosmic structure, since both would be produced with an energy density that implies $\Omega_0 \gg 1$. Studies of topological defects as the origin of cosmological density perturbations have therefore focused on cosmic strings and textures. In contrast to inflation, cosmic strings and textures in general generate non-Gaussian and isocurvature perturbations. However, the angular power spectrum of the temperature fluctuations in the cosmic microwave background predicted by these models does not agree with current observations (e.g. Pen et al., 1997), so that virtually all present-day models focus on inflation as the main mechanism to generate the primordial perturbations.

5
Gravitational Collapse and Collisionless Dynamics

Many objects in the present-day Universe, including galaxies and clusters of galaxies, have densities orders of magnitude higher than the average density of the Universe. These objects are thus in the highly nonlinear regime, where $\delta \gg 1$. To complete our description of structure formation in the Universe, we therefore need to go beyond perturbation growth in the linear and quasi-linear regimes, discussed in the previous chapter, and address the gravitational collapse of overdensities in the nonlinear regime.

In this chapter, we study the nonlinear gravitational collapse and dynamics of collisionless systems in which non-gravitational effects are negligible. In general, nonlinear gravitational dynamics is difficult to deal with analytically, and so in many applications computer simulations have to be used to follow the evolution in detail. However, if simple assumptions are made about the symmetry of the system, analytical models can still be constructed (§§5.1–5.3). Although these models are not expected to give accurate descriptions of the true nonlinear problem of gravitational collapse, they provide valuable insight into the complex processes involved. In §5.4 we describe the dynamics of collisionless equilibrium systems. These dynamical models describe the end states of the nonlinear gravitational collapse of a collisionless system, and are applicable to both galaxies and dark matter halos in a steady state. As such, these models are often used to model the observed kinematics of galaxies in an attempt to constrain their masses and their orbital structures. In §5.5 we present a description of the physical relaxation mechanisms that cause the collapsing perturbation to settle in an equilibrium configuration. As we will see, in general we cannot predict the structural and dynamical properties of a virialized object, even if its initial conditions are known. This is largely a reflection of the nonlinear dynamics involved, but also reflects our continuing lack of a full understanding of the various relaxation mechanisms at work. Finally, in §5.6, we use the models described in this chapter to understand how gravitational collapse proceeds in the cosmic density field.

5.1 Spherical Collapse Models

5.1.1 Spherical Collapse in a $\Lambda = 0$ Universe

In the absence of a cosmological constant, the radius r of a mass shell in a spherically symmetric density perturbation evolves according to the Newtonian equation,

$$\frac{\mathrm{d}^2 r}{\mathrm{d} t^2} = -\frac{GM}{r^2}, \tag{5.1}$$

where M is the mass *within* the mass shell. Before shell crossing, M is independent of t for a given mass shell, and so Eq. (5.1) can be integrated once to give

$$\frac{1}{2}\left(\frac{dr}{dt}\right)^2 - \frac{GM}{r} = \mathcal{E}, \tag{5.2}$$

where $\mathcal{E}$ is the specific energy of the mass shell. For $\mathcal{E} = 0$, the solution of this equation is particularly simple: $r = (9GM/2)^{1/3}t^{2/3}$. For $\mathcal{E} \neq 0$, the motion of a mass shell can be written in a parameterized form which depends on the sign of $\mathcal{E}$. Since the descriptions for $\mathcal{E} > 0$ and $\mathcal{E} < 0$ are parallel, we will only present the $\mathcal{E} < 0$ case for which the mass shell eventually collapses rather than expands forever. The motion of a mass shell with $\mathcal{E} < 0$ can be written as

$$r = A(1 - \cos\theta); \quad t = B(\theta - \sin\theta), \tag{5.3}$$

where A and B are two constants to be determined by the initial conditions for the mass shell in question. At early times when $\theta \ll 1$, we can expand Eq. (5.3) in powers of θ. Keeping the first two non-zero terms we have $r \approx A[\theta^2/2 - \theta^4/24]$ and $t \approx B[\theta^3/6 - \theta^5/120]$. Inserting these into Eqs. (5.1) and (5.2) and keeping the lowest non-zero orders we get

$$A^3 = GMB^2 \quad \text{and} \quad A = GM/(-2\mathcal{E}). \tag{5.4}$$

In order to specify A and B in terms of initial conditions, we assume that at an early time $t_i \ll 1$ the radius and the velocity of the mass shell are r_i and v_i, respectively. The specific energy of the mass shell (which is a constant during the evolution) is then $\mathcal{E} = v_i^2/2 - GM/r_i$. Inserting this into the second expression of Eq. (5.4) gives

$$\frac{r_i}{2A} = 1 - \frac{(v_i/H_i r_i)^2}{\Omega_i(1+\delta_i)}, \tag{5.5}$$

where H_i is the Hubble constant, $\Omega_i = \overline{\rho}(t_i)/\rho_{\rm crit}(t_i)$ is the cosmic density parameter at time t_i, and δ_i is the average mass overdensity within the mass shell, related to M and r_i by $(1+\delta_i)\overline{\rho}(t_i)(4\pi r_i^3/3) = M$. Since M is independent of t, the velocity of the mass shell can be written as

$$v_i = \frac{dr_i}{dt_i} = H_i r_i \left[1 - \frac{1}{3H_i t_i}\frac{\delta_i}{1+\delta_i}\frac{d\ln\delta_i}{d\ln t_i}\right]. \tag{5.6}$$

At sufficiently early times when $t_i \ll 1$, $\Omega_i \to 1$ and $\delta_i \to 0$. To first order in δ_i and $(\Omega_i^{-1} - 1)$, we have $H_i t_i \approx 2/3$, $\delta_i \propto t_i^{2/3}$, and $v_i/(H_i r_i) \approx 1 - \delta_i/3$. It then follows from Eqs. (5.4) and (5.5) that

$$A = \frac{1}{2}\frac{r_i}{[5\delta_i/3 + 1 - \Omega_i^{-1}]}; \quad B = \frac{3}{4}\frac{t_i}{[5\delta_i/3 + 1 - \Omega_i^{-1}]^{3/2}}. \tag{5.7}$$

Thus, the motion of a mass shell is completely specified by Eq. (5.3) in a given cosmology via the initial conditions on the radius r of the mass shell, and on the mean overdensity δ within it. Note that the mass shell reaches its maximum expansion at $\theta = \pi$. The radius and time at maximum expansion are

$$r_{\max} = 2A \quad \text{and} \quad t_{\max} = \pi B. \tag{5.8}$$

It is sometimes useful to specify the initial conditions in terms of quantities defined at the time of interest (say t). For example, a mass shell can also be specified by its Lagrangian radius $r_l \equiv [3M/4\pi\overline{\rho}(t)]^{1/3}$ and $\delta_l(t)$, the initial perturbation δ_i evolved to time t according to *linear* perturbation theory. For $t_i \ll t_0$ (where t_0 is the present time), the linear growth factor $g_i \approx 1$ (see §4.1.6) and so $\delta_i = \delta_l(t)a_i g_i/a_t g_t \approx (a_i/a_t)[\delta_l(t)/g_t]$, where a_t and g_t are, respectively, the scale factor and linear growth factor evaluated at time t. Since $(\Omega_i^{-1} - 1) = (\Omega_t^{-1} - 1)a_i/a_t$ for a matter dominated universe [which follows from Eqs. (3.79) and (3.77)], we can write Eq. (5.3) as

$$\frac{r}{r_l(t)} = \frac{1}{2} \frac{1 - \cos\theta}{[5\delta_l(t)/3g_t + (1 - \Omega_t^{-1})]}; \quad (5.9)$$

$$H_t t = \frac{1}{2\Omega_t^{1/2}} \frac{\theta - \sin\theta}{[5\delta_l(t)/3g_t + (1 - \Omega_t^{-1})]^{3/2}}, \quad (5.10)$$

where $H_t \equiv H(t)$, and we have used the fact that

$$r_l(t) \approx r_i a_t/a_i, \quad t_i^{-1} \approx \frac{3}{2}\Omega_t^{1/2} H_t(a_t/a_i)^{3/2}, \quad (5.11)$$

at $t_i \ll t_0$ when $\delta_i \ll 1$ and $H_i t_i \approx 2/3$. Eqs. (5.9) and (5.10) give a complete description of the mass shell in terms of $r_l(t)$, $\delta_l(t)$ and the cosmological quantities at time t (Ω_t and H_t). Since there is no shell crossing, the mean density within the mass shell at time t is just

$$\rho(t) = \overline{\rho}(t)[r_l(t)/r]^3. \quad (5.12)$$

In terms of $r_l(t)$ and $\delta_l(t)$, the maximum expansion radius $r_{\max}$ and time $t_{\max}$ can be written as

$$r_{\max}/r_l(t) = [5\delta_l(t)/3g_t + (1 - \Omega_t^{-1})]^{-1}, \quad (5.13)$$

and

$$H_t t_{\max} = \frac{\pi}{2\Omega_t^{1/2}} [5\delta_l(t)/3g_t + (1 - \Omega_t^{-1})]^{-3/2}. \quad (5.14)$$

For a mass shell turning around (i.e. reaching maximum expansion) at a time t_{ta}, its linear overdensity δ_l at t_{ta} needs to be

$$\delta_l(t_{\mathrm{ta}}) = \frac{3g(t_{\mathrm{ta}})}{5} \left\{ \left[\frac{\pi}{2\Omega^{1/2}(t_{\mathrm{ta}})H(t_{\mathrm{ta}})t_{\mathrm{ta}}}\right]^{2/3} - [1 - \Omega^{-1}(t_{\mathrm{ta}})] \right\}. \quad (5.15)$$

For $\Omega = 1$, this overdensity is

$$\delta_l(t_{\mathrm{ta}}) = \frac{3}{5}\left(\frac{3\pi}{4}\right)^{2/3} \approx 1.06, \quad (5.16)$$

and the real density at this time is

$$\rho(t_{\mathrm{ta}}) = \overline{\rho}(t_{\mathrm{ta}}) \left[\frac{r_l(t_{\mathrm{ta}})}{r_{\max}(t_{\mathrm{ta}})}\right]^3 = \left(\frac{3\pi}{4}\right)^2 \overline{\rho}(t_{\mathrm{ta}}) \approx 5.55\overline{\rho}(t_{\mathrm{ta}}). \quad (5.17)$$

According to Eqs. (5.9) and (5.10), the radius of a mass shell r becomes zero at $t = 2t_{\max}$ (corresponding to $\theta = 2\pi$), and it is usually said that the mass shell is collapsed by this time. Strictly speaking, however, Eqs. (5.9) and (5.10) are not valid for arbitrarily small r. As the mass shell turns around and begins to collapse, particles in the mass shell in question can cross the mass shells that were originally inside it, and consequently the mass enclosed by the mass shell is no longer constant, making the assumption of a constant M invalid. Indeed, as to be discussed in §5.2, by the time $t = 2t_{\max}$ all the mass shells initially enclosed by the mass shell that has just collapsed have crossed each other so many times that they form an extended, quasi-static structure called a virialized halo. Thus, we may identify $t_{\mathrm{col}} \equiv 2t_{\max}$ with the collapse time of the mass shell, even though the spherical collapse model described above is no longer valid by this time. From Eq. (5.14) we see that for a mass shell to collapse at time $t_{\rm col}$, the overdensity within it, linearly extrapolated to the collapse time, must be

$$\delta_{\rm c}(t_{\rm col}) = \frac{3g(t_{\rm col})}{5}\left\{\left[\frac{\pi}{\Omega^{1/2}(t_{\rm col})H(t_{\rm col})t_{\rm col}}\right]^{2/3} - \left[1 - \Omega^{-1}(t_{\rm col})\right]\right\}. \qquad (5.18)$$

Since $H(t)t$ and $g(t)$ depend only on $\Omega(t)$ in the matter dominated epoch (see §3.2.3 and §4.1.6), the critical overdensity for collapse, $\delta_{\rm c}(t_{\rm col})$, depends only on $\Omega(t_{\rm col})$. This critical overdensity can be approximated by

$$\delta_{\rm c}(t_{\rm col}) = \frac{3}{5}\left(\frac{3\pi}{2}\right)^{2/3}[\Omega(t_{\rm col})]^{0.0185} \approx 1.686[\Omega(t_{\rm col})]^{0.0185}, \qquad (5.19)$$

which is accurate to better than 1%. Note that the dependence on Ω is very weak.

5.1.2 Spherical Collapse in a Flat Universe with $\Lambda > 0$

In a universe with a non-zero cosmological constant, the motion of a mass shell in a spherically symmetric perturbation is given by

$$\frac{{\rm d}^2 r}{{\rm d}t^2} = -\frac{GM}{r^2} + \frac{\Lambda}{3}r, \qquad (5.20)$$

where we have used the fact that the cosmological constant contributes to the gravitational acceleration through an effective density $\rho + 3P/c^2 = -2\rho = -\Lambda c^2/(4\pi G)$. Integrating the above equation once we have

$$\frac{1}{2}\left(\frac{{\rm d}r}{{\rm d}t}\right)^2 - \frac{GM}{r} - \frac{\Lambda c^2}{6}r^2 = \mathscr{E}, \qquad (5.21)$$

where, as before, $\mathscr{E}$ is a constant. Before the mass shell reaches its maximum expansion, ${\rm d}r/{\rm d}t > 0$ and the solution of Eq. (5.21) can be written as

$$t = \int_0^r {\rm d}r\left[2\mathscr{E} + \frac{2GM}{r} + \frac{\Lambda c^2}{3}r^2\right]^{-1/2}. \qquad (5.22)$$

Suppose that the maximum-expansion radius of the mass shell is $r_{\max}$. Since ${\rm d}r/{\rm d}t = 0$ at this radius, we have $\mathscr{E} = -GM/r_{\max} - \Lambda c^2 r_{\max}^2/6$. Eq. (5.22) can then be written as

$$H_0 t = \left(\frac{\zeta}{\Omega_{\Lambda,0}}\right)^{1/2}\int_0^{r/r_{\max}}{\rm d}x\left[\frac{1}{x} - 1 + \zeta(x^2 - 1)\right]^{-1/2}, \qquad (5.23)$$

where

$$\zeta \equiv (\Lambda c^2 r_{\max}^3/6GM) < 1/2, \qquad (5.24)$$

with the inequality following from $\ddot{r} < 0$ at $r = r_{\max}$. After maximum expansion, the corresponding equation is

$$H_0(t - t_{\max}) = \left(\frac{\zeta}{\Omega_{\Lambda,0}}\right)^{1/2}\int_{r/r_{\max}}^1 {\rm d}x\left[1 - \frac{1}{x} - \zeta(x^2 - 1)\right]^{-1/2}, \qquad (5.25)$$

where

$$t_{\max} = \frac{1}{H_0}\left(\frac{\zeta}{\Omega_{\Lambda,0}}\right)^{1/2}\int_0^1 {\rm d}x\left[\frac{1}{x} - 1 + \zeta(x^2 - 1)\right]^{-1/2} \qquad (5.26)$$

is the time at maximum expansion. Thus, for given Λ and M, the three quantities, $r_{\max}$, ζ and $H_0 t_{\max}$, which are needed to specify the evolution given by Eqs. (5.23) and (5.25), are equivalent and only one of them needs to be determined through the initial conditions. At early times, when $r_i \ll r_{\max}$, Eq. (5.23) can be approximated by

$$H_0 t_i = \frac{2}{3}\left(\frac{\zeta}{\Omega_{\Lambda,0}}\right)^{1/2}\left(\frac{r_i}{r_{\max}}\right)^{3/2}\left[1+\frac{3}{10}(1+\zeta)\frac{r_i}{r_{\max}}\right], \qquad (5.27)$$

which follows from approximating the integrand in Eq. (5.23) by $x^{1/2}[1+(x+\zeta x-\zeta x^3)/2]$. For a flat universe with $\Omega_{m,0}+\Omega_{\Lambda,0}=1$, Eq. (3.99) gives $H_0 t_i = \frac{2}{3}\sqrt{\omega_i/\Omega_{\Lambda,0}}$ at $t_i \ll t_0$, where $\omega_i \equiv \Omega_\Lambda(t_i)/\Omega_m(t_i) = (\Omega_{\Lambda,0}/\Omega_{m,0})(1+z_i)^{-3} = \Omega_{m,i}^{-1}-1$. Since $r_i/r_{\max} \ll 1$, to first order Eq. (5.27) gives $r_i/r_{\max} \approx (\omega_i/\zeta)^{1/3}$. Replacing $r_i/r_{\max}$ in the bracket on the right-hand side of Eq. (5.27) by this first order relation, we get

$$\frac{r_i}{r_{\max}} \approx \left(\frac{\omega_i}{\zeta}\right)^{1/3}\left[1-\frac{1}{5}(1+\zeta)\left(\frac{\omega_i}{\zeta}\right)^{1/3}\right]. \qquad (5.28)$$

This equation, together with the relation $(1+\delta_i)\Omega_{m,i}\rho_{\rm crit}(t_i)(4\pi r_i^3/3) = M$ and the definition of ζ, gives

$$\delta_i = \frac{3}{5}(1+\zeta)\left(\frac{\omega_i}{\zeta}\right)^{1/3}. \qquad (5.29)$$

In terms of δ_0, which is δ_i linearly evolved to the present time, this relation is

$$\delta_0 = \frac{a_0 g_0}{a_i g_i}\delta_i = \frac{3}{5}g_0(1+\zeta)\left(\frac{\omega_0}{\zeta}\right)^{1/3}, \qquad (5.30)$$

where we have used that $\omega_i = \omega_0/(1+z_i)^3$ ($\omega_0 \equiv \Omega_{\Lambda,0}/\Omega_{m,0}$) and that $g_i \to 1$ for $a_i \ll a_0$. Eq. (5.30) relates ζ to the initial condition (δ_0, the initial overdensity within the mass shell evolved to the present time using linear theory) and is the relation we are seeking. If, as before, we assume that the collapse of a mass shell occurs at $t_{\rm col} = 2t_{\max}$, Eqs. (5.26) and (5.30) specify the relation between $t_{\rm col}$ and δ_0. At the time of collapse, the linear overdensity is

$$\delta_c(t_{\rm col}) = \frac{3}{5}g(t_{\rm col})(1+\zeta)\left[\frac{\omega(t_{\rm col})}{\zeta}\right]^{1/3}. \qquad (5.31)$$

The above relation can be approximated by

$$\delta_c(t_{\rm col}) = \frac{3}{5}\left(\frac{3\pi}{2}\right)^{2/3}[\Omega_m(t_{\rm col})]^{0.0055} \approx 1.686[\Omega_m(t_{\rm col})]^{0.0055}, \qquad (5.32)$$

which is accurate to better than 1%. As for cosmologies with zero cosmological constant, the dependence on Ω_m is extremely weak [see Eq. (5.19)]. To good approximation, therefore, $\delta_c(t_{\rm col}) \simeq 1.68$ for all realistic cosmologies.

5.1.3 Spherical Collapse with Shell Crossing

Because of its collisionless nature, a mass shell of collisionless particles in the spherical collapse model will oscillate about the center after the collapse, with an amplitude that may change with time. Gunn (1977) considered a simple model in which the oscillation amplitude of a mass shell is assumed to be a constant proportional to the radius of the mass shell at its first turnaround. Since the absolute value of the velocity of a mass shell is smaller at larger radius, the mass shell is expected to spend most time near its apocenter, and the total mass within a mass shell at the apocenter may be approximated by the original mass enclosed by the mass shell. In this case, the mean density within a radius r can be written as

$$\overline{\rho}(r) = \frac{3M}{4\pi r^3(M)}, \tag{5.33}$$

where $r(M) \propto r_{\rm ta}(M)$, with $r_{\rm ta}(M)$ the turnaround radius of the mass shell M. The density profile can then be obtained if we know how $r_{\rm ta}$ changes with M. As an example, consider a perturbation $\delta_{\rm i} \propto r_{\rm i}^{-3\varepsilon} \propto M^{-\varepsilon}$ in an Einstein–de Sitter universe. In this case, $r_{\rm ta} \propto r_{\rm i}/\delta_{\rm i} \propto M^{(\varepsilon+1/3)}$, and the density profile is

$$\rho(r) \propto r^{-\gamma} \quad \text{with} \quad \gamma = 9\varepsilon/(1+3\varepsilon). \tag{5.34}$$

For the special case in which the initial perturbation is associated with a point mass embedded in an Einstein–de Sitter background, $\varepsilon = 1$ and $\rho(r) \propto r^{-9/4}$. This solution was first obtained by Gunn & Gott (1972).

Unfortunately, the above treatment of shell crossing is not accurate. In general, the total mass within a mass shell at apocenter includes not only the particles initially enclosed by the mass shell, but also those shells which were initially outside it but have current radii smaller than its apocentric radius. Because of this additional mass, the apocentric radius in general changes with time, and so the density profile cannot be obtained simply by assuming the conservation of mass within individual mass shells. In the following section we describe a more general model that includes a more proper treatment of shell crossing.

5.2 Similarity Solutions for Spherical Collapse

5.2.1 Models with Radial Orbits

Consider an initial (spherical) density perturbation with density profile $\rho_{\rm i}(r_{\rm i})$, where $r_{\rm i}$ is the radius to the center at some fiducial time $t_{\rm i}$. The initial mass within a mass shell with radius $r_{\rm i}$ is

$$M_{\rm i}(r_{\rm i}) = 4\pi \int_0^{r_{\rm i}} \rho_{\rm i}(y) y^2 \, {\rm d}y. \tag{5.35}$$

At a later time $t > t_{\rm i}$, the radius of the mass shell with initial radius $r_{\rm i}$ [or with initial mass $M_{\rm i}(r_{\rm i})$] becomes $r(r_{\rm i},t)$, and the mass enclosed by it becomes $M(r,t)$. Assuming that all particles in the mass shell have purely radial orbits, the equation of motion of the mass shell is given by

$$\frac{{\rm d}^2 r}{{\rm d}t^2} = -\frac{GM(r,t)}{r^2}. \tag{5.36}$$

For simplicity we have assumed the cosmological constant to be zero. Before shell crossing, $M(r,t) = M_{\rm i}(r_{\rm i})$ is a constant, and the solution of this equation is the same as that discussed in §5.1. In general, the solution to the above equation has to be obtained numerically by following the time evolution of all individual mass shells. For a special set of problems where the collapse proceeds in a self-similar way, simpler solutions can still be found (e.g. Fillmore & Goldreich, 1984; Bertschinger, 1985). Before presenting these solutions, we caution that none of them are viable models for real halos since all are subject to strong non-radial instabilities which cause evolution from these initial conditions to produce strongly prolate, rather than spherical, systems (Carpintero & Muzzio, 1995; MacMillan et al., 2006). These similarity solutions nevertheless give useful insight into how halos grow.

In order for a problem to admit self-similar solutions, the time, t, has to be the only independent physical scale; all other characteristic scales are power laws of t. In such a case, the solution looks the same when physical quantities are expressed in terms of their characteristic scales. For a problem with spherical symmetry, such as the one considered here, this means that any physical quantity, $q(r,t)$, can be written in the form, $q(r,t) = Q(t)\mathcal{Q}[r/R(t)]$, where $R(t)$ and $Q(t)$ are the characteristic scales for the radius r and for the quantity q, respectively, both having power-law dependence on t, and $\mathcal{Q}$ is an arbitrary function of the normalized radius. When considering gravitational collapse in an expanding background, self-similarity requires the expansion of the universe to be scale-free [i.e. $a(t)$ is a power law of t], and so we need to assume an Einstein–de Sitter universe. In this case, the turnaround radius and time (which again refer to the radius and time at the *first* apocenter) are

$$\frac{r_{\max}}{r_i} = \frac{C_r}{\delta_i}; \quad \frac{t_{\max}}{t_i} = \left(\frac{C_t}{\delta_i}\right)^{3/2}, \quad (5.37)$$

where $C_r = 3/5$, $C_t = C_r(3\pi/4)^{2/3}$ and δ_i is the initial mean density contrast within the mass shell [see Eq. (5.8)]. In order for the problem to have self-similar solutions, we also need to assume the initial density perturbation to be scale-free. We therefore write

$$\delta_i \equiv \frac{\delta M_i}{M_i} = \left(\frac{M_i}{M_*}\right)^{-\varepsilon}, \quad (5.38)$$

where ε is a constant and M_* is a reference mass scale. Denoting the turnaround radius at time t by r_t and the mass within it by $M_t \equiv M(r_t,t)$, we find from Eqs. (5.37) and (5.38) that

$$\frac{M_t}{M_*} = \frac{1}{C_t^{1/\varepsilon}} \left(\frac{t}{t_i}\right)^{2/3\varepsilon}; \quad (5.39)$$

$$r_t = \frac{C_r}{C_t} \left[\frac{3M_t}{4\pi\overline{\rho}(t_i)}\right]^{1/3} \left(\frac{t}{t_i}\right)^{2/3}, \quad (5.40)$$

where $\overline{\rho}(t_i)$ is the background density at t_i. Under the assumption of self-similarity, the mass $M(r,t)$ must have the self-similar form

$$M(r,t) = M_t \mathcal{M}(r/r_t), \quad (5.41)$$

where $\mathcal{M}$ is a dimensionless mass profile. Since $M(r,t)$ contains all the mass in the mass shells with current radii smaller than $r(r_i,t)$, we have for $r < r_t$ that

$$\mathcal{M}(r/r_t) = \frac{1}{M_t} \int_0^{M_t} \mathcal{H}\left[r(r_i,t) - r(r_i',t)\right] dM_i', \quad (5.42)$$

where $\mathcal{H}(x)$ is the Heaviside step function: $\mathcal{H}(x) = 1$ for $x \geq 0$ and $\mathcal{H}(x) = 0$ otherwise. In terms of the dimensionless variables,

$$\lambda = r/r_{\max} \quad \text{and} \quad \tau = t/t_{\max}, \quad (5.43)$$

where $r_{\max}$ and $t_{\max}$ are the radius and time at the *first* apocenter, respectively, the equation of motion, Eq. (5.36), can be written as

$$\frac{d^2\lambda}{d\tau^2} = -\frac{\pi^2}{8} \frac{\tau^{2/3\varepsilon}}{\lambda^2} \mathcal{M}\left[\frac{\lambda}{\Lambda(\tau)}\right], \quad (5.44)$$

where

$$\mathcal{M}(x) = \frac{2}{3\varepsilon} \int_1^\infty \frac{dy}{y^{1+2/3\varepsilon}} \mathcal{H}\left[x - \frac{\lambda(y)}{\Lambda(y)}\right] \quad \text{and} \quad \Lambda(\tau) = \tau^{2/3+2/9\varepsilon}. \quad (5.45)$$

The boundary conditions are $\lambda = 1$ and $d\lambda/d\tau = 0$ at the turnaround time $\tau = 1$. Eq. (5.44) is independent of r_i, and so it applies to all mass shells that have turned around before time t.

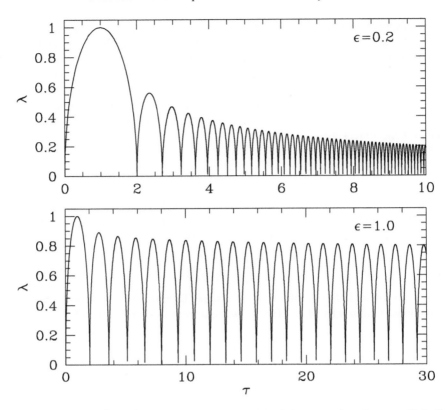

Fig. 5.1. The scaled radius $\lambda \equiv r/r_{\max}$ as a function of the scaled time $\tau \equiv t/t_{\max}$ in two self-similar radial orbit models with $\varepsilon = 0.2$ and 1.0, respectively. [After Fillmore & Goldreich (1984)]

This obviously is a consequence of the assumption of self-similarity that makes the problem scale-invariant. For a given ε, Eq. (5.44) can be integrated iteratively with an initial guess of the mass distribution $\mathcal{M}(x)$ (Fillmore & Goldreich, 1984). As an illustration, Fig. 5.1 shows λ as a function of τ for $\varepsilon = 0.2$ and 1.

As one sees from Fig. 5.1, each mass shell has an initial period of expansion which ends at the time of turnaround ($\tau = 1$). Thereafter, the mass shell oscillates around the center. Since $r_{\max}$ and $t_{\max}$ are fixed for a given mass shell, while the current turnaround radius r_t and time t increase with the passage of time, Fig. 5.1 shows that, for a given mass shell, the ratio between its apocentric radius r_a and the current turnaround radius r_t, as well as the ratio between its oscillation period and the current time t, both decrease with time. Consequently, a given mass shell becomes buried deeper and deeper in the halo of the collapsed material with the passage of time. During an oscillation period of a deeply buried mass shell, the change in the halo density profile is small – because the oscillation period is much smaller than the dynamical time scale of the halo – so that the oscillation is adiabatic. These properties allow us to examine the asymptotic behavior of the similarity solution.

The scale-free nature of the similarity solution suggests that the halo mass profile $M(r,t)$, and the time variation of the apocentric radius r_a, may both be approximated by power laws:

$$M(r,t) = \kappa(t) r^\alpha; \quad \frac{r_a}{r_{\max}} = \left(\frac{t}{t_{\max}}\right)^q. \tag{5.46}$$

Note that this mass profile corresponds to a density profile $\rho(r) \propto r^{-\gamma}$ with $\gamma = 3 - \alpha$. The remaining task is then to express α and q in terms of the index ε describing the initial perturbation. Inserting the first expression into Eq. (5.36) we obtain

$$\frac{d^2 r}{dt^2} = G\kappa(t) r^{\alpha-2}. \tag{5.47}$$

As discussed above, for orbits with apocentric radius $r_a \ll r_t$, $\kappa(t)$ can be treated as a constant over one orbital period. With this approximation, Eq. (5.47) can be integrated to give

$$\left(\frac{dr}{dt}\right)^2 \approx \frac{2G\kappa(t)}{\alpha-1} \left[r_a^{\alpha-1} - r^{\alpha-1}\right]. \tag{5.48}$$

The corresponding radial action of motion, which is an adiabatic invariant (see §11.1.3), is

$$J_r \equiv 4 \int_0^{r_a} \frac{dr}{dt} dr$$

$$\approx 4 \left[\frac{2G\kappa(t)}{\alpha-1}\right]^{1/2} r_a^{(\alpha+1)/2} \int_0^1 \left[1 - y^{\alpha-1}\right]^{1/2} dy. \tag{5.49}$$

Thus, if we write $\kappa(t) = ct^{-s}$, then $q = s/(\alpha+1)$. The value of s can be obtained by using the expressions of r_t and M_t given by Eqs. (5.39) and (5.40) in the relation $M_t = M(r_t, t) = ct^{-s}r_t^{\alpha}$ [see Eq. (5.46)]: $s = \frac{2}{3}(\alpha/3 + \alpha\varepsilon - 1)/\varepsilon$. It then follows that

$$q = \frac{2}{3\varepsilon(\alpha+1)} \left[\alpha/3 + \alpha\varepsilon - 1\right]. \tag{5.50}$$

The time a particle spends inside a radius r, $t \propto \int_0^r (dr/dt)^{-1} dr$, can be obtained from Eq. (5.48), and the fraction of time a particle with current apocentric radius r_a spends inside r is

$$P\left(\frac{r}{r_a}\right) = \begin{cases} \mathscr{T}\left(\frac{r}{r_a}\right)/\mathscr{T}(1) & (r \leq r_a) \\ 1 & (r > r_a), \end{cases} \tag{5.51}$$

where

$$\mathscr{T}(x) = \int_0^x \left[1 - y^{\alpha-1}\right]^{-1/2} dy. \tag{5.52}$$

It then follows from the first expression of Eq. (5.46) that

$$\left(\frac{r}{r_t}\right)^{\alpha} = \frac{M(r,t)}{M(r_t,t)} = \int_0^{M_t} \frac{dM_i}{M_t} P\left[\frac{r}{r_a(M_i,t)}\right]. \tag{5.53}$$

Using Eqs. (5.37), (5.38) and (5.46) to express r_i, $r_{\max}$ and $t_{\max}$ in terms of t_i, t and r_a, and making use of Eqs. (5.39) and (5.40), we can write Eq. (5.53) as

$$\left(\frac{r}{r_t}\right)^{\alpha-p} = \frac{1}{p} \int_{r/r_t}^{\infty} \frac{dy}{y^{1+p}} P(y), \tag{5.54}$$

where

$$p = \frac{6}{2 + 3(2-3q)\varepsilon}. \tag{5.55}$$

For $y \ll 1$, $P(y) \propto y$. Thus, if $p < 1$, then the integration in Eq. (5.54) converges to a constant value for $r/r_t \to 0$, giving $\alpha = p$. On the other hand, if $p > 1$, the integration diverges as $(r/r_t)^{1-p}$, which gives $\alpha = 1$. Inserting these results into Eqs. (5.50) and (5.55), we can finally express all the power indices in terms of ε:

$$\begin{aligned}&\alpha=1, \quad p=\tfrac{6}{(4+3\varepsilon)}\geq 1, \quad q=\tfrac{(3\varepsilon-2)}{9\varepsilon}, \quad \gamma=2, \quad &\text{for } \varepsilon\leq\tfrac{2}{3};\\ &\alpha=p, \quad p=\tfrac{3}{(1+3\varepsilon)}<1, \quad q=0, \quad \gamma=\tfrac{9\varepsilon}{(1+3\varepsilon)}, \quad &\text{for } \varepsilon>\tfrac{2}{3}.\end{aligned} \quad (5.56)$$

The physical distinction between the two scaling solutions can be understood by examining the initial specific binding energy of particles in different mass shells. Because a mass shell collapses when $\delta M/M \approx 1.68$, the mass accretion history implied by Eq. (5.38) is

$$M(t) \propto [D(t)]^{1/\varepsilon} \propto [H(t)]^{-2/3\varepsilon}, \qquad (5.57)$$

where $D(t)$ is the linear growth factor at time t and is related to the Hubble parameter, $H(t)$, as $D(t) \propto [H(t)]^{-2/3}$ in the Einstein–de Sitter universe considered here. The specific binding energy of a mass shell with turnaround time t is

$$E(M_i) \propto -\frac{M_i}{r_t} \propto -\frac{M_i \delta_i}{r_i} \propto -M_i^{2/3}\delta_i \propto -M_i^{2/3-\varepsilon} \propto -[H(z)]^{2(1-2/3\varepsilon)/3}. \qquad (5.58)$$

For $\varepsilon > 2/3$, mass shells with smaller M_i have higher (negative) binding energy. In this case, the mass in the inner halo is dominated by particles which have oscillated through the center many times, and so the mass profile is almost independent of time. In this case, the apocentric radius r_a is a fixed fraction of $r_{\max}$ (see the $\varepsilon = 1$ case in Fig. 5.1) so that $q=0$ and

$$M(r,t) \propto M_i(r_{\max}=r) \propto r^{3/(1+3\varepsilon)}, \qquad (5.59)$$

and the density distribution obeys Eq. (5.34). For $\varepsilon < 2/3$, on the other hand, mass shells with smaller M_i are less bound and the mass in the inner halo comes from particles whose apocenters spread throughout the halo. Consequently, $\alpha = 1$ corresponds to $\rho(r) \propto r^{-2}$, independent of the initial density profile. In this case, r_a decreases with time because the loosely bound particles in the inner part of the halo can lose energy to the mass shells that have recently collapsed.

5.2.2 Models Including Non-Radial Orbits

Self-similar models can also be constructed for spherical collapse with non-radial particle orbits. White & Zaritsky (1992) considered a model in which the specific angular momentum of a particle in a mass shell is specified by

$$\mathscr{L} = \mathscr{J}\sqrt{GM_{\max}r_{\max}}, \qquad (5.60)$$

where $r_{\max}$ is the turnaround radius, and $M_{\max}$ is the mass interior to $r_{\max}$ and is equal to the initial mass within the mass shell. If $\mathscr{J}$ is the same constant for all mass shells, then no new physical scale is introduced and the problem will still have self-similar solutions. Note that for a spherical system, $\mathscr{L}$ is conserved for a given mass shell. In the presence of angular momentum, the equation of motion for a mass shell becomes

$$\frac{d^2 r}{dt^2} = -\frac{GM(r,t)}{r^2} + \frac{\mathscr{L}^2}{r^3}, \qquad (5.61)$$

which is the same as Eq. (5.36) except for the centrifugal force term. Because of the centrifugal force, the radius r cannot reach zero; instead it oscillates between a pericenter r_p and an apocenter r_a, both being time-dependent. The equation corresponding to Eq. (5.44) is

$$\frac{d^2\lambda}{d\tau^2} = -\frac{\pi^2}{8}\frac{\tau^{2/3\varepsilon}}{\lambda^2}\mathscr{M}\left[\frac{\lambda}{\Lambda(\tau)}\right] + \frac{\pi^2}{8}\frac{\mathscr{J}^2}{\lambda^3}, \qquad (5.62)$$

which can be solved in the same way as Eq. (5.44).

5.2 Similarity Solutions for Spherical Collapse

To study the asymptotic behavior of the solution, we again write $M(r,t)$ and the apocentric radius r_a in the scale-free form of Eq. (5.46). Consider again mass shells with $r_a \ll r_t$ so that $\kappa(t)$ can be treated as a constant over one orbital period. In this case, Eq. (5.61) can be integrated once to give

$$\left(\frac{dr}{dt}\right)^2 \approx -\frac{2G\kappa(t)}{\alpha-1}r^{\alpha-1} - \frac{\mathscr{L}^2}{r^2} + 2E, \tag{5.63}$$

where E is a constant. Since $dr/dt = 0$ for both $r = r_a$ and $r = r_p$, it is straightforward to show that

$$2E = \frac{2G\kappa(t)}{\alpha-1}r_a^{\alpha-1} + \frac{\mathscr{L}^2}{r_a^2}; \quad \frac{2G\kappa(t)}{\alpha-1} = \mathscr{L}^2 \frac{r_p^{-2} - r_a^{-2}}{r_a^{\alpha-1} - r_p^{\alpha-1}}. \tag{5.64}$$

This in Eq. (5.63) gives

$$\frac{dr}{dt} = \frac{\mathscr{L}}{r_a}\left[(\zeta^{-2}-1)\frac{1-y^{\alpha-1}}{1-\zeta^{\alpha-1}} + (1-y^{-2})\right]^{1/2}, \tag{5.65}$$

where $y \equiv r/r_a$ and $\zeta \equiv r_p/r_a$. The radial action can then be written as

$$J_r \equiv 2\int_{r_p}^{r_a} \frac{dr}{dt} dr$$

$$\approx 2\mathscr{L} \int_\zeta^1 \left[(\zeta^{-2}-1)\frac{1-y^{\alpha-1}}{1-\zeta^{\alpha-1}} + (1-y^{-2})\right]^{1/2} dy, \tag{5.66}$$

which is again an adiabatic invariant. Since $\mathscr{L}$ is conserved, invariance of J_r implies that ζ is time independent. The conservation of $\mathscr{L}$ also implies that $GM_{\max}r_{\max} = GM(r_a,t)r_a = G\kappa(t)r_a^{\alpha+1} = (\mathscr{L}/\mathscr{J})^2$. Using these relations in the second equation of Eq. (5.64), we obtain $\mathscr{J}^2(\alpha-1) = 2(\zeta^2 - \zeta^{\alpha+1})/(1-\zeta^2)$. Since $\mathscr{J}$ is the same for all mass shells in the model considered here, this relation implies that ζ is the same for all mass shells.

The same arguments that led to Eq. (5.54) are still valid in the present case, and can therefore be used to obtain the asymptotic density profile for $r \to 0$. However, since $P(y) = 0$ for $y < \zeta$ (i.e. for $r < r_p$), the lower limit in the integration in Eq. (5.54) should be replaced by ζ for $r \ll \zeta r_t$. The right-hand side of Eq. (5.54) is then independent of r, implying that $\alpha = p$. This relation, together with Eqs. (5.50) and (5.55), implies that the asymptotic inner slope of the density profile is

$$\gamma = -\frac{d\ln\rho}{d\ln r} = 9\varepsilon/(1+3\varepsilon) \tag{5.67}$$

for all $\varepsilon > 0$ (Nusser, 2001). Note that for $\varepsilon \geq 2/3$, this solution is the same as that for the model assuming pure radial orbits. Because of the centrifugal force, the mixing of particles that leads to $\alpha = 1$ in the pure radial case is absent even for $0 < \varepsilon \leq 2/3$. Instead, the profile is determined by the initial conditions specified by ε. For large enough $\mathscr{J}$ we may expect these models to be free of the radial orbit instability which destroys the $\mathscr{J} = 0$ similarity solutions, but their stability has not yet been studied in detail.

If $\mathscr{J}$ is different for different mass shells, the problem in general does not admit self-similar solutions. In this case, one has to follow the time evolution of every mass shell separately. Lu et al. (2006) considered a model where the velocity of particles is isotropized during the collapse. One-dimensional simulations show that the asymptotic slope in this case is $\gamma = 9\varepsilon/(1+3\varepsilon)$ for $\varepsilon > 1/6$ and $\gamma = 1$ for $0 < \varepsilon \leq 1/6$. Fig. 5.2 shows the γ-ε relation in this model compared to the model assuming pure radial orbits and the model in which all particles have the same $\mathscr{J}$.

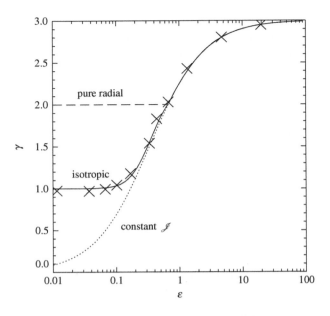

Fig. 5.2. The relation between the inner logarithmic slope, $\gamma \equiv -d\ln\rho/d\ln r = 3 - \alpha$, as a function of the exponent of the initial perturbation defined in Eq. (5.38). The long dashed curve shows the solution of the model with pure radial motion and the dotted curve shows the solution when all particles have the same $\mathcal{J}$. The solution with an isotropized inner velocity distribution is shown as the solid curve, and is compared with the result obtained directly from 1-D simulations (crosses). All the three curves overlap for $\varepsilon \geq 2/3$. [Adapted from Lu et al. (2006)]

One important feature of the γ-ε relation for the isotropized model is that $\gamma \approx 1$ for all $0 < \varepsilon \lesssim 1/6$. As shown in Eq. (5.58), for ε in this range, the specific energy of the accreted mass increases rapidly with time, with a time scale that is shorter than a Hubble time. In this case, not only can particles with small apocenters (low orbital energies) reach the inner part, but so can many particles with large apocenters (high orbital energies) and small angular momenta. Velocity isotropization mixes these orbits, resulting in $\gamma \approx 1$. For $\varepsilon > 1/6$ the gravitational potential is changing gradually and particles joining the inner halo all have a similar energy and orbital shape. The resulting profile can then be described by the self-similar solution assuming the same orbital shape for all particles. An inner logarithmic slope of $d\ln\rho/d\ln r \sim -1$ thus results from rapid collapse and orbit isotropization. Both conditions are required to produce such an inner slope. If the collapse is fast ($\varepsilon \lesssim 1/6$) but the velocity dispersion is not isotropic, the inner slope can be as shallow as 0 (for constant $\mathcal{J}$) and as steep as -2 [for pure radial orbits, though these models are unstable and evolve to a shallower inner slope (MacMillan et al., 2006)]. If the orbits are isotropized, but the mass accretion rate is small, i.e. $\varepsilon > 1$, the inner slope can be much steeper than -1. As we will see in Chapter 7, cold dark matter halos in numerical simulations typically have $\gamma = -d\ln\rho/d\ln r \simeq 1$ in the inner region.

5.3 Collapse of Homogeneous Ellipsoids

So far we have assumed that the collapse of density perturbations occurs under spherical symmetry, which is clearly an over-simplification. As a simple non-spherical model for gravitational collapse, let us consider the evolution of an overdense region that is spatially uniform inside an

5.3 Collapse of Homogeneous Ellipsoids

ellipsoidal volume. For simplicity, we assume that the ellipsoid consists of an ideal fluid of zero pressure. The equation of motion of a fluid element at the *comoving* coordinates **x** inside the ellipsoid can in general be written as

$$\frac{d\mathbf{v}}{dt} = -\frac{1}{a}\nabla\Phi(\mathbf{x}), \tag{5.68}$$

where **v** is the peculiar velocity of the fluid element, and the gravitational potential perturbation Φ obeys the Poisson equation

$$\nabla^2\Phi = 4\pi G\overline{\rho}_m(t)a^2\delta(t), \tag{5.69}$$

with $\delta(t)$ the overdensity of the ellipsoid at time t. In general, Φ can be separated into two parts,

$$\Phi = \Phi_{\text{int}} + \Phi_{\text{ext}}, \tag{5.70}$$

where Φ_{int} and Φ_{ext} are the potential perturbations due to matter interior and exterior to the ellipsoid, respectively. Note that $\nabla^2\Phi_{\text{ext}} = 0$ for **x** inside the ellipsoid. For simplicity, we work in the principal axes system of the ellipsoid. The potential perturbation due to the ellipsoid can then be written as

$$\Phi_{\text{int}}(\mathbf{x}) = \pi G a^2 \overline{\rho}_m(t)\delta(t) \sum_{i=1}^{3} \alpha_i x_i^2, \tag{5.71}$$

where

$$\alpha_i = X_1 X_2 X_3 \int_0^\infty dy (X_i^2 + y)^{-1} \prod_{j=1}^{3} (X_j^2 + y)^{-1/2}, \tag{5.72}$$

with $X_i(t)$ ($i = 1, 2, 3$) the comoving lengths of the principal axes at time t (e.g. Chandrasekhar, 1969). Since Φ_{int} obeys Eq. (5.69), it is straightforward to show that $\sum_i \alpha_i = 2$. Thus, the equation of motion (5.68) applied to the mass elements near the ends of the principal axes gives

$$\frac{d^2 X_i}{dt^2} + \frac{2\dot{a}}{a}\frac{dX_i}{dt} = -2\pi G\overline{\rho}_m \delta \alpha_i X_i - \frac{1}{a^2}\nabla_i \Phi_{\text{ext}}, \tag{5.73}$$

where δ is given by the mass of the ellipsoid, M, as

$$\frac{4\pi}{3}(1+\delta)\overline{\rho}_m a^3 X_1 X_2 X_3 = M. \tag{5.74}$$

If the external potential can be neglected, as is the case when the density contrast in the ellipsoid is high enough to dominate the dynamics, Eq. (5.73) can be integrated for a given initial condition. In this case, Eq. (5.73) is invariant under the transformation $x_i \to k_i x_i$, so that the ellipsoid remains ellipsoidal and homogeneous during the evolution.

Since Φ_{ext} is of the same order as Φ_{int} for $\delta \ll 1$ in a cosmological density field (see below), Eq. (5.73) with Φ_{ext} set to zero does not reproduce the linear evolution. In order to integrate the equation of motion from a linear initial condition, the external potential has to be taken into account properly. To find an expression for Φ_{ext} in the linear regime, we assume that the ellipsoid evolves from an initial Lagrangian sphere with radius r_0. According to the Zel'dovich approximation (see §4.1.8), the boundary of this sphere evolves into an ellipsoid with principal axes $X_i(t) = r_0[1 - \lambda_i D(t)]$ in the linear regime. Recall that λ_i ($i = 1, 2, 3$) are the eigenvalues of the deformation tensor, $\nabla\nabla\Phi_i/(4\pi G\overline{\rho}_m a^3)$, where Φ_i is the potential at some initial time t_i, related to the linear potential perturbation Φ_ℓ by $\Phi_\ell = (D/a)\Phi_i$. Thus, the principal axes of the ellipsoid are parallel to those of the deformation tensor in the Zel'dovich approximation.

Expanding Φ_ℓ in the neighborhood of the center of the ellipsoid up to the second order, we obtain

$$\Phi_\ell(\mathbf{x}) = \Phi_\ell(0) - \sum_i x_i \left(\frac{\partial \Phi_\ell}{\partial x_i}\right)_{\mathbf{x}=0} + \frac{1}{2}\sum_{i,j} x_i x_j \left(\frac{\partial^2 \Phi_\ell}{\partial x_i \partial x_j}\right)_{\mathbf{x}=0}. \tag{5.75}$$

Since the constant term has no effect on the motion and the linear gradient term represents the motion of the center of mass, we keep only the quadratic term which describes the tidal field around $\mathbf{x} = 0$. The linear potential can then be written as

$$\begin{aligned}\Phi_\ell(\mathbf{x}) &= \frac{1}{2}\sum_{i,j} T_{ij} x_i x_j \\ &= \Phi_{\ell,\text{int}}(\mathbf{x}) + \frac{1}{2}\sum_{i,j}\left(T'_{ij} - 2\pi G a^2 \overline{\rho}_\text{m} \delta\, \alpha'_i \delta_{ij}\right) x_i x_j, \end{aligned} \tag{5.76}$$

where $\alpha'_i = \alpha_i - 2/3$, and

$$T_{ij} = \frac{\partial^2 \Phi_\ell}{\partial x_i \partial x_j} \quad \text{and} \quad T'_{ij} = T_{ij} - \frac{1}{3}\delta_{ij}\nabla^2 \Phi_\ell \tag{5.77}$$

are the linear tidal tensor and its traceless part, respectively. Since to first order $\alpha'_i \propto \delta$, the term involving $\delta \alpha'_i$ in Eq. (5.76) is of second order in δ and can be ignored at early times. Hence, the external potential perturbation in the linear regime is

$$\Phi_{\ell,\text{ext}}(\mathbf{x}) = \frac{1}{2}\sum_{i,j} T'_{ij} x_i x_j, \tag{5.78}$$

which represents the entire traceless part of the total tide and in general cannot be neglected. In the principal-axis system, we can write

$$\Phi_{\ell,\text{ext}}(\mathbf{x}) = 2\pi G \overline{\rho}_\text{m} a^2 \sum_j \left(\lambda_j - \frac{\delta_\text{i}}{3}\right) D(t) x_j^2, \tag{5.79}$$

where δ_i is the initial linear overdensity of the ellipsoid. As a simple approximation, we may assume $\Phi_\text{ext} = \Phi_{\ell,\text{ext}}$ during the entire evolution of the ellipsoid. This assumption is reasonable, because by definition it reproduces the linear evolution and because once the ellipsoid turns around and begins to collapse, the nonlinear internal gravitational field will overcome the external tidal field so as to make the assumption regarding the external field irrelevant. The dynamical equations of X_j can then be written as

$$\frac{d^2 X_j}{dt^2} + \frac{2\dot{a}}{a}\frac{dX_j}{dt} = -4\pi G \overline{\rho}_\text{m} X_j \left[\frac{1}{2}\alpha_j \delta + D(t)\left(\lambda_j - \frac{1}{3}\delta_\text{i}\right)\right]. \tag{5.80}$$

In this case, the ellipsoid remains both ellipsoidal and homogeneous during the evolution. Combined with Eq. (5.74), Eq. (5.80) can be solved numerically for any given δ_i and λ_j in a given cosmology.

To proceed, we use that

$$\alpha_j \approx \frac{2X_h}{3X_j}, \tag{5.81}$$

where $X_h = 3/\sum_j X_j^{-1}$. This approximation is accurate to about 10% (White & Silk, 1979). Inserting this into Eq. (5.80) we obtain

$$a\frac{d}{dt}a^2\frac{dX_j}{dt} = -\frac{4\pi G}{3}\bar{\rho}_m a^3 [\delta X_h - \delta_i DX_j + 3\lambda_j DX_j]. \tag{5.82}$$

Since the two terms involving DX_j are expected to be important only in the linear regime when $X_j \sim X_j(t_i) \sim X_h$, this equation can formally be integrated twice to give

$$X_j(t) \approx X_j(t_i)[1 - D\lambda_j] - \frac{4\pi G}{3}\bar{\rho}_m a^3 \int \frac{dt}{a^2} \int \frac{dt}{a}(\delta - D\delta_i)X_h. \tag{5.83}$$

This solution has the desired asymptotic properties: in the linear regime, when $\delta = D\delta_i$, it is the same as that given by the Zel'dovich approximation; in the highly nonlinear regime where $\delta \gg D\delta_i$ it gives the solution of an ellipsoid with vanishing external tide. Note that the second term on the right-hand side of Eq. (5.83) does not depend on the index j, and so the comoving lengths of all three axes contract by the same amount relative to the Zel'dovich approximation. For a uniform sphere with the same initial overdensity δ_i, the solution given by Eq. (5.83) is exact for its comoving radius, $X(t)$. Thus, if we assume that the time-dependence of $X_h(\delta - D\delta_i)$ is independent of the shape of the perturbation, the last term of Eq. (5.83) can be replaced by $X(t) - X(t_i)(1 - D\delta_i/3)$, where $X(t_i) \approx X_h(t_i)$ is the initial radius of the uniform sphere. The solution (5.83) can then be written as

$$X_j(t) \approx X_j(t_i)[1 - D\lambda_j] - X_h(t_i)\left[1 - \frac{D\delta_i}{3} - \frac{a_e(t)}{a(t)}\right], \tag{5.84}$$

where $a_e(t)$ is the expansion factor in a universe with initial density $(1 + \delta_i)\bar{\rho}_m(t_i)$ at t_i. This solution is exact both for spherical collapse and in the linear regime, and is found to match the exact solution of Eq. (5.80) reasonably well up until the collapse of the shortest axis (White, 1996; Shen et al., 2006).

From the above description, we see that the collapse of a uniform ellipsoid can be specified by its initial overdensity $\delta_i = \lambda_1 + \lambda_2 + \lambda_3$, and the parameters e and p that characterize the asymmetry of the tidal field:

$$e \equiv \frac{\lambda_1 - \lambda_3}{2\delta_i}, \quad p \equiv \frac{\lambda_1 + \lambda_3 - 2\lambda_2}{2\delta_i}, \tag{5.85}$$

assuming $\lambda_1 \geq \lambda_2 \geq \lambda_3$. Thus defined, e (≥ 0) is a measure of the ellipticity in the (λ_1, λ_3) plane, and p determines the oblateness ($0 \leq p \leq e$) or prolateness ($0 \geq p \geq -e$) of the tidal ellipsoid. Oblate spheroids have $p = e$, prolate spheroids have $p = -e$ and spheres have $e = p = 0$. In general the shortest axis (the one parallel to λ_1) will be the first to collapse to zero. When this happens Eq. (5.80) is no longer valid. However, in order to study the collapse of the whole ellipsoid, we need to make additional assumptions so as to integrate Eq. (5.80) all the way to when the longest axis is considered to have collapsed. A common practice is to assume that all axes are frozen at the same radius equal to $(200)^{1/3}$ times the Lagrangian radius so that the mean density within it is about 200 times the mean density of the universe at the time of collapse. This choice is somewhat arbitrary but, by construction, reproduces the 'virial' density contrast from the spherical collapse model, as we will see in §5.4.4.

In order for a particular axis to collapse at a given time, the initial overdensity δ_i must be sufficiently high. For the ranges of e and p expected from cold dark matter models, a reasonable approximation to the critical linear overdensity for the longest axis to collapse can be obtained by solving

$$\frac{\delta_{ec}(e,p)}{\delta_{sc}} \approx 1 + \beta \left[5(e^2 \pm p^2)\frac{\delta_{ec}^2(e,p)}{\delta_{sc}^2}\right]^{\gamma} \tag{5.86}$$

for $\delta_{ec}(e,p)$, where $\beta \approx 0.47$, $\gamma \approx 0.615$, $\delta_{sc} = \delta_c$ is the critical overdensity for spherical collapse (i.e. for $e = p = 0$), and the plus (minus) sign is used if p is negative (positive). This relation has been obtained by Sheth et al. (2001) fitting the values of $\delta_{ec}(e,p)$ given by the ellipsoidal collapse model described above. Thus, if the collapse time of an ellipsoid is defined to be the time when its longest axis collapses, then for a given collapse time the required overdensity is higher for ellipsoidal collapse than for spherical collapse. In other words, for a given initial overdensity, collapse occurs later for an ellipsoidal perturbation than for a spherical perturbation. As shown by White (1996) and Shen et al. (2006), the opposite is true for the collapse along the shortest axis.

5.4 Collisionless Dynamics

In the previous sections we have described the gravitational collapse of density perturbations consisting of collisionless particles. The ultimate end state of such a collapse is a system in equilibrium, whose structure is governed by collisionless dynamics. Before discussing the physical relaxation mechanisms that cause the collapsing system to reach equilibrium, we first describe the dynamics of collisionless equilibrium. After demonstrating that galaxies and dark matter halos can be well approximated as collisionless systems, we describe the collisionless Boltzmann equations, the Jeans equations, the virial theorem, orbit theory, the Jeans theorem, and present some useful applications. A more detailed treatment of this topic can be found in Binney & Tremaine (1987, 2008).

5.4.1 Time Scales for Collisions

Consider a system of size r consisting of N bodies of radius r_p. The cross-section for a *direct* collision is $\sigma_d = \pi r_p^2$, and the mean free path of a particle is $\lambda = 1/(n\sigma_d)$ with $n = 3N/(4\pi r^3)$ the number density of particles. The collision time, defined as the characteristic time scale on which a particle experiences a direct collision, is then

$$t_{\text{direct}} = \frac{\lambda}{v} \simeq \left(\frac{r}{r_p}\right)^2 \frac{t_{\text{cross}}}{N}, \qquad (5.87)$$

where $t_{\text{cross}} = r/v$ is the crossing time, which is the average time it takes a particle to cross the system. As an example, consider a typical galaxy consisting of $N = 10^{10}$ stars, having $r \sim 10\,\text{kpc}$ and $v \sim 200\,\text{km}\,\text{s}^{-1}$. If r_p is roughly the radius of the Sun ($r_p = 6.9 \times 10^5\,\text{km}$), one finds that $t_{\text{direct}} \simeq 10^{21}\,\text{yrs}$, eleven orders of magnitude larger than the Hubble time, $t_H \sim 10^{10}\,\text{yrs}$. Therefore, direct collisions among stars are completely negligible in galaxies.

But what about encounters, also called indirect collisions? Consider the encounter of a particle of mass m with another particle of mass m'. The equation of motion of the vector $\mathbf{x}$ separating m and m' is given by

$$\left(\frac{mm'}{m+m'}\right)\ddot{\mathbf{x}} = -\frac{Gmm'}{r^2}\hat{\mathbf{x}}. \qquad (5.88)$$

Let $\mathbf{v}_0 = \dot{\mathbf{x}}_0$ be the relative initial velocity (at $t \to -\infty$) between the two particles with an impact parameter b, defined as the initial separation vector $\mathbf{x}_0$ projected in the direction perpendicular to $\mathbf{v}_0$. It can be shown that after the encounter, at $t \to \infty$, the changes in the velocity of m perpendicular and parallel to $\mathbf{v}_0$ are

$$|\delta v_{m,\perp}| = \frac{2m'bv_0^3}{G(m+m')^2}\left[1 + \frac{b^2v_0^4}{G^2(m+m')^2}\right]^{-1}, \qquad (5.89)$$

5.4 Collisionless Dynamics

and

$$|\delta v_{m,\|}| = \frac{2m' v_0}{m+m'} \left[1 + \frac{b^2 v_0^4}{G^2(m+m')^2}\right]^{-1}, \qquad (5.90)$$

respectively (see e.g. Binney & Tremaine, 2008). For simplicity, let us assume that all particles have the same mass m. From Eqs. (5.89) and (5.90), it can be seen that the change in the velocity δv becomes comparable to the approaching velocity v_0 itself if the impact parameter is of the order $b \sim b_1 \equiv Gm/v_0^2 \simeq r/N$. An encounter with $b \lesssim b_1$ is therefore called a close encounter, and the time scale for such a close encounter to occur follows from $t_{\rm direct}$ by simply replacing $r_{\rm p}$ with b_1:

$$t_{\rm close} = \left(\frac{r}{b_1}\right)^2 \frac{t_{\rm cross}}{N} \simeq N t_{\rm cross}. \qquad (5.91)$$

The mean separation between particles in a system of radius r is roughly $\overline{d} = (r^3/N)^{1/3} \sim (Gm/\sigma^2)(M/m)^{2/3}$, where M, r and $\sigma \sim (GM/r)^{1/2}$ are the mass, characteristic radius, and characteristic velocity dispersion of the system, respectively. Since the approaching velocity of a typical encounter is $v_0 \sim \sigma$ and m is in general much smaller than M, we have $\overline{d} \gg b_1$. Hence, typical encounters in a system with large N have $\delta v/v_0 \ll 1$. In this limit, the velocity perturbation in the perpendicular direction reduces to

$$|\delta v_\perp| \approx \frac{2Gm}{bv}. \qquad (5.92)$$

When a particle crosses the system once, the average number of encounters with impact parameter less than b is given by

$$\mathcal{N}(<b) = N \frac{\pi b^2}{\pi r^2} = N \left(\frac{b}{r}\right)^2. \qquad (5.93)$$

Differentiating with respect to b yields the total number of encounters per crossing with impact parameters in the range $(b, b+db)$:

$$\mathcal{N}(b) db = \frac{2N}{r^2} b \, db. \qquad (5.94)$$

Because the change in $\mathbf{v}$ is random, the cumulative $\delta v_\perp$ due to all these encounters is zero. However, the sum of the *square* of the velocity changes is non-zero:

$$\delta v_\perp^2 \approx \left(\frac{2Gm}{bv}\right)^2 \frac{2N}{r^2} b \, db. \qquad (5.95)$$

Integrating the above expression over the range $[b_1, r]$, we obtain that, for each crossing, the change in $v_\perp^2$ is

$$\Delta v_\perp^2 \equiv \int_{b_1}^{r} \delta v_\perp^2 \, db \approx 8N \left(\frac{Gm}{rv}\right)^2 \ln \Lambda, \qquad (5.96)$$

where $\ln \Lambda = \ln(r/b_1)$ is called the Coulomb logarithm. Using that $v^2 \sim GNm/r$, and that $\Lambda = r/b_1 \approx rv^2/(Gm) \approx N$, we have

$$\frac{\Delta v_\perp^2}{v^2} \sim \frac{8 \ln N}{N}. \qquad (5.97)$$

Thus it takes of the order of $N/(10 \ln N)$ crossings for $\Delta v_\perp^2$ to become comparable to v^2. This defines the so-called two-body relaxation time,

$$t_{\rm relax} = \frac{N}{10 \ln N} t_{\rm cross}. \qquad (5.98)$$

Thus, for all practical purposes, galaxies, as N-body systems with $N \gg 100$, can almost always be considered collisionless, with

$$t_{\text{direct}} \gg t_{\text{close}} \gg t_{\text{relax}} \gg t_{\text{H}} \gg t_{\text{cross}}. \tag{5.99}$$

This not only holds for galaxies, but even more so for the inner regions of their dark matter halos, as long as non-gravitational interactions among their constituent particles can be neglected. In the outer regions of halos orbital times become comparable to the Hubble time, i.e. $t_{\text{H}} \sim t_{\text{cross}}$.

5.4.2 Basic Dynamics

Consider a large number of particles moving under the influence of a smooth potential field $\Phi(\mathbf{x},t)$. At any time, a full description of the system is given by its distribution function (DF) $f(\mathbf{x},\mathbf{v},t)$, which describes the number density of particles in phase space $(\mathbf{x},\mathbf{v})$.[1] In particular, the density structure is given by $f(\mathbf{x},\mathbf{v},t)$ integrated over velocity:

$$\rho(\mathbf{x},t) = m \int f(\mathbf{x},\mathbf{v},t)\, \mathrm{d}^3\mathbf{v}, \tag{5.100}$$

where m is the particle mass, while the velocity structure is given by the velocity moments,

$$\langle v_i \rangle (\mathbf{x},t) = \frac{m}{\rho(\mathbf{x})} \int v_i f(\mathbf{x},\mathbf{v},t)\, \mathrm{d}^3\mathbf{v}, \tag{5.101}$$

$$\langle v_i v_j \rangle (\mathbf{x},t) = \frac{m}{\rho(\mathbf{x})} \int v_i v_j f(\mathbf{x},\mathbf{v},t)\, \mathrm{d}^3\mathbf{v}, \tag{5.102}$$

and so on.

As long as the particles are collisionless, and are neither created nor destroyed, the flow in phase space must conserve mass, which is expressed via the collisionless Boltzmann equation (CBE):

$$\frac{\mathrm{d}f}{\mathrm{d}t} = \frac{\partial f}{\partial t} + \sum_i v_i \frac{\partial f}{\partial x_i} - \sum_i \frac{\partial \Phi}{\partial x_i} \frac{\partial f}{\partial v_i} = 0, \tag{5.103}$$

with the subscript $i \in (1,2,3)$ denoting the three spatial components. Note that $\mathrm{d}f/\mathrm{d}t$ expresses the Lagrangian derivative along trajectories through phase space, and so the CBE implies that the phase-space density around a given particle remains constant.

The gravitational potential, $\Phi(\mathbf{x})$, is related to the mass density, $\rho(\mathbf{x})$, via the Poisson equation

$$\nabla^2 \Phi(\mathbf{x}) = 4\pi G \rho(\mathbf{x}). \tag{5.104}$$

For a spherical system, one can make use of Newton's theorems to obtain a general solution for $\Phi(r)$ given $\rho(r)$. According to Newton's first theorem, a body inside a spherical shell of matter experiences no net gravitational force from that shell, while Newton's second theorem states that the gravitational force on a body outside a closed spherical shell of mass M is the same as that of a point mass M at the center of the shell. Consequently, $\Phi(r)$ follows from splitting $\rho(r)$ in spherical shells, and adding the potentials of all these shells:

$$\Phi(r) = -4\pi G \left[\frac{1}{r} \int_0^r \rho(r')\, r'^2\, \mathrm{d}r' + \int_r^\infty \rho(r')\, r'\, \mathrm{d}r' \right]. \tag{5.105}$$

---

[1] Note, however, that this interpretation becomes ill defined when examining phase space on very small scales, as we will see in §5.5.

In the more general case, of an ellipsoidal body whose isodensity surfaces are concentric, tri-axial ellipsoids, and whose axis ratios are constant, the solution to the Poisson equation can be written as

$$\Phi(\mathbf{x}) = -\pi G \left(\frac{a_2 a_3}{a_1}\right) \int_0^\infty \frac{\mu(\infty) - \mu[s(\mathbf{x}, \tau)]}{\sqrt{(\tau + a_1^2)(\tau + a_2^2)(\tau + a_3^2)}} \, d\tau, \qquad (5.106)$$

where $a_1 \geq a_2 \geq a_3$ are the lengths of the three principal axes, assumed to be aligned with the Cartesian coordinate system (x_1, x_2, x_3),

$$\mu(s) \equiv \int_0^{s^2} \rho(s'^2) ds'^2, \qquad (5.107)$$

and

$$s(\mathbf{x}, \tau) = a_1^2 \sum_{i=1}^3 \frac{x_i^2}{\tau + a_i^2}. \qquad (5.108)$$

5.4.3 The Jeans Equations

In general, the complete solution of the CBE (5.103) is extremely difficult to obtain. Rather, it is common practice to consider the velocity moments of f. To do this, we proceed in a similar way as in §4.1.4.

The average value of a quantity Q in the neighborhood of $\mathbf{x}$ is

$$\langle Q \rangle(\mathbf{x}, t) \equiv \frac{1}{n(\mathbf{x}, t)} \int Q f \, d^3\mathbf{v}, \quad n(\mathbf{x}, t) \equiv \int f \, d^3\mathbf{v} = \frac{\rho(\mathbf{x}, t)}{m}. \qquad (5.109)$$

Multiplying Eq. (5.103) by Q and integrating over $\mathbf{v}$ we get

$$\frac{\partial}{\partial t}[n\langle Q \rangle] + \sum_i \frac{\partial}{\partial x_i}[n\langle Q v_i \rangle] + n \sum_i \frac{\partial \Phi}{\partial x_i} \left\langle \frac{\partial Q}{\partial v_i} \right\rangle = 0, \qquad (5.110)$$

where the last term on the left-hand side follows from partial integration and the assumption that $f = 0$ for $v^2 \to \infty$. Setting $Q = 1$ we obtain the continuity equation,

$$\frac{\partial n}{\partial t} + \sum_i \frac{\partial}{\partial x_i}[n\langle v_i \rangle] = 0, \qquad (5.111)$$

while setting $Q = v_j$ results in a set of three momentum equations (for $j = 1, 2, 3$)

$$\frac{\partial}{\partial t}[n\langle v_j \rangle] + \sum_i \frac{\partial}{\partial x_i}[n\langle v_i v_j \rangle] + n \frac{\partial \Phi}{\partial x_j} = 0. \qquad (5.112)$$

These momentum equations are the direct analogs of the Euler equations of fluid dynamics but for a collisionless system. Writing the velocity tensor $\langle v_j v_k \rangle$ as the sum of that due to coherent steaming motion and that due to random motion,

$$\langle v_j v_k \rangle = \langle v_j \rangle \langle v_k \rangle + \sigma_{jk}^2, \qquad (5.113)$$

and subtracting $v_j \times$ Eq. (5.111) from Eq. (5.112), we obtain the Jeans equations

$$\frac{\partial \langle v_j \rangle}{\partial t} + \sum_i \langle v_i \rangle \frac{\partial \langle v_j \rangle}{\partial x_i} = -\frac{1}{n} \sum_i \frac{\partial \left(n \sigma_{ij}^2\right)}{\partial x_i} - \frac{\partial \Phi}{\partial x_j} = 0. \qquad (5.114)$$

Note that $n\sigma_{ij}^2$ has a similar effect as the pressure in the Euler equation, except that it is now a tensor (called the stress tensor), rather than a scalar. It is a *local* quantity that quantifies the random motions of the particles relative to their local mean motion. Because σ_{ij}^2 is a symmetric second-rank tensor, one can always, locally, choose the coordinate system such that it is diagonalized, i.e. $\sigma_{ij} = \mathrm{diag}(\hat{\sigma}_{11}, \hat{\sigma}_{22}, \hat{\sigma}_{33})$, with $\hat{\sigma}_{ii}$ the eigenvalues of the stress tensor. An ellipsoid, whose principal axes are defined by the orthogonal eigenvectors of σ_{ij}^2 and have lengths proportional to $\hat{\sigma}_{ii}$, is called the velocity ellipsoid.

In general, the Jeans equations have nine unknowns (three streaming motion components $\langle v_i \rangle$ and six independent components of the stress tensor). With only three equations, this is clearly not a closed set. Note that adding higher-order moment equations of the CBE will not help in closing the set. Although this adds more equations, it also adds more unknowns, i.e. higher-order velocity moments such as $\langle v_i v_j v_k \rangle$, etc. Therefore, in order to be able to solve the Jeans equations, one has to make assumptions regarding the form of the stress tensor (see §5.4.7 below). For comparison, in fluid dynamics there are only five unknowns (three streaming motions, the isotropic pressure and the density). The three Euler equations, together with the continuity equation, therefore form a closed set once they are combined with an equation of state relating the pressure to the density. As a final warning, not all solutions to the Jeans equations can represent physical systems, since the solutions are not guaranteed to correspond to a phase-space DF with $f \geq 0$ for all $(\mathbf{x}, \mathbf{v})$.

5.4.4 The Virial Theorem

Using the moment equations of the CBE we can also obtain a relation between global parameters of a collisionless system. Multiplying Eq. (5.112) by mx_k and integrating over space we obtain

$$\int x_k \frac{\partial [\rho \langle v_j \rangle]}{\partial t} d^3\mathbf{x} = -\sum_i \int x_k \frac{\partial [\rho \langle v_i v_j \rangle]}{\partial x_i} d^3\mathbf{x} - \int \rho x_k \frac{\partial \Phi}{\partial x_j} d^3\mathbf{x}. \tag{5.115}$$

Using the theorem of divergence, we can write the first term on the right-hand side as

$$-\sum_i \int x_k \frac{\partial [\rho \langle v_i v_j \rangle]}{\partial x_i} d^3\mathbf{x} = 2K_{kj} + \Sigma_{kj}, \tag{5.116}$$

where

$$K_{jk} = \frac{1}{2} \int \rho \langle v_j v_k \rangle d^3\mathbf{x} \tag{5.117}$$

is the kinetic energy tensor, and

$$\Sigma_{jk} = -\sum_i \int x_k \rho \langle v_j v_i \rangle dS_i \tag{5.118}$$

is the surface pressure term. The last term on the right-hand side of Eq. (5.115) is the Chandrasekhar potential energy tensor, W_{jk}. Therefore

$$\int x_k \frac{\partial [\rho \langle v_j \rangle]}{\partial t} d^3\mathbf{x} = 2K_{jk} + W_{jk} + \Sigma_{jk}. \tag{5.119}$$

Using Eq. (5.113) we can write K_{jk} as

$$K_{jk} = T_{jk} + \Pi_{jk}/2, \tag{5.120}$$

where

$$T_{jk} = \frac{1}{2} \int \rho \langle v_j \rangle \langle v_k \rangle d^3\mathbf{x} \quad \text{and} \quad \Pi_{jk} = \int \rho \sigma_{jk}^2 d^3\mathbf{x}. \tag{5.121}$$

Note that the partial derivative on the left-hand side of Eq. (5.119) can be pulled outside the integral and then changed into a complete derivative (because the spatial part is integrated out). Using that T_{jk}, Π_{jk} and W_{jk} are symmetric with respect to their subscripts, Eq. (5.119) can be written as

$$\frac{1}{2}\frac{d}{dt}\int \rho \left[x_k\langle v_j\rangle + x_j\langle v_k\rangle\right] d^3\mathbf{x} = 2T_{jk} + \Pi_{jk} + W_{jk} + \Sigma_{jk}. \quad (5.122)$$

The left-hand side can be written in terms of the moment of inertia tensor,

$$I_{jk} = \int \rho x_j x_k d^3\mathbf{x}, \quad (5.123)$$

because

$$\frac{dI_{jk}}{dt} = \int \frac{\partial \rho}{\partial t} x_j x_k d^3\mathbf{x} = -\int \frac{\partial(\rho\langle v_i\rangle)}{\partial x_i} x_j x_k d^3\mathbf{x}$$

$$= \int \rho \left[x_k\langle v_j\rangle + x_j\langle v_k\rangle\right] d^3\mathbf{x}, \quad (5.124)$$

where we have used the continuity equation (5.111) in the second step. Finally, we obtain from Eq. (5.122) that

$$\frac{1}{2}\frac{d^2 I_{jk}}{dt^2} = 2T_{jk} + \Pi_{jk} + W_{jk} + \Sigma_{jk}. \quad (5.125)$$

This is the *tensor virial theorem*. Taking the trace on both sides of this equation gives

$$\frac{1}{2}\frac{d^2 I}{dt^2} = 2K + W + \Sigma, \quad (5.126)$$

where

$$I = \text{Tr}(I_{ij}) = \int \rho r^2 d^3\mathbf{x}, \quad K = \text{Tr}(K_{ij}) = \frac{1}{2}\int \rho\langle v^2\rangle d^3\mathbf{x}, \quad (5.127)$$

$$W = \text{Tr}(W_{ij}) = -\int \rho\mathbf{x}\cdot\nabla\Phi d^3\mathbf{x}, \quad \Sigma = \text{Tr}(\Sigma_{ij}) = -\int \rho\langle v^2\rangle \mathbf{x}\cdot d\mathbf{S}. \quad (5.128)$$

Note that I is the moment of inertia, K is the kinetic energy of the system, and Σ is the work done by the external pressure. The potential energy, W, is equal to the gravitational energy of the system only if any mass outside the surface S can be ignored in the computation of the potential. Eq. (5.126) is the important *scalar virial theorem*. Although it is derived for a system of collisionless particles, it also applies for an ideal fluid, because the Jeans equations on which the virial theorem is based have the same form as the Euler equations if the velocity dispersion is isotropic. From Eq. (5.126) we see that, if surface pressure can be neglected, a system whose kinetic energy is much larger (smaller) than the negative value of its potential energy tends to expand (contract). For a static system, $d^2 I/dt^2 = 0$, and

$$2K + W + \Sigma = 0, \quad (5.129)$$

giving a constraint on the global properties of any system in a static state. Finally, for $\Sigma = 0$ we have

$$E = -K = \frac{W}{2}, \quad (5.130)$$

where $E = K + W$ is the total energy.

Application to Spherical Collapse Consider the collapse of a uniform sphere of mass M. For simplicity, we assume that the cosmological constant $\Lambda = 0$. As we have seen in §5.1, all mass shells in such a sphere reach their maximum expansion at the same time, which we denote by t_max. Since the kinetic energy is zero at maximum expansion, the total energy of this sphere is $E = -3GM^2/5r_\text{max}$, where r_max is the radius of the outermost mass shell at maximum expansion. According to the spherical collapse model, the sphere will collapse to a point at $t = 2t_\text{max}$. However, since a collisionless system cannot dissipate its energy, during the collapse the gravitational potential energy has to be converted into kinetic energy of the mass shells, or, more realistically, into kinetic energy of the particles involved in the collapse. The sphere therefore eventually relaxes to a quasi-static structure supported by random motions. If the final object were uniform, its potential energy would be $W \approx -3GM^2/5r_{\rm vir}$, where $r_{\rm vir}$ is the (virial) radius of the object. The virial theorem, $W = 2E$, then implies that $r_{\rm vir} = r_{\max}/2$. It is usually assumed that virialization is achieved at $t_{\rm vir} = 2t_{\max}$. The mean overdensity within $r_{\rm vir}$ at $t_{\rm vir}$ is then given by

$$1 + \Delta_{\rm vir} = \frac{\rho(t_{\max})(r_{\max}/r_{\rm vir})^3}{\overline{\rho}(t_{\rm vir})} = \frac{\rho(t_{\max})}{\overline{\rho}(t_{\max})} \frac{\overline{\rho}(t_{\max})}{\overline{\rho}(2t_{\max})} \left(\frac{r_{\max}}{r_{\rm vir}}\right)^3, \quad (5.131)$$

where $\overline{\rho}(t)$ is the background density at t. If $\Omega_{m,0} = 1$, then according to Eq. (5.17) we have $\rho(t_{\max})/\overline{\rho}(t_{\max}) = 9\pi^2/16$, while $\overline{\rho}(t_{\max})/\overline{\rho}(2t_{\max}) = 4$, so that $\Delta_{\rm vir} = 18\pi^2 \approx 178$. If $\Omega_m \neq 1$ the situation is more complicated, because $\Delta_{\rm vir}$ depends both on $\Omega_m(t_{\rm vir})$ and $\Omega_m(t_{\rm vir}/2)$. A useful approximation is given by:

$$\Delta_{\rm vir} \approx (18\pi^2 + 60x - 32x^2)/\Omega_m(t_{\rm vir}), \quad (5.132)$$

where $x = \Omega_m(t_{\rm vir}) - 1$ (Bryan & Norman, 1998).

If $\Lambda \neq 0$, the cosmological constant also contributes to the gravitational energy (see §5.1.2). The virial theorem applied to the final halo (assumed to be a uniform sphere of radius $r_{\rm vir}$) is then

$$K = -\frac{W}{2} + W_\Lambda, \quad (5.133)$$

where $W = -3GM^2/5r_{\rm vir}$, $W_\Lambda = -c^2 \Lambda M r_{\rm vir}^2/10$, and the total energy is $E = K + W + W_\Lambda$. Since the total energy is equal to the potential energy at maximum expansion, we have $E = -3GM^2/5r_{\max} - c^2 \Lambda M r_{\max}^2/10$. It then follows that

$$4\zeta \left(\frac{r_{\rm vir}}{r_{\max}}\right)^3 - 2(1+\zeta)\frac{r_{\rm vir}}{r_{\max}} + 1 = 0, \quad (5.134)$$

where $\zeta = c^2 \Lambda r_{\max}^3/6GM$, which can be solved to get $r_{\rm vir}/r_{\max}$ in terms of ζ. For given $t_{\max}$, we can obtain ζ from Eq. (5.26), which gives the density of the sphere at $t_{\max}$: $\rho(t_{\max}) = 3M/4\pi r_{\max}^3 = \Lambda/8\pi G\zeta$. The density of the sphere at virialization, $\rho(t_{\rm vir}) = \rho(t_{\max})(r_{\max}/r_{\rm vir})^3$, can then be computed. If we again assume virialization to occur at $t_{\rm vir} = 2t_{\max}$, the overdensity at virialization is just $\Delta_{\rm vir} = \rho(t_{\rm vir})/\overline{\rho}(t_{\rm vir}) - 1$. For a flat universe with $\Omega_m + \Omega_\Lambda = 1$, a useful approximation is

$$\Delta_{\rm vir} \approx (18\pi^2 + 82x - 39x^2)/\Omega_m(t_{\rm vir}), \quad (5.135)$$

where as before $x = \Omega_m(t_{\rm vir}) - 1$ (Bryan & Norman, 1998).

This simple application of the virial theorem yields a rule of thumb for the average densities of virialized objects formed through gravitational collapse in a cosmological background. Although the values of $\Delta_{\rm vir}$ given by Eqs. (5.132) and (5.135) are widely used to define the extent of a virialized halo (see §7.5), it should be kept in mind that they are based on several oversimplified assumptions.

5.4.5 Orbit Theory

Since orbits are the building blocks of collisionless systems, we here present a brief, and very general, description of orbit theory. More elaborate treatments can be found in Goldstein (1980), Boccaletti & Pucacco (1996) and Binney & Tremaine (2008).

(a) Classical Mechanics In classical mechanics the equations of motion are best expressed in terms of generalized (i.e. 'unspecified') coordinates q_i and their conjugate momenta $p_i \equiv \partial \mathscr{L}/\partial \dot{q}_i$. Here $i = 1, ..., n$, with n the number of degrees of freedom (i.e. the dimensionality of configuration space), $\dot{q}_i$ is the time derivative of q_i, and $\mathscr{L} = \mathscr{L}(q_i, \dot{q}_i, t) = T - V$ is the

Lagrangian of the system, with T and V the kinetic and potential energy, respectively. The phase-space coordinates $(\mathbf{q},\mathbf{p})$ are canonical variables, which are defined to satisfy the fundamental Poisson bracket relations,

$$\{q_i,q_j\} = 0, \quad \{p_i,p_j\} = 0, \quad \{q_i,p_j\} = \delta_{ij}, \tag{5.136}$$

with δ_{ij} the Kronecker delta function. The Poisson bracket is defined such that for two functions $A(\mathbf{q},\mathbf{p})$ and $B(\mathbf{q},\mathbf{p})$ of the generalized coordinates and their conjugate momenta,

$$\{A,B\} \equiv \sum_{i=1}^{n} \left[\frac{\partial A}{\partial q_i} \frac{\partial B}{\partial p_i} - \frac{\partial A}{\partial p_i} \frac{\partial B}{\partial q_i} \right]. \tag{5.137}$$

The functions A and B are said to be in involution if their Poisson bracket vanishes, i.e. if $\{A,B\} = 0$.

In the Hamiltonian formalism of classical mechanics, the equations of motion reduce to

$$\frac{\partial \mathscr{H}}{\partial p_i} = \dot{q}_i, \quad \frac{\partial \mathscr{H}}{\partial q_i} = -\dot{p}_i, \tag{5.138}$$

where $\mathscr{H} = \mathscr{H}(\mathbf{q},\mathbf{p},t) = \sum p_i \dot{q}_i - \mathscr{L}(q_i,\dot{q}_i,t)$ is the system's Hamiltonian, which is equal to the total energy of the system in the case of a fixed potential. It is clear from Eq. (5.138) that if a generalized coordinate, say q_i, does not explicitly appear in the Hamiltonian, then the corresponding conjugate momentum, p_i, is a conserved quantity.

Since Hamilton's equations are first-order differential equations, one can determine $\mathbf{q}(t)$ and $\mathbf{p}(t)$ at any time t once the initial conditions $(\mathbf{q}_0, \mathbf{p}_0)$ are given. Therefore, through each point in phase-space passes a *unique* trajectory $\Gamma[\mathbf{q}(\mathbf{q}_0,\mathbf{p}_0,t),\mathbf{p}(\mathbf{q}_0,\mathbf{p}_0,t)]$: no two trajectories Γ_1 and Γ_2 can pass through the same $(\mathbf{q},\mathbf{p})$ unless $\Gamma_1 = \Gamma_2$. Note that an orbit, $\mathbf{q}(t)$, is simply the projection of $\Gamma(\mathbf{q},\mathbf{p},t)$ onto configuration space.

(b) Integrals of Motion Any function $I(\mathbf{q},\mathbf{p})$ of the phase-space coordinates $(\mathbf{q},\mathbf{p})$ *alone* (i.e. without an explicit time dependence) which is constant along an orbit is called an integral of motion. The class of integrals of motion that are one-valued functions of the phase-space coordinates are called isolating integrals of motion, and these are of great practical and theoretical importance. There also exists a class of non-isolating integrals of motion, which are multiple-valued functions of $(\mathbf{q},\mathbf{p})$ and have little practical value.

The importance of isolating integrals of motion for orbit theory follows from the fact that an isolating integral of motion reduces the dimensionality of the trajectory $\Gamma(t)$ by one. Thus, a trajectory in a dynamical system with n degrees of freedom and with k independent isolating integrals of motion is restricted to a $2n - k$ dimensional manifold in the $2n$-dimensional phase-space. Depending on the Hamiltonian of the dynamical system, k can be anywhere between 1 and $2n - 1$.

For a stationary, collisionless system, energy is a conserved quantity for each individual orbit, and is thus an isolating integral of motion [this integral is obviously isolating, since each point in the phase-space corresponds to a single value of energy $E = \mathscr{H}(\mathbf{q},\mathbf{p})$]. According to Noether's theorem, each conserved quantity corresponds to a symmetry of the Lagrangian. In the case of energy, this is the invariance of the Lagrangian under time translation. We can also use Noether's theorem to identify other isolating integrals of motion. For example, for a spherically symmetric system, there are three additional isolating integrals of motion associated with the three independent spatial rotations. Thus, all trajectories in a three-dimensional stationary spherical system are confined to two-dimensional manifolds, the orbital projections of which are planar rosettes. In the case of a stationary axisymmetric system, with the symmetry axis along the z direction, L_z is an isolating integral of motion (in addition to E). Note that the energy and angular momenta are special cases, in the sense that their dependence on the phase-space coordinates can be written down explicitly. These integrals are sometimes called classical integrals of motion. In general, however, there is no prescription to allow one to explicitly write down the $(\mathbf{q},\mathbf{p})$ dependence of an (isolating) integral of motion.

(c) Canonical Transformations and Action-Angle Variables A canonical transformation is a transformation $(\mathbf{q},\mathbf{p}) \to (\mathbf{Q},\mathbf{P})$ between two canonical coordinate systems that leaves the equations of motion invariant. If a canonical transformation exists so that the new Hamiltonian does not explicitly depend on the new coordinates Q_i, i.e. $\mathcal{H}(\mathbf{q},\mathbf{p}) \to \mathcal{H}'(\mathbf{P})$, then the system is said to be integrable. In such cases, Hamilton's equations of motion become

$$\frac{\partial \mathcal{H}'}{\partial Q_i} = -\dot{P}_i = 0, \quad \frac{\partial \mathcal{H}'}{\partial P_i} = -\dot{Q}_i, \tag{5.139}$$

which can be solved to give

$$Q_i(t) = \omega_i t + Q_i(0), \quad P_i = \text{constant}. \tag{5.140}$$

Here $\omega_i = \partial \mathcal{H}'/\partial P_i$ are frequencies, while $Q_i(0)$ are integration constants. The fact that the conjugate momenta P_i are constant implies that they are integrals of motion. Indeed, since the mapping $(\mathbf{q},\mathbf{p}) \to (\mathbf{Q},\mathbf{P})$ is one-to-one, they are isolating integrals of motion. This highlights an important aspect of dynamical systems: if a Hamiltonian system with n degrees of freedom is integrable, it admits (at least) n independent isolating integrals of motion in involution, and vice versa.

For a bound system, a particularly useful set of canonical variables are the action-angle variables $(\mathbf{J}, \Theta)$. These are characterized by the fact that increasing one of the angles Θ_i by 2π, while keeping all other action-angle variables fixed, brings one back to the same point in phase space. The momenta conjugate to these angle variables are called the actions and are defined by

$$J_i = \frac{1}{2\pi} \oint_{\gamma_i} \mathbf{p} \cdot d\mathbf{q}, \tag{5.141}$$

with γ_i the path along which Θ_i increases from 0 to 2π. If the system is integrable, then the actions are isolating integrals of motion in involution, $\mathcal{H} = \mathcal{H}(\mathbf{J})$, and $\Theta_i(t) = \omega_i t + \Theta_i(0)$, where $\omega_i(\mathbf{J})$ are called the fundamental frequencies.

(d) Orbit Classification If in a system with n degrees of freedom a particular orbit admits n independent isolating integrals of motion, the orbit is said to be regular, and its corresponding trajectory $\Gamma(t)$ is confined to a $2n - n = n$-dimensional manifold in phase-space. Topologically this manifold is called an invariant torus (or torus for short), and is uniquely specified by the n isolating integrals. The best set of isolating integrals are the actions J_i defined above, which specify the radii of the various cross-sections of the torus. In addition, the angles Θ_i uniquely label a phase-space point *on* the torus, and the orbit (i.e. the mapping of the trajectory $\Gamma(t)$ – which is confined to the surface of the torus – onto the configuration space) is said to be quasi-periodic with (fundamental) frequencies $\omega_i(\mathbf{J})$.

In general, the fundamental frequencies are incommensurable, i.e. their ratios cannot be expressed as ratios of integers. In this case, the trajectories will, given sufficiently long time, densely fill the entire surface of the torus. However, since $\omega_i = \omega_i(\mathbf{J})$, one can always find suitable values for J_i such that two or more of the n frequencies ω_i are commensurable, i.e. for which $l\omega_i = m\omega_j$, with l and m both integers. An orbit with commensurable frequencies is called a resonant orbit, and has a dimensionality that is one lower than that of the non-resonant regular orbits. This implies that there is an extra isolating integral of motion, namely $I_{n+1} = l\omega_i - m\omega_j$, which is in involution with all J_i. If all n frequencies are commensurable (i.e. if $\mathbf{n} \cdot \boldsymbol{\omega} = 0$, with $\mathbf{n}$ an n-dimensional non-zero integer vector), the orbit returns to exactly the same phase-space point after a period $T = 2\pi/\omega_k$, where k is the index of the smallest integer value of **n**. In this case the orbit is said to be closed.

In an integrable system, all orbits are regular, and phase-space is completely filled (one says 'foliated') with non-intersecting invariant tori. However, integrable Hamiltonians are extremely rare; mathematically speaking they form a set of measure zero in the space of all Hamiltonians. Thus, in all likelihood, galaxies and dark matter halos are non-integrable systems. In order to understand the orbital structure of non-integrable Hamiltonians, consider a small perturbation away from an integrable Hamiltonian. According to the Kolmogorov–Arnold–Moser (KAM) theorem, the tori corresponding to regular orbits with fundamental frequencies sufficiently incommensurable will survive a small perturbation; they retain their topology so that the motion in their vicinity remains quasi-periodic (i.e. regular) and confined to (slightly deformed) invariant tori (Lichtenberg & Lieberman, 1992). Resonant tori, and those with fundamental frequencies close to being commensurable, on the other hand, can be strongly affected by even a very small perturbation. A certain fraction of the resonant orbits are unstable, so that the perturbation causes them to be no longer confined to tori. They no longer admit n isolating integrals of motion in involution, and their trajectories move more-or-less randomly through a phase-space of dimensionality $2n - 1$, called the Arnold web (only bounded by the equipotential surface, since their energy is still an isolating integral of motion). Such orbits are called irregular or stochastic. They behave chaotically, in that trajectories initially very close to each other diverge exponentially. Because the majority of tori are very non-resonant, as rational numbers are rare compared to irrational ones, the KAM theorem implies that a small perturbation of an integrable system will leave most orbits regular. Hence, in non-integrable Hamiltonians, phase-space has a complicated structure of interleaved regular and stochastic regions.

Although most of phase-space is covered by the trajectories of non-resonant orbits, resonant tori still form a dense set in phase space, just like the rational numbers are dense on the real axis. Since one can always find a rational number in between two real numbers, there will always be a resonant torus in between any two tori. Since many of these will be unstable, they create many stochastic regions when the system is perturbed. However, as long as the resonance is of higher order (for example $l : m = 16 : 23$ rather than $1 : 2$) the corresponding stochastic regions are extremely small, and tightly bound by their surrounding tori. Since two trajectories cannot cross, an irregular orbit is bounded by its neighboring regular orbits, causing it to be almost confined to a n-dimensional manifold, i.e. it still behaves as if it has n isolating integrals of motion. However, if $n > 2$ a stochastic orbit may slip through a 'crack' between two confining tori, a process known as Arnold diffusion, causing the orbit to reveal its true stochastic nature. The stochastic regions also grow when the perturbations becomes larger. Eventually, stochastic regions associated with different unstable resonances will start to overlap, producing large interconnected regions of phase-space in which the motion is stochastic

In galactic dynamics, the KAM theorem of near-integrability is often used to the extreme, by assuming that irregular orbits can be ignored, and that the Hamiltonians associated with 'galaxy-like' potentials are integrable. Clearly the validity of this approximation depends on the fraction of phase-space that admits three isolating integrals of motion. For most potentials used in galactic dynamics this fraction is still very uncertain. Over the years, though, it has become clear that stochasticity may play an important role in governing the structure and evolution of galaxies. For example, Valluri & Merritt (1998) have shown that the growth of a supermassive black hole or a central density cusp in a triaxial system often destroys all existing isolating integrals of motion aside from the energy. Consequently, the central regions become highly stochastic, and evolve from being triaxial to being axisymmetric, which may play an important role in controlling the rate at which the central black hole can accrete mass.

5.4.6 The Jeans Theorem

The distribution function (DF) of a steady-state system with n spatial dimensions is a function of $2n$ phase-space coordinates $(\mathbf{x}, \mathbf{v})$. Obviously these are extremely complicated functions, whose interpretation is far from intuitive. However, one can simplify the interpretation and handling of the DF by making use of the concept of integrals of motion (see §5.4.5 above). Because of its definition, an integral of motion, I, obeys

$$\frac{dI}{dt} = \sum_i \frac{\partial I}{\partial x_i}\frac{dx_i}{dt} + \sum_i \frac{\partial I}{\partial v_i}\frac{dv_i}{dt} = \mathbf{v}\cdot\nabla I - \nabla\Phi\cdot\frac{\partial I}{\partial \mathbf{v}} = 0, \quad (5.142)$$

and so any integral of motion is a solution of the steady-state collisionless Boltzmann equation (CBE). Since any solution $f(\mathbf{x},\mathbf{v})$ of the steady-state CBE is itself an integral of motion (because $df/dt = 0$), all solutions to the steady-state CBE must depend on $(\mathbf{x},\mathbf{v})$ through the integrals of motion. These two statements are called the Jeans theorem, and they allow us to write $f(\mathbf{x},\mathbf{v})$ as $f(I_1,I_2,...,I_k)$.

As discussed in §5.4.5, the Hamiltonian of a dynamical system with n spatial dimensions is said to be integrable if it has m independent isolating integrals of motion, with $m \geq n$. The *strong* Jeans theorem states that the DF of a steady-state system in which (almost) all orbits are regular can be written as a function of the independent isolating integrals of motion. An equivalent statement is that the DF is constant on every invariant torus. Since the orbits in an integrable system are uniquely and completely specified by the values of the m independent isolating integrals of motion, such a DF basically describes the system in terms of its orbit building blocks, i.e. f indicates the distribution of particles or stars over all possible orbits.

In static spherical systems there are always four independent isolating integrals of motion, namely energy E, and the three components of the angular momentum vector $\mathbf{L}$, and each orbit is confined to a plane. There are two special cases, namely the point mass (Kepler potential) and the homogeneous sphere (harmonic potential), for which all orbits are closed ellipses. These two systems therefore have five independent isolating integrals of motion, the fifth being simply the angular phase of the line connecting the orbit's apo- and pericenter. Axisymmetric and triaxial systems with an integrable Hamiltonian always have three independent isolating integrals of motion: (E, L_z, I_3) in the case of axisymmetry and (E, I_2, I_3) in the case of triaxiality, where I_i indicates isolating integrals of motion that are non-classical. Based on the KAM theorem, these properties are, to good approximation, also valid for near-integrable systems.

5.4.7 Spherical Equilibrium Models

As we have seen above, spherical potentials in general have $f = f(E,\mathbf{L})$. The orbits are planar rosettes, which fill an annulus with pericenter radius r_- and apocenter radius r_+. From the symmetries of the individual orbits it is easy to see that any spherical, steady-state dynamical model must obey

$$\langle v_r \rangle = \langle v_\vartheta \rangle = 0, \quad \text{and} \quad \langle v_r v_\vartheta \rangle = \langle v_r v_\varphi \rangle = \langle v_\vartheta v_\varphi \rangle = 0, \quad (5.143)$$

which leaves four unknowns: $\langle v_\varphi \rangle$, $\langle v_r^2 \rangle$, $\langle v_\vartheta^2 \rangle$ and $\langle v_\varphi^2 \rangle$ (up to second order of the velocity moments). If one makes the assumption that the system is spherically symmetric in all its properties, then by symmetry the DF is independent of the direction of $\mathbf{L}$, i.e. $f(E,\mathbf{L}) \to f(E,L^2)$. Consequently, $\langle v_\varphi \rangle = 0$ and $\langle v_\vartheta^2 \rangle = \langle v_\varphi^2 \rangle$.

To obtain the corresponding steady-state Jeans equation, we use Eq. (5.112) in its vector form without the $\partial/\partial t$ term:

$$\nabla\cdot(\rho\langle\mathbf{v}\mathbf{v}\rangle) + n\nabla\Phi = 0. \quad (5.144)$$

In spherical coordinates, this equation becomes

$$\frac{1}{r^2}\frac{\partial}{\partial r}\left(r^2\langle v_r\mathbf{v}\rangle\rho\right) + \frac{1}{r\sin\vartheta}\frac{\partial}{\partial\vartheta}\left(\sin\vartheta\langle v_\vartheta\mathbf{v}\rangle\rho\right) + \frac{1}{r\sin\vartheta}\frac{\partial}{\partial\varphi}\left(\langle v_\varphi\mathbf{v}\rangle\rho\right) = -\rho\nabla\Phi. \quad (5.145)$$

Using the properties of unit vectors in spherical coordinates, and adopting the symmetries that hold for a system with $f = f(E, L^2)$, the only non-trivial equation is the radial part of Eq. (5.145),

$$\frac{1}{\rho}\frac{\mathrm{d}(\rho\langle v_r^2\rangle)}{\mathrm{d}r} + 2\beta\frac{\langle v_r^2\rangle}{r} = -\frac{\mathrm{d}\Phi}{\mathrm{d}r}, \quad (5.146)$$

where $\beta(r)$ is the anisotropy parameter, defined by

$$\beta(r) \equiv 1 - \langle v_\vartheta^2\rangle/\langle v_r^2\rangle. \quad (5.147)$$

With two unknowns, $\langle v_r^2\rangle$ and β, this Jeans equation cannot be solved without making additional assumptions. The special case $\beta = 0$ corresponds to an isotropic sphere for which $f = f(E)$. Models with $0 < \beta \leq 1$ are radially anisotropic. In the extreme case of $\beta = 1$ all orbits are radial and $f = g(E)\delta(L)$, with $g(E)$ some function of energy. Tangentially anisotropic models have $\beta < 0$, with $\beta = -\infty$ corresponding to a model in which all orbits are circular. In that case $f = g(E)\delta[L - L_{\max}(E)]$, where $L_{\max}(E)$ is the maximum angular momentum for energy E. Another special case is the one in which $\beta(r) = \beta$ is constant. As shown by Kent & Gunn (1982), such models have $f = g(E)L^{-2\beta}$.

For isotropic spheres ($\beta = 0$) it is possible to find the unique $f = f(E)$ corresponding to a given density distribution $\rho(r)$. For this it is convenient to define a new gravitational potential and a new energy:

$$\Psi \equiv -\Phi + \Phi_0 \quad \text{and} \quad \mathscr{E} \equiv -E + \Phi_0 = \Psi - v^2/2, \quad (5.148)$$

where Φ_0 is some constant. In practice, Φ_0 is chosen so that $f > 0$ for $\mathscr{E} > 0$ and $f = 0$ for $\mathscr{E} \leq 0$. Thus Φ_0 can be considered as the gravitational potential at the system's boundary (e.g. at infinity). The mass density at radius r can then be written as

$$\rho(r) = 4\pi \int_0^\Psi f(\mathscr{E})\sqrt{2(\Psi - \mathscr{E})}\,\mathrm{d}\mathscr{E}. \quad (5.149)$$

Since both $\rho(r)$ and $\Psi(r)$ are in general monotonic functions of r, we may consider ρ as a function of Ψ. Differentiating this function once with respect to Ψ, we obtain

$$\frac{1}{\pi\sqrt{8}}\frac{\mathrm{d}\rho}{\mathrm{d}\Psi} = \int_0^\Psi \frac{f(\mathscr{E})}{\sqrt{\Psi - \mathscr{E}}}\,\mathrm{d}\mathscr{E}. \quad (5.150)$$

This is an Abel integral equation for $f(\mathscr{E})$, which can be inverted to give

$$f(\mathscr{E}) = \frac{1}{\pi^2\sqrt{8}}\frac{\mathrm{d}}{\mathrm{d}\mathscr{E}}\int_0^\mathscr{E} \frac{\mathrm{d}\rho}{\mathrm{d}\Psi}\frac{\mathrm{d}\Psi}{\sqrt{\mathscr{E}-\Psi}}$$

$$= \frac{1}{\pi^2\sqrt{8}}\left[\int_0^\mathscr{E}\frac{\mathrm{d}^2\rho}{\mathrm{d}\Psi^2}\frac{\mathrm{d}\Psi}{\sqrt{\mathscr{E}-\Psi}} + \frac{1}{\sqrt{\mathscr{E}}}\left(\frac{\mathrm{d}\rho}{\mathrm{d}\Psi}\right)_{\Psi=0}\right]. \quad (5.151)$$

Thus, for any given spherically symmetric density distribution, one can always use Eq. (5.151) to find a DF, $f(\mathscr{E})$. However, in order for the solution to be physically meaningful, the DF must be nowhere negative. If this requirement is not satisfied, then the density profile cannot be represented by an equilibrium state with isotropic velocity dispersion. It can be shown that a spherically symmetric density profile can be represented by a DF $f(\mathscr{E})$ if, and only if, the integration in the first line of Eq. (5.151) is an increasing function of $\mathscr{E}$. In practice this requires $\rho(r)$ to drop off with radius sufficiently rapidly.

To end our discussion on spherical equilibrium models, we discuss three specific models that are often used to describe dark matter halos, galaxies, or globular clusters.

(a) The Isothermal Sphere

One of the simplest models that is often used to model spherical mass distributions is the isothermal sphere, whose DF is

$$f(\mathcal{E}) = \frac{(\rho_0/m)}{(2\pi\sigma^2)^{3/2}} \exp\left(\frac{\mathcal{E}}{\sigma^2}\right), \tag{5.152}$$

where $\sigma^2 = \langle v^2 \rangle /3$ is a constant. The corresponding density is $\rho(r) = \rho_0 \exp(\Psi/\sigma^2)$, which can be solved together with the Poisson equation,

$$\frac{1}{r^2}\frac{d}{dr}\left(r^2 \frac{d\Psi}{dr}\right) = -4\pi G \rho_0 \exp\left(\frac{\Psi}{\sigma^2}\right), \tag{5.153}$$

and the boundary condition $\rho(0) = \rho_0$ and $d\Psi/dr(0) = 0$. The resulting density profile is characterized by a *King radius*,

$$r_0 = \frac{3\sigma}{\sqrt{4\pi G \rho_0}}. \tag{5.154}$$

For $r \lesssim 2r_0$, the density profile can be approximated by

$$\rho(r) \approx \frac{\rho_0}{[1+(r/r_0)^2]^{3/2}}, \tag{5.155}$$

while at large radii ($r > 10 r_0$) the density profile approaches that of a singular isothermal sphere,

$$\rho(r) = \sigma^2/(2\pi G r^2). \tag{5.156}$$

Unless the density profile is truncated at an outer radius, the total mass of the isothermal sphere is therefore infinite. In order for this model to be applicable to systems with finite masses, some modifications have to be made.

(b) The King Model

One such modification is the King model, which is based on the DF

$$f(\mathcal{E}) = \begin{cases} (\rho_0/m)(2\pi\sigma^2)^{-3/2}\left(e^{\mathcal{E}/\sigma^2} - 1\right) & (\mathcal{E} > 0) \\ 0 & (\mathcal{E} \leq 0). \end{cases} \tag{5.157}$$

This distribution is similar to Eq. (5.152) except that no particles are allowed to have $\mathcal{E} \leq 0$. The density distribution corresponding to this DF is

$$\rho(r) = 4\pi \int f(\mathcal{E}) v^2 \, dv$$
$$= \rho_0 \left[e^{\Psi/\sigma^2} \mathrm{erf}\left(\frac{\sqrt{\Psi}}{\sigma}\right) - \sqrt{\frac{4\Psi}{\pi\sigma^2}}\left(1 + \frac{2\Psi}{3\sigma^2}\right)\right], \tag{5.158}$$

which, upon inserting into the Poisson equation, gives an equation for Ψ. This equation for Ψ can be integrated numerically with the boundary conditions $\Psi = \Psi_0$ and $d\Psi/dr = 0$ at $r = 0$. For a given Ψ_0/σ^2, the predicted density becomes zero at some radius r_t, which is referred to as the tidal radius. The ratio between the tidal radius and the core radius given by Eq. (5.154) defines the concentration of a King profile, $c \equiv \log(r_t/r_0)$.

(c) Double Power-Law Density Distributions

Dark matter halos and elliptical galaxies are often described as spheres with a double (broken) power-law density distribution given by

$$\rho(r) = \rho_0 \left(\frac{r}{r_0}\right)^{-\gamma}\left[1 + \left(\frac{r}{r_0}\right)^{\alpha}\right]^{(\gamma-\beta)/\alpha}. \tag{5.159}$$

5.4 Collisionless Dynamics

Table 5.1. Double power-law density profiles.

(α,β,γ)	Name	Reference
$(1,3,1)$	NFW profile	Navarro et al. (1997)
$(1,4,\gamma)$	Dehnen profile	Dehnen (1993)
$(1,4,1)$	Hernquist profile	Hernquist (1990)
$(1,4,2)$	Jaffe profile	Jaffe (1983)
$(2,2,0)$	Modified isothermal sphere	Sackett & Sparke (1990)
$(2,3,0)$	Modified Hubble profile	Binney & Tremaine (1987)
$(2,4,0)$	Perfect sphere	de Zeeuw (1985)
$(2,5,0)$	Plummer sphere	Plummer (1911)

At small radii $\rho \propto r^{-\gamma}$, while at large radii $\rho \propto r^{-\beta}$, and α determines the sharpness of the break. Table 5.1 lists a number of special cases that have often been used in the literature. The total mass corresponding to a density distribution of the form of Eq. (5.159) is given by

$$M = \frac{4\pi}{\alpha}\rho_0 r_0^3 \mathrm{B}\left(\frac{3-\gamma}{\alpha},\frac{\beta-3}{\alpha},1\right), \tag{5.160}$$

with

$$\mathrm{B}(a,b,x) = \int_0^x \mathrm{d}t\, t^{a-1}(1-t)^{b-1} \tag{5.161}$$

the incomplete Beta function. The corresponding gravitational potential follows from Eq. (5.105) and is given by

$$\Phi(r) = -\frac{4\pi G}{\alpha}\rho_0 r_0^2 \left[\frac{r_0}{r}\mathrm{B}\left(\frac{3-\gamma}{\alpha},\frac{\beta-3}{\alpha},\chi\right) + \mathrm{B}\left(\frac{\beta-2}{\alpha},\frac{2-\gamma}{\alpha},\chi\right)\right], \tag{5.162}$$

with

$$\chi = \frac{(r/r_0)^\alpha}{1+(r/r_0)^\alpha}. \tag{5.163}$$

In several cases, Eq. (5.162) reduces to an analytical function. For example, the potential of the Dehnen profile, which includes the Hernquist profile and Jaffe profile as special cases, reduces to

$$\Phi(r) = \frac{GM}{r_0} \times \begin{cases} -\frac{1}{2-\gamma}\left[1-\left(\frac{r}{r+r_0}\right)^{2-\gamma}\right] & \gamma \neq 2 \\ \ln\left(\frac{r}{r+r_0}\right) & \gamma = 2, \end{cases} \tag{5.164}$$

while that of the Plummer sphere can be written as

$$\Phi(r) = -\frac{GM}{\sqrt{r^2+r_0^2}}. \tag{5.165}$$

The Dehnen models are particularly useful, since many of their properties can be calculated analytically, including the intrinsic velocity dispersion for all real values of γ between 0 and 3, and, for $\gamma = 0, 1, 2$, the projected mass density and velocity dispersion as well (Dehnen, 1993; Tremaine et al., 1994). Furthermore, the Dehnen model with $\gamma = 3/2$ closely resembles the de Vaucouleurs $r^{1/4}$ profile often used to fit the surface brightness profiles of elliptical galaxies (see §2.3.2).

For an isotropic sphere the DF depends only on energy, $f = f(E)$, and can be uniquely determined from the density distribution using Eq. (5.151). For the Dehnen models it is convenient, for this purpose, to work in terms of the following dimensionless quantities:

$$\tilde{\rho} \equiv \frac{4\pi r_0^3}{(3-\gamma)M}\rho, \quad \tilde{\Psi} \equiv -\frac{r_0}{GM}\Phi, \quad \tilde{\mathcal{E}} \equiv -\frac{r_0}{GM}E. \tag{5.166}$$

The density profile (5.159) for $(\alpha, \beta) = (1,4)$ can then be expressed in terms of the potential as

$$\tilde{\rho}(\tilde{\Phi}) = (1-y)^4 y^{-\gamma}, \tag{5.167}$$

where

$$y(\tilde{\Psi}) = \begin{cases} [1-(2-\gamma)\tilde{\Psi}]^{1/(2-\gamma)} & (\gamma \neq 2) \\ \exp(-\tilde{\Psi}) & (\gamma = 2). \end{cases} \tag{5.168}$$

If the system has isotropic velocity dispersion, we can obtain its DF from Eq. (5.151):

$$f(\tilde{\mathcal{E}}) = \frac{(3-\gamma)}{2} f_0 \int_0^{\tilde{\mathcal{E}}} \frac{(1-y)^2[\gamma + 2y + (4-\gamma)y^2]}{y^{4-\gamma}\sqrt{\tilde{\mathcal{E}}-\tilde{\Psi}}} \, d\tilde{\Psi}, \tag{5.169}$$

where

$$f_0 \equiv \frac{M}{(2\pi^2 GMr_0)^{3/2}}. \tag{5.170}$$

For $\gamma = 1$ (the Hernquist model), the integration in Eq. (5.169) can be carried out explicitly, resulting in

$$f(\tilde{\mathcal{E}}) = 4f_0(1-\tilde{\mathcal{E}})^{-5/2}\left[3\sin^{-1}\sqrt{\tilde{\mathcal{E}}} + \sqrt{\tilde{\mathcal{E}}(1-\tilde{\mathcal{E}})}(1-2\tilde{\mathcal{E}})(8\tilde{\mathcal{E}}^2 - 8\tilde{\mathcal{E}} - 3)\right]. \tag{5.171}$$

For $\gamma = 2$ (the Jaffe model), the DF is

$$f(\tilde{\mathcal{E}}) = \sqrt{2}f_0\left[F_-\left(\sqrt{2\tilde{\mathcal{E}}}\right) - \sqrt{2}F_-\left(\sqrt{\tilde{\mathcal{E}}}\right) - \sqrt{2}F_+\left(\sqrt{\tilde{\mathcal{E}}}\right) + F_+\left(\sqrt{2\tilde{\mathcal{E}}}\right)\right], \tag{5.172}$$

where

$$F_\pm(x) = e^{\mp x^2}\int_0^x e^{\pm y^2}\, dy. \tag{5.173}$$

Fig. 5.3 shows some examples of DFs given by Eq. (5.169), together with the DF for a sphere that projects to a $r^{1/4}$ profile. Note that, as already alluded to above, the latter closely matches that of a Dehnen model with $\gamma \sim 3/2$. In particular, the asymptotic profile, $\rho(r) \propto r^{-4}$ for $r \to \infty$, gives a very close approximation to the $r^{1/4}$ profile at large radii (corresponding to $\tilde{\mathcal{E}} \to 0$).

5.4.8 Axisymmetric Equilibrium Models

In general, axisymmetric models in which most orbits are regular have a three-integral DF $f(E, L_z, I_3)$. It is customary to describe the orbits in cylindrical coordinates (R, ϕ, z), where the z-axis is the axis of symmetry. The three-dimensional motion can be considered as motion in the meridional plane (R, z), in an effective potential $\Phi_{\text{eff}}(R, z) = \Phi(R, z) + L_z^2/(2R^2)$, combined with a motion in the ϕ-direction given by the conservation of angular momentum $L_z = Rv_\phi$. Thus, these orbits have a net sense of rotation around the z-axis, and they lack a turning point in the ϕ direction. Such orbits are called short-axis tubes. From the symmetries of the individual orbits, it is evident that in this case

$$\langle v_R \rangle = \langle v_z \rangle = 0, \quad \text{and} \quad \langle v_R v_\phi \rangle = \langle v_z v_\phi \rangle = 0. \tag{5.174}$$

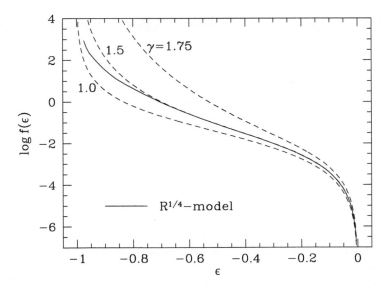

Fig. 5.3. Distribution function of the $r^{1/4}$ model (solid curve), compared with those corresponding to Dehnen models with $\gamma = 1$, 1.5 and 1.75, all plotted against $\varepsilon = -\mathscr{E}$. Units are chosen so that $M = G = r_0 = 1$. [After Hernquist (1990)]

Thus, general axisymmetric models have stress tensors with four non-zero components, and the velocity ellipsoid is not aligned with (R,ϕ,z). There are two non-trivial Jeans equations, given by

$$\frac{\partial(\rho\langle v_R^2\rangle)}{\partial R} + \frac{\partial(\rho\langle v_R v_z\rangle)}{\partial z} + \rho\left[\frac{\langle v_R^2\rangle - \langle v_\phi^2\rangle}{R} + \frac{\partial\Phi}{\partial R}\right] = 0; \quad (5.175)$$

$$\frac{\partial(\rho\langle v_R v_z\rangle)}{\partial R} + \frac{\partial(\rho\langle v_z^2\rangle)}{\partial z} + \rho\left[\frac{\langle v_R v_z\rangle}{R} + \frac{\partial\Phi}{\partial z}\right] = 0, \quad (5.176)$$

which clearly does not suffice to solve for the four unknowns.

For this reason, one often considers two-integral models with $f = f(E,L_z)$. In this case one has, in addition to Eq. (5.174), that

$$\langle v_R^2\rangle = \langle v_z^2\rangle, \quad \text{and} \quad \langle v_R v_z\rangle = 0, \quad (5.177)$$

(e.g. Binney & Tremaine, 2008), so that the Jeans equations reduce to

$$\frac{\partial(\rho\langle v_R^2\rangle)}{\partial R} + \rho\left[\frac{\langle v_R^2\rangle - \langle v_\phi^2\rangle}{R} + \frac{\partial\Phi}{\partial R}\right] = 0; \quad (5.178)$$

$$\frac{\partial(\rho\langle v_z^2\rangle)}{\partial z} + \rho\frac{\partial\Phi}{\partial z} = 0, \quad (5.179)$$

which is now a closed set for the two unknowns. Note, however, that the Jeans equations provide no information about how $\langle v_\phi^2\rangle$ splits in streaming and random motions. In practice one often follows Satoh (1980), and writes that

$$\langle v_\phi\rangle = k\sqrt{\langle v_\phi^2\rangle - \langle v_R^2\rangle}, \quad (5.180)$$

with k a free parameter. When $k = 1$ the azimuthal dispersion is $\sigma_\phi^2 = \langle v_\phi^2\rangle - \langle v_\phi\rangle^2 = \sigma_R^2 = \sigma_z^2$ everywhere. Models of this kind are referred to as oblate isotropic rotators.

For axisymmetric two-integral models, it is useful to split the DF into parts that are even and odd in L_z:

$$f(E,L_z) = f_+(E,L_z) + f_-(E,L_z), \tag{5.181}$$

with

$$f_\pm(E,L_z) \equiv \frac{1}{2}[f(E,L_z) \pm f(E,-L_z)]. \tag{5.182}$$

The density corresponding to the effective potential $\Psi(R,z) = -\Phi(R,z)$ depends only on the even part of the DF:

$$\rho = \frac{4\pi}{R}\int_0^\Psi dE \int_0^{R\sqrt{2(\Psi-E)}} f_+(E,L_z)dL_z. \tag{5.183}$$

For $L_z = 0$ the inverse of this equation is given analytically by Eq. (5.151), but the evaluation of $f_+(E,L_z)$ for other values of L_z from a given $\rho(R,z)$ is difficult. Lynden-Bell (1962), Hunter (1977a), and Dejonghe (1986) used different transformation methods to evaluate $f_+(E,L_z)$ for a number of models. However, since all these methods require the analytical knowledge of $\rho(R,\Psi)$, they are not widely applicable. A more general scheme has been provided by Hunter & Qian (1993) using a contour integral expression, which is the generalization of Eq. (5.151). This method has the advantage that it does not require explicit knowledge of $\rho(R,\Psi)$. Furthermore, the same method can also be used to obtain the odd part of the DF, $f_-(E,L_z)$, from the mean stellar streaming motion $\langle v_\phi(R,z)\rangle$ in the meridional plane.

Axisymmetric Power-Law Models A specific example of axisymmetric two-integral models is the set of power-law models introduced by Evans (1994). This is the largest known set of axisymmetric systems with simple and explicit two-integral DFs. Another attractive property of these models is that all the line-of-sight velocity moments are simple elementary functions of the coordinates on the sky, making them extremely useful for comparison with kinematic and photometric observations of elliptical galaxies.

The gravitational potential of the power-law models is given by

$$\Phi(R,z) = \begin{cases} \frac{v_0^2}{\gamma}R_c^\gamma(R_c^2+R^2+z^2/q^2)^{-\gamma/2} & (\gamma \neq 0) \\ \frac{v_0^2}{2}\ln(R_c^2+R^2+z^2/q^2) & (\gamma = 0), \end{cases} \tag{5.184}$$

where R_c is a core radius and q is the axis ratio of the spheroidal equipotentials. Oblate and prolate models have $q < 1$ and $q > 1$, respectively, while models with $\gamma > 0$ and $\gamma < 0$ correspond to falling and rising rotation curves, respectively. The density distribution that self-consistently generates the potential of Eq. (5.184) is

$$\rho(R,z) = \frac{v_0^2 R_c^\gamma}{4\pi G q^2}\frac{R_c^2(1+2q^2) + R^2(1-\gamma q^2) + z^2[2-(1+\gamma)q^{-2}]}{(R_c^2+R^2+z^2/q^2)^{\gamma/2+2}}. \tag{5.185}$$

In order for this density distribution to be everywhere positive requires that $q^2 \geq (1+\gamma)/2$. The even part of the DF that corresponds to this self-consistent model is given by

$$f_+(E,L_z) = \begin{cases} A_+L_z^2|E|^{4/\gamma-3/2} + B_+|E|^{4/\gamma-1/2} + C_+|E|^{2/\gamma-1/2} & (\gamma \neq 0) \\ A_+L_z^2 e^{-4E/v_0^2} + B_+ e^{-4E/v_0^2} + C_+ e^{-2E/v_0^2} & (\gamma = 0), \end{cases} \tag{5.186}$$

where the coefficients A_+, B_+, and C_+ depend on γ, q and R_c, and are given by Evans (1994). Evans & de Zeeuw (1994) have shown that the following simple choice for the mean streaming law leads to an odd part of the two-integral DF that is equally simple:

$$\langle v_\phi \rangle = k|1-q^2|^{1/2} v_0 \frac{RR_c^{\gamma/2}}{(R_c^2+R^2+z^2/q^2)^{\gamma/4+1/2}}, \tag{5.187}$$

where k is defined by Satoh's streaming law (5.180). Note that the use of the absolute value of $1-q^2$ ensures the streaming law to be applicable also to prolate models ($q > 1$), in which the streaming is along the major axis. The associated odd part of the DF is

$$f_-(E,L_z) = k|1-q^2|^{1/2} \begin{cases} A_- L_z^3 |E|^{5/\gamma-2} + B_- L_z |E|^{5/\gamma-1} + C_- L_z |E|^{3/\gamma-1} & (\gamma \neq 0) \\ A_- L_z^3 e^{-5E/v_0^2} + B_- L_z e^{-5E/v_0^2} + C_- L_z e^{-3E/v_0^2} & (\gamma = 0), \end{cases} \tag{5.188}$$

where the coefficients A_-, B_-, and C_- depend on γ, q and R_c, and are given in Evans & de Zeeuw (1994). The model is physical if, and only if, $f = f_+ + f_- \geq 0$, which limits k to the range $0 \leq k \leq k_{\max}(q,\gamma)$, where $k_{\max}$ is given explicitly in Evans & de Zeeuw (1994).

5.4.9 Triaxial Equilibrium Models

For triaxial systems, one generally has only one classical integral of motion, namely the energy. The other two isolating integrals of motion present in an integrable triaxial system are generally not known explicitly. An exception is the perfect spheroid, whose density distribution is given by

$$\rho(m) = \frac{\rho_0}{(1+m^2)^2}, \tag{5.189}$$

with

$$m^2 = \frac{x^2}{a^2} + \frac{y^2}{b^2} + \frac{z^2}{c^2} \quad (a \geq b \geq c). \tag{5.190}$$

This density distribution is stratified on similar concentric ellipsoids with semi-axes ma, mb and mc, and we have aligned the long and short axes with the x-axis and z-axis, respectively. Note that ρ has a constant density core, and falls off as m^{-4} at large distances. The total mass of the perfect spheroid is $M = \pi^2 abc\rho_0$.

The perfect spheroid is best studied in the ellipsoidal coordinate system (λ,μ,ν), which are the three roots for τ of

$$\frac{x^2}{\tau+\alpha} + \frac{y^2}{\tau+\beta} + \frac{z^2}{\tau+\gamma} = 1, \tag{5.191}$$

where (x,y,z) are Cartesian coordinates, $\alpha < \beta < \gamma$ are three arbitrary constants, and we take $\lambda > \mu > \nu$. The three roots (which depend on the three constants, α, β, γ) then satisfy $-\gamma \leq \nu \leq -\beta \leq \mu \leq -\alpha \leq \lambda$. Surfaces of constant λ are ellipsoids, of constant μ are hyperboloids of one sheet around the x-axis, and of constant ν are hyperboloids of two sheets (see de Zeeuw, 1985, for more detail).

In ellipsoidal coordinates, the potential of the perfect spheroid is given by

$$\Phi(\lambda,\mu,\nu) = \frac{F(\lambda)}{(\lambda-\mu)(\lambda-\nu)} + \frac{F(\mu)}{(\mu-\nu)(\mu-\lambda)} + \frac{F(\nu)}{(\nu-\lambda)(\nu-\mu)}, \tag{5.192}$$

where

$$F(\tau) = -\pi G\rho_0 abc(\tau+\alpha)(\tau+\gamma) \int_0^\infty \frac{\sqrt{s-\beta}}{(s-\alpha)(s-\gamma)} \frac{ds}{s+\tau} \tag{5.193}$$

(de Zeeuw, 1985). A potential of the form (5.192) is called a Stäckel potential, which has the desirable property that the equations of motion are separable in the ellipsoidal coordinates

(λ, μ, ν) so that each individual orbit can be considered as the sum of three motions, one in each coordinate (e.g. Weinacht, 1924).

A particular property of the perfect spheroid is that it admits three isolating integrals of motion (i.e. the Hamiltonian is integrable, and all orbits are regular) which can all be written down explicitly in terms of $\tau = (\lambda, \mu, \nu)$ and its conjugate momentum p_τ. This makes the perfect spheroid ideally suited to examine the characteristic orbital structure of triaxial systems with (near)-integrable Hamiltonians. In particular, the perfect spheroid admits four major orbit families:

- **Box orbits:** These orbits have turning points where the velocity $\mathbf{v} = 0$, and a particle on a box orbit oscillates (one says 'librates') back and forth in each coordinate, so that it does not have a fixed sense of rotation.
- **Inner and outer long-axis tubes:** These orbits have no turning point in ν, so that particles on these orbits have a fixed sense of rotation around the x-axis. The distinction between 'inner' and 'outer' depends on whether the orbit crosses the (y,z)-plane inside or outside the focal ellipse $y^2/(\beta - \alpha) + z^2/(\gamma - \alpha) = 1$, respectively.
- **Short-axis tubes:** These orbits have no turning point in μ, so that particles on these orbits have a fixed sense of rotation around the z-axis.

This orbital structure is generic for all triaxial mass models that are centrally concentrated, have a finite central density and a Stäckel potential.

Tube orbits are 'centrophobic', in that they avoid the center, while box orbits can pass arbitrarily close to the center of the potential. Since long-axis and short-axis tubes contribute angular momentum to the x- and z-axes, respectively, the *total* angular momentum vector of a triaxial system may point anywhere in the (x,z)-plane. Such a misalignment between the angular momentum vector and the eigenvectors of the moment of inertia is a characteristic property of a triaxial system.

Although the perfect spheroid is useful in that it gives valuable insight into the orbital structure of triaxial systems, it has a large constant density core. In general, however, galaxies and dark matter halos have cusps (i.e. inner density profiles with $d\ln\rho/d\ln r < 0$), which cannot be described by a Stäckel potential. As mentioned in §5.4.5, these cusps have a tendency to destroy the box orbits, giving rise to significant amounts of stochasticity in the central regions. Triaxial systems with realistic density distributions are therefore difficult to treat analytically, and one in general relies on numerical techniques to study their dynamical structure (see discussion in §13.1.4).

5.5 Collisionless Relaxation

In dynamics, relaxation is the process by which a system approaches equilibrium or by which it returns to equilibrium after a disturbance. So far in this chapter we have discussed the gravitational collapse of collisionless systems and the properties of virialized equilibrium systems. We now turn our attention to the relaxation mechanisms that operate during the collapse process and lead to these equilibrium configurations.

The range of equilibrium configurations accessible to a collisionless system is extremely large. Yet galaxies only seem to occupy a relatively small volume of this configuration space, as is evident from the fact that galaxies obey a number of relatively tight scaling relations, and have a fairly restricted variety of density profiles (see Chapter 2). Similarly as we will see in Chapter 7, dark matter halos as a class of objects are also remarkably homologous in terms of their structural and dynamical properties (at least in numerical simulations). This raises the question as to why collisionless systems settle in their particular equilibrium configuration. Is it because of the relaxation mechanism, or because of some aspects of the initial conditions?

For an isolated gas, it is well known that the collisions and interactions among the gas particles cause the system to relax to a Maxwellian distribution characterized by a single temperature. During this relaxation process most information regarding the initial conditions is erased. In the case of a collisionless system, however, the two-body relaxation time is much longer than the Hubble time (see §5.4.1), so that two-body interactions cannot be the main relaxation mechanism. In fact, a collisionless system obeys the collisionless Boltzmann equation (CBE) given by Eq. (5.103), which also remains valid during the gravitational collapse. So how, then, can we talk about relaxation, if the distribution function (DF) itself, which fully describes the system, remains constant ($df/dt = 0$)

In order to address this question, we need to take a closer look at the physical interpretation of the DF. In §5.4.2 we defined $f(\mathbf{x}, \mathbf{v})$ as the number density of particles. However, on infinitesimally small scales the distribution of particles becomes a sum of Dirac delta functions, so that the DF can no longer be considered as a smooth density field in phase space, and the CBE becomes ill-defined. This problem can be circumvented by interpreting f as a probability function, with $f(\mathbf{x}, \mathbf{v}) d\mathbf{x} d\mathbf{v}$ the probability of finding a particle in the phase-space volume $(\mathbf{x} \pm d\mathbf{x}/2, \mathbf{v} \pm d\mathbf{v}/2)$. Although this probability function is well-defined, with its evolution governed by the CBE, it is not directly measurable. Rather, an observer can only measure the numbers of stars in finite phase-space volume elements. The corresponding DF, $f_c(\mathbf{x}, \mathbf{v})$, is called the coarse-grained distribution function,[2] and corresponds to the average of $f(\mathbf{x}, \mathbf{v})$ in some phase-space volume element centered on $(\mathbf{x}, \mathbf{v})$. Note that, contrary to f, the coarse-grained DF does not obey the CBE. However, if $\partial f_c/\partial t = 0$, then the system will appear relaxed to an observer, as it does not reveal any observable evolution. Thus, as a working definition we may refer to relaxation as the process that drives the system towards a state with $\partial f_c/\partial t \simeq 0$.

There are four different relaxation mechanisms at work in gravitational N-body systems: phase mixing, chaotic mixing, violent relaxation and Landau damping. We now discuss each of these mechanisms in turn.

5.5.1 Phase Mixing

Consider two regular orbits with similar values for the actions J_i, i.e. whose tori are similar and close to each other in phase space (see §5.4.5). Since the fundamental frequencies are functions of the actions, they will also be similar for the two orbits. Now consider two trajectories on these two tori that initially have similar phases (i.e. the two trajectories are initially close to each other in phase space). Since the fundamental frequencies of the two trajectories are similar but not identical, the trajectories will separate. In particular, the phase difference in direction i will increase with time according to $\Delta\phi_i(t) = 2\pi(\Delta\omega_i)t$. Thus, points that are initially close to each other in phase space will separate linearly with time, a process called phase mixing. A simple example for a system with $n = 1$ degree of freedom is shown in Fig. 5.4. The points that initially occupy the same small region in phase space are sheared out, and eventually become tightly wound. Note that during this process the fine-grained DF remains constant. However, the coarse-grained DF, measured at the location of the initial phase-space region initially decreases with time, as more and more 'vacuum' (i.e. unoccupied phase-space volume) is mixed in. At some point, however, depending on the coarseness used to measure f_c, the coarse-grained DF will stop evolving. Hence, phase mixing is a relaxation process. Note, though, information is lost only through the coarse-graining. At the fine-grained level phase mixing (and indeed all relaxation

---

[2] To avoid confusion, f is sometimes called the fine-grained distribution function.

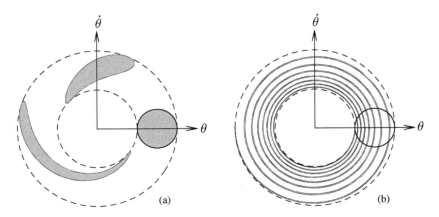

Fig. 5.4. An illustration of phase mixing. Here particles are assumed to orbit in a simple one-dimensional potential. Each dashed circle in phase space represents an orbit of given energy ($\theta^2 + \dot\theta^2 = $ const.). The shaded circle in the phase-space diagram on the left represents a group of particles that initially occupy a small region in phase space. Since 'orbits' with a larger amplitude are assumed to have longer periods, this particle cloud shears as it evolves (as in a), and eventually becomes tightly wound (as in b). [After Binney & Tremaine (1987)]

described purely by the coupled CBE and Poisson equations) is reversible and preserves all knowledge of the initial conditions.

An important corollary of phase mixing is that, for a system of fixed mass, the *maximum* of the coarse-grained DF, $f_{c,\max}$, cannot increase with time. This simply follows from the fact that, prior to reaching a relaxed state, phase mixing always continues to mix in more and more 'vacuum'. This consequence of phase mixing can be used to constrain the nature of the dark matter (e.g. Tremaine & Gunn, 1979) and the formation process of galaxies (see e.g. §13.3.3).

The time scale for phase mixing is not well defined. It basically depends on the range of (fundamental) frequencies covered by the group of particles under consideration. Note, however, that the time scale is never less than a dynamical time and can be much longer. In fact, for trajectories confined to the same torus the time scale is infinite, and no phase mixing occurs.

5.5.2 Chaotic Mixing

While regular orbits experience phase mixing, stochastic orbits experience chaotic mixing. As described in §5.4.5, a characteristic property of stochastic orbits is that two stochastic trajectories that are initially close to each other diverge exponentially. A group of stars that are initially close to each other in a stochastic region of phase space will mix by spreading through the 'Arnold web', the phase-space region accessible to every stochastic trajectory of given energy in a system with three degrees of freedom (see §5.4.5). After a sufficiently long time they will uniformly cover the Arnold web which implies that $\partial f_c / \partial t = 0$. As phase mixing, chaotic mixing thus smooths out (i.e. relaxes) the coarse-grained DF, but leaves the fine-grained DF invariant.

Consider two trajectories Γ_1 and Γ_2 that at some time t_0 pass close to some point $(\mathbf{x}, \mathbf{v})$ in phase space, and let $\Delta L_i(t)$ (with $i = 1, ..., 2n$) indicate the distance between Γ_1 and Γ_2 in phase-space coordinate i at time t. The Lyapunov exponents for $(\mathbf{x}, \mathbf{v})$ are defined as

$$\lambda_i = \lim_{t \to \infty} \frac{1}{t} \ln \left(\frac{\mathrm{d}L_i(t)}{\mathrm{d}L_i(t_0)} \right). \tag{5.194}$$

In a collisionless system

$$\sum_{i=1}^{2n} \lambda_i = 0, \tag{5.195}$$

which expresses the incompressibility of the flow (i.e. the fact that $df/dt = 0$). If the trajectory through $(\mathbf{x}, \mathbf{v})$ is regular then $L_i(t) \propto t$, so that $\lambda_i = 0$ for all i. However, if the trajectory is stochastic then $\lambda \equiv \max\{\lambda_i\} > 0$, which implies that two trajectories near $(\mathbf{x}, \mathbf{v})$ will diverge exponentially as $\delta\Gamma(t) \propto e^{\lambda t}$. The inverse of the largest Lyapunov exponent is called the Lyapunov time, and defines the characteristic e-folding time for chaotic mixing, which is typically much shorter than the time scale for phase mixing. Note, though, that the actual mixing rate of stochastic ensembles typically falls below the Lyapunov rate once the trajectories separate. This reflects the fact that stochastic regions are often bounded by regular tori, which can confine stochastic orbits for long periods of time to restricted parts of phase space. The time scale on which an orbit uniformly spreads over its accessible phase-space volume then becomes dependent on the efficiency of Arnold diffusion, which can be very low.

Note that, just like phase mixing, chaotic mixing is reversible at the fine-grained level, but that the exponential divergence of neighboring trajectories means that almost infinitely precise knowledge of the phase-space structure is required to undo its effects. Hence, chaotic mixing effectively erases knowledge of the initial conditions much more rapidly than phase mixing. On the other hand, while phase mixing operates in almost all collisionless systems, chaotic mixing is only important if a significant fraction of phase space is occupied by chaotic orbits.

5.5.3 Violent Relaxation

Phase mixing and chaotic mixing are two relaxation processes that occur even when the gravitational potential associated with the dynamical system is constant. During these relaxation processes the energies of the individual particles are conserved. However, the collapse or disturbance of a collisionless system is generally accompanied with changes of the gravitational potential, $\Phi(\mathbf{x},t)$, giving rise to an additional relaxation process.

Let $\varepsilon = \frac{1}{2}v^2 + \Phi$ be the specific energy for a given particle. Then

$$\begin{aligned}\frac{d\varepsilon}{dt} &= \frac{\partial \varepsilon}{\partial \mathbf{v}} \cdot \frac{d\mathbf{v}}{dt} + \frac{\partial \varepsilon}{\partial \Phi}\frac{d\Phi}{dt} = -\mathbf{v}\cdot\nabla\Phi + \frac{d\Phi}{dt}\\ &= -\mathbf{v}\cdot\nabla\Phi + \frac{\partial\Phi}{\partial t} + \mathbf{v}\cdot\nabla\Phi = \frac{\partial\Phi}{\partial t},\end{aligned} \tag{5.196}$$

where we have used that $(d\mathbf{v}/dt) = -\nabla\Phi$. Thus, a time-dependent potential of a collisionless system can induce a change in the energies of the particles involved (i.e. in a time-varying potential, energy is no longer an integral of motion). Exactly how the energy of a particle changes depends in a complex way on the initial position and energy of the particle: in fact, particles can both gain or lose energy, and some particles can even become unbound. Overall, the effect is to broaden the range of energies (see Fig. 5.5). In this respect, a time-varying potential provides an additional relaxation mechanism, which is called violent relaxation. Note that violent relaxation still obeys the CBE, $df/dt = 0$; however, unlike for a steady state system, the Eulerian time derivative $\partial f/\partial t \neq 0$. Furthermore, unlike phase mixing and chaotic mixing in a static potential, violent relaxation causes 'mixing' of particles also in binding energy.

As is evident from Eq. (5.196), $d\varepsilon/dt$ is independent of particle mass, so violent relaxation has no tendency to segregate particles of differing mass during the relaxation process. This is different from *collisional* relaxation, where the momentum exchange associated with the two-body gravitational enounters drives the system towards equipartition of kinetic energy. More massive particles tend to transfer energy to their lighter neighbors and so become more tightly

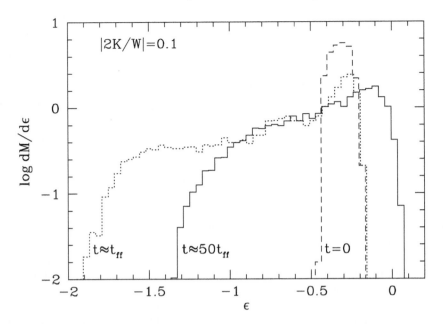

Fig. 5.5. The differential energy distributions of particles in a numerical simulation of the collapse of an N-body system that initially has a virial ratio $|2K/W| = 0.1$. Energies are measured in units of the final potential energy. Results are shown at three representative times, as indicated. Initially (at $t = 0$) the particle energies lie in a narrow range, which is rapidly broadened due to violent relaxation. After one free-fall time, $t_{\rm ff}$, particles have mainly gained (potential) energy; after 50 free-fall times, however, when the system has relaxed, there are both particles that have made a net gain in energy and particles that have made a net loss, and $dM/d\varepsilon$ peaks near the escape energy $\varepsilon = 0$. [Adapted from van Albada (1982)]

bound, sinking towards the center of the gravitational potential. This effect is known as mass segregation and is particularly important in the evolution of globular star clusters.

It is important to realize that a time-varying potential does not *guarantee* violent relaxation. Indeed, one can construct oscillating models exhibiting no tendency to relax by making sure that no mixing occurs. In this case, the energies of the individual particles change with time, but the relative distribution of energies is invariant (e.g. Sridhar, 1989). Although unrealistically artificial, this demonstrates that violent relaxation requires both a time-varying potential and mixing to occur simultaneously.

The time scale for violent relaxation can be defined as

$$t_{\rm vr} = \left\langle \frac{1}{\varepsilon^2} \left(\frac{\partial \Phi}{\partial t} \right)^2 \right\rangle^{-1/2}, \tag{5.197}$$

where the average $\langle \cdots \rangle$ is over all particles that make up the collective potential. As shown by Lynden-Bell (1967), this is approximately equal to the free-fall time[3] of the system, $t_{\rm ff} = (3\pi/32G\bar{\rho})^{1/2}$, with $\bar{\rho}$ the average density. This indicates that the relaxation process is very fast (hence 'violent').

Like phase mixing and chaotic mixing, violent relaxation tends to erase a system's memory of its initial conditions. However, the mixing associated with violent relaxation is self-limiting, because as soon as a system approaches *any* equilibrium state, the large-scale potential fluctuations which drive the evolution vanish. Mixing destroys the coherent motions required

---

[3] The free-fall time is the characteristic time that it would take for a body to collapse under its own gravitational attraction, if no other forces existed to oppose the collapse.

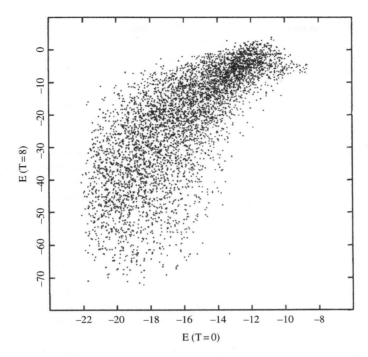

Fig. 5.6. Scatter plot of final against initial specific energies for particles in a cold, collisionless collapse. Note that the correlation is significant, indicating that violent relaxation has not completely erased the memory of the system's initial conditions. [Adapted from van Albada (1982)]

to maintain these variations, for example, in the later phases of the collapse of a system or the merger of two systems. As a result it is difficult to predict the extent to which the properties of the initial conditions are reflected in the final equilibrium state. Numerical simulations have shown that violent relaxation is, in general, never 'complete', in the sense that the final energies of particles are correlated with their initial values (White, 1978; van Albada, 1982; May & van Albada, 1984; see also Fig. 5.6) and the shape of the final system clearly remembers that of the initial conditions (Aarseth & Binney, 1978).

5.5.4 Landau Damping

The final relaxation process that plays an important role in collisionless dynamics is Landau damping, which is the damping of perturbations (waves) due to the (gravitational) interaction between the wave and the (background) particles.

To gain insight, it is useful to start by considering a fluid. A perturbation analysis of the fluid shows that if the perturbation has a wavelength $\lambda < \lambda_J$, with λ_J the Jeans length, then the perturbation is stable, and the wave propagates with a phase velocity $v_p = c_s(1 - \lambda^2/\lambda_J^2)^{1/2}$, where c_s is the sound speed. Note that larger waves move more slowly because self-gravity becomes more important. When $\lambda \geq \lambda_J$ the wave no longer propagates, but becomes a standing wave whose amplitude grows with time due to gravitational instability (see §4.1.3).

One can apply a similar perturbation analysis to self-gravitating collisionless systems (see §11.5.2). This yields a similar Jeans criterion, but with the 'sound speed' equal to the velocity dispersion of the particles, i.e. $c_s = \sigma$. As for a fluid, perturbations with $\lambda \geq \lambda_J$ are unstable and grow, while those with $\lambda < \lambda_J$ result in waves propagating with a group velocity $v_p \leq c_s = \sigma$.

However, while the fluid supports gravity-modified sound waves that are stable, the equivalent waves in gravitational collisionless systems experience Landau damping.

Consider a particle with $v \gg v_p$. At some point this particle will overtake the wave. As it 'falls' into the potential perturbation, it gains energy. But the energy gain is subsequently lost as the particle climbs out of the potential perturbation. Consequently, the particle experiences no net change in energy. The same holds for particles with $v \ll v_p$, which are overtaken by the wave. However, particles with $v \simeq v_p$ (i.e. that are near resonance with the wave) may experience a net gain or loss in energy.

To see this, consider a particle with a velocity that is slightly larger than v_p. Depending on the phase of the wave at its initial position, the particle will either be accelerated or decelerated by the gravitational potential perturbation associated with the wave. Those that are accelerated will move away from the resonance, while those that are decelerated approach the resonant velocity. The latter interact more effectively with the wave (will be decelerated for a longer time), and on average there will thus be a net transfer of energy from particles with $v \gtrsim v_p$ to the wave. Conversely, particles with $v \lesssim v_p$, on average gain net energy from the wave. In the case of a Maxwellian distribution, $f(\mathbf{v}) \propto \exp(-\frac{1}{2}v^2/\sigma^2)$, there are more particles in the latter class than in the former, causing a net transfer of energy from the wave to the particles so that the wave is damped.[4] Although collisionless systems in general do not have a Maxwellian velocity distribution, in almost all cases $\partial f/\partial v < 0$, so that there will still be damping, with a rate determined by the gradient of the velocity distribution function at the group velocity of the wave.

The main effect of Landau damping is to convert the energy in a wave, which is created due to a disturbance of the system (e.g. a close encounter with another system), into random motion of the particles in the system. Together with phase mixing, Landau damping is also responsible for limiting the effectiveness of violent relaxation, by damping the potential fluctuations that cause a mixing in particle binding energy.

5.5.5 The End State of Relaxation

A number of authors have tried to use the principles of statistical mechanics to predict the end state of a relaxed, collisionless system. Unfortunately, this approach has encountered a large number of formal difficulties, and little progress has been made. There is currently no clear understanding of how to proceed, and attention has mainly shifted to using numerical N-body simulations to study the end state of the various relaxation processes discussed above. Here we briefly mention the ideas behind the statistical mechanics approach, and we highlight its shortcomings. We then give a brief description of the results obtained using numerical simulations.

(a) Statistical Mechanics As discussed above, violent relaxation causes the energies of particles in a collisionless system to change. Therefore, if the mixing is very efficient or prolonged, one might expect the distribution of energies to approach a 'most probable' state, from a statistical mechanics point of view. After all, that is what happens to a collisional gas in which the collisions among molecules result in an efficient exchange of kinetic energies, and the most probable distribution is obtained by maximizing the entropy, leading to the well-known Maxwell–Boltzmann velocity distribution. One can carry out a similar calculation for collisionless systems, with the not very surprising result that the velocity distribution should be Maxwellian in this case as well (e.g. Ogorodnikov, 1965; Lynden-Bell, 1967). However, as shown in §5.4.7, the self-consistent model corresponding to such a DF is the isothermal sphere,

---

[4] For a mathematical treatment of Landau damping in collisionless systems, see Binney & Tremaine (2008).

which has infinite mass and energy. Clearly, this solution is therefore unphysical. In fact, if one defines the entropy of a collisionless system in analogy to a collisional gas as

$$S = -\int\int f \ln f \, d^3\mathbf{x} \, d^3\mathbf{v}, \tag{5.198}$$

it can be shown that, for a system of finite mass and energy, no $f(\mathbf{x},\mathbf{v})$ maximizes the entropy. One can always make S larger by making the system more centrally concentrated (see e.g. Tremaine et al., 1986). Physically this is due to the long-range nature of gravity, which allows bound particles to fill an infinitely large volume in phase space. In fact, the interpretation of S given by Eq. (5.198) as an entropy is also controversial. Clearly, if f is the fine-grained DF, the CBE guarantees that $dS/dt = 0$. This would imply that collisionless relaxation is adiabatic, in which case there is no evolution to a 'most probable' state. If, on the other hand, one uses the coarse-grained DF, $f_c(\mathbf{x},\mathbf{v},t)$ instead, then it has been argued that $S = S(t)$ is a non-decreasing function of time. Following Tolman (1938) and Tremaine et al. (1986) this argument goes as follows:

$$\begin{aligned}
S(t_f) &\equiv \left[-\int\int f_c \ln(f_c) \, d^3\mathbf{x} \, d^3\mathbf{v} \right]_{t_f} \\
&\geq \left[-\int\int f \ln(f) \, d^3\mathbf{x} \, d^3\mathbf{v} \right]_{t_f} \\
&= \left[-\int\int f \ln(f) \, d^3\mathbf{x} \, d^3\mathbf{v} \right]_{t_i} \\
&= \left[-\int\int f_c \ln(f_c) \, d^3\mathbf{x} \, d^3\mathbf{v} \right]_{t_i} \equiv S(t_i),
\end{aligned} \tag{5.199}$$

where t_i and t_f denote the initial and final times, respectively. The inequality in the second line is a result of the averaging implicit in the definition of f_c together with the fact that, for $C(f) = -f\ln(f)$, $\overline{C(f)} > C(\overline{f})$ for any set of unequal values of f. The equality in the third line of Eq. (5.199) holds because, as we have already seen, an entropy defined using Eq. (5.198) with the fine-grained DF is conserved. The equation in the final line of Eq. (5.199) reflects the assumption that, in the initial conditions, the fine-grained phase space density is a slowly varying function of phase-space coordinates so that $f_c = f$. Eq. (5.199) is often interpreted as implying that $S(t)$ is a non-decreasing function of time, in agreement with its interpretation as an entropy. However, all that Eq. (5.199) really says is that there is in general less information in the coarse-grained DF than in the fine-grained DF. It does not imply $S(t_2) \geq S(t_1)$ for $t_2 > t_1 > t_i$. In fact, we cannot say anything about the relative values of $S(t_1)$ and $S(t_2)$. Thus, even if Eq. (5.198) is used with the coarse-grained DF, it is not clear whether S can really be interpreted as an entropy. Furthermore, as noted by Tremaine et al. (1986), the relations in Eq. (5.199) still hold if one replaces $f \ln f$ by any convex function,[5] $C(f)$. The situation once again differs from that for gaseous systems, where the standard proof of Boltzmann's H theorem shows that Eq. (5.198) is the only possible definition for which entropy always increases. In the case of collisionless systems, there are infinitely many possible choices of $C(f)$. If the end state of relaxation is associated with the f_c that maximizes

$$S_c = -\int\int C(f_c) d^3\mathbf{x} \, d^3\mathbf{v}, \tag{5.200}$$

the solution is different for different choices of $C(f)$. Unless it is clear which convex function to use, this introduces a large amount of ambiguity.

---

[5] A twice differentiable function $f(x)$ is convex if and only if its second derivative is non-negative, i.e. $d^2f/dx^2 \geq 0$, for all x.

Yet another difficulty with using entropy-based arguments to describe the equilibrium end state of violent relaxation is the lack of reasons to believe that the system is maximally mixed, i.e. that the system has relaxed to the 'most probable' state. As we have shown above, violent relaxation is self-limiting: the relaxation process itself 'damps' the potential fluctuations that drive the mixing. Additional relaxation processes, such as phase mixing and chaotic mixing, are also unlikely to have proceeded to a maximally mixed state, simply because the outskirts of galaxies and dark matter halos are only a few dynamical times old.

(b) Numerical Simulations It is clear from the discussion above that statistical mechanics alone does not determine the relaxed equilibrium state of a collisionless system. This state must depend on the details of the collapse process and on the initial conditions that determine how efficient violent relaxation will be (i.e. how much mixing will occur). This has been confirmed with N-body simulations of the collapse and merger of collisionless systems (e.g. White, 1978; Aarseth & Binney, 1978; van Albada, 1982; May & van Albada, 1984), which show that violent relaxation does not continue to completion (i.e. mixing is not maximal), resulting in final particle energies and final shapes that correlate with the initial values (see Fig. 5.6).

Most importantly, the simulations show that the final density profile depends strongly on the initial conditions, in particular on the initial virial ratio $|2T/W|$. As shown in §5.4.4, ignoring the surface pressure a virialized system has a virial ratio of unity, so that $|2T/W|$ basically expresses how far away the initial system is from virial equilibrium. Since $T \propto M\sigma^2$ and $W \propto -GM^2/r = -MV_c^2$, with V_c the circular velocity at r, smaller values for the initial virial ratio $|2T/W| \propto (\sigma/V_c)^2$ indicate that the initial conditions are 'colder'. There are two ways of making the initial conditions colder: lowering the initial velocity dispersion of the particles, and making the initial conditions more clumpy.

The simulations show that, during the early stages of the collapse, the system rapidly contracts to a compact configuration (on a time scale comparable to the free-fall time $t_{\rm ff}$), with the collapse factor inversely proportional to $|2T/W|$. The initial collapse is followed by a series of expansion and contraction phases, during which the particles either gain or lose energy resulting in a final distribution of particle energies that is much broader than the initial distribution (Fig. 5.5). The time it takes for the system to settle in equilibrium depends on the initial virial ratio: colder initial conditions induce larger potential fluctuations, which take a longer time to be washed out. In addition, the larger potential fluctuations also cause a larger fraction of particles to be flung out to large radii, giving rise to a more extended halo which takes longer time to settle in equilibrium. The final equilibrium distribution extends well beyond the outer boundary of the initial configuration, with the central density comparable to the density of the system at the time when it first collapses (and thus higher for smaller values of the initial $|2T/W|$). The velocity field of the equilibrated structure is nearly isotropic in the inner regions, but dominated by radial orbits in the outer part. This is a natural consequence of the fact that the particles in the outskirts were launched there due to potential fluctuations that occurred in the central region, putting them on highly elongated orbits. Similar processes occur during mergers of collisionless systems and result in similar final structures.

As shown by van Albada (1982), collapse from initial conditions with $|2T/W| \sim 0.2$ produces a final, relaxed system whose density distribution, $\rho_{\rm f}(r)$, is well fit by the $r^{1/4}$-law profile [except, perhaps, near the center where $\rho_{\rm f}(r)$ falls below the $r^{1/4}$-law profile]. The fact that the surface brightness profiles of elliptical galaxies are well fitted by the $r^{1/4}$-law profile has therefore been interpreted as suggesting that elliptical galaxies may form through dissipationless collapse of a stellar system from cold initial conditions. However, as we will see in §13.2, this monolithic collapse model faces a number of problems, and it is currently believed that dissipation must play an important role during the formation of elliptical galaxies. A route where final assembly is via mergers appears more promising.

Another case in which cold initial conditions are believed to play an important role is the formation of cold dark matter (CDM) halos. As discussed in §3.3, the initial velocity dispersion of CDM particles is negligibly small. In addition, since there is power on all scales, the initial conditions for the gravitational collapse are also very clumpy. Hence, all CDM halos are expected to form out of initial conditions that are extremely cold, with an initial virial ratio close to zero. As discussed in §7.5, these initial conditions result in the collapse and formation of relaxed dark matter halos with a universal density profile that is similar to a Sérsic profile. Although it is tempting to link this universality to the similarity in the initial conditions, we are still lacking a clear understanding of why the relaxed state of a virialized dark matter halo takes on exactly this particular form (see discussion in §7.5.1).

5.6 Gravitational Collapse of the Cosmic Density Field

Now that we have described various models for the evolution of cosmic density perturbations, we can use these results to examine how gravitational collapse proceeds in the cosmic density field and what kind of structures are to be expected from such collapse. In this section we first use simple considerations to gain qualitative understanding of the problem, and then turn to N-body simulations to show how the various aspects of gravitational collapse are realized in detail.

5.6.1 Hierarchical Clustering

As we have seen in Chapter 4, in the linear regime the density perturbations $\delta(\mathbf{x},t)$ grow with time as $\delta(\mathbf{x},t) \propto D(t)$, and so the variance [defined in Eq. (4.271)] $\sigma^2(r;t) \propto D^2(t)$. For a power-law spectrum $P(k) \propto k^n$, $\sigma^2(r) \propto r^{-(n+3)}$ so that we can write

$$\sigma^2(r;t) = \left[\frac{r}{r^*(t)}\right]^{-(n+3)} = \left[\frac{M}{M^*(t)}\right]^{-(n+3)/3}, \qquad (5.201)$$

where

$$M^*(t) \propto [D(t)]^{6/(n+3)}, \text{ and } r^*(t) \propto [D(t)]^{2/(n+3)}, \qquad (5.202)$$

are the fiducial mass and length scales on which $\sigma = 1$ at time t. Since the critical linear overdensity of gravitational collapse is ~ 1.68 according to the spherical collapse model, we can imagine that nonlinear structures with mass $\sim M$ begin to form in significant numbers when the linear value of $\sigma(M;t)$ reaches ~ 1.68. Therefore, the time dependence of M^* can be used to understand how nonlinear structures develop with time for a given linear power spectrum. Since $D(t)$ increases with t, the mass scale of nonlinearity, $M^*(t)$, increases with t for $n > -3$. In this case, structure formation proceeds in a 'bottom-up' fashion, in the sense that smaller structures form prior to larger ones. For cold dark matter (CDM) models, the effective spectral index is larger than -3 over all length scales,[6] and so these models belong to the 'bottom-up' category. On the other hand, for hot dark matter (HDM) models where the effective spectral index $n < -3$ on scales smaller than the free-streaming length, the first structures to collapse are pancakes with sizes comparable to the free-streaming length. Smaller structures will then have to form via the fragmentation of larger ones, giving rise to a 'top-down' scenario of structure formation. For $n \sim -3$, we expect a lot of 'cross-talk' between large and small scales, as structures on all scales form roughly coevally.

In a 'bottom-up' scenario, the characteristic time, t_M, for the collapse of objects of mass M is given by

$$D(t_M) \propto M^{(n+3)/6} \quad [t_M \propto M^{(n+3)/4} \text{ for EdS}]. \qquad (5.203)$$

[6] This is true for the baryonic component only on scales larger than the Jeans length.

As shown in §5.4.4, the mean density of a spherical object that collapsed at time t is roughly proportional to the mean density of the universe at that time. Thus, the density, radius and velocity dispersion of a collapsed object scale with its mass as

$$\rho_M \propto \overline{\rho}(t_M) \propto [1+z(t_M)]^3 \quad [\propto M^{-(n+3)/2} \text{ for EdS}]; \tag{5.204}$$

$$r_M \propto (M/\rho_M)^{1/3} \quad [\propto M^{(n+5)/6} \text{ for EdS}]; \tag{5.205}$$

$$v_M^2 \propto GM/r_M \quad [\propto M^{(1-n)/6} \text{ for EdS}]. \tag{5.206}$$

It then follows that, for an EdS universe,

$$\rho_M \propto r_M^{-\gamma}, \quad \gamma = \frac{3n+9}{n+5}. \tag{5.207}$$

If $n > -3$, as required for hierarchical clustering, smaller objects formed earlier and have higher densities. At any given time t, the Universe contains collapsed objects of various masses up to some characteristic maximum scale $M^*(t)$. As time goes on, M^* increases, and larger and larger objects form by accretion and by the merger of smaller objects. For $n < 1$, v_M^2 increases with M, and so the typical specific binding energy increases as larger objects form. For $n > 1$, on the other hand, the binding energy of an object is dominated by that of its progenitors.

Based on these simple considerations, we can divide the evolution of the cosmic density field in three different regimes. First, on mass scales $M \gg M^*$, the density fluctuations are still in the linear regime, and their growth is governed by the linear perturbation theory described in Chapter 4. Second, on mass scales $M \ll M^*$, the density fluctuations have collapsed to form virialized objects with density profiles that are determined by their initial conditions and by the various relaxation processes described in §5.5. Finally, on intermediate mass scales ($M \sim M^*$), density fluctuations are in the quasi-linear regime, and, according to the theory described in §4.1.8 and §5.3, the structure is dominated by large-scale pancakes and filaments. At any given time, the evolved cosmic density field is therefore a complicated web (called the 'cosmic web') consisting of virialized halos arranged in a network of large-scale filaments and pancakes that surround large low-density regions (voids). As we will see below, these expected properties of gravitational collapse in the cosmic density field are well reflected in detailed numerical simulations.

5.6.2 Results from Numerical Simulations

Although the physics behind gravitational collapse is simple, in the sense that only gravitational interactions are involved, the evolution of the cosmic density field is in general complicated. This complexity arises because the initial density field contains perturbations over a wide range of scales, and nonlinear evolution couples structures of different scale (see §4.1.7). Thus, to follow the evolution of the cosmic density field in detail, one must use N-body simulations. The first step is to generate an initial density field with precisely the desired statistical properties. For a Gaussian field, which is what we are mostly interested in, this is relatively simple. One can start with an array of Fourier modes, each characterized by its wavevector $\mathbf{k}$, and assign each a random amplitude $|\delta_\mathbf{k}|$ and a random phase $\varphi_\mathbf{k}$ according to the distribution function described in §4.4.2. The linear overdensity field, $\delta(\mathbf{x})$, can then be obtained using fast Fourier transforms (FFTs), and set up in a simulation box using the techniques described in §C1.1.3 and §C1.1.4. Because both $|\delta_\mathbf{k}|$ and $\varphi_\mathbf{k}$ are random variables, the field $\delta(\mathbf{x})$ from a particular set of $(|\delta_\mathbf{k}|, \varphi_\mathbf{k})$ represents only one specific realization of the model in consideration. The perturbation fields will differ from one realization to another, but should all be equivalent in a

5.6 Gravitational Collapse of the Cosmic Density Field

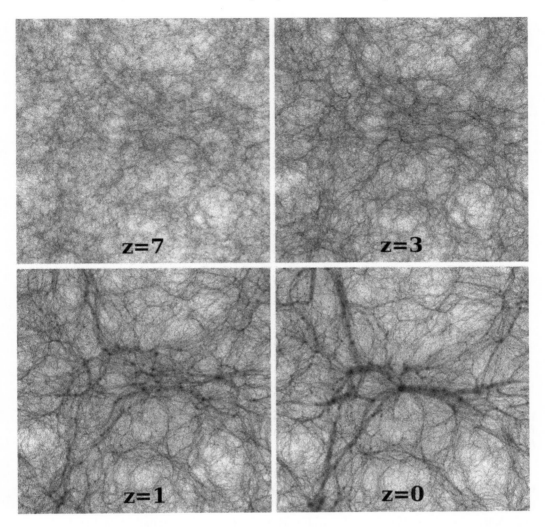

Fig. 5.7. Images of the dark matter distribution at four different redshifts ($z = 7, 3, 1$ and 0, as indicated) in a thin slice 140 Mpc on a side and 14 Mpc thick through a high resolution N-body simulation of structure formation in the ΛCDM cosmology. Note how a network of dense clumps (quasi-equilibrium halos), filaments and voids develops with time. [Kindly provided to us by M. Boylan-Kolchin, based on the Millenium-II simulation presented in Boylan-Kolchin et al. (2009)]

statistical sense. Because the density field must be sampled in a finite simulation box, there are also artificial effects due to the finite number of Fourier modes that is sampled. The realization to realization variance caused by these effects of finite volume is similar to the uncertainty in inferences from observational surveys due to their finite volume and is usually referred to as cosmic variance.

Once the initial conditions are set up, one of the algorithms described in §C1.1.1 can be chosen to evolve the density field forward in time. As an example, Fig. 5.7 shows the result of such a simulation for a ΛCDM model. As one can see, the structures indeed evolve in a hierarchical fashion, with smaller structures developing earlier and subsequently merging into larger and larger structures. The most prominent large-scale structures are highly flattened pancakes and

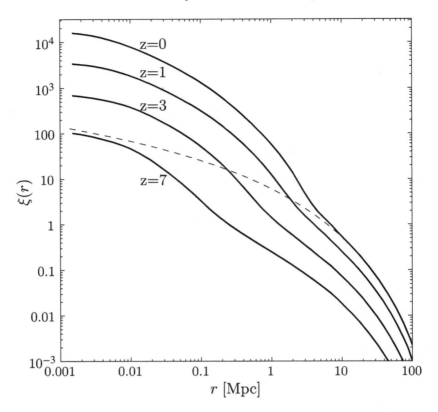

Fig. 5.8. Evolution of the two-point correlation function of dark matter in the ΛCDM simulation of Fig. 5.7 from redshift $z=7$ to the present time ($z=0$). The two-point correlation function corresponding to the *linear* power spectrum at $z=0$ is also shown.

highly elongated filaments, which are produced by the gravitational collapse of density perturbations in the quasi-linear regime. Relatively round clumps (dark matter halos) are found in high-density locations. Massive halos with masses $\sim M^*$ are preferentially found at the intersections of large-scale filaments, while halos with masses $\ll M^*$ are preferentially embedded in filaments. Since matter flows towards high-density regions, low-density regions expand, giving rise to a population of large voids.

Because of nonlinear evolution, the evolved density field has properties that are very different from those of the initial density field. In particular, the shape of the power spectrum has changed, as can be seen from Fig. 5.8, where the evolution of the two-point correlation function is shown. This behavior can be understood qualitatively as follows. On large scales where the structures are still in the linear regime, $\xi(x,t) \propto [D(t)]^2$, where $D(t)$ is the linear growth factor. On small scales, the clustering amplitude increases rapidly at the early stage of nonlinear evolution, and then evolves as a^3 due to the formation of virialized clumps whose characteristic densities change little while the background density decreases as $1/a^3$. Because of the rapid growth of ξ on small scales relative to its linear growth on large scales, $\xi(r)$ is steeper than its linear counterpart at the translinear scale (corresponding to $\xi \sim 1$), which itself increases with time due to hierarchical clustering.

Nonlinear gravitational collapse also changes other statistical properties of the density field. At late times the field is highly non-Gaussian, as can be seen from the large disparity between overdense and underdense regions. Because of this, the evolved density field is much more complicated than the initial linear density field, and can no longer be described completely by its power spectrum or two-point correlation function. We therefore need to develop further statistical tools to describe the properties of the evolved cosmic density field, which is the topic of the next chapter. Subsequently, in Chapter 7, we will focus on the properties of the high-density clumps, i.e. the virialized dark matter halos.

6
Probing the Cosmic Density Field

In the last two chapters we have described various models for the formation of structures in the cosmic density field. In this chapter we focus on how to test these models with observations. Since the cosmic density field is believed to be a random field generated by some random processes, model tests should be based on statistical properties of the cosmic density field, rather than on matching the predicted and observed fields point by point. Our task is therefore twofold. First, we need to develop statistical measures to characterize the cosmic density field. At any given time, the dynamical state of the cosmic density field is given by the mass distribution in space and the velocities of all mass elements. Thus, statistical characterizations of the cosmic density field are mostly based on the density and velocity fields of matter in the Universe. Such statistical characterizations are described in §§6.1–6.3, and models for the time evolution of some of these statistics are presented in §6.4. The second task is to find suitable observational probes of the cosmic density field. For many years, the distribution of galaxies has been used to infer the mass distribution in the Universe, based on the assumption that there is a simple and well-defined relation between the two. We describe various statistical measures of the galaxy distribution in §6.5. However, the assumption that galaxies trace mass cannot be true in detail, given that the total mass associated with galaxies is only a small fraction of the total mass in the Universe (see §2.10.2) and that galaxies of different properties have different spatial distributions (see §2.4.5). Because of this, other probes have been used. One such probe is the peculiar velocity distribution for galaxies. Since the large-scale structure of the Universe is believed to be produced by gravitational instability, the peculiar velocities produced by the gravitational field should be directly related to the underlying mass distribution. We describe statistical measures of the cosmic velocity field and their relations to the mass distribution in §§6.2–6.3. Gravitational lensing provides another very promising way to probe the cosmic mass distribution. The images of distant background sources (usually galaxies) are distorted by the foreground mass distribution near the line-of-sight, and we can reconstruct this foreground mass distribution by observing and modeling the distortion pattern. The principle behind such reconstructions is described in §6.6 along with some important applications. Finally, temperature fluctuations in the cosmic microwave background (CMB) provide another important avenue to probe the cosmic density field. As described in §3.5.2, the CMB photons were tightly coupled with the mass density field before decoupling, and so the temperature fluctuations in the CMB are expected to be closely related to the cosmic density field at the time of decoupling. In §6.7 we describe how to use the observed CMB anisotropy to probe the cosmic density field.

6.1 Large-Scale Mass Distribution

6.1.1 Correlation Functions

The properties of the cosmic density field can be described statistically by considering the moments of the distribution function $\mathscr{P}_x$ defined in Eq. (4.250). The two-point correlation function of the density perturbation field is defined as

$$\xi(x) = \langle \delta_1 \delta_2 \rangle, \tag{6.1}$$

where $x = |\mathbf{x}_1 - \mathbf{x}_2|$ is the separation in comoving units. Note that ξ depends only on the amplitude x of $\mathbf{x}$, which follows from the fact that the density perturbation field, $\delta(\mathbf{x})$, is, by assumption, a homogeneous and isotropic random field. Writing both δ_1 and δ_2 in terms of their Fourier transforms, we have

$$\xi(x) = \frac{1}{V_u} \sum_{\mathbf{k}} P(k) e^{i\mathbf{k}\cdot\mathbf{x}} = \frac{1}{(2\pi)^3} \int P(k) e^{i\mathbf{k}\cdot\mathbf{x}} d^3\mathbf{k}, \tag{6.2}$$

where

$$P(k) = V_u \langle |\delta_{\mathbf{k}}|^2 \rangle \tag{6.3}$$

is the power spectrum, and the second expression in the above equation follows from having $V_u \to \infty$ in our convention of Fourier transformations. The two-point correlation function and the power spectrum therefore form a Fourier transform pair. Carrying out the integration over the angle between $\mathbf{k}$ and $\mathbf{x}$ we have

$$\xi(x) = \frac{1}{2\pi^2} \int_0^\infty k^3 P(k) \frac{\sin kx}{kx} \frac{dk}{k} = \int_0^\infty \Delta^2(k) \frac{\sin kx}{kx} \frac{dk}{k}, \tag{6.4}$$

where

$$\Delta^2(k) \equiv \frac{k^3}{2\pi^2} P(k). \tag{6.5}$$

Similarly, the definition of the power spectrum gives

$$P(k) = 4\pi \int_0^\infty \xi(x) \frac{\sin kx}{kx} x^2 dx. \tag{6.6}$$

The volume average of $\xi(x)$ is defined as

$$\overline{\xi}(x) = \frac{3}{x^3} \int_0^x \xi(x') x'^2 dx', \tag{6.7}$$

which is related to the power spectrum by

$$\overline{\xi}(x) = \int_0^\infty \Delta^2(k) \left[\frac{3(\sin kx - kx \cos kx)}{(kx)^3} \right] \frac{dk}{k}. \tag{6.8}$$

The window function (the term in brackets) involved here dies off faster with increasing k than the term $(\sin kx/kx)$ in the expression of $\xi(x)$, so $\overline{\xi}(x)$ gives a cleaner measure of the power spectrum at $k \sim 1/x$ than does $\xi(x)$. A related quantity is the J_3 integral, defined as

$$J_3(x) = \int_0^x \xi(x') x'^2 dx' = \frac{1}{3} x^3 \overline{\xi}(x). \tag{6.9}$$

If $P(k) \propto k^n$ as $k \to 0$, then for $n > -3$ the main contribution to the integral of $\overline{\xi}(x)$ is from $k \lesssim 1/x$. Since the window function in $\overline{\xi}(x)$ is about unity for $k \lesssim 1/x$, we have

$$\overline{\xi}(x) \simeq \int_0^{1/x} \Delta^2(k) \frac{dk}{k} \propto x^{-(n+3)}, \tag{6.10}$$

and $J_3(x) \propto x^{-n}$. Thus, for $n > 0$ we have the following integral constraint on $\xi(x)$:

$$\int_0^\infty \xi(x) x^2 dx = 0. \tag{6.11}$$

Because $\xi(0) = \langle \delta^2(\mathbf{x}) \rangle > 0$, this constraint means that $\xi(x)$ must pass through zero at some large x.

The ℓ-point correlation function is defined as

$$\xi^{(\ell)}(\mathbf{x}_1,\mathbf{x}_2,\ldots,\mathbf{x}_\ell) \equiv \langle \delta_1 \delta_2 \ldots \delta_\ell \rangle. \tag{6.12}$$

In practice we are more interested in the connected ℓ-point function, which is obtained by subtracting from $\xi^{(\ell)}$ all the disconnected terms arising from lower-order correlations. For example, the connected three- and four-point correlation functions are

$$\begin{aligned}\zeta(\mathbf{x}_1,\mathbf{x}_2,\mathbf{x}_3) &\equiv \langle \delta_1 \delta_2 \delta_3 \rangle - \langle \delta_1 \rangle \langle \delta_2 \delta_3 \rangle \text{ (three terms)} - \langle \delta_1 \rangle \langle \delta_2 \rangle \langle \delta_3 \rangle \\ &= \langle \delta_1 \delta_2 \delta_3 \rangle;\end{aligned} \tag{6.13}$$

$$\begin{aligned}\eta(\mathbf{x}_1,\mathbf{x}_2,\mathbf{x}_3,\mathbf{x}_4) &\equiv \langle \delta_1 \delta_2 \delta_3 \delta_4 \rangle - \langle \delta_1 \rangle \langle \delta_2 \delta_3 \delta_4 \rangle \text{ (four terms)} - \langle \delta_1 \rangle \langle \delta_2 \rangle \langle \delta_3 \delta_4 \rangle \text{ (six terms)} \\ &\quad - \langle \delta_1 \delta_2 \rangle \langle \delta_3 \delta_4 \rangle \text{ (three terms)} - \langle \delta_1 \rangle \langle \delta_2 \rangle \langle \delta_3 \rangle \langle \delta_4 \rangle \\ &= \langle \delta_1 \delta_2 \delta_3 \delta_4 \rangle - \xi(x_{12}) \xi(x_{34}) \text{ (three terms)}.\end{aligned} \tag{6.14}$$

For a Gaussian random field, $\zeta = \eta = 0$, and so are all connected higher-order correlation functions. This highlights the advantage of working with the connected correlation functions: connected high-order correlation functions can be used to test whether a density perturbation field is Gaussian.

6.1.2 Particle Sampling and Bias

In many applications it is necessary to represent the cosmic density field by a set of mass particles. For example, if galaxies trace the mass distribution to some degree, the distribution of galaxies in space should be considered as a point set that represents the underlying mass distribution. In this subsection we first describe simple models to sample a continuous mass distribution with particles. We then define statistical measures based on such sampling, and discuss how the distribution of particles (tracers) may be biased relative to the original mass distribution.

(a) Poisson Sampling The simplest way to represent a continuous density field with particles is to divide the space into infinitesimal cells, and sample the density field in such a way that the number of particles in a particular cell has a Poisson distribution with a mean proportional to the mean density of the cell. Suppose that the volume of each cell, ΔV, is chosen so small that the probability for a cell to contain more than one particle is zero. Then the occupancy of an arbitrary cell ($\mathcal{N}_i$) has the following properties:

$$\mathcal{N}_i = \mathcal{N}_i^2 = \mathcal{N}_i^3 = \cdots. \tag{6.15}$$

Thus, to sample a given density field $\rho(\mathbf{x})$ by a Poisson process, we just need to specify $p^{(1)}(\mathbf{x})$, the probability for a cell located at $\mathbf{x}$ to contain one particle. Clearly this probability is equal to the average of $\mathcal{N}$ at $\mathbf{x}$:

$$p^{(1)}(\mathbf{x}) = \langle \mathcal{N}(\mathbf{x}) \rangle_\mathrm{P} = [1 + \delta(\mathbf{x})]\overline{n}\Delta V, \tag{6.16}$$

where $\overline{n} = \overline{\rho}/m$ (m being the mass of a particle) is the mean number density of particles, and $\langle \cdots \rangle_\mathrm{P}$ is the average over the Poisson distribution.

Now suppose we have an ensemble of realizations of a random density field. Since each of the realizations is represented by an independent set of particles according to the Poisson sampling, the ensemble average can be considered to be over both the realizations of the random field and the Poisson distribution. This can be understood by grouping the realizations into subgroups in such a way that the difference among the realizations within each subgroup is small. Averaging within a subgroup is therefore equivalent to that over the Poisson distribution, while averaging among the subgroups is equivalent to that over different realizations of the random field. This separation of an ensemble average into two steps is useful, because the Poisson sampling of the density at a point is independent of that at another and so the average over the Poisson distribution can be carried out first. For example, the ensemble average of the number of particles in a cell must be $\bar{n}\Delta V$, which follows from

$$\langle \mathcal{N} \rangle = \langle \langle \mathcal{N} \rangle_\mathrm{P} \rangle = \langle p^{(1)} \rangle = \bar{n}\Delta V. \tag{6.17}$$

Similarly,

$$\langle \mathcal{N}_i \mathcal{N}_j \rangle = \langle \langle \mathcal{N}_i \mathcal{N}_j \rangle_\mathrm{P} \rangle = \langle p^{(1)}(\mathbf{x}_i) p^{(1)}(\mathbf{x}_j) \rangle = (\bar{n}\Delta V)^2 [1 + \xi(x_{ij})], \tag{6.18}$$

where $x_{ij} \equiv |\mathbf{x}_i - \mathbf{x}_j| \neq 0$, and the second equation follows from the fact the Poisson samplings at different locations are independent of each other.

Using the assumption of ergodicity, the ensemble average, $\langle p^{(1)}(\mathbf{x}_i) p^{(1)}(\mathbf{x}_j) \rangle$, can be written in terms of a spatial average,

$$\left\langle p^{(1)}(\mathbf{x}_i) p^{(1)}(\mathbf{x}_i + \mathbf{x}) \right\rangle_{\mathbf{x}_i} = (\bar{n}\Delta V)^2 [1 + \xi(x)]. \tag{6.19}$$

The left-hand side is just the probability of finding a pair of particles from a randomly chosen pair of cells with comoving separation x, which is equal to $(\bar{n}\Delta V)^2$ for a random distribution. Hence, the two-point correlation function $\xi(x)$ is a measure of the excess in the number of pairs of particles with comoving separation x over a random sample with the same $\bar{n}$. The conditional probability for finding a pair of particles from a pair of cells that is separated by x, given that one cell contains a particle, is

$$\frac{\langle p^{(1)}(\mathbf{x}_i) p^{(1)}(\mathbf{x}_i + \mathbf{x}) \rangle_{\mathbf{x}_i}}{\langle p^{(1)}(\mathbf{x}_i) \rangle_{\mathbf{x}_i}} = (\bar{n}\Delta V)[1 + \xi(x)]. \tag{6.20}$$

This is just the mean number of neighbors each particle has in a volume element ΔV at a comoving distance x. The mean number of particles within a spherical shell that is centered on a particle and has a *proper* radius r (related to the comoving radius x by $r = ax$) and thickness dr is

$$dN(r) = 4\pi r^2 \bar{n}[1 + \xi(r)] dr, \tag{6.21}$$

and the mean number of neighbors within a proper distance r from a given particle is

$$N(r) = \frac{4\pi}{3} \bar{n} r^3 \left[1 + \overline{\xi}(r)\right]. \tag{6.22}$$

According to these arguments, $[1 + \xi(r)]$ as a function of r can be interpreted as the mean density profile around each particle, and $(4\pi/3)\bar{\rho} r^3 [1 + \overline{\xi}(r)]$ can be interpreted as the mean of the masses enclosed in spheres of radius r centered on particles. The distribution of galaxies can be viewed as a point process which samples the underlying mass distribution in some specified way. The two-point correlation function of galaxies, which can be estimated by counting the number of galaxy pairs as a function of their separation and comparing this with that given by a random sample, can therefore be used to infer the power spectrum of the mass distribution (see §6.5).

Similar to the two-point correlation function, higher-order correlation functions of a density field can also be interpreted in terms of the relationships among the particles that sample the density field. For example, the average number of triplets (with a given configuration) each particle can form is given by

$$\frac{\langle p^{(1)}(\mathbf{x}_1) p^{(1)}(\mathbf{x}_1 + \mathbf{x}_{12}) p^{(1)}(\mathbf{x}_1 + \mathbf{x}_{13}) \rangle_{\mathbf{x}_1}}{\langle p^{(1)}(\mathbf{x}) \rangle_{\mathbf{x}}} = (\bar{n}\Delta V)^2 h_3, \tag{6.23}$$

$$h_3 = 1 + \xi(x_{12}) + \xi(x_{13}) + \xi(x_{23}) + \zeta(x_{12}, x_{13}, x_{23}), \tag{6.24}$$

where $x_{23} = |\mathbf{x}_{12} - \mathbf{x}_{13}|$. Thus, h_3 is a measure of the excess in the number of particle triplets with a given configuration (specified by the lengths of the three sides, x_{12}, x_{13} and x_{23}) relative to a random sample with the same $\bar{n}$. Similarly,

$$\begin{aligned}h_4 = {} & 1+\xi(x_{12}) \text{ (6 terms)} + \xi(x_{12})\xi(x_{34}) \text{ (3 terms)} \\ & + \zeta(x_{12},x_{23},x_{31}) \text{ (4 terms)} + \eta\end{aligned} \quad (6.25)$$

is a measure of the excess in the number of particle quartets with a given configuration (specified by six variables that are needed to fix the relative positions of the four points) relative to a random sample with the same $\bar{n}$. These interpretations allow us to estimate the higher-order correlation functions of a point set in a straightforward way (see §6.5).

(b) Sampling Bias In the simple model discussed above, the sampling of the (smoothed) density field with particles (e.g. galaxies) is assumed to be a Poisson process with the mean number of particles at each point being proportional to the local mass density. This sampling is local, because the process at one point is independent of that at another, and is unbiased, because the density of particles at one point is proportional to the mass density. We have seen that in such a sampling the correlation functions of the point process are the same as those of the underlying density field. However, this sampling is only one among many possibilities. In general, the sampling may be non-local, in that the probability of having a sampling particle at one location may depend on the values of the density field at other locations. The sampling may even be non-Poissonian, in the sense that the probability of having a sampling particle at one location may depend on the presence of other sampling particles at or near this location. If we allow such freedom, many different models can be constructed to sample a mass density field with a point set, and it is not guaranteed that the correlation functions of the point set are the same as those of the underlying mass density field. As a simple example, if the sampling is still a Poisson process but with the mean proportional to $1+b\delta$ ($b>0$ is a constant, and the probability for δ to be $\leq -1/b$ is assumed to be small so that the mean is always positive) instead of to $1+\delta$, then the two-point correlation function of the particles will be b^2 times that of the underlying density field. In this case, the particles are biased tracers of the underlying density field. This simple example demonstrates an important point. If we want to use the spatial distribution of a population of objects (e.g. galaxies) to study the properties of the cosmic density field, we must understand *how* these objects trace the cosmic density field.

6.1.3 Mass Moments

Given a density perturbation field $\delta(\mathbf{x})$, we can filter it with some window function to get a smoothed field:

$$\delta(\mathbf{x};R) = \int \delta(\mathbf{x}')W(\mathbf{x}+\mathbf{x}';R)\,\mathrm{d}^3 x', \quad \int W(\mathbf{x};R)\,\mathrm{d}^3 x = 1, \quad (6.26)$$

where W is the window function, assumed to be spherically symmetric with characteristic radius R. The Fourier transform of $\delta(\mathbf{x};R)$ can be obtained from the convolution theorem:

$$\delta_{\mathbf{k}}(R) = \delta_{\mathbf{k}}\widetilde{W}(kR), \quad (6.27)$$

where $\widetilde{W}(kR)$ is the Fourier transform of the window function, and $\delta_{\mathbf{k}}$ is given by Eq. (4.25). For a Gaussian random field, $\delta(\mathbf{x};R)$ as a linear combination of $\delta(\mathbf{x})$ must obey the following Gaussian distribution:

$$\mathscr{P}(\delta;R)\,\mathrm{d}\delta = \frac{1}{\sqrt{2\pi}}\exp\left[-\frac{\delta^2}{2\sigma^2(R)}\right]\frac{\mathrm{d}\delta}{\sigma(R)}, \quad (6.28)$$

where the variance $\sigma^2(R)$ is related to the power spectrum by

$$\sigma^2(R) \equiv \langle \delta^2(\mathbf{x};R) \rangle = \int \Delta^2(k) \widetilde{W}^2(kR) \frac{dk}{k}. \tag{6.29}$$

More generally we can define a set of spectral moments weighted by powers of k^2:

$$\sigma_\ell^2(R) = \int k^{2\ell} \Delta^2(k) \widetilde{W}^2(kR) \frac{dk}{k}. \tag{6.30}$$

The variance σ^2 is just the zeroth moment: $\sigma^2(R) = \sigma_0^2(R)$.

The two most commonly used window functions are the top-hat window function:

$$W_{\text{th}}(\mathbf{x};R) = \left(\frac{4\pi}{3}R^3\right)^{-1} \begin{cases} 1 & (\text{for } |\mathbf{x}| \leq R) \\ 0 & (\text{for } |\mathbf{x}| > R), \end{cases} \tag{6.31}$$

$$\widetilde{W}_{\text{th}}(kR) = \frac{3(\sin kR - kR \cos kR)}{(kR)^3}; \tag{6.32}$$

and the Gaussian window function:

$$W_G(\mathbf{x};R) = \frac{1}{(2\pi)^{3/2} R^3} \exp\left(-\frac{|\mathbf{x}|^2}{2R^2}\right), \tag{6.33}$$

$$\widetilde{W}_G(kR) = \exp\left[-\frac{(kR)^2}{2}\right]. \tag{6.34}$$

The volumes of these two types of windows are related to the smoothing radius R as

$$V(R) = \begin{cases} 4\pi R^3/3 & (\text{for a top-hat}) \\ (2\pi)^{3/2} R^3 & (\text{for a Gaussian}). \end{cases} \tag{6.35}$$

Thus, for a given type of window, we can label its size by the mean mass contained in it: $M(R) \equiv \bar{\rho} V(R)$, where $\bar{\rho}$ is the mean density of the Universe at some given time. A particularly convenient convention is to refer to R as the comoving radius measured in present-day units. In this case

$$M(R) \equiv \bar{\rho}(t_0) V(R), \tag{6.36}$$

where $\bar{\rho}(t_0)$ is the mean mass density at the present time. In what follows, we will always use this convention.

Another useful window function is the k-space top-hat window, also called the sharp k-space window, which has the form:

$$\widetilde{W}_k(kR) = \begin{cases} 1 & (\text{for } kR \leq 1) \\ 0 & (\text{for } kR > 1). \end{cases} \tag{6.37}$$

This corresponds to a window in real space with the form:

$$W_k(\mathbf{x};R) = \frac{1}{2\pi^2 R^3} y^{-3} (\sin y - y \cos y), \quad (y \equiv |\mathbf{x}|/R). \tag{6.38}$$

This window function is particularly convenient in the analysis of Gaussian density fields, since the change in the field strength, $\Delta\delta$, due to a change in the window radius from $R \to R + \Delta R$, is independent of the original field, $\delta(\mathbf{x},R)$. However, the disadvantage here is that the integration of $W_k(\mathbf{x};R)$ over all space diverges, and so it is not straightforward to associate a well-defined volume (or mass) with the window. Formally, we may assign a volume to the window so that $W_k(0;R)V(R) = 1$, as is the case for both W_{th} and W_G. This gives $V(R) = 6\pi^2 R^3$ for W_k.

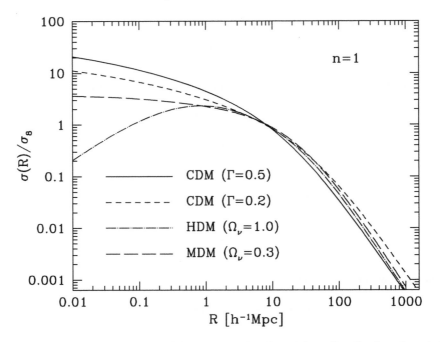

Fig. 6.1. The variance of the linear density field, σ, as a function of the radius, R, of top-hat windows in different cosmogonic models. The variance is normalized so that $\sigma_8 \equiv \sigma(8h^{-1}\text{Mpc}) = 1$.

For a power-law power spectrum, $P(k) = V_u A k^n$, Eq. (6.29) can easily be evaluated for a Gaussian window:

$$\sigma^2(R) = \frac{A}{4\pi^2} \Gamma\left(\frac{n+3}{2}\right) R^{-(n+3)} \propto R^{-(n+3)} \propto M^{-(n+3)/3}. \tag{6.39}$$

For a top-hat window, the above proportionality relations hold only for $-3 < n < 1$. The integration in Eq. (6.29) diverges in the limits $k \to 0$ and $k \to \infty$ for $n \leq -3$ and $n \geq 1$, respectively, because $\widetilde{W}^2(kR) \simeq 1$ for $k \ll R^{-1}$, and $\widetilde{W}^2(kR) \propto k^{-4}$ for $k \gg R^{-1}$. The constraint that $n > -3$ for small k (large length scale) is required because otherwise the perturbation amplitude would become larger and larger as the wavelength of the perturbation mode increases. The divergence at large k (short length scale) for $n \geq 1$ is due to the sharp edge of the top-hat window, which causes the Fourier transform of the window function to die off too slowly as $k \to \infty$ to suppress the small-scale power implied by $n \geq 1$. Since the mass variance in a top-hat window is a well-defined quantity and should be finite for any realistic density field, the above argument implies that the effective spectral index at $k \to \infty$ must be smaller than 1 for any realistic power spectrum. Fig. 6.1 shows $\sigma(R)$ versus R assuming top-hat windows for several power spectra described in §4.4.4. The meaning of $\sigma^2(R)$ can be understood in terms of variation in mass in randomly placed windows. If we denote the mass in a window centered at $\mathbf{x}$ as $M(\mathbf{x}; R)$, then the mass variance is

$$\sigma_M^2(R) \equiv \left\langle \left(\frac{M(\mathbf{x};R) - \overline{M}(R)}{\overline{M}(R)}\right)^2 \right\rangle_{\mathbf{x}} = \sigma^2(R), \tag{6.40}$$

where the last equality follows from ergodicity.

One can also define the ℓth order moment of the smoothed field:

$$\mu_\ell(R) \equiv \langle \delta^\ell(\mathbf{x};R) \rangle = \int \delta^\ell \mathscr{P}(\delta;R)\,\mathrm{d}\delta. \tag{6.41}$$

6.1 Large-Scale Mass Distribution

For a Gaussian field, this can be worked out explicitly:

$$\mu_\ell = \begin{cases} 0 & (\ell = \text{odd}) \\ \sigma^\ell(R)(\ell-1)!! & (\ell = \text{even}). \end{cases} \quad (6.42)$$

The connected part of the ℓth order moment, κ_ℓ, is defined by subtracting from μ_ℓ all disconnected terms arising from lower-order moments. For example, the third- and fourth-order moments, usually referred to as the skewness and kurtosis, are

$$\kappa_3 = \mu_3 - 3\kappa_1\kappa_2 - \kappa_1^3 = \mu_3; \quad (6.43)$$

$$\kappa_4 = \mu_4 - 3\kappa_2^2 - 4\kappa_1\kappa_3 - 6\kappa_1^2\kappa_2 - \kappa_1^4, \quad (6.44)$$

where we have used that $\kappa_1 \equiv \langle \delta \rangle = 0$. For Gaussian random fields it is easy to prove that

$$\kappa_\ell = 0 \quad \text{for all } \ell > 2. \quad (6.45)$$

Non-zero high-order connected moments therefore signify a non-Gaussian field. Similar to the variance, the high-order moments can also be interpreted in terms of the mass moments in randomly placed windows.

The mass moments are related to the correlation functions defined in §6.1.1. To find these relations, consider top-hat windows of volume V. Divide such a window into small subcells of volume ΔV. The mass in a subcell at a point $\mathbf{x}_i$ is therefore $M_i = \rho(\mathbf{x}_i)\Delta V$, and the mass in V is $M = \sum_i M_i$. It is then easy to show that

$$\langle M \rangle = \overline{\rho} V; \quad \langle M^2 \rangle = \overline{M}^2 + \frac{\overline{M}^2}{V^2}\int_V \xi(|\mathbf{x}_1 - \mathbf{x}_2|)\,\mathrm{d}^3\mathbf{x}_1\,\mathrm{d}^3\mathbf{x}_2, \quad (6.46)$$

and

$$\sigma^2 = \kappa_2 = \frac{1}{V^2}\int_V \xi(x_{12})\,\mathrm{d}V_1\,\mathrm{d}V_2. \quad (6.47)$$

Similarly, one can also prove that

$$\kappa_3 = \frac{1}{V^3}\int_V \zeta\,\mathrm{d}V_1\,\mathrm{d}V_2\,\mathrm{d}V_3; \quad \kappa_4 = \frac{1}{V^4}\int_V \eta\,\mathrm{d}V_1\,\mathrm{d}V_2\,\mathrm{d}V_3\,\mathrm{d}V_4. \quad (6.48)$$

If the density is sampled by particles so that the distribution of particle number in a window depends only on the mass within the window, the distribution function of the counts-in-cells of the particles is related to the distribution function of the density field $\mathscr{P}(\delta;R)$ by

$$f(N;V) = \int_{-\infty}^{\infty} \mathscr{P}(N|\delta)\mathscr{P}(\delta;R)\,\mathrm{d}\delta, \quad (6.49)$$

where $\mathscr{P}(N|\delta)$ is the probability of finding N particles in a window of radius R (corresponding to the volume V) and mean overdensity δ. In the special case where $\mathscr{P}(N|\delta)$ is a Poisson distribution,

$$f(N;V) = \int_{-\infty}^{\infty} \frac{e^{-\lambda}\lambda^N}{N!}\mathscr{P}(\delta;R)\,\mathrm{d}\delta, \quad (6.50)$$

where λ is related to δ. If we assume $\lambda = \langle N \rangle(1+\delta)$, where $\langle N \rangle = \overline{n}V$ is the mean number density of particles in a window, then it is easy to relate the central moments of $f(N;V)$ to the moments of the mass density field. For example,

$$\frac{\langle (N - \langle N \rangle)^2 \rangle}{\langle N \rangle^2} = \frac{1}{\langle N \rangle} + \sigma^2; \quad (6.51)$$

$$\frac{\langle (N-\langle N \rangle)^3 \rangle}{\langle N \rangle^3} = \frac{1}{\langle N \rangle^2} + \frac{3\sigma^2}{\langle N \rangle} + \kappa_3; \tag{6.52}$$

$$\frac{\langle (N-\langle N \rangle)^4 \rangle}{\langle N \rangle^4} = \frac{1}{\langle N \rangle^3} + \frac{3+7\sigma^2}{\langle N \rangle^2} + \frac{6(\sigma^2+\kappa_3)}{\langle N \rangle} + (3\sigma^2+\kappa_4). \tag{6.53}$$

We see that these central moments are the same as those defined in Eq. (6.41) for the smoothed field, except for the terms containing $\langle N \rangle$, which are due to the discrete sampling (shot noise) and go to zero as the sampling density approaches infinity (i.e. $\langle N \rangle \to \infty$).

It is instructive to derive the above expressions from the Poisson sampling discussed in §6.1.2. Divide a cell V into subcells with volume ΔV small enough so that the occupancy in each subcell satisfies $\langle \mathcal{N}_i \rangle = \langle \mathcal{N}_i^2 \rangle = \cdots$. The count in the volume V is therefore

$$N = \sum_i \mathcal{N}_i, \tag{6.54}$$

and the mean is

$$\langle N \rangle = \bar{n}V. \tag{6.55}$$

Using the properties of $\mathcal{N}_i$, it is easy to show that

$$\langle N^2 \rangle = \sum_i \langle \mathcal{N}_i^2 \rangle + \sum_{i \neq j} \langle \mathcal{N}_i \mathcal{N}_j \rangle$$
$$= \int \bar{n} \, dV + \int \bar{n}^2 [1 + \xi(x_{12})] \, dV_1 \, dV_2. \tag{6.56}$$

We leave it as an exercise for the reader to work out $\langle N^3 \rangle$ and $\langle N^4 \rangle$.

6.2 Large-Scale Velocity Field

In the linear regime, the peculiar velocities induced by density perturbations are proportional to the amplitude of the density fluctuations (see §4.1.6). At late times, when the growing mode dominates, we have

$$\mathbf{v}(\mathbf{x},t) = \sum_{\mathbf{k}} \mathbf{v}_{\mathbf{k}}(t) e^{i\mathbf{k}\cdot\mathbf{x}}, \quad \text{where} \quad \mathbf{v}_{\mathbf{k}}(t) = Haf(\Omega)\frac{i\mathbf{k}}{k^2}\delta_{\mathbf{k}}(t) \tag{6.57}$$

[see Eq. (4.77)]. Therefore, measurements of the linear velocity field provide direct constraints on the mass density field, $\delta_{\mathbf{k}}(t)$.

6.2.1 Bulk Motions and Velocity Correlation Functions

As with the density field, we can smooth the velocity field $\mathbf{v}(\mathbf{x})$ in spherical windows with a characteristic radius R:

$$\mathbf{v}(\mathbf{x};R) = \int \mathbf{v}(\mathbf{x}')W(\mathbf{x}+\mathbf{x}';R) \, d^3\mathbf{x}'. \tag{6.58}$$

The variance of the *bulk-motion* velocity is

$$\langle v^2(R) \rangle = \frac{V_u}{2\pi^2} \int (\mathbf{v}_{\mathbf{k}}^* \cdot \mathbf{v}_{\mathbf{k}}) \widetilde{W}^2(kR) k^2 dk, \tag{6.59}$$

where $\mathbf{v}_{\mathbf{k}}^*$ is the complex conjugate to $\mathbf{v}_{\mathbf{k}}$. Inserting Eq. (6.57) into this expression gives

$$\langle v^2(R) \rangle = \int_0^\infty \Delta_v^2(k) \widetilde{W}^2(kR) \frac{dk}{k}, \quad \Delta_v^2(k) = f^2(\Omega)(Ha)^2 k^{-2} \Delta^2(k). \tag{6.60}$$

Note that $\Delta_v^2(k)$ is proportional to $\Delta^2(k)$ weighted by k^{-2}, and so the velocity variance is sensitive to the mass density fluctuations on large scales. If $P(k) \propto k^n$, then for $n > -1$ the main contribution to the integral comes from $k \lesssim 1/R$. Since $\widetilde{W}(kR) \sim 1$ for $kR \lesssim 1$, we have

$$\langle v^2(R) \rangle \propto R^{-(n+1)}. \tag{6.61}$$

For $n \leq -1$ the peculiar velocity is dominated by large-scale fluctuations. Therefore realistic power spectra should have an effective spectral index $n > -1$ at $k \to 0$. For a spectrum similar to that predicted in a CDM universe, where the effective power index increases with scale from $n < -1$ to $n > -1$, the bulk motions are dominated by density fluctuations on scales where $n \simeq -1$.

We can also define the two-point correlation function of the velocity field as

$$\xi_{ij}^{(v)}(\mathbf{x}_1 - \mathbf{x}_2) \equiv \langle v_i(\mathbf{x}_1) v_j(\mathbf{x}_2) \rangle, \tag{6.62}$$

where v_i ($i = 1, 2, 3$) are the components of the velocity. Inserting Eq. (6.57) into the above expression we have

$$\xi_{ij}^{(v)}(\mathbf{x}) = (Hax)^2 f^2(\Omega_m) \int_0^\infty \Delta^2(k,t) \left[\delta_{ij} \frac{j_1(kx)}{(kx)^3} - \frac{x_i x_j}{x^2} \frac{j_2(kx)}{(kx)^2} \right] \frac{dk}{k}, \tag{6.63}$$

where $\mathbf{x} = \mathbf{x}_2 - \mathbf{x}_1$ and j_ℓ are spherical Bessel functions (Gorski, 1988). In terms of the mass correlation function,

$$\xi_{ij}^{(v)}(\mathbf{x}) = \frac{(Hax)^2 f^2(\Omega_m)}{2} \left[\left(\frac{J_3}{x^3} - \frac{J_5}{3x^5} + \frac{2K_2}{3x^2} \right) \delta_{ij} + \left(\frac{J_5}{x^5} - \frac{J_3}{x^3} \right) \frac{x_i x_j}{x^2} \right], \tag{6.64}$$

where

$$J_\ell = \int_0^x \xi(y) y^{\ell-1} \, dy; \quad K_\ell = \int_x^\infty \xi(y) y^{\ell-1} \, dy. \tag{6.65}$$

The correlation function $\xi_{ij}^{(v)}$ has been estimated by Groth et al. (1989) and Gorski et al. (1989) for galaxies with peculiar velocities estimated in a way described below.

6.2.2 Mass Density Reconstruction from the Velocity Field

In the gravitational instability scenario, the velocity of a test particle is induced directly by the underlying mass density field. Thus, if the peculiar velocities of galaxies are not severely affected by non-gravitational processes, the velocity field traced by galaxies can be used to infer the mass density fluctuations in the Universe.

By observing the spectra of galaxies one can determine their redshifts (i.e. their radial velocities). The radial velocity of a galaxy is partly due to the Hubble expansion and partly due to its peculiar motion. In order to separate these two components, one needs to know the distance (r, assumed to be in proper units) to the galaxy, so that the velocity due to the Hubble expansion, Hr, can be estimated. Two distance indicators have been used extensively to estimate the distances to galaxies. The first is based on the Tully–Fisher relation for spiral galaxies, which is an observational relation, $L \propto V_{max}^\alpha$ (with $\alpha \sim 3$), between the absolute luminosity, L, of a spiral galaxy and its maximum rotation velocity, V_{max} (see §2.3.3). The second is based on the diameter–velocity dispersion (D_n–σ) relation for elliptical galaxies, $D_n \propto \sigma_0^\beta$ (with $\beta \sim 1.2$), where D_n is the diameter of a galaxy at which its mean surface brightness drops to some fiducial value, and σ_0 is the central velocity dispersion of the galaxy (see §2.3.2). Since both V_{max} and σ_0 can be determined through spectroscopic observations, these relations allow us to estimate L (for spirals) and D_n

(for ellipticals). By comparing L so estimated with the apparent luminosity (or D_n with the angular diameter) of a galaxy, we can obtain the luminosity distance (or angular-diameter distance) of the galaxy from the formulae given in §3.2.6.

Given a sample of galaxies with redshifts (assumed to be small) and distances, we can estimate their peculiar velocities from

$$v_r(r_i) = cz_i - H_0 r_i, \tag{6.66}$$

and so the radial-velocity field is sampled at a set of points $\mathbf{r}_i = \{r_i, \vartheta_i, \varphi_i\}$ in space. From these sampled radial velocities, one can compute a smoothed radial-velocity field, $v_r(\mathbf{r})$, by smoothing the observed velocities in some pre-selected windows. The other two components of the velocity field can be recovered if we assume the velocity field to be curl-free so that it can be written as the gradient of a potential field,

$$\mathbf{v}(\mathbf{r}) = -\nabla_\mathbf{r} \Phi_v(\mathbf{r}). \tag{6.67}$$

As shown in §4.1.1, the curl of the peculiar velocity field dies off quickly with time in the linear regime. Vorticity is therefore negligible at late times even if it is present in the initial conditions. Thus, Eq. (6.67) is a good approximation as long as we are working in the linear and quasi-linear regimes. With this assumption, we have

$$\Phi_v(\mathbf{r}) = -\int_0^r v_r(r', \vartheta, \varphi)\, dr'. \tag{6.68}$$

The total velocity field can then be obtained from Eq. (6.67). With this technique, the mean (bulk-motion) velocities have been estimated for galaxies within spherical regions around the Local Group (e.g. Bertschinger et al., 1990). The mean bulk motion velocity is about 300–400 $\mathrm{km\,s^{-1}}$ within a spherical region of radius $\sim 50 h^{-1}\mathrm{Mpc}$.

In the linear and mildly nonlinear regimes, we can derive the underlying mass-density fluctuation field, $\delta(\mathbf{x})$, from the velocity field $\mathbf{v}(\mathbf{x})$ using the Zel'dovich approximation (§4.1.8):

$$\delta(\mathbf{x}) = \det\left[\mathbf{I} - \frac{D}{a\dot{D}}\frac{\partial \mathbf{v}(\mathbf{x})}{\partial \mathbf{x}}\right] - 1, \tag{6.69}$$

where $\mathbf{I}$ is the unit matrix. In the linear regime, the above equation reduces to

$$\delta(\mathbf{x}) = -\frac{\nabla \cdot \mathbf{v}(\mathbf{x})}{Haf(\Omega_m)}. \tag{6.70}$$

Fig. 6.2 shows an example of the mass density field derived from the peculiar velocities of galaxies in the local Universe. This density field is compared with the galaxy density field derived from

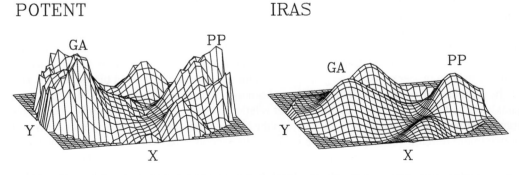

Fig. 6.2. The mass density field derived from galaxy peculiar velocities (left panel) in the local Universe is compared with the galaxy density field inferred from the IRAS galaxy survey in a similar region. [Adapted from Sigad et al. (1998) by permission of AAS]

the IRAS galaxy survey. As one can see, the reconstructed mass density field looks similar to the galaxy density field, suggesting that galaxies trace the mass distribution to some extent. If we assume the fluctuations in the galaxy distribution to be proportional to those in the mass distribution, i.e. $\delta_{\rm gal} = b\delta$ with b being a constant, then $\delta_{\rm gal}(\mathbf{x}) = -\nabla \cdot \mathbf{v}(\mathbf{x})b/Haf(\Omega_{\rm m})$. Thus, matching the reconstructed mass density field with the galaxy density field can be used to determine

$$\beta \equiv f(\Omega_{\rm m})/b \approx \Omega_{\rm m}^{0.6}/b. \tag{6.71}$$

Note that the Hubble constant H does not come in because physical distances inferred from the Hubble expansion are proportional to H^{-1}. The value of β estimated using this technique from data is about 0.5, but with large statistical uncertainties (see Dekel, 1994).

6.3 Clustering in Real Space and Redshift Space

Accurate distances to galaxies are difficult to obtain, because distances based on the Tully–Fisher or D_n–σ relations require careful observations of the kinematics and photometry of a large number of galaxies. More importantly, the uncertainties in such distance determinations are proportional to the distance itself, and rapidly become too large for the resulting peculiar velocity estimates to be useful as one moves away from the relatively local Universe. On the other hand, redshifts of galaxies are relatively easy to obtain, and currently almost a million galaxy redshifts are available from surveys such as the Sloan Digital Sky Survey (SDSS) and the 2-degree Galaxy Redshift Survey (2dFGRS). Since galaxy redshifts are not exact measures of distances, the galaxy distribution in redshift space is distorted with respect to the true distribution. In this section we show how to extract information regarding real space clustering and peculiar velocities from such redshift surveys.

6.3.1 Redshift Distortions

If we use redshift as distance, the inferred distance to a galaxy is related to its redshift, z, by

$$s \equiv cz. \tag{6.72}$$

This distance is usually referred to as the redshift distance of the galaxy. For convenience we will write distances in velocity units in this subsection. The true distance expressed in velocity units is

$$r \equiv H_0 d, \tag{6.73}$$

which we will refer to as the real distance of the galaxy. (For the moment we assume that all galaxies are local enough that a linear mean Hubble relation is appropriate.) s and r are related by

$$s = r + v_r, \tag{6.74}$$

where $v_r = \mathbf{v} \cdot \hat{\mathbf{r}}$ is the peculiar velocity along the line-of-sight. Peculiar velocities thus lead to redshift space distortions. Such distortions complicate the interpretation of galaxy clustering, but they also contain important additional information about the cosmic mass distribution, since the peculiar velocities are induced by this distribution which is itself correlated with galaxy positions.

To see how the pattern of galaxy clustering is distorted in redshift space, consider a simple spherical perturbation in an Einstein–de Sitter universe with initial overdensity profile $\bar\delta_{\rm i}(r) \propto r^{-\gamma}$ where $\gamma > 0$, and $\bar\delta_{\rm i}(r)$ is the mean overdensity within r. The evolution of each mass shell follows the spherical collapse model described in §5.1. For a large radius within which the overdensity is small, the expansion of the mass shell is decelerated but its peculiar velocity is still too small to compensate for the Hubble expansion. In redshift space the mass shell will then appear squashed

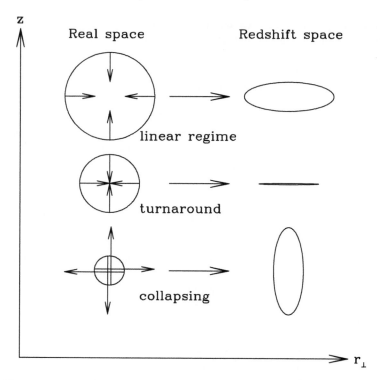

Fig. 6.3. An illustration of how peculiar velocities distort the galaxy distribution in redshift space in different regimes.

along the line-of-sight when observed from a distance much larger than its size. This effect is illustrated at the top of Fig. 6.3. A mass shell with linear overdensity $\delta_0 \sim 1$ is just turning around at the time it is observed, so its peculiar infall velocity is exactly equal to the Hubble expansion velocity across its radius. In redshift space this shell appears completely 'collapsed' to an observer at large distance, as shown in the middle panel of Fig. 6.3. Finally, a mass shell which has already turned around has a peculiar infall velocity which exceeds the Hubble expansion across its radius. If this infall velocity is less than twice the Hubble expansion velocity, the shell appears flattened along the line-of-sight, but with the nearer side having larger redshift distance than the farther side. At smaller radii the peculiar infall velocities of collapsing shells are much larger than the relevant Hubble velocites and are randomised by scattering effects. The structure then appears to be elongated along the line-of-sight in redshift space (a 'finger-of-God' pointing back to the observer). This is depicted in the bottom of Fig. 6.3. Clearly, redshift distortions have different observational consequences on different scales.

To examine such effects quantitatively, we sample the underlying density field by particles (galaxies) and denote the number density in real space by $n(\mathbf{r})$ and that in redshift space by $n^{(s)}(\mathbf{s})$. The conservation of particle number then implies that

$$n^{(s)}(s)\,\mathrm{d}^3\mathbf{s} = n(\mathbf{r})\,\mathrm{d}^3\mathbf{r}, \qquad (6.75)$$

or, in terms of density contrast,

$$\left[1 + \delta^{(s)}(\mathbf{s})\right]\mathrm{d}^3\mathbf{s} = [1 + \delta(\mathbf{r})]\,\mathrm{d}^3\mathbf{r}, \qquad (6.76)$$

6.3 Clustering in Real Space and Redshift Space

where we have used the fact that the mean number density is the same in both spaces. Since **s** and **r** are related by

$$\mathbf{s} = \mathbf{r} + v_r \hat{\mathbf{r}}, \tag{6.77}$$

we have

$$1 + \delta^{(s)}(\mathbf{s}) = \frac{r^2}{(r+v_r)^2}\left(1 + \frac{\partial v_r}{\partial r}\right)^{-1}[1+\delta(\mathbf{r})]. \tag{6.78}$$

Note that this equation is just a consequence of number conservation and applies in both linear and nonlinear regimes. Now let us apply this equation in the linear regime where $\delta(\mathbf{r}) \ll 1$. In this case the peculiar velocity field is related to the density field by

$$\delta(\mathbf{r}) = -\frac{\nabla_\mathbf{r} \cdot \mathbf{v}}{\beta}, \quad \text{where} \quad \nabla_\mathbf{r} \equiv \frac{\partial}{\partial \mathbf{r}}. \tag{6.79}$$

This is the same as Eq. (6.70), except that H is now absorbed into $r = Hd$ and that $f(\Omega_m)$ is replaced by $\beta \equiv f(\Omega_m)/b$ to take into account a possible bias of $n(\mathbf{r})$ relative to the underlying mass density field [i.e. we assume $\delta \equiv (n-\bar{n})/\bar{n}$ to be b times the mass density contrast]. Hence $|\partial v_r/\partial r| \ll 1$ in the linear regime. Assuming $|v_r| \ll r$ and keeping terms up to linear order, we can write Eq. (6.78) as

$$\delta^{(s)}(\mathbf{r}) = \delta(\mathbf{r}) - \left(\frac{\partial}{\partial r} + \frac{2}{r}\right)v_r. \tag{6.80}$$

In the linear regime, we may assume the curl of **v** to be zero, so that $\mathbf{v} = \nabla_\mathbf{r}\Phi_v$. We can then use Eq. (6.79) to formally write

$$v_r = -\beta \frac{\partial}{\partial r}\nabla_\mathbf{r}^{-2}\delta, \tag{6.81}$$

where $\nabla_\mathbf{r}^{-2}$ is the inverse Laplacian. It then follows that

$$\delta^{(s)}(\mathbf{r}) = \left[1 + \beta\left(\frac{\partial^2}{\partial r^2} + \frac{2}{r}\frac{\partial}{\partial r}\right)\nabla_\mathbf{r}^{-2}\right]\delta(\mathbf{r}). \tag{6.82}$$

Under the assumption that the scale sizes of perturbations in consideration are much smaller than their distances to us [the plane-parallel approximation originally proposed by Kaiser (1987)], we can expand the perturbations in a local Cartesian coordinate system. Choosing the z-axis along the radial direction and using the fact that the term containing $(2/r)$ in Eq. (6.82) can be neglected in the plane-parallel approximation, we can write Eq. (6.82) in Fourier space as

$$\delta_\mathbf{k}^{(s)} = \left(1 + \beta\mu_\mathbf{k}^2\right)\delta_\mathbf{k}, \quad \text{where} \quad \mu_\mathbf{k} \equiv k_z/k. \tag{6.83}$$

The power spectrum in redshift space is then related to that in real space by

$$P^{(s)}(\mathbf{k}) = \left(1 + \beta\mu_\mathbf{k}^2\right)^2 P(k). \tag{6.84}$$

In general we can expand $P^{(s)}(\mathbf{k})$ in harmonics of $\mu_\mathbf{k}$:

$$P^{(s)}(\mathbf{k}) = \sum_\ell \mathscr{P}_\ell(\mu_\mathbf{k})P_\ell^{(s)}(k), \quad P_\ell^{(s)}(k) = \frac{2\ell+1}{2}\int_{-1}^1 P^{(s)}(\mathbf{k})\mathscr{P}_\ell(\mu_\mathbf{k})\,\mathrm{d}\mu_\mathbf{k}, \tag{6.85}$$

where $\mathscr{P}_\ell$ is the ℓth Legendre polynomial. In the linear regime and under the plane-parallel approximation, $P^{(s)}(\mathbf{k})$ is given by Eq. (6.84) and the non-zero harmonics are

$$P_0^{(s)}(k) \equiv \left(1 + \frac{2}{3}\beta + \frac{1}{5}\beta^2\right)P(k), \tag{6.86}$$

$$P_2^{(s)}(k) = \left(\frac{4}{3}\beta + \frac{4}{7}\beta^2\right) P(k), \quad P_4^{(s)}(k) = \frac{8}{35}\beta^2 P(k). \tag{6.87}$$

Note that $P_0^{(s)}(k)$ is just $P^{(s)}(\mathbf{k})$ averaged over the position angle of $\mathbf{k}$:

$$P^{(s)}(k) \equiv \frac{1}{2}\int_{-1}^{1} P^{(s)}(\mathbf{k})\,d\mu_\mathbf{k} = P_0^{(s)}(k). \tag{6.88}$$

Thus, the power spectrum is enhanced in redshift space by a factor which depends only on β. The reason is simple: the infall peculiar velocity towards high density regions, which is proportional to β in the linear regime, makes the density contrast appear larger in redshift space.

The relations in Eqs. (6.86) and (6.87) can be used to estimate the value of β. From a redshift sample of galaxies we can estimate $P^{(s)}(\mathbf{k})$ as a function of $(k,\mu_\mathbf{k})$. The harmonics $P_0^{(s)}(k)$, $P_2^{(s)}(k)$ and $P_4^{(s)}(k)$ can then be estimated from Eq. (6.85), which allows the value of β to be obtained from the ratios between these harmonics. Applying this procedure to real galaxy samples gives $\beta \sim 0.5$ for optical galaxies and $\beta \sim 0.8$ for IRAS galaxies, indicating that optical and IRAS galaxies are differently biased. In practical applications the scheme described above can be improved. For example, one can relax the plane-parallel assumption by expanding perturbation quantities in spherical harmonics (e.g. Fisher et al., 1994b). One can also go beyond the linear regime by using either the Zel'dovich approximation or high-order perturbation theory to relate peculiar velocities with density fluctuations (e.g. Fisher & Nusser, 1996; Hatton & Cole, 1998).

6.3.2 Real-Space Correlation Functions

Redshift distortions can also be formulated in terms of the two-point correlation function. In redshift space the two-point correlation function is defined by

$$\xi^{(s)}(\mathbf{s}_1,\mathbf{s}_2) = \left\langle \delta^{(s)}(\mathbf{s}_1)\delta^{(s)}(\mathbf{s}_2) \right\rangle. \tag{6.89}$$

Because of peculiar velocities, this two-point correlation function is anisotropic and in general depends on both $\mathbf{s}_1$ and $\mathbf{s}_2$. However, under the plane-parallel assumption, the dependence is only through the value of $\mathbf{s} = \mathbf{s}_1 - \mathbf{s}_2$, and the orientation of $\mathbf{s}$ relative to the line-of-sight. In this case, we may write $\xi^{(s)}(\mathbf{s}_1,\mathbf{s}_2)$ as $\xi^{(s)}(r_\pi, r_p)$, where r_π and r_p are the (redshift-space) separations parallel and perpendicular to the line-of-sight, defined as

$$r_\pi = \frac{\mathbf{s}\cdot\mathbf{l}}{|\mathbf{l}|}, \quad r_p = \sqrt{s^2 - r_\pi^2}, \tag{6.90}$$

with $\mathbf{l} = (\mathbf{s}_1+\mathbf{s}_2)/2$. Equivalently, we can also write $\xi^{(s)}(\mathbf{s}_1,\mathbf{s}_2)$ as $\xi^{(s)}(s,\mu)$, where $\mu = r_\pi/s$. We can then expand $\xi^{(s)}(s,\mu)$ in harmonics of μ for given s. In the linear regime,

$$\xi^{(s)}(s,\mu) = \xi_0^{(s)}(s)\mathscr{P}_0(\mu) + \xi_2^{(s)}(s)\mathscr{P}_2(\mu) + \xi_4^{(s)}(s)\mathscr{P}_4(\mu), \tag{6.91}$$

where

$$\xi_0^{(s)}(s) = \left(1 + \frac{2}{3}\beta + \frac{1}{5}\beta^2\right)\xi(s), \tag{6.92}$$

$$\xi_2^{(s)}(s) = \left(\frac{4}{3}\beta + \frac{4}{7}\beta^2\right)\left[\xi(s) - \frac{3J_3(s)}{s^3}\right], \tag{6.93}$$

$$\xi_4^{(s)}(s) = \frac{8}{35}\beta^2\left[\xi(s) + \frac{15}{2}\frac{J_3(s)}{s^3} - \frac{35}{2}\frac{J_5(s)}{s^5}\right], \tag{6.94}$$

with $\xi(s)$ the real-space two-point correlation function evaluated at separation s, and J_ℓ defined in Eq. (6.65) (see Hamilton, 1992). These relations can also be used to infer the value of β. For example, from $\xi^{(s)}(r_\pi, r_p)$ one can determine the quadrupole-to-monopole ratio

$$q(s) = \frac{\xi_2^{(s)}(s)}{\frac{3}{s^3} \int_0^s \xi_0^{(s)}(s') s'^2 \, ds' - \xi_0^{(s)}(s)}, \qquad (6.95)$$

where

$$\xi_l^{(s)}(s) = \frac{2l+1}{2} \int_{-1}^{1} \xi^{(s)}(r_\pi, r_p) \mathscr{P}_l(\mu) \, d\mu. \qquad (6.96)$$

In the linear regime (i.e. at large s), we can use Eqs. (6.92)–(6.94) to write

$$q(s) = \frac{-\frac{4}{3}\beta - \frac{4}{7}\beta^2}{1 + \frac{2}{3}\beta + \frac{1}{5}\beta^2}. \qquad (6.97)$$

Thus, the value of the quadrupole-to-monopole ratio at large s can be used to infer β. An application of this method to the 2dFGRS yields $\beta \sim 0.5$ (Hawkins et al., 2003).

In order to get the real-space function ξ from $\xi^{(s)}(r_p, r_\pi)$, we define the projected correlation function

$$w(r_p) \equiv \int_{-\Delta s}^{\Delta s} \xi^{(s)}(r_p, r_\pi) \, dr_\pi. \qquad (6.98)$$

Note that $w(r_p)$ has the same units as r_π. If Δs is sufficiently large so as to include almost all correlated pairs, and if the density field is statistically isotropic in real space, we expect $w(r_p)$ to be related to the real-space correlation function $\xi(r)$ by

$$w(r_p) = 2 \int_0^\infty \xi\left[(r_p^2 + y^2)^{1/2}\right] dy = 2 \int_{r_p}^\infty \xi(r) \frac{r \, dr}{(r^2 - r_p^2)^{1/2}} \qquad (6.99)$$

in the plane-parallel approximation. Note that this relation does not require ξ to be small. The Abel integral in Eq. (6.99) can be inverted to give

$$\xi(r) = -\frac{1}{\pi} \int_r^\infty \frac{dw(y)}{dy} \frac{dy}{\sqrt{y^2 - r^2}}. \qquad (6.100)$$

If $w(r_p)$ can be modeled as a power law,

$$w(r_p) = A r_p^{1-\gamma}, \qquad (6.101)$$

then the real-space correlation function is also a power law,

$$\xi(r) = \left(\frac{r_0}{r}\right)^\gamma, \quad \text{with} \quad r_0^\gamma = \frac{A \Gamma(\gamma/2)}{\Gamma(1/2) \Gamma[(\gamma-1)/2]}, \qquad (6.102)$$

where $\Gamma(x)$ is the gamma function.

To (partly) recover the peculiar velocity field on nonlinear scales, one can model $\xi^{(s)}(r_p, r_\pi)$ as a convolution of the real-space correlation function, $\xi(r)$, with the distribution function of the relative velocity between a pair of particles at the separation in question:

$$\xi^{(s)}(r_p, r_\pi) = \int_{-\infty}^{\infty} \xi(r) f(u) \, du, \quad r \equiv \sqrt{r_p^2 + (r_\pi - u/H_0)^2}, \qquad (6.103)$$

where u is the projection of the relative (pairwise) peculiar velocity $\mathbf{v}_{12}$ *along the line-of-sight*, and $f(u)$ is its distribution function [normalized so that $\int f(u) \, du = 1$]. The form of $f(u)$ is not known a priori. Based on observations and theoretical considerations (e.g. Davis & Peebles,

1983; Fisher et al., 1994a; Diaferio & Geller, 1996; Sheth, 1996), an exponential form is usually adopted:

$$f(u) = \frac{1}{\sqrt{2}\,\sigma_u(r)} \exp\left[-\frac{\sqrt{2}}{\sigma_u(r)}|u - \langle u \rangle(r)|\right], \qquad (6.104)$$

where $\langle u \rangle(r)$ is the mean and $\sigma_u(r)$ the dispersion of u at the separation r. For small separations where the velocity dispersion σ_u is much larger than the systematic infall velocity $\langle u \rangle$, the redshift-space correlation function is smeared on velocity scale σ_u. In this case the Fourier transform of Eq. (6.103) gives

$$P^{(s)}(\mathbf{k}) = \tilde{f}(k\mu_\mathbf{k})P(k), \qquad (6.105)$$

where $\tilde{f}(k\mu_\mathbf{k})$ is the Fourier transform of $f(u)$. Assuming σ_u to be a constant, $\tilde{f}(k\mu_\mathbf{k})$ has the form

$$\tilde{f}(k\mu_\mathbf{k}) = \frac{1}{1 + (\sigma_u k\mu_\mathbf{k})^2/2}, \qquad (6.106)$$

for $f(u)$ given by Eq. (6.104).

If the data are sufficiently good, we can hope to recover both $\sigma_u(r)$ and $\langle u \rangle(r)$ from Eq. (6.103) together with the real-space correlation function $\xi(r)$ estimated from the projected correlation function. In practice, it is usually necessary to adopt a model for $\langle u \rangle(r)$. One such model often adopted has the form

$$\langle u \rangle(\mathbf{r}) = -\frac{r_\| H_0}{1 + (r/r_0)^\gamma}, \qquad (6.107)$$

where $r_\| = |r_\pi - u/H_0|$ is the separation in real space along the line-of-sight. This model is based on the self-similarity solution to be discussed in §6.4.2, and is a good approximation to the real infall pattern in CDM models (Jing et al., 1998).

6.4 Clustering Evolution

The observed structure in the present-day Universe is already nonlinear on scales smaller than $\sim 8h^{-1}$Mpc (§2.7). A theory of nonlinear gravitational clustering is thus required to study the properties of the cosmological density field on smaller scales. Nonlinear evolution substantially modifies the properties of the original linear density field. In particular, the evolved field will no longer be Gaussian even if the initial field was. In this section we study these modifications. Since our observations are primarily of the *evolved* field, it is necessary to understand these modifications in order to compare theory with observation.

6.4.1 Dynamics of Statistics

So far the evolution of the cosmic density field has been described in terms of the density contrast $\delta(\mathbf{x})$ (or its Fourier transform), and we have derived statistical measures based on the evolved density and velocity fields. Alternatively, one can also study the evolution of the cosmic density field by solving the dynamical equations for the statistical measures themselves. This approach is described extensively by Peebles (1980). In what follows we give a brief overview without going into detailed derivations.

(a) Pair Conservation Equation Consider an evolving point distribution (e.g. the galaxy distribution). At time t (when the scale factor is a), the average number of neighbors within comoving distance x of a particle is

$$N(x,t) = 4\pi \bar{n}(t) a^3 \int_0^x [1+\xi(y,a)] y^2 \, dy, \tag{6.108}$$

where $\bar{n}(t)$ is the mean particle number density at time t. This follows directly from the interpretation of ξ given in §6.1.1. Now suppose that the mean relative peculiar velocity of particle pairs with (comoving) separation x is $\langle v_{12}(x,a) \rangle$, where v_{12} is defined as $v_{12}(\mathbf{x}) \equiv [\mathbf{v}(\mathbf{y}) - \mathbf{v}(\mathbf{x}+\mathbf{y})] \cdot \hat{\mathbf{x}}$. The average particle flux through a spherical shell of radius x centered on a given particle is $4\pi \bar{n} a^2 x^2 [1+\xi(x,a)] \langle v_{12}(x,a) \rangle$. It then follows from mass conservation that

$$\frac{\partial N(x,t)}{\partial t} + 4\pi \bar{n} a^2 x^2 [1+\xi(x,a)] \langle v_{12}(x,a) \rangle = 0. \tag{6.109}$$

Since $\bar{n} a^3 = $ constant, we have,

$$\frac{\langle v_{12}(x,a) \rangle}{H(a)ax} = -\frac{1}{[1+\xi(x,a)]} \frac{a}{x^3} \frac{\partial}{\partial a} \int_0^x \xi(y,a) y^2 \, dy. \tag{6.110}$$

This equation is based on pair conservation and applies both in linear and nonlinear regimes. If $\xi(x,a)$ is known, the above equation can be used to predict the time evolution of the mean relative peculiar velocity for particle pairs at any given separation.

(b) Cosmic Energy Equation Since peculiar velocities are induced by the forces due to density fluctuations, the mean specific kinetic energy of particles must be related to their mean specific gravitational potential energy. The relation between the two is the Layzer–Irvine equation:

$$\frac{d}{dt}(K+W) + \frac{\dot{a}}{a}(2K+W) = 0, \tag{6.111}$$

where

$$K = \frac{1}{2} \langle v_1^2 \rangle, \quad W = -\frac{1}{2} G \bar{\rho} a^2 \int \frac{\xi(x)}{x} d^3 \mathbf{x}, \tag{6.112}$$

with $\langle v_1^2 \rangle$ the mean square of the peculiar velocities of particles. This equation can be derived as follows. Consider a set of mass particles, m_j ($j = 1, 2, \ldots$), with their phase-space coordinates labeled by $(\mathbf{x}_j, \mathbf{p}_j)$, where $\mathbf{x}$ denotes the comoving spatial coordinate, and $\mathbf{p} = a^2 \dot{\mathbf{x}}$ denotes the canonical momentum conjugate to $\mathbf{x}$ (see §4.1.4). The Hamiltonian of the system is then $H = M(K+W)$, where $M = \sum m_j$ is the total mass of the system. For fixed $\mathbf{x}_j$ and $\mathbf{p}_j$, we have $W \propto a^{-1}$ and $K = M^{-1} \sum p_j^2/(2 m_j a^2) \propto a^{-2}$. It is then easy to see that the energy equation, $dH/dt = \partial H/\partial t$ (where the partial derivative is for fixed $\mathbf{x}_j$ and $\mathbf{p}_j$), leads to Eq. (6.111). Writing Eq. (6.111) in the form

$$\frac{d}{da} a^2 \langle v_1^2 \rangle = \frac{\partial}{\partial \ln a} G \bar{\rho} a^3 \int_0^\infty \frac{\xi(x,a)}{x} d^3 \mathbf{x}, \tag{6.113}$$

and integrating once, we get

$$\langle v_1^2 \rangle = \frac{3}{2} \Omega_m(a) H^2(a) a^2 I_2(a) \left[1 - \frac{1}{a I_2(a)} \int_0^a I_2(a') \, da' \right], \tag{6.114}$$

where

$$I_2(a) \equiv \int_0^\infty \xi(x,a) x \, dx. \tag{6.115}$$

Note that $\langle v_1^2 \rangle$ is mass-weighted, because the number density of particles is proportional to the underlying mass density. This is different from the bulk-motion velocity described in §6.2.1, which is volume-weighted.

(c) Pairwise Peculiar Velocity Dispersions In Chapter IV of Peebles (1980) it is shown that the pairwise peculiar velocity $\mathbf{v}_{12} \equiv \mathbf{v}(\mathbf{x}_1) - \mathbf{v}(\mathbf{x}_2)$ obeys

$$\frac{\partial}{\partial t}(1+\xi)\langle v_{12}{}^i\rangle + \frac{\dot{a}}{a}(1+\xi)\langle v_{12}{}^i\rangle + \frac{1}{a}\sum_j \frac{\partial}{\partial x^j}(1+\xi)\langle v_{12}{}^i v_{12}{}^j\rangle$$

$$+ \frac{2Gm}{a^2}\frac{x^i}{x^3}(1+\xi) + 2G\overline{\rho}a\frac{x^i}{x^3}\int_0^x \xi\, d^3\mathbf{x}$$

$$+ 2G\overline{\rho}a\int \zeta(1,2,3)\frac{x_{13}{}^i}{x_{13}^3}d^3\mathbf{x}_3 \quad = 0, \quad (6.116)$$

where $x = |\mathbf{x}_2 - \mathbf{x}_1|$, $\zeta(1,2,3)$ is the three-point correlation function, and m is the mass of a particle. The fourth term describes the mutual attraction of a particle pair and can be neglected when we consider dark matter particles (for which m is small). It is convenient to write the velocity-dispersion tensor as

$$\langle v_{12}{}^i v_{12}{}^j\rangle = \left[\frac{2}{3}\langle v_1^2\rangle + \Sigma\right]\delta_{ij} + [\Pi - \Sigma]\frac{x^i x^j}{x^2}, \quad (6.117)$$

where Π and Σ represent the effects of correlated motion on the components of $\mathbf{v}_{12}$ that are parallel and perpendicular to $\mathbf{x} = \mathbf{x}_2 - \mathbf{x}_1$, respectively. Thus, the relative pairwise velocity projected along the line defined by the pair has a dispersion $\sigma_{12}^2 = \frac{2}{3}\langle v_1^2\rangle + \Pi - \langle v_{12}\rangle^2$. Assuming that $\xi(r \to \infty) = 0$ and $\Pi(r \to \infty) = 0$ and setting the i-axis to be along $\mathbf{x}_2 - \mathbf{x}_1$, one can solve for σ_{12}^2 by inserting Eq. (6.117) into Eq. (6.116) and integrating the latter over x^i from x to ∞ (Mo et al., 1997a). At small separations, where $\langle v_{12}\rangle = 0$ and the contribution from the three-point correlation (term containing ζ) dominates, one obtains

$$\sigma_{12}^2(r) = \frac{3\Omega_m H^2}{4\pi[1+\xi(r)]}\int_r^\infty \frac{dr}{r}\int d^3\mathbf{q}\frac{\mathbf{r}\cdot\mathbf{q}}{q^3}\zeta(r,q,|\mathbf{r}-\mathbf{q}|). \quad (6.118)$$

This is the cosmic virial theorem, and it applies on small scales where the structure is in a statistically static state. Note that Ω_m, H, ξ and ζ, and thus also σ_{12}, all depend on the scale factor a.

6.4.2 Self-Similar Gravitational Clustering

To see how gravitational clustering proceeds in the nonlinear regime, let us consider a simple example where the structure in the Universe grows in a self-similar way, namely the density field looks similar at different times once scaled by a time-dependent characteristic length scale. To achieve self-similar evolution, two conditions must be satisfied: (i) the background cosmology should not contain any characteristic scales, and (ii) the linear perturbation spectrum should be scale-free. These conditions are satisfied for a density field with power-law linear spectrum $[P(k) \propto k^n]$ in an Einstein–de Sitter universe. Since the nonlinear scale, $r^*(t)$, defined in Eq. (5.202) is the only characteristic scale involved in the problem, we expect that all statistical measures of the density field are independent of time when length scales are expressed in units of $r^*(t)$. For example, the two-point correlation function should scale as

$$\xi(x,t) = \xi(y) \text{ with } y \equiv x/r^*(t), \quad (6.119)$$

where x is the comoving separation, and

$$r^*(t) \propto a^\alpha \propto t^{2\alpha/3}, \quad \alpha = 2/(n+3). \quad (6.120)$$

Thus, under the assumption of self-similarity, the time evolution of ξ is completely determined once the functional form of ξ is obtained at an arbitrary time. It is easy to see that Eq. (6.119) applies in the linear regime; that it also applies in the nonlinear regime is a result of the self-similar assumption.

The conditions for self-similar clustering are not expected to hold rigorously in reality. As described in §4.4.4, all realistic models have perturbation spectra that contain characteristic scales at late times, even though the primordial spectrum is scale-free. Nevertheless, the idea of self-similar clustering provides a useful way to understand how nonlinear gravitational clustering proceeds. The growth of the nonlinear scale $r^*(t)$ with time and the structure of the self-similar solutions are exactly what is expected for idealized hierarchical clustering. Furthermore, in many models self-similar solutions are expected to hold approximately over limited ranges of length and time.

As an application, we can use the self-similar solution to construct a simple model for the slope of the two-point correlation function in the nonlinear regime. In terms of the scaled variable y, the pair conservation equation Eq. (6.110) can be written as

$$\frac{\langle v_{12}(x,a)\rangle}{H(a)ax} = \frac{\alpha}{[1+\xi(y)]}\left[\xi(y) - \frac{3}{y^3}\int_0^y \xi(y')y'^2\,dy'\right], \tag{6.121}$$

where x and y are related according to Eq. (6.119). If we approximate ξ as a power law, $\xi(y) \propto (y/y_0)^{-\gamma}$, then the above equation gives

$$\langle v_{12}\rangle(y) = -\frac{\alpha\gamma}{(3-\gamma)}\frac{Har^*y}{1+(y/y_0)^\gamma}. \tag{6.122}$$

For small separations where $\xi \gg 1$, it is reasonable to assume that the clumps of mass particles are bound and stable (stable clustering), i.e. there is no net streaming motion between particles in *physical* coordinates. In this case, $\langle \mathbf{u}_{12}\rangle = \dot{a}\mathbf{x}_{12} + \langle \mathbf{v}_{12}\rangle = 0$, or

$$\langle v_{12}(x,a)\rangle = -H(a)ax. \tag{6.123}$$

It then follows from Eq. (6.122) that $\gamma = 3/(1+\alpha)$, so that

$$\xi(y) \propto y^{-\gamma} \propto a^{(3-\gamma)}x^{-\gamma} \text{ with } \gamma = \frac{3n+9}{n+5}. \tag{6.124}$$

This result should be compared to that given by linear theory:

$$\xi(y) \propto y^{-\gamma} \propto a^2 x^{-\gamma} \text{ with } \gamma = 3+n. \tag{6.125}$$

We see that the nonlinear power spectrum resulting from an initial power-law spectrum with index n evolves self-similarly with an effective power index

$$n_{\text{eff}} \equiv \gamma - 3 = -\frac{6}{n+5}. \tag{6.126}$$

Note that for $-2 \leq n \leq +1$, the corresponding effective index only covers the relatively narrow range $-2 \leq n_{\text{eff}} \leq -1$.

It is sometimes useful to define the time evolution of the two-point correlation function in terms of the correlation amplitude at fixed *proper* separation $r = ax$. Such evolution is conventionally parameterized in the form

$$\xi(r,z) = \xi(r,0)(1+z)^{-(3+\varepsilon)}, \tag{6.127}$$

where z is the redshift at which the correlation function is measured, and ε parameterizes the rate of evolution. From Eqs. (6.124) and (6.125) we see that in an Einstein–de Sitter universe $\varepsilon = 0$ for stable clustering, and $\varepsilon = \gamma - 1$ for linear evolution. Another interesting case is when all particles are pasted on the expanding background (i.e. all peculiar velocities are zero). In this case $\xi(x,a)$ is independent of a for fixed x, so that $\xi(r,a) \propto a^\gamma r^{-\gamma}$ and $\varepsilon = \gamma - 3$.

6.4.3 Development of Non-Gaussian Features

As shown in §6.1.3, the high-order connected moments, κ_ℓ ($\ell > 2$), and the high-order connected correlation functions, ζ, η, etc., are all zero for a Gaussian density field. We saw in §4.4.2 that a Gaussian density field should remain Gaussian during linear evolution. However, non-Gaussian features can develop from an initially Gaussian density field due to nonlinear evolution, as can be understood from the high-order perturbation theory described in §4.1.7. For a density field which is initially Gaussian, $\langle [\delta^{(1)}]^3 \rangle = 0$ (see §4.1.7 for definitions). Thus, the lowest order contribution to the third moment, $\kappa_3 = \langle \delta^3 \rangle$, must have the form

$$\kappa_3 = 3 \left\langle [\delta^{(1)}]^2 \delta^{(2)} \right\rangle \propto \left\langle [\delta^{(1)}]^4 \right\rangle, \tag{6.128}$$

where we have used that $\delta^{(2)} \propto [\delta^{(1)}]^2$. Since $\delta^{(1)}$ has a Gaussian distribution, we have from Eq. (6.42) that $\langle [\delta^{(1)}]^4 \rangle = 3 \langle [\delta^{(1)}]^2 \rangle^2$. We can therefore write $\kappa_3 = S_3 \kappa_2^2$, where S_3 is some constant. Similar arguments can also be made for the higher-order moments. In general, we can write

$$\kappa_\ell = S_\ell \kappa_2^{\ell-1} \quad (\ell = 3, 4, 5, \ldots), \tag{6.129}$$

where the amplitudes S_ℓ are constant in time, and the values of S_ℓ can be calculated using the high-order perturbation theory outlined in §4.1.7.

For instance, in an Einstein–de Sitter universe where

$$\delta_{\mathbf{k}}^{(1)}(t) \propto a(t) \propto t^{2/3} \quad \text{and} \quad \mathbf{v}_{\mathbf{k}}^{(1)} = (i a \mathbf{k}/k^2) \dot{\delta}_{\mathbf{k}}^{(1)} \propto t^{-1/3},$$

Eq. (4.91) can be solved to second order to give

$$\delta^{(2)}(\mathbf{x}) = \sum_{\mathbf{k},\mathbf{k}'} \left[\frac{5}{7} - \frac{\mathbf{k} \cdot \mathbf{k}'}{k^2} + \frac{2}{7} \left(\frac{\mathbf{k} \cdot \mathbf{k}'}{kk'} \right)^2 \right] \delta_{\mathbf{k}}^{(1)} \delta_{\mathbf{k}'}^{(1)} e^{i(\mathbf{k}+\mathbf{k}') \cdot \mathbf{x}}. \tag{6.130}$$

With this expression, one finds that the skewness of the density fluctuation

$$\langle \delta^3 \rangle = 3 \left\langle [\delta^{(1)}]^2 \delta^{(2)} \right\rangle = \frac{34}{7} \langle \delta^2 \rangle^2, \quad \text{so that} \quad S_3 = \frac{34}{7}. \tag{6.131}$$

This result applies to the original field, $\delta(\mathbf{x})$, that is not smoothed. If, for example, $\delta(\mathbf{x})$ is smoothed with a top-hat window of radius R, then

$$\langle \delta^3(R) \rangle = \sum_{\mathbf{k},\mathbf{k}'} \left[\frac{5}{7} - \frac{\mathbf{k} \cdot \mathbf{k}'}{k^2} + \frac{2}{7} \left(\frac{\mathbf{k} \cdot \mathbf{k}'}{kk'} \right)^2 \right] P(k) P(k')$$
$$\times \widetilde{W}_{\text{th}}(kR) \widetilde{W}_{\text{th}}(k'R) \widetilde{W}_{\text{th}}(|\mathbf{k}+\mathbf{k}'|R), \tag{6.132}$$

which can be evaluated to give

$$\langle \delta^3(R) \rangle = \left(\frac{34}{7} + \gamma_1 \right) \langle \delta^2(R) \rangle^2, \quad \text{where} \quad \gamma_1 \equiv \frac{d \ln \sigma^2(R)}{d \ln R} \tag{6.133}$$

(Bernardeau, 1994). The extra term, γ_1, arises because the contribution to the mass moment at a given smoothing radius R comes from various mass scales. For instance, if $\sigma(R_0)$ (where R_0 corresponds to a mass $M = 4\pi \overline{\rho} R_0^3/3$) decreases with increasing M in the linear density field for R_0 around R (so that $\gamma_1 < 0$), then the contribution to the mass moments at R in the evolved density field from high-density regions should be smaller than that in the case where $\sigma(R_0)$ is independent of R_0. Consequently, the skewness of the smoothed field is reduced relative to the original field.

Similar calculations can be carried out for the higher-order moments of the smoothed field (Bernardeau, 1994). The values for the first few S_ℓ parameters are

$$S_3 = \frac{34}{7} + \gamma_1, \tag{6.134}$$

$$S_4 = \frac{60712}{1323} + \frac{62\gamma_1}{3} + \frac{7\gamma_1^2}{3} + \frac{2\gamma_2}{3}, \tag{6.135}$$

$$S_5 = \frac{200575880}{305613} + \frac{1847200\gamma_1}{3969} + \frac{6940\gamma_1^2}{63} + \frac{235\gamma_1^3}{27}$$
$$+ \frac{1490\gamma_2}{63} + \frac{50\gamma_1\gamma_2}{9} + \frac{10\gamma_3}{27}, \tag{6.136}$$

where

$$\gamma_n = \frac{d^n \ln \sigma^2(R)}{d(\ln R)^n}. \tag{6.137}$$

The results for the corresponding unsmoothed field can be obtained by setting $\gamma_n = 0$. The above results are derived assuming an Einstein–de Sitter universe, but the dependence on cosmology is found to be weak.

6.5 Galaxy Clustering

Since galaxies are the only population of cosmic objects that are abundant and are sufficiently bright to be observable at cosmological distances, they have long been used as tracers of the large-scale structure of the Universe. As discussed in §6.1.2, it is not guaranteed that galaxies are faithful tracers of the mass distribution. Nevertheless, it is generally believed that the relation between the mass and galaxy distributions is sufficiently simple that the study of galaxy clustering can provide useful information about the cosmic mass distribution. Furthermore, the study of galaxy clustering is also important for understanding galaxy formation, as it may provide clues about how galaxies influence each other as they form and at later times. In this section we describe how to characterize galaxy clustering statistically, and we present some results based on current observations.

To study the spatial clustering of galaxies, one usually starts with a galaxy sample for which sky positions and redshifts are listed for all members. Such a sample typically provides a non-uniform sampling of the true galaxy distribution in redshift space (and thus a distorted image of their distribution in real space) throughout a finite volume defined by the survey boundaries. Galaxies are much smaller than the scales normally of interest when studying galaxy clustering, and so can be considered as points in redshift space. The observational criteria which define real samples typically induce strong selection effects. For example, only the intrinsically brightest galaxies may be included at large distances whereas relatively faint systems may be included in the foreground. Such effects must be carefully taken into account when using real samples to study galaxy clustering.

Modeling of observational samples usually starts by defining a selection function which describes which galaxies are included in the sample. Mathematically, the selection function, which we denote by $S(\mathbf{x})$, is defined as the probability that a 'random' galaxy located near $\mathbf{x}$ (relative to us) is included in the sample. Clearly $S(\mathbf{x}) = 0$ if the direction of $\mathbf{x}$ lies outside the observed region of the sky. In general, $S(\mathbf{x})$ also varies strongly with $|\mathbf{x}|$ because of survey apparent magnitude limits. Thus the distribution of galaxies in real surveys has strong inhomogeneities resulting from survey construction which must be removed if the underlying physical clustering is to be measured properly. As a simple example, consider a *magnitude-limited* sample which selects all galaxies (within the observed sky area) with apparent magnitudes brighter than some limit $m_{\lim}$. A galaxy with apparent magnitude m, located at redshift z, has absolute magnitude $\mathcal{M} = m - 5\log[d_L(z)/\text{Mpc}] - 25$, where $d_L(z)$ is the luminosity distance corresponding to redshift z, and $\mathcal{M}$ is a logarithmic measure of luminosity L (see §2.1.1). The magnitude limit, $m_{\lim}$, therefore corresponds at redshift z to a luminosity limit,

$$\log(L_{\lim}/L_\odot) = 2\log[d_L(z)/\text{Mpc}] - 0.4(m_{\lim} - \mathcal{M}_\odot) + 10. \tag{6.138}$$

Only galaxies at z brighter than this luminosity limit are included in the sample. In this case the selection function can be written as

$$S(z) = \frac{\bar{n}(z)}{\bar{n}_0} = \frac{\int_{L_{\lim}(z)}^{\infty} \phi(L)\,dL}{\int_0^{\infty} \phi(L)\,dL}, \tag{6.139}$$

where $\phi(L)$ is the galaxy luminosity function, here assumed to be independent of spatial location. Once $S(z)$ is known, we can calculate the fraction of galaxies that is missed as a function of redshift. This can then be used to correct the clustering measurements for selection effects.

From a magnitude-limited sample, one can also construct a *volume-limited* subsample which is complete out to redshift z by taking all galaxies with luminosities brighter than $L_{\lim}(z)$. This is obtained simply by discarding all galaxies fainter than $L_{\lim}(z)$ or more distant than z. This results in a sample for which no further corrections for selection effects in the redshift direction are required. The shortcoming of this scheme is that it generally requires a large fraction of the entire survey to be discarded.

6.5.1 Correlation Analyses

(a) Two-Point Function The two-point correlation function of galaxies can be estimated from a redshift survey as

$$\xi(r) = \frac{DD(r)}{RR(r)} - 1, \tag{6.140}$$

where $DD(r)\Delta r$ is the observed number of galaxy–galaxy pairs with separations in the range $r \pm \Delta r/2$, and $RR(r)\Delta r$ is the expected number of such pairs in a 'random' (i.e. uniform, Poisson distributed) sample with the same number of objects filling the same volume and with the same selection function as the real sample. So defined, $\xi(r)$ is zero if galaxies are distributed as a uniform Poisson process; any deviation from zero indicates spatial clustering. This definition is motivated by the idea that galaxies can be treated as a Poisson sampling of a smooth underlying density field, as described in §6.1.1. Indeed, if the number density of galaxies at each spatial location is proportional to the underlying density field, then $\xi(r)$ as defined above will be the same as the two-point correlation function of the density field. Other estimators of $\xi(r)$ are possible. Three that are frequently used can be written symbolically as

$$\xi(r) = \frac{DD}{DR} - 1; \quad \xi(r) = \frac{DD \cdot RR}{DR^2} - 1; \quad \xi(r) = \frac{DD - 2DR + RR}{RR}, \tag{6.141}$$

where $DR(r)$ is the number of cross-pairs between the real and random samples. For real applications, the last two estimators are usually preferred because they are less affected by sample boundaries (e.g. Hamilton, 1993). In general, it is often advantageous to construct random samples that contain many more particles than there are galaxies in the real sample. In this case, RR has to be multiplied by $(N_g/N_r)^2$ and DR by N_g/N_r, where N_g and N_r are the number of galaxies in the real and the random samples, respectively.

Note that ξ so defined does not depend on the mean number density of galaxies in the sample, as long as the sample is a *fair* representation of the galaxy distribution in the Universe. Thus, if we have two (fair) samples with different mean densities, we can count the total number of pairs in the two samples separately, and estimate $\xi(r)$ from the total pair counts. Because of this, one can estimate $\xi(r)$ by counting pairs directly from a sample regardless of any selection effects, provided that these selection effects are taken into account when constructing random samples. This property of ξ also gives us the freedom to assign different weights to galaxies in different regions, as long as the clustering properties of galaxies are statistically the same in different regions. For instance, in a magnitude-limited sample, the number density of galaxies is higher in the nearby region than farther away, and the pair counts may be dominated by nearby galaxies. This is not desirable, because most of the weight is given to a relatively small volume which may not turn out to be typical. To overcome this problem, we may assign each galaxy a weight proportional to $1/S(z)$, where z is the redshift of the galaxy, so that all volume elements within the surveyed region have the same weight. The drawback of this weighting scheme is that it increases sampling noise, because the volume elements at larger redshifts are more sparsely populated. In practice, some compromise may be adopted. One possibility is to weight each galaxy pair of separation r at a mean redshift z by W^2, with

$$W(r;z) = \frac{1}{1 + 4\pi \bar{n}(z) J_3(r)}, \tag{6.142}$$

where J_3 is defined in Eq. (6.9). The motivation for this weighting scheme is that on average each galaxy at a redshift z has $N_c = 4\pi \bar{n}(z) J_3(r)$ correlated neighbors within a radius r (i.e. each correlated 'cluster' contains N_c galaxies). This formula has the desired property that it weights nearby regions (where N_c is large and so individual 'clusters' are oversampled) by volume and regions at large distances (where N_c is small) by number. The drawback is that the weights assigned to pairs depend on the correlation function to be estimated, and so an iterative procedure has to be used.

The weighting scheme described above has another serious problem. Because in reality galaxies with higher luminosity are more strongly correlated (see the left panel of Fig. 6.4), the

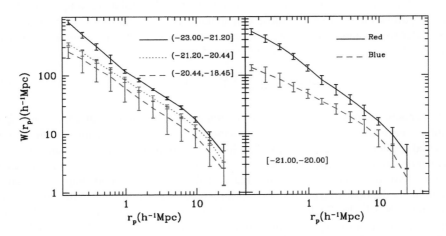

Fig. 6.4. The projected correlation function of galaxies in the SDSS. The left-hand panel shows the results for three separate luminosity bins. The corresponding ranges in $\mathcal{M}_r - 5\log h$ (here $\mathcal{M}_r$ is the SDSS r-band K-corrected to a redshift $z = 0.1$) are indicated in brackets. Note that brighter galaxies are more strongly clustered. The right-hand panel shows the results for red and blue galaxies with $-20 \geq \mathcal{M}_r - 5\log h > -21$; clearly red galaxies are more strongly clustered than blue galaxies. [Kindly provided by Y. Wang, based on data published in Wang et al. (2007c)]

correlation function estimate obtained from a galaxy sample can be affected by the weighting scheme adopted. For example, in an apparent-magnitude-limited sample, galaxies at higher redshift are on average brighter, and so a weighting scheme that gives more weight to galaxies at higher redshift would result in a correlation function with higher amplitude. Furthermore, since galaxy pairs with larger separations are more dominated by galaxies at larger distances (hence with larger luminosities), such a weighting scheme can also lead to an overestimate of the correlation amplitude on large scales relative to that on small scales, changing its slope. In order to avoid this problem, the simplest way is to use volume-limited samples, so that no weighting scheme is required. Alternatively, one may assume that the cross-correlation function between galaxies with luminosity L_1 and L_2, $\xi(r; L_1, L_2)$, is related to the correlation function of a reference population (e.g. L^* galaxies), $\xi(r; L^*)$, by $\xi(r; L_1, L_2) = b(L_1)b(L_2)\xi(r; L^*)$. One can then properly weight each galaxy according to the value of $b(L)$ appropriate for its luminosity so that all galaxies can be considered to sample the same correlation function (e.g. that of L^* galaxies) (e.g. Percival et al., 2004; Tegmark et al., 2004).

The error in the estimate of ξ is in general difficult to model. Ideally, we would like to estimate this error from the variance of an ensemble of independent realizations, each of which is statistically equivalent to the galaxy sample in question. Unfortunately, this is impractical, as we have only one realization, our Universe. In practice, one may use mock samples constructed from independent realizations of a realistic model of structure formation. The shortcoming here is that the result is model dependent. Because of this, 'internal' errors estimated directly from the data sample itself are still widely used.

One simple estimate is based on the assumption that the fluctuation in the number of independent pairs in a given bin of r has a Poisson distribution: $\delta N_{\mathrm{pair}} = N_{\mathrm{pair}}^{1/2}$. In this case, it is evident that the error in ξ is

$$\delta \xi(r) = \frac{1 + \xi(r)}{\sqrt{N_{\mathrm{pair}}(r)}}. \tag{6.143}$$

However, this method, which was adopted in many early studies, is found to significantly underestimate the true uncertainty in the data (e.g. Mo et al., 1992). Another class of 'internal' error estimators is based on dividing the data sample into a set of N similar subsamples in space. The simplest one, often referred to as the subsample method, is to use the variance among all the subsamples, under the assumption that each of them is an independent realization of the underlying distribution. The covariance matrix is then estimated from

$$C(\xi_i, \xi_j) = \frac{1}{N} \sum_{k=1}^{N} \left(\xi_i^k - \overline{\xi}_i\right)\left(\xi_j^k - \overline{\xi}_j\right), \tag{6.144}$$

where ξ_i^k denotes the measurement of ξ at separation r_i from the kth subsample, and $\overline{\xi}$ is the expectation value estimated independently from the N subsamples. The problem with this method is that in real applications the subsamples, each with a volume much smaller than the total sample, may not be mutually independent because of the existence of long-range modes in the density fluctuations. Two other commonly used estimators in this class are the bootstrap method and jackknife method. In the bootstrap method, one forms a set of N_{rs} resamplings of the original sample, each containing N galaxies (including duplicates) randomly picked from the original N galaxies with replacement (i.e. a galaxy is retained in the stack even if it has already been picked). Thus, although each resample consists of the same number of galaxies as the original sample, it will include some of the galaxies more than once, while others may not be included at all. In this case, the covariance matrix is estimated from

$$C(\xi_i, \xi_j) = \frac{1}{N_{\mathrm{rs}}} \sum_{k=1}^{N_{\mathrm{rs}}} \left(\xi_i^k - \overline{\xi}_i\right)\left(\xi_j^k - \overline{\xi}_j\right), \tag{6.145}$$

where $\overline{\xi}$ is the mean obtained from the N_{rs} resamplings. In practice for large samples, one chooses $N_{rs} \ll N$ but still large enough to provide a good measurement of the covariance matrix. In the jackknife method, one forms a set of N 'copies' of the original sample, each time leaving out one of the N galaxies. The covariance matrix is then estimated from

$$C(\xi_i, \xi_j) = \frac{N-1}{N} \sum_{k=1}^{N} \left(\xi_i^k - \overline{\xi}_i\right) \left(\xi_j^k - \overline{\xi}_j\right), \qquad (6.146)$$

where ξ_i^k is the measurement of ξ at separation r_i in the kth 'copy' and $\overline{\xi}$ is the mean of the N copies. However, for large samples the jackknife approach as outlined above is impractical, since excluding a single galaxy has almost no effect. An alternative is to divide the full sample into N_{rs} disjoint subsamples, each containing N/N_{rs} galaxies. One can then proceed as before, estimating a correlation function for N_{rs} jackknife 'copies' of the original sample obtained by leaving out one subsample at a time. A test of the performance of error estimators of this kind is given in Norberg et al. (2009).

The two-point correlation function of galaxies has been estimated from a variety of redshift surveys. In order to take care of redshift distortions, a common practice is to first estimate the two-dimensional function $\xi(r_\pi, r_p)$ (see §6.3.2), and then obtain the real-space correlation function from the projected function $w(r_p)$. For optically selected galaxies with $L \sim L^*$, the observed correlation function at $r \lesssim 20\,h^{-1}\mathrm{Mpc}$ can be approximated by a power law,

$$\xi(r) = \left(\frac{r}{r_0}\right)^{-\gamma}, \quad \text{with } r_0 \approx 5\,h^{-1}\mathrm{Mpc} \text{ and } \gamma \approx 1.7, \qquad (6.147)$$

as illustrated in Fig. 2.37 (see also Davis & Peebles, 1983; Jing et al., 1998; Norberg et al., 2002a; Zehavi et al., 2002). This indicates that the galaxy distribution becomes highly nonlinear on scales $\lesssim 5\,h^{-1}\mathrm{Mpc}$. As shown in Fig. 6.4, bright, red galaxies are more strongly clustered than faint, blue ones (see the discussion in §15.6).

(b) Three-Point Function The particle sampling of the density field described in §6.1.1 motivates the following definition of the three-point correlation function:

$$\zeta(r_{12}, r_{23}, r_{31}) = \frac{DDD}{RRR} - \xi(r_{12}) - \xi(r_{23}) - \xi(r_{31}) - 1, \qquad (6.148)$$

where $DDD(r_{12}, r_{23}, r_{31})$ is the number of triplets with their three sides covering the ranges $(r_{12}, r_{23}, r_{31}) \pm (dr_{12}, dr_{23}, dr_{31})/2$, and RRR is the corresponding number in the random sample. Largely for historical reasons (e.g. Groth & Peebles, 1977), the observed three-point correlation function of galaxies is usually written in the following 'hierarchical' form:

$$\zeta(r_1, r_2, r_3) = Q(r_1, r_2, r_3) \left[\xi(r_1)\xi(r_2) + \xi(r_2)\xi(r_3) + \xi(r_3)\xi(r_1)\right]. \qquad (6.149)$$

In general, the value of Q for galaxies depends on the shape of the triplets in question but only weakly on their size (e.g. Jing & Börner, 1998). For r_1, r_2 and r_3 near $\sim 1\,\mathrm{Mpc}$, Q is roughly a constant, $Q \sim 1$.

(c) Pairwise Peculiar Velocity Dispersion As discussed in §6.3.2, modeling the redshift-space distortions in the two-point correlation function allows one to obtain the pairwise peculiar velocity dispersion of galaxies σ_u. Recent estimates based on large redshift surveys give $\sigma_u \sim 500\,\mathrm{km\,s^{-1}}$ at $r_p \sim 1\,h^{-1}\mathrm{Mpc}$ (Jing et al., 1998; Hawkins et al., 2003; Zehavi et al., 2002). An accurate estimate of σ_u can provide important constraints on structure formation. Assuming the hierarchical form given by Eq. (6.149), the cosmic virial theorem (6.118) can be written as

$$\langle \sigma_{12}^2(r) \rangle^{1/2} \approx 850 Q^{1/2} \frac{\Omega_{m,0}^{1/2}}{b} \left(\frac{r_0}{5\,h^{-1}\mathrm{Mpc}}\right)^{\gamma/2} \left(\frac{r}{1\,h^{-1}\mathrm{Mpc}}\right)^{1-\gamma/2} \mathrm{km\,s^{-1}}. \qquad (6.150)$$

If we take $Q = 1$, the observed value of $\sigma_{12} \approx \sigma_u$ for galaxies at $1\,h^{-1}\mathrm{Mpc}$ would require $\Omega_{\mathrm{m},0}^{0.5}/b \sim 0.5$. This result, obtained from galaxy clustering on small scales, is consistent with the result, $\beta \equiv \Omega_{\mathrm{m},0}^{0.6}/b$, obtained from the large-scale velocity field (§6.2).

(d) Counts-in-Cells and Moments As shown in §6.1.3, the moments of the counts-in-cells for a point process are directly related to the correlation functions after correction for shot noise. Galaxy counts-in-cells can therefore be used to probe galaxy clustering. For a volume-limited sample, cell counts can be obtained by directly counting the number of galaxies in random cells of a given volume, and the moments of the counts can easily be calculated. Suppose that we have M random cells of volume V, and the number count in the ith cell is N_i. The distribution function of the counts-in-cells can be estimated through

$$f(N;V) = \frac{1}{M}\sum_{i=1}^{M}\delta_{NN_i}, \qquad (6.151)$$

where δ_{NN_i} equals 1 if $N_i = N$ and 0 otherwise. The ℓth moment can then be obtained:

$$\langle N^\ell(V)\rangle = \sum_{N=0}^{\infty} N^\ell f(N;V) = \frac{1}{M}\sum_{i=1}^{M} N_i^\ell. \qquad (6.152)$$

Using the definitions given in §6.1.3, we can then estimate the variance, skewness and other quantities.

Since the discreteness noise depends on the mean number density, $\bar{n}$, one has to be careful to take account of the change of $\bar{n}$ with redshift when using a magnitude-limited sample. One way to deal with this is to divide the sample into thick shells in such a way that the variation of the mean density due to the magnitude limit within each shell is negligible. Counts-in-cells can be carried out within each shell, and the moments of counts can then be corrected for shot noise using the mean density of the shell in consideration. Thus, each shell provides an estimate of the variance σ^2 (or the skewness κ_3, kurtosis κ_4, etc.), and the results from all independent shells can be combined to give an estimate through a likelihood analysis. Such an analysis for σ^2 has been performed by Efstathiou (1995) for a number of galaxy samples. Results for high-order moments can be found in Croton et al. (2004), for example.

6.5.2 Power Spectrum Analysis

Although the power spectrum is just the Fourier transform of the two-point correlation function (see §6.1), it is more advantageous to work with the power spectrum than with the two-point correlation function when studying galaxy clustering on large scales. The reason for this is that on large scales, where the density field is still in the linear regime, different Fourier modes evolve independently while the amplitude of the two-point correlation function is affected by many different modes. Consequently, power-spectrum amplitudes on large scales are less affected by small-scale structure than correlation function estimates, making observational results much easier to interpret.

Given a galaxy sample, it is straightforward to measure the power spectrum. Suppose that the volume of the sample is V_s, which is described by a window function $W(\mathbf{x})$ such that $W(\mathbf{x})$ is equal to 1 if $\mathbf{x} \in V_s$ and to 0 otherwise. Suppose that this volume is contained in a large box V_u on which the Universe is assumed to be periodic. We divide V_u into small cells of volume ΔV so that the galaxy occupation number in each cell, $\mathcal{N}_i$, is either 1 or 0. The observed galaxy density field can then be written as

$$n_o(\mathbf{x}) = \sum_j \mathcal{N}_j \delta^{\mathrm{D}}(\mathbf{x}-\mathbf{x}_j)W_j, \qquad (6.153)$$

where $W_j = W(\mathbf{x}_j)$ and $\delta^D(\mathbf{x})$ is the Dirac delta function. For a magnitude-limited sample, this density must be multiplied by the reciprocal of the selection function to correct for selection effects. The corrected density field is

$$n_c(\mathbf{x}) = A \sum_j \mathcal{N}_j S_j^{-1} \delta^D(\mathbf{x} - \mathbf{x}_j) W_j, \qquad (6.154)$$

where $S_j = S(\mathbf{x}_j)$, and the prefactor $A = \bar{n} V_s / \sum_j \mathcal{N}_j S_j^{-1} W_j$ is included so that $\int n_c(\mathbf{x}) d^3\mathbf{x} = \bar{n} V_s$, with $\bar{n}$ the mean number density of galaxies. Note that A is equal to 1 for the top-hat window considered here and is included in the equation to encompass other possible choices. Fourier transforming Eq. (6.154) with respect to $\mathbf{x}$, we have

$$\tilde{n}_c(\mathbf{k}) = \frac{A}{V_u} \sum_j \mathcal{N}_j S_j^{-1} W_j e^{-i\mathbf{k}\cdot\mathbf{x}_j}. \qquad (6.155)$$

The average of this is

$$\langle \tilde{n}_c(\mathbf{k}) \rangle = \bar{n} \widetilde{W}(\mathbf{k}), \quad \text{where} \quad \widetilde{W}(\mathbf{k}) = \frac{A}{V_u} \int W(\mathbf{x}) e^{-i\mathbf{k}\cdot\mathbf{x}} d^3\mathbf{x} \qquad (6.156)$$

is the Fourier transform of the window function. In obtaining the above relation, we have used that $\langle \mathcal{N}_j \rangle / S_j = \bar{n} \Delta V$. Similarly, using $\mathcal{N}_i^2 = \mathcal{N}_i$ and $\langle \mathcal{N}_i \mathcal{N}_j \rangle = \bar{n}^2 S_i S_j [1 + \xi(\mathbf{x}_i - \mathbf{x}_j)]$, we get

$$\langle \tilde{n}_c(\mathbf{k}) \tilde{n}_c^*(\mathbf{k}) \rangle = \frac{A^2}{V_u^2} \sum_{i,j} \langle \mathcal{N}_i \mathcal{N}_j \rangle S_i^{-1} S_j^{-1} W_i W_j e^{i\mathbf{k}\cdot(\mathbf{x}_i - \mathbf{x}_j)}$$

$$= \frac{A^2}{V_u^2} \sum_j \mathcal{N}_j (W_j / S_j)^2 + \bar{n}^2 |\widetilde{W}(\mathbf{k})|^2 + \frac{\bar{n}^2}{V_u} \sum_{\mathbf{k}'} |\widetilde{W}(\mathbf{k}')|^2 P(\mathbf{k} - \mathbf{k}'), \qquad (6.157)$$

where $P(\mathbf{k}) = \int \xi(\mathbf{x}) e^{-i\mathbf{k}\cdot\mathbf{x}} d^3\mathbf{x}$ is the power spectrum of the galaxy density field. If the survey volume is large in comparison to the wavelengths in consideration, $\widetilde{W}(\mathbf{k}')$ is sharply peaked at $\mathbf{k}' = 0$, and we can replace $P(\mathbf{k} - \mathbf{k}')$ by $P(\mathbf{k})$ in the above equation and pull it out of the summation. Rearranging, we get

$$P(\mathbf{k}) \approx \frac{V_u^2 \langle |\tilde{n}_c(\mathbf{k}) - \langle \tilde{n}_c(\mathbf{k}) \rangle|^2 \rangle - N_{\text{eff}}}{\bar{n}^2 V_u \sum_{\mathbf{k}'} |\widetilde{W}(\mathbf{k}')|^2}, \quad N_{\text{eff}} \equiv \frac{(\bar{n} V_s)^2 \sum_j \mathcal{N}_j (W_j S_j^{-1})^2}{(\sum_j \mathcal{N}_j W_j S_j^{-1})^2}. \qquad (6.158)$$

Note that the summations over j in N_{eff} can be replaced by those over all the galaxies in the sample with $\mathcal{N}_j$ set to 1. In a volume-limited sample, $S_i = 1$ and so $N_{\text{eff}} = \bar{n} V_s$. Eq. (6.158) shows that the true power spectrum is equal to the power spectrum of the corrected number density field, $n_c(\mathbf{x})$, with a subtraction of the shot noise, N_{eff}, and a deconvolution with the window function. The final power spectrum is usually binned in shells in $\mathbf{k}$-space:

$$P(k) = \frac{1}{V_k} \int_{V_k} P(\mathbf{k}') d^3\mathbf{k}', \qquad (6.159)$$

where V_k is the volume of the shell in $\mathbf{k}$-space.

Since the derivation of Eq. (6.158) is independent of the assumption that $W(\mathbf{x})$ is a top-hat, it is valid for any form of $W(\mathbf{x})$. As for the correlation functions, we may want to assign different weights to different regions to reduce the variance in the estimate. This can be done by using a window function that is inhomogeneous in space.

The power spectrum of galaxy clustering has been estimated for various redshift surveys (see Tegmark et al., 2004, for an example). Fig. 6.5 shows power spectrum estimates for optically selected galaxies compiled by Peacock (1997). The comparison with model predictions shows that a CDM model with $\Omega_{m,0} \sim 0.3$ matches the observed large-scale power spectrum, but an anti-bias on small scales is required to match the observed data.

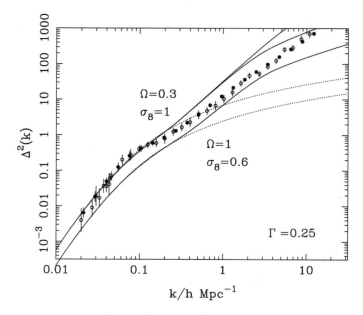

Fig. 6.5. Power spectrum estimates $\Delta^2(k) \equiv k^3 P(k)/(2\pi^2)$ for optically selected galaxies are compared to model predictions for the dark matter power spectrum in three cosmologies, all with $\Gamma = 0.25$: (1) $\Omega_{m,0} = 1$, $\Omega_{\Lambda,0} = 0$, $\sigma_8 = 0.6$ (the lowest solid curve); (2) $\Omega_{m,0} = 0.3$, $\Omega_{\Lambda,0} = 0$, $\sigma_8 = 1$ (the highest solid curve); (3) $\Omega_{m,0} = 0.3$, $\Omega_{\Lambda,0} = 0.7$, $\sigma_8 = 1$ (the middle solid curve). These curves all include a model for nonlinear evolution. The corresponding linear power spectrum for model (1) is the lower dotted curve, while the higher dotted curve shows that for models (2) and (3). [Adapted from Peacock (1997)]

6.5.3 Angular Correlation Function and Power Spectrum

Galaxy redshift samples are usually constructed from deep sky surveys that cover much larger volumes than the redshift samples themselves. Although only two-dimensional information about the galaxy distribution is available in such surveys, one may take advantage of their large volumes. For a two-dimensional sample, we can estimate the angular correlation function of galaxies, $w(\vartheta)$, from the definition:

$$\langle \nu(\hat{\mathbf{r}}_1)\nu(\hat{\mathbf{r}}_2)\rangle \, d\omega_1 \, d\omega_2 = \bar{\nu}^2 \left[1 + w(\vartheta)\right] d\omega_1 \, d\omega_2, \tag{6.160}$$

where $\nu(\hat{\mathbf{r}})d\omega$ is the number of galaxies within a solid angle $d\omega$ in the direction $\hat{\mathbf{r}}$, and $\bar{\nu}$ is the mean surface density. Suppose that the volume density of galaxies (assumed to be in comoving units) at redshift z and in the direction $\hat{\mathbf{r}}$ is $n(\hat{\mathbf{r}}, z)$. The surface density, which is the projection of the three-dimensional distribution on the sky, can be written as

$$\nu(\hat{\mathbf{r}}) = \int_0^\infty n(\hat{\mathbf{r}}, z) S(z) \frac{d^2V}{dz\,d\omega} \, dz, \tag{6.161}$$

where $S(z)$ is the selection function of the survey, and d^2V is the comoving volume element corresponding to dz and $d\omega$ at redshift z (see §3.2.6). Thus,

$$\bar{\nu} = \int_0^\infty \bar{n}(z) S(z) \frac{d^2V}{dz\,d\omega} \, dz, \tag{6.162}$$

$$w(\vartheta) = \frac{\bar{n}^2}{\bar{\nu}^2} \int_0^\infty \bar{n}(z_1)S(z_1)\bar{n}(z_2)S(z_2)\xi(r_{12},z) \frac{d^2V_1}{dz_1\,d\omega_1} \frac{d^2V_2}{dz_2\,d\omega_2} \, dz_1\,dz_2, \tag{6.163}$$

where $\xi(r_{12},z)$ is the spatial correlation function at redshift $z \equiv (z_1+z_2)/2$, r_{12} is the separation between $(\hat{\mathbf{r}}_1,z_1)$ and $(\hat{\mathbf{r}}_2,z_2)$, and $\cos\vartheta = \hat{\mathbf{r}}_1 \cdot \hat{\mathbf{r}}_2$. This is a general relation between the angular and spatial correlation functions, taking into account possible evolutions in $\bar{n}$ and ξ with redshift, as well as cosmological effects. If the survey is not very deep, so that the evolutionary and cosmological effects are all negligible, we can write

$$\bar{\nu} = \bar{n}\int_0^\infty S(r)r^2\,dr, \tag{6.164}$$

$$w(\vartheta) = \frac{1}{\bar{\nu}^2}\int_0^\infty S(r_1)S(r_2)\xi(r_{12})r_1^2 r_2^2\,dr_1\,dr_2, \tag{6.165}$$

where $r_{12}^2 = r_1^2 + r_2^2 - 2r_1 r_2 \cos\vartheta$. In the small-angle limit where ϑ is small and the mean distance of the pair, $x \equiv (r_1+r_2)/2$, is much larger than the value of $y \equiv r_1-r_2$, we can write $r_{12}^2 \approx y^2 + x^2\vartheta^2$ and

$$w(\vartheta) = \int_0^\infty x^4 S^2(x)\,dx \int_{-\infty}^\infty dy\,\xi[(y^2+x^2\vartheta^2)^{1/2}] \bigg/ \left[\int_0^\infty S(x)x^2\,dx\right]^2. \tag{6.166}$$

If $\xi(r)$ is a power law,

$$\xi(r) = A/r^\gamma, \tag{6.167}$$

the angular correlation function is also a power law,

$$w(\vartheta) = B/\vartheta^{\gamma-1}, \tag{6.168}$$

with the amplitude B related to A by

$$B = A\sqrt{\pi}\frac{\Gamma[(\gamma-1)/2]}{\Gamma(\gamma/2)}\int_0^\infty x^{5-\gamma}S^2(x)\,dx \bigg/ \left[\int_0^\infty x^2 S(x)\,dx\right]^2. \tag{6.169}$$

For given γ and A, the amplitude of the angular correlation function depends only on the shape of the selection function. For a magnitude-limited sample, the selection function can be written as $S(x) = f(x/d^*)$, where $d^* \propto 10^{0.2m_{\text{lim}}}$ (with m_{lim} the magnitude limit) is the characteristic depth of the sample, and f is a universal function (under the assumption that cosmological effects and K corrections are small). This form of $S(x)$ in Eq. (6.166) implies that for a given $\xi(r)$ the angular correlation function scales with the sample depth as

$$w(\vartheta;d^*) = \mathscr{W}(\vartheta d^*)/d^*, \tag{6.170}$$

where $\mathscr{W}$ is a scaling function.

The power spectrum can also be estimated for a two-dimensional sky survey of galaxies. Dividing the sky into small cells, $j = 1,2,\ldots$, so that the galaxy occupation number $\mathscr{N}_j$ is either 1 or 0, we can write the observed surface density field as

$$\nu_o(\hat{\mathbf{r}}) = \sum_j \mathscr{N}_j \delta^{(2)}(\hat{\mathbf{r}} - \hat{\mathbf{r}}_j)W_j, \tag{6.171}$$

where $W_j \equiv W(\hat{\mathbf{r}}_j)$ and $W(\hat{\mathbf{r}})$ is a window function specifying the sky coverage of the survey. Expanding $\nu_o(\hat{\mathbf{r}})$ in spherical harmonics,

$$\nu_o(\hat{\mathbf{r}}) = \sum_{\ell,m} a_{\ell m} Y_{\ell m}(\hat{\mathbf{r}}), \tag{6.172}$$

where $Y_{\ell m}(\hat{\mathbf{r}})$ is the spherical harmonic function calculated at the direction $\hat{\mathbf{r}} = (\vartheta,\varphi)$, we have

$$a_{\ell m} = \sum_j \mathscr{N}_j W_j Y_{\ell m}^*(\hat{\mathbf{r}}_j). \tag{6.173}$$

Following the same procedure as described in §6.5.2 and using the properties of the spherical harmonics, one obtains an expression similar to Eq. (6.158) for the angular power spectrum:

$$C_\ell \approx \frac{\langle |a_{\ell m} - \langle a_{\ell m}\rangle|^2\rangle}{\bar{\nu}^2 \langle J_{\ell m}\rangle} - \frac{1}{\bar{\nu}}, \quad (6.174)$$

where

$$\langle a_{\ell m}\rangle = \int W(\vartheta,\varphi) Y_{\ell m}(\vartheta,\varphi)\,\mathrm{d}\omega, \quad \langle J_{\ell m}\rangle = \int W(\vartheta,\varphi) |Y_{\ell m}(\vartheta,\varphi)|^2\,\mathrm{d}\omega, \quad (6.175)$$

and the approximation assumes that the size of the window is much larger than the angular scale corresponding to mode ℓ. The angular power spectrum C_ℓ is related to the angular correlation function by

$$w(\vartheta) = \frac{1}{4\pi}\sum_\ell (2\ell+1) C_\ell \mathscr{P}_\ell(\cos\vartheta), \quad (6.176)$$

where $\mathscr{P}_\ell$ is the Legendre function. Finally, using the relation between the angular and spatial correlation functions, one can relate C_ℓ to the spatial power spectrum $P(k)$:

$$C_\ell = \frac{2}{\pi(2\ell+1)}\int \mathrm{d}k k^2 P(k)\left[\int S(r) j_\ell(kr) r^2 \mathrm{d}r\right]^2 \bigg/ \left[\int S(r) r^2 \mathrm{d}r\right]^2, \quad (6.177)$$

where j_ℓ is the spherical Bessel function.

6.6 Gravitational Lensing

In Fourier space the Poisson equation (4.11) for the potential perturbations can be written in terms of the mass density fluctuations as

$$\Phi_{\mathbf{k}} = -\frac{3}{2}H^2 \Omega_\mathrm{m} a^2 \frac{\delta_{\mathbf{k}}}{k^2}. \quad (6.178)$$

The variance of potential fluctuations in windows of radius R can therefore be written as

$$\langle \Phi^2(R)\rangle = \int_0^\infty \Delta_\Phi^2(k)\widetilde{W}^2(kR)\frac{\mathrm{d}k}{k}; \quad \Delta_\Phi^2(k) = \left(\frac{3}{2}H^2 a^2 \Omega_\mathrm{m}\right)^2 k^{-4}\Delta^2(k). \quad (6.179)$$

Note that Eq. (6.178) does not require $\delta_{\mathbf{k}}$ to be small, so that the above relations apply in both linear and nonlinear regimes.

The potential fluctuations have direct observational consequences, because they affect the geodesics along which photons propagate, thereby distorting our images of distant objects. This effect is called gravitational lensing, because the situation is analogous to that of light deflected by an optical lens. Fig. 2.30 shows the gravitational lensing effects of a cluster. The arcs and arclets around the center of the cluster are strongly distorted images of background galaxies. The galaxy images in the outer region are also distorted, although less strongly. In this section, we examine how such gravitational lensing effects can be used to probe potential fluctuations, and hence the cosmic matter distribution.

6.6.1 Basic Equations

In the matter dominated epoch, a perturbed Robertson–Walker metric can be written as

$$\mathrm{d}s^2 = a^2(\tau)\left[(1+2\Phi/c^2)\mathrm{d}\tau^2 - (1-2\Phi/c^2)\mathrm{d}l^2\right], \quad (6.180)$$

$$\mathrm{d}l^2 \equiv \mathrm{d}\chi^2 + f_K^2(\chi)\mathrm{d}\omega^2; \quad \mathrm{d}\omega^2 = \mathrm{d}\vartheta^2 + \sin^2\vartheta\,\mathrm{d}\varphi^2, \quad (6.181)$$

where Φ is just the Newtonian potential (see Appendix A1.5), and $f_K(\chi)$ is given by Eq. (3.13). In an unperturbed universe where $\Phi = 0$, a photon travels along a null geodesic in the radial direction towards an observer at the origin, with the radial position of the photon given by the comoving distance $\chi(t) = \tau(t_0) - \tau(t)$. Potential fluctuations will change the photon trajectory, but as long as the perturbations are small, i.e. $\Phi \ll c^2$, the deviation of the photon trajectory from the unperturbed geodesic is small. In this limit, we can still use χ to label the perturbed photon path.

Without losing generality, let us consider the propagation of light rays confined to a narrow cone around the polar axis, $\omega \ll 1$. Close to the photon path at a comoving distance χ, we can construct a local Cartesian coordinate system so that $dl^2 = \delta_{ij} dx^i dx^j$ in the metric (6.180). In this frame, the photon geodesic equation (see Appendix A1.3) can be written to first order in Φ/c^2 as

$$\frac{d^2 \mathbf{x}}{d\tau^2} = -\frac{2}{c^2} \nabla \Phi. \tag{6.182}$$

This gives the motion of a light ray in the Newtonian potential Φ; the right-hand side is twice the Newtonian acceleration, because of relativistic effects. It is interesting to note that the above equation is the same as that for a light ray propagating in a medium with a refractive index, $n = 1 - 2\Phi/c^2 > 1$. This can be understood, since the 'effective speed of light' in the frame $(\mathbf{x}, \tau)$ is $dl/d\tau = c(1 + 2\Phi/c^2)$. Note also that the curvature term, f_K, does not show up explicitly in Eq. (6.182), because we are working in the local frame; it will come in when we relate the total deflection angle to distance (see below).

If we define $\mathbf{u} \equiv d\mathbf{x}/d\tau$, then $|\mathbf{u}| \approx c(1 + 2\Phi/c^2)$, and $\hat{\mathbf{u}} \equiv \mathbf{u}/|\mathbf{u}|$ represents the direction of light propagation. Using χ to replace τ as a label along the photon path, and to first order in Φ/c^2, we have from Eq. (6.182) that

$$\frac{d\hat{\mathbf{u}}}{d\chi} = -\frac{2}{c^2} \nabla_\perp \Phi, \tag{6.183}$$

where $\nabla_\perp$ denotes the gradient perpendicular to $\hat{\mathbf{u}}$. This equation gives the rate of change of the propagation direction at χ. Thus, the deflection as the light ray propagates from $\chi \to \chi + \delta\chi$ is $\delta\vec{\alpha}_d = -\frac{2}{c^2} \nabla_\perp \Phi \delta\chi$. Now suppose we have a light source located at a position specified by χ_S and $\mathbf{x}_{\perp,S}$, with the latter being the distance of the source from the unperturbed geodesic. As the light ray propagates from the source to the observer, it is deflected. Since the deflection $\delta\vec{\alpha}_d(\chi)$ at χ leads to a change in the image position at χ_S by $\delta\mathbf{x}_\perp(\chi) = f_K(\chi_S - \chi)\delta\vec{\alpha}_d(\chi)$, where $f_K(\chi_S - \chi)$ is the angular-diameter distance from χ_S to χ (in comoving units), the image position seen by the observer at the origin is

$$\mathbf{x}_{\perp,0} = \mathbf{x}_{\perp,S} - \frac{2}{c^2} \int_{\chi_S}^{0} f_K(\chi_S - \chi) \nabla_\perp \Phi(\chi) d\chi. \tag{6.184}$$

Using the relations $\mathbf{x}_{\perp,0} = \vec{\theta}_0 f_K(\chi_S)$, and $\mathbf{x}_{\perp,S} = \vec{\theta}_S f_K(\chi_S)$, where $\vec{\theta}_S$ and $\vec{\theta}_0$ are the angular positions of the source and image relative to that of the lens, respectively, we obtain

$$\vec{\theta}_S = \vec{\theta}_0 - \frac{2}{c^2} \int_0^{\chi_S} \frac{f_K(\chi_S - \chi)}{f_K(\chi_S)} \nabla_\perp \Phi(\chi) d\chi. \tag{6.185}$$

The above equation describes the mapping between the angular position $\vec{\theta}_S$ of a point in a source at $\chi = \chi_S$ and the angular position $\vec{\theta}_0$ of the corresponding point in its image (see Fig. 6.6). The local properties of this mapping are characterized by the Jacobian matrix:

$$A_{ij}(\vec{\theta}_0, \chi_S) \equiv \frac{\partial \theta_{Si}}{\partial \theta_{0j}} = \frac{1}{f_K(\chi_S)} \frac{\partial x_{Si}}{\partial \theta_{0j}}$$

$$= \delta_{ij} - \frac{2}{c^2} \sum_k \int_0^{\chi_h} g(\chi) \partial_i \partial_k \Phi(\mathbf{x}_\perp, \chi) A_{kj}(\vec{\theta}_0, \chi) d\chi, \tag{6.186}$$

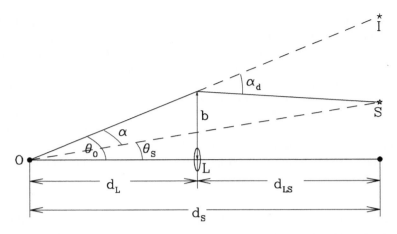

Fig. 6.6. This figure shows a common lensing configuration. The lens is at L, the source at S and the observer at O. Because of gravitational lensing, the image of the source will appear at I. The distances d_L, d_S and d_{LS} are the angular-diameter distances from the observer to the lens, from the observer to the source, and from the lens to the source, respectively.

where χ_h is the horizon radius, and

$$g(\chi) \equiv \frac{f_K(\chi_S - \chi) f_K(\chi)}{f_K(\chi_S)} \mathscr{H}(\chi_S - \chi), \qquad (6.187)$$

with $\mathscr{H}(x)$ the Heaviside step function. Because $\Phi \ll c^2$, we can replace A_{kj} in the integration term of Eq. (6.186) by δ_{kj} and write

$$A_{ij}(\vec{\theta}_0, \chi_S) = \delta_{ij} - \Psi_{,ij}\left[f_K(\chi)\vec{\theta}_0, \chi_S\right], \qquad (6.188)$$

$$\Psi(\mathbf{x}_\perp, \chi_S) = \frac{2}{c^2} \int_0^{\chi_h} d\chi\, g(\chi) \Phi(\mathbf{x}_\perp, \chi), \qquad (6.189)$$

where $\Psi_{,ij} \equiv \partial_i \partial_j \Psi$ is the shear tensor. The convergence κ and shear $\gamma = \gamma_1 + i\gamma_2$ at $\vec{\theta}_0$ are defined as

$$\kappa(\vec{\theta}_0) = \frac{1}{2}\left(\Psi_{,11} + \Psi_{,22}\right), \qquad (6.190)$$

$$\gamma(\vec{\theta}_0) = \frac{1}{2}\left(\Psi_{,11} - \Psi_{,22}\right) + i\Psi_{,12}. \qquad (6.191)$$

In terms of κ and γ, we can write the Jacobian matrix A_{ij} as

$$A(\vec{\theta}_0) = (1-\kappa)\begin{pmatrix} 1 & 0 \\ 0 & 1 \end{pmatrix} - \gamma \begin{pmatrix} \cos\varphi_\gamma & \sin\varphi_\gamma \\ \sin\varphi_\gamma & -\cos\varphi_\gamma \end{pmatrix}, \qquad (6.192)$$

where $\cos\varphi_\gamma \equiv \gamma_1/\gamma$, $\sin\varphi_\gamma \equiv \gamma_2/\gamma$. From this expression, we see that convergence causes an isotropic magnification of angular size in the neighborhood of $\vec{\theta}_0$, while shear produces anisotropy in the mapping. For a small circular source, its lensed image is an ellipse, with major and minor axes

$$a = (1-\kappa-\gamma)^{-1} \quad \text{and} \quad b = (1-\kappa+\gamma)^{-1}, \qquad (6.193)$$

and the magnification in area is

$$\mu \equiv \frac{\text{image area}}{\text{source area}} = \frac{1}{\det(A)} = \frac{1}{[(1-\kappa)^2 - \gamma^2]}. \quad (6.194)$$

As shown in §3.1.6, the apparent surface brightness of a source at a given redshift is independent of the properties of space-time. Gravitational light deflection therefore preserves surface brightness, and so μ defined above is also the magnification of the total flux from a lensed image. Since the Jacobian A is a function of $\vec{\theta}_0$, the equation $\det(A) = 0$ defines curves in the image plane. Such curves are called critical lines. On such lines the image is infinitely magnified ($\mu \to \infty$) and infinitely stretched ($a/b \to \infty$). The lines in the source plane which correspond to the critical lines are known as caustics.

In addition to image distortion and amplification, gravitational lensing also causes time delays. This comes from the fact that, in the presence of lensing, space is curved and the path length from source to observer is increased. The amount of the time delay can be obtained from Eq. (6.180):

$$\Delta t = \frac{2}{c^3} \int_0^{\chi_S} |\Phi(\chi)| \, d\chi. \quad (6.195)$$

6.6.2 Lensing by a Point Mass

To better understand the effects of gravitational lensing, we first consider a simple case where the lens is a point mass. In the neighborhood of the lens we can assume space-time to be Minkowskian with a (small) perturbation caused by the point mass, and so the gravitational potential is

$$\Phi(\mathbf{r}_\perp, z) = -\frac{GM}{(r_\perp^2 + z^2)^{1/2}} \quad (\mathbf{r} = a\mathbf{x}), \quad (6.196)$$

where M is the mass of the lens and z is the distance to the lens projected along the light ray. It then follows that

$$\nabla_{\mathbf{r}_\perp} \Phi(\mathbf{r}_\perp, z) = \frac{GM\mathbf{r}_\perp}{(r_\perp^2 + z^2)^{3/2}} \approx \frac{GM\mathbf{b}}{(b^2 + z^2)^{3/2}}, \quad (6.197)$$

where b is the impact parameter of the unperturbed light ray (see Fig. 6.6), and the approximation is valid because the deflection angle is small (since $\Phi \ll c^2$). The deflection occurs within $\Delta z \sim \pm b$ which, in virtually all applications of astrophysical interest, is much smaller than both the lens-to-source distance and the observer-to-lens distance. In this case, Eq. (6.185) can be written as

$$\vec{\theta}_S \equiv \vec{\theta}(\chi_S) = \vec{\theta}_0 - \frac{2}{c^2} \frac{f_K(\chi_S - \chi_L)}{f_K(\chi_S)} \int_0^{\chi_S} d\chi' \nabla_\perp \Phi(\chi'), \quad (6.198)$$

where $\vec{\theta}_S$ is the position of the source, and $\vec{\theta}_0$ the position of the image, on the sky (see Fig. 6.6). Since $f_K(\chi)$ is just the angular-diameter distance in comoving units (see §3.2.6), we can write

$$\vec{\theta}_S = \vec{\theta}_0 - \vec{\alpha}, \quad (6.199)$$

where

$$\vec{\alpha} = \frac{D_{LS}}{D_S} \vec{\alpha}_d = \frac{d_{LS}}{d_S} \vec{\alpha}_d, \quad (6.200)$$

and

$$\vec{\alpha}_d \equiv \frac{2}{c^2} \int \nabla_\perp \Phi d\chi = \frac{2}{c^2} \int \nabla_{\mathbf{r}_\perp} \Phi(b, z) \, dz \quad (6.201)$$

is the deflection angle (see Fig. 6.6). The distances D_{LS}, D_S and D_L are defined as

$$D_{LS} \equiv f_K(\chi_S - \chi_L), \quad D_S \equiv f_K(\chi_S), \quad D_L \equiv f_K(\chi_L). \tag{6.202}$$

These are *comoving* radial coordinates (angular-diameter distances in comoving units) and can be calculated using the formulae given in §3.2.6. They are related to the angular diameter distances d_{LS}, d_L and d_S according to $D_{LS} = d_{LS}(1+z_S)$, $D_L = d_L(1+z_L)$, and $D_S = d_S(1+z_S)$ with z_L and z_S the redshifts of the lens and source, respectively [see Eqs. (3.101) and (3.117)].

It is useful to define an effective lensing potential

$$\psi(\vec{\theta}_0) = \frac{d_{LS}}{d_L d_S} \frac{2}{c^2} \int \Phi(d_L \vec{\theta}_0, z) \, dz, \tag{6.203}$$

which is related to Ψ defined in Eq. (6.189) by

$$\psi = \Psi/D_L^2. \tag{6.204}$$

It then follows that

$$\vec{\alpha} = \nabla_{\vec{\theta}_0} \psi. \tag{6.205}$$

Eq. (6.199) is called the lens equation, and it is easy to see that it applies not only to point-mass lenses, but also to any other system in which the radial extent of the lens is much smaller than both d_{LS} and d_L. Such systems are called thin-lens systems. The thin-lens approximation is always valid in our discussion except when we consider the lensing effect of large-scale structure (see §6.6.4 below).

Under the same assumption, the time delay for an image at $\vec{\theta}_0$ can be written as

$$\Delta t(\vec{\theta}_0) = \frac{(1+z_L)}{c} \frac{d_L d_S}{d_{LS}} \left[\frac{1}{2} |\vec{\theta}_0 - \vec{\theta}_S|^2 - \psi(\vec{\theta}_0) \right]. \tag{6.206}$$

(a) Multiple Images and Einstein Rings For a point lens, the deflection angle given by Eq. (6.201) is

$$\alpha_d = \frac{4GM}{c^2 b}. \tag{6.207}$$

This in the lens equation (6.199) gives

$$\theta_S = \theta_0 - \frac{\theta_E^2}{\theta_0}, \tag{6.208}$$

where

$$\theta_E = \left[\frac{4GM}{c^2} \frac{d_{LS}}{d_L d_S} \right]^{1/2} \tag{6.209}$$

is the Einstein radius and we have set $b = d_L \theta_0$. If the source is exactly behind the lens, i.e. $\theta_S = 0$, the image is an Einstein ring with radius $\theta_0 = \theta_E$. In the more general case, there are two point images, each on one side of the source, with angular positions

$$\theta_{0\pm} = \frac{1}{2} \left(\theta_S \pm \sqrt{\theta_S^2 + 4\theta_E^2} \right). \tag{6.210}$$

The magnifications of the two images are

$$\mu_\pm = \left[1 - \left(\frac{\theta_E}{\theta_{0\pm}} \right)^4 \right]^{-1} = \frac{u^2 + 2}{2u\sqrt{u^2+4}} \pm \frac{1}{2}, \quad \text{with } u \equiv \frac{\theta_S}{\theta_E}. \tag{6.211}$$

6.6 Gravitational Lensing

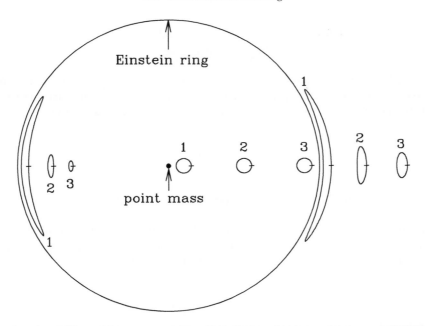

Fig. 6.7. This figure shows the images of extended sources (represented by the three small circles labeled 1, 2 and 3) lensed by a point lens. The images are labeled by the same number as their source. The small horizontal ticks on the sources and the images help to define the parities of the images relative to the sources: while all images to the right of the sources have the same parity as their source, those on the left have the opposite parity.

Note that μ_- is negative, which corresponds to an image with opposite parity to the source. Using Eq. (6.206), the time delay between the two images can also be calculated.

Clearly, the critical line for a point lens is the Einstein ring $\theta_0 = \theta_E$, and the caustic is a single point $\theta_S = 0$. When a source is on the caustic, its image is an Einstein ring. When the source moves away from the caustic, the ring breaks into two images (I_+ and I_-), each on one side of the lens. As the value of θ_S increases, image I_- moves towards the lens and becomes fainter and fainter, while image I_+ moves towards the source and tends towards a magnification of unity.

(b) Arcs If the source is extended and close in direction to the (point-mass) lens, its two images are two arcs with curvature radii approximately equal to the Einstein radius. This is demonstrated in Fig. 6.7 and can be easily understood by using the results for point sources described above.

6.6.3 Lensing by an Extended Object

For an extended lens, the lens equation (6.199) still applies as long as its radial thickness is much smaller than d_{LS} and d_L. In this case, the gravitational potential

$$\Phi(\mathbf{r}_\perp, z) = -\int \frac{G\rho(\mathbf{r}')\,\mathrm{d}^3\mathbf{r}'}{\left[(\mathbf{r}_\perp - \mathbf{r}'_\perp)^2 + (z-z')^2\right]^{1/2}}. \tag{6.212}$$

Under the thin-lens approximation, we can define a lens plane which is perpendicular to the line-of-sight at the lens position d_L. It then follows from Eq. (6.201) that the deflection angle at a point ξ on the lens plane is

$$\alpha_d(\xi) = \frac{4G}{c^2} \int \frac{(\xi - \xi')\Sigma(\xi')}{|\xi - \xi'|^2}\,\mathrm{d}^2\xi', \tag{6.213}$$

where
$$\Sigma(\xi) = \int \rho(\xi, z) \, dz \tag{6.214}$$

is the projected density of the lensing object onto the lens plane. In general, the deflection angle is a two-component vector. In the special case where $\Sigma(\xi)$ is circularly symmetric, we can choose the symmetric point to be the origin of the coordinate system (ξ_1, ξ_2) and write

$$\alpha_d = \frac{4GM(\xi)}{c^2 \xi}, \tag{6.215}$$

where
$$M(\xi) = 2\pi \int_0^\xi \Sigma(\xi') \xi' \, d\xi' \quad \text{and} \quad \xi = \sqrt{\xi_1^2 + \xi_2^2}. \tag{6.216}$$

This is the same as for the point-mass lens given by Eq. (6.207), except that the mass of the point is now replaced by the mass projected within a circle of radius ξ. If we define the mean surface density within ξ as $\overline{\Sigma}(\xi) \equiv M(\xi)/(\pi \xi^2)$, and a critical surface density

$$\Sigma_{\rm crit} = \frac{c^2}{4\pi G} \frac{d_S}{d_L d_{LS}}, \tag{6.217}$$

then the radius of the Einstein ring (the value of $\vec{\theta}_0$ corresponding to $\vec{\theta}_S = 0$) is given by

$$\overline{\Sigma}(\theta_E) = \Sigma_{\rm crit}. \tag{6.218}$$

Clearly, an Einstein ring only exists if the average surface density exceeds the critical density.

(a) Isothermal Spheres For a singular isothermal sphere the density profile is

$$\rho(r) = \frac{\sigma_v^2}{2\pi G r^2}, \tag{6.219}$$

where $\sigma_v = $ constant is the velocity dispersion of particles in the objects, and the surface density is

$$\Sigma(\xi) = \frac{\sigma_v^2}{2G\xi}. \tag{6.220}$$

It is then easy to see that the deflection angle is $\alpha_d = 4\pi \sigma_v^2/c^2$, independent of the impact parameter. For such a lens, the Einstein ring always exists [because $\Sigma(\xi)$ increases without limit as $\xi \to 0$], with a radius

$$\theta_E = 4\pi \frac{\sigma_v^2}{c^2} \frac{d_{LS}}{d_S} = \alpha. \tag{6.221}$$

Thus, for $\theta_S < \theta_E$, i.e. for a point source within the Einstein ring, its two images are at the positions

$$\theta_{0\pm} = \theta_S \pm \theta_E \tag{6.222}$$

on the two sides of the lens. The magnifications of the two images are

$$\mu_\pm = 1 \pm \frac{\theta_E}{\theta_S}. \tag{6.223}$$

For a point source outside the Einstein ring, i.e. $\theta_S > \theta_E$, there is only one image at $\theta_{0+} = \theta_S + \theta_E$. As for a point-mass lens, a singular isothermal sphere has the Einstein ring as its critical line and $\vec{\theta}_S = 0$ as its only caustic.

6.6 Gravitational Lensing

If the central cusp of a singular isothermal sphere is softened so that the surface density is

$$\Sigma(\xi) = \frac{\sigma_v^2}{2G[\xi^2 + \xi_c^2]^{1/2}}, \tag{6.224}$$

where ξ_c is a core radius of the lens, then the effective lensing potential is

$$\psi(\theta_0) = \frac{d_{LS}}{d_S} 4\pi \frac{\sigma_v^2}{c^2} \left(\theta_c^2 + \theta_0^2\right)^{1/2}, \tag{6.225}$$

where $\theta_c = \xi_c/d_L$, and the angle α can be written as

$$\alpha = 4\pi \frac{\sigma_v^2}{c^2} \frac{d_{LS}}{d_S} \frac{\theta_0}{\sqrt{\theta_c^2 + \theta_0^2}}. \tag{6.226}$$

Assuming that ξ_c is small enough for the Einstein ring to exist, then its radius is

$$\theta_E = \left[\left(\frac{4\pi \sigma_v^2}{c^2} \frac{d_{LS}}{d_S}\right)^2 - \theta_c^2\right]^{1/2}. \tag{6.227}$$

In this case, there are two critical lines, one being the Einstein ring given by $\theta_S = 0$, and the other being a circle given by $d\theta_S/d\theta_0 = 0$. The caustic corresponding to the second critical line is a circle, on which a source is also infinitely magnified.

(b) Mass inside a Giant Arc For a circularly symmetric lens, giant arcs are expected from sources close to the inner point-like caustic behind the center. In this case the radius of the arc is approximately the Einstein radius, and the mean surface density enclosed is approximately the critical surface density. Thus, the total mass enclosed by the arc can be estimated as

$$M(\theta_0) = \pi \Sigma_{\rm crit}(d_L \theta_0)^2 \approx 1.1 \times 10^{14} \, {\rm M}_\odot \left(\frac{\theta_0}{30''}\right)^2 \left(\frac{d_L}{d_{LS}}\right) \left(\frac{d_S}{10^3 \, {\rm Mpc}}\right), \tag{6.228}$$

where θ_0 is the angular radius of the arc. This technique has been applied to a number of clusters of galaxies. Mass-to-light ratios inferred are generally high, $\sim 200(M/L)_\odot$, implying that the masses of clusters are dominated by dark matter.

(c) Elliptical Objects If a lens is not circular, the mapping between the source plane and the image plane is more complicated and a wide variety of image configurations can occur. A simple non-circular lens is that given by an elliptical object (e.g. an elliptical galaxy). The effective lensing potential of elliptical galaxies with (close to) singular isothermal density distribution can be approximated by

$$\psi(\vec{\theta}_0) = \frac{d_{LS}}{d_S} 4\pi \frac{\sigma_v^2}{c^2} \left[\theta_c^2 + (1-\varepsilon)\theta_{01}^2 + (1+\varepsilon)\theta_{02}^2\right]^{1/2}, \tag{6.229}$$

where θ_c is an angular core radius and ε is the ellipticity (Blandford & Kochanek, 1987). Note that this potential reduces to that of a softened isothermal sphere [Eq. (6.225)] if we set $\varepsilon = 0$. Fig. 6.8 shows the critical lines (left) and caustics (right) of such an elliptical lens. The most prominent difference from that of a circular lens is that the point-like caustic at $\theta_S = 0$ now becomes diamond shaped. Fig. 6.8 also shows images of compact sources located at different positions relative to the caustics. Sources located close to the inner caustics can produce arc-like images while sources located elsewhere produce multiple images. If a source is extended, we can easily imagine that large arc-like or ring-like patterns can form as its lensed images merge.

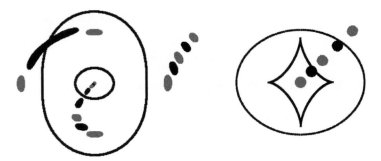

Fig. 6.8. The critical lines (left) and caustics (right) of an elliptical lens. The positions of five sources relative to the caustics and their images are also plotted. The source at the center of the caustic has four images symmetrically positioned relative to the critical lines. [Courtesy of M. Bartelmann]

6.6.4 Cosmic Shear

The principle of lensing by large-scale structures is the same as that described in §6.6.1. Suppose that there is a population of extended sources (galaxies) distributed over the sky. Because of the potential fluctuations caused by the large-scale structures in the Universe, the images of these galaxies are magnified and sheared. Thus, if one could somehow measure the change in the angular sizes and the shapes of these galaxies (relative to their unlensed images), one can obtain the convergence and shear at different positions in the $(\theta_{0,1}, \theta_{0,2})$ image plane. Clearly, the statistical properties of such shear (or convergence) fields are related to the underlying mass density field. The purpose of this subsection is to explore such relations.

Since different galaxies are located at different distances, a distribution function, $\mathscr{P}_\chi(\chi)\,\mathrm{d}\chi$, has to be used to describe the distribution of galaxies in radial distance. For a given image, the probability for its source to be at χ is given by $\mathscr{P}_\chi\,\mathrm{d}\chi$. The ensemble average (over $\mathscr{P}_\chi$) of the shear (or convergence) is then obtained by redefining $g(\chi)$ [Eq. (6.187)] as

$$g(\chi) = f_K(\chi) \int_\chi^{\chi_\mathrm{h}} \frac{f_K(\chi'-\chi)}{f_K(\chi')} \mathscr{P}_\chi(\chi')\,\mathrm{d}\chi'. \tag{6.230}$$

In practice, shear and convergence at a given point are obtained by averaging over the images of many galaxies in the vicinity of the point, so the quantities given by the redefined $g(\chi)$ are closely related to observations.

Take the shear field as an example. From Eqs. (6.191) and (6.189) we have

$$\gamma(\vec{\theta}_0) = \frac{1}{c^2} \int_0^{\chi_\mathrm{h}} \mathrm{d}\chi\, g(\chi) \left[\frac{\partial}{\partial x_1} + i\frac{\partial}{\partial x_2}\right]^2 \Phi(\mathbf{x}_\perp, \chi). \tag{6.231}$$

Since the potential fluctuations we are interested in have length scales much smaller than the curvature radius of the Universe, we can expand the potential field Φ in plane waves:

$$\Phi(\mathbf{x}_\perp, \chi) = \sum_\mathbf{k} \Phi_\mathbf{k} \exp\left(i\mathbf{k}_\perp \cdot \mathbf{x}_\perp + i k_\chi \chi\right). \tag{6.232}$$

This in Eq. (6.231) gives

$$\gamma(\vec{\theta}_0) = -\frac{1}{c^2} \int_0^{\chi_\mathrm{h}} \mathrm{d}\chi\, g(\chi) \sum_\mathbf{k} k_\perp^2 \Phi_\mathbf{k} e^{i k_\chi \chi} e^{i \mathbf{k}_\perp \cdot \mathbf{x}_\perp} \beta_k, \tag{6.233}$$

where $\beta_k = \exp(2i\varphi_k)$, with φ_k the azimuthal angle of $\mathbf{k}_\perp$, i.e. $\tan(\varphi_k) = k_2/k_1$. To first order, $\mathbf{x}_\perp$ in Eq. (6.233) can be replaced by $f_K(\chi)\vec{\theta}_0$.

6.6 Gravitational Lensing

One simple measure of the shear field is the variance of the average shear within some window. For a circular window of radius Θ, this average is

$$\bar{\gamma}(\Theta) = -\frac{1}{\pi\Theta^2} \int_0^\Theta d^2\vec{\theta}_0 \int_0^{\chi_h} d\chi\, g(\chi) \sum_{\mathbf{k}} k_\perp^2 \Phi_{\mathbf{k}} e^{ik_\chi\chi} \exp\left[if_K(\chi)\mathbf{k}_\perp \cdot \vec{\theta}_0\right] \beta_k$$

$$= \int_0^{\chi_h} d\chi\, g(\chi) \sum_{\mathbf{k}} k_\perp^2 \Phi_{\mathbf{k}} e^{ik_\chi\chi} \beta_k W_2[k_\perp f_K(\chi)\Theta], \qquad (6.234)$$

where $W_2(x) = 2J_1(x)/x$ with J_1 the Bessel function. Under the small angle approximation, which is valid in the limit where the source distance, χ_S, is larger than the largest-scale fluctuation we are concerned with, only the modes in the transverse direction contribute to the integral; the contribution from other modes are suppressed by the factor $\exp(ik_\chi\chi)$ owing to the cancellation of positive and negative fluctuations along the line-of-sight. This allows us to put $k_\perp \approx k$. With this approximation, the variance of $\bar{\gamma}(\Theta_0)$ can be written as

$$\langle \bar{\gamma}^2(\Theta_0)\rangle = \sum_{\mathbf{k}}\sum_{\mathbf{k'}} \int_0^{\chi_h} d\chi \int_0^{\chi_h} d\chi'\, g(\chi)g(\chi')e^{i(k_\chi\chi + k_{\chi'}\chi')}\beta_k\beta_{k'}$$

$$\times W_2[kf_K(\chi)\Theta_0]\, W_2[kf_K(\chi')\Theta_0]\, k^2 k'^2 \langle \Phi_{\mathbf{k}}\Phi_{\mathbf{k'}}\rangle. \qquad (6.235)$$

Using Eq. (6.178) and the definition of the power spectrum, we have

$$\langle \bar{\gamma}^2(\Theta_0)\rangle = \int_0^\infty \Delta_\gamma^2(k) W_2^2[kf_K(\chi)\Theta_0]\, \frac{dk}{k}; \qquad (6.236)$$

$$\Delta_\gamma^2(k) = \frac{9\pi}{4}\left(\frac{H_0}{c}\right)^4 \Omega_{m,0}^2 \int_0^{\chi_h} \frac{g^2(\chi)}{a^2(\chi)} \Delta^2(k;\chi)\, \frac{d\chi}{k}, \qquad (6.237)$$

where $a(\chi)$ is the expansion factor. Similarly the two-point correlation function of the shear field is

$$\xi^{(\gamma)}(\theta_0) \equiv \langle \gamma(0)\gamma^*(\vec{\theta}_0)\rangle$$

$$= \int_0^\infty \Delta_\gamma^2(k) J_0[kf_K(\chi)\Theta_0]\, \frac{dk}{k}. \qquad (6.238)$$

Similar statistics can be constructed for the convergence field. From Eq. (6.190) we have

$$\kappa(\vec{\theta}_0) = \frac{1}{c^2}\int_0^{\chi_h} d\chi\, g(\chi)\left[\frac{\partial^2}{\partial x_1^2} + \frac{\partial^2}{\partial x_2^2}\right]\Phi(\mathbf{x}_\perp,\chi). \qquad (6.239)$$

Adding in the bracket a term $\partial^2/\partial x_3^2$ (which cancels out upon the χ integration) and using the Poisson equation for Φ, we obtain

$$\kappa(\vec{\theta}_0) = \frac{3}{2}\left(\frac{H_0}{c}\right)^2 \Omega_{m,0} \int_0^{\chi_h} \frac{g(\chi)}{a(\chi)} \delta\left[f_K(\chi)\vec{\theta}_0,\chi\right] d\chi. \qquad (6.240)$$

So the convergence is just the density perturbation field $\delta(\mathbf{x})$ [in a radial window $g(\chi)/a(\chi)$] projected along the line-of-sight. Under the same assumptions as made for the shear field, the variance and the two-point correlation function (or power spectrum) of the convergence field can all be readily written in terms of the density perturbation spectrum.

In the weak-lensing regime, distortions of individual images are weak, typically a few percent, and the study of the lensing effects has to be based on the coherent patterns in the images of a large number of sources (usually galaxies). In order to obtain significant results, deep and relatively large imaging surveys are required. With such a survey, one can measure the quadrupole moment of individual galaxies:

$$Q_{ij} = \int x_i x_j w(x) I(x_1, x_2) \, dx_1 \, dx_2, \qquad (6.241)$$

where $I(x_1, x_2)$ is the surface brightness distribution of the image and $w(x)$ is a weighting function. The components of this moment can then be combined to form the ellipticities of the image:

$$\varepsilon_1 \equiv \frac{Q_{11} - Q_{22}}{Q_{11} + Q_{22}}; \quad \varepsilon_2 \equiv \frac{2Q_{12}}{Q_{11} + Q_{22}}. \qquad (6.242)$$

So defined, ε_1 describes compressions along the x_1 and x_2 axes, while ε_2 describes that along the two axes at $45°$ from the x_1 and x_2 axes. For a source with circular symmetry, $(\varepsilon_1, \varepsilon_2)$ is a direct measure of the local shear (γ_1, γ_2). Since galaxies in general are not intrinsically spherical, so that their images are elongated even in the absence of lensing, $(\varepsilon_1, \varepsilon_2)$ for a galaxy image is insufficient to estimate the local shear directly. However, under the assumption that the intrinsic ellipticities of the source galaxies are uncorrelated, the local shear can be estimated by averaging the ellipticities of many images in a small patch of the sky. The shear field over the survey area is then represented by many such patches.

To relate the observed shear field to the mass distribution, one also needs to know the source distribution, $\mathscr{P}_\chi(\chi)$. Since the sources are in general faint, spectroscopic redshifts are usually not available. A promising way to overcome this limitation is to use photometric redshifts, as described in §2.6.2.

Shear fields have now been observed around many clusters of galaxies, for example, Abell 2218 shown in Fig. 2.30. These shear fields can be used to estimate cluster masses. The mass-to-light ratios inferred from weak lensing are generally quite high, $\sim 400(M/L)_\odot$, and in agreement with other mass determinations (see Schneider, 2006, for a review). Weak lensing effects on large scales have also been successfully observed in a number of cosmic-shear surveys (see Fig. 2.38). These results can be used to infer the power spectrum of the mass density distribution, and the value of σ_8 (defined in §4.4.4) obtained from such analysis ranges from 0.7 to 1.0, assuming a ΛCDM model with $\Omega_{m,0} = 0.3$ and $\Omega_{\Lambda,0} = 0.7$ (e.g. Refregier, 2003). The existing measurements are limited primarily by statistics. These are expected to improve greatly in the near future by a number of planned surveys dedicated to cosmic shear measurements (see Refregier, 2003, for an overview).

6.7 Fluctuations in the Cosmic Microwave Background

The cosmic microwave background (CMB) is the radiation we receive from the early epoch ($t \sim 380{,}000$ years) when photons decoupled from baryons (see §3.5.2). Density fluctuations present at that time cause fluctuations in the CMB both through their coupling to the radiation field and through their perturbation of the space time metric. In this section, we describe how the observed properties of the CMB can be used to infer the properties of the cosmological density field.

6.7.1 Observational Quantities

As shown in §2.9, observations of the CMB provide us with microwave temperature maps of the sky. We can convert these into maps of the temperature fluctuations,

$$\frac{\Delta T}{T}(\hat{\mathbf{n}}) \equiv \frac{T(\hat{\mathbf{n}}) - \overline{T}}{\overline{T}}, \qquad (6.243)$$

where $\hat{\mathbf{n}} = (\vartheta, \varphi)$ is a direction on the sky and $\overline{T}$ is the mean temperature. Given an all-sky map, we can expand the temperature fluctuations in spherical harmonics,

$$\frac{\Delta T}{T}(\hat{\mathbf{n}}) = \sum_{\ell,m} a_{\ell m} Y_{\ell m}(\vartheta, \varphi). \tag{6.244}$$

Similar to the cosmological density field, the observed CMB sky should be considered as one realization of a cosmic random process. The expectation value of the square of the harmonic coefficients $a_{\ell m}$,

$$C_\ell = \langle |a_{\ell m}|^2 \rangle, \tag{6.245}$$

gives the power spectrum of the temperature fluctuations. Equivalently, we can define the autocorrelation function of the temperature fluctuations as

$$C(\vartheta) = \left\langle \frac{\Delta T}{T}(\hat{\mathbf{n}}_1) \frac{\Delta T}{T}(\hat{\mathbf{n}}_2) \right\rangle, \tag{6.246}$$

which is related to the power spectrum C_ℓ by

$$C(\vartheta) = \frac{1}{4\pi} \sum_\ell (2\ell + 1) C_\ell \mathscr{P}_\ell(\cos \vartheta) \qquad (\cos \vartheta = \hat{\mathbf{n}}_1 \cdot \hat{\mathbf{n}}_2), \tag{6.247}$$

with $\mathscr{P}_\ell$ the Legendre function. As we will see below, in the linear regime the CMB fluctuations are proportional to the density perturbations. Thus, if the linear cosmological density field is Gaussian, so also is the temperature fluctuation field. In this case, the power spectrum (or the autocorrelation function) provides a full statistical description of the temperature fluctuations. To test for departures from Gaussianity, one can use higher-order statistics, as in tests for non-Gaussianity of the density field (see §6.4.3).

The CMB radiation comes to us from the last scattering surface at redshift $z = z_{\text{dec}} \approx 1,100$ (see §3.5.2). A physical event at $z = z_{\text{dec}}$, characterized by a length scale l (in comoving units), therefore subtends an angle

$$\vartheta(l) = l/[d_{\text{A}}(z_{\text{dec}})(1 + z_{\text{dec}})], \tag{6.248}$$

where $d_{\text{A}}(z) = a_0 r(z)/(1+z)$ is the angular-diameter distance to redshift z. For $z \gg 1$, $r(z) \approx 2(H_0 a_0 \Omega_{\text{m},0})^{-1}$ in a universe with $\Omega_\Lambda = 0$, and $r(z) \approx 2(H_0^2 a_0^2 \Omega_{\text{m},0})^{-1/2}$ in a flat universe with $\Omega_{\text{m},0} + \Omega_{\Lambda,0} = 1$, so

$$\vartheta(l) \approx 0.6' \left(\frac{l}{h^{-1}\text{Mpc}} \right) \mathscr{K}(\Omega_{\text{m},0}), \quad \mathscr{K}(\Omega_{\text{m},0}) \equiv \begin{cases} \Omega_{\text{m},0}^{1/2} & (\Omega_{\text{m},0} + \Omega_{\Lambda,0} = 1) \\ \Omega_{\text{m},0} & (\Omega_{\Lambda,0} = 0). \end{cases} \tag{6.249}$$

Thus, for a given length scale, the corresponding angular scale depends on h, $\Omega_{\text{m},0}$ and $\Omega_{\Lambda,0}$. The Hubble radius (in proper units) at $z \gg 1$, $d_{\text{H}}(z) = cH^{-1}(z) \approx H_0^{-1}(\Omega_{\text{m},0} z)^{-1/2} z^{-1}$, subtends an angle

$$\vartheta_{\text{H}}(z) \approx 30° \frac{\Omega_{\text{m},0}^{-1/2} \mathscr{K}(\Omega_{\text{m},0})}{z^{1/2}}. \tag{6.250}$$

At decoupling, this defines an angular scale,

$$\vartheta_{\text{d}} \equiv \vartheta_{\text{H}}(z_{\text{dec}}) \approx 0.87° \, \Omega_{\text{m},0}^{-1/2} \mathscr{K}(\Omega_{\text{m},0}) \left(\frac{z_{\text{dec}}}{1,100} \right)^{-1/2}, \tag{6.251}$$

which depends on the curvature of the Universe, but not on the cosmological constant in a flat universe. On scales larger than ϑ_{d}, the observed temperature fluctuations are entirely due to super-horizon perturbations in space-time, while on scales smaller than ϑ_d evolutionary effects may be relevant. Since the angular scale corresponding to harmonic ℓ is roughly given by

$$\vartheta_\ell \approx \frac{\pi}{\ell}, \tag{6.252}$$

the Hubble radius at decoupling corresponds to $\ell \sim 200\Omega_{m,0}^{1/2}/\mathcal{H}(\Omega_{m,0})$.

6.7.2 Theoretical Expectations of Temperature Anisotropy

Given a cosmogonic model, it is straightforward to calculate the expected temperature fluctuations in the CMB. In fact, the temperature perturbation, at a space-time point $(\mathbf{x},t)$, for photons propagating in a direction $\hat{\mathbf{q}}$ is related to the radiation brightness Δ as

$$\Theta(\hat{\mathbf{q}},\mathbf{x},t) \equiv \frac{\Delta T}{T}(\hat{\mathbf{q}},\mathbf{x},t) = \frac{1}{4}\Delta(\hat{\mathbf{q}},\mathbf{x},t). \tag{6.253}$$

In §§4.2.4–4.2.6 we have described how $\Delta(\hat{\mathbf{q}},\mathbf{x},t)$ evolves in various models of structure formation. For example, we can use the linearized Boltzmann equation (4.170), which assumes a flat background, to write, for a perturbation mode $\mathbf{k}$,

$$(\Theta+\Psi)' + ik\mu(\Theta+\Psi) = (\Phi+\Psi)' + (\partial\Theta/\partial\tau)_C \quad (\mu \equiv \hat{\mathbf{q}}\cdot\hat{\mathbf{k}}), \tag{6.254}$$

where Φ and Ψ are the metric perturbations in the conformal Newtonian gauge (see §4.2.3), a prime denotes a derivative with respect to conformal time $\tau = \int dt/a(t)$, and the collisional term is

$$(\partial\Theta/\partial\tau)_C = \sigma_T n_e a \left(\delta_\gamma/4 + \mathbf{v}_e \cdot \hat{\mathbf{q}} - \Theta + \text{polarization term}\right). \tag{6.255}$$

The temperature fluctuations we observe today (at τ_0) are therefore related to those on the last-scattering surface (at τ_*) by

$$[\Theta+\Psi](k,\mu,\tau_0) = [\Theta+\Psi](k,\mu,\tau_*)e^{ik\mu(\tau_*-\tau_0)}$$
$$+ \int_{\tau_*}^{\tau_0} \left[(\Phi+\Psi)' + \left(\frac{\partial\Theta}{\partial\tau}\right)_C\right]_\tau e^{ik\mu(\tau-\tau_0)}d\tau, \tag{6.256}$$

where the exponential factors describe the phase shifts of the perturbation mode in question. The first term on the right-hand side is due to intrinsic fluctuations in the last-scattering surface, the term containing $(\Phi+\Psi)'$ is the contribution from the change of gravitational potential along the photon path, and the collisional term accounts for possible (non-gravitational) interactions of the CMB photons with baryons after decoupling. This equation can be integrated once the time dependence of the perturbation quantities is solved. The μ dependence can then be expanded in harmonics to obtain the angular power spectrum. Fig. 6.9 shows the predicted power spectrum C_ℓ from several models. In the following, we discuss how the patterns seen in C_ℓ depend on model parameters. For simplicity, we will adopt cold dark matter dominated models to illustrate the main physical effects. The details that are not covered here can be found in Hu & Dodelson (2002), for example. As we will see below, almost all cosmological parameters, such as $\Omega_{m,0}$, $\Omega_{\Lambda,0}$, $\Omega_{b,0}$, h, and $P(k)$, can affect the pattern of the predicted CMB anisotropies in one way or another (see Fig. 6.10 on page 309). It is therefore possible to determine all these parameters through accurate observations of this pattern.

(a) Large-Scale Fluctuations: Sachs–Wolfe Effect For angular scales $\vartheta \gg \vartheta_d$, the density perturbations responsible to the temperature fluctuations have $k \ll 2\pi a/ct_{\text{dec}}$, i.e. their scale sizes are much larger than the Hubble radius at z_{dec}. In this case, the collisional term can be neglected and, if we expand the μ dependence of $\Theta(k,\mu,\tau_*)$ in Legendre polynomials, only the lowest

6.7 Fluctuations in the Cosmic Microwave Background

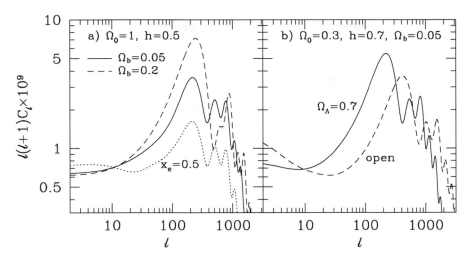

Fig. 6.9. The CMB power spectrum, C_ℓ, predicted by different CDM models. The result marked by $x_e = 0.5$ is for a model where the universe is re-ionized at $z \sim 100$, while the other models assume no re-ionization.

order components have non-negligible amplitudes [see Eqs. (4.178)–(4.182)]. Keeping only the first two components, we have

$$[\Theta + \Psi](k, \mu, \tau_0) = \left[\Psi + \frac{\delta_\gamma}{4} + \Theta_v\right](k, \tau_*)e^{ik\mu(\tau_* - \tau_0)} + \int_{\tau_*}^{\tau_0}(\Phi' + \Psi')e^{ik\mu(\tau - \tau_0)}d\tau, \quad (6.257)$$

where we have used that $\Theta_0 = \delta_\gamma/4$ and $\Theta_v \equiv -3i\Theta_1\mu = -i\theta_\gamma\mu/k = \mathbf{v}\cdot\hat{\mathbf{q}}$.

If Φ and Ψ do not depend on time explicitly (as is the case for linear perturbations in an Einstein–de Sitter universe) so that the second term on the right-hand side of Eq. (6.257) is zero, then the observed temperature fluctuations are caused by the intrinsic fluctuations of the photon density, $\delta_\gamma(\tau_*)$, at the last-scattering surface, the Doppler shift, $(\mathbf{v}\cdot\hat{\mathbf{q}})(\tau_*)$, due to the motion of the last-scattering surface, and the potential difference between the observer and the last scattering surface, $\Psi(\tau_*)$. The temperature fluctuations due to intrinsic and potential fluctuations together (which are sometimes referred to as the effective temperature) are called the Sachs–Wolfe effect. Notice that this decomposition of the Sachs–Wolfe effect into the intrinsic and potential parts is defined in the Newtonian gauge. If another gauge is used, the Sachs–Wolfe effect may correspond to a different combination of perturbation quantities.

Using the results given in §4.2, it is straightforward to relate the Sachs–Wolfe effect with initial perturbations. For simplicity, we assume the last-scattering surface to be a time when the Universe is already matter dominated. In this case, and for isentropic perturbations in the long wavelength limit, we have $\Psi(\mathbf{k}, \tau_*) = \Phi(\mathbf{k}, \tau_*) = 9A(\mathbf{k})/10$, where $A(\mathbf{k}) = \Psi(\mathbf{k}, \tau \to 0)$, $\delta_\gamma(\mathbf{k}, \tau_*) = -(8/3)\Psi(\mathbf{k}, \tau_*)$ and so

$$\Theta^{SW}(k, \mu, \tau_0) = \frac{1}{3}\Psi(\mathbf{k}, \tau_*)e^{ik\mu(\tau_* - \tau_0)}. \quad (6.258)$$

This is the most familiar form of the Sachs–Wolfe effect. Expanding $\Theta^{SW}(k, \mu, \tau_0)$ as

$$\Theta^{SW}(k, \mu, \tau_0) = \sum_\ell (-i)^\ell (2\ell+1)\Theta_\ell^{SW}(k, \tau_0)\mathscr{P}_\ell(\mu), \quad (6.259)$$

we have

$$\Theta_\ell^{SW}(k, \tau_0) = \frac{1}{3}\Psi(\mathbf{k}, \tau_*)j_\ell[k(\tau_0 - \tau_*)], \quad (6.260)$$

where j_ℓ are spherical Bessel functions. In order to link this to the CMB power spectrum, C_ℓ, we use Eq. (6.253) to write

$$\Theta(\hat{\mathbf{n}}, \mathbf{x}_0) = \sum_{\mathbf{k}} \sum_{\ell} (-i)^\ell (2\ell+1) \Theta_\ell^{SW}(k, \tau_0) \mathscr{P}_\ell(\mu) e^{i\mathbf{k}\cdot\mathbf{x}_0}. \tag{6.261}$$

Expanding the $\hat{\mathbf{n}}$ dependence on the left-hand side in harmonics [Eq. (6.244)] and using the fact that $\hat{\mathbf{k}} \cdot \hat{\mathbf{n}} = -\mu$, we obtain

$$C_\ell = \langle |a_{\ell m}|^2 \rangle = \frac{2V_u}{\pi} \int_0^\infty \langle |\Theta_\ell^{SW}(k, \tau_0)|^2 \rangle k^2 \, dk. \tag{6.262}$$

Inserting Eq. (6.260) into this equation and using $\tau_* \ll \tau_0$ we obtain

$$C_\ell = \frac{2V_u}{9\pi} \int_0^\infty \langle \Psi^2(\mathbf{k}, \tau_*) \rangle j_\ell^2(k\tau_0) k^2 \, dk \propto \int_0^\infty \frac{P_i(k)}{k^2} j_\ell^2(k\tau_0) \, dk, \tag{6.263}$$

where $P_i(k) = k^4 \langle |A(\mathbf{k})|^2 \rangle$ is the initial power spectrum. Once $P_i(k)$ is known, it is straightforward to calculate C_ℓ. For a power-law spectrum $P_i(k) \propto k^n$, the above equation gives

$$C_\ell = C_2 \frac{\Gamma\left[\ell + \left(\frac{n-1}{2}\right)\right] \Gamma\left[\frac{9-n}{2}\right]}{\Gamma\left[\ell + \left(\frac{5-n}{2}\right)\right] \Gamma\left[\frac{3+n}{2}\right]}. \tag{6.264}$$

In the special case of a scale-invariant power spectrum, $P_i(k) \propto k$, we have

$$C_\ell \propto \frac{1}{\ell(\ell+1)}, \tag{6.265}$$

and so $\ell(\ell+1)C_\ell$ is independent of ℓ. This explains the behavior of C_ℓ at low ℓ for an $\Omega_{m,0} = 1$ universe shown in Fig. 6.9.

(b) The Large-Scale Doppler Effect As mentioned above, the Θ_v term on the right-hand side of Eq. (6.257) arises from the motion of the last-scattering surface due to density perturbations. On large scale where the collisional term is negligible, we can obtain from Eq. (4.186) that

$$\theta_b = \frac{k^2}{a} \int a\Psi \, d\tau. \tag{6.266}$$

Assuming that the Universe is matter dominated at decoupling, and since the Universe is well approximated by an Einstein–de Sitter model at early epochs, we have $\Phi \sim$ constant, and $\mathbf{v}_b \cdot \hat{\mathbf{q}} \sim i(k\tau_*)\mu\Psi(\mathbf{k}, \tau_*)$. Thus temperature fluctuations due to motions in the last-scattering surface obey

$$\Theta^{\text{Doppler}} \sim k\tau_* \Theta^{SW}. \tag{6.267}$$

Thus, for a scale-invariant power spectrum where $\Theta^{SW} \propto \Psi$ is independent of scale, we have that Θ^{Doppler} scales with the angular size as $\propto \vartheta^{-1}$. In comparison to the Sachs–Wolfe effect, the Doppler effect therefore becomes significant only on scales smaller than the horizon size at decoupling.

Since the temperature fluctuations given by Eq. (6.257) are in the Newtonian gauge, in which the Universe is globally isotropic, they do not include any anisotropy due to the motion of the observer with respect to the CMB. As described in §2.9, such motion produces a strong dipole signal in the CMB map.

6.7 Fluctuations in the Cosmic Microwave Background

(c) The Integrated Sachs–Wolfe Effect If $\Phi + \Psi$ is time-dependent, there is an extra contribution to the temperature fluctuations due to the integral term in Eq. (6.257), which is called the integrated Sachs–Wolfe (ISW) effect. There are two possibilities to generate such an ISW effect in the linear regime. First, in an open universe, or in a flat universe with a cosmological constant, the linear gravitational potential decays with time (see §4.1.6). Because this kind of potential decay occurs only at later times when the curvature and/or cosmological constant become dynamically important, the effect on the CMB is usually referred to as the late ISW effect. Second, since the Universe is not completely matter dominated at the epoch of decoupling, the density perturbations grow with time more slowly than the scale factor [see Eq. (4.88)], which causes the potential to decay until the Universe becomes fully matter dominated. The temperature fluctuations caused by this potential decay are referred to as the early ISW effect because they are produced close to decoupling.

From Eq. (6.257), the contribution of the ISW effect can be written as

$$\Theta_\ell^{\rm ISW}(k,\tau_0) = \int_{\tau_*}^{\tau_0} (\Phi' + \Psi') j_\ell [k(\tau_0 - \tau)] \, d\tau. \tag{6.268}$$

Since the Bessel function j_ℓ peaks at $k(\tau_0 - \tau) \sim \ell$, and since potential evolution is important only for modes within the horizon, the largest effect is typically for modes with $\ell \sim \ell_c \sim (\tau_0 - \tau_c)/\tau_c$ (where τ_c is the time when the potential starts to evolve significantly), although all modes with $\ell > \ell_c$ can be affected. Thus, the early ISW effect is expected to peak roughly at the scale corresponding to the horizon size at decoupling, i.e. at $\ell \sim 200$ in a flat universe.

For a flat ΛCDM universe, the cosmological constant starts to dominate the energy content at a redshift $z \sim (\Omega_{\Lambda,0}/\Omega_{m,0})^{1/3} - 1$, which is about 0.3 for $\Omega_{m,0} = 0.3$. Thus, the late ISW effect is expected to peak roughly at the present horizon scale in this case, i.e. at very low ℓ. For an open universe, the curvature starts to be important at $z \sim \Omega_{m,0}^{-1} - 1$. If $\Omega_{m,0} \sim 0.3$, the late ISW effect is expected to be important again at relatively low ℓ.

(d) Acoustic Peaks On angular scales $\vartheta \lesssim \vartheta_{\rm d}$, the CMB can be affected by a number of non-gravitational effects. One important example is the acoustic oscillations of the baryon–radiation fluid in the pre-recombination era. At decoupling, the sound speed of the baryonic component is about

$$c_{\rm s} = \frac{c}{\sqrt{3(1+\mathscr{R})}}, \quad \mathscr{R} \equiv \frac{3\overline{\rho}_{\rm b}}{4\overline{\rho}_\gamma} \approx 27\Omega_{\rm b,0} h^2 \left(\frac{1+z_{\rm dec}}{1100}\right)^{-1}. \tag{6.269}$$

The Jeans length (see §4.1.3) thus corresponds to a comoving length scale

$$l_{\rm J} = \lambda_{\rm J}(1+z_{\rm dec}) \approx 268(\Omega_{\rm b,0} h^2)^{-1/2}(1+\mathscr{R})^{-1/2}\,{\rm Mpc}. \tag{6.270}$$

With $\Omega_{\rm b,0} h^2 \sim 0.02$, this scale is larger than the horizon size (in comoving units) at decoupling, $l_{\rm H}(z_{\rm dec}) \approx 200(\Omega_{\rm m,0} h^2)^{-1/2}\,{\rm Mpc}$. As a result, all baryonic perturbations that have come through the horizon before decoupling oscillate as acoustic waves in the tightly coupled photon–baryon fluid. Commensurabilities between the periods of these waves and the age of the universe at decoupling give rise to an oscillatory pattern in the CMB power spectrum at angular scales $\vartheta < \vartheta_{\rm c} \sim 1°$ (see Fig. 6.9).

The acoustic wave solution given in §4.1.6 can be used to understand the qualitative behavior of the acoustic pattern in the CMB. In the limit of tight coupling between photons and baryons, $\Theta_0 \equiv \delta_\gamma/4 = \delta_{\rm b}/3$, and we have from Eq. (4.82) that

$$\Theta_0(\tau) + \Psi = [\Theta_0(0) + (1+\mathscr{R})\Psi]\cos(kc_{\rm s}\tau) + \frac{1}{kc_{\rm s}}\Theta_0'(0)\sin(kc_{\rm s}\tau) - \mathscr{R}\Psi, \tag{6.271}$$

where $\Theta_0(0)$ and $\Theta_0'(0)$ are constants to be determined by initial conditions, and the perturbation quantities are in Fourier space. The above solution assumes the gravitational potential Ψ to be constant, which is true only in the matter dominated era in an Einstein–de Sitter universe. Note that Θ_0 is the temperature fluctuation due to the isotropic part of the perturbation in photon number density. The observed temperature fluctuation is, however, produced by the sum of the effective temperature $\Theta_0 + \Psi$ and the Doppler term due to the acoustic velocity,

$$\Theta_v(\tau) = -\frac{3\mu}{k}\Theta_0' = \sqrt{3}c_s[\Theta_0(0) + (1+\mathscr{R})\Psi]\sin(kc_s\tau) - \frac{\sqrt{3}}{k}\Theta_0'(0)\cos(kc_s\tau), \qquad (6.272)$$

where we have used the continuity equation $kv = -(3/4)\delta_\gamma'$ and neglected the time dependence of $\mathscr{R}$. The factor $\sqrt{3}$ is a result of averaging the radial components of randomly oriented velocities, which gives a factor of $1/\sqrt{3}$.

The form of the acoustic oscillation is governed by the two initial conditions, $\Theta_0(0)$ and $\Theta_0'(0)$. The cosine part in Eq. (6.271) represents the isentropic mode, since it is driven by initial metric perturbations; the sine part in Eq. (6.271) represents the isocurvature mode because it corresponds to zero initial metric perturbations. Since the acoustic solution is valid only after the perturbation has come through the horizon, $\Theta_0(0)$ and $\Theta_0'(0)$ should be the values appropriate to the epoch of horizon crossing, i.e. when $\tau \sim \tau_h(k) \equiv 1/(ck)$. Note that the phase of an acoustic wave at horizon crossing, $kc_s\tau_h(k) \sim 1$, is independent of k, and so the zero-point of the phase can be set consistently at horizon crossing.

Consider first the isentropic mode for which $\delta_S \equiv \frac{3}{4}\delta_\gamma - \delta_{\mathrm{dm}} = 0$. As we have seen above, the fluctuations in the effective temperature on super-horizon scales ($k\tau \to 0$) are due to the Sachs–Wolfe effect. Using the fact that $\Psi = \Phi$ are constant for super-horizon perturbations, and using Eq. (4.201) to relate δ_{dm} (in matter dominated case) or $\delta_\gamma = 4\Theta_0$ (in radiation dominated case) to Φ, we may write the initial condition as $(\Theta_0 + \Psi)(0) = \Psi/3$ for perturbations which come through the horizon in the matter dominated era, or $(\Theta_0 + \Psi)(0) = \Psi/2$ for perturbations which come through the horizon in the radiation dominated era. The acoustic waves at the decoupling epoch (assumed to be matter dominated) can therefore be written as

$$\Theta_0(\tau_*) + \Psi = \frac{\Psi}{3}(1+3\mathscr{R})\cos(kl_s) - \mathscr{R}\Psi, \qquad (6.273)$$

where l_s is the size of the sound horizon at decoupling:

$$l_s \equiv c_s(\tau_*)\tau_* \approx 100(\Omega_{\mathrm{m},0}h^2)^{-1/2}(1+\mathscr{R})^{-1/2}\,\mathrm{Mpc}. \qquad (6.274)$$

Note that there is a zero-point shift, $-\mathscr{R}\Psi$, in $\Theta_0(\tau_*) + \Psi$. The corresponding Doppler effect is

$$\Theta_v(\tau_*) = \frac{c_s}{\sqrt{3}}\Psi(1+3\mathscr{R})\sin(kl_s). \qquad (6.275)$$

Therefore, the amplitude of the temperature fluctuation given by a Fourier mode with wavenumber k is determined by Ψ, $\mathscr{R}$, and the phase kl_s of the acoustic wave at the last-scattering surface. If the initial perturbation spectrum does not oscillate significantly for $k \gtrsim 1/l_s$, then the modes of the effective temperature fluctuations with $k = m\pi/l_s$ ($m = 1, 2, \ldots$) have extremal amplitudes:

$$|\Theta_0(\tau_*) + \Psi| = \begin{cases} |\Psi|(1+6\mathscr{R})/3 & (m = \mathrm{odd}) \\ |\Psi|/3 & (m = \mathrm{even}). \end{cases} \qquad (6.276)$$

Hence the amplitude of an m = odd mode is enhanced by a factor of $1 + 6\mathscr{R}$ relative to that of an m = even mode. Since each k is associated with a characteristic scale on the last-scattering surface, these extrema correspond to peaks in the CMB power spectrum C_ℓ. The Doppler term $\Theta_v(\tau_*)$ due to acoustic velocity also produces peaks in the CMB, but with a phase shift of $\pi/2$ relative to $\Theta_0(\tau_*) + \Psi$. However, since $\mathscr{R}$ at decoupling is not much smaller than 1, the peaks given by Θ_v are lower than those given by $\Theta_0(\tau_*) + \Psi$ and so, in C_ℓ, they fill the valleys between the peaks of $\Theta_0(\tau_*) + \Psi$ rather than appearing as peaks. Unlike $\Theta_0(\tau_*) + \Psi$, the Doppler term has

6.7 Fluctuations in the Cosmic Microwave Background

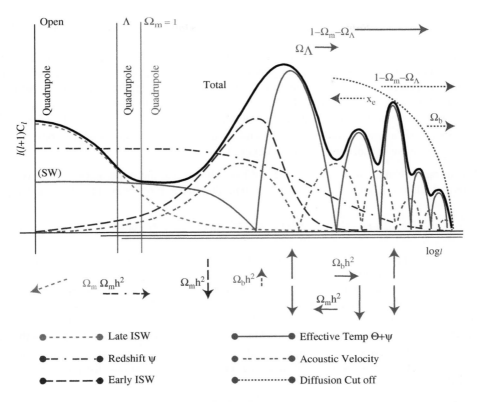

Fig. 6.10. An illustration of the various contributions of isentropic perturbations to the final CMB power spectrum and their dependences on model parameters (Ω_m, Ω_b and Ω_Λ are the density parameters in total matter, baryons, and the cosmological constant, respectively, all at $z = 0$). The arrows indicate the directions of changes led by the increase of a model parameter. For example, an increase in $\Omega_b h^2$ increases the heights of the first and third acoustic peaks, shifts the peaks to higher ℓ, and increases the peaks of acoustic velocity. [Courtesy of W. Hu; see Hu (1995)]

neither a zero-point shift nor disparity between odd and even modes. These features are depicted in Fig. 6.10.

The positions of the acoustic peaks in C_ℓ are determined by the physical size of the sound horizon at decoupling, l_s, and the angular-diameter distance of the last-scattering surface:

$$\ell_m \approx \frac{m\pi(1+z_{\rm dec})d_A(z_{\rm dec})}{l_s} \approx 200m(1+\mathcal{R})^{1/2}\frac{\Omega_{m,0}^{1/2}}{\mathcal{H}(\Omega_{m,0})}. \qquad (6.277)$$

Thus, the positions of the peaks depend significantly on $\Omega_{m,0}$ in models with $\Omega_\Lambda = 0$ but only weakly on $\Omega_{m,0}$ in a flat universe with $\Omega_{m,0} + \Omega_\Lambda = 1$. If $\Omega_{b,0}$ is large, ℓ_m may also increase significantly with $\Omega_{b,0}$ through the $\mathcal{R}$ dependence.

As shown by Eq. (6.276), the heights of odd peaks increase with $\Omega_{b,0}$ but even peaks are not affected by $\Omega_{b,0}$. Since the contributions from acoustic velocities increase with $\mathcal{R}$ as $(1+3\mathcal{R})/(1+\mathcal{R})^{1/2}$, the depths of the valleys depend also on $\Omega_{b,0}$. In addition, the heights of acoustic peaks can also be affected by the strength of initial perturbations on the relevant scales, and by $\Omega_{m,0}$. The dependence on the initial power spectrum is obvious, but that on $\Omega_{m,0}$ needs some explanation. In short, the $\Omega_{m,0}$ dependence comes from the fact that potentials are time dependent in realistic models. In fact, for perturbations inside the horizon (which is the case for acoustic waves), the time evolution in the mass component can induce potential evolution through the Poisson equation, and so potentials must respond to the acoustic oscillation in the radiation/baryon fluid. The strength of the response is expected to be larger if the acoustic perturbations contribute a larger part of the gravitational potential through their self-gravity. Since the change in potential is in resonance with the acoustic oscillation, this leads to a boost in the amplitude of the acoustic oscillation. In the isentropic case, the importance of the self-gravity of acoustic perturbations at a time t is proportional to $\overline{\rho}_\gamma/\overline{\rho}_m = a_{eq}/a(t) \propto (\Omega_{m,0}h^2)^{-1}(1+z)$. The boost is therefore larger for smaller $\Omega_{m,0}h^2$, and for acoustic perturbations which come through the horizon at higher z (i.e. for peaks with larger ℓ).

(e) Damping on Small Scales In §4.2.5 we have shown that the imperfect coupling between baryons and photons during decoupling causes damping in δ_γ on scales smaller than the damping scale

$$l_d \approx \frac{2\pi(1+z_{dec})}{\sqrt{6}} c\sqrt{t_c t_{dec}} \approx 4(\Omega_{m,0}h^2)^{-3/4} \text{Mpc}. \quad (6.278)$$

This damping suppresses the temperature fluctuations on scales smaller than a few arcminutes, corresponding to $\ell \gtrsim 2{,}000$ (see Fig. 6.9).

Another important damping of the temperature fluctuations on small scales comes from the fact that the last-scattering surface has a finite thickness. In §3.5.2 we have shown that the probability for the last scattering to happen at a particular redshift z, $\mathscr{P}(z)\,dz$, is approximately a Gaussian peaked at z_{dec} with a width $\Delta z \approx 80$. This width corresponds to a comoving length scale of $\sim 10(\Omega_{m,0}h^2)^{-1/2}$ Mpc and an angular scale of $\theta_{ls} \sim 0.1\Omega_{m,0}^{-1/2}$ degree. Thus, the observed temperature fluctuation is a superposition of the temperature fluctuations distributed within this finite width:

$$\Theta_{obs} = \int \Theta(z)\,\mathscr{P}(z)\,dz. \quad (6.279)$$

Clearly this leads to a smearing of the temperature fluctuations, Θ, on angular scales $\lesssim \theta_{ls}$, corresponding to $\ell \gtrsim 1{,}000$.

(f) Isocurvature Models Before leaving this subsection, let us very briefly describe the CMB anisotropy expected from isocurvature models. For isocurvature perturbations in the long wavelength limit, $\Psi(\mathbf{k},\tau_*) = I(\mathbf{k})/5$, $\delta_\gamma = 4\Psi$ [see Eqs. (4.229) and (4.230)], and so

$$\Theta^{SW}(k,\mu,\tau_0) = 2\Psi(k,\tau_*)e^{ik\mu(\tau_*-\tau_0)}. \quad (6.280)$$

Thus, the spectrum C_ℓ of the Sachs–Wolfe effect is related to the initial power spectrum, $P_i(k) = k^4 \langle |I(\mathbf{k})|^2 \rangle$, in the same way as in the isentropic case. However, for the same potential perturbations, the temperature fluctuations in an isocurvature model are about six times higher than in an isentropic model.

The acoustic peaks in the isocurvature case can be discussed in a way similar to the isentropic case, but there are several differences. First of all, potential perturbations in the isocurvature case are suppressed outside the horizon, because the density perturbations are set up initially to eliminate the curvature perturbations. Only near horizon crossing, when matter can be redistributed by causal processes, can the potential perturbations start to grow. However, because perturbations in the radiation are anticorrelated with the potential perturbations in dark matter before horizon crossing (see §4.2.6), the potential well initially corresponds to photon rarefaction rather than photon compression. As a perturbation approaches horizon crossing, photon pressure starts to resist the rarefaction, which causes the photon–baryon fluid to fall into the potential well of the dark matter, thereby increasing the depth of the potential well due to the self-gravity of the photon–baryon fluid. As the photon-baryon fluid reaches maximal compression, the pressure starts to push it back again, causing the potential well to decay. If the mass of the photon–baryon

6.7 Fluctuations in the Cosmic Microwave Background

fluid is significant, the potential change (now in resonance with the acoustic oscillation) boosts the amplitude of the acoustic oscillation. Since the oscillation starts from a state of weak compression (rather than maximal compression as in the isentropic case), the isocurvature initial condition drives the sine part of the acoustic solution (6.271), and the acoustic peaks are about $\pi/2$ out of phase with those in the isentropic case. As shown in Fig. 2.42, the acoustic wave pattern in the CMB power spectrum is now well constrained observationally, and it is very difficult to construct pure isocurvature models that match the observed CMB power spectrum. However, a mixed model, with dominant isentropic together with weak isocurvature perturbations, is still possible.

6.7.3 Thomson Scattering and Polarization of the Microwave Background

CMB photons are polarized by Thomson scattering of electrons as they decouple from the baryons and as they propagate from the last-scattering surface through the intergalactic medium to us. The differential cross-section per solid angle for Thomson scattering is

$$\frac{d\sigma}{d\omega} = \frac{3}{8\pi}\sigma_T |\hat{\mathbf{E}} \cdot \hat{\mathbf{E}}'|, \tag{6.281}$$

where $\hat{\mathbf{E}}'$ and $\hat{\mathbf{E}}$ are the incoming and outgoing directions of the electric field (the polarization vector). Since the polarization vector must be orthogonal to the propagation direction, incoming radiation that is polarized parallel to the outgoing direction cannot be scattered. For instance, consider incoming radiation in the $-\mathbf{x}$ direction, so that the polarization vector is $(0, E_y, E_z)$, which is scattered at a right angle into the $\mathbf{z}$ direction (see Fig. 6.11). The outgoing radiation must have a polarization vector $(0, E_y, 0)$. Thus, even if the incoming radiation is unpolarized, i.e. $|E_y| = |E_z|$, the outgoing radiation is linearly polarized in the $\mathbf{y}$ direction. Of course, the radiation observed along the $\mathbf{z}$ direction may be scattered from various directions. For example, the incoming radiation may be from the $-\mathbf{y}$ direction, $(E'_x, 0, E'_z)$, producing a polarization $(E'_x, 0, 0)$. If the radiation were completely isotropic, so that $|E'_x| = |E_y|$, the outgoing radiation would again

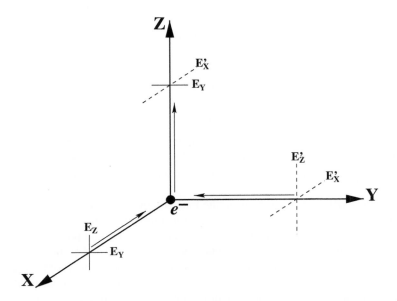

Fig. 6.11. An illustration of how Thomson scattering of a radiation field with quadruple anisotropy produces linear polarization. [After Hu & White (1997)]

be unpolarized. As we shall see, only a quadrupole anisotropy in the incoming radiation can generate a net linear polarization from Thomson scattering.

The polarization in a radiation field is usually described by the Stokes parameters, Q, U and V. In a coordinate system with the z axis chosen to be along the direction of propagation, $Q > 0$ ($Q < 0$) represents linear polarization along the x axis (y axis) and $U > 0$ ($U < 0$) represents the linear polarization along the axis which is at $45°$ ($-45°$) from the x axis. The parameter V is non-zero only if the radiation is circularly polarized, a phenomenon not expected in the CMB. With V neglected, the polarization dependent intensity can be written as

$$\mathscr{I}_{ij} = \begin{pmatrix} T+Q & U \\ U & T-Q \end{pmatrix}, \tag{6.282}$$

where T represents the unpolarized part. It is clear from this that Q measures the difference between the intensity along the x axis (1 axis) and that along the y axis (2 axis). Thus, according to Eq. (6.281), we can write the value of Q for light scattered in the z direction as

$$Q(\hat{\mathbf{z}}) = A \int d\omega_{\hat{\mathbf{n}}'} I(\hat{\mathbf{n}}') \sum_{j=1}^{2} \left(|\hat{\mathbf{x}} \cdot \hat{\mathbf{E}}'_j|^2 - |\hat{\mathbf{y}} \cdot \hat{\mathbf{E}}'_j|^2 \right), \tag{6.283}$$

where $I(\hat{\mathbf{n}}')$ is the intensity of the incoming radiation (assumed to be unpolarized here) from direction $\hat{\mathbf{n}}'$ [corresponding to (ϑ', φ') in spherical coordinates], the term containing the summation is the polarization produced by the radiation from this direction, and the integration is over all incoming directions. In the above equation, we have included a constant factor, A, to take into account the fact that only a fraction of the incoming photons are scattered because of the finite optical depth of Thomson scattering (see below). Writing the Cartesian components of $\hat{\mathbf{E}}'_1$ and $\hat{\mathbf{E}}'_2$ in terms of (ϑ', φ'), we obtain

$$Q(\hat{\mathbf{z}}) = -A \int d\omega_{\hat{\mathbf{n}}'} I(\hat{\mathbf{n}}') \sin^2 \vartheta' \cos(2\varphi'). \tag{6.284}$$

Using the fact that $U(\hat{\mathbf{z}})$ is related to $Q(\hat{\mathbf{z}})$ via a coordinate transformation, $\varphi' \to \varphi' - 45°$, we have

$$U(\hat{\mathbf{z}}) = -A \int d\omega_{\hat{\mathbf{n}}'} I(\hat{\mathbf{n}}') \sin^2 \vartheta' \sin(2\varphi'). \tag{6.285}$$

Note that

$$\sin^2 \vartheta' \cos(2\varphi') \propto Y_{2,2}(\vartheta', \varphi') + Y_{2,-2}(\vartheta', \varphi');$$

$$\sin^2 \vartheta' \sin(2\varphi') \propto Y_{2,2}(\vartheta', \varphi') - Y_{2,-2}(\vartheta', \varphi'),$$

where $Y_{\ell,m}$ are the spherical harmonics. Since spherical harmonics are orthogonal, we see that Q and U pick out the $\ell = 2$, $m = \pm 2$ harmonics in the incoming radiation I. Thus, Q and U are non-zero only if $I(\hat{\mathbf{n}}')$ contains a quadrupole component, as mentioned above and evident from Fig. 6.11.

In the case of the CMB, the fluctuations in the intensity of the incoming radiation are proportional to the temperature fluctuations. If we expand the temperature anisotropy $\Theta(\hat{\mathbf{n}})$ in terms of spherical harmonics as in Eq. (6.261), Q and U will pick out the Θ_2 component. We can then use the Boltzmann equations (4.178)–(4.182) to obtain Θ_2, and hence Q and U. In order to be consistent, we need to include in these equations the collisional term due to the Thomson scattering. In general, these equations have to be solved numerically (e.g. Bond & Efstathiou, 1984; Dodelson, 2003). However, qualitative understanding can be obtained with simplified assumptions. Since polarization is a secondary effect, we can neglect it in the incoming radiation so that the collisional term is given by Eq. (4.184). In this case, Θ_2 is the same as in the absence of Thomson scattering, and $4\pi A$ should be replaced by $(3/8\pi)\tau_T$ (with τ_T the optical depth of Thomson scattering) to account for the fact that only a fraction of the incoming photons are scattered. With these simplifications, the polarization parameters corresponding to the temperature fluctuations generated by a single perturbation mode $\mathbf{k}$ can be written as

$$Q(\hat{\mathbf{z}},\mathbf{k}) = \frac{3\tau_T}{8\pi}\sin^2\vartheta_\mathbf{k}\cos(2\varphi_\mathbf{k})\Theta_2(\mathbf{k}); \quad U(\hat{\mathbf{z}},\mathbf{k}) = \frac{3\tau_T}{8\pi}\sin^2\vartheta_\mathbf{k}\sin(2\varphi_\mathbf{k})\Theta_2(\mathbf{k}), \qquad (6.286)$$

where $(\vartheta_\mathbf{k},\varphi_\mathbf{k})$ specifies the direction of the wavevector $\mathbf{k}$, and $\Theta_2(\mathbf{k})$ is the quadrupole component produced by the perturbation mode $\mathbf{k}$. Note that Eqs. (6.284) and (6.285) correspond to the special case where the outgoing radiation is in the $\hat{\mathbf{z}}$ direction. They can be generalized to an arbitrary outgoing direction $\hat{\mathbf{n}}$ by replacing $\sin^2\vartheta_\mathbf{k}$ in the expressions of Q and U by $1-(\hat{\mathbf{n}}\cdot\hat{\mathbf{k}})^2$. Thus,

$$Q(\hat{\mathbf{n}},\mathbf{k}) = \frac{3\tau_T}{8\pi}[1-(\hat{\mathbf{n}}\cdot\hat{\mathbf{k}})^2]\Theta_2(\mathbf{k})\cos(2\varphi_\mathbf{k}); \quad U(\hat{\mathbf{n}},\mathbf{k}) = \frac{3\tau_T}{8\pi}[1-(\hat{\mathbf{n}}\cdot\hat{\mathbf{k}})^2]\Theta_2(\mathbf{k})\sin(2\varphi_\mathbf{k}). \qquad (6.287)$$

Finally, the polarization produced by all perturbation modes is given by

$$Q(\hat{\mathbf{n}}) = \sum_\mathbf{k} Q(\hat{\mathbf{n}},\mathbf{k}); \qquad U(\hat{\mathbf{n}}) = \sum_\mathbf{k} U(\hat{\mathbf{n}},\mathbf{k}). \qquad (6.288)$$

These can be expanded in spherical harmonics to obtain the polarization power spectra. For small sections of the sky, for which the sky's curvature can be neglected, the harmonic decompositions become two-dimensional Fourier transforms, $\tilde{Q}(\mathbf{l})$ and $\tilde{U}(\mathbf{l})$, where $\mathbf{l}$ is the two-dimensional wavevector, conjugate to the Cartesian coordinates (ξ_1,ξ_2). We can then define the E and B harmonics as

$$E(\mathbf{l}) = \tilde{Q}(\mathbf{l})\cos(2\varphi_\mathbf{l}) + \tilde{U}(\mathbf{l})\sin(2\varphi_\mathbf{l}); \qquad (6.289)$$

$$B(\mathbf{l}) = -\tilde{Q}(\mathbf{l})\sin(2\varphi_\mathbf{l}) + \tilde{U}(\mathbf{l})\cos(2\varphi_\mathbf{l}), \qquad (6.290)$$

where $\varphi_\mathbf{l}$ is the angle between $\mathbf{l}$ and the ξ_1 axis. Since $\tilde{Q}(\mathbf{l}) \propto \cos(2\varphi_\mathbf{l})$ and $\tilde{U}(\mathbf{l}) \propto \sin(2\varphi_\mathbf{l})$ for the scalar perturbations considered here, the B mode power spectrum must be zero in this case.

To gain some insight into the polarization spectrum expected from decoupling, let us examine the properties of Θ_2 using the corresponding Boltzmann equation. Since the incoming radiation is produced when photons are tightly coupled with baryons, we have $k\tau_c \ll 1$, where $\tau_c = 1/(a\sigma_T n_e)$. Including the collisional term and neglecting the high-order moment Δ_3 in Eq. (4.181), we obtain $\Theta_2' - (2/5)k\Theta_1 = (9/10)\Theta_2/\tau_c$. Since $|\Theta_2'| \ll |\Theta_2|/\tau_c$ in the tight coupling limit, we have $\Theta_2 \approx -(4/9)k\tau_c\Theta_1 \ll \Theta_1$. In the same limit, one can also show that $\Theta_\ell \propto k\tau_c\Theta_{\ell-1} \ll \Theta_{\ell-1}$ for $\ell > 2$, so that our neglect of Δ_3 in Eq. (4.181) is justified. The fact that $\Theta_2 \propto \Theta_1$ shows that the polarization power spectrum produced at decoupling should contain acoustic peaks similar to those produced by the acoustic velocity. There is, therefore, a phase shift of $\pi/2$ in the polarization peaks relative to temperature peaks. Since the amplitudes of the velocity peaks are comparable to those of the temperature peaks, the overall amplitude of the polarization power spectrum is expected to be smaller than the temperature power spectrum by a factor of $\sim k\tau_c$. Polarization power spectra with such characteristics have indeed been observed both through the Θ-E cross-spectrum and through the E-E power spectrum (e.g. Nolta et al., 2009; Pryke et al., 2009).

In the presence of gravitational waves, a tensor mode temperature quadrupole can be produced, which can also produce polarization through Thomson scattering. Since the temperature fluctuations produced by a given Fourier mode, $\mathbf{k}$, of tensor perturbations have an angular dependence relative to $\mathbf{k}$ that is different from that produced by scalar perturbations, the polarization generated by gravitational waves have a non-zero B mode. Since inflation generically predicts the production of gravitational waves in the early Universe, a detection of the B mode power spectrum would confirm the inflationary paradigm and provide important constraints on the potential of the inflaton (see Dodelson, 2003, for details).

6.7.4 Interaction between CMB Photons and Matter

CMB photons can interact with an ionized intergalactic medium (IGM) through free–free processes (free–free absorption and bremsstrahlung) and Compton scattering (see §B1.3). Both the free–free and the double-photon (radiative) Compton processes involve the creation of photons, while normal Compton scattering conserves the number of photons. This difference has important observable effects on the CMB. If the temperature of the IGM is higher than that of the radiation field, photons can gain energy from electrons through inverse Compton scattering. Since the number of photons is conserved in this process, the energy gain must result in a change in the spectral energy distribution of the photons. On the other hand, the free–free and double-photon Compton processes may thermalize the radiation field by creating more photons, leading to a blackbody distribution with a higher temperature. Whether the radiation field is thermalized with the IGM or a spectrum distortion is produced depends on how effective the thermalization processes are. In §3.5.4 we have seen that, since the baryon–photon ratio in the Universe is low, double-photon Compton scattering is the main thermalization process, and the time scale for this process is shorter than the Hubble time scale at $z \gtrsim 10^6$. Thus, any energy release to the IGM at $z \gtrsim 10^6$ would be thermalized quickly, but if the IGM was heated at lower redshift, significant spectral distortions in the CMB may be produced by inverse Compton processes.

(a) Spectral Distortion of the CMB by Hot Gas The starting point for calculating the spectral distortion due to the inverse Compton effect of a hot IGM is the Kompaneets equation (B1.67) described in §B1.3.6. In the limit that the electron temperature is much higher than the CMB temperature ($T_e \gg T_\gamma$), the term $\mathcal{N}(\mathcal{N}+1)$ in Eq. (B1.69) can be neglected, giving

$$\frac{\partial \mathcal{N}}{\partial y} \approx \frac{1}{x^2}\frac{\partial}{\partial x}\left[x^4 \frac{\partial \mathcal{N}}{\partial x}\right]. \tag{6.291}$$

This is a linear equation and can be solved by transforming it into a diffusion-type equation in (y,z) where $z \equiv \ln x + 3y$. The solution is

$$\mathcal{N}(x,y) = \frac{1}{\sqrt{4\pi y}} \int_0^\infty \frac{d\xi}{\xi} \mathcal{N}(\xi,0) \exp\left[-\frac{1}{4y}\left(3y + \ln\frac{x}{\xi}\right)^2\right] \tag{6.292}$$

(e.g. Bernstein & Dodelson, 1990), which can be calculated for a given initial spectrum $\mathcal{N}(x,0)$. From this we can determine the total number and energy of photons:

$$n_\gamma(y) = n_\gamma(0), \quad u_\gamma(y) = u_\gamma(0)e^{4y}. \tag{6.293}$$

Thus, while the number of photons is conserved, the photon energy is increased by a factor of e^{4y}.

Now suppose that the initial spectrum is Planckian: $\mathcal{N}(x,0) = (e^x - 1)^{-1}$. Since the total energy of photons is increased (because $T_e \gg T_\gamma$) while the total number is conserved, there must be some redistribution of photon energies. In the Rayleigh–Jeans tail, $\mathcal{N}(x,0) \approx x^{-2}$. In Eq. (6.292), this gives

$$\mathcal{N}(x,y) \approx x^{-1} e^{-2y} \quad (\text{at } x \ll 1). \tag{6.294}$$

Since $\mathcal{N} \propto T$, the change in the effective temperature in the Rayleigh–Jeans tail is

$$T(y) = T(0)e^{-2y}, \quad \text{or} \quad \frac{\delta T}{T} = -2y \text{ for } y \ll 1, \tag{6.295}$$

which corresponds to a decrement in the temperature. Since the total photon number is conserved, the analogous change in the Wien tail of the spectrum corresponds to an *increment* in the effective temperature.

Observations of the CMB radiation spectrum by COBE shows that the CMB has a Planck spectrum with $T = 2.725 \pm 0.002\,\text{K}$ (see §2.9). This result can be transformed into a limit on the Compton y parameter: $y \lesssim 1.5 \times 10^{-5}$, which, in turn, puts a stringent constraint on the possible existence of a hot IGM. Suppose that the IGM was heated to a high temperature at redshift z_1 and has evolved adiabatically to the present time t_0. If the IGM is highly ionized since z_1, then the number density and temperature of electrons evolve with redshift as $n_e(z) \propto (1+z)^3 n_0$ and $T_e(z) \propto (1+z)^2 T_0$ (assuming z is low enough so that Compton drag is negligible; see §3.1.5), where n_0 and T_0 are the values at the present time. Using the definition of y, we can write

$$y = \frac{c}{H_0} \frac{k_B T_0}{m_e c^2} \sigma_T n_0 \int_0^{z_1} \frac{(1+z)^4}{E(z)} dz$$

$$\approx 1.2 \times 10^{-3} \left(\frac{T_0}{10^8\,\text{K}}\right) \Omega_{\text{IGM}} h \int_0^{z_1} \frac{(1+z)^4}{E(z)} dz, \qquad (6.296)$$

where $E(z)$ is given by Eq. (3.75) and the second equation assumes the IGM to be completely ionized. Thus, only a tiny fraction of the IGM is allowed to be in the X-ray emitting phase.

(b) The Sunyaev–Zel'dovich Effect of X-Ray Clusters Clusters of galaxies are known to contain lots of X-ray emitting gas with temperatures of about $10^8\,\text{K}$ (see §2.5.1). Such gas can scatter the CMB photons and cause spectral distortions in the CMB radiation in the direction of the cluster. The net effect is similar to that in Eq. (6.295). In the Rayleigh–Jeans part of the spectrum, the effective temperature of the CMB is changed by

$$\frac{\delta T}{T} = -\frac{2\sigma_T}{m_e c^2} \int P_e(\ell) \, d\ell, \qquad (6.297)$$

where $P_e(\ell) = n_e k_B T_e$ is the electron pressure at a position ℓ along the line-of-sight in the cluster. This effect, called the Sunyaev–Zel'dovich (SZ) effect (Sunyaev & Zel'dovich, 1972), causes a 'dip' (decrement) in the CMB in the radio band (i.e. in the Rayleigh–Jeans part of the CMB radiation spectrum) with an angular size of about ℓ/d_A (where ℓ is the size, and d_A the angular-diameter distance, of the cluster), and has been observed for several clusters (e.g. Birkinshaw, 1999; Bonamente et al., 2006). The temperature T_e can in principle be measured from the X-ray spectrum of a cluster. Since the X-ray emission is produced by bremsstrahlung, the X-ray luminosity can be written as $L_X = A n_e^2 \ell^3 T_e^{1/2}$, where the proportionality A depends on the details of the density and temperature profiles of the cluster. Thus, with the assumption of spherical symmetry of the cluster, a measurement of $\delta T/T$ together with the measurements of the X-ray temperature and luminosity, can in principle be used to estimate the physical size of the cluster, ℓ. If the redshift of the cluster is low so that the curvature effect in the d_A-z relation can be neglected, a comparison of ℓ with the angular size of the cluster then determines the Hubble constant (e.g. Birkinshaw, 1999; Bonamente et al., 2006).

(c) Kinematic SZ Effect Peculiar motions of the hot intracluster gas lead to a Doppler shift of the scattered photons (see §B1.3.6), which can also change the spectrum of the CMB. For small optical depths, the change in the intensity of the CMB at frequencies near ν is given by

$$\frac{\delta I}{I} = \frac{\sigma_T}{c} \int d\ell \, v_{\text{pec}} n_e \left(\frac{xe^x}{e^x - 1}\right), \quad x = \frac{h_P \nu}{k_B T}, \qquad (6.298)$$

where $v_{\rm pec}$ is the peculiar velocity of the intracluster gas along the line-of-sight, and T is the temperature of the CMB. This equation is similar to the thermal effect, but with the replacement of $k_{\rm B} T_{\rm e}/m_{\rm e} c^2$ by $v_{\rm pec}/c$. In the Rayleigh–Jeans part of the spectrum, $x \ll 1$ and

$$\frac{\delta T}{T} \approx \frac{\delta I}{I} \approx \frac{\sigma_{\rm T}}{c} \int v_{\rm pec} n_{\rm e}\, {\rm d}\ell. \tag{6.299}$$

For a typical cluster, the kinematic SZ effect at the cluster center is of the order

$$\Delta T \sim 30 \left(\frac{n_{\rm e}}{3\times 10^{-3}\,{\rm cm}^{-3}}\right)\left(\frac{r_{\rm c}}{0.4{\rm Mpc}}\right)\left(\frac{v_{\rm pec}}{500\,{\rm km\,s}^{-1}}\right)\,\mu{\rm K}, \tag{6.300}$$

where $n_{\rm e}$ is the electron density in the core and $r_{\rm c}$ is the core radius. The spectral signature of this effect (i.e. its strength as a function of frequency) is indistinguishable from that of the CMB fluctuations induced at decoupling, and as a result it is difficult to separate the two effects.

(d) Thomson Scattering due to Re-ionization Suppose that the Universe is re-ionized at a redshift $z_{\rm ri}$ and that the intergalactic medium (IGM) remained fully ionized at $z < z_{\rm ri}$. The optical depth to Thomson scattering from us to a redshift $z < z_{\rm ri}$ is

$$\tau_{\rm T}(z) = \int_0^z \sigma_{\rm T} n_{\rm e}(z) c\, \frac{{\rm d}t}{{\rm d}z}\, {\rm d}z = \frac{c\sigma_{\rm T} n_{\rm e}(0)}{H_0} \int_0^z \frac{(1+z)^2}{E(z)}\, {\rm d}z. \tag{6.301}$$

For $z \gg \Omega_{{\rm m},0}^{-1}$ and $z \gg (\Omega_{\Lambda,0}/\Omega_{{\rm m},0})^{1/3}$, we have

$$\tau_{\rm T}(z) \approx \frac{H_0 c \sigma_{\rm T}}{4\pi G m_{\rm p}}\left(\frac{\Omega_{{\rm b},0}}{\Omega_{{\rm m},0}^{1/2}}\right) z^{3/2} \approx 0.017 h \left(\frac{\Omega_{{\rm b},0}}{\Omega_{{\rm m},0}^{1/2}}\right) z^{3/2}. \tag{6.302}$$

Thus, the re-ionization redshift, $z_{\rm ri}$, can be estimated by measuring the optical depth:

$$z_{\rm ri} \approx 15\, \Omega_{{\rm m},0}^{1/3} (\Omega_{{\rm b},0} h)^{-2/3} \tau_{\rm T}^{2/3}. \tag{6.303}$$

Because of the scattering, temperature fluctuations in the CMB will be smeared out. This effect is most significant on scales smaller than the Hubble radius at $z_{\rm ri}$. From Eq. (6.250), the angular scale corresponding to this Hubble radius is

$$\vartheta_{\rm H}(z_{\rm ri}) \approx \frac{30°}{z_{\rm ri}^{1/2}} \approx 7.5° (\Omega_{{\rm b},0} h)^{1/3} \Omega_{{\rm m},0}^{-1/6} \tau_{\rm T}^{-1/3} \quad (\text{for } \Omega_{{\rm m},0} + \Omega_{\Lambda,0} = 1). \tag{6.304}$$

Thus, if the Universe was re-ionized at an early epoch, so that $\tau_{\rm T}$ is significant, primordial anisotropy in the CMB can be significantly suppressed on angular scales smaller than a few degrees.

In order to obtain an accurate estimate of the primordial power spectrum from the observed CMB anisotropy, we need to take these re-ionization effects into account. To do this, it is important to measure the polarization spectrum of the CMB. Since Thomson scattering is polarization dependent, as described in §6.7.3, the effects of re-ionization can be probed by the polarization power spectrum on large scales. This polarization signal has indeed been observed by the Wilkinson Microwave Anisotropy Probe (WMAP) (e.g. Nolta et al., 2009), and is used together with the observed temperature anisotropy to constrain both cosmological parameters and the epoch of re-ionization (see below).

6.7.5 Constraints on Cosmological Parameters

As described in §2.9, there has been dramatic progress in our observational assessment of the CMB, starting with the preliminary detection of anisotropy on large angular scales by COBE, up

Table 6.1. Summary of the cosmological parameters of the ΛCDM model and the corresponding 68% intervals.

Parameter	WMAP5	WMAP5+BAO+SN
$100\Omega_{b,0}h^2$	2.273 ± 0.062	2.265 ± 0.059
$\Omega_{CDM}h^2$	0.1099 ± 0.0062	0.1143 ± 0.0034
$\Omega_{\Lambda,0}$	0.742 ± 0.030	0.721 ± 0.015
n_s	$0.963^{+0.014}_{-0.015}$	$0.960^{+0.014}_{-0.013}$
τ_T	0.087 ± 0.017	0.084 ± 0.016
$10^9 \times \Delta^2(k_0)$ [a]	2.41 ± 0.11	$2.457^{+0.092}_{-0.093}$
σ_8	0.796 ± 0.036	0.817 ± 0.026
H_0	$71.9^{+2.6}_{-2.7}$ km s^{-1} Mpc^{-1}	70.1 ± 1.3 km s^{-1} Mpc^{-1}
$\Omega_{b,0}$	0.0441 ± 0.0030	0.0462 ± 0.0015
Ω_{CDM}	0.214 ± 0.027	0.233 ± 0.013
$\Omega_{m,0}h^2$	0.1326 ± 0.0063	0.1369 ± 0.0037
z_{ri} [b]	11.0 ± 1.4	10.8 ± 1.4
t_0/Gyr [c]	13.69 ± 0.13	13.73 ± 0.12

[a] $k_0 = 0.002$ Mpc^{-1}. $\Delta^2(k) = k^3 P(k)/(2\pi^2)$
[b] 'Redshift of re-ionization', if the Universe was re-ionized instantaneously from the neutral state to the fully ionized state at z_{ri}
[c] The present-day age of the Universe

to accurate measurements of the acoustic peaks, damping tail and polarization. Since the physics behind these observable quantities is well understood, as described earlier in this section, the observational results can be used to put stringent constraints both on cosmology and on the initial conditions for structure formation. Table 6.1 is a summary of the model parameters obtained from the WMAP five-year data (WMAP5), assuming a ΛCDM model, i.e. that the curvature term, $\Omega_K \equiv 1 - \Omega_0$, is zero, and that the dark energy component has an equation of state with $w = -1$ (see Komatsu et al., 2009). The results shown in the second column are based on WMAP5 alone, while those in the third column are based on combining WMAP5 with data on the baryon acoustic oscillation (BAO) feature in the distribution of galaxies and on the redshift–distance relation of Type Ia supernovae (SN). As is evident, all model parameters of the ΛCDM cosmology have already been constrained to high accuracy.

CMB data alone cannot provide stringent constraints on the curvature of the Universe, because the patterns observed in the CMB power spectrum are angular and distances are required in order to interpret them in terms of physical processes. For a given high redshift, the corresponding distance depends not only on the curvature, but also on the expansion history of the Universe. The latter is determined by the energy content of the Universe and the Hubble constant at the present time. Thus, curvature effects on the CMB power spectrum are degenerate with those produced by the Hubble constant, and by the density and equation of state of the dark energy component. The degeneracy can be broken with an independent measure of the expansion history. For instance, one can achieve this by using one or more of the following distance scales: (i) the present-day Hubble constant obtained from, for example, the Hubble Key Project (Freedman et al., 2001); (ii) the luminosity distances provided by Type Ia supernovae (SN); (iii) the angular-diameter distances obtained from the BAO in the distribution of galaxies. The combination of WMAP5 with the current BAO and SN data gives $-0.0175 < \Omega_K < 0.0085$ and $-0.11 < 1+w < 0.14$ (both at 95% confidence level), in excellent agreement with the ΛCDM model.

With improved measurements from the next generation of CMB experiments, such as the Planck satellite, and from ground-based measurements, the determinations of cosmological parameters can be pushed to even higher accuracy (see Hu & Dodelson, 2002). These will not only give more stringent limits on models within the current ΛCDM paradigm, but also provide the opportunity to detect possible deviations from this paradigm, such as non-Gaussianity, non-adiabaticity of the primordial perturbations, or time dependence of the equation of state of dark energy. In addition, these observations may also provide a clear detection of the B mode polarization due to the gravitational waves predicted by the inflationary paradigm, and they may put stringent constraints on the re-ionization history of the Universe.

7
Formation and Structure of Dark Matter Halos

As we have seen in Chapter 2, there is ample evidence that galaxies reside in extended halos of dark matter. According to the current paradigm, these dark matter halos form through gravitational instability. As we have seen in Chapters 4 and 5, density perturbations grow linearly until they reach a critical density, after which they turn around from the expansion of the Universe and collapse to form virialized dark matter halos. These halos continue to grow in mass (and size), either by accreting material from their neighborhood or by merging with other halos. Some of these halos may survive as bound entities after merging into a bigger halo, thus giving rise to a population of subhalos. This process is illustrated in Fig. 7.1, which shows the formation of a dark matter halo in a numerical simulation of structure formation in a CDM cosmology. It shows how a small volume with small perturbations initially expands with the Universe. As time proceeds, small-scale perturbations grow and collapse to form small halos. At a later stage, these small halos merge together to form a single virialized dark matter halo with an ellipsoidal shape, which reveals some substructure in the form of dark matter subhalos.

In Chapter 6 we have described the overall statistical properties of the cosmic density field. In this chapter we focus on the statistical properties of the discrete halos, and on their internal structure. In particular, we will discuss the following topics:

- the mass function of dark matter halos (i.e. the number density of halos as a function of halo mass), its dependence on cosmology, and its evolution with redshift;
- the mass distribution of the progenitors of individual halos;
- the merger rate of dark matter halos;
- intrinsic properties of dark matter halos, such as density profile, shape and angular momentum;
- the mass function of dark matter subhalos, and its dependence on the mass of the host halo;
- the large-scale environment and spatial clustering of dark matter halos.

Clearly, since dark matter halos are the hosts of galaxies, these properties will have a direct link to the mass function, progenitor mass function, merger rate, clustering properties and internal properties of galaxies (see Chapter 15). As a result, understanding the structure and formation of dark matter halos plays a pivotal role in the understanding of the formation and evolution of galaxies.

In this chapter, we start with a description of peaks in the cosmic density field (§7.1). In the peak formalism, it is *assumed* that the material which will collapse to form nonlinear objects (i.e. dark matter halos) of given mass can be identified in the *initial* density field by first smoothing it with a filter of the appropriate scale and then locating all peaks above some threshold. The properties of peaks in a Gaussian random field can be analyzed in a mathematically rigorous way, but the formalism nevertheless has significant limitations. Although it predicts the number density of peaks as a function of peak height, it cannot be used to obtain the mass function of nonlinear objects (dark halos), nor does it provide a model for how dark matter halos grow with time. To remedy these limitations, an alternative formalism, built on an idea originally due

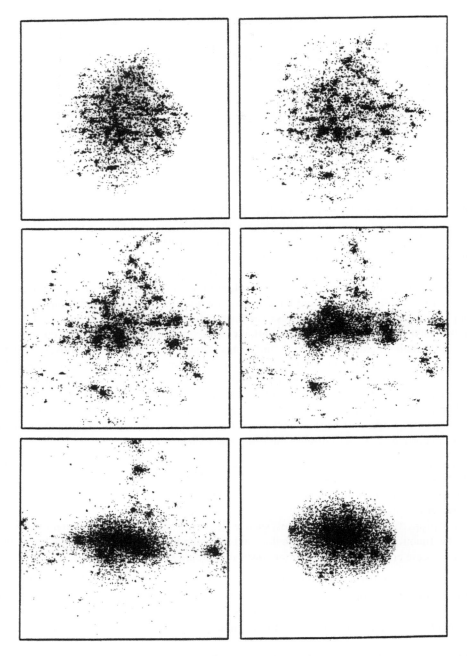

Fig. 7.1. The formation of a single dark matter halo in a numerical simulation of structure formation in a CDM cosmology. All panels correspond to the same physical size, and show the same 20,000 particles. At redshift $z = 0$ (last panel) these end up in a single halo. From left to right, and top to bottom, the panels correspond to redshifts of 3.5, 2.3, 1.5, 0.82, 0.35, and 0.0. [See White (1996)]

to Press & Schechter (1974), has been developed. Although mathematically less rigorous, this formalism, and its extensions, allow one to calculate many properties of the population of dark matter halos, such as their mass function (§7.2), the mass distribution of their progenitors, their merger rate (§7.3), and their clustering properties (§7.4). The analytic nature of this extended Press–Schechter formalism helps us to understand how the properties of the halo population are related to the cosmological framework. Its non-rigorous nature implies, however, that its predictions should always be checked using other methods, primarily numerical N-body simulations of cosmic structure formation. In §7.5 we describe the internal structure of dark matter halos, focusing on their density profiles, their shapes, their angular momenta, and their substructure (i.e. their subhalo populations). In a hierarchical model like CDM, most of the mass at late times is contained in dark matter halos. One can therefore describe the mass distribution by specifying the spatial distribution of dark matter halos and their internal density structure. This 'halo model' of dark matter clustering is the topic of §7.6.

7.1 Density Peaks

As we have seen in Chapter 4, in the linear regime the overdensity field grows as $\delta(\mathbf{x},t) \propto D(t)$, with $D(t)$ the linear growth rate. Thus for a given cosmology [which governs $D(t)$] we can determine the properties of the density field at any given time from those at some early time t_i. Can we extend this to the nonlinear regime in order, for example, to predict the distribution and masses of collapsed objects from $\delta(\mathbf{x},t_i)$ without following the nonlinear dynamics in detail? An object of mass M forms from an overdense region with volume $V = M/\bar{\rho}$ in the initial density field. It is tempting to assume that such a region will correspond to a peak in the density field after smoothing with window of characteristic scale $R \propto M^{1/3}$. This suggests that a study of the density peaks of the smoothed density field can shed some light on the number density and spatial distribution of collapsed dark matter halos.

7.1.1 Peak Number Density

Consider an overdensity field $\delta(\mathbf{x};R)$ which is a smoothed version of $\delta(\mathbf{x})$:

$$\delta(\mathbf{x};R) \equiv \int \delta(\mathbf{x}')W(\mathbf{x}+\mathbf{x}';R)\,\mathrm{d}^3\mathbf{x}', \tag{7.1}$$

where $W(\mathbf{x};R)$ is a window function of characteristic radius R. To be concise, we will drop the argument R and denote the smoothed field simply by $\delta(\mathbf{x})$. The field $\delta(\mathbf{x})$ generally contains many peaks (local maxima) and valleys (local minima). The spatial distribution of peaks can formally be written as

$$n_{\mathrm{pk}}(\mathbf{x}) = \sum_p \delta^{(\mathrm{D})}(\mathbf{x}-\mathbf{x}_p), \tag{7.2}$$

where the $\mathbf{x}_p$ are the peak positions. Since, by definition, the gradient of the smoothed field is zero at a peak, we have that $\eta(\mathbf{x}_p) \equiv \nabla \delta(\mathbf{x}_p) = 0$. Expanding $\eta(\mathbf{x})$ in the vicinity of a peak gives

$$\eta_i(\mathbf{x}) \approx \sum_j \zeta_{ij}(\mathbf{x}_p) \times (\mathbf{x}-\mathbf{x}_p)_j, \tag{7.3}$$

where

$$\begin{aligned}\zeta_{ij}(\mathbf{x}_p) &\equiv \nabla_i \nabla_j \delta(\mathbf{x}_p)\\ &= -\sum_{\mathbf{k}} k_i k_j \delta_{\mathbf{k}} e^{i\mathbf{k}\cdot\mathbf{x}_p}.\end{aligned} \tag{7.4}$$

Provided ζ_{ij} is non-singular at $\mathbf{x}_p$, we have that

$$(\mathbf{x} - \mathbf{x}_p)_i \approx \sum_j \zeta_{ij}^{-1}(\mathbf{x}_p)\eta_j(\mathbf{x}); \tag{7.5}$$

hence

$$\delta^{(D)}(\mathbf{x} - \mathbf{x}_p) = |\det \zeta(\mathbf{x}_p)| \, \delta^{(D)}[\eta(\mathbf{x})]. \tag{7.6}$$

Note that Eq. (7.6) is true not only for peaks but for all (extremal) points where $\eta = 0$. In order for an extremal point to be a peak, the 3×3 matrix ζ_{ij} has to be negative definite. If we denote the three eigenvalues of $-\zeta_{ij}$ as $\Lambda_1 \geq \Lambda_2 \geq \Lambda_3$, then the constraint for an extremum to be a peak is $\Lambda_3 > 0$. In addition, it is usually more useful to consider density peaks with some specific heights, and so we also put a constraint, $(\delta/\sigma) \geq \nu$ [where $\sigma = \langle \delta^2 \rangle$; see Eq. (6.29)], on the peak height. Inserting Eq. (7.6) into Eq. (7.2) and taking into account the constraints $\Lambda_3 > 0$, we can write the density distribution of peaks with $(\delta/\sigma) \geq \nu$ as

$$n_{\rm pk}(\geq \nu; \mathbf{x}) = \mathscr{H}(\delta/\sigma - \nu)|\Lambda_1 \Lambda_2 \Lambda_3|\mathscr{H}(\Lambda_3)\delta^{(D)}(\eta), \tag{7.7}$$

where $\mathscr{H}(x)$ is the Heaviside step function, and all field quantities are evaluated at $\mathbf{x}$. The ensemble average is

$$n_{\rm pk}(\geq \nu) = \left\langle \mathscr{H}(\delta/\sigma - \nu)|\Lambda_1 \Lambda_2 \Lambda_3|\mathscr{H}(\Lambda_3)\delta^{(D)}(\eta) \right\rangle, \tag{7.8}$$

or, in differential form,

$$\mathscr{N}_{\rm pk}(\nu)\, d\nu = \left\langle \delta^{(D)}(\delta/\sigma - \nu)|\Lambda_1 \Lambda_2 \Lambda_3|\mathscr{H}(\Lambda_3)\delta^{(D)}(\eta) \right\rangle d\nu. \tag{7.9}$$

The ensemble average in the above equations can be carried out once the distribution function $\mathscr{P}(\delta, \zeta_{ij}, \eta_i)$ is known. For a Gaussian random field, this distribution function is a multivariate Gaussian specified by the covariance matrix of the variates:

$$\langle \delta^2 \rangle = \sigma_0^2, \quad \langle \eta_i \eta_j \rangle = \frac{\sigma_1^2}{3}\delta_{ij}, \quad \langle \zeta_{ij}\zeta_{kl} \rangle = \frac{\sigma_2^2}{15}\left(\delta_{ij}\delta_{kl} + \delta_{ik}\delta_{jl} + \delta_{il}\delta_{jk}\right), \tag{7.10}$$

$$\langle \delta \eta_i \rangle = 0, \quad \langle \delta \zeta_{ij} \rangle = -\frac{\sigma_1^2}{3}\delta_{ij}, \quad \langle \eta_i \zeta_{jk} \rangle = 0, \tag{7.11}$$

where σ_0, σ_1, σ_2 are defined by Eq. (6.30). With all these, it is straightforward (though tedious) to perform the ensemble averages in the expressions of $n_{\rm pk}$ and $\mathscr{N}_{\rm pk}$ (see Bardeen et al., 1986). The result for the comoving differential peak density is

$$\mathscr{N}_{\rm pk}(\nu)\, d\nu = \frac{1}{(2\pi)^2 R_*^3} e^{-\nu^2/2} G(\gamma, \gamma\nu)\, d\nu, \tag{7.12}$$

where

$$R_* \equiv \sqrt{3}\frac{\sigma_1(R)}{\sigma_2(R)}, \quad \gamma \equiv \frac{\sigma_1^2(R)}{\sigma_2(R)\sigma_0(R)}, \tag{7.13}$$

and

$$G(\gamma, y) = \frac{1}{\sqrt{2\pi(1-\gamma^2)}} \int_0^\infty \exp\left[-\frac{(x-y)^2}{2(1-\gamma^2)}\right] f(x)\, dx, \tag{7.14}$$

with

$$f(x) = \frac{x^3 - 3x}{2}\left\{\text{erf}\left[\left(\frac{5}{2}\right)^{1/2}x\right] + \text{erf}\left[\left(\frac{5}{2}\right)^{1/2}\frac{x}{2}\right]\right\}$$
$$+ \left(\frac{2}{5\pi}\right)^{1/2}\left[\left(\frac{31x^2}{4} + \frac{8}{5}\right)e^{-5x^2/8} + \left(\frac{x^2}{2} - \frac{8}{5}\right)e^{-5x^2/2}\right]. \quad (7.15)$$

For high peaks, where $\nu \gg 1$, this reduces to

$$\mathcal{N}_{\text{pk}}(\nu)\,d\nu = \frac{(\sigma_2^2/3\sigma_0^2)^{3/2}}{(2\pi)^2}(\nu^3 - 3\nu)e^{-\nu^2/2}\,d\nu, \quad (7.16)$$

and

$$n_{\text{pk}}(\geq \nu) = \frac{(\sigma_2^2/3\sigma_0^2)^{3/2}}{(2\pi)^2}(\nu^2 - 1)e^{-\nu^2/2}. \quad (7.17)$$

Using the same principle, one can also calculate the number densities of minima and saddle points. The Euler characteristic, χ, of the overdensity field $\delta(\mathbf{x})$, defined as

$$\chi \equiv \text{number of maxima} + \text{number of mimima} - \text{number of saddles}, \quad (7.18)$$

can then be obtained. The genus of the field, which is the negative of the mean of χ per unit volume, is

$$\text{genus} = -n_\chi(\geq \nu) = 2\frac{(\sigma_2^2/3\sigma_0^2)^{3/2}}{(2\pi)^2}(1-\nu^2)e^{-\nu^2/2}. \quad (7.19)$$

Since this result does not depend on the assumption that the peaks are high, it applies to all Gaussian fields. Except for its normalization, the genus as a function of peak height ν is independent of the power spectrum. Thus, the shape of the genus curve (n_χ as a function of ν) can be used to test whether a density field is Gaussian in a way that is independent of the power spectrum (e.g. Gott et al., 1986).

7.1.2 Spatial Modulation of the Peak Number Density

So far we have described how to calculate the mean number density of peaks in the cosmic density field. In order to study their spatial distribution, we need to examine how the number density of peaks is modulated by large-scale density fluctuations.

Denote by $\delta_b(\mathbf{x})$ the background field obtained by smoothing the original field in large spherical windows of radius R_b. To avoid confusion, the smoothed field defining the density peaks will now be denoted by $\delta_s(\mathbf{x})$, and the corresponding window radius by R_s. Note that the window functions for δ_b and δ_s may have different forms, one being Gaussian and the other being top-hat, for example. The number density of peaks with height $(\delta_s/\sigma_s) = \nu_s$ in the δ_s field, given that the background field at the location of the peaks has height $(\delta_b/\sigma_b) = \nu_b$, can be written as

$$\mathcal{N}_{\text{pk}}(\nu_s|\nu_b)d\nu_s = \frac{\mathcal{N}_{\text{pk}}(\nu_s,\nu_b)\,d\nu_s\,d\nu_b}{\mathscr{P}(\nu_b)\,d\nu_b}. \quad (7.20)$$

Since peak conditions are not imposed on the δ_b field, the distribution function of the background field, $\mathscr{P}(\nu_b)$, is simply a Gaussian for an underlying Gaussian density field. The quantity $\mathcal{N}_{\text{pk}}(\nu_s,\nu_b)$ is the number density of peaks at locations where the background field has an amplitude $\nu_b \pm d\nu_b/2$, and can be calculated by imposing an extra constraint, $\delta^{(D)}(\delta_b/\sigma_b - \nu_b)$, on Eq. (7.9) for the number density of peaks. The remaining task is then to evaluate the multivariate Gaussian distribution function which now depends not only on ν_s, $(\zeta_{ij})_s$ and η_s, but also on ν_b.

In the limit $R_b \gg R_s$, $\mathcal{N}(\nu_s|\nu_b)$ can be written in a form similar to that of $\mathcal{N}(\nu)$ in Eq. (7.12):

$$\mathcal{N}_{pk}(\nu_s|\nu_b)\,d\nu_s = \frac{1}{(2\pi)^2 R_{*,s}^3} e^{-\nu_p^2/2} G(\gamma_p, \gamma_p \nu_p)\,d\nu_p, \tag{7.21}$$

where

$$\gamma_p = \frac{\gamma_s}{(1-\varepsilon^2)^{1/2}}, \quad \nu_p = \frac{\nu_s - \varepsilon\nu_b}{(1-\varepsilon^2)^{1/2}}, \quad \varepsilon \equiv \langle \nu_s \nu_b \rangle. \tag{7.22}$$

It can be shown that $\varepsilon = \sigma_b/\sigma_s$ for δ_b defined by a sharp-k filter (see §6.1.3). For a top-hat filter this holds only approximately (Bower, 1991). In the limit $R_b \gg R_s$, we have that $\varepsilon \to 0$ for a power spectrum with an effective power index larger than -3, and

$$\mathcal{N}_{pk}(\nu_s|\nu_b) \simeq \mathcal{N}_{pk}(\nu_p), \quad \nu_p = (\delta_s - \delta_b)/\sigma_s. \tag{7.23}$$

Thus, as long as $R_b \gg R_s$, the effect of the background field is just to shift the peak height from $\nu_s = \delta_s/\sigma_s$ to $(\delta_s - \delta_b)/\sigma_s$. This is called the peak-background split.

The modulation in the peak number density can be described by the overdensity of peaks,

$$\delta_{pk}^L(\nu_s|\nu_b) = \frac{\mathcal{N}(\nu_s|\nu_b)}{\mathcal{N}(\nu_s)} - 1, \tag{7.24}$$

where the superscript 'L' implies that the quantity is defined in Lagrangian coordinates. Assuming that $\delta_b \ll \delta_s$ (so that $\varepsilon\nu_b/\nu_s \ll 1$) and working to first order in $\varepsilon\nu_b/\nu_s$, we have

$$\delta_{pk}^L(\nu_s|\nu_b) = \frac{(\nu_s^2 - g_1)}{\delta_s}\delta_b, \quad \text{with} \quad g_1 \equiv \left.\frac{\partial \ln G(\gamma_s, y)}{\partial \ln y}\right|_{y=\gamma_s\nu_s}. \tag{7.25}$$

Since the number of peaks is conserved during the linear evolution of the density field, the enhancement factor taking into account the linear evolution is

$$\delta_{pk}(\nu_s|\nu_b) = \frac{\mathcal{N}(\nu_s|\nu_b)}{\mathcal{N}(\nu_s)}\frac{V_L}{V_E} - 1, \tag{7.26}$$

where $V_L \propto R_b^3$ is the background volume in the *unevolved* Lagrangian space, while V_E is that in the *evolved* Eulerian space. Since $V_L/V_E = (1 + \delta_b)$, the peak overdensity to first order becomes

$$\delta_{pk}(\nu_s|\nu_b) = b_{pk}(\delta_s; R_s)\delta_b, \tag{7.27}$$

where

$$b_{pk}(\delta_s; R_s) = 1 + \frac{(\nu_s^2 - g_1)}{\delta_s} \tag{7.28}$$

(Mo et al., 1997b). Thus, the overdensity of peaks is enhanced with respect to the background mass overdensity δ_b by a factor b_{pk} that depends both on the height ν_s and the size R_s (or mass) of the peaks. For high peaks, i.e. for $\nu_s \gg 1$, we have $b_{pk}(\delta_s; R_s) \approx \nu_s/\sigma_s$.

7.1.3 Correlation Function

Given the peak number density field, $n_{pk}(\mathbf{x})$, one can also calculate the spatial correlation function of peaks. For example, the two-point correlation function, defined as

$$\xi_{pk}^L(\mathbf{x}_1, \mathbf{x}_2) = \frac{\langle n_{pk}(\mathbf{x}_1)n_{pk}(\mathbf{x}_2)\rangle}{\langle n_{pk}(\mathbf{x})\rangle^2} - 1, \tag{7.29}$$

can be evaluated by working out the covariance matrix of δ, $\nabla\delta$ and $\nabla\nabla\delta$ at the two peak locations, $\mathbf{x}_1$ and $\mathbf{x}_2$. In the limit $(\nu/\sigma)^2 \xi(|\mathbf{x}_1 - \mathbf{x}_2|) \ll 1$, where ξ is the two-point correlation function of the linear density field $\delta(\mathbf{x})$, the two-point correlation function of the peaks with height ν reduces to

$$\xi_{\rm pk}^{\rm L}(r) = \frac{(\nu^2 - g_1)^2}{\delta^2} \xi(r). \tag{7.30}$$

Taking into account the change in the local peak number density due to the linear evolution of the background field, we have

$$\xi_{\rm pk}(r) = b_{\rm pk}^2(\delta, R)\xi(r), \tag{7.31}$$

where $b_{\rm pk}$ is given by Eq. (7.28). Compared to the two-point correlation function of the mass, $\xi(r)$, the two-point correlation function of the density peaks is enhanced by a bias parameter $b_{\rm pk}^2$. Thus, if a population of galaxies is associated with high peaks in the initial density field, they are expected to be more strongly clustered than the underlying matter.

Similar procedures can be applied to obtain the high-order correlation functions of density peaks (see Bardeen et al., 1986, for details).

7.1.4 Shapes of Density Peaks

In the neighborhood of a density peak, the density profile can be approximated by

$$\delta(\mathbf{x}) \approx \delta(0) - \frac{1}{2}\sum \Lambda_i x_i^2, \tag{7.32}$$

where we have chosen a coordinate system in which the origin is at the peak location and the three coordinate axes coincide with the three principal axes of the isodensity surface. With the density profile given by Eq. (7.32), an isodensity surface, $\delta(\mathbf{x}) = c$, defines a triaxial ellipsoid with semi-axes

$$a_i = \left(\frac{2[\delta(0) - c]}{\Lambda_i}\right)^{1/2}, \tag{7.33}$$

and whose shape is characterized by the parameters

$$e = \frac{\Lambda_1 - \Lambda_3}{2\sum \Lambda_i}, \quad p = \frac{\Lambda_1 - 2\Lambda_2 + \Lambda_3}{2\sum \Lambda_i}, \tag{7.34}$$

which are measures of the ellipticity and oblateness/prolateness of the triaxial ellipsoid (see §5.3). To describe the expected shapes of density peaks, one can start by calculating the conditional probability for e and p subject to the constraint that the peak has a given height $\nu \equiv \delta(0)/\sigma$ and a given value of $x \equiv (\sum_i \Lambda_i)/\sigma_2$:

$$d\mathscr{P} = \mathscr{P}(e, p|\nu, x)\, de\, dp. \tag{7.35}$$

For a Gaussian density field, this probability can be calculated analytically (Bardeen et al., 1986). It turns out that this probability is independent of ν and has the form

$$\mathscr{P}(e, p|x) = \frac{225\sqrt{5}}{\sqrt{2\pi}} \frac{x^8}{f(x)} \exp\left[-\frac{5}{2}x^2(3e^2 + p^2)\right] W(e, p), \tag{7.36}$$

where $f(x)$ is given by Eq. (7.15), and

$$W(e, p) = e(e^2 - p^2)(1 - 2p)\left[(1 + p)^2 - 9e^2\right]\chi(e, p), \tag{7.37}$$

$$\chi(e, p) = \begin{cases} 1 & (\text{if } 0 \leq e \leq 1/4 \text{ and } -e \leq p \leq e) \\ 1 & (\text{if } 1/4 \leq e \leq 1/2 \text{ and } -(1 - 3e) \leq p \leq e) \\ 0 & (\text{otherwise}). \end{cases} \tag{7.38}$$

The conditional probability for x for given peak height ν is independent of e and p:

$$\mathscr{P}(x|\nu)\,dx = \frac{1}{\sqrt{2\pi(1-\gamma^2)}}\exp\left[-\frac{(x-x_*)^2}{2(1-\gamma^2)}\right]\frac{f(x)}{G(\gamma,\gamma\nu)}\,dx, \qquad (7.39)$$

where $x_* \equiv \gamma\nu$. The distribution of e and p can then be obtained for given peak height ν using $\mathscr{P}(e,p|\nu) = \int \mathscr{P}(e,p|x)\mathscr{P}(x|\nu)\,dx$.

In the large x limit (i.e. for high peaks), both e and p are expected to be small, and we can approximate $\mathscr{P}(e,p|x)$ by a Gaussian

$$\mathscr{P}(e,p|x) \approx \mathscr{P}(e_{\rm m},p_{\rm m})\exp\left[-\frac{(e-e_{\rm m})^2}{2\sigma_e^2} - \frac{(p-p_{\rm m})^2}{2\sigma_p^2}\right], \qquad (7.40)$$

where the most probable values and dispersions are

$$e_{\rm m} = \frac{1}{(5x^2+6)^{1/2}}, \quad p_{\rm m} = \frac{30}{(5x^2+6)^2}, \quad \sigma_e = \frac{e_{\rm m}}{\sqrt{6}}, \quad \sigma_p = \frac{e_{\rm m}}{\sqrt{3}}. \qquad (7.41)$$

Hence, higher peaks tend to be more spherically symmetric. Note that this property applies to any Gaussian random field, independent of the power spectrum.

7.2 Halo Mass Function

The peak formalism described above allows one to calculate the abundance and clustering properties of density peaks of the smoothed overdensity field $\delta_{\rm s}(\mathbf{x};R)$ as a function of their height and of auxiliary properties such as their shapes. This should be related to the properties of objects of a given mass (corresponding to the smoothing radius R) forming at different times (corresponding to different peak heights). It is tempting to interpret the number density of peaks in terms of a number density of collapsed objects of mass $M \propto \bar{\rho}R^3$. However, there is a serious problem with this identification, because a mass element which is associated with a peak of $\delta_1(\mathbf{x}) = \delta_{\rm s}(\mathbf{x};R_1)$ can also be associated with a peak of $\delta_2(\mathbf{x}) = \delta_{\rm s}(\mathbf{x};R_2)$, where $R_2 > R_1$. Should such a mass element be considered part of an object of mass M_1 or of mass M_2? If $\delta_2 < \delta_1$ the mass element can be considered part of both. The overdensity will first reach (at time t_1) the critical value for collapse on scale R_1, and then (at time $t_2 > t_1$) on scale R_2. The situation reflects the fact that M_1 is the mass of a collapsed object at t_1 which merges to form a bigger object of mass M_2 at t_2, so that M_1 should no longer be considered as a separate object. In the opposite case, where $\delta_2 > \delta_1$, the mass element apparently can never be part of a collapsed object of mass M_1 but rather must be incorporated directly into a larger system of mass M_2. Such peaks of the field δ_1 should therefore be excluded when calculating the number density of collapsed objects of mass M_1 but should be considered parts of the collapsed objects of mass M_2. This difficulty is known as the 'cloud-in-cloud' problem, and it illustrates that deriving a mass function for collapsed objects using the peak formalism requires treating the statistics of peaks as R is varied. This turns out to be quite difficult to treat rigorously (see Bond & Myers, 1996, for a numerical approach).

What is required to predict the mass function of collapsed objects is a method to partition the linear density field $\delta(\mathbf{x})$ at some early time into a set of disjoint regions (patches) each of which will form a single collapsed object at some later time, and to calculate the statistical properties of this partition. In what follows we describe a simple, but less rigorous, formalism which can be used to compute the mass function of collapsed objects, as well as their mass assembly histories.

7.2.1 Press–Schechter Formalism

Consider the overdensity field $\delta(\mathbf{x},t)$, which in the linear regime evolves as $\delta(\mathbf{x},t) = \delta_0(\mathbf{x})D(t)$, where $\delta_0(\mathbf{x})$ is the overdensity field *linearly extrapolated* to the present time and $D(t)$ is the linear growth rate normalized to unity at the present (see §4.1.6). According to the spherical collapse model described in §5.1, regions with $\delta(\mathbf{x},t) > \delta_c \simeq 1.69$, or equivalently, with $\delta_0(\mathbf{x}) > \delta_c/D(t) \equiv \delta_c(t)$, will have collapsed to form virialized objects. In order to assign masses to these collapsed regions, Press & Schechter (1974) considered the smoothed density field

$$\delta_s(\mathbf{x};R) \equiv \int \delta_0(\mathbf{x}')W(\mathbf{x}+\mathbf{x}';R)\,\mathrm{d}^3\mathbf{x}', \tag{7.42}$$

where $W(\mathbf{x};R)$ is a window function of characteristic radius R corresponding to a mass $M = \gamma_f \bar{\rho} R^3$, with γ_f a filter-dependent shape parameter (e.g. $\gamma_f = 4\pi/3$ for a spherical top-hat filter; see §6.1.3). The ansatz of the Press–Schechter (PS) formalism is that the probability that $\delta_s > \delta_c(t)$ is the same as the fraction of mass elements that at time t are contained in halos with mass greater than M. If $\delta_0(\mathbf{x})$ is a Gaussian random field then so is $\delta_s(\mathbf{x})$, and the probability that $\delta_s > \delta_c(t)$ is given by

$$\mathscr{P}[>\delta_c(t)] = \frac{1}{\sqrt{2\pi}\sigma(M)} \int_{\delta_c(t)}^{\infty} \exp\left[-\frac{\delta_s^2}{2\sigma^2(M)}\right] \mathrm{d}\delta_s = \frac{1}{2}\,\mathrm{erfc}\left[\frac{\delta_c(t)}{\sqrt{2}\sigma(M)}\right]. \tag{7.43}$$

Here

$$\sigma^2(M) = \langle \delta_s^2(\mathbf{x};R)\rangle = \frac{1}{2\pi^2} \int_0^\infty P(k)\widetilde{W}^2(kR)\,k^2\,\mathrm{d}k \tag{7.44}$$

is the mass variance of the smoothed density field with $P(k)$ the power spectrum of the density perturbations, and $\widetilde{W}(kR)$ the Fourier transform of $W(\mathbf{x};R)$ (see §6.1.3).

According to the PS ansatz, the probability (7.43) is equal to $F(>M)$, the mass fraction of collapsed objects with mass greater than M. There is, however, a problem here. As $M \to 0$, then $\sigma(R) \to \infty$ (at least for a power-law spectrum with $n > -3$) and $\mathscr{P}[>\delta_c(t)] \to 1/2$. This would suggest that only half of the mass in the Universe is part of collapsed objects of any mass. This is a characteristic of linear theory, according to which only regions that are initially overdense can form collapsed objects. However, underdense regions can be enclosed within larger overdense regions, giving them a finite probability of being included in a larger collapsed object. This is another manifestation of the cloud-in-cloud problem mentioned above. Press & Schechter (1974) argued, without a proper demonstration, that the material in initially underdense regions will eventually be accreted by the collapsed objects, doubling their masses without changing the shape of the mass function. Press & Schechter therefore introduced a 'fudge factor' 2 and adopted $F(>M) = 2\mathscr{P}[>\delta_c(t)]$. This results in a number density of collapsed objects with masses in the range $M \to M+\mathrm{d}M$ given by

$$\begin{aligned} n(M,t)\,\mathrm{d}M &= \frac{\bar{\rho}}{M}\frac{\partial F(>M)}{\partial M}\,\mathrm{d}M = 2\frac{\bar{\rho}}{M}\frac{\partial \mathscr{P}[>\delta_c(t)]}{\partial \sigma}\left|\frac{\mathrm{d}\sigma}{\mathrm{d}M}\right|\mathrm{d}M \\ &= \sqrt{\frac{2}{\pi}}\frac{\bar{\rho}}{M^2}\frac{\delta_c}{\sigma}\exp\left(-\frac{\delta_c^2}{2\sigma^2}\right)\left|\frac{\mathrm{d}\ln\sigma}{\mathrm{d}\ln M}\right|\mathrm{d}M. \end{aligned} \tag{7.45}$$

This is known as the Press & Schechter (PS) mass function. Note that time enters Eq. (7.45) only through $\delta_c(t)$, and that mass enters through $\sigma(M)$ and its derivative. Upon defining the variable $\nu = \delta_c(t)/\sigma(M)$ the PS mass function can be written in a more compact form:

$$n(M,t)\,\mathrm{d}M = \frac{\bar{\rho}}{M^2}f_{\mathrm{PS}}(\nu)\left|\frac{\mathrm{d}\ln\nu}{\mathrm{d}\ln M}\right|\mathrm{d}M, \tag{7.46}$$

where

$$f_{\rm PS}(\nu) = \sqrt{\frac{2}{\pi}}\,\nu\,\exp(-\nu^2/2), \tag{7.47}$$

is the multiplicity function which gives the fraction of the mass associated with halos in a unit range of $\ln\nu$.

The PS formalism provides a useful way to understand how nonlinear structure develops in a hierarchical model. As one can see from Eq. (7.45), halos with mass M can only form in significant number when $\sigma(M) \gtrsim \delta_{\rm c}(t)$. If we define a characteristic mass $M^*(t)$ by

$$\sigma(M^*) = \delta_{\rm c}(t) = \delta_{\rm c}/D(t), \tag{7.48}$$

then only halos with $M \lesssim M^*$ can have formed in significant number at time t. Since, in hierarchical models, $D(t)$ increases with t and $\sigma(M)$ decreases with M, the characteristic mass increases with time. Thus, as time passes, more and more massive halos will start to form.

7.2.2 Excursion Set Derivation of the Press–Schechter Formula

(a) The Excursion Set of a Gaussian Density Field An alternative derivation of the halo mass function was obtained by Bond et al. (1991), using what is called the excursion set formalism, also known as the extended Press–Schechter (EPS) formalism.

In order to be concise, in what follows we drop the argument t of $\delta_{\rm c}(t)$, and adopt the shorthand notation

$$S \equiv \sigma^2(M) \quad \text{where} \quad M = \gamma_{\rm f}\overline{\rho}R^3. \tag{7.49}$$

We will use $S = S(M)$ as our mass variable, and since, in hierarchical models, S is a monotonically declining function of M, a larger value of S corresponds to a smaller mass. Each location $\mathbf{x}$ in the density field $\delta_0(\mathbf{x})$ corresponds to a trajectory $\delta_{\rm s}(S)$, which reflects the value of the density field at that location when smoothed with a filter of 'mass' S. In the limit $S \to 0$, which corresponds to $M \to \infty$, we have that $\delta_{\rm s} = 0$ for each $\mathbf{x}$. Increasing S corresponds to decreasing the filter mass, and $\delta_{\rm s}$ starts to wander away from zero.

In the excursion set formalism one adopts the sharp k-space filter for which $\gamma_{\rm f} = 6\pi^2$ (see §6.1.3). The smoothed field is then

$$\delta_{\rm s}(\mathbf{x};R) = \int {\rm d}^3\mathbf{k}\,\widetilde{W}_k(kR)\,\delta_{\mathbf{k},0}\,e^{i\mathbf{k}\cdot\mathbf{x}} = \int_{k<k_{\rm c}} {\rm d}^3\mathbf{k}\,\delta_{\mathbf{k},0}\,e^{i\mathbf{k}\cdot\mathbf{x}}, \tag{7.50}$$

where $k_{\rm c} = 1/R$ is the size of the top-hat in k-space, and $\delta_{\mathbf{k},0}$ are the Fourier modes of $\delta_0(\mathbf{x})$. The advantage of using this particular filter is that the change $\Delta\delta_{\rm s}$ corresponding to an increase from $k_{\rm c}$ to $k_{\rm c} + \Delta k_{\rm c}$ is a Gaussian random variable with variance

$$\langle(\Delta\delta_{\rm s})^2\rangle = \langle[\delta_{\rm s}(\mathbf{x};k_{\rm c}+\Delta k_{\rm c}) - \delta_{\rm s}(\mathbf{x};k_{\rm c})]^2\rangle = \frac{1}{2\pi^2}\int_{k=k_{\rm c}}^{k=k_{\rm c}+\Delta k_{\rm c}} P(k)\,k^2\,{\rm d}k = \sigma^2(k_{\rm c}+\Delta k_{\rm c}) - \sigma^2(k_{\rm c}), \tag{7.51}$$

where

$$\sigma^2(k_{\rm c}) = \frac{1}{2\pi^2}\int_{k<k_{\rm c}} P(k)\,k^2\,{\rm d}k, \tag{7.52}$$

[see Eq. (6.29)]. Thus the distribution of $\Delta\delta_{\rm s}$ is independent of the value of $\delta_{\rm s}(\mathbf{x};k_{\rm c})$. When $k_{\rm c}$ [or the associated $S = \sigma^2(k_{\rm c}) = \sigma^2(M)$ with $M = 6\pi^2\overline{\rho}k_{\rm c}^{-3}$] is increased, the value of $\delta_{\rm s}$ at a given point $\mathbf{x}$ executes a Markovian random walk. Note that for any other filter, the trajectory $\delta_{\rm s}(S)$ will not be Markovian.

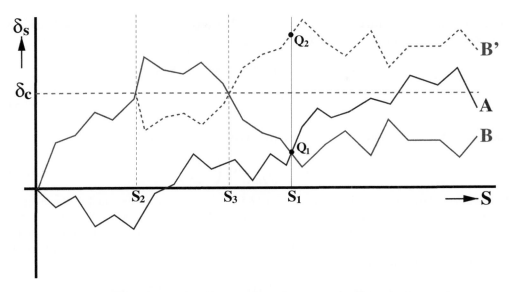

Fig. 7.2. A schematic representation of random walk trajectories in (S, δ_s) space. Each trajectory corresponds to a mass element in the initial (Gaussian) density field, and δ_s is the overdensity at the location of that mass element when the initial density field, linearly extrapolated to the present time, is smoothed with a sharp k-space filter of mass M given by $S = \sigma^2(M)$. The horizontal dashed line indicates the critical density for spherical collapse, $\delta_s = \delta_c$. Trajectory B' is obtained by mirroring trajectory B in $\delta_s = \delta_c$ for $S \geq S_2$, and, since the trajectories are Markovian, is equally likely as trajectory B.

As an illustration, Fig. 7.2 shows examples of two trajectories, A and B. Now consider a given mass scale M_1, corresponding to S_1, as indicated by the vertical line. According to the PS ansatz, the fraction of all trajectories having $\delta_s > \delta_c$ at S_1 is equal to the fraction of mass elements in collapsed objects with mass $M > M_1$. For trajectory B, which has $\delta_s < \delta_c$ at S_1, this mass element would not be part of a collapsed object with $M > M_1$ according to the PS ansatz. However, since it has $\delta_s > \delta_c$ over the interval $S_2 < S < S_3$, it should be part of a halo with $M > M_3 > M_1$ according to the same PS ansatz. Clearly the PS ansatz is not self-consistent. The problem is that it fails to account for the mass elements with trajectories such as B. A correction for this can be made rather easily by realizing that the trajectory from (S_2, δ_c) to (S_1, Q_1) is equally as likely as the trajectory B' (indicated by the dashed line) obtained by mirroring B for $S \geq S_2$ in $\delta_s = \delta_c$.[1] This implies that each trajectory such as B that is missed in 'counting' based on the PS ansatz corresponds to an equally likely trajectory, B', that passes through a point with $\delta_s > \delta_c$ at S_1. The fraction of mass in halos with $M > M_1$ is therefore given by *twice* the fraction of trajectories that crosses S_1 at $\delta_s > \delta_c$. This gives a natural explanation for the fudge factor 2 in the original PS formalism.

In order to derive the halo mass function from the excursion set formalism, we start by deriving the fraction of trajectories that have their *first upcrossing* of the barrier $\delta_s = \delta_c$ at $S > S_1$ (an example is trajectory A in Fig. 7.2). In the spirit of the PS formalism, we associate these trajectories with the mass elements in collapsed objects of masses $M < M_1$. Clearly, such a trajectory must have $\delta_s(S_1) < \delta_c$. However, this includes trajectories such as B, which have their first upcrossing of the barrier at some $S < S_1$. In order to exclude such trajectories, we once again use the fact that trajectories B and B' have equal probabilities. Indeed, *all* trajectories that pass through $(S, \delta_s) = (S_1, Q_1)$ but have $\delta_s > \delta_c$ for some $S < S_1$ have a mirror trajectory that

---

[1] The fact that B and B' are equally probable is a consequence of the trajectories of $\Delta \delta_s$ being Markovian random walks described by a Gaussian distribution with variance given by Eq. (7.51).

passes through $(S, \delta_s) = (S_1, Q_2)$, where $Q_2 = Q_1 + 2(\delta_c - Q_1) = 2\delta_c - Q_1$. Thus, the fraction of trajectories with a first upcrossing at $S > S_1$ is simply given by

$$F_{\rm FU}(>S_1) = \int_{-\infty}^{\delta_c} [\mathscr{P}(\delta_s, S_1) - \mathscr{P}(2\delta_c - \delta_s, S_1)] \, d\delta_s$$
$$= \int_{-\infty}^{\nu_1} \frac{dx}{\sqrt{2\pi}} e^{-x^2/2} - \int_{\nu_1}^{\infty} \frac{dx}{\sqrt{2\pi}} e^{-x^2/2}, \qquad (7.53)$$

where $\nu_1 \equiv \delta_c/\sqrt{S_1}$, and we have used that, for a Gaussian random field,

$$\mathscr{P}(\delta_s, S) d\delta_s = \frac{1}{\sqrt{2\pi S}} \exp\left(-\frac{\delta_s^2}{2S}\right) d\delta_s. \qquad (7.54)$$

As discussed above, the fraction given by Eq. (7.53) is considered to be equal to the fraction, $F(<M_1)$, of mass elements contained in halos with $M < M_1$. If $S \to \infty$ as $M \to 0$, *every* trajectory will cross the barrier δ_c at some point. Hence, each mass element in the Universe is expected to be in a collapsed object with some mass $M > 0$.[2] Consequently, $F(>M) = 1 - F(<M)$ and we obtain the halo mass function

$$n(M,t) \, dM = \frac{\overline{\rho}}{M} \frac{\partial F(>M)}{\partial M} dM = \frac{\overline{\rho}}{M} f_{\rm FU}(S, \delta_c) \left|\frac{dS}{dM}\right| dM, \qquad (7.55)$$

where

$$f_{\rm FU}(S, \delta_c) \, dS = \frac{\partial F_{\rm FU}}{\partial S} dS = \frac{1}{\sqrt{2\pi}} \frac{\delta_c}{S^{3/2}} \exp\left[-\frac{\delta_c^2}{2S}\right] dS, \qquad (7.56)$$

is the fraction of trajectories that have their first upcrossing in the range $(S, S+dS)$. Simply substituting Eq. (7.56) in Eq. (7.55) yields exactly the PS mass function (7.46), this time without having to include a fudge factor of 2.

The excursion set formalism described above can be cast into a more elegant form in terms of a diffusion equation. As described above, for a Gaussian random field, each step (characterized by a change $S \to S + \Delta S$) of the random walk is determined by the mean and variance of the change in δ_s, $\langle \Delta \delta_s \rangle \equiv \langle \Delta \delta_s | \delta_s \rangle$ and $\sigma_\Delta^2 \equiv \langle (\Delta \delta_s - \langle \Delta \delta_s \rangle)^2 | \delta_s \rangle$, through the Gaussian distribution function

$$P(\Delta \delta_s | \delta_s) = \frac{1}{\sqrt{2\pi} \sigma_\Delta} \exp\left[-\frac{(\Delta \delta_s - \langle \Delta \delta_s \rangle)^2}{2\sigma_\Delta^2}\right] d(\Delta \delta_s). \qquad (7.57)$$

In the limit $\Delta S \to 0$, the problem is equivalent to the diffusion of a 'particle' in δ_s-space with S being the 'time' variable. The probability $\Pi(\delta_s, S)$ for the 'particle' to lie between δ_s and $\delta_s + \Delta \delta_s$ when the variance of the density field (the 'time') is S obeys the diffusion-type equation

$$\frac{\partial \Pi}{\partial S} = -\frac{\partial (\mu \Pi)}{\partial \delta} + \frac{1}{2} \frac{\partial^2 (\Sigma^2 \Pi)}{\partial \delta^2}, \qquad (7.58)$$

where

$$\mu \equiv \lim_{\Delta S \to 0} \frac{\langle \Delta \delta_s | \delta_s \rangle}{\Delta S}, \quad \text{and} \quad \Sigma^2 \equiv \lim_{\Delta S \to 0} \frac{\langle (\Delta \delta_s)^2 | \delta_s \rangle}{\Delta S}, \qquad (7.59)$$

are the drift and variance parameters, respectively.

For sharp k-space filters, $\mu = 0$ and $\Sigma^2 = 1$, so that the diffusion equation reduces to

$$\frac{\partial \Pi}{\partial S} = \frac{1}{2} \frac{\partial^2 \Pi}{\partial \delta^2}. \qquad (7.60)$$

---

[2] In reality CDM has a small but non-zero Jeans mass and the effective n may be < -3 as $k \to \infty$. $\sigma^2(M)$ then approaches a finite value for $M \to 0$ and not *all* mass has to be in collapsed objects.

We wish to calculate the probability for 'particles' starting from $(\delta_s, S) = (0,0)$ to reach (δ_s, S) without exceeding a critical value δ_c at smaller S. This is equivalent to the solution of the diffusion equation with an absorbing barrier at $\delta_s = \delta_c$. For Eq. (7.60) (i.e. for sharp k-space filters) this has the following analytic solution:

$$\Pi(\delta_s, S) = \frac{1}{\sqrt{2\pi S}} \left\{ \exp\left(-\frac{\delta_s^2}{2S}\right) - \exp\left[-\frac{(2\delta_c - \delta_s)^2}{2S}\right] \right\} \qquad (7.61)$$

(Chandrasekhar, 1943). We thus have that

$$F_{\rm FU}(> S_1) = \int_{-\infty}^{\delta_c} \Pi(\delta_s, S_1) d\delta_s, \qquad (7.62)$$

which, upon substitution of Eq. (7.61), is identical to Eq. (7.53).

Casting the excursion set approach in terms of a diffusion problem allows one to see more clearly how to proceed when some of the assumptions in the model are relaxed. For example, if we use top-hat or Gaussian filters instead of the sharp k-space filters, Eq. (7.58) is still valid, except that the drift parameter μ and the variance parameter (the 'diffusion coefficient') Σ^2 should be evaluated for the new filter. Furthermore, if the critical density, δ_c, changes with S instead of being a constant, the problem is equivalent to diffusion in the presence of a moving absorbing barrier.

(b) Interpretation of the Excursion Set In deriving the halo mass function using the excursion set formalism, we have made the ansatz that the fraction of trajectories with a first upcrossing on a mass scale $(M, M+dM)$ is equal to the mass fraction of the Universe in collapsed objects with masses in this range. However, the following example shows that this cannot formally be true. Consider a uniform spherical region of mass M and radius R, with density equal to the critical density for collapse at some time t, embedded in a larger region with lower density. According to the spherical collapse model, all mass elements in this spherical region will be part of a collapsed halo of mass M at time t. Now consider an interior point $\mathbf{x}$ which is at a distance r from the center of the spherical region, and calculate the mean overdensities within spheres centered on this point. Clearly, the overdensity reaches the critical density only for spheres with radii smaller than $R - r$. Thus, the mass element at $\mathbf{x}$ has its first upcrossing at a mass scale $M(R-r)^3/R^3$, which is smaller than M unless $\mathbf{x}$ is at the center of M. This simple example suggests that the mass associated with the first upcrossing of a mass element is only a lower limit on the actual mass of the collapsed object to which the mass element belongs. Thus, the ansatz on which the EPS formalism is based cannot be correct for individual mass elements. Yet, as we will see in §7.2.4 below, the resulting mass function seems to be in good agreement with numerical simulations. This paradox is often interpreted as indicating that the excursion set formalism predicts, in a *statistical* sense, how much mass ends up in collapsed objects of a certain mass at a given time, but that it cannot be used to predict the halo mass in which a *particular* mass element ends up.

7.2.3 Spherical versus Ellipsoidal Dynamics

The Press–Schechter formalism described above is based on the assumption that the collapse of density perturbations is described by the spherical collapse model. This leads to a critical overdensity for collapse, δ_c, which is independent of halo mass and time. However, as we show below, the collapse of overdensities in a Gaussian density field is generically ellipsoidal rather than spherical. In this subsection we show how to incorporate ellipsoidal dynamics into the excursion set approach.

As we have seen in §5.3, initial gravitational collapse in a cosmological context is largely driven by the initial tidal field. For a Gaussian density field smoothed on some scale, $\delta_s(\mathbf{x})$, we can calculate at any point the tidal tensor (the second derivatives of the potential field) and its eigenvalues, λ_1, λ_2 and λ_3. Since the potential field is also Gaussian, the joint distribution of λ_1, λ_2 and λ_3 is a multivariate Gaussian specified by their covariance matrix. This distribution function was obtained by Doroshkevich (1970):

$$\mathscr{P}(\lambda_1, \lambda_2, \lambda_3) = \frac{675\sqrt{5}}{8\pi\sigma^6} \exp\left(-\frac{3I_1^2}{\sigma^2} + \frac{15I_2}{2\sigma^2}\right)(\lambda_1 - \lambda_2)(\lambda_2 - \lambda_3)(\lambda_1 - \lambda_3), \tag{7.63}$$

where $\sigma^2 = \langle \delta_s^2 \rangle$, $I_1 = \lambda_1 + \lambda_2 + \lambda_3 = \delta_s$ and $I_2 = \lambda_1\lambda_2 + \lambda_2\lambda_3 + \lambda_3\lambda_1$. Define the ellipticity and prolateness of the tidal field as

$$e \equiv (\lambda_1 - \lambda_3)/(2\delta_s), \quad p \equiv (\lambda_1 - 2\lambda_2 + \lambda_3)/(2\delta_s). \tag{7.64}$$

(Note that e and p here correspond to the tidal field and should not be confused with the corresponding quantities for the density field defined in §7.1.4.) We can then use Eq. (7.63) to write the distribution function for e and p given δ_s:

$$\mathscr{P}(e, p | \delta_s) = \frac{1125}{\sqrt{10\pi}} e(e^2 - p^2) \left(\frac{\delta_s}{\sigma}\right)^5 \exp\left[-\frac{5\delta_s^2}{2\sigma^2}(3e^2 + p^2)\right]. \tag{7.65}$$

For all e this distribution peaks at $p = 0$. For $p = 0$, the maximum occurs at $e_m = (\delta_s/\sigma)^{-1}/\sqrt{5}$, which suggests that regions with a higher δ_s/σ have a more spherical tidal tensor. Note, though, that even for $\delta_s/\sigma \sim 1$ the deviation from spherical symmetry can be fairly substantial, indicating that in general the collapse will be ellipsoidal.

In the ellipsoidal collapse model described in §5.3, the critical overdensity for collapse depends on e and p. Since both e and p are random fields, their values at a given point change stochastically as the size of the filter increases. Consequently, the critical density for collapse changes stochastically at each step of the random walk. In this case, the excursion set is given by the statistics of the random walk in (δ_s, e, p) space, which depends on the distribution of $(\Delta\delta_s, \Delta e, \Delta p)$ given (δ_s, e, p). Although this distribution can in principle be obtained for a Gaussian random field, the corresponding excursion set is rather complicated (e.g. Sandvik et al., 2007). In what follows, we give an oversimplified prescription, which nevertheless captures the essence of the ellipsoidal dynamics.

As discussed above, for a Gaussian density field the joint distribution of p and e peaks at $p = p_m = 0$ and $e = e_m = (\delta_s/\sigma)^{-1}/\sqrt{5}$. Thus, in order for regions *with the most probable e and p* to collapse and form a bound object at time t, their initial overdensity must be $\delta_{ec}(e_m, 0)$, where $\delta_{ec}(e, p)$ is the critical overdensity for collapse given by Eq. (5.86) with $\delta_{sc} = \delta_c(t)$. If we set δ_s on the right-hand side of the expression for e_m to be equal to this critical value, we can express e_m in terms of σ^2. Inserting this expression of e_m along with $p_m = 0$ into Eq. (5.86), one obtains the following relation between δ_{ec} and σ^2:

$$\delta_{ec}(\sigma, t) = \delta_c(t) \left\{ 1 + \beta \left[\frac{\sigma^2}{\sigma_*^2(t)}\right]^\gamma \right\}, \tag{7.66}$$

where $\sigma_*(t) \equiv \delta_c(t)$, $\beta \sim 0.47$ and $\gamma \sim 0.615$ (see Sheth et al., 2001). Although this relation is derived for perturbations with the most probable e and p, it may be used to approximate the average trend of δ_{ec} with σ, since $\mathscr{P}(e, p | \delta_s)$ is quite strongly peaked around (e_m, p_m) for typical objects (with $\sigma \sim 1$). Note that the perturbation power spectrum only enters in the relation between σ and the size of the filter (or the mass of the object), whereas the effects of cosmology only enter in $\delta_c(t) = \delta_c/D(t)$. Several other features of Eq. (7.66) are also worth noticing. For massive objects with $\sigma/\sigma_* \ll 1$, this equation suggests that $\delta_{ec}(\sigma, t) \approx \delta_c(t)$, so that the critical overdensity required for collapse at time t is approximately the same as that given by the spherical collapse model. The critical overdensity increases with σ, and so is larger for less massive objects. Thus, smaller objects are more strongly influenced by tides and therefore must have a larger internal density in order to be able to collapse.

Eq. (7.66) is useful because it allows one to include the effects of ellipsoidal collapse into the excursion set model in a straightforward way. All one needs to do is to use Eq. (7.66) to set the barrier in the (S, δ_s) plane, and the mass function associated with ellipsoidal collapse can then be obtained from the mass distribution of the first upcrossings of this barrier by independent random walks. For example, for a given model of structure formation, one can simulate an ensemble of independent unconstrained random walks, record the values of S (or mass M) at all first upcrossings of the mass-dependent ('moving') ellipsoidal collapse barrier, and estimate the halo mass function based on the distribution of the masses at these first upcrossings. To a good approximation, the resulting mass function can be written in the form (7.46), but with the multiplicity function $f_{PS}(\nu)$ replaced by

$$f_{EC}(\nu) = A\left(1 + \frac{1}{\tilde{\nu}^{2q}}\right) f_{PS}(\tilde{\nu}), \qquad (7.67)$$

where $\tilde{\nu} = 0.84\nu$ and $q = 0.3$. The normalization $A \approx 0.322$ is set by requiring that the integral of $f_{EC}(\nu)/\nu$ over all ν is unity, which means that all matter is in collapsed objects of some mass, however small (Sheth et al., 2001).

The great virtue of interpreting Eq. (7.66) as the 'moving' barrier is that, once the barrier shape is known, the entire excursion set formalism developed on the basis of spherical collapse dynamics (constant barrier) can be extended relatively straightforwardly. For simplicity, however, our descriptions in the following will be based on spherical collapse dynamics.

7.2.4 Tests of the Press–Schechter Formalism

Given the many crude assumptions underlying the Press–Schechter (PS) and extended PS (EPS) formalisms, it is important to test their predictions for the halo mass function against numerical simulations, which follow the growth and collapse of structures directly by solving the equations of motion for dark matter particles (see Appendix C). Until the end of the 1990s, most numerical simulations yielded halo abundances in reasonable agreement with the PS prediction (7.46) based on spherical collapse (see Monaco, 1998, for a detailed review). As higher resolution simulations became available, it became clear that this PS mass function overpredicts the abundance of sub-M^* halos, and underpredicts that of halos with $M > M^*$ (e.g. Governato et al., 1999; Sheth & Tormen, 1999). The PS mass function for ellipsoidal collapse, on the other hand, seems to fit the simulation results much better (see Fig. 7.3). In fact, the highest resolution simulations to date match the halo mass function based on Eq. (7.67) with discrepancies at the $\sim 20\%$ level (e.g. Jenkins et al., 2001; Warren et al., 2006).

Note, however, that these comparisons depend somewhat on how halos are identified in the simulations. Different authors often use different halo-finding algorithms, most of which have one or more free parameters. For example, the frequently used friends-of-friends (FOF) algorithm defines halos as structures whose particles are separated by distances less than a percolation parameter b, called the linking length, times the mean interparticle distance (Davis et al., 1985). A heuristic argument, based on spherical collapse, suggests that one should use $b \simeq 0.2$, for which the mean overdensity of a halo is ~ 180. Obviously, using different values for b results in different halo mass functions, which introduces some arbitrariness in the comparison between PS predictions and numerical simulations (e.g. White, 2002).

A more stringent test of the PS formalism is to use the excursion set to predict the formation sites and masses of individual halos, and to compare these predictions with N-body simulations. As already discussed in §7.2.2, the excursion set formalism is not expected to work on an object-by-object basis, as was confirmed by direct comparison with simulations (e.g. Bond

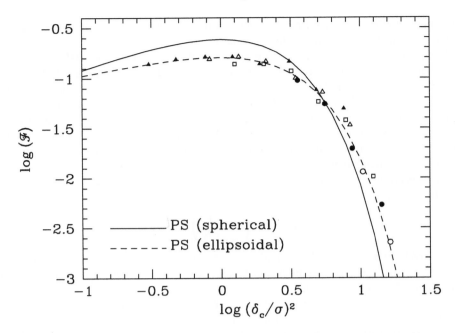

Fig. 7.3. The mass function of dark matter halos, here plotted as the logarithm of $\mathscr{F} \equiv (nM^2/\overline{\rho})|\mathrm{d}\ln M/\mathrm{d}\ln\sigma|$ versus the logarithm of $(\delta_c/\sigma)^2$. The results obtained from N-body simulations (symbols) are compared to the predictions of the Press–Schechter formalism for both spherical (solid curve) and ellipsoidal (dashed curve) collapse. Clearly, the latter is in much better agreement with the simulation results. [Based on data published in Sheth & Tormen (1999)]

et al., 1991). However, according to our earlier discussion, the mass at first upcrossing of a mass element should give a lower limit to the mass of the collapsed object to which the mass element belongs. For an ellipsoidal collapse barrier, this prediction was found to be consistent with N-body simulations by Sheth et al. (2001).

7.2.5 Number Density of Galaxy Clusters

As an example of one of the many applications of the PS formalism, consider the number density of clusters of galaxies, the most massive virialized objects in the Universe.

As we have seen in §2.5.1, the masses of rich clusters fall in the range 10^{14}–$10^{15}\,\mathrm{M}_\odot$, and are dominated by dark matter. Usually, this mass is estimated within the Abell radius, $r_\mathrm{A} = 1.5\,h^{-1}\mathrm{Mpc}$, and we write it as

$$M_\mathrm{A} = 7.8 \times 10^{14} m_\mathrm{A} h^{-1}\,\mathrm{M}_\odot, \qquad (7.68)$$

where m_A is a mass parameter, defined such that the mean overdensity within r_A is

$$\overline{\delta}(r_\mathrm{A}) \equiv \frac{3M_\mathrm{A}}{4\pi\overline{\rho}_0 r_\mathrm{A}^3} - 1 = 200 m_\mathrm{A}/\Omega_{\mathrm{m},0}. \qquad (7.69)$$

To obtain the number density of clusters from the PS formalism, we need to know how r_A is related to the total 'virial' mass M. Motivated by the spherical collapse model described in §5.1, we define M as

$$M = \frac{4\pi}{3} r_{200}^3 \overline{\rho}_0 \left[1 + \overline{\delta}(r_{200})\right], \qquad (7.70)$$

where
$$\bar{\delta}(r_{200}) = 200/\Omega_{m,0} \qquad (7.71)$$
is the average overdensity of the halo, and r_{200} is the corresponding halo radius. Under the assumption that the density profile of a cluster between r_A and r_{200} can be approximated as $\propto r^{-2}$, we have that
$$r_{200} = m_A^{1/2} r_A, \quad \text{and} \quad M \approx m_A^{1/2} M_A. \qquad (7.72)$$
The linear (Lagrangian) radius of the cluster is
$$R \equiv \left(\frac{3M}{4\pi\bar{\rho}_0}\right)^{1/3} \approx 1.1 \Omega_{m,0}^{-1/3} m_A^{1/2} r_8, \qquad (7.73)$$
where $r_8 \equiv 8h^{-1}$Mpc. From this we see that the mass range of rich clusters corresponds to a relatively small range in R around r_8. Within this range of R, we can parameterize the mass variance $\sigma(R)$ in top-hat windows as
$$\sigma(R) = \sigma_8 \left(\frac{R}{r_8}\right)^{-\beta}. \qquad (7.74)$$
For CDM-type power spectra with index $n = 1$, β can be approximated as $\beta \approx 0.6 + 0.8\Gamma$, where Γ is the shape parameter of a CDM power spectrum defined in §4.3. Using Eq. (7.46) we can then write the mean number of rich clusters in a sphere of radius r_8, with mass exceeding M_A, as
$$N(M_A) = \frac{4\pi r_8^3}{3} \int_{M_A}^{\infty} n(M,t_0) \, dM$$
$$\approx \frac{2}{\sqrt{\pi}} \left(\frac{\delta_c}{\sqrt{2}\sigma_8}\right)^{3/\beta} \int_{y_{\min}}^{\infty} y^{-3/\beta} \exp(-y^2) \, dy, \qquad (7.75)$$
where
$$y_{\min} \equiv \frac{\delta_c}{\sqrt{2}\sigma_8} \left[1.2 m_A \Omega_{m,0}^{-2/3}\right]^{\beta/2}. \qquad (7.76)$$
From this we see that the number density of clusters (with mass parameter m_A above some value) can be used to constrain σ_8 and $\Omega_{m,0}$. Since $N(M_A) \ll 1$, Eq. (7.75) determines the combination $(m_A \Omega_{m,0}^{-2/3})^{\beta/2}/\sigma_8$ accurately even if the cluster number density is uncertain. White et al. (1993a) estimated that $N(M_A) \approx 8.5 \times 10^{-3}$ for clusters with $M_A \approx 5 \times 10^{14} h^{-1} M_\odot$. This in Eq. (7.75) gives the following constraint on σ_8 and $\Omega_{m,0}$:
$$\sigma_8 \approx (0.5 - 0.6) \Omega_{m,0}^{-0.5}. \qquad (7.77)$$
Note that this result is very similar to that based on the cosmic virial theorem discussed in §6.5.1. This is easy to understand, because the main contribution to the small-scale velocity dispersion in the virial theorem comes from clusters of galaxies (Mo et al., 1997a). A more detailed analysis by Viana & Liddle (1996) gives the following results:
$$\sigma_8 = 0.60 \Omega_{m,0}^{-C(\Omega_{m,0})}, \qquad (7.78)$$
where
$$C(\Omega_{m,0}) = \begin{cases} 0.36 + 0.31\Omega_{m,0} - 0.28\Omega_{m,0}^2 & (\Omega_{m,0} \leq 1, \Omega_\Lambda = 0) \\ 0.59 - 0.16\Omega_{m,0} + 0.06\Omega_{m,0}^2 & (\Omega_{m,0} + \Omega_\Lambda = 1). \end{cases} \qquad (7.79)$$
This value of σ_8 can be used to normalize the power spectrum of the cosmological density perturbations (see §4.4.4). Such a normalization is usually referred to as cluster-abundance normalization.

7.3 Progenitor Distributions and Merger Trees

7.3.1 Progenitors of Dark Matter Halos

An advantage of the excursion set approach of the extended Press-Schechter (EPS) formalism is that it provides a neat way to calculate the properties of the progenitors which give rise to a given class of collapsed objects. For example, one can calculate the mass function at $z = 5$ of those halos (progenitors) which by $z = 0$ end up in a massive cluster-sized halo of mass $10^{15}\,\mathrm{M}_\odot$. Thus, the EPS formalism allows one to describe how a dark matter halo assembled its mass via mergers of smaller mass halos.

Consider a spherical region (a patch) of mass M_2, corresponding to a mass variance $S_2 = \sigma^2(M_2)$, with linear overdensity $\delta_2 \equiv \delta_c(t_2) = \delta_c/D(t_2)$ so that it forms a collapsed object at time t_2. We are interested in the fraction of M_2 that was in collapsed objects of a certain mass at an earlier time $t_1 < t_2$. To proceed, we adopt the same excursion-set approach as we did in deriving the Press–Schechter (PS) formula (see §7.2.2), and calculate the probability for a random walk originating at $(S, \delta_s) = (S_2, \delta_2)$ to execute a first upcrossing of the barrier $\delta_s = \delta_1 \equiv \delta_c(t_1)$ at $S = S_1$, corresponding to mass scale M_1. This is exactly the same problem as before except for a translation of the origin in the (S, δ_s) plane (see Fig. 7.2). Hence, the probability we want is given by

$$f_{\mathrm{FU}}(S_1, \delta_1 | S_2, \delta_2)\,\mathrm{d}S_1 = \frac{1}{\sqrt{2\pi}} \frac{\delta_1 - \delta_2}{(S_1 - S_2)^{3/2}} \exp\left[-\frac{(\delta_1 - \delta_2)^2}{2(S_1 - S_2)}\right]\mathrm{d}S_1, \qquad (7.80)$$

which follows from Eq. (7.56) upon replacing δ_c by $\delta_1 - \delta_2$ and S by $S_1 - S_2$. According to the interpretation of the excursion set, Eq. (7.80) gives the fraction of mass elements in (M_2, δ_2) patches that were in collapsed objects of mass M_1 at the earlier time t_1. Converting from mass weighting to number weighting, one obtains the average number of progenitors at t_1 in the mass interval $(M_1, M_1 + \mathrm{d}M_1)$ which by time t_2 have merged to form a halo of mass M_2:

$$\begin{aligned}n(M_1, t_1 | M_2, t_2)\,\mathrm{d}M_1 &= \frac{M_2}{M_1} f_{\mathrm{FU}}(S_1, \delta_1 | S_2, \delta_2)\left|\frac{\mathrm{d}S_1}{\mathrm{d}M_1}\right|\mathrm{d}M_1 \\ &= \frac{M_2}{M_1^2} f_{\mathrm{PS}}(\nu_{12})\left|\frac{\mathrm{d}\ln\nu_{12}}{\mathrm{d}\ln M_1}\right|\mathrm{d}M_1, \end{aligned}\qquad (7.81)$$

where $\nu_{12} \equiv (\delta_1 - \delta_2)/\sqrt{S_1 - S_2}$ and $f_{\mathrm{PS}}(\nu)$ is the PS multiplicity function given by Eq. (7.47).

As with the halo mass function, numerical simulations have been used to test the progenitor mass functions predicted by the EPS formalism. These show that although Eq. (7.81) is in good agreement with simulation results when $\Delta t = t_2 - t_1$ is small, it significantly underestimates the mass fraction in high mass progenitors for relatively large Δt (e.g. Somerville & Kolatt, 1999; van den Bosch, 2002b; Cole et al., 2008). Despite these shortcomings, the EPS formalism is often used since it allows for a fast computation of many interesting properties of dark matter halos, some of which we discuss below.

7.3.2 Halo Merger Trees

The conditional mass function derived above can be used to describe the statistical properties of the merger histories of dark matter halos. The merger history of a given dark matter halo can be represented pictorially by a merger tree such as that shown in Fig. 7.4. In this figure, time increases from top to bottom, and the widths of the branches of the tree represent the masses of the individual progenitors. If we start from an early time, the mass which ends up in a halo at the present time t_0 (the trunk of the tree) is distributed over many small branches; pairs of small branches merge into bigger ones as time goes on. Obviously, merger trees play a very

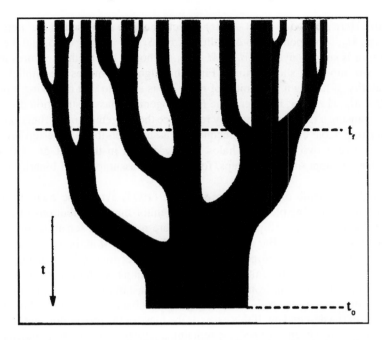

Fig. 7.4. Illustration of a merger tree, depicting the growth of a dark matter halo as the result of a series of mergers. Time increases from top to bottom, and the widths of the tree branches represent the masses of the individual halos. A horizontal slice through the tree, such as that at $t = t_\mathrm{f}$ gives the distribution of the masses of the progenitor halos at a given time. [Adapted from Lacey & Cole (1993)]

important role in hierarchical models of galaxy formation. In particular, they are the backbone of the semi-analytical models of galaxy formation discussed in §15.7.1.

In order to construct a merger tree of a halo, it is most convenient to start from the trunk and work upwards. Consider a halo with mass M at a time t (which can be the present time, or any other time). At a slightly earlier time $t - \Delta t$, the progenitor distribution, in a statistical sense, is given by Eq. (7.81). This can be used to draw a set of progenitor halos, after which the procedure is repeated for each of these progenitors, thus advancing upwards into the merger tree. In general, the construction of a set of progenitor masses for a given parent halo mass needs to obey two requirements. First of all, the *number* distribution of progenitor masses of many independent realizations needs to follow Eq. (7.81). Secondly, mass needs to be conserved, so that in each individual realization the sum of the progenitor masses is equal to the mass of the parent halo. In principle, this requirement for mass conservation implies that the probability for the mass of the nth progenitor needs to be conditional on the masses of the $n - 1$ progenitor halos already drawn. Unfortunately, these conditional probability functions are not derivable from the EPS formalism, so additional assumptions have to be made. This has resulted in the development of various different algorithms for constructing halo merger trees, each with its own pros and cons (Kauffmann & White, 1993; Sheth & Lemson, 1999; Somerville & Kolatt, 1999; Cole et al., 2000, 2008; Zhang et al., 2008; Neistein & Dekel, 2008).

One of the simplest merger trees is the so-called binary tree, in which the assumption is made that each merger involves exactly two progenitors (e.g. Lacey & Cole, 1993; Cole et al., 2000). In order to obtain the masses of the two progenitor halos of a halo of mass M at time t, one can proceed as follows. First draw a value for ΔS from the mass-weighted probability function

$$f_{\mathrm{FU}}(\Delta S, \Delta \delta)\, \mathrm{d}\Delta S = \frac{1}{\sqrt{2\pi}} \frac{\Delta \delta}{\Delta S^{3/2}} \exp\left[-\frac{\Delta \delta^2}{2\Delta S}\right] \mathrm{d}\Delta S, \qquad (7.82)$$

[see Eq. (7.80)]. Here $\Delta\delta = \delta_c(t - \Delta t) - \delta_c(t)$ reflects the time step used in the merger tree. The progenitor mass, M_p, corresponding to ΔS follows from $S(M_p) = S(M) + \Delta S$, and the mass of the second progenitor is simply given by $M'_p = M - M_p$. Although this method ensures mass conservation, it has two important shortcomings. First, this algorithm makes the implicit assumption that the probability of having a progenitor of mass M_p is equal to that of having a progenitor of mass $M'_p = M - M_p$. However, in general the EPS progenitor mass distribution does not respect this symmetry, causing a failure to accurately reproduce the progenitor mass functions, especially at earlier times (e.g. Cole et al., 2008). Second, the assumption that a timestep involves only a single merger between two progenitor halos is only correct in the limit $\Delta t \to 0$. In practice, however, one has to adopt finite time steps, for which the assumption of binarity is not strictly valid.

The tree-building scheme of Kauffmann & White (1993) avoids the assumption of binary mergers and ensures, in principle, that the EPS progenitor mass distributions of Eq. (7.80) (or the equivalent distributions for ellipsoidal collapse) are reproduced for all t_1 and t_2 and for all M_1 above some resolution limit. For each descendant halo mass M at time t_n one makes a library of N_{real} realizations of possible progenitor sets at the earlier time t_{n-1} by using Eq. (7.80) to calculate the total number of progenitors (over all realizations) with mass $M_p < M$ in each of a number of mass bins. Proceeding from most massive to least massive, these progenitors are then distributed at random among the N_{real} descendants with probability proportional to the remaining unassigned mass of the descendant, if this exceeds the mass of the progenitor, and zero otherwise. Once all progenitors have been assigned, the remaining descendant mass is assumed to have been accreted smoothly in objects below the resolution limit. Randomly chosen progenitor sets from such 'one-step' libraries can then be combined into trees giving the complete merger history of each final halo. As discussed by Zhang et al. (2008), considerable care is needed in programming this algorithm to avoid discreteness artifacts due to the mass binning.

A third method for constructing halo merger trees is the N-branch scheme with accretion of Somerville & Kolatt (1999). For each progenitor, a value for ΔS is drawn from Eq. (7.82), and its corresponding mass, M_p, is determined. For each new progenitor one checks whether the sum of the progenitor masses drawn thus far exceeds the mass of the parent, M. If this is the case the progenitor is rejected and a new progenitor mass is drawn. Any progenitor with $M_p < M_{min}$ is added to the mass component M_{acc} that is considered to be accreted onto the parent in a smooth fashion (i.e. the assembly history of such small progenitors is not followed). Here again M_{min} reflects the mass resolution chosen for the merger tree. This procedure is repeated until the total mass left, $M_{left} = M - M_{acc} - \sum M_p$, is less than M_{min}. This remaining mass is then assigned to M_{acc} and one moves on to the next time step. Although this algorithm, by construction, ensures exact mass conservation, and yields conditional mass functions that are in fair agreement with direct predictions from EPS theory (i.e. the method is self-consistent), it is not very rigorous. In particular, mass conservation is enforced 'by hand', by rejecting progenitor masses that overflow the mass budget. Consequently, the mass distribution of first-drawn progenitors differs from that for second-drawn progenitors, and so on. Somewhat fortunately, the sum of these distributions matches the distribution of Eq. (7.81) for all progenitors, but only if sufficiently small time steps $\Delta\delta$ are used.

These three tree-building algorithms were compared to each other, to the scheme of Cole et al. (2000), and to several new schemes by Zhang et al. (2008). Of the older schemes, only that of Kauffmann & White (1993) was found to reproduce the EPS conditional probability distributions over wide ranges of mass and time. This is not a sufficient condition for an algorithm to be realistic, however, since the EPS formalism itself is not rigorous and it does not specify many merger tree properties of interest. Further evaluation of tree-building schemes thus requires detailed comparison with simulations of structure formation. Trees extracted directly from such simulations can provide an alternative to EPS merger trees and have the advantages that they

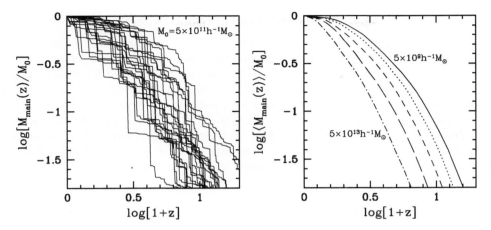

Fig. 7.5. Main progenitor histories obtained using EPS merger trees as described in van den Bosch (2002b). The left-hand panel shows 25 (normalized) main progenitor histories, $M_{\rm main}(z)/M_0$, for a halo with $M_0 = 5 \times 10^{11} h^{-1} \, M_\odot$ in an EdS Universe. The right-hand panel shows the *average* main progenitor histories, averaged over 1,000 realizations, for halos with five different masses: $5 \times 10^9 h^{-1} \, M_\odot$ (solid line), $5 \times 10^{10} h^{-1} \, M_\odot$ (dotted line), $5 \times 10^{11} h^{-1} \, M_\odot$ (short-dashed line), $5 \times 10^{12} h^{-1} \, M_\odot$ (long-dashed line), $5 \times 10^{13} h^{-1} \, M_\odot$ (dot-dashed line). Note that more massive halos assemble later.

do not depend on the EPS assumptions and that they automatically provide spatial and velocity information, but even here there are major limitations. High-resolution simulations are expensive, are time-consuming to process, and have limited resolution. In addition, there is no unambiguous definition in a simulation (or, presumably, in the real Universe) both for the boundary of a halo, and for the time when two halos should be considered to have merged. Different definitions give rise to different trees.

7.3.3 Main Progenitor Histories

The full merger history of any individual dark matter halo is a complex structure containing a lot of information. A useful subset of this information is provided by the main progenitor history, sometimes called the mass accretion history or mass assembly history, which restricts attention to the main 'trunk' of the merger tree. This main trunk is defined by following the branching of a merger tree back in time, and selecting at each branching point the most massive (main) progenitor. Note that with this definition, the main progenitor is not necessarily the most massive progenitor of its generation at a given time, even though it never accretes other halos more massive than itself.

The main progenitor history, $M_{\rm main}(t)$, can be obtained from merger trees constructed using either the EPS formalism (van den Bosch, 2002b) or numerical simulations (Wechsler et al., 2002; Zhao et al., 2008). The left-hand panel of Fig. 7.5 shows 25 examples of main progenitor histories as functions of redshift, $M_{\rm main}(z)$, normalized by the present-day halo mass, M_0. All histories correspond to a halo of $M_0 = 5 \times 10^{11} h^{-1} \, M_\odot$ in an EdS cosmology, and have been obtained using EPS merger trees constructed using the N-branching method described in the previous section. Note that these main progenitor histories reveal a large amount of scatter. The right-hand panel of Fig. 7.5 shows the *average* main progenitor histories, $\langle M_{\rm main}(z)/M_0 \rangle$, obtained by averaging 1,000 random realizations for a given M_0. Results are shown for five values of M_0, as indicated. As is evident, on average halos that are more massive assemble later. As we demonstrate below, this is a characteristic of hierarchical structure formation.

7.3.4 Halo Assembly and Formation Times

The concept of the main progenitor defined above allows one to define a useful characteristic time for the assembly of a dark matter halo. A convenient operational definition for the assembly time, t_a, of a halo of mass M_0 is the time when its main progenitor first reaches a mass $M_0/2$, i.e. $M_{\text{main}}(t_a) = M_0/2$. The probability distribution of t_a can be obtained as follows. From Eq. (7.80) the probability that a random mass element of a collapsed object of mass M_0 at time t_0 was part of a progenitor of mass M_1 at an earlier time $t_1 < t_0$ is $f_{\text{FU}}(S_1, \delta_1 | S_0, \delta_0)|dS_1/dM_1|dM_1$, and the average number of progenitors that each M_0 halo has is given by Eq. (7.81). However, since each object can have at most one progenitor with mass $M_0/2 < M_1 < M_0$, the probability that any particular object has a progenitor in this mass range, and thus an assembly time earlier than t_a, is given by

$$\mathscr{P}(<t_a|M_0,t_0) = \int_{M_0/2}^{M_0} \frac{M_0}{M_1} f_{\text{FU}}(S_1,\delta_1|S_0,\delta_0) \left|\frac{dS_1}{dM_1}\right| dM_1$$

$$= \int_{S_0}^{S_{\frac{1}{2}}} \frac{M(S_0)}{M(S_1)} f_{\text{FU}}(S_1,\delta_1|S_0,\delta_0) \, dS_1, \qquad (7.83)$$

where $S_{\frac{1}{2}} = S(M_0/2)$ and $t_1 = t_a$. Introducing the variables

$$\tilde{S} \equiv \frac{S_1 - S_0}{S_{\frac{1}{2}} - S_0}, \quad \text{and} \quad \tilde{\omega}(t_1) \equiv \frac{\delta_1 - \delta_0}{\sqrt{S_{\frac{1}{2}} - S_0}}, \qquad (7.84)$$

one can cast Eq. (7.83) in the form

$$\mathscr{P}(<t_a|M_0,t_0) = \mathscr{P}(>\tilde{\omega}_a|M_0,t_0)$$

$$= \frac{1}{\sqrt{2\pi}} \int_0^1 \frac{M(S_0)}{M(S_1)} \frac{\tilde{\omega}_a}{\tilde{S}^{3/2}} \exp\left[-\frac{\tilde{\omega}_a^2}{2\tilde{S}}\right] d\tilde{S}, \qquad (7.85)$$

where $\tilde{\omega}_a = \tilde{\omega}(t_a)$ (Lacey & Cole, 1993). For given M_0 and t_0, Eq. (7.85) gives the cumulative distribution of $\tilde{\omega}_a$ (and therefore of t_a). The advantage of using the variables $\tilde{S}$ and $\tilde{\omega}$ is that $P(>\tilde{\omega}_a)$ depends only very mildly on halo mass, M_0, and cosmology (this dependence is largely absorbed by the variables themselves). For example, for a power-law fluctuation spectrum with spectral index n, we can write

$$S(M) = \delta_c^2 \left[\frac{M}{M^*}\right]^{-\alpha}, \quad \text{with} \quad \alpha = \frac{n+3}{3}, \qquad (7.86)$$

(see §6.1.3), so that

$$\frac{M(S_0)}{M(S_1)} = \left[1 + (2^\alpha - 1)\tilde{S}\right]^{1/\alpha}. \qquad (7.87)$$

Upon substituting Eq. (7.87) in Eq. (7.85) it is clear that $\mathscr{P}(>\tilde{\omega}_a)$ is independent of M_0 and t_0 and with only weak dependence on the spectral index n.

The shaded histograms in Fig. 7.6 show the differential distributions, $d\mathscr{P}/d\tilde{\omega}_a$, for dark matter halos in the mass range $M_{\text{min}} < M < M_{\text{max}}$ (with M_{min} and M_{max} as indicated in each panel) in a flat ΛCDM cosmology with $\Omega_{m,0} = 0.3$, $n = 1$ and $\sigma_8 = 0.9$ obtained from an N-body simulation. For comparison, the solid curves show the corresponding predictions from the EPS formalism, obtained using Eq. (7.85) for halos with $M_0 = 8.4 \times 10^{11} h^{-1} \, \text{M}_\odot$ (left-hand panel) and $M_0 = 1.1 \times 10^{13} h^{-1} \, \text{M}_\odot$ (right-hand panel). As is evident, the EPS formalism predicts values for $\tilde{\omega}_a$ that are somewhat too low, corresponding to assembly times t_a that are too high; halos in N-body simulations assemble earlier than predicted by the EPS formalism when adopting the spherical collapse model to compute $\delta_c(t)$ (e.g. van den Bosch, 2002b; Neistein et al., 2006).

7.3 Progenitor Distributions and Merger Trees

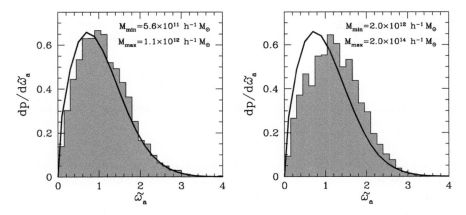

Fig. 7.6. The differential distribution of halo assembly times parameterized by the scaled variable $\tilde{\omega}_a$. The shaded histograms show the results for halos in the mass range $M_{\min} < M < M_{\max}$ in a flat ΛCDM cosmology with $\Omega_{m,0} = 0.3$, $n = 1$ and $\sigma_8 = 0.9$ obtained from an N-body simulation, while the solid curves show the corresponding distributions obtained from Eq. (7.85). Note that the assembly times in the simulation are offset from those obtained using the EPS formalism, and that the discrepancy is larger for more massive halos. [After van den Bosch (2002b)]

However, as shown by Giocoli et al. (2007), using the ellipsoidal collapse model instead yields assembly times in much better agreement with the simulation results.

For an EdS cosmology, $\delta_c(t) \propto 1+z$, where z is the redshift corresponding to time t. Using the definition of $\tilde{\omega}_a$, we obtain a simple approximate expression for the median value of the assembly redshift:

$$z_{f,m} = z_0 + \tilde{\omega}_{a,m} (1+z_0) \left\{ \sqrt{2^{(n+3)/3} - 1} \left[\frac{M_0}{M^*} \right]^{-(n+3)/6} \right\}. \tag{7.88}$$

Here z_0 is the redshift at which the halo of mass M_0 is identified, and $\tilde{\omega}_{a,m} \sim 1$ is the median value of $\tilde{\omega}_a$. From this we can see that, in a model with $n > -3$, halos of a higher mass assemble later, which is an important characteristic of hierarchical structure formation. Note also that, for a given power spectrum, the distribution of assembly times for halos with the same M/M^* only depends on cosmology through the linear growth rate $D(t)$. Since $D(t)$ has a weaker dependence on t in a cosmology with lower $\Omega_{m,0}$ (see §4.1.6), halos of a given mass typically assemble earlier in a lower-density universe.

In addition to a halo assembly time based on main progenitor growth, one can also define a characteristic formation time based on a halo's *entire* merger history. For example, following Navarro et al. (1996), one may define the formation time, t_f, of a halo of mass M_0 at time t_0, as the time when the sum of all progenitors with mass $M > M_{\min}$, hereafter $M_{\rm all}$, is equal to $M_0/2$. The minimum progenitor mass in this definition can be taken to be a fixed fraction of the final mass, as was done by Navarro et al. (1996), or as a fixed value, independent of the mass of the descendant, as was done by Neistein et al. (2006).

In principle, obtaining the formation history, $M_{\rm all}(t)$, of an individual halo requires the construction of an entire merger tree with a mass resolution $M \leq M_{\min}$. However, the *average* formation history of a halo of mass M_0, averaged over many merger trees, can be derived straightforwardly from the EPS formalism as

$$\langle M_{\rm all} \rangle(t) = \int_{M_{\min}}^{M_0} n(M,t|M_0,t_0) M \, dM, \tag{7.89}$$

where $n(M,t|M_0,t_0)$ is the mass function of progenitors at time t that have merged to form a halo of mass M_0 by time $t_0 > t$. Substituting Eq. (7.81) into the above equation yields

$$\frac{\langle M_{\rm all}\rangle(t)}{M_0(t_0)} = 1 - {\rm erf}\left[\frac{\delta_{\rm c}(t) - \delta_{\rm c}(t_0)}{\sqrt{2(S_{\rm min} - S_0)}}\right], \qquad (7.90)$$

with $S_{\rm min} = S(M_{\rm min})$. Defining the average formation time, $\bar{t}_{\rm f}$, according to $\langle M_{\rm all}\rangle(\bar{t}_{\rm f}) = M_0/2$, we obtain that

$$\delta_{\rm f} \equiv \delta_{\rm c}(\bar{t}_{\rm f}) = \delta_{\rm c}(t_0) + \beta\sqrt{S_{\rm min} - S_0}, \qquad (7.91)$$

where $\beta = \sqrt{2}/{\rm erf}(1/2) \simeq 0.6745$. In a hierarchical universe, S decreases with increasing halo mass, and $\delta_{\rm c}(t)$ decreases with time. We then find that $\bar{t}_{\rm f}$ increases with increasing halo mass if $M_{\rm min}$ is defined as a fixed fraction of M_0, but *decreases* with increasing halo mass if it is defined to be independent of M_0. Thus, although more massive halos, on average, form later, their progenitors typically cross any given mass threshold earlier.

This is particularly relevant for understanding the star-formation histories of galaxies. Since star formation is expected to take place in all halos above a given minimum mass (in which atomic cooling is expected to be efficient; see §8.4), and not only in the main progenitor halo, one expects that more massive halos host galaxies with older stellar populations (e.g. Li et al., 2008). As we will see in §13.5, this is consistent with observations.

7.3.5 Halo Merger Rates

From Eq. (7.80) we can obtain the reverse conditional probability

$$f_{\rm FU}(S_2,\delta_2|S_1,\delta_1)\,{\rm d}S_2 = \frac{f_{\rm FU}(S_1,\delta_1|S_2,\delta_2)\,f_{\rm FU}(S_2,\delta_2)}{f_{\rm FU}(S_1,\delta_1)}\,{\rm d}S_2$$

$$= \frac{1}{\sqrt{2\pi}}\frac{\delta_2(\delta_1-\delta_2)}{\delta_1}\left[\frac{S_1}{S_2(S_1-S_2)}\right]^{3/2}\exp\left[-\frac{(\delta_2 S_1 - \delta_1 S_2)^2}{2S_1 S_2(S_1-S_2)}\right]{\rm d}S_2, \quad (7.92)$$

where $S_i = \sigma^2(M_i)$ and $\delta_i = \delta_{\rm c}(t_i)$ with $S_1 > S_2$ and $\delta_1 > \delta_2$, and $f_{\rm FU}(S,\delta_{\rm c})$ is given by Eq. (7.56). According to the interpretation of the excursion set, this is the conditional probability that a halo of mass M_1 at time t_1 is incorporated into a halo with a mass between M_2 and $M_2 + {\rm d}M_2$ ($M_2 > M_1$) at a later time $t_2 > t_1$. If we set $M_2 = M_1 + \Delta M$ and $t_2 = t_1 + \Delta t$, then Eq. (7.92) gives the probability for the halo to gain a mass ΔM by merging or accretion in the time interval Δt. Thus, the rate at which a halo with mass M transits to a halo with mass between M and $M + \Delta M$ is given by

$$\mathscr{P}(\Delta M|M,t)\,{\rm d}\ln\Delta M\,{\rm d}\ln t = \frac{1}{\sqrt{2\pi}}\left[\frac{S_1}{(S_1-S_2)}\right]^{3/2}\exp\left[-\frac{\delta_{\rm c}^2(S_1-S_2)}{2S_1 S_2}\right]$$

$$\times \left|\frac{{\rm d}\ln\delta_{\rm c}}{{\rm d}\ln t}\right|\frac{\delta_{\rm c}}{\sqrt{S_2}}\left|\frac{{\rm d}\ln S_2}{{\rm d}\ln\Delta M}\right|{\rm d}\ln t\,{\rm d}\ln\Delta M, \qquad (7.93)$$

where $S_1 = \sigma^2(M)$ and $S_2 = \sigma^2(M+\Delta M)$. In any finite time interval Δt, the change in mass, ΔM, can be due to the cumulative effects of more than one merger. However, for an infinitesimal time interval ${\rm d}t$, the transition from M_1 to M_2 is most likely due a single merger event. Thus Eq. (7.93) gives the merger rate of a halo with mass M at time t with another halo with mass ΔM. Fig. 7.7 shows $\mathscr{P}(\Delta M|M,t)$ as a function of $\log(\Delta M/M)$ for power-law spectra with spectral index $n = -2$ and 0 in an EdS universe. In such models, ${\rm d}\ln\delta_{\rm c}(t)/{\rm d}\ln t = -2/3$, and $\mathscr{P}$ is independent of t if all masses are measured in units of $M^*(t)$. There are several interesting properties to notice. (i) For $\Delta M \ll M$, $\mathscr{P}$ is a power law of ΔM: $\mathscr{P} \propto (\Delta M)^{-1/2}$. This asymptote can be obtained

7.3 Progenitor Distributions and Merger Trees

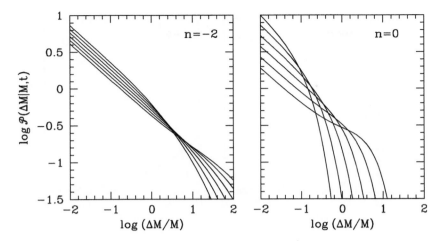

Fig. 7.7. The merger rate $\mathscr{P}(\Delta M|M,t)$ given by Eq. (7.93) as a function of $\log(\Delta M/M)$ for power-law spectra with spectral index $n = -2$ and 0 in an EdS universe. The six curves in each panel correspond to $M/M^* = 16, 8, 4, 2, 1$, and 0.5, where a larger M has a larger value of $\mathscr{P}$ at the low mass end of ΔM. [After Lacey & Cole (1993)]

directly from Eq. (7.93). Thus the number of merger events is dominated by minor mergers (those with $\Delta M \ll M$), whereas the mass accretion rate (i.e. the mass-weighted merger rate) is dominated by major mergers with $\Delta M \simeq M$. (ii) For $M \gg M^*$, mergers with halos of similar or larger masses are rare, while such mergers are more frequent for $M \ll M^*$. (iii) Since a spectrum with a more negative n has more power on large scales relative to that on small scales, major mergers (i.e. those with a large $\Delta M/M$) happen more frequently for a spectrum with a more negative spectral index.

7.3.6 Halo Survival Times

An interesting quantity that can be calculated from Eq. (7.92) is the survival time of a given dark matter halo. Let us consider all halos of mass M_1 at time t_1. The probability for such a halo to merge, by a time $t_2 > t_1$, into a new halo with mass larger than M_2 is

$$\mathscr{P}(S < S_2, \delta_2 | S_1, \delta_1) = \int_0^{S_2} f_{\rm FU}(S_2', \delta_2 | S_1, \delta_1) \, dS_2'. \tag{7.94}$$

This is equal to the probability that a halo makes a transition from $M < M_2$ to $M > M_2$ at some $t < t_2$, which we denote by $\mathscr{P}(S_2, t < t_2 | S_1, t_1)$. Inserting Eq. (7.92) into Eq. (7.94) we obtain

$$\mathscr{P}(S_2, t < t_2 | S_1, t_1) = \frac{1}{2}[1 - {\rm erf}(A)] \\ + \frac{1}{2}\frac{(\delta_1 - 2\delta_2)}{\delta_1} \exp\left[\frac{2\delta_2(\delta_1 - \delta_2)}{S_1}\right][1 - {\rm erf}(B)], \tag{7.95}$$

where $S_2 < S_1$, $\delta_2 < \delta_1$, and

$$A = \frac{S_1\delta_2 - S_2\delta_1}{\sqrt{2S_1 S_2(S_1 - S_2)}}; \qquad B = \frac{S_2(\delta_1 - 2\delta_2) + S_1\delta_2}{\sqrt{2S_1 S_2(S_1 - S_2)}}. \tag{7.96}$$

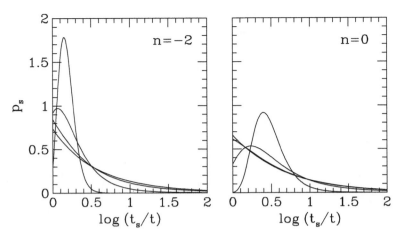

Fig. 7.8. The differential distribution of halo survival times calculated from Eq. (7.97) with $t_1 = t$, $t_2 = t_s$, $M_1 = M$, and $M_2 = 2M$, for power-law spectra with $n = -2$ (left panel) and 0 (right panel) in an EdS universe. The four curves in each panel correspond to $[M/M^*(t)]^{-(n+3)/6} = 3, 1, 0.3$, and 0.1 (from the most peaked curve to the most extended curve). [After Lacey & Cole (1993)]

The differential form of the distribution can be obtained by differentiating Eq. (7.95) with respect to t_2:

$$p_s(S_2,t_2|S_1,t_1)dt_2 \equiv \frac{\partial}{\partial t_2} \mathscr{P}(S_2,t<t_2|S_1,t_1)\,dt_2$$

$$= \frac{\delta_2}{\delta_1}\left[\frac{S_1}{S_2(S_1-S_2)}\right]^{1/2} \exp\left[\frac{2\delta_2(\delta_1-\delta_2)}{S_1}\right]$$

$$\times \left\{\left(\frac{2}{\pi}\right)^{1/2} \frac{-S_2(\delta_1-2\delta_2)-S_1(\delta_1-\delta_2)}{S_1} \exp(-B^2)\right.$$

$$\left. + \left[\frac{S_2(S_1-S_2)}{S_1}\right]^{1/2}\left[1-\frac{(\delta_1-2\delta_2)^2}{S_1}\right][1+\mathrm{erf}(-B)]\right\}d\ln\delta_2. \quad (7.97)$$

If we define a survival time t_s for a halo with mass M at time t as the cosmic time when the mass of the halo has doubled, the halo survival time distribution can be obtained by setting $t_2 = t_s$, $t_1 = t$, $S_1 = \sigma^2(M)$ and $S_2 = \sigma^2(2M)$ in Eqs. (7.95) and (7.97).

Fig. 7.8 shows the differential distribution of halo survival times for power-law spectra with $n = -2$ and 0 in an EdS universe. For $M \ll M^*(t)$, the distribution of t_s/t is very broad, indicating that some low-mass halos can survive for many Hubble times. In contrast, halos with $M \gg M^*(t)$ are unlikely to survive for more than a few Hubble times. For models with $n > -3$, the asymptotic behavior of the median survival time, $\hat{t}_s$, can be obtained from Eq. (7.95): $\hat{t}_s/t \to 2^{3/2}$ for $M \ll M^*(t)$ and $\hat{t}_s/t \to 2^{3\alpha/2}$ for $M \gg M^*(t)$, with α given by Eq. (7.86). Thus, for $n < 0$, the median survival time decreases with increasing halo mass. For a given M/M^*, the median survival time also decreases with decreasing n, reflecting the fact that a smaller value of n corresponds to stronger clustering power on large scales relative to small scales. Thus the survival time scales very similarly to the time since assembly, as given by Eq. (7.88).

It is interesting to compare the survival time interval $\Delta t_s \equiv t_s - t$ or the time since assembly $\Delta t_f \equiv t - t_f$ to the intrinsic dynamical time of a halo, $t_{\rm dyn}$. Only if the dynamical time is sufficiently short compared to these times will the halo spend much of its life in dynamical equilibrium. In the spherical collapse model discussed in §5.1, a mass shell reaches its maximum expansion at

$t_{\max}$ and collapses at $t_{\mathrm{col}} = 2t_{\max}$. This suggests that we take $t_{\max}$ as a rough estimate of the time a halo needs to reach equilibrium. Hence, we define $t_{\mathrm{dyn}} = t_{\max}$, and a halo that collapses at time t has $t_{\mathrm{dyn}} \sim t/2$. The power-law models discussed above yield $\Delta \hat{t}_{\mathrm{s}}$ and $\Delta \hat{t}_{\mathrm{f}}$ both $\sim 2t_{\mathrm{dyn}}$ for $M \ll M^*(t)$, and both $\sim [2^{(n+5)/2} - 2]t_{\mathrm{dyn}}$ for $M \gg M^*(t)$. Hence, for $n \lesssim -2$, most halos with $M \gg M^*$ spend little time in dynamical equilibrium. Although the exact fraction depends on the definition of t_{dyn}, it is clear that for the CDM-like power spectra discussed in §4.4.4, which all have an effective spectral index close to -2 on galactic scales, a large fraction of the galaxy-mass dark matter halos are expected to be dynamically young, especially at earlier times when the characteristic mass, $M^*(t)$, is smaller.

7.4 Spatial Clustering and Bias

7.4.1 Linear Bias and Correlation Function

The progenitor distribution of dark matter halos described in §7.3.1 can be extended to construct a model for their spatial clustering (Mo & White, 1996). To start with, recall that the average number of halos with mass M_1 identified at redshift z_1 that will merge into a larger halo with mass M_0 at redshift $z_0 < z_1$ is given by

$$N(1|0)\,dM_1 = \frac{M_0}{M_1} f(1|0) \left| \frac{dS_1}{dM_1} \right| dM_1, \qquad (7.98)$$

where

$$f(1|0)\,dS_1 \equiv \frac{1}{\sqrt{2\pi}} \frac{\delta_1 - \delta_0}{(S_1 - S_0)^{3/2}} \exp\left[-\frac{(\delta_1 - \delta_0)^2}{2(S_1 - S_0)} \right] dS_1. \qquad (7.99)$$

Eq. (7.98) is the same as Eq. (7.81) except for the change of notation. In deriving this equation it is not necessary that M_0 itself is a halo; this equation holds equally well even if M_0 is an uncollapsed spherical region. In this case, Eq. (7.98) can be interpreted as the average number of M_1 halos identified at redshift z_1 in a spherical region with comoving radius $R_0 \equiv (3M_0/4\pi \overline{\rho}_0)^{1/3}$ and with linear density contrast δ_0. Clearly this number has significant dependence on δ_0, and we quantify this by calculating the average overabundance of halos in such regions relative to the global mean halo abundance:

$$\delta_{\mathrm{h}}^{\mathrm{L}}(1|0) = \frac{N(1|0)}{n(M_1,z_1)V_{\mathrm{L}}} - 1, \quad \text{where} \quad V_{\mathrm{L}} \equiv \frac{4\pi}{3} R_0^3. \qquad (7.100)$$

This expression becomes particularly simple when $M_0 \gg M_1$ (so that $S_0 \ll S_1$) and $|\delta_0| \ll \delta_1$:

$$\delta_{\mathrm{h}}^{\mathrm{L}}(1|0) = \frac{\nu_1^2 - 1}{\delta_1} \delta_0, \quad \text{where} \quad \nu_1 = \frac{\delta_1}{\sqrt{S_1}}. \qquad (7.101)$$

Note that $\delta_1 = \delta_{\mathrm{c}}/D(t_1)$, and δ_0 is the linear density contrast linearly extrapolated to the present time. This expression gives the overabundance of halos without taking into account the dynamical contraction (or expansion) of the region (R_0, δ_0). Hence, it is the overabundance in *Lagrangian* space. Since by definition $N(1|0)$ is the number of halos associated with the mass of the spherical region (R_0, δ_0), the enhancement factor at time $t > t_1$ taking into account the overdensity of the region is

$$\delta_{\mathrm{h}}(1|0) = \frac{N(1|0)}{n(M_1,z_1)V_{\mathrm{L}}} \frac{V_{\mathrm{L}}}{V_{\mathrm{E}}} - 1, \qquad (7.102)$$

where V_E is the volume in the evolved *Eulerian* space. Using that $V_L/V_E = 1 + \delta(t)$, where δ is the average mass overdensity within the region in Eulerian space, we have

$$\delta_h(1|0) = \delta(t) + \frac{\nu_1^2 - 1}{\delta_1}\delta_0 + \frac{\nu_1^2 - 1}{\delta_1}\delta_0\delta(t). \tag{7.103}$$

In the linear regime where $\delta(t) \approx \delta_0 D(t) \ll 1$, we have

$$\delta_h(1|0) = b_h(M_1, \delta_1; t)\delta(t), \tag{7.104}$$

where

$$b_h(M_1, \delta_1; t) = 1 + \frac{1}{D(t)}\left(\frac{\nu_1^2 - 1}{\delta_1}\right) \tag{7.105}$$

is the bias factor at time t for halos identified at time t_1. Thus, the overabundance of halos is enhanced with respect to the underlying mass overdensity δ by a factor b_h, which depends on both their mass, M_1, and the time, t_1 (through δ_1), at which they are identified. If we define a characteristic mass at t_1 by $\sigma(M^*) = \delta_1$ [see Eq. (7.48)], we see that halos with masses $M_1 > M^*$ are biased ($b_h > 1$), while halos with $M_1 < M^*$ are anti-biased ($b_h < 1$), relative to the mass density field. Since b_h defined above is independent of δ and R, Eq. (7.104) is called the *linear bias* relation for halos. In general, as we will see in the next subsection, the halo bias may depend on both δ and R. In such cases we speak of nonlinear and scale-dependent bias, respectively.

As first shown by Jing (1998), the bias model described above, which is based on the spherical collapse model, suffers from similar inaccuracies as the original Press–Schechter mass function, and indeed the two discrepancies are closely related (see Sheth & Tormen, 1999). A more precise model for halo bias can be obtained from the ellipsoidal collapse model described in §7.2.3:

$$b_h(M_1, \delta_1, t) = 1 + \frac{1}{D(t)\delta_1}\left[\nu_1'^2 + b\nu_1'^{2(1-c)} - \frac{\nu_1'^{2c}/\sqrt{a}}{\nu'^{2c} + b(1-c)(1-c/2)}\right], \tag{7.106}$$

where $\nu_1' = \sqrt{a}\nu_1$, $a = 0.707$, $b = 0.5$ and $c = 0.6$ (Sheth et al., 2001). Numerical simulations show that this revision is substantially more accurate than its spherical counterpart, Eq. (7.105), especially for halos with $M < M^*$.

The bias relation (7.104) gives the average overabundance of halos in regions of radius R with mean mass density contrast δ. For a particular region, however, the overabundance depends not only on δ but also on other properties of the region. As a result, the relation between δ_h and δ is expected to be stochastic instead of deterministic. Taking this into account, we can write the linear bias relation between δ_h and δ at a point $\mathbf{r}$ as

$$\delta_h(1|0; \mathbf{r}, t) = b_h(M_1, \delta_1; t)\delta(\mathbf{r}, t) + \varepsilon(\mathbf{r}), \tag{7.107}$$

where ε is a stochastic term having zero mean: $\langle \varepsilon | \delta \rangle = 0$. This can be considered to be the overdensity field of halos of mass M_1 identified at time t_1. The two-point correlation function of halos can then be written as

$$\xi_h(M_1, \delta_1; r, t) = \langle \delta_h(1|0; \mathbf{r}_1, t)\delta_h(1|0; \mathbf{r}_1 + \mathbf{r}, t)\rangle_{\mathbf{r}_1}. \tag{7.108}$$

If the stochastic term is 'short-ranged' so that $\langle \varepsilon(\mathbf{r}_1)\varepsilon(\mathbf{r}_1 + \mathbf{r})\rangle = 0$ and $\langle \delta(\mathbf{r}_1, t)\varepsilon(\mathbf{r}_1 + \mathbf{r})\rangle = 0$, then

$$\xi_h(M_1, \delta_1; r, t) = b_h^2(M_1, \delta_1, t)\xi(r, t), \tag{7.109}$$

where $\xi(r) = \langle \delta(\mathbf{r}_1)\delta(\mathbf{r}_1 + \mathbf{r})\rangle$ is the autocorrelation function of mass.

We can rewrite Eq. (7.109) as

$$\xi_h(r, t) = \left[\sqrt{\xi_h^L(r)} + \sqrt{\xi(r, t)}\right]^2, \tag{7.110}$$

where

$$\xi_{\rm h}^{\rm L}(r) \equiv \langle \delta_{\rm h}^{\rm L}(1|0;{\bf r}_1)\delta_{\rm h}^{\rm L}(1|0;{\bf r}_1+{\bf r}) \rangle = D^2(t)\left[b_{\rm h}(M_1,\delta_1,t)-1\right]^2 \xi(r,t_0) \qquad (7.111)$$

is the autocorrelation function of halos in Lagrangian space, and $\xi(r,t_0) = \xi(r,t)/D^2(t)$ is the mass correlation function linearly extrapolated to the present time t_0. Note that the Lagrangian correlation function $\xi_{\rm h}^{\rm L}(r)$ is independent of t. Thus, for halos with $M_1 \gg M^*(t_1)$, $\xi_{\rm h}^{\rm L}$ dominates over ξ, and the correlation function is determined by the 'pattern' in the initial density field rather than by gravitational clustering of the underlying mass distribution. For low-mass halos that formed at some epoch t_1, however, the correlation function becomes more and more comparable to that of the underlying mass distribution as time proceeds, which follows from the fact that $D(t)$ increases monotonically with t. Note, though, that Eqs. (7.105) and (7.106) have to be interpreted with care when $t > t_1$, since it is only strictly valid when the halos of mass M_1 at time t_1 survive to time t. Clearly, this is not valid in a hierarchical Universe, in which halos of a given mass will merge to give rise to a population of more massive halos. In this sense, the bias relation (7.104) and the correlation function (7.109) are only valid in the case where $t = t_1$. However, galaxies which formed at the halo centers at t_1 may still remain distinct at t (i.e. they may survive as distinct objects even when their parent halos merge). The results for $t > t_1$ may then be relevant for galaxies, and suggest that galaxies formed at high redshifts tend to become more and more faithful tracers of the underlying mass distribution as gravitational clustering proceeds (provided that they survive as individual systems).

The bias model described above can be used to predict the two-point correlation function of galaxy clusters (Mo et al., 1996). According to the bias model described above, clusters, which are the most massive virialized objects in the Universe, are expected to be strongly biased with respect to the mass distribution. Because clusters of galaxies are separated by large distances, their two-point correlation function can only be measured reliably on scales larger than about $5\,h^{-1}{\rm Mpc}$, where the mass distribution is only mildly nonlinear. Hence, the linear bias model is expected to be a good approximation. Using Eqs. (7.74), (7.105) and (7.109), we can write the two-point correlation function of clusters, $\xi_{\rm cc}$, as

$$\left[\frac{\xi_{\rm cc}(r)}{\xi_{\rm N}(r)}\right]^{1/2} = \sigma_8\left(1-\frac{1}{\delta_{\rm c}}\right) + \frac{\delta_{\rm c}}{\sigma_8}\left(\frac{R_0}{r_8}\right)^{2\beta}, \qquad (7.112)$$

where $\xi_{\rm N}$ is the two-point correlation function of the underlying mass distribution for a linear power spectrum with $\sigma_8 = 1$. Using Eq. (7.73) we find that

$$\left[\frac{\xi_{\rm cc}}{\xi_{\rm N}}\right]^{1/2} = \sigma_8\left(1-\frac{1}{\delta_{\rm c}}\right) + 1.1^{2\beta}\delta_{\rm c} m_{\rm A}^\beta \left(\sigma_8 \Omega_{\rm m,0}^{2\beta/3}\right)^{-1}. \qquad (7.113)$$

Taking $\delta_{\rm c} \approx 1.69$, $m_{\rm A} = 1$, $\Gamma = 0.2$ (so that $\beta \approx 0.75$), and neglecting the first term on the right-hand side (which is small compared to the second term), we obtain

$$\frac{\xi_{\rm cc}}{\xi_{\rm N}} \approx 3.8 \left(\sigma_8 \Omega_{\rm m,0}^{0.5}\right)^{-2}. \qquad (7.114)$$

Thus $\xi_{\rm cc} \sim 10\xi_{\rm N}$ for $\sigma_8\Omega_{\rm m,0}^{0.5} \sim 0.6$ [see Eq. (7.77)]. Since $\xi_{\rm N}$ is approximately the two-point correlation function of normal galaxies, we see that CDM spectra with a shape parameter $\Gamma \simeq 0.2$, which is consistent with the observed large-scale clustering of galaxies (see §6.5), predict that the amplitude of the two-point correlation function of rich clusters is higher than that of normal galaxies by a factor of about 10, consistent with observational results (e.g. Bahcall & Soneira, 1983; Peacock & West, 1992; Yang et al., 2005c).

7.4.2 Assembly Bias

In the formalism described above, one-dimensional Markovian random walks are used to derive fomulae for halo bias in the spherical and ellipsoidal collapse models. In such EPS models, the trajectories corresponding to all mass elements in halos of mass M' at time t' pass through the same point in (S, δ) space, say (S', δ'). The large-scale environment of these halos depends on the trajectories at $S < S'$ (i.e. $M > M'$) where they must satisfy $\delta < \delta'$. Halo assembly histories, on the other hand, depend on the trajectories for $S > S'$. The Markov assumption then implies that the environments of halos of given mass should be independent of their assembly history (White, 1996).

Early numerical work suggested that this property does, in fact, hold for structure formation in CDM-like universes (Lemson & Kauffmann, 1999), but a more refined statistical analysis of the same data unambiguously demonstrated that halos in close pairs have slightly earlier assembly times than typical objects of the same mass (Sheth & Tormen, 2004). With the advent of much larger and higher resolution simulations, it became clear that low-mass halos ($M \ll M_*$) in the high-redshift tail of the formation-time distribution are *much* more strongly clustered than halos of similar mass in the low-redshift tail (Gao et al., 2005). This dependence of halo clustering on halo assembly history at fixed halo mass is known as assembly bias. Subsequent numerical work confirmed the original detection and showed that halo clustering at fixed mass depends not only on formation time, but independently on other halo properties such as concentration, spin and substructure content (Wechsler et al., 2006; Jing et al., 2007; Gao & White, 2007). The mass dependence of assembly bias turns out to be different when different halo properties are used to split the population. Thus the clustering of simulated halos depends not only on their mass but also, in a complex way, on their assembly history. This is inconsistent with the one-dimensional Markovian random walks of EPS theory.

There have been several theoretical studies of the origin of assembly bias. Wang et al. (2007a) found that low-mass halos with early assembly times tend to reside near massive halos, and suggested that truncation of accretion by tidal forces might induce the assembly bias effect. Maulbetsch et al. (2007) and Hahn et al. (2008) showed that many low-mass halos in high-density regions have indeed ceased accretion at recent times, while Ludlow et al. (2009) showed that in many cases this can be traced to the fact that the small halos have actually passed right through their larger neighbors, and were ejected again by three-body interactions with other substructures. Nevertheless, strong assembly bias is present even if all ejected halos are excluded (Wang et al., 2009), so additional effects must play an important role. Indeed, the dependence of clustering on other halo properties such as concentration, spin and substructure content seems inconsistent with production by this dynamical mechanism alone, and it is likely that a number of different mechanisms are at work. Halo formation models which extend the standard EPS models, either by increasing the dimensionality of the random walks or by relaxing their Markovian properties, have been considered by a number of authors but a clear understanding of assembly bias is still missing (Sandvik et al., 2007; Desjacques, 2008; Dalal et al., 2008).

7.4.3 Nonlinear and Stochastic Bias

From Eq. (7.103) we see that the bias relation between δ_h and δ is in general not linear: the last term on the right-hand side causes the bias relation to bend down for halos with $M_1 < M^*(t_1)$ and to bend up for halos with $M_1 > M^*(t_1)$. More accurately, we can use the spherical collapse model to relate δ_0 to δ. As shown in §5.1, the evolution of the radius R of a mass shell is completely determined by R_0 and δ_0, and so for a given R, $\delta \equiv (R_0/R)^3 - 1$ is completely determined by δ_0 (at least in the spherical collapse model). Fig. 7.9 shows the bias relation given by this model along with that obtained from N-body simulations. The mean bias relation is well described by

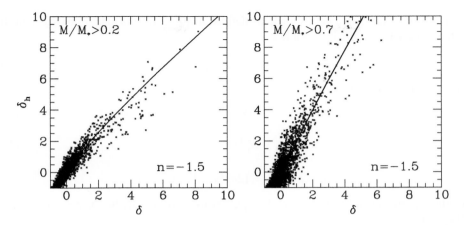

Fig. 7.9. The bias relation of dark matter halos. Both panels show the overdensity of dark matter halos, δ_h, versus the overdensity of mass, δ, in spherical windows with fixed radius $R = 0.08L$ (where L is the size of the simulation box). These results have been obtained from an N-body simulation of structure formation in a cosmology with a scale-free power spectrum with index $n = -1.5$. The two panels correspond to two different lower limits in halo mass, as indicated. The solid curves show the model predictions for the mean bias relation. [After Mo & White (1996)]

the model (see Mo & White, 1996). Clearly, the average bias deviates from linearity in the sense discussed above, and depends on the radius R. There is also scatter around the mean bias relation. Thus, the bias relation is nonlinear, scale-dependent and stochastic. In general we can write this relation as

$$\delta_h(\mathbf{x}) = \hat{\delta}_h + \varepsilon(\mathbf{x}), \quad \text{where} \quad \hat{\delta}_h \equiv b_h(M_1, \delta_1; \delta)\delta(\mathbf{x}). \tag{7.115}$$

This is the same as Eq. (7.107), except that we have allowed for the dependence of b_h on δ; the R dependence is always implicitly implied. Note that Eq. (7.115) is still not the most general form, because the bias relation may not be local, in the sense that $\delta_h(\mathbf{x})$ may depend not only on $\delta(\mathbf{x})$ but also on the properties of the density field at other points. For simplicity, though, we will restrict our discussion to local bias.

(a) Nonlinear Bias Let us first neglect the stochastic part and focus on the effect of nonlinearity in the bias relation on the statistics of the halo distribution relative to that of the mass. In this case δ_h is a deterministic function of δ: $\delta_h = \hat{\delta}_h(\delta)$. Expanding this function in a Taylor series gives

$$\delta_h = \hat{\delta}_h(\delta) = \sum_0^\infty \frac{b_k}{k!} \delta^k. \tag{7.116}$$

By definition, $\langle \delta_h \rangle = 0$, and so $b_0 = -\sum_{k=2}^\infty b_k \langle \delta^k \rangle/k!$ in general is non-zero. Thus, even the second-order moment of δ_h depends on the moments of δ to all orders. However, in the quasi-linear regime, where $\delta \lesssim 1$, the connected moments (defined in §6.1.3) of δ_h can be expressed explicitly in terms of those of δ. Keeping the lowest non-zero orders, we have for the first few moments that

$$\kappa_{2,h} = b_1^2 \kappa_2, \tag{7.117}$$

$$\kappa_{3,h} = b_1^3 \kappa_2^2 (S_3 + 3c_2), \tag{7.118}$$

$$\kappa_{4,h} = b_1^4 \kappa_2^3 (S_4 + 12c_2 S_3 + 4c_3 + 12c_2^2), \tag{7.119}$$

where $c_k \equiv b_k/b_1$, and S_j are the coefficients defined in §6.4.3 for the matter density field. If we define the same set of coefficients for halos,

$$\kappa_{\ell,h} = S_{\ell,h}\kappa_{2,h}^{\ell-1}, \tag{7.120}$$

then it is easy to show that

$$S_{2,h} = b_1^2, \tag{7.121}$$

$$S_{3,h} = b_1^{-1}(S_3 + 3c_2), \tag{7.122}$$

$$S_{4,h} = b_1^{-2}\left(S_4 + 12c_2 S_3 + 4c_3 + 12c_2^2\right). \tag{7.123}$$

These relations are general and apply to all objects whose overdensity is a deterministic function of δ (Fry & Gaztanaga, 1993).

For the bias relation given by Eq. (7.103), and using spherical models to relate δ_0 and δ, we can work out explicitly the coefficients b_k in Eq. (7.116) (see Mo et al., 1997b). For example, in an EdS universe,

$$b_1 = 1 + \frac{\nu_1^2 - 1}{\delta_1}, \tag{7.124}$$

$$b_2 = \frac{8}{21}\frac{\nu_1^2 - 1}{\delta_1} + \left(\frac{\nu_1}{\delta_1}\right)^2 (\nu_1^2 - 3). \tag{7.125}$$

These results apply to halos identified at t_1, but whose clustering properties are analyzed at the present time. The extension to cases where halos are identified at t_1 while clustering properties are analyzed at $t \geq t_1$ is to replace δ_1 by $\delta_c D(t)/D(t_1)$ in the above equations (but without replacing δ_1 in ν_1).

For $\delta_1 \gg 1$ and $\nu_1 \sim 1$, i.e. for halos with $M_1 \sim M^*$ identified at an early epoch, we have $b_1 = 1$ and $b_2 = 0$, so that $\kappa_{2,h} = \kappa_2$ and $S_{3,h} = S_3$. Indeed, it can be shown that in this limit $\kappa_{\ell,h} = \kappa_\ell$ for all $\ell \geq 2$, implying that halos have the same clustering properties as the mass. The reason for this is that the distribution of halos that are *not* strongly biased relative to the mass catches up quickly with the mass distribution due to the motions in the gravitational field of the underlying matter. In contrast, if $\nu_1 \gg \delta_1$ (i.e. for halos that are strongly biased), then $b_k = b_1^k$ for $k > 1$ and all $S_{\ell,h}$ are determined completely by the statistical properties of the initial density field, independent of S_ℓ and the dynamical evolution. In this case, we have

$$S_{3,h} = 3, \quad S_{4,h} = 16, \quad S_{5,h} = 125. \tag{7.126}$$

Finally, for halos with $\nu_1 = 1$, $S_{3,h} = S_3 - 6/\delta_1^2$, which is significantly smaller than S_3, unless δ_1 is large.

(b) Stochastic Bias In general the stochasticity of the bias relation is described by the conditional distribution, $\mathscr{P}(\varepsilon|\delta)\,\mathrm{d}\varepsilon$. With this, the distribution function of δ_h can be written as

$$\mathscr{P}(\delta_h) = \int \mathscr{P}(\varepsilon|\delta)\mathscr{P}(\delta)\,\mathrm{d}\delta, \quad \varepsilon = \delta_h - \hat{\delta}_h, \tag{7.127}$$

where $\mathscr{P}(\delta)$ is the distribution function of the mass overdensity δ. The moments of δ_h can then be written as

$$\langle \delta_h^\ell \rangle = \int \delta_h^\ell \mathscr{P}(\delta_h)\,\mathrm{d}\delta_h = \int \mathrm{d}\delta\,\mathscr{P}(\delta) \int (\hat{\delta}_h + \varepsilon)^\ell \mathscr{P}(\varepsilon|\delta)\,\mathrm{d}\varepsilon. \tag{7.128}$$

Evidently, the halo density field is completely specified by the mass density field once the mean bias relation $\hat{\delta}_h(\delta)$ and the distribution function $\mathscr{P}(\varepsilon|\delta)$ are known. For example, the second and third moments of δ_h are

$$\langle \delta_h^2 \rangle = \langle \hat{\delta}_h^2 \rangle + \langle \varepsilon^2 \rangle \tag{7.129}$$

and

$$\langle \delta_h^3 \rangle = \langle \hat{\delta}_h^3 \rangle + \langle \hat{\delta}_h \langle \varepsilon^2|\delta \rangle \rangle + \langle \varepsilon^3 \rangle, \tag{7.130}$$

respectively. Here

$$\langle \varepsilon^\ell \rangle \equiv \int \mathrm{d}\delta \, \mathscr{P}(\delta) \langle \varepsilon^\ell|\delta \rangle, \tag{7.131}$$

and we have used the fact that $\langle \varepsilon|\delta \rangle = 0$. In the quasi-linear regime, we have

$$S_{2,h} = b_1^2(1+\alpha), \quad \alpha \equiv \frac{\langle \varepsilon^2 \rangle}{b_1^2 \kappa_2}, \tag{7.132}$$

$$S_{3,h} = \frac{(S_3 + 3c_2)/b_1 + 3\left\langle \hat{\delta}_h \langle \varepsilon^2|\delta \rangle \right\rangle / b_1^4 + \alpha^2 S_{3,\varepsilon}}{(1+\alpha)^2}, \tag{7.133}$$

where $S_{3,\varepsilon} = \langle \varepsilon^3 \rangle / \langle \varepsilon^2 \rangle^2$ (e.g. Dekel & Lahav, 1999).

7.5 Internal Structure of Dark Matter Halos

The theory developed in the previous sections describes the abundance, formation history and clustering of dark matter halos as a function of their mass. To complete our description of the halo population, we now focus on their internal structure. In particular, we discuss their density profiles (§7.5.1), shapes (§7.5.2), substructure (§7.5.3), and angular momentum (§7.5.4).

7.5.1 Halo Density Profiles

To a first approximation, we can model a dark matter halo as a spherical object (deviations from sphericity will be discussed in the next subsection). In this case, the internal mass distribution is fully described by a density profile, $\rho(r)$. As discussed above, different halos have different formation histories, so we may expect a significant halo-to-halo variation in density profile. On the other hand, dark matter halos are highly nonlinear objects, and it may be that information regarding their formation histories has largely been erased by their nonlinear collapse. In the latter case, density profiles may be more closely related to the violent relaxation process than to initial conditions. Here we focus on theoretical predictions for the density profiles of dark matter halos; some observational constraints are discussed in §11.1.2.

In the similarity model of spherical collapse described in §5.2, if we start with an initial perturbation $\delta_i(r) \propto r^{-3\varepsilon}$, then the final profile will be $\rho(r) \propto r^{\gamma-3}$, where $\gamma = 1$ for $\varepsilon \leq 2/3$, and $\gamma = 3/(1+3\varepsilon)$ for $\varepsilon > 2/3$. According to the linear scaling relation in §5.6.1, the typical density profile around each particle in a linear density field with scale-free power spectrum, $P(k) \propto k^n$, is $\delta_i(r) \sim \xi(r) \propto r^{-(n+3)}$ [see the interpretation of $\xi(r)$ in §6.1.2]. The typical halo profile will then be

$$\rho(r) \propto \begin{cases} r^{-2} & \text{(for } n \leq -1\text{)} \\ r^{-(3n+9)/(n+4)} & \text{(for } n > -1\text{)}. \end{cases} \tag{7.134}$$

Over the range of interest, $-3 < n < 0$, this model thus predicts that virialized halos resemble isothermal spheres. The simplest plausible model is therefore to assume that dark matter halos are truncated singular isothermal spheres:

$$\rho(r) \propto r^{-2} \quad \text{for} \quad r \leq r_h. \tag{7.135}$$

We define the limiting radius of a dark matter halo, r_h, to be the radius within which the mean matter density is

$$\rho_h = \Delta_h \overline{\rho} = \Delta_h \rho_{\text{crit}} \Omega_m, \tag{7.136}$$

where $\overline{\rho}$ is the mean matter density of the Universe at the time in question, and ρ_{crit} is the corresponding critical density for closure. Under this assumption, we have

$$\rho(r) = \frac{V_h^2}{4\pi G r^2}, \quad r_h = \sqrt{\frac{200}{\Delta_h \Omega_m}} \frac{V_h}{10 H(t)}, \quad V_h = \sqrt{\frac{GM_h}{r_h}}, \tag{7.137}$$

where M_h is the mass of the halo (i.e. the mass within r_h), V_h is its circular velocity at r_h, and $H(t)$ is the Hubble constant at time t. It is common practice to refer to r_h and V_h as the virial radius and virial velocity.

Based on the results presented in §5.4.4, a physically motivated choice would be to take $\Delta_h = \Delta_{\text{vir}}$. However, since the criterion for virialization is not strict, other definitions are also in use in the literature. For example, some studies adopt $\rho_h = 200 \times \overline{\rho}$, so that $\Delta_h = 200$, while others use $\rho_h = 200 \times \rho_{\text{crit}}$, so that $\Delta_h = 200/\Omega_m$. Note that different definitions imply different relations among M_h, r_h and V_h. However, for a given density profile, it is straightforward to convert these quantities from one definition to another.

The isothermal model is, at best, an approximation. Many effects may cause deviations from the profile predicted by the simple similarity model. For example, (i) collapse may never reach an equilibrium state in the outer region of a dark matter halo, (ii) non-radial motion may be important, and (iii) mergers associated with the (hierarchical) formation of a halo may render the spherical-collapse model invalid. Unfortunately, the importance of such effects on the final density profile is difficult to model analytically, and we have to rely on numerical simulations to make progress.

Using high-resolution N-body simulations of structure formation in a CDM cosmogony, Navarro et al. (1996) showed that the density profiles of the simulated dark matter halos are shallower than r^{-2} at small radii and steeper at large radii. In fact, they found the density profiles to be well described by what has become known as the Navarro, Frenk & White (NFW) profile:

$$\rho(r) = \rho_{\text{crit}} \frac{\delta_{\text{char}}}{(r/r_s)(1 + r/r_s)^2} \quad \text{(NFW profile)}. \tag{7.138}$$

Here r_s is a scale radius, and δ_{char} is a characteristic overdensity. The logarithmic slope of the NFW profile changes gradually from -1 near the center to -3 at large radii, and only resembles that of an isothermal sphere at radii $r \sim r_s$. In a follow-up paper, Navarro et al. (1997) found Eq. (7.138) to be a good representation of the equilibrium density profiles of dark matter halos of all masses and in all CDM-like cosmogonies. Thus, halos formed by dissipationless hierarchical clustering seem to have a universal density profile. The enclosed mass of the NFW profile is given by

$$M(r) = 4\pi \rho_{\text{crit}} \delta_{\text{char}} r_s^3 \left[\ln(1 + cx) - \frac{cx}{1 + cx} \right], \tag{7.139}$$

where $x \equiv r/r_h$, and

$$c \equiv \frac{r_h}{r_s} \tag{7.140}$$

is the halo concentration parameter. Note that the total mass, M_h, of the halo is given by Eq. (7.139) with $x = 1$, and that this mass depends on the definition chosen for r_h. It is important to pay attention to this last point because several different definitions for the bounding radius of a halo are in common use, so that an object with given density profile will be assigned a different mass and concentration by different authors. Using Eqs. (7.138)–(7.140), we obtain a relation between characteristic overdensity and concentration parameter,

$$\delta_{\text{char}} = \frac{\Delta_h \Omega_m}{3} \frac{c^3}{\ln(1+c) - c/(1+c)}. \qquad (7.141)$$

Thus, for a given cosmology, the NFW profile is completely characterized by its mass, M, and its concentration parameter, c, or equivalently (and without any ambiguity about boundary definition) by r_s and δ_{char}. Navarro et al. (1997) showed that characteristic overdensity is closely related to formation time. Halos which form earlier are more concentrated. For a definition of formation time similar to that of Eq.(7.91), they found that the 'natural' relation $\delta_{\text{char}} \propto \Omega_{m,0}(1+z_f)^3$ describes how the overdensity of halos varies with their formation redshift both in the mean, as halo mass, linear power spectrum and cosmological density parameters are changed, and for individual halos of given mass in a given cosmology.

Subsequent N-body simulations explored in more detail this relation between concentration and formation history (Wechsler et al., 2002; Zhao et al., 2003b). Halos that have experienced a recent major merger typically have low concentrations, $c \sim 4$, while halos which have experienced a longer phase of relatively quiescent growth have larger concentrations (Zhao et al., 2003a, 2008). Physically, the central structure of a dark matter halo seems to be established through violent relaxation during a phase of rapid mergers, which leads to a 'universal' NFW profile with $c \sim 4$. Later accretion increases the mass and size of the halo without adding much material to its inner regions, thus increasing r_h while leaving r_s almost unchanged. Once the initial power spectrum and cosmology are chosen, the accretion history of a halo depends mainly on its mass relative to the characteristic mass M^* (see §7.3). In particular, more massive halos assemble later, and thus had their last major merger more recently. Consequently, concentration should be a decreasing function of halo mass, as found by the simulations (e.g. Navarro et al., 1997; Bullock et al., 2001b; Zhao et al., 2003a; Neto et al., 2007; Macciò et al., 2007). This has motivated a number of authors to develop simple models to estimate mean concentration as a function of halo mass for a given cosmology (Navarro et al., 1997; Bullock et al., 2001b; Eke et al., 2001; Macciò et al., 2008; Zhao et al., 2008). Although different in detail, these models all follow Navarro et al. (1997) in assuming that the characteristic density of a dark matter halo is related to the mean cosmic density at some characteristic epoch in the halo's history. For example, in the model of Zhao et al. (2008), dark matter halos are defined using $\Delta_h = \Delta_{\text{vir}}$, and the average concentration parameter of a halo of mass M at time t is given by a simple empirical formula:

$$c(M,t) = 4 \times \left\{ 1 + \left[\frac{t}{3.75 t_{0.04}(M,t)} \right]^{8.4} \right\}^{1/8}, \qquad (7.142)$$

where $t_{0.04}(M,t)$ is the time at which the main progenitor of the halo had gained 4% of its final mass M. For a given cosmological model, the value of $t_{0.04}(M,t)$ can be obtained by following the halo assembly history (§7.3.4). As shown in Zhao et al. (2008), this simple model accurately describes the mean concentrations of simulated halos as function of mass and redshift in a variety of cosmological models. At a given halo mass, there is significant scatter in concentration, reflecting the scatter in formation history. This scatter is reasonably well modeled by a log-normal distribution, with a variance $\sigma_{\log c} \simeq 0.12 \pm 0.02$ (e.g. Jing, 2000; Bullock et al., 2001b; Wechsler et al., 2002; Neto et al., 2007).

Although the NFW profile is widely used, numerical simulations of high resolution (Navarro et al., 2004; Hayashi & White, 2008; Gao et al., 2008; Springel et al., 2008) have shown that the spherically averaged density profiles of dark matter halos show small but systematic deviations from NFW form and, in the mean, are more accurately described by an Einasto (1965) profile,

$$\rho(r) = \rho_{-2} \exp\left[\frac{-2}{\alpha}\left\{\left(\frac{r}{r_{-2}}\right)^{\alpha} - 1\right\}\right] \quad (7.143)$$

[compare with Eq. (2.22) which gives a very similar formula for the projected density profile] with r_{-2} the radius at which the logarithmic slope of the density distribution is equal to -2 and $\rho_{-2} = \rho(r_{-2})$. The best-fit values for the index α typically span the range $0.12 \lesssim \alpha \lesssim 0.25$ and increase systematically with increasing mass (Hayashi & White, 2008; Gao et al., 2008). The Einasto profile has three free parameters, compared to only two for the NFW profile, so the fact that it fits simulation results better may seem unsurprising. However, fixing $\alpha = 0.17$ gives a two-parameter function which still fits mean halo profiles significantly better than the NFW form. Note that the systematic variation of α with halo mass demonstrates a small but significant deviation of mean density profiles from any 'universal' shape. A characteristic property of the Einasto profile is that its logarithmic slope is a power-law in radius:

$$\frac{d\ln\rho}{d\ln r} = -2\left(\frac{r}{r_{-2}}\right)^{\alpha}. \quad (7.144)$$

Thus, contrary to an NFW profile, which has a central r^{-1} cusp, the logarithmic slope of the Einasto profile continues to become shallower as $r \to 0$. Note, though, that the radius at which $d\ln\rho/d\ln r = -0.5$ is typically less than $10^{-3} r_{-2}$, which implies that in practice the profile (7.143) may still be considered 'cuspy' for most astrophysical applications. Note also that, depending on formation history details, individual dark halo profiles often differ much more from either NFW or Einasto form than the two fitting formulae differ from each other.

It is still unclear why halos formed through dissipationless hierarchical clustering have near-universal profiles. Quite different initial conditions give similar results, so the universality must result from relaxation processes in very general circumstances. As discussed earlier in this chapter, the essence of hierarchical clustering is that larger systems form through the mergers of smaller systems. Thus, the collapse leading to the formation of a dark matter halo is in general clumpy and chaotic, particularly during the early phases. This means that violent relaxation (discussed in §5.5) must play an important role. The ongoing accretion and stripping of small mass halos may also be important. On the other hand, even in the absence of mergers or pre-existing structure, cold collapse from asymmetric initial conditions produces objects with near-NFW density profiles, so hierarchical growth of structure is apparently not required (see van Albada, 1982; Huss et al., 1999; Wang & White, 2008). Although different explanations for the origin of the NFW profile have been proposed (see Syer & White, 1998; Dekel et al., 2003; Lu et al., 2006), none currently seems convincing.

7.5.2 Halo Shapes

(a) Theoretical Predictions As we have seen in §7.2.3, the collapse of overdensities in the cosmic density field is generically aspherical. Thus there is no reason to expect the resulting halos to be spherical. Even the earliest simulations of structure formation in a CDM universe emphasized that halos are substantially flattened (Davis et al., 1985). Subsequent work has shown that, to a good approximation, halo equidensity surfaces can be described by ellipsoids, each of which is characterized by the lengths of its axes ($a_1 \geq a_2 \geq a_3$). These axes can be used to specify the dimensionless shape parameters,

$$s = \frac{a_3}{a_1}, \quad q = \frac{a_2}{a_1}, \quad p = \frac{a_3}{a_2}, \qquad (7.145)$$

and/or the triaxiality parameter

$$T = \frac{a_1^2 - a_2^2}{a_1^2 - a_3^2} = \frac{1-q^2}{1-s^2} \qquad (7.146)$$

(Franx et al., 1991). Note that oblate and prolate shapes correspond to $T = 0$ and $T = 1$, respectively. The majority of CDM halos in numerical simulations have $0.5 \lesssim T \lesssim 0.85$, and $0.5 \lesssim s \lesssim 0.75$ (e.g. Jing & Suto, 2002; Bailin & Steinmetz, 2005; Kasun & Evrard, 2005; Allgood et al., 2006). In particular, less massive halos are more spherical, and halos of a given mass become flatter with increasing redshift. Using numerical simulations, Allgood et al. (2006) found that this mass and redshift dependence is well characterized by

$$\langle s \rangle (M, z) = (0.54 \pm 0.03) \left[\frac{M}{M^*(z)} \right]^{-0.050 \pm 0.003}, \qquad (7.147)$$

where $M^*(z)$ is the characteristic halo mass at redshift z [see Eq. (7.48)], and the scatter in s is $\sigma_s \sim 0.1$. The probability for p given s is found to be well fit by

$$\mathscr{P}(p|s) = \frac{3}{2(1-\tilde{s})} \left[1 - \left(\frac{2p - 1 - \tilde{s}}{1 - \tilde{s}} \right)^2 \right], \qquad (7.148)$$

with $\tilde{s} = \max[s, 0.55]$ (Jing & Suto, 2002; Allgood et al., 2006). Simulations suggest that the shape of a halo is tightly correlated with its merger history, with halos that assembled earlier being more spherical. In particular, halos that experienced a recent major merger have a tendency to be close to prolate, with the major axis reflecting the direction along which the last merger event occurred. In addition, as shown by Bailin & Steinmetz (2005), there is a strong tendency for the minor axes of halos to lie perpendicular to large-scale filaments, and a much weaker tendency for the major axes to lie along the filaments. This alignment is found to be stronger for more massive halos, and shows that the shapes of dark matter halos reflect the large-scale tidal field in which they are embedded.

(b) Observational Constraints A large number of different observational techniques have been used to determine the shapes of dark matter halos associated with galaxies and clusters. These include using the orbits of satellite galaxies as traced by their tidal streams (e.g. Helmi, 2004), the shapes of X-ray halos around galaxies and clusters (e.g. Buote & Canizares, 1992), gravitational lensing (Hoekstra et al., 2004), and the angular distribution of satellite galaxies (e.g. Wang et al., 2008a). Each of these methods has its own pros and cons, and often different studies obtain different results, even when the same method has been used. In addition, it is important to take into account that the halo shapes obtained from numerical simulations have typically ignored the presence of baryons. However, when baryons cool and accumulate at the centers of their dark matter halos, they tend to make the central regions of their halos more spherical (e.g. Kazantzidis et al., 2004). Unfortunately, the magnitude of this effect depends on how much gas cools out at the center of a halo, and on other details related to the formation of the central galaxy. Because of these uncertainties in both the data and the model predictions, it is still unclear whether the halo shapes inferred from observations are consistent with the predictions of CDM cosmology.

7.5.3 Halo Substructure

When a small halo merges with a significantly larger halo it becomes a subhalo orbiting within the potential well of its host. As it orbits, it is subjected to strong tidal forces from the host, which

Fig. 7.10. The density distribution of dark matter in a high-resolution N-body simulation. The image shows a pair of dark matter halos (each with a mass of $\sim 2 \times 10^{12} M_\odot$) at redshift $z = 0$, separated by about 1 Mpc. These two halos contain over 2,000 dark matter subhalos with a circular velocity larger than $10\,\mathrm{km\,s^{-1}}$. [Courtesy of B. Moore; see Moore (2001)]

cause it to lose mass (see §12.2). In addition, the orbit itself evolves, as the subhalo is subjected to dynamical friction (see §12.3) which causes it to lose energy and angular momentum to the dark matter particles of its host. Whether a subhalo survives as a self-bound entity depends on its mass (relative to that of its host), its density profile (which is related to its formation redshift) and its orbit.

Up until the end of the 1990s, numerical simulations of halo formation revealed little substructure. This was a result of the relatively small number of particles which could be used to represent them. With increasing computing power and better algorithms, ever greater numbers of particles and ever better resolution have become possible. It is now obvious that substantial amounts of substructure are expected (e.g. Moore et al., 1999a; Klypin et al., 1999; Diemand et al., 2007; Springel et al., 2008). An example is given in Fig. 7.10, which shows the distribution of dark matter in a pair of dark matter halos. Each clearly has a large number of subhalos.

Let m and M be the masses of the subhalo and host halo, respectively. At any given time, the subhalo mass function is well fitted by

$$\frac{dn}{d\ln(m/M)} = \frac{f_0}{\beta \Gamma(1-\gamma)} \left(\frac{m}{\beta M}\right)^{-\gamma} \exp\left[-\left(\frac{m}{\beta M}\right)\right]. \tag{7.149}$$

7.5 Internal Structure of Dark Matter Halos

Here βM (with $\beta < 1$) indicates a characteristic mass such that for $m \gg \beta M$ the subhalo mass function declines exponentially, and f_0 is the mean subhalo mass fraction, i.e.

$$f_0 = \frac{1}{M}\int_0^\infty m \frac{dn}{dm} dm = \int_0^\infty \frac{dn}{d\ln(m/M)} d\left(\frac{m}{M}\right). \tag{7.150}$$

Numerical simulations have shown that $\gamma \sim 0.9 \pm 0.1$ (De Lucia et al., 2004; Gao et al., 2004; Springel et al., 2008) while $0.1 \lesssim \beta \lesssim 0.5$ (van den Bosch et al., 2005; Shaw et al., 2006). These relatively large uncertainties are due in part to ambiguities in identifying and assigning masses to subhalos. To a given mass limit, subhalos are found to have a radial number density profile that is substantially less centrally concentrated than the dark matter density profile (e.g Springel et al., 2008). This arises because subhalos at smaller radii are subjected to stronger tidal forces, and are therefore more strongly truncated or disrupted. As a result, the subhalo mass fraction parameter f_0 is a strong function of the outer radius within which it is measured.

Consider a halo of mass M_0 at redshift z_0. Its population of subhalos corresponds to the halos that have been accreted by the halo's main progenitor (see §7.3.3) and that have survived tidal disruption. It is thus interesting to link the subhalo mass function to the mass function of halos accreted onto the main progenitor over its entire history, termed the unevolved subhalo mass function. Numerical simulations by Giocoli et al. (2008) show that the unevolved subhalo mass function is well fit by

$$\frac{dn}{d\ln(m/M_0)} = A\left(\frac{m}{fM_0}\right)^{-p} \exp\left[-\left(\frac{m}{fM_0}\right)^q\right], \tag{7.151}$$

with $A \simeq 0.345$, $f \simeq 0.43$, $p \simeq 0.8$ and $q \simeq 3$, with little dependence on halo mass or redshift. Note that in this formula m corresponds to the mass of the subhalo at the time it was accreted by the main progenitor. During its subsequent evolution it loses mass at a rate which can be written as

$$\frac{dm}{dt} = \frac{dm}{dr_{\rm tid}}\frac{dr_{\rm tid}}{dt} \tag{7.152}$$

where $r_{\rm tid}$ is its instantaneous tidal radius. We assume the latter to be the radius where the original (unstripped) density of the subhalo equals the density of its host's halo at its current orbital radius, $r_{\rm orb}$, i.e. $\rho_{\rm sub}(r_{\rm tid}) = \rho_{\rm host}(r_{\rm orb})$. If we assume that both $\rho_{\rm sub}(r)$ and $\rho_{\rm host}(r)$ are singular isothermal spheres, then

$$\frac{1}{r_{\rm tid}}\frac{dr_{\rm tid}}{dt} = \frac{1}{r_{\rm orb}}\frac{dr_{\rm orb}}{dt}, \tag{7.153}$$

so that both m and $r_{\rm tid}$ decrease in direct proportion to $r_{\rm orb}$. The evolution of $r_{\rm orb}$ is governed by dynamical friction, and, assuming a circular orbit, is given by

$$\frac{dr_{\rm orb}}{dt} = -0.428 \frac{Gm}{V_c\, r_{\rm orb}} \ln\Lambda, \tag{7.154}$$

with V_c the circular velocity of the host halo (see §12.3). To leading order, the Coulomb logarithm $\ln\Lambda$ is just a function of the mass ratio m/M, which is a constant (equal to v_c^3/V_c^3 where v_c is the circular velocity of the subhalo) under the physically reasonable assumption that M should be taken as the mass interior to the satellite's orbit. In this very simple model the satellite's mass and the radius of its (circular) orbit thus decrease linearly with time:

$$\frac{m}{m_{\rm i}} = \frac{r_{\rm orb}}{r_{\rm orb,i}} = 1 - 0.428\ln\Lambda \frac{v_c^3 t}{V_c^2 r_{\rm orb,i}}, \tag{7.155}$$

where $m_{\rm i}$ and $r_{\rm orb,i}$ are the initial values of m and $r_{\rm orb}$ respectively. Since we expect $\ln\Lambda$ to be of the order of a few and $r_{\rm orb,i} \sim r_{\rm vir}$, the virial radius of the host halo, this predicts that newly accreted subhalos should merge completely with their host after of order $0.1(V_c/v_c)^3 \sim 0.1 M/m_i$ times the initial orbital periods, where M is the total mass of the host. Thus infalling subalos with mass greater than a few percent of that of the main object are predicted to merge rapidly, whereas substantially lower mass objects will survive for long times with little stripping.

High-resolution numerical simulations show that while this conclusion is qualitatively correct, Eq. (7.155) is a poor description of the typical mass loss behavior (Diemand et al., 2007). Most accreted halos fall in on highly elongated orbits and lose a large fraction of their mass at first pericenter. The most massive and lowest concentration objects continue to lose mass thereafter and are often completely disrupted, but less massive and more concentrated objects often stabilize on their new orbit at their reduced mass. Thereafter they evolve rather little either in mass or in orbit. The detailed behavior depends significantly on the structure of both the host halo and the infalling satellite, as well as on the eccentricity of their relative orbit. Regardless of this, it is clear that, given the universality of the unevolved subhalo mass function, host halos assembling earlier (i.e. accreting their subhalos at higher redshifts) will end up with less mass in subhalos, simply because there has been more time for mass loss to operate. As we have seen in §7.3.3, in a CDM cosmogony, halos that are more massive assemble later, and are thus expected to have a larger subhalo mass fraction, f_0, than lower mass objects. This is indeed confirmed by simulations, which show that, to a good approximation, the (evolved) subhalo mass function is given by Eq. (7.149) with f_0 an increasing function of halo mass, and with γ and β roughly constant (Giocoli et al., 2008).

7.5.4 Angular Momentum

(a) Halo Spin Parameters Another important property of a dark matter halo is its angular momentum. As originally pointed by Hoyle (1949) and first demonstrated explicitly using numerical simulation by Efstathiou & Jones (1979), asymmetric collapse in an expanding universe produces objects with significant angular momentum. This is traditionally parameterized through a dimensionless spin parameter,

$$\lambda = \frac{J|E|^{1/2}}{GM^{5/2}}, \tag{7.156}$$

where J, E and M are the total angular momentum, energy and mass of the halo, respectively. For an isolated system, all these quantities are conserved during dissipationless gravitational evolution, and so, therefore, is λ itself. The spin parameter thus defined is roughly the square root of the ratio between the rotational and the total energy of the system, and so characterizes the overall importance of angular momentum relative to random motion.

The energy of a spherical dark matter halo is easily obtained from the virial theorem, $E = -K$, assuming all particles to be on circular orbits:

$$E = -4\pi \int_0^{r_h} \frac{\rho(r)V_c^2(r)}{2} r^2 dr \equiv -\frac{M_h V_h^2}{2} F_E, \tag{7.157}$$

where $V_h = V_c(r_h)$ is the circular velocity at r_h and F_E is a parameter that depends on the halo's density distribution; $F_E = 1$ for a singular isothermal sphere, while for a NFW halo with a concentration parameter c,

$$F_E = \frac{c}{2} \frac{[1 - 1/(1+c)^2 - 2\ln(1+c)/(1+c)]}{[c/(1+c) - \ln(1+c)]^2}. \tag{7.158}$$

In the literature one often finds an alternative definition of the spin parameter which avoids the need to calculate the halo energy explicitly:

$$\lambda' = \frac{J}{\sqrt{2}MV_h r_h}. \tag{7.159}$$

This spin parameter is related to that defined by Eq. (7.156) through $\lambda' = \lambda F_E^{-1/2}$.

Numerical simulations have shown that the spin parameter distribution for halos formed by dissipationless hierarchical clustering is well fit by a log-normal distribution,

$$p(\lambda)\,d\lambda = \frac{1}{\sqrt{2\pi}\sigma_{\ln\lambda}} \exp\left[-\frac{\ln^2(\lambda/\overline{\lambda})}{2\sigma_{\ln\lambda}^2}\right] \frac{d\lambda}{\lambda}, \tag{7.160}$$

with $\overline{\lambda} \approx 0.035$ and $\sigma_{\ln\lambda} \approx 0.5$. Detailed simulations show that the median and width of this log-normal distribution depend only weakly on halo mass, redshift and cosmology (e.g. Barnes & Efstathiou, 1987; Warren et al., 1992; Lemson & Kauffmann, 1999; Bullock et al., 2001a; Macciò et al., 2007). At all halo masses there is a marked tendency for halos with higher spin to be in denser regions and thus to be more strongly clustered (Gao & White, 2007). Note that the distribution (7.160) is fairly broad, spanning roughly a factor of 5 between the 10th and 90th percentiles. The fact that the median spin parameter is so small indicates that dark matter halos are mainly supported by random motions of their particles rather than by rotation. For comparison, as we will see in §11.1.4, the spin parameter of a self-gravitating, rotationally supported disk is ~ 0.4, roughly an order of magnitude larger than that of a dark matter halo.

(b) Tidal Torque Theory As first suggested by Hoyle (1949), the spin of a protogalaxy may arise from the tidal field of its neighboring structure. This has prompted detailed studies into angular momentum growth within the framework of the gravitational instability picture (Peebles, 1969; Doroshkevich, 1970). Here we follow the analysis of White (1984), which gives a consistent description of the growth of the angular momentum in the quasi-linear regime.

Consider the material that ends up as part of a virialized dark matter halo. Let the *Lagrangian* region it occupies in the early universe be V_L. The angular momentum of this material at early times (well before collapse) can then be written as

$$\mathbf{J} = \int_{V_L} d^3\mathbf{x}_i \overline{\rho}_m a^3 (a\mathbf{x} - a\overline{\mathbf{x}}) \times \mathbf{v} = \overline{\rho}_m a^5 \int_{V_L} d^3\mathbf{x}_i (\mathbf{x} - \overline{\mathbf{x}}) \times \dot{\mathbf{x}}, \tag{7.161}$$

where $\overline{\mathbf{x}}$ is the barycenter of the volume, a is the scale factor of the universe, and an overdot denotes derivative with respect to t. Using the Zel'dovich approximation described in §4.1.8, this can be written to lowest order in the perturbation amplitude as

$$\mathbf{J} = -\overline{\rho}_m a^5 b \int_{V_L} d^3\mathbf{x}_i (\mathbf{x}_i - \overline{\mathbf{x}}_i) \times \nabla\Phi_i, \tag{7.162}$$

where $b(t) = D(t)/4\pi G\overline{\rho}_m a^3$ with $D(t)$ the linear growth rate. Using the divergence (Gauss') theorem, this expression can be converted into an integral over the surface, Σ_L, of V_L:

$$\mathbf{J} = -\overline{\rho}_m a^5 b \int_{\Sigma_L} \Phi_i(\mathbf{x}_i)(\mathbf{x}_i - \overline{\mathbf{x}}_i) \times d\mathbf{S}. \tag{7.163}$$

Thus $\mathbf{J}$ vanishes to first order if V_L is spherical or if Σ_L is an equipotential surface of Φ_i. If we assume that $\nabla\Phi_i$ is smooth enough so that it can be expanded in a Taylor series around $\overline{\mathbf{x}}_i$,

$$\nabla\Phi_i|_{\mathbf{x}_i} = \nabla\Phi_i|_{\overline{\mathbf{x}}_i} + (\mathbf{x}_i - \overline{\mathbf{x}}_i) \cdot (\nabla\nabla\Phi_i)|_{\overline{\mathbf{x}}_i}, \tag{7.164}$$

then the volume integral form of $\mathbf{J}$ gives

$$J_i(t) = -a^2 b \sum_{j,k,l} \varepsilon_{ijk} T_{jl} I_{lk}, \tag{7.165}$$

where ε_{ijk} is the completely antisymmetric tensor with $\varepsilon_{123} = 1$,

$$T_{jl} = \nabla_j \nabla_l \Phi_i|_{\bar{\mathbf{x}}_i} \tag{7.166}$$

is the tidal tensor at $\bar{\mathbf{x}}_i$, and

$$I_{lk} = \int_{V_L} (x_{i,l} - \bar{x}_{i,l})(x_{i,k} - \bar{x}_{i,k})\bar{\rho}_m a^3 \, d^3\mathbf{x}_i \tag{7.167}$$

is the inertia tensor of the matter in V_L. Both T_{jl} and I_{lk} are evaluated at the fiducial time t_i. As one can see, if the principal axes of T_{ij} and I_{ij} are different, $\mathbf{J}$ is non-zero because the tidal field couples to the quadrupole generated by the boundary of V_L.

To gain more insight into the problem, let us approximate the boundary of V_L by an ellipsoid with principal axes $R_1 \geq R_2 \geq R_3$. In the principal axis system of the ellipsoid, the inertia tensor is

$$I_{ij} = \delta_{ij}\mu_j; \quad \mu_1 = \frac{M}{5}R_1^2, \quad \mu_2 = \mu_1\left(\frac{R_2}{R_1}\right)^2, \quad \mu_3 = \mu_1\left(\frac{R_3}{R_1}\right)^2, \tag{7.168}$$

where M is the mass of the ellipsoid. In this same coordinate system, T_{ij} can be written as

$$T_{ij} = -4\pi G\bar{\rho}_m a^3 \sum_{\alpha=1}^{3} \lambda_\alpha A_{\alpha,i} A_{\alpha,j}, \tag{7.169}$$

where λ_α ($\alpha = 1, 2, 3$) are the eigenvalues of $\nabla\nabla\Phi_i/4\pi G\bar{\rho}_m a^3$, and $A_{\alpha,i}$ is the cosine of the angle between the αth principal axis of $\nabla\nabla\Phi_i$ and the ith coordinate axis. Thus the value of $\mathbf{J}$ can be written as

$$J(t) = \frac{1}{5}\left[\frac{2G}{H^2(t)\Omega_m(t)}\right]^{2/3} \dot{D}(t)\bar{\delta}_i M^{5/3} \mathscr{G}, \tag{7.170}$$

where

$$\mathscr{G} = \left(\frac{R_1^2}{R_2 R_3}\right)^{2/3} \frac{\lambda_1}{\lambda_1 + \lambda_2 + \lambda_3}\left[\sum_{\alpha=1}^{3}\frac{\lambda_\alpha^2}{\lambda_1^2}\sum_{j>k}A_{\alpha,j}^2 A_{\alpha,k}^2\left(\frac{\mu_j - \mu_k}{\mu_1}\right)^2\right]^{1/2}, \tag{7.171}$$

with $\bar{\delta}_i$ the initial density contrast of the ellipsoid. We can now see clearly how J depends on various parameters. The 'geometrical' factor $\mathscr{G}$ is determined by the shape of the ellipsoid (given by R_2/R_1 and R_3/R_2), the shape of the tidal field (given by λ_2/λ_1 and λ_3/λ_1), and the position angles between them. For given $\mathscr{G}$, the angular momentum J increases with the mass of the ellipsoid as $M^{5/3}$ and with the initial density contrast $\bar{\delta}_i$ linearly. The time dependence of J is given by the time dependence of $H^2(t)\Omega_m(t)$ and $\dot{D}(t)$. For an EdS universe $\Omega_m(t) = 1$, $H(t) \propto t^{-1}$ and $D(t) \propto t^{2/3}$, so that $J \propto t$, in good agreement with numerical simulations (White, 1984).

The angular momentum growth stops once a protogalaxy separates from the overall expansion and starts to collapse. As an approximation, the final angular momentum of the protogalaxy may therefore be estimated as the value of J predicted by Eq. (7.170) at the time t_f when $D(t_f)\bar{\delta}_i = 1$. According to the Zel'dovich approximation, this is approximately the time when the object collapses to form a pancake. The final angular momentum can then be written as

$$J_f = \frac{(2G)^{2/3}}{5}[\Omega_m(t_f)]^{-0.07}[H(t_f)]^{-1/3}M^{5/3}\mathscr{G}, \tag{7.172}$$

where we have used the approximation $a\dot{D}/\dot{a}D = \Omega_m^{0.6}$ (see §4.1.6). Since $H(t_f) \sim t_f^{-1}$, this equation shows that the angular momentum of a protogalaxy depends strongly on its mass, weakly on its time of collapse, and almost not at all on Ω_m at that time.

The linear theory described above gives us some idea about the acquisition of angular momentum during the early stages of collapse of dark matter halos in the cosmological density field. It should be realized, however, that this angular momentum may not correspond to the final angular momentum of a dark matter halo, because a dark matter halo may acquire significant amounts of angular momentum during the late stages of nonlinear collapse and due to mergers with other halos (e.g. Maller et al., 2002; Vitvitska et al., 2002). In fact, numerical simulations show that the linear angular momentum is a relatively poor predictor of the final angular momentum (e.g. White, 1984; Sugerman et al., 2000; Porciani et al., 2002a,b).

Numerical simulations show that the *direction* of the angular momentum vector is strongly aligned with the minor axis of the halo, with a median misalignment angle of $\sim 25°$ (Bailin & Steinmetz, 2005). On larger scales, the angular momentum vectors of dark matter halos embedded in mildly nonlinear, two-dimensional sheets have a strong tendency to be parallel to the sheet (Hahn et al., 2007). Another form of alignment is that between the angular momentum vectors of neighboring dark matter halos. This can be expressed via the spin–spin correlation function

$$\xi_{\rm spin}(r) \equiv \langle \mathbf{J}_1(\mathbf{x}) \cdot \mathbf{J}_2(\mathbf{x}+\mathbf{r}) \rangle, \tag{7.173}$$

where $\mathbf{J}_1$ and $\mathbf{J}_2$ are the angular momentum vectors of two halos separated by a distance $r = |\mathbf{r}|$. Using numerical simulations, Hahn et al. (2007) find that there is a weak tendency for massive halos ($M \gtrsim 5 \times 10^{12} h^{-1} \rm M_\odot$) to have their angular momentum vectors antiparallel to those of halos within a distance of a few Mpc. On larger scales, and for less massive halos, however, $\xi_{\rm spin}(r)$ is consistent with being zero. These various alignments are not only important for understanding the angular momentum acquisition of galaxies, but may also cause a systematic contamination in weak lensing maps of cosmic shear (e.g. Hirata & Seljak, 2004). Unfortunately, the origin of all these alignment effects is still poorly understood, and much work is needed in this direction.

(c) Internal Angular Momentum Distribution The spin parameter defined in Eq. (7.156) describes the total angular momentum of a dark matter halo, but contains no information regarding the *distribution* of angular momentum within the halo. As we will see in §11.4.1 this specific angular momentum distribution is an important ingredient for modeling the mass distribution of disk galaxies.

Using numerical simulations of structure formation, Bullock et al. (2001a) measured the specific angular momentum distributions of dark matter halos, and found that they can be adequately fit by

$$P(\mathcal{J}) = \begin{cases} \frac{\mu \mathcal{J}_0}{(\mathcal{J}_0 + \mathcal{J})^2} & \text{if } \mathcal{J} \geq 0 \\ 0 & \text{if } \mathcal{J} < 0. \end{cases} \tag{7.174}$$

Here $\mathcal{J}$ is the specific angular momentum in the direction of the total angular momentum vector of the halo (and can thus be negative), $\mathcal{J}_0$ is a characteristic value of $\mathcal{J}$, and μ is a shape parameter. This distribution is normalized, so that

$$M(<\mathcal{J}) = \begin{cases} M_{\rm vir} \frac{\mu \mathcal{J}}{\mathcal{J}_0 + \mathcal{J}} & \text{if } \mathcal{J} \geq 0 \\ 0 & \text{if } \mathcal{J} < 0, \end{cases} \tag{7.175}$$

with $M(<\mathcal{J})$ the halo mass with specific angular momentum less than $\mathcal{J}$. The distribution has a maximum specific angular momentum, $\mathcal{J}_{\max}$, specified by $M(<\mathcal{J}) = M_{\rm vir}$, which gives $\mathcal{J}_{\max} = \mathcal{J}_0/(\mu - 1)$. The total specific angular momentum of a halo with this distribution is given by

$$\mathcal{J}_{\rm tot} = \int_0^{\mathcal{J}_{\max}} \mathcal{J} P(\mathcal{J}) \, {\rm d}\mathcal{J} = \zeta \mathcal{J}_{\max}, \tag{7.176}$$

with

$$\zeta = 1 - \mu \left[1 - (\mu - 1) \ln \left(\frac{\mu}{\mu - 1} \right) \right]. \qquad (7.177)$$

This $\mathscr{J}_{\text{tot}}$ can be related to the halo spin parameter using Eq. (7.156), according to which $\mathscr{J}_{\text{tot}} = \sqrt{2} \lambda r_{\text{vir}} V_{\text{vir}} F_{\text{E}}^{-1/2}$. Thus the pair (λ, μ) completely specifies the angular momentum distribution of a dark matter halo. As shown by Bullock et al. (2001a), μ has a log-normal distribution with mean $\overline{\mu} \simeq 1.25$ and scatter $\sigma_{\ln \mu} \simeq 0.4$.

Bullock et al. (2001a) also measured the angular momentum distribution in spherical shells, and found that the specific angular momentum in a shell at radius r is roughly given by

$$\mathscr{J}(r) \propto r^\alpha \quad \text{with} \quad \alpha \simeq 1.1 \pm 0.3 \qquad (7.178)$$

(see also Barnes & Efstathiou, 1987; Frenk et al., 1988). Alternatively, one can also express this radial distribution as a function of the mass enclosed by the shell, M, which yields

$$\mathscr{J}(M) \propto M^\beta \quad \text{with} \quad \beta \simeq 1.3 \pm 0.3. \qquad (7.179)$$

7.6 The Halo Model of Dark Matter Clustering

In a hierarchical cosmogony with a density perturbation power spectrum that increases monotonically and without bound as the mass scale decreases, all dark matter particles are expected to reside in dark matter halos. This suggests that one can describe the dark matter distribution in terms of its halo building blocks: on small scales the density field is related to the density distribution of individual halos, which, as we have seen in §7.5.1, have roughly a universal density profile. On large scales it reflects the spatial distribution of halos of different masses. To some extent, this is an old idea. One of the earliest models for galaxy clustering considered a density field composed of randomly distributed independent clumps with some universal density profile (Neyman et al., 1953). Now that we have accurate models for the mass function, density profile and spatial correlation of dark matter halos, it is possible to use this halo model to give an accurate description of the dark matter distribution. Here we present the main framework; a more thorough description can be found in Cooray & Sheth (2002).

For simplicity, we assume that the density profile of a halo depends only on its mass, so that we can write the density profile of a halo with mass M as

$$\rho(\mathbf{x}) = M u(\mathbf{x}|M), \qquad (7.180)$$

where $\mathbf{x}$ is the position relative to the halo center, and the profile function $u(\mathbf{x}|M)$ is normalized, i.e. $\int u(\mathbf{x}|M) d^3 \mathbf{x} = 1$. Now, imagine that space is divided into many small volumes, ΔV_i ($i = 1, 2, \ldots$), which are so small that none of them contains more than one halo center. The occupation number in the ith volume $\mathscr{N}_i$ is therefore either 1 or 0, and so $\mathscr{N}_i = \mathscr{N}_i^2 = \mathscr{N}_i^3 = \cdots$.

In terms of the occupation number $\mathscr{N}_i$, the density field of dark matter can formally be written as

$$\rho(\mathbf{x}) = \sum_i \mathscr{N}_i M_i u(\mathbf{x} - \mathbf{x}_i | M_i), \qquad (7.181)$$

where M_i is the mass of the halo whose center is in ΔV_i. Note that the ensemble average $\langle \mathscr{N}_i M_i u(\mathbf{x} - \mathbf{x}_i | M_i) \rangle$ is equal to $\int dM\, n(M) M \Delta V_i u(\mathbf{x} - \mathbf{x}_i | M)$, where $n(M)$ is the halo mass function. Thus

7.6 The Halo Model of Dark Matter Clustering

$$\langle \rho(\mathbf{x}) \rangle = \int dM\, M\, n(M) \sum_i \Delta V_i\, u(\mathbf{x} - \mathbf{x}_i | M)$$

$$= \int dM\, M\, n(M) \int d^3 x'\, u(\mathbf{x} - \mathbf{x}' | M)$$

$$= \int dM\, M\, n(M) = \overline{\rho}, \qquad (7.182)$$

where $\overline{\rho}$ is the mean mass density of the Universe. Similarly, the two-point correlation function of the density field is

$$\langle \rho(\mathbf{x}_1) \rho(\mathbf{x}_2) \rangle = \sum_{i,j} \langle \mathcal{N}_i M_i \mathcal{N}_j M_j u(\mathbf{x}_1 - \mathbf{x}_i | M_i) u(\mathbf{x}_2 - \mathbf{x}_j | M_j) \rangle. \qquad (7.183)$$

We can divide the summation $\sum_{i,j}$ into two parts, one with $i = j$ and the other with $i \neq j$. The first part describes the case where the two contributions to the correlation are from the same halo, and is called the '1-halo term'. The second part represents the case where the two contributions are from different halos, and is called the '2-halo term'. For $i = j$ the ensemble average in the above equation can be simplified by using $\mathcal{N}_i \mathcal{N}_j = \mathcal{N}_i^2 = \mathcal{N}_i$. For $i \neq j$, the ensemble average can be written as

$$\int dM_1 M_1 n(M_1) \int dM_2 M_2 n(M_2) [1 + \xi_{hh}(\mathbf{x}_i - \mathbf{x}_j | M_1, M_2)] u(\mathbf{x}_1 - \mathbf{x}_i | M_1) u(\mathbf{x}_2 - \mathbf{x}_j | M_2), \qquad (7.184)$$

where $\xi_{hh}(\mathbf{x}_i - \mathbf{x}_j | M_1, M_2)$ is the two-point cross-correlation function between halos of masses M_1 and M_2. It is then easy to show that the two-point correlation function for the density fluctuation field, $\delta(\mathbf{x}) \equiv \rho(\mathbf{x})/\overline{\rho} - 1$, is

$$\xi(r) \equiv \langle \delta(\mathbf{x}_1) \delta(\mathbf{x}_2) \rangle = \xi^{1h}(r) + \xi^{2h}(r) \quad (r \equiv |\mathbf{x}_1 - \mathbf{x}_2|), \qquad (7.185)$$

where

$$\xi^{1h}(r) = \frac{1}{\overline{\rho}^2} \int dM\, M^2 n(M) \int d^3 y\, u(\mathbf{y} - \mathbf{x}_1 | M) u(\mathbf{y} - \mathbf{x}_2 | M), \qquad (7.186)$$

$$\xi^{2h}(r) = \frac{1}{\overline{\rho}^2} \int dM_1 M_1 n(M_1) \int dM_2 M_2 n(M_2) \int d^3 x \int d^3 x'$$
$$\times u(\mathbf{x}_1 - \mathbf{x} | M_1) u(\mathbf{x}_2 - \mathbf{x}' | M_2) \xi_{hh}(\mathbf{x} - \mathbf{x}' | M_1, M_2). \qquad (7.187)$$

Because the model correlation function involves convolution, it is more convenient to work in Fourier space. The Fourier transform of the density field defined in Eq. (7.181) is

$$\rho(\mathbf{k}) = \sum_i \mathcal{N}_i M_i \tilde{u}(\mathbf{k} | M_i) e^{-i \mathbf{k} \cdot \mathbf{x}_i}. \qquad (7.188)$$

Here $\tilde{u}(\mathbf{k} | M_i)$ is the Fourier transform of the density profile, which for a spherically symmetric profile truncated at the virial radius r_h is given by

$$\tilde{u}(k | M) = \frac{4\pi}{M} \int_0^{r_h} \rho(r | M) \frac{\sin kr}{kr} r^2\, dr. \qquad (7.189)$$

Using the properties of $\mathcal{N}_i$, it is straightforward to show that the power spectrum of the density field is

$$P(k) \equiv V_u \langle |\delta_\mathbf{k}|^2 \rangle = P^{1h}(k) + P^{2h}(k), \qquad (7.190)$$

where

$$P^{1h}(k) = \frac{1}{\overline{\rho}^2} \int dM\, M^2 n(M) |\tilde{u}(k | M)|^2, \qquad (7.191)$$

$$P^{2h}(k) = \frac{1}{\bar{\rho}^2} \int dM_1 \, M_1 n(M_1) \tilde{u}(k|M_1) \int dM_2 \, M_2 n(M_2) \tilde{u}(k|M_2) P_{hh}(k|M_1, M_2), \quad (7.192)$$

with $P_{hh}(k|M_1,M_2)$ the cross-power spectrum of halos of masses M_1 and M_2. As expected, the one-halo terms of $\xi(r)$ and $P(k)$ are completely determined by the density profiles of individual halos and the halo mass function. The dependence on the halo profile in the two-halo terms is due to the finite sizes of individual halos. If the scale in consideration is much larger than the sizes of individual halos, only the two-halo term is important. In this case, we can replace $u(\mathbf{y}|M)$ in Eq. (7.187) by a delta function and $\tilde{u}(\mathbf{k}|M)$ in Eq. (7.192) by 1, and so the two-halo terms become independent of the density profiles of individual halos.

As we have seen in §7.4.1, dark matter halos are biased with respect to the mass distribution. On large scales, this can be expressed via the linear halo bias parameter $b_h(M)$, which allows us to write

$$P_{hh}(k|M_1, M_2) = b_h(M_1) b_h(M_2) P_{\text{lin}}(k), \quad (7.193)$$

with $P_{\text{lin}}(k)$ the linear power spectrum.[3] Substituting Eq. (7.193) into Eq. (7.192) yields

$$P^{2h}(k) = P_{\text{lin}}(k) \left[\frac{1}{\bar{\rho}} \int dM \, M \, n(M) \, b_h(M) \, \tilde{u}(k|M) \right]^2. \quad (7.194)$$

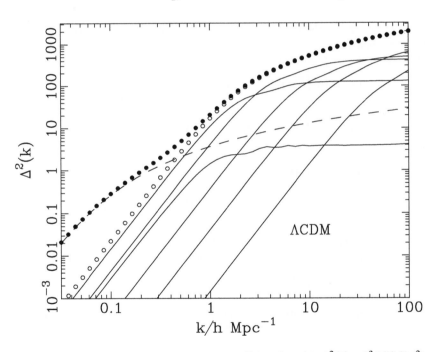

Fig. 7.11. The decomposition of the mass power spectrum [here plotted as $\Delta^2(k) \equiv k^3 P(k)/2\pi^2$] according to the halo model, for a flat cosmology with $\Omega_{m,0} = 0.3$, $\Omega_{\Lambda,0} = 0.7$, $h = 0.7$, and $\sigma_8 = 0.8$. The dashed line shows the linear power spectrum and the open circles show the predicted one-halo contribution. Summing up the two yields the total power spectrum represented by the solid points. The solid lines show the contribution of different mass ranges to the one-halo term: each line covers one dex in halo mass, starting at 10^{10}–$10^{11} h^{-1} \text{M}_\odot$ and ending at 10^{15}–$10^{16} h^{-1} \text{M}_\odot$. The more massive halos have larger radii and hence contribute to the power spectrum on progressively larger scales. Most of the quasi-linear power is contributed by halos with masses close to $M^* \sim 10^{13} h^{-1} \text{M}_\odot$. [Courtesy of J. Peacock; see Peacock (2003)]

---

[3] The discussion for the two-point correlation function is the same and is omitted here.

Since the distribution of matter is by definition unbiased with respect to itself, we have that

$$\frac{1}{\bar{\rho}} \int dM\, M\, n(M)\, b_\mathrm{h}(M) = 1. \tag{7.195}$$

Thus, on large scales (small k), where $\tilde{u}(k|M) = 1$, we have that $P^{2\mathrm{h}}(k) = P_\mathrm{lin}(k)$. On small scales, Eq. (7.194) is no longer an accurate approximation of the true two-halo term, due to the facts that (i) Eq. (7.193) only applies to large scales that are still in the linear regime, and (ii) it does not account for halo exclusion (halos are spatially exclusive). Fortunately, high accuracy for $P_{2\mathrm{h}}(k)$ on small scales is not really needed, since the one-halo term, associated with the internal structure of the dark matter halos, clearly dominates for large k. Although this halo model is obviously an approximation, in practice it works extremely well (Ma & Fry, 2000a,b; Smith et al., 2003). In addition, it provides a novel way to understand the features in the nonlinear power spectrum (see Fig. 7.11). According to the ideas presented here, the flat small-scale spectrum arises because halos have central density cusps proportional to r^{-1}, but not much steeper. The sharp fall in power at smaller k reflects the cutoff at the virial radii of the halos that dominate the correlation signal.

Similar derivations can also be carried out for higher-order correlation functions, or for the correlation function and power spectrum in redshift space (Cooray & Sheth, 2002).

8
Formation and Evolution of Gaseous Halos

So far we have concentrated on the formation of structure under the influence of gravity alone. However, since the astronomical objects we are able to observe directly are made of baryons and electrons, the role of gas-dynamical and radiative processes must also be taken into account in order to understand how the structures we observe form and evolve. As demonstrated in §4.1.3, since baryons and dark matter are expected to be well mixed initially, the density perturbation fields of the baryons, δ_b, and dark matter, δ_{dm}, are expected to be equal in the linear regime, except for perturbations on scales smaller than or comparable to the Jeans length of the gas. In this chapter, we examine the role of gas-dynamical and radiative processes for the evolution of structures in the highly nonlinear regime. We start in §8.1 with a brief description of the basic dissipational processes. §8.2 describes the structure of gas in hydrostatic equilibrium within dark matter halos. The formation of gaseous halos in the absence of cooling and heating is discussed in §8.3, while §8.4 focuses on the impact of cooling. §8.5 describes several thermal and hydrodynamical instabilities of cooling gas, and §8.6 discusses the evolution of gaseous halos in the presence of energy sources. §8.7 gives a summary of the current status of numerical studies of the formation and structure of gaseous halos, while §8.8 discusses observations of gaseous halos associated with clusters and galaxies.

8.1 Basic Fluid Dynamics and Radiative Processes

8.1.1 Basic Equations

In many problems to be discussed below, the gas component can be approximated as an ideal fluid, which means that we can neglect heat conduction and viscous stress in the fluid equations (see §B1.2). Written in physical coordinates, $\mathbf{r}$, the continuity, Euler, and energy equations are:

$$\frac{\partial \rho}{\partial t} + \nabla \cdot (\rho \mathbf{v}) = 0, \tag{8.1}$$

$$\frac{\partial \mathbf{v}}{\partial t} + (\mathbf{v} \cdot \nabla) \mathbf{v} = -\left(\nabla \Phi + \frac{\nabla P}{\rho}\right), \tag{8.2}$$

$$\frac{\partial}{\partial t}\left[\rho\left(\frac{v^2}{2} + \mathscr{E}\right)\right] + \nabla \cdot \left[\rho\left(\frac{v^2}{2} + \frac{P}{\rho} + \mathscr{E}\right)\mathbf{v}\right] - \rho \mathbf{v} \cdot \nabla \Phi = \mathscr{H} - \mathscr{C}. \tag{8.3}$$

Here ρ, $\mathbf{v}$, P, $\mathscr{E}$ are the density, velocity, pressure and specific internal energy of the fluid, respectively, and $\mathscr{H}$ and $\mathscr{C}$ are the heating and cooling rates per unit volume. For an ideal gas with an adiabatic index γ (also called the ratio of specific heats), we have $P = \rho(\gamma - 1)\mathscr{E}$. Eq. (8.3) can then be replaced by the following entropy equation:

8.1 Basic Fluid Dynamics and Radiative Processes

$$\frac{P}{\gamma-1}\left(\frac{\partial}{\partial t}+\mathbf{v}\cdot\nabla\right)\ln\left(\frac{P}{\rho^\gamma}\right) = \mathcal{H}-\mathcal{C}, \tag{8.4}$$

(see §B1.2). The gravitational potential Φ satisfies the Poisson equation

$$\nabla^2\Phi = 4\pi G\rho_{\text{tot}}, \tag{8.5}$$

where ρ_{tot} is the total mass density of the universe. If the Universe contains a collisionless dark matter component in addition to the baryonic fluid, ρ_{tot} should also include ρ_{dm}, the evolution of which is governed by the collisionless Boltzmann equation (5.103).

As one can see, in order to study the evolution of the baryonic component using the ideal-fluid approximation, one has to deal with processes that can heat or cool the baryonic gas. In what follows, we describe briefly some of these processes; a more detailed description is given in Appendix B.

8.1.2 Compton Cooling

When photons of low energy $h_P\nu$ pass through a thermal gas of non-relativistic electrons with temperature T_e ($h_P\nu \ll m_e c^2$; $k_B T_e \ll m_e c^2$), photons and electrons exchange energy due to Compton scattering (see §B1.3). If the radiation field is a thermal background with temperature $T_\gamma \ll T_e$, the net effect is for electrons to lose energy to the radiation field, causing the gas to cool. The cooling rate per unit volume is equal to the rate of increase in the energy density of the radiation field, allowing us to write

$$\mathcal{C}_{\text{Comp}} = \frac{du_\gamma}{dt} = \frac{4k_B T_e}{m_e c^2}c\sigma_T n_e a_r T_\gamma^4, \tag{8.6}$$

where the second equality follows from Eq. (B1.77). For a fully ionized gas of primordial composition ($Y_p \simeq 1/4$ so that $n_{\text{He}} \simeq n_H/12$; see §3.4.3), the energy content is $(3/2)kT_e \times (27/14)$ per electron, where the factor $27/14$ is the number of particles per electron. Thus, the gas can cool against the radiation field on a time scale

$$t_{\text{Comp}} \approx \frac{3k_B T_e n_e}{\mathcal{C}_{\text{Comp}}} = \frac{3m_e c}{4\sigma_T a_r T_\gamma^4}, \tag{8.7}$$

provided $T_e \gg T_\gamma$. Note that this time scale is independent of the density and temperature of the gas.

An important application is the cooling against the cosmic microwave background (CMB), for which the temperature changes with redshift as $T_\gamma \approx 2.73(1+z)$ K. We can approximate the age of the Universe at redshift $z \geq \Omega_{m,0}^{-1} - 1$ by $t \approx 6.7 \times 10^9 \Omega_{m,0}^{-1/2} h^{-1}(1+z)^{-3/2}$ yr (see §3.2.5). It is then easy to show that

$$\frac{t_{\text{Comp}}}{t} \approx 350\Omega_{m,0}^{1/2} h(1+z)^{-5/2}. \tag{8.8}$$

For $\Omega_{m,0} = 0.3$ and $h = 0.7$, this gives $t_{\text{Comp}}/t = 1$ at $z \sim 6$. Hence, Compton cooling against the CMB is only important for gas at high redshifts.

8.1.3 Radiative Cooling

The primary cooling processes relevant for structure formation are two-body radiative processes, in which gas loses energy through radiation as a result of two-body interactions. For our purposes, processes involving three bodies or more can be ignored, since the gas densities involved are too low. At temperatures above 10^6 K, primordial gas (composed of hydrogen and helium) is almost entirely ionized, and above a few $\times 10^7$ K, enriched gas (which contains also heavier elements) is fully ionized as well. The only significant radiative cooling at such high temperatures is bremsstrahlung due to the acceleration of electrons as they encounter atomic nuclei. For an optically thin gas, the volume cooling rate is related to the bremsstrahlung (free–free) emissivity $\varepsilon_{\rm ff}$ [see Eq. (B1.63)] through

$$\mathscr{C}_{\rm ff} = \int \varepsilon_{\rm ff}(\nu)\,{\rm d}\nu$$
$$\approx 1.4 \times 10^{-23} T_8^{1/2} \left(\frac{n_{\rm e}}{\rm cm^{-3}}\right)^2 {\rm erg\,s^{-1}\,cm^{-3}}, \qquad (8.9)$$

where $T_8 \equiv T/(10^8\,{\rm K})$, and the second expression assumes a charge number of unity and $n_{\rm i} \sim n_{\rm e}$, valid for a completely ionized hydrogen gas. When species other than hydrogen are present, the total cooling rate is the sum over all species of ions. For example, for a primordial gas, with $n_{\rm He} \simeq n_{\rm H}/12$ the above cooling rate needs to be multiplied by a factor $16/14$.

At lower temperatures, several other processes are important. The first is collisional ionization, in which atoms (including partially ionized ones) are ionized by collisions with electrons, removing from the gas an amount of kinetic energy equal to the ionization threshold. The second is recombination, in which an electron recombines with an ion, emitting a photon. The third is collisional excitation, in which atoms are first excited by collisions with electrons and then decay radiatively to the ground state. All these processes depend strongly on T, in the first and second cases because the recombination and (collisional) ionization rates are temperature sensitive, and in the third case because the ion abundance depends strongly on temperature.

With the assumption that the gas is optically thin to the emitted photons, the cooling rates due to these various processes can be calculated for a gas at a given temperature, and the details of such calculation are described in §B1.4. In practice, the cooling rate of the cosmic gas (which is rich in hydrogen) is usually represented by a cooling function defined as

$$\Lambda(T) \equiv \frac{\mathscr{C}}{n_{\rm H}^2}, \qquad (8.10)$$

where $\mathscr{C}$ is the total cooling rate per unit volume and $n_{\rm H}$ is the number density of hydrogen atoms (both neutral and ionized). Thus defined, Λ is independent of gas density for an optically thin gas. Note that other normalizations, such as $\mathscr{C}/n_{\rm e}^2$, are also in use in the literature, which may lead to slightly different values of the cooling function. Fig. 8.1 shows the cooling function as a function of T for gases with three different metallicities. In all cases, the cooling rate is dominated by bremsstrahlung at the high-temperature end, where $\Lambda \propto T^{1/2}$. For a primordial gas with $T < 10^{5.5}\,{\rm K}$, a large fraction of the electrons are bound to their atoms, and the dominant cooling process is collisional excitation followed by radiative de-excitation; the peaks in the cooling function at $\sim 15{,}000\,{\rm K}$ and $\sim 10^5\,{\rm K}$ are due to collisionally excited electronic levels of H^0 and He^+, respectively. For an enriched gas, there is an even stronger peak at $T = 10^5\,{\rm K}$ due to the collisionally excited levels of ions of oxygen, carbon, nitrogen, etc. In an enriched gas, the cooling function is also enhanced at $\sim 10^6\,{\rm K}$ by other common elements, noticeably neon, iron and silicon.

At temperatures below $10^4\,{\rm K}$, most of the electrons have recombined and cooling due to collisional excitation drops precipitously. In this case, radiative cooling is still possible, albeit at a much reduced rate, if the gas is enriched. Here collisions with neutral hydrogen atoms and with the few free electrons left can excite the fine structure levels of low ions, such as OI, OII, OIII and CII. If molecules (H_2, CO, etc.) are present in the gas, collisional excitation of their rotational/vibrational levels can also contribute to gas cooling at low temperature. Detailed descriptions about these cooling processes can be found in Dalgarno & McCray (1972) and

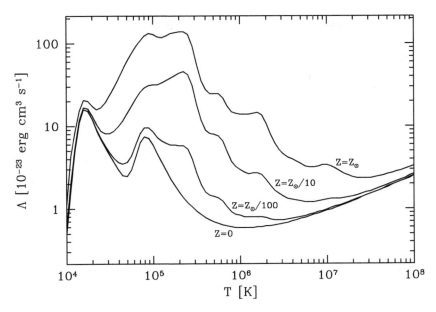

Fig. 8.1. Cooling functions for primordial ($Z = 0$) gas (assuming $n_{He}/n_H = 1/12$), and for gases with metallicities $Z/Z_\odot = 0.01$, 0.1 and 1.0, as indicated. [Based on data published in Sutherland & Dopita (1993)]

Spitzer (1978), for example. In the case of primordial gas, cooling at temperatures below 10^4 K is only possible if significant amounts of molecular hydrogen can form in the gas. In the absence of dust grains, as is the case for a primordial gas, the formation of H_2 has to proceed via gas-phase reactions such as $H^0 + e \to H^- + \gamma$ followed first by $H^- + H^0 \to H_2 + e$ and/or $H^+ + H^0 \to H_2^+ + \gamma$ and then by $H_2^+ + H^0 \to H_2 + H^+$. The cross-sections of all these reactions are reasonably well known, allowing the corresponding cooling function to be calculated (e.g. Abel et al., 1997; Galli & Palla, 1998).

It should be emphasized that the cooling functions shown in Fig. 8.1 assume ionization equilibrium, i.e. that the densities of all ions are equal to their equilibrium values. This is only expected to be applicable if the time scales for the radiative processes in question are much shorter than the hydrodynamical time scales of the gas. This may not be the case in very dilute gas (where the reaction rates are very low) or in shocks (where the hydrodynamical times are short). For gas that is not in ionization equilibrium, the cooling rates have to be calculated using non-equilibrium densities obtained by solving the time-dependent ionization equation (see §B1.3).

8.1.4 Photoionization Heating

In addition to collisional ionization, an atom can also be ionized by absorbing a photon, a process called photoionization (see §B1.3). Thus, the presence of an ionizing radiation field can change the population densities of ions, thereby changing the cooling rate of the gas. In addition, photoionization can also heat the gas, through a process called photoionization heating. When an ionizing photon with energy $h_P \nu$ ionizes an electron from an atom with threshold frequency ν_i (i.e. whose ionization threshold is $h_P \nu_i$), the surplus energy, $h_P(\nu - \nu_i)$, is transformed into the kinetic energy of the electron. In a static state, photoionization is balanced by recombination. However, the loss of energy due to recombination is smaller than the gain from photoionization, because the recombination rate is in general higher for lower-energy electrons, causing a net heating. The photoionization heating rate per unit volume is expected to be proportional to the

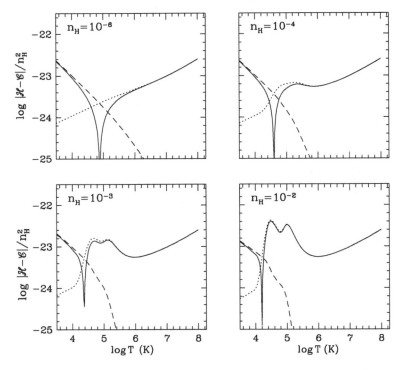

Fig. 8.2. Net cooling rates as a function of temperature for a gas of primordial composition in ionization equilibrium with a UV radiation background of intensity $J(\nu) = 10^{-22}(\nu_H/\nu)\,\mathrm{erg\,s^{-1}\,cm^{-2}\,sr^{-1}\,Hz^{-1}}$. Results are shown for four different hydrogen number densities $n_H = 10^{-6}$, 10^{-4}, 10^{-3}, and $10^{-2}\,\mathrm{cm^{-3}}$, as indicated. In each panel, the dotted line shows the cooling rate and the dashed line shows the rate of heating by photoionization. The solid curve shows the absolute value of the net cooling rate (in $\mathrm{erg\,cm^3\,s^{-1}}$); heating dominates at low temperatures and cooling at high temperatures. [Adapted from Weinberg et al. (1997a) by permission of AAS]

intensity of the ionizing radiation field, $J(\nu)$, the density of the ions that can be ionized by the radiation field, n, and the photoionization cross-section, σ_{phot}, and can be written as

$$\mathcal{H} = \sum_i n_i \varepsilon_i, \qquad (8.11)$$

where

$$\varepsilon_i = \int_{\nu_i}^{\infty} \frac{4\pi J(\nu)}{h_P \nu} \sigma_{\mathrm{phot},i}(\nu)\, (h_P \nu - h_P \nu_i)\, d\nu \qquad (8.12)$$

is the mean kinetic energy involved in photoionizing species 'i'. Note that Eq. (8.11) does not include the energy loss due to recombinations, which is included in the recombination cooling.

As an illustration, Fig. 8.2 shows the net cooling functions for primordial gas in ionization equilibrium at various densities and exposed to a UV radiation background with intensity $J(\nu) = 10^{-22}(\nu_H/\nu)\,\mathrm{erg\,s^{-1}\,cm^{-2}\,sr^{-1}\,Hz^{-1}}$. The intensity adopted here is roughly that expected from the emission of quasars and young galaxies at $z \sim 3$. As one can see, heating dominates over cooling at $T < 10^4\,\mathrm{K}$ to $10^5\,\mathrm{K}$, depending on the density of the gas. Note also that, for very dilute gas, there is very little cooling due to radiative de-excitation (i.e. the peaks in the cooling function due to H^0 and He^+ disappear).

8.2 Hydrostatic Equilibrium

For gas in hydrostatic equilibrium, the gravitational forces are balanced by pressure gradients, so that

$$\nabla P(\mathbf{r}) = -\rho(\mathbf{r})\nabla\Phi(\mathbf{r}), \tag{8.13}$$

where Φ is the gravitational potential which satisfies the Poisson equation

$$\nabla^2\Phi = 4\pi G(\rho_{dm}+\rho). \tag{8.14}$$

Here we have explicitly included a possible contribution to the gravitational potential due to dark matter. As is clear from the above equations, the isopotential surfaces are the same as the isobaric surfaces. Under the assumption of spherical symmetry and for an ideal gas, the pressure and potential gradients are

$$\frac{dP}{dr} = \frac{d(k_B T\rho/\mu m_p)}{dr}, \quad \frac{d\Phi}{dr} = \frac{GM(r)}{r^2}, \tag{8.15}$$

where $M(r)$ is the total mass (gas plus dark matter) within radius r, $\mu = \rho/(nm_p)$ is the mean molecular weight of the gas, m_p is the proton mass, ρ is the mass density, and n is the number density of particles. The hydrostatic equation can then be written as

$$M(r) = -\frac{k_B T(r)r}{\mu m_p G}\left[\frac{d\ln\rho}{d\ln r} + \frac{d\ln T}{d\ln r}\right]. \tag{8.16}$$

Thus, the hydrostatic equation provides an estimate of the *total* mass of a halo within some radius r from measurements of the density and temperature profiles, $\rho(r)$ and $T(r)$, of the gas. This is in fact the standard method for estimating the mass of X-ray clusters, for which both $\rho(r)$ and $T(r)$ can be obtained from their X-ray emissions (see §8.8). Note, however, that the total pressure may have a substantial contribution, P_{nt}, due to non-thermal turbulence, magnetic fields and/or cosmic rays. In that case Eq. (8.16) needs to be replaced by

$$M(r) = -\frac{k_B T(r)r}{\mu m_p G}\left[\frac{d\ln\rho}{d\ln r} + \frac{d\ln T}{d\ln r} + \frac{P_{nt}}{P_{th}}\frac{d\ln P_{nt}}{d\ln r}\right], \tag{8.17}$$

where P_{th} is the thermal pressure. Since P_{nt} is in general difficult to measure, and since real clusters are not exactly spherical, the masses inferred with this method can still have significant uncertainties (e.g. Evrard et al., 1996; Rasia et al., 2006).

8.2.1 Gas Density Profile

In general, the assumption of hydrostatic equilibrium alone is not sufficient to determine the density distribution of the gas, because the dependence of the temperature on radius remains unspecified. In order to make progress, one typically either assumes that the gas has a polytropic equation of state, or that it has a constant temperature $T(r) = T$. We now discuss these two cases in turn.

In the case of a polytropic gas, the equation of state is given by $P = A\rho^\Gamma$, where A is a constant, and $\Gamma = \mathrm{const.}$ is the polytropic index (see §B1.1). If $\Gamma > 1$, the hydrostatic equation implies that

$$k_B T(\mathbf{r}) = \frac{(1-\Gamma)}{\Gamma}\mu m_p \Phi(\mathbf{r}), \tag{8.18}$$

where we have assumed that both ρ and Φ vanish at large distances. In this case, the gas temperature exactly follows the gravitational potential field. Note that, since $P \propto T\rho \propto \rho^\Gamma$, the gravitational potential field also completely specifies the density and pressure profiles of the gas.

Another example for which the hydrostatic equation can be solved is an isothermal[1] sphere, for which T is independent of r. In this case the hydrostatic equation can be written as

$$\rho(r) = \rho_0 \exp\left(-\frac{\Phi}{c_T^2}\right), \quad c_T^2 \equiv \frac{k_B T}{\mu m_p}, \tag{8.19}$$

and the Poisson equation gives

$$\frac{1}{r^2}\frac{d}{dr}\left(r^2 \frac{d\Phi}{dr}\right) = 4\pi G \left[\rho_{\rm dm}(r) + \rho_0 \exp\left(-\frac{\Phi}{c_T^2}\right)\right]. \tag{8.20}$$

This equation can be solved for given $\rho_{\rm dm}(r)$. In particular, if $\rho_{\rm dm} = 0$, the above equation reduces to the Lane–Emden equation. If we assume that $\rho(r)$ is a power law, there is a simple solution:

$$\Phi(r) = \frac{2k_B T}{\mu m_p} \ln\frac{r}{r_0}, \quad \rho(r) = \frac{2k_B T}{\mu m_p}\frac{1}{4\pi G r^2}, \quad M(r) = \frac{2k_B T}{\mu m_p}\frac{r}{G}, \tag{8.21}$$

with r_0 defined by $\Phi(r_0) = 0$. Defining the circular velocity of the gaseous sphere as $V_c \equiv [GM(r)/r]^{1/2}$, we have

$$V_c^2 = \frac{2k_B T}{\mu m_p}, \quad \rho(r) = \frac{V_c^2}{4\pi G r^2}, \quad M(r) = \frac{V_c^2 r}{G}. \tag{8.22}$$

These are the properties of a singular isothermal sphere. It is clear from Eq. (8.20) that if $\rho_{\rm dm}$ has a $1/r^2$ profile, the above solution still holds, except that ρ and M are now the total density and total mass, respectively.

The singular isothermal sphere solution, although simple, is unphysical, because the density diverges at $r = 0$. To obtain a physical solution, we need to impose physical boundary conditions on the problem. There are two such conditions: $\rho(r=0) = \rho_0 = $ constant and $(d\Phi/dr)(r=0) = 0$. The second condition follows from the fact that the force at the center is zero. The solution of the Lane–Emden equation with these boundary conditions can be obtained numerically. The resulting density profile can be well approximated by the King profile,

$$\rho(r) = \frac{\rho_0}{[1+(r/r_0)^2]^{3/2}}, \quad r_0 = \frac{3 c_T}{\sqrt{4\pi G \rho_0}}, \tag{8.23}$$

for $r \lesssim 2r_0$. For $r \gtrsim 10 r_0$, $\rho(r)$ approaches that of a singular isothermal sphere (see §5.4.7).

In general, if both gas and dark matter are in static equilibrium within the same potential well $\Phi(\mathbf{r})$, then

$$\frac{1}{\rho_{\rm gas}}\nabla_i P = \frac{1}{\rho_{\rm dm}}\sum_j \nabla_j \left[\rho_{\rm dm}\langle v_j v_i\rangle\right]. \tag{8.24}$$

This relation follows from substituting the gravitational term $\nabla\Phi$ in Eq. (8.13) with that appearing in

$$\sum_j \nabla_j \left[\rho_{\rm dm}\langle v_j v_i\rangle\right] = -\rho_{\rm dm}\nabla_i\Phi, \tag{8.25}$$

which are the momentum equations in static form [see Eq. (5.112)]. If the velocity distribution of dark matter particles is isotropic, i.e. $\langle v_i v_j\rangle = \sigma^2 \delta_{ij}$, and if the halo is isothermal so that σ^2 and T are independent of $\mathbf{r}$, we obtain

$$\rho_{\rm gas}(r) \propto [\rho_{\rm dm}(r)]^\beta, \tag{8.26}$$

---

[1] Here 'isothermal' refers to the fact that T is constant in space. This should not be confused with an isothermal equation of state, for which T is constant in time during a thermodynamic process.

where
$$\beta \equiv \frac{\mu m_p \sigma^2}{k_B T}. \tag{8.27}$$

Since the specific kinetic energy of the dark matter is equal to $\langle v^2 \rangle/2 = 3\sigma^2/2$, and the specific internal energy of the gas is equal to $\mathscr{E} = (\gamma - 1)^{-1}(k_B T/\mu m_p)$, for a monatomic gas with $\gamma = 5/3$ the value of β is equal to the ratio of specific energies of the dark matter and gas.

As a final example, consider the NFW dark halo profile discussed in §7.5.1,
$$\rho_{dm}(r) = \rho_{crit} \frac{\delta_{char}}{(r/r_s)(1 + r/r_s)^2}. \tag{8.28}$$

If we again make the assumption that the gas temperature profile is constant, $T(r) = T$, and we assume in addition that the self-gravity of the gas is negligible (i.e. $\rho \ll \rho_{dm}$), then hydrostatic equilibrium implies a gas density profile
$$\rho(r) = \rho_0 e^{-b} \left(1 + \frac{r}{r_s}\right)^{br_s/r}, \tag{8.29}$$

where $\rho_0 = \rho(0)$, and
$$b = 4\pi G \rho_{crit} \delta_{char} r_s^2 \frac{\mu m_p}{k_B T}. \tag{8.30}$$

Thus, unlike the dark matter profile, the gas profile $\rho(r)$ is finite at $r = 0$ with a core radius that depends on b and r_s. In reality, though, neither is the gas temperature profile isothermal, nor can the self-gravity of the gas be completely neglected, and one has to solve Eqs. (8.13) and (8.14) numerically. More detailed discussion of the density and temperature profiles of gas in NFW halos can be found in Suto et al. (1998) and Komatsu & Seljak (2001).

8.2.2 Convective Instability

The hydrostatic equation (8.13) specifies the condition of mechanical equilibrium for gas in a gravitational potential. However, the equilibrium state can be convectively unstable, depending on the distribution and thermal state of the gas. To obtain the criterion for convective instability, consider a small blob of gas initially in thermal and mechanical equilibrium with an ambient medium. For simplicity, consider a one-dimensional case and assume that the gravitational force points in the $-z$ direction. Suppose that the blob is displaced adiabatically along the $-z$ direction by an infinitesimal amount. To maintain pressure equilibrium with the ambient gas, the change in the density of the blob must be
$$(d\rho)_{blob} = \left(\frac{\partial \rho}{\partial P}\right)_S dP. \tag{8.31}$$

In the new place, the density of the ambient gas is different from that in the old place by an amount that can be written as
$$\left(\frac{d\rho}{dz}\right)_{ambient} dz = \left(\frac{\partial \rho}{\partial P}\right)_S dP + \left(\frac{\partial \rho}{\partial S}\right)_P dS, \tag{8.32}$$

where dP and dS are the differences of the pressure and specific entropy of the ambient medium between the two places. If the increase in ρ_{blob} is faster than the increase in $\rho_{ambient}$, the blob will sink further along the $-z$ axis and the perturbation is amplified. Thus, convective instability will occur if the ambient medium satisfies
$$\left(\frac{\partial \rho}{\partial S}\right)_P \frac{dS}{dz} > 0. \tag{8.33}$$

Using Maxwell's relation, $(\partial V/\partial S)_P = (\partial T/\partial p)_S$, we see that

$$(\partial \rho/\partial S)_P \propto -\rho^2 (\partial V/\partial S)_P = -\rho^2 (\partial T/\partial P)_S < 0, \qquad (8.34)$$

where we have used the fact that the gas temperature increases if it is adiabatically compressed. We then obtain Schwarzschild's criterion for convective instability:

$$dS/dz < 0, \qquad (8.35)$$

i.e. the specific entropy of the ambient medium increases in the direction of gravity.

For an ideal gas, $dS = c_P dT/T - dP/(\rho T)$, where c_P is the specific heat at constant pressure. The mechanical equilibrium condition gives $dP/dz = -g\rho$, where g is the gravitational acceleration. The condition (8.35) can therefore be written as

$$\frac{dT}{dz} < \frac{T}{P}\left(1 - \frac{c_V}{c_P}\right)\frac{dP}{dz} = \left(\frac{dT}{dz}\right)_S = -\frac{g}{c_P}, \qquad (8.36)$$

implying that convective instability occurs whenever the temperature gradient along the gravitational field is sufficiently high.

8.2.3 Virial Theorem Applied to a Gaseous Halo

As shown in §5.4.4, the virial theorem provides a constraint on the global properties of any dynamical system:

$$\frac{1}{2}\frac{d^2 I}{dt^2} = 2K + W + \Sigma, \qquad (8.37)$$

where I is the momentum of inertia, K is the kinetic energy, W is the gravitational energy, and Σ is the work done by the surface pressure. For a static system, $d^2 I/dt^2 = 0$ and the (scalar) virial theorem reduces to

$$2K + W + \Sigma = 0. \qquad (8.38)$$

As an application of the virial theorem, consider a uniform cloud of monatomic gas ($\gamma = 5/3$) with constant temperature T. The static virial theorem applied to such a system gives

$$2 \times \frac{3 M_{\rm gas} k_B T}{2\mu m_p} - \frac{3 G M_{\rm gas} M}{5 r_{\rm cl}} - 4\pi r_{\rm cl}^3 P_{\rm ext} = 0, \qquad (8.39)$$

where $r_{\rm cl}$ is the radius of the cloud, $M_{\rm gas}$ and M are, respectively, the gas and total masses within $r_{\rm cl}$, and $P_{\rm ext}$ is the external pressure. If $P_{\rm ext} = 0$, the virial theorem defines a virial temperature,

$$T_{\rm vir} = \frac{\mu m_p G M}{5 k_B r_{\rm cl}} = \frac{\mu m_p}{5 k_B} V_c^2, \qquad (8.40)$$

with V_c the virial velocity, defined as the circular velocity at $r_{\rm cl}$. This virial temperature is the equilibrium temperature for an isothermal cloud with uniform density.

If the external pressure is zero, Eq. (8.39) also defines a mass for the gas,

$$M_J = \left(\frac{3}{4\pi\rho_{\rm gas}}\right)^{1/2} \left(\frac{5 f_{\rm gas} k_B T}{\mu m_p G}\right)^{3/2}, \quad \text{where} \quad f_{\rm gas} \equiv \frac{M_{\rm gas}}{M}, \qquad (8.41)$$

which is similar to the Jeans mass for gravitational instability. If $M_{\rm gas} > M_J$, then $d^2 I/dt^2 < 0$ and the system contracts. On the other hand, if $M_{\rm gas} < M_J$, then $d^2 I/dt^2 > 0$ and the system expands with time. As is evident from Eq. (8.39), for a self-gravitating gas cloud with constant T, the contraction is catastrophic, because d^2I/dt^2 becomes more negative as r decreases. This is also true for any gas cloud with polytropic index $\Gamma < 4/3$, because $T \propto \rho^{\Gamma-1} \propto r_{\rm cl}^{-3(\Gamma-1)}$.

In the presence of external pressure, the gas cloud is compressed and instability can occur if the external pressure is too large. Consider an isothermal, self-gravitating gas cloud ($f_{\rm gas} = 1$). From Eq. (8.39) we see that compression will occur whenever

$$P_{\rm ext} > P_{\rm crit}(r) = \frac{3M_{\rm gas}}{4\pi r^3} \left(\frac{k_B T}{\mu m_p} - \frac{GM_{\rm gas}}{5r} \right). \tag{8.42}$$

This critical, external pressure is a function of the radius of the gas cloud, r, and has a maximum

$$P_{\rm max} \approx 3.15 \left(\frac{k_B T}{\mu m_p} \right)^4 \frac{1}{G^3 M_{\rm gas}^2} \quad \text{at} \quad r = r_{\rm max} \equiv \frac{4G\mu m_p M_{\rm gas}}{15 k_B T}. \tag{8.43}$$

If $P_{\rm ext}$ exceeds $P_{\rm max}$, no equilibrium is possible as the contraction can never bring the cloud onto the $P_{\rm crit}(r)$ curve. Hence, the cloud will collapse. If $P_{\rm crit} < P_{\rm ext} < P_{\rm max}$, the outcome depends on whether the radius of the cloud is smaller or larger than $r_{\rm max}$. If $r > r_{\rm max}$, then the cloud will contract until it reaches a radius at which $P_{\rm crit}(r) = P_{\rm ext}$ and it stops contracting. If $r \leq r_{\rm max}$, however, contraction can never re-establish equilibrium, and the system will collapse.

A gas cloud can be stabilized against gravitational collapse in several ways. If there is turbulent motion of the gas in the cloud, a kinetic energy term should be added to K, and the cloud becomes more stable. Similarly, the existence of a magnetic field, which adds a positive pressure term $3 \int P_B d^3\mathbf{x}$ (with P_B the pressure of the magnetic field) to the right-hand side of Eq. (8.37), also acts to stabilize the cloud. Furthermore, if the gas cloud is in the potential well of a static dark matter halo, it may also be stabilized if the halo density profile is sufficiently shallow. Suppose the halo density profile is $\rho_{\rm dm} \propto r^{-\alpha}$. The gravitational interaction between the gas cloud (assumed to be uniform) and the halo adds a term proportional to $-M_{\rm gas} r^{2-\alpha}$ to the virial, d^2I/dt^2. If $\alpha < 2$, the virial increases as the cloud contracts (r decreases), and the collapse of the gas is stabilized. On the other hand, if $\alpha > 2$ so that the added term decreases with decreasing r, the gas cloud is actually destabilized by the presence of the dark matter halo.

We have to be cautious when applying the virial theorem to a singular isothermal sphere, because the system is not finite (M goes to infinity as $r \to \infty$). However, we can truncate such a halo at some radius r_v and apply the virial theorem to the truncated halo. In order for the truncated halo to be in its original equilibrium state, we can imagine that there is an external pressure,

$$P_{\rm ext} = \rho_{\rm gas}(r_v) \frac{k_B T}{\mu m_p}, \tag{8.44}$$

acting on the outer surface. Since the gravitational energy of the truncated halo is $W = -V_c^2 M_{\rm gas}$ (where $V_c = \sqrt{GM/r_v}$ is the halo circular velocity), the virial theorem gives

$$T_{\rm vir} = \frac{\mu m_p}{2k_B} V_c^2 \simeq 3.6 \times 10^5 \, {\rm K} \left(\frac{V_c}{100 \, {\rm km \, s^{-1}}} \right)^2, \tag{8.45}$$

which is similar to the virial temperature defined in Eq. (8.40) except for a factor $5/2$. Although the virial temperature is a useful, and often used, concept in galaxy formation, realistic dark matter halos are not isothermal spheres (see §7.5.1), so that in general the gas inside the virial radius will have a radial temperature profile. In addition, we have so far ignored radiative cooling, which can have an important impact on the temperature and radial distribution of the gas inside a dark matter halo (see §8.4).

8.3 The Formation of Hot Gaseous Halos

The properties of gaseous halos that can form in the cosmic density field are determined by cosmological initial and boundary conditions, together with gravitational, gas-dynamical and radiative processes. As discussed in §4.1.3, the Jeans mass in the baryonic component after recombination is about $10^6\,M_\odot$, much smaller than the typical mass of galaxies. Thus, the baryonic component is expected to cluster in the same way as collisionless dark matter particles during the early evolutionary stages, when the perturbations on galaxy scales are still in the linear regime. However, once overdense regions become nonlinear and start to collapse, the baryonic gas associated with the overdensity can be shocked and heated, while collisionless dark mater particles cannot. Consequently, in a collapsed object the distribution of the gas component may deviate from that of the dark matter. In this section we describe how gravitational collapse leads to the formation of gaseous halos.

8.3.1 Accretion Shocks

According to Eq. (8.4), the specific entropy of a fluid element is conserved in the absence of net heating and net cooling. However, this is only true when the fluid remains smooth during the evolution so that the process is reversible. Entropy can be generated whenever there is an abrupt change in the fluid properties so that the flow process becomes irreversible, as is the case across a shock front. Shock waves can be produced when an obstructive body moves through a gaseous medium with supersonic speed. This is what happens, for example, when supernova explosions drive gas clouds (shells) into the interstellar medium. More relevant to our discussion in the following are the accretion shocks produced when cold gas is accreted into a gravitational potential well, such as a dark matter halo. In this subsection, we first consider the general properties of gas near a shock front, which we then use to understand how accretion shocks produced by dark matter halos result in the formation of gaseous halos.

(a) Shock Front To exemplify the essence of how the state of gas is affected by a shock, let us idealize the problem by a planar shock propagating at a constant speed $v_{\rm sh}$ through a uniform medium (see Fig. 8.3). The motion is one-dimensional, and we assume $v_{\rm sh}$ to be along the $-x$ axis. For an observer moving with the shock wave, the fluid is steady. The properties in the upstream (pre-shock fluid) and downstream (post-shock fluid) are therefore related by the

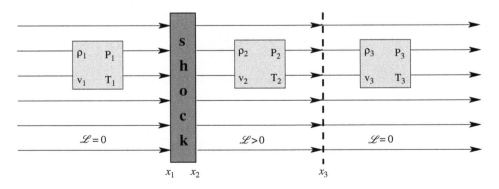

Fig. 8.3. An illustration of different regions of a planar shock. The supersonic flow in the region $x < x_1$ is shocked between x_1 and x_2 (i.e. the shock has a finite, though small, width), after which it becomes a hot subsonic flow. In between x_2 and x_3 the gas is out of thermal equilibrium resulting in net cooling ($\mathscr{L} > 0$). At $x > x_3$ the gas has cooled, reached a new thermal equilibrium, and continues to flow subsonically. The arrows indicate the direction (but not the speed) of the flow.

8.3 The Formation of Hot Gaseous Halos

steady-state fluid equations. For an ideal gas with small mean free path (so that a fluid treatment is valid, and viscosity and heat conduction can be neglected) and for which radiative cooling is unimportant, the continuity, Euler, and energy equations give the following jump conditions for the density ρ, pressure P and velocity v:

$$\rho_2 v_2 = \rho_1 v_1, \tag{8.46}$$

$$\rho_2 v_2^2 + P_2 = \rho_1 v_1^2 + P_1, \tag{8.47}$$

$$\frac{1}{2} v_2^2 + \frac{P_2}{\rho_2} + \mathscr{E}_2 = \frac{1}{2} v_1^2 + \frac{P_1}{\rho_1} + \mathscr{E}_1, \tag{8.48}$$

where subscripts '1' and '2' denote quantities for the upstream and downstream gas, respectively. Note that $v_1 = -v_{\rm sh}$. These jump conditions, usually referred to as the Rankine–Hugoniot jump conditions, can be written in more useful forms in terms of the Mach number of the upstream gas $\widehat{M}_1 \equiv v_1/c_{s,1}$ (where $c_s^2 = \gamma P/\rho$ is the sound speed, and we use the symbol $\widehat{M}$ in order to avoid confusion with mass):

$$\frac{\rho_2}{\rho_1} = \frac{v_1}{v_2} = \left[\frac{1}{\widehat{M}_1^2} + \frac{\gamma-1}{\gamma+1} \left(1 - \frac{1}{\widehat{M}_1^2} \right) \right]^{-1}, \tag{8.49}$$

$$\frac{P_2}{P_1} = \frac{2\gamma}{\gamma+1} \widehat{M}_1^2 - \frac{\gamma-1}{\gamma+1}, \tag{8.50}$$

$$\frac{T_2}{T_1} = \frac{P_2}{P_1} \frac{\rho_1}{\rho_2} = \frac{\gamma-1}{\gamma+1} \left[\frac{2}{\gamma+1} \left(\gamma \widehat{M}_1^2 - \frac{1}{\widehat{M}_1^2} \right) + \frac{4\gamma}{\gamma-1} - \frac{\gamma-1}{\gamma+1} \right], \tag{8.51}$$

where we have used that, for an ideal gas,

$$P = \rho(\gamma-1)\mathscr{E} = \frac{k_{\rm B} T}{\mu m_{\rm p}} \rho. \tag{8.52}$$

Thus if $\widehat{M}_1 > 1$, gas is compressed ($\rho_2 > \rho_1$ and $P_2 > P_1$), decelerated ($v_2 < v_1$) and heated ($T_2 > T_1$) by the shock. In the case of a strong shock, $\widehat{M}_1 \gg 1$ and the density ratio ρ_2/ρ_1 tends to a finite value $(\gamma+1)/(\gamma-1)$ (= 4 for $\gamma = 5/3$), while the temperature and pressure of the downstream gas diverge as $\widehat{M}_1^2$. In fact, for the downstream temperature we can write $T_2 = [2(\gamma-1)/(\gamma+1)^2] (\mu m_{\rm p}/k_{\rm B}) v_1^2$ (= $3\mu m_{\rm p} v_1^2/16 k_{\rm B}$ for $\gamma = 5/3$).

The compression and heating will generally push the post-shock gas out of thermal equilibrium and the gas has to cool in order to reach a new equilibrium. Consider the layout of Fig. 8.3, in which the gas is shocked between x_1 and x_2 (i.e. realistic shocks have a finite width, typically comparable to the mean free path of the gas particles). The pre-shocked gas is supposed to be in thermal equilibrium so that there are no radiative losses [$\mathscr{L} = 0$, with $\mathscr{L}(\rho,T) = (\mathscr{C} - \mathscr{H})/\rho$ the net specific cooling function]. If we continue to ignore radiative losses in the shock ($x_1 < x < x_2$), then the conditions of the post-shocked gas at $x = x_2$ are given by the adiabatic jump conditions discussed above. Behind the shock, however, the gas, which is now moving subsonically ($v_2 < c_{s,2}$), has been heated and compressed [with $\rho_2/\rho_1 \leq (\gamma+1)/(\gamma-1)$] and is therefore no longer in thermal equilibrium. For a steady flow we now have the conditions:

$$\rho v = \rho_2 v_2, \tag{8.53}$$

$$\rho v^2 + P = \rho_2 v_2^2 + P_2. \tag{8.54}$$

These two equations, together with the time-independent form of the energy equation (8.3), give

$$\rho v \frac{d}{dx}\left[\frac{1}{\gamma-1}\frac{P}{\rho}\right] = -P\frac{dv}{dx} - \rho \mathscr{L}. \qquad (8.55)$$

Using that $dP = -\rho v\, dv$, which follows from Eqs. (8.53)–(8.54), the above equation can be written as

$$\frac{1}{\gamma-1}[c_s^2 - v^2]\frac{dv}{dx} = -\mathscr{L}(\rho,T). \qquad (8.56)$$

Supplemented with an equation of state, these equations can be solved for $x > x_2$, subject to the boundary conditions at $x = x_2$ given by the adiabatic jump conditions (8.49)–(8.51). Qualitative results can be obtained even without solving the equations. For post-shock gas, $v < c_s$, and since $\mathscr{L} > 0$ (cooling), we see that v decreases with x. Since the mass flux ρv has to remain constant, the decrease in velocity implies that the gas is compressed (i.e. $d\rho/dx > 0$). Similarly, we see from Eq. (8.54) that the pressure will also increase with x. The temperature at x is $T \propto P/\rho = (A-v)v$, where $A = (\rho v^2 + P)/(\rho v)$ is a constant. Thus, as v decreases with x, the temperature will also decrease along the flow. This continues until the gas comes back to thermal equilibrium ($\mathscr{L} = 0$) at some location $x = x_3$.

In some special cases, one can obtain the properties of the gas in the new equilibrium state (i.e. at $x > x_3$) by directly matching the conditions at x_1 and x_3, without having to be concerned with the evolution in the intermediate region. An example is the case with $T_3 = T_1$ (this is known as the isothermal shock). In this case, the equation of state implies that $P_3/\rho_3 = P_1/\rho_1 = c_s^{1/2}$, which together with

$$\rho_3 v_3 = \rho_1 v_1, \quad \text{and} \quad \rho_3 v_3^2 + P_3 = \rho_1 v_1^2 + P_1, \qquad (8.57)$$

can be used to obtain that $v_3 = v_1/\widehat{M}_1^2$ and $\rho_3 = \rho_1 \widehat{M}_1^2$. Note that, contrary to the non-radiative shock, the downstream density is now no longer limited as $\widehat{M}_1^2 \to \infty$. Hence, an isothermal shock can create an unlimited amount of compression and deceleration for $\widehat{M}_1^2 \to \infty$.

(b) Heating by Accretion Shocks During the formation of a dark matter halo from a density perturbation, the gas initially associated with the perturbation also collapses in the gravitational potential well of the halo. However, unlike dark matter particles, shell crossing is not allowed for the gas component, and so the gas associated with a mass shell will eventually be stopped and shocked by the gaseous structure that has already collapsed. If radiative cooling is negligible, the shocked gas will remain hot, forming a hot gaseous halo in hydrostatic equilibrium in the potential well of the dark matter halo.

We can use the Rankine–Hugoniot jump conditions described above to relate the properties of the shocked gas to those of the pre-shock gas. If the infall velocity of the accreted gas is $v_{\rm in}$ and the shocked gas has zero velocity (i.e. the kinetic energy of the infalling gas is completely thermalized), then the upstream velocity $v_1 = v_{\rm in} + v_{\rm sh}$ and the downstream velocity $v_2 = v_{\rm sh}$, with $v_{\rm sh}$ the velocity of the shock front. In this case, the post-shock gas temperature, T_2, can be written as

$$\frac{k_B T_2}{\mu m_p} = \frac{v_{\rm in}^2}{16\gamma}\left[2\gamma\left(1+\sqrt{1+\varepsilon}\right)^2 - (\gamma-1)\varepsilon\right]\left[\frac{2\varepsilon}{(1+\sqrt{1+\varepsilon})^2} + (\gamma-1)\right], \qquad (8.58)$$

where

$$\varepsilon \equiv \frac{\gamma}{v_{\rm in}^2}\left(\frac{4}{\gamma+1}\right)^2 \frac{k_B T_1}{\mu m_p}, \qquad (8.59)$$

while the pre- and post-shock densities are related by

$$\frac{\rho_2}{\rho_1} = \frac{\gamma+1}{2(\gamma-1)}\left(1-\frac{T_1}{T_2}\right) + \left[\frac{1}{4}\left(\frac{\gamma+1}{\gamma-1}\right)^2\left(1-\frac{T_1}{T_2}\right)^2 + \frac{T_1}{T_2}\right]^{1/2}. \quad (8.60)$$

Once $v_{\rm in}$ is given, the above equations can be used to determine ρ_2 and T_2 in terms of ρ_1 and T_1. Note that if $v_1^2 \gg k_{\rm B} T_1/(\mu m_{\rm p})$ (i.e. $\varepsilon \to 0$ and $T_1/T_2 \to 0$), these relations reduce to those for strong shocks, with

$$T_2 = (\gamma-1)T_{\rm vir}\left(\frac{v_{\rm in}}{V_{\rm c}}\right)^2 \quad \text{and} \quad \frac{\rho_2}{\rho_1} = \frac{\gamma+1}{\gamma-1}, \quad (8.61)$$

where we have used Eq. (8.45). Thus, if $v_{\rm in}$ is comparable to the circular velocity at the radius of the shock, $V_{\rm c}$, the temperature of the shocked gas will be comparable to the virial temperature of a truncated isothermal sphere.

In general, one can write the velocity of a mass shell of gas as

$$\frac{1}{2}v_{\rm in}^2 = \frac{1}{2}v_{\rm ff}^2 + \Delta W - \frac{c_{\rm s1}^2}{\gamma-1}\left[1-\left(\frac{\rho_{\rm ta}}{\rho_{\rm sh}}\right)^{\gamma-1}\right], \quad (8.62)$$

where $c_{\rm s1}$ is the sound speed of the pre-shock gas, $\rho_{\rm ta}$ and $\rho_{\rm sh}$ are the gas densities at the turn-around radius, $r_{\rm ta}$, and the shock radius, $r_{\rm sh}$, respectively, and

$$\frac{1}{2}v_{\rm ff}^2 \equiv \frac{GM}{r_{\rm sh}} - \frac{GM}{r_{\rm ta}} \quad (8.63)$$

is the free-fall energy of the gas shell within which the mass is M. The term ΔW in Eq. (8.62) together with $v_{\rm ff}^2/2$ gives the total work done by the gravitational potential on the gas shell from $r_{\rm ta}$ to $r_{\rm sh}$. In the absence of shell crossing $\Delta W = 0$; however, this is not a realistic assumption in the presence of dark matter (see §8.3.3), so that $\Delta W \neq 0$ in general. The last term on the right-hand side of Eq. (8.62) reflects the change in the internal energy of the gas between the time of turnaround and the time just before it is shocked, which arises from the fact that the gas shell is compressed when moving to a smaller radius. If $r_{\rm ta} = 2r_{\rm sh}$, which is the case if the shock is located at the virial radius (see §5.4.4), then we have that $v_{\rm ff}^2 = GM/r_{\rm sh} = V_{\rm c}^2$. Thus, if we could ignore shell crossing and the change in the internal energy of the gas when it moves from $r_{\rm ta}$ to $r_{\rm sh}$, then the temperature of the post-shocked gas would be $T_2 = (\gamma-1)T_{\rm vir}$. In the more realistic case, including shell crossing for the dark matter, $v_{\rm in} \lesssim V_{\rm c}$ (see Tozzi & Norman, 2001) and the temperature of the post-shocked gas will be somewhat lower.

8.3.2 Self-Similar Collapse of Collisional Gas

To gain insight into the formation of gaseous halos in the cosmic density field, we first consider a simple self-similar model with spherical symmetry. In order for the problem to admit a self-similar solution, we must assume that (i) the background is an EdS universe, so that the expansion of the universe is scale-free ($a \propto t^{2/3}$); (ii) the initial perturbation is a power law of the mass scale; (iii) the gas is initially cold and radiative cooling is negligible, so that no characteristic scale is introduced through the gas component.

Let us first consider the case in which the Universe contains only collisional gas. This assumption simplifies the discussion, and will be relaxed later by including a collisionless dark matter component. Under the assumptions made above, the only characteristic length scale at any given time t is the turnaround radius $r_{\rm ta}(t)$. The problem then admits the following set of self-similar solutions:

$$\rho(r,t) = \overline{\rho}(t)\mathscr{D}(\lambda), \quad P(r,t) = \overline{\rho}(t)\left(\frac{r_{\rm ta}}{t}\right)^2 \mathscr{P}(\lambda), \tag{8.64}$$

$$v(r,t) = \frac{r_{\rm ta}}{t}\mathscr{V}(\lambda), \quad T(r,t) = \frac{\mu m_{\rm p}}{k_{\rm B}}\frac{P(r,t)}{\rho(r,t)} = \frac{\mu m_{\rm p}}{k_{\rm B}}\left(\frac{r_{\rm ta}}{t}\right)^2 \mathscr{T}(\lambda), \tag{8.65}$$

$$M(r,t) = \frac{4\pi}{3}\overline{\rho}(t)r_{\rm ta}^3 \mathscr{M}(\lambda). \tag{8.66}$$

Here $\lambda \equiv r/r_{\rm ta}$ is a dimensionless radius, ρ, P, v and T are the density, pressure, velocity and temperature of the (ideal) fluid, $M(r,t)$ is the mass within radius r, and $\overline{\rho}(t)$ is the mean density of the universe at time t.

For a power-law initial perturbation, $\delta_{\rm i} \propto M^{-\varepsilon} \propto r_{\rm i}^{-3\varepsilon}$ (with $r_{\rm i}$ the initial radius enclosing mass M), the turnaround radius is a power law of t:

$$r_{\rm ta} \propto t^\eta, \quad \text{with} \quad \eta = \frac{2(1+3\varepsilon)}{9\varepsilon}, \tag{8.67}$$

which follows from Eq. (5.8): $r_{\rm ta} = r_{\rm max} \propto r_{\rm i}/\delta_{\rm i} \propto \delta_{\rm i}^{-1-1/3\varepsilon}$, $t_{\rm ta} = t_{\rm max} \propto \delta_{\rm i}^{-3/2}$. Inserting the similarity solutions into the fluid equations,

$$\frac{\partial \rho}{\partial t} + \frac{1}{r^2}\frac{\partial}{\partial r}(r^2 \rho v) = 0, \tag{8.68}$$

$$\frac{\partial v}{\partial t} + v\frac{\partial v}{\partial r} = -\frac{1}{\rho}\frac{\partial P}{\partial r} - \frac{GM(r,t)}{r^2}, \tag{8.69}$$

$$\left(\frac{\partial}{\partial t} + v\frac{\partial}{\partial r}\right)\ln\left(\frac{P}{\rho^\gamma}\right) = 0, \tag{8.70}$$

$$\frac{\partial M}{\partial r} = 4\pi r^2 \rho, \tag{8.71}$$

we obtain

$$(\mathscr{V} - \eta\lambda)\mathscr{D}' + \mathscr{D}\mathscr{V}' + \frac{2}{\lambda}\mathscr{D}\mathscr{V} - 2\mathscr{D} = 0, \tag{8.72}$$

$$(\mathscr{V} - \eta\lambda)\mathscr{V}' + (\eta-1)\mathscr{V} = -\frac{\mathscr{P}'}{\mathscr{D}} - \frac{2}{9}\frac{\mathscr{M}}{\lambda^2}, \tag{8.73}$$

$$(\mathscr{V} - \eta\lambda)\left(\frac{\mathscr{P}'}{\mathscr{P}} - \gamma\frac{\mathscr{D}'}{\mathscr{D}}\right) = 2(2-\eta) - 2\gamma, \tag{8.74}$$

$$\mathscr{M}' = 3\lambda^2 \mathscr{D}, \tag{8.75}$$

where a prime denotes a derivative with respect to λ. The fluid equations are thus reduced to a set of ordinary differential equations under the assumption of self-similarity.

Before being shocked, the gas is assumed to be cold and its mass shells follow the spherical collapse model (see §5.1), so that

$$\lambda = \sin^2\left(\frac{\theta}{2}\right)\frac{\pi}{\theta - \sin\theta}, \quad \mathscr{V} = \frac{\pi}{2}\cot\left(\frac{\theta}{2}\right), \tag{8.76}$$

$$\mathscr{D} = \frac{1}{3}\left(\frac{3\pi}{4}\right)^2 \frac{1}{\lambda(\lambda-\mathscr{V})\sin^2(\theta/2)}, \quad \mathscr{M} = \left(\frac{3\pi}{4}\right)^2 \frac{\pi}{\theta - \sin\theta}, \tag{8.77}$$

8.3 The Formation of Hot Gaseous Halos

where $\theta \in [0, 2\pi]$. Since r_{ta} is the only length scale, the shock radius at time t must have the form

$$r_{sh}(t) = \lambda_{sh} r_{ta}(t), \qquad (8.78)$$

where λ_{sh} is a constant. This shock radius separates the cold gas in spherical infall from the gas that has been shocked. The above equation therefore implies that the shocked region becomes larger with the passage of time. Applying the Rankine–Hugoniot jump conditions (see §8.3.1) for strong shocks across r_{sh}, we obtain

$$\mathscr{D}_2 = \frac{\gamma+1}{\gamma-1}\mathscr{D}_1, \quad \mathscr{V}_2 = \eta\lambda_{sh} + \frac{\gamma-1}{\gamma+1}(\mathscr{V}_1 - \eta\lambda_{sh}), \qquad (8.79)$$

$$\mathscr{P}_2 = \frac{2}{\gamma+1}\mathscr{D}_1(\mathscr{V}_1 - \eta\lambda_{sh})^2, \quad \mathscr{M}_2 = \mathscr{M}_1, \qquad (8.80)$$

where subscripts '1' and '2' denote quantities of the pre-shock ($r \gtrsim r_{sh}$) and post-shock ($r \lesssim r_{sh}$) gas, respectively. Note that $\mathscr{P}_1 = 0$ because the gas is cold before being shocked. The values of $\mathscr{D}_1$, $\mathscr{V}_1$ and $\mathscr{M}_1$ are given by Eqs. (8.76)–(8.77) with $\theta = \theta_{sh}$, where θ_{sh} corresponds to $\lambda = \lambda_{sh}$. With these boundary conditions, Eqs. (8.72)–(8.75) can be integrated from $\lambda = \lambda_{sh}$ inwards to $\lambda = 0$. To completely specify the solution, we need to know the value of λ_{sh}. This is given by the inner boundary condition: $\mathscr{M} = \mathscr{V} = 0$ at $\lambda = 0$.

Fig. 8.4 shows the solutions for initial perturbations with $\varepsilon = 2/3$ ($\eta = 1$) and $\gamma = 5/3$. In this case $\lambda_{sh} \approx 0.29$. As one can see, the velocity decrement across the shock is quite large, indicating that the deceleration due to the large post-shock pressure is very effective. The post-shock kinetic energy is much smaller than the thermal and gravitational energies, implying that the pre-shock kinetic energy associated with gas infall is effectively transformed into thermal energy by the accretion shock. The mean density contrast within the shock radius is $\mathscr{D}(\lambda_{sh}) \sim 100$. For small

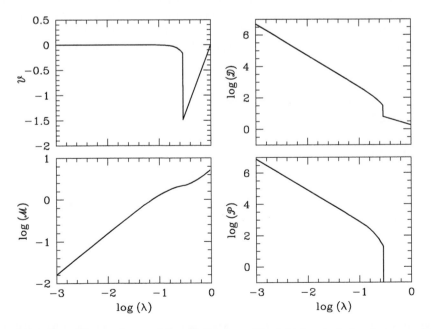

Fig. 8.4. Self-similar collapse of a $\gamma = 5/3$ collisional gas. The initial perturbation is assumed to have a profile with $\varepsilon = 2/3$. The accretion shock is located at $\lambda = \lambda_{sh} = 0.29$. [After Bertschinger (1985)]

λ, the solutions are power laws, which can be understood as follows. Deep in the shocked region, gas is in a hydrostatic state and $\mathscr{V} = 0$. This in Eqs. (8.72)–(8.75) gives

$$\mathscr{D}(\lambda) \propto \lambda^{-2/\eta}, \qquad \mathscr{P}(\lambda) \propto \lambda^{2(\eta-2)/\eta}, \qquad (8.81)$$

$$\mathscr{T}(\lambda) = \frac{\mathscr{P}(\lambda)}{\mathscr{D}(\lambda)} \propto \lambda^{2-2/\eta}, \qquad \mathscr{M}(\lambda) \propto \lambda^{3-2/\eta}. \qquad (8.82)$$

For $\eta = 1$, we have $\mathscr{D}(\lambda) \propto \mathscr{P}(\lambda) \propto \lambda^{-2}$, $\mathscr{M}(\lambda) \propto \lambda$, $\mathscr{T}(\lambda) =$ constant, and an isothermal sphere is produced. For secondary infall onto a seed perturbation, $\varepsilon = 1$ ($\eta = 8/9$) and $\lambda_{\rm sh} = 0.339$ for gas with $\gamma = 5/3$.

It is interesting to see how a mass shell collapses and settles into the final static state by following the trajectory of a mass shell. In terms of $\xi \equiv \ln(t/t_{\rm ta})$ (where $t_{\rm ta}$ is the time when the mass shell in question turns around) and the dimensionless radius $\lambda \equiv r/[r_{\rm ta}(t)]$, the equation of motion can be written as

$$\frac{d^2\lambda}{d\xi^2} + (2\eta - 1)\frac{d\lambda}{d\xi} + \eta(\eta - 1)\lambda = -\frac{2}{9}\frac{\mathscr{M}}{\lambda^2} - \frac{1}{\mathscr{D}}\frac{d\mathscr{P}}{d\lambda}, \qquad (8.83)$$

where $\mathscr{D}(\lambda)$, $\mathscr{P}(\lambda)$ and $\mathscr{M}(\lambda)$ are the similarity solutions discussed above. The boundary conditions are $\lambda(\xi = 0) = 1$ and $(d\lambda/d\xi)(\xi = 0) = -\eta$. For given η and γ, Eq. (8.83) can be solved. The left-hand panel of Fig. 8.5 shows $r/[r_{\rm ta}(t_{\rm ta})] = \lambda \exp(\eta\xi)$ as a function of $t/t_{\rm ta}$ for models with $\eta = 1$ and for various values of the adiabatic index γ in the range $[4/3, 5/3]$. The mass shell first expands from its initial radius at $t = 0$ to a maximum expansion radius at $t = t_{\rm ta}$, then contracts by some factor before it is shocked at a time $t \sim 2t_{\rm ta}$, and eventually settles into a radius which is about a constant fraction of its maximum expansion radius. This behavior of a gas shell is different from that of a collisionless mass shell which oscillates at $t > 2t_{\rm ta}$ (see Fig. 5.1), and the difference is obviously due to the fact that shell crossing occurs for collisionless particles but is not allowed for a collisional gas. The final radius into which a gas shell settles depends on the value of γ, because $P \propto \rho^\gamma$ so that gas with a larger value of γ is 'stiffer' and more difficult to compress. For $\gamma \leq 4/3$, the final radius approaches zero at $t \gg t_{\rm ta}$, i.e. the gas collapses catastrophically, because it is too 'soft' to balance its self-gravity. Thus, in the presence of effective cooling, so that the effective γ is low, a self-gravitating gas cloud is violently unstable (see §8.4.4).

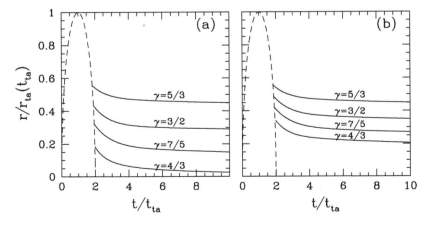

Fig. 8.5. The trajectories of gas shells (with given γ) in the similarity solution. The initial perturbation is assumed to have a profile with $\eta = 1$. Panel (a) shows results for a pure collisional gas, while panel (b) shows results for collisional gas in the potential of a dominating collisionless component. [After Bertschinger (1985)]

8.3.3 The Impact of a Collisionless Component

Since similarity solutions exist for both collisional and collisionless collapse (see §5.2), such solutions should also exist for any combination of the two, provided that the total density adds up to $\Omega = 1$ (recall that similarity solutions only exist for an Einstein–de Sitter universe). In this subsection, we consider a model in which the cosmic density in collisional particles (baryons) is much smaller than unity, i.e. the Universe is dominated by CDM. We also assume that the collisional and collisionless components are well mixed initially, so that the initial perturbations in the two components are the same.

Since $\Omega_b \ll \Omega_{CDM}$, we can neglect the gravitational force of the collisional gas, which simplifies the problem. In this case, the fluid equations for the collisional gas have the same forms as in Eqs. (8.68)–(8.71) and (8.72)–(8.75), except that M in Eq. (8.71) and $\mathcal{M}$ in Eq. (8.75) have to be replaced by M_{CDM} and $\mathcal{M}_{CDM}$. Under this assumption, the motions of the collisionless particles are not affected by the collisional gas, so that $\mathcal{M}_{CDM}(\lambda)$ is the same as that given by the similarity solutions in §5.2. Eqs. (8.72)–(8.75) can then be integrated from $\lambda = \lambda_{sh}$ to $\lambda = 0$ using the post-shock boundary conditions given by the jump conditions (8.79) and (8.79). To specify the jump conditions, we need to obtain the pre-shock quantities $\mathcal{D}_1$, $\mathcal{P}_1$, $\mathcal{V}_1$ and $\mathcal{M}_1$ near the shock front. The situation here is different from that of a purely collisional collapse in which there is no shell crossing so that all mass shells outside the shock radius are in spherical infall and their motion follows exactly the spherical infall model. In the present case, however, some collisionless particles outside the shock radius are not in their initial infall phase but instead have already passed through the center once or several times. In this case, the total mass of collisionless particles inside the shock radius is not equal to the mass initially enclosed by this mass shell, and consequently, the pre-shock boundary conditions are not the same as those given by the spherical infall model.

Since the pre-shock collisional gas is pressureless, its motion is the same as that of the collisionless mass shells that turn around the *first* time. We can then use the similarity solutions of collisionless particles discussed in §5.2 to fix the pre-shock boundary conditions. The dimensionless trajectory $\lambda(\xi)$ for the collisionless mass shells (see Fig. 5.1) can be inverted to obtain ξ_{sh}, the value of ξ at which $\lambda = \lambda_{sh}$, for a mass shell that turns around the first time. The pre-shock velocity is therefore

$$\mathcal{V}_1 = \frac{d\lambda}{d\xi}(\lambda_{sh}) + \eta \lambda_{sh}. \tag{8.84}$$

Since the total mass of collisional gas within the shock radius is equal to the initial gas mass enclosed when the mass shell turned around, we have

$$M = \frac{4\pi}{3} \mathcal{M}_{ta} \overline{\rho}(t_{ta}) r_{ta}^3(t_{ta}), \tag{8.85}$$

where $\mathcal{M}_{ta} = (3\pi/4)^2$ is the density contrast at turnaround [see Eq. (5.17)], and

$$\mathcal{M}_1 = \mathcal{M}_{ta} \exp[-(3\eta - 2)\xi_{sh}]. \tag{8.86}$$

The pre-shock density and pressure are

$$\mathcal{D}_1 = \frac{1}{\lambda_{sh}^2} \frac{d\mathcal{M}}{d\lambda}(\lambda_{sh}) = -\frac{2}{9} \frac{\mathcal{M}_1}{\lambda_{sh}^2} \left[\frac{d\lambda}{d\xi}(\lambda_{sh})\right]^{-1}, \quad \text{and} \quad \mathcal{P}_1 = 0. \tag{8.87}$$

These boundary conditions, together with the inner boundary conditions, $\mathcal{M} = \mathcal{V} = 0$ at $\lambda = 0$, can be used to integrate the fluid equations to obtain $\mathcal{D}(\lambda)$, $\mathcal{P}(\lambda)$, $\mathcal{V}(\lambda)$ and $\mathcal{M}(\lambda)$ for the post-shock gas.

The effect of including the collisionless component is most clearly seen in the trajectories of the collisional gas shown in the right-hand panel of Fig. 8.5. These trajectories are obtained for the same models as in the left-hand panel but now with the inclusion of dark matter [i.e. with the replacement $\mathscr{M} \to \mathscr{M}_{CDM}$ in Eq. (8.83)]. For $\gamma = 5/3$ the result with $\Omega_b \to 0$ is almost identical to that of the pure collisional model (with $\Omega_b = 1$), indicating that the results obtained in §8.3.2 are applicable to any combination of Ω_b and Ω_{CDM} provided that $\Omega_b + \Omega_{CDM} = 1$. For $\gamma = 4/3$, however, there is a marked difference: instead of going to zero, as is the case without dark matter, the mass shells now settle towards a finite radius. The reason for this is that the gravitational force is now dominated by dark matter, and the gas can reach hydrostatic equilibrium by adjusting its pressure gradient.

8.3.4 More General Models of Spherical Collapse

If the assumption of self-similarity is not imposed, spherical collapse models can be constructed to incorporate general accretion histories of dark matter halos, the possibility that the gas to be accreted is not cold, and the effect of clumpiness in the accreted gas. In general, the solutions to such problems can only be obtained through solving the full time-dependent fluid equations (e.g. Lu & Mo, 2007). Here we use simple considerations to demonstrate these effects on the structure of gaseous halos.

The gas accretion history in a halo may be specified by the gas mass within a mass shell, M_{gas}, as a function of its turnaround time, t_{ta}. Since the shock radius, r_{sh}, is roughly a constant fraction (e.g. a half) of the turnaround radius, r_{ta}, we can use Eqs. (8.58), (8.60) and (8.62) to calculate the post-shock specific entropy of the gas shell:

$$S(M_{gas}) = \frac{P(M_{gas})}{\rho^\gamma(M_{gas})} = \frac{k_B}{\mu m_p} \frac{T(M_{gas})}{\rho^{\gamma-1}(M_{gas})}. \tag{8.88}$$

If radiative cooling is negligible in the shocked gas, the value of S is conserved for each mass shell. The entropy profile, $S(M_{gas})$, can then be used to determine the density and pressure profiles of the gas in the dark matter halo. For example, if the gravitational field is dominated by dark matter, the structure of the shocked gas can be obtained by solving the two equations,

$$\frac{dP}{dM_{gas}} = \frac{GM_{CDM}}{4\pi r^4} \quad \text{and} \quad \frac{dr}{dM_{gas}} = \frac{1}{4\pi \rho_{gas} r^2}, \tag{8.89}$$

together with Eq. (8.88).

We can use the equations derived above to gain some insight into the formation of gaseous halos in the cosmic density field. If the pre-shock gas is cold, the characteristic gas density within a mass shell is roughly proportional to the mean density of the universe at the time when the mass shell turns around. If the gravitational potential is dominated by a dark matter halo, the temperature of the post-shock gas is typically the virial temperature of the halo and therefore increases as the dark matter potential well deepens. Thus, the gas that is accreted earlier has lower temperature but higher density, and hence lower specific entropy. Since the low-entropy gas must settle into the inner part of the halo (otherwise the gaseous halo would be convectively unstable, see §8.2.2), the gas density has to increase rapidly towards the halo center in order for the low-entropy gas to have a sufficiently high pressure gradient to counterbalance gravitational compression. On the other hand, if the gas has a significant amount of initial entropy and radiative cooling is negligible, there must be an entropy floor below which no gas can exist. In this case, the gas in the inner part of the halo, for which the entropy generated by the accretion shock is small compared to the initial entropy, should have roughly constant specific entropy. This makes the gas more difficult to compress, and consequently, the gas distribution develops a core within which the gas density changes only slowly with radius. The size of the core is roughly the radius at which the entropy generated by the accretion shock is comparable to the initial entropy. Outside the core, where the entropy generated by accretion shocks is larger than the initial entropy, the gas distribution is expected to be similar to that in a cold collapse.

If the pre-shock gas is clumpy so that part of the gas has a density higher than the average, the post-shock density will be higher for the denser component, although the post-shock temperature is density-independent. Consequently, the specific entropy generated by the shock will be lower for the denser component. Convection will cause the lower-entropy gas to sink deeper in the halo before it reaches hydrostatic equilibrium.

8.4 Radiative Cooling in Gaseous Halos

In this section we investigate how cooling impacts on the formation and evolution of gaseous halos. We start in §8.4.1 by computing the radiative cooling times for uniform clouds, which gives some idea about the importance of radiative cooling as a function of halo mass. In §8.4.2 we consider the more realistic case of gas halos with radial density distributions, and we will see that such halos tend to cool from the inside out, giving rise to a cooling wave propagating outward with time. In §8.4.3 we discuss self-similar solutions for such cooling waves, and we end in §8.4.4 with a discussion of the impact of cooling on the formation of gaseous halos during the spherical collapse of the associated dark matter halo.

8.4.1 Radiative Cooling Time Scales for Uniform Clouds

Gas in a halo can lose energy through radiative cooling, thereby changing its thermodynamic properties and structure. In this subsection, we examine when radiative cooling is expected to be important. Hydrostatic equilibrium can be established and maintained only in systems where radiative cooling is negligible or balanced by heating. For systems where radiative cooling is important, it must be included as part of the dynamical evolution.

As a simple example, consider the cooling time for a uniform spherical cloud in virial equilibrium. Let the total mass of the cloud be M, and assume that a fraction $f_{\rm gas}$ of this mass is gas and the rest is dark matter. In the absence of external pressure, we obtain from Eq. (8.39) that

$$\frac{3k_{\rm B}T}{\mu m_{\rm p}} = \frac{3}{5}\frac{GM}{r} = \frac{3}{5}\frac{GM_{\rm gas}}{f_{\rm gas}r}. \tag{8.90}$$

Solving for $M_{\rm gas}$ gives

$$M_{\rm gas} \approx 8.4 \times 10^{12} T_6^{3/2} f_{\rm gas}^{3/2} n_{-3}^{-1/2} \, {\rm M}_\odot, \tag{8.91}$$

where we have used that $\mu \simeq 0.59$ for a fully ionized gas of primordial composition ($Y_{\rm p} \simeq 1/4$). The temperature is written as $T = 10^6 T_6$ K and the mean particle density $n = \rho_{\rm gas}/(\mu m_{\rm p}) = 10^{-3} n_{-3} \, {\rm cm}^{-3}$. It is useful to express this formula in terms of the overdensity $\delta \equiv \rho/\overline{\rho} - 1$ and the cosmological parameters H_0 and $\Omega_{\rm m,0}$. Then, for a fully ionized gas,

$$n_{-3} \approx 1.9 \times 10^{-2} f_{\rm gas} (1+\delta)(\Omega_{\rm m,0}h^2)(1+z)^3, \tag{8.92}$$

and thus

$$M_{\rm gas} \approx 6.1 \times 10^{13} T_6^{3/2} f_{\rm gas} (1+\delta)^{-1/2}(\Omega_{\rm m,0}h^2)^{-1/2}(1+z)^{-3/2} \, {\rm M}_\odot. \tag{8.93}$$

Since a newly collapsed object has an overdensity $\delta \sim 200$, we see, for example, that a protogalaxy with a gas mass of $10^{11} \, {\rm M}_\odot$ at redshift 3 in a universe with $\Omega_{\rm m,0}h^2 = 0.15$ and $f_{\rm gas} = 0.15$

will have a virial temperature of 0.6×10^6 K. The cooling time for gas at temperature T and density n as a result of radiative cooling can be written as

$$t_{\rm cool} \equiv \frac{\rho \mathscr{E}}{\mathscr{C}} = \frac{3 n k_B T}{2 n_{\rm H}^2 \Lambda(T)} \approx 3.3 \times 10^9 \frac{T_6}{n_{-3} \Lambda_{-23}(T)} \,{\rm yr}, \qquad (8.94)$$

where $\mathscr{E}$ is the internal energy per unit mass, $n_{\rm H}$ is the number density of hydrogen atoms [$n_{\rm H} = (12/27)n$ for a fully ionized primordial gas], and we have assumed an ideal gas with adiabatic index $\gamma = 5/3$. Note that we have written the cooling function as $\Lambda = 10^{-23}\Lambda_{-23}\,{\rm erg\,cm^3\,s^{-1}}$. The $10^{11}\,{\rm M}_\odot$ protogalaxy forming at $z=3$ has $n_{-3} \approx 5.5$ and $\Lambda_{-23} \approx 0.5$ assuming primordial gas (see Fig. 8.1), which implies a cooling time of $t_{\rm cool} \sim 7.4 \times 10^8$ yr. This is roughly twice as long as its free-fall time,

$$t_{\rm ff} = \sqrt{\frac{3\pi}{32 G \rho}} = \sqrt{\frac{3\pi f_{\rm gas}}{32 G n \mu m_p}} \approx 2.1 \times 10^9 f_{\rm gas}^{1/2} n_{-3}^{-1/2}\,{\rm yr}, \qquad (8.95)$$

or $t_{\rm ff} \approx 3.5 \times 10^8$ yr for the case we are considering. Note that for a given temperature $t_{\rm cool} \propto (1+z)^{-3}$ and $t_{\rm ff} \propto (1+z)^{-3/2}$ so that cooling is more effective at higher redshifts. Note also that, in the absence of a strong UV background, protogalaxies with $10^4\,{\rm K} < T_{\rm vir} < 10^5\,{\rm K}$ (where $\Lambda_{-23} \sim 10$) have $t_{\rm cool}$ much smaller than $t_{\rm ff}$. Thus, the gas in small halos at high redshifts is expected to cool effectively and on short time scales.

Fig. 8.6 shows the locus of $t_{\rm cool} = t_{\rm ff}$ in the n–T plane, which separates clouds that can cool effectively ($t_{\rm cool} \ll t_{\rm ff}$) from those that cannot. Superposed on the diagram are the loci of constant $M_{\rm gas}$ and the densities n that correspond to overdensities of $\delta = 200$ at redshifts $z = 0, 1, ..., 5$. As one can see, over this entire redshift range halos with (primordial) gas masses larger than about $10^{11}\,{\rm M}_\odot$ cannot cool effectively. This mass increases to $\sim 10^{12}\,{\rm M}_\odot$ if the gas has solar

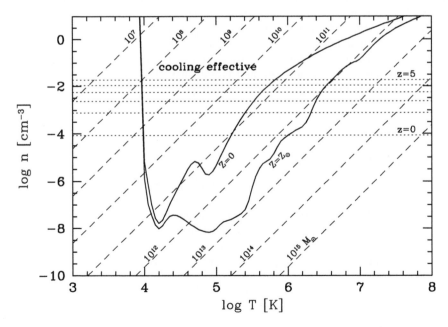

Fig. 8.6. Cooling diagram showing the locus of $t_{\rm cool} = t_{\rm ff}$ in the n–T plane. The upper and lower curves correspond to gas with zero and solar metallicity, respectively. The tilted dashed lines are lines of constant total mass (in ${\rm M}_\odot$), while the horizontal dotted lines show the gas densities expected for virialized halos ($\delta = 200$) at different redshifts. All calculations assume $f_{\rm gas} = 0.15$, $\Omega_{m,0} = 0.3$, and $h = 0.7$. Cooling is effective for clouds with n and T above the locus.

metallicity. This is why clusters and groups of galaxies at the present time usually contain large amounts of hot gas. The critical mass for effective cooling is similar to that of the most massive galaxies, suggesting that the physics of cooling plays an important role in limiting the mass of galaxies (e.g. Rees & Ostriker, 1977; White & Rees, 1978; Blumenthal et al., 1984). Note, though, that this argument applies only to galaxies that form directly from gas cooling in dark matter halos. If galaxies form via mergers of smaller galaxies, which may be an important formation channel, especially for giant ellipticals (see Chapter 13), this cooling argument may not be directly relevant. Since major mergers between dark matter halos are believed to reheat the gas to the virial temperature of the merged halo, a more meaningful comparison is that between the cooling time and the time since the last major merger, $t_{\rm lmm}$. The loci of $t_{\rm cool} = t_{\rm lmm}$ may be significantly different from the $t_{\rm cool} = t_{\rm ff}$ shown in Fig. 8.6, especially for low mass (low virial temperature) halos, which may not have experienced a major merger in the last couple of Gyr. We also caution that the results presented here are based on the oversimplified assumption that the gas has a uniform and spherical distribution.

In a hierarchical model of structure formation, smaller halos are expected to form earlier. The fraction of mass in halos with masses exceeding M [corresponding to linear mass variance $\sigma(M)$], $F(>M)$, can be estimated from the Press–Schechter formalism (see §7.2.1),

$$F(>M) = {\rm erfc}\left[\frac{\delta_{\rm c}(t)}{\sqrt{2}\sigma(M)}\right], \qquad (8.96)$$

which, for a given redshift z, goes to unity if M is sufficiently small. Thus, a large fraction of the mass in the universe is expected to have collapsed into small halos at high redshift. Since gas cooling is effective in small halos at high redshifts, this suggests that a (very) large fraction of the gas in the universe can cool at high redshifts. If nothing prevents the cooled gas from forming stars, then we run into an *overcooling problem*, because it would mean that the stellar mass density $\Omega_{\star,0} \simeq \Omega_{\rm b,0}$, which, as we have seen in §2.10.2, is in violent disagreement with the data (White & Rees, 1978). As discussed in §8.1.4, a strong UV background may significantly suppress gas cooling in halos with $T_{\rm vir} \lesssim 10^5$ K, and thus help to alleviate the overcooling problem. Furthermore, energy feedback from supernovae (see §8.6.1) and active galactic nuclei (see §14.4) may also prevent gas from cooling too fast in dark matter halos.

8.4.2 Evolution of the Cooling Radius

The cooling times derived above are based on the assumption that the gaseous halos have a uniform density. In reality, the gaseous halos have a density distribution $\rho(\mathbf{x})$, and the cooling time will be a local quantity. In a spherically symmetric gaseous system, the cooling time at radius r can be defined as

$$t_{\rm cool}(r) = \frac{3n(r)k_{\rm B}T(r)}{2n_{\rm H}^2(r)\Lambda(T)}, \qquad (8.97)$$

which is the same as Eq. (8.94) except that the temperature, T, and the densities, n and $n_{\rm H}$, now all refer to their local values at radius r. Radiative cooling at r is important if $t_{\rm cool}(r)$ is shorter than the age t of the system.

To understand how radiative cooling proceeds in a gaseous halo, let us model the *adiabatic* density and pressure profiles by power-law forms:

$$\rho_{\rm ad}(r) = \rho_0 \left(\frac{r}{r_0}\right)^{-\alpha}, \quad P_{\rm ad}(r) = P_0 \left(\frac{r}{r_0}\right)^{-\beta}. \qquad (8.98)$$

For an ideal gas, the temperature profile is then

$$T_{\rm ad}(r) = T_0 \left(\frac{r}{r_0}\right)^{\alpha-\beta}, \quad \text{with} \quad T_0 = \frac{\mu m_{\rm p}}{k_{\rm B}} \frac{P_0}{\rho_0}. \tag{8.99}$$

Piecewisely, the cooling function may be written as a power law of T,

$$\Lambda(T) = \Lambda_0 \left(\frac{T}{T_0}\right)^\nu, \tag{8.100}$$

and for a gas with cosmic composition, $\nu \in (-1,0)$ in the temperature range $10^5\,{\rm K} < T < 10^7\,{\rm K}$ (see §8.1.3). The cooling time can then be written as

$$t_{\rm cool} = t_0 \left(\frac{r}{r_0}\right)^{1/\tau}, \tag{8.101}$$

where

$$t_0 \equiv \frac{3}{2}\frac{k_{\rm B} T_0}{\Lambda_0} \left(\frac{\mu m_{\rm p}}{\rho_0}\right)\left(\frac{n}{n_{\rm H}}\right)^2, \quad \tau \equiv [\alpha + (\alpha-\beta)(1-\nu)]^{-1}. \tag{8.102}$$

If we define the cooling radius, $r_{\rm cool}$, as the radius at which the cooling time is equal to the age, t, we have

$$r_{\rm cool}(t) = r_0 \left(\frac{t}{t_0}\right)^\tau. \tag{8.103}$$

An estimate of the rate at which gas cools out in the halo then follows from

$$\dot{M}_{\rm cool}(t) = 4\pi \rho(r_{\rm cool}) r_{\rm cool}^2 \frac{dr_{\rm cool}}{dt}$$
$$= \frac{4\pi \rho_0 r_0^3}{t_0}\tau \left(\frac{t}{t_0}\right)^{\tau(3-\alpha)-1}, \tag{8.104}$$

which implies that

$$M_{\rm cool}(t) = \begin{cases} \frac{4\pi \rho_0 r_0^3}{3-\alpha}\left(\frac{t}{t_0}\right)^{\tau(3-\alpha)} & \text{if } \alpha \neq 3 \\ \frac{4\pi \rho_0 r_0^3}{3+(3-\beta)(1-\nu)} \ln\left(\frac{t}{t_0}\right) & \text{if } \alpha = 3. \end{cases} \tag{8.105}$$

For an isothermal sphere, $\alpha = \beta = 2$, so that $r_{\rm cool} \propto M_{\rm cool} \propto t^{1/2}$. Thus, in this case, the cooling region expands with the passage of time. Since no mass transport is involved with this expansion, the propagation of the cooling radius with time may be considered as a cooling wave in the gaseous halo.

Note, however, that this description of the cooling rate is quite crude at best (see also the discussion in §8.7.2). Because of cooling, the properties of the gas in the cooling region will differ from those given by the adiabatic model. In particular, once gas starts to cool out, the remaining hot gas tends to re-establish a new hydrostatic equilibrium, which in turn will have an impact on the subsequent cooling rates. In the remainder of this section we examine these issues in more detail, starting with a self-similar solution.

8.4.3 Self-Similar Cooling Waves

Consider the case in which a halo of gas and dark matter has already formed. We assume that the initial density and pressure profiles have the power law forms of Eq. (8.98), and that the gas is initially in hydrostatic equilibrium, so that the gravitational mass follows from Eq. (8.16):

$$M(r) = M_0 \left(\frac{r}{r_0}\right)^{\alpha-\beta+1}, \quad M_0 \equiv \beta \frac{r_0 P_0}{G\rho_0}. \tag{8.106}$$

For simplicity, we also assume the gravitational mass distribution to be static, a reasonable approximation for gas in a virialized dark matter halo that dominates gravitationally. The fluid equations for the gas are the same as those given by Eqs. (8.68)–(8.71), except that $M(r,t)$ in Eq. (8.69) should now be replaced by $M(r)$ given in Eq. (8.106) and that an energy loss term should be included in the entropy equation:

$$\left(\frac{\partial}{\partial t} + v\frac{\partial}{\partial r}\right)\ln\left(\frac{P}{\rho^\gamma}\right) = -\frac{(\gamma-1)}{P}\mathscr{C}, \tag{8.107}$$

where $\mathscr{C} = n_H^2 \Lambda(T)$ is the energy loss rate per unit volume due to radiative cooling.

The evolution of the cooling gas is expected to be self-similar if the gas flow is characterized by a single length scale. In the presence of radiative cooling, the cooling radius given by Eq. (8.103) is such a length scale, and we denote the characteristic density, pressure and velocity that can be derived from the cooling radius as $\rho_a(r_{\rm cool})$, $P_{\rm ad}(r_{\rm cool})$ and $v_{\rm cool} \equiv dr_{\rm cool}/dt$, respectively. However, $r_{\rm cool}$ is not the only length scale in the problem, as the sound speed at $r_{\rm cool}$ times the age, $c_s t$, is another length scale which in general is independent of the cooling radius (physically this represents the scale over which hydrodynamical perturbations produced by the cooling can have propagated through the system). Thus, the problem only admits a similarity solution under some special conditions.

In order to obtain a similarity solution, we define dimensionless fluid variables in terms of their characteristic values:

$$r = r_{\rm cool}(t)\lambda, \quad \rho = \rho_{\rm ad}(r_{\rm cool})\mathscr{D}(\lambda), \tag{8.108}$$

$$P = P_{\rm ad}(r_{\rm cool})\mathscr{P}(\lambda), \quad v = v_{\rm cool}\mathscr{V}(\lambda). \tag{8.109}$$

In terms of these variables, the fluid equations can be written as

$$\left(\frac{\mathscr{V}-\lambda}{\mathscr{V}}\right)\frac{d\ln\mathscr{D}}{d\ln\lambda} + \frac{d\ln\mathscr{V}}{d\ln\lambda} + 2 - \frac{\alpha\lambda}{\mathscr{V}} = 0, \tag{8.110}$$

$$\frac{d\ln\mathscr{P}}{d\ln\lambda} + \beta\frac{\mathscr{D}}{\mathscr{P}}\lambda^{\alpha-\beta} = -\varsigma\gamma\frac{\mathscr{D}}{\mathscr{P}}\mathscr{V}^2\left[\left(\frac{\mathscr{V}-\lambda}{\mathscr{V}}\right)\frac{d\ln\mathscr{V}}{d\ln\lambda} + \frac{(\tau-1)}{\tau}\frac{\lambda}{\mathscr{V}}\right], \tag{8.111}$$

$$\left(1 - \frac{\mathscr{V}}{\lambda}\right)\frac{d}{d\ln\lambda}\ln(\mathscr{P}\mathscr{D}^{-\gamma}) = \gamma\alpha - \beta + \frac{1}{\tau}\mathscr{D}^{2-\nu}\mathscr{P}^{\nu-1}, \tag{8.112}$$

where

$$\varsigma = \left[\frac{v_{\rm cool}}{c_s(r_{\rm cool})}\right]^2. \tag{8.113}$$

In general ς is a function of time, so that the dimensionless fluid variables depend on both λ and t. Only when $\varsigma\mathscr{V}^2 \ll |d\ln\mathscr{P}/d\ln\lambda| \sim 1$, so that the right-hand side of Eq. (8.111) can be neglected, does the problem admit a similarity solution. For $\alpha = \beta = 2$ and $\gamma = 5/3$, ς can be written as

$$\varsigma = \frac{3}{20}\frac{\mu m_p}{k_B T_0}\left(\frac{r_{\rm cool}}{t}\right)^2$$

$$\approx 1.7 \times 10^{-3} h(1+z)^{3/2}\left(\frac{f_{\rm gas}}{0.1}\right)\left(\frac{T_0}{10^6\,{\rm K}}\right)^{-1}\left(\frac{\Lambda_0}{10^{-23}\,{\rm erg\,cm^3\,s^{-1}}}\right), \tag{8.114}$$

where $r_{\rm cool}$ is given by Eq. (8.103), and we have used Eqs. (8.99) and (3.96) plus the fact that $c_s^2 = \gamma P/\rho$. This shows that $\varsigma \ll 1$ at redshifts $z \ll 20$. As we will see below, the similarity solution gives that $\mathcal{V} \sim \lambda^{-1/2}$. We can therefore define a limiting value for λ from $\varsigma \mathcal{V}^2 = 1$,

$$\lambda_\ell \sim \varsigma, \qquad (8.115)$$

so that the similarity model is valid for $\lambda > \lambda_\ell$.

The outer boundary conditions are given by the requirement that, as $\lambda \to \infty$, the gas state must approach the static solution without cooling,

$$\mathcal{D}\lambda^\alpha \to 1, \quad \mathcal{P}\lambda^\beta \to 1, \quad \mathcal{V} \to 0 \quad \text{for } \lambda \gg 1. \qquad (8.116)$$

Because the assumption of self-similarity breaks down at $\lambda < \lambda_\ell$, one has to solve the time-dependent fluid equations for the fluid at $\lambda < \lambda_\ell$ in order to set the inner boundary conditions for the similarity solutions. Bertschinger (1989) discussed in detail how to achieve this, and found that the only physically interesting self-similarity solution is the one with $t_{\rm cool} \approx t_{\rm flow} \equiv -r/v$ at $\lambda \ll 1$. In terms of the dimensionless fluid variables, the cooling time can be written as $t_{\rm cool} = t\mathcal{D}^{\nu-2}\mathcal{P}^{1-\nu}$, and so the condition $t_{\rm cool} = t_{\rm flow}$ leads to

$$\mathcal{V} = -\lambda\tau^{-1}\mathcal{D}^{2-\nu}\mathcal{P}^{\nu-1}. \qquad (8.117)$$

For $\lambda \ll 1$, we have $t_{\rm flow} \sim t_{\rm cool} \ll t$, and $-\mathcal{V} = r/(v_{\rm cool}t) = \lambda r_{\rm cool}/(v_{\rm cool}t) \ll \lambda$. Eq. (8.117) can be combined with Eqs. (8.110)–(8.112) to give the similarity solution. If $\beta = 0$ or $\alpha - \beta > 3/(3-\nu)$, the density, temperature and inflow velocity are all power laws of λ, with slopes

$$\frac{d\ln\mathcal{D}}{d\ln\lambda} = -\frac{3}{3-\nu}, \quad \frac{d\ln(\mathcal{P}/\mathcal{D})}{d\ln\lambda} = \frac{3}{3-\nu}, \quad \frac{d\ln\mathcal{V}}{d\ln\lambda} = \frac{2\nu-3}{3-\nu}. \qquad (8.118)$$

If $\beta \neq 0$ and $\alpha - \beta < 3/(3-\lambda)$, then

$$\frac{d\ln\mathcal{D}}{d\ln\lambda} = \frac{-3 + (\alpha-\beta)(1-\nu)}{2}, \quad \frac{d\ln(\mathcal{P}/\mathcal{D})}{d\ln\lambda} = \alpha-\beta, \qquad (8.119)$$

$$\frac{d\ln\mathcal{V}}{d\ln\lambda} = \frac{-1 - (\alpha-\beta)(1-\nu)}{2}. \qquad (8.120)$$

In the special case of an isothermal halo, $\alpha = \beta = 2$ and this model yields

$$\mathcal{D} \propto \lambda^{-3/2}, \quad \frac{\mathcal{P}}{\mathcal{D}} = \frac{4}{3}, \quad \mathcal{V} \propto \lambda^{-1/2}, \qquad (8.121)$$

and

$$T(r) = T(r_{\rm cool})\frac{\mathcal{P}}{\mathcal{D}} = \frac{4}{3}T(r_{\rm cool}). \qquad (8.122)$$

In this case, the temperature within the cooling region is actually higher than that at the cooling radius, because the gas is compressed as it cools and flows inwards. The central mass inflow rate is somewhat lower than in the simple model of Eq. (8.104).

8.4.4 Spherical Collapse with Cooling

The self-similar cooling wave solution discussed in the previous section is only valid for a static halo where the gas is initially (i.e. prior to the onset of cooling) in hydrostatic equilibrium. In reality, however, dark matter halos continue to accrete new material (both dark matter and gas), which introduces another scale length to the problem, namely the shock radius $r_{\rm sh}(t)$. Consequently the problem no longer admits a self-similar solution. In order to investigate what happens to a system in which both the cooling radius and shock radius evolve with time, one generally needs to resort to numerical simulations. However, some useful results can still be obtained from simple considerations.

An important time scale in the problem is the time when the cooling radius $r_{\rm cool}$ is equal to the shock radius $r_{\rm sh}$ (White & Frenk, 1991). For a singular isothermal sphere with circular velocity V_c, the cooling radius $r_{\rm cool} \propto \Lambda^{1/2}(T) t^{1/2}$ [which follows from Eq. (8.97) and the fact that $n \propto r^{-2}$ for an isothermal sphere], and the shock radius $r_{\rm sh} \propto r_{\rm ta} \propto V_c t$ (which follows from the fact that $r_{\rm ta}$ is proportional to the virial radius). The time at which $r_{\rm cool} = r_{\rm sh}$ therefore scales as

$$t_{\rm crit} \propto \Lambda(T)/V_c^2. \tag{8.123}$$

Since $r_{\rm sh}$ increases with time faster than $r_{\rm cool}$, we expect the adiabatic collapse model described in §8.3 to be valid only at $t \gg t_{\rm crit}$, i.e. when $r_{\rm cool}$ is much smaller than $r_{\rm sh}$. In this case, the gas between $r_{\rm cool}$ and $r_{\rm sh}$ is approximately in hydrostatic equilibrium with the dark matter halo, and the cooling wave propagates in a manner similar to that described in the previous subsection. At earlier times, however, when $t \ll t_{\rm crit}$, gas can cool as soon as it is shocked. In this regime, neither the adiabatic collapse model nor the self-similar cooling wave model applies. In fact, if cooling is effective, the equation of state of the accreted gas may become so soft that pressure support of the gas is negligible. In this case, the accreted gas will be in free-fall and no accretion shock is produced: the gas reaches the center via a cold flow.[2] This is sometimes referred to as the cold mode of gas accretion, to discriminate it from the hot mode, in which the gas is heated to the virial temperature due to an accretion shock at the outskirts of the halo. Since $t_{\rm crit} \propto V_c^{-2}$, less massive halos have larger $t_{\rm crit}$, and one thus expects that halos with M smaller than a certain critical mass, $M_{\rm crit}$, accrete their gas via the cold mode, while more massive halos experience hot mode accretion. We now proceed with a crude derivation of this critical mass scale.

As shown by Birnboim & Dekel (2003), in the presence of cooling the post-shock gas is gravitationally stable as long as

$$\gamma_{\rm eff} \equiv \frac{d \ln P}{d \ln \rho} = \frac{\dot{P}}{P} \frac{\rho}{\dot{\rho}} > \frac{2\gamma}{\gamma + \frac{2}{3}}, \tag{8.124}$$

where an upper dot denotes a time derivative following a comoving volume element. Note that in the adiabatic case $\gamma_{\rm eff} = \gamma$, so that the stability criterion reduces to $\gamma > 4/3$, in agreement with the criterion we derived in §8.2.3 using the virial theorem. Using that

$$\dot{P} = (\gamma - 1)\left[\rho \dot{\mathscr{E}} + \dot{\rho} \mathscr{E}\right], \tag{8.125}$$

and that

$$\dot{\mathscr{E}} = -P\dot{V} - \mathscr{L} = \frac{P\dot{\rho}}{\rho^2} - \mathscr{L} \tag{8.126}$$

(with V the specific volume), the effective adiabatic index can be written as

$$\gamma_{\rm eff} = \gamma - \frac{\rho}{\dot{\rho}} \frac{\mathscr{L}}{\mathscr{E}}. \tag{8.127}$$

We now apply the stability criterion (8.124) to the post-shock gas, whose density, velocity and pressure are related to those of the pre-shock gas by the Rankine–Hugoniot jump conditions for a strong shock:

$$\rho_2 = \frac{\gamma+1}{\gamma-1}\rho_1, \quad v_2 = \frac{\gamma-1}{\gamma+1}v_1, \quad P_2 = \frac{2\rho_1 v_1^2}{\gamma+1}, \tag{8.128}$$

where subscripts '1' and '2' refer to the pre- and post-shock conditions, respectively, and we have assumed that the velocity of the shock $v_{\rm sh} = 0$, adequate for an accretion shock whose post-shock

---

[2] Note, though, that at the center of the potential well, the gas may experience a shock due to the convergence of the flow.

conditions are marginally stable. In order to proceed, we make the assumption that the (radial) velocities are proportional to the radius (as in the Hubble flow): $v(r) = v_2(r/r_{\rm sh})$. As shown by Birnboim & Dekel (2003) this assumption is valid close to the accretion shock, and allows us to write

$$\frac{\dot{\rho}}{\rho} = -\nabla \cdot \mathbf{v} = -\frac{1}{r^2}\frac{\partial}{\partial r}(r^2 v) = -\frac{3v_2}{r_{\rm sh}}. \tag{8.129}$$

Combining Eqs. (8.127)–(8.129), and using that

$$\mathscr{L} = \frac{\rho_2 \Lambda(T)}{\mu^2 m_{\rm p}^2}\left(\frac{n_{\rm H}}{n}\right)^2, \tag{8.130}$$

the stability criterion (8.124) for a monatomic gas with $\gamma = 5/3$ reduces to

$$\Lambda(T_2) < \Lambda_{\rm crit} = 0.022 \frac{m_{\rm p}^2 |v_1|^3}{\rho_1 r_{\rm sh}}, \tag{8.131}$$

where we have adopted $n_{\rm H}/n = 12/27$ as appropriate for a fully ionized gas of primordial composition.

In order to translate this into constraints on the properties of dark matter halos, we define virialized dark matter halos as having an average density equal to $\delta_{\rm vir}$ times the critical density, i.e.

$$\frac{3M_{\rm vir}}{4\pi r_{\rm vir}^3} = \delta_{\rm vir}\rho_{\rm crit}(z). \tag{8.132}$$

In addition, we make the assumptions that the density of the pre-shock gas is equal to the average baryonic density of the universe, $\rho_1 = \overline{\rho}_{\rm b} = \Omega_{\rm b,0}(3H_0^2/8\pi G)(1+z)^3$, and that the accretion velocity is equal to the virial velocity of the halo, $v_1 = V_{\rm vir} = \sqrt{GM_{\rm vir}/r_{\rm vir}}$. Assuming an Eds universe, it then follows that

$$\Lambda_{\rm crit} \sim 61.8 \times 10^{-23}\,{\rm erg\,cm^3\,s^{-1}}\left(\frac{T_{\rm vir}}{10^6\,{\rm K}}\right)\left(\frac{\delta_{\rm vir}}{100}\right)^{1/2}\left(\frac{\Omega_{\rm b,0}h^2}{0.024}\right)^{-1}\left(\frac{h}{0.7}\right)(1+z)^{-3/2}, \tag{8.133}$$

where we have used that $r_{\rm sh} = r_{\rm vir}$, and

$$T_{\rm vir} = \frac{\mu m_{\rm p}}{2k_{\rm B}}V_{\rm vir}^2 \sim 7.5 \times 10^5\,{\rm K}\left(\frac{M_{\rm vir}}{10^{12}h^{-1}{\rm M}_\odot}\right)^{2/3}\left(\frac{\delta_{\rm vir}}{100}\right)^{1/3}(1+z) \tag{8.134}$$

[see Eq. (8.45)].

The shaded area in Fig. 8.7 shows $\Lambda_{\rm crit}(T)$ for halos in the redshift range $0 \leq z \leq 3$. Overplotted are the same cooling curves as in Fig. 8.1. On the upper axis we have translated the (virial) temperature into a virial mass, using Eq. (8.134). As can be seen, halos with $M_{\rm vir} \lesssim 10^{10}h^{-1}{\rm M}_\odot$ will always accrete their gas in the cold mode, while those with $M_{\rm vir} \gtrsim 10^{12}h^{-1}{\rm M}_\odot$ will have a stable accretion shock close to the virial radius, causing hot mode accretion. Whether halos in the intermediate mass range accrete gas in the cold mode or in the hot mode depends on the redshift and the metallicity of the accreted gas. Note that we have made a number of crude assumptions. In particular, in reality the infall velocity is likely to be somewhat lower than the virial velocity of the halo (see discussion in §8.3.1), and the density of the pre-shock gas may be significantly higher than that of the universal baryon density. Both these trends will lower $\Lambda_{\rm crit}$, and thus increase the corresponding $M_{\rm crit}$. Nevertheless, detailed hydrodynamical simulations by Birnboim & Dekel (2003) and Kereš et al. (2005) have shown that $10^{11}h^{-1}{\rm M}_\odot \lesssim M_{\rm crit} \lesssim 10^{12}h^{-1}{\rm M}_\odot$, with a remarkably weak dependence on redshift, in rough agreement with our simple predictions. As we will see in Chapter 15, this critical mass is tantalizingly close to the scale at which the mass-to-light ratios of dark matter halos reveal a pronounced minimum, suggesting that the cold-mode to

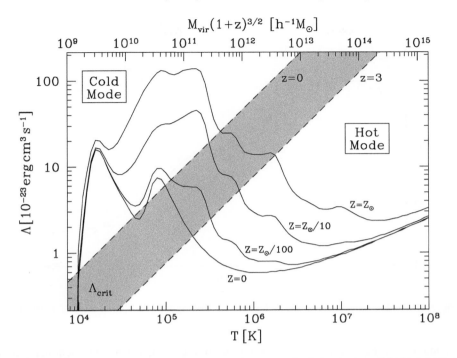

Fig. 8.7. Same as Fig. 8.1, except that now we also indicate $\Lambda_{\rm crit}$ for halos in the redshift range $0 \leq z \leq 3$ (shaded area). Halos with $\Lambda > \Lambda_{\rm crit}$ accrete their gas in the cold mode, while those with $\Lambda < \Lambda_{\rm crit}$ experience hot mode accretion.

hot-mode transition may have an important impact on galaxy formation (see e.g. White & Frenk, 1991; Cattaneo et al., 2006; Birnboim et al., 2007)

8.5 Thermal and Hydrodynamical Instabilities of Cooling Gas

The preceding section shows that gas can cool through radiative processes, but the models considered do not tell us anything about the state of the cooling gas. Is the cooling gas in a single phase, or in a multi-phase with gas clouds of different temperatures coexisting? In this section we show that, because of thermal instability, a multi-phase medium is likely to develop in a cooling gas. We also describe other processes that can affect the properties of a multi-phase medium.

8.5.1 Thermal Instability

Consider a gas in thermal equilibrium. Since cooling balances heating, the net cooling–heating rate (per unit mass) must be zero, i.e. $\mathscr{L} \equiv (\mathscr{C} - \mathscr{H})/\rho = 0$. As we have seen earlier in this chapter, for a given chemical composition, the rate $\mathscr{L}$ is typically a function of both the temperature, T, and density, ρ, of the gas. The condition for cooling–heating balance therefore defines a locus in the ρ–T plane, as shown schematically in Fig. 8.8. Gas above this locus has $\mathscr{L} > 0$ (net cooling) because the temperature of the gas is higher than the equilibrium temperature, while gas below this locus has $\mathscr{L} < 0$ (net heating).

The locus is expected to have the basic shape shown in Fig. 8.8, which can be understood as follows. As the temperature of the gas approaches $\sim 10^4\,{\rm K}$ from a lower value, the ground and low-excitation states of common elements, such as hydrogen and helium, begin to be collisionally

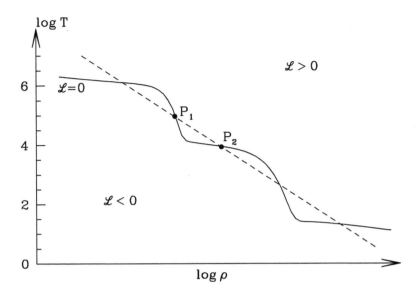

Fig. 8.8. The locus of $\mathscr{L} = 0$ in the n–T plane, illustrating the principle of thermal instability.

excited, and the subsequent radiative de-excitations of the excited states represent net cooling of the gas. For a Maxwell–Boltzmann distribution, the fraction of particles with energy exceeding the threshold energy, ΔE, required for collisional excitations increases with temperature as $\exp(-\Delta E/k_B T)$, and so the cooling rate per unit mass changes rapidly as the temperature changes. Thus, in order to keep a balance between cooling and heating, a large change in ρ is required to counterbalance a small change in T, which explains the almost flat locus at $T \sim 10^4$ K. As the temperature exceeds 10^4 K, the fraction of particles with energy exceeding the threshold energy approaches unity and cannot be increased any further by increasing T. In this case, a small change in ρ requires a large change in T in order to keep a cooling–heating balance, which gives rise to the almost vertical locus for T above 10^4 K. When T approaches $\sim 10^6$ K, the inner shells of heavy elements, such as oxygen and iron, can be collisionally excited, and the cooling becomes dominated by the radiative de-excitations of these excited states. The situation is then similar to that for $T \sim 10^4$ K. Similarly, at temperatures below 10^2 K, gas cooling is dominated by the radiative de-excitation of the excited rotational levels of molecules, and the locus once again depends strongly on T with only a weak dependence on ρ.

Now suppose that we have a static, homogeneous gas in thermal equilibrium, and that a blob of gas is perturbed away from the equilibrium state in such a way that its pressure is kept equal to that of the surrounding medium. The assumption of pressure equilibrium is valid if the thermal time scale is much longer than the time required for acoustic waves to travel across the blob. If we neglect any change of molecular weight in the perturbation, the locus of constant pressure is given by $\rho T = $ constant, which is indicated by the dashed line in Fig. 8.8. The intersections of this line with the locus $\mathscr{L}(\rho, T) = 0$ represent the possible equilibrium states corresponding to the given pressure. As one can see from Fig. 8.8, these equilibrium states can be divided into two classes, those with $(\partial \ln T/\partial \ln \rho)_{\mathscr{L}} > -1$, and those with $(\partial \ln T/\partial \ln \rho)_{\mathscr{L}} < -1$. Let us first consider an isobaric perturbation from an equilibrium state in the former class (e.g. point P_2 in Fig. 8.8). When the gas blob is perturbed along the isobaric line from this equilibrium state to a higher temperature, it enters a region where $\mathscr{L} > 0$, i.e. with net cooling, and the blob will cool back to its unperturbed temperature. Similarly, if the blob is perturbed to a lower temperature, it enters a region with $\mathscr{L} < 0$, and the blob will heat up and expand back towards P_2. Clearly, state

P_2 is stable against thermal instability. As the reader can easily verify, the situation is exactly the opposite for an equilibrium state in the latter class (e.g. point P_1 in Fig. 8.8), indicating that such an equilibrium state is unstable against isobaric perturbations.

From the above argument, it is easy to see that, for an isobaric medium, the criterion for thermal instability can be written as

$$\left(\frac{\partial \mathscr{L}}{\partial T}\right)_P < 0. \tag{8.135}$$

To derive a more general instability criterion, we start with the specific entropy of the gas:

$$dS = \frac{dQ}{T} = -\frac{\mathscr{L}\,dt}{T}. \tag{8.136}$$

Now suppose that the entropy of a gas blob is perturbed from the mean value by an amount δS in such a way that some thermodynamic quantity A is kept fixed ($\delta A = 0$). The rate of change in δS can be written as

$$\frac{d\delta S}{dt} = \delta\left(\frac{dS}{dt}\right) = -\delta\left(\frac{\mathscr{L}}{T}\right). \tag{8.137}$$

Clearly, if δS and $\delta(\mathscr{L}/T)$ have the same sign, the entropy will return to the background value (stability). On the other hand, if δS and $\delta(\mathscr{L}/T)$ have opposite signs, the entropy perturbation will grow with time (instability). Thus, the instability criterion can be written as

$$\left[\frac{\partial}{\partial S}\left(\frac{\mathscr{L}}{T}\right)\right]_A < 0. \tag{8.138}$$

This relation was derived by Balbus (1986) and, for a gas in which heating balances cooling (i.e. $\mathscr{L} = 0$), reduces to

$$\left(\frac{\partial \mathscr{L}}{\partial S}\right)_A < 0, \tag{8.139}$$

which is known as the Field criterion (Field, 1965).

Because of thermal instability, we expect that gas cooling generally leads to a multi-phase medium, with most gas in the thermally stable phases. From Fig. 8.8, we see that these phases have temperatures at $T \sim 10^6$ K, $\sim 10^4$ K, and ~ 10–100 K. Indeed, the thermal instability is the basic idea behind the multi-phase models for the interstellar medium (e.g. Field et al., 1969; McKee & Ostriker, 1977).

In addition, the thermal instability may also cause cloud fragmentation in the cooling flows associated with hot gaseous halos, which may have potentially important implications for galaxy formation (e.g. Mo & Miralda-Escudé, 1996; Maller & Bullock, 2004). There is, however, a complication when the thermal instability criterion is applied to gas that is vertically stratified in a gravitational field. In this case, the specific entropy of the gas must decrease in the direction of the gravitational field to prevent convective instability (see §8.2.2). A slightly overdense, cold gas blob that has a specific entropy slightly lower than its surrounding medium will then fall in the gravitational field until it encounters material of the same specific entropy and buoyancy starts to push it back. Consequently the gas blob will oscillate vertically around the location where the material in the unperturbed ambient medium has the same specific entropy. This causes a drastic reduction of the growth rate of thermal instabilities in the linear regime (Balbus & Soker, 1989). However, if the initial perturbations in the gas density are sufficiently large, a multi-phase medium may still develop, simply because the cooling rates at $T \sim 10^6$ K, $\sim 10^4$ K, and ~ 10–100 K are relatively low compared to those at other temperatures.

8.5.2 Hydrodynamical Instabilities

(a) Kelvin–Helmholtz Instability For a multi-phase medium in a gravitational field, the cold phase may be subject to several dynamical instabilities that can destroy cold clouds. If the mass of a cold cloud, $M_{\rm cl}$, exceeds the Jeans mass, $M_{\rm J}$, it is subject to collapse under its own gravity. If $M_{\rm cl} < M_{\rm J}$, the cloud can remain in pressure equilibrium with the hot phase, but will be accelerated with respect to the surrounding medium by buoyancy (due to the stratification by gravity).

As a cold, dense cloud moves through a hot, tenuous medium, the interface between the two phases is subject to the growth of Kelvin–Helmholtz instabilities. In the ideal case where two incompressible fluids slide past one another at a flat interface and there is no gravity perpendicular to the interface, the growth rate in the linear regime, obtained from the dispersion relation of the fluid equations, is

$$\omega = \frac{(\rho_{\rm h}\rho_{\rm c})^{1/2} v}{\rho_{\rm c}+\rho_{\rm h}} k \qquad (8.140)$$

(e.g. Drazin & Reid, 1981), where $\rho_{\rm h}$ and $\rho_{\rm c}$ are the densities of the hot and cold media, respectively, k is the wavenumber of the perturbation and v is the relative velocity. If we set $k = 2\pi/R_{\rm c}$, where $R_{\rm c}$ is the cloud radius, the characteristic time scale for instability growth is

$$\tau_{\rm KH} \equiv \frac{2\pi}{\omega} = \frac{R_{\rm c}}{v} \frac{(2+\delta)}{\sqrt{1+\delta}}, \qquad (8.141)$$

with $\delta = (\rho_{\rm c}/\rho_{\rm h}) - 1$ the overdensity of the cold cloud with respect to the hot medium. Note that this equation is only valid in the linear regime ($\delta \ll 1$), where we thus have that $\tau_{\rm KH} \simeq 2R_{\rm c}/v$ to good approximation. If the hot phase is in hydrostatic equilibrium with a gravitational potential, then the final speed of the cold clouds is expected to be roughly the sound speed of the hot medium. In this case, if the cold cloud is in pressure equilibrium with the medium, then $\tau_{\rm KH} \simeq (R_{\rm c}/c_{\rm s,c})[(2+\delta)/(1+\delta)]$, with $c_{\rm s,c}$ the sound speed in the cloud. Thus, pressure confined cold clouds moving at the sound speed of a hot medium are disrupted by the Kelvin–Helmholtz instability on a time scale comparable to their internal dynamical time, $\tau_{\rm dyn} \equiv R_{\rm c}/c_{\rm s,c}$. In principle the clouds may be stabilized by their self-gravity ($\delta \gg 1$), but this basically implies that the cloud mass exceeds the critical mass for the onset of gravitational instability (Murray et al., 1993). Alternatively, the cold clouds may be stabilized by magnetic fields (e.g. Malagoli et al., 1996), or by radiative cooling at the cloud–medium interface (Vietri et al., 1997).

(b) Rayleigh–Taylor Instability Another hydrodynamic instability in a multi-phase medium is the Rayleigh–Taylor instability, which occurs when a heavy fluid rests on top of a light fluid in an effective gravitational field. Such a situation can arise when a cold cloud moves through a tenuous medium and is decelerated by the wind (ram pressure). The linear growth rate for this instability is given by

$$\omega^2 = \left(\frac{\rho_{\rm c}-\rho_{\rm h}}{\rho_{\rm c}+\rho_{\rm h}}\right) gk = gk \frac{\delta}{(2+\delta)}, \qquad (8.142)$$

where g is the gravitational acceleration of the cloud due to the wind (e.g. Drazin & Reid, 1981). If we assume that the gas in the wind transfers all its momentum to the cloud upon impact, then

$$g \sim \rho_{\rm h} \pi R_{\rm c}^2 v^2 / M_{\rm c}, \qquad (8.143)$$

and the characteristic time scale for instability growth is

$$\tau_{\rm RT} \equiv \frac{2\pi}{\omega} \sim \frac{R_{\rm c}}{v} \left[\frac{(2+\delta)(1+\delta)}{\delta}\right]^{1/2}, \qquad (8.144)$$

which, for $\delta \sim 1$, is similar to the characteristic time scale for the Kelvin–Helmholtz instability [see Eq. (8.141)]. However, if $\delta \ll 1$ then $\tau_{\rm KH} \ll \tau_{\rm RT}$ and the Rayleigh–Taylor instability is of little importance.

8.5.3 Heat Conduction

In a multi-phase medium, cold clouds also suffer evaporation because of conductive heat input from the hot medium. The energy flux due to heat conduction is proportional to the temperature gradient and is given by:

$$\mathbf{F}_{\rm cond} = -\mathcal{K}\nabla T, \tag{8.145}$$

where $\mathcal{K}$ is the coefficient of thermal conductivity (see §B1.2). To derive an expression for $\mathcal{K}$, let us consider an idealized case with a temperature gradient only in the z–direction. In this case the energy flux through a unit area perpendicular to the z-axis is

$$F_{\rm cond} \propto nv\ell \frac{\partial \varepsilon}{\partial z} = nv\ell c_V \frac{\partial T}{\partial z}, \tag{8.146}$$

where n is the number density of the gas, v and ε are the average velocity and average energy per particle, respectively, $\ell = 1/n\sigma$ is the mean free path, and c_V is the heat capacity per particle. Thus, the conductivity can be written as

$$\mathcal{K} \propto nv\ell c_V \propto T^{1/2} c_V/\sigma. \tag{8.147}$$

In a neutral gas, σ is independent of T and so $\mathcal{K} \propto T^{1/2}$; for a fully ionized gas $\sigma \propto T^{-2}$ and so $\mathcal{K} \propto T^{5/2}$ (Spitzer, 1962). Numerically,

$$\mathcal{K} = \begin{cases} 2.5 \times 10^5 T_4^{1/2} & (10^{-2} < T_4 < 1) \\ 6.0 \times 10^3 T_4^{5/2} & (T_4 > 1), \end{cases} \tag{8.148}$$

where $T_4 = T/10^4\,{\rm K}$ and $[\mathcal{K}] = {\rm erg\,cm^{-1}\,s^{-1}\,K^{-1}}$.

Heat conduction adds a term $\nabla \cdot \mathbf{F}_{\rm cond}$ to the left-hand side of the energy equation (8.3). In a steady state including radiative cooling, we therefore have

$$\nabla \cdot \left[\rho \mathbf{v}\left(\frac{v^2}{2} + \frac{P}{\rho} + \mathscr{E}\right)\right] + \nabla \cdot \mathbf{F}_{\rm cond} = \mathscr{H} - \mathscr{C}. \tag{8.149}$$

This equation can be solved to obtain the structure of a conduction front under various assumptions. For a spherical cloud with radius r_c embedded in a hot gas that has density n_h and temperature T_h far from the cloud, under the assumption that radiative cooling and magnetic fields can be neglected, the classical mass-loss (evaporation) rate of the cloud is

$$\dot{M} = 16\pi\mu m_p r_c \mathcal{K}(T_h)/(25 k_B) \tag{8.150}$$

(Cowie & McKee, 1977). The time scale for evaporation is then

$$t_{\rm ev} = \frac{M_c}{\dot{M}} \approx 1.5 \times 10^8 \left(\frac{\bar{n}_c}{10^{-3}\,{\rm cm}^{-3}}\right)\left(\frac{r_c}{1\,{\rm kpc}}\right)^2 \left(\frac{T_h}{10^6\,{\rm K}}\right)^{-5/2}\,{\rm yr}, \tag{8.151}$$

where $\bar{n}_c$ is the average number density of the cloud.

If radiative cooling is important, one can define a critical radius at which radiative losses balance conductive heating. For $T_h \gtrsim 10^5\,{\rm K}$, an approximation to this critical radius is

$$r_{\rm crit} \approx 0.16 \left(\frac{\bar{n}_c}{10^{-3}\,{\rm cm}^{-3}}\right)^{-1}\left(\frac{T_h}{10^6\,{\rm K}}\right)^2\,{\rm kpc} \tag{8.152}$$

(McKee & Cowie, 1977). Clouds smaller than this critical radius evaporate, while those larger than this condense.

The classical thermal conduction is based on the assumption that the mean free path is much shorter than the temperature scale height, $l \ll T/|\nabla T|$. When the mean free path becomes comparable to or larger than the temperature scale height, the heat flux is no longer equal to that given by Eq. (8.145). In this case, the heat flux is said to be saturated. The degree of saturation is described by a saturation parameter which, for a spherical cloud embedded in a hot gas, can be written as

$$\sigma_{\rm ev} = 4.2 \times 10^{-3} \left(\frac{\bar{n}_{\rm c}}{10^{-3}\,{\rm cm}^{-3}}\right)^{-1} \left(\frac{r_{\rm c}}{1\,{\rm kpc}}\right)^{-1} \left(\frac{T_{\rm h}}{10^6\,{\rm K}}\right)^2, \tag{8.153}$$

and evaporation is saturated when $\sigma_{\rm ev} > 1$.

From Eqs. (8.152) and (8.153), we see that clouds are stabilized by radiative cooling against evaporation if

$$\sigma_{\rm ev} \lesssim 0.026. \tag{8.154}$$

Saturated evaporation of spherical clouds in a hot gas is discussed in detail by Cowie & McKee (1977).

8.6 Evolution of Gaseous Halos with Energy Sources

So far we have focused on the formation of gaseous halos in which gas is only heated by gravitational accretion shocks. However, in the presence of energy sources, gas can also be heated through non-gravitational processes, such as radiation from stars and AGN, stellar explosions and stellar winds. If such heating is important, the properties of the gaseous halos can be significantly affected.

In the presence of source terms, gas dynamics is described by the following set of mass, momentum and energy conservation equations:

$$\frac{\partial \rho}{\partial t} + \nabla \cdot (\rho \mathbf{v}) = S_{\rm m}, \tag{8.155}$$

$$\frac{\partial (\rho \mathbf{v})}{\partial t} + \sum_i \nabla_i (\rho v_i \mathbf{v}) + \nabla P + \rho \nabla \Phi = \mathbf{S}_{\rm mom}, \tag{8.156}$$

$$\frac{\partial}{\partial t}\left[\rho\left(\frac{v^2}{2}+\mathscr{E}\right)\right] + \nabla \cdot \left[\rho\left(\frac{v^2}{2}+\frac{P}{\rho}+\mathscr{E}\right)\mathbf{v}\right] - \rho \mathbf{v} \cdot \nabla \Phi = S_{\rm e}, \tag{8.157}$$

where $S_{\rm m}$, $\mathbf{S}_{\rm mom}$ and $S_{\rm e}$ are the changes per unit time in mass density, momentum density, and energy density due to the sources. Note that $\mathbf{S}_{\rm mom} = S_{\rm m}\mathbf{v}_{\rm inj}$, with $\mathbf{v}_{\rm inj}$ the local mean velocity of the injected material, and that

$$S_{\rm e} = \mathscr{H} - \mathscr{C} + S_{\rm m}\left(\Phi + \frac{3}{2}\theta_{\rm inj} + \frac{1}{2}v_{\rm inj}^2\right), \tag{8.158}$$

where $\mathscr{C} = n_{\rm H}^2 \Lambda$ is the cooling rate per unit volume, $\mathscr{H}$ is the (radiative) heating rate per unit volume, and $\theta_{\rm inj} \equiv k_{\rm B} T_{\rm inj}/\mu m_{\rm p}$ is the local mean thermal velocity of injected material. If all three source terms are known, the above set of equations can be solved together with the Poisson equation for the gravitational potential Φ. In §10.5.2 we show how the source terms can be computed in the case of stellar feedback.

8.6.1 Blast Waves

In some applications, the energy source may be approximated by a blast wave, in which a large amount of energy is released locally during a short period of time, and the disturbance produced in the medium propagates as a shock wave. Here we describe a simple model for such blast waves, and consider its application to supernova explosions in the interstellar medium.

(a) Self-Similar Model Consider an instantaneous release of energy ε_0 from a source of negligible size in a uniform medium with density ρ_0. We assume that the medium can be described as an ideal fluid, and idealize the problem by assuming that the mass ejected from the explosion is negligible. This assumption is valid when the mass swept by the expanding shock wave is much larger than the initial mass of the ejecta. In the energy-conserving phase of the evolution, i.e. when the shock wave is at a stage where the total radiated energy is much smaller than ε_0, the only time scale involved is t, the age of the explosion, and the only length scale is $r_{\rm sh}(t)$, the radius of the shock wave at time t. In this case, we expect the problem to admit 'self-similar' solutions, meaning that any dimensional quantity $Q(r,t)$ at radius r (from the center of the explosion) and time t can be written as $Q(r,t) = Q_{\rm ch}\mathcal{Q}(r/R)$, where $Q_{\rm ch}$ is the characteristic value of Q, obtained by combining $r_{\rm sh}$, t and ρ_0 (or ε_0). For instance, the gas density, velocity and pressure at (r,t) can be written, respectively, in the following forms:

$$\rho(r,t) = \rho_0 \mathcal{D}(\lambda), \quad v(r,t) = \frac{r_{\rm sh}}{t}\mathcal{V}(\lambda), \quad P(r,t) = \rho_0 \left(\frac{r_{\rm sh}}{t}\right)^2 \mathcal{P}(\lambda), \tag{8.159}$$

where

$$\lambda \equiv r/r_{\rm sh}. \tag{8.160}$$

An important property of such solutions is that, once $r_{\rm sh}(t)$ is known, the evolution is completely determined by the forms of the single-variable functions $\mathcal{D}(\lambda)$, $\mathcal{V}(\lambda)$ and $\mathcal{P}(\lambda)$. Since the evolution of $r_{\rm sh}(t)$ should be completely determined by ε_0 and ρ_0, and since no combination of t, ε_0 and ρ_0 can give a dimensionless quantity, the most general form of $r_{\rm sh}(t)$ is

$$r_{\rm sh}(t) = At^\eta \varepsilon_0^\alpha \rho_0^\beta, \tag{8.161}$$

where A is a dimensionless constant. As the dimensions on both sides of the above equation must be equal, we have $\eta = 2\alpha = -2\beta = 2/5$, and so

$$r_{\rm sh}(t) = A\left(\frac{\varepsilon_0}{\rho_0}\right)^{1/5} t^{2/5}. \tag{8.162}$$

Thus, under the assumption of self-similarity, the time dependence of $r_{\rm sh}$ is completely determined by dimensionality considerations. The expansion speed of the shock front is therefore

$$v_{\rm sh} \equiv \frac{dr_{\rm sh}}{dt} = \frac{2}{5}A\left(\frac{\varepsilon_0}{\rho_0}\right)^{1/5} t^{-3/5}, \tag{8.163}$$

which shows that the shock becomes weaker as it expands. Once $v_{\rm sh}$ is reduced to a level comparable to the sound speed of the ambient medium, the shock disappears.

The above dimensionality analysis can be extended to cases where ρ_0 is a power law of r, and ε_0 is a power law of t:

$$\rho_0 \propto (r/r_0)^a, \quad \varepsilon_0 \propto (t/t_0)^b. \tag{8.164}$$

In this case,

$$r_{\rm sh}(t) \propto t^{(b+2)/(5+a)}. \tag{8.165}$$

The value of A in Eq. (8.162) and the forms of $\mathscr{D}$, $\mathscr{V}$ and $\mathscr{P}$ can all be obtained by solving the fluid equations. In spherical symmetry, the mass, momentum and energy equations are

$$\frac{\partial \rho}{\partial t} + \frac{1}{r^2}\frac{\partial}{\partial r}(r^2 \rho v) = 0, \tag{8.166}$$

$$\frac{\partial v}{\partial t} + v\frac{\partial v}{\partial r} = -\frac{1}{\rho}\frac{\partial P}{\partial r}, \tag{8.167}$$

$$\frac{\partial}{\partial t}\left[\rho\left(\frac{1}{2}v^2 + \mathscr{E}\right)\right] + \frac{1}{r^2}\frac{\partial}{\partial r}\left[r^2 \rho v\left(\frac{1}{2}v^2 + \frac{P}{\rho} + \mathscr{E}\right)\right] = 0, \tag{8.168}$$

where we have ignored any heating or cooling. As we have seen in §8.1.1, the last equation can be replaced by the following entropy equation,

$$\left(\frac{\partial}{\partial t} + v\frac{\partial}{\partial r}\right)\ln\left(P\rho^{-\gamma}\right) = 0. \tag{8.169}$$

Under the assumption of similarity, all quantities depend on r and t only through the combination $\lambda = r/r_{\rm sh}(t)$, and Eqs. (8.166), (8.167) and (8.169) are reduced to the following set of ordinary differential equations:

$$(\mathscr{V} - \eta\lambda)D' + \mathscr{D}\mathscr{V}' + \frac{2}{\lambda}\mathscr{D}\mathscr{V} = 0, \tag{8.170}$$

$$(\mathscr{V} - \eta\lambda)\mathscr{V}' + (\beta - 1)\mathscr{V} = -\frac{\mathscr{P}'}{\mathscr{D}}, \tag{8.171}$$

$$(\mathscr{V} - \eta\lambda)\left(\frac{\mathscr{P}'}{\mathscr{P}} - \gamma\frac{\mathscr{D}'}{\mathscr{D}}\right) = 2(2-\eta) - 2\gamma, \tag{8.172}$$

where a prime denotes a derivative with respect to λ, and $\eta = 2/5$ is the power of the time dependence of $r_{\rm sh}(t)$. These equations can be integrated from $\lambda = 1$ to $\lambda = 0$ subject to the jump conditions (8.49) and (8.50) at $\lambda = 1$. Note that these jump conditions are obtained by an observer moving with the shock wave. For an observer at rest with the ambient medium, $v_1 = -v_{\rm sh}$, $v_2 = v_{\rm b} - v_{\rm sh}$, where $v_{\rm b}$ is the rest-frame velocity of the flow just behind the shock front. For strong shocks, the jump conditions are

$$\mathscr{D}(1) = \frac{\gamma+1}{\gamma-1}, \quad \mathscr{V}(1) = \frac{2\eta}{(\gamma+1)}, \quad \mathscr{P}(1) = \frac{2\eta^2}{(\gamma+1)}. \tag{8.173}$$

An extra condition is required in order to specify the constant A in Eq. (8.162). This condition can be obtained by integrating the energy equation (8.168) over the entire space. Since $rv = 0$ at both $r = 0$ and $r = \infty$, only the first term in Eq. (8.168) contributes to the integration:

$$\frac{d}{dt}\int_0^\infty \left(\frac{P}{\gamma-1} + \frac{1}{2}\rho v^2\right) 4\pi r^2\, dr = 0. \tag{8.174}$$

Outside the shock radius, $r_{\rm sh}$, the velocity $v = 0$ and the pressure is a constant P_0. The integration from $r_{\rm sh}$ to infinity is then equal to

$$-\frac{P_0}{\gamma-1} 4\pi r_{\rm sh}^2 \frac{dr_{\rm sh}}{dt}, \tag{8.175}$$

which can be ignored if the ambient medium is cold so that $P_0/\rho_0 = k_{\rm B}T_0/\mu_0 m_{\rm p} \ll v^2$. Thus,

$$\int_0^{r_{\rm sh}}\left(\frac{P}{\gamma-1} + \frac{1}{2}\rho v^2\right) 4\pi r^2\, dr = \varepsilon_0, \tag{8.176}$$

8.6 Evolution of Gaseous Halos with Energy Sources

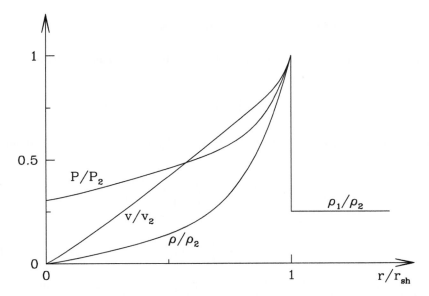

Fig. 8.9. The density, pressure and velocity profiles of a self-similar blast wave. All quantities are normalized to unity directly behind the shock.

or, in terms of our dimensionless variables,

$$\int_0^A \left[\mathscr{P}(\lambda) + \mathscr{D}(\lambda)\mathscr{V}^2(\lambda)\right] \lambda^4 \, d\lambda = \frac{25(\gamma^2 - 1)}{32\pi}. \tag{8.177}$$

For $\gamma = 5/3$, this gives $A \approx 1.15$. The corresponding density, pressure and velocity profiles are shown in Fig. 8.9.

(b) Applications to Supernova Remnants With the above results, we can now discuss in more detail the different evolutionary stages of a blast wave. As a concrete example, we consider supernova remnants, which are produced by supernova explosions at the late evolutionary stages of relatively massive stars (see §10.2.3 for details). The typical energy output of such an explosion is about 10^{51} erg, which is released within a matter of seconds. Almost all this energy is initially in the form of kinetic energy of the ejecta; the integrated photon luminosity of a supernova is a factor ~ 100 smaller. However, as we will see below, the supernova ejecta creates a shock that becomes radiative, and by the time the supernova blast wave fades away, almost 95% of its energy has been radiated away.

The evolution of a supernova blast wave consists of four well-defined stages. In the first stage the ejected mass exceeds the swept-up ambient mass, and to lowest order the ejecta undergo free expansion. Once the mass of the swept-up material becomes comparable to the mass of the ejecta, the blast wave enters the adiabatic (or Sedov) phase, in which the evolution is self-similar as long as the ambient medium is homogeneous. Eventually the radiative losses from the interior of the blast wave become significant, and the supernova remnant enters the third, radiative stage of its evolution. Finally, once the interior pressure becomes comparable to that of the ambient medium, the supernova remnant merges with the ISM. We now describe this evolution in more detail.

At the beginning of a supernova explosion, when the gas swept up by the shock is still smaller than the mass of the ejecta, M_{ejecta}, the remnant is in free expansion, with a constant velocity v_{sh}, and its radius increases as $r_{\text{sh}} = v_{\text{sh}} t$. The free expansion continues until the mass swept by the shock is comparable to M_{ejecta}:

$$M_{\rm sh} = \frac{4\pi}{3} r_{\rm sh}^3 \rho_1 = M_{\rm ejecta}. \tag{8.178}$$

Since the total explosion energy is $\varepsilon_0 = (1/2) M_{\rm ejecta} v_{\rm sh}^2$, the free-expansion phase terminates at a time

$$t_{\rm f} \approx 189 \left(\frac{M_{\rm ejecta}}{M_\odot}\right)^{5/6} \left(\frac{n_{\rm H}}{\rm cm^{-3}}\right)^{-1/3} \varepsilon_{51}^{-1/2} \, {\rm yr}, \tag{8.179}$$

where $\varepsilon_{51} \equiv \varepsilon_0/10^{51}$ erg, and $n_{\rm H}$ is the number density of hydrogen nuclei in the medium. At this time the size and velocity of the remnant are

$$r_{\rm sh} \approx 1.9 \left(\frac{n_{\rm H}}{\rm cm^{-3}}\right)^{-1/3} \left(\frac{M_{\rm ejecta}}{M_\odot}\right)^{1/3} {\rm pc}, \tag{8.180}$$

$$v_{\rm sh} \approx 10^4 \left(\frac{M_{\rm ejecta}}{M_\odot}\right)^{-1/2} \varepsilon_{51}^{1/2} \, {\rm km\, s^{-1}}. \tag{8.181}$$

At $t > t_{\rm f}$, the mass swept by the shell is larger than $M_{\rm ejecta}$ and so the expansion of the shell is decelerated. As long as radiative energy loss is negligible, i.e. in the adiabatic (or Sedov) phase, the evolution of the remnant is given by the similarity solution discussed above [Eqs. (8.162) and (8.163)], with

$$r_{\rm sh} \approx 32 \left(\frac{n_{\rm H}}{\rm cm^{-3}}\right)^{-1/5} \varepsilon_{51}^{1/5} \left(\frac{t}{10^5 \, {\rm yr}}\right)^{2/5} {\rm pc}, \tag{8.182}$$

$$v_{\rm sh} \approx 124 \left(\frac{n_{\rm H}}{\rm cm^{-3}}\right)^{-1/5} \varepsilon_{51}^{1/5} \left(\frac{t}{10^5 \, {\rm yr}}\right)^{-3/5} {\rm km\, s^{-1}}. \tag{8.183}$$

As a crude approximation, this phase ends when the radiative energy loss is about half of the thermal energy, or a quarter of the total energy:

$$f_{\rm loss}(t) \equiv \frac{\varepsilon_{\rm loss}}{\varepsilon_0} = \frac{1}{\varepsilon_0} \int_0^t dt' \int_0^{r_{\rm sh}} 4\pi r^2 \, n_{\rm H}^2(r) \, \Lambda(r) \, dr \approx \frac{1}{4}. \tag{8.184}$$

To gain some insight into the problem, let us approximate the cooling function by a power-law function of temperature:

$$\Lambda(T) = \Lambda_0 (T/T_0)^\nu. \tag{8.185}$$

For gas with solar metallicity, $\nu \approx -2/3$ in the temperature range $10^5 \, {\rm K} < T < 10^7 \, {\rm K}$ (see Fig. 8.1), and $\Lambda_0 \approx 2 \times 10^{-23} \, {\rm erg\, cm^3\, s^{-1}}$ for $T_0 = 10^7$ K. Since post-shock gas is the densest and coolest just behind the shock, we expect that most of the energy loss occurs there. The post-shock temperature near $r_{\rm sh}$ is

$$T_2 = \frac{\mu m_{\rm p}}{k_{\rm B}} \frac{P_2}{\rho_2} = \left(\frac{2}{5}\right)^2 \frac{2(\gamma-1)}{(\gamma+1)^2} \frac{\mu m_{\rm p}}{k_{\rm B}} \left(\frac{r_{\rm sh}}{t}\right)^2, \tag{8.186}$$

where we have used the self-similar solutions (8.159) and (8.173). Inserting this and Eq. (8.185) into Eq. (8.184), and using $\gamma = 5/3$, we have

$$f_{\rm loss}(t) = \frac{1}{\varepsilon_0} \int_0^t dt' \left[\frac{3}{100} \frac{\mu m_{\rm p}}{k_{\rm B}} \left(\frac{r_{\rm sh}}{t'}\right)^2\right]^\nu \int_0^{r_{\rm sh}} 4\pi r^2 \Lambda_0 n_{\rm H}^2(r) \, dr. \tag{8.187}$$

Replacing $n_H^2(r)$ by its rms value within r_{sh}, $\overline{n_H^2} \approx \langle n_H^2 \rangle = \zeta n_{H,1}^2$, where $n_{H,1}$ is the number density of hydrogen in the pre-shock gas, and $\zeta \approx 2.29$ for the adiabatic (Sedov) solution, we finally obtain

$$f_{\text{loss}}(t) = \frac{4\pi}{3} \frac{\Lambda_0 \zeta}{(1-2\nu)} \frac{n_{H,1}^2}{\varepsilon_0} \left(\frac{3\mu m_p}{100 k_B T_0}\right)^\nu \left(\frac{\varepsilon_0 A^5}{\rho_1}\right)^{(3+2\nu)/5} t^{(11-6\nu)/5}. \tag{8.188}$$

If we define t_{rad} by $f_{\text{loss}}(t_{\text{rad}}) = 1/4$, then for fully ionized gas with $n_{He} = n_H/12$,

$$t_{\text{rad}} \approx 4.3 \times 10^4 \left(\frac{n_H}{\text{cm}^{-3}}\right)^{-5/9} \varepsilon_{51}^{2/9} \text{ yrs}, \tag{8.189}$$

$$r_{sh}(t_{\text{rad}}) \approx 23 \left(\frac{n_H}{\text{cm}^{-3}}\right)^{-19/45} \varepsilon_{51}^{13/45} \text{ pc}, \tag{8.190}$$

$$v_{sh}(t_{\text{rad}}) \approx 200 \left(\frac{n_H}{\text{cm}^{-3}}\right)^{2/15} \varepsilon_{51}^{1/15} \text{ km s}^{-1}, \tag{8.191}$$

where we have used the values of A, ν, T_0 and Λ_0 given above. Note that $v_{sh}(t_{\text{rad}})$ is almost independent of ε_{51} and n_H.

For $t > t_{\text{rad}}$, the adiabatic model is no longer valid. The problem should then be solved by adding the cooling term, $-(\gamma-1)\mathscr{C}/P$, to the right-hand side of the energy equation (8.169). This in general introduces extra scales and so the problem no longer admits self-similar solutions.

At $t = t_{\text{sp}} \gg t_{\text{rad}}$, however, simple arguments can again be used to give some useful results. During this late stage, the pressure inside the shock is negligible, and the remnant resembles a 'snowplow', in which ambient gas is swept up by the *inertia* of the moving shell. In this case, the evolution of the remnant is governed by momentum conservation: $M_{sh} v_{sh} = $ constant. Since $M_{sh} \propto r_{sh}^3$ (most gas is swept up), we have $r_{sh}/t \propto v_{sh} \propto r_{sh}^{-3}$, and thus $r_{sh} \propto t^{1/4}$. Numerical calculations give a somewhat different result, $r_{sh} \propto t^{0.31}$. This is due to the fact that the internal pressure is not entirely negligible, which slightly increases the momentum (Chevalier, 1974).

We have seen that a supernova explosion can accelerate the gas in its surrounding. An interesting question is: what fraction of the total explosion energy is transformed into kinetic energy of the gas (the rest is radiated away). The final kinetic energy in the remnant is given by the mass and velocity at the time when it fades into the interstellar medium: $E_{\text{kin}} = (1/2) M_{\text{fade}} v_{\text{fade}}^2$. If we denote the mass and velocity at the onset of the 'snowplow' phase by M_{sp} and v_{sp}, then

$$f_{\text{kin}} \equiv \frac{E_{\text{kin}}}{E} = \left(\frac{M_{\text{sp}} v_{\text{sp}}^2}{2E}\right) \left(\frac{M_{\text{fade}} v_{\text{fade}}^2}{M_{\text{sp}} v_{\text{sp}}^2}\right) \approx \frac{1}{2} \frac{v_{\text{fade}}}{v_{\text{sp}}}, \tag{8.192}$$

where we have used that momentum is conserved during the 'snowplow' phase, and assumed that another quarter of the total energy is radiated away between t_{rad} and t_{sp}. If we take $v_{\text{fade}} = 10 \text{ km s}^{-1}$ (the typical velocity dispersion of the ISM) and $v_{\text{sp}} = 100 \text{ km s}^{-1}$ (half the value at t_{rad}), then $f_{\text{kin}} \sim 0.05$. Although this number is fairly uncertain, it is clear that only a relatively small fraction of the total explosion energy is ultimately transferred to kinetic energy.

(c) Supernova Heating So far we have only considered the effect of a single blast wave on the surrounding gas. In reality, there may be several or many explosions confined to a small region. In this case, individual shocks can overlap and be thermalized, thereby heating the gas. How much of the initial explosive energy can be thermalized depends on the time scale of thermalization relative to the cooling time scale. Suppose that the thermalized energy is a fraction f_{th} of the explosive energy ε_0. In this case the volume heating rate due to explosions can formally be written as

$$\mathscr{H} = \frac{\dot{\rho}_{\text{inj}}}{\mu m_p} \times \frac{3}{2} k_B T_{\text{inj}}, \tag{8.193}$$

where

$$\dot{\rho}_{\rm inj} = M_{\rm ejecta} \dot{n}_{\rm blast} \tag{8.194}$$

is the mass injection rate per unit volume (with $\dot{n}_{\rm blast}$ the rate of explosions per unit volume), and

$$T_{\rm inj} = \frac{2}{3} \frac{(f_{\rm th} \varepsilon_0 / k_{\rm B})}{(M_{\rm ejecta}/\mu m_{\rm p})} \tag{8.195}$$

is the effective temperature of the ejected material.

The value of $f_{\rm th}$ may be estimated by assuming that thermalization occurs at a time when the volume filling factor of the supernova remnants approaches unity. The volume filling factor of supernova remnants can be defined as

$$\mathscr{P}_{\rm SNR}(t) = \int_0^t \nu_{\rm SN}(t') V_{\rm SNR}(t-t') \, dt', \tag{8.196}$$

where $\nu_{\rm SN}(t')$ is the supernova rate per unit volume at time t' and $V_{\rm SNR}(t-t')$ is the volume of a supernova remnant at the age $t-t'$. With the model described above, one can estimate $\mathscr{P}_{\rm SNR}$ as a function of t for given $\nu_{\rm SN}(t)$, and thus identify the time $t_{\rm ov}$ at which $\mathscr{P}_{\rm SNR} = 1$. The value of $f_{\rm th}$ then follows from

$$f_{\rm th} = \frac{\int_0^{t_1} \nu_{\rm SN}(t') \varepsilon_{\rm SNR}(t_{\rm ov} - t') \, dt'}{10^{51} \, {\rm erg} \times \int_0^{t_1} \nu_{\rm SN}(t') \, dt'}. \tag{8.197}$$

Here $\varepsilon_{\rm SNR}(t_{\rm ov} - t')$ is the sum of the kinetic and thermal energy of a supernova remnant at an age of $t_{\rm ov} - t'$, and we have assumed that each supernova releases a total of 10^{51} ergs. If $\nu_{\rm SN}(t)$ is very high, such as in starbursts, the individual bubbles can start to overlap before the supernova remnants have reached their radiative stages, and $f_{\rm th}$ can be close to unity. If, on the other hand, $\nu_{\rm SN}(t)$ is low, then bubbles only overlap when the supernova remnants are already in the snowplow stage and have already radiated away most of their energy. In this case $f_{\rm th} \sim f_{\rm kin} \sim 0.05$.

8.6.2 Winds and Wind-Driven Bubbles

In some cases, the energy injection from a source occurs over an extended period of time, rather than instantaneously. Examples in this category include stellar winds driven by the radiation pressure of stars, and galactic winds driven by multiple supernova explosions associated with extended periods of star formation. In these cases, the blast wave model, in which energy injection is assumed to be an explosion, is no longer valid.

In order to understand how a long-lasting wind propagates and interacts with its surrounding medium, let us consider the following idealized case. Suppose that at time $t = 0$ a point source begins to blow a spherically symmetric wind with some terminal velocity $v_{\rm w}$ and mass-loss rate $dM_{\rm w}/dt$. The power of the wind is given by the mechanical luminosity,

$$L_{\rm w} = \frac{1}{2} \frac{dM_{\rm w}}{dt} v_{\rm w}^2. \tag{8.198}$$

For a steady wind, both $v_{\rm w}$ and $dM_{\rm w}/dt$ are constant, but in general they may depend on time. We assume the wind to be cold, so that the sound speed in the wind is much smaller than the terminal velocity, and that the source is embedded in an ambient cold medium of constant density, ρ_0. Throughout we also adopt $\gamma = 5/3$. Our task is to obtain the structure of the interaction between the wind and the ambient medium.

Without going into great detail, we may use our knowledge about shock waves to infer that the dynamical system in consideration should consist of four distinct zones (Weaver et al., 1977):

- **Region I** ($r < r_1$): the hypersonic wind. In this region the wind is propagating with the terminal velocity v_w.
- **Region II** ($r_1 < r < r_c$): hot, shocked wind. This is material that is part of the wind, but that has been shocked.
- **Region III** ($r_c < r < r_2$): a shell of shocked interstellar gas. This is material from the ambient medium that has been shocked by the wind.
- **Region IV** ($r > r_2$): the ambient medium which has not yet been affected by the wind.

Thus, in this case there are two shocks, one at r_1 and the other at r_2.

During the early stages of the evolution, radiative losses are everywhere negligible. In this case, if the wind is assumed to be steady, so that L_w is a constant, the only dimensionless variable composed of L_w, ρ_0, the radial coordinate r and time t is

$$\lambda = r/r_2(t), \tag{8.199}$$

where

$$r_2(t) = A \left(\frac{L_{\text{wind}}}{\rho_0} \right)^{1/5} t^{3/5}, \tag{8.200}$$

with A a constant of order unity whose exact value remains to be determined. With the same argument as made in §8.6.1, we expect that the problem admits self-similar solutions, and so all quantities should depend on r and t only through λ. In this case, the fluid equations to be solved reduce to the set of ordinary differential equations (8.170)–(8.172) with $\eta = 3/5$. With the jump conditions given in Eq. (8.173), this set of equations can be integrated numerically (see Weaver et al., 1977) to obtain the structure of the gas distribution inside r_2. Unlike the blast wave solution, the gas density drops rapidly to zero at a radius $r_c = 0.86 r_2$, where there is a contact discontinuity separating the swept-up gas from the shocked wind. At this radius the velocity is $v(r_c) = 0.86 v_2$, where $v_2 = dr_2/dt$, and the pressure is $P(r_c) = 0.59 \rho_0 v_2^2$. Note that $v(r_c)$ is the velocity of the gas at $r = r_c$, which should not be confused with $\dot{r}_c$, the velocity with which the contact discontinuity propagates. Since region III admits a self-similar solution with $r_c = 0.86 r_2$, we have that $\dot{r}_c = v_2$.

In region II, the gas is approximately isobaric, because the temperature of the shocked wind is so high that the time for a sound wave to cross the region is much smaller than the age of the system. The density in this region is roughly uniform. Both the gas density and pressure can be estimated from the jump conditions at r_1:

$$\rho(r_1^+) \sim 4\rho(r_1^-), \quad P \sim (3/4)\rho(r_1^-) v_w^2, \tag{8.201}$$

where $\rho(r_1^-) = (dM_w/dt)/(4\pi r_1^2 v_w)$ is the density of the freely propagating wind just interior to the shock radius r_1. Note that r_1 may depend on time, and so both the density and pressure are time dependent. In order to determine r_1, we use the adiabatic condition $d(P/\rho^\gamma)/dt = 0$. Since $P \sim P(r_c) \propto t^{-4/5}$, we have

$$\frac{1}{r^2} \frac{\partial (r^2 v)}{\partial r} = -\frac{1}{\rho} \frac{d\rho}{dt} = \frac{12}{25 t}. \tag{8.202}$$

With the boundary condition that $v(r_c) = 3 r_c / 5t$, the solution of the above equation is

$$v(r,t) = \frac{11}{25} \frac{r_c^3}{r^2 t} + \frac{4}{25} \frac{r}{t}. \tag{8.203}$$

On the other hand, in the region $r_1 \leq r \ll r_c$, the flow is nearly steady, and so $v \sim (v_w/4)(r_1/r)^2$, where we have used the jump condition $v(R_1^+) = v_w/4$. Matching this velocity with the above solution and using the fact that $r_1 \ll r_c$, we get $r_1 \sim (44/25)^{1/2} r_c^{3/2}/(v_w t)^{1/2}$. Inserting this into the expression of P in Eq. (8.201) and matching the pressure thus obtained with $P(r_c)$, we obtain $A \approx 0.87$. Note that $r_2/r_1 \propto t^{1/5}$, so that the shock at r_2 propagates faster than that at r_1.

At some later stage, the cooling in the swept-up gas becomes important so that it collapses into a thin shell, while the cooling of the shocked wind is still negligible. Thus, the system then consists of a thin expanding shell enclosing and driven by a hot bubble, whose internal energy is much larger than its kinetic energy. At this stage, the time for sound waves to cross the hot bubble is still small compared to the age, so that the entire region is approximately isobaric. Since the volume of region I is much smaller than that inside r_2, the pressure is related to the internal energy simply by

$$\varepsilon = \frac{2}{3} \times \frac{4\pi}{3} r_2^3 P. \tag{8.204}$$

Assuming the shell to be infinitesimally thin and the mass in the bubble negligible, the momentum equation for the shell can be written as

$$\frac{d}{dt}\left(\frac{4\pi}{3} r_2^3 \rho_0 \frac{dr_2}{dt}\right) = 4\pi r_2^2 P, \tag{8.205}$$

and the energy balance for the hot region is

$$\frac{d\varepsilon}{dt} = L_w - P dV = L_w - 4\pi r_2^2 P \frac{dr_2}{dt}. \tag{8.206}$$

These equations have the following solution:

$$\varepsilon = \frac{5}{11} L_w t; \quad r_2 = \left(\frac{125}{154\pi}\right)^{1/5} \left(\frac{L_w}{\rho_0}\right)^{1/5} t^{3/5}. \tag{8.207}$$

More complicated models for wind-driven bubbles can be constructed by including radiative cooling in the bubble (which becomes important at late stages of the wind), the evaporation of gas from the swept-up shell to the bubble, and the impact of the gravitational potential in which the wind is embedded (e.g. Weaver et al., 1977; Ostriker & McKee, 1988).

8.6.3 Supernova Feedback and Galaxy Formation

In general, the full set of fluid equations (8.155)–(8.157) has to be solved, in order to study how a wind is generated by energy and mass sources, and how it evolves with time. Here we consider a simple model to demonstrate qualitatively how supernova explosions may drive galactic winds and affect star formation in galaxies. The wind is assumed to be spherical and steady, propagating in a static, spherically symmetric potential, Φ. Under these assumptions, the fluid equations (8.155)–(8.157) can be combined to give two first-order differential equations for the fluid velocity, v, and the adiabatic sound speed of the gas, w:

$$\frac{r}{v^2}\frac{dv^2}{dr} = \frac{-w^2}{2\pi(v^2 - w^2)}\left[4\pi\left(2 - \frac{V_c^2}{w^2}\right) + \frac{2}{3}A - \left(\frac{4}{3} + \frac{w_i^2}{v^2}\right)B\right], \tag{8.208}$$

$$\frac{r}{w^2}\frac{dw^2}{dr} = \frac{-v^2}{6\pi(v^2 - w^2)}\left[4\pi\left(2 - \frac{V_c^2}{v^2}\right) + \left(\frac{5}{3} - \frac{w^2}{v^2}\right)A \right.$$
$$\left. + \left(\frac{w^2}{v^2} - \frac{5w_i^2}{2v^2} - \frac{3w^4}{2v^4} + \frac{3w_i^2 w^2}{2v^4} - \frac{5}{6}\right)B\right], \tag{8.209}$$

where $V_c^2 \equiv r(d\Phi/dr)$ specifies the shape of the gravitational potential well, and the injected gas is assumed to have an initial *isothermal* sound speed, $w_i \equiv (k_B T_i/\mu m_p)^{1/2}$, with T_i the initial temperature of the injected gas. The quantities A and B are given by

8.6 Evolution of Gaseous Halos with Energy Sources

$$A = \frac{\dot{M}\Lambda(T)n_H^2}{r(\rho w v)^2}, \quad B = \frac{\dot{\rho}_{\text{inj}}\dot{M}}{r(\rho w)^2}, \quad \dot{M} = 4\pi \int_0^r \dot{\rho}_{\text{inj}}(r') r'^2 \, dr', \quad (8.210)$$

where $\dot{\rho}_{\text{inj}}$ is the mass injection rate per unit volume. As one can see, the sonic point, where $v = w$, is a critical point at which $v^2(r)$ and $w^2(r)$ are not smooth functions of r. Note also that Eqs. (8.208) and (8.209) are invariant under the transformation $v \to -v$, so that they describe both outflow and inflow (depending on the boundary conditions).

In some applications, the source term, $\dot{\rho}_{\text{inj}}$, is non-zero only within a confined region near the center of the potential well (i.e. that of a dark matter halo), where stars form. In such cases, one can separate a halo into an inner heating base and an outer region. The equations describing the flow in the outer region are Eqs. (8.208) and (8.209) with $B = 0$. For supersonic flows, which are most relevant to the large-scale outflows from galactic systems, the boundary conditions can be set at the radius r_1 where v^2 is slightly above w^2. If radiative cooling is negligible in the flow, so that $A = 0$, the properties of the flow are determined by the gas temperature relative to the halo gravitational potential at the heating base. Depending on whether the sound speed near r_1, w_1, is bigger or smaller than $V_c(r_1)/\sqrt{2}$, the outflow results in either a galactic wind or a hot corona. Numerical integrations of Eqs. (8.208) and (8.209) show that the wind can reach a bulk velocity of about $\sqrt{2.5}w_1$ (Efstathiou, 2000). Thus, if $w_1 \gtrsim V_{\text{esc}}/\sqrt{2.5}$, where V_{esc} is the escape velocity from the central part of the halo, the wind may escape; otherwise a hot corona is produced. The importance of radiative cooling is specified by $A(r_1)$, which is roughly the ratio between the flow time (r/v) and the cooling time $(\rho w^2/\Lambda n_H^2)$ near r_1. If $A(r_1) \gg 1$ so that cooling is effective, the outgoing gas can cool and, via thermal and hydrodynamical instabilities (see §8.5), may form cold clouds that either leave the galaxy (if $w_1 \gtrsim V_{\text{esc}}/\sqrt{2.5}$), or fall back (if $w_1 < V_{\text{esc}}/\sqrt{2.5}$) (see Wang, 1995).

The gas temperature at the heating base is determined primarily by the intensity of the supernova heating, which depends on the supernova rate per unit volume, and the cooling rate of the shocked gas. If these rates were independent of halo mass so that the temperature at the heating base does not depend strongly on V_c, then the ratio w_1/V_{esc} would be larger for less massive halos, making outflows easier to generate in halos of lower mass. In order to quantify the conditions for gas removal from dark matter halos, Dekel & Silk (1986) considered a simple model in which the star-formation rate (the mass that turns into stars per unit time) in a galaxy is assumed to be $\dot{M}_\star = M_g/(\varepsilon_{\text{SF}} t_{\text{ff}})$, where $M_g = f_g M$ is the mass of cold gas in the galaxy, ε_{SF} is a constant specifying the star formation efficiency, and $t_{\text{ff}} \equiv [3\pi/(32G\rho)]^{-1/2}$ is the free-fall time, with ρ the mass density. With the assumption that the star-formation rate is a constant, the number of supernova remnants at time t can be written as $N_{\text{SN}}(t) = \mu_{\text{SN}}\dot{M}_\star t$, where μ_{SN} is the number of supernovae corresponding to a unit mass of stars that have formed. The total energy in the supernova remnants can then be written as

$$E(t) = \mu_{\text{SN}}\dot{M}_\star \varepsilon_0 t_{\text{rad}} \mathscr{F}(t), \quad (8.211)$$

where

$$\mathscr{F}(t) \equiv \frac{1}{t_{\text{rad}}} \int_0^t \frac{\varepsilon_{\text{SNR}}(\tau)}{\varepsilon_0} d\tau, \quad (8.212)$$

with $\varepsilon_{\text{SNR}}(\tau)$ the total energy of a supernova remnant at the age τ. In the adiabatic phase, $\varepsilon_{\text{SNR}}(\tau) = (1 - f_{\text{loss}})\varepsilon_0$ with f_{loss} defined in Eq. (8.184). At later times, the energy content of a supernova remnant is roughly $\varepsilon_{\text{SNR}}(\tau) \sim 0.22\varepsilon_0 [r_{\text{sh}}(\tau)/r_{\text{sh}}(t_{\text{rad}})]^{-2}$ (Cox, 1972). Dekel & Silk (1986) argued that gas can be removed from a dark matter halo if the total energy of the supernova remnants at time t_{ov}, when $\mathscr{P}_{\text{SNR}} \sim 1$, is larger than the binding energy of the gas: $E(t_{\text{ov}}) > \frac{1}{2}M_g V_c^2$. This defines a critical value for V_c:

$$V_{\rm crit} = [2\varepsilon_0 \mathscr{F}(t_{\rm ov})(t_{\rm rad}/t_{\rm ff})(\mu_{\rm SN}/\varepsilon_{\rm SF})]^{1/2}, \qquad (8.213)$$

so that gas removal occurs in halos with $V_c < V_{\rm crit}$. Using the results for the evolution of supernova remnants described above, Dekel & Silk (1986) found $V_{\rm crit} \sim 100\,{\rm km\,s^{-1}}$.

If the energy input from star formation is equal to the binding energy of the cold gas, the star-formation rate, $\dot{M}_\star$, is given by $\mathscr{E}_0 \dot{M}_\star = (\dot{M}_{\rm g} - \dot{M}_\star) V_c^2/2$, where $\mathscr{E}_0$ measures the energy feedback per unit mass of formed stars, and the right-hand side is a crude estimate of the binding energy of the cold gas. Solving for $\dot{M}_\star$ we obtain

$$\dot{M}_\star = \frac{\dot{M}_{\rm g}}{1 + (V_0/V_c)^2}, \qquad (8.214)$$

where $V_0^2 = 2\mathscr{E}_0$. Based on the discussion presented above, we have $V_0 \sim V_{\rm crit}$. In this simple model, the star-formation efficiency in halos with $V_c \ll V_{\rm crit}$ is reduced by a factor proportional to V_c^2. As we will see in Chapter 15, it is necessary to suppress the efficiency of star formation in low-mass halos in order to explain the observed galaxy luminosity function at the faint end in the CDM scenario of galaxy formation. Star formation feedback through supernova explosions provides an appealing mechanism. Unfortunately, the details regarding this feedback mechanism have yet to be quantified. It is still unclear how effective the energy feedback from star formation is coupled to the gas. Some numerical simulations show that the coupling is rather poor so that much of the feedback energy can escape from a galaxy without affecting the bulk of the gas (e.g. Mac Low & Ferrara, 1999a). Furthermore, the evolution of supernova remnants in real star forming regions is expected to be much more complicated than that given by the simple model described above, so that the fraction of supernova energy available to drive a potential galactic wind is also uncertain.

8.7 Results from Numerical Simulations

In the preceding sections we have discussed various processes that can affect the formation of gaseous halos. Our discussion so far has been largely based on idealized models with various simplifications, and so the results obtained can only serve as approximations. As we have seen in Chapter 7, dark matter halos do not grow by smooth, spherical accretion in a homogeneous background. Rather, the formation is hierarchical, involving numerous mergers, making the formation of gaseous halos a complex, clumpy, non-spherical process. In addition, because halos are highly nonlinear systems, the effects of different processes may be coupled, implying that the reliability of model predictions depends on modeling all relevant processes accurately. This typically requires sophisticated numerical simulations. In this section we give a brief summary of the current status of numerical investigations regarding the formation and evolution of gaseous halos. A brief description of hydrodynamic simulation techniques is given in Appendix C.

8.7.1 Three-Dimensional Collapse without Radiative Cooling

The simplest case is that of a non-radiative gas without heating (other than shock heating), cooling, or star formation. Such gas is often (incorrectly) referred to as 'adiabatic'. These assumption may be valid for clusters of galaxies, because cooling is expected to be unimportant over the bulk of such massive objects (see §8.4.1) and the energy output from star formation contributes only a small fraction of the total potential energy of the intracluster gas. Numerical simulations with a small fraction ($\sim 10\%$) of gas that is initially cold and has the same initial distribution as the dominant dark matter show that by and large the gas follows the dark matter during the early (linear) stages of gravitational clustering. As soon as the structures become nonlinear, and start to merge, a complex network of shocks develop. These shocks heat the gas by thermalizing the kinetic energy of the gas as it falls into the gravitational potential wells provided by the underlying dark matter. Since the collisionless dark matter does not experience these shocks, the gas and dark matter become somewhat segregated on small scales, with the gas distribution being less centrally concentrated. In addition, the gas pressure is isotropic, while the equivalent for the dark matter, the velocity dispersion of the dark matter particles, can be significantly anisotropic. Consequently, the gas distribution ends up being somewhat rounder than that of the dark matter. In simulations, the temperature distribution in the inner parts typically has an approximately flat profile, which begins to decline beyond roughly half the virial radius, reaching $\sim 30\%$ of its central value at the virial radius (e.g. Evrard, 1990; Navarro & White, 1993; Bryan et al., 1994; Frenk et al., 1999).

As shown in §8.2, for an isothermal gas in hydrostatic equilibrium in the potential well of a dark matter halo, the gas density, $\rho_{\rm gas}$, is related to the dark matter density, $\rho_{\rm dm}$, via $\rho_{\rm gas} \propto \rho_{\rm dm}^{\beta}$, where $\beta = \mu m_{\rm p} \sigma^2 / k_{\rm B} T$ is the ratio of specific energies of the dark matter and the gas. The value of β is a measure of the degree of thermalization of the gas. If the gas and the dark matter particles have the same specific energy, then $\beta = 1$. The values of β estimated from simulations of CDM models are typically 1.1–1.2, suggesting that gravitational infall onto collapsed objects is thermalized effectively. In addition to this thermal energy, the mergers associated with the formation of the halo disturb the dynamical state of the gas, resulting in bulk motions whose kinetic energy is $\sim 15\%$ of the thermal energy of the gas (e.g. Bartelmann & Steinmetz, 1996; Frenk et al., 1999).

8.7.2 Three-Dimensional Collapse with Radiative Cooling

In semi-analytical models for galaxy formation, the rate at which gas cools is often estimated using the method described in §8.4.2. In this model, originally proposed by White & Frenk (1991), the cooling radius is defined as the radius at which the cooling time equals the age of the universe, and the cooling rate then follows from the assumed gas density profile, $\rho_{\rm gas}(r,t)$, and the rate at which the cooling radius moves outward, as given by Eq. (8.104).

Using hydrodynamical simulations of hot gas with radiative cooling in *static* dark matter halos, Viola et al. (2008) have shown that the simple cooling rates obtained from Eq. (8.104) in general underpredict the actual cooling rates, especially during the early stages. This is because Eq. (8.104) does not account for the fact that the gas distribution can readjust itself once gas starts to cool out. In particular, once gas starts to cool in the center, the outer gas starts to flow inwards. This increases the density of the gas, which subsequently decreases its cooling time.

In hydrodynamical cosmological simulations with radiative cooling, the total amount of gas that cools depends strongly on the resolution of the simulation. This is because radiative cooling in dark matter halos is dominated by two-body processes whose rates depend crucially on the gas density. In general, simulations of higher resolution predict larger mass fractions of cold gas, which is a consequence of the overcooling problem eluded to in §8.4.1. High-resolution simulations in general overpredict the stellar mass function, especially at the high-mass end (e.g Davé et al., 2001). In addition, as discussed in §11.2.6, overcooling also causes (disk) galaxies to be much too compact, a problem known as the angular momentum catastrophe. It is generally believed that both these problems reflect the need to include heating mechanisms into the simulations.

As discussed in §8.6, gas in small halos may be heated by energy feedback from supernova explosions. Since the cosmological simulations cannot resolve the physical scales of individual supernova explosions, this feedback mechanism can only be implemented by modeling its impact on the resolution scale of the simulations. Unfortunately, the details of supernova feedback are still poorly understood. In particular, it is still unclear what fraction of the supernova energy goes into generating bulk motions, what fraction goes into heating the diffuse high-pressure gas component, and what fraction is radiated away by dense clouds in the immediate surroundings of the supernova event. Various methods have been developed to implement supernova feedback into the simulations, often with significantly different outcomes (e.g. Thacker & Couchman, 2000; Springel & Hernquist, 2003; Oppenheimer & Davé, 2006).

8.8 Observational Tests

The gaseous halos discussed in the previous sections can be probed observationally by their emission and absorption properties. If the temperature of the gas exceeds $\sim 10^6$ K it can be detected via X-ray emission due to thermal bremsstrahlung. At lower temperatures, diffuse halo gas can be probed either by line emission or via its absorption of background sources.

In this section we discuss some observational aspects of gaseous halos. After a detailed discussion of the X-ray emission from clusters and groups, we briefly discuss the gaseous halos around isolated elliptical and spiral galaxies.

8.8.1 X-ray Clusters and Groups

Rich clusters of galaxies are observed to be strong X-ray sources, with X-ray luminosities $L_X = 10^{43}$–10^{45} erg s^{-1}. The X-ray spectra show that this emission originates mainly from the thermal bremsstrahlung of hot, diffuse plasma characterized by a temperature of 10^7–10^8 K and a gas density of 10^{-4}–10^{-2} cm^{-3}. The implied mass of this intracluster medium (ICM) is typically an order of magnitude larger than the total stellar mass in the member galaxies, making it the dominant baryonic component in clusters.

(a) Hydrostatics and the β-model If the ICM is in hydrostatic equilibrium, its density and temperature profiles can be used to infer the total dynamical mass. The sound speed of the ICM is $c_s \simeq \sqrt{k_B T/\mu m_p} \simeq 1{,}000$ km s^{-1}, which implies a sound crossing time of $\tau_s = 2R_A/c_s \simeq 4 \times 10^9$ yr. Here $R_A = 1.5h^{-1}$ Mpc is the Abell radius and we have adopted $h = 0.7$. Thus, unless clusters formed a long time ago, they are not necessarily expected to be in perfect hydrostatic equilibrium. Indeed, many clusters show strong asymmetries and substructure in their X-ray emission, indicating that they are still in the process of formation. Such clusters are generally discarded when using X-ray data to infer dynamical masses. In the following discussion we therefore focus only on clusters that appear smooth and symmetrical, and hence relaxed, in their X-ray emission.

For given density and temperature distributions, $\rho_{\rm gas}({\bf x})$ and $T({\bf x})$, the bremsstrahlung emissivity can be calculated at each position ${\bf x} \equiv (r, {\bf R})$ in the cluster, where r is the radial coordinate along the line-of-sight, and ${\bf R}$ labels the position in the perpendicular plane. The X-ray surface brightness at a position ${\bf R}$ on the sky can then be obtained by integrating the emissivity along the line-of-sight:

$$S_X({\bf R}) = \frac{1}{4\pi} \int d\nu w(\nu) \int dr \frac{\varepsilon_{\rm ff}[r, {\bf R}; \nu(1+z)]}{(1+z)^3}, \tag{8.215}$$

where $w(\nu)$ is a response function of the passband, z is the redshift of the source, and $\varepsilon_{\rm ff} \propto \rho_{\rm gas}^2 T^{-0.5}$ is the (free–free) bremsstrahlung emissivity (see §B1.3). The surface brightness is usually averaged in circular annuli centered on the cluster center. The resulting distribution is then fit to some theoretical profile. If the cluster is spherically symmetric, the observed surface brightness profile provides a constraint on the combination of $\rho_{\rm gas}(r)$ and $T(r)$. If X-ray spectroscopy

8.8 Observational Tests

is available, one can determine the temperature profile $T(r)$, and thus also the gas density profile $\rho_{\rm gas}(r)$. This can then be used to infer the total mass of the cluster from the hydrostatic equation (8.16).

A simple model commonly used to fit cluster X-ray data assumes that the inner cluster approximates an isothermal sphere with total mass distribution given roughly by Eq. (8.23):

$$\rho_{\rm m}(r) = \rho_{\rm m}(0)\left[1+(r/r_0)^2\right]^{-3/2}, \quad (8.216)$$

where r_0 is a core radius. As we have seen in §8.2, this profile is valid only for the inner region ($r \lesssim 2r_0$) of a true isothermal sphere. From Eq. (8.26) we see that the inner profile of an isothermal gas in hydrostatic equilibrium in such a cluster is approximately

$$\rho_{\rm gas}(r) = \rho_{\rm gas}(0)\left[1+(r/r_0)^2\right]^{-3\beta/2}, \quad (8.217)$$

where β is the ratio of the specific energies of the dark matter and the gas [see Eq. (8.27)]. The corresponding X-ray surface-brightness profile is then

$$S_{\rm X}(R) = S_0\left[1+\left(\frac{R}{r_0}\right)^2\right]^{-3\beta+1/2}, \quad (8.218)$$

where $S_0 \propto \rho_{\rm gas}^2(0)$ is the central surface brightness. Since S_0, r_0 and β can all be determined from the X-ray data (see Fig. 8.10), one can obtain the gas mass and the total mass by integrating $\rho_{\rm gas}(r)$ and $\rho_{\rm m}(r)$ over r. This kind of analysis shows that clusters have gas mass fractions of

$$f_{\rm gas} = \frac{M_{\rm gas}}{M_{\rm total}} \simeq (0.06 \pm 0.01)h^{-3/2} \quad (8.219)$$

(e.g. White et al., 1993b; Evrard, 1997; Ettori & Fabian, 1999; Allen et al., 2002). The dependence on h can be understood as follows. The total X-ray luminosity $L_{\rm X} \propto \rho_{\rm gas}^2(0)r_0^3$ while the observed values of $L_{\rm X}$ and r_0 scale with h as h^{-2} and h^{-1}, and so $M_{\rm gas} \propto \rho_{\rm gas}(0)r_0^3 \propto h^{-5/2}$. The

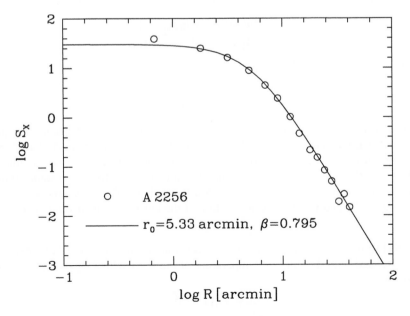

Fig. 8.10. The observed X-ray surface brightness for the cluster A 2256 and the fit by the model (8.218). [Based on data published in Henry et al. (1993)].

observed total cluster mass $M_{\rm total}$ scales as h^{-1}. The mass fraction of stars in rich clusters (estimated from the optical luminosities of the cluster galaxies) is about 1%, which is significantly smaller than that of the X-ray gas. Since the matter content of rich clusters is thought to provide a fair sample of the matter content of the Universe as a whole, the observed ratio of the baryonic to total mass in clusters should closely match the universal baryon fraction $\Omega_{\rm b,0}/\Omega_{\rm m,0}$. Adopting the cosmic nucleosynthesis value $\Omega_{\rm b,0}h^2 = 0.0205 \pm 0.0018$ (O'Meara et al., 2001, see also §3.4.4), the inferred gas mass fraction of Eq. (8.219) implies that $\Omega_{\rm m,0} \sim (0.25 \pm 0.05)h^{-1/2}$, in good agreement with other determinations of the universal matter density (see §2.10).

The value of β obtained by fitting $S_X(R)$ is typically $\beta_{\rm fit} \sim 0.7$. This would imply that the gas is hotter than the dark matter. However, direct spectroscopic measurements from the optical velocity dispersion of galaxies and the temperature of the ICM give $\beta_{\rm spec} = 0.9$–1.2. This discrepancy in the value of β obtained by the two methods is known as the β-discrepancy. In order to understand its origin, recall that the model considered above is based on the assumption that both the gas and the galaxies are in equilibrium with the cluster potential. Under the assumption of spherical symmetry, hydrostatic equilibrium implies that the enclosed mass of the cluster obeys Eq. (8.16), while virial equilibrium implies that

$$M(r) = -\frac{r\sigma_r^2}{G}\left[\frac{d\ln\rho_{\rm gal}}{d\ln r} + \frac{d\ln\sigma_r^2}{d\ln r}\right], \qquad (8.220)$$

(see §13.1.4). Here σ_r is the radial velocity dispersion of the galaxies and we have assumed that the galaxies have an isotropic velocity distribution. Combining Eqs. (8.16) and (8.220) we obtain that

$$\beta \equiv \frac{\mu m_{\rm p}\sigma_r^2}{k_{\rm B}T} = \frac{d\ln\rho_{\rm gas}/d\ln r + d\ln T/d\ln r}{d\ln\rho_{\rm gal}/d\ln r + d\ln\sigma_r^2/d\ln r}. \qquad (8.221)$$

Adopting isothermal dark matter and gas distributions, a standard assumption which is incorrect in detail is most real clusters, one has that $d\ln T/d\ln r = d\ln\sigma_r^2/d\ln r = 0$, and the line-of-sight velocity dispersion of the galaxies, $\sigma_{\rm gal} = \sigma_r$. This implies that

$$\beta_{\rm spec} \equiv \frac{\mu m_{\rm p}\sigma_{\rm gal}^2}{k_{\rm B}T} = \frac{d\ln\rho_{\rm gas}/d\ln r}{d\ln\rho_{\rm gal}/d\ln r}. \qquad (8.222)$$

Comparing Eqs. (8.216) and (8.217), we see that

$$\beta_{\rm fit} = \frac{d\ln\rho_{\rm gas}/d\ln r}{d\ln\rho_{\rm dm}/d\ln r}. \qquad (8.223)$$

Assuming isothermality, $\beta_{\rm spec} = \beta_{\rm fit}$ if only the dark matter and galaxy distributions follow the King profile (8.23). Bahcall & Lubin (1994) show that this is not the case in detail, and argue that this is the main reason for the β-discrepancy. Other possible explanations include that clusters are not perfectly isothermal and/or isotropic, that they are not in perfect hydrostatic equilibrium, that there is additional pressure support from non-thermal turbulence, magnetic fields and/or cosmic-rays (see §8.2), and that the thermalization of the cluster gas is incomplete (e.g. Evrard, 1990).

(b) Scaling Relations Observationally, it has been found that the X-ray luminosity of clusters, L_X, is correlated with the inferred temperature of the ICM. This L_X–T relation is typically parameterized as

$$L_X \propto T^\xi (1+z)^\zeta. \qquad (8.224)$$

Observations indicate that $\xi \sim 3$ and $\zeta \sim 0$ with scatter in L_X at fixed T of approximately 0.1 dex (White et al., 1997; Markevitch, 1998).

If clusters are formed from dark matter and cold intergalactic gas in a self-similar fashion (Kaiser, 1986), so that the gas and dark matter distributions are similar for different clusters, then, for a cluster with radius R,

$$L_X \propto \mathscr{C} R^3 \propto M\rho T^{1/2}, \tag{8.225}$$

where we have used that $\mathscr{C} = n_H^2 \Lambda(T) \propto \rho^2 T^{1/2}$ for thermal bremsstrahlung. Assuming that the X-ray gas is isothermal and in hydrostatic equilibrium, and using the fact that the average density of dark matter halos is independent of halo mass (see §5.4.4), we have

$$M \propto \frac{V_c^3}{GH(z)} \simeq \frac{1}{GH(z)} \left(\frac{k_B T}{\mu m_p}\right)^{3/2} \propto (1+z)^{-3/2} T^{3/2} \text{ (for EdS)}, \tag{8.226}$$

where we have used Eq. (8.45). Substitution in Eq. (8.225) yields

$$L_X \propto T^2 (1+z)^{3/2}. \tag{8.227}$$

This T dependence is shallower than observed, suggesting that non-gravitational heating may be important. For example, if the gas in a cluster had been pre-heated to a high entropy level before the cluster was assembled, the gas density in the central region of the cluster can be reduced, which has the effect of decreasing the X-ray luminosity of the cluster. Since for a given initial entropy the effect is larger in lower-mass clusters, preheating can steepen the L_X–T relation. An alternative explanation for the discrepancy is that cooling causes a deviation from self-similarity. Cooling preferentially removes the low entropy gas from the hot phase, causing a net increase in the entropy (and temperature) of the remaining hot gas. Since the cooling efficiency depends on halo mass, cooling may also modify the L_X–T relation (e.g. Nagai et al., 2007).

(c) Cooling Flows The X-ray emission from clusters indicates that the ICM continues to lose energy via thermal bremsstrahlung. As we have seen in §8.4.1, the rate at which the ICM cools is in general low, with the average cooling time longer than the Hubble time, $t_H \propto H_0^{-1}$. However, since the hot gas is roughly in hydrostatic equilibrium within the gravitational potential of the cluster, its density in general increases towards the center. As the cooling time $t_{\rm cool} \propto T^{1/2} \rho_{\rm gas}^{-1}$, which follows from Eq. (8.94) together with the fact that for bremsstrahlung $\Lambda \propto T^{1/2}$, the cooling time typically becomes shorter at smaller cluster-centric radii. The densities and temperatures of the ICM inferred from X-ray data indicate that the cooling times in the central ~ 100 kpc of most clusters are shorter than 10^{10} yr (e.g. Mushotzky, 1993; Fabian, 1994). Consequently, in the absence of a balancing heat source, a slow, subsonic inflow of gas is expected to develop in the central regions. This flow takes place because the cooling reduces the temperature and thus the pressure of the gas. In order to support the weight of the overlying gas, the central gas has to increase its density, which it does by flowing inward. This, essentially cooling-driven flow, is called a cooling flow.

The mass deposition rate, $\dot{M}$ (i.e. the rate at which gas cools in the cooling flow), can be estimated from the X-ray image of a cluster. Consider a small blob of gas (with temperature T_1 and volume V_1) which has cooled down and moved inwards by a distance Δr from its original radius, where it had a temperature $T \gg T_1$ and occupied a volume V. Since the pressure in the inner region is higher than that in the outer region, $V \gg V_1$. Now imagine that this blob is brought back to its original place and state. The energy required is the sum of the thermal energy originally in the gas, the PdV work required to inflate the blob and the gravitational energy. In the reverse process, this same amount of energy is radiated away. If the gravitational potential is flat in the central region, or if a gas blob can only cover a short distance before it cools down, only the thermal energy and PdV work are important. In this case, the radiated luminosity within a radius r should be equal to the rate of change of enthalpy within r, implying that

$$\dot{M}(<r) = \frac{2\mu m_{\rm p}}{5k_{\rm B}T}L(<r),\tag{8.228}$$

with $L(<r)$ the luminosity of the cooling gas inside radius r. The inferred mass deposition rates are typically of the order of 50–100 $M_\odot\,{\rm yr}^{-1}$, while some clusters even have $\dot{M} \gtrsim 1000\,M_\odot\,{\rm yr}^{-1}$.

If all the gas were to flow to the center, so that $\dot{M}(<r)$ is a delta function, the X-ray surface-brightness profile would also be a delta-function. This is inconsistent with the data, which indicates that although the X-ray surface brightness profiles of cooling-flow clusters are sharply peaked, they are well resolved. The inference is that, to good approximation, $\dot{M}(<r) \propto r$, and thus that, in the absence of other effects, cold gas should be deposited over relatively large volumes in the central regions of clusters, while the rest of the gas remains hot and continues to flow inwards. Such a multi-phase cooling flow could form naturally out of density inhomogeneities due to the thermal instability discussed in §8.5.1 (e.g. Nulsen, 1986), although it is not entirely clear how effective the thermal instability is in the presence of gravitationally induced convective motion (e.g. Hattori & Habe, 1990).

Key evidence that the gas in the central regions of cooling flow clusters is indeed cooling to lower temperatures comes from X-ray spectroscopy, which shows that temperature drops of up to a factor of three are common in the central $\sim 100\,{\rm kpc}$ of cooling flow clusters. However, the actual mass deposition rates have been, and still are, heavily debated. The inferred mass deposition rates would imply that more than $10^{11}\,M_\odot$ of cold gas is being deposited within less than a Gyr beyond 100 kpc but no further. Although some evidence for warm and cold molecular gas has been found in the central regions of cooling flow clusters (e.g. Jaffe & Bremer, 1997; Edge et al., 2002), it remains unclear whether this is actually associated with the cooling flow. What is clear, though, is that the majority of the gas that is cooling out cannot be forming stars, as this would make the central cluster galaxies much bluer than observed. Although many central galaxies in cooling flow clusters have excess blue light indicative of recent star formation (e.g. Crawford et al., 1999), the inferred star-formation rates are one to two orders of magnitude lower than the mass deposition rates inferred from the X-ray data. Furthermore, high-resolution spectroscopy with the XMM-Newton satellite has revealed a surprising absence of X-ray spectral signatures of gas with temperatures below 1–2 keV (e.g. David et al., 2001; Peterson et al., 2003). Hence, it seems that the gas is cooling at a high rate to about one third of the mean temperature beyond 100 kpc but then vanishes.

The mass deposition rates are based on the assumption that there is no heating in addition to the gravitational heating associated with the flow. However, this appears to be an oversimplification. In many some clusters observations have revealed the presence of X-ray cavities with diffuse radio emission (e.g. Böhringer et al., 1993; McNamara et al., 2000; Fabian et al., 2002). These cavities are inflated by jets launched from AGN associated with one or more radio galaxies located near the center of the cluster's potential well (see §14.4.2), and may well be a significant source of heating that can offset the radiative losses (e.g. Birzan et al., 2004). An additional heating source may be thermal conduction (e.g. Narayan & Medvedev, 2001). The central temperature gradients observed in the ICM may induce heat transport from the outside in via conduction. The level of conductivity may be sufficient for this process to play an important role, provided that it is not strongly suppressed by magnetic fields.

8.8.2 Gaseous Halos around Elliptical Galaxies

Many elliptical galaxies are observed to be X-ray sources, with X-ray luminosities $L_{\rm X} \sim 10^{39}$–$10^{42}\,{\rm erg\,s}^{-1}$. This X-ray emission usually has a diffuse component due to hot gas and a discrete component due to stellar X-ray binaries. The X-ray luminosities of elliptical galaxies are correlated with their optical luminosities, although with a huge amount of scatter (close to two orders of magnitude). For low luminosity ($L_B \lesssim 3 \times 10^9 L_\odot$) ellipticals, $L_X \propto L_B$, consistent with the X-ray emission being dominated by X-ray binaries. For more luminous ellipticals, however, $L_X \propto L_B^2$, indicating that the hot diffuse gas is dominating the X-ray emissivity (e.g. Eskridge et al., 1995; O'Sullivan et al., 2001). The diffuse gas is characterized by temperatures $k_B T \sim 0.5$–1 keV, lower than that from X-ray binaries ($k_B T > 3$ keV), and in general has a smooth distribution, indicating rough hydrostatic equilibrium with the gravitational potential. Consequently, X-ray measurements can be used to probe the dark matter halos of elliptical galaxies (see §13.1.5).

Originally it was believed that the hot gaseous atmospheres around ellipticals have formed due to the accumulation of normal mass loss from a dominant population of old stars, which is supported by the $L_X \propto L_B^2$ scaling (e.g. Tsai & Mathews, 1995; Brighenti & Mathews, 1996). The mass ejection rate for a typical old stellar population is

$$\dot{M}_{\rm inj} \sim 1.6 \left(\frac{M_\star}{10^{12} M_\odot}\right) \left(\frac{t}{t_0}\right)^{-1.3} M_\odot \, {\rm yr}^{-1}, \qquad (8.229)$$

with $M_\star$ the stellar mass and t_0 the age of the galaxy, which is sufficient to account for the amounts of hot gas observed. In this picture, the gaseous envelopes expelled from normally evolving stars collide with the ambient gas, shock and dissipate the orbital energy of the parent stars. This would imply that the temperature of the gas is similar to the virial temperature of the stars, i.e. $T = \mu m_p \sigma^2/k_B \sim 3 \times 10^6$ K, with σ the velocity dispersion of the stars. However, with more detailed X-ray data becoming available, it was found that the gas has significantly higher temperatures, with $\beta = \mu m_p \sigma^2/k_B T = 0.6 \pm 0.1$ (e.g. Davis & White, 1996). This cannot be explained by assuming that there is significant additional heating due to supernovae of type Ia,[3] since this would dramatically overpredict the iron abundances of the gas. Rather, it is now generally accepted that a significant fraction of the hot gas must have an external origin: either it is gas left over from the formation process, or it is gas recently accreted from the ambient cosmic flow. In this case the temperature of the gas reflects the virial temperature of the dark matter halo, which is higher than that of the stellar component.

Similar to the case for clusters discussed above, the X-ray emission from the hot gas around elliptical galaxies has been interpreted in terms of a cooling flow. The inferred mass deposition rates [Eq. (8.228)] are of the order of $\dot{M} \sim 1.5 M_\odot \, {\rm yr}^{-1}$ for a typical X-ray luminosity of $L_X = 5 \times 10^{41}$ erg s^{-1}. This is comparable to the mass ejection rate expected from stellar mass loss [Eq. (8.229)], indicating that the amount of hot gas can remain more or less constant. However, as with clusters, there is a serious problem with this simple cooling flow model. There is no observational evidence for the amounts of mass expected to have cooled at the centers of these ellipticals. In particular, the amounts are an order of magnitude larger than the masses of their central black holes (see §13.1.6). Multi-phase cooling models, in which the cold gas is assumed to deposit over a large range of radii, also do not seem to work. In particular, as for clusters, intermediate (multi-phase) temperatures, which are an essential outcome of radiative cooling and mass deposition, are not observed in the X-ray spectra of galactic scale flows (e.g. Xu et al., 2002). As mentioned above, the most popular explanation for this cooling flow puzzle is that the hot gas is being heated by conduction and/or by a central AGN. This is supported by the fact that many bright ellipticals have extended non-thermal radio emission, and by high resolution images of the Chandra satellite, which show that the central regions of the hot atmospheres are highly disturbed. Unfortunately, we still lack a detailed understanding of how exactly these heating mechanisms operate (see Mathews & Brighenti, 2003, for a review).

---

[3] As discussed in §10.2.3, the rate of these supernovae is still very uncertain, allowing for significant amounts of freedom in the models.

8.8.3 Gaseous Halos around Spiral Galaxies

Diffuse X-ray emission is also detected in some spiral galaxies. Such X-ray emission is generally soft, reflecting the low temperature (a few $\times 10^6$ K) of the gas, and has relatively low luminosity, $L_X < 10^{40}\,\mathrm{erg\,s^{-1}}$. In almost all cases this X-ray emission is concentrated close to the galactic disks, suggesting that the emission is mainly due to star-formation activity in the disk or due to an active galactic nucleus. There is currently no convincing evidence for the general existence of large X-ray halos around normal spiral galaxies. This is in sharp contrast to simple theoretical expectations. As discussed in Chapter 11, disk galaxies are believed to form out of gas that cools inside extended dark matter halos. During this formation process, the material that ends up in the disk has to get rid of its binding energy. At the virial temperature of galaxy halos, the dominant cooling mechanism is thermal bremsstrahlung, and it thus seems likely that a large fraction of this binding energy is radiated away as X-ray emission. Under the assumption that the mass of the present disk, M_{disk}, has been built up via a constant rate of accretion over the age of the Universe, $t_0 \simeq 13\,\mathrm{Gyr}$, a simple (but crude) estimate for the X-ray luminosity is given by

$$L_X \simeq \frac{\frac{1}{2} M_{\mathrm{disk}} V_{\mathrm{esc}}^2}{t_0}. \tag{8.230}$$

Here $V_{\mathrm{esc}} = V_c \sqrt{2\ln(r_{\mathrm{vir}}/r_{\mathrm{disk}}) + 2}$ is the velocity required for the material at the characteristic disk radius, r_{disk}, to escape the halo, which is assumed to have a constant circular velocity, $V_c(r) = V_c$, and a virial radius, r_{vir}. Substituting in values characteristic for the Milky Way, we obtain $L_X \sim 5 \times 10^{41}\,\mathrm{erg\,s^{-1}}$. Using more realistic models that follow the actual formation of a disk galaxy predicts present-day X-ray luminosities for massive disk galaxies that are roughly a factor two higher (White & Frenk, 1991). This is approximately an order of magnitude higher than the typical upper limit for the diffuse X-ray emission around present-day disk galaxies (Benson et al., 2000).

There are numerous possible solutions for this discrepancy. One possibility is that the gas is simply too diffuse to emit X-rays efficiently. However, this is difficult to reconcile with our picture of hierarchical formation, since it basically implies that no gas is cooling at the present time. Hence disk galaxies must have assembled their mass at high redshifts and evolved essentially as closed systems thereafter, which seems inconsistent with the fact that disk galaxies have young stellar populations. Alternatively, the disk may have formed via cold mode accretion (see §8.4.4), in which the gas radiates at least part of its binding energy at much lower temperatures. Hydrodynamical simulations by Fardal et al. (2001) indeed suggest that most of the cooling radiation is actually emitted by gas at $T < 20{,}000\,\mathrm{K}$ in the Lyα line. Finally, it is important to realize that not all binding energy has to be radiated away. Part of it may actually be transferred to the dark matter. This could happen, for instance, if part of the disk material assembles out of massive clouds that lose energy to the dark matter via dynamical friction.

9
Star Formation in Galaxies

By and large, a galaxy is observed through, and defined by, its stellar content. Hence, any theory of galaxy formation has to address the question of how stars form. As we have seen in the previous chapter, the baryonic gas in galaxy-sized halos can cool within a time that is shorter than the age of the halo. Consequently the gas is expected to lose pressure support and to flow towards the center of the halo potential well, causing its density to increase. Once its density exceeds that of the dark matter in the central part of the halo, the cooling gas becomes self-gravitating and collapses under its own gravity. As we have seen in the previous chapter, in the presence of efficient cooling, self-gravitating gas is unstable and can collapse catastrophically. Ultimately, this cooling process may lead to the formation of dense, cold gas clouds within which star formation can occur. In this chapter we take a closer look at the actual process of star formation. The main questions to be addressed are:

(i) When does large-scale star formation occur in a galaxy?
(ii) What are the main processes that drive star formation?
(iii) What is the rate at which stars form in a gas cloud?
(iv) What mass fraction of a gas cloud can be converted into stars?
(v) What is the initial mass function (IMF), which describes the mass distribution of stars at birth?

The answers to these questions play an important role in our treatment of galaxy formation and evolution. In fact, if we know the star-formation history of a galaxy, which describes the total mass in stars formed per unit time, and if we know the IMF, then we can use stellar evolution models, which describe how stars of different masses evolve with time (see Chapter 10), to predict the luminosity and color of the galaxy as a function of time.

In the Milky Way, and other nearby galaxies that can be observed with high spatial resolution, it is found that all star formation takes place in dense molecular clouds (e.g. Blitz, 1993; Williams et al., 2000). In addition, observations of CO emission from starburst galaxies show that they are associated with large amounts of molecular gas (10^8–$10^{10}\,M_\odot$) confined to small volumes with sizes $\lesssim 2$ kpc. Although the corresponding molecular gas densities are about 10–100 times higher than in the inner 1 kpc region of the Milky Way, and starbursts have very different star-formation properties from normal spirals, it is clear that their star-formation rates are also driven by the availability of dense molecular gas. Based on these observations, it is now generally accepted that the overall star-formation rate (SFR) of a galaxy is determined by its ability to form dense molecular clouds. Therefore, the topic of star formation in galaxies can be divided into two broad parts: 'microphysics' and 'macrophysics'. The former deals with the formation of individual stars in dense molecular clouds (often called 'cores'), while the latter deals with the formation and structure of molecular clouds in galaxies. The presentation in this chapter is based on these considerations. In §9.1 we start with a description of the properties of molecular clouds, while their formation is discussed in §9.2. In §9.3 we start looking into the microphysics and discuss mechanisms that are believed to control the star-formation efficiency (SFE) of individual molecular clouds. In §9.4 we give a brief description of the formation of individual stars.

A complete understanding of star formation in a cosmological framework is extremely challenging. The typical mass and density of gas in a galaxy-sized halo are $\sim 10^{11}\,M_\odot$ and $\sim 10^{-24}\,{\rm g\,cm^{-3}}$, respectively, while those for a typical star are $\sim 1\,M_\odot$ and $\sim 1\,{\rm g\,cm^{-3}}$. Thus, a complete description of the formation of individual stars spans some 11 orders of magnitude in mass and 24 orders of magnitude in density. Extremely diverse physical processes can be involved in the problem, many of which are still poorly understood at the present time. However, for our purpose of studying galaxy formation and evolution, it is sufficient to understand the properties of the stellar population in a volume that is comparable to that of a galaxy, and so progress may still be made without going into the details of each individual process. The question we want to address is how the *global* properties of star formation, averaged over a large volume of gas, depend on the *global* properties of the gas, such as mass, density, temperature and chemical composition. Since the time and spatial scales relevant for the formation of individual stars are much smaller than the scales involved in the formation and evolution of galaxies, it is possible to describe star formation in galaxies with some statistical relations between the SFR and the global properties of the star-forming gas. These star-formation 'laws', which are the topic of §9.5, can be calibrated empirically and may even give us some insight into the physical processes underlying star formation.

In §9.6 we give a detailed description of the observational constraints on and physical origin of the IMF, and we end this chapter in §9.7 with a discussion of the formation of population III stars. The reader interested in more in-depth discussions on star formation is referred to the excellent reviews by Shu et al. (1987), Mac Low & Klessen (2004), Elmegreen & Scalo (2004), and McKee & Ostriker (2007).

9.1 Giant Molecular Clouds: The Sites of Star Formation

Detailed information regarding the fine structure of the interstellar medium (ISM) is available only for our own Galaxy. However, since the spatial and time scales for star formation are typically much smaller than those of a galaxy, it is likely that the relationship between star formation and the fine structure of the ISM does not vary too much from galaxy to galaxy, so that the results obtained for the Milky Way may also help us to understand star formation in other galaxies.

The ISM of the Milky Way is contained in a highly flattened disk-like structure of predominantly cold gas (see §2.3.4). This cold gas is dominated by neutral hydrogen, but a significant fraction, $\sim 20\%$, is in molecular form, mainly in molecular hydrogen (H_2). In addition, a small fraction (about 1%) of the ISM is in dust grains. The molecular fraction is found to increase with gas density: it increases from almost zero in the diffuse ISM to almost 100% in the dense regions near the Galactic center.

Detailed observations of the molecular gas in the Milky Way show that it is highly clumpy; virtually all molecular gas is distributed over giant molecular clouds, which themselves reveal large amounts of substructure. This section gives an overview of the observed properties and dynamical state of these molecular clouds.

9.1.1 Observed Properties

The largest molecular structures considered to be single objects are giant molecular clouds (GMCs), which have masses of 10^5–$10^6\,M_\odot$ and extend over a few tens of parsecs. The corresponding densities are of the order of $n_{H_2} \simeq 100$–$500\,{\rm cm^{-3}}$ ($\rho \sim 10^{-21}\,{\rm g\,cm^{-3}}$). Emission line observations of GMCs reveal clumps and filaments on all scales accessible by present-day telescopes, ranging from molecular 'clumps' with masses in the range $\sim 10^2$–$10^4\,M_\odot$, sizes $\sim$ 1–10 pc, and densities of $n_{H_2} \sim 10^2$–$10^4\,\mathrm{cm}^{-3}$, to 'cores' with masses ~ 0.1–$10\,M_\odot$ and densities $n_{H_2} > 10^5\,\mathrm{cm}^{-3}$. In what follows, we use the term 'molecular clouds' to refer to this ensemble of GMCs, clumps and cores. Note that all this terminology is somewhat arbitrary, and not fixed; different authors may use different terms to refer to the hierarchy of substructures within GMCs.

Star formation is found to occur only in the most massive clumps (forming star clusters) and cores (forming single stars, which is why they are also often called 'protostellar cores'). This suggests that only a small fraction of the total GMC mass ends up forming stars (see Hunter, 1992, for a review). As we will see below, understanding why the overall star-formation efficiency of GMCs is so low is one of the most important questions for our understanding of star formation.

The observed mass distribution of GMCs and their subclumps is found to be a power law over a wide range in masses, with a relatively sharp cutoff:

$$\frac{dN}{d\ln M} = N_u \left(\frac{M}{M_u}\right)^{-\zeta} \quad (M \lesssim M_u), \tag{9.1}$$

where $0.3 < \zeta < 0.9$ and $M_u \sim 5 \times 10^6\,M_\odot$ (e.g. Williams & McKee, 1997; Heithausen et al., 1998; Rosolowsky, 2005). The observed mass distribution of cores is significantly steeper, with a power law index $0.9 < \zeta < 1.5$ at $M \gtrsim 1\,M_\odot$ (e.g. Testi & Sargent, 1998; Motte et al., 1998). In addition, there are indications that the core mass function starts to decline below a mass $M \lesssim 1\,M_\odot$ (Motte et al., 1998; Reid & Wilson, 2006). This core mass distribution is similar to the IMF of young stars (see §9.6), supporting the notion that dense molecular cores are directly linked to the formation of individual stars.

The temperature of GMCs, inferred from molecular line ratios, is typically about 10 K. In fact, except for regions heated by UV radiation from massive stars, GMCs and their subclumps and cores, which span about three orders of magnitude in gas densities, are remarkably homogeneous in temperature. As shown by Goldsmith & Langer (1978), this is consistent with cosmic rays being the main source of heating.

Although GMCs in general have very complex structures, they obey a number of relatively well-defined scaling relations. In the mass range $10^2\,M_\odot \lesssim M \lesssim 10^6\,M_\odot$, the masses and velocity widths of individual molecular clouds and clumps are observed to be tightly correlated with their radii:

$$M \propto R^2, \quad \Delta v \propto R^{1/2} \propto \rho^{-1/2} \tag{9.2}$$

(Larson, 1981; Myers & Fuller, 1992; Heyer & Brunt, 2004). Except for the protostellar cores, the observed linewidths (about $10\,\mathrm{km\,s}^{-1}$ on the scale of a GMC) are much larger than what is expected from simple thermal broadening ($\sim 0.2\,\mathrm{km\,s}^{-1}$ for a temperature of 10 K). This indicates that GMCs and molecular clumps have high levels of supersonic turbulence. As discussed below, this may have important ramifications for the process of star formation.

Finally, GMCs show a strong spatial correlation with young star clusters with ages $\lesssim 10^7$ yr, but little correlation with older star clusters. This indicates that the typical lifetime of a GMC is $\sim 10^7$ yr, much shorter than the typical age of a galaxy (e.g. Leisawitz et al., 1989; Ballesteros-Paredes et al., 1999; Fukui et al., 1999).

9.1.2 Dynamical State

Given their relatively short lifetimes, and given that stars form within them, it is important to investigate the dynamical state of GMCs: are they collapsing, expanding, or in some form of equilibrium? To start with, let us assume that GMCs are self-gravitating, homogeneous, isothermal spheres of gas, and ignore, for the moment, any potential turbulence and/or magnetic fields. In this highly idealized case, a GMC will collapse under its own gravity if its mass exceeds the thermal Jeans mass,

$$M_{\rm J} = \frac{\pi^{5/2}}{6} \frac{c_{\rm s}^3}{(G^3\rho)^{1/2}} \simeq 40\,{\rm M_\odot} \left(\frac{c_{\rm s}}{0.2\,{\rm km\,s^{-1}}}\right)^3 \left(\frac{n_{\rm H_2}}{100\,{\rm cm^{-3}}}\right)^{-1/2} \quad (9.3)$$

(see §4.1.3), where we have adopted a mean molecular mass of $3.4 \times 10^{-24}\,{\rm g}$, appropriate for molecular hydrogen. For an isothermal sphere in pressure equilibrium with its surrounding, which may be a more appropriate description for molecular clumps and cores, the equivalent to the Jeans mass is called the Bonnor–Ebert mass,

$$M_{\rm BE} = 1.182 \frac{c_{\rm s}^3}{(G^3\rho)^{1/2}} = 1.182 \frac{c_{\rm s}^4}{(G^3 P_{\rm th})^{1/2}}, \quad (9.4)$$

where $P_{\rm th} = \rho c_{\rm s}^2$ is the surface pressure (Bonnor, 1956; Ebert, 1957). Clouds with $M > M_{\rm BE}$ cannot be prevented from collapsing by thermal pressure alone. Note that the Bonnor–Ebert mass is very similar to the thermal Jeans mass. Whereas protostellar cores have masses comparable to the Bonnor–Ebert mass, suggesting that they are roughly supported against gravitational collapse by thermal pressure (in the absence of cooling), GMCs and molecular clumps both have masses that exceed the thermal Jeans mass (or the Bonnor–Ebert mass) by orders of magnitude. Hence, in the absence of any additional pressure forces, they should collapse (and form stars) on a free-fall time

$$\tau_{\rm ff} = \left(\frac{3\pi}{32G\rho}\right)^{1/2} \simeq 3.6 \times 10^6\,{\rm yr} \left(\frac{n_{\rm H_2}}{100\,{\rm cm^{-3}}}\right)^{-1/2}. \quad (9.5)$$

This is significantly shorter than their inferred lifetimes, which indicates that GMCs and their subclumps must be supported against gravitational collapse by some non-thermal pressure.

As mentioned above, observations suggest that GMCs and molecular clumps are characterized by substantial amounts of turbulent motion. If the turbulence is isotropic, one can replace the sound speed in Eqs. (9.3) and (9.4) by an effective sound speed, $c_{\rm s,eff}^2 = c_{\rm s}^2 + \sigma^2$, where σ is the one-dimensional mean square velocity due to turbulent motion (e.g. Chandrasekhar, 1951). Thus, a GMC with a mass of $10^6\,{\rm M_\odot}$ requires $\sigma \sim 6\,{\rm km\,s^{-1}}$ in order to be stable against gravitational collapse, in rough agreement with the values inferred from the observed molecular line widths. This suggests that GMCs, and their subclumps, may be supported by turbulence.

There is another source of non-thermal pressure that is potentially important: magnetic fields. Equating the potential energy of a cloud with the magnetic energy yields a characteristic mass

$$M_\Phi \equiv \frac{5^{3/2}}{48\pi^2} \frac{B^3}{G^{3/2}\rho^2} \simeq 1.6 \times 10^5\,{\rm M_\odot} \left(\frac{n_{\rm H_2}}{100\,{\rm cm^{-3}}}\right)^{-2} \left(\frac{B}{30\,\mu{\rm G}}\right)^3, \quad (9.6)$$

where the magnetic field, B, is assumed to be uniform across the cloud (Spitzer, 1968); for different field configurations M_Φ may be somewhat different (e.g. McKee et al., 1993). Magnetic fields alone cannot prevent gravitational collapse in clouds with $M > M_\Phi$, which are said to be magnetically supercritical. In order for GMCs to be magnetically subcritical (i.e. $M < M_\Phi$), magnetic fields of the order of 10–100 μG are required. Unfortunately, accurate measurements of magnetic field strengths in molecular clouds are extremely difficult to obtain. In the few cases where measurements have been obtained, they seem to suggest that molecular clouds are magnetically supercritical, but only marginally so (Crutcher, 1999; Bourke et al., 2001).

Thus, both turbulence and magnetic fields may play an important role in governing the structure and evolution of molecular clouds, and hence in regulating the overall star-formation efficiency. This has given rise to two competing theories of star formation, which we discuss in more detail in §9.3.

9.2 The Formation of Giant Molecular Clouds

As indicated in §9.1.1, the typical lifetime of a GMC is estimated to be of the order of only $\sim 10^7$ yr, based on the fact that GMCs show little correlation with stars older than this. At the same time, observations show that it is rare to find any GMC that is not forming stars. This suggests that star formation in a GMC starts as soon as the cloud has formed. Altogether, this indicates that, at least in a normal spiral galaxy like the Milky Way, the rate of large-scale star formation is essentially proportional to the formation rate of GMCs. Hence, understanding the formation of GMCs is an essential ingredient of our understanding of star formation.

9.2.1 The Formation of Molecular Hydrogen

We start our investigation by examining the conditions under which molecular gas can form abundantly. By far the most abundant molecule in interstellar space is molecular hydrogen (H_2), whose abundance is set by its formation and destruction processes. The main formation mechanism of H_2 in the ISM is recombination of pairs of adsorbed hydrogen atoms on the surfaces of dust grains (e.g. Gould & Salpeter, 1963; Hollenbach & Salpeter, 1971). For a mass ratio of atomic hydrogen gas to dust of ~ 100, typical of what is observed in low-density clouds in the Milky Way, the time scale for H_2 formation on dust grains is

$$t_{\rm form} = 1.5 \times 10^7 \,{\rm yr} \left(\frac{n}{100\,{\rm cm}^{-3}}\right)^{-1}, \tag{9.7}$$

(e.g. Hollenbach et al., 1971). This process is more efficient than H_2-formation via gas phase reactions (see §B1.3) by many orders of magnitude.

The main destruction process is photodissociation. Without going into details (see e.g. Kouchi et al., 1997), the photodissociation rate of H_2 is $k_{\rm pd} \simeq 5 \times 10^{-11} {\rm s}^{-1}$ (corresponding to a typical lifetime of a H_2 molecule of ~ 600 yr) in the unattenuated interstellar radiation field (Stecher & Williams, 1967). However, inside an interstellar cloud the intensity of the ambient radiation field will be diminished due to continuum attenuation by dust grains and line attenuation of H_2 itself (self-shielding): H_2 molecules at the edge of the cloud absorb all photons at certain wavelengths, so that molecules deeper in the cloud 'see' virtually no photons at all, and are not dissociated. This self-shielding is very efficient, resulting in a fairly sudden HI-to-H_2 transition (e.g. Hollenbach et al., 1971; Federman et al., 1979).

Elmegreen (1993) has shown that the mass ratio of molecular to atomic hydrogen, $R_{\rm mol} \equiv n_{H_2}/n_{\rm HI}$, is expected to depend sensitively on the local pressure and radiation field. Using detailed calculations of the formation and destruction rates of molecular hydrogen, including self-shielding and dust extinction, Elmegreen finds that

$$R_{\rm mol} \propto P_{\rm ext}^{2.2} j^{-1}, \tag{9.8}$$

with $P_{\rm ext}$ the external pressure and j the radiation intensity. If we make the simple assumption that in a disk consisting of gas and stars j is proportional to the star-formation rate per unit surface density, $\dot{\Sigma}_\star$, and we use that $\dot{\Sigma}_\star \propto \Sigma_{H_2} \simeq R_{\rm mol}\Sigma_{\rm gas}$ (see §9.5.2 below),[1] we thus expect

---

[1] Here we have assumed that $R_{\rm mol} \lesssim 0.1$, so that $f_{\rm mol} \equiv \rho(H_2)/\rho_{\rm gas} \simeq R_{\rm mol}$.

that $R_{\mathrm{mol}} \propto P_{\mathrm{ext}}^{1.1} \Sigma_{\mathrm{gas}}^{-0.5}$. A numerical solution to the equation of hydrostatic equilibrium for a disk consisting of gas and stars suggests that the total mid-plane pressure is given, to within 10%, by

$$P_{\mathrm{ext}} \approx \frac{\pi}{2} G \Sigma_{\mathrm{gas}} \left[\Sigma_{\mathrm{gas}} + \frac{\sigma_{\mathrm{gas}}}{\sigma_{\star,z}} \Sigma_\star \right] \quad (9.9)$$

(Elmegreen, 1989). Here Σ_{gas} and $\Sigma_\star$ are the surface densities of gas and stars, respectively, σ_{gas} is the velocity dispersion of the gas, and $\sigma_{\star,z}$ is the velocity dispersion of stars in the vertical direction. Thus, in the outer regions of disk galaxies, where $\Sigma_\star \ll \Sigma_{\mathrm{gas}}$, we expect that $R_{\mathrm{mol}} \propto P_{\mathrm{ext}}^{0.85}$. This is in excellent agreement with observations, which show that the mass ratio of molecular to atomic hydrogen in disk galaxies is tightly correlated with the ambient pressure [estimated using Eq. (9.9)], following a power law behavior $R_{\mathrm{mol}} \propto P_{\mathrm{ext}}^\alpha$, with α in the range 0.8–0.92 (Wong & Blitz, 2002; Blitz & Rosolowsky, 2006; Leroy et al., 2008). Further support comes from the simulations of Robertson & Kravtsov (2008), which suggest $\alpha \sim 0.9$.

9.2.2 Cloud Formation

In this subsection we describe a number of processes that have been linked to GMC formation. These can be roughly divided in two different 'classes'. In one scenario it is the actual cooling, and the corresponding molecule formation, that triggers gravitational collapse, and hence the formation of GMCs. In the other scenario, cooling and molecule formation is a consequence of compression or collapse in turbulent or gravitationally unstable gas. Both seem plausible, and currently it is still unclear which of the following processes represents the main channel for GMC formation. Most likely, they all play some role, and it may even be that different mechanisms give rise to different modes of star formation.

(a) Thermal Instability The thermal instability, which has been discussed in §8.5.1, is one of the basic elements in some of the most influential models of the ISM (Field et al., 1969; McKee & Ostriker, 1977). Its application to the ISM naturally leads to the standard picture of a three-phase medium, in which a hot phase with temperature $T \sim 10^6$ K, a warm phase with $T \sim 10^4$ K, and a cold phase with $T \sim 10^2$ K, coexist in rough pressure equilibrium (McKee & Ostriker, 1977; see also §8.5.1). Because of the large range in temperatures, the gas density in such a medium can cover a large range, consistent with observations of the ISM. In order to maintain pressure equilibrium, the transition to the cold phase coincides with a large increase in density, and hence a large increase in the molecular fraction. This raises the possibility that GMCs are simply the cold gas parcels in the ISM, with their formation being a direct consequence of thermal instability. However, this does not seem to be the case: although the thermal instability may result in the formation of molecular gas, it does not by itself result in the formation of GMCs, which seem to have an internal pressure that is roughly an order of magnitude higher than that of the warm ISM. An additional mechanism seems to be required to compress the molecular gas to GMC densities.

(b) Disk Gravitational Instability The fact that GMCs seem to be out of pressure equilibrium with their surroundings suggests that they are gravitationally bound objects (or, as we will see below, that they are transient structures). Consequently, a logical formation mechanism for GMCs is gravitational instability. We have already encountered the Jeans criterion, according to which a perturbations will grow if its self-gravity overpowers the internal pressure. However, in a disk galaxy, pressure is not the only restoring force. Unless the galaxy's circular velocity scales as $V_c \propto R^{-1}$, conservation of angular momentum forces a perturbation to rotate around its own center. The resulting Coriolis force provides support against collapse. In §11.5.2 we use perturbation theory to show that perturbations in a rotationally supported disk are unstable against gravitational collapse if the Toomre parameter

$$Q \equiv \frac{c_s \kappa}{\pi G \Sigma} < 1, \qquad (9.10)$$

where

$$\kappa = \sqrt{2} \left(\frac{V_c^2}{R^2} + \frac{V_c}{R} \frac{dV_c}{dR} \right)^{1/2} \qquad (9.11)$$

is the epicycle frequency (see §11.1.5) and Σ is the surface density of the disk. This criterion is known as the Toomre stability criterion (Toomre, 1964). If locally Q dips slightly below unity, only perturbations with a critical wavelength,

$$\lambda_{\text{crit}} = \frac{2\pi^2 G \Sigma}{\kappa^2}, \qquad (9.12)$$

become unstable against gravitational collapse. In a typical disk galaxy, λ_{crit} is of the order of 1 kpc, which corresponds to a mass

$$M_{\text{crit}} = \pi \left(\frac{\lambda_{\text{crit}}}{2} \right)^2 \Sigma \simeq 2.4 \times 10^7 \, \text{M}_\odot \left(\frac{\Sigma}{30 \, \text{M}_\odot \, \text{pc}^{-2}} \right), \qquad (9.13)$$

much larger than that of a GMC. For perturbations on the scale of GMCs and smaller (i.e. $M < 10^6 \, \text{M}_\odot$) to be unstable requires values for Q as low as 0.1. In the warm ISM ($T \sim 10^4 \, \text{K}$, $c_s \sim 10 \, \text{km s}^{-1}$) such low values of Q are rarely encountered. However, gas in the cold phase has a sound speed $c_s \sim 0.2 \, \text{km s}^{-1}$; hence, wherever the ISM conditions favor the formation of a cold phase, the ISM is likely to become gravitationally unstable on a wide range of scales. This suggests that the thermal instability discussed above may *trigger* gravitational instability, and hence the formation of GMCs (Schaye, 2004).

Alternatively, GMCs can also form if some mechanism can cause perturbations on the scale of M_{crit} to fragment.

(c) Turbulence In the picture discussed above, GMCs are assumed to be self-gravitating objects in quasi-equilibrium that form due to the combination of thermal and gravitational instability. This ignores, however, the presence of magnetic fields and turbulence, which, as discussed in §9.1.2, are believed to play an important role in governing the dynamical state of a GMC.

In fact, turbulence is present not only on the scale of GMCs. As discussed in more detail in §9.3.2, there are numerous processes that appear to be driving turbulence on larger scales as well. Indeed, many of the processes of GMC formation discussed in this section may themselve induce strong motions in the ISM, and hence be a source of turbulence. Consequently, it has been suggested that GMCs may actually be transient objects forming (and dissolving) at converging points in the larger-scale turbulent flow (Heitsch et al., 2005; Vázquez-Semadeni et al., 2006; Ballesteros-Paredes et al., 2007). In this scenario, the post-shock gas in the stagnation region represents the nascent GMC, which never reaches a state of virial equilibrium, and will be dispersed again on a time scale over which the turbulent flows change direction. Numerical simulations show that shock compression in a self-gravitating, turbulent gas can produce gas densities sufficiently high to form H_2 on a time scale of about 10^6 years (Koyama & Inutsuka, 2000; Glover & Mac Low, 2007). In particular, Hennebelle & Pérault (1999) have shown that shock compressions associated with converging turbulent flows can trigger thermal instability, resulting in further compression of the gas and a rapid formation of molecular hydrogen.

(d) Parker Instability Another process that can trigger the growth and formation of massive clouds from the diffuse ISM is the Parker instability (Parker, 1966). Briefly, this instability, also known as the magnetic buoyancy instability, works as follows. Consider a uniform disk of gas which is coupled to a magnetic field that is parallel to the disk. Suppose that the disk is gravitationally stratified in the vertical direction, and is in dynamical equilibrium under the balance of gravity and pressure (thermal and magnetic). Now consider a small perturbation which causes the field lines to rise in certain parts of the disk and to sink in others. Because of gravity, the gas loaded onto the field lines tends to slide off the peaks and sink into the valleys. The increase of mass load in the valleys makes them sink further, while the magnetic pressure causes the peaks to rise as their mass load decreases. Consequently, the initial perturbation is amplified, causing the production of density fluctuations in an initially uniform disk. The characteristic scale for the Parker instability is $\sim 4\pi H$, where H is the scale height of the diffuse component of the disk (Shu, 1974). For the Milky Way, where $H \sim 150$ pc, this scale is about 1–2 kpc. Numerical simulations show that the density contrast generated by the Parker instability is generally of order unity before the instability saturates (e.g. Kim et al., 2002). This implies that the Parker instability on its own may not be able to drive collapse on large scales. Nevertheless, it may trigger gravitational instability in a marginally unstable disk and/or induce strong motions in the medium, thereby acting as a source of turbulence on large scales.

(e) Spiral Arms Observations show that most of the molecular gas in spiral galaxies is concentrated in the spiral arms, suggesting that these structures promote the formation of GMCs. As we shall see in §11.6, grand-design spiral arms are believed to be spiral density waves. Without depending on the details of the formation mechanism, interstellar gas swept by such a spiral density wave is compressed, which promotes both thermal and gravitational instability, and hence GMC formation. In addition, spiral shocks are potentially a major source of turbulence in the interstellar medium (e.g. Kim et al., 2006), which may further promote the formation of GMCs, both within the spiral arms themselves as well as downstream from the arms in trailing gaseous spurs (e.g. Kim & Ostriker, 2002).

Many disk galaxies reveal small, flocculent arm fragments that have a limited radial and angular extent, unlike the grand-design spirals. These arm fragments also seem to reveal an enhancement in the local SFR, but its origin is believed to be different from that in the case of the grand-design spirals. Arm fragments are simply *local* density enhancements (probably created due to gravitational instability of the disk) that have been sheared into spiral structures due to the differential rotation of the disk. Thus, whereas the star formation in a grand-design spiral is *triggered* by the spiral density wave, in the case of arm fragments the star formation is triggered by a local density enhancement and subsequently sheared into a spiral pattern (see §11.6 for more details).

(f) Galaxy Interactions and Mergers Gas in galaxies can also be disturbed by interactions with nearby galaxies or by merging with another galaxy. For instance, gas can be compressed by shocks as two galaxies merge, and be energized as the energy in the merger is converted into turbulent motion. However, as shown by numerical simulations, the most dramatic effect of galaxy interactions is to cause the gas in a galaxy to rapidly lose angular momentum and to flow towards the center of the galaxy. In simulations of mergers of gaseous disk galaxies of comparable masses, an exceedingly high gas surface density, $\sim 10^4 \, M_\odot \, \mathrm{pc}^{-2}$, can be produced in the central region with a size of only about ~ 100 pc, making the gas capable of forming large quantities of molecular gas to support a high SFR. Indeed, starburst galaxies with star-formation rates as high as 10–$100 \, M_\odot \, \mathrm{yr}^{-1}$ are commonly observed to be associated with gas concentrations in the central regions of interacting or merging systems (see §12.4.3 for a more detailed discussion).

9.3 What Controls the Star-Formation Efficiency

Galaxies can be seen as gravitational potential wells containing gas that has been able to radiatively cool within a Hubble time. In the absence of any hindrance, this gas should collapse gravitationally to form stars on a free-fall time. Clearly this is not what has happened: disk galaxies are many free-fall times old, yet contain a large reservoir of gas and are still forming stars at the present day. The time scale for star formation, also called the gas consumption time, is defined as the ratio between the total gas mass available for star formation and the rate at which gas turns into stars, $\tau_{SF} \equiv M_{gas}/\dot{M}_{gas}$. For spiral galaxies of different morphological types $\tau_{SF} \simeq (1-5) \times 10^9$ yr (e.g. Kennicutt, 1998a), an order of magnitude larger than any other relevant dynamical time scale of the gas. Note, though, that in starburst galaxies the star-formation time scales are in the range $\sim 10^7$–10^8 yr, comparable to the dynamical times of the cold gas in these galaxies, and much shorter than those in normal disk galaxies.

The most important question for our understanding of star formation is what causes the overall star-formation efficiency (SFE) in normal disk galaxies to be so low. Part of the answer is that angular momentum prevents the gas from collapsing all the way to the center of the potential well and acquire very high densities. In fact, this probably underlies, at least in part, the difference between disk galaxies and starbursts; in the latter some process – for example tides associated with a merger or close encounter – has removed the angular momentum of the gas, allowing it to accumulate at the center and to form stars with high efficiency. However, angular momentum is not the entire story. The ISM densities in present-day disk galaxies still imply local free-fall times of $\sim 10^8$ yr, much shorter than the gas consumption time scale.

Since star formation in the Milky Way is observed to exclusively occur in GMCs, the overall SFE is a product of the efficiency with which individual GMCs form stars and the efficiency with which a galaxy can form GMCs. The latter efficiency has already been addressed in the previous section. Here we focus on what determines the SFE within individual GMCs. A useful way to parameterize the efficiency with which a GMC is forming stars, is to compare its gas consumption time scale, $\tau_{SF,GMC}$, to its free-fall time, $\tau_{ff,GMC}$. Observations show that $\varepsilon_{SF,GMC} \equiv \tau_{ff,GMC}/\tau_{SF,GMC} \sim 0.002$ (see §9.5.2 below). In what follows we discuss three processes that have been invoked to explain why the SFE of GMCs is so low.

9.3.1 Magnetic Fields

As discussed in §9.1.2, magnetic fields may be an important source of pressure support against the gravitational collapse of molecular clouds. Therefore, one possibility is that the SFE of GMCs is regulated by magnetic fields.

As long as the ionization level is sufficiently high for the field to be frozen to the matter, the magnetic flux, $\Phi = \pi R^2 B$, through a cloud of size R, as well as the critical magnetic mass, M_Φ, remains constant. This implies that if a cloud is subcritical to start with, it remains subcritical even if the cloud is compressed. In other words, in order for a subcritical cloud to be able to collapse and form stars, it needs to be able to reduce its magnetic flux. As first pointed out by Mestel & Spitzer (1956), the magnetic support can diminish through a process called ambipolar diffusion. For a cloud containing both ionized and neutral particles, the neutral population is only indirectly coupled to the B field through collisions with the ionized population. Once the ionized fraction is sufficiently low, the collisions between neutral particles and ions become insufficient to couple the neutral gas to the magnetic field, and the neutral particles can diffuse through the magnetic field, reducing the magnetic flux in the gas.

The ambipolar diffusion time scale, τ_{ad}, can be derived by considering the relative drift velocity of neutrals and ions (Spitzer, 1968). As shown by Elmegreen (1979), on the scale of a GMC,

$$\tau_{\rm ad} \simeq 1.1 \times 10^8 \, {\rm yr} \left(\frac{n}{100 \, {\rm cm}^{-3}}\right)^{1/2} \left(\frac{R}{10 \, {\rm pc}}\right)^{3/2} \left(\frac{B}{30 \mu {\rm G}}\right)^{-1}, \quad (9.14)$$

where we have used an ionization fraction of

$$x = 2.75 \times 10^{-8} \left(\frac{n}{10^5 \, {\rm cm}^{-3}}\right)^{-1/2}, \quad (9.15)$$

which is valid for reasonable ionization rates of neutral molecules due to cosmic rays. Thus, if GMCs are magnetically subcritical, their star-formation time scale will be comparable to the ambipolar diffusion time scale, which implies a star-formation efficiency per free-fall time of

$$\varepsilon_{\rm SF, GMC} \equiv \tau_{\rm ff, GMC}/\tau_{\rm ad} \sim 0.004, \quad (9.16)$$

similar to the observed value.

This picture of star formation being regulated by magnetic support was generally considered as the standard theory of (low-mass) star formation during the 1980s (see Shu et al., 1987, for a detailed review). However, in the 1990s it became clear that it suffers from a number of severe observational and theoretical shortcomings. Most importantly, observations of magnetic field strengths suggest that virtually all clouds are consistent with being magnetically supercritical, or at most marginally critical (Crutcher, 1999; Bourke et al., 2001), and therefore unable to regulate star formation. In addition, star-formation time scales of the order of $\tau_{\rm ad} \sim 10^8$ yr are inconsistent with the inferred lifetime of GMCs of $\sim 10^7$ yr; in particular, the observed age spread in star clusters is comparable to the dynamical time, more than an order of magnitude lower than $\tau_{\rm ad}$ (e.g. Hodapp & Deane, 1993; Hillenbrand & Hartmann, 1998). Several additional problems for this picture of star formation are discussed in Mac Low & Klessen (2004).

9.3.2 Supersonic Turbulence

Turbulence is ubiquitous in the ISM over a large range of scales, from that comparable to the host galaxy all the way down to that of the massive clumps in molecular clouds. Based on these observations, and on the results from numerical simulations and theoretical modeling, a new paradigm has emerged, in which star formation is regulated by supersonic turbulence, rather than by magnetic fields (e.g. Vázquez-Semadeni et al., 2003; Krumholz & McKee, 2005; Li et al., 2006b; Vázquez-Semadeni et al., 2007; Padoan et al., 2007; Glover & Mac Low, 2007).

Although a turbulent medium in general has a very complicated and irregular structure, a key property of turbulence is that it possesses orders in the correlation of fluid variables, which are insensitive to the details of the driving mechanism. For instance, between the driving and dissipation scales, the power spectrum of the velocity field can typically be approximated by a power law, $P_v(k) \propto k^{-n}$, so that the velocity dispersion on a scale l is $\sigma_v(l) \propto l^q$, with $q = (n-3)/2$. In the limit of negligible compression, i.e. the turbulence has Mach number $\mathcal{M} \ll 1$, the classical Kolmogorov theory gives $n = 11/3$, corresponding to $q = 1/3$. In the opposite limit, i.e. for highly compressible (supersonic) turbulence, which is more relevant to interstellar media, one has $n = 4$ and $q = 1/2$. This scaling is very similar to the velocity structure observed for molecular clouds [see Eq. (9.2)].

In a self-gravitating turbulent medium, the collapse of gas clouds can be affected by turbulence in two different ways. First, because turbulent motion increases the effective velocity dispersion of the gas, turbulence can delay or suppress gravitational collapse. Second, since gas can be swept and compressed by shocks produced by the supersonic flow in a turbulent medium, turbulence can promote gravitational collapse by increasing the gas density. For an isothermal gas with density ρ, the post-shock density is $\rho' = \mathcal{M}^2 \rho$ (assuming strong shocks), where $\mathcal{M} \gg 1$ is the Mach number of the shock (see §8.3.1). In this case, the effective Jeans mass, M_J, defined in Eq. (9.3), decreases by a factor of $\mathcal{M}$. The combined effect of turbulent motion and shock compression is therefore to change the effective Jeans mass to $M_J \propto (\sigma_{th}^2 + \sigma_{nt}^2)^{3/2}/(\mathcal{M}\rho^{1/2})$, where σ_{th} and σ_{nt} are the thermal velocity dispersion and the nonthermal (turbulent) velocity dispersion, respectively. On scales comparable to the driving scale, the typical shock velocity is of the same order as σ_{nt} and is much larger than σ_{th}, and so $M_J \propto \sigma_{nt}^2$. In this case, the net effect of turbulence is to increase M_J, i.e. to suppress gravitational collapse. On the other hand, since σ_{nt} is expected to decrease with scale [see Eq. (9.2)], small, dense clouds may have σ_{nt} much smaller than the shock velocity. In this case, the net effect of turbulence is to promote gravitational collapse. Thus, turbulence may suppress gravitational collapse globally, but can promote gravitational collapse locally.

Numerical models of highly compressible, self-gravitating turbulence show that clumps can form over large ranges in both mass and density. For an isothermal gas, appropriate for a GMC, the volume-weighted probability distribution function (i.e. the fraction of volume as a function of gas density) is roughly log-normal,

$$p(\ln x) d\ln x = \frac{1}{\sqrt{2\pi \sigma_{\ln x}^2}} \exp\left[-\frac{(\ln x - \langle \ln x \rangle)^2}{2\sigma_{\ln x}^2}\right] d\ln x, \tag{9.17}$$

where $x \equiv n/n_0$ (with n_0 the average density), and $\langle \ln x \rangle = -\sigma_{\ln x}^2/2$ (e.g. Li et al., 2004). Most of these structures will be redispersed by turbulence. However, the densest regions may become self-gravitating cores and collapse further to form stars. In this picture, at any given time, efficient star formation is expected to occur only in the small fraction of the gas clouds that get compressed sufficiently to collapse, while the overall SFE remains low because of the support by turbulence. As we will see in §9.6.2, the collapse of the densest regions given by the above distribution function also leads to a core-mass function that has the same characteristics as the observed IMF of stars. All in all, the scenario of star formation based on supersonic turbulence is quite successful in explaining many of the observed properties of the ISM (see reviews by Mac Low & Klessen, 2004; Elmegreen & Scalo, 2004; McKee & Ostriker, 2007).

The important question for this scenario is what drives the turbulence. An obvious candidate is the actual formation mechanism of galaxies. According to our current understanding of galaxy formation, the cold gas in a star-forming galaxy is assembled hierarchically through the cooling and accretion of gas in merging dark matter halos. This process is expected to generate high levels of turbulence in the gas. Indeed, cosmological numerical simulations of galaxy disk formation show that gravitational instabilities combined with shear flows and tidal interactions can generate high levels of turbulence in the ISM on large scales (e.g. Kravtsov, 2003). The large-scale turbulence may then cascade down in scales, producing a supersonic turbulent medium similar to the observed ISM. However, using numerical simulations, it has become clear that supersonic turbulence in molecular clouds typically decays in less than a free-fall time, regardless of whether they are magnetized or not (Stone et al., 1998; Mac Low et al., 1998; Padoan & Nordlund, 1999). Hence, some driving mechanism must operate to maintain the observed levels of turbulence. Broadly speaking, one can distinguish two different modes of energy injection: external and internal. In the external mode, the GMC taps energy from the turbulence in the diffuse ISM, which is believed to be driven by gravitational, thermal and magnetodynamical instabilities, spiral density waves, as well as supernova explosions associated with massive young star clusters. As discussed in §9.2.2, this large-scale turbulence may play an important role in forming GMCs. In addition, it may also supply GMCs with high levels of turbulence at formation. However, it may be difficult for these processes to continue to pump energy into GMCs during their lifetime: the density contrast between molecular clouds and the ambient medium may simply cause energy to be reflected from the clouds, rather than being transmitted into them. Internal energy injection, from protostellar outflows, stellar winds and ionizing radiation from newly formed stars therefore seems a more likely candidate for maintaining high levels of turbulence within GMCs, although a detailed understanding is still lacking (see the reviews by Mac Low & Klessen, 2004; McKee & Ostriker, 2007).

9.3.3 Self-Regulation

A third mechanism that may be important for regulating the overall SFE in GMCs is self-regulation. Here it is the actual process of star formation (and evolution) itself which destroys the molecular clouds, and hence regulates the SFE. In particular, feedback from protostellar winds is believed to play an important role in regulating the overall SFE of protostellar cores (e.g. Matzner & McKee, 2000). In addition, stellar feedback may even cause the disruption of an entire GMC. As mentioned earlier, the typical lifetime of a GMC is $\sim 10^7$ yr, much shorter than the typical age of a galaxy. It is believed that GMCs are destroyed by the energy feedback from massive (OB) stars that form in them. Possible mechanisms include the formation and expansion of HII regions (due to the ionization and heating by UV photons from massive stars), stellar winds and supernova explosions. The energy outputs of these modes are comparable over the lifetime of a young star cluster, but detailed evaluations of the overall efficiency suggest that the dominant destruction mechanism is probably photoevaporation by HII regions (e.g. Williams & McKee, 1997; Krumholz et al., 2006). Note, though, that if GMCs are transient objects formed by converging flows in a turbulent medium, they may also be dispersed by the same turbulence, without having to invoke stellar feedback (e.g. Ballesteros-Paredes et al., 1999; Hartmann et al., 2001).

In addition to controlling or terminating the star formation in individual GMCs and protostellar cores, self-regulation may also operate on larger scales. In particular, star formation may heat the ISM, increasing the velocity dispersion of the gas, which in turn may make the disk stable against gravitational collapse. This type of self-regulation has been invoked by Silk (1997) in order to explain why disk galaxies seem to have a Toomre parameter $Q \sim 1$ over the extent of the disk in which star formation is prevalent. The argument goes as follows. If $Q \ll 1$ (i.e. the disk is very unstable), star formation in the disk would proceed rapidly, which enhances the heating of the gas by the energy feedback from stars, thereby increasing the value of Q; if $Q \gg 1$, on the other hand, the SFR would be insufficient to prevent the gas from cooling, and the value of Q must decrease.

Finally, stellar feedback may not only be important for terminating or preventing star formation; it may also *promote* it under certain circumstances. For instance, star formation might spread through an individual molecular cloud or even through an entire galaxy as a wave of gravitational collapse propagating through the ISM, produced by the compression of gas associated with supernova shocks, stellar winds or ionization fronts. Such scenarios are usually called 'induced' star formation, because star formation in one place is induced by the feedback of star-formation activity in other places. In this case, the environments for star formation, i.e. the density and velocity structure of the ISM, are actually produced by stars themselves. Even if the large-scale star formation environments are produced by the other processes described above, compression by energy feedback from nearby stars may play an important role in triggering star formation in pre-existing clouds (e.g. Elmegreen, 2002). Currently, it is still unclear how important this 'induced' process is in comparison with the 'spontaneous' process, in which star formation is triggered by a process other than star formation itself.

9.4 The Formation of Individual Stars

Individual stars are believed to form in the high-density cores of GMCs. Based on our current understanding, the process that leads to the formation of stars in the dense cores contains the following basic elements:

(i) The collapse of dense cores, which leads to the development of highly stratified density profiles within which protostars form.
(ii) The growth of protostars by the accretion of infalling material from the original cloud core. The material with high specific angular momentum is expected to be accreted first into a disk and then transported inwards by viscous processes.
(iii) The development of protostellar feedback (such as outflow and radiation) that tends to disperse the gas around a protostar, reducing or even terminating further accretion.
(iv) The contraction of protostars to form pre-main-sequence stars.

The details about this formation process are still poorly understood. Here we give a very brief account of the basic ideas that have been proposed.

The formation of individual stars is usually divided into two parts according to whether the formation time scale is longer or shorter than the Kelvin–Helmholtz time, $t_{KH} = GM_\star^2/(RL)$, where $M_\star$, R, and L are the mass, radius and luminosity of the star, respectively. This is the time needed to radiate away the gravitational potential energy. The dividing line is roughly at a mass of $\sim 8\,M_\odot$. Protostars with masses below this value form in a time shorter than t_{KH} and have luminosities dominated by accretion; protostars with masses above this value form on a time scale longer than t_{KH} and their luminosities are dominated by nuclear burning while accreting gas. Consequently, powerful feedback effects due to the radiation pressure and photoionization generated by the nuclear reactions can play an important role in the evolution of massive protostars, but not so much in the evolution of low-mass protostars. Furthermore, while low-mass stars are expected to form in relatively isolated environments, high-mass protostars tend to form in clusters so that interactions with other stars and protostars may play an important role.

9.4.1 The Formation of Low-Mass Stars

(a) The Collapse of Molecular Cores Early models of the collapse of low-mass molecular cores are based on the assumption that molecular cores are isolated, spherical objects that evolve under the interplay between self-gravity and pressure gradients. Based on observations (see §9.1.1), such a core has a typical density $\rho \sim 10^{-19}\,\mathrm{g\,cm^{-3}}$, a typical effective temperature $T \approx 10\,\mathrm{K}$, and is composed of H_2, neutral He, heavy elements and dust grains. Magnetic fields and rotation are usually neglected in these models, but the effects of small-scale turbulence can be included as an effective pressure. In the beginning of the collapse, the density is still low so that the gas is optically thin to the infrared photons emitted by the dust grains. In this case, the cooling by the dust grains can effectively dissipate the gravitational energy associated with the collapse, so that the temperature of the gas remains at about $10\,\mathrm{K}$. Thus, the collapse of a molecular core in the early stage is roughly isothermal.

The collapse of a self-gravitating isothermal sphere is well studied (see §8.3). Without depending on the details of the initial density profile, the collapse of an isothermal sphere in general leads to a final profile with $\rho(r) \propto r^{-2}$. Thus, if we start with an uniform sphere, the density in the inner region increases with time until the $1/r^2$ profile is established at smaller and smaller radius (see Fig. 9.1). The time scale for the collapse is roughly the free-fall time corresponding to the initial density, which is about $4 \times 10^5\,\mathrm{yr}$ for a core with $\rho = 10^{-19}\,\mathrm{g\,cm^{-3}}$.

Once the density in the central region reaches a level of $\sim 10^{-13}\,\mathrm{g\,cm^{-3}}$, the gas there is no longer transparent to the infrared radiation from the dust grains. The photons produced are

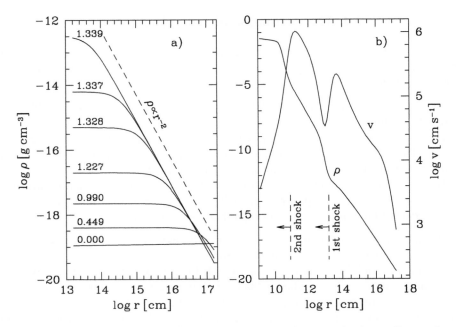

Fig. 9.1. The collapse of a protostellar cloud. The left-hand panel shows the density profile at various times during the collapse of an isothermal sphere that has an uniform density distribution at the initial time $t = 0$. Each curve is labeled with the evolutionary time in units of 10^{13} s. The right-hand panel shows the density and velocity distributions just after the formation of an initial stellar core. The positions of the first and second shock fronts are marked by the two vertical lines. Within these two radii one finds the first and second (stellar) cores. [After Larson (1969)]

then trapped by the gas, and the collapse becomes adiabatic. In this case, the optically thick gas is heated effectively by the gravitational energy associated with the contraction, and the rapid increase of the pressure in the gas slows down the gravitational contraction near the center, producing an adiabatic core and an accretion shock exterior to it. This shock front is referred to as the 'first shock front' and is shown schematically in Fig. 9.1 as '1st shock'. The corresponding adiabatic core is called the 'first core'. As the adiabatic core gains mass and is compressed by gravity, its temperature increases. Once the temperature in the inner region of the core reaches $\sim 2,000$ K, dissociation of H_2 begins. Since such dissociation consumes heat, which reduces the thermal pressure, a second phase of collapse ensues in the core center, which erodes the 'first core' and causes the 'first shock front' to disappear. This second phase of collapse continues until the dissociation of H_2 is completed at $\rho \sim 10^{-3}\,{\rm g\,cm^{-3}}$ and $T \sim 2 \times 10^4$ K. Thereafter, a 'second hydrostatic core' (called stellar core) develops near the center of the first core because of the resumed adiabatic collapse and the associated accretion shock. The place where the 'second core' develops is marked in Fig. 9.1 as the '2nd shock'.

Since much material is still in the isothermal phase at the early stage of the formation of the second (stellar) core, the accretion of material will continue. Detailed modeling shows that the protostellar mass accretion rates are $\dot{M} \sim 30 c_s^3/G$ (Hunter, 1977b). Since the temperature of the gas is set by H_2 dissociation and changes little, the mass accretion rate is roughly constant. The accretion phase will eventually be terminated by stellar winds once nuclear reactions set off in the stellar core.

(b) The Effects of Rotation and Magnetic Fields The model of protostellar collapse described above is obviously an idealization of what happens in reality. Since protostellar clouds in general have angular momentum, rotation may play an important role in protostellar collapse (e.g.

Bodenheimer, 1995). Detailed calculations show that the collapse of a rotating protostar generally leads to the formation of a central object surrounded by a rotationally supported gaseous disk. Therefore, high angular-momentum material is expected to be accreted first into the disk and then transported inwards as the disk material loses angular momentum due to gravitational and magnetic stresses, and the viscosity of the gas. Because of the rotational support, the accretion rate into the central object is in general lower than that given by the non-rotating model. If the disk is sufficiently massive, it may become gravitationally unstable and fragment to form a planetary system or a binary stellar system.

In the presence of magnetic fields, the magnetic pressure and tension act to dilute the gravitational force, thereby inhibiting the contraction of the core. Since the contraction is less inhibited along the magnetic field lines, the collapse in general leads to the formation of a disk-like structure (which is not supported by rotation). The accretion into the central part of the core is then determined by the interplay between gravity, gas pressure, magnetic force, and ambipolar diffusion. If the magnetic field is sufficiently strong (i.e. for magnetically subcritical clouds), the evolution may reach a quasi-static state in which the accretion is regulated by ambipolar diffusion at a roughly constant rate. On the other hand, for supercritical cores the evolution is expected to be dynamical.

Magnetic fields can effectively remove angular momentum from the infalling gas as long as the neutral and ionized components of the gas are well coupled and the collapse time scale is longer than the magneto-dynamical time scale. The interaction between the magnetic fields and the rotation of the gas can also generate winds and outflows from a protostar. The magnetic fields that are anchored on the disk and extend to large distance may be twisted by the differential rotation of the gas, forming a toroidal magnetic field with a significant component in the azimuthal direction of the disk and with a large magnetic pressure gradient along the rotation axis. This magnetic pressure gradient together with the compression by gravity, rotation and magnetic tension, can drive winds from the disk. In the inner part of the wind, where the magnetic pinch (Lorentz force due to the toroidal field) is important, a jet-like structure may be produced (e.g. Pudritz et al., 2007). Once formed, the strong protostellar outflows may clear up the surrounding gas, eventually terminating the gas accretion onto the protostar (Matzner & McKee, 2000).

(c) The Birth of a Young Star With the above discussion, we are now in a position to discuss how a young, low-mass star forms in the dense core of a molecular cloud. The structure of such a protostar in the accretion phase is depicted in Fig. 9.2. The innermost part is the stellar core, which continues to grow by accretion. As the accreted material is shocked near the surface of the core, almost all of its kinetic energy is dissipated into heat, producing a luminosity

$$L = \frac{GM_c}{R_c}\dot{M}, \qquad (9.18)$$

where M_c and R_c are the instantaneous mass and radius of the stellar core, respectively. Since the gas close to the shock front has a temperature $T \gg 2000\,\text{K}$ (for the H_2 dissociation to be completed), there is an 'opacity gap' exterior to the shock front, where the temperature is above $1500\,\text{K}$ and dust grains are evaporated. Surrounding this 'opacity gap' is a dust envelope which is optically thick to UV photons. The UV photons radiated near the shock front travel freely through the 'opacity gap' into the dust envelope, where they are processed by the dust into infrared photons. For an observer located outside of the system, there is a dust photosphere from which the infrared photons can be observed without being further processed. At this stage, the system will therefore be observed as an infrared source, with a luminosity approximately given by Eq. (9.18). The embedded infrared sources observed in molecular clouds, such as Taurus, are believed to be protostars in this phase.

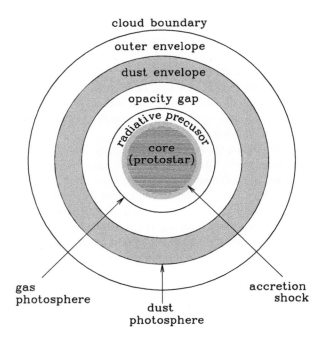

Fig. 9.2. The structure of a theoretical protostar in the phase of accreting material from the collapsing envelope. [After Stahler et al. (1980)]

Once the gas accretion is terminated, the optical depth of the dust envelope becomes negligible. At this stage, an observer outside the system can observe photons propagating directly from the gas photosphere just behind the shock front. For the model considered by Stahler et al. (1980), where the accretion with a rate of $\dot{M} = 10^{-5}\,M_\odot\,\mathrm{yr}^{-1}$ is assumed to be terminated at $M_c = 1\,M_\odot$, the photosphere at the time of accretion termination is found to have a radius of $\sim 4.7\,R_\odot$, an effective temperature of $\sim 7300\,\mathrm{K}$, and a (shock plus interior) luminosity of $\sim 66\,L_\odot$. However, prior to this, when the mass of the stellar core reached $\sim 0.3\,M_\odot$, the ignition of deuterium has already begun to occur. Thus, immediately after the accretion is terminated, the shock layer in the outer part of the stellar core rapidly readjusts itself to the conditions appropriate for a stellar photosphere. A young star is then born. For the case considered by Stahler et al., the young star has an effective temperature of $4250\,\mathrm{K}$ and a luminosity of $6.39\,L_\odot$. The location of a young star in the luminosity–effective-temperature plane at the time of birth defines a 'birth point', and the locus of the birth points for young stars of different masses defines the birthline of new born stars. Such birthlines have been calculated by Palla & Stahler (1992). After birth, a young star evolves along a pre-main-sequence track until hydrogen is ignited in its center (Iben, 1965a). The star then enters the main sequence and evolves further according to the stellar evolution model discussed in the next chapter.

9.4.2 The Formation of Massive Stars

Our understanding about how massive stars form is still poor at the present time. The standard assumption is that massive stars form from the collapse of massive self-gravitating molecular cores. The early stage of the core collapse is expected to be similar to that of low-mass cores, except that the internal turbulence may play a more important role. As in the low-mass case, the core collapse is likely to form a rotationally supported disk because of angular momentum, and the disk gas is transported inwards to feed the protostar by gravitational/magnetic stresses and gas viscosity.

An additional complication in the case of high-mass protostars is that nuclear burning can set in long before it acquires its final mass. Consequently, the radiation pressure and photoionization generated by the nuclear burning can play an important role in the formation of massive stars. The radiation pressure is due to the absorption of UV photons by the dust grains, and the balance between the star's gravity and radiation pressure defines an Eddington luminosity,

$$L_{E,d} = \frac{4\pi c G M_\star}{\kappa_d}, \tag{9.19}$$

where $M_\star$ is the stellar mass, κ_d the dust opacity per unit mass, and c is the speed of light. Assuming $\kappa_d = 8\,\mathrm{cm}^2\,\mathrm{g}^{-1}$, a value suitable for the gas around a protostar, the Eddington luminosity $L_{E,d} \sim 1600(M_\star/M_\odot)\,L_\odot$, which is lower than the luminosity of a massive star with $M_\star \gtrsim 13\,M_\odot$. Thus, for a star more massive than $\sim 13\,M_\odot$, the radiation pressure can in principle prevent the growth of the star before it reaches its final mass. This indicates that there must be some processes through which radiation pressure can be overcome so that massive stars can grow to their final masses. A number of mechanisms have been proposed. As already mentioned, it is likely that the accretion is through a disk, and the strong ram pressure associated with the disk accretion can overcome the radiation pressure. It is also possible that the beaming effect by the accretion disk can redirect the radiation to the polar regions, so that the radiation pressure on the infalling gas from other regions is reduced. Finally, the gas being accreted may be subjected to radiation-driven Rayleigh–Taylor instability and form a highly inhomogeneous flow, so that the radiation can escape through the low-density regions without affecting much the accretion of high-density clumps.

The UV photons generated by the protostellar nuclear burning can also ionize the surrounding gas through photoionization. The associated HII regions can provide strong feedback on the gas infall and accretion, and may play an important role in defining the final mass of a massive star. Unfortunately, the details of this process are yet to be worked out.

Finally, most stars, particularly the massive ones, are born in clusters, and so the evolution of a protostar may be influenced by the interaction with other protostars residing in the same cluster. The tidal force from other objects may truncate the disk of a protostar, thereby terminating the accretion onto the protostar. In a dense environment, close encounters and collisions between different protostars can occur, changing the orbits of the protostars, and even ejecting some of the low-mass members from the system. The interaction among protostars in a cluster may also cause segregation according to stellar mass, with more massive stars sinking deeper into the central part of the cluster.

9.5 Empirical Star-Formation Laws

It is common to characterize the star-formation rate (SFR) in a galaxy, either globally or locally, in terms of the mass in stars formed per unit time per unit area; $\dot{\Sigma}_\star = \dot{M}_\star/\text{area}$. A related quantity is the gas consumption time $\tau_{SF} = \Sigma_{gas}/\dot{\Sigma}_\star$. Ideally, we would like to have a star-formation law derived from first principles that describes $\dot{\Sigma}_\star$ as a function of the relevant physical conditions of the ISM (e.g. density, temperature, and metallicity of the gas, radiation field, magnetic field strength, etc.). Unfortunately, our understanding of the physical processes involved is still very limited. In order to make progress, numerous investigators have therefore used observations to determine *empirical* star-formation laws, i.e. empirical scaling relations between $\dot{\Sigma}_\star$ and certain ISM properties. Such scaling relations are not only useful in that they can be used to include star formation in analytical models or numerical simulations of galaxy formation; they may also give us useful insight into the physics underlying star formation.

Before discussing the results, it is important to stress that all empirical star-formation laws suffer from systematic uncertainties. For example, as discussed in detail in §10.3.8, inferring a star-formation rate from observations is far from trivial, and is often hampered by dust extinction and uncertainties regarding the initial mass function (see also §9.6). Another important uncertainty concerns the molecular gas densities. The surface density of molecular hydrogen, Σ_{H_2}, is typically determined from the integrated CO intensity, I_{CO}, using a constant CO-to-H_2 conversion factor, $X_{CO} \equiv N(H_2)/I_{CO}$. Although it is common practice to assume that X_{CO} is constant – typically at $X_{CO} \sim 2 \times 10^{20}\,\mathrm{cm}^{-2}/(K\,\mathrm{km\,s}^{-1})$ – there are strong indications that it may actually be a function of metallicity and radiation field (e.g. Israel, 1997). This may have a strong impact on the inferred Σ_{H_2}, especially in dwarf and low surface brightness galaxies, where X_{CO} may be substantially higher, causing Σ_{H_2} to be severely underestimated (e.g. Leroy et al., 2008). Another potential concern is that CO may be a more direct tracer of star formation than of H_2. This could come about if the energy input from young stars is responsible for exciting the CO emission (e.g. Dopita & Ryder, 1994).

9.5.1 The Kennicutt–Schmidt Law

Since the most obvious requirement for star formation is the presence of gas, it is only logical to investigate the relation between $\dot{\Sigma}_\star$ and the surface density of gas, Σ_{gas} (be it in atomic form, molecular form, or the sum of both). A power-law relation of the form

$$\dot{\Sigma}_\star \propto \Sigma_{\mathrm{gas}}^N, \tag{9.20}$$

is known as the Schmidt law of star formation, after Schmidt (1959) who found that such a relation with $N \sim 2$ could adequately account for the observed distributions of HI and stars perpendicular to the Galactic plane.[2]

Numerous observational studies of star formation in normal spiral galaxies have shown that the Schmidt law is a surprisingly good description of the global star-formation rates (averaged over the entire star-forming disk) in these galaxies, with the best-fit values of N typically in the range from 1 to 2. Kennicutt (1998b) extended such an analysis by also including starburst galaxies, and found that these systems follow the same power-law relation between $\dot{\Sigma}_\star$ and the *total* surface densities of gas (atomic plus molecular) as normal spiral galaxies (see left-hand panel of Fig. 9.3). As is evident, the simple Schmidt law gives a surprisingly good parameterization of the global star-formation rate over a large range of surface densities, from the most gas-poor spiral disks to the cores of the most luminous starburst galaxies. The best fit to the observational data gives

$$\dot{\Sigma}_\star = (2.5 \pm 0.7) \times 10^{-4} \left(\frac{\Sigma_{\mathrm{gas}}}{M_\odot\,\mathrm{pc}^{-2}} \right)^{1.4 \pm 0.15} M_\odot\,\mathrm{yr}^{-1}\,\mathrm{kpc}^{-2} \tag{9.21}$$

(Kennicutt, 1998b).[3]

This Kennicutt–Schmidt law for star formation is often interpreted as indicating that the star-formation rate is controlled by the self-gravity of the gas. In that case, the rate of star formation will be proportional to the gas mass divided by the time scale for gravitational collapse. For a gas cloud with mean density $\bar{\rho}$, the free-fall time $\tau_{\mathrm{ff}} \propto 1/\sqrt{\bar{\rho}}$ so that

$$\dot{\rho}_\star = \varepsilon_{\mathrm{SF}} \frac{\rho_{\mathrm{gas}}}{\tau_{\mathrm{ff}}} \propto \rho_{\mathrm{gas}}^{1.5}, \tag{9.22}$$

---

[2] To be precise, Schmidt (1959) used volume densities, rather than surface densities, but as long as the scale height of the disk is roughly constant, the two laws are equivalent.
[3] Note that Kennicutt (1998b) defined $\Sigma_{\mathrm{gas}} = \Sigma_{HI} + \Sigma_{H_2}$, which does not account for helium.

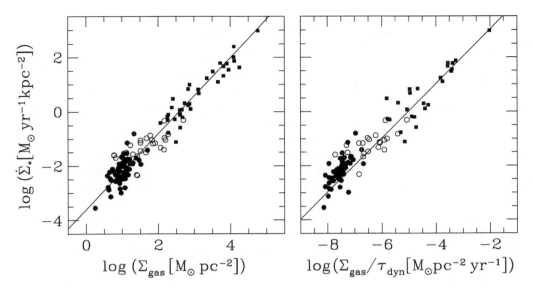

Fig. 9.3. Global star-formation rate per unit area as function of the surface density of the total (atomic plus molecular) gas, Σ_{gas} (left-hand panel) and as function of Σ_{gas} divided by the dynamical time, τ_{dyn} (right-hand panel). Results are shown for normal disk galaxies (filled circles), starbursts (squares), and for the centers of the normal disk galaxies (open circles). The solid line in the left-hand panel corresponds to the Kennicutt–Schmidt law of Eq. (9.21), while that in the right-hand panel corresponds to the star-formation law of Eq. (9.23). [Based on data published in Kennicutt (1998b)]

where ε_{SF} is the ratio of the free-fall time of the gas divided by the gas consumption time, which is a measure for the efficiency of star formation. If all galaxies have approximately the same scale height, this implies that $\dot{\Sigma}_\star \propto \Sigma_{gas}^{1.5}$, in good agreement with the empirical Kennicutt–Schmidt law. The problem with this interpretation, however, is that the efficiency parameter $\varepsilon_{SF} \ll 1$, which suggests that self-gravity of the gas cannot be the entire story. Either the star-formation time scale really is τ_{ff}, but only a small fraction ε_{SF} of the gas participates in star formation, or the actual star-formation time scale is $\tau_{ff}/\varepsilon_{SF}$. In either case, some additional physics is required to explain the origin of the empirical Kennicutt–Schmidt law (see discussion in §§9.2 and 9.3).

In addition to the Schmidt law, Kennicutt (1998b) has shown that the data reveals an equally tight correlation between $\dot{\Sigma}_\star$ and Σ_{gas}/τ_{dyn} (shown in the right-hand panel of Fig. 9.3). Here τ_{dyn} is defined somewhat arbitrarily as $2\pi R/V_{rot}(R)$, the orbital time at the outer radius R of the relevant star-forming region. The best fit obtained by Kennicutt (1998b) is

$$\dot{\Sigma}_\star \approx 0.017 \Sigma_{gas} \Omega, \tag{9.23}$$

with Ω the circular frequency [see Eq. (11.47)]. This implies that about 10% of the available gas is consumed by star formation per orbital time (i.e. $\tau_{SF} \propto \tau_{dyn}$). As demonstrated by Silk (1997), a star-formation law of this form follows naturally from models in which star formation is self-regulated such that disks maintain $Q \sim 1$. However, this is not the only possible explanation. For example, a law of the form $\dot{\Sigma}_\star \propto \Sigma_{gas}\Omega$ can also be reproduced in models in which the SFR is governed by the rate of collisions between gravitationally bound clouds (Tan, 2000), or in models in which spiral arms play an important role in triggering star formation (e.g. Wyse & Silk, 1989).

9.5.2 Local Star-Formation Laws

Clearly, with two empirical star-formation laws that fit the data equally well, and with multiple models that seem able to reproduce these laws, there is little hope that we can gain true insight into the underlying physics of star formation. The main culprit is the fact that these global star-formation laws are based on quantities averaged over the entire star-forming disk. Although useful for modeling galaxy evolution, furthering our physical understanding of star formation requires *local* star-formation laws, which correlate star-formation rates, gas densities and orbital time scales that are measured on much smaller scales. This has only recently become feasible with the advent of better telescopes and detectors.

Using high-resolution data to measure $\dot{\Sigma}_\star$, $\Sigma_{\rm gas}$ and Ω as function of galactocentric radius, it has become clear that although there is a roughly linear trend between star-formation efficiency and orbital frequency *among* galaxies, in agreement with Eq. (9.23), no strong relation exists *within* galaxies (Wong & Blitz, 2002; Leroy et al., 2008). Hence, the orbital time seems to have little direct impact on the local star-formation efficiency.

Fitting individual Schmidt laws to individual galaxies, for which $\dot{\Sigma}_\star$ and $\Sigma_{\rm gas}$ are determined either in azimuthally averaged rings or on a pixel-by-pixel basis, yields best-fit values for $N = {\rm d}\log\dot{\Sigma}_\star/{\rm d}\log\Sigma_{\rm gas}$ that cover the entire range from $1 < N < 3$ (e.g. Bigiel et al., 2008). The main reason for this large amount of scatter is the fact that a single power-law is actually a poor fit to the *local* relation between $\dot{\Sigma}_\star$ and $\Sigma_{\rm gas}$. As first noted by Kennicutt (1998b) and Martin & Kennicutt (2001), based on azimuthally averaged radial profiles, the $\dot{\Sigma}_\star$–$\Sigma_{\rm gas}$ relation reveals a pronounced break at low surface densities of the gas.[4] This is also evident in the right-hand panel of Fig. 9.4, where the grayscale is proportional to the number of individual resolution elements (which have a typical scale of $\sim 750\,{\rm pc}$) with the corresponding values of $\dot{\Sigma}_\star$ and $\Sigma_{\rm gas}$ in a sample of 18 nearby galaxies studied by Bigiel et al. (2008).

The origin of this relatively abrupt truncation in star formation is discussed in §9.5.3 below. Here we focus on what regulates $\dot{\Sigma}_\star$ in regions where star formation is prevalent. Important insight comes from examining how $\dot{\Sigma}_\star$ correlates with atomic and molecular gas separately. The left-hand panel of Fig. 9.4 shows that $\dot{\Sigma}_\star$ is only poorly correlated with $\Sigma_{\rm HI}$: $\dot{\Sigma}_\star$ covers three orders of magnitude over less than one order of magnitude in $\Sigma_{\rm HI}$. Hence, $\Sigma_{\rm HI}$ cannot be used to predict the local star-formation rate surface density. Interestingly, as first noticed by Martin & Kennicutt (2001) and Wong & Blitz (2002), the surface density of atomic hydrogen seems to saturate at $\Sigma_{\rm HI} \sim 10\,{\rm M}_\odot\,{\rm pc}^{-2}$ (corresponding to a column density of $N_{\rm HI} \sim 10^{21}$ atoms cm^{-2}). This could result from a tendency for HI to convert to H$_2$ at higher column densities due to self-shielding.

The middle panel of Fig. 9.4 shows that there is a relatively well-defined Schmidt law for the molecular gas, which is well fit by

$$\dot{\Sigma}_\star = (7\pm 3)\times 10^{-4}\left(\frac{\Sigma_{\rm H_2}}{{\rm M}_\odot\,{\rm pc}^{-2}}\right)^{1.0\pm 0.2}\,{\rm M}_\odot\,{\rm yr}^{-1}\,{\rm kpc}^{-2}, \qquad (9.24)$$

where the relatively large uncertainty on the normalization reflects the pixel-to-pixel variation (Bigiel et al., 2008). The fact that the power-law index of the Schmidt law is close to unity implies a roughly constant star-formation efficiency of $\dot{\Sigma}_\star/\Sigma_{\rm H_2} = (7\pm 3)\times 10^{-10}\,{\rm yr}^{-1}$, equivalent to a molecular gas consumption time of $\sim 1.5 \times 10^9$ yr (see also Wong & Blitz, 2002; Leroy et al., 2008). This suggests that the Schmidt law between $\dot{\Sigma}_\star$ and the total gas surface density is actually an admixture of two 'laws', the first governing the transformation of atomic to molecular gas, and the second describing the formation of stars from molecular gas. This implies that we can write

---

[4] This surface density threshold did not introduce any scatter in the Kennicutt–Schmidt law of Eq. (9.21), since $\dot{\Sigma}_\star$ and $\Sigma_{\rm gas}$ were only averaged over the extent of the disk that is forming stars.

9.5 Empirical Star-Formation Laws

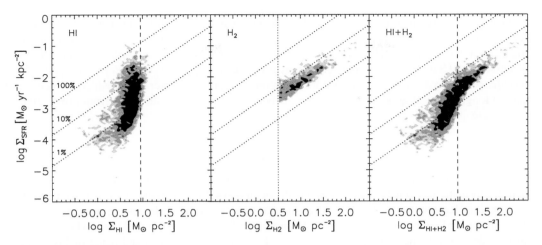

Fig. 9.4. Local star-formation rate per unit area (measured on a scale of $\sim 750\,\mathrm{pc}$) as a function of the local atomic gas density (left-hand panel), molecular gas density (middle panel), and total gas density (right-hand panel). The grayscale is proportional to the number of independent data points (resolution elements) obtained from a sample of 18 nearby galaxies. The diagonal dotted lines show lines of constant star-formation efficiency, indicating the level of $\dot{\Sigma}_\star$ needed to consume 1%, 10% and 100% of the gas reservoir (including helium) in $10^8\,\mathrm{yr}$. Dashed vertical lines in the panels on the left and right show the surface density at which HI saturates. The dotted vertical line in the middle panel indicates the typical sensitivity of the CO data used to infer the molecular gas densities. [Kindly provided by F. Bigiel, based on data published in Bigiel et al. (2008)]

$$N \equiv \frac{\mathrm{d}\log\dot{\Sigma}_\star}{\mathrm{d}\log\Sigma_{\mathrm{gas}}} = 1 + \frac{\mathrm{d}\log f_{\mathrm{mol}}}{\mathrm{d}\log\Sigma_{\mathrm{gas}}}, \tag{9.25}$$

where $f_{\mathrm{mol}} = \Sigma_{\mathrm{H_2}}/\Sigma_{\mathrm{gas}}$ and we have used that $\mathrm{d}\log\dot{\Sigma}_\star/\mathrm{d}\log\Sigma_{\mathrm{H_2}} \simeq 1$. Using that $R_{\mathrm{mol}} \equiv \Sigma_{\mathrm{H_2}}/\Sigma_{\mathrm{HI}} = f_{\mathrm{mol}}/(1 - f_{\mathrm{mol}}) \propto P_{\mathrm{ext}}^\alpha$, with $\alpha \simeq 0.85 \pm 0.05$ (see §9.2.1), and defining $\nu \equiv \mathrm{d}\log P_{\mathrm{ext}}/\mathrm{d}\log\Sigma_{\mathrm{gas}}$ we can write this as

$$N = 1 + \frac{\alpha\nu}{1 + R_{\mathrm{mol}}}. \tag{9.26}$$

This allows us to understand the local Schmidt law in the right-hand panel of Fig. 9.4: in the central regions of disk galaxies, where the gas is largely molecular, R_{mol} is large resulting in $N \sim 1$. In the outer regions, the molecular fraction decreases (due to a decline in P_{ext}, see §9.2.1), causing N to increase towards $1 + \alpha\nu$. Using that $1 \leq \nu \leq 2$, which follows from Eq. (9.9), we thus find that N can become as large as 2.7 in the outskirts where $\Sigma_{\mathrm{gas}} \gg \Sigma_\star$. This is in good agreement with the main trend in the observed relation between $\dot{\Sigma}_\star$ and Σ_{gas}.

Given that observations in the Milky Way and other nearby galaxies show that star formation is exclusively restricted to GMCs, the fact that the SFE of molecular gas appears to be constant suggests that individual GMCs have a roughly constant SFE, with no significant dependence on their formation mechanism or environment. Furthermore, since Rosolowsky (2005) has shown that the mass function of GMCs depends on environment, the SFE of GMCs cannot be a strong function of cloud mass either. The typical mean densities within GMCs imply a typical free-fall time of $\tau_{\mathrm{ff,GMC}} \sim 4 \times 10^6\,\mathrm{yr}$. Compared to the empirical molecular gas consumption time, we infer that

$$\varepsilon_{\mathrm{SF,GMC}} \equiv \frac{\tau_{\mathrm{ff,GMC}}}{\tau_{\mathrm{SF,GMC}}} \sim 0.002. \tag{9.27}$$

In fact, using different tracers of the molecular gas, which probe different densities, the star-formation efficiency per free fall time is found to vary only weakly with density, except for the highest density, protostellar cores, for which ε_{SF} is found to be of the order of 0.1−0.3. As discussed in §9.3 it is currently believed that the low value of $\varepsilon_{SF,GMC}$ is an outcome of supersonic turbulence. The reason why ε_{SF} is still less than unity in protostellar cores, in which turbulence can no longer prevent gravitational collapse, is believed to be due to feedback from protostellar winds (see §9.4.1).

9.5.3 Star-Formation Thresholds

The observation that HI in disk galaxies typically extends well beyond the optical disk suggests that star formation is somehow suppressed in the outer disks. Indeed, as already mentioned above, observations have shown that the local star-formation rates in disk galaxies are truncated rather abruptly beyond a few disk scale lengths. In particular, $\dot{\Sigma}_*$ is found to drop sharply whenever the surface densities of the cold gas drop below $\sim 10 M_\odot \, \mathrm{pc}^{-2}$ (see right-hand panel of Fig. 9.4). The physical origin of these star-formation thresholds is still heavily debated. In what follows we discuss three of the most popular explanations.

(a) Gravitational Instability As originally suggested by Quirk (1972), if the origin of the empirical Schmidt law is indeed related to gravitational instability, it is expected to break down when the surface density of the gas falls below the critical surface density

$$\Sigma_{\mathrm{crit}} = \frac{c_s \kappa}{\pi G}, \qquad (9.28)$$

which is the surface density for which the Toomre parameter Q (see Eq. [9.10]) equals unity. After all, in this picture stars only form where the disk is unstable to gravitational collapse, and gas with $\Sigma_{\mathrm{gas}} < \Sigma_{\mathrm{crit}}$ is locally stable (i.e. $Q > 1$). Kennicutt (1998b) tested this hypothesis using the radial distribution of HII regions to trace the star-formation rate. He adopted a constant velocity dispersion for the gas of $\sigma_{\mathrm{gas}} = 6 \, \mathrm{km \, s}^{-1}$, and found that the truncation occurs at a threshold density $\Sigma_{\mathrm{th}} = \alpha_Q \Sigma_{\mathrm{crit}}$, with $\alpha_Q \simeq 0.5$.[5] A similar result was obtained by Martin & Kennicutt (2001) who found $\alpha_Q = 0.53$ from a sample of 26 disk galaxies with well-defined thresholds. These results have been used to argue that star formation, and in particular the threshold, is governed by gravitational instability.

However, this interpretation faces a number of problems. First of all, it remains to be understood why the average value of α_Q is not equal to unity, as one would naively expect. Secondly, the scatter in α_Q is found to be relatively large: Martin & Kennicutt (2001) and Boissier et al. (2003) quote a galaxy-to-galaxy scatter of 0.2 and 0.33, respectively. Hunter et al. (1998), using a sample of low surface brightness irregular galaxies, obtained $\alpha_Q \simeq 0.25$, and suggest that α_Q may actually vary systematically along the Hubble sequence (see also Leroy et al., 2008). And finally, a number of disk galaxies have disks that are subcritical throughout (i.e. have $\Sigma < \Sigma_{\mathrm{crit}}$ over their entire extent), yet reveal widespread star formation (e.g. Martin & Kennicutt, 2001; Wong & Blitz, 2002; Boissier et al., 2003; Leroy et al., 2008).

Part of the scatter may simply be due to uncertainties regarding the actual sound speed of the gas (which is often assumed to be constant), and regarding the CO-to-H_2 conversion factor, X_{CO}. As argued by Martin & Kennicutt (2001), the fact that α_Q is not exactly equal to unity, is expected from the fact that Toomre's stability criterion only applies to infinitesimally thin gas disks. Real galaxies, however, contain both gas and stars, and the system can be unstable, even when both the stellar and gaseous components individually meet their requirements for stability

---

[5] Actually, Kennicutt (1989) quoted a value $\alpha_Q = 0.67$, but used a constant of 3.36 rather than π in the definition of Σ_{crit}, and did not correct for the fact that $\sigma_{\mathrm{gas}} = c_s / \gamma^{1/2}$, with $\gamma = 5/3$ the adiabatic index (Schaye, 2004).

(e.g. Jog & Solomon, 1984). Wang & Silk (1994) have shown that one can roughly account for the stellar disk by multiplying the critical surface density of Eq. (9.28) with a factor

$$\mathscr{F} = \left[1 + \frac{\Sigma_\star \sigma_{\rm gas}}{\Sigma_{\rm gas} \sigma_\star}\right]^{-1} \tag{9.29}$$

where $\Sigma_\star$ and $\sigma_\star$ are the surface density and velocity dispersion of the stars, respectively. This implies that the values of α_Q have to be multiplied by a factor $1/\mathscr{F}$. Using a sample of 16 disk galaxies, Boissier et al. (2003) obtained an average value for $\mathscr{F}$ of ~ 0.7. Hence, taking account of the contribution of stars alleviates (but does not completely remove) the disagreement between the observed values of α_Q and the theoretical expectation. In addition, it also alleviates the problem with some galaxies appearing to be subcritical throughout. On the other hand, as shown by Boissier et al. (2003), correcting α_Q for the stellar contribution also increases its scatter and still leaves low surface brightness galaxies having a smaller mean value of α_Q than high surface brightness disks.

(b) Shear Instability In light of such difficulties, Hunter et al. (1998) suggested that rather than the Coriolis force, it is the shear due to differential rotation that prevents perturbations from growing. Following Elmegreen (1993), they argued that in the presence of magnetic fields, angular momentum can easily be transferred from the clouds, rendering the Toomre stability criterion, which effectively assumes angular momentum conservation, inappropriate. In that case, what is more important is whether the perturbation can grow in the presence of shear, and the critical surface density for perturbation growth becomes

$$\Sigma_{{\rm crit},A} = \frac{2.5 c_{\rm s} A}{\pi G}. \tag{9.30}$$

Here

$$A = -\frac{R}{2}\frac{d\Omega}{dR} = \frac{\Omega}{2}(1-\beta), \tag{9.31}$$

is Oort's constant, which describes the local shear rate, and the second equality follows from $\Omega = V_{\rm c}/R$ with $\beta = d\log V_{\rm c}/d\log R$. Note that this critical surface density is similar to that of Eq. (9.28), but with the epicycle frequency κ replaced by $2.5A$.[6] In fact,

$$\frac{\Sigma_{{\rm crit},A}}{\Sigma_{\rm crit}} = \frac{2.5}{2\sqrt{2}}\frac{(1-\beta)}{(1+\beta)^{1/2}}. \tag{9.32}$$

Thus, for flat rotation curves ($\beta = 0$), both critical surface densities are the same to within 12%. However, in the inner regions of low surface brightness galaxies, which reveal close to solid-body rotation, $\beta \sim 1$ and $\Sigma_{{\rm crit},A} \ll \Sigma_{\rm crit}$. Thus, although these regions are prevented from forming stars according to Toomre's stability criterion, they are still able to form stars according to the shear criterion (shear vanishes for solid-body rotation). As shown by Hunter et al. (1998), the parameter $\alpha_A \equiv \Sigma_{\rm th}/\Sigma_{{\rm crit},A}$ is much closer to unity than α_Q, suggesting that cloud formation in (irregular) galaxies may involve more of a competition between self-gravity and shear than between self-gravity and coriolis force (see also Leroy et al., 2008).

(c) Ability to Form Molecular Hydrogen Another alternative explanation for the star-formation thresholds is that they reflect the ability to form a cold neutral medium, rather than large-scale gravitational instability or cloud destruction by shear. Skillman (1996) suggested that the threshold for star formation occurs at a constant column density of $N_{\rm HI} \simeq 10^{21}\,{\rm cm}^{-2}$, and reflects the ability of atomic hydrogen to convert to molecular hydrogen due to self-shielding.

---

[6] The factor 2.5 originates from the requirement that perturbations must grow by at least a factor of 100 during the time allowed by shear, and is somewhat arbitrary.

This has support from the fact that the surface density of atomic hydrogen is seen to saturate at this level (see left-hand panel of Fig. 9.4). Along similar lines, Elmegreen & Parravano (1994) suggested that the SFE in the outer parts of disk galaxies drops because the pressure becomes too low to allow a cold phase to form. Schaye (2004) combined this concept with gravitational instability, and argued that the formation of a cold phase lowers the velocity dispersion of the gas, which in turn triggers gravitational instability.

9.6 The Initial Mass Function

The properties of stellar populations in stellar systems depend not only on the rate and efficiency of star formation, but also on what kind of stars are being formed. Since the properties and evolution of individual stars are primarily determined by their masses (as we will see in the next chapter), the initial mass function (IMF), i.e. the mass spectrum with which stars form, is another important property of star formation.

The IMF, $\phi(m)$, is defined so that $\phi(m)\,dm$ is the relative number of stars *born* with masses in the range $m \pm dm/2$. The IMF is assumed to be a continuous function, and we adopt the normalization

$$\int_{m_\ell}^{m_u} m\phi(m)\,dm = 1\,M_\odot, \tag{9.33}$$

where m_ℓ and m_u are the lower and upper mass limits for stars.[7] So normalized, $\phi(m)\,dm$ is the number of stars *born* with masses in the range $m \pm dm/2$ for every $M_\odot$ of newly formed stars. Thus, for a total mass, $M_\star$, of newly formed stars, the total number and total mass of stars with masses in the range $m \pm dm/2$ are given by

$$dN(m) = \frac{M_\star}{M_\odot}\phi(m)\,dm \quad \text{and} \quad dM(m) = \frac{M_\star}{M_\odot}m\phi(m)\,dm, \tag{9.34}$$

respectively. Since the central temperatures of stellar objects with masses $0.08\,M_\odot$ are too low for hydrogen fusion to take place, and stars with masses greater than about $100\,M_\odot$ are unstable against radiation pressure (see §10.2), one normally adopts $m_\ell \simeq 0.08\,M_\odot$ and $m_u \simeq 100\,M_\odot$. The logarithmic IMF is defined as $\xi(m)\,d\log m = \phi(m)dm$, and is related to $\phi(m)$ through

$$\xi(m) = \ln(10)m\phi(m). \tag{9.35}$$

It is sometimes convenient to use the logarithmic slopes of $\phi(m)$ and $\xi(m)$,

$$b(m) \equiv -\frac{d\log\phi}{d\log m} \quad \text{and} \quad \beta(m) \equiv -\frac{d\log\xi}{d\log m} = b-1, \tag{9.36}$$

to characterize the shape of the IMF.

In general, the form of the IMF may vary within a galaxy, and from galaxy to galaxy. Any such variation will undoubtedly complicate the description of the IMF. However, as we will see below, observational results for the Milky Way seem to suggest that the IMF has roughly the same basic form independent of the location in the Galaxy. Consequently, it is often assumed that the IMF is universal, not only within the Milky Way, but also from galaxy to galaxy and for galaxies at different redshifts. We caution, though, that current observational constraints on the IMF still carry large uncertainties (see below) and cannot rule out the possibility of a systematic change of the IMF with the physical conditions (i.e. temperature, density, metallicity) of the ISM out of which the stars form. Furthermore, as we will see in §9.6.2, the idea of the IMF being universal is also a challenge from a theoretical point of view.

---

[7] Note that different normalizations are in use in the literature. For instance, some authors use $\int \phi(m)dm = 1$ so that $\int m\phi(m)dm$ is equal to the average mass of the stars being formed.

9.6 The Initial Mass Function

9.6.1 Observational Constraints

For most stars, the mass is not a direct observable, but has to be estimated from its (apparent) luminosity. Consequently, an observational derivation of the IMF consists of the following steps. First, one measures the apparent magnitudes and distances of all stars in a particular volume down to some magnitude limit. Since this requires individual stars to be resolved, this is only possible for volumes in the Milky Way or in some nearby galaxies. Second, the observed apparent magnitudes are converted into absolute magnitudes, $\mathcal{M}$, using their distances, and one derives the luminosity function, $\Phi(\mathcal{M})$, defined as the number density of stars as a function of their absolute magnitude. Third, the luminosity function of stars is transformed into a mass function, using the mass–luminosity relation of stars, $m(\mathcal{M})$, which can be obtained from the stellar evolution models to be discussed in Chapter 10. Finally, the IMF is obtained from the mass function, once a correction is made for the fact that some stars born earlier may have already died. This final step depends on the star-formation history, $\psi(t)$, in the volume in consideration. If the IMF is independent of time, the luminosity function, Φ_0, corrected for the fact that some stars may have already died is given by

$$\Phi_0(\mathcal{M}) = \Phi(\mathcal{M}) \times \frac{\int_0^{t_0} \psi(t)\,\mathrm{d}t}{\int_{t_0-\tau_{\mathrm{MS}}(\mathcal{M})}^{t_0} \psi(t)\,\mathrm{d}t} \qquad (9.37)$$

for $\tau_{\mathrm{MS}}(\mathcal{M}) < t_0$, and as $\Phi_0(\mathcal{M}) = \Phi(\mathcal{M})$ for $\tau_{\mathrm{MS}}(\mathcal{M}) \geq t_0$. Here t_0 is the present age of the Universe, and $\tau_{\mathrm{MS}}(\mathcal{M})$ is the main-sequence lifetime for a star of magnitude $\mathcal{M}$. This expression follows from the fact that for stars with $\tau_{\mathrm{MS}} < t_0$, only the fraction with ages younger than τ_{MS} can be observed. The IMF then follows from

$$\phi(m) \propto \frac{\mathrm{d}\mathcal{M}}{\mathrm{d}m} \Phi_0[\mathcal{M}(m)]. \qquad (9.38)$$

Although this procedure may appear straightforward, considerable uncertainties are involved in each of the steps. To convert the observed apparent magnitude of a star to its intrinsic luminosity, one needs to know the distance of the star and the amount of extinction by intervening dust. The distances of stars can be determined from trigonometric, spectroscopic and photometric parallaxes (see §2.1.3), but reliable measurements are only possible for stars in a relatively small volume around the solar neighborhood. Additional uncertainties arise from the conversion from absolute magnitude to stellar mass, because the relation between the two is not always one-to-one, especially for massive stars. Finally, the determination of the star-formation history, $\psi(t)$, from observables carries a large uncertainty (see §10.3.9), which impacts on the accuracy of the correction for the stars that have already died. Consequently, there are still large uncertainties in the IMF inferred from observations, as we will see below.

The IMF in our Galaxy has been estimated by a number of investigators. In the solar neighborhood (which is dominated by disk stars), the first determination by Salpeter (1955) gave

$$\phi(m)\mathrm{d}m \propto m^{-b}\mathrm{d}m \quad \text{with} \quad b = 2.35 \quad \text{[Salpeter IMF]} \qquad (9.39)$$

for stars in the mass range $0.4\,\mathrm{M}_\odot \leq m \leq 10\,\mathrm{M}_\odot$. More recent determinations suggest that the IMF deviates from a pure power law, becoming flatter at the lowest mass end and steeper at the highest mass end. Miller & Scalo (1979) approximated the observed IMF by a log-normal form,

$$\xi(x) = a_0 - a_1 x - a_2 x^2 \quad \text{with} \quad x \equiv \log(m/\mathrm{M}_\odot) \quad \text{[Miller–Scalo IMF]}, \qquad (9.40)$$

where $(a_0, a_1, a_2) = (1.53, 0.96, 0.47)$. Scalo (1986) compiled the Milky Way field star IMF based on a large number of references, and the result is shown in Fig. 9.5. The Scalo IMF can be represented by the following broken power law for $m > 0.2\,\mathrm{M}_\odot$:

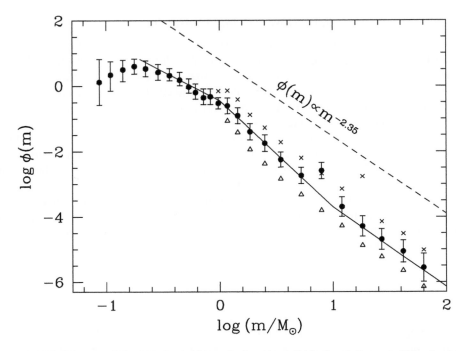

Fig. 9.5. The IMF from Scalo (1986) for field stars in the solar neighborhood. For $m > 1\,M_\odot$ the three sets of points represent results assuming different ratios of the current star-formation rate to the average over the lifetime of the solar neighborhood. The solid lines show the broken power law (9.41), while the dashed line shows the shape of the Salpeter IMF. [Based on data published in Scalo (1986)]

$$\phi(m) \propto \begin{cases} m^{-2.45} & (m > 10\,M_\odot) \\ m^{-3.25} & (1\,M_\odot < m < 10\,M_\odot) \\ m^{-1.80} & (0.2\,M_\odot < m < 1\,M_\odot). \end{cases} \quad \text{[Scalo IMF]} \quad (9.41)$$

At $m > 1\,M_\odot$ the Scalo IMF is similar to the Salpeter IMF, but it contains some features at lower masses, most noticeably the shoulder at $m \sim 1\,M_\odot$ and the turnover at $m \lesssim 0.2\,M_\odot$. A determination based on field stars in the solar neighborhood by Kroupa (2002) gives an IMF

$$\phi(m) \propto \begin{cases} m^{-2.7} & (1.0\,M_\odot < m < 100\,M_\odot) \\ m^{-2.3} & (0.5\,M_\odot < m < 1.0\,M_\odot) \\ m^{-1.3} & (0.08\,M_\odot < m < 0.5\,M_\odot) \\ m^{-0.3} & (0.01\,M_\odot < m < 0.08\,M_\odot). \end{cases} \quad \text{[Kroupa IMF]} \quad (9.42)$$

This IMF has a similar shape as the Salpeter IMF for $m > 0.5\,M_\odot$ but flattens successively at lower masses. Chabrier (2003) presented estimates of the IMFs for different stellar components of the Galaxy, such as disk stars, bulge stars and stars in young and globular clusters, and found that these IMFs have similar forms. For disk stars, the result is

$$\xi(m) \propto \begin{cases} m^{-1.35} & (m > 1.0\,M_\odot) \\ \exp\{-[\log(m/0.2\,M_\odot)]^2/0.6\} & (m < 1.0\,M_\odot) \end{cases} \quad \text{[Chabrier IMF]} \quad (9.43)$$

(see Chabrier, 2005).

For $m \gtrsim 1\,M_\odot$, all these IMFs roughly follow a power law, similar to the original Salpeter IMF. However, at smaller masses there are significant differences among different estimates (see e.g. Fardal et al., 2007, for a more complete list of IMFs).

Beyond the Milky Way, only the brightest stars in relatively nearby galaxies can be resolved. Consequently, for galaxies other than the Milky Way, only the massive end of the IMF can be determined. The current status is that, within the large uncertainties involved, the inferred IMF appears to be similar for galaxies with different properties, supporting the possibility that the IMF may be universal (see Hunter, 1992, Elmegreen, 2008, for details).

9.6.2 Theoretical Models

Theoretical models that describe the origin of the IMF from first principles require detailed understanding of the formation of individual stars within GMCs. Since our understanding of star formation is still rather limited, we currently lack a detailed understanding of the origin of the IMF. Nevertheless, numerous models have been proposed in the literature. Below we briefly describe some of the ideas that have been proposed. Other examples can be found in the review by Zinnecker et al. (1993) and Elmegreen (2008).

(a) Self-Regulated Accretion In this class of models, the masses of stars are assumed to be limited by how much matter they can accrete during their formation. In the models considered by Silk (1995) and Adams & Fatuzzo (1996) it is assumed that each gas clump produces only a single star. Since the mass and density of a clump are both correlated with the velocity dispersion within the clump (see §9.1), the mass of the star is assumed to be determined primarily by the velocity dispersion, Δv. The stellar mass spectrum can then be written as

$$\frac{dN}{dm} = \frac{dN}{dM_{cl}} \frac{dM_{cl}}{d(\Delta v)} \frac{d(\Delta v)}{dm}. \tag{9.44}$$

Using the observed scalings for molecular clouds, $dN/dM_{cl} \propto M_{cl}^{-p}$ ($p \approx 1.5$), and $M_{cl} \propto (\Delta v)^q$ ($q \approx 4$), we can formally write the IMF as

$$\phi(m) \propto m^{-b}, \quad \text{with} \quad b = q(p-1)/\eta + 1, \tag{9.45}$$

where the mass of a star formed in a cloud is assumed to be related to the cloud velocity dispersion as

$$m \propto (\Delta v)^\eta. \tag{9.46}$$

The task is then to work out the relation between m and Δv to obtain η. Assuming that the final mass of a star is determined by the condition that the wind from the star is strong enough to reverse the infall onto the star, Adams & Fatuzzo (1996) obtained $\eta = 11/6$, which, according to Eq. (9.45), yields an IMF slope $b \approx 2.1$, not too different from what is observed. Note that this result is based on the assumption that the wind is spherically symmetric, while observations show that stellar winds are typically collimated. Hence, it is unclear to what extent the results based on a spherical model are applicable.

(b) Hierarchical Accretion Observations indicate that star-forming clouds consist of self-similar networks of filaments over a wide range of scales (e.g. Scalo, 1990). If stars form by accretion in such a self-similar structure, and if their masses are related to the masses of the fractal where they form, a power-law IMF can be produced. This idea of star formation has been considered by Larson (1992), Elmegreen (1997) and others. In a self-similar fractal structure, the number density of subclusters with a typical size l can be written as

$$\frac{dN}{d\ln l} \propto l^{-D}, \tag{9.47}$$

where D is the fractal dimension of the cloud boundary. If the mass of a subcluster is related to its size as $M \propto l^\gamma$, the mass function of the subclusters is

$$\frac{dN}{d\ln M} \propto M^{-D/\gamma}. \tag{9.48}$$

Assuming that the mass of a star is proportional to that of the subcluster in which it forms, the IMF has a logarithmic slope $b = 1 + D/\gamma$. Observations show that the boundary of molecular clouds can be described as fractals with a dimension of about 1.3, which suggests that the surfaces of molecular clouds have a fractal dimension ~ 2.3. If the three-dimensional structure of a cloud is very open and is represented by the structure near the cloud surface, then the fractal dimension of a cloud is $D \sim 2.3$, which gives an IMF slope $b = 1 + 2.3/\gamma$. Thus, $b = 3.3$ for $\gamma \sim 1$ (i.e. for filament-like subclusters), and $b = 1.8$ for $\gamma \sim 3$ (i.e. for three-dimensional subclusters). These slopes bracket that of the Salpeter IMF, suggesting that the origin of the IMF may indeed be related to the geometry of the molecular ISM.

(c) IMF Based on the Core Mass Function Padoan & Nordlund (2002) proposed an IMF model based on the mass distribution of molecular cloud cores expected in a turbulent medium. They argued that, since the strength of compression in a shock must be related to the pre-shock spatial scale, the mass function of dense cores in the GMCs generated by supersonic turbulence with a power-law spectrum, $|\Delta v|(l) \propto l^q$ (which is the rms velocity on scale l), should have a mass function

$$\frac{dN}{d\ln m} \propto m^{-3/(3-2q)}. \tag{9.49}$$

For $q = 0.5$ (as expected for supersonic turbulence), this mass function is a power law with index -1.5. Since only cores with sufficiently high density are able to collapse to form stars, the IMF of stars should be related to the gravitationally bound fraction of the mass function. For a given gas density n, the thermal Jeans mass (which is similar to the Bonnor–Ebert mass) is

$$m_J = m_{J,0} \left(\frac{n}{n_0}\right)^{-1/2}, \quad \text{where} \quad m_{J,0} \approx 1.2 \times \left(\frac{T}{10\,\mathrm{K}}\right)^{3/2} \left(\frac{n_0}{1000\,\mathrm{cm}^{-3}}\right)^{1/2} \mathrm{M}_\odot \tag{9.50}$$

is the Jean mass corresponding to a fiducial density n_0. If the volume-weighted density probability distribution function (PDF) is $p(x)\,d\ln x$ (where $x \equiv n/n_0$), then the mass-weighted PDF is $xp(x)\,d\ln x$. Thus, for gas with a constant temperature, and assuming that the distribution function of the average density of cores of a given mass is the same as the overall density distribution function, the distribution function of local Jeans mass is $P(y)\,dy = 2y^{-2}p[x(y)]\,dy$, where $y \equiv m_J/m_{J,0} = x^{-1/2}$. For cores with a given mass m, only the fraction with local Jeans mass $m_J < m$, $f(m) = \int_0^{m/m_{J,0}} P(y)\,dy$, can collapse. Hence, the mass function of the cores that can collapse is

$$\frac{dN}{d\ln m} \propto f(m) m^{-3/(3-2q)}. \tag{9.51}$$

If $p(x)$ is log-normal (see §9.3.2), then $f(m) \sim 1$ for $m \gg m_{J,0}$, and $f(m) \sim 0$ for $m \ll m_{J,0}$. Therefore, the resulting core mass function is a power law, $dN/dm \propto m^{-3/(3-2q)-1}$ ($\propto m^{-2.5}$ assuming $q = 0.5$), at $m \gg m_{J,0}$, peaks at $m \sim m_{J,0}$, and decreases to 0 as $m \to 0$. Such a core mass function has the same characteristics as the observed IMF of stars. Thus, as long as the mass of the star that forms in a core is proportional to the core mass, the observed IMF can be reproduced.

(d) (Non)-Universality of the IMF Before leaving this section, let us go back to the question whether or not the IMF is universal. As pointed out at the beginning of this section, current observations based on local star-forming galaxies are consistent with the notion that the IMF is universal on scales much larger than individual star-forming regions. However, the uncertainties in the data are still large. In addition, there is evidence for local variations of the IMF in different star-forming regions. For instance, Herbig (1962) noted that star-forming regions such as Taurus contain no stars more massive than $\sim 2\,M_\odot$, whereas the more massive Orion cloud contains both high- and low-mass stars. Subsequently, Solomon et al. (1985) found that, when molecular clouds are divided into cold ($< 10\,\rm K$) and warm ($> 20\,\rm K$) populations, small and cold clouds in general do not contain massive stars with spectral type earlier than late B stars, while warm clouds, which tend to be the largest and the most massive, are found to be associated with HII regions produced by massive stars. These observational results may have important implications. For instance, starburst galaxies are observed to be associated with large amounts of concentrated gas, which is likely to enhance the formation of large molecular clouds relative to that in normal spirals. The 'bimodal' star formation suggested by local star-forming molecular clouds would then mean that a larger fraction of massive stars might form in starburst galaxies than in normal galaxies. If this is the case, the IMF appropriate for starburst galaxies should be 'top heavy'. Unfortunately, it is unclear to what extent the results based on the small number of local star-forming clouds can be generalized, and our knowledge about the ISM in starburst galaxies is still limited. Hence, it remains to be seen whether the IMF in starburst galaxies is significantly different from that in normal spiral galaxies.

The ability of a gas cloud to collapse and fragment is determined by the local thermal Jeans mass; fragmentation can occur only when the cloud mass is larger than the Jeans mass (e.g. Larson, 2005). Since the local Jeans mass is determined by the density and temperature of the gas, the ability to fragment depends on how efficiently the gas can cool. If the cooling time is shorter than the dynamical time, the cloud can cool and fragment as it collapses, otherwise the collapse will come to a halt as the gas is heated up by the compression accompanying the collapse. The radiative cooling rate of a cloud depends on the chemical composition of the gas, and so the balance between radiative cooling and heating by adiabatic compression defines a critical metallicity, $Z_{\rm crit}$. Only gas clouds with metallicity $Z > Z_{\rm crit}$ can fragment to form low-mass systems. This suggests that the IMF may depend on the metallicity of the gas out of which the stars form: while star formation in gas with $Z > Z_{\rm crit}$ may yield a normal IMF, the stars that form in gas with $Z < Z_{\rm crit}$ may have an IMF that is more top-heavy. Theoretical modeling and numerical simulations including atomic and molecular cooling processes show that $10^{-4} \lesssim Z_{\rm crit} \lesssim 10^{-3}$ (e.g. Santoro & Shull, 2006; Smith & Sigurdsson, 2007).[8] Since most of the stars observed in galaxies have metallicities much higher than $Z_{\rm crit}$, this metallicity effect may not be important. However, these results are still premature, and it may well be that the mass function of star-forming clouds changes systematically with metallicity even at $Z > Z_{\rm crit}$. In addition, since high-redshift galaxies are expected to have lower metallicities than present-day galaxies, this may also imply an evolution in the IMF with redshift, as discussed below.

In addition to metallicity, the ambient radiation field may also play an important role in shaping the IMF. For example, far-infrared radiation produced by heated dust grains in starburst regions, and the cosmic microwave background, whose temperature increases with redshift as $2.73(1+z)\,\rm K$, may substantially raise the minimum temperature and increase the Jeans mass in star-forming clouds. This may then shift the mass scale of the IMF upwards at high z and in starburst regions, thereby producing a top-heavy IMF (e.g. Larson, 2005).

---

[8] If dust grains manage to form in the gas, this critical metallicity can be much lower, $Z_{\rm crit} \sim 10^{-6} Z_\odot$ (Schneider et al., 2006).

9.7 The Formation of Population III Stars

Theoretically, if we believe that the structures in the Universe grew through hierarchical clustering, the first objects that went nonlinear and collapsed are expected to have masses much smaller than that of a typical galaxy. Since the gas in these systems has primordial composition, the first generation of stars that managed to form in them are expected to be extremely metal poor. These stars are referred to as Population III stars, to distinguish them from disk (Population I) and halo (Population II) stars.

How could Population III stars have formed? In a hierarchical scenario, such as the CDM model, low-mass dark matter halos already start to collapse at high redshift. The virial temperature of a CDM halo (see §8.2.3) is related to its mass, M, and assembly redshift, $z_{\rm vir}$, by

$$T_{\rm vir} \approx 442 \Omega_{m,0}^{1/3} \left(\frac{M}{10^4 h^{-1} \, {\rm M}_\odot} \right)^{2/3} \left(\frac{1+z_{\rm vir}}{100} \right) \, {\rm K}, \qquad (9.52)$$

where we assumed that the average density of dark matter halos is 200 times the critical density, and that $H(z) = H_0 \Omega_{m,0}^{1/2} (1+z)^{3/2}$ for $z \gg 1$. The virial temperature of a dark matter halo is a measure of the specific binding energy of the halo material, and so the halo can only trap baryonic gas with temperatures $T_{\rm gas} \lesssim T_{\rm vir}$. At redshifts $z \gtrsim 200$, when the effect of Compton scattering is important (see §3.5.3), the temperature of the intergalactic medium (IGM) is

$$T_{\rm gas} = T_\gamma = 273 \left(\frac{1+z}{100} \right) \, {\rm K}. \qquad (9.53)$$

In this case, only halos with $M \gtrsim 10^4 \, {\rm M}_\odot$ can trap significant amounts of baryonic gas. At lower redshifts, when the temperature of the IGM decreases with time faster than T_γ, halos with lower masses can start to trap baryonic gas. The trapped gas might initially be heated to the virial temperature of the halo by shocks accompanying the gravitational collapse. In the absence of radiative cooling, the gas would stay in hydrostatic equilibrium with the dark matter halo and nothing interesting would happen. However, if the gas can cool, it contracts within the halo, becomes gravitationally unstable, and eventually fragments to form stars.

Hence, whether or not stars can form in these early small halos depends on whether or not gas can cool in them. As shown in §8.1.3, for a H–He plasma the only significant radiative cooling at temperatures $< 10^4 \, {\rm K}$ is due to molecular hydrogen. In order for the cooling to be effective, the cooling time scale must be shorter than the Hubble time. This gives a lower limit on the fraction of H_2 molecules in the gas.[9] With the H_2 cooling rate given in §B1.3, it is straightforward to calculate this limit for given gas density and temperature. This required fraction of H_2 should be compared to the fraction expected from the formation of H_2 molecules in the gas cloud. Given the rates of the reactions leading to the formation or destruction of H_2, the expected fraction can also be calculated. Since the formation of H_2 proceeds through reactions that involve both electrons and ionized hydrogen atoms, the expected fraction of H_2 depends not only on the density and temperature of the gas, but also on its ionization fraction.

The requirement that the H_2 fraction should be sufficiently high puts a constraint on the gas temperature and density, or, equivalently, on the virial temperature $T_{\rm vir}$ (or the mass) and assembly redshift $z_{\rm vir}$ of the dark halo containing the gas – note that dark matter halos that assembled

---

[9] Note that primordial gas is free of dust, which implies that the formation of H_2 must proceed via the slow gas-phase reactions given in §B1.3.

9.7 The Formation of Population III Stars

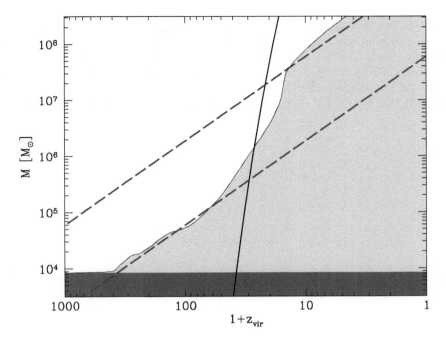

Fig. 9.6. The minimum halo mass (thin solid curve) within which H_2 cooling can lead to gas collapse, as a function of the redshift at which the halo is assembled (note that the halo density increases with z_{vir}). The plot shown is for a CDM model with $\Omega_{m,0} = 1$, $\Omega_{b,0} = 0.06$, and $h = 0.5$. The dashed lines correspond to $T_{vir} = 10^4$ K (upper) and $T_{vir} = 10^3$ K (lower). The dark shaded region is that in which no radiative cooling can be effective, because T_{vir} would be lower than the CMB temperature. The solid line corresponds to 3σ peaks, i.e. $\nu(M) \equiv \delta_c/\sigma(M;z) = 3$, assuming $\sigma_8 = 0.7$, where $\delta_c \approx 1.69$ is the critical overdensity for spherical collapse, $\sigma(M)$ and σ_8 are defined in §4.4.4. [Adapted from Tegmark et al. (1997) by permission of AAS]

earlier had higher characteristic density. Fig. 9.6 shows the minimum mass of halos within which H_2 cooling is sufficiently effective to lead to gas collapse. The minimum halo mass is between $\sim 10^4 \, M_\odot$ (for $z_{vir} \gtrsim 100$) and $\sim 10^8 \, M_\odot$ (for $z_{vir} \lesssim 10$), and the corresponding virial temperature is in the range 10^3–10^4 K. For a CDM-type spectrum, such halos begin to form in large number at $z \lesssim 30$.

There are still a large number of uncertainties in the scenario discussed above. For instance, in the presence of UV photons, H_2 could be photodissociated, so that H_2 cooling could be quenched. In particular, H_2 can be dissociated by photons with an energy below the Lyman limit (Stecher & Williams, 1967). Such photons can penetrate a cloud with high HI column density (because of their low interaction cross-sections with atomic H and He), and can effectively destroy H_2 in a virialized halo or in a collapsing cloud, even if the UV flux is low. Thus, only a small fraction of the gas in small halos is expected to be able to form stars before the H_2 cooling is quenched by the UV flux from the stars. Further star formation may then have to wait until the system (halo plus gas) has grown sufficiently massive that atomic cooling becomes efficient, resulting in a sufficiently large gas contraction that the gas clouds become self-shielding for H_2 molecules (e.g. Haiman et al., 1996). However, even if the gas in a halo is able to cool, it is unclear what fraction of the cooled gas can actually form stars before the rest of the gas is ejected by the feedback effects of the stars already formed.

As discussed in §9.6.2, because of the low metallicities, the radiative cooling time scale of the gas from which the first stars form may be longer than the time scale of the gravitational collapse, and so the gas clouds may not be able to fragment to form low-mass systems (e.g. Santoro & Shull, 2006; Smith & Sigurdsson, 2007). Consequently, the collapse of the gas clouds may lead to the formation of massive stars, resulting in an initial mass function (IMF) that is more top-heavy than for stars of Populations I and II (e.g. Abel et al., 1998; Bromm & Larson, 2004).

10
Stellar Populations and Chemical Evolution

Following up on the previous chapter, in which we discussed star formation, we now address how individual stars evolve with time. As we will see below, most stars, during most of their evolutionary histories, can be considered as spherically symmetric objects with a constant mass that are in hydrostatic equilibrium. Under these conditions, the evolution of a star is almost completely determined by its mass and chemical composition through a set of ordinary differential equations that describe the structure of the star. In this chapter, we start with a brief description of the basic concepts of the theory of stellar evolution. A detailed description is beyond the scope of this book, but can be found in many monographs and textbooks on this subject (e.g. Schwarzschild, 1965; Clayton, 1983; Kippenhahn & Weigert, 1990). We then use the theory of stellar evolution to predict the properties of individual stars (e.g. spectrum, luminosity, metal production, etc.) at different evolutionary stages, and discuss how these results can be synthesized to make predictions regarding the evolution of populations of stars (e.g. galaxies). Finally, we discuss how the evolution of stars affects the chemical evolution of galaxies.

10.1 The Basic Concepts of Stellar Evolution

A stellar evolution model generally starts with two basic assumptions: (i) stars are in hydrostatic equilibrium, and (ii) stars are spherically symmetric. Under these two assumptions, the structure of a star can be described by a set of equations in which all physical quantities depend only on the distance to the center of the star. Of course stars are not completely spherical; as a gaseous body a star can be flattened by rotation. The importance of rotation can be estimated as follows. Consider an element of mass m near the equator of a star. The centrifugal force on the mass element due to rotation is $m\omega^2 R$, where ω is the angular velocity of the rotation and R is the equatorial radius of the star. This force will not cause significant departure from spherical symmetry if it is much smaller than the gravitational force, GMm/R^2, i.e. if $\omega^2 R^3/GM \ll 1$. For the Sun, $\omega \approx 2.5 \times 10^{-6}\,\mathrm{s}^{-1}$, $\omega^2 R^3/GM \approx 2 \times 10^{-16}$, and so flattening due to rotation is completely negligible. In general, normal stars can be considered to be spherical to very high accuracy. To see whether the assumption of hydrostatic equilibrium is valid, consider the acceleration of a mass element at a radius r in a spherical star:

$$\ddot{r} = -\frac{GM}{r^2} - \frac{1}{\rho(r)} \frac{\partial P}{\partial r}. \tag{10.1}$$

Here the partial derivative of P is used because the pressure is a function of both radius r and time t. We can define two time scales based on the ratio between the radius of the star, R, and the acceleration $|\ddot{r}|$. The first is a gravitational dynamical time scale, $t_{\rm dyn} = (G\rho)^{-1/2}$, which corresponds to setting the pressure gradient in Eq. (10.1) to zero. The second is a hydrodynamical time scale, $t_{\rm hydro} = R/c_{\rm s}$ (with $c_{\rm s}$ the sound speed of the stellar gas), which corresponds to setting the gravitational term in Eq. (10.1) to zero. For the Sun with $\rho \sim 1\,\mathrm{g\,cm^{-3}}$, $R \sim 7 \times 10^{10}\,\mathrm{cm}$ and mean temperature $T \sim 7000\,\mathrm{K}$, these time scales are both ~ 1 hour. Since they characterize how rapidly gravity and sound waves allow a perturbed star to adjust to a new equilibrium configuration, they should be compared to the time scales for processes that can cause the star to deviate from its equilibrium state.

Let us first consider the thermal time scale of a star, which can be defined as the ratio between the total thermal energy of a star, E_{th}, and the energy loss rate at its surface, L: $t_{th} = E_{th}/L$. This time scale characterizes the rate at which a star changes its structure by radiating its thermal energy. For the Sun, $t_{th} \sim 10^7$ yr. Since this is much longer than both t_{dyn} and t_{hydro}, the structure of a normal star can adjust quickly to a new configuration as it radiates. As we will see below, the main energy source in a normal star is nuclear reactions and, for a star in equilibrium, the energy generation rate is equal to its luminosity. We can therefore define the time scale for nuclear energy generation as the ratio between the total energy resource in a star and its current luminosity, L: $t_{nuc} \sim \eta M c^2/L$, where $\eta \sim 10^{-3}$ is the efficiency at which nuclear reactions convert the rest mass of a star into radiation (see §10.1.3). For the Sun, this time scale is about 10^{10} yr, much longer than any of the other time scales described above. Thus, for a normal star like the Sun, $t_{nuc} \gg t_{th} \gg t_{dyn}$. This implies that the evolution of a star is largely determined by nuclear reactions, and that thermal and mechanical equilibrium can be achieved to high accuracy during the evolution. This conclusion is valid for almost all stars; the only exceptions are variable stars and stars in their late evolutionary stages, where the luminosity of a star can change significantly over a time scale comparable to the dynamical time scale.

10.1.1 Basic Equations of Stellar Structure

Under the assumption of spherical symmetry and hydrostatic equilibrium, the equation governing the structure of a star is the hydrostatic equation,

$$\frac{dP(r)}{dr} = -\frac{GM(r)\rho(r)}{r^2}, \tag{10.2}$$

where $P(r)$, $\rho(r)$ are the pressure and density at the radius r, and $M(r)$ is the total mass within r,

$$\frac{dM(r)}{dr} = 4\pi r^2 \rho(r). \tag{10.3}$$

These are two differential equations for three quantities, P, M and ρ. In order to complete the description, another independent relation among these quantities is needed. The equation of state of the stellar material can provide such a relation. In general, the equation of state relates the local pressure P to the local density ρ, temperature T, and mass fractions $\{X_i\}$ of elements $i = 1, 2, \cdots, n$,[1] and can formally be written as

$$P = P(\rho, T, \{X_i\}), \tag{10.4}$$

where the exact form of $P(\rho, T, \{X_i\})$ has yet to be specified. Now that we have introduced two new quantities, T and $\{X_i\}$, we need at least two new equations to complete the description. To get an equation for the temperature, we use the fact that a temperature gradient in a star causes energy transport. In general, there are three different modes of the energy transport: (i) heat conduction, which transfers energy as electrons from hotter regions collide and exchange energy with electrons from cooler regions; (ii) radiation, which transfers energy as photons propagate from hotter regions to cooler regions; and (iii) convection, which transfers energy as material is convected between hot and cold regions. For the moment, let us consider radiative stars in which

---

[1] Note that the mass fractions of hydrogen and helium are usually denoted by X and Y, respectively.

10.1 The Basic Concepts of Stellar Evolution

energy is transported mainly by photon diffusion. In this case, we can write the energy flux F ($[F] = \text{erg s}^{-1}\text{cm}^{-2}$) in terms of the temperature gradient and an energy transport coefficient λ as

$$F(r) = -\lambda \frac{dT}{dr}. \tag{10.5}$$

In practice, we work in terms of a quantity called opacity, which is related to λ by

$$\kappa \equiv \frac{4 a_r c T^3}{3\rho\lambda}, \tag{10.6}$$

where a_r is the radiation constant, c is the speed of light, and $[\kappa] = \text{cm}^2\text{g}^{-1}$. In terms of κ, the temperature gradient can be written as

$$\frac{dT}{dr} = -\frac{3\kappa L \rho}{16\pi a_r c r^2 T^3}, \tag{10.7}$$

where $L \equiv 4\pi r^2 F$ is the luminosity ($[L] = \text{erg s}^{-1}$) at radius r. To complete this equation, the form of κ has to be specified. In general κ is a function of ρ, T and $\{X_i\}$, and at the moment we denote this function as

$$\kappa = \kappa(\rho, T, \{X_i\}). \tag{10.8}$$

Now that a new quantity, L, is introduced, we need an equation for L as well. Suppose that the energy release rate at r is ε ($[\varepsilon] = \text{erg g}^{-1}$), then

$$\frac{dL}{dr} = 4\pi r^2 \rho \varepsilon. \tag{10.9}$$

To complete this equation we need to know ε as a function of ρ, T and $\{X_i\}$. At the moment we assume that this function can be obtained and write it as

$$\varepsilon = \varepsilon(\rho, T, \{X_i\}). \tag{10.10}$$

Eqs. (10.2), (10.3), (10.7), and (10.9) are the four structure equations for the five quantities ρ, M, P, T, L. Together with the equation of state, they provide a complete description of a radiative star with given composition $\{X_i\}$, provided that P, κ and ε are known functions of $(\rho, T, \{X_i\})$.

The boundary conditions for the structure of a star can be set as follows:

$$(M, L) = (0, 0) \text{ at } r = 0, \text{ and } (P, T) = (P_s, T_s) \text{ at } r = r_s, \tag{10.11}$$

where r_s is the radius of the star. Since by definition the luminosity of a star is related to its radius and effective temperature by

$$L = 4\pi r_s^2 \sigma_{SB} T_{\text{eff}}^4, \tag{10.12}$$

where $\sigma_{SB} = a_r c/4$ is the Stefan–Boltzmann constant, we can choose $T_s = T_{\text{eff}}$ so that $L(r) = L$ at $r = r_s$. The value of P_s can be set by the condition of hydrostatic equilibrium for the stellar atmosphere:

$$\frac{dP}{dr} = -\rho g, \tag{10.13}$$

where g is the gravitational acceleration. Since the mass in the atmosphere is in general much smaller than the mass within it, the pressure on the stellar surface (i.e. at the bottom of the atmosphere) can be written as

$$P(r_s) = GM_s \int_{r_s}^{\infty} \rho(r) \frac{dr}{r^2}. \tag{10.14}$$

We relate this pressure to the optical depth of the atmosphere,

$$\tau(r_s) = \int_{r_s}^{\infty} \kappa(r)\rho(r)\,dr. \tag{10.15}$$

In general, $\rho(r)$ falls rapidly as r increases and the main contribution to $P(r_s)$ and $\tau(r_s)$ comes from layers near r_s. We can therefore write $P(r_s)$ and $\tau(r_s)$ in the following approximate forms:

$$P(r_s) \approx \frac{GM_s}{r_s^2} \int_{r_s}^{\infty} \rho(r)\,dr, \quad \tau(r_s) \approx \kappa(r_s) \int_{r_s}^{\infty} \rho(r)\,dr. \tag{10.16}$$

It then follows that

$$P(r_s)\kappa(r_s) \approx \frac{GM_s}{r_s^2}\tau(r_s). \tag{10.17}$$

Under the 'gray-atmosphere' approximation, i.e. the opacity is independent of photon frequency (see Shu, 1991b), the temperature is related to the optical depth as

$$T^4(r) = \frac{1}{2}T_{\rm eff}^4\left[1 + \frac{3}{2}\tau(r)\right], \tag{10.18}$$

and so $\tau(r_s) = 2/3$ [because $T(r_s) = T_{\rm eff}$]. With all these relations, we can finally write the boundary conditions on the surface of a star as

$$T_s = T_{\rm eff}, \quad P_s = \frac{2}{3}\frac{GM_s}{r_s^2}\frac{1}{\kappa(r_s)}. \tag{10.19}$$

Since the temperature and pressure on the surface of a star are generally much smaller than those in the stellar interior, the outer boundary conditions may be replaced by the so-called 'zero boundary conditions':

$$T_s = 0, \quad P_s = 0. \tag{10.20}$$

For stars whose outermost layers are in radiative equilibrium, these zero boundary conditions provide a good approximation to the actual boundary conditions.

Theoretically, the mass of a star is chosen to be a constant, while the radius is determined only after the model calculation. It is therefore more convenient to write the structure equations using $M(r)$ (instead of r) as the independent variable, and set the boundary conditions at M_s. In real applications, there is no problem with this change of variable, because the relation between r and $M(r)$ is almost always monotonic. In terms of $M(r)$, the four stellar structure equations, for a star of mass M_s, can be written as

$$\frac{dr}{dM} = \frac{1}{4\pi r^2 \rho(r)}, \tag{10.21}$$

$$\frac{dP}{dM} = -\frac{GM}{4\pi r^4}, \tag{10.22}$$

$$\frac{dL}{dM} = \varepsilon, \tag{10.23}$$

$$\frac{dT}{dM} = -\frac{3\kappa L}{64\pi^2 a_r c r^4 T^3}, \tag{10.24}$$

and the boundary conditions are

$$(r, L) = (0, 0) \text{ at } M = 0, \quad (P, T) = (P_s, T_s) \text{ at } M = M_s. \tag{10.25}$$

10.1.2 Stellar Evolution

The stellar structure equations given above are all static equations without any explicit time dependence. But stars do evolve with time. As nuclear reactions proceed in the central region of a star and energy is radiated away from its surface, both the structure and chemical composition of the star can change with time. Do we then need a complete set of *dynamical* equations in order to describe such evolution? The answer is no and the reason is the following. As we will see later in this chapter, nuclear reactions are the ultimate source that drive the evolution of a star. Since in a normal star $t_{\rm dyn} \ll t_{\rm th} \ll t_{\rm nuc}$ as discussed above, any deviation from dynamical and thermal equilibrium caused by nuclear reactions can be compensated very quickly, so that at any time the star can be considered to be in an equilibrium state governed by the static structure equations. The evolution comes in only because at each time step the star must adjust its structure and thermal state to a new chemical composition determined by the ongoing nuclear reactions. If there is no bulk motion of material in a star (i.e. convection is negligible), the change of chemical composition is localized and is given by the chemical evolution equation,

$$\frac{\partial X_i}{\partial t} = f_i(\rho, T, \{X_j\}), \tag{10.26}$$

where the time derivative is for a fixed element of mass, and the form on the right-hand side takes into account the fact that the rates of nuclear reactions in general depend on ρ, T and chemical composition. It is this change in chemical composition that drives stellar evolution. The procedure for solving the evolution of a star is then clear. We start with an initial assumption on the composition as a function of r or $M(r)$ at some time t_0, and solve the structure equations under the boundary conditions. With the newly obtained ρ, T and the old $\{X_i\}$ we can calculate, for each radius in the star, the new composition at a slightly later time $t = t_0 + \delta t$:

$$X_i(t) = X_i(t_0) + f_i[\rho(t), T(t), \{X_j\}(t_0)]\delta t. \tag{10.27}$$

With this modified chemical composition, the stellar structure equations can be solved once again. This process can be repeated until a final time is reached.

10.1.3 Equation of State, Opacity, and Energy Production

The above discussion sets up the basic framework for calculating the evolution of a radiative star. To complete the program, we must specify the equation of state, $P(\rho, T, \{X_i\})$, the opacity, $\kappa(\rho, T, \{X_i\})$, and the energy production rate, $\varepsilon(\rho, T, \{X_i\})$. In this subsection, we give a very brief description about how to achieve this goal.

(a) Equation of State The pressure in a star consists of two components, the gas pressure $P_{\rm gas}$ and the radiation pressure $P_{\rm rad}$. For a blackbody, which is a good approximation for stellar material, the radiation pressure is

$$P_{\rm rad} = \frac{1}{3} a_{\rm r} T^4. \tag{10.28}$$

Since the mean separation of gas particles in a normal star is much larger than the typical size of atoms, stellar material can be considered as an ideal gas, which allows us to write the gas pressure as

$$P_{\rm gas} = n k_{\rm B} T, \tag{10.29}$$

where n is the number density of particles (i.e. atoms, ions and electrons). In terms of the mean molecular weight,

$$\mu \equiv \rho/(n m_{\rm p}), \tag{10.30}$$

where m_p is the mass of a proton, Eq. (10.29) can be written in the form

$$P_{\text{gas}} = \frac{k_B T}{\mu m_p} \rho. \tag{10.31}$$

To express P_{gas} in the form of $P(\rho, T, \{X_i\})$, we need to express μ in terms of ρ, T and chemical composition. The particle number density in a gas depends on the ionization states of different elements at the density and temperature in consideration. For stellar material where the temperature is so high that all elements are highly ionized, simple approximations to μ can be made. Since a fully ionized atom of charge number Q_i consists of $(Q_i + 1)$ particles, one nucleus and Q_i electrons, the number density contributed by such an element is $X_i(Q_i+1)\rho/(M_i m_p)$, where X_i is the mass abundance of element i and M_i is its mass number in units of m_p. The total number density can therefore be written as

$$n = \frac{\rho}{m_p} \sum_i \frac{(Q_i+1)X_i}{M_i} \approx \frac{1}{4}(6X+Y+2)\frac{\rho}{m_p}, \tag{10.32}$$

where $X \equiv X_H$ and $Y \equiv X_{He}$. The approximation uses the facts that $M_i \approx 2(Q_i+1)$ for elements heavier than helium and that, by definition, $\sum X_i = 1$. Since heavier elements are far less abundant than hydrogen and helium this approximation is very accurate, and allows us to write

$$\mu = 4/(6X+Y+2). \tag{10.33}$$

If the metallicity $Z \equiv 1 - X - Y$ is very small, to good approximation we can write

$$\mu = 4/(5X+3). \tag{10.34}$$

Under a similar approximation, the number density of free electrons can be written as

$$n_e = \frac{\rho}{m_p}\sum_i \frac{Q_i X_i}{M_i} \approx \frac{1}{2}(1+X)\frac{\rho}{m_p}. \tag{10.35}$$

The ideal gas law will break down when the ionized gas becomes so dense that the pressure arising from Pauli's exclusion principle is no longer negligible. This principle states that no more than one fermion of any given type (e.g. electron) can occupy a given quantum state, and so a gas of fermions can show strong resistance to compression when the effect of the exclusion becomes important. For free electrons (or other fermions) in an ionized gas, Pauli's exclusion principle is related to Heisenberg's principle of uncertainty, $\delta x \delta p \geq h_P/4\pi$, where δx is the uncertainty in position and δp is the uncertainty in the corresponding momentum. Thus, if a gas is highly compressed so that gas particles are closely packed in space, then the electrons will have momentum higher than that predicted by the classical kinetic theory, and the gas will have pressure higher than that given by the classical gas law. A gas in which Pauli's exclusion principle is important is called a degenerate gas. Since degeneracy pressure is proportional to phase-space density, and since ions have higher momentum than electrons for a given temperature, the degeneracy pressure from ions is in general much smaller than that from electrons if their number densities are similar. The equation of state for a degenerate gas can be found in e.g. Kippenhahn & Weigert (1990).

(b) Opacity Photons emitted in the stellar interior can be absorbed and scattered before they reach the surface. The opacity of stellar material, κ, in Eq. (10.7) is a measure of the resistance of the material to the passage of photons. To find an expression for κ, we need to know how photons interact with stellar material (ionized gas). The main interactions are:

- Compton or Thomson scattering: Photons can be scattered by ions or free electrons, which changes the directions of photons and slows down the net rate of energy transport.

- Free–free absorption: A photon can be absorbed by a free electron, giving its total energy $h_P \nu$ to the electron.
- Bound–bound absorption: A photon can be absorbed by an atom or an ion, moving an electron from one bound orbit to another bound orbit of higher energy. This process is responsible for the observed spectral absorption lines of stars, but does not contribute much to the opacity in the deep interior of stars where the atoms are highly ionized so that the number of bound states is small.
- Bound–free absorption: A photon can be absorbed by an atom or an ion, removing one electron from its bound orbit. Again, because of the rarity of bound electrons in a star's interior, the contribution of this process to the opacity is small.

The cross-sections of these processes can in principle be calculated from quantum theory or obtained from physical experiments (see §B1.3). The total opacity can then be obtained by summing up the contributions from all different processes.

For given density, temperature and chemical composition, the stellar opacity can be different for photons with different frequencies. The opacity in Eq. (10.7) should therefore be a proper average of the frequency-dependent opacity, κ_ν, over frequency ν. To find such an average, we start by writing Eq. (10.7) as $L = (4\pi r^2 c/3\rho\kappa)(\mathrm{d}B/\mathrm{d}r)$, where $B = a_\mathrm{r} T^4$ is the energy density of radiation. In general, we can write the luminosity within a unit frequency interval near ν as

$$L_\nu = \frac{4\pi r^2 c}{3\rho} \frac{1}{\kappa_\nu} \frac{\mathrm{d}B_\nu}{\mathrm{d}r}, \tag{10.36}$$

where B_ν is the Planck function. Since $L = \int L_\nu \, \mathrm{d}\nu$, $B(T) = \int B_\nu(T) \, \mathrm{d}\nu$ and since $\mathrm{d}B_\nu/\mathrm{d}r = (\mathrm{d}B_\nu/\mathrm{d}T)(\mathrm{d}T/\mathrm{d}r)$ (with $\mathrm{d}T/\mathrm{d}r$ independent of ν), integrating both sides of Eq. (10.36) over ν leads to

$$L = \frac{4\pi r^2 c}{3\rho \kappa} \frac{\mathrm{d}B}{\mathrm{d}r}, \tag{10.37}$$

where

$$\kappa \equiv \left(\int_0^\infty \frac{\mathrm{d}B_\nu}{\mathrm{d}T} \, \mathrm{d}\nu \right) \bigg/ \left(\int_0^\infty \frac{1}{\kappa_\nu} \frac{\mathrm{d}B_\nu}{\mathrm{d}T} \, \mathrm{d}\nu \right) \tag{10.38}$$

is known as the Rosseland mean opacity, which can be calculated for given T, ρ and $\{X_i\}$.

In practice, the calculation of stellar opacity is very involved, because we have to take into account all important atoms and ions. It is beyond the scope of this book to describe such calculations in detail. There are specialized research groups who make detailed calculations of κ and publish and update their results in opacity tables (e.g. Seaton et al., 1994). For a given chemical composition, the effective opacity is a function of temperature and density, and it is found that κ is in general low for both very high and very low temperatures. When the temperature is very high, most photons have very high energy and are not absorbed effectively, because the absorption cross-sections generally decrease with photon energy (see §B1.3). At very low temperatures, on the other hand, many atoms are neutral and few electrons are available to scatter photons or to take part in the free–free absorption. The effective opacity can be approximated by power laws in specific ranges of temperature and density. We may write these power laws as

$$\kappa = \kappa_0 \rho^{\alpha-1} T^{3-\beta}, \tag{10.39}$$

where κ_0 is a constant which depends on the chemical composition, and the values of κ_0, α and β depend on the particular ranges of T and ρ in question.

At high temperatures, where the main contribution to the opacity is the scattering of photons by free electrons, the opacity is given by

$$\kappa_e = \frac{n_e \sigma_T}{\rho} = \frac{4\pi}{3} \frac{q_e^4}{c^4 m_e^2 m_H} (1+X) \approx 0.2(1+X), \tag{10.40}$$

where σ_T is the Thomson cross-section. At somewhat lower temperatures, where light elements like hydrogen and helium are fully ionized while heavier elements are partially ionized, the main sources of opacity are free–free (ff) and bound–free (bf) processes. In this case the opacity can be approximated by:

$$\kappa_{\rm ff} \propto \rho(X+Y)(1+X)T^{-7/2}, \quad \kappa_{\rm bf} \propto (1-X-Y)(1+X)\rho T^{-7/2}. \tag{10.41}$$

Here the factor $(1+X)$ comes from the dependence of the opacity on the electron density given by Eq. (10.35); the factor $(X+Y)$ comes from the fact that the main contribution to $\kappa_{\rm ff} \propto \sum Q_i^2 n_i$ is from H and He; and the factor $(1-X-Y)$ comes from the assumption that H and He are fully ionized so that only heavier elements have bound electrons to produce bound–free absorption. At even lower temperature ($T \lesssim 10^4$ K), hydrogen becomes partially ionized. Because of the nature of the Saha equation (see §B1.3), the number density of free electrons is a very rapidly increasing function of T in this case. Since both the free–free and bound–free absorption rates increase with n_e, the opacity in this temperature range increases very rapidly with temperature:

$$\kappa \sim \kappa_0 \rho^{1/2} T^{10}. \tag{10.42}$$

(c) The Effect of Convection One of the most uncertain aspects in stellar evolution is the treatment of convection. According to Schwarzschild's criterion (§8.2.2), convection occurs when the specific entropy of the stellar gas increases in the direction of gravity, i.e. when the temperature gradient required at a radius r for photons to carry the heat flux is larger than the temperature gradient implied by a constant specific entropy in the neighborhood of r. In this case, blobs of hot gas in the stellar interior move upwards (i.e. away from the center of the star) and are replaced by blobs of cooler gas that move downwards. This is a very efficient process for transporting energy and material, and so it can greatly affect the evolution of a star in which convection does take place. Unfortunately, this process is very difficult to describe rigorously. A standard phenomenological model commonly used is the so called 'mixing-length' theory, which assumes that a convecting blob typically travels one mixing length $l_{\rm mix}$ before dissolving into the ambient medium. The mixing length is usually represented by a dimensionless factor, $\alpha_{\rm mix} \equiv l_{\rm mix}/H$, where $H = [d(\ln P)/dr]^{-1}$ is the pressure scale height in the stellar interior. In general, the results of stellar evolution models are quite sensitive to the assumed value of $\alpha_{\rm mix}$.

Because of convection, metals that are produced near the bottom of the convective zone can be effectively transported to the top. In practice a convective zone cannot have a sharp boundary, and so we expect that some metals can be overshot into the region above the top of the convective zone. Such convective overshoot can affect stellar evolution. As with convection itself, convective overshoot is very difficult to quantify. In model calculations, the effect is usually parameterized by a dimensionless parameter, $\alpha_{\rm os} = l_{\rm os}/H$, which describes the size of the overshoot region.

In the absence of a rigorous theory, the values of $\alpha_{\rm mix}$ and $\alpha_{\rm os}$ have to be calibrated so that model calculations best match the observed properties of the Sun and other well-observed stars. Unfortunately, it remains unclear how accurate such calibration is for other stars. Detailed modeling shows that main-sequence stars (which are powered by burning H into He in their cores, see next section) contain only small convective zones, and so their structures are not affected significantly by uncertainties related to convection. In contrast, stars in late evolutionary stages can develop large convective zones, and model predictions for such stars are still very uncertain at the present.

(d) Stellar Nucleosynthesis and Energy Production Our Sun is currently emitting with a luminosity $L_\odot \approx 3.9 \times 10^{33}\,\mathrm{erg\,s^{-1}}$. Based on radioactive dating of rocks on Earth, it can be inferred that the Sun has been shining with such a luminosity for at least about 5×10^9 yr. The total energy that has been released by the Sun is therefore about 6×10^{50} erg, or about a fraction of 3×10^{-4} of its present total mass. An obvious question is, what is the basic energy source for the Sun? The gravitational energy, $GM_\odot^2/R_\odot$, can sustain the Sun at its present luminosity for a period of $GM_\odot^2/R_\odot L_\odot \sim 3 \times 10^7$ yr, much too short compared to the age of 5×10^9 yr. The total thermal energy, which is about half of the gravitational energy, according to the virial theorem, is also far too small to supply the energy. Chemical reactions associated with the burning of material can transform a fraction of $\sim 10^{-10}$ of the rest mass into heat. This fraction is much smaller than the required fraction mentioned above, so that chemical energy is also insufficient to power the Sun. The only energy source that can convert rest mass into energy with an efficiency larger than 3×10^{-4} is nuclear reactions. Today we are quite sure that normal stars like the Sun are all powered by nuclear fusion, a process by which light elements synthesize to form heavier ones. The simplest example is the fusion of four hydrogen nuclei to form a helium nucleus, $4\mathrm{H} \to \mathrm{He}$. Since the rest mass of He is $3.97m_p$, the rest mass that is converted into energy in this reaction is therefore $4m_p - 3.97m_p = 0.03m_p$, or about 7×10^{-3} of the original mass, $4m_p$. This efficiency is large enough to power the Sun for a period of about 8×10^{10} yr. In fact, as we will see below, the structure and luminosity of the Sun will change drastically once it uses up about 13% of its total hydrogen. Thus the fusion of H to He can sustain the Sun at its present state for about 9×10^9 yr, or about twice its present age.

Nuclear fusion not only provides the main energy source in stars, but also is the main process for synthesizing the heavy elements observed in the Universe. As we have seen in §3.4, cosmological primordial nucleosynthesis is effective only in producing light elements, in particular He and Li. All heavier elements are believed to have been synthesized in the interiors of stars.

The binding energy per nucleon in a nucleus of mass m with Q protons and N neutrons is defined as

$$\Delta E(Q,N) = \frac{[Qm_p + Nm_n - m(Q,N)]c^2}{Q+N}. \tag{10.43}$$

For a helium nucleus, $\Delta E \approx 0.007 m_p c^2$. Thus, when four hydrogen nuclei fuse into a helium nucleus, the binding energy per nucleon increases. From the point of view of thermodynamics, it is therefore more advantageous for hydrogen to fuse into helium than to remain as free protons. Fig. 10.1 shows the binding energy per nucleon for various atomic nuclei. We see that there is a general trend for the binding energy per nucleon to become larger as we go to heavier elements, but the trend reverses for elements heavier than ^{56}Fe. This behavior can be explained as a result of the balance between the net attraction of the strong nuclear force and the repulsion of the positively charged protons. As long as the nucleus is not too big, the short-range nuclear force tends to win over the long-range electric force, causing the binding energy per nucleon to increase when a nucleon (proton or neutron) is added to the nucleus. However, since the nuclear density is nearly constant, heavier nuclei are also physically bigger. The increase in the binding energy per nucleon with the number of nucleons saturates with iron-56; further addition of nucleons actually reduces the binding energy per nucleon. This is the reason why very heavy nuclei, such as uranium, are unstable.

If it is more advantageous for nucleons to be synthesized into heavy elements such as iron, why, then, is most of the cosmic material still in the form of light elements such as H and He? The answer to this question is that nuclear fusion reactions can occur only under certain conditions. Since atomic nuclei are positively charged, the two (or more) nuclei in a fusion reaction tend to

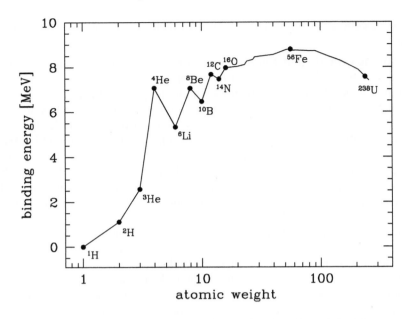

Fig. 10.1. Binding energy per nucleon as a function of atomic weight.

repel each other due to the Coulomb force. In order for the components to come close enough so that the attractive nuclear force becomes dominant, they have to move rapidly towards each other to overcome the Coulomb barrier. Thus, nucleosynthesis in the Universe is expected to be efficient only in places where gas density and temperature are both high so that many high-velocity collisions among nuclei can occur. Stellar interiors are such places. Since the nuclear reactions involved are driven by thermal motions of nuclei, they belong to the category of thermonuclear reactions.

To have a closer look at how thermonuclear reactions depend on temperature, let us consider the collision of two nuclei of charges q_i and q_j, and masses m_i and m_j, with a relative speed v at large separation. Based on classical mechanics, the closest separation the two nuclei can achieve is $r = 2q_i q_j / mv^2$, where $m = m_i m_j/(m_i + m_j)$ is the reduced mass. However, the two particles *can* penetrate the Coulomb barrier due to quantum tunneling. The probability of this penetration is $\mathscr{P}_r \propto \exp(-2\pi^2 r/\lambda)$, where $\lambda = h_P/mv$ is the de Broglie wavelength. For a classical gas with temperature T, the relative velocity v of particles obeys the Maxwell–Boltzmann distribution, $\mathscr{P}_v \propto \exp(-mv^2/2k_B T)$. The probability for a fusion to take place is therefore proportional to $\mathscr{P}_r \mathscr{P}_v$. For a given T, this probability is maximized at $v = (4\pi^2 q_i q_j k_B T / m h_P)^{1/3}$, with a maximum value

$$\mathscr{P}_{\max} \sim \exp\left[-\left(\frac{T_0}{T}\right)^{1/3}\right], \quad (10.44)$$

where $T_0 \equiv (3/2)^3 (4\pi^2 q_i q_j / h_P)^2 (m/k_B)$. Thus, in order for the fusion reaction to proceed at a significant rate, the temperature T must be $\gtrsim T_0$. The temperature required is in general higher for heavier nuclei, because of the larger values of masses and charges. One might therefore expect that, as a star evolves and its internal temperature rises, light elements will be converted successively into heavier and heavier elements, until all the material in the star is converted into elements in the neighborhood of iron in the periodic table. However, this process is slow, because the temperature in a stellar interior does not change much during most of its lifetime. This is the reason why most baryonic material in the Universe is still in the form of H and He. In the following, we examine in some detail the most important nuclear reactions under the conditions applicable to stellar interiors.

There are two important channels for converting hydrogen into helium. The first is through the pp chain and the second is through the CNO cycle. The pp chain starts with two hydrogen nuclei (i.e. two protons) to form a deuteron (D) which then captures another proton to form a ^{3}He:

$$p+p \rightarrow D+\bar{e}+\nu_e, \quad D+p \rightarrow {}^3He+\gamma. \tag{10.45}$$

The ^{3}He so created is converted into ^{4}He either through the collision of two ^{3}He,

$$^3He+{}^3He \rightarrow {}^4He+2p, \tag{10.46}$$

or, in the presence of ^{4}He, through

$$^3He+{}^4He \rightarrow {}^7Be+\gamma, \quad {}^7Be+e \rightarrow {}^7Li+\nu_e, \quad {}^7Li+p \rightarrow 2\,{}^4He. \tag{10.47}$$

The CNO cycle involves the following set of reactions:

$$^{12}C+p \rightarrow {}^{13}N+\gamma, \quad {}^{13}N \rightarrow {}^{13}C+\bar{e}+\nu_e, \quad {}^{13}C+p \rightarrow {}^{14}N+\gamma,$$

$$^{14}N+p \rightarrow {}^{15}O+\gamma, \quad {}^{15}O \rightarrow {}^{15}N+\bar{e}+\nu_e, \quad {}^{15}N+p \rightarrow {}^{12}C+{}^4He. \tag{10.48}$$

The net effect of the CNO cycle is to convert four protons into a helium nucleus, with C, N, and O nuclei acting only as catalysts. Note that the CNO cycle requires the pre-existence of carbon.

The rate with which the pp chain and the CNO cycle convert H into He is proportional to the square of the density (because the reactions involved are two-body in nature) and increases rapidly with temperature. The CNO cycle is more sensitive to temperature because the elements involved are heavier. The rate of energy production from each reaction is equal to the reaction rate times the energy released from each conversion and can be calculated as a function of T. The resulting specific energy production rate, ε, is in general a smooth function of T, and can be represented as a piecewise power law of T:

$$\varepsilon = \varepsilon_0 \rho T^\eta. \tag{10.49}$$

In the temperature range where the pp chain is important,

$$\varepsilon \propto X_H^2 \rho T^4 \quad \text{(pp chain)}, \tag{10.50}$$

while in the temperature range where the CNO cycle dominates,

$$\varepsilon \propto X_H X_C \rho T^{17} \quad \text{(CNO cycle)}. \tag{10.51}$$

Since thermonuclear reactions need high temperature to proceed, there must be systems in which the temperature is never high enough to ignite H-burning. Such stars are called black dwarfs, because they radiate very little. As we will see in §10.2, this minimum temperature corresponds to a minimum mass, $M_H \approx 0.08 M_\odot$. Although stars less massive than this cannot burn hydrogen, they can burn deuterium which has a lower ignition temperature. A star powered by deuterium burning is called a brown dwarf.

For systems with masses larger than M_H, we expect the conversion of H to He to first occur in the central region of a star where the temperature is the highest. There will then be a time when most hydrogen in the central region is exhausted. When this happens the star will contract as the thermal energy keeps on flowing out. According to the virial theorem, such contraction will lead to an increase in the temperature in the central part of the star. The contraction continues until the next important nuclear reaction is able to generate sufficient thermal energy to halt the contraction. As can be inferred from Eq. (10.44), when the temperature reaches $\sim 10^8$ K, fusion of helium begins to synthesize the next stable nuclei, ^{12}C, through the following reactions:

$$^4\text{He} + ^4\text{He} \to ^8\text{Be} + \gamma, \quad ^8\text{Be} + ^4\text{He} \to ^{12}\text{C} + \gamma. \tag{10.52}$$

This process is called the triple-α process because three α particles (^{4}He) are involved. Since ^{8}Be is unstable and can decay back to two ^{4}He in a short time, only small amounts of ^{12}C would be produced if the second reaction occurred with a 'normal' rate. This is known as the 'bottleneck' for the synthesis of heavy elements. A breakthrough in this 'bottleneck' problem occurred when it was realized that the second reaction in (10.52) can proceed through a fast (resonant) channel with the production of an excited state ^{12}C*, which then de-excites to the ground level ^{12}C (Hoyle, 1954). Once ^{12}C is produced, the synthesis of heavy elements can proceed further.

In our following discussion, two groups of elements are particularly important. One group is called the α-elements, which includes ^{16}O, ^{20}Ne, ^{24}Mg, ^{28}Si, ^{32}S, ^{36}A, ^{40}Ca, and it is named after the fact that all the elements in this group can be formed by adding α-particles to ^{12}C. The synthesis of these elements is either through the capture of α-particles in reactions such as

$$^{12}\text{C} + ^4\text{He} \to ^{16}\text{O} + \gamma \quad \text{and} \quad ^{20}\text{Ne} + ^4\text{He} \to ^{24}\text{Mg} + \gamma, \tag{10.53}$$

or through the burning of C and O in reactions like

$$^{12}\text{C} + ^{12}\text{C} \to ^{20}\text{Ne} + ^4\text{He} \quad \text{and} \quad ^{16}\text{O} + ^{16}\text{O} \to ^{28}\text{Si} + ^4\text{He}. \tag{10.54}$$

Since these elements can be synthesized in a star that starts out with pure hydrogen and helium, their abundances in a star are essentially independent of the initial metallicity. The other group, called the iron-peak elements, includes all elements with atomic numbers in the range $40 < A < 65$, i.e. Sc, Ti, V, Cr, Mn, Fe, Co, Ni and Cu. Because of their high nuclear charges, iron-peak elements are formed late in a star's life, when its core becomes extremely hot.

Elements that can be synthesized directly in stars of zero initial metallicity are referred to as primary elements, while elements that can form only in the presence of primary elements are called secondary elements. An important secondary element is N, which is produced in the CNO cycle from pre-existing C and O. Note that the carbon produced later in a star *cannot* directly capture a proton to form N, because very few protons are present by the time the carbon nuclei are synthesized.

10.1.4 Scaling Relations

Once P, κ and ε are known as functions of ρ, T and chemical composition, the stellar evolution equations can be integrated. In general, these equations have to be solved numerically. With certain assumptions, however, we can find some simple scaling relations that can help us to understand how the properties of a star change with its mass and chemical composition.

One example of this kind is the homologous model of stellar structure.[2] In this model, stellar opacity and energy generation rate are assumed to have the power-law forms given in the last subsection, radiation pressure and convection are assumed to be negligible, and the chemical composition is assumed to be the same everywhere in the stellar interior (so that μ, κ_0 and ε_0 are constant throughout the interior of a star). All stars with the same κ_0, ε_0 and μ, but having different masses form a homologous sequence. In such a sequence, M_s is the only independent quantity; characteristic values of all other quantities are determined by M_s. Thus, if we use the fractional mass $m \equiv M/M_s$ as the independent variable, each of the other dependent quantities in units of its characteristic value should be the same function of m without an explicit dependence

---
[2] Another example are the so-called polytropic models, which will not be discussed in this book (see e.g. Kippenhahn & Weigert, 1990).

on M_s. The solutions to the structure equations for homologous stars can therefore be written in the following form:

$$r = r_0 \tilde{r}(m), \quad \rho = \rho_0 \tilde{\rho}(m), \quad L = L_0 \tilde{L}(m), \quad T = T_0 \tilde{T}(m), \quad P = P_0 \tilde{P}(m), \qquad (10.55)$$

where a subscript '0' denotes the characteristic value of a quantity, and the tilded functions depend only on m. We can choose $r_0 = r_s$, the radius of the star. Using the argument given above, r_s should depend only on M_s. With M_s as the characteristic mass and r_s as the characteristic radius, we can define the following characteristic values for ρ, P and T:

$$\rho_0 = \frac{M_s}{r_s^3}, \quad P_0 = \frac{GM_s^2}{r_s^4}, \quad T_0 = \frac{\mu m_p}{k_B} \frac{GM_s}{r_s}. \qquad (10.56)$$

Using Eqs. (10.23) and (10.49) we can also define a characteristic luminosity,

$$L_0 = \varepsilon_0 M_s \rho_0 T_0^\eta = \varepsilon_0 \left(\frac{G\mu m_p}{k_B}\right)^\eta \frac{M_s^{2+\eta}}{r_s^{3+\eta}}. \qquad (10.57)$$

Note that all these characteristic quantities are power laws of both M_s and r_s. To get an expression for r_s as a function of M_s, we insert the homologous solutions (10.55) into Eq. (10.24). With the use of Eq. (10.39), we get

$$\frac{d\tilde{T}}{dm} = \frac{3\kappa_0 \varepsilon_0}{64\pi^2 ac} \left(\frac{G\mu m_p}{k_B}\right)^\eta \frac{1}{G^{\beta+1}} \frac{\tilde{\rho}\tilde{L}}{\tilde{r}^4 \tilde{T}^\beta} \frac{M_s^{1+\eta+\alpha-\beta}}{r_s^{3+\eta+3\alpha-\beta}}. \qquad (10.58)$$

Since this equation should have the same form for all M_s, $r_s(M_s)$ must have the form

$$r_s(M_s) \propto M_s^q \quad \text{with} \quad q = \frac{1+\eta+\alpha-\beta}{3+\eta+3\alpha-\beta}. \qquad (10.59)$$

Inserting this into Eqs. (10.56)–(10.57), we find that

$$\rho_0(M_s) \propto M_s^{1-3q}, \quad T_0(M_s) \propto M_s^{1-q}, \quad L_0(M_s) \propto M_s^{2-3q+(1-q)\eta}. \qquad (10.60)$$

For main-sequence (MS) stars, which are burning hydrogen into helium in their centers, $\alpha \approx 2$ and $\beta \approx 6.5$ for near-solar mass stars in which the main source of opacity is free–free and bound–free absorptions, while $\alpha = 1$ and $\beta = 3$ for more massive stars in which electron scattering is the main source of opacity. If we take $\eta = 4$ (for the pp chain), or $\eta = 17$ (for the CNO chain), then the homologous model predicts a luminosity–mass relation, $L_s \propto L_0 \propto M_s^a$, with $a \approx 5$ for near-solar mass stars and $a \approx 3$ for more massive ones. These results are qualitatively in agreement with detailed model calculations, as well as with observational results. For example, detailed model calculations by Bressan et al. (1993) give

$$\frac{L_{MS}}{L_\odot} = \begin{cases} 81(M_s/M_\odot)^{2.14} & \text{(for } M_s \gtrsim 20 M_\odot) \\ 1.78(M_s/M_\odot)^{3.5} & \text{(for } 2 M_\odot < M_s \lesssim 20 M_\odot) \\ 0.75(M_s/M_\odot)^{4.8} & \text{(for } M_s \lesssim 2 M_\odot). \end{cases} \qquad (10.61)$$

From the definition of the effective temperature in Eq. (10.12), we can also obtain a luminosity-effective temperature relation,

$$L_s \propto T_{\text{eff}}^b, \quad \text{where} \quad b = \frac{4[2-3q+(1-q)\eta]}{2-5q+(1-q)\eta}. \qquad (10.62)$$

For the values of α, β and η given above, the value of b falls in the range from 4.1 (for low-mass stars) to 8.6 (for high-mass stars). These results are also qualitatively in agreement with detailed calculations of main-sequence stars. Stars on a homologous sequence therefore lie approximately on a straight line in the theoretical H-R diagram [plot of $\log(L_s)$ versus $\log(T_{\text{eff}})$]. This is the reason why the locus of main-sequence stars in the observed H-R diagram (see Fig. 2.6) is approximately a straight line.

In reality, stars on the main sequence may have different chemical compositions. The simple homologous model can also help to understand how the properties of a star of a given mass change with chemical composition. The dependence on composition is through that of κ_0, ε_0 and μ. From Eq. (10.58) we see that, for a given mass, M_s, the radius scales with κ_0, ε_0 and μ as

$$r_s \propto \left(\kappa_0 \varepsilon_0 \mu^{\eta-\beta-1}\right)^p \quad \text{where} \quad p = \frac{1}{3+\eta+3\alpha-\beta}. \tag{10.63}$$

Inserting this into Eq. (10.60) and Eq. (10.12), we obtain

$$L_s = L_0 \tilde{L}(1) \propto \varepsilon_0 \mu^\eta \left(\kappa_0 \varepsilon_0 \mu^{\eta-\beta-1}\right)^{-p(3+\eta)}, \tag{10.64}$$

and

$$T_{\text{eff}} \propto \varepsilon_0^{1/4} \mu^{\eta/4} \left(\kappa_0 \varepsilon_0 \mu^{\eta-\beta-1}\right)^{-p(5+\eta)/4}. \tag{10.65}$$

As an example, let us consider a case where

$$\kappa \propto Z(1+X)\rho T^{-7/2} \quad \text{and} \quad \varepsilon \propto X^2 \rho T^4, \tag{10.66}$$

so that

$$\alpha = 2, \quad \beta = 13/2, \quad \eta = 4 \quad \text{and} \quad \kappa_0 \propto Z(1+X), \quad \varepsilon_0 \propto X^2. \tag{10.67}$$

These forms of κ and ε are reasonable approximations for the main-sequence stars with masses $\sim 1 M_\odot$. If Z is small so that $\mu = 4/(3+5X)$ to good approximation, the scalings of r_s, L_s and T_{eff} with chemical composition become

$$r_s \propto X^{4/13} (1+X)^{2/13} (5X+3)^{7/13} Z^{2/13}, \tag{10.68}$$

$$L_s \propto X^{-2/13} (1+X)^{-14/13} (5X+3)^{-101/13} Z^{-14/13}, \tag{10.69}$$

$$T_{\text{eff}} \propto X^{-5/26} (1+X)^{-9/26} (5X+3)^{-115/52} Z^{-9/26}. \tag{10.70}$$

Thus, r_s increases, while both L_s and T_{eff} decrease, with increasing X and Z.

10.1.5 Main-Sequence Lifetimes

As shown above, the luminosity of a main-sequence (hydrogen-burning) star scales with mass roughly as $L_s \propto M_s^4$. Since a star converts roughly a fixed fraction of its rest mass into radiation before it leaves the main sequence, the main-sequence lifetime scales with mass as

$$t_{\text{H}\to\text{He}} \propto M_s/L_s \propto M_s^{-3}. \tag{10.71}$$

For a star like the Sun, the main-sequence lifetime is $\sim 10^{10}$ yr, comparable to the age of the Universe. According to the scaling given above, a star with a mass of about $10 M_\odot$ is expected to have a main-sequence lifetime of only $\sim 10^7$ yr, much shorter than the age of the Universe. For a given mass, the main-sequence lifetime is expected to be shorter for stars with lower metallicity, because L_s increases with decreasing Z. Detailed stellar evolution models can be used to calculate the main-sequence lifetimes as a function of initial mass and initial metallicity. In the mass range between $0.08 M_\odot$ to $100 M_\odot$, and for a metallicity not very different from $Z_\odot$, the main-sequence lifetime can conveniently be approximated by the following formula:

$$t_{\text{MS}} = \frac{2.5 \times 10^3 + 6.7 \times 10^2 m^{2.5} + m^{4.5}}{3.3 \times 10^{-2} m^{1.5} + 3.5 \times 10^{-1} m^{4.5}} \text{Myr}, \tag{10.72}$$

where m is the mass of the star in units of $M_\odot$ (e.g. Schaller et al., 1992).

The fact that stars of different masses have different main-sequence lifetimes and that massive stars have main-sequence lifetimes much shorter than the age of the Universe has important implications for galaxy evolution. For example, when we observe a galaxy today (i.e. at redshift $z = 0$), we are observing the light from the stars that have evolved to the present time. Hence, the main-sequence stars with $M_s \sim M_\odot$ observed today include all stars of such masses that have formed during the past 9×10^9 yr, while the main-sequence stars with $M_s \sim 10 M_\odot$ observed today are only those that have formed during the past 10^7 yr. Consequently, the stellar population observed from a galaxy depends strongly on its star-formation history. Put differently, it is possible to learn about the star-formation history of a galaxy by studying its stellar population. Furthermore, as we will see later in this chapter, massive stars in their post-main-sequence evolution can eject material and energy into the interstellar medium either as stellar winds or through supernova explosions. The fact that massive stars have relatively short lifetimes implies that the chemical composition and thermal state of the gas in a galaxy can be affected almost simultaneously as star formation proceeds.

10.2 Stellar Evolutionary Tracks

As shown above, the evolution of a star is almost completely determined by its initial mass and chemical composition. Thus, for given initial mass and metallicity, a stellar evolution model should, in principle, yield all the properties of a star at any time, t, after its birth. The two most important properties of a star are its luminosity, L, and its effective temperature, T_{eff}, and so the evolution of a star can be represented conveniently by its evolutionary track in the T_{eff}–L plane. Since T_{eff} and L are related to the color and absolute magnitude of a star, such evolutionary tracks provide the basis for the understanding of observed H-R diagrams, such as the one shown in Fig. 2.6. Indeed, the prominent branches heavily populated by stars in a H-R diagram can be considered as stellar 'traffic jams' where stars evolve slowly and many tracks run close to each other.

With the stellar evolution model described above, one can calculate the evolutionary track for any given initial mass and metallicity, and standard computer programs are available for such calculations (see Girardi et al., 2000; Yi et al., 2003, and references therein). In what follows, we present a brief discussion of the most important results obtained from such calculations.

10.2.1 Pre-Main-Sequence Evolution

In §9.4.1 we saw that a young star is born at a late stage of protostellar collapse on the birthline where the star is in a state of quasi-equilibrium and optically visible due to deuterium burning. The star then quickly evolves onto the main sequence. The evolution between the birthline and the zero-age main sequence is usually referred to as pre-main-sequence evolution. The evolutionary tracks can be obtained using the stellar evolution model outlined above (e.g. Iben, 1965b; D'Antona & Mazzitelli, 1994), and Fig. 10.2 shows some examples of such tracks. For low-mass stars, the pre-main-sequence evolution is characterized by almost vertical tracks (called Hayashi tracks), because such stars, with relatively low surface temperatures, become almost entirely convective when they begin to burn deuterium. Stars with masses $0.08 M_\odot \lesssim M_s \lesssim 0.3 M_\odot$ remain convective as they contract towards the main sequence, and their pre-main-sequence tracks remain vertical. More massive stars, however, become radiative before reaching the main sequence, and the late stage of their pre-main-sequence evolution is characterized by contraction with nearly constant luminosity (see §10.1.4). Stars with masses smaller than $\sim 0.08 M_\odot$ cannot ignite hydrogen burning in their centers, and so they shine only during the deuterium-burning

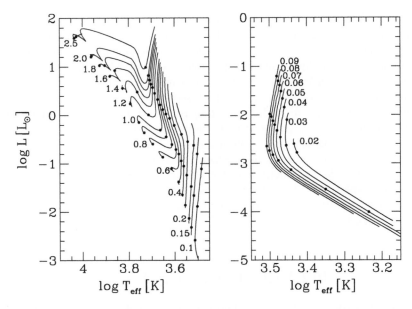

Fig. 10.2. The pre-main-sequence evolutionary tracks of stars with solar metallicity. The positions reached by stars after 1, 10 and 100 Myr from birth are marked by dots. Left and right panels show results for stars with main-sequence masses (in solar units) in the range [0.1, 2.5] and [0.02, 0.09], respectively. [Based on data published in D'Antona & Mazzitelli (1994)]

phase as brown dwarfs. As a brown dwarf runs out of deuterium fuel, it shrinks until it is supported by the degeneracy pressure of electrons, eventually becoming a black dwarf.

The energy released during the pre-main-sequence phase of a star is mainly from gravitational contraction and deuterium-burning, and is generally much smaller than the energy released during the main-sequence and post-main-sequence evolution. Because of this, pre-main-sequence stars are usually neglected in modeling the spectra of galaxies.

10.2.2 Post-Main-Sequence Evolution

Once a star reaches its zero-age main sequence, it will stay on the main sequence for a time interval given by the main-sequence lifetime, t_{MS}, (§10.1.5) until the hydrogen in its core is burned into helium. During its main-sequence lifetime the star moves little in the T_{eff}-L plane, because the luminosity of a main-sequence star is limited by the photon-diffusion rate, which depends primarily on the initial mass (see §10.1.4). The amount of H converted into He at the end of the main-sequence lifetime is $Lt_{MS}/0.007c^2$, which is about 13% of the total hydrogen mass for a star like the Sun. The properties of stars on the main sequence have been discussed in the last section. Here we examine how stars evolve after their main-sequence lifetimes. As one can see from Fig. 10.3, where the post-main-sequence evolutionary tracks are shown for various initial masses, the evolution is qualitatively different for low- and high-mass stars.

(a) Low-Mass Stars Even after hydrogen has been exhausted in the core of a star, energy transport will continue in its interior. Since the core is still too cold to burn He, there is no thermal energy to counterbalance the energy leak, and so the core contracts. As it contracts, the core is heated up (according to the virial theorem), and so is the gas layer above the He core. Since hydrogen has a lower ignition temperature than helium, the hydrogen in the shell just above the He core will burn first as the temperature increases, while the He core itself still remains

10.2 Stellar Evolutionary Tracks

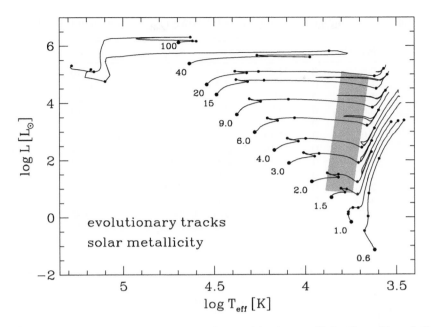

Fig. 10.3. Post-main-sequence evolutionary tracks of stars with solar metallicity. To avoid confusion, tracks for stars with masses smaller than $2\,M_\odot$ are terminated at their He flash. The six points on each track mark the positions reached by the star after the times from the zero-age main sequence listed in Table 10.1. The shaded area is the instability strip within which various variable stars are located. [Based on data published in Girardi et al. (2000)]

dormant. The H shell burning is quite effective in generating energy. But as long as the star is not convective, the amount of energy that can leak from the star is limited by photon diffusion and is roughly constant (see §10.1.4), with the extra heat causing the star to expand. Since the luminosity scales with the radius and the effective temperature as $L \propto r_s^2 T_{\rm eff}^4$, the star evolves more-or-less horizontally in the $T_{\rm eff}$-L plane to the right (Fig. 10.4), turning into a subgiant. However, as the temperature in the outer envelope drops, the temperature gradient between the outer and inner regions increases, and the star eventually becomes fully convective. When this happens the effective temperature cannot decrease any further and the star starts to ascend almost vertically in the $T_{\rm eff}$-L plane to the red giant branch (RGB). At this stage, the envelope of the star has been distended so much that substantial mass-loss (stellar wind driven by the radiation pressure) may occur. Meanwhile, the He core becomes hotter and hotter as it contracts and as more He ash is added to it from the H-burning shell. When the core temperature reaches $\sim 10^8$ K, helium begins to burn into carbon through the triple-α process; some oxygen is also synthesized as some of the ^{12}C nuclei capture a ^{4}He. The He-burning heats the core and causes it to expand, lowering the gravity on the H-burning shell and thereby reducing the strength of shell H burning. Consequently, the star reaches the tip of the red-giant branch and then settles onto the horizontal branch (HB) where the star is mainly powered by He burning in its core. If the envelope were peeled off the star, the He core would behave like a He main-sequence star. Stars on the horizontal branch are thus expected to have relatively high effective temperatures, but the exact value of $T_{\rm eff}$ for a star depends on how much mass the star had lost during its red-giant phase.

The core He burning will come to a halt once most of the helium nuclei in the core are synthesized into carbon and oxygen. At this stage, the star contains a C/O core, an inner He-burning shell and an outer H-burning shell. As the core temperature is still too low for carbon to burn,

Table 10.1. Stellar-evolution times (in 10 Myr) at the points in Fig. 10.3.

$M_s/M_\odot$	τ_1	τ_2	τ_3	τ_4	τ_5	τ_6
0.6	0	7142	7660	7801	7837	7840
1.0	0	732.5	887.4	1101	1137	1225
1.5	0	251.9	257.5	258.4	262.9	282.5
2.0	0	107.5	109.7	109.9	110.8	113.1
3.0	0	36.25	36.99	37.04	37.20	37.44
4.0	0	17.90	18.22	18.24	18.30	18.37
6.0	0	7.046	7.148	7.156	7.167	7.214
9.0	0	2.945	2.986	2.989	2.991	2.995
15.0	0	1.237	1.250	1.251	1.252	1.253
20.0	0	0.8462	0.8547	0.8552	0.8559	0.8565
40.0	0	0.4430	0.4567	0.4574	0.5115	0.5147
100.0	0	0.1267	0.1975	0.2964	0.3383	0.3402

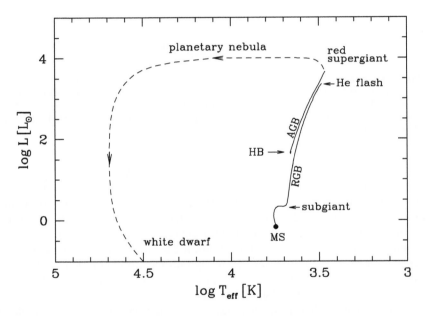

Fig. 10.4. An illustration of the complete evolutionary track of low-mass stars.

the C core contracts. The He-burning shell outside the C core and the H-burning shell outside the He-burning shell also contract in this phase because of the loss of pressure support. Such contraction enhances the double-shell burning, causing the star to ascend the asymptotic giant branch (AGB) in a way similar to what happened during the RGB phase. During the AGB phase, the envelope of the star can be greatly distended, and a large amount of mass loss can occur. When the star reaches the top of the AGB, where it becomes a red supergiant, it can lose all of its blanketing hydrogen layer. The star then makes a sudden move leftwards in the H-R diagram as it peels away its relatively cold mantle. At this stage, the intense ionizing radiation from the star may cause the ejected H envelope to shine as a fluorescent planetary nebula. The He burning will gradually die off, leaving behind a C/O core with a mass in the range 0.55–$0.6 M_\odot$ (independent of the initial mass). Because of the small mass, further nuclear burning cannot be ignited in the core. The star will then contract until it is supported by the degeneracy pressure of electrons, becoming a white dwarf.

The evolutionary track discussed above is valid for all stars with initial masses $M_s \lesssim M_{up} \approx 8\,M_\odot$. However, stars with masses $M_s \lesssim M_{He-f} \approx 2.3\,M_\odot$ experience an additional event, called the helium flash. For stars with masses as low as this, the He core must contract a lot before it reaches the ignition temperature of He burning. Consequently, the core becomes supported by the degeneracy pressure of electrons before He is ignited. Since the core continues to grow in mass, its temperature increases until He is ignited. Because the pressure of degenerate matter does not depend on its temperature, the corresponding release of nuclear energy does not result in an expansion of the core. Instead the nuclear reactions cause the temperature of the core to increase, which leads to an increase in the rate of energy production, thereby giving rise to a thermal runaway instability called the helium flash. The drastic and rapid increase in temperature ultimately removes the degeneracy, causing the core to expand and the He burning to continue in a stable fashion. Therefore, after the helium flash the core contains an ordinary helium plasma that burns into carbon on the HB in a way similar to the core He burning described above.

(b) Massive Stars For stars with initial masses $M_s \gtrsim M_{up} \approx 8\,M_\odot$, core-H exhaustion, He-core contraction, shell-H burning, and core-He ignition proceed in much the same way as for low-mass stars, except that the helium flash is absent. However, because of the high mass, core ignition and shell burning can proceed successively to heavier elements. During this process, a star in general moves to a higher T_{eff} after a core ignition, and to a lower T_{eff} during core contraction. Since a massive star can maintain radiative equilibrium during the evolution, its luminosity changes only little. As a result, the evolutionary tracks of such stars appear more-or-less horizontal in the H-R diagram (see Fig. 10.3).

At some stage, all the material in the core may be converted into iron-peak elements that cannot be ignited any more. The burning in the core will then cease, while Si, O, Ne, C, He, and H continue to burn in shells at successively larger radii. When this happens, the core contracts until it becomes supported by the degeneracy pressure of electrons. Since the shell-burning outside the core keeps dropping ash onto the core, the degenerate core will grow until it reaches Chandrasekhar's mass limit, $M \approx 1.4\,M_\odot$, beyond which the iron-core can no longer be supported by the degeneracy pressure of electrons and must collapse further. This collapse can heat the core to such a high temperature ($\gtrsim 10^9$ K) that iron nuclei can be photodisintegrated into α-particles (^{4}He nuclei). Since thermal energy is used in the photodisintegration, the core will collapse further and be heated to an even higher temperature. The α-particles produced will then also be photodisintegrated into protons and neutrons, draining more thermal energy from the core. The core then undergoes a phase of catastrophic contraction, and its density can become so high that almost all electrons are squeezed into protons to form neutrons. If the mass is smaller than $\sim 3\,M_\odot$, the core will be a neutron star supported by the degeneracy pressure of neutrons; otherwise the core will collapse further to form a black hole. The collapse of the iron-core is so violent that a supernova explosion may take place, as we will discuss in the following.

(c) Radiation in the Giant Phase As a star evolves off the MS, it becomes redder, and if its mass is $\lesssim 2\,M_\odot$, it also becomes brighter. With the stellar evolution models described above, one can in principle estimate the total energy that is radiated in the giant phase as compared to that in the MS phase. It turns out that, for a star with an initial mass $< 1.5\,M_\odot$, the total energy radiated in the giant phase is actually larger than that in the MS phase. Since the MS lifetime of a $1.5\,M_\odot$ star is about 2.5 Gyr, the luminosity of a starburst with an age longer than this is therefore expected to be dominated by giant-branch stars rather than by dwarfs on their MS. This has important implications. For instance, for a stellar population older than a few Gyr, such as that in an elliptical galaxy, the spectrum is expected to resemble that of giant-branch stars, as we shall see in §10.3.

(d) Variable Stars An important feature in the H-R diagram is the instability strip, an almost vertical strip containing various variable stars (Fig. 10.3). The origin of this variability is a vibrational instability of the outer envelope of the star. As gas clouds, all stars oscillate to some degree. The oscillations in normal stars are in general quite small, because they are excited by random fluctuations. In contrast, the oscillations of stars in the instability strip may have much larger amplitudes, as they are excited by global instabilities. Most of the large-amplitude oscillations of stars are generated by the so called κ mechanism. The physics underlying this mechanism is that, in some special states, the stellar material can become more opaque when it is heated, as is the case for many stars in the instability strip. To see how this property is related to the generation of oscillations in the stellar structure, let us consider a layer of such material that has lost support against gravity and is moving inwards. As it shrinks, the layer is compressed and heated up, and therefore becomes more opaque. Since it is now more difficult for photons to diffuse through the layer, heat will build up below it. The rising pressure below the layer will eventually halt the contraction and push it outwards. As the layer expands, its temperature drops and it becomes more transparent to radiation. This increased transparency allows radiation in the inner region to diffuse outwards more freely, thereby decreasing the pressure support to the layer. As the layer loses pressure support, it falls back, and the cycle repeats. During such a cycle, part of the energy generated below the layer is used to push the layer outwards. An initial small oscillation will then be amplified after each cycle.

As a star oscillates, its atmosphere moves in and out, leading to pulsations of luminosity with time. The pulsation period, $\mathscr{P}$, is of the order of the sound-crossing time of the star, and it can be shown that

$$\mathscr{P} \sim Q \left(\frac{\overline{\rho}}{\overline{\rho}_\odot} \right)^{-1/2}, \qquad (10.73)$$

where $\overline{\rho}$ is the mean density of the star, and $Q \sim (G\overline{\rho}_\odot)^{-1/2} \sim 1\,\mathrm{hr}$ is the pulsation constant. Because the instability strip is roughly vertical in the T_eff-L plane, all stars in the strip have approximately the same effective temperature. Since for a given T_eff the mean density of a star can be related to its luminosity, the above relation between $\mathscr{P}$ and $\overline{\rho}$ implies a period–luminosity relation. In general, the mean density of a star decreases with increasing luminosity, so that bright stars have longer pulsation periods. A special class of variable stars are the Cepheids, which have masses $M_\mathrm{s} \gtrsim 5\,\mathrm{M}_\odot$, and which reveal a regular period–luminosity relation. This can be used to determine the Cepheid's distance by simply measuring its period and apparent magnitude (see §2.1.3).

10.2.3 Supernova Progenitors and Rates

At the end of their lifetimes some stars will explode as supernovae. These are observed as objects that radiate very intensely (with a luminosity $\sim 10^{10}\,\mathrm{L}_\odot$) for a period of a few weeks. The emission and absorption lines observed from a supernova in general show strong Doppler shifts produced by the explosion-driven motions of the gas associated with the supernova, with the expanding velocities typically in the range $2{,}000$–$10{,}000\,\mathrm{km\,s}^{-1}$. Supernovae are classified into Type Ia, Type Ib and Type II by the presence of hydrogen. Type II supernovae are distinguished from the other two types by the existence of hydrogen lines in their spectra. The distinction between Type Ia and Type Ib is that the former has significant Si^+ absorption in its spectrum while the latter does not. Type II and Ib are rare in early-type galaxies, while Type Ia are observed in all types of galaxies (Table 10.2). In spiral galaxies, Type II and Ib

Table 10.2. Supernova rates in h^2SNU, where 1 SNU equals one supernova per century per $10^{10} L_{\odot B}$. [Data taken from Cappellaro et al. (1997)]

	E	S0	S0a-Sa	Sab-Sb	Sbc-Sc	Scd-Sd	others
Type Ia	0.23	0.32	0.44	0.30	0.41	0.43	0.48
Type Ib	≤ 0.05	≤ 0.07	0.28	0.16	0.36	0.07	0.39
Type II	≤ 0.07	≤ 0.07	0.28	0.94	1.33	2.17	0.73
All	0.23	0.32	1.01	1.40	2.11	2.66	1.60

supernovae are preferentially found in spiral arms where young (OB) stars form, whereas Type Ia do not reveal such a preference. These observational results suggest that the progenitors of Type II and Ib supernovae are probably short-lived massive stars, while the progenitors of Type Ia are long-lived low mass stars. As we will see in the following, these expectations are supported by the theory of stellar evolution.

(a) Type Ia Supernovae As discussed earlier in this section, a star with initial mass $M_s \lesssim M_{up} \approx 8 M_\odot$ evolves into a C/O white dwarf with a mass $\sim 0.6 M_\odot$. If such a white dwarf is part of a close binary system with a red giant or another white dwarf, it can accrete material from the companion. When the mass of the white dwarf reaches about $1.38 M_\odot$, carbon is ignited in the center of the degenerate C/O core (Woosley, 1990). This ignition produces a burning wave in the star, converting about half of the C and O into iron-peak elements in about one second, giving rise to a Type Ia supernova. The light curve and spectrum of the supernova depend sensitively on how fast the burning wave propagates through the star. The best agreement with observations is achieved in the carbon-deflagration (C-deflagration) model in which the deflagration wave propagates with a speed slightly smaller than the sound speed (Nomoto et al., 1984; Thielemann et al., 1993). In this model, about $0.6 M_\odot$ of ^{56}Ni is produced, which powers the light curve of the supernova by the radioactive decay of ^{56}Ni to ^{56}Co and then to ^{56}Fe. Because of the large nuclear energy release, the star is completely disrupted, returning all its mass to the interstellar medium.

Consider a binary of mass $M_B = M_1 + M_2$, with $M_2 \leq M_1$. In order for the primary, the initially more massive star, to result in a Type Ia supernova we have that $M_{B,\min} \leq M_B \leq M_{B,\max}$, with $M_{B,\min} \sim 3 M_\odot$ and $M_{B,\max} \sim 16 M_\odot$. The upper limit comes from the requirement that the mass of each component of the binary cannot exceed $8 M_\odot$, the maximum mass for producing a C/O white dwarf. The lower mass limit is more uncertain; it is taken to be $3 M_\odot$ to ensure that the primary star is massive enough that the C/O white dwarf eventually reaches the Chandrasekhar limit by accretion from the companion.

The specific rate of Type Ia supernovae (i.e. the number of supernova explosions per unit mass per unit time) depends on the formation rate of stars in the relevant mass range, as well as on the fraction of such stars that end up in close binary systems. The specific rate of Type Ia supernovae in a galaxy at time t can be written as

$$R_{Ia}(t) = \int_{M_{B,\min}/2}^{M_{B,\max}/2} dM_1 \, \frac{\phi(M_1)}{M_\odot} \int_{M_{\min,2}}^{M_1} dM_2 \, f_B(M_1, M_2) \psi(t - \tau_2). \tag{10.74}$$

Here $M_{\min,2} = \max[0, M_{B,\min} - M_1]$, $\phi(M)$ is the IMF normalized according to Eq. (9.33), $\psi(t)$ is the specific star-formation rate (i.e. the star-formation rate per unit mass) at t, and $\tau_2 = \tau(M_2)$ is the main-sequence lifetime of the companion star (which signals the time when the companion starts to overflow its Roche lobe, resulting in a transfer of mass to the primary). The function $f_B(M_1, M_2)$ indicates the fraction of stars of mass M_1 that have a binary companion of mass M_2

at a separation sufficiently close to allow for a Type Ia supernova. Unfortunately, this function is quite uncertain at the present, and one has to resort to simple empirical models to proceed.

Since the progenitors of Type Ia supernovae (C/O white dwarfs with a narrow mass range) have quite uniform properties, the luminosities of Type Ia supernovae are found to be within a narrow range around $L_B = 9.6 \times 10^9 \, L_\odot$. Because of this uniformity, Type Ia supernovae can be used as standard candles to measure cosmic distances (see §2.1.3).

(b) Type II Supernovae Type II supernovae are assumed to originate from stars with an initial mass $M_s \gtrsim M_{up} \approx 8 \, M_\odot$. As discussed above, such a massive star can develop an iron core during its late evolution that collapses catastrophically. Whether the energy released by this implosion results in a Type II supernova depends critically on the exact fraction of this energy (of the order of only $\sim 1\%$) that is transferred to the stellar envelope. If all massive stars can produce Type II supernovae, the rate is simply given by

$$R_{II}(t) = \int_{M_{up}}^{M_u} \frac{\phi(M)}{M_\odot} \psi(t - \tau_M) \, dM, \quad (10.75)$$

where $M_u \sim 100 \, M_\odot$ is the mass of the most massive stars. Since the lifetimes, τ_M, for these massive stars are short, we may neglect them in the expression for $R_{II}(t)$, and write

$$\frac{R_{II}(t)}{\psi(t)} \simeq \int_{M_{up}}^{M_u} \frac{\phi(M)}{M_\odot} \, dM. \quad (10.76)$$

For a Salpeter IMF, $R_{II}/\psi \sim 0.01 \, M_\odot^{-1}$, which implies that about one Type II supernova is produced for every $100 \, M_\odot$ of new stars formed. This rate is roughly consistent with observations (see Table 10.2).

10.3 Stellar Population Synthesis

In §10.2 we have shown that the stellar evolution models can be used to predict the evolutionary tracks of stars in the theoretical H-R (T_{eff}-L) diagram. In this section, we describe how one can assign a spectrum to a model star of given age and metallicity, and how the spectra of individual stars can be synthesized to predict the spectrum of an entire stellar population.

10.3.1 Stellar Spectra

(a) Empirical Spectra There are two different approaches to obtain the spectrum of a star with given L, T_{eff} and metallicity. The first is empirical. Here one uses a sample of nearby stars with measured (bolometric) absolute magnitudes, M_{bol}, effective temperatures, T_{eff}, and metallicities, Z, for which an accurate spectrum is available. The spectrum of a star with given M_{bol}, T_{eff} and Z can then be obtained by simple interpolation (or extrapolation) of the available data. In practice, the observed stars are characterized by their spectral types and luminosity classes (see §2.2). Since the spectral energy distribution (SED) of a star is roughly Planckian, there is a tight correlation between T_{eff} and spectral type. The luminosity class is a measure of the envelope size and atmospheric density of the star, and is therefore correlated with the surface gravity, $g \equiv GM_s/r_s^2$, where M_s and R_s are the mass and radius of the star. Thus, an interpolation in the T_{eff}-g plane can be used to predict the spectrum for a model star with given T_{eff} and g. Libraries of observational stellar spectra for such purposes are available in the literature (e.g. Gunn & Stryker, 1983; Pickles, 1998; Le Borgne et al., 2003). Fig.2.5 in Chapter 2 shows examples of empirical stellar spectra for a number of typical spectral types. As one can see, hot O and B stars emit most of their light in the UV, while cold K and M stars emit mostly in the optical and near-infrared. Spectral features, such as the Lyman break at 912Å in the spectra of OB stars, and the 4,000Å break in the spectra of redder stars, are clearly visible.

This empirical method is limited by the fact that accurate spectra are only available for stars in the solar neighborhood, which only sample a limited range of metallicities and abundance ratios. Since extrapolation is not very reliable, another approach is required to predict the SEDs for stars with metallicities and abundance ratios that differ significantly from those in the solar neighborhood.

(b) Theoretical Spectra An additional approach to modeling stellar spectra, which, in principle, is unhindered by available data, is to use theoretical models of stellar atmospheres. As one can see from Eq. (10.19), the temperature at the bottom of a stellar atmosphere is $T_{\rm eff}$, while the pressure there can be determined by the surface gravity, g. Assuming the atmosphere to be in both thermal and hydrostatic equilibrium, its density, temperature and ionization structures can all be determined once $T_{\rm eff}$, g and the chemical composition are known. The specific intensity of the radiation emerging from the surface of the atmosphere can then be obtained by following the radiative transfer through the atmosphere and using the constraint that the total energy emitted per unit time must be equal to the luminosity of the star. A detailed discussion of this problem is beyond the scope of this book, and we refer the reader to Mihalas (1978) for a comprehensive description. In practice, sophisticated stellar-atmosphere codes have been developed to predict the spectra of stars covering wide ranges of metallicity, effective temperature and surface gravity (e.g. Kurucz, 1992; Bessell et al., 1998; Allard & Hauschildt, 1995).

10.3.2 Spectral Synthesis

Consider now an ensemble of stars (e.g. a galaxy). The luminosity (at some given wavelength λ) of the galaxy at any time t can formally be written as

$$L_\lambda(t) = \int_0^t \mathscr{L}_\lambda^{\rm (cp)}(t-t')\Psi(t')\,{\rm d}t', \qquad (10.77)$$

where $\Psi(t)$ is the star-formation rate (i.e. the mass that turns into stars per unit time), and $\mathscr{L}_\lambda^{\rm (cp)}(\tau)$ is the luminosity per unit stellar mass of all stars of a coeval population (cp) of age τ. It is easy to see that $\mathscr{L}_\lambda^{\rm (cp)}$ can be written as

$$\mathscr{L}_\lambda^{\rm (cp)}(\tau) = \int \mathscr{L}_\lambda(m,\tau)\frac{\phi(m)}{{\rm M}_\odot}\,{\rm d}m, \qquad (10.78)$$

where m is the mass of a star, $\phi(m)$ is the IMF normalized according to Eq. (9.33), and $\mathscr{L}_\lambda(m,\tau)$ is the luminosity of a star with initial mass m at age τ. As discussed above, the quantity $\mathscr{L}_\lambda(m,\tau)$ can be obtained from stellar evolution models. Thus, once the star-formation rate, $\Psi(t)$, and the IMF, $\phi(m)$, are specified, the SED at any given time can be obtained by summing the spectra of coeval populations with different ages. Therefore, the spectra of coeval populations, i.e. $\mathscr{L}_\lambda^{\rm (cp)}$ as a function of τ, play a pivotal role in spectral synthesis modeling.

Fig. 10.5 shows the SEDs for coeval stellar populations at different ages. These distributions can be understood qualitatively using the results of stellar evolution and stellar spectra. At an age younger than $\sim 10^7$ yr, the spectrum is almost completely dominated by the blue main-sequence stars, which have strong emission in the UV due to their high effective temperatures. At an age of about 10^7 yr, the most massive stars have already evolved off the main sequence and become red supergiants, which causes a drop in UV flux and a rise in near-infrared flux in the synthesized spectrum. From a few times 10^8 yr to about 1×10^9 yr, the AGB stars maintain a relatively high flux in the near-infrared, while the UV flux continues to drop as

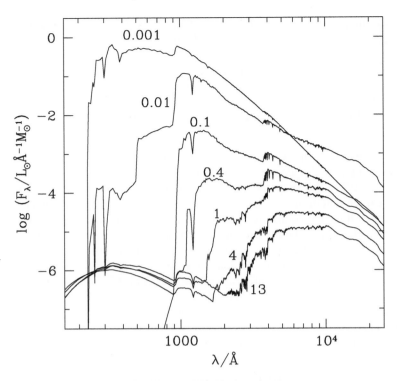

Fig. 10.5. The predicted spectra of a coeval stellar population at ages 0.001, 0.01, 0.1, 0.4, 1, 4, and 13 Gyr. The model assumes solar metallicity and a Salpeter IMF. [Based on data kindly provided by S. Charlot]

stars of lower masses evolve off the main sequence. After a few times 10^9 yr, stars on the red giant branch take over as the main contributors of the near-infrared flux. The noticeable rise in the far-UV flux after about 3–10 Gyr is produced by low-mass stars in their post-AGB phase.

With the spectral energy distribution, we can obtain the broad-band luminosities and colors for a galaxy using the photometric systems described in §2.1.1.

10.3.3 Passive Evolution

As a simple application of spectral synthesis modeling, consider a star-formation history that is given by a simple delta function $\Psi(t) \propto \delta(t - t_{\rm form})$, so that $L_\lambda(t - t_{\rm form}) = \mathscr{L}_\lambda^{\rm (cp)}(t - t_{\rm form})$. The evolution of such a coeval stellar population is often called passive evolution, and, as we will see in Chapter 13, is a reasonable approximation for the evolution of elliptical galaxies.

Fig. 10.6 shows how the broad-band luminosities and colors of a coeval stellar population evolve with age. As one can see, the broad-band luminosities of such a population all decrease with age, and the decrease is more rapid for a bluer band. This is mainly due to the disappearance of bright main-sequence stars and supergiants during the course of evolution. The galaxy also becomes redder with the passage of time, as more and more stars evolve off the main sequence. As we have seen in §10.1.5, stars with masses larger than $1.25 \, {\rm M}_\odot$ have lifetimes shorter than 4 Gyr. Thus, any coeval population older than ~ 4 Gyr contains only stars with masses $\lesssim 1 \, {\rm M}_\odot$. Such stars emit most of their light when they are on the giant branch in the H-R diagram, and so the integrated light from an old coeval population is dominated by giant-branch stars. Furthermore, because the giant branch is located in the red part

10.3 Stellar Population Synthesis

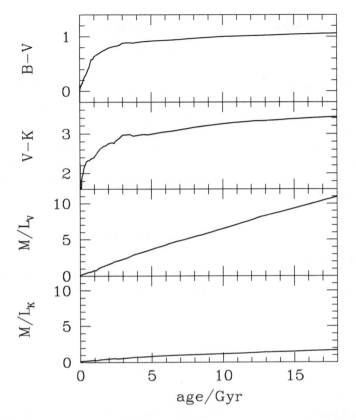

Fig. 10.6. The predicted $B-V$ and $V-K$ colors and mass-to-light ratios in the V and K bands for a coeval population of stars as functions of age. The model assumes solar metallicity and a Salpeter IMF. [Based on data kindly provided by S. Charlot]

of the H-R diagram, and since the effective temperature of the giant branch only depends weakly on the masses of the corresponding stars, old coeval populations have a red color that changes very slowly with time (see the upper two panels of Fig.10.6). Since the time that a star with $m < 1.25 \, M_\odot$ spends on the giant branch is much shorter than its main-sequence lifetime, the integrated luminosity is determined by the number of stars that evolve off the main sequence onto the giant branch. Suppose that the mass of the stars which evolve onto the giant branch at time t is $m_{\rm GB}(t)$ and that each star emits a total energy $E_{\rm GB}(M_{\rm GB})$ when on the giant branch. Then the luminosity of the stellar population can approximately be written as

$$L \propto E_{\rm GB}(m_{\rm GB}) \phi(m_{\rm GB}) \left| \frac{dm_{\rm GB}}{dt} \right|, \qquad (10.79)$$

with $\phi(m)$ the IMF. In the neighborhood of $m = 1 \, M_\odot$, the main-sequence lifetime of a star is related to its mass by $t_{\rm MS} \approx 10 (m/M_\odot)^{-3}$ Gyr (see §10.1.5), and so

$$m_{\rm GB}(t) \approx \left(\frac{t}{10 \, {\rm Gyr}} \right)^{-1/3} M_\odot. \qquad (10.80)$$

Inserting this into Eq. (10.79), and assuming a power-law IMF, $\phi(m) \propto m^{-b}$, one obtains that

$$\frac{d\ln L}{d\ln t} = \frac{b}{3} - \frac{1}{3}\left(4 + \frac{d\ln E_{GB}}{d\ln m_{GB}}\right). \tag{10.81}$$

Stellar evolution models indicate that $0 \lesssim d\ln E_{GB}/d\ln m_{GB} \lesssim 1$, and we thus see that the luminosity of an old, passively evolving stellar population will be declining with time as long as $b < 4$. For a Salpeter IMF, which has $b = 2.35$, Eq. (10.81) implies that $L \propto (t - t_{\rm form})^{-\gamma}$, with $0.55 \lesssim \gamma \lesssim 0.88$. More detailed calculations show that the value of γ not only depends on the IMF, but also on the metallicity of the stellar population and on the photometric passband (as is evident from the lower two panels of Fig. 10.6). In the B-band, the stellar population models of Worthey (1994), Vazdekis et al. (1996), and Bruzual & Charlot (2003) give $0.86 < \gamma_B < 1.0$ for a Salpeter IMF and $-0.5 < [{\rm Fe/H}] < 0.5$.

10.3.4 Spectral Features

(a) Spectral Discontinuities As one can see in Fig. 10.5, the synthesized spectra of stellar populations reveal several pronounced discontinuities, such as the Lyman break at 912 Å and the D(4000) break at 4000 Å. The strengths of these breaks can be quantified as follows:

$$D(912) = \int_{1000}^{1100} F(\lambda)\,d\lambda \bigg/ \int_{800}^{900} F(\lambda)\,d\lambda; \tag{10.82}$$

$$D(4000) = \int_{4050}^{4250} F(\lambda)\,d\lambda \bigg/ \int_{3750}^{3950} F(\lambda)\,d\lambda, \tag{10.83}$$

where the integration limits are in units of Å (Bruzual A., 1983). As one can inspect from Fig. 10.5, the amplitude of the Lyman break for a coeval population of stars increases rapidly at an age of 10^7 yrs when the most massive stars start to evolve off the main sequence. At an age of about 10^9, substantial numbers of UV photons are generated again, as the main-sequence turnoff mass drops below $2.5\,M_\odot$ and the evolved stars spend more time in the post-AGB phase (see §10.2.2). For a galaxy in which stars have formed over an extended period of time, rather than in a single burst, the change in the amplitude of the Lyman break is significantly smaller because of the dispersion in stellar age. Thus, a large amplitude of the Lyman break indicates that star formation occurred recently, and on a short time scale. Unfortunately, a sharp break at 912 Å in the spectrum of a real galaxy can also be produced by absorption due to neutral hydrogen in the galaxy or the intergalactic medium. This complicates the interpretation of the Lyman break in terms of the star-formation history of the galaxy in consideration.

The strength of the 4000 Å break for a coeval population is controlled mainly by the most massive stars on the main sequence, and increases almost monotonically with time from about 10^7 yr to about 10 Gyr. The amplitude of this break can therefore be used to indicate the mean stellar age of a galaxy, although it also depends significantly on metallicity.

(b) Spectral Indices Based on the theoretical spectra obtained from the spectral synthesis model described above, one can define several spectral indices that are sensitive to the abundance of certain elements and to the age of the stellar population. An example is the Lick indices introduced by Faber et al. (1985). The properties of some of the spectral indices, as given in Worthey et al. (1994) and Worthey & Ottaviani (1997), are listed in Table 10.3. Each index is defined by a central band of width $\Delta\lambda_0$ and two side bands. The side-band intensity, I_s, is defined to be the mean intensity over the two side bands, while the central-band intensity, I_c, is defined to be the mean over the central band. Some indices, such as Hβ, Hδ_A, Mgb, Fe$_1$, Fe$_2$, G and Na are measured in terms of an equivalent width defined as $W = (1 - I_c/I_s)\Delta\lambda_0$. Others, such as Mg$_1$,

Table 10.3. A subset of the spectral Lick indices.

Index	Feature	Central band [Å]	First side band [Å]	Second side band [Å]
G4300	CH	4282.625–4317.625	4267.625–4283.875	4320.125–4336.375
Mgb	Mgb	5160.625–5192.625	5142.625–5161.375	5191.375–5206.375
Fe5270	Fe^0, Ca^0	5245.650–5285.650	5233.150–5248.150	5285.650–5318.150
Fe5335	Fe^0, Cr^0, Ca^0	5312.125–5352.125	5304.625–5315.875	5353.375–5363.375
Na	Na D	5878.625–5911.125	5862.375–5877.375	5923.875–5949.875
Hβ	Hβ	4847.875–4876.625	4827.825–4847.875	4876.625–4891.625
Hγ_A	Hγ	4319.75–4363.50	4283.50–4319.75	4367.25–4419.75
Hδ_A	Hδ	4083.50–4122.25	4041.60–4079.75	4128.50–4161.00
Mg$_1$	MgH	5069.125–5134.125	4895.125–4957.625	5301.125–5366.125
Mg$_2$	MgH, Mgb	5154.125–5196.625	4895.125–4957.625	5301.125–5366.125
TiO$_1$	TiO	5938.375–5995.875	5818.375–5850.875	6040.375–6105.375
TiO$_2$	TiO	6191.375–6273.875	6068.375–6143.375	6374.375–6416.875

Mg$_2$, TiO$_1$ and TiO$_2$ are measured in terms of a magnitude defined as $-2.5\log(1 - I_c/I_s)$. The hope is that a systematic study of such indices and other spectral features can help to understand the properties of the underlying stellar population, such as age and metallicity.

10.3.5 Age–Metallicity Degeneracy

The main goal of the spectral evolution models described above is to use the observed spectra of galaxies to constrain their age and metallicity, or, more precisely, their star-formation history and element abundances. Unfortunately, there is a fundamental limitation to this, which arises from a degeneracy between stellar age and metallicity. As shown in §10.1.4, the evolution of a star with a given initial mass in the L-$T_{\rm eff}$ diagram depends on its chemical composition: stars with higher metallicities evolve faster. As a result, a population of young stars with a relatively high metallicity has a SED that looks very similar to that of an older population with lower metallicity. Detailed population synthesis calculations show that two stellar populations with the same $\tau Z^{3/2}$ (where τ is the age and Z is the metallicity) have virtually identical colors. In order to break this age–metallicity degeneracy one has to use additional information that probes the fine structure of the SED. The spectral indices defined above are ideally suited for this. The strength of the Balmer lines, such as Hβ and Hγ, are more sensitive to stellar age, while the metal line strengths, such as Mgb, Fe$_1$ and Fe$_2$, are more sensitive to metallicity. Therefore, measurements of these line indices can be used to break the age–metallicity degeneracy (see e.g. Worthey, 1994).

10.3.6 K and E Corrections

Because of the expansion of the Universe, the observed SED of a galaxy is redshifted with respect to the rest-frame SED. Thus, when we use a given waveband to image galaxies at different redshifts, we are in fact looking at these galaxies in different rest-frame wavebands. Consequently, even two identical galaxies will appear to have different absolute luminosities in the same observational waveband, if they have different redshifts. Therefore, in order to allow for meaningful comparisons between galaxies at different redshifts, one needs to correct their magnitudes to a common rest-frame waveband. Such corrections are called K corrections.

Consider a galaxy whose rest-frame SED is given by $L(\nu_e)$, which is defined so that $L(\nu_e)d\nu_e$ is the energy emitted by the galaxy in the frequency range $\nu_e \to \nu_e + d\nu_e$ per unit time, and denote the observed flux by $f(\nu_o)d\nu_o$. In the absence of absorption, energy conservation implies that

$$f(\nu_o)d\nu_o = \frac{L(\nu_e)d\nu_e}{4\pi d_L^2} \quad \text{where} \quad \nu_o = \nu_e(1+z)^{-1}, \tag{10.84}$$

or

$$f(\nu_o) = \frac{L(\nu_e)}{L(\nu_o)}(1+z)\frac{L(\nu_o)}{4\pi d_L^2}, \tag{10.85}$$

with $d_L = d_L(z)$ the luminosity distance out to redshift z (see §3.1.6). Consider a photometric waveband 'j' centered at ν_o. Then, in terms of magnitudes, we have that the observed apparent magnitude, m_j, is related to the absolute magnitude $\mathcal{M}_j$ that corresponds to the *rest-frame* waveband 'j', as

$$m_j = \mathcal{M}_j + 5\log\left[\frac{d_L(z)}{10\,\text{pc}}\right] + K_j(z), \tag{10.86}$$

where $K_j(z)$ is the corresponding K correction, defined by

$$K_j(z) = -2.5\log(1+z) - 2.5\log\left[\frac{L(\nu_e)}{L(\nu_o)}\right]. \tag{10.87}$$

The first term on the right-hand side of this equation represents the stretch of the waveband due to redshift, i.e. from $(d\nu_e/d\nu_o)$, while the second term represents the difference between the intrinsic spectral energies at $\nu = \nu_e$ and $\nu = \nu_o$. If we neglect the evolution in galaxy spectra and if we assume that galaxies of the same class (e.g. Hubble type) all have a similar SED, we can obtain the form of $L(\nu_e)$ for a class by observing the spectra of local galaxies for which redshift effects are negligible. Such spectra can then be used as templates to calculate the K corrections for the galaxies of the same class at higher redshift (e.g. Coleman et al., 1980).

Because stars evolve, the intrinsic SED of a galaxy will change with time (redshift), even if no new stars form. Such evolution is called passive evolution. Therefore, in order to check whether or not galaxies evolve with time only passively, one needs to correct the magnitudes of galaxies observed at different redshifts. Such corrections are called E corrections. Formally we can write the E correction as

$$E_j(z) = -2.5\log\left[\frac{L(\nu_e,t)}{L(\nu_e,t_0)}\right], \tag{10.88}$$

where $L(\nu_e,t)$ is the SED at time $t = t(z)$, while $L(\nu_e,t_0)$ is that at the present time. Such E corrections can be obtained using the spectral synthesis models described above. In addition to correcting for passive evolution, one can also use this technique to E-correct the magnitudes for any other star-formation history, if needed.

10.3.7 Emission and Absorption by the Interstellar Medium

Real galaxies are not made solely of stars. Rather, the stars are embedded in an interstellar medium (ISM) consisting of gas (both hot and cold) and dust. In order to interpret the observed spectra of galaxies, spectral synthesis models based purely on stellar spectra are insufficient. We also need to understand how the stellar spectra are attenuated by the ISM. Such attenuations are expected to be more important in gas-rich galaxies, such as present-day late-type galaxies and young galaxies at high redshift, but the details depend on the physical state of the ISM. Observations show that the ISM in local late-type galaxies in general has a very complex structure, making it extremely difficult to accurately model the ISM and its interactions with the interstellar radiation field. Extensive discussions about the properties of the ISM and related astrophysics can be found in Kaplan & Pikelner (1970), Spitzer (1978), and Osterbrock (1989). In this subsection, we select two topics that are most relevant for the interpretation of galaxy spectra: the emission of HII regions and the emission and extinction due to dust.

(a) Emission from HII Regions HII regions in galaxies are usually observed to be associated with OB stars whose UV radiation is believed to cause the photoionization of the surrounding ISM. In a static state, the number of ionizing photons emitted per unit time, $\dot{N}_{\rm ion}$, has to be balanced by the recombination rate, so that

$$\dot{N}_{\rm ion} = \alpha_{\rm B} \mathcal{N}_{\rm p} \mathcal{N}_{\rm e} V, \tag{10.89}$$

where V is the volume of the HII region, and the medium is assumed to be uniform with proton density $\mathcal{N}_{\rm p}$ and electron density $\mathcal{N}_{\rm e}$. The recombination coefficient $\alpha_{\rm B}$ in the above equation refers to that of Case B (see §B1.3) for the reasons to be explained below.

In a HII region, a hydrogen nucleus can recombine with a free electron to form a hydrogen atom. If the capture is to an excited state, the atom will cascade to the ground state, emitting line photons. Direct recombination to the ground state produces a Lyman continuum photon. If the optical depth of the HII region is small for the recombination photons (Case A), they will escape the HII region without causing further ionizations or excitations. In the opposite case (Case B), in which the optical depth of the HII region is large, the recombination photons will be absorbed, causing further ionizations or excitations of the gas. For real HII regions, Case B may be a better approximation, although the real situations must always be somewhere between the two. Under Case B, recombination to the ground state must always proceed through the emission of a Balmer series photon (see Fig. B1.1) and a transition from the $n=2$ level to the ground state, possibly (but not necessarily) accompanied by some lower-series photons (Pα, etc.). The decay from the $n=2$ state to the ground state is either through $2P \to 1S$ with the emission of a Lyα photon, or through $2S \to 1S$. The latter process must be accompanied by the emission of two continuum photons, in order to conserve angular momentum. Since the two-photon process (which is forbidden to the first order) is not very effective, a Lyα line should still be produced by recombination, even for HII regions that are optically thick to Lyα photons.

The emissivity of a given recombination line is given by

$$\varepsilon_{nn'} = \frac{h_{\rm P} \nu_{nn'}}{4\pi} \sum_{L=0}^{n-1} \sum_{L'=L\pm 1} \mathcal{N}_{nL} A_{nL,n'L'}, \tag{10.90}$$

where $\mathcal{N}_{nL}$ is the population density of the state with principal quantum number n and angular quantum number L, $A_{nL,n'L'}$ is the spontaneous transition coefficient from level (nL) to level $(n'L')$, and the second summation takes account of the selection rule $\Delta L = \pm 1$ that applies for a one-electron system. It is convenient to define an effective recombination coefficient for the level in consideration:

$$\mathcal{N}_{\rm p} \mathcal{N}_{\rm e} \alpha_{nn'}^{\rm eff} = \sum_{L=0}^{n-1} \sum_{L'=L\pm 1} \mathcal{N}_{nL} A_{nL,n'L'} = \frac{4\pi \varepsilon_{nn'}}{h_{\rm P} \nu_{nn'}}. \tag{10.91}$$

In order to obtain the emissivity (or equivalently, the effective recombination coefficient), we need to know the population density of the ionization state in question. In general, we can write

$$\mathcal{N}_{nL} = b_{nL}(2L+1)\left(\frac{h_{\rm P}^2}{2\pi m k_{\rm B} T}\right)^{3/2} e^{E_n/k_{\rm B} T} \mathcal{N}_{\rm p} \mathcal{N}_{\rm e}, \tag{10.92}$$

where E_n is the energy at level n, and $b_{nL} = 1$ if the system is in thermodynamic equilibrium. If the gas temperature, T, is different from that required to maintain thermodynamical equilibrium among different levels, the value of b_{nL} has to be obtained from the equation of statistical equilibrium:

$$\mathcal{N}_{\rm p} \mathcal{N}_{\rm e} \alpha_{nL}(T) + \sum_{n'>n}^{\infty} \sum_{L'} \mathcal{N}_{n'L'} A_{n'L',nL} = \mathcal{N}_{nL} \sum_{n'=n_0}^{n-1} \sum_{L'} A_{nL,n'L'}, \tag{10.93}$$

where α_{nL} is the recombination coefficient to the level nL. The summation on the right-hand side starts from level 1 ($n_0 = 1$) for Case A, and from level 2 ($n_0 = 2$) for Case B. Inserting Eq. (10.92) into the above equation we obtain

$$\frac{\alpha_{nL}}{(2L+1)} \left(\frac{2\pi m k_B T}{h_P^2}\right)^{3/2} e^{-\frac{E_n}{k_B T}} + \sum_{n'>n} \sum_{L'} b_{n'L'} A_{n'L',nL} = b_{nL} \sum_{n'=n_0}^{n-1} \sum_{L'} A_{nL,n'L'}. \quad (10.94)$$

This equation expresses b_{nL} in terms of $b_{n'L'}$ with $n' > n$. Since the population density must change continuously from the discrete states to the continuum state in an ionized gas of temperature T, we expect the population of a very high level ($n \to \infty$, $E_n \to 0$) to be approximately that at thermodynamic equilibrium. The above equation can then be solved downward in n for b_{nL}, starting from some high n where $b_{nL} \approx 1$. Inserting these solutions into Eq. (10.92) to get $\mathcal{N}_{nL}$ and using Eq. (10.91) one obtains $\alpha_{nn'}^{\text{eff}}$ for a given gas temperature T. With this procedure, the emissivity for any given recombination line can be obtained (see Osterbrock, 1989).

If the HII region is optically thin to the line in question, the line luminosity is simply

$$L_{nn'} \equiv 4\pi \varepsilon_{nn'} V = h_P \nu_{nn'} V \mathcal{N}_p \mathcal{N}_e \alpha_{nn'}^{\text{eff}} = h_P \nu_{nn'} \left(\frac{\alpha_{nn'}^{\text{eff}}}{\alpha_B}\right) \dot{N}_{\text{ion}}, \quad (10.95)$$

where the last equation follows from Eq. (10.89). For example, for $T \sim 10^4$ K and an electron density of $\mathcal{N}_e < 10^6 \text{ cm}^{-3}$, the luminosities in the Hα and Hβ lines are

$$L_{H\alpha} = 0.450 \, h_P \nu_{H\alpha} \dot{N}_{\text{ion}}, \quad (10.96)$$

$$L_{H\beta} = 0.117 \, h_P \nu_{H\beta} \dot{N}_{\text{ion}}. \quad (10.97)$$

Because recombination to the ground state in Case B must proceed through the $n = 2$ level, the Lyα emission is determined by the difference between the Case B recombination coefficient α_B and the effective recombination coefficient to $2S$. For $T \sim 10^4$ K and $\mathcal{N}_e < 10^4 \text{ cm}^{-3}$, $\alpha_{2S}^{\text{eff}} = 0.838 \times 10^{-13} \text{ cm}^3 \text{ s}^{-1}$, so that

$$L_{\text{Ly}\alpha} = 0.676 \, h_P \nu_{\text{Ly}\alpha} \dot{N}_{\text{ion}}. \quad (10.98)$$

Thus, by measuring the luminosity of a HII region in a recombination line, one can in principle infer the rate $\dot{N}_{\text{ion}}$ which, in turn, can be used to infer the number of OB stars that generate the ionizing photons. Since OB stars are short lived, their number is directly related to their birth rate. Eq. (10.95) therefore provides the principle for deriving the star-formation rate in a star-forming region from its recombination line strengths (see §10.3.8 below).

The above description serves as a simple example for how to calculate the emission line spectra of gaseous nebulae. A more extensive treatment can be found in Osterbrock (1989).

(b) Dust Extinction Most of the dust in galaxies is believed to be produced in the envelopes of AGB stars and injected into the ISM through stellar winds, although supernovae may also make a significant contribution. The presence of dust in the ISM of our own Galaxy can be inferred from the extinction and polarization of starlight (see e.g. Mathis, 1990). In general, the strength of dust extinction increases with decreasing wavelength in the range from the infrared to the UV, and so dust extinction causes the spectrum of a source to become redder. By measuring the reddening of stars with known (intrinsic) spectra, one can in principle infer the amount of dust in the ISM of our Galaxy. In practice, this also requires knowledge of the extinction law, which describes the variation of extinction with wavelength. For galaxies in which the individual stars cannot be resolved, one may infer the presence of dust either by studying the reddening in the composite spectrum, or from the infrared emission produced by the dust.

Because of interstellar dust, the spectrum we observe from a galaxy is not the original stellar spectrum. Such extinction must be taken into account when interpreting an observed spectrum in terms of stellar populations. In general, the strength of dust extinction along a given line-of-sight can be described by the optical depth τ_λ:

$$I_\lambda = I_{\lambda 0} e^{-\tau_\lambda}, \tag{10.99}$$

where $I_{\lambda 0}$ is the intensity at the wavelength λ that would be observed in the absence of extinction, and I_λ is the observed intensity. Conventionally, dust extinction is expressed in terms of an empirical extinction law,

$$k(\lambda) \equiv \frac{A_\lambda}{E(B-V)} \equiv R_V \frac{A_\lambda}{A_V}, \tag{10.100}$$

where $A_\lambda = (2.5 \log e)\tau_\lambda$ is the change in magnitude at wavelength λ due to extinction, $E(B-V) \equiv A_B - A_V$ is the color excess measured between the B and V bands, and $R_V \equiv A_V/E(B-V)$. Note that sometimes the extinction law is defined as

$$k'(\lambda) \equiv \frac{E(\lambda - V)}{E(B-V)} = k(\lambda) - R_V. \tag{10.101}$$

The advantage of working with R_V and $k(\lambda)$ or $k'(\lambda)$ is that they are insensitive to the total amount of dust along a line-of-sight, which is often expressed in terms of the dust-to-gas ratio,

$$\xi \equiv \tau_V/N_{21} \approx 0.921 R_V E(B-V)/N_{21}, \tag{10.102}$$

where N_{21} is the column density of neutral gas (HI and H_2) along the line-of-sight in units of $10^{21} \mathrm{cm}^{-2}$. With the above formulation, the dust extinction, A_λ, along a line-of-sight can be determined from the observed neutral hydrogen column density, once the dust-to-gas ratio, the value of R_V, and the form of the extinction law $k(\lambda)$ are known.

The extinction law depends on the physical properties (i.e. sizes and chemical composition) of the dust grains, and can be different from galaxy to galaxy, or even, within a given galaxy, from location to location. Empirical determinations of the wavelength dependence of interstellar extinction have been carried out for the Milky Way, the LMC and the SMC (see Fig. 10.7). They are quite similar for $\lambda \gtrsim 2{,}600\,\text{Å}$, but there is a marked difference at shorter wavelengths. Note that the curve for the Milky Way has a strong feature at $2{,}175\,\text{Å}$ (believed to be produced by graphite dust grains), which is weak for the LMC and almost absent for the SMC. The corresponding values of R_V are ~ 3.1 for the MW and the LMC, and ~ 2.7 for the SMC, although the scatter among different sight-lines is quite large (e.g. Cardelli et al., 1989). These differences indicate that the dust compositions in these galaxies are different.

In order to accurately model the dust extinction of a galaxy, one needs to know how dust grains and stars are distributed in the galaxy. Statistically, dust extinction may be described by the distribution of the optical depths, $\mathscr{P}(\tau_\lambda)\,d\tau_\lambda$, which gives the probability that a sight-line to a star has an optical depth in the range $\tau_\lambda \pm d\tau_\lambda/2$. If the τ_λ distribution is the same for stars of different types, the transmission function for the galaxy can be written as

$$T_\lambda \equiv \frac{L_\lambda}{L_{\lambda 0}} = \int_0^\infty d\tau_\lambda \mathscr{P}(\tau_\lambda) e^{-\tau_\lambda}, \tag{10.103}$$

where L_λ is the observed luminosity, and $L_{\lambda 0}$ is the luminosity in the absence of dust extinction (Charlot & Fall, 2000). If we know the extinction curve, we can infer $\mathscr{P}(\tau_\lambda)$ from the dust column density distribution $\mathscr{P}(N_\mathrm{d})$ which, for a given dust-to-gas ratio, is equivalent to the column density distribution of neutral gas. The dust column density distribution can be obtained if the dust distribution is known. For example, if the dust distribution is a uniform screen in front of all stars, then

$$\mathscr{P}(N_\mathrm{d}) = \delta(N_\mathrm{d} - N_\mathrm{d}^{\mathrm{sc}}), \tag{10.104}$$

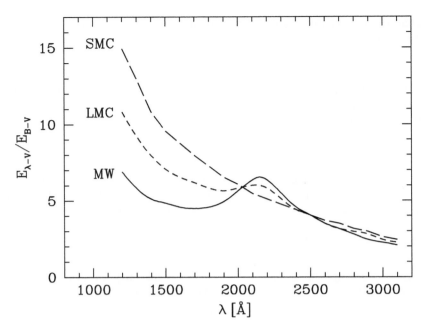

Fig. 10.7. The extinction laws, defined in terms of the ratio of color excesses, for the Milky Way (MW), the Large Magellanic Cloud (LMC) and the Small Magellanic Cloud (SMC). These are based on data published in Seaton (1979) and Howarth (1983) in the case of the MW, in Koornneef & Code (1981) in the case of the LMC, and in Bouchet et al. (1985) in the case of the SMC.

where δ is the Dirac delta function and N_d^{sc} is the dust column density of the screen. If dust grains have the same properties along all lines-of-sight, the transmission function is just

$$T_\lambda = \exp(-\tau_\lambda^{sc}), \qquad (10.105)$$

where τ_λ^{sc} is the optical depth of the screen.

If stars of different types have different distributions relative to the dust, the dust column density distribution and transmission function should be calculated separately for each type. This can be understood using the following simple example. Suppose that we have two groups of stars, each containing the same number of stars that are identical within the group but have different properties from the other group. Suppose that the dust distribution is an opaque screen and half of the total stars are in front of the screen and the other half behind it. The dust column density is therefore either zero or infinity, with equal probability. If stars in the two groups have the same distribution relative to the screen, then half of the stars in each group will be observed. If, on the other hand, one group is in front of the screen while the other is behind it, we only observe stars in one group. The resulting spectra will obviously be different in these two cases, although there is no change in the overall dust column density distribution. Although this particular example is obviously unrealistic, it may well be that young stars are located in regions with more dust than old stars, clearly complicating the treatment of dust extinction.

For a population of galaxies which, in a statistical sense, have identical compositions and distributions of dust, one may derive an effective extinction law that can be applied without having to model the detailed dust distribution. Such an analysis has been carried out for local starburst galaxies by Calzetti et al. (1994). For this particular population of galaxies, the UV continuum can be approximated by a power law, $F(\lambda) \propto \lambda^\beta$, and the power index β is found to be linearly correlated with

$$\tau_B \equiv \ln\left[\frac{F(H\alpha)/F(H\beta)}{2.86}\right], \qquad (10.106)$$

where $F(H\alpha)$ and $F(H\beta)$ are the intensities of the Hα and Hβ emission lines. In the absence of dust extinction, $F(H\alpha)/F(H\beta) \approx 2.86$, so that τ_B is a measure of the amount of dust extinction. Calzetti et al. divided their galaxies into several groups according to the value of τ_B. They used the group with the lowest value of τ_B as the 'unreddened' template, and examined the average spectrum in each of the other groups relative to the template spectrum. For each group this resulted in an effective extinction law given by

$$Q_n(\lambda) = \frac{\tau_n(\lambda)}{\tau_{B,n} - \tau_{B,1}}, \quad \tau_n(\lambda) \equiv -\ln\left[\frac{F_n(\lambda)}{F_1(\lambda)}\right], \qquad (10.107)$$

where a subscript 'n' refers to a certain group, with $n = 1$ the template group. Defined in this way, the effective extinction law is a measure of $\tau(\lambda)/\tau_B$. Once τ_B is observed for a given galaxy (either directly or using the τ_B-β relation), the effective extinction law can be used to estimate the amount of dust extinction.

So far we have described dust extinction in terms of empirical models. However, there have also been numerous efforts to understand dust extinction laws in terms of the physical properties of the dust. Detailed observations of dust absorption in the Milky Way, from the near-infrared to the UV, have revealed interesting spectral features in the extinction curve (see Fig. 10.8). For example, the broad peak around 2,175 Å is believed to be caused by graphite dust grains, while the features at 10 μm and 18 μm are best explained as due to silicate dust grains. Consequently, many models of dust absorption in galaxies have focused on these two types of dust grains, and the corresponding absorption cross-sections have been calculated as a function of grain size under various approximations (e.g. Mathis et al., 1977; Draine & Lee, 1984). A theoretical extinction curve can then be derived from these cross-sections together with an assumption regarding the size distribution of the dust grains and an assumption about the relative abundances of the two types of dust grains. The size distribution is usually assumed to be a power law, $dn/da \propto a^{-3.5}$ with a in the range from 0.005 μm to 0.25 μm. The relative abundance has to be tuned to match the observed extinction law. In the case of the Milky Way, this suggests roughly equal abundances of graphite and silicate grains (Fig. 10.8), while successively lower fractions of graphites are required for the LMC and the SMC.

(c) Dust Emission Dust grains are heated when absorbing photons. The temperature of the interstellar dust therefore depends on the local radiation field, and ranges from $\sim$ 20–40 K in the diffuse ISM to $\sim$ 100–500 K in star-forming regions. Temperatures as high as $\sim$ 1,000 K may also be possible but only for small grains ($a < 100$ Å) that are transiently heated by the absorption of single photons. Thus, interstellar dust is expected to radiate in the wavelength range from mid- to far-infrared. The SED of the dust emission is commonly expressed in terms of the dust emission curve, which is defined to be the ratio between the emission intensity along a line-of-sight and the corresponding hydrogen column density. Such an emission curve, obtained from the local ISM, is shown in the lower panel of Fig. 10.8. The observed dust emission curve can be understood in terms of three main components. Big grains with $a > 100$ Å, which are large enough to maintain thermal equilibrium with the local radiation field, have a SED that can be approximated by that of a gray body at a given temperature. Such grains dominate the emission at the far-infrared ($\lambda \gtrsim 50$ μm). Small grains with $a \lesssim 100$ Å, which can transiently be heated by absorption of single photons to a temperature as high as $\sim$ 1,000 K (Guhathakurta & Draine, 1989), can make a significant contribution to the emission in the mid-infrared ($\lambda \sim 20$–50 μm). In addition to the normal dust grains, polycyclic aromatic hydrocarbons (PAHs), which are transiently heated by UV photons, dominate the emission in the wavelength range $\lambda \sim 3$–10 μm, producing the strong bands at 3, 6, 8 and 11 μm due to the vibrational modes of C–C and C–H bonds (e.g. Desert et al.,

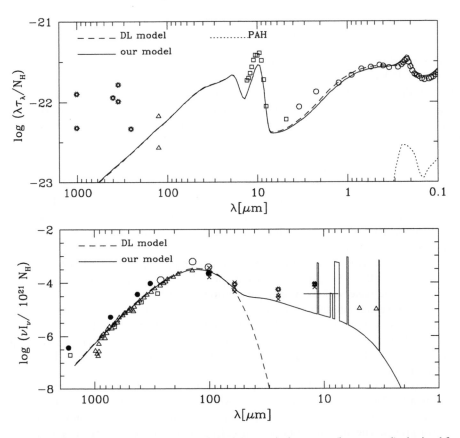

Fig. 10.8. The dust extinction curve (upper panel) and dust emission curve (lower panel) obtained from the local diffuse ISM. The solid curves are theoretical predictions by Silva et al. (1998) assuming three dust components: large grains in thermal equilibrium with the local radiation field, small grains which can be transiently heated, and polycyclic aromatic hydrocarbons (PAHs). The dashed curves are the predictions given by Draine & Lee (1984) assuming only the first component. [Adapted from Silva et al. (1998) by permission of AAS]

1990). The dust emission is assumed to be the sum of the SEDs of these three components, and the parameters of the model are usually tuned to fit the extinction and emission properties of the local diffuse ISM (Fig. 10.8).

10.3.8 Star-Formation Diagnostics

The population synthesis models described above can be used to translate certain properties of a galaxy's SED into a star-formation rate (SFR). In this subsection we discuss the most important diagnostics that are used to probe the star-formation rates of galaxies (see Kennicutt, 1998a, for more detailed discussion). As we will see, each of these diagnostics has its own pros and cons. In general, more accurate results are obtained by using a combination of various diagnostics, although this is observationally expensive, and often difficult for practical reasons.

(a) UV Continuum As we have seen in §10.1.5, massive stars with masses $M_s \gtrsim 5\,M_\odot$ have main-sequence lifetimes $\lesssim 10^8\,\mathrm{yr}$, much shorter than the typical age of a galaxy. Therefore, the number of massive stars in a galaxy is directly proportional to its current star-formation rate

(SFR). Since massive stars emit most of their energy as UV continuum radiation, the UV luminosity of a galaxy can in principle be used as a diagnostic of its current SFR, as long as the radiation manages to reach the observer without being absorbed. The optimal UV wavelengths to probe a galaxy's SFR are in the range 1250–2500Å, short enough to minimize the contamination due to older stellar populations, and long enough to prevent significant absorption due to hydrogen along the line-of-sight. Note, however, that these wavelengths are only accessible from the ground for galaxies in the redshift range $1 \lesssim z \lesssim 5$, because of the transmission of the atmosphere and the cosmological redshift. For lower redshift galaxies, space telescopes have to be used instead.

The conversion from the UV luminosity in the above mentioned wavelength range to a SFR can be derived from spectral synthesis models. Since the UV light is dominated by stars with $M_s \gtrsim 5\,\mathrm{M}_\odot$, while the total mass of newly born stars is dominated by stars with lower masses, the conversion from UV luminosity to SFR is sensitive to the assumed form of the IMF. For a Salpeter IMF, the SFR (for stars in the mass range 0.1–$100\,\mathrm{M}_\odot$) is related to the UV luminosity as

$$\frac{\dot{M}_\star}{\mathrm{M}_\odot\,\mathrm{yr}^{-1}} \approx 1.4 \times 10^{-28} \left(\frac{L_{\mathrm{UV}}}{\mathrm{erg\,s}^{-1}\,\mathrm{Hz}^{-1}}\right), \tag{10.108}$$

where L_{UV} is the luminosity within the wavelength range 1500–2800Å. This calibration is valid for galaxies in which the star-formation time scale is $\sim 10^8\,\mathrm{yr}$ or longer. For a burst with duration $< 10^8\,\mathrm{yr}$, Eq. (10.108) may underestimate the SFR, because the star-formation time scale is overestimated. In this case a re-calibration of the $\dot{M}_\star$-L_{UV} relation is required.

In addition to the dependence on the unknown form of the IMF, the main drawback of the UV diagnostic is its sensitivity to dust extinction. In particular, the dust distribution in real galaxies can be quite clumpy, and it may well be that star formation preferentially occurs in highly obscured regions. In that case the total UV luminosity may be dominated by the radiation coming from the older population (mainly blue horizontal branch and post-AGB stars) and thus be a poor diagnostic of the actual SFR. This potential correlation between the local amount of extinction and the local star-formation rate makes any correction for extinction extremely problematic.

(b) Nebular Emission Lines The interstellar medium in the neighborhood of young, massive stars is ionized by the Lyman continuum photons produced by these stars, thereby giving rise to HII regions. The recombination of this ionized gas produces hydrogen emission lines, which can be used as a SFR diagnostic because their flux is proportional to the Lyman continuum flux produced by the young, massive stars.

The nebular line most commonly used is the Hα line of hydrogen, but other lines, such as Hβ, Pα, Pβ, Brα, and Brγ can also be used. Only stars with masses $M_s \gtrsim 10\,\mathrm{M}_\odot$ (lifetimes $\lesssim 2 \times 10^7\,\mathrm{yr}$) contribute significantly to the Lyman continuum luminosity, and so the strength of a nebular emission line is almost an instantaneous measure of the SFR. The conversion from nebular-line strength to SFR can be obtained from spectral synthesis models. For solar abundances and a Salpeter IMF (for stars in the mass range 0.1–$100\,\mathrm{M}_\odot$), the conversion between Hα luminosity and SFR is given by

$$\frac{\dot{M}_\star}{\mathrm{M}_\odot\,\mathrm{yr}^{-1}} \approx 7.9 \times 10^{-42} \left(\frac{L_{\mathrm{H}\alpha}}{\mathrm{erg\,s}^{-1}}\right). \tag{10.109}$$

Because the Hα luminosity is a measure of the formation rate of massive stars with $M_s \gtrsim 10\,\mathrm{M}_\odot$, the derived SFR is again sensitive to the assumed IMF. In addition, dust extinction may also cause systematic errors, because the efficiency of dust absorption at the Hα wavelength may not be negligible. If the Hβ line is also measured, one may use Eq. (10.106) to correct for dust extinction. Alternatively, one can also use recombination lines in the infrared, which are less affected by dust extinction, although these lines are usually much weaker.

(c) Forbidden Lines For galaxies with redshifts $z \gtrsim 0.5$, the Hα emission line is redshifted out of the optical waveband. Probing their SFR using emission lines therefore requires another strong emission line blueward of Hα. The strongest emission feature in the blue is the [OII]λ3727 forbidden-line doublet. Unfortunately, the luminosities of forbidden lines depend not only on the local radiation field, but also on the ionization state and metallicity of the interstellar medium. Consequently, the flux in forbidden lines is not directly related to the UV luminosity produced by massive young stars. Despite of this, it is found that the [OII] emission is sufficiently well behaved that it can be calibrated empirically (through Hα at $z \sim 0.5$) to trace the SFRs in galaxies out to $z \sim 1.6$ (at higher redshifts even [OII] shifts out of the optical). Such calibrations give

$$\frac{\dot{M}_\star}{\mathrm{M}_\odot \, \mathrm{yr}^{-1}} \approx 1.4 \times 10^{-41} \left(\frac{L_{\mathrm{OII}}}{\mathrm{erg \, s}^{-1}} \right). \tag{10.110}$$

Since this relation is calibrated through Hα, it adds all the uncertainties of the $L_{\mathrm{H}\alpha}$-$\dot{M}_\star$ relation to its own substantial intrinsic uncertainties.

(d) Far-Infrared Continuum Typically the ISM associated with a star-forming region can be quite dusty, so that a significant fraction of the UV luminosity produced by massive young stars can be absorbed. This radiation heats the dust and is subsequently re-emitted in the far-infrared (FIR). Since the absorption efficiency of dust is strongly peaked in the UV, the FIR luminosity of a galaxy is often used as an additional SFR diagnostic. Unfortunately, this diagnostic suffers from various uncertainties as well. First of all, the conversion from an observed FIR luminosity to a UV luminosity depends on the opacity of the dust. If the dust is not optically thick, one needs to specify the fraction of UV photons that can escape unextincted. In principle, this problem can be avoided by combining the FIR flux with a measurement of the UV flux, if available. Secondly, observations show that the FIR spectra of galaxies contain both a 'warm' component ($\overline{\lambda} \sim 60\,\mu\mathrm{m}$) associated with dust around star-forming regions, and a cooler and more extended 'cirrus' component ($\overline{\lambda} \gtrsim 100\,\mu\mathrm{m}$) in which dust is heated not only by the UV radiation from massive young stars but also by the visible light from old stars. Thus, the cirrus component has to be excluded from the FIR luminosity in order to estimate the SFR. In general, these two uncertainties are difficult to quantify.

However, the FIR emission should be a good measure of the SFR for the population of starburst galaxies, which, by definition, have a significant fraction of their total stellar mass generated within a short period around the time of observation. In these systems, the FIR emission is therefore predominantly caused by young stars. Furthermore, starbursts are in general very dusty, and a very large fraction of the UV photons radiated by massive young stars are absorbed and re-emitted in the FIR. For starbursts with ages in the range of 10–100 Myr, stellar population synthesis models that assume a standard Salpeter IMF predict the following relation between SFR and FIR luminosity:

$$\frac{\dot{M}_\star}{\mathrm{M}_\odot \, \mathrm{yr}^{-1}} \approx 4.5 \times 10^{-44} \left(\frac{L_{\mathrm{FIR}}}{\mathrm{erg \, s}^{-1}} \right) \quad \text{(for starbursts)}, \tag{10.111}$$

where L_{FIR} is the luminosity over the wavelength range 8–1000 μm.

10.3.9 Estimating Stellar Masses and Star-Formation Histories of Galaxies

The population synthesis model described above can be used to predict the SEDs of galaxies from their star-formation histories. For the purpose of understanding galaxy formation and evolution, the inverse problem is actually more interesting: how can one infer the physical properties of galaxies (e.g. stellar masses and star-formation histories) from quantities that are directly observable? As we have seen above, the observable quantities, such as the luminosity and spectrum of a galaxy, are the convolutions of the star-formation history, initial mass function (IMF),

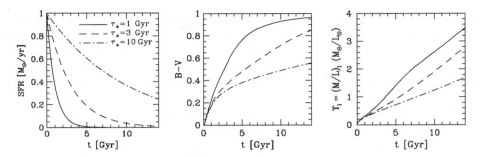

Fig. 10.9. The left-hand panel shows exponential star-formation histories of the form (10.112) for three different values of the characteristic time scale for star formation, $\tau_\star$, as indicated. The middle panel shows the evolution of the corresponding $B-V$ colors, obtained using the spectral evolution models of Bruzual & Charlot (2003), assuming a Salpeter IMF. The right-hand panel shows the corresponding evolution of the I-band mass-to-light ratios, Υ_I. [Kindly provided by Y. Lu]

dust extinction, and other quantities that characterize the formation and evolution of the galaxy. Hence, it should not come as a surprise that the inverse problem is highly degenerate. Consequently, careful analysis is required in order to identify the observables that are best suited for deriving the quantities of interest. As seen above, some emission line strengths and the luminosity in the UV and FIR can be used as diagnostics for the *current* star-formation rate in a galaxy. Here we describe how the stellar population synthesis models described above can be used to develop methods to derive the stellar masses and star-formation *histories* of galaxies from observations.

Using a population synthesis model and assuming various star formation histories, Bell & de Jong (2001) have shown that the stellar mass-to-light ratio, Υ, of a galaxy is tightly correlated with its color in the optical bands, but depends only moderately on the star-formation history. This is evident from Fig. 10.9, which shows the $B-V$ colors and I band mass-to-light ratios as function of time for stellar populations that have star-formation histories of the form

$$\Psi(t) = \frac{1}{\tau_\star} \exp\left(-\frac{t}{\tau_\star}\right), \qquad (10.112)$$

for three different values of the characteristic time scale for star formation, $\tau_\star$, as indicated. The middle panel of Fig. 10.9 shows that a given $B-V$ color can correspond to stellar populations with different combinations of t and $\tau_\star$. For example, for $\tau_\star = 1, 3$ and $10\,\mathrm{Gyr}$, a color $B-V = 0.5$ corresponds to population ages of $t = 2.7, 5.2$ and $10.8\,\mathrm{Gyr}$, respectively (for the Salpeter IMF adopted here). However, as is evident from the right-hand panel of Fig. 10.9, the I band mass-to-light ratios of these three different populations are remarkably similar at $\Upsilon_I \sim 1.0 \pm 0.3\,\mathrm{M}_\odot/\mathrm{L}_\odot$. As shown by Bell & de Jong (2001), the Υ color relation is well approximated by $\log \Upsilon_X = a_X + b_X \times \mathtt{color}$, where a_X and b_X are two constants which depend both on the photometric band in question, X, and on the adopted color. The slope, b_X, is rather insensitive to the assumed IMF and dust reddening. The normalization, a_X, though, depends critically on the shape of the IMF at the low-mass end. This is due to the fact that low-mass stars contribute significantly to the mass, but insignificantly to the luminosity and color of a stellar system. These results suggest that the optical broad-band colors may provide a reliable estimate of the stellar mass-to-light ratio, and hence the stellar mass of a galaxy, once the normalization of the Υ color relation is fixed either by adopting an IMF or by independent measurements of the stellar masses. Indeed, ignoring the uncertainty on the IMF, the typical error of the stellar mass-to-light ratio predicted with this method is about 0.1 dex in optical bands, and about 0.1–0.2 dex in the near-infrared. Values for a_X and b_X for different photometric bands, different colors, and different IMFs are presented in Bell et al. (2003b).

Based on stellar population synthesis models, Kauffmann et al. (2003b) find that the two stellar absorption-line indices, $D_n(4000)$, which are defined in a way similar to $D(4000)$ [see Eq. (10.83)] but with narrower bands (3850–3950 Å, 4000–4100 Å), and the Balmer absorption-line index $H\delta_A$ (defined in Table 10.3) can be used to constrain star-formation histories, dust attenuation and stellar masses of galaxies. The strength of the 4000 Å break is an indicator of the mean stellar age of a galaxy, as described in §10.3.4, while strong $H\delta$ absorption lines arise in galaxies that experienced a burst of star formation that ended about 0.1–1 Gyr ago. Together, these two indices allow one to constrain the mean stellar ages of galaxies and the fractional stellar mass formed in bursts over the past few Gyr. Since the overall intrinsic SED of a galaxy with a given total stellar mass is predominantly determined by the mean stellar age and the recent burst strength, a comparison of the predicted SED with the broad-band photometry then yields estimates of dust attenuation and stellar mass. For a given IMF, the typical 95% confidence range for the estimated stellar masses is ±40% (see Kauffmann et al., 2003b).

Finally, there have also been attempts to use the full spectrum to constrain the star-formation histories and stellar masses of individual galaxies (e.g. Panter et al., 2007). In such analyses, the star-formation history of a galaxy is either parameterized by some simple functional forms, or sampled in broad time bins. The observed spectrum is then fitted to the population-synthesized spectrum to infer the star-formation history and the stellar mass. As for the other two methods described above, this method is also limited by the uncertainties in the stellar population model, in the choice of the IMF, and in the treatment of the dust attenuation.

10.4 Chemical Evolution of Galaxies

In §10.2.3 we have seen that stars may return substantial fractions of their initial masses to the ISM at the end of their lifetimes through stellar winds and supernova explosions. Since the returned material is enriched in metals, this process can change the metallicity of the interstellar gas, thereby affecting the properties of the ISM. In this section, we first examine the chemical production of individual stars, and then use the results to model the chemical enrichment of galaxies.

10.4.1 Stellar Chemical Production

The mass ejection from a star of initial mass M_s (and some initial chemical composition) can be described by its lifetime $\tau(M_s)$ and by $M_j^{(\mathrm{ej})}(M_s)$, the total mass of element j ejected from the star, where the dependence of these quantities on the initial chemical composition is implicitly implied. The values of $\tau(M_s)$ and $M_j^{(\mathrm{ej})}(M_s)$ can, in principle, be calculated from stellar evolution models. However, it should be pointed out that this part of the theory involves modeling stars during their late evolutionary stages and some of the results are still fairly uncertain.

(a) Metal Production by Low-mass Stars Stars with initial masses $M_s \lesssim 8\,\mathrm{M}_\odot$ end up as C/O white dwarfs after their AGB phases (see §10.2). These white-dwarf remnants have masses in the range 0.4–1.4 $\mathrm{M}_\odot$ (with a typical mass $\sim 0.6\,\mathrm{M}_\odot$), and so the progenitor stars return most of their initial mass to the ISM. Observations show that a low-mass star can lose mass rapidly once it reaches the red-giant tip of the AGB, where mass loss is driven by stellar winds generated by radiation pressure from the star. Since metals (C, O and α elements) are synthesized by He-burning in the stellar core, one might think that these metals are locked up in the white dwarf so that low mass stars can only enrich the ISM with He. However, because of convection, metals produced below the He-burning shell can be brought up to the envelope by a process called dredge-up. In fact, the existence of carbon stars (most or them are M giants) suggests

Table 10.4. The ejected masses (in $M_\odot$) from winds and Type II SNe for stars with solar metallicity. [Based on data published in Marigo et al. (1996) and Portinari et al. (1998)]

M_s	$M_{ej}^{(wind)}$	$M_{ej}^{(total)}$	He	C	N	O	Ne	Mg	Fe
0.99	0.43	0.43	0.13	0.001	0.001	0.004			
1.49	0.91	0.91	0.27	0.003	0.002	0.009			
2.00	1.40	1.40	0.43	0.011	0.004	0.014			
3.00	2.34	2.34	0.73	0.014	0.007	0.022			
4.00	3.14	3.14	0.96	0.014	0.010	0.029			
6.00	4.70	4.70	1.58	0.017	0.019	0.044			
7.00	5.70	5.70	2.15	0.019	0.027	0.050			
9.00	0.32	7.69	2.99	0.074	0.040	0.178	0.021	0.007	0.057
12.0	0.53	10.56	3.95	0.170	0.055	0.712	0.125	0.028	0.121
15.0	0.78	13.13	4.70	0.268	0.067	1.27	0.177	0.036	0.098
20.0	1.90	17.89	6.01	0.281	0.089	2.76	0.217	0.069	0.186
30.0	17.11	22.82	8.28	0.310	0.128	3.08	0.504	0.124	0.022
40.0	34.00	37.94	19.05	3.03	0.211	2.06	0.417	0.039	0.133
60.0	52.65	57.91	32.14	4.93	0.343	2.66	0.474	0.056	0.234
100.0	91.89	97.88	55.48	11.40	0.682	4.28	1.07	0.130	0.251

that this dredge-up process can even bring metals all the way to the surface of an AGB star. Unfortunately, because of the uncertainties in the treatment of convection, theoretical predictions for the amount of metals pumped into the ISM by these stars remain uncertain. Observationally, the abundances of He, C, N, O in the ejected material from AGB stars can be estimated from the spectra of planetary nebulae in the Milky Way and the Magellanic Clouds. In order to determine the metal production (i.e. the mass in metals that was newly synthesized in the progenitor star) from such observations, however, one has to estimate the difference between the abundances of the ejecta and those of the ISM in the direct vicinity of the planetary nebula (whose abundances are presumably similar to those of the gas out of which the star originally formed). This is no easy task, because the difference is in general quite small. At the moment, chemical production by low-mass stars is still treated in a semi-analytical fashion, and calculations for stars with initial masses $\lesssim 8 M_\odot$ have been made by Renzini & Voli (1981) and Marigo et al. (1996), among others (see also Table 10.4).

(b) Metal Production by Massive Stars Massive stars, with masses $M_s \gtrsim 8 M_\odot$, enrich the ISM with metals via both stellar winds and their final explosions as core-collapse (Type II) SNe (e.g. Portinari et al., 1998). The structure of a massive star just prior to core collapse is relatively straightforward to calculate; it consists of an inert iron core surrounded from inside out by shells within which Si, O, Ne, C, He, and H are burning. When a Type II SN explodes, these shells of enriched material may be ejected by shock waves. However, the exact amount of material that can be blown away is unclear, as it is very hard to determine accurately the small fraction ($\sim 1\%$) of the explosive energy that is transferred to the stellar envelope. Equally uncertain is the change of chemical composition during and after the core collapse. When the iron core collapses, the shocks generated by the explosion may induce a burst of nuclear reactions in the shocked layer; in particular, a substantial amount of Si (which is burning in a shell just above the core) may be burnt into iron-peak elements. In the collapsing core itself, the temperature can become so high ($> 10^9$ K) that iron is photodisintegrated into α particles (He nuclei) and neutrons. These neutrons may collide with the iron-peak elements in the core and convert many of them into r-process elements (ones that are formed via rapid neutron capture). As a result, only a small amount of Fe may be ejected into the ISM by a Type II SN. Detailed observations of SN 1987A (which is a Type II) support these basic ideas (e.g. McCray, 1993), although many uncertainties remain at the quantitative level. Despite this, theoretical models have been used to predict the chemical yields from massive stars (e.g. Maeder, 1992; Woosley & Weaver, 1995; Thielemann et al., 1996). The lower part of Table 10.4 lists the total mass ejected by both stellar winds and core-collapse supernovae for stars with initial masses $M_s \gtrsim 8 M_\odot$. Note that stars in the mass range $8 M_\odot \lesssim M_s \lesssim 20 M_\odot$ contribute to the chemical enrichment mainly through Type II SNe, while more massive stars contribute mainly via their stellar winds. It should be kept in mind, though, that there are still substantial differences among the results obtained from different models.

(c) Metal Production by Type Ia SNe As discussed in §10.2.3, the progenitor of a Type Ia SN is a C/O white dwarf, which accretes material from a close companion, reaches the Chandrasekhar limiting mass for a C/O dwarf ($1.4 M_\odot$), and explodes. In the carbon-deflagration model (§10.2.3), calculations by Nomoto et al. (1984) and Thielemann et al. (1993) show that the deflagration wave can convert C and O very quickly into iron-peak elements in the inner region of a star. In particular, about $0.6 M_\odot$ of ^{56}Ni is produced, which is sufficient to power the light curve of a Type I SN by the radioactive decays of ^{56}Ni (and ^{56}Co) into ^{56}Fe. A substantial amount of α elements is predicted to be synthesized in the outer layers by the decaying deflagration wave, which is consistent with the observed spectra of Type Ia SNe near the peaks of their light curves. In the models considered by Thielemann et al. (1993), a Type Ia SN typically ejects about $1.38 M_\odot$ of material, of which $\sim 0.75 M_\odot$ is Fe, $\sim 0.05 M_\odot$ is C, $\sim 0.14 M_\odot$ is O, $\sim 0.005 M_\odot$ is Ne, $\sim 0.01 M_\odot$ is Mg, $\sim 0.15 M_\odot$ is Si, and $\sim 0.09 M_\odot$ is S. Typically, a Type Ia SN produces about 5–10 times as much Fe as does a Type II SN.

10.4.2 The Closed-Box Model

Having discussed the chemical production from individual stars, we now describe how to model the chemical enrichment in galaxies.

Consider a system consisting of gas and stars (e.g. a galaxy or part of a galaxy). We denote the masses in gas and stars by $M_{\rm gas}(t)$ and $M_\star(t)$, respectively, where we have made it explicit that both may change with time. In the so-called closed-box model, one assumes that there is no mass flow into or out of the system, so that the total mass $M_{\rm tot} = M_{\rm gas}(t) + M_\star(t)$ is a constant. The evolution of $M_{\rm gas}(t)$ can be written as

$$\frac{dM_{\rm gas}(t)}{dt} = -\Psi(t) + \mathscr{E}(t), \quad (10.113)$$

where $\Psi(t)$ is the star-formation rate of the system, and $\mathscr{E}(t)$ is the rate at which the stars return mass to the gas phase by stellar winds and supernovae (hereafter called the return rate). If we use m to denote the initial mass of a star, and $\phi(m)$ to denote the IMF [normalized according to Eq. (9.33)], we can write

$$\mathscr{E}(t) = \int_{m_\tau}^{\infty} m f_{\rm M}^{\rm (ej)}(m) \Psi(t - \tau_m) \frac{\phi(m)}{M_\odot} dm, \quad (10.114)$$

where m_τ is the mass of stars whose lifetime is τ, and $f_{\rm M}^{\rm (ej)}(m)$ is the fraction of a star's initial mass that is ejected:

$$f_{\rm M}^{\rm (ej)}(m) = \sum_j f_{{\rm M},j}^{\rm (ej)}(m) = 1 - \frac{m_{\rm rem}}{m}, \quad (10.115)$$

with the summation over all elements j, and $m_{\rm rem}$ the mass of the stellar remnant. If we further assume that the system is a one-zone system, so that the material ejected from the stars is uniformly mixed with the gas, then the time evolution of metals can be written as

$$\frac{d(ZM_{\text{gas}})}{dt} = -Z\Psi(t) + \mathscr{E}_Z(t), \tag{10.116}$$

where Z is the metallicity of the gas, defined by $Z \equiv M_Z/M_{\text{gas}}$, and

$$\mathscr{E}_Z(t) = \int_{m_\tau}^{\infty} m f_Z^{(\text{ej})}(m) \Psi(t - \tau_m) \frac{\phi(m)}{\text{M}_\odot} dm, \tag{10.117}$$

with

$$f_Z^{(\text{ej})}(m) = \sum_{j \in \text{metals}} f_{\text{M},j}^{(\text{ej})}(m). \tag{10.118}$$

Eqs. (10.113) and (10.116) can be solved once m_τ, $f_{\text{M},j}^{(\text{ej})}$, $\Psi(t)$, and $\phi(m)$ are known.

Under the instantaneous recycling approximation, all stars with masses above some limit m_{lim} are assumed to have negligible lifetimes, while those with $m < m_{\text{lim}}$ are assumed to live forever. With this approximation, the lower integration limit m_τ in Eqs. (10.114) and (10.117) can be replaced by m_{lim}, and $\Psi(t - \tau_m)$ by $\Psi(t)$. The return rate can then be written as $\mathscr{E}(t) = \Psi(t)\mathscr{R}$, with

$$\mathscr{R} = \int_{m_{\text{lim}}}^{\infty} m f_{\text{M}}^{(\text{ej})}(m) \frac{\phi(m)}{\text{M}_\odot} dm \tag{10.119}$$

the return fraction. The metal yield, y_Z, is defined as the ratio between the mass of *newly produced* metals (i.e. those metals produced in the star via nucleosynthesis) and the total mass which remains locked up in the remnant:

$$y_Z = \frac{1}{1 - \mathscr{R}} \int_{m_{\text{lim}}}^{\infty} m f_Z^{(\text{p})}(m) \frac{\phi(m)}{\text{M}_\odot} dm, \tag{10.120}$$

where $f_Z^{(\text{p})}(m)$ is the ratio between the mass of newly produced metals that have been ejected and the star's initial mass m. Note that

$$f_Z^{(\text{ej})}(m) = f_Z^{(\text{p})}(m) + f_{\text{M}}^{(\text{ej})}(m) Z(t - \tau_m), \tag{10.121}$$

where the second term on the right-hand side describes the mass fraction of metals that were already present in the star at formation and have been returned to the ISM due to stellar winds and/or supernovae. Using the instantaneous recycling approximation, $\mathscr{E}_Z(t)$ can then be written as

$$\mathscr{E}_Z(t) = \Psi(t) \left[\mathscr{R} Z(t) + y_Z(1 - \mathscr{R}) \right]. \tag{10.122}$$

Inserting the expressions for $\mathscr{E}(t)$ and $\mathscr{E}_Z(t)$ into Eqs. (10.113) and (10.116) and assuming y_Z to be independent of Z, we obtain the following solution for the evolution of the metallicity:

$$Z(t) = Z(0) + y_Z \ln \left[\frac{M_{\text{gas}}(0)}{M_{\text{gas}}(t)} \right], \tag{10.123}$$

where $M_{\text{gas}}(0)$ and $Z(0)$ are, respectively, the gas mass and metallicity at the starting time $t = 0$. If $M_{\text{gas}}(0) = M_{\text{tot}}$, the above solution can be written as

$$Z(t) = Z(0) - y_Z \ln \left[\mu(t) \right] \quad \text{with} \quad \mu(t) \equiv M_{\text{gas}}(t)/M_{\text{tot}}. \tag{10.124}$$

In this simple model, the metallicity of the gas is entirely determined by the metal yield and the instantaneous gas mass fraction. It is easy to see that this solution also applies to individual elements:

$$X_i(t) = X_i(0) - y_i \ln \left[\mu(t) \right], \tag{10.125}$$

where X_i is the mass fraction of the gas in element i and y_i is the yield of this element.

Finally, the above equations can also be used to determine the metallicity distribution of the stars. In particular, the mass of stars with metallicity $<Z$ is

$$M_\star(<Z) = M_{\rm tot} - M_{\rm gas}(t) = M_{\rm tot}\left\{1 - \exp\left[-\frac{Z-Z(0)}{y_Z}\right]\right\}. \tag{10.126}$$

10.4.3 Models with Inflow and Outflow

In the presence of gas flow, the total mass of the system changes with time according to

$$\frac{dM_{\rm tot}}{dt} = \mathscr{A}(t) - \mathscr{W}(t), \tag{10.127}$$

where $\mathscr{A}(t)$ and $\mathscr{W}(t)$ are the inflow and outflow rates of gas mass, respectively. The equation for the gas mass now becomes

$$\frac{dM_{\rm gas}}{dt} = -\Psi(t) + \mathscr{E}(t) + \mathscr{A}(t) - \mathscr{W}(t). \tag{10.128}$$

Under the assumption of uniform mixing of the gas in the system, the equation for the metallicity is

$$\frac{d(ZM_{\rm gas})}{dt} = -Z\Psi(t) + \mathscr{E}_Z(t) + Z_A\mathscr{A}(t) - Z(t)\mathscr{W}(t), \tag{10.129}$$

where Z_A is the metallicity of the inflowing gas. These equations can be solved once $\mathscr{A}(t)$ and $\mathscr{W}(t)$ are known.

In the special case where $\mathscr{A}(t) = 0$ and $\mathscr{W}(t) = \alpha(1-\mathscr{R})\Psi(t)$ (i.e. gas inflow is negligible while the gas outflow rate is proportional to the star-formation rate through a constant $\alpha > 0$), the instantaneous recycling approximation gives

$$Z(t) = Z(0) + \frac{y_Z}{1+\alpha}\ln\left[\frac{M_{\rm gas}(0)}{M_{\rm gas}(t)}\right]. \tag{10.130}$$

The quantity $y_Z/(1+\alpha)$ is called the effective yield, and its difference with respect to y_Z is a measure for the impact of the outflow (see also §11.8.2). Note that the effective yield is smaller than the true, nucleosynthetic yield, simply because some of the newly processed metals are now ejected from the system by the outflow. Since the total mass of the system decreases with time as $M_{\rm tot}(t) = M_{\rm tot}(0) - \alpha M_\star(t)$, the mass contained in stars with metallicity $<Z$ is

$$M_\star(<Z) = M_{\rm tot}(t) - M_{\rm gas}(t)$$
$$= \frac{M_{\rm tot}(0)}{1+\alpha}\left\{1 - \exp\left[-\frac{Z-Z(0)}{y_Z/(1+\alpha)}\right]\right\}. \tag{10.131}$$

In the case where $\mathscr{W}(t) = 0$, Eq. (10.129) can, with the use of Eqs. (10.127) and (10.128), be cast into the following form:

$$\frac{dZ}{du} + Z = Z_A + y_Z\left(1 - \frac{d\ln M_{\rm gas}}{du}\right), \quad \text{where } u \equiv \int\frac{dM_{\rm tot}}{M_{\rm gas}}. \tag{10.132}$$

The general solution is

$$Z = Z_A + y_Z\left(1 - Ce^{-u} - e^{-u}\int_0^u e^{u'}\frac{d\ln M_{\rm gas}}{du'}du'\right), \tag{10.133}$$

with C a constant. In the special case where the infall rate is equal to the rate at which mass is locked up in stars [$\mathscr{A}(t) = \Psi(t) - \mathscr{E}(t)$], so that $M_{\rm gas}$ is constant, the solution reduces to

$$Z = Z_A + y_Z\left\{1 - \exp\left[1 - \frac{M_{\rm tot}(t)}{M_{\rm gas}}\right]\right\}, \quad (M_{\rm gas} = \text{constant}). \tag{10.134}$$

In this model, $Z \approx Z_A + y_Z$ once M_{tot} becomes much larger than M_{gas}, and the mass in stars with metallicity $< Z$ is

$$M_\star(<Z) = M_{tot}(t) - M_{gas} = -M_{gas} \ln\left(1 - \frac{Z - Z_A}{y_Z}\right). \qquad (10.135)$$

Note that the above solutions are obtained under the assumption that the yield y_Z is independent of Z. Elements for which this assumption holds are called primary elements, while those for which the yield depends on the composition of the progenitor star are called secondary elements. Unfortunately, many elements, including important ones such as nitrogen, display secondary behavior (Talbot & Arnett, 1974). In general, therefore, the solutions for $Z(t)$ derived above have to be modified in order to account for secondary behavior of the elements.

10.4.4 Abundance Ratios

As discussed in §10.4.1, Type Ia SNe are the main producers of iron, while α elements, such as O, Ne, Mg, Si and S, are produced by both Type Ia and Type II SNe. Since the progenitors of Type II SNe are massive stars ($\gtrsim 8\,M_\odot$) that have lifetimes shorter than $\sim 10^7$ yr, while the progenitors of Type Ia SNe are mostly stars with masses about $3\,M_\odot$ that have lifetimes $\gtrsim 10^8$ yr, there is a time delay of at least $\sim 10^8$ yr between the onset of Type Ia supernova explosions and that of Type II for a coeval population of stars. Consequently, there is also a delay of Fe production relative to that of α elements. Here we discuss the observational consequences of this delay, and how these can be used to constrain the time scale for star formation in a galaxy.

If we consider a coeval stellar population as a closed box, the metallicity of the ISM (e.g. [Fe/H]) must increase with time, as more and more metals are injected by supernova explosions. In the early stage where only Type II supernova explosions can occur, the abundance ratio [α/Fe] must be high and has roughly a constant value determined by the enrichment pattern of Type II SNe. After about 10^8 yr when Type Ia SNe start to make a significant contribution to the chemical enrichment, the abundance ratio [α/Fe] must decrease with time (or with the total metallicity, [Fe/H]). As the age reaches about 10^9 yr when the enrichment of the ISM is dominated by Type Ia SNe, the abundance ratio [α/Fe] remains at a low constant level determined by the enrichment pattern of Type Ia supernovae. The chemical enrichment by a coeval stellar population should therefore have a pattern similar to that shown in Fig. 10.10.

One can also use the time delay between Type Ia and Type II SNe to constrain the star-formation history by observing the metallicities and abundance ratios of low mass stars (rather than the ISM). Low-mass stars, such as F and G stars, have lifetimes comparable to the age of the Universe. These stars, therefore, sample the entire star-formation history of a galaxy. If all stars formed over a short period of time ($\lesssim 10^8$ yr), then the gas out of which these stars formed could only have been enriched by Type II SNe. Consequently, all stars are expected to have an overabundance in α elements relative to Fe. On the other hand, if star formation lasted for a long period, so that the enrichment by Type Ia SNe is important, then the abundance of Fe is expected to be more enhanced for stars with younger ages. By studying the metallicity and [α/Fe] ratio of low-mass stars in a galaxy in detail, it is thus possible to constrain the time scale over which star formation took place. Unfortunately, there is a complication here: the expected [α/Fe] ratio depends not only on the star-formation history, but also on the form of the IMF that determines the mass spectrum of stars at formation. For instance, if the IMF in a galaxy is skewed to massive stars (i.e. the IMF is top-heavy), the fraction of stars that end up as Type II supernovae is enhanced, leading to a higher [α/Fe] ratio than that predicted by the standard IMF. For a stellar population with a Salpeter IMF and a star-formation history given

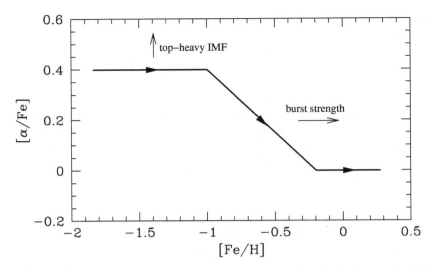

Fig. 10.10. A schematic of the chemical enrichment pattern of the ISM for a single coeval burst of star formation. Time advances along the thick curve as indicated by the arrows. The thin arrows indicate the impact of making the IMF more top-heavy and of increasing the strength of the burst.

by a Gaussian with a FWHM Δt, closed-box evolution gives the following relation between Δt and $[\alpha/\text{Fe}]$:

$$\log\left(\frac{\Delta t}{\text{Gyr}}\right) \approx 1.2 - 6[\alpha/\text{Fe}] \tag{10.136}$$

(Thomas et al., 2005).

10.5 Stellar Energetic Feedback

In broad terms, stars can output energy in three different channels: radiation, neutrino emission, and mass flow. The energy output in radiation can be obtained by integrating the luminosity of a star over its lifetime. Typically, a star of solar mass emits about $10^{51.5}$ erg in radiation, while the corresponding numbers for $10\,\text{M}_\odot$ and $100\,\text{M}_\odot$ stars are $\sim 10^{52.5}$ erg and $\sim 10^{53.5}$ erg, respectively. Note that the numbers quoted above do not include contributions from supernova explosions.

Of particular interest is the kinetic energy that is loaded in supernova ejecta and stellar winds, because such energy may be effectively injected into the ISM, affecting the gas properties and star formation in galaxies. In this section we examine how much mass-loaded kinetic energy is expected from individual stars. We also present models to describe the effects of such energy injections on the properties of the ISM.

10.5.1 Mass-Loaded Kinetic Energy from Stars

(a) Stellar Winds As we have seen in §10.2, some stars can lose substantial amounts of mass via stellar winds during their late evolutionary stages. Such mass loss can be calculated from stellar evolution theory, as described in §10.4.1, and some results obtained from such calculations are given in Table 10.4. Conventionally, a stellar wind is specified by its mass-loss

Table 10.5. Supernova mass ejection (in $M_\odot$) and kinetic energy (in 10^{51} erg).

SN	1987A	1969L	1980K	1993J	1991T	1989B	1992A	1991bg
Type	SNII-S	SNII-P	SNII-L	SNII/I	SNIa	SNIa	SNIa	SNIa
$M_{\rm ej}/M_\odot$	15	17	2.2	3.26	1.1	0.87	0.8	0.7
E_{51}	1.7	1.7	1.0	1.7	1.48	1.18	1.0	0.69

rate, $\dot{M}$, and its terminal velocity, v_∞. The kinetic energy of the wind, sometimes called the wind 'luminosity', is

$$L_{\rm wind} = \frac{1}{2}\dot{M}v_\infty^2. \tag{10.137}$$

There have been many efforts to relate both $\dot{M}$ and v_∞ to the global properties of stars, such as luminosity, effective temperature, mass and radius, from both observational and theoretical considerations (see Chiosi & Maeder, 1986, for a review). For example, based on about 300 stars of all spectral types and luminosities, Waldron (1984) suggests the following relations:

$$\log \dot{M} = 1.07\log(L/L_\odot) + 1.77\log(R/R_\odot) - 14.30, \tag{10.138}$$

$$\log v_\infty = 1.77\log(T_{\rm eff}/T_\odot) + 1.97, \tag{10.139}$$

$$\log L_{\rm wind} = 1.96\log(L/L_\odot) + 25.13, \tag{10.140}$$

where $\dot{M}$ is in $M_\odot\,{\rm yr}^{-1}$, v_∞ in $\rm km\,s^{-1}$ and $L_{\rm wind}$ in $\rm erg\,s^{-1}$. Unfortunately, the scatter in such relations is in general quite large. From these relations we see that the typical wind velocity for supergiant OB stars is about $2000\,\rm km\,s^{-1}$, and the corresponding kinetic energy is $E_{\rm kin} \sim 4\times 10^{50}(M_{\rm wind}/10\,M_\odot)(v_\infty/2000\,{\rm km\,s^{-1}})^2$ erg. As we will see below, this energy is comparable to the kinetic energy output from a supernova.

(b) Supernova Explosions For supernova explosions, the typical mass-loaded kinetic energy can be estimated by modeling in detail supernovae observed in the local Universe. Table 10.5 list some of the results (adapted from Arnett, 1996). As one can see, the mass-loaded kinetic energies are quite similar for Type Ia and Type II SNe. As an approximation we may write

$$E_{\rm SN} = E_{51} \times 10^{51}\,{\rm erg}, \quad \text{with} \quad E_{51} \sim 1.0. \tag{10.141}$$

For Type Ia SNe, the ejecta masses are in a small range around $M_{\rm ejecta} = 1\,M_\odot$, as expected from the fact that their progenitors are C/O white dwarfs near the Chandrasekhar mass limit ($1.4\,M_\odot$) and that such SNe do not leave any remnants behind. In this case, the initial velocity of the ejecta is $\sim 10^4\,\rm km\,s^{-1}$. In contrast, the ejecta mass of Type II SNe can change substantially from object to object, as their progenitors can cover a large mass range and the remnant mass depends on the details of the properties of the progenitor (see §10.2.3). In this case, the initial velocity is difficult to predict. Fortunately, in applications to galaxy formation and evolution, only the total kinetic energy is relevant, because here the energy in the ejecta is transferred into a gas component that has mass much larger than that of the ejecta themselves.

10.5.2 Gas Dynamics Including Stellar Feedback

In §8.6 we have seen that a general description of the evolution of gaseous halos in terms of the fluid equations needs to include the source terms that describe the gas consumption due to star formation, as well as the energy and mass injection due to stellar evolution. Assuming that the injected material is well mixed with the interstellar gas so that the total gas can be considered as a single fluid, the fluid equations are given by Eqs. (8.155)–(8.157). Here we describe the source terms S_m, $\mathbf{S}_\mathrm{mom}$, S_e, which are the changes per unit time in mass density, momentum density, and energy density, due to the consumption of gas by star formation and the injection of gas by stellar winds and supernovae. For simplicity, we consider scales that are much larger than individual stars, so that the velocities of the injected material average out. In other words, we assume that the kinetic energy of the injected material has already been thermalized, which allows us to set $\mathbf{S}_\mathrm{mom} = 0$.

Let $\psi(\mathbf{x},t)$ describe the local *specific* star-formation rate (i.e. the star-formation rate per unit mass), then for a given IMF, $\phi(m)$, we can write

$$S_\mathrm{m}(\mathbf{x},t) = \rho(\mathbf{x},t)\left[-\psi(\mathbf{x},t) + \varepsilon(\mathbf{x},t)\right], \tag{10.142}$$

where

$$\varepsilon(\mathbf{x},t) = \int_0^\infty \int_0^t m f_\mathrm{M}^{(\mathrm{ej})}(m,t') \frac{\phi(m)}{M_\odot} \psi(\mathbf{x}, t-t')\, dm\, dt' \tag{10.143}$$

is the local *specific* rate of mass feedback, with $f_\mathrm{M}^{(\mathrm{ej})}(m,t)dt$ the fraction of the star's initial mass m that is returned to the ISM in the time interval $[t, t+dt]$. If we assume that most of the mass is ejected from a star at the end of its lifetime, then $f_\mathrm{M}^{(\mathrm{ej})}(m,t) \approx \delta(t - \tau_m) f_\mathrm{M}^{(\mathrm{ej})}(m)$, with τ_m the lifetime for a star of mass m, and

$$\varepsilon(\mathbf{x},t) = \int_0^\infty m f_\mathrm{M}^{(\mathrm{ej})}(m) \frac{\phi(m)}{M_\odot} \psi(\mathbf{x}, t - \tau_m)\, dm. \tag{10.144}$$

Similarly, we can write

$$S_\mathrm{e}(\mathbf{x},t) = \rho(\mathbf{x},t) \int_0^\infty m f_\mathrm{E}^{(\mathrm{ej})}(m) \frac{\phi(m)}{M_\odot} \psi(\mathbf{x}, t - \tau_m)\, dm, \tag{10.145}$$

where $f_\mathrm{E}^{(\mathrm{ej})}(m)$ is the energy *per unit mass* that is ejected by a star with mass m at the end of its life. For supernova explosions where $f_\mathrm{E}^{(\mathrm{ej})}$ is independent of m, the rate of energy injection per volume reduces to

$$S_\mathrm{e}(\mathbf{x},t) = \rho(\mathbf{x},t)\left[R_\mathrm{Ia}(\mathbf{x}) + R_\mathrm{II}(\mathbf{x})\right] E_\mathrm{SN}, \tag{10.146}$$

where $R_\mathrm{Ia}(\mathbf{x})$ and $R_\mathrm{II}(\mathbf{x})$ are the local *specific* rates for Type Ia and Type II SNe, respectively (see §10.2.3).

For a given IMF, we can use the mass return discussed in §10.4.1, the supernova rates discussed in §10.2.3, and the energy feedback discussed in the last subsection, to calculate both S_m and S_e. It can be shown that stellar mass loss provides most of the feedback mass but little energy, while supernova explosions provide most the kinetic energy but relatively little mass.

11
Disk Galaxies

In the previous chapters we have discussed the various physical ingredients that play a role in galaxy formation, from the growth, collapse and virialization of dark matter halos, to the formation of stars out of the baryonic material that cools within these halos. In this chapter we combine all this information to examine the structure and formation of disk galaxies.

As we have seen in Chapter 2, disk galaxies in general consist of a disk component made up of stars, dust and cold gas (both atomic and molecular), a central bulge component, a stellar halo, and a dark halo. The disk itself reveals spiral arms and often – in roughly half of all cases – a central bar component. Any successful theory for the formation of disk galaxies has to be able to account for all these components. In addition, it should also be able to explain a variety of observational facts (see §2.3.3), the most important of which are:

- Brighter disks are, on average, larger, redder, rotate faster, and have a smaller gas mass fraction.
- Disk galaxies have flat rotation curves.
- The surface brightness profiles of disks are close to exponential.
- The outer parts of disks are generally bluer, and of lower metallicity than the inner parts.

We start in §11.1 with a description of simple mass models for disks, and discuss how one can infer the dynamical masses of disk and halo from the observed kinematics. In addition, we also give a brief description of the angular momentum of disk galaxies and show how this can be used to infer the presence of an extended dark matter halo. In §11.2 we present various models, of increasing complexity, for the formation of disk galaxies, and use them to investigate the origin of the observed distribution of disk sizes. The scaling relations of disk galaxies and their origin are discussed in §11.3. In §11.4 we examine various possible explanations for the fact that disks have exponential surface brightness profiles. Disk instabilities and their role in star formation, in driving secular evolution, and in creating spiral arms are discussed in §§11.5 and 11.6. Finally, the stellar populations and chemical properties of disk galaxies are described in §§11.7 and 11.8, respectively.

Throughout this chapter we will use the terms disk galaxy and spiral galaxy without distinction.

11.1 Mass Components and Angular Momentum

Before addressing the formation of disk galaxies, we take a closer look at their dynamical properties. After a brief overview of realistic potential-density pairs for disks, we discuss how the kinematics of disk galaxies can be used to infer their dynamical masses and their angular momentum distributions, and demonstrate that both of these indicate the presence of an extended dark halo.

11.1.1 Disk Models

(a) Potential-Density Pairs As we have seen in §2.3.3, the disks of spiral galaxies typically have exponential surface brightness distributions. It is therefore common to model the disk as an infinitesimally thin, exponential disk, with surface density distribution

$$\Sigma(R) = \Sigma_0 \exp(-R/R_{\rm d}), \tag{11.1}$$

where $R_{\rm d}$ is the scale length of the disk. The total mass of the disk is $M_{\rm d} = 2\pi\Sigma_0 R_{\rm d}^2$. As originally shown by Toomre (1963), the potential of an infinitesimally thin disk with surface density distribution $\Sigma(R)$ can be written as

$$\Phi(R,z) = -2\pi G \int_0^\infty J_0(kR)\tilde\Sigma(k) e^{-k|z|} dk, \tag{11.2}$$

with $J_0(x)$ the cylindrical Bessel function of order zero, and

$$\tilde f(k) = \int_0^\infty J_0(kx) f(x) x\, dx \tag{11.3}$$

the Hankel transform of $f(x)$, also known as the Fourier–Bessel transform. For the thin exponential disk given by Eq. (11.1) this reduces to

$$\Phi(R,z) = -2\pi G\Sigma_0 R_{\rm d}^2 \int_0^\infty \frac{J_0(kR) e^{-k|z|}}{[1+(kR_{\rm d})^2]^{3/2}}\, dk. \tag{11.4}$$

Although the infinitesimally thin, exponential disk is used extensively for modeling rotation curves (see §11.1.2) and disk formation (see §11.2), it is generally inadequate for more detailed dynamical models. After all, realistic disks have a non-zero thickness. In principle it is straightforward to construct thick disk models, as long as the density distribution is separable, i.e. can be written as the product of a function of R and a function of z. Consider an infinitesimally thin disk, whose surface density is $\rho(R,z) = \Sigma(R)\delta(z)$, and with corresponding potential $\Phi(R,z) = g(R,z)$. Here Σ and g are arbitrary functions that are related via the Poisson equation. One can thicken this disk by replacing the Dirac delta function, $\delta(z)$, by any other function $h(z)$ that describes the vertical density distribution. Since potentials are additive, the potential of the thick disk simply follows from summing the potentials of an infinite number of infinitesimally thin disks, properly weighted by $h(z)$:

$$\Phi(R,z) = \int_{-\infty}^\infty g(R, z-z') h(z')\, dz'. \tag{11.5}$$

For Eq. (11.2) this gives

$$\Phi(R,z) = -2\pi G \int_0^\infty dk\, J_0(kR) \int_{-\infty}^\infty dz'\, \tilde\rho(k,z') e^{-k|z-z'|}, \tag{11.6}$$

with $\tilde\rho(k,z)$ the Hankel transform of $\rho(R,z)$ in the R variable.

As discussed in §2.3.3, the vertical density distribution of spiral galaxies is often well fit by an exponential profile, so that

$$\rho(R,z) = \rho_0 \exp(-R/R_{\rm d})\exp(-|z|/z_{\rm d}). \tag{11.7}$$

The face-on projection of this density distribution is given by Eq. (11.1), with $\Sigma_0 = 2\rho_0 z_{\rm d}$. For this particular case, the potential given by Eq. (11.6) reduces to a single integration:

$$\Phi(R,z) = -2\pi G\Sigma_0 R_{\rm d}^2 \int_0^\infty dk\, \frac{J_0(kR)}{[1+(kR_{\rm d})^2]^{3/2}}\, \frac{e^{-k|z|} - (kz_{\rm d}) e^{-|z|/z_{\rm d}}}{1-(kz_{\rm d})^2} \tag{11.8}$$

(Kuijken & Gilmore, 1989b). A downside of this double exponential disk is its unphysical characteristic that the derivative $\partial \rho/\partial z$ is discontinuous at $z=0$. Furthermore, the oscillatory behavior of the Bessel function in the integrand makes the numerical evaluation of the potential and its derivatives tedious.

An alternative, better behaved model for a thick disk is the exponential spheroid. This model, like the double-exponential disk, has an exponential surface brightness along both the radial and vertical directions. Consider a sphere with projected surface brightness given by Eq. (11.1). The corresponding density distribution follows from the Abel integral equations,

$$\Sigma(R) = 2\int_R^\infty \frac{\rho(r) r\, dr}{\sqrt{r^2-R^2}} \quad \text{and} \quad \rho(r) = -\frac{1}{\pi}\int_r^\infty \frac{d\Sigma(R)}{dR}\frac{dR}{\sqrt{R^2-r^2}}, \quad (11.9)$$

which give

$$\rho(r) = \rho_0 K_0(r/R_d), \quad (11.10)$$

with $\rho_0 = \Sigma_0/\pi R_d$, and K_0 the modified Bessel function. Replacing r by the spheroidal coordinate $m = \sqrt{R^2 + z^2/q_d^2}$ we obtain an axisymmetric model that has exponential surface brightness when projected along any axis, namely

$$\Sigma_i(x,y) = \frac{q_d}{q_d'}\Sigma_0 \exp(-R^*/R_d). \quad (11.11)$$

Here $R^* = \sqrt{x^2 + y^2/q_d'^2}$, and q_d' is the projected axis ratio related to the intrinsic flattening, q_d, and the inclination angle, i, through[1]

$$q_d'^2 = \cos^2 i + q_d^2 \sin^2 i. \quad (11.12)$$

When $q_d \ll 1$ the model represents exponential disks of non-zero thickness, and with total mass $M_d = 2\pi^2 q_d \rho_0 R_d^3$. The potential of the exponential spheroid is

$$\Phi(R,z) = \frac{GM_d}{\pi R_d^2}\int_0^\infty \frac{t K_1(t/R_d)\, d\tau}{(\tau+1)\sqrt{\tau+q_d^2}}, \quad (11.13)$$

with

$$t = \sqrt{\frac{R^2}{\tau+1} + \frac{z^2}{\tau+q_d^2}} \quad (11.14)$$

(van den Bosch & de Zeeuw, 1996). Thus, the potential can also be written as a single integral, similar to the case of the double exponential model in Eq. (11.7). However, the numerical evaluation of $\Phi(R,z)$ is much simpler here, because the modified Bessel function, $K_1(x)$, decays similar to an exponential and is positive definite. Furthermore, it also has the advantage that the derivative $\partial\rho/\partial z$ is continuous at $z=0$. A downside of the exponential spheroid disk is that its density distribution is not separable, so that it cannot be written in the form $\rho(R,z) = f(R)g(z)$. This means that the characteristic scale height of the exponential spheroid is not independent of R, in contrast to observations (see §2.3.3).

---

[1] Note that there is a degeneracy between inclination angle and intrinsic flattening: any combination (q_d, i) that obeys Eq. (11.12) will project to exactly the same surface brightness distribution.

(b) The Isothermal Sheet A simple model for the vertical structure of disk galaxies, which has been used extensively, is the isothermal sheet originally proposed by Spitzer (1942). This model assumes that the vertical density distribution of a disk is *locally* isothermal, so that the distribution function of stars in the z direction is Maxwellian,

$$f = \frac{1}{\sqrt{2\pi}\sigma_z} \exp\left(-\frac{E_z}{\sigma_z^2}\right). \tag{11.15}$$

Here σ_z is the velocity dispersion of stars in the vertical direction, and $E_z \equiv v_z^2/2 + \Phi(R,z)$, with $\Phi(R,z)$ the gravitational potential, is the energy associated with the motion in the z direction. For a very flattened system, $\Phi(R,z)$ satisfies the Poisson equation, $\partial^2 \Phi/\partial z^2 = 4\pi G \rho(R,z)$. Since the density is equal to the integral of the distribution function over velocity space, we can write the Poisson equation as

$$\frac{d^2\phi}{d\zeta^2} = \frac{1}{2}e^{-\phi}, \tag{11.16}$$

where

$$\phi \equiv \frac{\Phi}{\sigma_z^2}, \quad \zeta \equiv \frac{z}{z_d}, \quad z_d \equiv \frac{\sigma_z}{\sqrt{8\pi G \rho(R,0)}}, \tag{11.17}$$

with z_d the vertical scale height of the disk (see §2.3.3). With the boundary conditions that both ϕ and its derivative $d\phi/d\zeta$ are zero at $z=0$, the density of the disk can be solved for, yielding

$$\rho(R,z) = \rho(R,0)\,\mathrm{sech}^2(z/2z_d). \tag{11.18}$$

The mid-plane density is related to the surface density of the disk by

$$\rho(R,0) = \frac{\Sigma(R)}{4z_d}, \tag{11.19}$$

so that the disk scale height is related to $\sigma_z(R)$ and $\Sigma(R)$ by

$$z_d = \frac{\sigma_z^2(R)}{2\pi G \Sigma(R)}. \tag{11.20}$$

The scale height is generally independent of R, and is approximately a constant fraction of the radial scale length R_d (e.g. van der Kruit & Searle, 1981). If we write $z_d = \varsigma R_d$, then for a locally isothermal disk with exponential R profile,

$$\sigma_z^2(R) = \frac{\varsigma G M_d}{R_d} \exp\left(-\frac{R}{R_d}\right). \tag{11.21}$$

Thus, for an isothermal sheet one can estimate the disk mass from a measurement of the vertical velocity dispersion, σ_z, and the radial scale length, R_d, as long as one adopts a value for ς (e.g. Bottema, 1993). Since $\sigma_z(R)$ is best measured in nearly face-on systems, for which the information regarding z_d is lost, this method clearly suffers from the uncertainty in ς, which is observed to have a substantial amount of scatter.

11.1.2 Rotation Curves

As we have seen in §2.3.3 the observed rotation curves of disk galaxies, $V_{\rm rot}(R)$, are flat on large scales, suggesting that the dominant matter component of a disk galaxy is an extended dark halo. Here we describe how the rotation curves of disk galaxies can be used to infer the matter distribution of this dark halo. This is important since, as we have seen in Chapter 7, if baryon effects are neglected, CDM cosmologies predict that dark matter halos have a near-universal density distribution accurately described by a NFW profile. In particular, they should have a central cusp with density scaling as $\rho \propto r^{-1}$. Disk rotation curves can be used to test these predictions.

It is generally assumed that the observed rotation velocities of a disk galaxy, after correcting for inclination effects, reflect the circular velocities (i.e. that deviations from perfectly circular orbits can be neglected). In what follows we consider a model in which the galaxy consists of an axisymmetric disk and a spherical dark matter halo. The circular velocity in the disk plane follows from balancing the centrifugal force with the gravitational force, and is given by

$$V_c^2(R) = R|\mathbf{F}_R|, \tag{11.22}$$

with R the cylindrical radius and $\mathbf{F}_R(R)$ the radial force per unit mass (acceleration) in the plane of the disk ($z = 0$). Since this radial force is the sum of the forces due to the disk and the halo, we can split the circular velocity into a disk component, $V_{c,d}$, and a halo component, $V_{c,h}$:

$$V_c^2(R) = V_{c,d}^2(R) + V_{c,h}^2(R). \tag{11.23}$$

For a spherical system the circular velocity is given by

$$V_c^2(r) = r\frac{d\Phi}{dr} = \frac{GM(r)}{r}, \tag{11.24}$$

where $M(r)$ is the total mass within a sphere of radius r. For the special cases where the mass distribution is a point mass, a singular isothermal sphere (SIS) [with density profile $\rho(r) \propto r^{-2}$], or a uniform sphere with constant density, the circular velocities obey:

$$V_c(r) \propto \begin{cases} r^{-1/2} & \text{for point mass} \\ \text{constant} & \text{for SIS} \\ r & \text{for uniform sphere.} \end{cases} \tag{11.25}$$

For the more realistic case of a dark matter halo with a NFW profile [Eq. (7.138)], the circular velocity is

$$\frac{V_{c,h}(r)}{V_{\rm vir}} = \left[\frac{1}{x}\frac{\ln(1+cx) - cx/(1+cx)}{\ln(1+c) - c/(1+c)}\right]^{1/2}, \tag{11.26}$$

where $x = r/r_{\rm vir}$, c is the halo concentration parameter, and $V_{\rm vir} = \sqrt{GM_{\rm vir}/r_{\rm vir}}$ is the circular velocity at the virial radius. This circular velocity curve reaches a maximum

$$V_{\rm max} \approx 0.465\sqrt{\frac{c}{\ln(1+c) - c/(1+c)}}\, V_{\rm vir} \tag{11.27}$$

at a radius $r \simeq (2.163/c)\, r_{\rm vir}$.

For an axisymmetric disk, the circular velocity in the plane of the disk is given by

$$V_c^2(R) = R\left(\frac{\partial \Phi}{\partial R}\right)_{z=0}. \tag{11.28}$$

Using Eq. (11.2) we obtain, for an infinitesimally thin disk, that

$$V_{c,d}^2(R) = 2\pi GR \int_0^\infty J_1(kR)\tilde{\Sigma}(k)k\,dk. \tag{11.29}$$

For a thin exponential disk given by Eq. (11.1), this reduces to

$$V_{c,d}^2(R) = 4\pi G\Sigma_0 R_d y^2 \left[I_0(y)K_0(y) - I_1(y)K_1(y)\right], \tag{11.30}$$

where $y = R/(2R_d)$, and I_n and K_n are modified Bessel functions of the first and second kinds, respectively. This circular velocity curve reaches a maximum at $R \approx 2.16R_d$, while at large radii ($R \gg R_d$) it approaches the Keplerian form $V_{c,d} \propto R^{-1/2}$ expected for a point mass.

For the exponential spheroid given by Eq. (11.10) the circular velocity in the equatorial plane ($z = 0$) follows directly from differentiating the potential [Eq. (11.13)], and is given by

$$V_{c,d}^2(R) = 2\pi G \rho_0 q_d R^2 \int_0^\infty \frac{K_0(R/R_\tau) d\tau}{(\tau+1)\sqrt{\tau+q_d^2}}, \qquad (11.31)$$

where $R_\tau = R_d \sqrt{\tau+1}$. More generally, the circular velocity of a disk of non-zero thickness can be obtained from the radial force in the plane of the disk and Eq. (11.22). Using the Hankel transform one obtains

$$F_R(R) = 8G \int_0^\infty du \int_0^\infty dz [\mathcal{K}(\rho) - \mathcal{E}(\rho)] \left(\frac{u}{\rho R}\right)^{1/2} \frac{\partial \rho(u,z)}{\partial u}, \qquad (11.32)$$

where $\mathcal{K}(x)$ and $\mathcal{E}(x)$ are the complete elliptic integrals of the first and second kind, $\rho = t - \sqrt{t^2-1}$ and $t = (R^2+u^2+z^2)/2Ru$ (Casertano, 1983). In general, as long as the vertical scale height of a disk is sufficiently small compared to its radial scale length, the actual vertical density distribution of the disk has only a small effect on the resulting rotation curve. For most practical purposes it therefore suffices to assume the disk to be infinitesimally thin.

As is evident from Eq. (11.30), the circular velocity of an infinitesimally thin, exponential disk is entirely determined by its scale length, R_d, and its central surface density, Σ_0. Under the assumption that the disk is made entirely of stars with a homogeneous stellar population, R_d is simply identical to the scale length of the stellar light, while $\Sigma_0 = \Upsilon I_0$, with Υ the stellar mass-to-light ratio and I_0 the central surface brightness. In this case, the observables $V_{\rm rot}(R)$, I_0 and R_d contain all the information required to compute the enclosed mass of the halo

$$M_h(R) = \frac{R[V_{\rm rot}^2(R) - V_{c,d}^2(R)]}{G}, \qquad (11.33)$$

for a given choice of Υ. In general, disks also contain a bulge component and a significant amount of cold gas. One must therefore account for contributions from these components also, which is straightforward once their density distributions are known. For example, the circular velocity of an (infinitesimally thin) gas disk simply follows from the observed HI surface density and Eq. (11.29).

Unfortunately, the stellar mass-to-light ratios of disk galaxies are not well constrained, and Υ is generally treated as a free parameter. One can obtain a strict upper limit on Υ from the fact that $V_{c,d} \propto \Upsilon^{1/2}$ and the constraint that $V_{c,d}(R) \leq V_{\rm rot}(R)$ at all R. A model in which the disk is assumed to have this maximal mass-to-light ratio, $\Upsilon_{\max}$, is called a maximal disk model, and results in a lower limit on $M_h(r)$. The upper limit on $M_h(r)$ simply follows from ignoring the self-gravity of the disk (i.e. setting $\Upsilon = 0$).[2] In most cases, the difference between these upper and lower limits on the enclosed halo mass are very substantial, at least at small radii. This so-called disk–halo degeneracy severely impedes our ability to obtain a unique rotation-curve decomposition. This is illustrated in Fig. 11.1, which shows the observed rotation curve of NGC 2403 together with two different decompositions that fit the data equally well. In the left-hand panel, a maximal disk model is used, which gives an R band stellar mass-to-light ratio of $\Upsilon_R = 3.1(M/L_R)_\odot$, in combination with a dark matter halo that has a central core: $\rho \propto r^{-\alpha}$ with $\alpha = 0$. In the right-hand panel the halo is assumed to follow a NFW profile (for which $\alpha = 1$), which results in a best-fit $\Upsilon_R = 1.1(M/L_R)_\odot$.

The impact of this disk–halo degeneracy can be minimized by focusing on low surface brightness (LSB) galaxies and dwarf galaxies. As it turns out, these systems have extremely high values of $\Upsilon_{\max}$, clearly in conflict with any realistic stellar population. For any reasonable value

---

[2] One may further limit the range of Υ by using constraints from stellar population models, although these are often model dependent (see §10.3).

11.1 Mass Components and Angular Momentum

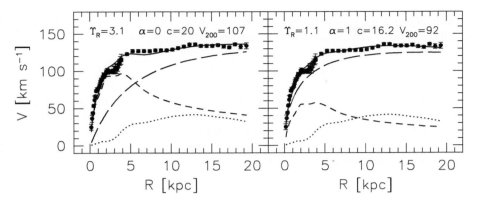

Fig. 11.1. The rotation curve of NGC 2403 (solid dots with error bars) and two decompositions in contributions from the stellar disk (short-dashed lines), the gas disk (dotted lines) and the dark matter halo (long-dashed lines). In the left panel, a maximal disk with $\Upsilon_R = 3.1(M/L_R)_\odot$ is used together with a dark matter halo with a central density core, while the decomposition in the right panel uses a NFW halo and a submaximal disk with $\Upsilon_R = 1.1(M/L_R)_\odot$. Both decompositions fit the observations equally well, illustrating the disk–halo degeneracy. [After Dutton et al. (2005); courtesy of A. Dutton]

of the stellar mass-to-light ratio, they are dominated by dark matter at all radii, making them ideally suited for inferring the density distribution of the dark matter halo, and for testing the prediction that $\alpha \equiv d\ln\rho/d\ln r \sim -1$ at small radii. Many dwarf and LSB galaxies are found to have rotation curves rising like $V_{\rm rot}(R) \propto R$ in the inner regions, suggesting that $M_{\rm h}(r) \propto r^3$, hence that the dark matter halo has a constant density core ($\alpha = 0$) as opposed to a r^{-1} cusp. Although this seems inconsistent with the predictions of CDM cosmologies, there are a number of caveats. First of all, because the rotation curves are measured with finite resolution, the observed rotation curve may be shallower than the true rotation curve (e.g. Begeman, 1989; van den Bosch et al., 2000). Secondly, a large fraction of disk galaxies have central bars. In these systems, as well as in those where the halo is triaxial rather than spherical, the orbits may deviate significantly from circular, making it difficult to interpret $V_{\rm rot}(R)$ (e.g. Hayashi & Navarro, 2006). Thirdly, even for a fixed stellar mass-to-light ratio one can still trade off the halo concentration against the slope of the central cusp and obtain equally good fits to the data. This cusp-core degeneracy further hampers an accurate determination of α (van den Bosch & Swaters, 2001). Finally, when a disk forms in the center of a dark matter halo, its gravity can modify the dark matter distribution. As a result, the distribution inferred from a rotation-curve decomposition may differ from that expected from dissipationless gravitational collapse. This should be taken into account when modeling disk–halo systems (see §11.1.3 below). Because of all these uncertainties, the question of whether the observed rotation curves of disk galaxies are consistent with a CDM cosmology is still hotly debated. It is clear, though, that the rotation curves of LSB and dwarf galaxies pose a potentially severe challenge to the CDM paradigm.

11.1.3 Adiabatic Contraction

As already alluded to above, when baryonic gas cools and concentrates in the center of a dark matter halo, the halo structure may be modified due to the gravitational action of the baryons. In general, it is difficult to accurately model this action of the disk on the halo, because the details depend on the exact formation history of the disk–halo system. However, if the growth of the disk in its halo is so slow that the potential of the system changes only a little during the typical orbital period of a dark matter particle, the system adjusts itself adiabatically, which means that its final state is independent of the path taken. As shown in §5.4.5, under adiabatic evolution, the action defined through a canonical coordinate, q_i, and its conjugate momentum, p_i,

$$J_i = \frac{1}{2\pi} \oint p_i \, dq_i, \tag{11.34}$$

is a conserved quantity, called an adiabatic invariant. For the idealized case of a spherical halo in which all dark matter particles move on circular orbits, the adiabatic invariant reduces to the specific angular momentum, $rV(r)$. If, in addition, the distribution of the baryons has spherical symmetry, this reduces further to $rM(r)$, with $M(r)$ the mass enclosed within radius r. This means that a particle initially at a mean radius, r_i, will end up at a mean radius, r_f, given by

$$r_f M_f(r_f) = r_i M_i(r_i), \tag{11.35}$$

where $M_i(r_i)$ is the initial mass profile, and $M_f(r_f)$ is the total final mass within r_f. Eq. (11.35) is often used to model the contraction of a dark matter halo in response to the formation of a disk galaxy at its center. In this case the final mass is the sum of the dark matter mass inside the initial radius r_i plus the mass contributed by the disk. Hence

$$M_f(r_f) = M_d(r_f) + (1 - m_d) M_i(r_i), \tag{11.36}$$

where m_d is the fraction of the total mass in the disk, and $M_d(r_f)$ is the disk mass within radius r_f. For given $M_d(r)$ and $M_i(r)$, the above two equations can be used to solve for r_f as a function of r_i, thereby giving the final, total mass profile $M_f(r_f)$ and the modified halo mass profile $M_f(r_f) - M_d(r_f)$. For realistic values of m_d, it turns out that the effect of adiabatic contraction can be very substantial, causing the circular velocity curve of the halo to change by a large amount. As we will see in §11.3 this has important implications for interpreting the Tully–Fisher relation.

Note that $rM(r)$ is only an adiabatic invariant for a spherical system in which all particles are on circular orbits. This clearly is unrealistic. First of all, in realistic dark matter halos the particles typically move on highly eccentric orbits (Ghigna et al., 1998). Taking this into account reduces the effect of adiabatic contraction (Gnedin et al., 2004). Secondly, disks are not spherical. Therefore, in principle one should use $rV(r)$ as the adiabatic invariant, rather than $rM(r)$. The use of the latter typically tends to increase the effect of adiabatic contraction (Dutton et al., 2007). Somewhat fortuitously, the two effects partially cancel, so that the prescription given above predicts contracted dark matter density profiles that are in fair agreement with, though somewhat more concentrated than, results from numerical simulations (e.g. Barnes & White, 1984; Abadi et al., 2009).

11.1.4 Disk Angular Momentum

One of the most important properties of spiral galaxies is that their disks are supported by rotation. The importance of rotation in a system is usually described by the spin parameter,

$$\lambda = \frac{J |E|^{1/2}}{G M^{5/2}}, \tag{11.37}$$

where M, J, and E are the mass, angular momentum, and total energy of the system, respectively. For an isolated exponential disk (i.e. without any dark matter halo), the rotation curve is given by Eq. (11.30) and the total angular momentum is

$$J_d = 2\pi \int_0^\infty V_c(R) \Sigma(R) R^2 \, dR \approx 1.11 G^{1/2} M_d^{3/2} R_d^{1/2}. \tag{11.38}$$

According to the virial theorem $E = -K$, where K is the total kinetic energy of the disk, so that

$$E = -2\pi \int_0^\infty \frac{V_c^2(R)}{2} \Sigma(R) R dR \approx -0.147 G M_d^2 R_d^{-1}. \qquad (11.39)$$

Substituting Eqs. (11.38) and (11.39) into Eq. (11.37) then yields a spin parameter

$$\lambda \approx 0.425 \quad \text{for an isolated exponential disk.} \qquad (11.40)$$

As we have seen in §7.5.4, tidal torques give protogalaxies an amount of angular momentum that corresponds to a spin parameter in the range $0.01 \lesssim \lambda \lesssim 0.1$, with a median near 0.035. Comparing this to the spin parameter of an isolated exponential disk yields an estimate of the collapse time of the gas cloud that forms the disk. Consider a self-gravitating gas cloud containing *no* dark matter. As the cloud radiates and shrinks, both its mass M and its angular momentum $\mathbf{J}$ are conserved, but its binding energy, $-E$, increases in inverse proportion to its size R. As a result, the spin parameter of the cloud scales as

$$\lambda = \lambda_i (R/R_i)^{-1/2}. \qquad (11.41)$$

To increase the spin parameter from an initial value $\lambda_i \sim 0.05$ to the value $\lambda \sim 0.425$ characteristic of a centrifugally supported exponential disk thus requires a contraction factor of $R_i/R \sim 70$. Consider the disk of a spiral galaxy like the Milky Way, with a mass of $M \sim 5 \times 10^{10} M_\odot$ and a radius $R \sim 10\,\text{kpc}$. The inferred radius of the initial cloud is then $R_i \sim 700\,\text{kpc}$, which implies a free-fall time of $t_{\text{ff}} = \sqrt{3\pi/32 G \rho} \simeq 4.3 \times 10^{10}\,\text{yr}$. This is much longer than the age of the Universe, and thus rules out the possibility that disk galaxies could have formed from pure gas clouds with spin parameters expected from tidal torque theory.

The situation is quite different if the gas contracts inside a massive dark halo. Consider a halo with a circular velocity independent of radius (i.e. $\rho \propto r^{-2}$), and assume that the gas cools and flows inward conserving its specific angular momentum. This implies that the gas at radius R in the disk came from an initial radius in the halo given by $R_i = R(V_c/V_{\text{rot},i})$, where $V_{\text{rot},i}$ is the typical initial rotation velocity of the gas at radius R_i. As we have seen in §7.5.4, the specific angular momentum scales roughly with radius as $\mathcal{J} \propto r$. In our singular isothermal halo, this implies that the average rotation velocity of the dark matter at radius r is given by $V_{\text{rot}}(r) = \eta V_c$, with η a constant whose value is related to λ. For a singular isothermal sphere, truncated at large radii so that its mass is finite, one finds that $\lambda = 2^{-3/2}\eta$, so that a spin parameter of $\lambda = 0.05$ corresponds to $\eta \simeq 0.14$. If the gas inside a dark matter halo has the same specific angular momentum distribution as the dark matter, which is a reasonable assumption given that both components have experienced the same tidal torques, one finds that the gas only needs to collapse by a factor of $1/\eta = 1/0.14 \approx 7$ in order to bring its rotation speed up to V_c and make it reach centrifugal equilibrium in the potential well of the dark matter halo. Thus the presence of the dark halo has reduced the collapse factor of the gas by a full order of magnitude. Since the free-fall time in an isothermal potential $t_{\text{ff}} \propto R_i^{3/2}$, the gas can settle in centrifugal equilibrium in $\sim 1.4 \times 10^9\,\text{yr}$. Therefore, in the presence of dark halos, there is plenty of time to form large disks. As first noted by Efstathiou & Silk (1983), this simple argument gives strong support for the presence of extended dark halos, independent of the usual dynamical arguments.

11.1.5 Orbits in Disk Galaxies

Orbits in an axisymmetric potential conserve both energy and the angular momentum in the direction of the symmetry axis (hereafter the z direction). Solving Newton's equation of motion, $d^2\mathbf{r}/dt^2 = -\nabla \Phi$, in cylindrical coordinates (R, ϕ, z) yields

$$\ddot{R} - R\dot{\phi}^2 = -\frac{\partial \Phi}{\partial R}; \tag{11.42}$$

$$\frac{d}{dt}(R^2\dot{\phi}) = 0; \tag{11.43}$$

$$\ddot{z} = -\frac{\partial \Phi}{\partial z}, \tag{11.44}$$

where a dot denotes a time derivative. The second equation expresses the conservation of the component of angular momentum about the z axis, $L_z = R^2\dot{\phi}$, while the other two equations describe the coupled oscillations in the R and z directions. We can reduce these equations of motion to

$$\ddot{R} = -\frac{\partial \Phi_{\rm eff}}{\partial R} \quad \text{and} \quad \ddot{z} = -\frac{\partial \Phi_{\rm eff}}{\partial z}, \tag{11.45}$$

with $\Phi_{\rm eff}(R,z) = \Phi(R,z) + (L_z^2/2R^2)$ the effective potential. The L_z^2/R^2 term serves as a centrifugal barrier, only allowing orbits with $L_z = 0$ to be close to the symmetry axis. This allows us to reduce the three-dimensional motion to two-dimensional motion in the meridional plane (R,z), which rotates non-uniformly around the symmetry axis with $\dot{\phi} = L_z/R^2$.

The effective potential has a minimum at $(R,z) = (R_{\rm g}, 0)$, called the guiding center, where

$$\frac{\partial \Phi_{\rm eff}}{\partial R} = \frac{\partial \Phi}{\partial R} - \frac{L_z^2}{R^3} = 0. \tag{11.46}$$

The radius $R_{\rm g}$ corresponds to the radius of a circular orbit with energy $E = \Phi_{\rm eff}$ and with circular frequency

$$\Omega(R) = \frac{V_{\rm c}(R)}{R} = \left[\frac{1}{R}\left(\frac{\partial \Phi}{\partial R}\right)_{(R_{\rm g},0)}\right]^{1/2} = \frac{L_z}{R^2}. \tag{11.47}$$

Defining $x = R - R_{\rm g}$ and expanding $\Phi_{\rm eff}$ in a Taylor series around the point $(x,z) = (0,0)$, and keeping only terms up to second order, the equations of motion in the meridional plane become

$$\ddot{x} = -\kappa^2 x \quad \text{and} \quad \ddot{z} = -\nu^2 z, \tag{11.48}$$

where κ and ν are the epicyclic frequency and vertical frequency, respectively, defined by

$$\kappa^2 \equiv \left(\frac{\partial^2 \Phi_{\rm eff}}{\partial R^2}\right)_{(R_{\rm g},0)} \quad \text{and} \quad \nu^2 \equiv \left(\frac{\partial^2 \Phi_{\rm eff}}{\partial z^2}\right)_{(R_{\rm g},0)}. \tag{11.49}$$

Thus, when the third and higher-order terms of the Taylor expansion of $\Phi_{\rm eff}$ can be neglected the x and z motions reduce to simple harmonic oscillations. This is called the epicyclic approximation, and allows the motion of stars in disk potentials to be characterized by three frequencies: Ω, κ and ν. Using the definition of the effective potential, it is straightforward to show that the epicyclic and circular frequencies are related through

$$\kappa = \sqrt{2}\Omega\left[1 + \frac{d\log V_{\rm c}}{d\log R}\right]^{1/2} = \left[R\frac{d\Omega^2}{dR} + 4\Omega^2\right]^{1/2}. \tag{11.50}$$

For realistic galactic potentials $\Omega < \kappa < 2\Omega$, where the limits correspond to a homogeneous mass distribution ($\kappa = 2\Omega$) and the Kepler potential ($\kappa = \Omega$). Therefore, in general, the period of the radial oscillation is incommensurable with the orbital period, so that the orbit does not form a closed figure in an inertial frame. Rather, the motion in (R,ϕ) can be described as retrograde motion on an ellipse, whose guiding center is in prograde motion around the center of the system.

When is the epicyclic approximation valid? First consider the z motion: the equation of motion, $\ddot{z} = -\nu^2 z$ with constant ν, implies a constant density in the z direction. Hence, the epicyclic approximation is valid only when $\rho(z)$ is roughly constant. This is only approximately true very close to the equatorial plane. In the radial direction realize that the Taylor expansion is only accurate when R is sufficiently close to R_g. Hence, the epicyclic approximation is only valid for small oscillations around the guiding center; i.e. for orbits with an angular momentum close to that of the corresponding circular orbit.

11.2 The Formation of Disk Galaxies

We now turn to models for the formation of disk galaxies within the CDM framework outlined in the previous chapters. This means that we consider disk galaxies to be embedded in extended dark matter halos. In this section we assume that the formation and growth of the disk is a slow, adiabatic process. This implies that the end state is independent of the exact formation history, so that we can focus on static models of the end states. Under the assumption that disks are in centrifugal equilibrium, we can then link the structural properties of the disks to those of their dark matter halos.

Obviously, with a static model one cannot model the actual star-formation histories of the disk galaxies. This requires more 'dynamic' models that follow the actual formation of the disk galaxies starting from high redshifts. In §11.2.5 we will briefly describe such models. Note, however, that the star-formation history mainly governs the final mass-to-light ratio of the stars, which we consider a free parameter in the static models discussed below.

11.2.1 General Discussion

Disk galaxies are highly flattened systems supported by rotation. A natural way to form such an object is through the dissipational collapse of a gas cloud with some initial angular momentum. Consider a gas cloud for which radiative cooling is very effective (this will generally be the case provided the cloud is sufficiently dense and has a temperature $T > 10^4$ K; see §8.4). Such a cloud will radiate away its binding energy and contract, causing it to approach a state in which its energy is as low as possible. In the absence of any interactions with other mass components (e.g. a dark matter halo), the cloud will conserve its angular momentum, simply because the radiation field from cooling is roughly isotropic and so should not carry away much angular momentum. The preferred end state is a rotating disk, since in such a configuration the angular momenta of all mass elements point in the same direction. In fact, for a given total angular momentum, J, the state of lowest energy is the one in which all but an infinitesimal fraction (with mass ΔM) of the gas collapses into a 'black hole', while the infinitesimal part is on a Keplerian orbit with a large radius R given by $J = \Delta M (GMR)^{1/2}$, where M is the mass of the cloud. This structure is, of course, very different from a real disk. The explanation for this paradoxical discrepancy is that although the lowest energy state is preferred by energy considerations, its realization requires a very effective transfer of angular momentum from the inside of the disk to the outside. As we will see in §11.4.2, the efficiency of this kind of angular-momentum transfer depends on the effective viscosity of the disk material. In fact, in the absence of viscosity or non-axisymmetric structure, each mass element of the cloud will conserve its own specific angular momentum, so that the end state is a disk with surface density directly related to the initial angular momentum distribution of the cloud. Thus, the fact that disk galaxies contain extended disks suggests that these effects are not extremely efficient. Nevertheless, as we will see in §11.4.2, viscosity may still play an important role in determining the density distribution of the final disk.

11.2.2 Non-Self-Gravitating Disks in Isothermal Spheres

As an illustration, let us consider an idealized case in which we (i) ignore the self-gravity of the disk, and (ii) assume that the dark matter halo has the density distribution of a singular, isothermal sphere: $\rho(r) = V_{\rm vir}^2/(4\pi G r^2)$. Here $V_{\rm vir}$ is the circular velocity at the virial radius $r_{\rm vir}$ defined so that the average density within it is $\Delta_{\rm vir}$ times the critical density for closure. The value of $\Delta_{\rm vir}$ depends on both cosmology and redshift, and follows from the spherical collapse model described in Chapter 5. For a flat ΛCDM cosmology with $\Omega_{m,0} = 0.3$ one has that $\Delta_{\rm vir} \simeq 100$ at $z = 0$.

Since the critical density for closure at redshift z is given by $\rho_{\rm crit} = 3H^2(z)/(8\pi G)$, with $H(z)$ the Hubble parameter, the virial radius and virial mass are related to the virial velocity as

$$r_{\rm vir} = \sqrt{\frac{2}{\Delta_{\rm vir}(z)}} \frac{V_{\rm vir}}{H(z)} \quad \text{and} \quad M_{\rm vir} = \sqrt{\frac{2}{\Delta_{\rm vir}(z)}} \frac{V_{\rm vir}^3}{GH(z)}. \tag{11.51}$$

Note that these relations hold independent of the density profile of the dark matter halo; they follow directly from the definition of the virial radius. The redshift dependence of the halo properties are determined by both $H(z)$ and $\Delta_{\rm vir}(z)$.

If we assume that the mass which settles into the disk is a fixed fraction, $m_{\rm d}$, of the halo mass, then the disk mass is

$$M_{\rm d} \approx 1.3 \times 10^{11} h^{-1} {\rm M}_\odot \left(\frac{m_{\rm d}}{0.05}\right) \left(\frac{V_{\rm vir}}{200\,{\rm km\,s^{-1}}}\right)^3 \mathscr{D}^{-1}(z), \tag{11.52}$$

where we have used the shorthand notation

$$\mathscr{D}(z) \equiv \left[\frac{\Delta_{\rm vir}(z)}{100}\right]^{1/2} \left[\frac{H(z)}{H_0}\right]. \tag{11.53}$$

For simplicity we assume the disk to be infinitesimally thin and to have an exponential surface density distribution given by Eq. (11.1). If the gravitational effect of the disk is neglected, its rotation curve is flat at the level $V_{\rm vir}$ and its angular momentum is just

$$J_{\rm d} = 2\pi \int_0^\infty V_{\rm c}(R)\Sigma(R)R^2 dR = 2M_{\rm d} R_{\rm d} V_{\rm vir}, \tag{11.54}$$

where $R_{\rm d}$ is the disk scale length. We assume this angular momentum to be a fraction $j_{\rm d}$ of that of the halo, i.e.

$$J_{\rm d} = j_{\rm d} J_{\rm vir}, \tag{11.55}$$

and we relate $J_{\rm vir}$ to the spin parameter λ of the halo through the definition given by Eq. (11.37). Eqs. (11.54)–(11.55) then imply that

$$R_{\rm d} = \frac{\lambda G M_{\rm vir}^{3/2}}{2 V_{\rm vir} |E|^{1/2}} \left(\frac{j_{\rm d}}{m_{\rm d}}\right). \tag{11.56}$$

The total energy of a truncated singular isothermal sphere is easily obtained from the virial theorem by assuming all particles to be on circular orbits:

$$E = -\frac{M_{\rm vir} V_{\rm vir}^2}{2}. \tag{11.57}$$

On inserting this into Eq. (11.56), and using Eqs. (11.51) and (11.52), we obtain

$$R_{\rm d} = \frac{1}{\sqrt{2}} \left(\frac{j_{\rm d}}{m_{\rm d}}\right) \lambda r_{\rm vir}$$

$$\approx 10 h^{-1} {\rm kpc} \left(\frac{j_{\rm d}}{m_{\rm d}}\right) \left(\frac{\lambda}{0.05}\right) \left(\frac{V_{\rm vir}}{200\,{\rm km\,s^{-1}}}\right) \mathscr{D}^{-1}(z), \tag{11.58}$$

and

$$\Sigma_0 \approx 207 h\, M_\odot\, \text{pc}^{-2} \left(\frac{m_d}{0.05}\right)\left(\frac{j_d}{m_d}\right)^{-2}\left(\frac{\lambda}{0.05}\right)^{-2}\left(\frac{V_{\text{vir}}}{200\,\text{km s}^{-1}}\right)\mathscr{D}(z). \quad (11.59)$$

Eqs. (11.52), (11.58) and (11.59) relate the disk properties to the properties of the dark matter halo (V_{vir} and λ), and to the quantities $m_d = M_d/M_{\text{vir}}$ and $j_d = J_d/J_{\text{vir}}$. The latter two quantities depend on the details of the processes by which the disk has formed, such as the efficiencies of cooling and feedback. It is common practice to assume that $j_d = m_d$, i.e. that the specific angular momentum of the material forming the disk is the same as that of the halo. This is motivated by the fact that the baryons experience the same tidal forces as the dark matter. However, as we will see below, there are several processes that can make j_d significantly different from m_d, even if the baryons and dark matter originally started out with the same specific angular momentum.

How do the predictions of this naive model compare to observations? For the Milky Way, $V_{\text{rot}} \simeq 220\,\text{km s}^{-1}$, $M_d \simeq 5 \times 10^{10}\,M_\odot$, and $R_d \simeq 3.5\,\text{kpc}$. Using $h = 0.7$, and assuming that $V_{\text{rot}} = V_{\text{vir}}$, this implies that $m_d \sim 0.01$ and $\lambda \sim 0.011$. To put this in perspective we write m_d as the product of the universal baryon fraction, $f_{\text{bar}} \equiv \Omega_{b,0}/\Omega_{m,0}$, and the galaxy formation efficiency parameter, ε_{gf}, which describes what fraction of the baryonic mass inside the halo ultimately ends up in the disk. For the ΛCDM cosmology considered here, $f_{\text{bar}} \simeq 0.17$ (see §2.10.1) so that $m_d = 0.01$ implies that $\varepsilon_{\text{gf}} \sim 0.06$. Since baryons and dark matter are well mixed prior to the onset of cooling, this means that only $\sim 6\%$ of the baryons originally associated with the Lagrangian volume out of which the halo formed have ended up in the disk. The remaining 94% either never cooled down, or have been expelled from the disk by some feedback mechanism. This poses a serious problem, because feedback is not expected to be very effective in halos with V_{vir} as large as $\sim 200\,\text{km s}^{-1}$. As for the spin parameter, a value of 0.011 is relatively rare: as we have seen in §7.5.4, dark matter halos have spin parameters that follow a log-normal distribution [Eq. (7.160)] with a median $\overline{\lambda} \approx 0.035$ and a standard deviation $\sigma_{\ln \lambda} = 0.5$. This implies that $< 3\%$ of all halos have $\lambda < 0.011$. Alternatively, one can form a disk with $R_d \simeq 3.5\,\text{kpc}$ in a halo with $\lambda = 0.05$ if $j_d \sim 0.2 m_d$ (i.e. the disk material has a lower specific angular momentum than the dark matter halo). This may come about if the disk grows preferentially out of the low angular momentum material, or if the baryons transfer a significant amount of their angular momentum to the dark matter halo during the cooling process.

There is another problem for this simple model. Using the redshift dependence of the Hubble constant, we find that, for the ΛCDM cosmology adopted here, a disk galaxy at $z = 1$ will have a scale length about 70% smaller than a disk galaxy in a halo of the same λ and V_{vir} at $z = 0$. Since simulations show that the distribution of halo spin parameters does not evolve with redshift, this is inconsistent with the observation that disks at $z \sim 1$ are only marginally smaller than present-day disk galaxies of the same stellar mass (e.g. Barden et al., 2005).

11.2.3 Self-Gravitating Disks in Halos with Realistic Profiles

The naive, idealized model discussed above has three serious problems: it predicts that the disk mass fraction is much smaller than the universal baryon fraction, that disks form in halos with very low spin parameters, and that disk sizes evolve more rapidly with redshift than observed. We now turn to a more realistic model, in which we account for the self-gravity of the disk, and use more realistic density profiles for the dark matter halos.

Consider a halo with some unperturbed density profile

$$\rho(r) = \frac{1}{4\pi r^2}\frac{dM(r)}{dr}, \quad (11.60)$$

where $M(r)$ is the halo mass within radius r. As before, we define the limiting radius of a virialized halo, r_{vir}, to be the radius within which the mean density is $\Delta_{\text{vir}}\rho_{\text{crit}}$. The total energy of the halo will be different from that for an isothermal sphere. Without losing generality, we write

$$E = -\frac{M_{\rm vir} V_{\rm vir}^2}{2} F_{\rm E}, \quad (11.61)$$

where $F_{\rm E}$ is a factor which depends on the exact form of $\rho(r)$ (see §7.5.4). For a NFW halo it is given by Eq. (7.158).

As before, we assume that the baryons ending up in the disk initially have the same specific angular momentum as the dark matter. If we continue to assume that the disk material is moving on perfectly circular orbits, so that $V_{\rm rot}(R) = V_{\rm c}(R)$, the total angular momentum of the disk can be written as

$$J_{\rm d} = 2 M_{\rm d} R_{\rm d} V_{\rm vir} F_{\rm R}, \quad (11.62)$$

where

$$F_{\rm R} = \frac{1}{2} \int_0^{r_{\rm vir}/R_{\rm d}} u^2 e^{-u} \frac{V_{\rm c}(u R_{\rm d})}{V_{\rm vir}} \, {\rm d}u. \quad (11.63)$$

Note that $V_{\rm vir} = V_{\rm c}(r_{\rm vir})$ is unaffected by disk formation. In practice we can set the upper limit of integration to infinity because the disk surface density drops exponentially and $r_{\rm vir} \gg R_{\rm d}$. Substituting Eq. (11.61) into Eq. (11.37) and writing $J_{\rm d} = j_{\rm d} J_{\rm vir}$, we obtain

$$R_{\rm d} = \frac{1}{\sqrt{2}} \left(\frac{j_{\rm d}}{m_{\rm d}} \right) \lambda r_{\rm vir} F_{\rm R}^{-1} F_{\rm E}^{-1/2}. \quad (11.64)$$

Since the computation of $F_{\rm R}$ requires knowledge of both $R_{\rm d}$ and the circular velocity curve $V_{\rm c}(r)$, the above set of equations has to be solved iteratively. For example, one can start with $F_{\rm R} = 1$, use Eq. (11.64) to obtain $R_{\rm d}$, and use Eq. (11.63) to update $F_{\rm R}$. Due to the self-gravity of the disk, $V_{\rm c}(r)$ is the sum in quadrature of the contributions from the disk and from the dark matter halo modified by adiabatic contraction:

$$V_{\rm c}^2(r) = V_{\rm c,d}^2(r) + \frac{G M_{\rm h,ac}(r)}{r}. \quad (11.65)$$

Here $V_{\rm c,d}(r)$ is the circular velocity due to the exponential disk alone [see Eq. (11.30)] and $M_{\rm h,ac}(r)$ is the mass profile of the contracted dark matter halo, which can be obtained using the procedure outlined in §11.1.3.

The self-gravity of the disk, and the inclusion of adiabatic contraction, strongly boost the circular velocity at a few disk scale lengths. This implies that a disk galaxy of a given $V_{\rm rot}$ resides in a less massive halo than in the naive model presented in the previous section. This in turn implies that a given disk mass requires a larger value of $m_{\rm d}$. In addition, since a less massive halo also has a smaller virial radius, a certain disk scale length corresponds to a larger value of λ. Detailed calculations, in which halos are modeled as spheres with an NFW density distribution (before adiabatic contraction), show that one can obtain a disk galaxy like the Milky Way with $m_{\rm d} \sim 0.05$ and $\lambda \sim 0.05$ (Mo et al., 1998). This alleviates two of the problems for the idealized model discussed in the previous section. In fact, the model described here can reproduce the observed distribution of disk sizes from the distribution of halo spin parameters given by Eq. (7.160), provided that $j_{\rm d}/m_{\rm d} \sim 1$. This implies that baryons have to largely conserve their specific angular momentum when cooling.

As shown by Somerville et al. (2008a), the model presented here also predicts a significantly weaker redshift evolution of disk sizes, in agreement with observations. This is owing to the fact that NFW halos are expected to be less concentrated at higher redshifts (see §7.5.1). Everything else being equal, a less concentrated halo results in a disk with a larger scale length. This trend largely counteracts the fact that halos of given mass have smaller virial radii at larger redshifts, leading to a much weaker redshift evolution of $R_d(M_d, z)$.

Thus, all in all, the model outlined here is remarkably successful in explaining the observed sizes of disk galaxies in a ΛCDM cosmology.

11.2.4 Including a Bulge Component

The majority of disk galaxies also contain a spheroidal bulge component, which has so far not been taken into account. In §11.5.4 we describe how the formation of bulges may be related to disk instabilities. Here we ignore the details of the formation process itself, and simply show how a bulge component can be included in the modeling scheme outlined above.

When the galaxy consists of a bulge and disk its baryonic mass can be written as

$$M_{\rm gal} \equiv M_{\rm d} + M_{\rm b} \equiv m_{\rm gal} M_{\rm vir}, \qquad (11.66)$$

with M_d and M_b the masses of the disk and bulge, respectively. We define the bulge-to-disk mass ratio as $\Theta \equiv M_b/M_d$ so that

$$M_{\rm d} = \frac{1}{1+\Theta} m_{\rm gal} M_{\rm vir}, \quad M_{\rm b} = \frac{\Theta}{1+\Theta} m_{\rm gal} M_{\rm vir}. \qquad (11.67)$$

In addition, we write the total angular momentum of the baryons, out of which the disk and bulge form, as

$$J_{\rm gal} \equiv j_{\rm gal} J_{\rm vir}. \qquad (11.68)$$

To proceed, let $J_b = j_b J_{\rm vir}$ indicate the original angular momentum of the baryonic material out of which the bulge forms. We assume that the bulge formation process transfers this angular momentum to the disk and the halo, so that the final angular momentum of the bulge is zero. If we define f_t as the fraction of J_b that is transferred to the disk (the rest being transferred to the dark matter halo), the final angular momentum of the disk is equal to

$$J_{\rm d} = \left[j_{\rm gal} - (1-f_{\rm t}) j_{\rm b} \right] J_{\rm vir}. \qquad (11.69)$$

Using the same procedure as before, one then obtains

$$R_{\rm d} = \frac{1}{\sqrt{2}} \left(\frac{j_{\rm gal}}{m_{\rm gal}} \right) \left[1 + (1-f_{\rm j})\Theta \right] \lambda r_{\rm vir} F_{\rm R}^{-1} F_{\rm E}^{-1/2}, \qquad (11.70)$$

where

$$f_{\rm j} = (1-f_{\rm t}) \left(\frac{J_{\rm b}/M_{\rm b}}{J_{\rm gal}/M_{\rm gal}} \right) \qquad (11.71)$$

expresses the ratio between the specific angular momentum lost to the halo due to bulge formation and the total specific angular momentum of the material out of which the disk plus bulge have formed. Note that this is exactly the same as Eq. (11.64) except for the factor $1 + (1-f_j)\Theta$, which describes how bulge formation impacts on the scale length of the disk. A value of $f_j > 1$ causes the disk size to decrease, while $f_j < 1$ causes an increase in R_d. The transition does not occur exactly at $f_j = 1$, however, because the presence of a bulge component also modifies the circular velocity curve and thus F_R. Numerical simulations indicate that bulge formation causes an increase in disk scale lengths (Debattista et al., 2006), suggesting that $f_j < 1$.

11.2.5 Disk Assembly

The 'static' models discussed above are based on the ansatz that disk formation is a slow, adiabatic process. Under the assumption that disks are in centrifugal equilibrium, such a static model is sufficient to predict the structural properties of disk galaxies. However, it does not suffice to describe the star-formation history of the disk galaxy, or to make predictions regarding radial age and/or metallicity gradients. This requires more 'dynamic' models that follow the actual assembly of the disk over time. Here we briefly outline how such models can be constructed.

Consider a disk with surface density $\Sigma(R,t)$ that at time t is embedded in a halo with a virial mass $M_{\rm vir}(t)$. Suppose that the gas accretion rate of the disk is equal to $\dot{M}_{\rm d}(t)$, and that this newly accreted material has a specific angular momentum distribution $P(\mathcal{J},t){\rm d}\mathcal{J}$. If the gas settles in the disk conserving its specific angular momentum we have that

$$2\pi\dot{\Sigma}(R,t)R{\rm d}R = \dot{M}_{\rm d}(t)P(\mathcal{J},t){\rm d}\mathcal{J}, \qquad (11.72)$$

where $\mathcal{J} = RV_{\rm c}(R,t)$ with $V_{\rm c}(R,t)$ the circular velocity curve of the disk–halo system at time t. This implies that

$$\dot{\Sigma}(R,t) = \frac{\dot{M}_{\rm d}(t)}{2\pi R^2}P(\mathcal{J},t)RV_{\rm c}(R,t)\left[1+\frac{\partial \ln V_{\rm c}(R,t)}{\partial \ln R}\right]. \qquad (11.73)$$

For a given density distribution of the dark matter halo, $V_{\rm c}(R,t)$ can be obtained from $\Sigma(R,t)$ (with or without adiabatic contraction), so that the growth of the disk is determined once $\dot{M}_{\rm d}(t)$ and $P(\mathcal{J},t)$ are given.

The disk accretion rate, $\dot{M}_{\rm d}(t)$, is set by the rate at which the dark matter halo accretes mass, $\dot{M}_{\rm vir}(t)$, and the rate at which the baryons in the halo can radiate away their binding energy through radiative cooling. The rate $\dot{M}_{\rm vir}(t)$ can be obtained from numerical simulations, or from merger trees constructed using the extended Press–Schechter formalism (see §7.3). Since baryons and dark matter are well mixed prior to cooling, the baryonic material accreted by the dark matter halo at time $t_{\rm acc}$ is simply given by $\dot{M}_{\rm bar}(t_{\rm acc}) = f_{\rm bar}\dot{M}_{\rm vir}(t_{\rm acc})$, with $f_{\rm bar} \equiv \Omega_{\rm b,0}/\Omega_{\rm m,0}$ the universal baryon fraction. If the gas is assumed to be heated to the virial temperature upon accretion, and if some assumptions regarding the density distribution of the gas are made, the cooling time of the gas, $t_{\rm cool}(t_{\rm acc})$, can be computed. Here we have made it implicit that material accreted at different times may have different cooling times. Since the material newly accreted into the halo takes at least a free-fall time, $t_{\rm ff}(t_{\rm acc})$, to reach the disk, we have that $\dot{M}_{\rm d}(t) = f_{\rm bar}\dot{M}_{\rm vir}(t')$, where t' is the root of $t' + \max[t_{\rm cool}(t'),t_{\rm ff}(t')] = t$.

Modeling the specific angular momentum distribution of the newly accreted material is more uncertain. Either one makes oversimplified assumptions, for example that the newly accreted material resides in a thin shell in solid-body rotation (e.g. van den Bosch, 2001), or one measures $P(\mathcal{J},t)$ from numerical hydrodynamical simulations.

In general the disk is made up of both stars and cold gas: $\Sigma(R,t) = \Sigma_\star(R,t) + \Sigma_{\rm gas}(R,t)$. One can solve for the evolution of these separate components once Eq. (11.73) is supplemented with prescriptions for star formation and/or supernova feedback. Let $\dot{\Sigma}_\star(R,t)$ denote the star-formation rate at radius R at time t, and assume that supernova feedback expels cold gas from the disk at a rate $\dot{\Sigma}_{\rm fb}(R,t) = \varepsilon_{\rm fb}\dot{\Sigma}_\star(R,t)$, with $\varepsilon_{\rm fb}$ an efficiency parameter. Then the evolution of $\Sigma_{\rm gas}(R,t)$ is simply given by $\dot{\Sigma}_{\rm gas}(R,t) = \dot{\Sigma}(R,t) - (1 - \mathcal{R} - \varepsilon_{\rm fb})\dot{\Sigma}_\star(R,t)$, where $\mathcal{R}$ is the instantaneous recycling fraction, defined as the ratio between the mass instantaneously returned to the ISM via stellar winds/supernovae and the mass of stars that have formed (see §10.4.2). Within such a scheme it is straightforward to include chemical evolution, and to compute the metallicities and ages of the stars as function of radius, R, and time, t. Combined with a stellar population synthesis model one can then compute the surface brightness distribution of the stellar disk in any photometric band. Various authors have presented this kind of model, which shows that disk galaxies are expected to form from the inside-out (e.g. Kauffmann, 1996a; Cole et al., 2000; Firmani & Avila-Reese, 2000; van den Bosch, 2002a; Stringer & Benson, 2007; Dutton & van den Bosch, 2009).

11.2.6 Numerical Simulations of Disk Formation

The first simulations of galaxy formation using SPH techniques (see Appendix C) were published in Katz & Gunn (1991) and Katz (1992). This work considered the collapse from a uniform, uniformly rotating, initially expanding spherical state on which small-scale irregularities were imposed at about the level predicted in a CDM universe. The simulations considered a mix of 90% dark matter and 10% gas, and included radiative cooling and, in some cases, star formation and feedback. It was shown that, in models with only a moderate amount of initial irregularity, the gas settles to a centrifugally supported disk before making a substantial number of stars, with the structure of the disks encouragingly similar to that of real disks. On the other hand, in simulations with a high degree of initial irregularity, the gas cooled off and made stars in subclumps before the main collapse of the system, and the final stellar configuration was ellipsoidal in shape.

The first attempt to carry out SPH simulations of galaxy formation in its proper cosmological context was that of Navarro & Benz (1991). These authors carried out a few simulations in a universe with $\Omega_{m,0} = 1$, $\Omega_{b,0} = 0.1$, and with an initial perturbation power spectrum $P(k) \propto k^{-1}$. Their resolution was too poor to study the internal structure of the 'galaxies' which formed, but they did note an important process: as dark halos merge to form larger objects, the gaseous cores at their centers also merge to make a larger 'galaxy'. However, during this process the cores transfer most of their orbital angular momentum to the surrounding dark matter. This means that the resulting disks are far more compact than they would be if they had the same specific angular momentum as the halo in which they are embedded. Much higher resolution simulations of disk formation in a CDM universe were carried out by Navarro & White (1994), and confirmed that most angular momentum of the disk material is lost to the dark matter during the highly inhomogeneous assembly process. The reason for this is that the gas cools quickly and collapses in dense subunits within the halos. A combination of dynamical friction (see §12.3) and gravitational torques then transfers most of the orbital angular momentum of the baryons to the dark matter. This problem has become known as the angular momentum catastrophe.

Three possible solutions to this angular momentum catastrophe have been discussed in the literature:

(i) The simulations do not have adequate numerical resolution. When disk formation is simulated with poor mass and spatial resolution, numerical viscosity (present in SPH to describe shocks; see Appendix C) and numerical two-body relaxation introduced by noise in the global potential due to small particle numbers reduce the angular momentum of the disk by artificial dissipation and by transfer to the halo component, respectively (e.g. Governato et al., 2004).
(ii) The treatment of feedback associated with the formation of stars is inadequate. Strong feedback can prevent the gas from cooling in small subunits, so that the gas remains hot and has a spatial distribution similar to that of the dark matter. This will reduce the amount of angular momentum transfer from gas to dark matter, resulting in more extended disks (e.g. Navarro & Benz, 1991; Brook et al., 2004a).
(iii) Dark matter is warm rather than cold. In this case, free streaming of dark matter particles will cause a reduction of the abundance of low mass halos. Having fewer low-mass halos in which the gas can cool has a similar effect as feedback, in that it keeps the gas more extended (e.g. Sommer-Larsen & Dolgov, 2001).

In recent years high-resolution numerical simulations with more realistic treatments of feedback, sometimes even with a multi-phase treatment of the ISM, have been performed (e.g. Okamoto et al., 2005; Robertson et al., 2006a; Scannapieco et al., 2008). Although such simulations have been able to produce more realistic disk galaxies, they also demonstrate that it is extremely difficult to produce disk galaxies without a substantial bulge component. It remains to be seen whether the situation will improve with even higher resolution simulations and with more sophisticated feedback prescriptions, or whether this failure signals the need for a more fundamental modification of the underlying cosmogony.

11.3 The Origin of Disk Galaxy Scaling Relations

As shown in §2.3.3, disk galaxies obey scaling relations between their luminosity, scale length and rotation velocity. Fig. 11.2 shows these relations for a large sample of disk galaxies compiled by Courteau et al. (2007). The rotation velocities are measured from Hα and HI rotation curves and basically reflect the maxima of the rotation curves. The upper-left panel shows the scaling relation between rotation velocity and luminosity, i.e. the Tully–Fisher (TF) relation. In the I band used here, the data can be described by

$$L_{\rm d} = 2.9 \times 10^{10} h^{-2} L_\odot \left(\frac{V_{\rm obs}}{200\,{\rm km\,s^{-1}}} \right)^{3.44}, \qquad (11.74)$$

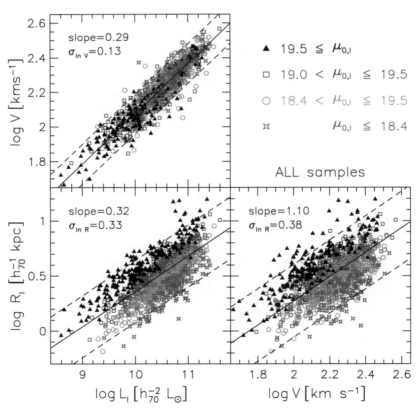

Fig. 11.2. The scaling relations between rotation velocity V, luminosity L, and scale length R. Different symbols represent different bins in central surface brightness, as indicated (all photometry is in the I band). Solid lines show the best-fit linear relations between $\log V$, $\log L$ and $\log R$. The 2σ observed scatter is indicated by the dashed lines, and the slope and scatter are listed in each panel. [Adapted from Courteau et al. (2007) by permission of AAS]

with a scatter in $\ln(V_{\rm obs})$ at fixed L of $\sigma_{\ln V} = 0.13$, uncorrelated with the disk surface brightness. The lower two panels show that disks are larger for brighter galaxies ($R_{\rm d} \propto L^{0.32}$) and for galaxies with larger rotation velocities ($R_{\rm d} \propto V_{\rm obs}^{1.1}$), although the amounts of scatter in these relations are significantly larger than in the TF relation.

The slopes of these scaling relations can vary substantially from one photometric band to another. For example, the slope $\partial \ln L / \partial \ln V$ of the TF relation ranges from ~ 2.5 in the B band to ~ 4 in the K band. Such variation reflects the fact that brighter (disk) galaxies are redder (see §2.4.3). In addition to using luminosity, one can also use stellar mass or total baryonic mass (sum of stellar mass plus cold gas mass) to obtain a baryonic TF relation. Such a relation is found to have a slope $\partial \ln M_{\rm bar}/\partial \ln V \simeq 3.5$ (Bell & de Jong, 2001), similar to that for the I band shown in Fig. 11.2.

Below we investigate the extent to which the disk-formation models described in the previous section can account for the observed scaling relations of disk galaxies.

(a) Self-Gravitating Disks The tight TF relation implies a close relation between the total gravitational mass (related to $V_{\rm obs}$) and the total amount of stars (related to $L_{\rm d}$). One might think that this can be achieved if the rotation curve of a disk is dominated by the luminous mass. However, this is true only if all disks with the same mass have the same density distribution. To see this more clearly, consider self-gravitating disks (i.e. disks without dark matter halos) with exponential surface density profiles. From Eq. (11.30) we see that the rotation curves of such disks peak at a radius $R \approx 2.16 R_{\rm d}$. It then follows that $V_{\rm max}^2 \approx 2.5 G \Sigma_0 R_{\rm d}$. Since $V_{\rm obs} \simeq V_{\rm max}$, and assuming a disk mass-to-light ratio $\Upsilon_{\rm d} \equiv M_{\rm d}/L_{\rm d}$, this relation can be written as

$$L_{\rm d} = 8.7 \times 10^{11} \, {\rm L}_\odot \left(\frac{V_{\rm obs}}{200\,{\rm km\,s^{-1}}}\right)^4 \left(\frac{I_0}{100\,{\rm L}_\odot\,{\rm pc}^{-2}}\right)^{-1} \left(\frac{\Upsilon_{\rm d}}{{\rm M}_\odot/{\rm L}_\odot}\right)^{-2}, \quad (11.75)$$

where $I_0 \equiv \Sigma_0/\Upsilon_{\rm d}$ is the disk central luminosity density. This relation looks like a TF relation with a slope $\partial \ln L/\partial \ln V = 4$. However, there is a problem: the scatter in this TF relation is proportional to the scatter in I_0, in conflict with the data that shows no correlation between the TF scatter and I_0. Thus, unless $\Upsilon_{\rm d} \propto I_0^{-1/2}$ with very little scatter, the observed TF relation is inconsistent with a picture in which the disk rotation is completely dominated by baryonic matter. Given that the rotation curves of disk galaxies are flat out to large radii, this should not come as a surprise.

(b) Non-Self-Gravitating Disks in Isothermal Spheres As another extreme, let us assume disk gravity to be negligible, so that disk rotation curves are determined entirely by their dark matter halos. For simplicity we model these halos as singular isothermal spheres. This is the same assumption as used in §11.2.2, and so the Tully–Fisher relation follows directly from Eq. (11.52):

$$L_{\rm d} = A \left(\frac{V_{\rm obs}}{200\,{\rm km\,s^{-1}}}\right)^3, \quad (11.76)$$

where $V_{\rm obs} = V_{\rm vir}$, and

$$A \approx 1.3 \times 10^{11} h^{-1} {\rm L}_\odot \left(\frac{m_{\rm d}}{0.05}\right) \left(\frac{\Upsilon_{\rm d}}{{\rm M}_\odot/{\rm L}_\odot}\right)^{-1} \mathscr{D}^{-1}(z). \quad (11.77)$$

Contrary to Eq. (11.75), the scatter in this TF relation is expected to be independent of surface brightness, in agreement with observational data. In addition, this relation matches both the slope and the zero-point of the observed I band TF relation given by Eq. (11.74), provided that

$$\left(\frac{m_{\rm d}}{\Upsilon_{\rm d}}\right) \simeq 0.016 \left(\frac{V_{\rm obs}}{200\,{\rm km\,s^{-1}}}\right)^{0.44}, \quad (11.78)$$

where we have adopted $h = 0.7$, and $\Upsilon_{\rm d}$ now represents the mass-to-light ratio of the disk in the I band (in solar units). To put this in perspective, we write $\Upsilon_{\rm d} = \Upsilon_\star/\varepsilon_{\rm sf}$, where $\Upsilon_\star$ is the stellar mass-to-light ratio and ε_{sf} is the star-formation efficiency describing the fraction of the disk mass that ends up in the form of stars. For a disk with $V_{obs} \sim 200\,\mathrm{km\,s^{-1}}$, such as our Milky Way, $\varepsilon_{sf} \sim 0.9$ and $\Upsilon_\star \sim 2\,(M/L)_\odot$ in the I band. From Eq. (11.78) we then infer that $m_d \sim 0.035$, corresponding to a galaxy formation efficiency of $\varepsilon_{gf} = m_d/(\Omega_{b,0}/\Omega_{m,0}) \simeq 0.2$; i.e. only about 20% of the baryons in the halo of a Milky-Way-like galaxy have ended up in the disk. The challenge is to find a physical mechanism (e.g. feedback) that can explain such a low efficiency while keeping the galaxy-to-galaxy scatter in m_d/Υ_d sufficiently small to match the observed TF scatter. Finally, the (weak) dependence of m_d/Υ_d on V_{obs} indicates that either m_d increases with rotation velocity for fixed Υ_d, or Υ_d decreases with rotation velocity for fixed m_d. Both are likely: the former is qualitatively consistent with supernova feedback, which is expected to be more effective in less massive halos (see §8.6.3), while the latter is consistent with the fact that ε_{sf} is larger in more massive systems (i.e. more massive disk galaxies have lower gas mass fractions; see §2.3.3).

(c) Self-Gravitating Disks in Halos with Realistic Profiles The above model, in which the self-gravity of the disk is negligible, and the halo is a singular isothermal sphere, seems to be able to explain all aspects of the observed TF relation, as long as the scatter in m_d/Υ_d is sufficiently small. In particular, it can produce a TF relation whose scatter is uncorrelated with the surface brightness of the disk, as is observed. However, as seen in §11.2.3, in realistic models dark matter halos are not well described by singular isothermal spheres, and the self-gravity of disks is usually not negligible.

The main effect of using realistic dark matter halos and including the self-gravity of the disk (and the associated adiabatic contraction of the halo) is that V_{obs} is no longer equal to V_{vir}. This means that Eq. (11.78) must be replaced by

$$\left(\frac{m_d}{\Upsilon_d}\right) \simeq 0.016 \left(\frac{V_{obs}}{V_{vir}}\right)^3 \left(\frac{V_{obs}}{200\,\mathrm{km\,s^{-1}}}\right)^{0.44}. \tag{11.79}$$

For realistic models with a NFW dark matter halo $1.2 \lesssim V_{obs}/V_{vir} \lesssim 1.8$, depending on how concentrated the disk and halo are. This has an important impact. For example, consider an average value of $V_{obs}/V_{vir} = 1.5$. Adopting $\varepsilon_{sf} \sim 0.9$ and $\Upsilon_\star \sim 2\,(M/L)_\odot$, as above, we obtain that $\varepsilon_{gf} \simeq 0.7$ for a disk galaxy with $V_{obs} \sim 200\,\mathrm{km\,s^{-1}}$. This significantly reduces the requirements for strong feedback compared to the model without self-gravity (which, as shown above, requires $\varepsilon_{gf} \simeq 0.2$ to match the TF zero-point). Another important advantage of taking the disk self-gravity into account is that it reduces the sensitivity of the TF relation to the scatter in m_d, simply because a more massive disk also implies a higher rotation velocity V_{obs}. In fact, it has been demonstrated that the scatter in m_d moves disk galaxies mainly *along* the TF relation, rather than perpendicular to it (Mo & Mao, 2000; Steinmetz & Navarro, 2000).

However, there is an important complication. The extra factor of $(V_{obs}/V_{vir})^3$ in Eq. (11.79) also *adds* scatter to the TF relation. First of all, halos of a fixed V_{vir} have a significant amount of scatter in halo concentrations (see §7.5.1), which translates into scatter in V_{obs}/V_{vir}. Secondly, the scatter in λ also translates into scatter in V_{obs}/V_{vir}, simply because a more concentrated disk (associated with a lower spin parameter) will have a larger V_{obs}/V_{vir}. This is a problem, because the surface density of the disk $\Sigma_0 \propto \lambda^{-2}$ [see Eq. (11.59)], and the TF scatter thus becomes correlated with the disk surface brightness whenever the self-gravity of the disk becomes important (i.e. V_{obs}/V_{vir} has a large value), in conflict with the data.

There is yet another complication: an increase in V_{obs}/V_{vir} also means that a disk galaxy of a given V_{obs} on average lives in a less massive halo. This has two effects: it causes the virial radius of the halo to be smaller, and it requires a larger value of m_d to match the TF zero-point. As can be seen from Eq. (11.64), $R_d \propto r_{vir}/m_d$. Therefore, increasing the value of V_{obs}/V_{vir} causes the scale length of the disks to become smaller. Matching the observed zero-point of the scaling relation between R_d and L (or between R_d and V_{obs}) then requires that disk galaxies form in halos with larger spin parameters. These together illustrate that the scaling relations shown in Fig. 11.2 are intimately related, and that a successful model must be able to reproduce all three of them simultaneously.

Detailed models that take all these effects into account have been presented by van den Bosch (2000) and Dutton et al. (2007). These studies show that it is extremely difficult to construct a model that can match all scaling relations simultaneously. The main culprit is that the models predict values for V_{obs}/V_{vir} that are too large. This, in turn, is due to the fact that NFW halos are predicted to be relatively concentrated in the ΛCDM concordance cosmology, and that adiabatic contraction is very effective. At the present it is unclear whether this implies a problem for the CDM framework, or rather indicates that one of the assumptions on which these models are based is incorrect.

11.4 The Origin of Exponential Disks

As we have seen in §2.3.3, the observed surface brightness profiles of galactic disks are well described by an exponential form, both in the radial and in the vertical direction. An important challenge for any theory of disk formation is to understand the origin of these exponential profiles. Broadly speaking two different concepts have been considered to explain why the radial surface brightness profiles are exponential: (i) the surface density distribution of the disk galaxy reflects the specific angular momentum distribution of the proto-galaxy, and (ii) the surface density distribution is a result of disk viscosity. Unfortunately, as we will see below, both of these models have problems, so that currently we do not have a clearly viable theory.

The vertical density distribution of disk galaxies turns out to reveal a wealth of structure that serves as a fossil record of their formation history. After discussing models for the origin of the radial density distribution of disk galaxies, we therefore also give a brief overview of the vertical structure of disk galaxies and sketch some of the main ideas regarding its origin.

11.4.1 Disks from Relic Angular Momentum Distribution

The disk formation models discussed in §11.2 are based on the assumption that the total specific angular momenta of the disks are the same as those of their halos (at least when $m_d/j_d = 1$ is assumed). We refer to this as the weak form of angular momentum conservation. We now consider a model in which there is no angular momentum redistribution in the gas as it flows to form a disk (Mestel, 1963; Freeman, 1970; Fall & Efstathiou, 1980). In this case, the angular momentum of each mass element is conserved and the initial angular momentum distribution of the gas is the same as that of the resulting disk galaxy. We refer to this as the strong form of angular momentum conservation.

In §7.5.4 we have seen that the specific angular momentum distribution of dark matter halos is well fit by

$$P(\mathcal{J}) = \frac{\mu \mathcal{J}_0}{(\mathcal{J}_0 + \mathcal{J})^2}, \qquad (11.80)$$

where μ is a free shape parameter, and $\mathcal{J}_0$ a parameter related to the total specific angular momentum by $\mathcal{J}_0 = \mathcal{J}_{tot}(\mu - 1)/\zeta$ with $\zeta = \zeta(\mu)$ given by Eq. (7.177).

In what follows we assume that the gas has the same specific angular momentum distribution $P(\mathcal{J})$ as the dark matter. Non-radiative hydrodynamical simulations have shown that this is a reasonable assumption (e.g. van den Bosch et al., 2002; Chen et al., 2003; Sharma & Steinmetz, 2005). If all the gas cools to form a disk, and if the strong form of angular momentum conservation is valid, the mass distribution of the disk obeys

$$\frac{M_\mathrm{d}(r)}{M_\mathrm{d}} = \frac{M_\mathrm{h}(<\mathcal{J})}{M_\mathrm{vir}}, \qquad (11.81)$$

where $\mathcal{J}$ and R are related according to $\mathcal{J} = RV_\mathrm{c}(R)$. Since

$$\Sigma_\mathrm{d}(R) = \frac{1}{2\pi}\frac{1}{R}\frac{\mathrm{d}M_\mathrm{d}(R)}{\mathrm{d}R}, \qquad (11.82)$$

one has that

$$\Sigma_\mathrm{d}(R) = \frac{M_\mathrm{d}}{2\pi R^2} P(\mathcal{J}) RV_\mathrm{c}(R) \left[1 + \frac{\partial \ln V_\mathrm{c}}{\partial \ln R}\right] \qquad (11.83)$$

[see Eq. (11.73)].

For a non-self-gravitating disk embedded in a singular isothermal sphere, $V_\mathrm{c}(R) = V_\mathrm{c}$ and Eq. (11.83) reduces to

$$\Sigma_\mathrm{d}(R) = \mu \frac{M_\mathrm{d}}{2\pi R_\mathrm{d}'^2} \left(\frac{R}{R_\mathrm{d}'}\right)^{-1} \left(1 + \frac{R}{R_\mathrm{d}'}\right)^{-2} \qquad (11.84)$$

where

$$R_\mathrm{d}' \equiv \frac{\mathcal{J}_0}{V_\mathrm{c}} = \sqrt{2}(\mu - 1)\zeta^{-1}\lambda r_\mathrm{vir}. \qquad (11.85)$$

Here we have used a prime to indicate that this disk scale length should not be confused with that of an exponential disk. In fact, Eq. (11.84) is a double power law rather than an exponential. It changes from $\Sigma_\mathrm{d} \propto R^{-1}$ at small radii ($R \ll R_\mathrm{d}'$) to $\Sigma_\mathrm{d} \propto R^{-3}$ at large radii ($R \gg R_\mathrm{d}'$), so that its decline is faster (slower) than exponential at small (large) radii.

More detailed calculations, including the self-gravity of the disk and using realistic NFW density distributions for the dark matter, paint a similar picture: $P(\mathcal{J})$ is not consistent with an exponential surface density distribution of disk galaxies. This is illustrated in Fig. 11.3, where a comparison is made between the specific angular momentum distributions of dark matter halos and those of three observed disk galaxies, obtained from their surface density profiles and rotation curves. The differences are very pronounced: disk galaxies lack both low and high specific angular momentum material compared to the distribution given by Eq. (11.80). This represents another angular momentum problem for disk galaxy formation, in addition to the angular

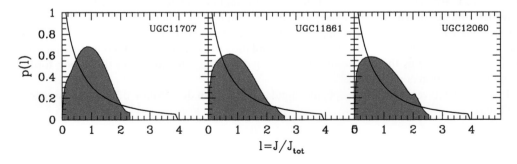

Fig. 11.3. The shaded areas indicate the normalized angular momentum distributions $p(l)$ (where $l = \mathcal{J}/\mathcal{J}_\mathrm{tot}$) of three disk galaxies, obtained from their observed density distributions and rotation curves. The solid lines show the typical, normalized angular momentum distribution of dark matter halos, given by Eq. (11.80) with $\mu = 1.25$. [After van den Bosch et al. (2001)]

momentum catastrophe discussed in §11.2.6: even if disks form conserving their detailed specific angular momentum distribution, which, as we have seen in §11.2.6, is inconsistent with hydrodynamical simulations, the resulting disks do not have realistic surface density profiles (Bullock et al., 2001a; van den Bosch et al., 2001; Sharma & Steinmetz, 2005).

Within the standard framework for disk formation outlined above, there are three possible solutions to this problem:

(i) The low specific angular momentum material is somehow transformed into a bulge component. Although this can result in disks with exponential surface brightness distributions (van den Bosch, 2001), it still leaves open the question how to form bulge-less, exponential disks.

(ii) Disks only form out of a special subset of the entire baryonic matter associated with the dark halo. Indeed, as we have seen in §11.3 the zero-point of the observed Tully–Fisher relation suggests that only a relatively small fraction of the baryonic material ends up in the disk. If, for example, feedback preferentially expels low angular momentum material from the disk, it could solve the problem at hand (e.g. Dutton & van den Bosch, 2009).

(iii) Somehow the specific angular momentum of the disk material is redistributed during the formation or evolution of the disk, so that its final $P(\mathscr{J})$ corresponds to an exponential disk. One mechanism to transfer angular momentum is disk viscosity, which we discuss next. Another would be gravitational torques during disk assembly.

11.4.2 Viscous Disks

An alternative explanation for the exponential density distribution of disk galaxies is angular momentum transport. In this case, the gas that forms a galactic disk is assumed to settle in a centrifugally supported disk with some arbitrary surface-density profile. In general, the disk material rotates differentially, and transport processes can therefore cause angular momentum to be exchanged between material at different radii. If gas at radius R loses angular momentum in this manner, it sinks deeper into the gravitational potential well, while the material which absorbs its angular momentum moves outwards. This redistribution of mass and angular momentum causes $\Sigma(R)$ to evolve with time, and the final density distribution of the disk will reflect dissipational processes during its formation and evolution, rather than the initial conditions.

The evolution of viscous, star-forming disks has been discussed in some detail by Lin & Pringle (1987b). The gas surface density, $\Sigma(R,t)$, is assumed to evolve into a stellar surface density, $\Sigma_\star(R,t)$, according to the star-formation law

$$\frac{\partial \Sigma_\star}{\partial t} = \frac{\Sigma(R,t)}{t_\star}, \qquad (11.86)$$

where $t_\star$ is the star-formation time scale whose form has to be specified. The evolution of the gas surface density is governed by the standard viscous accretion disk equations of continuity and conservation of angular momentum supplemented with a sink term due to the conversion of gas to stars. These equations are

$$\frac{\partial \Sigma}{\partial t} + \frac{1}{R}\frac{\partial}{\partial R}(\Sigma R v_R) = -\frac{\Sigma}{t_\star}, \qquad (11.87)$$

and

$$\frac{\partial}{\partial t}(\Sigma R^2 \Omega) + \frac{1}{R}\frac{\partial}{\partial R}(\Sigma R^3 \Omega v_R) = \frac{1}{R}\frac{\partial}{\partial R}\left(\nu \Sigma R^3 \frac{\partial \Omega}{\partial R}\right) - R^2 \Omega \frac{\Sigma}{t_\star} \qquad (11.88)$$

(see Pringle, 1981, for details). Here v_R is the radial velocity of the gas, $\Omega(R) = V_c(R)/R$ is the circular frequency, and ν is the kinetic viscosity. Eliminating v_R between Eq. (11.87) and Eq. (11.88) assuming $\partial \Omega/\partial t = 0$ yields

$$\frac{\partial \Sigma}{\partial t} = -\frac{1}{R}\frac{\partial}{\partial R}\left\{\frac{(\partial/\partial R)\left[\nu \Sigma R^3 (\partial \Omega/\partial R)\right]}{(\partial/\partial R)(R^2 \Omega)}\right\} - \frac{\Sigma}{t_\star}. \tag{11.89}$$

Without the star-formation term, the equation for $\Sigma(R,t)$ is of diffusion type. In this case, as the gas diffuses inwards the angular momentum is transported outwards in order to conserve the total angular momentum of the gas. The time scale of this viscous transport is of the order $t_\nu \approx R^2/\nu$. In the presence of star formation, the radial diffusion of a gas parcel is halted when it is turned into stars, giving rise to a final stellar distribution given by

$$\Sigma_\star(R) = \int_0^\infty \frac{\Sigma(R,t)}{t_\star}\,dt. \tag{11.90}$$

If $t_\star \ll t_\nu$, the redistribution of gas is negligible and $\Sigma_\star(R) \sim \Sigma(R,0)$. On the other hand, if $t_\star \gg t_\nu$, the effect of star formation on the gas redistribution can be neglected, and the disk rapidly evolves towards its ultimate endpoint in which all the mass is at the origin and all the angular momentum at infinity (see §11.2.1). However, Lin & Pringle (1987b) found that if $t_\star \approx t_\nu$ then the resulting stellar disk has an exponential profile quite independent of the initial gas surface density and of the assumed viscosity prescription. Although the dominant physical process responsible for the viscosity in disk galaxies is poorly constrained, the condition that $t_\star \approx t_\nu$ seems quite natural. If, for example, viscosity results from interactions between molecular clouds then it is likely that enhanced viscosity is associated with enhanced star formation through dissipation and/or aggregation of clouds. Furthermore, since angular momentum may be transferred by non-axisymmetric self-gravitating modes (Lin & Pringle, 1987a), such as spiral arms, both star formation and viscous dissipation may be linked to the importance of self-gravity in the disk.

Viscous evolution therefore seems a promising mechanism to explain the exponential surface brightness profiles of disk galaxies. However, since viscosity transports mass inwards, a requirement is that the initial gas profile, prior to the onset of evolution, is less centrally concentrated than an exponential. As we have seen above, this is inconsistent with the initial surface density profile predicted from the halo angular momentum distribution, which is actually more centrally concentrated than an exponential profile. Thus, although viscosity may play an important role in redistributing disk material, and in particular in setting the outer profile of the gas distribution (e.g. Olivier et al., 1991), it only aggravates the problem of the excess of low angular momentum material outlined above.

11.4.3 The Vertical Structure of Disk Galaxies

As mentioned in §2.3.3, galactic disks typically consist of two separate components: a thin disk and a thick disk. Here we briefly discuss the origin of this vertical structure.

(a) Thin Disks As discussed in §11.1.1 the vertical structure of thin disks is often modeled as an isothermal sheet. This is motivated by considerations of simplicity. However, since the stellar components in galaxies are collisionless, there is no reason to expect them to be isothermal. Indeed, as we have seen in §2.3.3, in many cases an exponential z profile is a better description of the vertical density distribution in galaxy disks than is a sech^2-distribution. Kinematic studies of stars in the solar neighborhood have shown that the velocity dispersion of stars increases systematically with their age. Defining the heating rate $\alpha_i = d\ln\sigma_i/d\ln t$, with σ_i the velocity dispersion in the i direction and t the age of the stars, observational data

11.4 The Origin of Exponential Disks

show that $\alpha_R \simeq 0.31$, $\alpha_\phi \simeq 0.34$ and $\alpha_z \simeq 0.47$ (Wielen, 1977; Nordström et al., 2004). It is generally assumed that this age dependence is due to some heating mechanism: stars are born close to the disk mid-plane with a small velocity dispersion, but the dispersion is subsequently increased by one or more heating mechanisms. In this case, the density distribution of the disk in the vertical direction is a result of both the star-formation rate and heating rate of the disk. For example, as shown by Just et al. (1996), one can obtain an exponential vertical profile by having a declining star-formation rate combined with a constant heating rate.

Spitzer & Schwarzschild (1953) suggested that disk stars could be heated by scattering against massive gas clouds orbiting in the disk. Because of differential rotation, a cloud with orbital radius R_c overtakes stars with guiding center radii $R > R_c$ and is overtaken by stars with $R < R_c$. Typically, stars gain energy from the encounters with the cloud because of their much smaller masses, and so increase their peculiar motion relative to the rotating disk. Although the discovery of a population of giant molecular clouds (GMCs) with masses in the proposed range ($M > 10^5 \, M_\odot$) was originally considered a major success for this heating model, more detailed calculations by Lacey (1984) showed that the heating rate is too low ($\alpha_z \lesssim 0.25$), while the resulting ratio σ_z/σ_R is too high. These results have subsequently been confirmed with numerical simulations (Villumsen, 1985; Hänninen & Flynn, 2002). An alternative heating mechanism, proposed by Goldreich & Lynden-Bell (1965) and Barbanis & Woltjer (1967), is heating by *transient* spiral arms. As discussed in more detail in §11.6, spiral arms are associated with spiral-shaped potential perturbations. In order for these perturbations to cause a net heating of the disk stars, it is crucial that the amplitude of the perturbations evolve with time, on a time scale sufficiently small compared to the time it takes for a star to cross the perturbation (see §7.5 in Binney & Tremaine, 1987). In this case the heating is somewhat similar to violent relaxation. As we discuss in §11.6, the flocculent spiral arm fragments seen in many disk galaxies are believed to be transient phenomena that form and dissappear on an orbital time scale. Unlike the grand-design spirals, which are believed to be quasi-stationary density waves, these structures may thus be responsible for a significant heating of disk stars. Indeed, numerical simulations by Carlberg & Sellwood (1985) and De Simone et al. (2004) have shown that heating by transient spirals is likely to be the dominant in-plane heating mechanism. However, since the spiral arms have characteristic sizes much larger than the scale height of the young stellar population, they are unable to cause a significant increase in σ_z. Jenkins & Binney (1990), therefore, considered a hybrid model, in which the main source of heating is transient spirals, while the GMCs are responsible for scattering the stars out of the disk plane. Although they were able to reproduce the observed ratio σ_z/σ_R, the resulting heating rate, $\alpha_z \sim 0.3$, is still too low compared to the data.

Two additional heating mechanisms that have been discussed in the literature are minor mergers with small satellite galaxies (Tóth & Ostriker, 1992) and heating by a population of massive black holes with masses of the order of $10^6 \, M_\odot$ that populate the dark halo surrounding the disk (Lacey & Ostriker, 1985). The former is likely to play a role in hierarchical structure formation models, and will be discussed in more detail in §12.4.4. The latter, although potentially capable of explaining the observed heating rates, lacks a natural explanation for the origin of the black holes. In addition, it has problems because massive black holes passing through the disk can accrete gas, giving rise to an unobserved population of high proper motion X-ray sources, and because microlensing observations of the Galactic bulge and the LMC exclude a dominant halo population of such objects.

In summary, it is generally accepted that the vertical structure of thin disks is produced by heating of disk stars that originally formed near the mid-plane with a small velocity dispersion. Although there is no shortage of possible heating mechanisms, it is still unclear which, if any, of these dominates. In addition, the fact that the scale heights of disk galaxies are observed to be remarkably independent of the galactocentric distance suggests that the heating rate is equally efficient over the entire disk. A natural explanation for this is still lacking.

(b) Thick Disks Thick disks were first discovered by Burstein (1979) from deep photometry of five edge-on S0 galaxies. Burstein noted faint envelopes of excess surface brightness at large distances from the mid-plane, which cannot be described by the regular sech^2 law of Eq. (2.31). Of course, the excess alone may simply reflect that the sech^2 form is not an appropriate fitting function, rather than indicating a separate disk component. The subsequent discovery of a thick disk in the Milky Way by Gilmore & Reid (1983) quickly settled this issue. With photometric and spectroscopic observations of individual thick disk stars it soon became clear that the thick disk, at least in the Milky Way, is a truly distinct component. Its stars are old ($\gtrsim 10-12$ Gyr), relatively metal poor ($-2.2 \lesssim$ [Fe/H] $\lesssim -0.3$ with a mean of ~ -0.6) and are more enhanced in α elements than thin disk stars of the same [Fe/H]. The old stellar age indicates that the thick disk may serve as a fossil record of physical processes during the early stages of the formation of the Milky Way.

Nowadays it is clear that thick disks are ubiquitous. They are found in essentially all edge-on disk galaxies of all Hubble types, indicating that they are a generic by-product of galaxy formation. As discussed in §2.3.3, extragalactic thick disks are almost always more radially extended, older and more metal poor than the more massive embedded thin disk, while their vertical scale heights are ~ 3 times larger. Of the three cases studied kinematically, two rotate slower than the thin disk, lagging behind by $\sim 20\%-50\%$, while in the other case there is evidence for counter-rotation (Yoachim & Dalcanton, 2005).

One can envisage a number of different formation scenarios for thick disks, which we categorize according to the birth place of the thick disk stars: (i) in the thin disk, (ii) in situ in the thick disk, or (iii) externally. In case (i) the stars must have been heated vertically by some mechanism, while either a substantial fraction stays behind or a new thin disk forms by gas inflow following the heating event. Potential candidates for heating the disk are mergers or close encounters with relatively massive satellite galaxies or dark matter subhalos (e.g. Quinn et al., 1993; Velazquez & White, 1999; Kazantzidis et al., 2008). A potential problem for this thickening scenario is that the event that heated the thin disk must have happened about 10–12 Gyr ago (i.e. at $z \sim 2$). Since the thick disk is at least as large as the present-day thin disk, this requires the thin disk at $z \sim 2$ to have a size comparable to that today, which is difficult to reconcile with the standard model for disk formation discussed in §11.2. In case (ii) the thick disk stars may form during a slow, pressure-supported collapse, or during a burst of intense star formation at the early stages of galaxy formation when clumps of gas coalesce to form the thin disk (Brook et al., 2004b). The former is unlikely as it fails to explain the lack of a large, intermediate age population with a scale height in between that of the thin and thick disks. Finally, in case (iii), the thick disk stars may have formed in small satellite galaxies, which deposited their stars into the thick disk when they were disrupted by the tidal field of the thin disk (Abadi et al., 2003). The main argument against this scenario is that none of the satellite galaxies of the Milky Way, not even the LMC, has metallicities and abundance patterns comparable to that of the thick disk. This indicates that if the thick disk is made up primarily of satellite debris, these satellites must have been very different from those that survived. Note, however, that this is not a very strong argument, since the satellites that formed the thick disk (and probably also the stellar halo) are likely to have been accreted a long time ago. It is not at all unlikely that these early satellites had different metallicities and abundance patterns from the present-day population of survivors. Currently it is unclear which of these various scenarios is the dominant mechanism for forming thick disks. It is possible that all the processes occur and contribute to some extent.

11.5 Disk Instabilities

Thus far we have considered models for the formation of disk galaxies without examining whether the resulting disks are stable or not. However, instabilities may play a very important role in transforming and regulating the properties of (disk) galaxies. In what follows we distinguish between *global* stability and *local* stability. The former can cause a significant transformation of the overall disk structure, and plays an important role in our understanding of the global properties of disk galaxies. Whenever a disk galaxy is globally unstable it will evolve towards a new, stable configuration, erasing information about the initial conditions under which the system formed. Local stability, on the other hand, controls whether or not perturbations much smaller than the size of the disk can grow. Since star formation requires the fragmentation and collapse of gas clouds, local instability is likely to be a necessary condition for star formation in disk galaxies.

11.5.1 Basic Equations

The analysis of disk instability presented here is based on standard first-order perturbation theory. Starting from the basic dynamical equations, we first write each quantity as a sum of its unperturbed value and a (small) perturbation. Next we keep only terms that are first order in the small perturbations to get a set of linear equations. After expanding the perturbations in terms of the eigenmodes (e.g. Fourier expansion), we determine the dispersion relation and investigate the unstable modes.

Consider a thin, self-gravitating disk (i.e. for the moment we ignore a possible dark halo). For simplicity we assume the disk to be gaseous so that fluid dynamics apply. We also assume the unperturbed disk to be axisymmetric so that it is convenient to work with cylindrical coordinates (R, ϕ, z), where $z = 0$ corresponds to the disk plane. Neglecting the thickness of the disk, the continuity equation can be written as

$$\frac{\partial \Sigma}{\partial t} + \frac{1}{R}\frac{\partial}{\partial R}(\Sigma R v_R) + \frac{1}{R}\frac{\partial}{\partial \phi}(\Sigma v_\phi) = 0, \tag{11.91}$$

where Σ is the surface density. The Euler equations in cylindrical coordinates are

$$\frac{\partial v_R}{\partial t} + v_R \frac{\partial v_R}{\partial R} + \frac{v_\phi}{R}\frac{\partial v_R}{\partial \phi} - \frac{v_\phi^2}{R} = -\frac{\partial}{\partial R}(\Phi + h), \tag{11.92}$$

$$\frac{\partial v_\phi}{\partial t} + v_R \frac{\partial v_\phi}{\partial R} + \frac{v_\phi}{R}\frac{\partial v_\phi}{\partial \phi} + \frac{v_\phi v_R}{R} = -\frac{1}{R}\frac{\partial}{\partial \phi}(\Phi + h), \tag{11.93}$$

where Φ is the gravitational potential at $z = 0$, related to Σ via the Poisson equation

$$\nabla^2 \Phi = 4\pi G \Sigma \delta(z), \tag{11.94}$$

with $\delta(z)$ the Dirac delta function. The quantity h introduced in the Euler equations represents the pressure force and its expression can be derived as follows. The pressure term should have appeared on the right-hand side of Eq. (11.92) as $-\rho^{-1}(\partial P/\partial R)$ and on the right-hand side of Eq. (11.93) as $-(\rho R)^{-1}(\partial P/\partial \phi)$, where ρ is the *volume* density and P is the pressure. Since we do not care about the vertical structure of the disk, we may assume the disk to be uniform in the z direction. In this case, using the definition of the sound speed, $c_s^2 = \partial P/\partial \rho$ (notice that c_s is defined for the unperturbed disk and so it depends only on R), we have $\rho^{-1}(\partial P/\partial R) = c_s^2 \partial \ln \Sigma/\partial R$ and $\rho^{-1}(\partial P/\partial \phi) = c_s^2 \partial \ln \Sigma/\partial \phi$. It then follows that h is related to Σ by

$$dh = c_s^2 d\ln\Sigma. \tag{11.95}$$

Now we write $\Sigma = \Sigma_0 + \Sigma_1$, $v_R = v_{R0} + v_{R1} = v_{R1}$, $v_\phi = v_{\phi 0} + v_{\phi 1}$, $\Phi = \Phi_0 + \Phi_1$ and $h = h_0 + h_1$, where the subscripts '0' and '1' refer to unperturbed and perturbed quantities, respectively. Keeping only terms to the first order in the perturbations, Eqs. (11.91)–(11.94) become

$$\frac{\partial \Sigma_1}{\partial t} + \frac{1}{R}\frac{\partial}{\partial R}(\Sigma_0 R v_{R1}) + \Omega \frac{\partial \Sigma_1}{\partial \phi} + \frac{\Sigma_0}{R}\frac{\partial v_{\phi 1}}{\partial \phi} = 0, \tag{11.96}$$

$$\frac{\partial v_{R1}}{\partial t} + \Omega \frac{\partial v_{R1}}{\partial \phi} - 2\Omega v_{\phi 1} = -\frac{\partial}{\partial R}(\Phi_1 + h_1), \tag{11.97}$$

$$\frac{\partial v_{\phi 1}}{\partial t} + \Omega \frac{\partial v_{\phi 1}}{\partial \phi} + \frac{\kappa^2}{2\Omega} v_{R1} = -\frac{1}{R}\frac{\partial}{\partial \phi}(\Phi_1 + h_1), \tag{11.98}$$

$$\nabla^2 \Phi_1 = 4\pi G \Sigma_1 \delta(z). \tag{11.99}$$

In the above equations, $h_1 = c_s^2 \Sigma_1 / \Sigma_0$, which follows from the linearized version of Eq. (11.95), and $\Omega(R)$ and $\kappa(R)$ are, respectively, the circular frequency and epicyclic frequency of the unperturbed disk at radius R (see §11.1.5).

In general, we can expand a perturbation in the form

$$Q_1 = \sum Q_a(R) e^{-i(\omega t - m\phi)}, \tag{11.100}$$

where Q_1 denotes a perturbation quantity (e.g. Σ_1), and the summation is over all modes a, each of which is characterized by an angular frequency, ω, and its azimuthal wavenumber $m > 0$. Throughout this section we consider it understood that, since Q_1 is a physical quantity, we need to take the real part of Eq. (11.100). Substituting such solutions into Eqs. (11.96)–(11.98) and rearranging, we have

$$i(m\Omega - \omega)\Sigma_a + \frac{1}{R}\frac{d}{dR}(\Sigma_0 R v_{Ra}) + \frac{im\Sigma_0}{R} v_{\phi a} = 0, \tag{11.101}$$

$$v_{Ra} = -\frac{i}{\varpi}\left[(m\Omega - \omega)\frac{d}{dR}(\Phi_a + h_a) + \frac{2m\Omega}{R}(\Phi_a + h_a)\right], \tag{11.102}$$

$$v_{\phi a} = \frac{1}{\varpi}\left[\frac{\kappa^2}{2\Omega}\frac{d}{dR}(\Phi_a + h_a) + \frac{m(m\Omega - \omega)}{R}(\Phi_a + h_a)\right], \tag{11.103}$$

where $h_a = c_s^2 \Sigma_a / \Sigma_0$, and

$$\varpi \equiv \kappa^2 - (m\Omega - \omega)^2. \tag{11.104}$$

This set of equations is complete for Σ_a, v_{Ra} and $v_{\phi a}$ once Φ_a is related to Σ_a via the Poisson equation.

The Poisson equation for a thin disk can be solved easily for Fourier modes. To proceed, we write Eq. (11.99) as

$$\nabla_{\mathbf{x}}^2 \Phi_1 + \frac{\partial^2 \Phi_1}{\partial z^2} = 4\pi G \Sigma_1 \delta(z), \tag{11.105}$$

where $\nabla_{\mathbf{x}}$ denotes derivatives in the disk plane. Consider a plane-wave perturbation on the disk: $\Sigma_1 = \tilde{\Sigma}_1 \exp[i(\mathbf{k}\cdot\mathbf{x} - \omega t)]$. Since $\nabla^2 \Phi_1 = 0$ for $z \neq 0$ and since Φ_1 must be of the form $\Phi_1(\mathbf{x}, z = 0, t) = \tilde{\Phi}_1 \exp[i(\mathbf{k}\cdot\mathbf{x} - \omega t)]$ on the disk plane ($z = 0$), it can be shown that the only continuous function that satisfies both conditions has the form $\Phi_1(\mathbf{x}, z, t) = \tilde{\Phi}_1 \exp[i(\mathbf{k}\cdot\mathbf{x} - \omega t)] \exp(-|kz|)$. Inserting this solution into Eq. (11.105) gives

$$-k^2 e^{-|kz|}\tilde{\Phi}_1 + \tilde{\Phi}_1 \frac{\partial^2 e^{-|kz|}}{\partial z^2} = 4\pi G \tilde{\Sigma}_1 \delta(z), \tag{11.106}$$

which can be solved to give

$$\tilde{\Phi}_1 = -\frac{2\pi G}{|k|}\tilde{\Sigma}_1. \tag{11.107}$$

Note, however, that this relation is only valid for Fourier modes which, in general, are not the eigenmodes for axisymmetric disks.

11.5.2 Local Instability

If the characteristic size of the perturbation is much smaller than the size of the disk, the stability analysis (one now speaks of local stability) can be simplified. To proceed, let us write a perturbation mode as

$$\Sigma_1(R,\phi,t) = A(R,t)\exp\{i[m\phi + f(R,t)]\}, \tag{11.108}$$

where $f(R,t)$ is the shape function, and $A(R,t)$ is a slowly varying function of R that sets the amplitude of the density wave. For a fixed t, $m\phi + f(R,t) =$ constant defines a curve (in the disk) where the phase of the perturbation mode is the same. The curves defined by $m\phi + f(R,t) = 2n\pi$ ($n = 0, \pm 1, \ldots$) therefore correspond to the peaks of the density waves and delineate the ridges of spiral density arms. The radial separation between adjacent arms at a given azimuth is ΔR, where $|f(R+\Delta R,t) - f(R,t)| = 2\pi$. Under the assumption of tight-winding $\Delta R \ll R$, and one has $f(R+\Delta R,t) \approx f(R,t) + (\partial f/\partial R)\Delta R$ so that $\partial f/\partial R = 2\pi/\Delta R$. In this case, we can write the perturbation in the neighborhood of a point (R_0, ϕ_0) as

$$\Sigma_1(R,\phi,t) \approx \Sigma_a \exp[ik(R_0,t)(R-R_0)], \tag{11.109}$$

where

$$\Sigma_a = A(R_0,t)\exp\{i[m\phi_0 + f(R_0,t)]\}, \tag{11.110}$$

$$k(R_0,t) \equiv \left[\frac{\partial f(R,t)}{\partial R}\right]_{R_0} = \frac{2\pi}{\Delta R}. \tag{11.111}$$

In the above expression we have neglected the variations with angle ϕ, because they are much slower than the radial variations for tightly wound waves. Under this assumption, a spiral density wave closely resembles a plane wave in the R direction, with wave-vector $k\mathbf{e}_R$ and wavelength ΔR. In this case, the potential perturbation Φ_a is related to Σ_a by $\Phi_a = -2\pi G\Sigma_a/|k|$ [see Eq. (11.107)]. Inserting such solutions into Eqs. (11.101)–(11.103) and using the fact that $kR \gg 1$ to neglect all small terms, we obtain

$$(m\Omega - \omega)\Sigma_a + k\Sigma_0 v_{Ra} = 0, \tag{11.112}$$

$$v_{Ra} = \frac{(m\Omega - \omega)k(\Phi_a + h_a)}{\varpi}, \quad v_{\phi a} = \frac{i\kappa^2 k(\Phi_a + h_a)}{2\Omega\varpi}. \tag{11.113}$$

Inserting the expression of v_{Ra} into Eq. (11.112) and using $h_a = c_s^2 \Sigma_a/\Sigma_0$, we finally get the following dispersion relation for a gaseous disk in the tight-winding approximation:

$$(m\Omega - \omega)^2 = \kappa^2 - 2\pi G\Sigma_0|k| + k^2 c_s^2. \tag{11.114}$$

For axisymmetric perturbations (i.e. $m = 0$), the dispersion relation reduces to

$$\omega^2 = \kappa^2 - 2\pi G\Sigma_0|k| + k^2 c_s^2. \tag{11.115}$$

Note that without the κ^2 term this is simply the Jeans criterion for gravitational instability (see §4.1.3) in a thin disk. The extra κ term is owing to the Coriolis force associated with the disk's rotation, and which provides a centrifugal force that combines with the pressure to resist the growth of perturbations. We can use Eq. (11.115) to investigate whether the disk is stable against perturbations with $m = 0$: modes with $\omega^2 < 0$ grow exponentially with time and so are unstable, while those with $\omega^2 > 0$ are stable. To proceed, we define the following two dimensionless parameters,

$$Q \equiv \frac{c_s \kappa}{\pi G \Sigma_0} \quad \text{and} \quad \lambda_{\text{crit}} \equiv \frac{4\pi^2 G \Sigma_0}{\kappa^2}, \tag{11.116}$$

which allows us to write the dispersion relation as

$$\omega^2 = \frac{4\pi^2 G \Sigma_0}{\lambda_{\text{crit}}} \left[1 - \frac{\lambda_{\text{crit}}}{\lambda} + \frac{Q^2}{4} \left(\frac{\lambda_{\text{crit}}}{\lambda} \right)^2 \right], \tag{11.117}$$

where $\lambda \equiv 2\pi/|k|$. As is evident from this equation, λ_{crit} is the largest unstable wavelength for a zero-pressure ($c_s = 0$) disk. The line separating stable and unstable modes is given by

$$Q(\lambda) = 2\sqrt{\frac{\lambda}{\lambda_{\text{crit}}} \left(1 - \frac{\lambda}{\lambda_{\text{crit}}} \right)}; \tag{11.118}$$

a disk with a Q value larger than $Q(\lambda)$ is stable for the perturbation mode with wavelength λ and vice versa. The function $Q(\lambda)$ has a maximum $Q_{\text{max}} = 1$ at $\lambda = \lambda_{\text{crit}}/2$. Thus, a disk is stable against all local perturbations if $Q > 1$, and as the Q value decreases, the mode that first becomes unstable has $\lambda = \lambda_{\text{crit}}/2$. In summary, for fluid disks,

$$Q \equiv \frac{c_s \kappa}{\pi G \Sigma_0} > 1 \quad \text{(for a stable disk)}, \tag{11.119}$$

and

$$\lambda = 0.5 \times \frac{4\pi^2 G \Sigma_0}{\kappa^2} \quad \text{(most unstable)}. \tag{11.120}$$

A similar derivation can be used to derive the dispersion relation for a *stellar* disk, in which the effective 'pressure' is due to the random motions of the stars. For a stellar disk the equivalent of Eq. (11.114) is given by

$$(m\Omega - \omega)^2 = \kappa^2 - 2\pi G \Sigma_0 |k| \mathscr{F} \left(\frac{\omega - m\Omega}{\kappa}, \frac{k^2 \sigma_R^2}{\kappa^2} \right), \tag{11.121}$$

where σ_R is the radial velocity dispersion, and

$$\mathscr{F}(s, \chi) = \frac{(1 - s^2)}{\sin(s\pi)} \int_0^\pi e^{-(1+\cos\tau)\chi} \sin(s\tau) \sin(\tau) \, d\tau. \tag{11.122}$$

The derivation of this relation was given independently by Kalnajs (1965) and Lin & Shu (1966), and can also be found in Binney & Tremaine (1987). In terms of λ_{crit} defined in Eq. (11.116) and $Q \equiv \sigma_R \kappa/(\pi G \Sigma_0)$, the line of neutral instability, $\omega = 0$, can be written as

$$\frac{\lambda_{\text{crit}}}{\lambda} \mathscr{F} \left(0, Q^2 \frac{\lambda_{\text{crit}}^2}{\lambda^2} \right) = 1, \tag{11.123}$$

which defines a curve $Q(\lambda)$. As before, a stellar disk is stable (unstable) against a perturbation mode with wavelength λ if Q is larger (smaller) than $Q(\lambda)$. The function $Q(\lambda)$ has a maximum value $Q_{\text{max}} \approx 3.36/\pi$ at $\lambda \approx 0.55 \lambda_{\text{crit}}$. Thus, for stellar disks,

$$Q \equiv \frac{\sigma_R \kappa}{3.36 G \Sigma_0} > 1 \quad \text{(for a stable disk)}, \tag{11.124}$$

and
$$\lambda = 0.55 \times \frac{4\pi^2 G \Sigma_0}{\kappa^2} \quad \text{(most unstable)}. \tag{11.125}$$

The inequalities (11.119) and (11.124) are known as Toomre's stability criterion (Toomre, 1964), and Q, which is slightly different for gaseous and stellar disks, is sometimes called the Toomre parameter. Note that the Toomre stability criterion is derived for axisymmetric perturbations with wavelengths that are much smaller than the size of the disk. It is independent of the form of the surface density profile, $\Sigma_0(R)$, indicating that it applies *locally*. Gas disks with $Q \lesssim 1$ can fragment into clumps with typical sizes $\sim \lambda_{\text{crit}}/2$, and it is believed that this plays an important role in modulating star formation in disk galaxies (see §9.2.2). As we will see in §11.6, the Toomre stability criterion may also play an important role in the formation of spiral arm fragments.

11.5.3 Global Instability

We now turn to global stability, and investigate whether disks are stable against perturbations with wavelengths that are comparable to the disk size. Obviously, the tight-winding approximation is not valid for such perturbations. In fact, whether or not a disk is globally stable depends on the global properties of the disk, such as its density distribution $\Sigma_0(R)$ and circular frequency, $\Omega_0(R)$, and it is not possible to write down a universal dispersion relation or stability criterion. However, in a few simple cases the dispersion relation can be worked out analytically, which may provide some insight into the problem.

One such case is the Maclaurin disk, which is a *fluid* disk with an (unperturbed) surface density of the form
$$\Sigma_0(R) = \begin{cases} \Sigma_c \left(1 - R^2/a^2\right)^{1/2} & (R \leq a) \\ 0 & (R > a), \end{cases} \tag{11.126}$$

with a the radius of the disk. The potential in the plane of the disk is
$$\Phi_0(R) = \frac{1}{2}\Omega_0^2 R^2 + \text{constant}, \quad \text{with} \quad \Omega_0^2 = \frac{\pi^2 G \Sigma_c}{2a}, \tag{11.127}$$

which implies a constant circular frequency of Ω_0. In other words, the circular velocity curve is that of a solid body.

For simplicity we assume that the equation of state is $P = K\Sigma^3$, with K constant, and that the pressure only acts in the plane of the disk. The resulting equilibrium angular frequency, Ω, follows from the Euler equation (11.92) by setting $v_R = 0$ (i.e. in equilibrium there can be no net radial motion):
$$\Omega^2 = \Omega_0^2 - \frac{3K\Sigma_c^2}{a^2}. \tag{11.128}$$

The perturbations on such a disk can be expanded in the following modes:
$$P_\ell^m(\xi)e^{im\phi} \quad \text{with} \quad \xi \equiv (1 - R^2/a^2)^{1/2}, \tag{11.129}$$

where P_ℓ^m is an associated Legendre function and, by its definition, $0 \leq m \leq \ell$. Since physical quantities correspond to $\xi \geq 0$, only modes with $\ell - m =$ even are required in the expansion. On inserting such solutions into Eqs. (11.101)–(11.103) one can obtain the dispersion relation for perturbations on Maclaurin disks (Takahara, 1976; see also Binney & Tremaine, 1987):
$$(\omega - m\Omega)^3 - (\omega - m\Omega)\left\{4\Omega^2 + \left(\ell^2 + \ell - m^2\right)\left[\Omega_0^2(1 - g_{\ell m}) - \Omega^2\right]\right\}$$
$$+ 2m\Omega\left[\Omega_0^2(1 - g_{\ell m}) - \Omega^2\right] = 0, \tag{11.130}$$

where

$$g_{\ell m} = \frac{(\ell+m)!(\ell-m)!}{2^{(2\ell-1)}\left[\left(\frac{\ell+m}{2}\right)!\left(\frac{\ell-m}{2}\right)!\right]^2}. \tag{11.131}$$

The $\ell = m = 0$ mode is unphysical, as it does not conserve mass, while the $\ell = m = 1$ mode corresponds to a mere translation of the disk. Therefore, the first modes corresponding to real perturbations of the disk are the $\ell = 2$ modes which have $m = 0$ or 2. The $m = 0$ mode corresponds to an axisymmetric pulsation (expansion and contraction) of the disk, and is sometimes called the breathing mode. For $\ell = 2$ and $m = 2$ the dispersion relation reduces to

$$\omega = \Omega \pm \sqrt{\Omega_0^2/2 - \Omega^2}, \tag{11.132}$$

and the mode is dynamically unstable (i.e. ω has an imaginary part) if

$$\Omega^2 > \Omega_0^2/2. \tag{11.133}$$

The density perturbation represented by this mode has the form

$$\Sigma_1 = \frac{3R^2}{a\sqrt{a^2 - R^2}} \cos(2\phi - \omega t), \tag{11.134}$$

which corresponds to a rotating, elliptical deformation of the disk. Since this looks similar to the bars observed in disk galaxies, this mode is called the bar mode, and the corresponding instability is known as the bar instability.

One can stabilize a disk against the bar instability by increasing its pressure (i.e. the random motion of the gas particles). To see this, we note that the sound speed, c_s, in a Maclaurin disk is given by

$$c_s^2 = \left(\frac{dP}{d\Sigma}\right)_0 = 3K\Sigma_c^2 \xi^2 = \left(\Omega_0^2 - \Omega^2\right) a^2 \xi^2, \tag{11.135}$$

where the later equality follows from Eq. (11.128). Using the definitions of Σ_0 and Ω_0 we have

$$\frac{c_s \Omega_0}{G\Sigma_0} = \frac{\pi^2}{2}\sqrt{1 - \frac{\Omega^2}{\Omega_0^2}}. \tag{11.136}$$

Eq. (11.133) then implies that stability requires

$$\frac{c_s \Omega_0}{G\Sigma_0} \geq \frac{\pi^2}{2^{3/2}}. \tag{11.137}$$

This suggests a useful criterion for bar instability based on the energies of the disk. Let T be the rotational kinetic energy, Π the kinetic energy in random motion and W the self-gravitational energy. These energies are related by the virial theorem: $T + \Pi/2 = -W/2$ (see §5.4.4). Thus, the importance of the random motions, which act to stabilize the disk, can be characterized by the following ratios:

$$t \equiv \frac{T}{|W|} \quad \text{or} \quad \frac{\Pi}{T} = \frac{1}{t} - 2. \tag{11.138}$$

Since $\Pi/T > 0$, we have that $0 \leq t \leq 1/2$. For a Maclaurin disk we have $t = \Omega^2/2\Omega_0^2$ and so

$$t < 1/4 \quad \text{or} \quad \frac{\Pi}{T} > 2 \tag{11.139}$$

for disk stability.

11.5 Disk Instabilities

This instability criterion also applies for disks with a finite thickness. For example, for Maclaurin spheroids, which are uniform spheroids of incompressible fluid (with density ρ) flattened by uniform rotation, the rotation speed Ω is related to the eccentricity of the surface, e, by

$$\frac{\Omega^2}{2\pi G\rho} = \frac{\sqrt{1-e^2}}{e^3}\left[(3-2e^2)\sin^{-1}e - 3e\sqrt{1-e^2}\right]. \tag{11.140}$$

Such spheroids are unstable to the bar mode if $e > 0.9529$, corresponding to $t > 0.2738$ (Chandrasekhar, 1969; Binney & Tremaine, 1987).

A similar analysis can also be carried out for stellar disks. A disk with the same surface density and potential as the Maclaurin disk, but consisting of stars rather than a fluid, is called a Kalnajs disk after A. Kalnajs, who analyzed their stability (Kalnajs, 1972b). As for a Maclaurin disk the mean angular frequency is independent of radius, but the pressure is now provided by the (isotropic) random motion of the stars in the disk plane with a velocity dispersion

$$\langle v_x^2\rangle = \langle v_y^2\rangle = \frac{a^2}{3}\left(\Omega_0^2 - \Omega^2\right)\left(1 - R^2/a^2\right). \tag{11.141}$$

In this case, disks are stable against bar instability if $\Omega^2/\Omega_0^2 < 125/486$, or in terms of energy

$$t < \frac{125}{972} \approx 0.1284 \quad \text{or} \quad \frac{\Pi}{T} > 5.776. \tag{11.142}$$

Comparing this result with Eq. (11.139) we see that fluid disks are substantially more stable than stellar disks.

Using N-body simulations of differentially rotating stellar disks, Ostriker & Peebles (1973) found that their model disks are stable against the bar mode if $t < 0.14 \pm 0.02$ or $\Pi/T \gtrsim 5$. These conditions are fairly similar to those for the Kalnajs disk. In the solar neighborhood, the random velocity of stars is about $60\,\mathrm{km\,s^{-1}}$ while the rotation velocity is about $220\,\mathrm{km\,s^{-1}}$. If these values are typical for the whole disk, then $\Pi/T \sim 0.15$, much smaller than the value required for the stability of stellar disks. Ostriker and Peebles therefore argued that there must be some unseen component of matter with $\Pi/T \gg 1$ to stabilize the disk. Although there are other possibilities, this stability argument gives additional support for the existence of massive, dark halos around disk galaxies.

Efstathiou et al. (1982) suggested an alternative criterion for bar stability, which has the advantage of being expressible in terms of quantities that are easier to obtain observationally. Based on N-body simulations of exponential disks embedded in a variety of halos, they found that stellar disks are stable against the bar mode as long as

$$\varepsilon_\mathrm{m} \equiv \frac{V_\mathrm{max}}{(GM_\mathrm{d}/R_\mathrm{d})^{1/2}} \gtrsim 1.1, \tag{11.143}$$

where V_max is the maximum rotation velocity of the disk. The corresponding stability threshold for gaseous disks is $\varepsilon_\mathrm{m} \gtrsim 0.9$ (Christodoulou et al., 1995). The parameter ε_m measures the importance of the self-gravity of the disk. An isolated exponential disk has $\varepsilon_\mathrm{m} \approx 0.63$, and is therefore unstable. Embedding the disk in an extended halo increases V_max and can thus stabilize the disk against the bar mode.

As discussed in §2.3.3, more than half of all disk galaxies in the local Universe possess bars with no strong dependence on the type (Sa, Sb or Sc). Given that the bar mode instability can lead to the formation of such structures, it is tempting to associate galactic bars with the global instability of disk galaxies. However, many barred disk galaxies have a central bulge component, and their rotation curves indicate that they are embedded in a massive halo. Both the bulge and the halo tend to stabilize the disk. Indeed, a simple application of the Ostriker–Peebles criterion or Eq. (11.143) suggests that these barred galaxies should be stable against the bar mode. Either the stability criteria presented here are not entirely correct, or the bar mode is induced by some external processes. For example, numerical simulations show that the encounter of a disk galaxy with another galaxy can produce bar-like structure in an otherwise stable disk (e.g. Noguchi, 1987). Given that galaxy interactions are quite frequent in hierarchical models, it is likely that some of the observed bars are produced by this process.

11.5.4 Secular Evolution

The instabilities discussed above may cause a disk galaxy to change its mass and angular momentum distribution. Such a relatively slow evolution, largely decoupled from the cosmological framework (i.e. also operating on galaxies in isolation), is called secular evolution. The main driver of secular evolution is the presence of global instabilities, such as the bar mode, spiral arms (to be discussed in §11.6) and the bending mode instability to be discussed below. It has become clear in recent years that secular evolution may play an important role in the evolution of disk galaxies. This is important, as it implies that the structural properties of disk galaxies may be more closely related to internal evolutionary processes than to the properties of their dark matter halos, as envisioned in the 'standard picture' outlined in §11.2. Whereas the latter allows us to make clear predictions for the sizes and rotation curves of disk galaxies, the former involves highly nonlinear processes that are still poorly understood. Below we describe the main mechanisms that play a role during secular evolution. More details can be found in Sellwood & Wilkinson (1993), Binney & Tremaine (2008), and the references given below.

(a) Resonance Coupling Disk galaxies unstable to the bar mode discussed above develop a bar in their central region. This bar is a highly flattened, triaxial structure whose figure rotates as a solid body with an angular frequency Ω_p, called the pattern speed. If bars are weak they may be considered as non-axisymmetric perturbations of an otherwise axisymmetric disk potential. Ignoring the z direction, and defining (R,ϕ) as the plane corotating with the bar, we have that

$$\Phi(R,\phi) = \Phi_0(R) + \Phi_1(R,\phi) \quad \text{with} \quad |\Phi_1/\Phi_0| \ll 1. \tag{11.144}$$

In the epicyclic approximation (see §11.1.5), the motion of a star in the unperturbed potential, $\Phi_0(R)$, is that of an epicycle with frequency $\kappa(R)$ around a guiding center which rotates around the center of the galaxy with frequency $\Omega(R)$. In the presence of Φ_1 the movement of the guiding center with respect to the frame corotating with the perturbation (i.e. the bar) is given by $\phi(t) = [\Omega(R) - \Omega_p]t$. In general Φ_1 will have m-fold symmetry ($m=2$ in the case of a bar or a two-armed spiral) so that the guiding center finds itself at effectively the same location in the (R,ϕ) plane with frequency $\Omega_d \equiv m[\Omega(R) - \Omega_p]$. Therefore, the motion in the R direction becomes that of a harmonic oscillator with natural frequency $\kappa(R)$ that is *driven* by a frequency Ω_d. At several R the natural and driven frequencies are in resonance. These resonances play an important role, because a star on an orbit at or near a resonance can be strongly perturbed, even if the perturbation Φ_1 is weak. For $m=2$ the most important resonances are the corotation resonance (CR) where $\Omega_d = 0$ [corresponding to $\Omega(R) = \Omega_p$], and the Lindblad resonances, where $\Omega_d = \pm\kappa(R)$. The plus sign corresponds to the inner Lindblad resonance (ILR), where $\Omega(R) - \kappa(R)/2 = \Omega_p$, and the minus sign corresponds to the outer Lindblad resonance (OLR), where $\Omega(R) + \kappa(R)/2 = \Omega_p$. Any perturbation with a non-zero pattern speed will introduce one CR, one OLR and, depending on the circular velocity curve either zero, one or two ILRs.

Resonances play two important roles. First of all, they delineate the radial regimes in which orbits have a particular orientation. As is common for driven oscillators, the response changes sign across a resonance. This means that the phase of the orbital orientation changes in phase by $\pi/2$ across a Lindblad resonance or across CR. In the case of a bar, the orbits are aligned with the bar between the ILR and CR, while they are perpendicular to the bar between CR and the OLR and inside of the ILR. Since the bar has to be built from stellar orbits, its shape reflects that of its orbital building blocks, and the bar can therefore not extend beyond CR. Secondly, resonances cause an exchange of angular momentum between the bar and the stars, and thus drive secular evolution of the disk. In general, the bar will transfer angular momentum to the outer disk, causing an increase in the disk scale length (Debattista et al., 2006). In fact, as shown by Hohl (1971), the angular momentum transfer even has the tendency to make non-exponential disks more exponential. In this respect, secular evolution induced by bars may be another mechanism to explain the exponential profile of disk galaxies, in addition to the mechanisms discussed in §11.4.

Bars can also exchange angular momentum via resonance coupling with the halo, causing the bar to slow down. As the decreasing bar pattern speed sweeps across a resonance with some halo orbits, their angular momenta may be substantially changed. Halo particles may either gain or lose angular momentum as they cross a resonance, and to first order there is no net loss or gain. However, to second order in the perturbing potential there is usually a net gain in angular momentum by the halo particles, leading to a friction-like drag on the bar even in a perfectly collisionless system. The bias arises because the number density of halo particles is usually a decreasing function of angular momentum, leading to an excess of gainers over losers (Tremaine & Weinberg, 1984; Weinberg, 1985). The slowdown of the bar typically causes CR to move to larger radii, hence resulting in an increase of the ratio between the corotation radius, $R_{\rm CR}$, and the radial extent of the bar, $R_{\rm bar}$. Observed bars have $R_{\rm CR}/R_{\rm bar} \sim 1.2 \pm 0.2$ (Aguerri et al., 2003), similar to the typical ratio with which they form due to the bar instability (at least in numerical simulations). This suggests that bars cannot have lost much angular momentum, and has been used to argue against a dense dark matter halo in the central regions of at least some disk galaxies (Debattista & Sellwood, 2000). However, many uncertainties remain, and it is currently unclear to what extent the properties of observed bars are consistent with the CDM prediction that halos should have dense, central cusps.

(b) Response of the Gas Bars also interact with gas in the disk. There are two reasons why gas reacts differently to the presence of a bar than stars. First of all, unlike stellar orbits, gas streamlines cannot cross, i.e. the gas must have a unique streaming velocity at each point in the flow. In a steady system, gas must therefore move on non-intersecting closed orbits, which, in the case of an axisymmetric disk, are circular orbits. However, in the presence of a perturbation, such as a bar or spiral arms, it may happen that closed orbits can no longer be nested without intersections. This is particularly true close to resonances, where orbit orientations change dramatically. In addition, many of the periodic orbits inside a (strong) bar are self-intersecting (i.e. they have loops). At any intersection of streamlines, shocks cause the gas to dissipate energy. Thus, just like a damped oscillator, the gas oscillations will have a phase lag with respect to the perturbation. This, in turn, causes a net torque between the perturbation and the gas, resulting in an exchange of energy and angular momentum between them. The net effect is that, in general, the gas is driven away from the CR resonance towards the OLR and ILR, where it can form ring-like structures and may even form stars. In nature, rings are observed to be quite ubiquitous in disk galaxies (e.g. Buta, 1986), and it is generally believed that these are produced by radial flows induced by bars and spiral arms.

The accumulation of gas in the central region of the galaxy, either in the very center or at the ILR, may cause a central starburst, may feed an AGN, and may even result in the destruction of the bar. The latter is due to the fact that a strong central mass concentration can scatter stars that come close to the center, causing their orbits to become chaotic. Since chaotic orbits tend to have more nearly round shapes (when averaged over long times), the bar weakens as more and more bar-supporting regular orbits become chaotic. In addition, the growth of a central mass concentration changes the potential, hence the orbital frequencies and the locations of the various resonances. Meanwhile, the gas loses its angular momentum to the bar when it 'falls' in, causing the bar to speed up (i.e. Ω_p increases). Both effects cause the resonances, and in particular the ILR, to move to different radii. Since the orbits within the ILR do not have the orientation to support the bar, the motion of the ILR adds to the weakening and possible destruction of the bar. Although early simulations suggested that a central mass concentration with a mass of only a few percent of that of the disk could destroy a bar (Friedli, 1994; Norman et al., 1996), more recent simulations of higher resolution seem to suggest that bars are in general remarkably resilient, and that the central mass concentration required for bar destruction is of the order of 20% of the total disk mass (Shen & Sellwood, 2004).

(c) The Bending Instability and Bulge Formation In §11.5.2 and §11.5.3 we focused on perturbations in the plane of the disk. However, one can also have perturbations out of the disk plane. Consider a perturbation that bends the disk in the vertical z direction. When a star follows the bend out of the unperturbed disk plane it experiences a destabilizing centrifugal force that opposes the stabilizing self-gravity of the disk. Toomre (1966) showed that an infinitesimally thin sheet is unstable to such a bending instability[3] if the stars have large random motions in the plane of the disk. Using a normal-mode analysis of a slab of finite thickness, Araki (1985) found that bending (at all wavelengths) is stabilized when the ratio of vertical to horizontal velocity dispersion exceeds ~ 0.293. As this is significantly smaller than the value observed in the solar neighborhood, the typical inference is that the bending mode is not important for disk galaxies. However, when a bar forms in the disk, it changes the stellar orbits, making them more elongated and aligned with the bar. This effectively increases the radial velocity dispersion, which may make the bar unstable to the bending mode.

Indeed, numerical simulations have shown that bars typically suffer a bending mode instability shortly after their formation, which displaces the central part in one vertical (z) direction and the outer part of the bar to the opposite direction (e.g. Combes et al., 1990a; Raha et al., 1991). After its maximum distortion the bend settles back to the galactic plane, cascading the energy in the bending mode to smaller scales. This increases the velocity dispersion in the vertical direction, thickening the bar. The effect is most pronounced towards the ends of the bar where the displacements from the plane are the largest. Consequently, the resulting bar has a distinctive peanut shape when viewed edge-on, very similar in appearance to the 'boxy bulges' (i.e. bulges with boxy isophotes) seen in many nearly edge-on disk galaxies. Thus secular evolution may be responsible for the formation of some bulges. Note, however, that these boxy peanut shaped bulges in reality are still bars (i.e. they are tumbling triaxial systems with a non-zero pattern speed). Although it was originally speculated that the bending mode could cause the bar to dissolve, high-resolution simulations have shown that bars virtually always survive the bending instability (e.g. Debattista et al., 2006). Indeed, kinematic studies of observed boxy peanut shaped bulges have shown clearly that they are bars (Kuijken & Merrifield, 1995; Bureau & Freeman, 1999).

Bulge components that are built entirely out of the disk by secular processes are sometimes called pseudo-bulges to distinguish them from the classical bulges, which are believed to have an origin similar to elliptical galaxies (see §13.6.1). In addition to having a boxy peanut shape when seen edge-on, pseudo-bulges have surface brightness profiles that are close to exponential, are more rotation dominated and more strongly flattened than classical bulges, and often contain relatively young stars (e.g. Kormendy & Kennicutt, 2004). Pseudo-bulges are found

---

[3] This instability is also sometimes referred to as buckling instability and is similar to the firehose instability in plasma physics.

predominantly in late-type spirals (Sc and Sd), while Sa and Sb type spirals mainly have classical bulges.

11.6 The Formation of Spiral Arms

Many disk galaxies reveal spiral structure, ranging from the 'grand-design' spirals that can be traced over large parts of the disk, to small, flocculent arm fragments that have a limited radial and angular extent (see §2.3.3). Given their prevalence, understanding the origin and nature of spiral arms is an integrable part of understanding the structure and formation of disk galaxies. Unfortunately, many of the details related to spiral structure are still unclear. It is likely that no unique explanation exists for all spiral arms; their nature and formation may well differ from one system to another. After a brief description of spiral morphology, we summarize the main ideas behind some of the more popular models.

(a) Morphology of Spirals Spiral arms are clearly associated with enhanced star formation, as is evident from the fact that HII regions and molecular gas are often the best tracers of spiral structure. Because of the young stars, spiral arms are more pronounced in bluer wavebands. However, they are also evident in near-infrared wavebands, indicating that they are also present in old stars. Evidently, spiral structure is not a phenomenon restricted to the gas, but involves the entire matter density of the disk. Hence, spiral structure has to be considered as a true density perturbation of the underlying disk. For a sinusoidal variation with the azimuthal angle ϕ, the surface density of a disk with m arms can be written as

$$\Sigma(R,\phi) = \Sigma_0(R) + \Sigma_1(R)\cos[m\phi + f(R)]. \tag{11.145}$$

Here $\Sigma_0(R)$ is the surface density of the unperturbed, axisymmetric disk, $\Sigma_1(R)$ is the radial amplitude of the perturbation, and $f(R)$ is the shape function describing the form of the spiral. The azimuthal profile of spiral arms is often much sharper than sinusoidal (especially for younger stars), in which case a Fourier expansion to higher order harmonics can be used to describe $\Sigma(R,\phi)$. The mathematical form for $f(R)$ most often used, which has been shown to give a reasonable, but certainly not perfect, description of the shapes of observed spirals, is the logarithmic spiral, $f(R) = f_0 \ln(R) + \phi_0$, where f_0 is a constant describing how tightly the spiral pattern is wound. This is more conveniently expressed by the pitch angle, i, which at any given point (R,ϕ) along the spiral is defined as the angle between the local tangent of the spiral and the circle of radius R, i.e.

$$\tan i = m \left| R \frac{\partial f}{\partial R} \right|^{-1}. \tag{11.146}$$

The logarithmic spiral is special in that its pitch angle $i = \arctan(m/f_0)$ is constant; in general i is a function of radius. Observed pitch angles range from $\sim 10°$ for tightly wound early-type spirals to $\sim 30°$ for late-type spirals (Garcia Gomez & Athanassoula, 1993). Actually, as discussed in §2.3.1, the openness of spiral arms is one of the basic classification criteria for disk galaxies.

(b) Arm Fragments The most basic reason for the formation of spiral arms is the differential rotation of disks. As an example, consider a disk with a constant rotational velocity, $V_{\rm rot}(R) = V = $ constant (as is the case for a light disk embedded in a singular isothermal sphere). Since the angular velocity, $\Omega = V/R$, is R-dependent, the material initially (at a time $t = 0$) distributed on the line with a constant azimuthal angle $\phi = \phi_0$ will be sheared into a spiral curve $\phi(R,t) = \phi_0 + Vt/R$ at time t. Thus, if an extended perturbation develops on the disk, it will be sheared into a spiral arm (see Fig. 11.4).

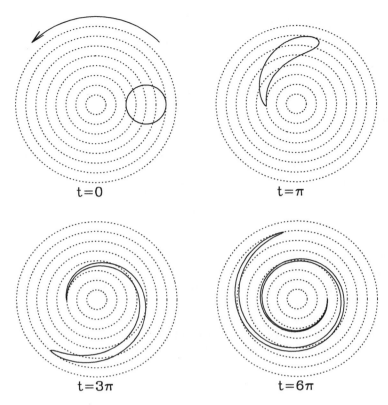

Fig. 11.4. The creation of a spiral arm. A circular patch at $t = 0$ is sheared into a spiral arm due to the differential rotation in the direction indicated by the arrow.

The multi-armed patterns seen in galaxies like the one shown in the middle panel of Fig. 2.7 are believed to be produced by patches of new stars formed through local disk instability and sheared by disk differential rotation. In agreement with this explanation, the characteristic sizes of the arm fragments observed in many external multi-armed disk galaxies are comparable to $\lambda_{\rm crit}/2$, with $\lambda_{\rm crit}$ the most unstable wavelength (see §11.5.2). With the passage of time, an arm fragment is more and more sheared while the brightest young stars (which have short lifetimes) die, leading to gradual dissolution of the arm fragment. Meanwhile new arm fragments may form in other places where the disk becomes locally unstable. Thus, flocculent arm fragments are believed to be short-lived, transient phenomena related to *local* disk instability. Such arms are called material arms as they are always made up of the same material from which they formed.

(c) Spiral Density Waves Grand-design spirals are believed to be of a different nature than the arm fragments discussed above. The reason is that material arms tend to wind up quickly due to differential rotation, so that long arms, such as those in grand-design spirals, would be much more strongly wound (i.e. have much smaller pitch angles) than observed. To avoid this winding problem, Lin & Shu (1964) proposed that grand-design spiral arms are produced by spiral-shaped density waves propagating through the disk (see also Lin & Shu, 1966). Similar to a bar, the spiral structure is a quasi-stationary, wave-like density perturbation with a pattern speed $\Omega_{\rm p}$. Since stars and interstellar gas clouds in the perturbed disk have angular speeds Ω that differ from $\Omega_{\rm p}$ (except near corotation), they move in and out of the spiral pattern. Therefore, like waves on the surface of the ocean, the material making up a spiral arm changes constantly as the arms propagate on the disk, and the arm is not sheared by differential rotation. Furthermore,

11.6 The Formation of Spiral Arms

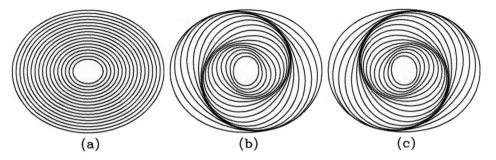

Fig. 11.5. Self-consistent construction of density perturbations in a disk. Due to a non-axisymmetric $m = 2$ perturbation periodic orbits become elliptical. In the case of a bar-like perturbation (left panel) the orbits align along the bar, while in the case of a spiral-like perturbation (middle and right panels), the orbit orientation changes as function of radius. Note how the crowding of the orbits gives rise to a self-consistent reproduction of the induced spiral pattern. [After Kalnajs (1973)]

disk material is compressed when swept by a spiral density wave. If the interstellar gas is cold, its response to a spiral wave can produce narrow gaseous arms of enhanced density, where young stars may form. Hence, spiral density waves can naturally explain why star formation occurs predominantly along the spiral arms (see §9.2.2).

This density-wave theory faces two challenges. First of all, in order to consider it as a viable theory for the observed grand-design spiral arms, one needs to demonstrate that a self-consistent model can be constructed. Similar to the situation with a bar, this means that one has to find a set of orbits in the perturbed potential that, once stacked together, can reproduce the density distribution of the perturbed disk. Secondly, one needs to find a mechanism to create a spiral-shaped density perturbation that is sufficiently long-lived. Although the details cannot be covered here (see Binney & Tremaine, 1987), we briefly discuss the current ideas.

Consider an unperturbed stellar disk where all stars are on circular orbits with angular velocity $\Omega(R)$. If the disk is subjected to a global $\ell = m = 2$ mode perturbation the circular orbits are deformed into elliptical orbits (see §11.5.3). In general the perturbation will have a non-zero pattern speed Ω_p, which causes the orientation of the elliptical orbits to rotate with the same frequency. In the case of a bar-like perturbation all orbits between the ILR and CR will have the same phase as the bar, so that the bar perturbation can be constructed self-consistently (see the left panel of Fig. 11.5). In the case of a spiral-shaped perturbation, the orientation of the elliptical (periodic) orbits tend to align with the perturbing potential. As can be seen from Fig. 11.5, this causes a crowding of the orbits along a two-armed spiral pattern that more or less follows the shape of the $m = 2$ density perturbation, and has a pattern speed equal to Ω_p. This demonstrates that two-armed spiral density waves can be constructed self-consistently, and nicely explains why the spiral pattern is also visible in the light emitted by the old stars. The real situation is a bit more complicated than what is shown in Fig. 11.5. A general phenomenon in driven oscillators is that the orientation of the response changes at resonance crossing. In the case of a bar this explains why the bar cannot cross CR. In the case of a spiral arm, the 4 : 1 resonance [i.e. where $\Omega(R) - \kappa(R)/4 = \Omega_p$][4] plays an important role, as it causes a phase-shift in the orbit orientation of $\pi/4$. Consequently, the orbits can only support the spiral structure between the ILR and the 4 : 1 resonance, and the spiral is not expected to extent much beyond the 4 : 1 resonance (Contopoulos & Grosbol, 1986), at least if grand-design spirals are long-lived features. In many of the best known examples (e.g. M 51 and M 81) the structure appears to be a direct consequence

---

[4] The 4 : 1 resonance is basically the first ultra-harmonic of the ILR and is located in between the ILR and CR.

of an encounter with a companion, and so may not, in fact, be long-lived [see Binney & Tremaine (1987) for further discussion].

In the case of a bar, the formation mechanism is the bar-mode instability discussed in §11.5.3. It thus seems likely that the spiral density waves are also associated with some disk instability. Unfortunately, deriving the dispersion relation for a general spiral wave is extremely complicated, mainly because the tight-winding approximation used in §11.5.2 only holds for spirals with a small pitch angle. In addition, the waves are traveling waves, because the disk has a finite extent so that any realistic wave on the disk must be described by a wave packet. The group velocity of a wave packet is $v_g = d\omega/dk$ which is non-zero in a dispersive medium [i.e. a medium in which $\omega(k)$ is not linear in k]. Because of this non-zero group velocity, the wave packet will move radially throughout the disk. At resonances, the waves can be reflected, transmitted or absorbed. In particular, as shown by Mark (1974), waves can be absorbed at the Lindblad resonances. To ensure long-lived spiral arms, a mechanism is therefore required either to prevent this absorption or to amplify the waves. Several of these mechanisms have been proposed (Lin, 1970; Mark, 1976; Toomre, 1981) and it is likely that they all play some role in the creation of spiral density waves (and bars).

(d) Bar-Driven Spiral Arms In many barred spirals, the spiral arms appear to start at the two ends of the bar. This suggests that the bar and spiral pattern have the same pattern speed and thus are related. Using hydrodynamical simulations, Sanders & Huntley (1976) studied the response of gas to a bar perturbation. They considered a uniform, axisymmetric gas disk (with no self-gravity) subjected to an additional potential due to a rigidly rotating bar. They found that the gas eventually settled into a steady state with prominent trailing spiral structure. This happens because of gas viscosity; simulations in which the gas disk was replaced by a collisionless disk of test particles did not give rise to any spiral structure.

Although this suggests that bars can indeed induce spiral arms without the help of spiral density waves, the model cannot explain why many grand-design spiral patterns are also seen in old stars. It is likely that spirals in barred galaxies are associated with a spiral density wave, probably with a pattern speed different from that of the bar. Indeed, Sellwood & Sparke (1988) have shown that, in numerical simulations, multiple pattern speeds are quite common in disk galaxies, with the spiral structure typically having a much lower pattern speed than the bar. This implies a more or less random distribution of the phase difference between the bar and the start of the arms, which seems to be in conflict with observations. However, as pointed out by Sellwood & Sparke (1988), contour plots of the non-axisymmetric density in their simulations show that the spiral arms appear to the eye to be joined to the ends of the bar for most of the beat frequency. This suggests that the observed correlation between bars and spirals might simply be an illusion.

11.7 Stellar Population Properties

Important insight regarding the formation and evolution of disk galaxies can be obtained from the properties of their stellar populations, such as ages and metallicities. In general, the stellar populations of disk galaxies are far more complicated than those of elliptical galaxies. Ellipticals have little ongoing star formation, with spectral properties (e.g. colors and absorption line indices) that can be well fit by a simple single-age stellar population (see §13.5). Disk galaxies, on the other hand, typically have significant ongoing star formation, and their star-formation histories, often 'modeled' as exponentially declining functions of time [see Eq. (10.112)], are clearly inconsistent with a single burst. In addition, disk galaxies are often dusty, so that broad-band colors have to be carefully corrected for extinction effects. To complicate the matter even further, the various components of a disk galaxy (thin disk, thick disk, bulge, stellar halo) all seem to have different stellar populations. Typically the spheroidal components (bulge and stellar halo) are dominated by old stars, while the disk component is populated by stars of various ages and with a broad range in metallicities. In addition, thick disk stars are older and more metal poor than thin disk stars (Seth et al., 2005). In what follows, we focus on the disk components.

11.7.1 Global Trends

As discussed in §10.3.8, there are various diagnostics to determine the current star-formation rate of a (disk) galaxy. However, determining the entire past star-formation history (SFH) is far from trivial. We can get some insight, though, by comparing the current star-formation rates of disk galaxies with their stellar masses (which reflect their integrated SFHs). A typical disk galaxy has a star-formation rate per unit stellar mass, known as the specific star-formation rate (SSFR), of a few $10^{-10}\,\mathrm{yr}^{-1}$. For a galaxy like the Milky Way, this translates into a star-formation rate of a few $\mathrm{M}_\odot\,\mathrm{yr}^{-1}$. One can characterize the relative importance of current star formation in a galaxy using the 'birthrate', b, defined as the ratio between the current star-formation rate and the average star-formation rate in the past: $b = \mathrm{SFR}/\langle\mathrm{SFR}\rangle$. Since $\langle\mathrm{SFR}\rangle = M_\star/t_0$, with $M_\star$ the stellar mass of the galaxy and $t_0 \simeq 10^{10}\,\mathrm{yr}$ the age of the Universe, we have that $b = \mathrm{SSFR} \times t_0$. The left-hand panel of Fig. 11.6 shows the birthrate for disk galaxies in the SDSS as a function of absolute magnitude. On average, brighter disk galaxies have a lower birthrate. The brightest disk galaxies have $b \sim 1$ (albeit with a large amount of scatter), indicating that their current star-formation rate is comparable to their average past star-formation rate. At the faint end, however, $b \sim 3$, implying that a large fraction of their stars have formed relatively recently.

The right-hand panel of Fig. 11.6 shows that the birthrate is fairly tightly correlated with the $^{0.1}(g-r)$ color. This suggests that some information about the SFHs of (disk) galaxies can be obtained from broad-band photometry. Since broad-band photometry is relatively easy to obtain, one can obtain measurements for large, statistically significant samples. The color–magnitude relation of disk galaxies is both steeper and broader than for early-type galaxies. A significant part of this broadness is likely due to dust extinction combined with inclination effects (more inclined galaxies are more extincted and hence redder). Indeed, using a dust extinction corrected

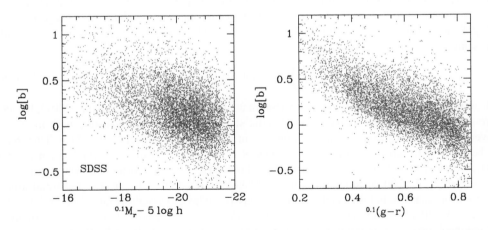

Fig. 11.6. The birthrate $b = \mathrm{SSFR} \times 10^{10}\,\mathrm{yr}$ as function of the absolute magnitude in the r band, K-corrected to $z = 0.1$ (left-hand panel) and as function of the $^{0.1}(g-r)$ color for disk galaxies in the SDSS (right-hand panel). Here, following Strateva et al. (2001), disk galaxies are (crudely) defined as galaxies with a concentration index $c = r_{90}/r_{50} \leq 2.6$, with r_{90} and r_{50} the radii that contain 90 and 50 percent of the Petrosian r band flux. The specific star-formation rates are derived using the stellar masses and star-formation rates obtained by Kauffmann et al. (2003b) and Brinchmann et al. (2004), respectively.

color–magnitude relation for edge-on spirals, Peletier & de Grijs (1998) obtain a tight color–magnitude relation with a steeper slope than for early-type galaxies. Using stellar population models, they conclude that this is most naturally explained as indicating trends in both age and metallicity, with fainter spirals having both younger ages and lower metallicities than brighter spirals.

However, as shown by Worthey (1994), the technique of using broad-band colors to probe the stellar populations and SFHs of galaxies suffers from a strong age-metallicity degeneracy: the spectra of composite stellar populations are virtually identical if the percentage change in age and metallicity, Z, follows $\Delta\ln(\text{age})/\Delta\ln Z \sim -3/2$ (see §10.3.5). This degeneracy can be partially broken using near-infrared photometry (e.g. H or K band) in addition to optical colors (de Jong, 1996b; Cardiel et al., 2003). Essentially, optical colors are sensitive to the position of the main-sequence turn-off, whereas near-infrared colors are more sensitive to the properties of the red giants and AGB stars. Hence, using a combination of optical and near-infrared colors one can differentiate between the effects of metallicity and the ratio of young ($\lesssim 2\,\text{Gyr}$) to old stars – the latter reflects the average population age or birthrate parameter. Defining the average age of a stellar population as

$$\langle A \rangle = t_0 - \frac{\int_0^{t_0} t\,\Psi(t)\,dt}{\int_0^{t_0} \Psi(t)\,dt}, \tag{11.147}$$

with $\Psi(t)$ the SFH, the use of optical and near-infrared broad-band photometry has revealed strong correlations of age and metallicity with Hubble type, rotation velocity, luminosity, gas mass fraction and surface brightness, in the sense that earlier-type, faster rotating, more luminous, higher surface brightness and gas-poorer disk galaxies are older and more metal-rich (e.g. de Jong, 1996b; Bell & de Jong, 2000; MacArthur et al., 2004). Several authors have argued that the SFH of a disk galaxy is primarily driven by surface density, which may reflect a local density dependence of the star-formation law, with total stellar mass seemingly a less important parameter (e.g. Bell & de Jong, 2000; Kauffmann et al., 2003a). Total stellar mass does correlate significantly with metallicity (see §2.4.4), suggesting that mass-dependent feedback may be an important process in the chemical evolution of disk galaxies.

More recently, observations in the near UV with the Galaxy Evolution Explorer (GALEX), which allow for an independent determination of star-formation rates, have also suggested that disk galaxies occupy a reasonably tight locus in the SSFR vs. stellar mass plane that is well represented by

$$\log\left(\frac{\text{SSFR}}{\text{yr}^{-1}}\right) = -0.36\log\left(\frac{M_\star}{M_\odot}\right) - 6.4, \tag{11.148}$$

with an intrinsic scatter of only ~ 0.5 dex (Salim et al., 2007; Schiminovich et al., 2007). Hence, more massive disk galaxies have lower specific star-formation rates, consistent with the trend between the birthrates and absolute magnitudes in the right-hand panel of Fig. 11.6. This trend can be difficult to reproduce in simple models of (disk) galaxy formation, which often predict inverted trends, namely more massive (disk) galaxies have higher SSFRs (e.g. van den Bosch, 2002a; Bell et al., 2003a; Somerville et al., 2008b). In these models the inverted trend arises because more massive halos (which host more massive galaxies) assemble later (see §7.3.4). The discrepancy with observation indicates that a more detailed treatment of accretion, star-formation and feedback processes is required, and indeed models which include such effects and attempt to fit the galaxy population as a whole do typically produce birthrate trends similar to those observed (e.g. Croton et al., 2006; Bower et al., 2006).

Arguably the best way to constrain the star-formation histories of (disk) galaxies is to measure the SFR-$M_\star$ relation as a function of redshift. Using multi-wavelength data, Noeske et al. (2007b) have shown that disk galaxies out to $z \sim 1$ follow a narrow relation between SFR and stellar mass,

11.7 Stellar Population Properties

with slope and scatter in agreement with that at $z=0$ [Eq. (11.148)], but with a normalization that evolves with redshift: disk galaxies at higher redshifts have higher SSFRs. This suggests that the evolution in the cosmic star-formation history (see §2.6.8) is not primarily driven by evolution in the frequency of (merger-induced) starbursts, but rather reflects a decline in the typical SSFRs of 'normal' star-forming (disk) galaxies (see also §15.4.2). Although the data are consistent with a simple model of gradual gas exhaustion, in which more massive galaxies experience an earlier onset of star formation (Noeske et al., 2007a), exactly how such a behavior comes about in a hierarchical model with continuing merging, gas accretion and various feedback modes is still not clear.

11.7.2 Color Gradients

Fig. 11.7 shows the $B-K$ colors as a function of R band surface brightness for a sample of 86 face-on spiral galaxies. There is a clear trend that disks are bluer at larger radii (lower surface brightness). Because of these color gradients, disks appear larger in bluer bands, so that color gradients also manifest themselves as systematic changes of disk scale length with waveband (e.g. MacArthur et al., 2003).

The two most straightforward explanations for the color gradients are radial gradients in the stellar populations and radial variations in the amount of reddening due to dust extinction. The gradients in stellar populations, in turn, may be due to gradients in the IMF, in age, and/or in metallicity. Using realistic 3D radiative transfer modeling, de Jong (1996b) and Kuchinski et al. (1998) found that reddening by dust extinction is unlikely to be the major cause of the observed color gradients: although dust is a likely contributor, especially in more massive disk galaxies, the tentative consensus to date is that stellar population gradients dominate (see also Byun et al., 1994; MacArthur et al., 2004).

If the color gradients are related to stellar population gradients, and if we ignore the possibility that the IMF may vary systematically with radius, then the outer parts of disk galaxies need to be younger and/or more metal-poor. There is evidence that both trends are present. Comparing

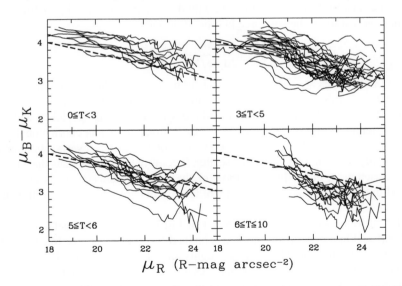

Fig. 11.7. Color gradients in disk galaxies. This plot shows the $B-K$ color of disk galaxies as a function of the azimuthally averaged R band surface brightness. Galaxies are divided into four morphological classes according to the de Vaucouleurs' T type (see §2.3.1 for definition). The reference line in each panel has a $B-K$ gradient of 0.14 magnitudes per unit surface brightness. [Adapted from de Jong (1996b)]

the observed optical and near-IR color gradients to stellar population synthesis models, Bell & de Jong (2000) inferred average gradients in age and metallicity per K-band disk scale length of -0.79 ± 0.08 Gyr and -0.14 ± 0.02 dex, respectively.[5] Although the metallicity gradients are less well measured and more sensitive to dust effects than age gradients, they are consistent with the trends seen in the gas metallicity and discussed in §11.8.2 below. In addition, the age gradients have additional support from the fact that the scale length of current star formation, as measured by the Hα flux, is larger than that of the underlying older stellar population (e.g. Ryder & Dopita, 1994). This indicates that the fraction of young stars increases with galactocentric radius, which would come about if star formation in a disk proceeds in an inside-out fashion. In that respect it is reassuring that such an inside-out formation scenario is a natural prediction of the standard formation models presented in §11.2.5.

11.8 Chemical Evolution of Disk Galaxies

The chemical properties of a galaxy reflect the amount of gas that has been reprocessed by stars and exchanged with its environment over the galaxy's lifetime. As such the metallicities and abundances of galaxies, and of stars within them, serve as a fossil record of their formation history. Before we describe some global relations for the population of disk galaxies, we focus on the solar neighborhood where the ages and metal abundances of individual stars can be measured.

11.8.1 The Solar Neighborhood

In the solar neighborhood the available data on the chemical composition are much larger than for any other system, allowing a detailed investigation of its chemical evolution. The most important observational constraints are the age–metallicity relation traced by the iron abundance of long-lived stars, the metallicity distribution of long-lived G-dwarfs, and the abundance ratios of various elements, in particular oxygen and iron. In principle, an assumed one-to-one relation between age and metallicity can be combined with the observed metallicity distribution to infer the star-formation history of the solar neighborhood through

$$\frac{dM_\star}{dt} = \frac{dM_\star}{d[\text{Fe}/\text{H}]} \frac{d[\text{Fe}/\text{H}]}{dt}. \tag{11.149}$$

Unfortunately, stellar ages are much harder to determine than stellar metallicities, and the age–metallicity relation is still poorly constrained. Using a sample of 189 F-dwarfs, Edvardsson et al. (1993) obtained a clear trend of increasing metallicity with decreasing age, albeit with substantial scatter. Such a trend is naturally predicted if star formation causes a net enrichment of the ISM. More recently, however, Nordström et al. (2004), using 462 F- and G-dwarfs, obtained an almost flat age–metallicity relation with very large scatter. If confirmed, this suggests that star formation causes little enrichment, either because most of the newly produced metals escape the solar neighborhood due to outflows, or because there is a significant amount of infall of metal-poor gas to dilute the ISM.

The shaded histogram in the left-hand panel of Fig. 11.8 shows the metallicity distribution of 287 G-dwarfs within 25 pc from the Sun, obtained by Rocha-Pinto & Maciel (1996). Note that the distribution is fairly broad, covering roughly an order of magnitude from $[\text{Fe}/\text{H}] \simeq -0.7$ ($Z \sim 0.2 Z_\odot$) to $[\text{Fe}/\text{H}] \simeq 0.3$ ($Z \sim 2 Z_\odot$). What does this tell us about the chemical evolution of the solar neighborhood? We can gain some insight by using the simple chemical evolution models described in §10.4. In the solar neighborhood the baryonic surface density is $\sim 50 \, \text{M}_\odot \, \text{pc}^{-2}$, of

---

[5] As pointed out by MacArthur et al. (2004), the age and metallicity gradients are typically steeper in the inner parts of the galaxies, so that they cannot be fully described by a single number.

11.8 Chemical Evolution of Disk Galaxies

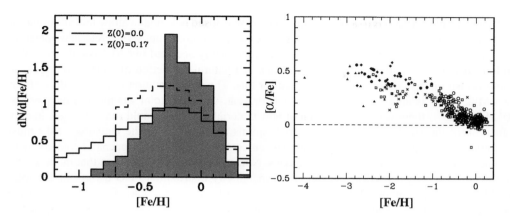

Fig. 11.8. *Left panel*: The shaded histogram shows the metallicity distribution of G-dwarfs in the solar neighborhood (data taken from Rocha-Pinto & Maciel, 1996). For comparison, we also show two predictions of closed-box models with initial metallicities of $Z(0) = 0$ and $Z(0) = 0.17$, as indicated. *Right panel*: Abundance ratios of stars as a function of metallicity, based on data published in Edvardsson et al. (1993, open circles) and Wheeler et al. (1989, other symbols).

which $\sim 20\%$ is in the form of gas (Kuijken & Gilmore, 1989a; Boissier & Prantzos, 1999). In the closed-box model, this means that $M_{\rm gas}(t_0)/M_{\rm gas}(0) \approx 0.2$ at the present time t_0. The metallicity of the local ISM is similar to that of the Sun, so that $Z(t_0) \approx Z_\odot$. Inserting these numbers into Eq. (10.124) and assuming that the gas started out with a primordial composition [i.e. $Z(0) = 0$] and that the instantaneous recycling approximation holds, we obtain a metal yield $y_Z \approx 0.62 Z_\odot \approx 0.012$. Using Eq. (10.126) we can then predict the metallicity distribution of long-lived stars, which is plotted in the left-hand panel of Fig.11.8 as a solid histogram. The model prediction is clearly inconsistent with the data. The closed-box model with $Z(0) = 0$ predicts that $\sim 28\%$ of the long-lived stars have metallicities [Fe/H] < -0.7, in violent disagreement with the data, which shows that only $\sim 3\%$ of the G-dwarfs have such low metallicities. This discrepancy is known as the G-dwarf problem.

An obvious way to reduce the number of low metallicity stars is to assume a non-zero initial metallicity, $Z(0)$. It is likely that the gas out of which the thin disk stars in the Milky Way formed was pre-enriched to appreciable levels. Both the spheroid (bulge plus stellar halo) and the thick disk of the Milky Way seem to have formed on a relatively short time scale about 10 Gyr ago. The metals produced and expelled during this epoch may well have been mixed with the surrounding gas, polluting the material for the subsequent formation of the thin disk. As a rough estimate, assume that the mass of metals produced per unit stellar mass is the same for all components of the Milky Way. The initial metallicity of the thin disk material can then be written as

$$Z(0) = \frac{M_{\star,\rm old}}{M_{\star,\rm old} + M_{\star,\rm d}} \overline{Z}(t_0), \quad (11.150)$$

where $\overline{Z}(t_0)$ is the mean metallicity of the old components at the present time. For the Milky Way, the stellar mass ratio between the disk and the 'old' component (spheroid plus thick disk) is $M_{\star,\rm d}/M_{\star,\rm old} \sim 5$, and $\overline{Z}(t_0) \sim Z_\odot$, yielding $Z(0) \sim 0.17 Z_\odot$. Substituting this in Eq. (10.124) and adopting again $M_{\rm gas}(t_0)/M_{\rm gas}(0) = 0.2$ results in a metal yield $y_Z \approx 0.52 Z_\odot \approx 0.010$. The corresponding metallicity distribution of long-lived stars given by Eq. (10.126) is indicated as the dashed histogram in the left-hand panel of Fig.11.8. Although it predicts a smaller number of low-metallicity stars than the model with $Z(0) = 0$, the resulting metallicity distribution is still a poor fit to the data, indicating that the solution to the G-dwarf problem is not as simple as merely raising the initial metallicity.

A more viable solution is to abandon the concept of a closed box, and to allow for inflow and/or outflow. As seen in §7.3, dark matter halos are expected to continuously accrete new material. As long as the baryonic material associated with the newly-accreted material is able to cool, a prolonged inflow of new gas onto disk galaxies is a natural prediction of CDM cosmogonies. There is also indirect observational evidence for this. The star-formation rate in the disk of our own Milky Way is about a few solar masses per year, while the total gas mass is only $\sim 5 \times 10^9 \, M_\odot$. Thus, if there were no accretion of new material, our disk would run out of gas in only a fraction of the Hubble time. Since gas-rich spirals are common in the Universe, disks must continuously be replenished with infalling gas, unless the present time happens to coincide with the end of the era of star forming disk galaxies. Numerous studies have shown that models with infall can accurately reproduce the observed metallicity distribution of G-dwarfs in the solar neighborhood (e.g. Larson, 1976; Sommer-Larsen, 1991; Prantzos, 2008). The details regarding the exact infall rate depend on the adopted star-formation rate, but in general the infall rate has to be a decreasing function of time.

Another important constraint on the chemical evolution of the solar neighborhood comes from the abundance ratios of various elements. The general discussion in §10.4.4 shows that the abundance ratio between the α-elements and Fe in stars with long lifetimes can be used to constrain the star-formation history of a galaxy. The right-hand panel of Fig. 11.8 shows the ratio $[\alpha/\text{Fe}]$ as a function of $[\text{Fe/H}]$, obtained by Edvardsson et al. (1993). Note that $[\alpha/\text{Fe}]$ is approximately constant for F and G stars with $[\text{Fe/H}] \lesssim -1$ but decreases with increasing metallicity for larger $[\text{Fe/H}]$. This behavior can be explained if stars with $[\text{Fe/H}] < -1$ formed on a relatively short time scale ($\lesssim 10^9$ yr) before the onset of the enrichment by Type Ia SNe, while stars with higher metallicity formed over a longer time scale (see discussion in §10.4.4).

It is clear from the above discussion that the combined data on the ages, metallicities and abundances of (long-lived) stars in the solar neighborhood put important constraints on the star-formation history, and on the amounts of inflow and outflow. Currently, the main problem is not the lack of successful models, but rather the multiplicity of them, and the uncertainties in the data (i.e. the age–metallicity relation discussed above). Larger data sets are required in order to be able to discriminate between these models.

11.8.2 Global Relations

(a) The Metallicity–Luminosity Relation In §2.4.4 we have shown that more massive (more luminous) galaxies have a higher gas-phase metallicity, something that holds for galaxies of all types. The upper-left panel of Fig. 11.9 shows the metallicity–luminosity relation for a sample of 70 disk galaxies (regular spirals and irregulars). There are a number of possible explanations for this relation. Since in the simple closed-box model the metallicity is directly related to the gas mass fraction, one possibility is that the metallicity–luminosity relation simply reflects that less massive disk galaxies have a larger gas mass fraction (see upper-right panel of Fig. 11.9), either because they are younger or because their star formation is less efficient. Alternatively, the metallicity–luminosity relation may reflect the impact of inflow and/or outflow. If the infall rate is larger than the star-formation rate, the accreted metal-poor gas will dilute the ISM faster than it can be enriched by evolving stars, thus causing the metallicity to drop. Similarly, one can lower the metallicity via outflows, but only if the material in the outflow has a higher metallicity than the ISM. Thus, inflow and/or outflow can explain the observed metallicity–luminosity relation provided that their efficiency is higher in lower mass galaxies. Both inflow and outflow are likely to occur: inflow because in CDM cosmogonies dark matter halos continue to accrete new mass, and outflow because this seems required to explain various observations (see Chapter 15). Although there is no obvious reason why low-mass galaxies should have a higher inflow rate than their massive counterparts, outflows are naturally expected

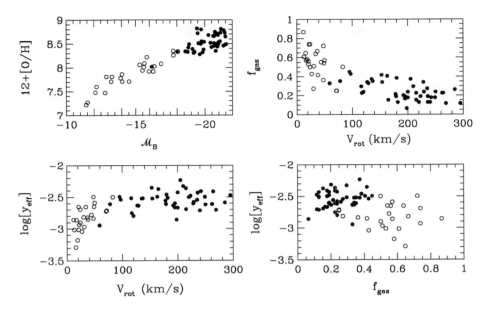

Fig. 11.9. The upper left panel shows the metallicity–luminosity relation for a sample of 70 disk galaxies. Solid and open circles correspond to spiral galaxies and irregulars, respectively. The upper right panel shows the relation between the gas mass fractions, $f_{\rm gas}$, and the rotation velocities, $V_{\rm rot}$. The lower panels show the effective yields as function of $V_{\rm rot}$ and $f_{\rm gas}$. [Based on data published in Pilyugin et al. (2004) and kindly provided by J. Dalcanton]

to be more efficient in less massive systems, simply because their potential wells are shallower (see §8.6.3).

We can discriminate between some of these ideas using the so-called effective yield. As we have seen in §10.4.2, under the assumption of closed-box evolution and using the instantaneous recycling approximation, the metallicity of the gas, Z, is a simple function of the gas mass fraction, $f_{\rm gas}$, and the true nucleosynthetic yield, y_Z:

$$Z = y_Z \ln(1/f_{\rm gas}), \qquad (11.151)$$

where we have assumed that the gas starts out with zero metallicity. The *effective* yield is defined as

$$y_{\rm eff} = \frac{Z}{\ln(1/f_{\rm gas})}. \qquad (11.152)$$

where Z is now the *observed* metallicity of a galaxy. Thus, if the galaxy has evolved as a closed box, then $y_{\rm eff} = y_Z$. As shown by Edmunds (1990), in the case of inflow or outflow one always has $y_{\rm eff} \leq y_Z$.[6] Thus, the effective yield is an observationally determined quantity that can be used to diagnose the importance of inflows and/or outflows. The lower panels of Fig. 11.9 show the relationships between $y_{\rm eff}$, $f_{\rm gas}$ and the rotation velocities, $V_{\rm rot}$. Compared to massive spirals, the effective yield is reduced by a factor of several in low-mass galaxies ($V_{\rm rot} \lesssim 40\,{\rm km\,s^{-1}}$), all of which are relatively gas rich ($f_{\rm gas} > 0.3$).

If the true nucleosynthetic yield is roughly constant among galaxies, then this indicates that low-mass disk galaxies do not evolve as a closed box, so that the metallicity–luminosity relation is not just a consequence of variations in the gas mass fraction along the luminosity sequence.

---

[6] The only exception is the accretion of gas with a metallicity comparable to or higher than that of the system itself. Since this is highly unlikely, we will not consider this case here.

Rather, the observed relation between $y_{\rm eff}$ and $V_{\rm rot}$ suggests that inflow and/or outflow is important, and that their efficiency depends on the rotation velocity (i.e. the depth of the gravitational potential well) of the galaxy. As demonstrated by Dalcanton (2007), the only mechanism that can explain the extremely low effective yields for low mass disk galaxies is metal-enriched outflows (i.e. outflows with a metallicity larger than that of the gas). To see this, note that $Z = M_Z/M_{\rm gas}$, with M_Z the mass in metals (in the gas phase), and that $f_{\rm gas} = M_{\rm gas}/(M_\star + M_{\rm gas})$. When $f_{\rm gas}$ is large, a simple Taylor series expansion of Eq. (11.152) yields

$$y_{\rm eff} \approx \frac{M_Z}{M_\star} \quad \text{for large } f_{\rm gas}. \tag{11.153}$$

This makes it immediately clear that in galaxies with a high gas mass fraction the accretion of new gas does not have a significant impact on the effective yield. Although it will lower the metallicity, Z, it will increase $f_{\rm gas}$ as well, so that $y_{\rm eff}$ is unaffected.

In the case of outflows, consider a galaxy with an initial gas mass $M_{\rm gas,i}$ and initial metallicity $Z_{\rm i}$ that experiences an impulsive outflow of gas with mass $\Delta M_{\rm gas} = f_{\rm out} M_{\rm gas,i}$ and with metallicity $Z_{\rm out} = xZ_{\rm i}$. The effective yield immediately after the outflow is then given by

$$y_{\rm eff,f} = y_{\rm eff,i} \frac{1 - x f_{\rm out}}{1 - f_{\rm out}} \frac{\ln(f_{\rm gas,i})}{\ln(f_{\rm gas,f})}, \tag{11.154}$$

where $f_{\rm gas,i}$ and $f_{\rm gas,f}$ are the gas mass fractions directly prior to and directly after the outflow event, respectively, which are related according to

$$\frac{f_{\rm gas,f}}{f_{\rm gas,i}} = \frac{1 - f_{\rm out}}{1 - f_{\rm out} f_{\rm gas,i}}. \tag{11.155}$$

In the case of an unenriched outflow ($x = 1$) the final effective yield can only be significantly lower than the initial one if the outflow removes nearly the entire ISM. Since the galaxies with a small $y_{\rm eff}$ are gas rich, outflows of unenriched gas cannot produce their low effective yields. However, outflows consisting primarily of escaped SN ejecta (so that $x \gg 1$) are extremely efficient at reducing the effective yield.

Thus, the data suggest that low-mass, gas-rich galaxies have experienced significant metal-enriched outflows. Note, however, that this does not mean that massive, gas-poor galaxies have not experienced similar metal-enriched outflows, or that inflows are not important. Rather, these two processes simply do not cause a significant change of the effective yield in these galaxies.

(b) Metallicity Gradients The study of metal abundances in HII regions in the disks of spiral galaxies has revealed the existence of metallicity gradients. In almost all cases the outer parts are found to have lower metallicities than the central regions. In particular, in the case of the Milky Way,

$$\Delta[{\rm O/H}]/\Delta R = -(0.07 \pm 0.01)\,{\rm dex\,kpc}^{-1} \tag{11.156}$$

(Smartt & Rolleston, 1997). Late-type galaxies have somewhat steeper gradients than the more luminous early-types, although this dependence disappears when the gradients are expressed in units of the disk scale length. In addition, barred galaxies are found to have shallower gradients than non-barred galaxies of the same luminosity (Vila-Costas & Edmunds, 1992; Zaritsky et al., 1994).

From a theoretical point of view, the existence of metallicity gradients is not surprising, and there are numerous explanations for their origin. Even within the simple closed-box chemical evolution model, metallicity gradients are a fairly natural outcome. To see this, consider a gas disk that forms stars according to a Schmidt-type star formation law (see §9.5.1),

$$\frac{d\Sigma_{\rm gas}}{dt} = -A\Sigma_{\rm gas}^n, \tag{11.157}$$

11.8 Chemical Evolution of Disk Galaxies

with $\Sigma_{\rm gas}$ the gas surface density, and A a constant. Simple integration gives

$$\ln\left(\frac{\Sigma_{\rm gas}}{\Sigma_{\rm tot}}\right) = -At \quad (n=1); \tag{11.158}$$

$$\frac{\Sigma_{\rm gas}}{\Sigma_{\rm tot}} = \left[1 + (n-1)At\Sigma_{\rm tot}^{n-1}\right]^{-1/(n-1)} \quad (n \neq 1), \tag{11.159}$$

where $\Sigma_{\rm tot}$ is the total surface density, taken to be equal to $\Sigma_{\rm gas}$ at $t=0$. Using Eq. (11.151) the metallicity is then given by

$$Z = \begin{cases} y_Z At & (n=1) \\ \frac{y_Z}{n-1}\ln\left[1 + (n-1)A\Sigma_{\rm tot}^{(n-1)}t\right] & (n \neq 1). \end{cases} \tag{11.160}$$

Since $d\Sigma_{\rm tot}/dR < 0$, this shows that simple closed-box evolution naturally yields a metallicity gradient with $d\log Z/dR < 0$ as long as star formation follows a Schmidt law with $n > 1$. However, as shown by Phillipps & Edmunds (1991), the amplitude of the resulting metallicity gradient is too small compared to observations. One way to steepen the metallicity gradients is to consider inflow. As discussed in §11.2.5, disk galaxies are expected to grow from the inside-out in the standard paradigm. This implies that the outer disks form later so that t is shorter, which naturally results in lower metallicities, hence steeper gradients. Indeed, metallicity gradients obtained by analytical models and numerical simulations that specifically model the inside-out formation via inflow match well observational results (e.g. Steinmetz & Mueller, 1994; Prantzos & Boissier, 2000).

Another process that may play an important role in producing metallicity gradients is radial gas flows. If gas moves inward, while the long-lived stars stay at the radii where they were born, a metallicity gradient may be produced in the stellar population (Lacey & Fall, 1985). Such radial gas flows may occur due to viscosity (see §11.4.2) or due to the torques from bars and/or spiral arms (see §11.5.4). On the other hand, gas flows can also wash out existing gradients by mixing gas of different metallicities, and it is generally believed that mixing due to bar-induced gas flows is responsible for the smaller metallicity gradients observed in barred galaxies in comparison to those in unbarred galaxies.

12

Galaxy Interactions and Transformations

So far we have treated galaxies as isolated, non-interacting systems. However, in the hierarchical scenario of structure formation, galaxies and their associated dark matter halos undergo frequent interactions with each other. As we have seen in §7.3.6, a large fraction of dark matter halos are expected to be dynamically young (i.e. to have experienced a merger event in their recent history). In fact, as shown by Li et al. (2007), each halo, independent of its mass, experiences about three major mergers (defined as mergers with a progenitor mass ratio larger than $1/3$) after its main progenitor has acquired 1% of its present-day mass. Hence, galaxies and their associated dark matter halos cannot be considered isolated 'island universes', but are constantly influenced by gravitational interactions with other systems. These interactions may have dramatic impact on the morphologies and star-formation histories of galaxies, making the study of their nature and frequency an important part of galaxy formation and evolution.

Consider a body S which has an encounter with a perturber P with impact parameter b and initial velocity v_∞ (in the limit of infinite, initial separation between S and P). Let q be a particle (e.g. a star) in S, at a distance $\mathbf{r}(t)$ from the center of S, and let $\mathbf{R}(t)$ be the position vector of P from S (see Fig. 12.1 for an illustration). Since the gravitational force due to P is not uniform over the body of S, the particle q experiences a tidal force per unit mass

$$\mathbf{F}_{\rm tid}(\mathbf{r}) = -\nabla\Phi_{\rm P}(|\mathbf{R}-\mathbf{r}|) + \nabla\Phi_{\rm P}(\mathbf{R}), \tag{12.1}$$

with $\Phi_{\rm P}$ the gravitational potential of P. Hence, because of the encounter, the rate at which particle q gains energy per unit mass is

$$\frac{dE_q}{dt} = \mathbf{v}\cdot\mathbf{F}_{\rm tid}(\mathbf{r}), \tag{12.2}$$

with $\mathbf{v}$ the velocity of q with respect to the center of S.

Similar to the way the Moon gives rise to oceanic tides on the Earth, the gravitational interaction between S and P enhances the gravitational multipole moments of both bodies, which in turn may cause a backreaction on their orbit. Let $\tau_{\rm tide}$ be the time for the tide to rise and $\tau_{\rm enc} \simeq R_{\rm max}/V$ the time of the encounter, with V the relative velocity of the two bodies, and $R_{\rm max} = \max[R_0, R_{\rm S}, R_{\rm P}]$ with $R_{\rm S}$ and $R_{\rm P}$ the (characteristic) radii of S and P, and R_0 the minimum distance of the encounter. If $\tau_{\rm enc} \gg \tau_{\rm tide}$ then the time scale for the internal structures of the deformable bodies to adjust themselves is much shorter than the time scale on which the tides change (due to the change of the relative position and orientation between S and P). Consequently, the effects of the encounter during approach and departure cancel each other (the deformations are adiabatic), and there is no net transfer of energy. However, if $\tau_{\rm enc} \lesssim \tau_{\rm tide}$ the response of the bodies lags behind the instantaneous magnitude and direction of the force, causing a backreaction on the orbit. The net effect in this case is a transfer of orbital energy to internal energy of the two bodies, causing an increase in their mutual binding energy. Under

12.1 High-Speed Encounters

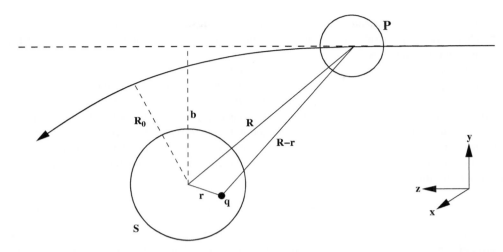

Fig. 12.1. Schematic illustration of an encounter with impact parameter b between a system S and its perturber P.

certain conditions, this may cause the two bodies to become gravitationally bound to each other, in which case we speak of gravitational capture.

In the case of an encounter between two stars, $\tau_{\rm tide}$ depends on the effective viscosity of the stars. If sufficiently large, the deformations caused by the encounter induce oscillations in the two stars, which are gradually damped by the viscosity and the corresponding energy is transformed into heat. In the case of interest to us, the two bodies are collisionless systems (dark matter halos and/or galaxies), and $\tau_{\rm tide} \sim R/\sigma$, with R the characteristic size of the system and σ its internal velocity dispersion. This implies that the situation $\tau_{\rm enc} \gg \tau_{\rm tide}$, which translates into $V \ll (R_{\rm max}/R)\sigma$, does not occur. After all, $R_{\rm max} \geq R$ by definition, and one always has that $V \gtrsim \sigma$ because of the gravitational acceleration during the encounter. Therefore, an encounter between two collisionless systems almost always leads to an increase of their internal energies. In general, numerical simulations are required to determine how much energy transfer takes place. However, in the case of high-speed encounters with $V \gg \sigma$, which we discuss in §12.1, the amount of energy transfer can be calculated analytically. This energy deposition gives rise to modifications of the internal structure of the two systems. In particular, certain stars or dark matter particles may gain sufficient kinetic energy to become unbound. Hence, tidal interactions during encounters may induce mass loss, which is discussed in §12.2. When a massive object moves through a dark matter halo, its gravitational interaction with the constituent particles of the halo induces a dynamical friction force on the object, which is the topic of §12.3. If the encounter velocity between two bodies is not too large, the encounter may result in gravitational capture and ultimately cause the two bodies to merge with each other. This merging process is discussed in §12.4. Finally, in §12.5 we discuss a number of galaxy transformation processes operating in clusters, which may be responsible for the various correlations between galaxy properties and their environments.

12.1 High-Speed Encounters

In general, an encounter between two collisionless systems is extremely complicated, and one typically has to resort to numerical simulations to investigate its outcome. However, in the limiting case where the encounter velocity is much larger than the internal velocity dispersion of the perturbed system the change in the internal energy can be approximated analytically. Such high-speed encounters play an important role in galaxy clusters, where the velocity dispersion of the cluster ($\sigma_{\rm cluster} \sim 1000 {\rm km\,s^{-1}}$) is significantly larger than the internal dispersions of the individual member galaxies.

Consider the encounter between S and P illustrated in Fig. 12.1. Let v_∞ be the initial velocity of P with respect to S when their separation is large. In the large-v_∞ limit the tidal forces due to P act on a time scale that is much shorter than the dynamical time of S, and the encounter is said to be impulsive. This means that we may consider q to be stationary with respect to the center of S during the encounter, only experiencing a change $\Delta \mathbf{v}$ in its velocity. In this impulse approximation, the potential energy of q before and after the encounter is the same (i.e. the density distribution of S remains unchanged during the encounter), so that the change in the total energy per unit mass of a particle of S is given by

$$\Delta E = \frac{1}{2}(\mathbf{v} + \Delta \mathbf{v})^2 - \frac{1}{2}\mathbf{v}^2 = \mathbf{v} \cdot \Delta \mathbf{v} + \frac{1}{2}|\Delta \mathbf{v}|^2. \tag{12.3}$$

We are interested in computing $\Delta E_{\rm S}$, obtained by integrating ΔE over the entire system S. Because of symmetry, the integral of the first term on the right-hand side of Eq. (12.3) is typically equal to zero, so that

$$\Delta E_{\rm S} = \frac{1}{2} \int |\Delta \mathbf{v}(\mathbf{r})|^2 \rho(\mathbf{r}) d^3 \mathbf{r}. \tag{12.4}$$

In the large-v_∞ limit the distance of closest approach $R_0 \to b$, and the velocity of P with respect to S is $\mathbf{v}_{\rm P}(t) \simeq v_\infty \mathbf{e}_z \equiv v_{\rm P} \mathbf{e}_z$. Hence, with a proper choice of time origin, $\mathbf{R}(t) = (0, b, v_{\rm P} t)$. Let $\mathbf{r}$ be the position vector of a particle in S and let R be the instantaneous distance between S and P. In the distant encounter approximation, where $b \gg \max[R_{\rm S}, R_{\rm P}]$, the perturber may be considered a point mass. Hence, the potential at $\mathbf{r}$ due to P is

$$\Phi_{\rm P}(\mathbf{r}) = -\frac{GM_{\rm P}}{|\mathbf{r} - \mathbf{R}|}. \tag{12.5}$$

Using the series expansion of $(1+x)^{-1/2}$ and the fact that $|\mathbf{r} - \mathbf{R}| = \sqrt{R^2 - 2rR\cos\phi + r^2}$, with ϕ the angle between $\mathbf{r}$ and $\mathbf{R}$, we obtain

$$\Phi_{\rm P}(\mathbf{r}) = -\frac{GM_{\rm P}}{R} - \frac{GM_{\rm P} r}{R^2}\cos\phi - \frac{GM_{\rm P} r^2}{R^3}\left(\frac{3}{2}\cos^2\phi - \frac{1}{2}\right) + \mathcal{O}\left[(r/R)^3\right]. \tag{12.6}$$

Dropping the higher-order terms in the series expansion of the potential constitutes the tidal approximation. Encounters for which both the tidal approximation and the impulsive approximation are valid are often called tidal shocks. The first term on the right-hand side of Eq. (12.6) is a constant and does not yield any force. The second term yields a uniform acceleration $GM_{\rm P}/R^2$ directed towards P, and describes how the center of mass of S changes its velocity. Since we want to calculate the velocity change of a particle with respect to the center of S, this term is not of interest to us. The third term of Eq. (12.6) corresponds to the tidal force per unit mass. In order to calculate this force, it is more convenient to work in a coordinate frame, (x', y', z'), with the origin at the center of S, with the x' axis pointing towards the instantaneous position of P, and with y' in the yz plane. It is then easy to show that the tidal force (per unit mass) at $\mathbf{r}$ is

$$\mathbf{F}_{\rm tid}(\mathbf{r}) = \frac{GM_{\rm P}}{R^3}(2x', -y', -z'), \tag{12.7}$$

with (x', y', z') the coordinates of $\mathbf{r}$ in the primed system. Transforming to the (x, y, z) coordinate system, and integrating $\mathbf{F}_{\rm tid} = d\mathbf{v}/dt$ over time, yields the cumulative change in velocity with respect to the center of S,

$$\Delta \mathbf{v} = \frac{2GM_{\rm P}}{v_p b^2}(-x, y, 0). \tag{12.8}$$

Substituting Eq. (12.8) in Eq. (12.4), and assuming spherical symmetry for S, so that $\langle x^2 \rangle = \langle y^2 \rangle = \langle r^2 \rangle/3$, we finally obtain the tidal heating of S due to a high-speed encounter with P:

$$\Delta E_S = \frac{4}{3} G^2 M_S \left(\frac{M_P}{v_P} \right)^2 \frac{\langle r^2 \rangle}{b^4}. \tag{12.9}$$

As shown by Aguilar & White (1985), this derivation, which is originally due to Spitzer (1958), is surprisingly accurate for encounters with $b \gtrsim 5 \max[R_P, R_S]$, even for relatively slow encounters with $v_\infty \simeq \sigma_S$. Note that Eq. (12.9) is only valid for distant encounters in which P may be considered a point mass. In the case of encounters with $b \lesssim R_P$, the internal structure of P has to be taken into account. As shown in Gnedin et al. (1999), in the case of a spherical perturber, this implies multiplying the right-hand side of Eq. (12.9) with a factor

$$f(b) = \frac{1}{2} \left[(3J_0 - J_1 - I_0)^2 + (2I_0 - I_1 - 3J_0 + J_1)^2 + I_0^2 \right]. \tag{12.10}$$

Here

$$I_0(b) = \int_1^\infty \frac{M_P(br)}{M_P} \frac{dr}{r^2(r^2-1)^{1/2}}, \tag{12.11}$$

$$J_0(b) = \int_1^\infty \frac{M_P(br)}{M_P} \frac{dr}{r^4(r^2-1)^{1/2}}, \tag{12.12}$$

$$I_1(b) = b \frac{dI_0}{db} \quad \text{and} \quad J_1(b) = b \frac{dJ_0}{db} \tag{12.13}$$

with $M_P(r)$ the mass of P enclosed within a radius r. In general, $f(b)$ decreases from unity in the large-b limit to zero when $b \to 0$. Note that $\Delta E_S \propto b^4$, indicating that close encounters are far more important than distant encounters. On the other hand, $f(b)$ rapidly decreases with b once $b \lesssim R_P$. Consequently, encounters with impact parameters $b \sim R_P$ are the ones that have the strongest impact on S.

In the impulse approximation, the encounter only changes the kinetic energy of a system, but leaves its potential energy intact. Consequently, after the encounter the system is no longer in virial equilibrium, and has to undergo a relaxation process in order to settle to a new virial equilibrium. Let the initial kinetic and total energies of S be K_S and E_S, respectively. According to the virial theorem (5.130) we have that $E_S = -K_S$. Due to the encounter, $E_S \to E_S + \Delta E_S$ and, since all this energy is invested in the internal kinematics of S, we also have that $K_S \to K_S + \Delta E_S$. After S has relaxed to a new virial equilibrium, $K_S = -(E_S + \Delta E_S)$. Thus, the relaxation process decreases the kinetic energy by $2\Delta E_S$. This energy is transferred to potential energy, which becomes less negative, implying that tidal shocks cause systems to expand.

Note that the net effect of pumping energy into the system is therefore a *decrease* of its kinetic energy (i.e. the system gets 'colder'). This is a consequence of the negative specific heat of self-gravitating systems. By analogy with the particles in an ideal gas, the kinetic energy in an N-body system of equal point masses can be assigned a mean 'temperature':

$$\frac{1}{2} N m \langle v^2 \rangle = \frac{3}{2} N k_B \langle T \rangle. \tag{12.14}$$

Here k_B is Boltzmann's constant, and $\langle v^2 \rangle$ and $\langle T \rangle$ are the mean velocity dispersion and mean temperature, respectively. According to the virial theorem we have that

$$E = -K = -\frac{3}{2} N k_B \langle T \rangle, \tag{12.15}$$

which allows us to define the heat capacity of the system as

$$C \equiv \frac{dE}{d\langle T \rangle} = -\frac{3}{2} N k_B. \qquad (12.16)$$

Note that C is always negative, so that a system becomes hotter when it loses energy. This is a characteristic, and somewhat counter-intuitive, property of all systems in which the dominant forces are gravitational. This includes the Sun, where the stability of nuclear burning is a consequence of $C < 0$: If the reaction rates become too high, the excess energy input into the core makes the core expand and cool. This makes the reaction rates drop, bringing the system back to equilibrium.

12.2 Tidal Stripping

In the previous section we have seen how tidal shocks due to high-speed encounters heat a collisionless system, which eventually results in expansion possibly associated with mass loss. We now examine how tidal forces impact a collisionless system in the more general case (i.e. no longer restricting ourselves to high-speed encounters). As we will see, even in a static configuration tidal forces can strip material from the outer parts of a collisionless system, which is generally known as tidal stripping. After introducing the concept of tidal radius, we briefly discuss how this tidally stripped material gives rise to tidal streams and tails.

12.2.1 Tidal Radius

Let us start our exploration with a subject mass m (hereafter the satellite) on a circular orbit of radius R in the potential well of a point mass M. In this case the tidal field across m is static. Although this implies that there are no tidal shocks, the subject mass can still suffer mass loss since particles in the outer parts may experience tidal forces that exceed their binding forces. The gravitational attraction of M causes the center of the subject mass to experience an acceleration GM/R^2. If the satellite is spherical and has a radius r, then different accelerations, $GM/(R+r)^2$ and $GM/(R-r)^2$, will be felt by the material at the nearest and farthest ends of the satellite from M, respectively. Assuming that $r \ll R$, the difference between the accelerations at these points and that at the center has the magnitude $2GMr/R^3$. If this tidal acceleration exceeds the binding force per unit mass, Gm/r^2, the material at a distance r from the center of m will be stripped away from the subject mass. This defines a critical radius (the tidal radius): $r_t = (m/2M)^{1/3} R$. If the radius of a subject mass is larger than its tidal radius r_t, it will experience mass loss due to tidal stripping. This simple derivation, however, ignores the fact that the subject mass also experiences a centrifugal force associated with its (circular) motion around the center of M. This can be accounted for using a more rigorous derivation based on the restricted three-body problem, resulting in a tidal radius:

$$r_t = \left[\frac{m/M}{(3+m/M)} \right]^{1/3} R \qquad (12.17)$$

(see Binney & Tremaine, 1987).

Note that Eq. (12.17) is based on the assumption that M can be approximated as a point mass, so that it is only valid if R is large compared to the size of M. In addition, we have made the assumption that the subject mass is on a circular orbit. In many cases, however, we will be concerned with the tidal stripping experienced by a subject mass on an eccentric orbit *within* a host system of mass M (e.g. a satellite galaxy orbiting within the dark matter halo associated with the Milky Way). In general, if the subject mass is on a non-circular orbit, its tidal radius cannot be rigorously defined. However, following King (1962), we can define an approximate tidal radius as the distance from the center of m at which a point on the line connecting the centers of m and M experiences zero acceleration with respect to the center of m when the subject mass has its pericentric passage. Let R be the distance between the centers of m and M. Then, during its pericentric passage the subject mass experiences an acceleration

$$\ddot{R} = -\left.\frac{d\Phi_h}{dR}\right|_{R_0} + R_0 \Omega^2, \qquad (12.18)$$

where R_0 is the pericentric distance, Ω is the angular speed, and Φ_h is the gravitational potential of the host mass. Now consider a particle 'p' that is part of the subject mass, located a distance R_p from the center of the host mass along the line connecting m and M. This particle's acceleration is

$$\ddot{R}_p = -\left.\frac{d\Phi_h}{dR}\right|_{R_p} - \left.\frac{d\Phi_s}{dr}\right|_r + R_p \Omega^2, \qquad (12.19)$$

where $r = |R_p - R|$ is the distance of p from the center of m, and Φ_s is the gravitational potential of the subject mass. Hence, the relative acceleration of p with respect to the center of m can be approximated as

$$\ddot{r} = \left(-\left.\frac{d^2\Phi_h}{dR^2}\right|_{R_0} - \frac{1}{r}\left.\frac{d\Phi_s}{dr}\right|_r + \Omega^2\right) r. \qquad (12.20)$$

Using that $d\Phi_i/dr = GM_i(r)/r^2$, with $M_i(r)$ the mass of spherical object i enclosed within radius r, and solving for $\ddot{r} = 0$, we obtain that the tidal radius r_t is given by the solution to

$$r_t = \left[\frac{m(r_t)/M(R_0)}{2 + \frac{\Omega^2 R_0^3}{GM(R_0)} - \left.\frac{d\ln M}{d\ln R}\right|_{R_0}}\right]^{1/3} R_0. \qquad (12.21)$$

This equation gives an approximation for the tidal radius of a subject mass m moving on an orbit with pericenter R_0 in or around a host system of mass M. In the limit where both m and M are point masses traveling at separation R_0 in a circular orbit around their mutual center of mass, $\Omega^2 = G(M+m)/R_0^3$, and Eq. (12.21) reduces to Eq. (12.17).

It should be pointed out, however, that even Eq. (12.21) is only a crude approximation. First of all, as already alluded to above, in the case of non-circular orbits the concept of a tidal radius is not well defined. In fact, one might argue that a pericentric passage is better described by the impulse approximation, i.e. the subject mass experiences a tidal shock, even though the encounter is not necessarily a high-speed encounter. Secondly, even in the case of point masses, the two-dimensional surface along which $\ddot{r} = 0$ is not spherical, and so cannot be characterized by a single radius. And finally, in the derivation of Eq. (12.21) we have ignored the orbital motion of particles within the subject mass. This, among other effects, gives rise to scatter in the circular frequencies Ω, and effectively introduces some 'thickness' to the shell of particles for which the internal and tidal forces balance. Despite these shortcomings, Eqs. (12.17) and (12.21) are often used to model tidal stripping of satellite galaxies, globular clusters and/or dark matter subhalos (e.g. Johnston, 1998; Taylor & Babul, 2001; Zentner & Bullock, 2003).

12.2.2 Tidal Streams and Tails

We now turn our attention to the fate of tidally stripped material. We consider two specific cases: the stripping of a relatively small satellite system orbiting in a larger host system, and the tidal stripping accompanying the merger between two disk galaxies of comparable masses.

(a) Tidal Stripping of Satellite Galaxies Consider a collisionless system S orbiting inside the (external) gravitational potential of a host system P (i.e. a satellite galaxy orbiting within the dark matter halo of its host galaxy). Let $(\bar{\mathbf{J}}, \bar{\Theta})$ be the average action-angle variables of the particles that make up S. Then, if the Hamiltonian of P is integrable, the equations of motion for S are

$$J_i(t) = J_i(0) = \text{constant},$$
$$\Theta_i(t) = \frac{\partial \mathcal{H}}{\partial J_i} t + \Theta_i(0), \qquad (12.22)$$

with $i = 1, 2, 3$ and $\mathcal{H} = \mathcal{H}(\mathbf{J})$ the Hamiltonian (see §5.4.5). Since the constituent particles of S have similar positions and velocities (phase-space coordinates) with respect to P, they also have similar values for $(\mathbf{J}, \Theta)$. In particular, if the distribution function of the ensemble of S is a multivariate Gaussian in configuration and velocity space, then the distribution function in action-angle variables is also a multivariate Gaussian (Helmi & White, 1999). Hence, the particles of S that are stripped due to the tidal forces of P will have action-angle variables that are similar to those of the particles that remain bound to S, and thus to $(\bar{\mathbf{J}}, \bar{\Theta})$. Since the actions are integrals of motion, the differences $\Delta J_i = J_i - \bar{J}_i$, where J_i corresponds to the action of a tidally stripped particle, remain constant. However, the differences $\Delta \Theta_i = \Theta_i - \bar{\Theta}_i$ evolve with time. Using Eq. (12.22) and expanding $\Delta \Theta_i$ to first order in $J_k - \bar{J}_k$, we have that

$$\Delta \Theta_i(t) = \Delta \Theta_i(t_{\text{strip}}) + \left. \frac{\partial^2 \mathcal{H}}{\partial J_i \partial J_k} \right|_{\mathbf{J}} (J_k - \bar{J}_k)(t - t_{\text{strip}}), \qquad (12.23)$$

where t_{strip} is the time when the particle is stripped from S. Since the phase-space trajectories $\Gamma(t)$ of objects with similar actions are confined to similar invariant tori (see §5.4.5), the stripped particles and S will be on similar orbits [which are the projections of $\Gamma(t)$ in configuration space]. However, since $|\Delta \Theta|$ increases with time, the differences in the orbital phases grow with time. Thus, stripped stars move on roughly the same orbit as S, but, depending on the sign of $J_k - \bar{J}_k$, either trail or lead S in terms of their orbital phases. As is evident from Eq. (12.23), the rate at which the stripped stars disperse along the orbit of S is determined by the initial spread in actions and by the (eigenvalues of the) Hessian of the Hamiltonian (Tremaine, 1999; Helmi & White, 1999). Usually one of the eigenvalues is significantly larger than the other two, so that the stripped particles form a relatively thin structure in configuration space, called a tidal stream or tidal tail.[1] Smaller systems S have a smaller spread in actions and therefore disperse more slowly (i.e. produce shorter streams). The width of the tidal stream is mainly governed by the values of the two non-dominant eigenvalues of the Hessian.

The most well-known example of tidal streams is the Magellanic stream, which is a stream of neutral hydrogen stripped from the Magellanic clouds, and extending more than 100 degrees on the sky (Wannier & Wrixon, 1972; Mathewson et al., 1974). Although gaseous streams may also be produced by stripping due to ram-pressure (for instance if the Magellanic clouds have traversed the HI disk of the Milky Way), the fact that the Magellanic stream also contains a leading stream indicates that tides are the dominant mechanism responsible for its formation (Putman et al., 1998). Other examples of tidal streams in the halo of the Milky Way are those associated with the Sagittarius dwarf galaxy (Newberg et al., 2002) and with the globular cluster Pal 5 (Odenkirchen et al., 2002).

Tidal streams are powerful diagnostics. First of all, they can be used to constrain the gravitational potential of their host system (e.g. Murai & Fujimoto, 1980; Johnston et al., 1999). This is easy to understand from the fact that the orbital energy, $E = \frac{1}{2}v^2 + \Phi(\mathbf{x})$, is an integral of motion. Hence, all stars along the stream should have similar values of E. This principle, together with

---

[1] The term 'tidal tail' is usually reserved to refer to the structures formed by tidally stripped stars in major mergers discussed below.

measurements of the positions and velocities of the stream stars, puts constraints on $\Phi(\mathbf{x})$. In fact, even without any kinematic data valuable constraints can be obtained. For example, in a spherical potential, orbits (and thus also tidal streams) are confined to a plane. Consequently, the stream stars will be located on a great circle on the sky. This principle applied to the tidal stream associated with the Sagittarius dwarf galaxy has been used to argue that the halo of the Milky Way is close to spherical (Ibata et al., 2001). However, current data is still consistent with a Galactic dark matter halo that is either oblate or prolate, with minor-to-major density axis ratios as low as 0.6 within the region probed by the orbit of the Sagittarius dwarf (Helmi, 2004).

A second application of tidal streams is to constrain the hierarchical formation history of the Milky Way halo. Even after a satellite is completely disrupted, the tidal stream survives. Although phase mixing may no longer allow one to detect the stream in configuration space, the stream is still detectable as a pronounced peak in action-space, because the distribution function $f = f(\mathbf{J})$ is invariant and unaffected by tidal stripping or tidal destruction. Thus, if (i) one has measured the phase-space coordinates of individual halo stars, (ii) the potential of the Milky Way system is known, so that the actions of individual stars can be computed from their phase-space coordinates, (iii) the Hamiltonian of the Milky Way system is (near-)integrable, and (iv) the Milky Way system has not experienced significant violent relaxation (i.e. its potential has either remained constant or evolved adiabatically), then one can in principle identify individual satellite systems, even if they have been tidally disrupted a long time ago. Such information can put constraints on the hierarchical formation of the Milky Way system. Although it is unlikely that all of the above requirements are met, this archaeological approach to constraining galaxy formation has recently become an extremely active field of research, largely due to the explosion of new data provided by the SDSS (see Helmi, 2008, for a review)

(b) The Formation of Tidal Tails in Mergers In addition to the tidal streams produced when satellite galaxies orbiting their hosts are stripped, tidal tails are also observed in merging (disk) galaxies. Fig. 2.10 shows the merging galaxy pair NGC 4038 and NGC 4039, also known as the Antennae, which reveals two prominent tails stretching a length over 100 kpc from end to end. In a seminal paper, Toomre & Toomre (1972) showed that such tail structures are tidal relics of close encounters (or mergers) between two disk galaxies. The tails are narrow because they originate from dynamically cold disks; mergers between (dynamically hot) spheroids do not produce narrow tidal tails.

Toomre & Toomre (1972) also showed that prograde encounters, in which the orbital angular momentum is aligned with the spin vectors of the initial disk galaxies, result in far more prominent tidal tails than retrograde encounters, in which the orbital angular momentum is anti-aligned with the disks' spin vectors. This is easy to understand. If we follow Toomre & Toomre (1972), and consider the disk galaxies to consist of point masses with disks made of massless test-particles, the orbital frequency of a ring of radius r is $\omega_{\rm ring} = \sqrt{GM/r^3}$. The angular velocity of the line joining the two massive particles at pericenter is

$$\omega_{\rm orb} = \left[\frac{2GM(1+e)}{R_0^3}\right]^{1/2}, \qquad (12.24)$$

where e is the eccentricity of the orbit and R_0 is the minimum separation of the two massive particles. If $\omega_{\rm orb} = \omega_{\rm ring}$, i.e. for a ring with radius $r = R_0/[2(1+e)]^{2/3}$, and if the encounter is prograde, then a test particle in the ring is in resonance with the tidal acceleration – it is continuously pulled either inwards or outwards, depending on its orbital phase. Consequently, the ring responds violently to the encounter. If, on the other hand, the encounter is retrograde, then test particles will be pulled alternatively inwards and outwards, with a high frequency, and the net effect of the tidal acceleration is small.

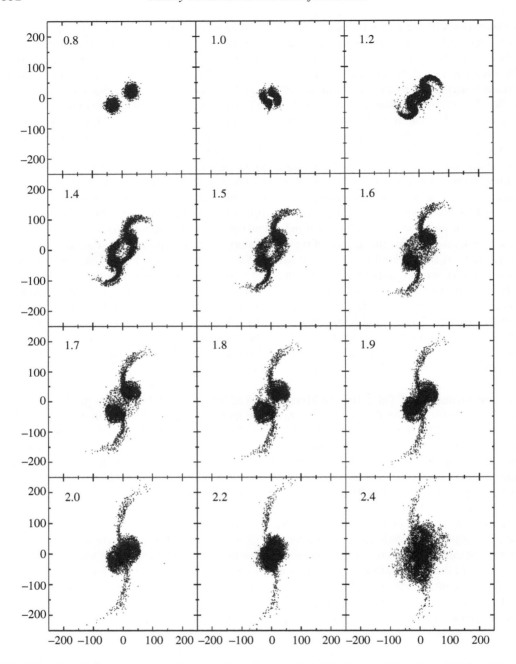

Fig. 12.2. Snapshots that show the time evolution of a numerical N-body simulation of a prograde merger between two disk galaxies that are embedded in dark matter halos. The snapshots show the disk particles ('stars') only, and clearly reveal the formation of two prominent tidal tails. The length units labeling the axes are given in h^{-1} kpc, and the numbers in the upper left corner of each panel indicate the elapsed time since the start of the simulation in units of $9.8 \times 10^8 h^{-1}$ yr. [Adapted from Springel & White (1999)]

In order to gain insight, the simulations of Toomre & Toomre (1972) were deliberately oversimplified: they considered disks consisting of massless particles orbiting around point masses (i.e. they ignored the self-gravity of the disks) and the disks were not embedded in dark matter halos. Nevertheless, the main conclusions reached by the Toomres have since been confirmed by increasingly larger numerical simulations that include self-gravity of the disks as well as dark matter halos (e.g. Gerhard, 1981; Farouki & Shapiro, 1982; Barnes, 1988; Hernquist, 1992). Fig. 12.2 shows an example of such a numerical simulation. It shows snapshots of a prograde merger between two disk galaxies embedded in dark matter halos. Note that the formation of the two prominent tidal tails carries away part of the energy and angular momentum of the initial orbit, causing the progenitors to merge.

As shown by Dubinski et al. (1996), increasing the mass of the dark matter halo with respect to that of the disk makes the tidal tails shorter and less massive. This simply reflects the fact that more massive halos have deeper potential wells and higher encounter velocities. As a consequence, the duration and overall strength of the tidal perturbations are smaller, and the perturbed material cannot as easily climb out of the deeper potential well (White, 1982). Based on their numerical simulations, Dubinski et al. (1996) argued that well-known merger candidates such as the Antennae could not have originated from encounters in which the progenitor galaxies have halo-to-disk mass ratios larger than 10:1. However, subsequent studies by Springel & White (1999) and Dubinski et al. (1999) have shown that the halo-to-disk mass ratio is not a sufficient condition to determine whether extended tidal tails can develop or not. Rather, Mo et al. (1998) suggested using the ratio

$$\mathscr{E} \equiv \frac{V_{\text{esc}}^2(2R_{\text{d}})}{V_{\text{c}}^2(2R_{\text{d}})}, \qquad (12.25)$$

where R_{d} is the disk scale length, and V_{esc} and V_{c} are the escape velocity and circular velocity, respectively. Thus, $\mathscr{E}$ compares the depth of the potential well ($\sim V_{\text{esc}}^2$) with the specific kinetic energy of the disk material ($\sim V_{\text{c}}^2$). The reason for evaluating this ratio at $2R_{\text{d}}$ is that this corresponds roughly to the half mass radius of an exponential disk. The simulation results of Springel & White (1999) and Dubinski et al. (1999) show that $\mathscr{E}$ is indeed a suitable indicator of a disk's tail-making ability; prominent tidal tails are produced for $\mathscr{E} \lesssim 6$ (assuming prograde mergers). Thus, the fact that prominent tidal tails such as those in the Antennae have been observed implies that (at least some) disk galaxies must have dark matter halos for which $\mathscr{E} \lesssim 6$.

12.3 Dynamical Friction

When an object of mass M_{S} (hereafter the subject mass) moves through a large collisionless system whose constituent particles (the field particles) have mass $m \ll M_{\text{S}}$, it experiences a drag force, called dynamical friction, which transfers energy and momentum from the subject mass to the field particles. Intuitively, this can be understood from the fact that two-body encounters cause particles to exchange energies in such a way that the system evolves towards thermodynamic equilibrium. Thus, in a system with multiple populations, each with a different particle mass m_i, two-body encounters drive the system towards equipartition, in which the mean kinetic energy per particle is locally the same for each population: $m_1 \langle v_1^2 \rangle = m_2 \langle v_2^2 \rangle = m_i \langle v_i^2 \rangle$. Since $M_{\text{S}} \gg m$ and particles at the same radius in inhomogeneous self-gravitating systems tend to have similar orbital velocities, the subject mass usually has a much larger kinetic energy than the typical field particles it encounters, producing a net tendency for it to lose energy and momentum. An alternative but equivalent way to think about dynamical friction is that the moving subject mass perturbs the distribution of field particles causing a trailing enhancement (or 'wake') in their

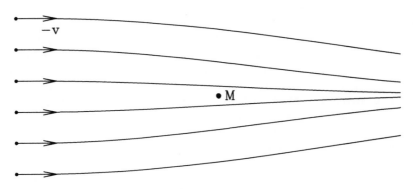

Fig. 12.3. As a massive object M moves through a sea of particles, the particles passing by are accelerated towards the object. As a result, the particle number density behind the object is higher than that in front of it, and the net effect is a drag force (dynamical friction) on the object.

density. The gravitational force of this wake on the subject mass M_S then slows it down (see Fig. 12.3).

In order to obtain an analytical expression for the dynamical friction force, we follow the original derivation due to Chandrasekhar (1943) in which dynamical friction is considered as a sum of uncorrelated two-body encounters between the subject mass and individual field particles. The principle here is that the momentum lost by the subject mass must be equal to the momentum gained by the field particles. Let us first consider a single encounter under the assumption that both the subject mass and the field particles are point masses. As shown in §5.4.1, when a point mass M_S moving with velocity $\mathbf{v}_S$ has an encounter with impact parameter b with a point mass m moving with velocity $\mathbf{v}_m$, the components of its velocity in the directions perpendicular and parallel to the initial encounter velocity change by

$$|\Delta \mathbf{v}_{s,\perp}| = \frac{2mbv_\infty^3}{G(M_S+m)^2}\left[1+\frac{b^2 v_\infty^4}{G^2(M_S+m)^2}\right]^{-1}, \qquad (12.26)$$

and

$$|\Delta \mathbf{v}_{s,\parallel}| = \frac{2mv_\infty}{(M_S+m)}\left[1+\frac{b^2 v_\infty^4}{G^2(M_S+m)^2}\right]^{-1}, \qquad (12.27)$$

where $v_\infty = |\mathbf{v}_\infty| = |\mathbf{v}_m - \mathbf{v}_S|$ is the initial relative velocity of the encounter. Note that these equations ignore the external potential due to the other field particles. The derivation that follows is therefore only strictly valid for a scenario in which the unperturbed orbits are straight lines, i.e. in which the distribution of field particles is infinite and homogeneous.

When moving through such a sea of field particles, the subject mass experiences many encounters with different impact parameters b and different encounter velocities $\mathbf{v}_m - \mathbf{v}_S$. The cumulative effect of all these encounters results in a total velocity change per unit time of

$$\left(\frac{d\mathbf{v}_S}{dt}\right)_i = \int\int \Delta \mathbf{v}_{s,i}(b,\mathbf{v}_m)\frac{d\mathcal{N}_{\rm enc}}{db\,d^3\mathbf{v}_m\,dt}\,db\,d^3\mathbf{v}_m, \qquad (12.28)$$

where i indicates either perpendicular or parallel to $\mathbf{v}_\infty$ and $\mathcal{N}_{\rm enc}$ is the number of encounters. If the phase-space distribution function of the field particles is given by $f(\mathbf{x},\mathbf{v})$, the number density of particles in the velocity-space volume $d^3\mathbf{v}$ is $f(\mathbf{x},\mathbf{v})d^3\mathbf{v}$. As mentioned above, our derivation assumes that the number density distribution of field particles is homogeneous, so that $f(\mathbf{x},\mathbf{v}) = f(\mathbf{v})$. Obviously, in this case the net cumulative effect on $\mathbf{v}_{s,\perp}$ is zero, and in what follows we focus on $d\mathbf{v}_{s,\parallel}/dt$. The number of encounters in a time interval Δt with impact parameters between b and $b + db$ is then simply the number density of particles times the volume of an annulus with inner radius b, outer radius $b + db$ and length $|\mathbf{v}_m - \mathbf{v}_S| \times \Delta t$. Thus

$$\frac{d\mathcal{N}_{enc}}{db\, d^3\mathbf{v}_m\, dt} = 2\pi b\, |\mathbf{v}_m - \mathbf{v}_S|\, f(\mathbf{v}_m). \tag{12.29}$$

Substituting Eqs. (12.27) and (12.29) in Eq. (12.28), and integrating the impact parameter over the range $0 < b < b_{max}$ yields

$$\frac{d\mathbf{v}_S}{dt} = \left(\frac{d\mathbf{v}_S}{dt}\right)_{\parallel} = 4\pi G^2 (M_S + m) m \int d^3\mathbf{v}_m\, f(\mathbf{v}_m) \ln \Lambda\, \frac{\mathbf{v}_m - \mathbf{v}_S}{|\mathbf{v}_m - \mathbf{v}_S|^3}. \tag{12.30}$$

Here

$$\ln \Lambda \equiv \frac{1}{2} \ln\left[1 + \left(\frac{b_{max}}{b_{90}}\right)^2\right] \simeq \ln\left(\frac{b_{max}}{b_{90}}\right), \tag{12.31}$$

is the Coulomb logarithm, with $b_{90} \equiv G(M_S + m)/v_\infty^2$ the impact parameter for which the deflection angle of the reduced particle of the encounter is equal to 90° (see Binney & Tremaine, 2008). This Coulomb logarithm basically describes the cut-off at the maximum impact parameter, b_{max}, as is required because Eq. (12.28) diverges logarithmically for large impact parameters (i.e. most of the contribution to $d\mathbf{v}_S/dt$ comes from encounters with large impact parameters). The second equality of Eq. (12.31) holds as long as $b_{max} \gg b_{90}$, which is virtually always the case. Note that the Coulomb logarithm depends on the encounter velocity $v_\infty = |\mathbf{v}_m - \mathbf{v}_S|$, and therefore has to be integrated over velocity space. In practice, however, since Λ is typically large, we do not make a large error by replacing v_∞ in the definition of $\ln \Lambda$ by a typical encounter speed $\bar{v}_{enc}$. We can then take the Coulomb logarithm outside of the integral.

In order to evaluate the integral over velocity space in Eq. (12.30), we use that

$$\nabla_\mathbf{x} \left(\frac{1}{|\mathbf{x}' - \mathbf{x}|}\right) = \frac{\mathbf{x}' - \mathbf{x}}{|\mathbf{x}' - \mathbf{x}|^3} \tag{12.32}$$

to write

$$\int d^3\mathbf{v}_m\, f(\mathbf{v}_m)\, \frac{\mathbf{v}_m - \mathbf{v}_S}{|\mathbf{v}_m - \mathbf{v}_S|^3} = \nabla H(\mathbf{v}_S), \tag{12.33}$$

where

$$H(\mathbf{v}_S) = \int d^3\mathbf{v}_m\, \frac{f(\mathbf{v}_m)}{|\mathbf{v}_m - \mathbf{v}_S|} \tag{12.34}$$

is known as the first Rosenbluth potential (e.g. Rosenbluth et al., 1957). In terms of Legendre polynomials, the inverse 'distance' between the vectors $\mathbf{v}_m$ and $\mathbf{v}_S$ can be written as

$$\frac{1}{|\mathbf{v}_m - \mathbf{v}_S|} = \sum_{l=0}^{\infty} \frac{v_-^l}{v_+^{l+1}} P_l(\cos\theta), \tag{12.35}$$

where $v_- = \min(|\mathbf{v}_m|, |\mathbf{v}_S|)$, $v_+ = \max(|\mathbf{v}_m|, |\mathbf{v}_S|)$ and θ is the angle between the two velocity vectors. If we now assume that $f(\mathbf{v}_m)$ is isotropic, so that it only depends on $v_m = |\mathbf{v}_m|$, symmetry dictates that the Rosenbluth potential only depends on $v_S = |\mathbf{v}_S|$. With the help of Eq. (12.35) we then have that

$$H(v_S) = 2\pi \sum_{l=0}^{\infty} \int_0^\infty dv_m\, \frac{v_m^2\, v_-^l}{v_+^{l+1}} f(v_m) \int_0^\pi d\theta\, \sin\theta\, P_l(\cos\theta). \tag{12.36}$$

Using that $\int_{-1}^{1} P_l(x)\, dx = 2\delta_{l0}$, this reduces to

$$H(v_S) = 4\pi \left[\frac{1}{v_S} \int_0^{v_S} f(v_m)\, v_m^2\, dv_m + \int_{v_S}^\infty f(v_m)\, v_m\, dv_m\right]. \tag{12.37}$$

Under the assumption of isotropy

$$\nabla H(\mathbf{v}_S) = \frac{\partial H}{\partial v_S} \frac{\mathbf{v}_S}{v_S}, \qquad (12.38)$$

so that we finally obtain that

$$\int d^3 \mathbf{v}_m f(\mathbf{v}_m) \frac{\mathbf{v}_m - \mathbf{v}_S}{|\mathbf{v}_m - \mathbf{v}_S|^3} = -4\pi \frac{\mathbf{v}_S}{v_S^3} \int_0^{v_S} f(v_m) v_m^2 dv_m. \qquad (12.39)$$

Substituting Eq. (12.39) into Eq. (12.30), and taking $M_S \gg m$, we finally obtain the dynamical friction force

$$\begin{aligned}\mathbf{F}_{df} = M_S \frac{d\mathbf{v}_S}{dt} &= -16\pi^2 G^2 M_S^2 m \ln\Lambda \left[\int_0^{v_S} f(v_m) v_m^2 dv_m \right] \frac{\mathbf{v}_S}{v_S^3} \\ &= -4\pi \left(\frac{GM_S}{v_S} \right)^2 \ln\Lambda \, \rho(<v_S) \frac{\mathbf{v}_S}{v_S}, \end{aligned} \qquad (12.40)$$

where $\rho(<v_S)$ is the density of field particles with speeds less than v_S. This is known as the Chandrasekhar dynamical friction formula. Similar to the frictional drag in fluid mechanics, dynamical friction exerts a force always pointing in the direction opposite to the motion. However, contrary to hydrodynamic friction, which always increases in strength when the velocity increases, the drag due to dynamical friction has a more complicated velocity dependence. In the low v_S limit, $f(v_m)$ in Eq. (12.40) may be replaced by $f(0)$, resulting in $F_{df} \propto v_S$, similar to hydrodynamic friction. However, at the high v_S limit, $\rho(<v_S)$ becomes independent of v_S so that $F_{df} \propto v_S^{-2}$.

Note that F_{df} is independent of the mass m of the field particles, at least in the limit $M_S \gg m$. Hence, Chandrasekhar's dynamical friction formula is also valid for a background field with a distribution of particle masses. Note also that $F_{df} \propto M_S^2$. This can be understood by considering the deflection of the field particles as illustrated in Fig. 12.3. Because of the gravitational focusing, a density wake is created downstream from the subject mass. Since the mass of the wake is proportional to M_S, the gravitational force of the wake on the subject mass is proportional to M_S^2.

12.3.1 Orbital Decay

Dynamical friction causes a relatively massive object orbiting in a background host system to lose energy and angular momentum to the 'field' particles of the host. Consequently, the orbit of the massive object decays with time, transporting it towards the center of the host's potential well. Thus, dynamical friction causes mass segregation, with more massive particles typically residing deeper in the potential well. This is in contrast to violent relaxation which does not separate particles according to their individual masses. In this section we use Chandrasekhar's dynamical friction formula to estimate orbital decay rates. These play an important role in galaxy formation and evolution.

(a) Circular Orbits Consider a subject mass on a circular orbit in a spherical, singular isothermal host halo with density distribution $\rho(r) = V_c^2/(4\pi G r^2)$, where V_c is the circular velocity which is independent of r. If we assume that the constituent 'field' particles all have mass m and follow a locally Maxwellian velocity distribution, the phase-space distribution function is

$$f(\mathbf{x}, \mathbf{v}) = f(r, v) = \frac{\rho(r)}{m} \frac{1}{(2\pi\sigma^2)^{3/2}} \exp\left(-\frac{v^2}{2\sigma^2} \right), \qquad (12.41)$$

where the velocity dispersion $\sigma = V_c/\sqrt{2}$. Consequently, in the limit $M_S \gg m$, dynamical friction results in a deceleration

$$\frac{d v_S}{dt} = -4\pi \frac{G^2 M_S}{v_S^2} \ln\Lambda \, \rho(r) \left\{ \mathrm{erf}\left(\frac{v_S}{V_c}\right) - \frac{2}{\sqrt{\pi}} \frac{v_S}{V_c} \exp\left[-\left(\frac{v_S}{V_c}\right)^2\right] \right\} \frac{\mathbf{v}_S}{v_S}. \quad (12.42)$$

Using that on circular orbits in a singular isothermal sphere $v_S = V_c$ and that the dynamical friction force is tangential, the rate at which the subject mass loses its specific angular momentum $L_S = r v_S$ is

$$\frac{d L_S}{dt} = r \frac{d v_S}{dt} = -0.428 \ln\Lambda \frac{G M_S}{r}. \quad (12.43)$$

Since the circular speed is independent of radius, the subject mass continues to orbit at speed V_c as it spirals inward, so that the radius of the orbit, r, changes with time as

$$r \frac{dr}{dt} = -0.428 \ln\Lambda \frac{G M_S}{V_c}. \quad (12.44)$$

For an initial orbit with radius r_i, the dynamical friction time (i.e. the time it takes for the orbit to decay to zero radius) is then

$$t_{\mathrm{df}} = \frac{1.17}{\ln\Lambda} \frac{r_i^2 V_c}{G M_S} = \frac{1.17}{\ln\Lambda} \left(\frac{r_i}{r_h}\right)^2 \left(\frac{M_h}{M_S}\right) \frac{r_h}{V_c}, \quad (12.45)$$

with r_h and M_h the radius and mass of the host system.

If we set the maximum impact parameter, $b_{\max}$, equal to the size of the host system, r_h, and we use that $\bar{v}_{\mathrm{enc}} \simeq V_c = \sqrt{G M_h / r_h}$, the Coulomb logarithm [Eq. (12.31)] becomes

$$\ln\Lambda \simeq \ln\left(\frac{M_h}{M_S}\right), \quad (12.46)$$

which is the form that is often used in estimates of orbital decay rates. If $r_i \sim r_h$, and for a dark matter halo in an EdS universe, we thus obtain

$$t_{\mathrm{df}} \approx \frac{1.17}{\ln(M_h/M_S)} \left(\frac{M_h}{M_S}\right) \frac{1}{10 H(z)}, \quad (12.47)$$

where we have used that, to good approximation, $r_h/V_c \approx 1/10 H(z)$ [see Eq. (7.137)]. Thus, the dynamical friction decay time from the edge of a halo to the center is longer than the age of the Universe for $M_h/M_S \gtrsim 15$: only the most massive subhalos and satellite galaxies in a dark matter halo are expected to be substantially segregated by mass.

(b) Eccentric Orbits Now consider the orbital decay of an eccentric orbit, whose eccentricity is defined as

$$e = \frac{r_+ - r_-}{r_+ + r_-}, \quad (12.48)$$

with r_+ and r_- the apo- and pericenter of the orbit, respectively. For an orbit with energy E and angular momentum L (both per unit mass) in a spherical potential, r_- and r_+ are the roots for r of

$$\frac{1}{r^2} + \frac{2[\Phi(r) - E]}{L^2} = 0 \quad (12.49)$$

(e.g. Binney & Tremaine, 1987). For a singular isothermal sphere, whose potential is given by $\Phi(r) = V_c^2 \ln(r/r_0)$, the maximum angular momentum for a given energy E is given by $L_c(E) = r_c(E) V_c$, with

$$r_c(E) = r_0 \exp\left[\frac{E - V_c^2/2}{V_c^2}\right], \quad (12.50)$$

the radius of a circular orbit with energy E. It is useful to express the angular momentum of an eccentric orbit in terms of the orbital circularity,

$$\eta \equiv \frac{L}{L_c(E)}, \tag{12.51}$$

which is a monotonically decreasing function of the orbital eccentricity (i.e. $d\eta/de < 0$). Circular orbits have $\eta = 1$ and $e = 0$, while purely radial orbits have $\eta = 0$ and $e = 1$.

Dynamical friction transfers both energy and angular momentum from the subject mass M to the field particles of mass m making up the host system. This causes a change in the orbital circularity given by

$$\begin{aligned}\frac{d\eta}{dt} &= \frac{1}{L_c(E)}\frac{dL}{dt} - \frac{L}{L_c^2(E)}\frac{\partial L_c(E)}{\partial E}\frac{dE}{dt} \\ &= \eta\left[\frac{1}{L}\frac{dL}{dt} - \frac{1}{V_c^2}\frac{dE}{dt}\right],\end{aligned} \tag{12.52}$$

where we have used Eq. (12.50). Using that $L = rv_\perp$, with $v_\perp$ the velocity in the direction perpendicular to the radial vector, we have that

$$\frac{dE}{dt} = v\frac{dv}{dt} \quad \text{and} \quad \frac{dL}{dt} = \frac{L}{v}\frac{dv}{dt}, \tag{12.53}$$

where dv/dt is the frictional deceleration given by Eq. (12.40). Substituting Eq. (12.53) in Eq. (12.52) we have that

$$\frac{de}{dt} = \frac{\eta}{v}\frac{de}{d\eta}\left[1 - \left(\frac{v}{V_c}\right)^2\right]\frac{dv}{dt}. \tag{12.54}$$

At pericenter $v > V_c$, and since $\eta > 0$, $de/d\eta < 0$ and $dv/dt < 0$ we thus have that $de/dt < 0$. However, at apocenter $v < V_c$ so that $de/dt > 0$. Thus, while dynamical friction causes an orbit to become more circular near pericenter, it causes an increase of the orbit's eccentricity near apocenter. Numerical simulations of the orbital decay of a solid body in a spherical halo with a realistic density distribution, show that the effects at apo- and pericenter almost cancel each other, so that there is virtually no *net* evolution in the orbital eccentricity (van den Bosch et al., 1999). The same simulations also show that the dynamical friction time scales with the orbit's circularity as $t_{df} \propto \eta^{0.53}$, indicating that more eccentric orbits decay more rapidly. These results agree quite well with those obtained by direct integration of Chandrasekhar's formula along eccentric orbits (White, 1976b)

(c) Orbital Decay in the Presence of Mass Loss In the above derivation of the dynamical friction time we have assumed that the subject mass M_S is constant. However, unless the subject mass is a compact object (e.g. a black hole), the tidal forces due to the host system, with potential Φ_h, will cause mass loss and thus a decrease of M_S with time (see §12.2). To estimate how tidal stripping impacts on the dynamical friction time, assume that the host system is a singular isothermal sphere with circular velocity V_h and mass profile $M_h(r) = V_h^2 r/G$. As above, we refer to the mass inside a radius r_h as the total host mass M_h. Now consider a subject mass, which has the density distribution of a singular isothermal sphere with circular velocity V_S. If located at a radius r within the host system, we assume that its mass distribution is truncated at its tidal radius

$$r_t = \left[\frac{M_S(r_t)}{2M_h(r)}\right]^{1/3} r, \tag{12.55}$$

(see §12.2.1). Hence, at radius r the subject mass is equal to

$$M_S(r) = \frac{V_S^2 r_t(r)}{G} = \frac{r}{\sqrt{2} G} \frac{V_S^3}{V_h}. \tag{12.56}$$

Substituting Eq. (12.56) in Eq. (12.44), and assuming that the subject mass starts on an initial circular orbit of radius r_i, we obtain a dynamical friction time

$$\tilde{t}_{df} = \frac{3.30}{\ln \Lambda} \left(\frac{r_i}{r_h}\right) \left(\frac{V_h}{V_S}\right)^3 \frac{r_h}{V_h}, \tag{12.57}$$

where the tilde is to indicate that this dynamical friction time accounts for mass loss due to tidal stripping. Since $(V_h/V_S)^3 \simeq (M_h/M_S)$, which follows from the fact that, according to the spherical collapse model, all virialized structures have similar densities, a comparison with the dynamical friction time in the absence of mass loss, given by Eq. (12.45), shows that $\tilde{t}_{df} \simeq 2.8 t_{df}$ for $r_i = r_h$. Thus, in the presence of mass loss due to tidal stripping, the dynamical friction time is expected to be almost three times as long as without mass loss. Note that we have assumed that the Coulomb logarithm is a constant along the orbit, and independent of whether mass loss occurs or not. Since $\ln \Lambda \sim \ln(M_h/M_S)$, the appropriate value depends on the stripping and also on which region of the host halo should be used to estimate M_h. Nevertheless, numerical simulations show that, for a satellite system with an initial mass ratio of $M_S/M_h = 0.1$, mass loss causes an increase in the dynamical friction time by a factor of 2–3 (Colpi et al., 1999; Boylan-Kolchin et al., 2008; Jiang et al., 2008), in good agreement with our estimate above. In addition, these simulations show that in the presence of mass loss the dynamical friction time scales with the orbit's circularity as $t_{df} \propto \eta^s$ with $s \sim 0.3$–0.4. This dependence is considerably weaker than in the absence of mass loss, reflecting the fact that stripping is more effective on orbits of small pericenter and partially counterbalances the enhanced friction.

12.3.2 The Validity of Chandrasekhar's Formula

Although the Chandrasekhar dynamical friction formula has often been used to estimate the decay times of satellite galaxies, globular clusters and supermassive black holes, it is important to be aware of its shortcomings. Its derivation is based on the following three assumptions:

(i) the subject mass and the field particles are point masses;
(ii) the self-gravity of the field particles can be ignored;
(iii) the distribution of field particles is infinite, homogeneous and isotropic.

None of these assumptions is realistic. In what follows we briefly describe how dynamical friction changes when the above assumptions are relaxed.

(a) Dynamical Friction on Extended Subject Mass In deriving Eq. (12.40) we have made the assumption that the subject mass M_S is a point mass. In general, though, we are interested in the dynamical friction experienced by extended objects, such as globular clusters, satellite galaxies or dark matter subhalos. When the impact parameter becomes smaller than or comparable to the size of the subject mass, Eq. (12.27) is no longer valid. Rather, as shown by White (1976a), in the case of an extended body Eq. (12.27) becomes

$$|\Delta v_{S,\parallel}| = 2 \frac{G^2 b^2 m}{(M_S + m) v_\infty^3} I_S^2(b), \tag{12.58}$$

with

$$I_S(b) = \int_b^\infty \frac{M_S(r) \, dr}{r^2 (r^2 - b^2)^{1/2}}. \tag{12.59}$$

Substituting Eq. (12.58) in Eq. (12.28), and repeating the same analysis as above, we obtain the same dynamical friction force as in Eq. (12.40), but with a Coulomb logarithm given by

$$\ln \Lambda = \frac{1}{M_S^2} \int_0^{b_{\max}} I_S^2(b)\, b^3\, db. \qquad (12.60)$$

Let r_t be the tidal radius of the subject mass, defined so that $M_S(r)$ is equal to the total mass M_S for $r \geq r_t$. Then we can rewrite Eq. (12.60) as

$$\ln \Lambda = \ln\left(\frac{b_{\max}}{k r_t}\right) \quad \text{with} \quad \ln\left(\frac{1}{k}\right) = \frac{1}{M_S^2} \int_0^{r_t} I_S^2(b)\, b^3\, db. \qquad (12.61)$$

This demonstrates that the Chandrasekhar dynamical friction formula can also be used for extended sources by simply replacing b_{90} in the Coulomb logarithm of Eq. (12.31) with $k r_t$. For realistic density distributions $0.1 \lesssim k \lesssim 0.3$ (White, 1976a).

(b) Self-Gravity of the Field Particles The derivation of Chandrasekhar's dynamical friction formula neglects the self-gravity of the host system, i.e. it only considers the interaction of the field particles with the subject mass, but not with one another. A more sophisticated treatment of dynamical friction, which does take the self-gravity of the background field particles into account, requires the linear response theory developed by Marochnik (1968) and Kalnajs (1972a). Rather than considering a sequence of uncorrelated two-body interactions, in linear response theory the subject mass is regarded as a moving, external potential which gives rise to a response density in the host system. Dynamical friction manifests itself as the gravitational force due to this response density, as obtained from the Poisson equation. An important difference between this derivation of dynamical friction and that due to Chandrasekhar is that in the latter case dynamical friction is considered a purely *local* effect. This is evident from the fact that in Chandrasekhar's formula the dynamical friction force is proportional to the local density $\rho(r)$ [see Eq (12.42)]. According to the linear response theory, however, dynamical friction is a global effect, arising from a global perturbation of the potential of the host system (e.g. Weinberg, 1986, 1989; Colpi, 1998). A clear demonstration that dynamical friction is not a purely local phenomenon comes from the fact that Eq. (12.40) predicts that a subject mass orbiting beyond the outer edge of a finite host system experiences no dynamical friction (i.e. the local density is zero). This, however, is inconsistent with numerical simulations, which clearly show that even in such a case the subject mass loses momentum due to the gravitational backreaction of the response density (e.g. Lin & Tremaine, 1983).

(c) Inhomogeneous Background and the Coulomb Logarithm Arguably the most problematic aspect of Chandrasekhar's dynamical friction formula is the introduction of the Coulomb logarithm, $\ln \Lambda$. It originates from the introduction of a maximum impact parameter, $b_{\max}$, required to avoid divergence. This divergence, however, arises because we made the (unrealistic) assumption of a homogeneous and infinite medium. In general we will be concerned with dynamical friction operating on a subject mass orbiting in a host system of mass $M_h \gg M_S$ (e.g. a galaxy in a dark matter halo). In this case, a logical value for $b_{\max}$ is the size of the host system, for which the Coulomb logarithm reduces to the form of Eq. (12.46). Note, however, that the derivation of Chandrasekhar's dynamical friction formula is based on Eq. (12.27), which is only valid for an infinite, homogeneous medium: in a finite host system the orbits will not be straight lines.

Despite this inconsistency, numerical simulations and linear response calculations have shown that Chandrasekhar's dynamical friction formula gives a reasonably accurate description of the dynamical friction experienced by a subject mass orbiting within a finite, inhomogeneous host system, provided that (i) the entire orbit lies inside the host system, (ii) mass loss is taken into account by replacing M_S with $M_S(t)$, (iii) the subject mass $M_S \lesssim M_h/10$, and (iv) the Coulomb logarithm is treated as a free parameter (e.g. Cora et al., 1997; Colpi et al., 1999; Velazquez &

White, 1999; Fujii et al., 2006). The latter implies that one can in general find a value for $\ln\Lambda$ for which integration of the equation of motion,

$$\frac{d\mathbf{v}_S}{dt} = -\nabla\Phi_h(\mathbf{r}) - 4\pi\frac{G^2 M_s \rho(<v_S)}{v_S^2}\ln\Lambda\frac{\mathbf{v}_S}{v_S} \qquad (12.62)$$

(with Φ_h the gravitational potential of the host system), yields a fairly accurate description of the orbital evolution. Unfortunately, the 'best-fit' value for $\ln\Lambda$ depends on the orbit and on the density distribution of the satellite. Currently, there is still no detailed understanding of how the Coulomb logarithm depends on these parameters, so that different semi-analytical models typically make different assumptions. Given the shortcomings of the Chandrasekhar dynamical friction formula, it is clear that the choice for $\ln\Lambda$ remains somewhat arbitrary.

Finally, we emphasize that in an inhomogeneous background density, there will be a net dynamical friction force in the direction perpendicular to $\mathbf{v}_S$, i.e. the velocity changes $\Delta\mathbf{v}_{s,\perp}$ of individual encounters [Eq. (12.26)] no longer sum to zero. In particular, any density gradient in the direction perpendicular to the motion of M_S induces a non-zero force (sometimes called the inhomogeneous dynamical friction force). For realistic density distributions, this inhomogeneous friction force is typically an order of magnitude smaller than the homogeneous friction force, and can therefore be ignored for most practical purposes (see e.g. Just & Peñarrubia, 2005).

12.4 Galaxy Merging

If the orbital energy is sufficiently low, close encounters between two systems can lead to a merger. In the hierarchical scenario of structure formation, mergers play an extremely important role in the assembly of galaxies and dark matter halos. After discussing the criteria under which mergers take place, we briefly discuss the demographics of mergers, the connection between mergers, starbursts and AGN activity, and the heating of galactic disks due to the accretion of satellite galaxies.

12.4.1 Criterion for Mergers

In order to examine the conditions under which an encounter can result in a merger, we closely follow Binney & Tremaine (1987) and consider a simple case in which the two galaxies are identical, non-rotating and spherical. Suppose that each galaxy has mass M and median radius $r_{\rm med}$. The internal mean-square velocity is $\langle v^2 \rangle = aGM/r_{\rm med}$, where a is a parameter of order unity that depends on the density distribution of the galaxy. Such an encounter is completely specified by $E_{\rm orb}$ (the orbital energy per unit mass) and L (the orbital angular momentum per unit mass) in units of the characteristic values derived from $\langle v^2 \rangle$ and $r_{\rm med}$:

$$\hat{E} \equiv \frac{E_{\rm orb}}{(1/2)\langle v^2\rangle} \quad \text{and} \quad \hat{L} \equiv \frac{L}{\langle v^2\rangle^{1/2} r_{\rm med}}. \qquad (12.63)$$

With this, each encounter is associated with a point in the $(\hat{E},\hat{L})$ plane which can be divided into different regions, as shown in Fig. 12.4. The line of parabolic orbits ($\hat{E}=0$) separates bound (elliptic) orbits, for which $\hat{E}<0$, from unbound (hyperbolic) orbits, for which $\hat{E}>0$. For bound orbits with a fixed $\hat{E}$, the largest angular momentum corresponds to a circular orbit. The corresponding locus $\hat{L}=\hat{L}_{\rm circ}(\hat{E})$ is indicated as a solid curve; no orbits can exist above this curve.

In principle, any bound orbit will eventually lead to a merger because the tidal interaction between the two galaxies always transfers orbital energy into internal energy. However, if the angular momentum is high and if the orbital energy is not low enough, the merger will not happen in a Hubble time. Numerical simulations show that $\hat{L} \lesssim 4$ is required for mergers from

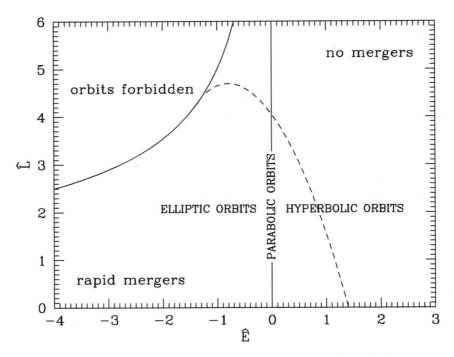

Fig. 12.4. Merging criteria for two spherical galaxies with the same mass. Orbits in the upper-left region are forbidden, because for a given orbital energy the largest possible angular momentum is that of a circular orbit (indicated by the solid curve). Encounters with orbital energy and/or angular momentum that are too high cannot lead to a merger. Mildly hyperbolic orbits can lead to a merger if the orbital angular momentum is sufficiently low. Mergers occur within a few galaxy dynamical times of first pericentric passage for encounters lying below and to the left of the dashed line. [After Binney & Tremaine (1987)]

parabolic encounters to happen within a few galactic dynamical times. Since tidal interactions can effectively drain orbital energy during a close encounter, mergers can also happen between two galaxies on an initially unbound orbit as long as $\hat{L}$ is sufficiently small. Numerical simulations of head-on encounters ($\hat{L} = 0$) show that mergers can still happen within a relatively small number of galaxy dynamical times for $\hat{E} \lesssim 1.4$. The dashed curve in Fig. 12.4 roughly delineates the region where the time between pericentric passage and merging is short enough to be relevant for real systems.

There are two important conclusions to draw from Fig. 12.4. First of all, mergers are only expected to be effective when $\hat{E} \lesssim 1$. If the galaxies are moving in a system with velocity dispersion σ, then the specific orbital energy for a typical encounter is $E_{\rm orb} \sim \sigma^2$. Thus, mergers are only effective in systems with a velocity dispersion smaller than or comparable to the internal velocities of the orbiting galaxies. Hence galaxy mergers can occur effectively in groups of galaxies, but not in rich clusters.[2] Secondly, for given orbital energy and angular momentum, mergers are expected to be more effective for more extended objects (i.e. with larger $r_{\rm med}$ and thus smaller $\hat{L}$). This means that it is crucial to take account of the fact that galaxies are believed to reside in extended dark matter halos. When two such systems have an encounter, their extended halos may merge to form a common halo even if the two systems are on a mildly hyperbolic orbit. The two galaxies themselves, however, being more compact than the halos, may not merge at the same time as their halos, but only later (if at all) when their distance and relative velocity at pericenter become comparable to their radii and internal velocities. Thus, the two galaxies are expected to

---

[2] Mergers onto the central cluster galaxy are an important exception here; see §12.5.2.

orbit in a common halo until dynamical friction and tidal interactions have removed sufficient orbital energy for the galaxies to merge.

12.4.2 Merger Demographics

Contrary to the cases of high-speed encounters and dynamical friction, for which reasonably accurate analytical descriptions are available, mergers between systems of comparable mass typically cannot be treated analytically. When two systems merge, their orbital energy is transferred to the internal energy of the merger product. In addition, some of the orbital energy can be carried away by material ejected from the progenitors (for instance in the form of tidal tails). In the case of galaxies embedded in extended dark matter halos, a significant fraction of the orbital energy of the stellar components can be transferred to the dark matter by dynamical friction. Due to the strong tidal perturbations and the exchange of energy between components, the system needs to settle to a new (virial) equilibrium after merging, which it does via violent relaxation, phase mixing and Landau damping of global modes (see Barnes, 1998, for a detailed review). As discussed in §5.5, the outcome of such relaxation is difficult to predict theoretically, and one has to resort to numerical simulations.

The first self-consistent three-dimensional simulations of encounters of galaxies, with and without rotation, were carried out by White (1978, 1979). These considered the collisions of two (or four) spherical, self-gravitating galaxies of equal mass on parabolic orbits, and demonstrated that interpenetrating encounters from such orbits generally lead to mergers within a few dynamical times, and that the structure of the merger remnants converges towards a density profile similar to that of elliptical galaxies. Since the early 1980s, N-body merger simulations of ever increasing numerical resolution have been performed. Gerhard (1981), Farouki & Shapiro (1982), Negroponte & White (1983), Barnes (1988), and Hernquist (1992) all carried out fully self-consistent merger simulations of two equal mass stellar disks embedded in extended dark matter halos. These simulations showed that mergers between disk galaxies of roughly equal mass produce remnants that resemble elliptical galaxies. This is also supported by observations. For example, Fig. 12.5 shows the heavily distorted galaxy NGC 7252, which reveals multiple

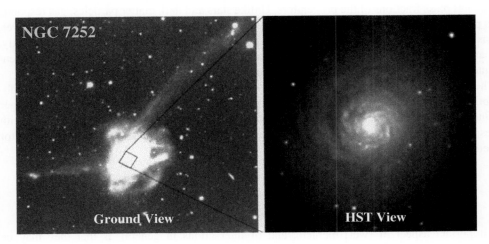

Fig. 12.5. The galaxy NGC 7252, whose distorted morphology, including several tidal tails, leaves little doubt that this is a remnant of a relatively major merger that occurred not too long ago. The right-hand panel shows a view of the central region obtained with the HST. It reveals the presence of a spiral structure, which formed out of the gas that has fallen towards the center during the merger. This gas has accumulated in a disk, and is forming stars. [Left Credit: F. Schweizer, taken with 4-meter telescope at the Cerro Tololo Inter-American Observatory. Right Credit: B. Whitmore and NASA]

tidal features and fine structure. There is little doubt that this is a system that has recently undergone a merger, as numerical simulations of mergers between two disk galaxies can reproduce various properties of this system in detail (e.g. Borne & Richstone, 1991; Hibbard & Mihos, 1995). The main part of NGC 7252 appears to have already relaxed towards an equilibrium state with properties similar to those of elliptical galaxies: its light profile is well fitted by an $R^{1/4}$ law (Schweizer, 1982) and it even obeys the Faber–Jackson relation (Lake & Dressler, 1986). These findings suggest that some, if not all, elliptical galaxies are the remnants of mergers between two or more disk galaxies. This idea, originally due to Toomre & Toomre (1972), will be explored in more detail in §13.2.2, where we also present a more detailed comparison of the properties of merger remnants in numerical simulations with the observed properties of elliptical galaxies.

A large number of more recent studies based on high-resolution numerical simulations (e.g. Barnes, 1992; Hernquist, 1992, 1993; Barnes & Hernquist, 1996; Dubinski et al., 1996; Springel & White, 1999; Naab & Burkert, 2003; Boylan-Kolchin et al., 2005; Cox et al., 2006) have made clear that the structure of the remnant of a merger between two galaxies depends primarily on four properties:

- The progenitor mass ratio $q \equiv M_1/M_2$, where $M_1 \geq M_2$. If $q \lesssim 4$ ($q \gtrsim 4$) one speaks of a major (minor) merger.[3] In major mergers violent relaxation plays an important role during the relaxation of the merger remnant, and the remnant typically has little resemblance to its progenitors. In minor mergers, on the other hand, phase mixing and/or Landau damping dominate, and the merger is less destructive. Consequently, the remnant of a minor merger often resembles its most massive progenitor.
- The morphologies of the progenitors (disks or spheroids). Galactic disks are fragile, and are therefore relatively easy to destroy, especially when q is small. Disks that accrete small satellites (i.e. in the minor merger regime with $q \gtrsim 10$) typically survive the merger event but can undergo considerable thickening (see §12.4.4 below). As discussed in §12.2.2, mergers that involve one or more disk galaxies tend to create tidal tails, which are absent in mergers between two spheroids.
- The gas mass fractions of the progenitors. Unlike stars and dark matter particles, gas responds to pressure forces as well as gravity and can lose energy through radiative cooling. Moreover, gas flows develop shocks whereas streams of stars can freely interpenetrate. Consequently, mergers between gas-rich progenitors (often called 'wet' mergers) can have a very different outcome from mergers between gas-poor progenitors ('dry' mergers).
- The orbital properties. The orbital energy and angular momentum not only determine the probability for a merger to occur (as discussed above), but also have an impact on the merger outcome. For example, as discussed in §12.2.2, the relative orientation of the orbital spin with respect to the intrinsic spins of the progenitors (prograde or retrograde) is an important factor determining the prominence of tidal tails.

12.4.3 The Connection between Mergers, Starbursts and AGN

An important aspect of (wet) mergers, and of interactions in general, is that they may be responsible for triggering (nuclear) starbursts and AGN activity. Numerical simulations of mergers and encounters between gas-rich disks show that the tidal perturbations can cause the disks to become globally unstable and to develop pronounced bars. Since the gas and stars do not have the same response to the tidal force, the gaseous and stellar bars generally

---

[3] Note that the exact value at which one distinguishes major from minor mergers is somewhat arbitrary.

have different phases (see §11.5.4). This phase difference gives rise to torques that can effectively remove angular momentum from the gas. As a result, the gas flows towards the central region and eventually forms a dense gas concentration at the center of the merger remnant (e.g. Negroponte & White, 1983; Noguchi, 1988; Combes et al., 1990b; Mihos & Hernquist, 1996).

The surface density of this central gas can easily be several orders of magnitude higher than the typical surface density of gas in (undisturbed) disk galaxies. A simple application of the Schmidt law of star formation (see §9.5.1) suggests that this gas may form stars at an extremely high rate, giving rise to a nuclear starburst (see also §2.3.7). In addition, if some of the nuclear gas can continue to lose angular momentum, it may be able to fuel or even form a central black hole, resulting in an active galactic nucleus (e.g. Hernquist, 1989; Barnes & Hernquist, 1991).

There is strong observational support that some starbursts and active galaxies are triggered by mergers and/or encounters between gas-rich galaxies. Using integrated colors, Larson & Tinsley (1978) showed that many interacting and merging galaxies have undergone short bursts ($\lesssim 10^8$ yr) of star formation involving up to $\sim 5\%$ of their total luminous mass. All ULIRGs, which are starbursting galaxies with far-infrared luminosities exceeding $10^{12} L_\odot$ (see §2.3.7), show clear signs of recent or continuing interactions, such as shells, tidal tails and complex velocity fields (Joseph & Wright, 1985). In addition, most ULIRGs have dominant non-thermal optical emission lines, indicative of an embedded active galactic nucleus (Sanders et al., 1988). Many bright radio galaxies also reveal structural peculiarities indicative of a recent interaction. Together with the absence of nearby companions, this suggests that some radio galaxies may have resulted from mergers of disk galaxies (e.g. Heckman et al., 1986).

Although many details are still poorly understood, it is clear from these and other observations that the presence of gas in a merger between two galaxies can result in enhanced star formation and/or AGN activity. Important outstanding questions are what fraction of the present-day stars formed in merger-induced starbursts (see e.g. Somerville et al., 2001) and what is the role of feedback in regulating and terminating the starburst and AGN activity (see e.g. Springel et al., 2005b).

12.4.4 Minor Mergers and Disk Heating

As discussed above, because of the highly nonlinear effects involved, mergers between galaxies and/or dark matter halos are best studied using numerical simulations. In the minor merger limit of large mass ratios $q = M_1/M_2$, though, the most massive progenitor will only be mildly disturbed, and it is still possible to use simple approximations to estimate the impact of the merger event[4] on the structural and kinematical properties of the massive progenitor. As an example, we discuss the (vertical) heating of a galactic disk due to the accretion of a low mass satellite galaxy.

Consider a satellite system of mass M_s (e.g. a dark matter subhalo, with or without an embedded satellite galaxy) orbiting within a halo of mass $M_h \gg M_s$ that hosts a central disk galaxy of mass M_d. As we have seen in §12.3, the subhalo experiences a drag force due to dynamical friction by the halo–disk particles and loses orbital energy. Because of energy conservation, this energy loss of the satellite must be equal to the energy pumped into the halo and the disk. As a result, both the halo and the disk will be heated up. We expect the ratio of the energies transferred to the disk and the halo to be of the order of the mass ratio of the energy sinks, i.e.

$$\eta \equiv \frac{\Delta E_h}{\Delta E_d} \sim \frac{M_h}{M_d}, \qquad (12.64)$$

which is also borne out by a more careful analysis (see Tóth & Ostriker, 1992).

---

[4] Note that in the case of a large mass ratio one also often speaks of an accretion event.

For simplicity, we assume that the disk plus halo system can be described by a singular isothermal sphere, whose potential is given by $\Phi_h(r) = V_c^2 \ln(r)$, with $V_c^2 = GM_{\rm vir}/r_{\rm vir}$ the circular velocity of the system, $M_{\rm vir} = M_h + M_d$, and $r_{\rm vir}$ the halo's virial radius. Since the rotation curve of the disk–halo system is flat (consistent with observations), the velocity of the satellite does not change significantly as it spirals towards the center. Hence, the change in the energy of the sinking satellite comes only from the change in potential energy. If the satellite starts out at the halo virial radius, $r_{\rm vir}$, and is (instantaneously) dissolved (due to tidal forces) at a radius $r_{\rm diss}$, the energy deposited into the disk is

$$\Delta E_d = m_d \Delta E_s = \mathscr{R} m_d M_s V_c^2, \tag{12.65}$$

with $m_d \equiv M_d/M_{\rm vir} = 1/(1+\eta)$ and

$$\mathscr{R} = \ln\left(\frac{r_{\rm vir}}{r_{\rm diss}}\right). \tag{12.66}$$

The actual mechanism by which the accreting satellite deposits its energy into the disk is resonant excitation, either of individual stellar orbits (direct heating; see Spitzer, 1958) or of collective modes (both vertical bending waves and horizontal density waves). These waves subsequently experience Landau damping (see §5.5.4) so that the energy deposited into the disk is typically distributed over large scales (Sellwood et al., 1998). We therefore do not make a large error if we simply assume that ΔE_d is distributed equally among all disk stars, so that the energy deposited into the disk per unit area is

$$\Delta e_d = \frac{\Sigma_d(R)}{M_d} \Delta E_d = m_d \mathscr{R} \Sigma_d(R) \frac{M_s}{M_d} V_c^2. \tag{12.67}$$

The disk heating energy will be shared by the energy increases in the vertical and planar directions. Since we are mainly interested in the effect on the disk thickness, we are only concerned with the former. For simplicity we assume that the energy is shared equally among all directions, so that $\Delta e_{d,z} = \Delta e_d/3$. As shown by Tóth & Ostriker (1992), this is a reasonable approximation.

Part of $\Delta e_{d,z}$ is stored in kinetic energy in the vertical direction, while the rest is stored in potential energy as the disk thickens. To estimate the changes in disk thickness and in stellar velocity dispersion, we use the virial theorem. A small part of the disk can be regarded as a one-dimensional system in the external potential of the halo, and the virial theorem can be applied to a unit surface area. Since the local gravitational potential balancing the local vertical velocity dispersion, σ_z, is dominated by disk self-gravity (rather than by halo gravity), we can ignore the potential energy associated with the disk–halo gravity. For a thin disk, the thickening may be treated locally as a one-dimensional problem in the vertical direction, and only the z dependence of the disk potential at a given radius matters. Setting the zero-point of the potential (i.e. that corresponding to zero thickness) to be zero for each R independently, the virial theorem can be written as

$$2t_d - w_{dd} \approx 0 \tag{12.68}$$

(Tóth & Ostriker, 1992). Here

$$t_d \equiv \frac{1}{2}\Sigma_d(R)\sigma_z^2(R), \tag{12.69}$$

is the kinetic energy in the vertical direction per unit area, and

$$w_{dd} \equiv \frac{1}{2}\int_{-\infty}^{\infty} \Phi_d(R,z)\rho_d(R,z)\,dz = \alpha G \Sigma_d^2(R) z_d \tag{12.70}$$

is the potential energy per unit area associated with the disk self-gravity, with Φ_d and ρ_d the potential and density of the disk, z_d the disk scale height, and α a geometrical factor of order unity. Using that $e_{d,z} \approx t_d + w_{dd}$, we have that $\Delta e_{d,z} = (3/2)\Delta w_{dd}$. Assuming that $\Sigma_d(R)$ is unaffected by the heating, and using Eq. (12.67), we obtain that

$$\frac{\Delta z_d}{z_d}(R) = \frac{1}{3}\frac{\Delta \sigma_z^2}{\sigma_z^2}(R) = \frac{2\mathscr{R}}{9} m_d \frac{M_s}{M_d}\left[\frac{V_c}{\sigma_z(R)}\right]^2, \quad (12.71)$$

where we have used that $z_d = \sigma_z^2(R)/[\alpha G \Sigma_d(R)]$, which follows from the virial theorem. To gain further insight, we assume that the initial disk has a density distribution

$$\rho_d(R,z) = \rho_0 \exp(-R/R_d)\operatorname{sech}^2(z/2z_d). \quad (12.72)$$

As shown in §11.1.1, the velocity dispersion in the z direction is then given by Eq. (11.21), which allows us to rewrite Eq. (12.71) as

$$\frac{\Delta z_d}{z_d}(R) = \frac{2}{9}\frac{\lambda}{\sqrt{2}}\ln\left(\frac{\sqrt{2}}{\lambda}\right)\frac{R_d}{z_d}\frac{M_s}{M_d}\exp(R/R_d), \quad (12.73)$$

where we have used that $R_d/r_{\rm vir} = \lambda/\sqrt{2}$, with λ the spin parameter of the disk (see §11.2.2), and we have assumed that $r_{\rm diss} \simeq R_d$. As is evident from Eq. (12.73), disk thickening is more effective (i) for more massive satellites, (ii) at larger radii, (iii) for disks with a larger spin parameter (i.e. with a larger disk scale length and lower surface density), and (iv) for disks that are initially thinner (i.e. have a larger value of R_d/z_d). The latter implies that disk thickening saturates, i.e. that the accretion of subsequent satellites of similar mass has a smaller effect on the scale height of the disk. As a specific example, for $\lambda = 0.05$ and $z_d/R_d = 0.1$, which are typical values, Eq. (12.73) reduces to

$$\frac{\Delta z_d}{z_d}(R = 2R_d) \simeq 1.4 \frac{M_s}{M_d}. \quad (12.74)$$

Thus, the accretion of satellites with $M_s \lesssim M_d/10$ only causes a relatively small amount of heating. However, if the disk accretes a satellite with a mass that is comparable to that of the disk itself, its scale height at $2R_d$ more than doubles. As argued by Tóth & Ostriker (1992), the fact that the disk of our Milky Way has a thin component with a scale height of $\sim 0.3\,\mathrm{kpc}$ at the solar radius ($R_\odot \simeq 8\,\mathrm{kpc} \simeq 2.3 R_d$), seems to suggest that it cannot have accreted more than a few percent of its stellar mass inside the solar radius within the last 5 Gyr.

Although the analytical treatment presented above is based on numerous assumptions and oversimplifications, detailed numerical simulations have largely confirmed its predictions, both qualitatively and quantitatively (e.g. Walker et al., 1996; Huang & Carlberg, 1997; Velazquez & White, 1999; Kazantzidis et al., 2008; Purcell et al., 2009). In addition, these simulations have shown that the accretion of satellite systems produces distinctive morphological features in the disk, including bars, warps, and low surface brightness features similar to those detected in the Milky Way (e.g. Newberg et al., 2002; Jurić et al., 2008) and the outskirts of M31 (e.g. Ferguson et al., 2002). Furthermore, the thickened disks in the simulations have structural and kinematical properties similar to those of the thick disk in the Milky Way (Villalobos & Helmi, 2008). Depending on the density and orbit of the satellite, the core of the satellite may survive the merger and sink towards the galactic center to form a bulge-like entity (e.g. Tremaine et al., 1975; Aguerri et al., 2001). Alternatively, if it is tidally disrupted before it reaches the center, its stars may contribute to the stellar component of the (thickened) disk (e.g. Abadi et al., 2003; Peñarrubia et al., 2006). Since thick disks, bars, warps and low surface brightness features are ubiquitous among disk galaxies, it seems likely that minor mergers play an important role in galaxy evolution along the Hubble sequence.

However, the disk heating associated with the accretion of satellite systems may also signal a potential problem for the hierarchical model of structure formation. As we have seen in §7.5.3, dark matter halos are expected to host large populations of subhalos (with or without satellite galaxies). Given that each of these subhalos may cause some amount of disk heating, this raises the question whether thin disks are expected to survive in a CDM cosmology. If the total heating rate due to the cumulative effect of the entire population of subhalos expected in the host halo of a disk galaxy is large, the ubiquity of observed thin disks may indicate a serious problem for the CDM paradigm.

In order to compute the net heating rate, we first compute the heating rate due to a single satellite. The time scale on which a satellite system transfers (part of) its orbital energy to the disk is the dynamical friction time, t_{df}, given by Eq. (12.45). Thus we have that

$$\frac{dE_d}{dt}(M_s) = \frac{\Delta E_d}{t_{df}} \simeq 8.5 \mathscr{R} m_d \ln \Lambda \frac{GM_s^2}{r_{vir}} H(z), \qquad (12.75)$$

where we have used Eq. (12.65) and the fact that, to good approximation, $r_h/V_c \approx 1/10H(z)$ [see Eq. (7.137)]. Thus, we have that the *total* heating rate in a host halo of mass M_h, due to the cumulative effect of *all* dark matter subhalos, is

$$\frac{dE_d}{dt} \propto \int n(M_s|M_h) M_s^2 \, dM_s, \qquad (12.76)$$

with $n(M_s|M_h)$ the mass function of dark matter subhalos that are accreted by a host halo of mass M_h (i.e. the unevolved subhalo mass function; see §7.5.3). For CDM cosmologies $n(M_s|M_h) \propto M_s^{-\gamma}$, with $\gamma \simeq 1.8$. Hence, the heating rate will be dominated by the few most massive satellite systems, making disk heating a stochastic process. Using numerical simulations of structure formation, Stewart et al. (2008) have shown that $\sim 70\%$ of Milky Way sized dark matter halos with a mass of $\sim 10^{12} M_\odot$ have accreted a system of mass $M_s \simeq 10^{11} M_\odot \simeq 3M_d$ in the last 10 Gyr. Since such a merger event largely destroys a thin disk (Purcell et al., 2009), the only recourse for explaining the ubiquity of thin disks in a CDM universe is that the disks have a significant gas component. Since gas can dissipate, it can have a stabilizing effect on the stellar disk, and it can even reform a thin disk after the heating event (e.g. Robertson et al., 2006a; Moster et al., 2009). Detailed hydrodynamical simulations of disk formation and evolution are required to investigate whether or not the CDM paradigm is consistent with the existence of a large population of (thin) disk galaxies.

12.5 Transformation of Galaxies in Clusters

As discussed in §2.4.5, denser environments host larger fractions of galaxies morphologically classified as early types. In addition, galaxies in denser environments are on average redder, less gas-rich, and have lower specific star-formation rates. Although far from unambiguous, this strong environment dependence is often interpreted as indicating that galaxies undergo transformations (e.g. late type → early type, star forming → passive) once they enter or become part of a denser environment. Clusters of galaxies are the largest virialized structures in the Universe, with masses of about 10^{14}–$10^{15} M_\odot$, and velocity dispersions of about $1,000 \, \mathrm{km \, s^{-1}}$, and the environments with the highest number densities of galaxies. Hence, galaxy interactions are frequent, making clusters the ideal environments to look for possible transformation processes.

In this section we take a closer look at various processes that operate in clusters of galaxies, and that may be responsible for transforming star-forming disk galaxies into passive spheroids. Roughly speaking, cluster galaxies can be affected by the cluster environment in three different ways: (i) tidal interactions with other cluster members and with the cluster potential, (ii) dynamical friction, which causes the galaxy to slowly make its way to the cluster center, and

(iii) interactions with the hot, X-ray emitting intracluster medium (ICM) that is known to permeate clusters (see §8.8.1). In what follows we discuss each of these processes in turn, focusing on their potential impact on the properties of the galaxies.

12.5.1 Galaxy Harassment

In a cluster of galaxies, the typical velocity of a galaxy is of the order of the velocity dispersion of the cluster, which is much larger than the internal velocity dispersion of the galaxy. Thus, the encounters of a galaxy with other member galaxies and cluster substructures are all high-speed in nature. As we have seen in §12.1, during a high-speed encounter, a colliding galaxy is impulsively heated (i.e. its internal energy is increased). As a result, the perturbed galaxy becomes less bound, and more vulnerable to disruptions by further encounters and by tidal interactions with the global cluster potential. The cumulative effect of multiple high-speed impulsive encounters is generally referred to as galaxy harassment. Early studies of this process focused on how it modifies the structure of elliptical galaxies in clusters and whether it can account for the extended outer envelopes of central cluster galaxies (Richstone, 1976; Aguilar & White, 1985). A particularly interesting result by Aguilar & White (1986) was that the de Vaucouleurs surface brightness profile appears invariant to harrassment, i.e. rapid encounters between galaxies with this structure can cause substantial mass loss but the profile of the final object is still well fit by the $r^{1/4}$ law; there is no tendency for harrassment to produce a tidal cut-off in the density profiles of cluster galaxies.

The effects of harassment on disk galaxies falling into clusters was first simulated by Farouki & Shapiro (1981), who showed that disks can be almost entirely destroyed by one or two passages through the cluster. These results were confirmed and extended using much larger simulations by Moore et al. (1996) and Moore et al. (1998b) who highlighted the damage caused to the fragile disks of late-type (Sc-Sd) spiral galaxies. If such galaxies experience several close encounters with relatively massive cluster members, they may lose very substantial amounts of mass as impulsive heating pushes stars onto unbound orbits. The disk stars that remain bound to the galaxy are also heated, causing a transformation of the (dynamically cold) disk into a spheroidal component closely resembling a dwarf elliptical. Since dwarf ellipticals are ubiquitous in clusters, it may well be that they are the remnants of the disk galaxies that have experienced such harassment. This is consistent with the fact that the galaxy population of clusters is observed to have evolved rapidly over the past few billion years: at redshifts $z \sim 0.4$, clusters contain a large population of star-forming galaxies, many of which are disturbed and show evidence for multiple bursts of star formation (see §2.5.1). This population of 'Butcher–Oemler' galaxies is almost entirely absent from clusters at $z \sim 0$ (Butcher & Oemler, 1978). As discussed in §13.6.2, however, this scenario is made less plausible by the fact that many dwarf ellipticals show little or no rotation, while the objects formed in numerical simulations by harassing disk galaxies always seem to preserve significant amounts of rotation.

Although harassment may have a strong impact on the morphology of late-type (Sc-Sd) disk galaxies, which typically have relatively low surface density disks, numerical simulations indicate that it has little impact on the more compact early-type (Sa-Sb) disk galaxies (Moore et al., 1999b). This is easy to understand from the fact that the dynamical time in denser, more compact disks is shorter. In Sa and Sb galaxies the orbital time within a couple of disk scale lengths is short enough for the disk to respond adiabatically to the high-speed encounters experienced in clusters. In addition, since denser systems have a smaller fraction of their mass located beyond the tidal radius, they are also less susceptible to tidal stripping. Tidal shocks therefore can neither remove large amounts of material from early-type disk galaxies, nor transform them into spheroids. Nevertheless harassment can still significantly heat the disks and drive disk instabilities that funnel gas into the central regions. Combined with ram-pressure stripping (see §12.5.3), it is thus plausible that harassment can transform Sa-Sb galaxies into S0 galaxies.

12.5.2 Galactic Cannibalism

As discussed in §12.4.1, galaxies in clusters are unlikely to merge since their encounter speed is typically much larger than their internal velocity dispersion. However, there is one important exception: because of dynamical friction, galaxies lose energy and momentum which causes them to 'sink' towards the center of the potential well. If the dynamical friction time is sufficiently short, the galaxy will ultimately reach the cluster center, where it will merge with the central galaxy already residing there. Hence, a central cluster galaxy may accrete satellite galaxies, a process called galactic cannibalism.

Roughly speaking, a satellite galaxy will be cannibalized by its central galaxy when dynamical friction can bring it to the center of the potential well within a Hubble time $1/H(z)$. From Eq. (12.45) we see that the critical radius for this to occur to a satellite galaxy of mass M_S in an isothermal host halo of mass M_h is

$$r_{\rm crit} \sim 0.1 r_{\rm h} \left(\frac{\ln \Lambda}{10}\right)^{1/2} \left(\frac{M_S}{10^{-4} M_{\rm h}}\right)^{1/2}, \qquad (12.77)$$

where we have used $V_{\rm h}/r_{\rm h} = 10 H(z)$ [see Eq. (7.137)]. Thus, although it is preferentially the massive satellites that will be cannibalized, in the central regions of a cluster even low-mass galaxies can be accreted by the central galaxy within a Hubble time.

Galactic cannibalism has two important effects: it causes the central galaxy to increase in mass, and it causes a depletion of massive satellite galaxies, for which the dynamical friction time is the shortest. Consequently, cannibalism causes an increase of the magnitude difference, $\Delta \mathcal{M}_{12}$, between the brightest and second brightest member of a cluster. If galactic cannibalism is the main mechanism regulating the luminosity of the central galaxy, the magnitude gap $\Delta \mathcal{M}_{12}$ can thus be used as a measure for the dynamical age of the cluster: older systems will have a larger magnitude gap.

As discussed in §2.5.1, the central cluster galaxy is typically also the brightest cluster galaxy and often has an extraordinarily diffuse and extended outer envelope, in which case it is called a cD galaxy (see Oemler, 1976; Schombert, 1986, for detailed photometric properties). Numerous studies have suggested that these cD galaxies are the product of galactic cannibalism (e.g. Ostriker & Tremaine, 1975; White, 1976b; Hausman & Ostriker, 1978; Malumuth & Richstone, 1984). This would explain not only their large masses, but also their diffuse envelopes, which, in this picture, consist of material tidally stripped from the cannibalized galaxies as they spiral into the cluster center (Gallagher & Ostriker, 1972; Richstone, 1976). Some of the assumptions made in these early models were poorly chosen and led to overestimates of the efficiency of the process. For example, most studies ignored mass loss from the satellite galaxies due to tidal stripping [as does Eq. (12.77)]. As we have seen in §12.3.1, this can increase dynamical friction times by a factor of several, a correction that results in current cD growth rates too low to explain the observed luminosities (Merritt, 1985), but in excellent agreement with the rates inferred from the fraction of cD galaxies harboring multiple nuclei (Lauer, 1988; Merrifield & Kent, 1991; Blakeslee & Tonry, 1992). This is not fatal, since clusters have been growing at the same time as their central galaxies, and merging onto the central object should have occurred more rapidly in lower mass progenitors than in today's cluster. As noted by Merritt (1984), this implies that cannibalism must be followed throughout the hierarchical growth of the cluster, rather than just in the present-day system. The merger tree techniques discussed in §7.3 make it possible to solve this problem in detail in a CDM cosmology. Current results show that the cannabalism model provides a detailed explanation for most properties of the cluster/cD population (Aragon-Salamanca et al., 1998; De Lucia & Blaizot, 2007), although there is still some tension with observational results suggesting that cDs built up most of their mass at earlier times than the model predicts (Collins et al., 2009).

12.5.3 Ram-Pressure Stripping

When a galaxy moves through the intracluster medium (ICM), its gas component experiences a ram pressure, just like one feels wind drag when cycling. As first discussed by Gunn & Gott (1972), if the ram pressure is sufficiently strong, it may strip the gas initially associated with the galaxy.

Consider a disk galaxy of radius R_d moving through an ICM of density ρ_{ICM} with velocity V. For simplicity, assume that the velocity vector of the disk galaxy is pointing perpendicular to the disk, so that it experiences a face-on wind. The amount of ICM material swept per unit time by the disk is $\pi R_d^2 \rho_{ICM} V$. If we assume that the wind is stopped by the interstellar medium (ISM) of the disk, then the momentum transferred to the disk per unit time is $\pi R_d^2 \rho_{ICM} V^2$, corresponding to a ram pressure $P_{ram} = \rho_{ICM} V^2$. If this pressure exceeds the force per unit area that binds the ISM to the disk, the gas will be stripped. To estimate the binding force, assume that the mean surface density of interstellar gas is Σ_{ISM} and that the mean mass density (usually dominated by stars) of the disk is $\Sigma_\star$. The gravitational field of the disk is approximately $2\pi G \Sigma_\star$ in the disk and so the gravitational force per unit area on the interstellar gas is $2\pi G \Sigma_\star \Sigma_{ISM}$. Hence, we expect that stripping occurs if

$$\rho_{ICM} > \frac{2\pi G \Sigma_\star \Sigma_{ISM}}{V^2}. \tag{12.78}$$

To put this in perspective, consider a disk similar to that of the Milky Way, with a stellar mass of $5 \times 10^{10} M_\odot$ and an ISM mass of $5 \times 10^9 M_\odot$, both spread over a disk of radius 10 kpc. Suppose that this disk is moving at a speed $V = 1,000 \, \text{km s}^{-1}$ (the typical velocity dispersion of galaxies in rich clusters) with respect to an ICM. Eq. (12.78) then gives $\rho_{ICM} > 4.6 \times 10^{-27} \, \text{g cm}^{-3}$ as the condition for ram-pressure stripping to occur. The typical density of the ICM in clusters is $\sim 10^{-3}$ particles per cubic centimeter (see §8.8.1), or $\sim 10^{-27} \, \text{g cm}^{-3}$, of the same order as required for ram-pressure stripping to be effective. In general, a disk galaxy will be on an eccentric orbit along which its velocity and the ICM density change as function of time. Hence, whether condition (12.78) is satisfied or not depends on time. In general, though, since the surface densities of stars and gas decline as a function of galactocentric distance, one can typically identify a radius in the disk beyond which ram-pressure stripping is efficient.

When a spiral galaxy loses most of its interstellar gas, its potential for future star formation is greatly reduced. Ram-pressure stripping is therefore often invoked to explain why dense environments, such as clusters, reveal a clear deficit of gas-rich, star-forming galaxies (see §2.4.5). With (most of) its interstellar medium removed, and with star formation quenched, the resulting disk galaxy may look like a S0 galaxy. Thus, ram-pressure stripping may explain why clusters contain a larger fraction of S0 galaxies than the field.

However, the importance of ram-pressure stripping for transforming spirals into S0s, and for quenching star formation, is still a matter of debate. Although there is ample observational evidence that ram-pressure stripping is occurring (see van Gorkom, 2004, for a review), in almost all cases only the gas at relatively large galactocentric radii is being stripped. This is consistent with numerical, hydrodynamical simulations, which show that if a Milky-Way-like galaxy falls through the center of the Coma cluster, only about 80% of its gas mass is stripped; the inner 20% survives a plunge through the densest region of the ICM (Abadi et al., 1999). Furthermore, it is not even clear that ram-pressure stripping necessarily results in a reduction of the galaxy's star-formation rate. The remaining non-stripped gas may actually be compressed by the ram pressure, giving rise to enhanced star formation in the disk (Dressler & Gunn, 1983; Gavazzi et al., 1995), or the stripped gas may remain bound to the galaxy, fall back and induce a later starburst (Vollmer et al., 2001). Somewhat surprisingly, the effect of ram-pressure stripping seems to be enhanced if the gas disk is porous, so that part of the wind can stream through the holes. Although this results in less direct momentum transfer to the gas disk, streaming through the holes can ablate their edges through turbulent viscous stripping, and can prevent the stripped gas from falling back. Numerical simulations suggest that ram-pressure stripping may in some cases strip porous disks of their entire gas reservoir within $\sim 10^8$ yr (Quilis et al., 2000), resulting in rapid truncation of their star formation. However, it should be noted that the apparent HI holes in real disk galaxies may not be empty, but rather filled with molecular gas. In this case, the flow cannot pass through the disk, and ram-pressure stripping remains inefficient in removing the inner material.

12.5.4 Strangulation

The ram-pressure stripping discussed above may strip a galaxy of its entire cold gas reservoir, causing an abrupt quenching of its star formation. However, if only the outer parts of a galaxy's gas disk are stripped, star formation may continue until all fuel is exhausted. In normal field spirals, the gas consumption time scale is typically only a few Gyrs (see §9.3). Their lifetimes can be significantly extended only if they continue to accrete fresh gas. Galaxies are believed to be surrounded not only by halos of dark matter, but also by substantial amounts of hot/warm gas. This consists in part of gas that is just falling onto the system for the first time, in part of gas which has already fallen in and been shocked to high temperature, but has not yet cooled, and in part of gas that has been reheated and expelled from the galaxy by feedback processes. The infall and cooling of this material can replenish the ISM as it is consumed by star formation, allowing the galaxy to continue forming stars over long time scales. Hence, this extended gas component can be regarded as a reservoir of fuel for future star formation.

Since this gas reservoir is only relatively loosely bound to the galaxy, it is fairly easily stripped off, either by tides or by ram pressure. Hence, it is to be expected that a large fraction, if not all, of this halo gas is stripped off from a galaxy after it is accreted into a larger system such as a cluster. This is called strangulation, and results in a fairly gradual decline of the galaxy's star-formation rate as it slowly runs out of fuel (Larson et al., 1980).

It is well established that galaxies in the field form stars at rates several times higher than systems of similar luminosity in the denser environments associated with groups and clusters (e.g. Dressler et al., 1985; Balogh et al., 1997). Although this is partly a reflection of the well-known morphology–density relation (see §2.4.5), in that ellipticals and S0 galaxies, which have lower specific star-formation rates than spirals, are more abundant in denser environments, even galaxies of given stellar mass and given internal structure show a strong correlation between star-formation rate and environment density (Kauffmann et al., 2004). Late-type disk galaxies in clusters clearly have less gas and form stars at lower rates than those in the field (e.g. van den Bergh, 1976; Balogh et al., 1999). As argued by several studies, strangulation seems to be the main mechanism responsible for this environment dependence of the specific star-formation rates (e.g. Balogh et al., 2000; Balogh & Morris, 2000; van den Bosch et al., 2008a; Weinmann et al., 2009).

Most current models for the evolution of the galaxy population and its environment dependence include strangulation by assuming that a galaxy's gas reservoir is completely and instantaneously stripped when it is accreted onto a bigger host system (e.g. Kauffmann et al., 1993; Baugh et al., 1996; Springel et al., 2001a). In combination with merging, strangulation can reproduce most of the observed trends of star-formation activity and morphology with stellar mass and environment. Nevertheless, a detailed comparison shows that current implementations predict too many faint galaxies in clusters, and, in particular, much too high a fraction of red satellite galaxies (Weinmann et al., 2006a; Baldry et al., 2006; Kimm et al., 2009). This has become known as the over-quenching problem. Extending the time scale on which the gas reservoir is stripped to ~ 2 Gyr reproduces the color distribution of satellite galaxies considerably better (Wang et al., 2007b; Kang & van den Bosch, 2008; Font et al., 2008). Such delayed stripping is also suggested by hydrodynamical simulations (McCarthy et al., 2008), which show that up to $\sim 30\%$ of the initial hot gas can remain bound to a satellite galaxy even 10 Gyr after it has been accreted.

13
Elliptical Galaxies

Elliptical galaxies are smooth, roundish stellar systems which contain little cold gas or dust. Until the 1970s, they were thought to be relaxed, oblate systems, flattened by rotation and with close-to isotropic velocity dispersions. The same was assumed for the bulges of disk galaxies. Nowadays it is known that elliptical galaxies are far more complex systems than their outwardly bland and symmetrical morphologies seem to suggest. Based on their kinematics and photometry, ellipticals can be roughly divided into three different classes. The bright end ($\mathcal{M}_B \lesssim -20.5$) is dominated by systems with little rotation, typically boxy isophotes, and relatively shallow central surface brightness profiles. Ellipticals of intermediate luminosity ($-20.5 \lesssim \mathcal{M}_B \lesssim -18$), on the other hand, seem to be supported by rotation, have disky isophotes and steep central surface brightness cusps. At the faint end ($\mathcal{M}_B \gtrsim -18$), the majority of dwarf ellipticals (dEs) and dwarf spheroidals (dSphs) reveal no or very little rotation, and have roughly exponential surface brightness profiles.

Despite these clear differences, elliptical galaxies as a class conform to a number of tight, well-defined scaling relations. As we have seen in §2.3.2, they occupy a two-dimensional plane (the fundamental plane) in three-dimensional parameter space of size, velocity dispersion and surface brightness. In addition, the colors and metallicities of elliptical galaxies are tightly correlated with their luminosities, in that more luminous galaxies are both redder and more metal rich. All in all, elliptical galaxies therefore form a remarkably homogeneous class that is clearly distinct from the disk population described in Chapter 11. We do stress, however, that S0 galaxies seem to represent a transition class, linking the early-type spirals to the disky ellipticals, in a smooth sequence of decreasing disk-to-bulge (or disk-to-spheroid) ratio.

In this chapter we use the various ingredients discussed in the previous chapters to investigate possible formation scenarios for elliptical galaxies. We start in §13.1 with a description of models that can be used to describe their photometric and kinematic structure. In §13.2 we discuss two constrasting scenarios for the formation of elliptical galaxies, which we compare to various observational tests and constraints in §13.3. The fundamental plane, together with its physical interpretation, is the topic of §13.4. The stellar populations and chemical evolution of elliptical galaxies are discussed in §13.5. Throughout this chapter we use the term 'elliptical galaxies' to refer mainly to relatively bright early-type galaxies with $\mathcal{M}_B \lesssim -18$. The dEs and dSphs, which make up the faint members of the early-type family, will be discussed separately in §13.6.

13.1 Structure and Dynamics

Elliptical galaxies are prime examples of collisionless systems (of stars). Consequently, their structure and dynamics are completely determined by the phase-space distribution function. After a description of how the (projected) observables are related to the intrinsic quantities, we give a

13.1 Structure and Dynamics

brief overview of the observed structure and kinematics of elliptical galaxies. Subsequently, we describe two techniques to model the dynamics of elliptical galaxies, give an overview of the evidence for the presence of dark halos and supermassive black holes in elliptical galaxies, and discuss the internal shape of elliptical galaxies and its relation to the dynamical structure.

13.1.1 Observables

The observables of a galaxy, such as the surface brightness, rotation velocity and velocity dispersion at a position (x,y) on the sky are line-of-sight projections of the corresponding three-dimensional quantities. Let $\rho(\mathbf{x})$ and $\nu(\mathbf{x})$ be the three-dimensional distributions of stellar mass and light, respectively, which are related via the stellar mass-to-light ratio $\Upsilon(\mathbf{x})$. In what follows we will ignore any spatial variations in the stellar populations (see §13.5) and simply assume a constant stellar mass-to-light ratio, $\Upsilon(\mathbf{x}) = \Upsilon$. The surface brightness at a position (x,y) on the sky is then given by

$$I(x,y) = \frac{\Sigma(x,y)}{\Upsilon} = \frac{1}{\Upsilon}\int \rho(\mathbf{x})\,dz, \tag{13.1}$$

where $\Sigma(x,y)$ is the projected surface mass density, and z is the distance along the line-of-sight. Similarly, the normalized line-of-sight velocity distribution (LOSVD) is given by

$$\mathscr{L}(x,y,v_z) = \frac{1}{\Sigma(x,y)} \int\int\int f(\mathbf{x},\mathbf{v})\,dv_x\,dv_y\,dz, \tag{13.2}$$

with $f(\mathbf{x},\mathbf{v})$ the phase-space distribution function. Observationally, the LOSVDs can be obtained from spectroscopy, by analyzing the profiles of absorption lines that arise in the atmospheres of stars along the line-of-sight.

In general, the LOSVDs are not perfectly Gaussian, and it has become customary to parameterize the LOSVD with a Gauss–Hermite series

$$\mathscr{L}(v) = \frac{\alpha(w)}{\sigma}\left[1 + \sum_{j=3}^{N} h_j H_j(w)\right], \tag{13.3}$$

where

$$\alpha(w) \equiv \frac{1}{\sqrt{2\pi}}e^{-w^2/2}, \quad w \equiv (v-V)/\sigma. \tag{13.4}$$

Here v is the line-of-sight velocity, V and σ are the mean and dispersion of the best-fit Gaussian, and H_j is the Hermite polynomial of degree j. The Gauss–Hermite moments h_j ($j \geq 3$) measure deviations of the LOSVD from the best-fit Gaussian, and are related to the jth velocity moments of the LOSVD (see van der Marel & Franx, 1993). These in turn are related to the intrinsic velocity moments according to

$$\left\langle v_{\mathrm{los}}^j \right\rangle(x,y) = \frac{1}{I(x,y)}\int \nu(\mathbf{x})\left\langle v_z^j\right\rangle(\mathbf{x})\,dz. \tag{13.5}$$

In general, the Gauss–Hermite moments h_j ($j \geq 3$) are quite small ($|h_j| \lesssim 0.1$), indicating that the LOSVDs do not depart dramatically from a Gaussian.

For a spherical system there is a unique relation between $\Sigma(R)\langle v_{\mathrm{los}}^j\rangle(R)$ and $\rho(r)\langle v_z^j\rangle(r)$, or, equivalently, between $I(R)\langle v_{\mathrm{los}}^j\rangle(R)$ and $\nu(r)\langle v_z^j\rangle(r)$, given by the Abel transformation:

$$\Sigma(R)\langle v_{\mathrm{los}}^j\rangle(R) = 2\int_R^\infty \rho(r)\langle v_z^j\rangle(r)\frac{r\,dr}{\sqrt{r^2-R^2}}, \tag{13.6}$$

$$\rho(r)\langle v_z^j\rangle(r) = -\frac{1}{\pi}\int_r^\infty \frac{d}{dR}\left[\Sigma(R)\langle v_{\mathrm{los}}^j\rangle(R)\right]\frac{dR}{\sqrt{R^2-r^2}}. \tag{13.7}$$

This analytical deprojection is used mainly to obtain the three-dimensional luminosity density $\nu(r)$ from the observed, two-dimensional surface brightness $I(R)$, which corresponds to setting $j = 0$. In practice, however, the observational data contains noise and a direct differentiation of the data is generally too noisy to be useful. Therefore, one normally either fits $I(R)$ with an analytical fitting function and uses Eq. (13.7) to infer (a smooth version of) $\nu(r)$, or adopts a functional form for $\nu(r)$ and uses Eq. (13.6) to constrain the free parameters by comparing $I(R)$ to the observed surface brightness profile.

In the case of an axisymmetric system, the deprojection of the observed surface brightness distribution, $I(x,y)$, to a three-dimensional luminosity density, $\nu(R,z)$, is indeterminate unless the galaxy is seen edge-on (in which case, the Abel transformation can be used). In all other cases, and for triaxial systems, no unique deprojection is possible. In practice this problem is often circumvented by making ad hoc assumptions about the density distribution. However, one can always redistribute the mass (or light) and still project to *exactly* the same surface density (or brightness). In fact, any density distribution whose Fourier transform is only non-zero inside a cone with half-opening angle θ and aligned with the symmetry axis of the Fourier transform of the density distribution (the so-called 'cone of ignorance') projects to zero surface density when seen under any inclination angle $i < 90° - \theta$ (Rybicki, 1987; Gerhard & Binney, 1996). Such density distributions, sometimes called konus densities, have regions with both positive and negative density, and basically describe how the matter can be redistributed without affecting the *projected* mass distribution. As such, they quantify the degeneracy in the deprojection. However, not all hope is lost: although the addition of a konus density to a mass model does not affect the surface densities, it does affect the dynamical structure (also in projection) so that the corresponding degeneracy can in principle be broken using kinematic data (Romanowsky & Kochanek, 1997; van den Bosch, 1997).

13.1.2 Photometric Properties

As discussed in §2.3.2, the surface brightness of elliptical galaxies is typically specified by a radial surface brightness profile, $I(R)$, and by the radial dependency of the shapes of the isophotes (ellipticity, position angle and diskiness). Bright ellipticals ($\mathcal{M}_B \lesssim -20.5$) have relatively low ellipticities ($\varepsilon \lesssim 0.3$) and often reveal boxy isophotes. Ellipticals with $-20.5 \lesssim \mathcal{M}_B \lesssim -18$, on the other hand, span a wider range in ellipticities ($\varepsilon \lesssim 0.7$), and typically have disky isophotes. This diskiness is often interpreted as due to an embedded stellar disk (Rix & White, 1990; Scorza & Bender, 1995). Disky ellipticals, therefore, seem to form a continuous sequence in the Hubble diagram from S0s to galaxies with smaller disk-to-bulge ratios (e.g. Kormendy & Bender, 1996).

A significant fraction of the boxy ellipticals show isophote twisting, in that the position angle of the isophote changes as a function of radius. This is impossible for an intrinsically axisymmetric system, but occurs naturally if one projects a triaxial system with varying axis ratios (e.g. Stark, 1977). This is one of several reasons why bright ellipticals are generally believed to be triaxial. Disky ellipticals, on the other hand, only rarely reveal isophote twisting, consistent with them being axisymmetric.

The dichotomy between disky and boxy ellipticals is reinforced by their nuclear properties: high-resolution images from the Hubble Space Telescope have shown that the central regions of disky ellipticals typically have steep cusps, while many boxy ellipticals have gently rising inner luminosity profiles or central cores (Ferrarese et al., 1994; Lauer et al., 1995; Rest et al., 2001). If the surface brightness profiles, $I(R)$, are inverted to obtain the three-dimensional luminosity density profiles, $\nu(r)$, the distribution of the inner logarithmic slope of these profiles, $S \equiv \mathrm{d}\ln\nu/\mathrm{d}\ln r$, appears bimodal, with most galaxies having S near -0.9 or -1.8. Furthermore,

13.1 Structure and Dynamics

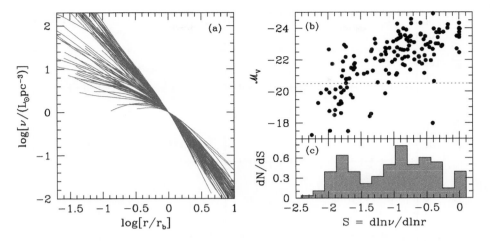

Fig. 13.1. Panel (a) shows luminosity density profiles of 86 elliptical galaxies obtained by deprojecting their surface brightness profiles observed with the HST. Panel (b) shows the correlation between the inner cusp slopes $S \equiv \mathrm{d}\ln\nu/\mathrm{d}\ln r$ (measured at $r = 0.1$ arcsec) and the absolute magnitude of the galaxy, while the distribution of cusp slopes, $\mathrm{d}N/\mathrm{d}S$, is shown in panel (c). Note that bright and faint ellipticals typically have shallow and steep cusps, respectively. [Based on data published in Lauer et al. (2007) and kindly provided by K. Gebhardt]

there is a correlation with luminosity, in the sense that brighter ellipticals tend to have shallower cusps (see Fig. 13.1).

A number of recent studies have argued that the transition at $\mathcal{M}_B \sim -20.5$ between disky ellipticals with cusps and boxy ellipticals with cores is too smooth to be characterized as a true dichotomy. Indeed, it is already apparent from Fig. 13.1 that there is no clear feature at $\mathcal{M}_B \sim -20.5$. Côté et al. (2007) have argued that the bimodality in the distribution of cusp slopes S in the lower-right panel of Fig. 13.1 could simply reflect sample selection effects. Using a more complete sample of early-types in the Fornax and Virgo clusters, Ferrarese et al. (2006b) find a smooth distribution of central cusp slopes with no sign of bimodality. The lack of a clear dichotomy or bimodality is also supported by Pasquali et al. (2007), who have shown that the fraction of early-type galaxies with disky isophotes is a smooth function of absolute magnitude, given by

$$f_{\mathrm{disky}}(\mathcal{M}_B) = (0.61 \pm 0.02) + (0.17 \pm 0.03)\left[\mathcal{M}_B - 5\log h + 20\right], \tag{13.8}$$

with no sign of any specific 'transition' scale. Thus, although there is no doubt that bright and faint ellipticals have different properties, whether or not there is any dichotomy or bimodality is still under debate.

13.1.3 Kinematic Properties

The kinematics of elliptical galaxies have to be determined from absorption line spectroscopy, requiring sensitive spectrographs and detectors. As a result, the first kinematic data for elliptical galaxies became available only in the 1970s. Before this, it was generally believed that both ellipticals and the bulges of disk galaxies are oblate systems with near-isotropic velocity dispersions and are flattened by rotation. This belief was overturned when Bertola & Capaccioli (1975) and Illingworth (1977) obtained kinematics for a number of elliptical galaxies and showed that their rotation velocities are much too small to explain their observed flattening, which must therefore reflect anisotropic velocity dispersions rather than rotation. As more and more data became available, it became clear that the dichotomy present in the photometric properties of ellipticals is also

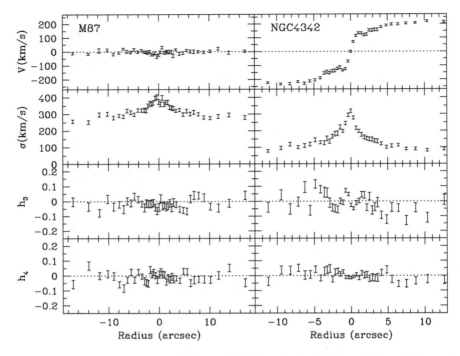

Fig. 13.2. Major axis kinematics of M 87 (left) and NGC 4342 (right). [Based on data published in van der Marel (1994) and van den Bosch et al. (1998)]

reflected in their kinematics (Davies et al., 1983). While bright, boxy ellipticals are slow rotators, supported by anisotropic velocity dispersions, the fainter, disky ellipticals typically have much higher rotation velocities, often consistent with them being purely rotationally flattened (see Fig. 2.16 and §13.1.7 below). As first discussed by Binney (1976), it is then more natural to assume that bright ellipticals, as a class, are triaxial, rather than axisymmetric. At the faint end ($\mathcal{M}_B \gtrsim -18$), spatially resolved kinematics have revealed an intriguing diversity among the population of dEs and dSphs: while the majority have no detectable rotation, roughly 20% have kinematics consistent with rotation producing their flattening (e.g. Bender et al., 1991; De Rijcke et al., 2001; Geha et al., 2002).

Fig. 13.2 shows the major axis kinematics of two typical examples: M87, one of the brightest ellipticals in the Virgo cluster, and NGC 4342, a much fainter E/S0 galaxy (with disky isophotes). M87 reveals absolutely no rotation, indicating that its modest flattening ($\varepsilon \sim 0.1$) must be due to anisotropy. NGC 4342, on the other hand, has a rotation curve fairly reminiscent of that of spiral galaxies, and its flattening is consistent with being due to the centrifugal forces generated by rotation. Both galaxies have a velocity dispersion profile that is characteristic of the kinematics of early-type galaxies: it shows a pronounced peak at the center, and becomes flat at larger radii. Finally, the Gauss–Hermite moments h_3 and h_4 are close to zero, indicating that the LOSVDs are close to Gaussian. Often, though, galaxies with substantial rotation tend to have significantly skewed LOSVDs, in the sense that h_3 has the opposite sign to the rotation velocity V. Such skewed profiles are most easily explained by a (thin) disk component embedded in a spheroid with little or no rotation (e.g. Cinzano & van der Marel, 1994; van den Bosch & de Zeeuw, 1996), and therefore provides further support for the idea that disky ellipticals are two-component systems similar to S0s, but with a larger bulge-to-disk ratio.

For an oblate rotator, the kinematic axis, defined as the axis along which the observed rotation velocity is zero, is aligned with the minor axis of the photometry. A number of elliptical galaxies, however, are observed to have rotation along both the major and the minor axes (e.g. Franx et al., 1991), so that their kinematic and photometric axes do not coincide. The magnitude of this misalignment is typically characterized by the kinematic misalignment angle

$$\Psi = |\theta_{\rm kin} - \theta_{\rm phot}|, \tag{13.9}$$

with $\theta_{\rm kin}$ and $\theta_{\rm phot}$ the position angles of the projected kinematic axis and the photometric minor axis, respectively. When integral field spectroscopy is available, the kinematic axis can be determined directly from the two-dimensional velocity field, $V(x,y)$. Alternatively, if the only kinematic data that is available is long-slit spectroscopy along the major and minor axes, it is customary to estimate the kinematic misalignment angle using $\Psi = \tan^{-1}(v_{\rm min}/v_{\rm maj})$, with $v_{\rm min}$ and $v_{\rm maj}$ the maximum rotation velocities along the minor and major axes, respectively (Franx et al., 1989a). In general, kinematic misalignment is another signature of triaxiality, and can be caused by two effects. First, for nearly all viewing angles the projected surface brightness of a triaxial galaxy has an apparent minor axis that is misaligned from its projected short axis. If the mean streaming is around the intrinsic short axis, an observer will generally measure rotation along both projected axes. Secondly, the total angular momentum of a triaxial galaxy can be *intrinsically* misaligned, and lie anywhere in the plane containing the long and short axes (e.g. Levison, 1987). This reflects the fact that triaxial potentials support orbits with a net sense of rotation both around the short axis and around the long axis (Statler, 1987). Kinematic misalignments are more common among the bright, velocity-dispersion supported ellipticals, consistent with their general shapes being triaxial. Rapidly rotating ellipticals, on the other hand, rarely reveal kinematic misalignment, consistent with them being axisymmetric.

Finally, the velocity fields of $\sim 25\%$ of all ellipticals have a kinematically decoupled core (KDC) whose angular momentum vector is misaligned with respect to that of the bulk of the galaxy. In extreme cases, the core can even be counter-rotating with respect to the outer regions. KDCs are usually attributed to dynamically distinct subsystems that are the remnants of accreted companions. However, kinematic twists can also result from the projection of the major families of circulating orbits in a triaxial potential, without the core being a separate dynamical subsystem (Statler, 1991). The photometry and velocity dispersion profiles of ellipticals with KDCs do not appear particularly unusual, and most data suggest a smooth variation of stellar population properties between the KDC and their direct surroundings (Forbes et al., 1995; Carollo et al., 1997). Thus, it remains unclear whether KDCs originate from accreted material or merely reflect projection effects in a triaxial system.

13.1.4 Dynamical Modeling

Modeling the dynamics of elliptical galaxies has two main goals. First of all, comparing models to kinematic data helps to constrain the mass-to-light ratio, which holds information regarding both stellar populations and the presence of dark matter or supermassive black holes. Secondly, the models can be used to constrain the orbital structure, or equivalently, the phase-space distribution function, in the hope that this can be used to discriminate between different formation scenarios. In §11.1 we have seen that, for disk galaxies, it is relatively straightforward to determine the mass profile from the observed rotation curves, and thus to infer the presence of dark matter halos. This reflects the relatively simple dynamics of an almost two-dimensional, rotationally supported configuration. In elliptical galaxies, however, the situation is more complicated. In the first place, observations only provide the stellar distribution in one of the three velocity coordinates and two of the three position coordinates. As we saw in §13.1.1, this can lead to degeneracy. Secondly, ellipsoidal structures support a much richer variety of orbits than thin, rotationally supported disks. In what follows we briefly describe two commonly used modeling techniques. A more general description of collisionless dynamics can be found in §5.4.

(a) Jeans Models One way to constrain the mass distribution from the observed kinematics is to solve the Jeans equations relating the gravitational potential to various intrinsic velocity moments, which are in turn related to the observed velocity moments via Eq. (13.5). As discussed in §5.4.3, however, the Jeans equations do not in general form a closed set, and certain assumptions have to be made in order to proceed.

As an example, consider a spherical system for which one has observed the surface brightness profile, $I(R)$, and the projected velocity dispersion profile, $\sigma_p(R)$, with R the projected radius. The only non-trivial Jeans equation for a spherical system is

$$\frac{1}{\rho}\frac{d(\rho\langle v_r^2\rangle)}{dr} + 2\beta\frac{\langle v_r^2\rangle}{r} = -\frac{d\Phi}{dr}, \qquad (13.10)$$

where $\beta(r)$ is the anisotropy parameter, defined by

$$\beta(r) \equiv 1 - \langle v_\vartheta^2\rangle/\langle v_r^2\rangle. \qquad (13.11)$$

Since $d\Phi/dr = GM_{\rm tot}(r)/r^2$, where $M_{\rm tot}(r)$ is the total enclosed mass within a radius r, we can rewrite the Jeans equation as

$$M_{\rm tot}(r) = -\frac{\langle v_r^2\rangle r}{G}\left[\frac{d\ln\rho}{d\ln r} + \frac{d\ln\langle v_r^2\rangle}{d\ln r} + 2\beta\right]. \qquad (13.12)$$

Thus, the mass profile depends both on $\langle v_r^2\rangle(r)$ and on the anisotropy profile $\beta(r)$. Since the line-of-sight velocity is given by $v_{\rm los} = v_r \sin\alpha - v_\vartheta \cos\alpha$, with $\cos\alpha = R/r$, we have that

$$\sigma_p^2(R) = \frac{2}{I(R)}\int_R^\infty \left(1 - \beta\frac{R^2}{r^2}\right)\frac{\nu\langle v_r^2\rangle r\,dr}{\sqrt{r^2 - R^2}}. \qquad (13.13)$$

The observable $\sigma_p^2(R)$ thus depends on both $\langle v_r^2\rangle(r)$ and $\beta(r)$, and there is no unique solution for $M_{\rm tot}(r)$. In the absence of additional information, this lack of uniqueness, often referred to as the mass-anisotropy degeneracy, prevents a determination of the mass-to-light ratio profile, even for spherical galaxies. Since systems with the same mass profile but a different velocity anisotropy have different orbital structures, one can in principle break this degeneracy, at least partially, by including higher-order velocity moments (see e.g. van der Marel & Franx, 1993). However, one then needs to consider higher-order Jeans equations in order to interpret these higher-order velocity moments of the LOSVD. As discussed in §5.4.3, such higher-order Jeans equations involve new unknowns, so that one still has to make ad hoc assumptions in order to close the set of equations.

In axisymmetric systems the situation is similar. One can only solve the Jeans equations under certain idealized assumptions. For axisymmetric systems one typically assumes that $f = f(E, L_z)$, which is equivalent to assuming that the system is isotropic in the meridional plane, i.e. $\sigma_R = \sigma_z$. Although this allows for a solution of the Jeans equations (e.g. Hunter, 1977a; Binney et al., 1990), there is no reason to believe that $f = f(E, L_z)$ should hold for real systems. Furthermore, even if a particular Jeans model is found to match the data perfectly, it is rarely unique and it may be unphysical, i.e. its distribution function may not obey $f \geq 0$ everywhere.

(b) Schwarzschild Orbit-Superposition Models One can improve upon the Jeans modeling described above by using the Schwarzschild orbit-superposition technique. This leads to completely numerical, three-integral models that are fully general, and, by construction, always obey $f \geq 0$. The method is computationally expensive, however, and has only been viable for accurate modeling of high-quality data since the late 1990s.

The method, first introduced by Schwarzschild (1979), works as follows. One starts by assuming a stellar mass-to-light ratio, $\Upsilon(\mathbf{x})$, which is used to transform the deprojected light distribution, $\nu(\mathbf{x})$, into a stellar mass distribution, $\rho(\mathbf{x})$. Next integrate the Poisson equation to find the corresponding potential, $\Phi(\mathbf{x})$, and add the contribution, $\Phi_{\text{ext}}(\mathbf{x})$, due to additional components such as a dark halo or a supermassive black hole. Then calculate a large number of orbits by numerical integration of the equations of motion, and compute each orbit's contribution to each of the observational constraints. For example, suppose that one has measured the surface brightness and LOSVDs on a grid (i, j) of pixels on the sky. Then, the contribution of orbit k to the surface density at (i, j) is proportional to the time the projected orbit appears in pixel (i, j), while its contribution to the LOSVD is simply its distribution of line-of-sight velocities during those periods. What remains to be done is to find the linear combination of orbit weights, w_k, that best matches the set of observations. If no acceptable fit can be obtained then the assumed $\Upsilon(\mathbf{x})$ and/or $\Phi_{\text{ext}}(\mathbf{x})$ can be excluded. Thus, this modeling technique can be used to constrain the mass profiles, and to test for the presence of dark halos and/or supermassive black holes.

The integrals of motion do not appear explicitly in this scheme, but the resulting orbital weights effectively describe the distribution of tracers as a function of them. Since each (regular) orbit is uniquely characterized by three independent, isolating integrals of motion (see §5.4.6), Schwarzschild's method provides a numerical representation of the distribution function which depends implicitly on these three integrals. For further details about this technique see, for example, Cretton et al. (1999) and references therein.

Schwarzschild (1979) originally used this orbit-superposition scheme to prove the existence of triaxial equilibrium models that resemble real ellipticals. Richstone (1980) used the method to show that many different orbit combinations can reproduce the same mass model. More recently, the method has mainly been used to investigate the existence of massive black holes at the centers of elliptical galaxies, and of dark halos surrounding them (e.g. Rix et al., 1997; Cretton & van den Bosch, 1999; Gebhardt et al., 2003), as well as to constrain the orbit structure of ellipticals by comparison with kinematic data including higher-order LOSVD moments (e.g. Cappellari et al., 2007).

13.1.5 Evidence for Dark Halos

As discussed in Chapter 11, the HI rotation curves of spiral galaxies provide strong evidence that they are embedded in extended dark halos. According to the standard paradigm for galaxy formation described in this book, elliptical galaxies should reside in similar dark halos. However, finding direct, dynamical evidence for dark halos around elliptical galaxies has proven difficult because of the lack of suitable and easily interpretable tracers at large radii. In addition, modeling the stellar dynamics of ellipticals is far more complicated than for the cold HI disks of spirals because one must solve simultaneously for the potential and for the orbit populations, given the observed kinematics. Here we describe a few techniques that have been used to probe the gravitational potentials of elliptical galaxies at large radii.

One possibility is to use the stellar kinematics, obtained from absorption line spectroscopy of the integrated stellar light. However, since the surface brightness of elliptical galaxies drops rapidly with radius, it is difficult to obtain reliable measurements much beyond one effective radius. To date, stellar kinematics have been measured out to $\sim 2R_e$ for only about a dozen cases. These typically show that the line-of-sight velocity dispersion profile is roughly constant beyond $\sim 1R_e$. Although consistent with a mass-to-light ratio profile that increases outward, as expected if the system is embedded in a dark halo, a constant $\sigma_p(R)$ can also signal a velocity distribution that becomes more tangentially anisotropic with increasing radius. As mentioned in §13.1.4 above, this mass-anisotropy degeneracy can be broken using measurements of the shapes of the velocity profiles beyond the second moments (i.e. via the Gauss–Hermite moments).

Comparing such data with dynamical models has indicated that, in general, the mass-to-light ratios increase with radius, consistent with the presence of dark halos, although there are also cases where the data is consistent with a constant M/L all the way out to $2R_e$ (e.g. Rix et al., 1997; Kronawitter et al., 2000). The inferred mass-to-light ratios in the central regions are higher than the stellar mass-to-light ratios inferred from stellar population models, and indicate that, on average, ~ 30 percent of the total mass within one R_e is dark matter (e.g. Gerhard et al., 2001; Cappellari et al., 2006).

An alternative way to probe the mass distribution at larger radii is to use discrete dynamical tracers, such as globular clusters, planetary nebulae or satellite galaxies. Although this has the advantage that the tracers can be observed out to large galactocentric radii, dynamical modeling of the observed radial velocities of a discrete tracer population is beset by the same mass-anisotropy degeneracy that plagues the interpretation of the integrated light measurements. Since the available number of tracers is relatively small, it is difficult to obtain accurate measurements of the higher-order moments of the LOSVD and so to constrain the velocity anisotropy. As a result, these studies have often produced ambiguous results (e.g. Ciardullo et al., 1993; Romanowsky et al., 2003; Dekel et al., 2005).

In addition to dynamical modeling, there are two other techniques that have been used to probe the potentials of elliptical galaxies at large radii: X-ray mapping and gravitational lensing. As discussed in §2.3.2, bright ellipticals are often surrounded by extended X-ray emitting coronae of hot gas. As we have seen in §8.2.1, under the assumption that the X-ray emitting gas is in hydrostatic equilibrium, the total gravitational mass enclosed within a radius r is given by Eq. (8.16). A comparison with Eq. (13.12) shows that this hydrostatic equation is similar to the stellar dynamical equivalent, but with the stellar velocity dispersion $\langle v_r^2 \rangle$ replaced by the gas temperature T and with $\beta(r) = 0$. For a few bright ellipticals, both the temperature, $T(r)$, and the density, $\rho(r)$, of the gas can be determined from X-ray measurements, and Eq. (8.16) can be solved for the total mass profile $M(r)$. In all cases the mass-to-light ratios thus derived reach values of $\sim 100 M_\odot/L_\odot$ on scales of ~ 100 kpc, providing strong evidence for the presence of dark halos (e.g. Forman et al., 1985; Mushotzky et al., 1994). Obviously, this method does not suffer from the mass-anisotropy degeneracy that hampers mass determinations based on kinematics, but as discussed in §8.2, there are other problems due to the fact that the pressure can have significant and poorly known non-thermal contributions due to turbulence, magnetic fields and relativistic particles (cosmic rays).

Another powerful method to probe the total gravitational mass of elliptical galaxies, independent of their kinematic structure, is provided by gravitational lensing (see §6.6). In particular, multiple image configurations yield information regarding the mass distribution of the lens, first and foremost a precise and robust measurement of the mass projected within the Einstein radius (see §6.6.2). Except for the central galaxies of rich clusters, the latter is usually smaller than the photometric effective radius of the galaxy (Bolton et al., 2008). Although only a small fraction of all ellipticals are lenses, this technique provides clear evidence that the mass distributions of elliptical galaxies are more extended than their stellar components, either from the analysis of individual systems or by considering statistical ensembles (e.g. Maoz & Rix, 1993; Kochanek, 1995; Rusin & Ma, 2001). Furthermore, when lensing data is combined with stellar kinematics, the mass-anisotropy degeneracy can be broken. This has been used to put constraints on the actual density distribution of the dark matter halos surrounding ellipticals (e.g. Treu & Koopmans, 2004).

13.1.6 Evidence for Supermassive Black Holes

Considerable evidence suggests that the energetic processes in active galaxies and quasars are due to the accretion of matter onto supermassive black holes (SMBHs) with masses in the range

13.1 Structure and Dynamics

$10^6 \, M_\odot \lesssim M_{BH} \lesssim 10^9 \, M_\odot$ (see Chapter 14). Since quasars were far more numerous at $z \sim 2$ than at the present, many normal (quiescent) galaxies should harbor such a dead quasar engine in their nucleus. Consequently, there has been a large effort to search for SMBHs in the centers of (nearby) galaxies through kinematic studies. In particular, one tries to find evidence for an increase in the mass-to-light ratio towards the center to values that cannot be explained by normal stellar populations. Although this may indicate the presence of a SMBH, it may also be due to the presence of a dense cluster of non-luminous objects, such as brown dwarfs or stellar remnants (white dwarfs or neutron stars). If the inferred mass-to-light ratio is sufficiently high, and restricted to a sufficiently small region, these alternatives can be ruled out by the fact that these dense clusters would have lifetimes that are significantly shorter than the age of the galaxy (e.g. Maoz, 1998). Currently, the only case in which we can convincingly rule out all alternatives in favor of a SMBH is the Milky Way (see §14.2.8). Relying on Occam's razor, it is then reasonable to assume that all central supermassive dark objects are SMBHs.

A massive black hole (BH) at the center of a galaxy only significantly influences the dynamics inside a radius of influence,

$$r_{BH} = \frac{GM_{BH}}{\sigma^2} = 10.8 \, \text{pc} \left(\frac{M_{BH}}{10^8 \, M_\odot} \right) \left(\frac{\sigma}{200 \, \text{km s}^{-1}} \right)^{-2}, \tag{13.14}$$

where σ is the characteristic velocity dispersion of the stars near the center of the galaxy. Clearly, finding evidence for SMBHs requires kinematic observations with high spatial resolution, which is why spectroscopy with the HST has played a major role in this field.

As already mentioned in §13.1.3, many elliptical galaxies have velocity dispersion profiles that rise steeply towards the center. Fitting simple isotropic models to these data often indicates a central rise of the mass-to-light ratio, in agreement with the presence of a SMBH. However, this interpretation is ambiguous because of the same mass-anisotropy degeneracy that plagues the search for massive halos. In particular, a central rise in $\sigma_p(R)$ can occur in a system without a SMBH if the central region becomes radially anisotropic (e.g. Binney & Mamon, 1982). As in the case of dark halos, this ambiguity can be largely removed by measuring the entire LOSVD (or at least some higher-order moments, such as the Gauss–Hermite moments h_3 and h_4). Such data, combined with sophisticated orbit superposition models, have provided strong (though not conclusive) evidence for SMBHs in a few dozen cases of nearby elliptical galaxies. Two examples are M87 and NGC 4342, whose kinematics are shown in Fig. 13.2. The inferred SMBH masses for these two cases are $\sim 3 \times 10^9 \, M_\odot$ and $\sim 3 \times 10^8 \, M_\odot$, respectively (van der Marel, 1994; Cretton & van den Bosch, 1999).

In addition to stellar kinematics, in some cases the presence of a SMBH is inferred from the kinematics of small disks of ionized gas (e.g. Harms et al., 1994; van der Marel & van den Bosch, 1998). Although the rotational velocities of gas disks in principle yield fairly direct measures of the enclosed mass, gas is very responsive to non-gravitational forces, so that the inferred kinematics have to be interpreted with care.

Over the last decade stellar- and gas-dynamical studies have indicated the presence of a SMBH in an ever-increasing number of galaxies (see also §14.2.8), and it is now generally accepted that every relatively massive spheroid, whether an elliptical or a bulge, contains a SMBH at its center (see Kormendy & Richstone, 1995; Ferrarese & Ford, 2005, for comprehensive reviews). This has permitted an examination of the correlation between BH mass and the properties of the host galaxy. It was soon realized that M_{BH} is well correlated with the luminosity of the spheroid. An even tighter correlation exists between the BH mass and the velocity dispersion of the spheroid (typically measured inside an aperture with radius proportional to R_e), which is given by

$$M_{BH} = (1.3 \pm 0.2) \times 10^8 \, M_\odot \left(\frac{\sigma_e}{200 \, \text{km s}^{-1}} \right)^\gamma \tag{13.15}$$

(see Fig. 2.17). The value of γ ranges from 3.75 (Gebhardt et al., 2000) to 4.8 (Ferrarese & Merritt, 2000) depending on the sample used and on the exact definition of σ_e (see discussion in Tremaine et al., 2002). The intrinsic dispersion in Eq. (13.15) is no larger than 0.3 dex. Equally tight relations have been obtained between the BH mass and the stellar mass of the spheroid (Häring & Rix, 2004) and between the BH mass and the Sérsic index of the light distribution of the spheroid (Graham et al., 2001). These BH demographics strongly suggest that the formation of the SMBH is tightly entwined with that of (the spheroidal component of) its host galaxy.

13.1.7 Shapes

(a) **Rotational Support versus Anisotropy** The tensor virial theorem (§5.4.4) applied to a steady-state stellar system is

$$2T_{ij} + \Pi_{ij} + W_{ij} = 0, \tag{13.16}$$

where

$$T_{ij} = \frac{1}{2} \int \rho \langle v_i \rangle \langle v_j \rangle \, d^3\mathbf{x}, \tag{13.17}$$

$$\Pi_{ij} = \int \rho \, \sigma_{ij}^2 d^3\mathbf{x} \quad \text{with} \quad \sigma_{ij}^2 = \langle v_i v_j \rangle - \langle v_i \rangle \langle v_j \rangle, \tag{13.18}$$

$$W_{ij} = -\int \rho \, x_i \frac{\partial \Phi}{\partial x_j} d^3\mathbf{x}. \tag{13.19}$$

For simplicity, let us consider an axisymmetric system that rotates about its symmetry axis (the z axis) and is seen edge-on (with the x axis being the line-of-sight). For such a system, $W_{xx} = W_{yy}$ and $W_{ij} = 0$ for $i \neq j$, and similar relations also hold for T_{ij} and Π_{ij}. Eq. (13.16) then reduces to

$$2T_{xx} + \Pi_{xx} + W_{xx} = 0; \quad 2T_{zz} + \Pi_{zz} + W_{zz} = 0. \tag{13.20}$$

If the only streaming motion is rotation about the z axis, then $T_{zz} = 0$ and

$$2T_{xx} = \frac{1}{2} \int \rho \langle v_\phi \rangle^2 d^3\mathbf{x} = \frac{1}{2} M v_0^2, \tag{13.21}$$

where $\langle v_\phi \rangle^2 = \langle v_x \rangle^2 + \langle v_y \rangle^2$, M is the mass of the system, and v_0 is the mass-weighted rotation velocity. Similarly, we can write

$$\Pi_{xx} = M\sigma_0^2; \quad \Pi_{zz} = (1-\delta)M\sigma_0^2, \tag{13.22}$$

where $\sigma_0^2 \equiv (1/M) \int \rho \, \sigma_{xx}^2 d^3\mathbf{x}$ is the mass-weighted velocity dispersion along the line-of-sight, and $\delta \equiv 1 - \Pi_{zz}/\Pi_{xx} < 1$ is a measure of the anisotropy in the velocity dispersion. It then follows, from the ratio between the two expressions in Eq. (13.20), that

$$\frac{W_{xx}}{W_{zz}} = \frac{1}{1-\delta}\left(1 + \frac{1}{2}\frac{v_0^2}{\sigma_0^2}\right). \tag{13.23}$$

As shown by Roberts (1962), for systems stratified on similar coaxial oblate ellipsoids, W_{xx}/W_{zz} depends only on the ellipticity ε. In particular, for an oblate system

$$\frac{W_{xx}}{W_{zz}} = \frac{1}{2}\left[\frac{\arcsin e - e\sqrt{1-e^2}}{e\sqrt{1-e^2} - (1-e^2)\arcsin e}\right], \tag{13.24}$$

with $e = \sqrt{1-(1-\varepsilon)^2}$. It is then obvious from Eq. (13.23) that a stellar system can be flattened (i.e. $\varepsilon > 0$) either by rotation (i.e. $v_0^2 > 0$) or by anisotropic velocity dispersion (i.e. $\delta > 0$).

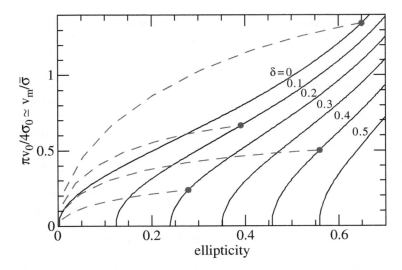

Fig. 13.3. The anisotropy diagram, showing the relation between the kinematic parameter $v_m/\overline{\sigma} \simeq \pi v_0/4\sigma_0$ and the ellipticity ε. The solid lines indicate the relation obtained from the tensor virial theorem for systems stratified on similar coaxial oblate ellipsoids. Results are shown for different values of the anisotropy parameter δ, as indicated. The dashed lines show how systems with edge-on quantities specified by the solid dots move through the anisotropy diagram if their inclination angles decrease towards face-on. [After Binney & Tremaine (1987)]

If we are able to measure ε, v_0 and σ_0 from observations, Eqs. (13.23) and (13.24) enable us to determine the anisotropy parameter δ. Long-slit spectroscopy typically yields $V_{\rm rot}(R)$ and $\sigma(R)$ measured along the projected major axis of the elliptical. It is customary to identify σ_0 with $\overline{\sigma}$, the mean velocity dispersion interior to half the effective radius, and v_0 with $4v_m/\pi$, with v_m the maximum rotation velocity. Although neither of these identifications has a rigorous basis, detailed dynamical models show that they are sufficiently accurate for most practical purposes (Binney, 2005).[1] Fig. 13.3 shows the relation between ε and $\pi v_0/4\sigma_0 \simeq v_m/\overline{\sigma}$ for different values of the anisotropy parameter δ, obtained using Eqs. (13.23) and (13.24). For an oblate system with isotropic velocity dispersion ($\delta = 0$), this relation is approximated by

$$\frac{v_m}{\overline{\sigma}} \approx \sqrt{\frac{\varepsilon}{1-\varepsilon}}, \qquad (13.25)$$

to good accuracy.

An important complication arises from the fact that the relation between ε and $v_m/\overline{\sigma}$ derived above is only (approximately) valid for systems observed close to edge-on. In general, a galaxy is observed under an inclination angle i, so that the observed values of ε and $v_m/\overline{\sigma}$ differ from those appearing in the above equations. It is easy to show that the values observed under an inclination angle i are related to those observed edge-on ($i = 90°$) according to

$$\varepsilon_i = 1 - \sqrt{(1-\varepsilon_{i=90°})^2 \sin^2 i + \cos^2 i}, \qquad (13.26)$$

and

$$\left(\frac{v_m}{\overline{\sigma}}\right)_i = \left(\frac{v_m}{\overline{\sigma}}\right)_{i=90°} \frac{\sin i}{\sqrt{1 - \delta \cos^2 i}}. \qquad (13.27)$$

---

[1] With integral-field spectroscopy one can in principle do better, and measure kinematic quantities that are more directly related to v_0 and σ_0 (e.g. Cappellari et al., 2007).

The dashed lines in Fig. 13.3 show how galaxies with edge-on quantities specified by the solid dots move through the anisotropy diagram if their inclination angles decrease towards face-on. This illustrates that these inclination effects can be very substantial and need to be corrected for with the use of the inverse of Eqs. (13.26) and (13.27). Unfortunately, in general the inclination angle under which an elliptical galaxy is observed is unknown, and so no reliable correction can be applied. A common practice is to ignore the inclination and to simply assume that the galaxy has been observed edge-on. As is apparent from Fig. 13.3, this results in an underestimate of the inferred anisotropy parameter δ, which should be kept in mind when interpreting anisotropy diagrams such as that shown in Fig. 2.16.

Kinematic studies of several dozen ellipticals and S0 galaxies have shown that the majority of bright ellipticals ($\mathscr{M}_B \lesssim -20.5$) are slow rotators, with $v_{\rm m}/\overline{\sigma} < \sqrt{\varepsilon/(1-\varepsilon)}$. They typically have only modest ellipticities ($\varepsilon \lesssim 0.3$), and their velocity dispersion has anisotropy that spans the range $0 < \delta \lesssim 0.3$. On the other hand, fainter ellipticals typically have much higher values of $v_{\rm m}/\overline{\sigma}$. Early studies indicated that their rotation velocities are consistent with being isotropic, oblate rotators. However, more recent studies based on integral field spectroscopy have shown that the majority of the fast-rotating, low luminosity ellipticals have velocity anisotropy, δ, that deviates strongly from zero, reaching values as high as 0.5 (Emsellem et al., 2007; Cappellari et al., 2007).

(b) Intrinsic Shapes The above analysis of the flattening of ellipticals is based on the observed, projected flattening. More stringent constraints on the formation of elliptical galaxies come from their intrinsic three-dimensional shapes.

As discussed in §13.1.1, in general the deprojection of an observed surface brightness distribution to the three-dimensional luminosity density is not unique. It is often *assumed* that the mass and luminosity density are stratified on concentric ellipsoids with semimajor axes $a \geq b \geq c$. When the axis ratios $\beta \equiv b/a$ and $\gamma = c/a$ are constant for a given galaxy ($0 \leq \gamma \leq \beta \leq 1$), the projected surface brightness is stratified on ellipses of constant projected axis ratio, q, and there is no isophote twist. In this case, q is a function of four variables; the intrinsic axis ratios β and γ as well as two viewing angles, ϑ and φ. Clearly, even in this idealized case of concentric ellipsoids with constant axis ratios one cannot determine a unique probability distribution $\psi(\beta, \gamma)$ from the observed probability distribution $\phi(q)$.

The situation simplifies if all ellipticals are either pure oblate ($\beta = 1$) or pure prolate ($\beta = \gamma$). In both cases, there are only two unknowns per galaxy, namely γ and the inclination angle i (where $i = 0°$ corresponds to face-on and end-on projections for oblate and prolate systems, respectively). Under the realistic assumption of randomly distributed orientations, the probability distribution of inclination angles is simply given by $p(i)\,di = \sin i\, di$, and the probability density of intrinsic flattening, $\psi(\gamma)$, can be uniquely inferred from the observed $\phi(q)$, and vice versa. The relation between γ, q, and i is given by

$$\gamma^2 \sin^2 i + \cos^2 i = \begin{cases} q^2 & \text{(oblate)} \\ q^{-2} & \text{(prolate)}. \end{cases} \qquad (13.28)$$

Using that

$$\phi(q) = \int_0^1 p(q|\gamma)\psi(\gamma)\,d\gamma, \qquad (13.29)$$

where

$$p(q|\gamma)\,dq = p(i[q]|\gamma)\,|dq/di|^{-1}\,dq, \qquad (13.30)$$

and the fact that the orientations are distributed randomly, the relation (13.28) between γ, q, and i yields

$$\phi(q) = \begin{cases} q \int_0^q \frac{\psi(\gamma) d\gamma}{\sqrt{(1-\gamma^2)(q^2-\gamma^2)}} & \text{(oblate)} \\ q^{-2} \int_0^q \frac{\psi(\gamma) \gamma^2 d\gamma}{\sqrt{(1-\gamma^2)(q^2-\gamma^2)}} & \text{(prolate).} \end{cases} \quad (13.31)$$

As shown by Fall & Frenk (1983), the inversion of these integral equations can be performed analytically, resulting in

$$\psi(\gamma) = \frac{2}{\pi}\sqrt{1-\gamma^2} \begin{cases} \frac{1}{\gamma} \int_0^\gamma \frac{d\phi}{dq} \frac{q dq}{\sqrt{\gamma^2-q^2}} & \text{(oblate)} \\ \frac{1}{\gamma^3} \int_0^\gamma \frac{d(q^3\phi)}{dq} \frac{q dq}{\sqrt{\gamma^2-q^2}} & \text{(prolate).} \end{cases} \quad (13.32)$$

Various studies have used this approach to infer $\psi(\gamma)$ from the observed $\phi(q)$ of elliptical galaxies (e.g. Fasano & Vio, 1991; Lambas et al., 1992; Tremblay & Merritt, 1995). In all cases, the inferred $\psi(\gamma)$ are negative at the high-γ end, indicating that the observed distribution of ellipticities is inconsistent with ellipticals as a class being either pure oblate or pure prolate.

Although the data are consistent with a mixed model, with partly oblate and partly prolate models, a more likely alternative is that at least some fraction of the ellipticals is triaxial. As discussed in §13.1.2 and §13.1.3 this is supported by various photometric and kinematic properties. Tremblay & Merritt (1996) presented evidence that elliptical galaxies have a bimodal distribution of shapes: while low-luminosity ellipticals are consistent with being oblate spheroids, bright ellipticals are only consistent with a triaxial intrinsic shape that is rounder on average than low luminosity ellipticals. A similar conclusion was reached by Franx et al. (1991) from their analysis of the distribution of kinematic misalignment angles. Although these results are supported by various other observations, we caution that the data can still be reproduced by a wide variety of intrinsic shape distributions. Furthermore, a proper interpretation of the data is complicated by the fact that the ellipticities and kinematic alignment angles are often functions of radius (e.g. Arnold et al., 1994; Emsellem et al., 2004) so that it is generally impossible to characterize the full shape of a single elliptical galaxy with only one or two parameters.

13.2 The Formation of Elliptical Galaxies

The fact that the dynamical structure of elliptical galaxies is less ordered than that of disk galaxies immediately suggests that their formation was more violent. Indeed, all theories for the formation of elliptical galaxies assume that violent relaxation played an important role during some stage of the formation process. This is in part a consequence of the discovery, already discussed in §5.5, that violent relaxation either during collapse from cold and asymmetric initial conditions or following a merger between objects of similar mass produces ellipsoidal systems with density profiles similar to those of elliptical galaxies. Before the current structure formation paradigm was established, there were two competing scenarios for the formation of elliptical galaxies:

(i) **The monolithic collapse scenario:** In this scenario elliptical galaxies form on a short time scale through collapse and virialization from idealized 'uncollapsed' initial conditions whose prior evolution is not considered. If the star-formation time scale is short compared to the free-fall time scale the collapse is effectively dissipationless. If the two time scales are comparable, then radiative energy losses are important and one speaks of dissipational collapse. The main characteristic of this scenario is that the stars form simultaneously with the assembly of the final galaxy.

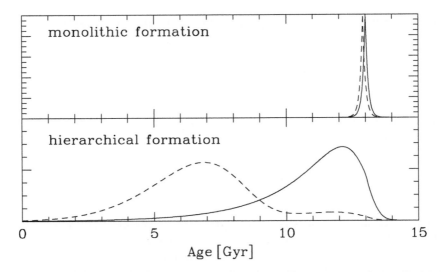

Fig. 13.4. A schematic illustration highlighting the differences between the traditional monolithic collapse (upper panel) and merger (lower panel) scenarios for the formation of the elliptical galaxy population. Solid curves show the mean star-formation history, while dashed curves show the mean rate of increase in the baryonic mass of the largest progenitor. In the monolithic collapse scenario, stars are assumed to form in a single short burst (typically more than 10 Gyr ago), and to assemble into the final galaxy during or shortly after the burst. In the merger scenario, however, the stars are formed over an extended period of time in two or more pre-existing galaxies, which only merge into the final elliptical at a (sometimes much) later time.

(ii) **The merger scenario:** In this scenario, an elliptical forms when two or more pre-existing and fully formed galaxies merge together. The main differences with respect to the monolithic collapse scenario is that formation of the stars occurs before, and effectivly independently of, the assembly of the final galaxy.

Neither of these scenarios is a good representation of how ellipticals are expected to form in the current ΛCDM structure formation paradigm, where mergers of dark matter halos occur continually across a very wide mass range and galactic star formation continues at a significant level from the epoch of re-ionization right down to the present day. In this paradigm both the assembly of galaxies and the formation of stars within them must be considered as processes rather than events. Nevertheless, for elliptical galaxies in particular, it is useful to distinguish the formation history of the stellar population from the assembly history of this baryonic material into a single object. Fig. 13.4 contrasts these in a schematic representation of the expectations for an ensemble of ellipticals in the two toy scenarios discussed above. We now discuss each of them individually before explaining why models for elliptical formation within the ΛCDM hierarchical clustering paradigm need to include aspects of both.

13.2.1 The Monolithic Collapse Scenario

In the monolithic collapse scenario elliptical galaxies form in a single burst of intense star formation at high redshift, which is coincident with their collapse to equilibrium and is followed by passive evolution of their stellar populations to the present day (Partridge & Peebles, 1967; Larson, 1975). This scenario was inspired by the fact that elliptical galaxies appear to be a remarkably homogeneous class of objects with uniformly old stellar populations. The morphology and size of the final object depend critically on when star formation occurs relative to collapse, and, in particular, on whether substantial radiative energy losses can increase the binding energy of the system before the stars form. In the dissipationless extreme, all the gas associated with the object is turned into stars either prior to the collapse or during its early stages. The collapse then effectively conserves energy and according to the spherical collapse model, the final system has an average density that is ~ 200 times that of the average density of the universe at the time of collapse (see §5.4.4). Given the observed sizes and masses of elliptical galaxies [see Eq. (13.38) below] this implies that they must all have formed at redshifts greater than 20. This is quite incompatible with our current understanding of the star-formation history of the Universe according to which only a very small fraction of all stars formed before $z = 6$ (see §2.6.8).

A second conflict between this dissipationless collapse scenario and our current understanding of cosmic structure comes from the fact that violent relaxation does not differentiate between stars and dark matter and so cannot separate them. This is incompatible with the observation that the visible regions of ellipticals contain rather little dark matter, but are surrounded by dark matter halos with masses at least 10 times the stellar mass of the galaxy and sizes which are well over an order of magnitude greater (see §13.1.5). Clearly any collapse model should apply to the *total* mass associated with the virialized system, but it then needs to explain why the stars occupy a very small region at the center of the final object. The dissipationless collapse simulations of van Albada (1982) (see §5.5.3), although able to produce convincing $R^{1/4}$ profiles, are clearly not a viable description for the formation of real ellipticals.

The situation is different if substantial dissipation can occur during collapse, requiring a star-formation time scale which is comparable to or somewhat longer than the collapse time scale. The gas can then segregate from the dark matter at the center of the potential well before turning into stars, and the final galaxy can end up with a binding energy that is substantially greater than, and a characteristic time scale which is much shorter than, those which characterized the protogalaxy before it collapsed and which continue to characterize the dark halo of the final system. During such an extended collapse, stars form at later times out of gas that was enriched in metals by earlier generations. This can result in metallicity gradients similar to those seen in many real ellipticals (e.g. Larson, 1974a).

Although clearly a more promising picture than dissipationless monolithic collapse, several issues remain. As we show below in §13.3.2 the observed properties of ellipticals lead to inferred collapse factors of their baryonic component which are comparable to those inferred for the material collecting in galaxy disks (see §11.2). Why then does the gas end up in a rotationally supported configuration in one case but not in the other? This would require ellipticals either to form from protogalaxies with substantially lower initial angular momentum, or for their formation process to be much more efficient at transferring angular momentum from baryons to dark matter. This is not implausible; the distribution of spin parameters of dark matter haloes is broad (see §7.5.4) and galaxy formation simulations often reveal efficient angular momentum transfer to the extent that it turns out to be much easier to make spheroid-dominated systems like ellipticals than disk-dominated systems like late-type spirals (e.g. Katz & Gunn, 1991; Katz, 1992, and the discussion in §11.2.6).

A more difficult problem for the monolithic collapse model comes from its principal assumption that the final assembly of an elliptical galaxy occurs simultaneously with the formation of the bulk of its stars over a relatively short time interval, perhaps a Gyr or two. The great majority of normal elliptical galaxies appear to have old stellar populations with a mean age of 10 Gyr or more, implying that the formation events should have occurred at $z \gtrsim 2$ and that the galaxies should have evolved passively thereafter. However, the total mass density in relatively massive, passively evolving galaxies appears to be a factor of 3–4 lower at $z \sim 1$ than it is today (Bell et al., 2004; Brown et al., 2007; Faber et al., 2007; Taylor et al., 2009, see also §2.6.3). Thus at least 70% of present-day ellipticals were either still forming stars at $z = 1$ or had not yet been assembled. In either case this contradicts the simple monolithic collapse hypothesis.

Additional evidence against purely passive evolution of ellipticals since high redshifts comes from recent observations that show strong evolution in their sizes: passive galaxies at $z \gtrsim 1.5$ are much smaller than present-day ellipticals of the same stellar mass (e.g. Daddi et al., 2005; Trujillo et al., 2006; van Dokkum et al., 2008; van der Wel et al., 2008).

Theoretical difficulties with the monolithic collapse scenario arise because its initial conditions are not well motivated and are difficult to reconcile with the cold dark matter paradigm. Why was there no significant star formation before the onset of collapse? Why did all star formation cease immediately after collapse? How do we relate the formation of ellipticals to that of other galaxies and other kinds of structure in order to understand why ellipticals are predominantly the most massive galaxies and are preferentially found in clusters? Such questions can be addressed within the hierarchical structure formation picture, where mergers appear a more natural explanation for many of the systematic properties of ellipticals. Thus while monolithic collapse is the simplest toy model for elliptical formation, it does not appear realistic and we therefore turn now to a discussion of mergers.

13.2.2 The Merger Scenario

The merger hypothesis was introduced by Toomre (1977) as an alternative to the monolithic collapse picture discussed above. In its extreme form, this hypothesis assumes that star formation only occurs in galactic disks and that *all* ellipticals formed by mergers of disk galaxies. The first assumption seems at least plausible, since star formation in the local Universe is restricted almost entirely to the disks of spiral and irregular galaxies and to starbursts where the fuel came from pre-existing disks (as in a merging system). Whether the second assumption is consistent with observation is less obvious. There is no doubt that mergers do occur. The outstanding isssues are: (i) whether mergers of observed galaxies (not necessarily $z=0$ galaxies) can produce remnants that resemble present-day ellipticals; and (ii) whether the merger rate as a function of progenitor properties and environment, when integrated over redshift, can reproduce the $z=0$ population, i.e. the observed abundance of elliptical galaxies as a function of stellar mass, age, metallicity, size, velocity dispersion and environment. These two aspects of the problem have typically been addressed separately. We discuss the first in this subsection and the second in the next.

Early simulations of disk mergers considered stellar disks embedded in relatively low-mass dark matter halos (Gerhard, 1981; Farouki & Shapiro, 1982; Negroponte & White, 1983; Barnes, 1988). In such mergers, the orbital angular momentum of the merging galaxies is largely converted to internal spin (some is carried away by tidal tails), producing remnants that rotate too fast to represent slowly rotating massive ellipticals. However, as pointed out by Barnes (1988), in simulations where the disks are embedded in extended dark matter halos, the final remnants are found to rotate more slowly, in qualitative agreement with observations of luminous ellipticals. The reason for this is that dynamical friction is able to transfer the orbital angular momentum of the two disks to the more weakly bound dark matter halo before the disks finally merge (see §12.3). As a result, the material in the tightly bound region of the remnant rotates only slowly, with a typical ratio between rotation speed and velocity dispersion of $v_0/\sigma_0 \sim 0.2$. In addition, the remnants display a variety of isophotal shapes (both disky and boxy), have luminosity profiles that are well fit by $R^{1/4}$ laws over a large radial range, and in projection have ellipticities that span a broad range with a peak near $\varepsilon \simeq 0.4$ (e.g. Hernquist, 1992; Barnes, 1992). The high ellipticities and low rotation speeds demonstrate that remnant shapes typically reflect an anisotropic velocity dispersion rather than rotation.

Although some of these aspects agree well with the observed properties of bright ellipticals, observationally bright ellipticals are always found to have relatively small ellipticities $\varepsilon \lesssim 0.3$, inconsistent with the simulations which suggest that the majority should have

$\varepsilon > 0.3$. In addition, the simulations cannot account for ellipticals of intermediate luminosity ($-18 \gtrsim \mathcal{M}_B \gtrsim -20.5$), which typically have values of v_0/σ_0 similar to those expected for oblate, isotropic rotators. Finally, the remnants produced by mergers of purely stellar disks are insufficiently centrally concentrated, with constant density cores that are much larger than those observed in bright ellipticals.

One possible reason for the discrepancy is that simulations have tended to focus on equal mass mergers. As shown by Naab & Burkert (2003), mergers between progenitors with mass ratios of the order of 1:3 produce remnants that are significantly different from 1:1 mergers. In particular, 1:3 mergers result in ellipticals that are predominantly disky and have kinematics more similar to oblate rotators (see also Naab et al., 1999; Bendo & Barnes, 2000). Although this is in good agreement with the observed properties of ellipticals of intermediate luminosity, it seems unlikely that mass ratio is the main parameter underlying the dichotomy among ellipticals. This would require more massive ellipticals to have similar mass progenitors more frequently than less massive ellipticals, which seems a priori surprising (Khochfar & Burkert, 2003). A similar suggestion from Weil & Hernquist (1996) is that the luminosity dependence of elliptical properties could be explained by considering multiple mergers. It is quite likely that multiple mergers occur, especially in (compact) groups of galaxies, but simulations of multiple mergers have produced remnants with even more extended cores than in the single-merger case. This arises from the fact that the phase-space densities of the progenitor disks (which were designed to mimic present-day disks) were not as high as those in the central regions of many elliptical galaxies (see §13.3.3 below). The discrepancy could be alleviated by adding dense bulges or gas components to the progenitors.

Hydrodynamical simulations have demonstrated that tidal perturbations during mergers cause much of the gas to concentrate near the center of the remnant (see discussion in §12.4.3). Simulations that include prescriptions for star formation show that this gas may be transformed into stars in bursts accompanying the first passage of the two galaxies and their final merger (Mihos & Hernquist, 1996), giving rise to a more concentrated remnant which agrees better with the structure of observed ellipticals. This process also has convincing observational support, since strongly star-bursting galaxies (i.e. galaxies with a greatly elevated star-formation rate) almost always have highly distorted morphologies suggestive of a recent merger (see §2.3.7). Since most disk galaxies have non-negligible gas mass fractions, it seems natural that the progenitor disks should have a gas component. In particular, since most ellipticals have old stellar populations, any merger involving disks is likely to have happened at relatively high redshift, where such galaxies are expected to have higher gas fractions than at the present time.

Using a large suite of simulations of mergers between equal-mass disk galaxies, both with and without gas, Cox et al. (2006) showed that dissipational merger remnants differ substantially from their dissipationless counterparts. In particular, gas dissipation and the associated star formation result in remnants that are more compact, rounder, have higher central velocity dispersions and rotation velocities, and are much closer to isotropic than without dissipation. Consequently, as shown in Fig. 13.5, dissipationless and dissipational remnants occupy different regions in the diagram of v_0/σ_0 vs. ellipticity: while dissipationless mergers only occupy the region with low v_0/σ_0 and extend to high ellipticities, dissipational mergers span a much larger range of rotation properties, including remnants that reside close to the oblate isotropic rotator line of Eq. (13.25). As is evident from Fig. 13.5, dissipational mergers are a much better match to the observational data (see also Burkert et al., 2008). In addition, such mergers also yield much smaller kinematic misalignments than their dissipationless counterparts, in better agreement with observation.

On the other hand, dissipational mergers have difficulty producing remnants that are slow rotators and boxy, and so comparable to the majority of luminous ellipticals. In particular, the presence of a dissipative component causes the production of a steep central cusp which is not

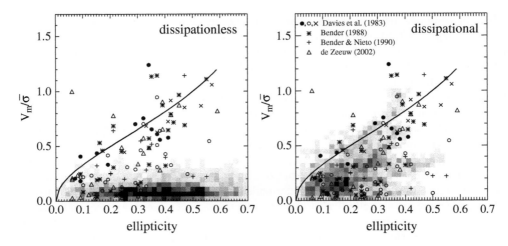

Fig. 13.5. The anisotropy diagram of $v_m/\bar{\sigma}$ vs. ellipticity for dissipationless (left-hand panel) and dissipational (right-hand panel) merger remnants obtained from a large suite of hydrodynamical simulations (gray scale). Overplotted, with different symbols, are data for observed ellipticals from Davies et al. (1983), Bender (1988), Bender & Nieto (1990), and de Zeeuw et al. (2002). Note that the dissipational simulations yield remnants in much better agreement with these data. The solid line in both plots corresponds to Eq. (13.25), and indicates the expected relation for an oblate isotropic rotator. [Adapted from Cox et al. (2006) by permission of AAS]

only inconsistent with the shallow cusps seen in massive ellipticals (see §13.1.2), but also destabilizes the box orbits responsible for the boxiness of the isophotes (Gerhard & Binney, 1985; Merritt & Fridman, 1996). It has therefore been suggested that massive ellipticals are the remnants of 'dissipationless' (also called 'dry') mergers of elliptical progenitors. This is supported by numerical simulations, which seem to indicate that the remnants of dry mergers between elliptical progenitors *of roughly equal mass* are typically slow rotating, boxy ellipticals (e.g. Khochfar & Burkert, 2005; Naab et al., 2006; Cox et al., 2006) that lie on the fundamental plane (Nipoti et al., 2003; Boylan-Kolchin et al., 2006; Robertson et al., 2006b).[2] If both progenitors contain a supermassive black hole in their nuclei, a black hole binary may form in the central region of the merger remnant. The orbital decay of such a binary can produce a core in the remnant by pumping energy into the orbits of the tightly bound stars that made up the cusps of the progenitors (e.g. Begelman et al., 1980; Ebisuzaki et al., 1991). As a rule of thumb, the total ejected mass of stars is comparable to that of the black hole binary (see Gualandris & Merritt, 2007, and references therein), which seems to be sufficient to explain the observed cores in massive ellipticals (e.g. Faber et al., 1997; Milosavljević et al., 2002). Hence, dry mergers appear to be a viable mechanism for the formation of massive ellipticals. The obvious question is why massive ellipticals apparently form via dissipationless ('dry') mergers, while their less massive counterparts seem to require dissipational ('wet') mergers. This is part of the second class of issues which we referred to at the beginning of this subsection and which we address in the next.

To summarize, the merger scenario seems able to explain a large number of observable properties of individual elliptical galaxies, as long as dissipation plays an important role at the low-mass end, while the more massive ellipticals are mainly the product of dry mergers. Further support for this conclusion is presented in §13.3.

---

[2] Dry mergers between disky progenitors with a mass ratio significantly different from unity typically yield disky remnants.

13.2.3 Hierarchical Merging and the Elliptical Population

Given that mergers between objects similar to observed $z > 0$ galaxies appear able to produce remnants with structural properties spanning the full range seen in present-day ellipticals, the critical question becomes whether real merger rates are, in fact, compatible with production of the current *population* of elliptical galaxies (i.e. their abundance as a function of stellar mass, stellar age, metallicity, size, velocity dispersion and environment) from the observed *population* of higher redshift galaxies. This is a complex question. Since it is very difficult to measure merger rates observationally to the necessary precision (see §13.3.6 below), it has so far been addressed primarily from a theoretical point of view. The question can be phrased as follows: Assume that structure grows hierarchically with galaxies condensing at the center of dark matter halos. If this process is tuned to reproduce the observed galaxy population at each redshift, does the merger rate predicted in a particular CDM cosmology then vary with progenitor properties and redshift as needed to produce the current elliptical galaxy population, given the simulation results for individual galaxy mergers reviewed in the previous subsection?

The first serious attempt to address this question was that of Kauffmann et al. (1993). These authors modeled the statistics of the build-up of dark matter halos in a CDM universe using merger trees of the kind described in §7.3. Within each halo they assumed gas to cool off onto a central disk at rates given by a model similar to those of §8.4 and to turn into stars according to a prescription similar to that of §9.5.1, modified by supernova feedback using the simple model discussed in §8.6.3. Adopting plausible efficiencies for star formation and feedback, they showed that these assumptions result in predicted galaxy luminosity functions in crude agreement with data, both at $z = 0$ and at higher redshift (see also Kauffmann et al., 1994). In conjunction with a dynamical friction model similar to that of §12.3, the halo merger trees then allowed merger rates to be calculated as a function of galaxy properties and redshift. Assuming that elliptical galaxies are produced *only* from those major mergers that are unable to regrow a disk at later times, Kauffmann et al. (1993) showed that $z = 0$ 'ellipticals' are predicted to dominate the population of high-mass galaxies, to be an increasingly small fraction of the population at lower masses, to have red colors, and to be found predominantly in clusters. These predicted trends all agreed quite well with observation.

Later work showed that cluster ellipticals are expected to have uniformly old stellar populations in this model, but that 'field' ellipticals (i.e. objects embedded in lower mass halos) should be significantly younger (Kauffmann, 1996b; Baugh et al., 1996). In addition, more massive ellipticals were predicted to be systematically younger. Both trends disagree with observation. As discussed in §13.5.1 below, real massive ellipticals appear to have *older* stellar populations than lower mass galaxies, and the variation of mean stellar age with environment is weaker than predicted. The resolution of this problem may be related to another difficulty of these early CDM models, namely their inability to produce a galaxy luminosity function that cuts off as sharply as observed at the bright end. This cut-off appears to require a process which suppresses star formation in massive systems without forming new stars (i.e. other than supernova feedback). As discussed in §14.4, feedback from an active galactic nucleus (AGN) is a candidate for such a process. An extension of the above models to include AGN feedback can not only reproduce the observed luminosity function cut-off, but also reverse the predicted trend between stellar mass and mean stellar age, and substantially reduce the predicted correlation between mean stellar age and environment, leading to much better agreement with the observed population of ellipticals (De Lucia et al., 2006, see also discussion in §15.5.1). Such hierarchical formation models also predict the gas content of merging galaxies (i.e. the fraction of 'wet' versus 'dry' mergers) to vary with remnant mass in the way needed to explain the luminosity dependence of the structural properties of elliptical galaxies highlighted in the previous subsection (Khochfar & Burkert, 2003; Kang et al., 2007).

13.3 Observational Tests and Constraints

The main areas where observations can test the theories for the formation of elliptical galaxies, discussed above, are the following:

(i) Star-formation history: while quasi-monolithic collapse assumes the stars of each elliptical to form in a single burst, hierarchical merging predicts star formation in several different sites and over a more extended period, possibly with merger-induced starbursts.
(ii) Assembly history: for quasi-monolithic collapse the assembly of an elliptical is coeval with the formation of its stars, while in the merger scenario, most of the stars form in progenitor galaxies well before assembly of the final elliptical. In addition, assembly may involve multiple mergers, and so occur over an extended period of time.
(iii) Progenitor properties: for quasi-monolithic collapse, the immediate progenitor of an elliptical galaxy is a single starbursting gas cloud, while in the hierarchical merging picture ellipticals have diverse progenitors (spirals, ellipticals, irregulars, etc.).

Observations of (the evolution of) the structural, kinematic, spectral and chemical properties of elliptical galaxies can clearly test these ideas in some detail. Here we focus predominantly on the structural and kinematical properties. Constraints from spectral and chemical properties will be discussed in §13.5 below.

13.3.1 Evolution of the Number Density of Ellipticals

An important difference between quasi-monolithic collapse and assembly through hierarchical merging concerns their predictions for how and when ellipticals are finally assembled. The first scenario requires the present population of ellipticals to be present with essentially no change in comoving number density out to the relatively high redshift when collapse occurred, whereas the second predicts that the comoving number density above any given stellar mass should decrease with increasing redshift. Note, however, that the rate of this decrease is quite dependent on cosmology, since this sets the merger rate for dark matter halos (see §7.3.5).

Early attempts to apply this test suggested strong evolution in the observed comoving number density of bright and/or red galaxies – presumably ellipticals (e.g. Kauffmann et al., 1996; Zepf, 1997; Kauffmann & Charlot, 1998b; Aragon-Salamanca et al., 1998). These analyses were primarily based on an assumed Einstein–de Sitter cosmology, and later studies with larger samples and the now favored ΛCDM gave more ambiguous results – the switch in cosmology weakens both the evolution inferred from the observational data and the evolution predicted by the models. As noted in §13.2.1, recent work seems to have reached a consensus that the total stellar mass in red, passively evolving galaxies is a factor of several lower at $z \sim 1$ than it is today (Bell et al., 2004; Brown et al., 2007; Faber et al., 2007; Taylor et al., 2009). Indeed, the total stellar mass in galaxies of *all types* at redshifts $1.3 < z < 2$ seems to be factor of several below the current value for ellipticals if one integrates down to individual stellar masses of a few $10^{10} M_\odot$ (Marchesini et al., 2008). At face value, this appears a definitive exclusion of quasi-monolithic collapse at $z \gtrsim 2$. However, the situation is less clear for the most massive galaxies. For example, a number of papers have argued that the abundance of the most massive ellipticals changes little if at all back to $z \sim 1$ (e.g. Daddi et al., 2000; Cimatti et al., 2002b; Collins et al., 2009). This may reflect the difficulty of reliably establishing the abundance of the most massive (and thus rarest) galaxies, as well as in measuring their masses accurately. At the highest masses both observational and theoretical abundances are extremely sensitive to mass, so that the comparison between the two becomes very sensitive both to theoretical assumptions and to observational error in the measured masses (e.g. Bower et al., 2006; Kitzbichler & White, 2007).

13.3.2 The Sizes of Elliptical Galaxies

As we have seen in §11.2, the standard formation paradigm allows us to make predictions for the sizes of galaxy disks. This reflects the fact that disk formation is assumed (though not proven) to be a near-adiabatic process governed by angular momentum conservation. The situation is far more complicated for elliptical galaxies, because of the highly nonlinear, and non-adiabatic processes involved. Nevertheless, with some idealizations one can still make simple predictions (e.g. Shen et al., 2003; Boylan-Kolchin et al., 2005). In general, the size of an equilibrium galaxy is related to its mass and binding energy via the virial theorem (see §5.4.4), which states that $E = W/2$. Here

$$W = -4\pi G \int_0^\infty \rho_s(r) \left[M_s(r) + M_h(r) \right] r \, dr, \tag{13.33}$$

is the potential energy of the stellar system, and we have assumed spherical symmetry. The subscripts s and h refer to stars and the dark matter halo, respectively. The potential energy thus consists of a term that describes the self-energy of the stellar system plus a cross-term that describes the interaction energy of the stellar system with the dark matter halo. In general, we can cast Eq. (13.33) in the form

$$W = -\zeta \frac{G M_s^2}{r_e}, \tag{13.34}$$

with r_e some characteristic radius of the stellar system, and ζ a form factor that depends on the density distributions of the stars and dark matter. In the absence of a dark matter halo, a Hernquist sphere has $\zeta \simeq 0.303$, if r_e is defined as the effective radius, while realistic models *with* a dark matter halo typically have $\zeta \sim 0.6 \pm 0.1$ (Boylan-Kolchin et al., 2005).

In the case of the monolithic collapse scenario, consider a perturbation consisting of both gas and dark matter with a homogeneous density distribution, and having a total mass of $M_{\rm vir}$ at turn-around. The initial energy of the gas at turnaround is then simply

$$E_i = W_i = -\frac{3}{5} \frac{G M_{\rm gas}^2}{f_{\rm gas} r_t}, \tag{13.35}$$

where $f_{\rm gas} = M_{\rm gas}/M_{\rm vir}$ and r_t is the turnaround radius. Suppose that, while the dark matter collapses and virializes, the gas dissipates its binding energy (due to radiative cooling) until it is instantaneously turned into stars at a time when the absolute value of its binding energy has increased to $|E_f| = \eta |E_i|$. From that point on the stellar system experiences dissipationless, gravitational collapse resulting in a virialized stellar system, embedded in a virialized dark matter halo. According to the virial theorem we then have that

$$E_f = \frac{W_f}{2} = -\zeta_f \frac{G M_{\rm gas}^2}{2 r_e}, \tag{13.36}$$

where we have assumed that all the gas is turned into stars. Relating this to the initial binding energy of the gas, and using that $\zeta_f \sim 0.6$, we obtain that

$$\eta \simeq f_{\rm gas} \frac{r_{\rm vir}}{r_e}, \tag{13.37}$$

where $r_{\rm vir}$ is the virial radius of the final dark matter halo, which we have taken to be half the turnaround radius (see §5.4.4). Using the Sloan Digital Sky Survey, Shen et al. (2003)

found that the effective radii of elliptical galaxies are related to their stellar masses, $M_\star$, according to[3]

$$r_e = 0.98 h^{-1}\,\text{kpc} \left(\frac{M_\star}{10^{10} h^{-1}\,M_\odot}\right)^{0.56}. \tag{13.38}$$

Substituting this into Eq. (13.37) and assuming that the average density of a dark matter halo is 100 times the critical density for closure (see §5.4.4), we finally obtain that

$$\eta = 6.9 \left(\frac{f_{\text{gas}}}{0.15}\right)^{0.44} \left(\frac{M_{\text{vir}}}{10^{12} h^{-1}\,M_\odot}\right)^{-0.23}. \tag{13.39}$$

Thus, in the monolithic collapse scenario the binding energy of the gas has to become more negative by a factor ~ 7 before it forms stars, in order to explain the observed sizes of elliptical galaxies.

In the merger scenario, the size of the remnant depends on those of the progenitors as well as on the amount of dissipation and star formation during the merger. The latter processes depend strongly on the gas content of the progenitors and on the details of the merger. Simulations which follow gas, stars, dark matter, star formation and feedback from supernovae are normally required to make realistic predictions for the sizes of merger remnants. In special circumstances, however, simple predictions are still possible. In particular, in the case of a merger between two galaxies with no gas (a 'dissipationless' or 'dry' merger), dissipation and star formation do not play a role, and one can predict the size of the remnant from the sizes and masses of the progenitors. Consider two virialized, gas-poor galaxies with stellar masses M_1 and M_2 and effective radii r_1 and r_2, which merge to form a new galaxy with stellar mass M_f and effective radius r_f. We assume that initially both progenitors reside in their own dark matter halos (both are two-component systems), and are on an orbit with energy

$$E_{\text{orb}} = -f_{\text{orb}} \frac{G M_1 M_2}{(r_1 + r_2)}. \tag{13.40}$$

Here, for convenience, we have introduced f_{orb} to express the orbital energy in units of the orbital energy of two point masses, M_1 and M_2, on circular orbits with separation $r_1 + r_2$. Note that $f_{\text{orb}} = 0$ for a parabolic orbit, while elliptic (bound) and hyperbolic (unbound) orbits have $f_{\text{orb}} > 0$ and $f_{\text{orb}} < 0$, respectively (see §12.4.1). Once the merger remnant has established virial equilibrium, the total energy of its stars is given by

$$E_f = -\zeta_f \frac{G(M_1 + M_2 - M_{\text{ej}})^2}{2 r_f} + E_{\text{ej}} + E_{\text{tr}}. \tag{13.41}$$

Here $E_{\text{ej}} = \frac{1}{2} M_{\text{ej}} v_{\text{ej}}^2$ is the kinetic energy of the stellar mass, M_{ej}, that has been ejected from the remnant (due to the violent potential fluctuations) with an average ejection speed v_{ej} at infinity, and E_{tr} is the energy that has been transferred from the stars to the dark matter. Energy conservation allows us to write

$$\zeta_f \frac{(M_1 + M_2 - M_{\text{ej}})^2}{r_f} = \zeta_1 \frac{M_1^2}{r_1} + \zeta_2 \frac{M_2^2}{r_2} + 2(f_{\text{orb}} + f_{\text{ej}} + f_{\text{tr}}) \frac{M_1 M_2}{r_1 + r_2}, \tag{13.42}$$

where f_{tr} and f_{ej} are defined by

$$E_{\text{tr}} = f_{\text{tr}} \frac{G M_1 M_2}{(r_1 + r_2)} \quad \text{and} \quad E_{\text{ej}} = f_{\text{ej}} \frac{G M_1 M_2}{(r_1 + r_2)}. \tag{13.43}$$

---

[3] Here we ignore the fact that observationally the effective radii are obtained from the projected surface brightness, which may differ somewhat from those obtained from the 3D luminosity density.

While, by definition, $f_{\rm ej} \geq 0$, the value of $f_{\rm tr}$ can be either positive or negative. The former corresponds to a net transfer of energy from the stars to the dark matter (due, for instance, to dynamical friction), while a negative value of $f_{\rm tr}$ indicates that energy has been transferred from the dark matter to the stars (due to violent relaxation). For brevity, in what follows we write the results in terms of $f_{\rm tot} \equiv f_{\rm orb} + f_{\rm tr} + f_{\rm ej}$. In addition, we assume homology, so that $\zeta_{\rm f} = \zeta_1 = \zeta_2 \equiv \zeta$. Numerical simulations suggest that this is a reasonable assumption, and that homology is fairly well preserved during dissipationless mergers (Boylan-Kolchin et al., 2005).

If we assume that the progenitors follow a (narrow) radius–mass relation of the form $r \propto M^\alpha$, then in terms of the progenitor mass ratio $q \equiv M_1/M_2$ (with $q \geq 1$), we can rewrite Eq. (13.42) as

$$\frac{r_{\rm f}}{r_1} = (1+q)^2 \left[q^2 + q^\alpha + \frac{2 f_{\rm tot}}{\zeta} \frac{q}{1+q^{-\alpha}} \right]^{-1} \left(1 - \frac{M_{\rm ej}}{M_1 + M_2} \right)^2, \quad (13.44)$$

which reduces to

$$\frac{r_{\rm f}}{r_1} = \frac{4}{2 + f_{\rm tot}/\zeta} \left(1 - \frac{M_{\rm ej}}{M_1 + M_2} \right)^2 \quad (13.45)$$

for an equal mass merger ($q = 1$). To gain some insight, consider a few idealized cases. In the case of single-component systems (i.e. elliptical galaxies without dark matter halos), we have that $f_{\rm tr} = 0$. If in addition we ignore mass ejection ($f_{\rm ej} = 0$) and adopt parabolic orbits ($f_{\rm orb} = 0$), then equal mass mergers result in $r_{\rm f} = 2r_1$. If some mass is ejected ($f_{\rm ej} > 0$), then $r_{\rm f} < 2r_1$, because the ejected mass carries away binding energy, leaving a more strongly bound remnant. In the absence of mass loss, equal mass mergers that start from elliptic and hyperbolic orbits result in $r_{\rm f} < 2r_1$ and $r_{\rm f} > 2r_1$, respectively. In the case of two component systems (i.e. elliptical galaxies embedded in extended dark matter halos), the situation is more complicated. In particular, since $f_{\rm tr}$ can be both positive and negative, it is difficult to make a priori predictions. Numerical simulations indicate that the ejected mass is typically sufficiently small that it can be neglected, and that equal mass mergers between two-component systems with initial parabolic orbits typically have $r_{\rm f}/r_1$ close to 2. This implies that $f_{\rm tr} \sim 0$, i.e. there is little net transfer of energy from the stellar component to the dark matter (Nipoti et al., 2003; Boylan-Kolchin et al., 2005; Robertson et al., 2006b). On the other hand, Boylan-Kolchin et al. (2005) have shown that realistic orbits typically have $f_{\rm orb} > 0$, with $f_{\rm orb} \sim 0.1$ for the 'most probable' orbit. Equal mass mergers starting from such orbits have $f_{\rm tr} \sim 0.4$ (i.e. dynamical friction causes a significant transfer of energy from stars to dark matter), so that $r_{\rm f}/r_1 \sim 1.4$.

It is interesting to see whether the remnants of dry mergers have the same radius–mass relation, $r \propto M^\alpha$, as the progenitors. Since the merger causes a transition $M \to (1+q^{-1})M$ in mass, we need $r_{\rm f}/r_1 = (1+q^{-1})^\alpha$ in order for the remnant to obey the same radius–mass relation. Using Eq. (13.44), and ignoring mass loss, the condition for conserving the radius–mass relation can be written as

$$f_{\rm tot} = \frac{\zeta}{2} \frac{q^\alpha + 1}{q} \left[(1+q)^{2-\alpha} - q^{2-\alpha} - 1 \right]. \quad (13.46)$$

Setting $\alpha = 0.56$ (Shen et al., 2003) and $\zeta = 0.6$ we find $f_{\rm tot}(q) \sim 0.43$, with remarkably weak dependence on q: it changes by no more than 2% over the entire range of q from 1 to 100. This is in reasonable agreement with the value of $f_{\rm tot}$ obtained by Boylan-Kolchin et al. (2005) for the most probable orbit, suggesting that dry mergers roughly preserve the observed radius–mass relation.

Finally, we emphasize that these results are only valid for dissipationless mergers. If the progenitor galaxies have a significant gas reservoir, dissipation can occur, which in general will result in a remnant with a smaller $r_{\rm f}$ than that of the equivalent remnant in a dissipationless merger.

13.3.3 Phase-Space Density Constraints

As discussed in §5.5.1 the maximum value of the coarse-grained phase-space density of a collisionless system cannot increase during its evolution. If giant elliptical galaxies are produced by collisionless mergers of stellar disks, then their maximum phase-space densities should not exceed those of their disk progenitors.

The phase-space density in the central region of a galaxy can be estimated by dividing the central mass density by the volume occupied by an ellipsoid with its axes equal to the three velocity dispersions. As a reasonable approximation, we assume that elliptical galaxies can be described by an isothermal sphere with the boundary conditions $\rho(0) = \rho_0$ and $d\Psi/dr(0) = 0$. As we have seen in §5.4.7, the core region of such a system has a density distribution given by

$$\rho(r) \approx \frac{\rho_c}{[1+(r/r_c)^2]^{3/2}}, \tag{13.47}$$

where

$$r_c = \frac{3\sigma}{\sqrt{4\pi G \rho_c}}, \tag{13.48}$$

is the King radius, at which the projected surface density drops to half of the central value, with σ the one-dimensional velocity dispersion of the system. The central phase-space density can then be written as

$$f_0 = \frac{27}{16\pi^2} \frac{1}{G\sigma_0 r_c^2} = 39.5 \, M_\odot \, pc^{-3} (km \, s^{-1})^{-3} \left(\frac{\sigma}{km \, s^{-1}}\right)^{-1} \left(\frac{r_c}{pc}\right)^{-2}. \tag{13.49}$$

Note that both σ and r_c can be obtained observationally.

For disk galaxies, we model the vertical structure as that of an isothermal sheet (see §11.1.1). The central velocity dispersion in the vertical direction can then be written as

$$\sigma_z^2(0) = 2\pi G \Sigma_0 z_d \quad \text{with} \quad z_d = \varsigma R_d, \tag{13.50}$$

where the vertical scale height, z_d, is assumed to be proportional to the disk scale length. Assuming further that the dispersions of the other two velocity components are proportional to σ_z,

$$\sigma_\phi = k_\phi \sigma_z, \quad \sigma_R = k_R \sigma_z, \tag{13.51}$$

the central phase-space density of a disk can be written as

$$f_c = \frac{3\rho_0}{4\pi \sigma_z(0) \sigma_\phi(0) \sigma_R(0)} = \frac{\Sigma_0}{4R_d} \frac{1}{(2\pi G \Sigma_0 R_d)^{3/2}} \frac{3}{4\pi \varsigma^{5/2} k_R k_\phi}. \tag{13.52}$$

For numerical values we take $\varsigma = 0.1$ (Bottema, 1997), $k_\phi = \sqrt{2}$ and $k_R = 2$. The values of k_R and k_ϕ are motivated by observations of our local disk and the flat rotation curves of disk galaxies (see Carlberg, 1986). Fig. 13.6 compares the observed central phase-space densities for disks with those for elliptical galaxies. As one can see, the central phase-space densities of ellipticals with $\mathcal{M}_B \gtrsim -22$ are more than three orders of magnitude higher than those of the observed disks. This result has been used to argue against the merger scenario for the formation of elliptical galaxies.

This argument has two important limitations, however. First of all, as shown in §11.2, disks that formed at high redshifts are expected to be smaller and denser than their present-day counterparts. It may therefore be possible that these high-redshift disks are dense enough to form the elliptical galaxies observed today. Second, the central phase-space density of an elliptical galaxy is associated with a small fraction of the total stellar mass at system center and is much higher than the values typical for the main body of the galaxy. An important question, therefore, is whether the main bodies of elliptical galaxies could be produced by stellar mergers, even if their

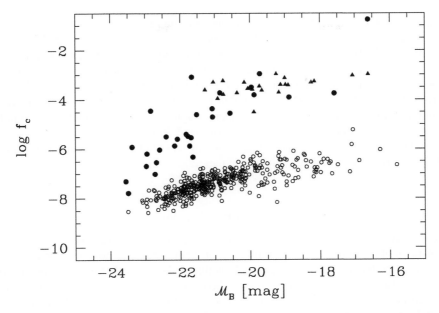

Fig. 13.6. The central phase-space densities (in arbitrary units) of elliptical galaxies (solid symbols, with triangles showing lower limits) compared to those of spiral galaxies (open circles). [After Mao & Mo (1998)]

central parts are not. These two issues have been discussed by Mao & Mo (1998). By defining 'average' phase-space densities for galaxies, they found that the main body of giant ellipticals can be produced by mergers of high-redshift disks. However, high-redshift disks are still not dense enough to produce the high central phase-space densities of (some) low-luminosity ellipticals. Using numerical simulations and analytical arguments, Hernquist et al. (1993) estimated that this discrepancy is confined to the inner $\sim 15\%$ of the stellar mass.

One obvious way to alleviate this problem is to assume that the progenitors had dense stellar bulges (Barnes, 1992). Another is to circumvent the phase-space constraints by assuming that the progenitors contained significant amounts of gas. Since gas can lose energy through radiation before it turns into stars, it can easily produce enhanced central phase-space densities. Hernquist et al. (1993) estimated that dissipation and star formation can result in sufficiently high phase-space densities if the progenitors had a gas mass fraction of $\sim 25-30\%$. Subsequent hydrodynamical simulations of disk galaxy mergers have clearly demonstrated that merging promotes the accumulation of large quantities of gas in the central regions. If converted into stars, this gas can give rise to dense cusps/cores in reasonable agreement with observation (e.g. Mihos & Hernquist, 1996; Cox et al., 2006).

13.3.4 The Specific Frequency of Globular Clusters

Elliptical galaxies are found to have more globular clusters per unit stellar mass than spirals (see Harris, 1991, for a review). In terms of the specific frequency, S_N, defined as the number of globular clusters divided by the galaxy luminosity in units of $\mathcal{M}_V = -15$, ellipticals in rich clusters have $S_N \sim 5$, field ellipticals have $S_N \sim 3$, and spirals have $S_N \sim 0.5-1$. This difference in globular cluster frequency appears to be a serious problem for the merger scenario: if elliptical galaxies are the merger remnants of spirals, they should have the same S_N as spirals. However, this argument applies only to mergers of pure stellar disks. If gas is involved in the mergers, the situation may be different. In fact, violent mergers often reveal large populations of newly forming, massive young star clusters, with masses and sizes consistent with globular clusters (e.g. Holtzman et al., 1992; Schweizer et al., 1996; Whitmore et al., 1999, see also Fig. 2.10). Furthermore, it has become clear that globular clusters have a bimodal color distribution, with a blue (metal-poor) population and a distinct red (metal-rich) population (e.g. Zepf & Ashman, 1993; Forbes et al., 1997). A consistent interpretation of the data appears to be that the metal-poor population formed at high redshifts in protogalactic fragments, while the metal-rich population formed during merger-induced starbursts (e.g. Ashman & Zepf, 1992; Beasley et al., 2002). In this picture, the high specific frequency of globular clusters in elliptical galaxies may simply be a relic of their formation by major, dissipational mergers.

13.3.5 Merging Signatures

If many elliptical galaxies indeed formed via major mergers, we would expect to find observational signatures of these events in some fraction of them. The most obvious sign of a (recent) merger is a clearly distorted morphology, such as that of NGC 7252 (Fig. 12.5) or the Antennae (Fig. 2.10). In many cases these peculiar galaxies reveal pronounced tidal tails, which, as we have seen in §12.2.2, are produced in the merger of two stellar disks with similar masses. Unfortunately, since violent relaxation lasts only a few dynamical times, the strongly distorted morphology is visible only for a fraction of the Hubble time. It the relevant time scale can be estimated, this signature may be used to determine the instantaneous merger rate (see below), but it cannot be used to test whether a specific elliptical indeed formed through a major merger. Tidal tails are somewhat longer-lived, but they are typically low surface brightness features that are difficult to detect, especially at high redshift. Moreover, not all merger configurations produce prominent tidal tails.

Mergers can also produce 'shells', 'ripples' and 'plumes' in the outer parts of the remnants which can survive for longer periods. Such fine structures are due to material that was originally in the outer disk(s) of the progenitor(s) and will mix into the remnant long after the inner region has relaxed (e.g. Hernquist & Spergel, 1992). Shells can also form in minor mergers when a dwarf galaxy disrupts on a relatively radial orbit and 'phase-wraps' its dynamically cold material around the larger galaxy it fell into (Quinn, 1984).

Schweizer et al. (1990) defined an empirical index to quantify the significance of fine structure:

$$\Sigma \equiv S + \log(1+n) + J + B + X, \tag{13.53}$$

where S is a visual measure of the most prominent ripple on a scale of 0 to 3, n is the number of detected ripples, J is the number of optical 'jets', B is a visual estimate of the maximum boxiness of the galaxy's isophotes on a scale from 0 to 4, and X equals 0 or 1 depending on whether or not the image of the galaxy shows an X-shaped structure. A survey by Seitzer & Schweizer (1990) showed that more than half of all ellipticals, and at least one third of all S0s, posses significant fine structure. The link to mergers (major and minor) is supported by observations that show a correlation between the presence and strength of fine structure and other merger signatures, such as tidal tails, multiple nuclei and young stellar populations (e.g. Malin & Carter, 1983; Schweizer & Seitzer, 1992). It is important to stress, however, that these fine structures form only when the merger involves at least one dynamically cold progenitor (disk or dwarf galaxy); mergers between two dynamically hot systems (i.e. between two ellipticals) do not produce shells and ripples, simply because the intrinsic velocity dispersion is too high. Thus, whereas the presence of fine-structure is a strong indication for a merger origin, the lack of fine-structure does not necessarily imply that the system has not undergone a major merger in its not-too-distant past.

13.3.6 Merger Rates

What is the rate of major mergers as a function of redshift and of galaxy properties? Obviously this is an important question, whose answer determines the abundance and properties of merger remnants, and hence, under the hypothesis that all ellipticals form via major mergers, the properties of the present-day elliptical population. We reviewed theoretical estimates of these rates in §13.2.3; we now turn to direct observational estimates.

Two different techniques have been used to probe the redshift evolution of the galaxy merger rate, $\dot{n}_{\rm mrg}$, defined as the number of merger events per unit time per unit comoving volume. The first method estimates the fraction of galaxies that are in the process of merging, $f_{\rm mrg}$, as function of redshift. This merger fraction is related to the merger rate according to

$$\dot{n}_{\rm mrg}(z) = \frac{f_{\rm mrg}(z) \, n_{\rm gal}(z)}{\tau_{\rm mrg}(z)}, \tag{13.54}$$

where $n_{\rm gal}$ is the comoving number density of the galaxy population probed, and $\tau_{\rm mrg}$ is the time scale over which the merger can be identified as such (which is proportional to the dynamical time). Unfortunately, because of cosmological surface brightness dimming (see §3.1.6), most of the merger signatures discussed in the previous section are too diffuse to be identified at high redshift, and one must typically resort to estimating the incidence of strongly disturbed galaxies as a function of redshift. With the high spatial resolution provided by the HST, it soon became clear that the number density of peculiar (i.e. strongly distorted) galaxies increases strongly with increasing redshift (e.g. Driver et al., 1995; Glazebrook et al., 1995; Brinchmann et al., 1998). Unfortunately, it has also become clear that not all galaxies with disturbed morphologies are undergoing a (major) merger; in some cases the distortions merely reflect individual star-formation regions within the galaxies. Because of this, results for the evolution of the merger fraction obtained using this technique are still very uncertain. If the evolution is parameterized as

$$f_{\rm mrg} \propto (1+z)^m, \tag{13.55}$$

the best-fit values for m claimed by different studies cover the entire range $0 \lesssim m \lesssim 4$ (e.g. Le Fèvre et al., 2000; Conselice et al., 2003; Lotz et al., 2008). In addition, the time scale $\tau_{\rm mrg}(z)$ and its dependence on redshift cannot be estimated reliably, leading to another major uncertainty when merger fraction is converted to merger rate.

An alternative is to infer the merger rate from the number density of close galaxy pairs (typically with separations less than ~ 20–$50\,{\rm kpc}$). Under the assumption that these pairs will merge on a time scale $\tau_{\rm df}$ due to dynamical friction, the merger rate is related to the pair fraction, $f_{\rm pair}$, according to

$$\dot{n}_{\rm mrg}(z) = \frac{1}{2} \frac{f_{\rm pair}(z) \, n_{\rm gal}(z)}{\tau_{\rm df}(z)}, \tag{13.56}$$

where the factor $1/2$ takes account of the fact that a galaxy pair creates a single merger. An advantage of this method over the direct identification of mergers is that it simultaneously provides statistics on the progenitor properties, such as mass ratios and morphologies. The downside is that one has to be aware of projection effects, and that evolution in the number density of close pairs does not necessarily reflect evolution in the merger rate (see Berrier et al., 2006). Furthermore, the characteristic time scale, $\tau_{\rm df}$, required to convert the number density of pairs into a merger rate has an uncertainty of at least a factor two (but see Kitzbichler & White, 2008). Parameterizing the redshift dependence of $f_{\rm pair}$ with the same power-law dependence as in Eq. (13.55), different studies once again have found values for m spanning the entire range $0 \lesssim m \lesssim 4$ (e.g. Patton et al., 2002; Lin et al., 2004a; Bell et al., 2006b).

What do these merger rates imply for the idea that all ellipticals have formed by major mergers? The number density of merger remnants at time t_0, $n_{\rm rem}(t_0)$, can be (approximately) estimated from the number density of merger candidates at t_0, $n_{\rm mrg}(t_0)$, using

$$\frac{n_{\rm rem}(t_0)}{n_{\rm mrg}(t_0)} \approx \frac{\int_0^{t_0} \dot{n}_{\rm mrg}(t)\,dt}{\int_{t_0-\tau_{\rm mrg}}^{t_0} \dot{n}_{\rm mrg}(t)\,dt}. \quad (13.57)$$

If we make the simplifying assumptions that $n_{\rm gal}$ does not evolve with redshift, and that $H_0 t = \frac{2}{3}(1+z)^{-3/2}$ (as for an Einstein–de Sitter cosmology; see §3.2.3), we can write the present-day ratio between the number of merger remnants and that of ongoing mergers as

$$\frac{n_{\rm rem}}{n_{\rm mrg}} = \frac{(1+z_{\rm f})^{m-3/2} - 1}{(m-3/2) H_0 \tau_{\rm mrg}}, \quad (13.58)$$

where $z_{\rm f}$ is the typical redshift from which present-day merger remnants began to form, and we have assumed that $H_0 \tau_{\rm mrg} \ll 1$. As an illustration, let us take $H_0 = 70\,{\rm km\,s^{-1}\,Mpc^{-1}}$, $\tau_{\rm mrg} = 5 \times 10^8$ yr, and $z_{\rm f} = 3$. It then follows from Eq. (13.58) that $15 \lesssim n_{\rm rem}/n_{\rm mrg} \lesssim 350$ for m in the range $0 < m < 4$. From the *New General Catalog*, Toomre (1977) found that 11 out of about 4,000 galaxies show tidal tails and other indications of mergers, while ~ 800 are ellipticals. Under the assumption that all ellipticals are merger remnants the data thus suggests that $n_{\rm rem}/n_{\rm mrg} \sim 70$. Given the huge uncertainties on m, on $z_{\rm f}$, and even on $\tau_{\rm mrg}$, all we can conclude at this stage is that the data are not inconsistent with the hypothesis that all elliptical galaxies formed via major mergers.

13.4 The Fundamental Plane of Elliptical Galaxies

As we have seen in §2.3.2, elliptical galaxies obey a tight scaling relation between central velocity dispersion, σ_0, effective radius, $R_{\rm e}$, and $\langle I \rangle_{\rm e}$, the average surface brightness within $R_{\rm e}$ (Djorgovski & Davis, 1987; Dressler et al., 1987). This so-called fundamental plane (hereafter FP; see Fig. 2.18) is generally written in the form

$$\log R_{\rm e} = a \log \sigma_0 + b \log \langle I \rangle_{\rm e} + {\rm constant}. \quad (13.59)$$

Typically the best-fit parameters range from $a \sim 1.2$ in the blue to $a \sim 1.5$ in the near-infrared, while $b \simeq -0.8$ with only very weak dependence on the photometric band (e.g. Jørgensen et al., 1996; Pahre et al., 1998; Colless et al., 2001b; Bernardi et al., 2003b). The most striking observational property of the FP is its small and nearly constant thickness: the distribution of $\log R_{\rm e}$ around the best-fit FP (at fixed σ_0 and $\langle I \rangle_{\rm e}$) has a measured rms corresponding to scatter in $R_{\rm e}$ of only 15–20%.

The origin of the FP is usually interpreted in terms of the virial theorem (§5.4.4):

$$\frac{GM}{\langle R \rangle} = \langle v^2 \rangle, \quad (13.60)$$

where M is the mass of the system, $\langle R \rangle$ is an average radius so defined that the left-hand-side is the absolute value of the mean potential energy per unit mass, and $\langle v^2 \rangle$ is the average squared velocity so that $\langle v^2 \rangle/2$ is the mean kinetic energy per unit mass. The virial relation given above can be expressed in terms of the observables of an elliptical galaxy. In particular, we write $R_{\rm e} = k_R \langle R \rangle$, and $\sigma_0 = k_V \sqrt{\langle v^2 \rangle}$, where k_R and k_V are dimensionless quantities which depend on the density profile and orbital structure of the galaxy, respectively. Using that the luminosity of the galaxy is $L = 2\pi \langle I \rangle_e R_e^2$ (which follows from the definition of $\langle I \rangle_e$), we obtain the following relation between R_e, σ_0 and $\langle I \rangle_e$:

$$R_e = \mathscr{C}_R \sigma_0^2 \langle I \rangle_e^{-1} \left(\frac{M}{L} \right)^{-1}, \quad \mathscr{C}_R \equiv \frac{1}{2\pi G k_R k_V^2}. \quad (13.61)$$

If elliptical galaxies are homologous, so that they have self-similar density and orbital distributions, $\mathscr{C}_R$ has the same numerical value for all ellipticals. If, in addition, all ellipticals have the same (M/L), then the virial relation defines a FP with $a = 2$ and $b = -1$. The deviation of the observed a and b from these virial predictions is called the 'tilt' of the fundamental plane, and indicates that (M/L) and/or $\mathscr{C}_R$ have power-law dependence on $\langle I \rangle_e$, σ_0 and/or R_e. Furthermore, the thinness of the observed FP indicates that such power-law dependence itself cannot have much scatter, which puts tight constraints on models for the formation and evolution of elliptical galaxies.

Given the observed diversity in the structural and kinematic properties among elliptical galaxies, it is clear that they do not form a perfectly homologous class. Furthermore, the color–magnitude relation clearly indicates that there are also systematic variations in stellar populations, and thus in the mass-to-light ratios. This is further supported by the fact that the value of the FP slope, a, depends on the photometric band in which the FP is defined. The challenge is to determine which of these effects is most important for setting the tilt of the FP, and to understand why the scatter in the FP is so small. Numerous studies have estimated the contribution of different effects to the origin of the tilt in the FP, sometimes with contradictory results, and there is still no clear consensus.

One extreme possibility is that the entire tilt is due to non-homology. As discussed in §2.3.2 and §13.1.3, bright galaxies are supported by velocity dispersion, while low-luminosity ellipticals have kinematics consistent with them being rotationally supported. However, using detailed dynamical models it has been shown that the variations of rotational support and/or velocity anisotropy from galaxy to galaxy contribute very little to the observed tilt (Ciotti et al., 1996; Lanzoni & Ciotti, 2003). On the other hand, the surface brightness profiles of ellipticals are well fit by a Sérsic profile, with a Sérsic index n that systematically increases with increasing luminosity (see Fig. 2.13). Several studies have argued that this non-homology is responsible for a significant part of the observed tilt (e.g. Graham & Colless, 1997; Trujillo et al., 2004), although others claim that non-homology cannot produce more than $\sim 15\%$ of the observed tilt (e.g. Cappellari et al., 2006).

The other extreme is to assume that ellipticals are homologous, and that the tilt is entirely due to systematic variations in the mass-to-light ratios. The observed FP then implies a change of (M/L) with L and $\langle I \rangle_e$ given by

$$(M/L) \propto L^{(2-a)/2a} \langle I \rangle_e^{-1/2 - (1+2b)/a}. \quad (13.62)$$

For example, the results of Jørgensen et al. (1996), who obtained $a = 1.24 \pm 0.07$ and $b = -0.82 \pm 0.02$ in the optical, imply that

$$(M/L) \propto L^{0.31} \langle I \rangle_e^{0.02} \quad \text{(in optical)}, \quad (13.63)$$

and thus that the mass-to-light ratio increases with increasing luminosity, with only very mild dependence on the average surface brightness.

Using that $(M/L) = \Upsilon (M/M_\star)$, with Υ the stellar mass-to-light ratio and $M_\star$ the stellar mass, it is clear that there are two effects that can contribute to variations in (M/L): (i) variations in the stellar populations (ages and/or metallicities), and (ii) variations in the contribution and/or concentration of the dark matter to the total, dynamical mass.[4] Comparing spatially resolved

---

[4] Note that, strictly speaking, variations in $M/M_\star$ are also a manifestation of non-homology.

kinematic data of relatively small samples of nearby ellipticals with detailed dynamical models, a number of studies have found that $(M/L) \propto M^\gamma$ with $\gamma \simeq 0.3 \pm 0.1$ (van der Marel, 1991; Magorrian et al., 1998; Gerhard et al., 2001; Cappellari et al., 2006). This corresponds to $(M/L) \propto L^\alpha$ with $\alpha = \gamma/(1-\gamma) \simeq 0.4 \pm 0.2$, and is thus well in the range needed to explain the tilt of the FP. Although the variation of the FP slope with wavelength is strong evidence for systematic variations in the stellar content, numerous studies have suggested that the expected change in Υ alone cannot explain the observed tilt (Prugniel & Simien, 1996; Pahre et al., 1998). Rather, what is needed is an additional, systematic change in the dark matter content of ellipticals, in the sense that $M/M_\star$ increases with the mass (or luminosity) of the galaxy (e.g. Cappellari et al., 2006).

13.4.1 The Fundamental Plane in the Merger Scenario

An important question is whether a thin FP relation, with the correct tilt, can be produced in the merger scenario discussed in §13.2.2. Using a large suite of hundreds of merger simulations, Robertson et al. (2006a) have demonstrated that a successful reproduction of the FP relation requires the presence of gas in the progenitor galaxies. Whereas dissipationless mergers of disks produce remnants that occupy a FP similar to that expected from the virial relation (i.e. $a \simeq 2$, $b \simeq -1$), mergers between disks with gas fractions $f_{\rm gas} \gtrsim 30\%$ results in remnants that occupy a tilted FP, similar to what is observed. In their simulations, this tilt arises mainly from systematic variations in $M/M_\star$. During the merger a large fraction of the gas is collisionally heated to high temperatures. In massive systems this hot gas becomes X-ray luminous (and thus provides an explanation for the X-ray halos observed around massive ellipticals), but does not cool efficiently. In low mass systems, on the other hand, the collisionally heated gas cools rapidly and forms new stars at the center of the merger remnant, causing a significant decrease in $M/M_\star$. Thus, the halo mass dependence of the cooling efficiency (see §8.4.1) induces a mass dependence of $M/M_\star$, hence a tilt in the FP.

Observations have shown that at $z \lesssim 1$ a significant fraction of the luminous elliptical galaxies undergo mergers with other ellipticals (van Dokkum, 2005; Bell et al., 2006a; McIntosh et al., 2008). These mergers are typically dissipationless (i.e. 'dry' mergers). This raises the question of whether such mergers will change the tilt and destroy the thinness of the observed FP. Numerical simulations suggest that dry mergers between spheroidal galaxies do, in fact, roughly maintain the FP tilt (e.g. Nipoti et al., 2003; Boylan-Kolchin et al., 2005; Robertson et al., 2006a). This can be understood from the fact that violent relaxation tends to mix dark matter and stars (i.e. it reverses the impact of dissipation which segregates baryons, and thus the stars, from dark matter). The net effect is that the dark matter fraction interior to the final $R_{\rm e}$ is larger than the dark matter fraction interior to the initial $R_{\rm e}$. This implies a modification of k_R, which, according to the simulations, is sufficient to maintain the tilt of the FP.

13.4.2 Projections and Rotations of the Fundamental Plane

(a) The κ Coordinates Instead of using σ_0, $\langle I \rangle_{\rm e}$, and $R_{\rm e}$, one can use some orthogonal combinations of them to label the parameter space of the FP. Algebraically, this corresponds to a rotation of the coordinate axes. A particularly useful example, proposed by Bender et al. (1992), is based on the parameters

$$\kappa_1 \equiv \left(\log \sigma_0^2 + \log R_{\rm e}\right)/\sqrt{2}, \tag{13.64}$$

$$\kappa_2 \equiv \left(\log \sigma_0^2 + 2\log \langle I \rangle_{\rm e} - \log R_{\rm e}\right)/\sqrt{6}, \tag{13.65}$$

$$\kappa_3 \equiv \left(\log \sigma_0^2 - \log \langle I \rangle_{\rm e} - \log R_{\rm e}\right)/\sqrt{3}, \tag{13.66}$$

where σ_0 is in km s^{-1}, R_e in kpc, and $\langle I \rangle_e$ in $L_\odot/\mathrm{pc}^2$. The three-dimensional parameter space $(\kappa_1, \kappa_2, \kappa_3)$ is called κ space, and has the advantage that the physical interpretation of the FP relation is more transparent. In particular, the parameter κ_1 is a measure of the mass because $\kappa_1 \propto \log(\sigma_0^2 R_e)$, κ_3 is a measure of the mass-to-light ratio because $\kappa_3 \propto \log(\sigma_0^2 R_e/\langle I \rangle_e R_e^2)$, and κ_2 is proportional to the mass-to-light ratio times $\langle I \rangle_e^3$, which is required so that it is orthogonal to κ_1 and κ_3. The κ_1-κ_2 projection is very close to a face-on projection of the FP, while the κ_1-κ_3 projection shows the FP nearly edge-on. In fact, if ellipticals are homologous and $(M/L) \propto M^\gamma$, the virial theorem (13.60) implies a FP relation given by $\kappa_3 = \sqrt{2/3}\,\gamma \kappa_1 + \mathrm{constant}$.

(b) The Faber–Jackson Relation Before it was realized that elliptical galaxies obey a FP relation, Faber & Jackson (1976) had noticed that the luminosity of an elliptical galaxy is correlated with its central velocity dispersion:

$$L \propto \sigma_0^\beta \quad \text{with} \quad \beta \sim 4. \tag{13.67}$$

Note that this Faber–Jackson relation is similar to the Tully–Fisher relation for spiral galaxies discussed in §11.3. Since $L \propto \langle I \rangle_e R_e^2$, the Faber–Jackson relation implies a FP relation with $a = \beta/2 \sim 2$ and $b = -1/2$. These values of a and b are very different from the best-fit values of the observed FP relation, indicating that the Faber–Jackson relation is far from an edge-on projection of the FP, which also explains why its scatter is considerably larger than that in the FP relation.

(c) The D_n-σ Relation Rather than expressing the size of an elliptical galaxy via its effective radius, R_e, that encloses half of the total light, one can also use the radius at which the average, enclosed surface brightness reaches a fixed value. One such example is the parameter D_n, which is defined as the diameter within which the mean surface brightness drops to some fiducial value $\langle I \rangle_n$, i.e.

$$\langle I \rangle_n = \frac{2\pi \int_0^{D_n/2} I(R)\,R\,\mathrm{d}R}{\frac{1}{4}\pi D_n^2}. \tag{13.68}$$

If, for simplicity, we write the surface brightness profiles of elliptical galaxies as a single power law $I(R) \propto \langle I \rangle_e (R/R_e)^{-\xi}$, we have that

$$D_n \propto R_e \langle I \rangle_e^{1/\xi}. \tag{13.69}$$

Since $\xi > 0$ this implies that higher surface brightness galaxies have a larger ratio of D_n/R_e (as long as ξ is roughly the same for all galaxies). Using Eq. (13.69) to eliminate R_e from Eq. (13.59), we obtain the FP relation expressed in terms of D_n:

$$D_n \propto \sigma_0^a \langle I \rangle_e^{b+1/\xi}. \tag{13.70}$$

Since in general ellipticals do not have pure power-law surface brightness profiles, the characteristic value of ξ depends on the choice of the fiducial value of $\langle I \rangle_n$. As shown by Dressler et al. (1987), the value of $b + 1/\xi$ can be made insignificantly small by adopting $\langle I \rangle_n = 20.75$ mag arcsec^{-2} in the B band, so that the FP relation reduces to

$$D_n \propto \sigma_0^a \quad \text{with} \quad a \sim 1.2. \tag{13.71}$$

This is called the D_n-σ relation, and is basically another edge-on projection of the FP: for a given σ_0, the scatter in D_n is typically 15% from galaxy to galaxy.

(d) The Kormendy Relation In Fig. 2.14 we have seen that more luminous ellipticals are larger, and have a lower characteristic surface brightness. Consequently, there is also an anticorrelation between size and surface brightness. In the optical

$$\langle I \rangle_e = R_e^\nu \quad \text{with} \quad \nu \sim -1.3 \pm 0.1 \tag{13.72}$$

(e.g. Kormendy, 1977; Bernardi et al., 2003a). This relation, which is a projection of the FP along the σ_0 direction, is known as the Kormendy relation.

An anticorrelation between surface brightness and size is a fairly natural prediction of the merger scenario. Consider an equal mass merger between two identical progenitors of mass M_i and radius r_i, that results in a remnant of mass $M_f = 2M_i$ and radius $r_f = \mathcal{R} r_i$. Assuming that $(M/L) \propto M^\gamma$, and using that $\langle I \rangle_e \propto L/r^2$, then under the additional assumption of homology the surface brightness of the final remnant will be $2^{1-\gamma} \mathcal{R}^{-2}$ times that of the progenitors. Consequently,

$$\nu = (1-\gamma)\frac{\ln 2}{\ln \mathcal{R}} - 2. \qquad (13.73)$$

If we adopt $\gamma = 0.3$, in rough agreement with what is required to explain the tilt of the FP (see above), then an anticorrelation between $\langle I \rangle_e$ and R_e will arise as long as $\mathcal{R} > 1.27$. As we have seen in §13.3.2, dry (dissipationless) mergers typically have $\mathcal{R} \sim 1.4$ for their most probable orbit, which would leave the observed size–mass relation of elliptical galaxies intact. However, this value of $\mathcal{R}$ implies $\nu \sim -0.6$, inconsistent with the observed Kormendy relation, which requires $\mathcal{R} \sim 2$ instead. Apparently, dry mergers cannot simultaneously explain the observed size–mass and size–surface-brightness relations (e.g. Nipoti et al., 2003).

13.5 Stellar Population Properties

In the previous sections we have focused on the structural and kinematic properties of elliptical galaxies. We now turn our attention to the properties of their stellar populations. In general, a stellar population is characterized by its star-formation history (SFH), $\psi(t)$, an initial mass function (IMF), $\phi(m)$, and its history of chemical enrichment. In the case of a closed box, the latter is completely determined by $\psi(t)$ and $\phi(m)$ (see §10.4.2), although in the more general case with inflow and/or outflow the various element abundances may differ strongly from that of a closed box.

In what follows we will make the assumption of a universal IMF, and discuss various constraints on the SFHs of elliptical galaxies. We will distinguish between two different approaches: the archaeological approach, in which one tries to infer $\psi(t)$ from the observable properties of present-day ellipticals, and the evolutionary approach, where $\psi(t)$ is constrained using the observed properties of ellipticals as a function of redshift. We will discuss both approaches, and their results, in turn.

13.5.1 Archaeological Records

(a) Color–Magnitude Relation With only very few exceptions, early-type galaxies (ellipticals and S0s) have red colors. More importantly, the colors of early-type galaxies are tightly correlated with their luminosities. Fig. 13.7 shows the color–magnitude relation for ellipticals and S0s in the Coma Cluster. Note that more luminous early-types are redder, and that the overall scatter in the color–magnitude relation is remarkably small (the rms scatter is ~ 0.05 mag, of which more than half can be accounted for by observational errors).

Unfortunately, an interpretation of this color–magnitude relation in terms of the physical properties of the underlying stellar populations is severely impeded by the age–metallicity degeneracy discussed in §10.3.5. If we provisionally make the naive assumption that elliptical galaxies all have the same metallicity, the slope of the color–magnitude relation implies that more massive early-types are older. As shown in §10.3.3, the colors of a passively evolving, single-age stellar population (SSP) younger than about 5 Gyr depend significantly on its age. For example,

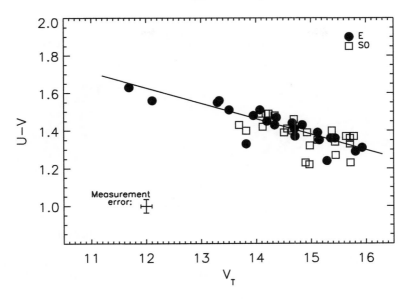

Fig. 13.7. The color–magnitude relation for early-type galaxies in the Coma Cluster. [Adapted from Bower et al. (1992)]

the $(U-V)$ color of a coeval stellar population formed 5 Gyr ago changes at a rate of about 0.05 magnitude per Gyr. Unless all early-types formed during a very narrow epoch in cosmic time, the small amount of scatter therefore implies that ellipticals must have formed the bulk of their stars more than ~ 10 Gyr ago (see Fig.10.6). However, as already alluded to, this interpretation is severely hampered by the age–metallicity degeneracy, in that redder colors can also indicate a more metal-rich stellar population. Thus, the slope of the color–magnitude relation may simply indicate that more massive early-types are more enriched in metals, while the small scatter may come about if stellar populations that formed earlier were less enriched in metals. Using additional data that can help to break the age–metallicity degeneracy (see below), it has become clear that more massive early-types are both older and more metal rich. Nowadays, most studies seem to agree that the slope of the color–magnitude relation is primarily driven by metallicity rather than age (e.g. Kodama & Arimoto, 1997; Graves et al., 2009), while the small amount of scatter mainly reflects that early-type galaxies have old, passively evolving stellar populations.

(b) Absorption Line Indices More detailed insight into the physical properties of the stellar populations of elliptical galaxies comes from their absorption line indices, such as the Lick indices discussed in §10.3.4. The main advantage of this approach is that it allows one to break the age–metallicity degeneracy that plagues the broad-band colors discussed above. Based on spectral synthesis modelling, it is found that some indices (e.g. G4300 and Balmer indices such as Hβ and Hδ) are more sensitive to age, while others (e.g. Mgb, Mg$_2$, Fe5270, Fe5335) are more indicative of the population's metallicity. Thus, by measuring these different indices, which is fairly straightforward and only requires spectra of moderate resolution, and comparing their values with predictions from models, one can determine both the (luminosity-weighted) age and the metallicity of the stellar population. Until recently a major problem with this approach has been that different metal line indices sometimes yielded different metallicities, [Z/H]. This problem was due to the fact that most models assumed solar abundance ratios for the various elements, while in reality giant ellipticals are typically enhanced in α elements (relative to iron) when compared to stars in the solar neighborhood (e.g. Worthey et al., 1992).

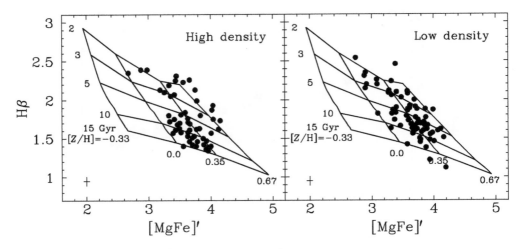

Fig. 13.8. The Lick indices Hβ vs. [MgFe]' for a sample of ellipticals, lenticulars and cD galaxies with overplotted a grid of SSP models of Thomas et al. (2003) for metallicities of [Z/H] = 0.0, 0.35 and 0.67, and ages of 2, 3, 5, 10 and 15 Gyr. The median 1σ error bars of the data points are shown in the lower-left corner. Results are shown separately for galaxies in high- and low-density environments (left and right panels, respectively). [Adapted from Thomas et al. (2005) by permission of AAS]

Using more sophisticated stellar population models that allow for non-solar abundance ratios (e.g. Trager et al., 2000b; Proctor & Sansom, 2002; Thomas et al., 2003), it has recently become possible to determine reliable ages, metallicities and α enhancements from Lick indices such as Hβ, Mgb, Fe5270 and Fe5335. In particular, the indices Hβ and

$$[\text{MgFe}]' \equiv \sqrt{\text{Mg}b \times [0.72\text{Fe}5270 + 0.28\text{Fe}5335]} \tag{13.74}$$

can be used to determine the ages and metallicities of single-age stellar populations with virtually no dependence on [α/Fe] (Thomas et al., 2005). Fig. 13.8 shows Hβ versus [MgFe]' for 124 early-type galaxies (ellipticals, lenticulars and cD galaxies). A comparison with the SSP models of Thomas et al. (2003), shown as the grid, indicates that their stellar populations are not as homogeneous as traditionally thought. In particular, elliptical galaxies seem to span a wide range of ages, from about 2 to 15 Gyr, but a small range of metallicities. In addition, there is a weak indication that ellipticals in high-density environments (groups and clusters) have somewhat older stellar populations than those in low density environments. Note, however, that these age estimates are heavily weighted towards young stellar populations. Detailed spectral-synthesis calculations show that a population with an age of 1 Gyr has Balmer absorption lines that are about 5 times stronger, and a luminosity that is more than 6 times higher, than a population of 15 Gyr with the same mass. Thus, even a few percent (by mass) of a young population can dominate the age estimate. Indeed, present data favor such a 'frosting' model, in which the low apparent SSP ages are produced by adding a small frosting of younger stars (typically $\sim$ 10–20 percent in mass) to an older 'base' population of 10-15 Gyr (de Jong & Davies, 1997; Trager et al., 2000a).

Observations show that various Lick indices of early-type galaxies are correlated with their velocity dispersions, σ (and thus also with the luminosity or stellar mass). In particular, Mgb and $\langle\text{Fe}\rangle = 0.5(\text{Fe}5270 + \text{Fe}5335)$ increase with increasing σ, while Hβ decreases with it (e.g. Terlevich et al., 1981; Bender et al., 1993; Kuntschner et al., 2001; Bernardi et al., 2003c). For ellipticals with stellar masses $M_\star \gtrsim 10^{10} M_\odot$, Thomas et al. (2005) inferred the following scaling relations:

13.5 Stellar Population Properties

$$\log(t/\text{Gyr}) = +0.46(+0.17) + 0.24(0.32)\log(\sigma/\text{km s}^{-1})$$
$$[Z/H] = -1.06(-1.04) + 0.55(0.57)\log(\sigma/\text{km s}^{-1}) \quad (13.75)$$
$$[\alpha/\text{Fe}] = -0.42(-0.42) + 0.28(0.28)\log(\sigma/\text{km s}^{-1})$$

where the values outside (inside) of the brackets correspond to high (low) density environments (see also Nelan et al., 2005). The observed scatter in these relations implies an intrinsic scatter of about 20% in age, ~ 0.1 dex in [Z/H] and ~ 0.05 dex in [α/Fe] at fixed velocity dispersion. Thus, more massive ellipticals are older, more metal enriched and more enhanced in α elements. As discussed in §10.4, the latter implies that more massive ellipticals have formed their stars on a shorter time scale (at least if ellipticals have a universal IMF). Furthermore, ellipticals in low-density environments are, on average, ~ 2 Gyr younger than their counterparts in high-density environments. Note that these ages refer to the old populations; about 15% of early-type galaxies have an additional, younger frosting population with an age between 1 and 2 Gyr. This fraction rapidly increases with decreasing stellar mass.

13.5.2 Evolutionary Probes

Rather than trying to infer the star-formation histories of elliptical galaxies from their present day archaeological records, one can also trace their evolution by studying their properties as function of redshift. Although this has the advantage of being more direct, the limited spatial resolution makes it difficult to accurately determine the morphologies of high redshift galaxies and to select clean samples of a fixed morphological type. In particular, the increasing population of blue galaxies with increasing redshift (the Butcher–Oemler effect) may introduce a systematically greater contamination at higher redshifts. Fortunately, the superior imaging capability of the HST has made it possible to identify elliptical galaxies out to redshifts $z \sim 1$ based purely on their morphologies (e.g. Driver et al., 1995; Abraham et al., 1996). Together with redshift measurements, these observations can be used to directly probe the evolution of elliptical galaxies.

Using a sample of 94 luminous ellipticals in nine clusters at redshifts $0.17 < z < 1.21$, Schade et al. (1997) found that the absolute magnitudes of elliptical galaxies of fixed sizes increase with redshift roughly as $\Delta \mathcal{M}_B \sim (-1.05 \pm 0.25)\Delta z$. High redshift ellipticals are also found to be bluer than their low redshift counterparts (e.g. Aragon-Salamanca et al., 1993; Rakos & Schombert, 1995), but they still obey a tight color–magnitude relation with the same intrinsic scatter and slope as at $z = 0$ (Ellis et al., 1997; Stanford et al., 1998). All these results are consistent with a picture in which elliptical galaxies in clusters formed their stars at high redshifts ($z > 2$) in a well-synchronized fashion, and evolved passively thereafter. Furthermore, the fact that the slope of the color–magnitude relation shows no significant change out to $z \sim 1$ supports the belief that it arises from a correlation between galaxy mass and metallicity, rather than age.

With modern 10-meter class telescopes it has also become possible to measure kinematics of (elliptical) galaxies out to redshifts $z \sim 1$. Together with HST photometry, these kinematic data make it possible to study the fundamental plane of elliptical galaxies at high redshifts. Several studies have shown that cluster ellipticals out to $z \sim 1$ satisfy a tight FP relation with a slope and scatter that are similar to those at $z = 0$ (e.g. van Dokkum & Franx, 1996; Kelson et al., 1997). The zero-point of the FP, however, is found to evolve with redshift in a way consistent with an evolution of the mass-to-light ratios given by $\Delta \log(M/L_B) \propto -0.5z$ (van Dokkum et al., 1998). Once again, these results indicate that elliptical galaxies formed their stars early ($z > 2$) and have been evolving passively ever since.

Finally, Ziegler & Bender (1997) used medium-resolution spectroscopy of elliptical galaxies in three clusters at $z \sim 0.37$ to measure absorption line indices. They found a clear relation between the line strength of Mgb and the galaxy's velocity dispersion, σ, similar to that of nearby ellipticals. However, at a given σ the Mgb line strength of the distant ellipticals is significantly lower than the mean value of the nearby sample. This difference can be fully attributed to the lower age of the distant galaxies and is in excellent agreement with a passively evolving population that formed the bulk of its stars at $z > 2$.

Thus, there seems to be good overall agreement, both among different evolutionary probes and with the archaeological records, that elliptical galaxies in clusters formed their stars in a relatively short time at high redshifts.

13.5.3 Color and Metallicity Gradients

In addition to the global properties discussed above, the stellar populations of elliptical galaxies also reveal radial gradients within individual galaxies. As discussed in §2.3.2, the outskirts of elliptical galaxies are typically bluer than their central regions. In the absence of dust, these color gradients indicate that the central regions are older and/or more metal rich than the outer regions.

Radial gradients have also been observed in various absorption line indices. Using a small sample of 13 elliptical and lenticular galaxies, Davies et al. (1993) obtained average gradients of $\Delta \mathrm{Mg}_2/\Delta \log r = -0.059 \pm 0.022$ mag and $\Delta \langle \mathrm{Fe} \rangle / \Delta \log r = -0.38 \pm 0.26$Å. From a comparison with SSP models, these authors inferred an average metallicity gradient of $\Delta [\mathrm{Fe/H}]/\Delta \log r = -0.2 \pm 0.1$, which largely accounts for the color gradients mentioned above. Similar metallicity gradients were obtained by Trager et al. (2000b), who used population models which allow for non-solar abundance ratios. In addition, they found that stellar ages are on average $\sim 25\%$ *older* in the outer parts, while the amount of α enhancement is nearly constant with radius. Two-dimensional maps of absorption line indices across the images of elliptical galaxies show a wide variety of behaviors, suggesting that there are a variety of formation paths which can lead to quite different metallicity structures (Kuntschner et al., 2006).

13.5.4 Implications for the Formation of Elliptical Galaxies

As discussed above, the stellar populations of elliptical galaxies are characterized by the following properties:

(i) The evolution of the stellar population of elliptical galaxies is roughly consistent with the passive evolution of a SSP.
(ii) More massive ellipticals form their stars earlier and over a shorter time scale.
(iii) Ellipticals in dense environments form their stars earlier than their counterparts of the same stellar mass in low-density environments.
(iv) More massive ellipticals are more enriched in metals.
(v) The inner parts of elliptical galaxies are more enriched in metals, and slightly younger, than their outer parts.

We now discuss whether these observational facts favor a single, major formation event (a quasi-monolithic collapse), or formation in diverse sites and assembly over an extended period, as expected for hierarchical merging.

(a) The Monolithic Collapse Scenario In the monolithic collapse scenario, elliptical galaxies are assumed to form the majority of their stars in a single burst of star formation and to evolve primarily passively thereafter. This is a particularly simple model which naturally accounts for some of the characteristic properties of the star-formation histories mentioned above. On the other hand, this scenario lacks a natural explanation for the mass and environment dependencies of the observed star-formation histories, and it is inconsistent with the observed abundances and sizes of $z \sim 1$ ellipticals which appears several times smaller than today.

If star formation were to proceed to completion according to a closed-box model (see §10.4.2), then the observed trend of metallicity with velocity dispersion would be difficult to establish. On the other hand, in the outflow model discussed in §10.4.3, star formation stops when the gas is exhausted, i.e. when $M_\star = M_{\text{tot}} = M_{\text{tot}}(0)/(1+\alpha)$. The mean stellar metallicity at this final stage is

$$\overline{Z}_\star = \frac{1+\alpha}{M_{\text{tot}}(0)} \int_0^\infty Z M_\star(Z)\,\mathrm{d}Z = \frac{yz}{1+\alpha}, \qquad (13.76)$$

where the second equation follows from differentiating Eq. (10.131). Thus, the average stellar metallicity is lower for systems where the gas outflow caused by star formation is more effective (i.e. where α has a larger value). Hence, what is required is a mechanism that can naturally explain a more pronounced outflow in less massive galaxies. As first postulated by Mathews & Baker (1971), the ISM of an elliptical galaxy can be heated significantly by supernova explosions and be driven out of the galaxy in a galactic wind once the thermal energy of the gas exceeds its gravitational binding energy. Larson (1974b) noted that the binding energy per unit gas mass is higher in more massive galaxies, and that more massive galaxies can therefore retain their gas for a longer period so as to attain a higher metallicity (see also Dekel & Silk, 1986). Once specific assumptions are made regarding the star-formation law and the density profiles of stars, gas and dark matter, one can work out the details of the galactic wind (see §8.6), and, in combination with the chemical evolution models described in §10.4.3, predict the chemical composition of the final galaxy. Such calculations have been carried out by a number of authors (e.g. Arimoto & Yoshii, 1987; Matteucci, 1992; Gibson, 1997), who have shown that monolithic collapse, in combination with this classic wind model, is able to explain the observed relation between metallicity and velocity dispersion.

As we have seen in §13.3.2, in order for the monolithic collapse scenario to be consistent with the observed sizes of elliptical galaxies, the original collapse needs to be extremely dissipative. This dissipation naturally leads to the formation of radial gradients in the metallicities and ages of the resulting stellar populations. The idea is that stars begin to form everywhere during the collapse and remain in orbit with little net inward motion, whereas the gas continues to sink towards the center due to dissipation, and is enriched by the evolving stars. Consequently, the stars that form late at the center of the galaxy are more metal-enriched, and somewhat younger, than those that formed earlier in the outskirts. Detailed calculations of this mechanism have typically produced metallicity gradients $\Delta[Z/H]/\Delta\log r \sim -1$, much steeper than observed (e.g. Larson, 1974a, 1975). To alleviate this problem, Carlberg (1984) suggested that the pressure associated with the energy output from supernova explosions during the collapse may reduce the gas inflow rate. This reduces the metallicity gradient to $\Delta[Z/H]/\Delta\log r \sim -0.5$, which is still somewhat too steep compared to the data. In addition, galactic winds may also *create* radial gradients. Since the escape velocity of a galaxy is lower at larger radii, a galactic wind is expected to occur earlier in the outer regions of a galaxy. As the star-formation rate is expected to be substantially reduced after the onset of the wind, stars in the outer region will have formed from gas which is less enriched by the evolving stars and thus have lower metallicity. Detailed calculations of this kind of wind models show that this effect by itself is already sufficient to reproduce the metallicity gradients observed (e.g. Martinelli et al., 1998). Thus, overall monolithic collapse models appear to produce metallicity gradients that are too steep.

(b) The Hierarchical Merger Scenario In the hierarchical merger scenario, the assembly of elliptical galaxies involves the merging of two or more progenitor galaxies. Since these progenitors have evolved largely independently of each other, and may have formed their stars over extended periods of time, and since merging is an ongoing process, one expects the resulting ellipticals to contain stars of different ages. At first sight, this seems inconsistent with the data which favors passive evolution of a SSP. However, there are two caveats. First of all, if most mergers happen early, and there is little or no subsequent star formation, the resulting stellar population will be virtually indistinguishable from that of an old SSP, and thus consistent with the data (which in any case often require at least a 'frosting' of younger stars). Secondly, the data suffer from what is called progenitor bias (van Dokkum & Franx, 1996); if the progenitors of some present-day ellipticals were spirals, these would not be included in observed samples of high-redshift ellipticals, causing evolution to appear more 'passive' than it actually is.

Detailed galaxy formation models within the hierarchical CDM framework show that, in spite of their violent formation history, the predicted scatter in the color–magnitude relation of cluster ellipticals is remarkably small, in reasonable agreement with observations (Kauffmann, 1996b; Baugh et al., 1996; Kauffmann & Charlot, 1998a; De Lucia et al., 2006). This demonstrates that the apparent 'homogeneity' of the stellar population properties of cluster ellipticals is not inconsistent with the merger scenario. In addition, these hierarchical models also predict that ellipticals in dense environments form their stars earlier than their counterparts in low-density environments, in qualitative agreement with the data. This is a natural outcome of the hierarchical scenario, because present-day ellipticals in clusters form from the highest peaks in the primordial density field, leading to an earlier onset of the collapse of their dark matter halos and to more rapid mergers.

Despite these successes, early models for the formation of (elliptical) galaxies in a hierarchical universe all predicted that more massive ellipticals should have younger stellar populations, in clear conflict with the (more recent) data. As we have seen in §7.3.4, a characteristic feature of hierarchical models is that more massive halos assemble later. Thus, if the star-formation history of an elliptical galaxy were to trace its assembly history, as is to some extent the case in these early models, the resulting age–mass trend is inverted with respect to the data. This is why the observed relation between age and mass is often called 'antihierarchical'. However, as illustrated in Fig. 13.4, in the hierarchical merger scenario the star-formation history and assembly history of an elliptical galaxy can be very different. Although more massive haloes *assemble* later, their progenitors *form* earlier. Since stars are expected to form whenever the progenitors form, and not only when the progenitors assemble to form the final object, it is more natural for the hierarchical scenario to predict that more massive galaxies have older stellar populations (Neistein et al., 2006; Li et al., 2008).

There are two reasons why the early models predicted the opposite trend. First of all, they considered an Einstein–de Sitter cosmology, in which there is far more merger activity at late times than in the currently popular ΛCDM cosmologies. Secondly, it has become clear that in order to be successful, the models need to include a mechanism that can prevent star formation at late times. One such mechanism, which has received a lot of attention in recent years, is feedback from an active galactic nucleus (AGN). Although the details of such a feedback mechanism are still poorly understood (see discussion in §14.4), modern models often incorporate it in such a way that it prevents hot gas from cooling in massive halos (Springel et al., 2005c; Monaco & Fontanot, 2005; Bower et al., 2006; Croton et al., 2006). As shown by De Lucia et al. (2006), once AGN feedback is included, and if one adopts a ΛCDM concordance cosmology, the models can nicely reproduce the observed trend between mass and stellar age. In addition, they even predict that more massive ellipticals formed their stars over a shorter time scale, in good qualitative agreement with observation.

In addition to AGN feedback, these models also include a description for galactic winds (assumed to be powered by supernovae). Since more massive ellipticals typically have more massive progenitors, galactic winds have the same impact as in the monolithic collapse model, and their efficiency can be tuned to yield a mass–metallicity relation that matches the observational data (e.g. Kauffmann & Charlot, 1998a; De Lucia et al., 2006).

The age and metallicity gradients observed in elliptical galaxies would be difficult to explain in the hierarchical formation scenario if mergers would completely randomize the material in their progenitors, i.e. if the final binding energies of the stars would be uncorrelated with their binding energies in the progenitors. However, as discussed in §5.5, violent relaxation never proceeds to completion. Consequently, any gradient present in the progenitors will be retained to some degree in the merger remnant. Using numerical simulations, White (1980) showed that any pre-existing gradients are diluted by roughly a factor of two after three merger events. Furthermore, the merger process itself may also *create* metallicity and/or age gradients. If galactic winds occurred in the progenitors, stars in more massive progenitors are expected to be more enriched in metals. Since stars from the more massive progenitor have a tendency to end up at smaller galactocentric radii in the merger remnant, unequal mass mergers can create metallicity gradients even if these are not present in the progenitors. Furthermore, as discussed in §13.2.2, mergers may drive gas to the centers of their remnants to ignite a starburst, thus giving rise to, or enhancing, an age and/or metallicity gradient.

13.6 Bulges, Dwarf Ellipticals and Dwarf Spheroidals

Thus far our discussion has mainly focused on elliptical galaxies with $\mathcal{M}_B \lesssim -18$. These form the bright end of a much larger family of dynamically hot, spheroidal systems which encompasses, in addition to elliptical galaxies, the bulges of spiral and S0 galaxies, as well as dwarf ellipticals (dEs) and dwarf spheroidals (dSphs).

Fig. 13.9 shows the distribution of various classes of dynamically hot galaxies in the κ-space (see §13.4.2a). Note that we have split the elliptical population into bright ellipticals ($\mathcal{M}_B \leq -20.5$) and those with intermediate luminosities ($-20.5 < \mathcal{M}_B \leq -18.5$). This roughly represents a separation of boxy, velocity-dispersion supported systems from disky, rotation supported systems. The (κ_1, κ_3) plane (upper panel) is an almost edge-on view of the fundamental plane (FP). Except for the dSphs, all systems seem to roughly obey the same FP. Recall that both the brightest ellipticals and the dEs are (mainly) supported by anisotropic velocity dispersion, while the galaxies with intermediate luminosities are mainly rotationally flattened. Yet, they all roughly populate the same FP, although the dEs tend to have a higher FP zeropoint than the overall population. As discussed in Bender et al. (1992), this result most likely indicates that dEs have somewhat higher mass-to-light ratios than their more massive counterparts. This trend continues to lower κ_1, i.e. towards the dSphs, where galaxies lie well above the FP relation defined by the overall population. Indeed, dSphs are known to have extremely high mass-to-light ratios, and their dynamics are almost completely dominated by the dark matter halos in which they reside (e.g. Mateo, 1998).

The lower panel of Fig. 13.9 shows the distribution of dynamically hot galaxies in the (κ_1, κ_2) plane, which is an almost face-on view of the FP. Note that the ellipticals (both bright and intermediate) and bulges form a single smooth sequence in this plane. In fact, as shown in §2.3.3, ellipticals of intermediate luminosity and (massive) bulges share many characteristics and seem to form a smooth, continuous sequence in bulge-to-disk ratio, suggesting that they may have been formed by the same mechanism. On the other hand, the dynamically hot dwarf galaxies (dEs and dSphs) populate a sequence in the (κ_1, κ_2) space almost perpendicular to that populated by the ellipticals and bulges. This is clearly due to the fact that the surface brightness of dwarf systems increases with increasing luminosity, while ellipticals and bulges show the opposite trend (see §2.3.2a).

If it were not for the bulges, one could argue that the ellipticals of intermediate luminosity actually occupy the same sequence as the dwarf galaxies, and that it is only the brightest ellipticals that seem to deviate from this sequence (see also the discussion in §2.3.5). However, once

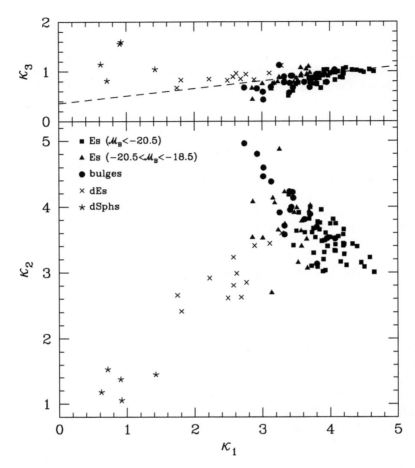

Fig. 13.9. The distribution of dynamically hot galaxies in κ space. The upper and lower panels show the edge-on and face-on views of the (fundamental) plane populated by these systems. Galaxies are split in five subclasses as indicated. Note that dwarf galaxies (dEs + dSphs) occupy a different locus in the face-on projection (κ_1, κ_2) than that occupied by ellipticals and bulges. The dashed line in the upper panel indicates the FP of Virgo ellipticals ($\kappa_3 = 0.15\kappa_1 + 0.36$) and is shown for comparison. [Based on data published in Bender et al. (1992)]

the bulges are taken into account, there seems to be an indication that dwarf galaxies occupy a sequence distinct from that of the other dynamically hot systems. We do caution, however, that the κ parameters of the bulges may be significantly more uncertain than those for the other systems due to the larger than average measurement errors associated with the bulge–disk decomposition (see §2.3.3b).

13.6.1 The Formation of Galactic Bulges

As briefly mentioned in §2.3.3, massive bulges, which are predominantly found in S0 and Sa galaxies, share many properties with ellipticals of intermediate luminosities (see Wyse et al., 1997, for a review). In particular, massive bulges are consistent with being flattened by rotation (Fig. 2.16), and the best-fit Sérsic parameter for their surface brightness profiles scales with luminosity in the same way as for ellipticals (Fig. 2.13). In addition, massive bulges and ellipticals obey similar color–magnitude relations (Balcells & Peletier, 1994), similar metallicity–luminosity relations (Jablonka et al., 1996), similar fundamental plane relations (see Fig. 13.9) and the same $M_{\rm BH}$-σ relation (see §13.1.6). All this suggests that massive bulges form in the same way as ellipticals of intermediate luminosity (i.e. most likely via the merging of gas-rich progenitor galaxies). Whether they end up as bulges or as (disky) ellipticals just depends on the size and mass of the disk that manages to survive the merger, or that manages to grow after the merger due to the accretion of new gas (e.g. Kauffmann et al., 1993; Baugh et al., 1996). In this picture, early-type disk galaxies, S0s and disky ellipticals are all part of the same family, and represent a sequence of decreasing disk-to-bulge ratio. As described in §13.2.2, such formation of bulges via major mergers fits nicely in the framework of hierarchical structure formation. However, it does not provide a natural explanation of why the effective radii of bulges are always of the order of 20% of the disk scale length (see §2.3.3), nor does it explain why the colors of bulge and central disk are so similar. It remains to be seen whether or not these observations pose a serious challenge to this picture.

It is often assumed that small bulges, which are predominantly found in late-type spiral galaxies, form in a different fashion than their more massive counterparts. This is motivated by the fact that the properties of small bulges are more reminiscent of disks than of ellipticals. Their surface brightness profiles are close to exponential, and their ratios of $v_{\rm m}/\sigma$ are significantly higher than expected for an isotropic oblate rotator (Kormendy & Kennicutt, 2004). As shown in Fig. 13.3, this latter property indicates that the system must be highly flattened intrinsically (i.e. disk-like). Furthermore, in late-type disk galaxies seen close to edge-on, the bulge component often has a boxy or peanut shape. These bulges are often called 'pseudo-bulges', to differentiate them from the 'classical' bulges that are more elliptical-like. It is generally believed that pseudo-bulges form via secular evolution from the disk component. As discussed in §11.5.4, bars and spiral arms can rearrange angular momentum in disk galaxies, which may result in the formation of a central mass concentration. When this central disk material is kinematically heated in the vertical direction, for example via the bending instability, it may result in the formation of a peanut-shaped bulge.

A merit of this scenario is that it explains naturally why the colors of bulges are very similar to those of the inner disk. Support also comes from numerical simulations, which show that angular momentum redistribution in disk galaxies may indeed result in the formation of bulges with close to exponential surface brightness profiles, and with bulge-to-disk size ratios that are in good agreement with observations (e.g. Pfenniger & Friedli, 1991; Debattista et al., 2006). The simulations even show that the angular momentum transfer can create outer breaks in the surface density distributions of the disks as seen in real disk galaxies (§2.3.3). However, it should be emphasized that, although secular evolution can indeed transfer mass towards the center, there is no reason to believe that disks should form as pure exponentials. Indeed, as discussed in §11.4, it may well be that disks form already with an 'excess' of low angular momentum material in the central part. Hence, pseudo-bulges may well be a direct outcome of disk formation, without the additional need for secular evolution (e.g. van den Bosch, 1998).

A final process that may play a role in bulge building is the accretion of satellite galaxies and/or globular clusters. As discussed in Chapter 12, dynamical friction causes satellite galaxies and globular clusters to transfer their orbital energy and angular momentum to the dark matter, and thus to sink to the center of the potential well. If sufficiently dense, they can survive complete tidal disruption before reaching the center, thus giving rise to a bulge component, or adding mass to a pre-existing bulge. In addition to being sufficiently dense, they also need to have masses in a fairly restricted range to form a bulge. If the mass of the satellite is too low, the dynamical friction time scale will be too long to bring it to the center. If too massive, the satellite will destroy the disk component, which basically puts the accretion in the major-merger regime. Numerical simulations show that when the mass of the satellite is of the order of 10% of the disk, it can sink to the center of the potential well in a few Gyr (e.g. Walker et al., 1996). Aguerri et al. (2001) have shown that the accretion of multiple satellite galaxies may even create a relation between the bulge mass and the best-fit Sérsic profile n that is in good agreement with the observed relation. In addition to building a bulge, this process also causes a significant modification of the disk component. Dynamical friction can transport part of the orbital energy and angular momentum of the satellite(s) to the disk stars, causing an increase of both the scale length and scale height. The formation of the bulge may thus be intimately coupled to the creation of the thick disk. Whether or not this mechanism can explain the fact that the bulge-to-disk size ratio is roughly independent of Hubble type remains to be seen.

To summarize, there are multiple processes that may be responsible for the formation of bulges. It is almost certain that each of these processes is at work; what remains unclear, however, is what their relative importance is as a function of the various properties of the bulges, and whether such a hybrid scenario for bulge formation can match the data.

13.6.2 The Formation of Dwarf Ellipticals

It is tempting, and natural, to envision that dEs form in the same way as their more massive counterparts, namely via the merger scenario discussed in §13.2.2. The only difference with respect to the regular ellipticals is then simply that the progenitors of dEs are dIrrs rather than the more luminous spirals. In this scenario, the lower metallicities and lower surface densities of dEs compared to regular ellipticals simply reflect the lower metallicities and lower surface densities of their progenitor galaxies, which in turn can be explained by assuming that the feedback from supernova results in a relatively low efficiency of star formation and chemical enrichment in low-mass halos (Larson, 1974b; Dekel & Silk, 1986; Vader, 1986; Yoshii & Arimoto, 1987; Dekel & Woo, 2003, see also §8.6.3).

A prediction of this scenario is that the majority of dEs should reside in low-mass halos, and so should be less strongly clustered than more massive ellipticals. This is in violent disagreement with data from large galaxy redshift surveys, which show that faint red galaxies are as strongly clustered as the most massive ellipticals (e.g. Norberg et al., 2002a). Indeed, dEs are preferentially found in dense cluster environments to the extent that there are few, if any, examples of isolated dEs (Binggeli et al., 1987; Wang et al., 2008c). Another problem with the simple merger scenario is that, as mentioned in §13.1.3, the majority of dEs have kinematics more akin to that of bright ellipticals rather than of ellipticals of intermediate luminosity: they are supported by anisotropic velocity dispersion rather than rotation. It is unclear why mergers between two dIrrs, which are gas rich, would result in velocity-dispersion supported systems. After all, as discussed in §13.2.2, simulations of mergers between gas-rich disk galaxies typically result in a disky elliptical with significant rotation.

Because dEs seem to be restricted to massive clusters, it is often suggested that they are the remnants of a population of progenitor galaxies that are transformed into dEs via cluster-specific processes. Indeed, the spatial and velocity distributions of cluster dEs support the notion that they have been accreted into the cluster fairly recently (e.g. Bothun & Mould, 1988; Conselice et al., 2001). In what follows we describe three different formation scenarios for dEs and dSphs in clusters.

(a) Gas Stripping (dIrr → dE/dSph) In this scenario, originally proposed by Lin & Faber (1983), it is assumed that the progenitors of dEs and dSphs are dIrrs that had their gas stripped due to ram pressure by the intracluster medium after they were accreted into the cluster. This scenario is supported by recent observations which indicate that a significant fraction of (relatively bright) dEs contain (or are) stellar disks (Lisker et al., 2006, and references therein). This scenario can also accommodate the stellar population properties of dEs, and yields a natural explanation for why they have surface brightness profiles that are close to exponential. However, it faces two important challenges. First of all, since gas stripping should not have a dramatic impact on the kinematics of the stars, this model predicts that dEs and dSphs should be supported by rotation, in clear contradiction with the observed kinematics (see §13.1.3). Secondly, since it is unlikely that ram-pressure stripping promotes the formation of globular clusters, dEs are expected to have a similar specific frequency of globulars, S_N, as dIrrs. However, as shown by Durrell et al. (1996) and Miller et al. (1998), dEs have $S_N = 5.2 \pm 1.1$, comparable to that of more massive ellipticals but much larger than the typical value, $S_N < 1$, for spirals and dIrrs (see §13.3.4). The removal of gas can result in a quenching of star formation, and thus in a (passive) fading of the galaxy, which boosts S_N under the conservation of the number of globulars. However, the amount of fading expected is not sufficient to explain the high values of S_N observed.

(b) Galaxy Harassment (S $\rightarrow$ dE/dSph) As discussed in §12.5.1, high-speed impulsive encounters with other cluster galaxies can transform a low surface-brightness disk galaxy into a dE or dSph. An interesting aspect of this harassment scenario is that the progenitor population of dEs might be the blue Butcher–Oemler disk galaxies seen in clusters at moderate redshifts (Moore et al., 1996, see also §2.5.1). Although this process tends to increase the velocity dispersion of the harassed system, simulations suggest that a significant fraction of the progenitor's rotation is preserved. Hence, it seems unlikely that harassment can account for the large fraction of dEs that show absolutely no sign of rotation. In addition, this scenario has the same problem as the gas-stripping scenario in explaining the high specific frequency of globular clusters, unless their formation is somehow promoted by the harassment.

(c) Tidal Stripping (E $\rightarrow$ dE/dSph) In this scenario dEs and dSphs are the remnants of more massive ellipticals that have been tidally stripped by the gravitational potential of the cluster and its massive member galaxies. Of all scenarios, this is perhaps the least likely one, as it faces two important problems. First of all, it would predict that the dwarfs have metallicities and chemical abundances similar to those of more massive ellipticals, which is clearly inconsistent with the data (except for special cases like M 32). Secondly, since tidal stripping should not impact the nuclear black hole, the dEs should have SMBHs with masses that would place them far off the M_{BH}-σ relation. This is clearly in contradiction with kinematic data, which excludes that dEs harbor SMBHs more massive than $\sim 10^7 \, M_\odot$ (Geha et al., 2002).

To summarize, there is currently no single formation model for dwarf ellipticals that can explain all their observed properties. It may well be that there are multiple formation channels leading to the formation of dEs and/or dSphs. For example, there are indications that nucleated and non-nucleated dEs have different spatial distributions in clusters (Ferguson & Sandage, 1989; Lisker et al., 2007) and different S_N (Miller et al., 1998). On the other hand, nucleated and non-nucleated dEs are remarkably similar in terms of their position in the FP, their morphologies, and their stellar populations. Hence, how exactly dwarf ellipticals have formed is still an unsolved problem.

14
Active Galaxies

Our discussion so far has mainly focused on the formation and evolution of 'normal' galaxies, the ones whose emission is dominated by stars. Since stellar atmospheres are basically in hydrodynamical and thermal equilibrium, this emission is predominantly thermal radiation whose spectrum is roughly a sum of the Planck spectra corresponding to the temperatures of all the individual stars in the galaxy. Since the temperatures of stars only cover a relatively narrow range, $3,000\,\text{K} \lesssim T \lesssim 40,000\,\text{K}$, and since the Planck spectrum for a given temperature is fairly narrow, the spectrum of a normal galaxy is largely confined to the wavelength range between $\sim 4,000\text{Å}$ and $\sim 20,000\text{Å}$. If the galaxy is actively forming stars and dusty, the emission from young hot stars can extend this range to smaller wavelengths, while thermal emission from the dust (heated by these young stars) extends it to the far-infrared.

A small, but important, fraction of all galaxies have a spectral energy distribution that is much broader than what is expected from a collection of stars, gas and dust. They typically emit over the full wavelength range from the radio to the X-ray, suggesting that the radiation is non-thermal. In addition, the optical/UV parts of their spectra often reveal numerous strong and very broad emission lines. These galaxies are referred to as active galaxies, and examples include Seyfert galaxies, radio galaxies, and quasars. The non-thermal emission of active galaxies in general emanates from a very small central region, often less than a few parsec across, which is called the active galactic nucleus (AGN).[1] The amazing aspect of an AGN is that, despite its extremely small size, the (non-thermal) luminosity can exceed that of the host galaxy, sometimes by as much as a factor of a thousand.

Understanding the properties and formation of active galaxies is an important part of galaxy formation. First of all, active galaxies form an important population of galaxies, and so any theory of galaxy formation should also address the formation of AGN. Secondly, as discussed in §14.2 below, it is believed that AGN are powered by matter accreting onto a supermassive black hole (SMBH). The observed correlation between the masses of SMBHs and the masses of their host galaxies, shown in Fig. 2.17, strongly suggests that the formation of SMBHs is closely connected to galaxy formation. Furthermore, the fact that virtually all spheroids are found to harbor a SMBH suggests that many, if not all, normal galaxies may have experienced an active phase in their past. Finally, AGN are powerful energy sources, and their energy feedback may have important impact on the intergalactic medium as well as on the formation and evolution of galaxies. The effects of such feedback must be taken into account in any theory of galaxy formation and evolution.

In §14.1 we start with a summary of the properties of different classes of active galaxies. Theoretical models, based on the paradigm that AGN are powered by SMBHs, are described in §14.2, together with a brief description of the various emission mechanisms. In §14.3, the evolution

---

[1] The term AGN is also often used to denote active galaxies, although strictly speaking an active galaxy also includes the host galaxy.

Table 14.1. Local number densities.

Type of object	Number density [Mpc^{-3}]
Field galaxies	10^{-1}
Luminous spirals	10^{-2}
Seyfert galaxies	10^{-4}
Radio galaxies	10^{-6}
QSOs	10^{-7}
Radio-loud quasars	10^{-9}

of the AGN population is discussed in connection to our current ideas of galaxy formation. Finally, in §14.4, we discuss the feedback effects of the AGN population on galaxy formation. The presentation in this chapter serves only as a brief introduction to the topic of active galaxies, emphasizing the connections to galaxy formation and evolution. Readers who want to learn more about AGN and the related physics are referred to the excellent text books and monographs by Robson (1996), Krolik (1998) and Kembhavi & Narlikar (1999).

14.1 The Population of Active Galactic Nuclei

Roughly speaking, an object is defined to be an AGN if one or more of the following properties are observed:

(i) a compact nuclear region much brighter than a region of the same size in a normal galaxy;
(ii) non-stellar (non-thermal) continuum emission;
(iii) strong emission lines;
(iv) variability in continuum emission and/or in emission lines on relatively short time scales.

The observed AGN population is classified into subgroups according to their observational properties. As with the classification of galaxies, some elements of this classification are mainly historical and do not necessarily differentiate the underlying physical processes. In the following, we give a brief overview of the different classes of AGN. Table 14.1 gives a rough idea of the number densities of the objects we are concerned with. Clearly, the AGN population only makes up a relatively minor fraction of all galaxies observed at the present time. However, as we will see, it is believed that AGN have relatively short lifetimes, and it is well possible that many, if not all galaxies, have gone through one or more AGN phases. This is supported by the fact that virtually all spheroids seem to harbor a SMBH.

(a) Seyfert Galaxies Seyfert galaxies, named after their discoverer Carl Seyfert (1943), are active galaxies characterized by spiral-like morphologies with bright star-like nuclei. Spectroscopic observations show that the spectra of Seyfert nuclei have non-thermal continua and contain strong and broad emission lines of high excitation (see Fig. 14.1). Moreover, in many Seyfert galaxies, the radiation from the nuclei can vary by a factor of more than two in less than a year, indicating that the sizes of these nuclei are smaller than one light year across (see §14.2.8).

Seyfert galaxies are subdivided into two categories, Seyfert 1 and Seyfert 2, according to the widths of their emission lines. In Seyfert 1 galaxies, the permitted lines, mainly from hydrogen, are very broad with full widths at half maxima (FWHM) corresponding to velocities in the range 1000–5000 km s^{-1}. The forbidden lines (see §14.2.4 for a definition), such as [OIII], are much narrower, with FWHM corresponding to velocities of a few hundred km s^{-1}. In Seyfert 2 galaxies, on the other hand, both permitted and forbidden lines have narrow velocity widths, typically

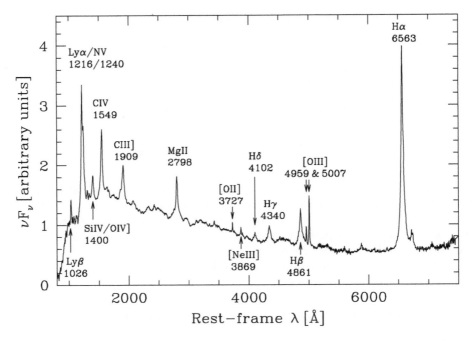

Fig. 14.1. A composite spectrum of QSOs revealing the typical non-thermal continuum and various emission lines. Lines in brackets are forbidden lines, those in semibrackets are semiforbidden lines, and lines without brackets are permitted lines. Note that the permitted lines are much broader than the forbidden lines, as is typical for QSOs and Seyfert 1 galaxies. [Courtesy of C. Foltz and P. Hewett, based on an extension of the data published in Francis et al. (1991)]

of a few hundred km s^{-1}. In general, Seyfert 1 galaxies have stronger non-stellar continua and stronger hard X-ray emission than Seyfert 2 galaxies.

The linewidths reflect the velocities of the emitting gas clouds. The large difference in the FWHM between the permitted and forbidden lines in Seyfert 1 galaxies indicate that these two types of emission lines are probably produced in different regions, while in Seyfert 2 galaxies both permitted and forbidden lines are likely to emanate from the same region. In general, very broad lines, with FWHM larger than 10^3 km s^{-1}, are said to come from a broad-line region, while lines with smaller FWHM are said to arise from a narrow-line region. As we will see §14.2.7, the lack of broad emission lines in Seyfert 2 does not necessarily imply the absence of a broad-line region; it may merely reflect that this region is blocked from our view.

A class of active galaxies related to Seyferts are the LINERs (low-ionization nuclear emission line regions). These objects have properties similar to those of Seyfert 2 galaxies, except that their forbidden lines tend to arise from less ionized atoms. LINERs were initially thought of as a separate class of AGN, but are nowadays most often considered as a low-luminosity extension of Seyferts.

(b) Radio Galaxies Radio galaxies are a class of active galaxies characterized by relatively strong radio emission. Since normal spirals often have weak radio emission (mainly due to supernova remnants), with power $P_{1.4\text{GHz}} \lesssim 2 \times 10^{23}$ W Hz^{-1} at ~ 1.4 GHz, radio galaxies are commonly defined as galaxies with $P_{1.4\text{GHz}}$ larger than this. Radio galaxies were initially identified in radio surveys, such as the well-known third survey carried out at Cambridge (the 3C catalog; see Edge et al., 1959). Over time almost all 3C objects have been unambiguously identified with optical sources, which has revealed that almost all host galaxies of radio AGN are ellipticals.

14.1 The Population of Active Galactic Nuclei

Fig. 14.2. Radio image of 3C175 (a prototypical FR II radio galaxy) at 4.9 GHz (see Bridle et al., 1994). The source has a redshift $z = 0.768$, and the overall linear size of the image is $212\,h^{-1}$ kpc. The source shows double lobes with prominent hot spots, a narrow jet, but no counterjet. [NASA/courtesy of nasaimages.org]

The spectra of the optical sources associated with radio galaxies typically reveal strong emission lines. Similar to Seyfert galaxies, one distinguishes radio galaxies between broad-line radio galaxies (BLRGs) and narrow-line radio galaxies (NLRGs). In principle BLRGs and NLRGs can be considered as radio-loud Seyfert 1 and Seyfert 2 galaxies, respectively, albeit with a different morphology of the host galaxy.

In addition to the classification in BLRGs and NLRGs based on their optical spectra, radio galaxies are also distinguished based on their radio morphology. Radio maps of powerful radio galaxies typically reveal a double-lobed structure extending to several hundred kiloparsecs or even megaparsecs from the central nucleus (Fig. 14.2) that coincides with the center of the host galaxy. Often jet-like structures are observed to stretch from the compact core towards the lobes, suggesting that these jets are responsible for transporting energy from the core out into the radio lobes. The jets are rarely symmetric: often only one jet is observed, and in sources with two jets, one is usually significantly brighter than the other. For weak radio galaxies, the radio emission has a more irregular and more concentrated structure. Based on radio morphology Fanaroff & Riley (1974) divided radio galaxies into two subgroups. In class FR I, the distance between the two most intense spots on either side of the nucleus (called 'hot spots') is smaller than half the overall source size, while in class FR II this distance is larger. It turns out that class FR II are powerful radio sources, with $P_{1.4\text{GHz}} \gtrsim 4 \times 10^{25}\,\text{W}\,\text{Hz}^{-1}$, while class FR I typically have weaker radio power.

On the basis of their radio spectral properties, radio galaxies are also subdivided into 'steep-spectrum' and 'flat-spectrum' sources. This is done by fitting the flux at $\sim 1\,\text{GHz}$ with a power law, $F_\nu \propto \nu^{-\alpha}$, and the division is made at $\alpha = 0.4$. Flat-spectrum sources tend to be compact and show observable variability, while steep-spectrum sources are usually extended.

(c) Quasars and QSOs The name quasar (from Quasi-Stellar Radio Source) was originally used for the optical identifications of compact radio sources with Seyfert-like spectra. The radio characteristics of quasars are similar to those of powerful radio sources, but their optical images are unresolved (at $\theta \sim 1''$), luminous ($M_B \lesssim -21.5 + 5\log h$) nuclei that are unusually blue and often variable.

Radio observations of high spatial resolution have shown that flat-spectrum quasars often contain very compact nuclei ($\sim 10^{-3}$ arcsec). Some of these have elongated structure (jets) consisting of two or more closely separated components whose relative motion to each other can be detected on time scales of a few years. In some cases, the apparent velocity of separation in the transverse direction is inferred to exceed the speed of light. Such 'superluminal' motion is likely due to relativistic jets lying almost along the line-of-sight (see §14.2.5).

The extremely blue colors of quasars means they can also be detected using optical photometry only, without the need to first detect them in the radio. This color-selection technique has proved very successful, and resulted in the detection of many intrinsically bright, high-redshift sources with broad emission lines. It came somewhat as a surprise when many of these sources turned out to be invisible in the radio. Originally, the term quasi-stellar object (QSO) was used to refer to these radio-quiet quasar analogs, which were found to outnumber the radio-loud quasars by a factor 10 to 100. Because of their similarities in optical properties, it has become common practice to use the terms 'QSO' and 'quasar' without distinction. When referring to their radio properties, the adjectives 'radio-quiet' and 'radio-loud' are used, giving rise to confusing nomenclature such as 'radio-loud QSOs' and 'radio-quiet quasars'. At the risk of confusing the reader, we also use this muddled nomenclature throughout this book.

The optical spectrum of a quasar is very similar to that of a Seyfert 1 galaxy. Formally these two classes of AGN are separated at an absolute magnitude of $\mathcal{M}_B = -21.5 + 5\log h$, although this distinction is mainly historical, as there seem to be no fundamental differences between quasars and Seyfert 1 galaxies, other than their AGN luminosities. QSOs are the most luminous AGNs, with luminosities as high as a thousand times that of an L^* galaxy. Consequently, being outshined by the QSO, their host galaxies are extremely difficult to detect. Nevertheless, largely due to the unprecedented spatial resolution achievable with the HST, host galaxies of several low-redshift QSOs have now been observed. Although in many cases the images are not sufficiently clear to reveal the details about their morphologies, they suggest that the host galaxies of low-redshift QSOs have diverse properties; while some appear to be normal spirals and ellipticals, others are strongly disturbed or interacting systems (Bahcall et al., 1997).

(d) BL Lac Objects and OVVs A special subclass of quasars are the optically violently variables (or OVVs for short), which are characterized by very strong and rapid optical variability. The optical flux of OVVs can vary by a significant fraction in less than one day, suggesting that the regions from which the optical emission emanates are less than 1 light-day across. In addition to their strong variability, OVVs are also characterized by a relatively strong polarization of the optical light, typically at the level of a few percent (compared to $\lesssim 1\%$ for regular quasars). OVVs are also variable at other wavelengths, with the time scale of variability typically becoming smaller towards shorter wavelengths.

A class of objects closely related to OVVs are the BL Lac objects, named after their prototypical source BL Lacertae, originally believed to be a variable star (which explains its stellar name). Like OVVs, BL Lac objects are strong radio emitters, highly variable in optical and X-ray emission, and with strongly polarized radio and optical emission. But unlike OVVs, BL Lac objects reveal no emission lines; their spectra are featureless power laws. The similarity of their radio properties with flat-spectrum radio quasars suggests that BL Lac objects are probably associated with elliptical galaxies. Indeed, in a number of cases, faint surrounding nebulosity is detected with a surface brightness profile similar to that of an elliptical galaxy. However, there are also cases where the profile of the nebulosity is disk-like.

Although the distinction between BL Lac objects and OVVs is that the latter reveal broad emission lines which are absent in the former, this difference may well be a reflection of the intrinsic variability of these sources: the emission lines may be visible when the continuum is relatively faint, but become undetectable if the continuum is bright. Consequently, BL Lac objects and OVVs are often grouped together in a single class known as blazars.

14.2 The Supermassive Black Hole Paradigm

In terms of energetics, an AGN is extraordinary in that it emits a large amount of energy from a very small region. An obvious question is how this energy is generated. The small size of the emission region and the large amount of energy output suggest that the central engine must be compact and have relatively large mass. It is now generally believed that this central engine is a supermassive black hole (SMBH), an idea originally proposed by Salpeter (1964), Zel'dovich & Novikov (1964) and Lynden-Bell (1969). In this section, we describe the current paradigm for understanding the properties of AGN. In addition to the central engine, the black hole (BH), the standard AGN model also assumes the existence of other components, such as broad and narrow line regions (BLR and NLR), an accretion disk, and jets (see Fig. 14.3). The roles of these components for interpreting the observational properties of AGN are discussed in the following subsections. Observational support for the existence of SMBHs is presented in §14.2.8.

14.2.1 The Central Engine

In the SMBH paradigm, an AGN is assumed to be powered by a SMBH accreting gas, and the energy source is the gravitational potential of the central black hole. To comprehend the energy scale involved with such accretion, consider a simple spherical model in which a central source with luminosity L is surrounded by gas with density distribution $\rho(r)$. The flux at a radius r from the source is $L/(4\pi r^2)$ and the radiation pressure is

$$P_{\rm rad}(r) = \frac{L}{4\pi r^2 c}. \tag{14.1}$$

If the gas is ionized, the pressure force on a unit volume of gas due to the scattering of photons by electrons is

$$\mathbf{F}_{\rm rad} = \sigma_{\rm T} P_{\rm rad}(r) n_{\rm e}(r) \hat{\mathbf{r}}, \tag{14.2}$$

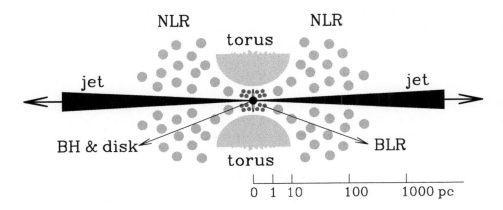

Fig. 14.3. Different components of an AGN in the standard paradigm.

where σ_T is the Thomson scattering cross-section, and $n_e(r)$ is the electron density at radius r. In order for the gas not to be dispersed quickly, this pressure force must be smaller than the gravitational force on the gas, i.e.

$$|\mathbf{F}_{\rm rad}| \le F_{\rm grav} = \frac{GM_{\rm BH}\rho(r)}{r^2}. \tag{14.3}$$

This defines a maximum luminosity (the Eddington luminosity) for a given black hole $M_{\rm BH}$:

$$L_{\rm Edd} \equiv \frac{4\pi Gcm_{\rm p}}{\sigma_T} M_{\rm BH} \approx 1.28 \times 10^{46} M_8 \,\text{erg s}^{-1} \quad (M_8 \equiv M_{\rm BH}/10^8\,M_\odot), \tag{14.4}$$

with $m_{\rm p}$ the proton mass. This Eddington luminosity is the largest possible luminosity that can be achieved by spherical accretion. The above relation can be inverted to give a minimum central mass required to achieve a given luminosity:

$$M_{\rm Edd} = 8 \times 10^7 L_{46}\,M_\odot, \quad [L_{46} \equiv L/(10^{46}\,\text{erg s}^{-1})]. \tag{14.5}$$

From these simple considerations, we see that a bright quasar with a luminosity $L \sim 10^{46}\,\text{erg s}^{-1}$ may be powered by a black hole with a mass $M_{\rm BH} \sim 10^8\,M_\odot$.

What kind of mass accretion rate is required in order to power an AGN? To answer this question, we note that, under the assumption that the luminosity is powered by the gravitational potential of the central BH, the accretion luminosity can be written as

$$L = \frac{GM_{\rm BH}}{r}\dot{M}_{\rm BH}, \tag{14.6}$$

where $\dot{M}_{\rm BH}$ is the mass accretion rate, i.e. the rate at which mass crosses a shell of radius r. From this we see that the efficiency at which the rest mass of accreted material is converted into radiation is

$$\varepsilon_{\rm r} \equiv \frac{L}{\dot{M}_{\rm BH}c^2} = \frac{1}{2}\frac{r_{\rm S}}{r}, \tag{14.7}$$

where

$$r_{\rm S} = \frac{2GM_{\rm BH}}{c^2} \approx 10^{-2} M_8 \text{ light-days} \tag{14.8}$$

is the Schwarzschild radius of a black hole with mass $M_{\rm BH}$. As we shall see below, the bulk of the continuum radiation (which is dominated by photons with frequencies around the blue bump; see Fig. 14.4) originates from $r \sim 5 r_{\rm S}$. It then follows that $\varepsilon_{\rm r} \sim 0.1$. This is a very high efficiency, much higher than the efficiency with which hydrogen is burned into helium, which is only ~ 0.007. With this efficiency, an accretion rate of $\dot{M} \sim 2\,M_\odot\,\text{yr}^{-1}$ is required to power a bright quasar with luminosity $L = 10^{46}\,\text{erg s}^{-1}$. In particular, the Eddington luminosity defined in Eq. (14.4) corresponds to a mass accretion rate

$$\dot{M}_{\rm Edd} = \frac{L_{\rm Edd}}{\varepsilon_{\rm r}c^2} \approx 2.2 M_8 \left(\frac{\varepsilon_{\rm r}}{0.1}\right)^{-1}\,M_\odot\,\text{yr}^{-1}, \tag{14.9}$$

which is the highest possible accretion rate in the simple spherical model. Super-Eddington accretion is possible if the mass accretion is non-spherical, e.g. primarily in the equatorial plane of a disk while the radiation escapes from the polar zones.

14.2.2 Accretion Disks

Since the gas to be accreted by a SMBH in general has angular momentum, the accretion is most likely through a Keplerian disk. If the accretion disk is axisymmetric and thin, the structure of the disk is described by its surface density $\Sigma(R,t)$, where R is the radius on the disk. The time evolution of $\Sigma(R,t)$ is governed by the conservations of mass and angular momentum. For a constant accretion rate, i.e. $2\pi \int \dot{\Sigma} R \, dR = \dot{M}_{BH}$ = constant, conservation of angular momentum gives

$$\frac{\partial \Sigma}{\partial t} = -\frac{1}{R}\frac{\partial}{\partial R}\left\{\frac{(\partial/\partial R)\left[\nu_k \Sigma R^3 (d\omega/dR)\right]}{(d/dR)(R^2 \omega)}\right\}, \quad (14.10)$$

where ν_k is the kinetic viscosity, and ω is the rotation velocity of the gas (see §11.4.2). The viscosity is thought to be due to turbulence or magnetic stress, but the details are still poorly understood. Usually a simple prescription proposed by Shakura & Sunyaev (1973) is adopted (e.g. Pringle, 1981), in which the effective viscosity is assumed to be a constant, α, times the total internal energy density in the disk. Integrating the above equation twice with respect to R, we get

$$2\pi \nu_k \Sigma R^3 \omega' = A - \dot{M}_{BH} R^2 \omega, \quad (14.11)$$

where a prime denotes derivative with respect to R, and A is independent of R. For a Keplerian disk, $\omega = (GM_{BH}/R^3)^{1/2}$, and

$$\nu_k \Sigma = \frac{\dot{M}_{BH}}{3\pi}\left(1 - \sqrt{\frac{R_{in}}{R}}\right), \quad (14.12)$$

where R_{in} is an inner radius of the accretion disk.

In order to predict the emission properties of an accretion disk, we need to know its temperature structure. Consider a mass element within a ring between R and $R+dR$. Because of viscosity and differential rotation, there is angular momentum flow at both the inner radius R and the outer radius $R+dR$. The angular-momentum transfer rate at a radius R due to viscosity is $\mathcal{G} = 2\pi R^3 \nu_k \Sigma \omega'$, which corresponds to an energy transfer rate $\mathcal{G}\omega$. Thus, the total work done on the material in the ring is $(\mathcal{G}\omega)' dR$, and the work done on a unit area is

$$W = \frac{(\mathcal{G}\omega)' \Delta R}{2\pi R \Delta R} = \frac{\mathcal{G}'\omega + \mathcal{G}\omega'}{2\pi R}. \quad (14.13)$$

The term with ω' reflects the change of the rotation velocity of the ring due to angular momentum transfer, while the term containing $\mathcal{G}'$ describes the energy dissipation in the ring.

Assuming that all the dissipated energy is radiated away at the radius where it is produced and that the disk is optically thick, so that the dissipated energy is completely thermalized, the temperature at R is then given by

$$2\sigma_{SB} T^4(R) = \frac{3GM_{BH}\dot{M}_{BH}}{4\pi R^3}\left(1 - \sqrt{\frac{R_{in}}{R}}\right), \quad (14.14)$$

where σ_{SB} is the Stefan–Boltzmann constant, and the factor 2 on the left-hand side is due to the fact that the dissipated energy is radiated away from both faces of the disk. The temperature peaks at $R = (7/6)^2 R_{in}$, while the energy emitted in rings with fixed ΔR peaks at $R = (5/4)^2 R_{in}$. The total luminosity is $L = GM_{BH}\dot{M}_{BH}/R_{in}$, and half of the energy is radiated within a radius of $\sim 8 R_{in}$. Clearly, most emission emanates from the inner disk where the temperature is also the highest. If $R_{in} \sim r_S$, then $L \sim \dot{M}_{BH} c^2/4$.

Outside R_{in}, the temperature is

$$T(R) \sim \left(\frac{3GM_{BH}\dot{M}_{BH}}{8\pi\sigma_{SB} r_S^3}\right)^{1/4}\left(\frac{R}{r_S}\right)^{-3/4} \approx 6.3 \times 10^5 \left(\frac{\dot{M}_{BH}}{\dot{M}_{Edd}}\right)^{1/4} M_8^{-1/4}\left(\frac{R}{r_S}\right)^{-3/4} \text{ K}, \quad (14.15)$$

where $\varepsilon_r = 0.1$ is assumed. For a disk around a $M_{BH} = 10^8 \, M_\odot$ black hole accreting at the Eddington limit, the thermal radiation at a radius not much larger than $R_{in} \sim r_S$ peaks at a frequency

$\nu \sim 10^{15}$ - 10^{16} Hz. This frequency is similar to that of the blue bump often observed in quasar continua. It is therefore believed that the blue bump is produced by the accretion disk. Because of the change of temperature with radius R, the expected spectrum from an accretion disk at low frequency is actually very different from a thermal spectrum. In the Rayleigh–Jeans limit ($h_P \nu \ll k_B T$), where the Planck function $B_\nu(T) \propto \nu^2 T$, the spectrum is

$$F_\nu \propto \nu^2 \int T(R) R dR \propto \nu^{1/3}, \qquad (14.16)$$

where the second relation uses $T \propto R^{-3/4}$, and the integration is from $R_{\rm in}$ to an outer radius given by $k_B T(R) = h_P \nu$, beyond which the Rayleigh–Jeans approximation fails.

At small radii, the assumption of a geometrically thin disk is not valid, as the inner disk can be puffed up either by radiation pressure from the central source or due to insufficient cooling of the accreted gas. In both cases, a geometrically thick disk may form in the inner part of an accretion disk. Detailed modeling shows that such a thick disk has a torus-like structure, which may provide a funnel to collimate the radiation produced in the central region. The pressure of such collimated radiation may help to produce a collimated jet of material along the funnel, as seen in some AGN (see Rees, 1984, and references therein).

14.2.3 Continuum Emission

An AGN typically has a very broad spectrum, ranging from the radio all the way to gamma rays (see Fig. 14.4). To a very rough approximation, the overall spectral energy distribution (SED) of an AGN can be described by a power law, $F_\nu \propto \nu^{-\alpha}$, with $0 \lesssim \alpha \lesssim 1$. A closer examination

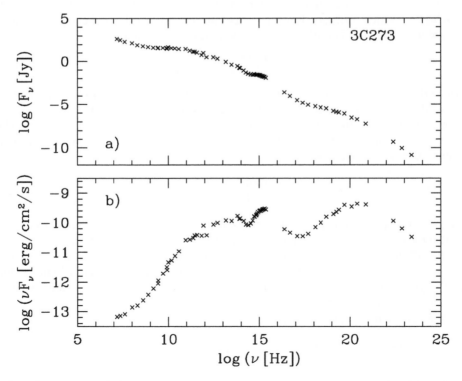

Fig. 14.4. The spectrum of 3C273, F_ν (upper panel) and νF_ν (lower panel), showing the large-scale power-law behavior as well as several pronounced bumps. [Based on data presented in Lichti et al. (1995)]

reveals that the SED contains depressions and bumps, with the blue bump around $\nu \sim 10^{15}$–10^{16} Hz and the broad bump around $\nu \sim 10^{20}$–10^{21} Hz being the most prominent.

The broad energy range observed in an AGN spectrum suggests that a variety of emission mechanisms is involved. It is generally believed that relativistic electrons play a crucial role here. To see this, define a 'brightness temperature' for a radiation field in terms of its specific intensity J_ν:

$$T_{\rm b} \equiv \frac{c^2 J_\nu}{2 k_{\rm B} \nu^2}. \tag{14.17}$$

Inspecting the Planck function $B_\nu(T)$ we see that the brightness temperature is the lowest possible temperature to produce the specific intensity if the radiation field were thermal. For compact radio sources, the values of $T_{\rm b}$ can often reach 10^{11} K, but the amount of gamma-ray emission corresponding to such high temperatures is not observed. This indicates that the radio emission is non-thermal. On the other hand, the electrons responsible for the radio emission must each have an energy of the order $k_{\rm B} T_{\rm b}$. This is much higher than the rest-mass energy of an electron, $m_{\rm e} c^2 \approx 5.8 \times 10^9 k_{\rm B}$ K, and so the electrons involved must be highly relativistic.

Relativistic electrons in AGN are believed to be generated by the first-order Fermi acceleration in shocks that are likely to form from supersonic flows near the central SMBH. Fermi acceleration is the acceleration that charged particles undergo when they are reflected by a moving interstellar magnetic field (Fermi, 1949). In shocks, this process is particularly effective: a charged particle ahead of the shock front can pass through the shock and then be scattered by magnetic inhomogeneities behind the shock. The particle gains energy from this bounce and flies back across the shock, where it can be scattered by magnetic inhomogeneities ahead of the shock. This enables the particle to bounce back and forth again and again, gaining energy each time. Because the mean energy gain depends only linearly on the shock velocity, this process is called first-order Fermi acceleration, and typically results in a power-law energy distribution for the accelerated particles (Blandford & Eichler, 1987; Jones & Ellison, 1991). The motion of relativistic electrons with a power-law energy distribution in a magnetic field produces synchrotron radiation with a power-law spectrum that can cover many decades in frequency in the radio band. In addition, photons with energies all the way up to the X-ray band can also be generated directly from relativistic electrons through the inverse Compton process. In what follows we examine in more detail the continuum to be expected from these two processes.

(a) Synchrotron Emission When a charged particle is accelerated, it radiates photons. In the instantaneous rest-frame of the charge, the total emitted power is given by Larmor's formula,

$$P_{\rm S} = \frac{2q^2}{3c^3} |\dot{\bf v}|^2, \tag{14.18}$$

where q is the charge of the particle and $\dot{\bf v}$ is the acceleration in the instantaneous rest-frame. To obtain the emitted power in the observer's frame, we seek a form of $P_{\rm S}$ which is invariant under the Lorentz transformation. In special relativity, the Lorentz-invariant space-time interval is

$$ds^2 = \eta_{\mu\nu} dx^\mu dx^\nu = c^2 dt^2 - dx^i dx_i = c^2 \gamma^{-2} dt^2, \tag{14.19}$$

where $\eta_{\mu\nu}$ is the Minkowski metric (see Appendix A), and

$$\gamma = (1 - \beta^2)^{-1/2} \quad \text{with} \quad \beta = v/c \tag{14.20}$$

is the Lorentz factor. From the four-coordinate x^μ one can form a four-velocity, $U^\nu = c\, dx^\mu/ds$, and a four-acceleration,

$$a^\mu \equiv c \frac{dU^\mu}{ds} = \gamma \frac{dU^\mu}{dt}. \tag{14.21}$$

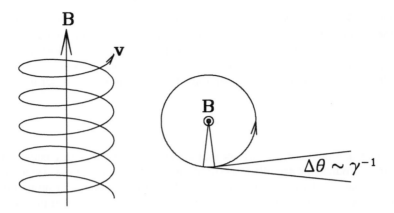

Fig. 14.5. Synchrotron radiation is generated by relativistic electrons spiraling in a magnetic field. The radiation is beamed into a forward cone with opening angle $\Delta\theta \sim \gamma^{-1}$.

One can then form a Lorentz scalar,

$$P_S = \frac{2q^2}{3c^3} a^\mu a_\mu, \tag{14.22}$$

which reduces to Larmor's formula in the instantaneous rest-frame of the charge. The power P_S in the observer's frame can be obtained by inserting the expression of a^μ into Eq. (14.22):

$$P_S = \frac{2q^2}{3c} \left\{ \left[\gamma \frac{d}{dt} \left(\gamma \frac{\mathbf{v}}{c} \right) \right]^2 - \left(\gamma \frac{d\gamma}{dt} \right)^2 \right\}. \tag{14.23}$$

Using $\gamma^2 = (1 - v^2/c^2)$ and $d\gamma/dt = \gamma^3(\dot{\mathbf{v}} \cdot \mathbf{v}/c^2)$, we can write

$$\begin{aligned} P_S &= \frac{2q^2}{3c^3} \gamma^6 \left[\left(\dot{\mathbf{v}} \cdot \frac{\mathbf{v}}{c} \right)^2 + \left(1 - \frac{v^2}{c^2} \right) |\dot{\mathbf{v}}|^2 \right] \\ &= \frac{2q^2}{3c^3} \gamma^6 \left(|\dot{\mathbf{v}}|^2 - |\dot{\mathbf{v}} \times \mathbf{v}/c|^2 \right). \end{aligned} \tag{14.24}$$

For synchrotron radiation created by a relativistic electron spiraling in a static magnetic field $\mathbf{B}$ (Fig. 14.5), $\dot{\mathbf{v}} = \omega_B \times \mathbf{v}$, where the gyrotation frequency of the electron is

$$\omega_B = \omega_L/\gamma, \tag{14.25}$$

with $\omega_L \equiv q_e \mathbf{B}/(m_e c)$ the Larmor angular frequency. It then follows that

$$P_S = 2\sigma_T c \gamma^2 \beta^2 \mathscr{E}_B \sin^2 \alpha, \tag{14.26}$$

where α is the angle between $\mathbf{v}$ and $\mathbf{B}$, σ_T is the Thomson cross-section, and $\mathscr{E}_B \equiv B^2/8\pi$ is the energy density of the static magnetic field. For an isotropic distribution of relativistic electrons, the average power is

$$\langle P_S \rangle = 2\sigma_T c \gamma^2 \beta^2 \mathscr{E}_B \langle \sin^2 \alpha \rangle = \frac{4}{3} \sigma_T c \gamma^2 \beta^2 \mathscr{E}_B. \tag{14.27}$$

The synchrotron emission from a relativistic electron is concentrated within a small beam, with an opening angle $\Delta\theta \sim \gamma^{-1}$, aligned with the instantaneous velocity of the electron (Fig. 14.5). Thus, as an electron is spiraling in the magnetic field, the radiation signals received by an observer are short pulses. The width of each pulse and the interval between successive pulses can be estimated as follows. The time interval for the electron to cover the angle $\Delta\theta$ is

$\Delta t_{\rm em} = \Delta\theta/\omega_B \sim \gamma^{-1}/\omega_B$. For an observer in the direction $\hat{\bf n}$, the time interval of being swept by the beam is

$$\Delta t_{\rm obs} \approx \left(\frac{{\rm d}\tau}{{\rm d}t}\right)^{-1} \Delta t_{\rm em} = (1 - \boldsymbol{\beta}\cdot\hat{\bf n})\Delta t_{\rm em}, \tag{14.28}$$

where $\boldsymbol{\beta} = {\bf v}/c$, and ${\rm d}\tau$ is the proper time interval for the electron. For $\gamma \gg 1$ we obtain

$$\Delta t_{\rm obs} \sim \left[1 - \beta + (\Delta\theta)^2/2\right]\Delta t_{\rm em} \sim \frac{1}{2\gamma^3\omega_B}. \tag{14.29}$$

The time interval between successive pulses is the same as the time of gyration $2\pi/\omega_B$. Thus, the observer receives radiation pulses of width $\sim \gamma^{-3}/2\omega_B$ separated with intervals $\sim 2\pi/\omega_B$. If we were to Fourier transform the signal, we would see significant power at frequencies all the way from the fundamental frequency $\nu_B = \omega_B/2\pi \propto 1/\gamma$ to the high harmonic

$$\nu_{\rm c} \equiv \frac{\omega_B}{2\pi}\gamma^3 = \frac{\omega_L}{2\pi}\gamma^2 \propto \gamma^2. \tag{14.30}$$

For given B, the power spectrum depends only on γ, and we can formally write the angle-averaged power spectrum as

$$P_{\rm S}(\nu;\gamma){\rm d}\nu = \frac{4}{3}\sigma_{\rm T}c\gamma^2\beta^2\mathscr{E}_B\phi(\nu;\gamma){\rm d}\nu, \tag{14.31}$$

where $\phi(\nu;\gamma)$ is the spectrum shape normalized so that $\int \phi(\nu;\gamma){\rm d}\nu = 1$.

If the electrons in a cosmic radio source have some distribution in their energy $m_ec^2\gamma$, i.e. the number density of electrons per unit Lorentz factor is given by some distribution function $\mathscr{N}(\gamma){\rm d}\gamma$, we can integrate over the distribution to get the total volume emissivity:

$$\varepsilon_\nu = \int P_{\rm S}(\nu;\gamma)\mathscr{N}(\gamma){\rm d}\gamma. \tag{14.32}$$

Based on the discussion given above, we may assume a power-law form for $\mathscr{N}(\gamma)$:

$$\mathscr{N}(\gamma){\rm d}\gamma = K\gamma^{-p}{\rm d}\gamma. \tag{14.33}$$

As an approximation, we treat the spectral emissivity of a single particle by a delta-function centered on the highest harmonic frequency $\nu_{\rm c}$, i.e. we write $\phi(\nu;\gamma) = \delta(\nu - \nu_{\rm c})$. This is a good approximation as long as the width of ϕ in ν is much narrower than the frequency range covered by the electron energy distribution. Assuming further that the main contribution to the total emissivity is from electrons with $\gamma \gg 1$ so that $\beta = (1 - 1/\gamma^2)^{1/2} \sim 1$, we can write the volume emissivity as

$$\varepsilon_\nu \approx \int_1^\infty P_{\rm S}(\nu;\gamma)\mathscr{N}(\gamma){\rm d}\gamma \sim \frac{2}{3}c\sigma_{\rm T}K\mathscr{E}_B\nu_{\rm L}^{-1}\left(\frac{\nu}{\nu_{\rm L}}\right)^{-s}, \tag{14.34}$$

where $s = (p - 1)/2$ is the spectral index. The observed spectral index for extended radio sources is $s \sim 0.7$. The corresponding index of the electron energy distribution is $p \sim 2.5$, consistent with what is expected from first-order Fermi acceleration in shocks.

The above result is valid only when the sources are optically thin so that the photons created by the synchrotron process are emitted without being absorbed and re-emitted by the plasma. However, this may not be true at low frequencies where emitted photons can be effectively absorbed by the electrons. The effect of self-absorption on the spectral shape can be obtained using the radiative transfer equation:

$$\frac{{\rm d}J_\nu}{{\rm d}x} = \frac{\varepsilon_\nu}{4\pi} - \kappa_\nu J_\nu, \tag{14.35}$$

where J_ν is the specific intensity, $dx = c\,dt$ is the proper length along the path of the light ray, and κ_ν is the opacity (the fractional decrease of the specific intensity due to the absorption over a unit of x). Detailed calculations of synchrotron absorption (see Shu, 1991b) show that the opacity of self-absorption scales as

$$\kappa_\nu \propto n_e B^{(2+p)/2} \nu^{-(4+p)/2}, \qquad (14.36)$$

where p is the index in Eq. (14.33). Since p is expected to be larger than -4, a dense source may become optically thick at low frequencies. For a source with uniform properties, the radiative transfer equation can be solved to give

$$J_\nu = S_\nu \left(1 - e^{-\kappa_\nu x}\right), \quad S_\nu \equiv \frac{\varepsilon_\nu}{4\pi \kappa_\nu} \sim \left(\frac{m_e^3 c}{q_e}\right)^{1/2} B^{-1/2} \nu^{5/2}. \qquad (14.37)$$

It then follows that

$$J_\nu = \begin{cases} [\varepsilon_\nu x / 4\pi] \propto \nu^{-s} & (\text{for } \kappa_\nu x \ll 1) \\ S_\nu \propto \nu^{5/2} & (\text{for } \kappa_\nu x \gg 1). \end{cases} \qquad (14.38)$$

The spectrum is thus peaked at a frequency given by $\kappa_\nu x \sim 1$, with the peak frequency determined by the magnetic field strength B and the electron column density $n_e x$ of the source.

(b) Inverse Compton Scattering Another important radiation mechanism involving relativistic electrons is the production of high-energy photons through the scattering of low-energy ones by high-energy electrons. The net result of this process is the emission of radiative energy from electrons, and it arises because electrons moving in a radiation field are accelerated by the electromagnetic fields (see §B1.3.6). The radiation power in this process is given by Eq. (14.22) with the four-acceleration given by

$$a^\mu = c \frac{dU^\mu}{ds}, \quad \frac{dU^\mu}{ds} = \frac{q_e}{m_e} F^{\mu\nu} U_\nu, \qquad (14.39)$$

where $F^{\mu\nu} = \partial^\mu A^\nu - \partial^\nu A^\mu$. It then follows that the angle-averaged emission rate per electron is

$$\langle P_{IC} \rangle = \frac{\sigma_T c \gamma^2}{4\pi} \left\langle (\mathbf{E} + \boldsymbol{\beta} \times \mathbf{B})^2 - (\mathbf{E} \cdot \boldsymbol{\beta})^2 \right\rangle = \frac{4}{3} \sigma_T c \gamma^2 \beta^2 \mathscr{E}_{\text{rad}}, \qquad (14.40)$$

where $\mathscr{E}_{\text{rad}}$ is the energy density of the radiation field. Since the number of photons is conserved in the inverse Compton process, on average each collision upshifts the photon frequency as

$$\nu_0 \to \langle \nu \rangle = \frac{4}{3} \gamma^2 \nu_0 \quad (\gamma \gg 1), \qquad (14.41)$$

where ν_0 is the frequency of the seed photons. The photon frequency is therefore boosted by a factor of $\sim \gamma^2$, which is large for highly relativistic electrons with $\gamma \gg 1$. Thus, as long as relativistic electrons are available, photons generated by the synchrotron process can be scattered into the optical regime and beyond by the inverse Compton process. Note that the mean frequency $\langle \nu \rangle$ given above has the same γ dependence as ν_c in Eq. (14.30) for the synchrotron emission. If the frequency distribution around $\langle \nu \rangle$ is not very broad for a given γ, the spectral shape of the inverse Compton emission is determined by $\mathscr{N}(\gamma)\,d\gamma$, the energy distribution of the relativistic electrons.

The importance of Compton losses relative to synchrotron losses is given by the ratio

$$\frac{\langle P_{IC} \rangle}{\langle P_S \rangle} = \frac{\mathscr{E}_{\text{rad}}}{\mathscr{E}_B} = \frac{(4\pi/c) \int J_\nu\, d\nu}{B^2/8\pi}, \qquad (14.42)$$

where J_ν is the specific intensity of the seed radiation field for the inverse Compton process. To get an order-of-magnitude estimate, we write $\int J_\nu\, d\nu = J_{\nu_*} \nu_*$, where ν_* is some average frequency. If the seed radiation field is provided by the synchrotron photons, and if ν_* is high enough so that the optical depth of the source is small, we can write

$$J_{\nu_*} = \frac{2\nu_*^2}{c^2} k_B T_b \sim \left(\frac{m_e^3 c}{q_e}\right)^{1/2} B^{-1/2} \nu_*^{5/2}, \quad (14.43)$$

where T_b is the brightness temperature defined in Eq. (14.17). It then follows that

$$\frac{\langle P_{IC}\rangle}{\langle P_S\rangle} \sim \left(\frac{k_B T_b}{m_e c^2}\right)^5 \frac{\sigma_T^{1/2} \nu_*}{c} \sim \left(\frac{T_b}{10^{12}\,\mathrm{K}}\right)^5 \left(\frac{\nu_*}{1\,\mathrm{GHz}}\right). \quad (14.44)$$

Thus, the power of inverse-Compton emission exceeds that of synchrotron emission if $T_b > 10^{12}$ K. When this occurs, the cooling of the electrons by radiation is catastrophic, because the inverse-Compton photons can themselves be scattered by the relativistic electrons, enhancing the inverse-Compton emission power even further. Thus, the brightness temperature of the synchrotron emission must in general be below 10^{12} K in order for the electrons to remain relativistic.

To determine the values of T_b for AGN, we need to know the sizes of the regions from which the radio emission originates. For a given luminosity, the smaller the size, the higher the brightness temperature. Currently, the most stringent limit comes from estimates based on source variability, $l \sim c\Delta t$. Based on the observed time scale of variability, the inferred values of T_b often exceed 10^{12} K, indicating that the real size of the emission region must be larger than that implied by the observed variability time scale. This can be achieved if relativistic motion is involved in the source, so that time dilation causes the observed time scale of variability to be compressed relative to the real time scale of variability.

14.2.4 Emission Lines

As mentioned in §14.1, an important characteristic of AGN spectra is the presence of strong emission lines produced by the transitions of excited atoms. In this subsection we briefly describe how studies of these emission lines can be used to infer both the physical conditions of the emitting gas (such as density, temperature, chemical composition, ionization state and turbulent motion) and the properties of the ionization sources. A more detailed description can be found in Osterbrock (1989).

The observed emission lines are generally divided into two categories, permitted lines and forbidden (or improbable) lines, according to the rate of spontaneous transition between the energy levels responsible for the emission. Lines with an intermediate spontaneous transition probability are called semiforbidden lines. Permitted lines are associated with transitions allowed by the electric-dipole selection rules, while forbidden lines are associated with transitions with zero dipole component but non-zero high-order components. Forbidden lines are denoted by square brackets, such as the [OIII] lines of doubly ionized oxygen, and semiforbidden lines are designated with a single square bracket, such as CIII]. In general, forbidden transitions have lower probability than permitted ones. Because of the low transition probability, forbidden lines are not expected from gas with high density, where relevant ions can be removed quickly from the excited state by collisions with electrons before they have a chance to make the forbidden transition. On the other hand, permitted lines are expected in both high- and low-density gas. To see this more clearly, let us consider a simple example in which an ion has only two levels, 1 and 2, with $E_1 < E_2$. The equation for the balance between excitation and de-excitation is

$$n_e n_1 P_{12} = n_e n_2 P_{21} + n_2 A_{21}, \quad (14.45)$$

where n_1 and n_2 are the population densities of states 1 and 2, respectively, n_e is the electron density, A_{21} the Einstein coefficient of spontaneous emission, P_{12} the probability for the ion to undergo a transition from state 1 to state 2 in a unit time due to collisions with electrons, and P_{21} the corresponding probability from state 2 to state 1 (see Appendix B for details). The relative population of the two states is then given by

$$\frac{n_2}{n_1} = \frac{n_e P_{12}}{A_{21}} \frac{1}{1 + n_e P_{21}/A_{21}}. \tag{14.46}$$

Since the line emission is produced mainly by spontaneous transitions, the line luminosity per unit volume due to collisional excitation is

$$\mathscr{L}_c = n_2 A_{21} h_P \nu_{12} = \frac{n_e n_1 P_{12} h_P \nu_{12}}{1 + n_e P_{21}/A_{21}}. \tag{14.47}$$

It then follows that

$$\mathscr{L}_c = \begin{cases} n_e n_1 P_{12} h_P \nu_{12} & (\text{if } A_{21} \gg n_e P_{21}) \\ n_1 (P_{12}/P_{21}) A_{21} h_P \nu_{12} & (\text{if } A_{21} \ll n_e P_{21}). \end{cases} \tag{14.48}$$

In low density regions where $n_e \ll A_{21}/P_{21}$, $\mathscr{L}_c$ is independent of A_{21} and so both the forbidden and permitted lines can be produced with significant strengths. On the other hand, in high density regions where $n_e \gg A_{21}/P_{21}$ for the forbidden line but not for the permitted line, the strength of the forbidden line is reduced by a factor of $A_{21}/(n_e P_{21})$. Consequently, the relative strength of forbidden lines with respect to permitted lines can be used as a probe into the density of the emitting gas. The most important diagnostic lines are the forbidden [OIII]$\lambda\lambda(4959, 5007)$Å doublet and the semiforbidden CIII]$\lambda 1909$Å. The critical electron density, A_{21}/P_{21}, is $\sim 10^6 \text{ cm}^{-3}$ for [OIII] and $\sim 10^{10} \text{ cm}^{-3}$ for CIII], while for permitted lines it is much higher.

As mentioned in §14.1, if we divide the emission lines in AGN spectra into two classes according to their linewidth, broad lines with FWHM corresponding to Doppler-broadening velocities $\gtrsim 10^3 \text{ km s}^{-1}$ are almost exclusively permitted lines, while narrow lines with velocities $\sim 10^2 \text{ km s}^{-1}$ can be either permitted lines or forbidden lines. Using the forbidden-line diagnostics discussed above, we infer that broad lines are produced in the innermost emission-line regions where the gas densities and velocities are high, while narrow lines emanate from more extended regions characterized by lower gas densities and velocities (see also §14.2.6).

To interpret an observed emission-line spectrum in detail, we must know how the emitting gas is ionized. In general, gas can be ionized by collisional processes and by photoionization, depending on the density and temperature of the gas, and on the local intensity of ionizing radiation. With the principles described in §B1.3, it is straightforward to calculate the number densities of various species for a gas with given density, temperature and chemical composition. The volume emissivity of an emission line, which is proportional to $n_m A_{mn} h_P \nu_{mn}$ (where m and n are the two energy levels responsible for the line emission), can then be obtained. The line profile can also be related to the temperature and turbulent motion of the gas, as described in detail in §16.4. With such a procedure, one can in principle fit an observed emission-line spectrum by tuning the conditions in the emitting gas and the properties of the ionizing source.

One important application of emission-line spectra is to use line ratios to separate AGN from star-forming galaxies whose spectra also contain emission lines due to the HII regions generated by young massive stars (see §10.3.7). As far as emission lines are concerned, the main difference between an AGN and a star-forming galaxy is in the level of ionization and temperature of the emitting gas, both expected to be higher in an AGN due to the stronger UV flux involved. In addition, the radiation field in an AGN is also 'harder', i.e. contains a larger fraction of high-energy photons. These photons have a larger mean free path in a neutral medium, and can therefore penetrate farther to give rise to a larger partially ionized zone. Detailed photoionization modeling (Osterbrock, 1989) shows that the optical line ratio, [OIII]$\lambda 5007/\text{H}\beta$, is a good indicator of the mean level of ionization and temperature, while [OI]$\lambda 6300/\text{H}\alpha$, [SII]$\lambda\lambda 6717, 6731/\text{H}\alpha$ and

14.2 The Supermassive Black Hole Paradigm

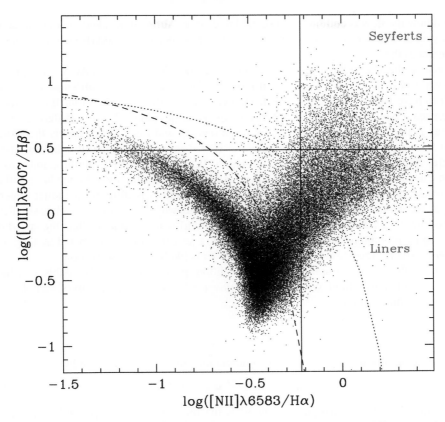

Fig. 14.6. The emission-line flux ratio [OIII]$\lambda 5007$/Hβ versus the ratio [NII]$\lambda 6583$/Hα for a galaxy sample constructed from the SDSS. A diagram of emission-line ratios like this is often called a BPT diagram, after Baldwin et al. (1981) who demonstrated its usefulness for separating AGN from normal star-forming galaxies. The dashed curve represents the demarcation line of pure star formation defined by Kauffmann et al. (2003a) and the dotted line is the extreme starburst demarcation line of Kewley et al. (2001). Seyfert galaxies are often defined to have [OIII]/H$\beta > 3$ and [NII]/H$\alpha > 0.6$, and LINERs to have [OIII]/H$\beta < 3$ and [NII]/H$\alpha > 0.6$. [Adapted from Kauffmann et al. (2003a)]

[NII]$\lambda 6583$/Hα are sensitive to the relative importance of a large partially ionized zone produced by high-energy photoionization. Fig. 14.6 shows an example of how one can use these line ratios to separate AGN from star-forming galaxies. Typically, systems with both high [OIII]$\lambda 5007$/Hβ and high [NII]$\lambda 6583$/Hα are likely to be ionized by an AGN (e.g. Baldwin et al., 1981).

14.2.5 Jets, Superluminal Motion and Beaming

(a) Jets and Collimation In broad terms, jets are well-collimated outflows of material. As we have seen in §14.1, in many radio galaxies, the radio emission is not isotropic but shows an elongated structure with a length that is many times larger than the width (opening angle $\lesssim 0.1$). Such structure is called a radio jet (see Fig. 14.2 for an example). With high-resolution observations, the structure of a radio jet can be resolved down to sub-parsec scales, and the results are remarkable. In some cases, a thin jet (with width of $\sim 1\,\mathrm{pc}$) is observed to extend continuously from the innermost parsec-scale region of a galaxy to a distance up to several hundreds of kiloparsecs where it eventually runs into a flaring lobe. Clearly, the radio lobes are fueled by the narrow, core-powered jets. The question is how the material in the core is funneled into an out-going flow, and how the flow is collimated to such high degree. Although jets and their collimation are still not well understood, numerous interesting ideas have been proposed (Begelman et al., 1984, and references therein).

Suppose we have a well-collimated jet flow somehow generated by an AGN. The question is then how the flow is kept collimated as it moves to large distances. We all know that it is very difficult to have well-collimated smoke from a chimney in the presence of air flow (wind), but we often see a well-collimated thrust from a supersonic jet plane. The key difference in these two cases is that the gas in the thrust has large density and high speed, and is therefore less affected by the medium in which it is moving. However, as long as the flow is subsonic, it is in general difficult to sustain a good collimation. If the jet is diffuse, perturbations from the medium can easily change its course; if the jet is dense, its pressure over the medium will cause it to disperse sideways before it propagates far. The situation is different if the jet material is moving at supersonic speed. In this case, the speed of sideways dispersion (which is smaller than or of the same order as the sound speed of the medium) is much smaller than the speed of the jet flow, and collimation can be sustained. Indeed, a jet can remain collimated as long as it is supersonic, which suggests that a well-collimated jet is likely a supersonic flow.

Supersonic flows arise naturally under the conditions present in AGN. To see this, consider the behavior of a relativistic fluid, whose motion is governed by the conservations of particle number and energy-momentum tensor:

$$\frac{\partial}{\partial x^\mu} J_\mu = 0; \qquad \frac{\partial}{\partial x^\nu} T^{\mu\nu} = 0. \tag{14.49}$$

Here the particle flux, J^μ, and the energy–momentum tensor, $T^{\mu\nu}$, are given by

$$J^\mu = nU^\mu, \qquad T^{\mu\nu} = wU^\mu U^\nu + Pg^{\mu\nu}. \tag{14.50}$$

In these equations, $U^\mu = (dx^\mu/d\tau) = \gamma(c, \mathbf{v})$ is the four-velocity, $w \equiv \rho + P/c^2$, ρ and P are the gas density and pressure, and n is the proper number density of gas particles. The $\mu = 0$ component of the energy–momentum equation gives

$$\frac{d\gamma^2 w}{dt} = \frac{1}{c^2}\frac{\partial P}{\partial t} - \gamma^2 w \nabla \cdot \mathbf{v}, \tag{14.51}$$

which, together with the conservation equation of particle number, yields

$$\frac{d}{dt}\left(\frac{\gamma w}{n}\right) = \frac{\dot P}{\gamma nc^2}. \tag{14.52}$$

For a steady flow, this equation gives $\gamma w/n = $ constant. The application of this steady-flow equation to an ultra-relativistic fluid (for which $w = 4P/c^2 \propto n^{4/3}$) gives

$$\gamma = (P_0/P)^{1/4}, \tag{14.53}$$

where P_0 is the pressure at the origin of the flow. Thus, a relativistic flow can be generated from an internally relativistic fluid by a significant decrease in the pressure along the flow. For a steady jet flow, Eq. (14.51) reduces to

$$\int \gamma^2 w \nabla \cdot \mathbf{v} \, d^3 \mathbf{x} = \gamma^2 w v A = Q, \tag{14.54}$$

where Q is a constant, A is the cross-section of the jet at a location where the flow velocity is v, and the volumic enthalpy is w. Since $w = 4P/c^2$ and $P = P_0 \gamma^4$, we get

$$A = \left(\frac{Q}{4P_0 c}\right)\frac{1}{y(1-y)^{1/2}}, \tag{14.55}$$

14.2 The Supermassive Black Hole Paradigm

with $y \equiv (P/P_0)^{1/2}$. Thus if the pressure decreases monotonically along the jet, the cross-section decreases with decreasing y initially, reaches a minimum at $y = 2/3$, and then flares out at $y < 2/3$. At the point of minimum cross-section, the flow has a speed $v = c/\sqrt{3}$, identical to the sound speed in an ultra-relativistic fluid, and the pressure drops by a factor of only $4/9$. The fluid is subsonic at $y > 2/3$ and supersonic at $y < 2/3$. If the pressure drops with increasing distance as $P \propto r^{-\alpha}$, the solid angle subtended by the jet is

$$A/r^2 \propto y^{-1} \propto r^{(\alpha-4)/2}. \tag{14.56}$$

This indicates that a relativistic supersonic jet can remain collimated if $\alpha < 4$. The simple consideration given here suggests that it is relatively easy to generate a relativistic jet and keep it collimated. What is needed is an internally relativistic fluid, a relatively small gradient of pressure in the medium that confines the jet, and an initial flow in some preferred direction. This mechanism was first considered by Blandford & Rees (1974).

The simple version of the Blandford–Rees model has a problem, though. From Eq. (14.55) we see that the jet cross-section A should be large at its origin ($y \sim 1$), while observations show that jets originate from sub-pc scales. For a given Q, the cross-section can be made smaller by increasing the pressure P_0, but this would require the existence of a hot gas in the nucleus and predict extremely strong X-ray emission, which is not observed. Thus, although the Blandford–Rees mechanism may play a role in jet collimation at large radii, the processes involved in the production and initial collimation of the jet may be more complicated.

Jets may form naturally in the funnels of a radiation-supported torus (Fig. 14.3). However, the material in jets formed in this way can only be accelerated to a small Lorentz factor ($\gamma \lesssim 3$), even if the jets consist of pair-plasma and the sources radiate at super-Eddington luminosities (see Sikora & Wilson, 1981). The values of γ are even smaller if the jets consist of normal plasma. These values of γ are too small to explain the observed superluminal motions (see below). Furthermore, such a mechanism is not relevant for sources radiating at sub-Eddington luminosities.

As shown in §14.2.3, strong magnetic fields must exist to generate the synchrotron emission from AGN. It is therefore possible that magnetohydrodynamic processes play an important role in the production and collimation of jets (Blandford & Payne, 1982; Camenzind, 1990). Because of the coupling between the magnetic field and the plasma, the magnetic field lines wind up as material is accreted onto a differentially rotating disk, producing a strong magnetic field along the rotation axis. Outflows generated by, for example, radiation pressure in the central region may then stream along the rotation axis, producing a collimated jet.

(b) Superluminal Motion Given that the material in jets is moving at relativistic velocities, how does this relativistic motion appear to an observer? To see this, consider a blob of material ejected from an AGN with a velocity $v \equiv \beta c$ towards an observer at an angle θ relative to the line-of-sight. For simplicity, we assume the AGN to be at rest with respect to the observer. Suppose that at some time, t_1, the blob is at point P_1 with a distance r_1 from the observer. The light signal emitted by the blob from this point is then observed at $t_{\text{obs},1} = t_1 + r_1/c$. Suppose that after a small time interval Δt the blob moves to another point P_2 at a distance r_2 from the observer. The light signal emitted by the blob at this point will then be observed at a time $t_{\text{obs},2} = t_1 + \Delta t + r_2/c$. Simple geometrical consideration shows that $r_2 = r_1 - v\Delta t \cos\theta$ (for small Δt). Thus for the observer the time taken for the blob to move from P_1 to P_2 is

$$\Delta t_{\text{obs}} = t_{\text{obs},2} - t_{\text{obs},1} = \Delta t \left(1 - \beta \cos\theta\right). \tag{14.57}$$

The transverse distance between P_1 and P_2 is $v\Delta t \sin\theta$ and so, to the observer, the apparent transverse velocity of the blob, $v_\perp$, is given by

$$\frac{v_\perp}{c} = \frac{\beta \sin\theta}{1 - \beta \cos\theta}. \tag{14.58}$$

Clearly, if $\beta = v/c$ is close to 1, $v_\perp$ can exceed the speed of light provided that θ is sufficiently small. For example, if $\beta = 0.99$ and $\theta = 5°$, then $v_\perp = 6.27c$. This kind of motion with an apparent velocity exceeding the speed of light is called superluminal motion. High-resolution observations of radio jets over long time scales (several years) can detect the motion of bright spots in a jet relative to the core. In many radio-loud quasars, the inferred apparent velocities range from $\sim 3h^{-1}c$ to $\sim 10h^{-1}c$. The discussion given above shows that such superluminal motion arises naturally from relativistic motion of the material in jets.

(c) Relativistic Beaming Another observational effect of relativistic jets is beaming. Consider a blob of material in a jet which is moving towards us with a velocity, v, in a direction, θ, relative to the sightline to the AGN. The observed frequency, ν, is related to the frequency in the rest-frame of the blob, ν', by

$$\nu = \mathscr{D}\nu', \tag{14.59}$$

where

$$\mathscr{D} = \frac{1}{\gamma(1 - \beta\cos\theta)} \tag{14.60}$$

is the Doppler boosting factor. Suppose that the observed specific intensity of the blob is $J(\nu)$. It can be shown (see Rybicki & Lightman, 1979) that $J(\nu)/\nu^3$ is a Lorentz invariant, and so the observed intensity is related to the intensity in the rest frame of the blob by

$$J(\nu) = \mathscr{D}^3 J'(\nu'). \tag{14.61}$$

For an optically thin source, the observed flux is proportional to the source intensity: $F_\nu \propto J(\nu)$. If the source has a power-law spectrum $J'(\nu') \propto \nu'^{-\alpha}$, then

$$F_\nu = D^{3+\alpha} F'_\nu. \tag{14.62}$$

Clearly, for a given rest-frame flux, F'_ν, the observed flux depends on θ: it is higher if the jet is moving towards us and lower if moving away from us. For two blobs with the same intrinsic properties but moving in opposite directions, the flux ratio between the blobs moving in (towards us) and out (away from us) is

$$\frac{F_{\nu,\text{in}}}{F_{\nu,\text{out}}} = \left(\frac{1 + \beta\cos\theta}{1 - \beta\cos\theta}\right)^{3+\alpha}. \tag{14.63}$$

The boost is somewhat different for a jet that can be considered to be a number of unresolved blobs. In this case, the total observed flux is an integration over the jet,

$$F_\nu = \int J(\nu)\,\mathrm{d}\Omega\,\mathrm{d}l = \int \varepsilon(\nu)\frac{\mathrm{d}a\,\mathrm{d}l}{d_A^2}, \tag{14.64}$$

where $\varepsilon(\nu)$ is the emissivity, $\mathrm{d}\Omega$ is the element of solid angle, $\mathrm{d}a$ is the area element perpendicular to the line-of-sight, and d_A is the angular-diameter distance to the source. The emissivity transforms as $\varepsilon(\nu) = \mathscr{D}^2 \varepsilon'(\nu')$, and so

$$F_\nu = \frac{\mathscr{D}^{2+\alpha}}{d_A^2} \int \varepsilon'(\nu)\,\mathrm{d}a\,\mathrm{d}l = \mathscr{D}^{2+\alpha} F'_\nu. \tag{14.65}$$

Thus, for two identical jets propagating in opposite directions, we have

$$\frac{F_{\nu,\text{in}}}{F_{\nu,\text{out}}} = \left(\frac{1+\beta\cos\theta}{1-\beta\cos\theta}\right)^{2+\alpha}. \tag{14.66}$$

14.2 The Supermassive Black Hole Paradigm

If $\beta \sim 1$ and θ is small, the flux from the incoming jet is much higher than that of the outgoing one. This explains why the observed jets in powerful radio galaxies tend to be one-sided (see Fig. 14.2).

14.2.6 Emission-Line Regions and Obscuring Torus

(a) Broad-Line Region The optical spectra of Seyfert 1 galaxies and quasars contain strong emission lines with velocity widths $\sigma_v > 1{,}000 \,\mathrm{km\,s^{-1}}$. If these velocity widths are due to the gravitational motion in the vicinity of the central black hole, the sizes of the broad-line regions should be of the order

$$R \sim \frac{GM}{\sigma_v^2} \lesssim 0.5 M_8 \left(\frac{\sigma_v}{1{,}000\,\mathrm{km\,s^{-1}}}\right)^{-2} \mathrm{pc}. \tag{14.67}$$

This argument suggests that the broad lines are produced in a small inner region surrounding the accretion disk. Direct imaging of the broad-line region has not yet been possible, but the observed variabilities of broad lines suggest a typical size of $\lesssim 1$ light-year ($\sim 0.3\,\mathrm{pc}$), consistent with that based on the velocity width. Broad forbidden lines (such as [OIII]) are not observed, indicating that the electron density of the gas producing the broad lines is very high (see §14.2.4). Detailed photoionization models of the broad-line spectra give densities as high as $10^{10}\,\mathrm{cm^{-3}}$ and gas temperatures $\sim 10^4\,\mathrm{K}$.

(b) Narrow-Line Region The optical spectra of many AGN reveal strong narrow lines with velocity widths $\sigma_v \sim 100\,\mathrm{km\,s^{-1}}$. From Eq. (14.67) we see that such lines may be produced in a region of size $\sim 50\,\mathrm{pc}$ around the central engine. Because of their relatively large sizes, narrow-line regions in nearby AGN can be spatially resolved even with ground-based observations. The observed sizes range from $\sim 10\,\mathrm{pc}$ to $\sim 100\,\mathrm{pc}$, again in good agreement with the sizes inferred from the velocity widths. Since forbidden [OIII] lines are usually observed, the density of the narrow-line gas must be $\lesssim 10^6\,\mathrm{cm^{-3}}$. Detailed modeling of the narrow-line spectra yields densities in the range $(10^4\text{–}10^6)\,\mathrm{cm^{-3}}$. This relatively large range is also required within individual AGN to explain the observed line ratios, indicating that the gas within a narrow-line region is highly inhomogeneous. Indeed, high-resolution HST imaging of narrow-line regions of some Seyfert 2 galaxies shows very clumpy and distorted media rather than smooth gas distributions (Capetti et al., 1996; Bennert et al., 2002). The gas is mainly photoionized by the central source, although collisional ionization may also be involved in some cases. The temperature of the gas is in the range $(1\text{–}2) \times 10^4\,\mathrm{K}$.

(c) Obscuring Torus One defining difference between Seyfert 1 and Seyfert 2 galaxies is the presence of strong broad lines in the spectra of Seyfert 1 versus the absence of such lines in the spectra of Seyfert 2. It was therefore a surprise to find that the spectrum of NGC1068 (a classical Seyfert 2 galaxy) shows broad emission lines in its polarized light (Antonucci & Miller, 1985). The simplest mechanism for producing polarized light is by the reflection from a 'mirror' consisting of dust grains or electrons. It is thus likely that NGC1068 contains a Seyfert 1 nucleus (i.e. a broad-line region) surrounded by an obscuring torus (or just obscuring clouds, see Fig. 14.3), which obscures the broad-line region from our direct view but whose emission is reflected to us by dust and electrons above the obscuring torus. The torus must be inside the narrow-line region in order for the narrow-line emission not to be obscured. In order to provide sufficient extinction to hide the broad-line region, the torus must contain a sufficient amount of material to make it opaque to the broad-line emission. Furthermore, since strong X-ray and UV continua are observed in Seyfert 1 but not in Seyfert 2, the torus should also be opaque to these continuum photons. The torus must therefore have a large column density of absorbing gas (e.g. Krolik & Begelman, 1988). In addition to NGC1068, several other Seyfert 2 galaxies have now been observed in polarized light, again revealing broad emission line regions (e.g. Tran, 1995a). Furthermore, using optical interferometry, it has recently become possible to directly image the nuclear regions of AGN on the scale of the expected torus. These confirm the presence of a geometrically thick, torus-like dust distribution (Tristram et al., 2007; Raban et al., 2009). All these observations support the idea that Seyfert 1 and 2 galaxies are the same type of objects, but observed at different angles (Tran, 1995b).

14.2.7 The Idea of Unification

Although different classes of AGN appear quite differently, many of them have properties in common. For example, radio-quiet quasars (QSOs) and radio-loud quasars have very different radio properties, but their emission line properties are very similar. An obvious question is whether different classes of AGN are intrinsically different or whether they are intrinsically similar but only *appear* different because we observe them at different angles. Given that the basic structure of an AGN is axisymmetric rather than spherically symmetric, similar objects may appear different if their orientations relative to the line-of-sight are different. The question is whether such 'geometrical' effect is sufficient to explain the observed diversity of the AGN population. As discussed above, in the case of Seyfert galaxies, there is substantial evidence to support the notion that Seyfert 1 and 2 galaxies are intrinsically the same objects, and this may indicate that other classes of AGN are also similar to Seyfert galaxies except for their orientation with respect to the observer. Although this idea of unification is widely accepted, many open issues remain (see Antonucci, 1993; Urry & Padovani, 1995). In what follows we give a brief overview of some of the basic ideas related to AGN unification.

As described in §14.1, radio quasars are observed both as compact core-dominated radio sources and as extended double-lobed radio sources. Observations of faint extended regions around compact quasars show that they have luminosities comparable to those of the lobes in the extended sources. It is therefore possible that these two types of quasars belong to the same population, with compact core-dominated quasars being the double-lobed ones seen end-on. A detailed model for this unification was developed by Orr & Browne (1982). This idea can be tested observationally, because relativistic beaming is expected to be more important for the core-dominated component from the relativistic jet than for the extended lobes. The importance of relativistic beaming can be characterized by the ratio

$$\mathscr{R} = F_{\nu,\text{cc}}(\theta)/F_{\nu,\text{ec}}, \qquad (14.68)$$

where $F_{\nu,\text{cc}}$ is the flux from the compact component of a source, while $F_{\nu,\text{ec}}$ is that from the extended component. Because of beaming, the flux of the compact component depends on the angle, θ, between the jet and the line-of-sight, and this dependence is given by Eq. (14.66). We can then write

$$\mathscr{R} = \frac{\mathscr{R}_\perp}{2}\left[(1-\beta\cos\theta)^{-(2+\alpha)} + (1+\beta\cos\theta)^{-(2+\alpha)}\right], \qquad (14.69)$$

with $\mathscr{R}_\perp \equiv \mathscr{R}(\theta = \pi/2)$ and α the power-law index of Eq. (14.62). The first term on the right-hand side is the contribution of the incoming jet, while the second term is that of the outgoing one. If the idea of unification is correct, then the distribution of $\mathscr{R}$ should follow that given by a random distribution of source orientations. Unfortunately, observational results are not yet conclusive (see Murphy et al., 1993; Urry & Padovani, 1995). A related unification scheme was proposed by Blandford & Rees (1978) for blazars and double-lobed radio sources. In this scheme, a blazar is a radio source seen almost exactly along the jet, and the observed rapid variability and high degree of polarization of blazars are attributed to the relativistic flow in the jet.

Along this line of reasoning, it has been suggested that all radio sources form a continuous sequence in the viewing angle, θ. As θ changes from $\pi/2$ to 0, a radio source would appear successively as a radio galaxy, an extended (lobe-dominated) quasar, a compact (core-dominated) quasar, and a blazar. There is some observational support for this unification scheme (see Urry & Padovani, 1995), but more work is needed for a decisive conclusion.

14.2.8 Observational Tests for Supermassive Black Holes

The SMBH paradigm is by far the most successful model for interpreting the observed properties of AGN. Obviously, an ultimate test of this paradigm is to prove the existence of SMBHs observationally. By definition SMBHs cannot be seen directly. Instead, evidence for their existence has to come from the impact of SMBHs on their direct surroundings. One of the strongest motivations for considering SMBHs as the engines of AGN is the fact that AGN are extremely luminous (they can outshine their entire host galaxy) and confined to extremely small regions. Strong limits on the sizes of the emission regions come from variability arguments. The fact that some AGN reveal large changes in luminosity on time scales as short as $\Delta t \sim 1$ hour, suggests that the emission emanates from a region not much larger than $\sim c\Delta t$ (i.e. one light-hour across). Although SMBHs with their associated accretion disks are the only known sources that could plausibly explain these phenomena, this in itself constitutes no proof. Unambiguous evidence for the existence of a SMBH requires the detection of relativistic motion in a region the size of a few Schwarzschild radii.

Although galaxies with AGN are the obvious places to look for evidence for the existence of SMBHs, the fact that quasars were much more numerous at $z \sim 2$ than at the present suggests that dead quasar engines (i.e. SMBHs with little or no accretion) should be hiding in many nearby, quiescent galaxies. As discussed in §13.1.6, the kinematics of stars and gas in the central regions of spheroidal galaxies clearly indicate that their nuclei contain massive dark objects, with masses comparable to those of SMBHs required to power AGN. Although suggestive, these kinematics are typically measured on scales comparable to the radius of influence of the SMBH, r_{BH}, given by Eq. (13.14). For a typical galaxy, this corresponds to $r_{BH} \sim 10^6 r_S$. Consequently, these data only prove the existence of a massive dark object, but not necessarily a SMBH. Alternatives are dense clusters of stellar remnants (e.g. neutron stars), or more exotic objects such as balls of heavy fermions (sterile neutrinos, gravitinos or axinos) held up by degeneracy pressure (e.g. Tsiklauri & Viollier, 1998) or boson stars, which are hypothetical stars made of bosons (e.g. Torres et al., 2000). Ruling against these alternatives and in favor of a SMBH requires the detection of higher velocities on smaller scales.

In a few cases, special circumstances allow us to probe the kinematics at much smaller scales. The most impressive example is our own Milky Way, where the radial velocities, proper motions and accelerations of individual stars can be measured. This has provided strong evidence for the presence of a central massive object with $M \sim 3 \times 10^6 M_\odot$ within a radius of $\lesssim 10$ light-hours centered on the radio source Sgr A* (Genzel et al., 2000; Schödel et al., 2003; Ghez et al., 2005). This implies a mass density of $> 3 \times 10^{19} M_\odot \, pc^{-3}$, much higher than that of any known astronomical object, leaving a SMBH as the most likely option. In the galaxy NGC 4258, water masers have been detected, whose kinematics reveal perfect Keplerian motion around a central mass of $\sim 3.6 \times 10^7 M_\odot$ within a radius $< 0.13 \, pc$ (Miyoshi et al., 1995). As for the Milky Way, it is difficult to explain these observations by any object other than a SMBH. Finally, the extreme broadness and asymmetry of the fluorescent Fe K line (at ~ 6 keV) detected in the active galaxy MCG-6-30-15 suggests relativistic motions ($\sim 10^5 \, km \, s^{-1}$) of the gas. The most likely explanation is that the line arises from an accretion disk in a region between three and ten Schwarzschild radii from a SMBH (Tanaka et al., 1995).

Another method to estimate the masses of SMBHs in AGN is reverberation mapping (e.g. Blandford & McKee, 1982; Peterson, 1993). This method uses the intrinsic flux variability of the UV/optical continuum generated in the accretion disk surrounding the SMBH and the time delay in the response of the broad emission lines to the change in the continuum flux. The time delay, τ, is a measure of the light crossing time of the broad-line region, and so provides a measure of its size: $r \sim c\tau$. Meanwhile, the emission linewidth, Δv, reflects the velocity dispersion of the emission-line clouds in the gravitational potential well which is dominated by the SMBH. Thus, the mass of the SMBH can be estimated through $M_{\rm BH} = fc\tau(\Delta v)^2/G$, where f is a dimensionless factor of order unity that depends on the kinematics, geometry, and inclination of the AGN. So far this method has been used to measure black hole masses in a few dozen AGN, and the reverberation mass measurements are consistent with other mass estimates within a factor of a few (e.g. Bentz et al., 2009).

To summarize, in a restricted few cases the evidence for the existence of SMBHs is extremely strong, in particular in our own Milky Way. Although there is still no direct proof for the existence of an event horizon, virtually all other alternatives are currently ruled out. Unless the Milky Way happens to be a special place, Occam's razor therefore strongly suggests that the nuclei of many, if not all, (spheroidal) galaxies indeed host SMBHs that can power AGN activity.

14.3 The Formation and Evolution of AGN

In the SMBH paradigm described above, the two most important conditions for producing an AGN are (1) the existence of a central SMBH, and (2) a sufficient amount of gas to fuel the nucleus. Thus, in order to understand the formation of AGN, we must understand how SMBHs form and which mechanisms are responsible for transporting gas towards the center of the host galaxy to feed the black hole. In this section, we describe possible scenarios for the formation of AGN. We demonstrate that not only can the formation of AGN be understood in the general framework of galaxy formation, but AGN can also have significant impact on the formation and evolution of galaxies. Finally, we will show how to test the scenario of AGN formation by studying the redshift evolution of the AGN population in connection to other populations of galaxies.

14.3.1 The Growth of Supermassive Black Holes and the Fueling of AGN

One important fact that any theory of AGN formation must take into account is that quasars are observed at redshifts up to ~ 7. As shown in §3.2.5, the cosmic time at such a redshift is about 0.5 Gyr, and so the time scale for the formation of a SMBH must be shorter than 0.5 Gyr. Since black holes are believed to form in collapsed objects, another relevant time scale is the free-fall time scale of a virialized halo, $t_{\rm ff} \sim r_{\rm vir}/V_{\rm c} \sim 1/[10H(z)]$ (with $r_{\rm vir}$ the virial radius and $V_{\rm c}$ the circular velocity of the halo), which is about 5×10^7 yr at $z = 7$. These two time scales should be compared to the time scale for the growth of a SMBH. If the growth of a SMBH is through radiative accretion, the mass accretion rate can be written as

$$\dot{M}_{\rm BH} = \frac{L}{\varepsilon_{\rm r} c^2} = \left(\frac{L}{L_{\rm Edd}}\right) \frac{M_{\rm BH}}{\varepsilon_{\rm r} t_{\rm Edd}}, \tag{14.70}$$

where the first equation follows from the definition of $\varepsilon_{\rm r}$ in Eq. (14.7), $L_{\rm Edd}$ is the Eddington luminosity defined in Eq. (14.4), and

$$t_{\rm Edd} \equiv \frac{\sigma_{\rm T} c}{4\pi G m_{\rm p}} \approx 4.4 \times 10^8 \, {\rm yr} \tag{14.71}$$

is the Eddington time. If both $L/L_{\rm Edd}$ and $\varepsilon_{\rm r}$ are independent of time, we get

$$M_{\rm BH}(t) = M_{\rm BH,0} \exp\left(\frac{t}{t_{\rm BH}}\right), \tag{14.72}$$

14.3 The Formation and Evolution of AGN

where $M_{\rm BH,0}$ is the mass of the black hole at an initial time $t=0$, and

$$t_{\rm BH} = \left(\frac{L}{L_{\rm Edd}}\right)^{-1} \varepsilon_{\rm r} t_{\rm Edd} \approx 4.4 \times 10^7 \left(\frac{\varepsilon_{\rm r}}{0.1}\right) \left(\frac{L}{L_{\rm Edd}}\right)^{-1} \text{ yr} \qquad (14.73)$$

is the time scale for the black hole mass to increase by a factor of e.

The time required for a black hole to reach $M_{\rm BH} \sim 10^8 \, {\rm M}_\odot$ depends on the seed mass $M_{\rm BH,0}$ which, in turn, depends on how the black hole was initially created. In broad terms, there are three possibilities for the initial creation of a black hole (Rees, 1984): (i) the collapse of an isolated massive star, (ii) the merger and accretion of neutron stars, and (iii) the collapse of a gas cloud. The first two possibilities are almost certain to happen. As discussed in §10.1.5, massive stars with initial mass $\gtrsim 20 \, {\rm M}_\odot$ have lifetimes $\sim 10^7 \, {\rm yr}$. At the end of their lifes, these stars either collapse directly into black holes, or produce core-collapse supernovae with black holes or neutron stars as the remnants. Even if the remnants are neutron stars, some of them may end up in close binaries, and a black hole can form when some dissipative processes (e.g. gravitational radiation) brings them together. Black holes that form through such a channel should have stellar masses, and so $M_{\rm BH,0}$ is of the order of $10 \, {\rm M}_\odot$. However, supermassive Population III stars, with masses of a few hundred solar masses, may be able to form at $z \sim 10$–20 when the gas in the Universe is too metal poor to cool down to a temperature that allows gravitational instability on smaller mass scales (see §9.7). Such stars may collapse directly to form black holes with masses $\sim 100 \, {\rm M}_\odot$. The third possibility is uncertain: it is unclear how often the collapse of a gas cloud can lead directly to the formation of a black hole; it is also difficult to estimate the initial mass of a black hole that can form through such a collapse (see Koushiappas et al., 2004, and references therein)

If $M_{\rm BH,0}$ is of the order of $100 \, {\rm M}_\odot$, we see from Eq. (14.72) that about 14 e-foldings are required to reach a mass $M_{\rm BH} = 10^8 \, {\rm M}_\odot$. The time interval required is then

$$t \approx 14 t_{\rm BH} \approx 6 \times 10^8 \left(\frac{\varepsilon_{\rm r}}{0.1}\right) \left(\frac{L}{L_{\rm Edd}}\right)^{-1} \text{ yr}. \qquad (14.74)$$

If $L \lesssim L_{\rm Edd}$ and $\varepsilon_{\rm r} \sim 1$, this would signal a problem for growing black holes to the required mass within a Hubble time at $z \sim 7$. In order to achieve a sufficiently high growth rate, either the gas accretion is at a super-Eddington rate (so that $L \gg L_{\rm Edd}$), or the radiative efficiency is low (so that $\varepsilon_{\rm r} \ll 0.1$). Super-Eddington accretion can be achieved if the accretion is very anisotropic, so that the outgoing radiation is well separated from the infalling material. A smaller $\varepsilon_{\rm r}$ can lead to higher accretion rates, allowing the black hole to accrete the required mass in a shorter period of time and yet to radiate at the Eddington limit. However, the value of $\varepsilon_{\rm r}$ cannot be much smaller than 0.1, at least for the last few e-foldings before the AGN is observed, as it is constrained by the radiative properties of the accretion disk (see §14.2.2).

Despite the many uncertainties, the formation of a SMBH generically requires a considerable amount of gas to be funneled into the center of a dark matter halo on a short time scale. Indeed, even with $\varepsilon_{\rm r} \sim 0.1$ and $L \sim L_{\rm Edd}$, the time scale $t_{\rm BH}$ is shorter than the free-fall time scale of a dark matter halo at $z \lesssim 5$. Once a SMBH has managed to form in the center of a host galaxy, whether or not it can be observed as a bright AGN again depends on whether or not it is fed with gas at a sufficiently high rate. The typical rate required to power a black hole at the Eddington luminosity is that given by Eq. (14.9), or $\sim 2 \, {\rm M}_\odot \, {\rm yr}^{-1}$ for a $10^8 \, {\rm M}_\odot$ black hole. Note that the fueling of a bright AGN leads to further growth of the black hole mass, and the time scale of the growth is $M_{\rm BH}/\dot{M}_{\rm BH} \sim \varepsilon_{\rm r} t_{\rm E} (L/L_{\rm Edd})^{-1}$. This time scale is comparable to that for the growth of a SMBH accreting at the Eddington limit, indicating that the growth of a SMBH and the fueling of a bright AGN go hand in hand. Hence, the key step towards understanding the formation of an AGN is therefore to identify the mechanism(s) that can effectively bring gas into the center of the host galaxy.

The amount of mass involved in both the formation of a SMBH and the fueling of a bright AGN is only a small fraction of the total mass of the host galaxy. So, in principle, there is no difficulty in having a sufficiently large gas reservoir. The real problem is how to funnel the required amount of gas into a very small region. Let us assume first that the gas reservoir is cold, so that the pressure of the gas does not resist accretion. Even in this case, gas accretion would be impossible if the gas has a significant amount of angular momentum. The specific angular momentum of gas clouds on a orbit of radius R in the potential well of mass M is $j \sim (GMR)^{1/2}$. For a parcel of gas in a typical galaxy, $M \sim 10^{11} \, M_\odot$ and $R \sim 10 \, \text{kpc}$. If this parcel is to be moved to a location within $\sim 0.1 \, \text{pc}$ of a $10^8 \, M_\odot$ central black hole, its specific angular momentum must be reduced by a factor of 10^4. This clearly indicates that the formation of a bright AGN must be connected with events in which the gas in the host galaxy can lose its angular momentum very effectively.

As discussed in §12.4.3, one such process is gravitational interaction with other galaxies. Tidal forces during galaxy encounters can cause otherwise stable disks to develop bars, and gas in such barred disks, subjected to strong gravitational torques, can lose angular momentum very effectively and flow into the central region. Numerical simulations show that the merger of a pair of galaxy-sized gaseous disks can produce a central cloud with mass $\sim 10^9 \, M_\odot$ and size $\sim 100 \, \text{pc}$ (Barnes & Hernquist, 1996). The detailed structure of such clouds is not yet resolved by current simulations, and it is unclear whether the gas can indeed be funneled directly into the central parsec region (see Phinney, 1994, and references therein).

Significant gas inflow may also be generated in minor mergers in which a bar instability is induced in a disk by the perturbation from a smaller satellite galaxy. Although such interactions may be less effective than major mergers in bringing a large amount of gas into the central region of a galaxy, they are more common. In some cases, a disk may become unstable spontaneously and form a bar-like structure. This may occur as a result of the interaction between the disk and other components of the galaxy, or when the disk becomes globally unstable during the formation process. It is therefore also possible that some AGN are fueled without involving strong interaction with other galaxies.

Fig. 14.7 is a schematic chart showing how various paths of gas evolution in galaxies can lead to the formation of SMBHs and to the fueling of AGN. The mechanisms involved are plausible, but the details have yet to be worked out in order to determine which processes are the most relevant.

One prediction of the merger scenario of AGN formation is that bright AGN powered by large accretion rates should be preferentially observed in interacting systems. There is some evidence that this may indeed be the case for quasars at low redshifts. Early suggestions that quasars have close companions are supported by recent observations of quasar host galaxies out to $z \sim 1$ (e.g. Lacy et al., 2002). Tidal tails and other signatures of violent interaction are also observed in nearby host galaxies (Bahcall et al., 1997). Many of these systems involve small satellites, presumably because such encounters are more common. Unfortunately, the signatures of interactions are usually faint, hence difficult to observe at high redshifts (see §13.3.5). However, in the hierarchical scenario of structure formation it is expected that at high redshift galaxy mergers are more frequent, and that the merging galaxies are more gas rich. Both are favorable for fast gas accretion onto a central black hole. Thus, if the merger scenario is correct, and if SMBHs had managed to form at high redshift, we would expect to find more bright AGN at higher redshifts. As we will see in §14.3.2, the comoving number density of quasars peaks around $z \sim 2$–3 at a value more than 100 times higher than that in the local Universe. Since the total number of SMBHs is not expected to decrease with time, the drop at lower redshift is likely due to a decline of systems in which a large amount of gas can be funneled to feed the central SMBHs. The drop in the number density at higher z is also expected, because SMBHs need time to grow and because there were not as many galaxy-sized objects to host bright AGN at $z \gg 3$.

14.3 The Formation and Evolution of AGN

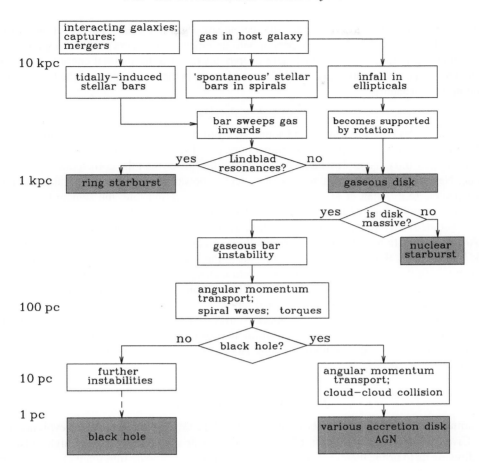

Fig. 14.7. Flow-chart diagram showing how SMBHs may have formed and how AGN may be fueled in a host galaxy. [After Shlosman et al. (1990)]

SMBHs more massive than $10^{9.5} M_\odot$ are rare (Marconi et al., 2004), and so a SMBH cannot accrete at the Eddington limit continuously over its entire lifetime. This is perhaps not surprising, given that AGN can release huge amounts of energy, both in radiation and in kinetic forms, into its surrounding. Such energy feedback can heat the gas in the host galaxy or even drive the gas out of it, thereby quenching further gas accretion. It is therefore possible that the growth of a SMBH is self-regulated, so that significant growth only occurs during the active (quasar) phase(s) of the AGN, while the mass remains more or less constant during the long dormant phase(s). The growth of a SMBH will eventually stop if the host galaxy runs out of cold gas to feed the black hole. Mergers among black holes may also play a role in the growth of SMBHs, but their importance is limited by the merger frequency of massive galaxies (e.g. Yoo et al., 2007).

Most of the SMBHs at low redshifts are observed to be hosted by early-type galaxies, such as ellipticals and the bulges of early-type spirals. This is consistent with the formation scenario for SMBHs outlined above, because the formation of early-type galaxies involves many of the same processes – galaxy mergers, galaxy interactions, disk instability, etc (see §13.2). These black holes are not observed as bright AGN at the present time, because they are starved of gas. Early-type galaxies do contain large amounts of gas, but it is hot and difficult for the SMBHs to accrete. In a hot medium, a SMBH can still accrete some gas through Bondi accretion (Bondi, 1952). Bondi accretion is spherical accretion onto an object, and occurs within a region where the gravitational potential of the black hole overcomes the specific thermal energy of the gas. This defines an accretion radius, $r_A \sim GM_{\rm BH}/c_{\rm s}^2$, where $c_{\rm s}$ is the sound speed of the hot gas. The accretion rate is roughly

$$\dot{M}_{\rm BH} \sim 4\pi r_A^2 \rho_A c_{\rm s}(r_A), \qquad (14.75)$$

where ρ_A and $c_{\rm s}(r_A)$ are the density and sound speed of the gas at r_A, respectively. Assuming a black hole mass $M_{\rm BH} = 10^9 \, {\rm M}_\odot$, a gas temperature $T = 1\,{\rm keV}$ and a gas density $n = 0.1\,{\rm cm}^{-3}$, we have $\dot{M}_{\rm BH} \sim 0.04\,{\rm M}_\odot\,{\rm yr}^{-1}$, which is about 500 times lower than the corresponding Eddington rate. Nevertheless, such a rate may be sufficient to power many of the radio galaxies in the local Universe (e.g. Allen et al., 2006). At such a low accretion rate, the standard geometrically thin, optically thick disk model described in §14.2.2 may not be valid. Instead, the disk may be geometrically thick and optically thin, as in the advection dominated accretion flow (ADAF) and advection dominated inflow–outflow solution (ADIOS) models (Narayan & Yi, 1994; Blandford & Begelman, 1999). In this case, the accretion is expected to be radiatively inefficient, with $\varepsilon_{\rm r} \ll 0.1$, and much of the accretion energy is expected to be in strong gas outflows, which may explain why many massive elliptical galaxies are radio sources that have low radiative efficiencies but possess radio jets and lobes.

As we will see in §14.4, the energy feedback from AGN not only limits the growth of SMBHs, but can also have important impact on the formation of galaxies in their surroundings. Hence, the study of AGN is not just for understanding AGN per se, but is an inseparable part of galaxy formation.

14.3.2 AGN Demographics

The formation scenario discussed above predicts that the AGN population must be evolving with time. Given the intrinsic brightness of QSOs, they can be studied out to very high redshifts, thus allowing us to probe their evolution directly. Such studies provide important clues about the formation and evolution of the AGN population, and its connection to galaxies and dark matter halos.

(a) Luminosity Function and Number Density Information about the evolution of the AGN population can be obtained by studying how the luminosity function of AGN varies with redshift. Fig. 14.8 shows the luminosity function of QSOs in six redshift intervals covering the range $0.4 < z < 2.1$ (Croom et al., 2004). The observed luminosity functions are well fitted by the following form:

$$\phi(L,z)\,{\rm d}L = \phi^*(z) \left\{ \left[\frac{L}{L^*(z)} \right]^{\beta_1} + \left[\frac{L}{L^*(z)} \right]^{\beta_2} \right\}^{-1} \frac{{\rm d}L}{L^*(z)}, \qquad (14.76)$$

where β_1 and β_2 are two constant indices and $L^*(z)$ is a characteristic luminosity. Assuming a flat universe with $\Omega_{\rm m,0} = 0.3$, the best fit gives $\beta_1 = 3.9$ and $\beta_2 = 1.5$. The data are consistent with pure luminosity evolution in which ϕ^* is independent of z but $L^*(z) \propto L^*(0)(1+z)^k$ with $k \sim 3.45$. Results based on the SDSS QSO survey, which covers the redshift range from ~ 0.5 up to about 6 (Richards et al., 2006; Fan et al., 2004), show that the overall number density $\phi^*(z)$ is quite independent of z over the whole redshift range, but the characteristic luminosity peaks at $z \sim 2$ and declines rapidly towards higher z.

To see more directly how the number density of AGN changes with redshift, one can calculate the number density of AGN of a given luminosity as a function of redshift. Fig. 14.9 shows the results based on a combination of a number of large QSO surveys (see Hopkins et al., 2007). The

14.3 The Formation and Evolution of AGN

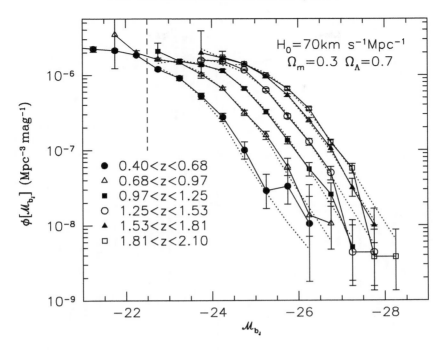

Fig. 14.8. The luminosity function of QSOs in six redshift intervals covering the range $0.4 < z < 2.1$ determined from the 2dF QSO Redshift Survey (Croom et al., 2004). The dotted lines show the best fit to the data by the double power-law model given in Eq. (14.76). The dashed vertical line denotes the separation between QSOs and Seyfert galaxies. [Adapted from Croom et al. (2004)]

data clearly show that the number density of bright QSOs peaks at $z \sim 2$–3, and declines rapidly towards both the lower- and higher-redshift ends. The peak appears to shift to lower redshift for fainter QSOs.

(b) Spatial Clustering With the advent of large QSO surveys, it is now possible to determine the spatial correlation function for QSOs at different redshifts. Results based on the 2dF QSO survey (e.g. Porciani et al., 2004; Croom et al., 2005) show that the amplitude of the two-point correlation function increases with redshift, in contrast to the amplitude of the correlation function of mass in the ΛCDM model, which decreases with redshift. This implies that the linear bias factor of QSOs, which measures the ratio between the correlation amplitude of the QSOs and that of the mass on large scales, increases rapidly with redshift, with QSOs at higher redshift being more strongly biased with respect to the mass.

Assuming that each QSO is associated with a dark matter halo in a ΛCDM universe, the bias parameter for QSOs should reflect that of the dark matter halos in which they reside (see §15.6). As shown in §7.4, at a given redshift, the bias parameter of dark matter halos is determined by their masses. If the halos which can host QSOs have the same clustering properties as the total halo population, the observed bias parameter for QSOs can be used to estimate the typical mass of the host halos. The results of Porciani et al. (2004) and Croom et al. (2005) show that the host halo mass is roughly $10^{12.5}$–$10^{13} h^{-1} M_\odot$ (assuming the ΛCDM concordance cosmology), quite independent of redshift over the range $0.4 < z < 2.5$. At lower redshifts the number density of QSOs is too low to reliably measure their clustering properties. However, as discussed in §14.1, the optical properties of Seyferts, whose clustering properties *can* be measured at low redshifts ($z \sim 0.1$), resemble those of QSOs except that their luminosities are lower. Interestingly, as shown by Wake et al. (2004) and Li et al. (2006a), Seyfert galaxies also preferentially reside in halos

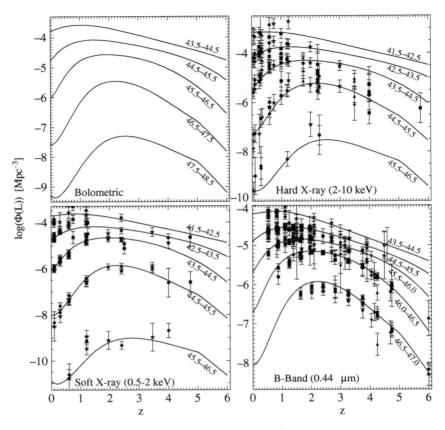

Fig. 14.9. The comoving space density as a function of redshift for QSOs of different luminosities. [Adapted from Hopkins et al. (2007) by permission of AAS]

with masses in the range 10^{12}–$10^{13} h^{-1} M_\odot$. A similar result was obtained by Pasquali et al. (2009) using a large galaxy group catalogue. It thus seems that halos of this mass range are the typical hosts of optical AGN, and that their activity is stronger at higher redshifts (giving rise to QSOs) than at low redshifts (where the AGN are typically Seyfert galaxies and LINERS). This is consistent with the fact that the peak in the number density–redshift relation shifts to smaller redshifts for fainter QSOs (see Fig. 14.9).

(c) Implications for AGN Formation At any redshift, the number density of halos that can host QSOs (i.e. halos with masses $\sim 10^{12.5} h^{-1} M_\odot$) is much larger than the number density of bright QSOs observed at the same redshift. Clearly, not all halos with the right mass host a bright AGN at any time. It is believed that halos of the right mass have relatively short 'duty cycles', so that each of them hosts a bright QSO only for a brief period of time. If the average duty time of a halo is t_Q, then the probability to observe a halo during the duty cycle at redshift z is roughly the ratio $t_Q/t(z)$, where $t(z)$ is the age of the Universe at redshift z. This ratio should be equal to the ratio between the number density of QSOs and that of dark matter halos capable of hosting QSOs. Using the observed number density of QSOs and the abundance of dark matter halos with masses $\gtrsim 10^{12.5} h^{-1} M_\odot$ predicted by the ΛCDM model, t_Q is estimated to be of the order of 10^7–10^8 yr (Porciani et al., 2004; Croom et al., 2005).[2] This time scale is much shorter than the

---

[2] Note that the data cannot discriminate between a single activity phase of duration t_Q, and many shorter bursts with a *combined* duration of t_Q.

age of the Universe at any redshift at which QSOs are observed. Because of this, the number density of bright quasars observed at a given redshift is directly related to the rate at which QSOs are activated at that redshift. The results shown in Fig. 14.9 therefore indicate that bright QSOs started to form at $z \gtrsim 6$, with the activation rate peaking at $z \sim 2$–3 and declining rapidly at lower redshifts. The question is: what determines such redshift-dependence of the activation rate?

If QSOs are hosted by relatively massive dark matter halos, as indicated by the clustering data described above, the number density of QSOs should depend on the number density of dark matter halos available to host them. In the ΛCDM model, the number density of dark matter halos with $M \sim 10^{12.5} h^{-1} \mathrm{M_\odot}$ increases rapidly with time at $z > 3$. It is thus possible that the rapid increase of QSO number density with decreasing redshift seen at $z > 3$ is dictated directly by the increase of the number of massive halos. At $z \lesssim 3$, on the other hand, the number density of $M \sim 10^{12.5} h^{-1} \mathrm{M_\odot}$ halos keeps on increasing with time, albeit at a slower rate, in contrast to the rapid decrease in the number density of QSOs. Hence, there must be some processes that regulate the formation of bright AGN in massive halos at $z \lesssim 3$ in such a way that the fraction of SMBHs in a bright AGN phase decreases.

If quasar activity were triggered by galaxy interactions, the decline of the number density of bright quasars at $z \lesssim 3$ may reflect a decline in the rate of major mergers. The observed merger rate of galaxies increases with redshift and is usually approximated by a power law, $(1+z)^m$ with $m > 0$. If $m \sim 4$, this effect alone would mean a drop by a factor of ~ 100 in the number density of quasars from $z = 2$ to $z = 0$, consistent with the data shown in Fig. 14.9. Unfortunately, the value of m is still very uncertain (see §13.3.6). In addition to a drop in the merger rate, there are a number of other plausible effects that may also have played a role. First of all, galaxies are less gas-rich at lower redshift, and so a stronger (and rarer) interaction may be required in order to funnel a large amount of gas into the halo center to feed a bright AGN. Secondly, as discussed below, AGN in their bright phases can output large amounts of energy into their surroundings, thereby suppressing the gas accretion by the black hole. It is possible that such feedback is more effective at lower redshift when the gas density in the host galaxies is lower. Finally, at low redshift many galaxies that host SMBHs may have merged into massive systems (clusters and rich groups) where the gas is too hot to be accreted by the SMBH. Indeed, as shown by Pasquali et al. (2009), optical AGN become rare in halos with $M \gtrsim 10^{14} h^{-1} \mathrm{M_\odot}$; the AGN hosted by such high-mass halos are preferentially radio galaxies.

14.3.3 Outstanding Questions

Although there has been a tremendous increase in the amount of data, many open questions regarding AGN demographics remain. For example, what is so special about halos of $10^{12.5} h^{-1} \mathrm{M_\odot}$, i.e. why do AGN appear to be absent in lower mass halos, or equivalently in galaxies with stellar masses below 10^{10} M., as seems to be the case, at least at the present day (Kauffmann et al., 2003a; Pasquali et al., 2009). Is it mainly because the AGN in these halos are too faint to be detected with current surveys, or because these halos do not host (sufficiently massive) black holes, or because they do not have sufficient gas accretion to feed the black holes? In the first case one would expect a correlation between AGN luminosity and halo mass, which would then mean that fainter QSOs are less strongly clustered. Currently, there is no convincing evidence that the clustering strength of QSOs depends strongly on their luminosities (Croom et al., 2005; Myers et al., 2007).

Another outstanding question is what determines an AGN to be radio-quiet or radio-loud. There are indications that this may be related to the environment in which the active galaxy is located. Radio-loud AGN in the local Universe, which are mainly hosted by elliptical galaxies, are found to be located preferentially in central galaxies of groups and clusters as opposed to other galaxies of the same (stellar) mass (e.g. Best et al., 2007; Pasquali et al., 2009). This indicates that the conditions for triggering a radio AGN may be very different from those for triggering an optical AGN. The optical AGN phenomenon apparently requires a galaxy with a young stellar population, and therefore seems to be linked with the availability of the supply of cold gas that not only serves as the fuel for ongoing star formation, but also feeds the central SMBH. In contrast, radio AGN are triggered in dense environments (i.e. massive halos) where large amounts of cold gas are not available. However, exactly how radio jets develop in such environments is still unclear. It may be that the radio mode predominantly develops in massive halos because they contain large amounts of hot gas. It is unclear, though, whether the hot halo is *required* to form the jet in the first place, or whether the hot corona merely acts as a working surface for the jet to dissipate its energy (e.g. Kauffmann et al., 2008).

In recent years, it has become clear that the presence of a hot corona, combined with the absence of a significant cold gas supply in the host galaxy, may have an important impact on the radio loudness of the AGN. At low redshift, the majority of radio galaxies are low-power FR I sources residing mainly in elliptical galaxies with very little ongoing star formation and with weak emission lines (e.g. Ledlow & Owen, 1995). X-ray studies suggest that the active nuclei in these radio galaxies are not radiatively efficient (Hardcastle et al., 2007). There is almost never any evidence for a heavily absorbed nuclear X-ray component, indicating that these systems may lack the classical accretion disk and obscuring torus. One possibility is that the absence of large amounts of cold gas result in radiatively inefficient, advection dominated accretion flows (e.g. Narayan & Yi, 1994). Alternatively, the SMBH may accrete the hot gas directly via Bondi accretion (see §14.3.1). The latter is supported by a tight observed correlation between the Bondi accretion rate, inferred from the estimated SMBH mass and the observed temperature and density profile of the gas, and the power emerging from these systems in relativistic jets (Allen et al., 2006).

In contrast, the more powerful FR II radio galaxies often show strong emission lines with strengths tightly correlated with their radio fluxes (e.g Baum et al., 1995). This suggests that both the optical and radio emission are due to radiatively efficient accretion via an accretion disk, similar to the accretion mode of an optical AGN. Hence, it seems that the difference between FR I and FR II radio galaxies is in their gas accretion mode. Powerful radio sources only develop when a sufficient supply of cold gas is available so that a radiatively efficient accretion disk can develop. On the other hand, when only hot gas is available, the accretion is radiatively inefficient, and almost all the accretion energy is channeled into the jets, giving rise to a radio galaxy of relatively low power (see Körding et al., 2006, for an interesting analogy with X-ray binaries). Since galaxies are expected to be more gas rich at higher redshifts, this picture gives a natural explanation for the fact that the space density of FR II sources increases strongly with increasing redshift (e.g. Laing et al., 1983; Dunlop & Peacock, 1990).

14.4 AGN and Galaxy Formation

As shown above, an AGN can release a huge amount of energy during its lifetime. In general we may express the power of the energy output as

$$\frac{dE}{dt} = \varepsilon \dot{M}_{\rm BH} c^2, \tag{14.77}$$

where $\varepsilon = \varepsilon_{\rm r} + \varepsilon_{\rm m}$ is an efficiency factor, and we use $\varepsilon_{\rm r}$ and $\varepsilon_{\rm m}$ to denote the radiative and mechanical efficiencies, respectively. Integrating the power over the history of an AGN, we get the total energy output,

$$E = \bar{\varepsilon} M_{\rm BH} c^2, \tag{14.78}$$

where $\bar{\varepsilon}$ is the mean efficiency. In order to see whether the energy feedback from AGN can have significant impact on galaxy formation and evolution, compare the total energy output from an AGN with the binding energy of its host galaxy. Suppose that the mass and velocity dispersion of the host galaxy (assumed to be an early-type galaxy) are $M_{\rm gal}$ and σ, respectively. According to the virial theorem, the gravitational binding energy is roughly $W \sim -M_{\rm gal}\sigma^2$. It then follows that

$$\frac{E}{|W|} \sim \frac{\bar{\varepsilon} M_{\rm BH}}{M_{\rm gal}} \left(\frac{c}{\sigma}\right)^2. \qquad (14.79)$$

According to the observational result shown in Fig. 2.17, $M_{\rm BH}/M_{\rm gal} \sim 10^{-3}$. Thus, for a massive galaxy with $\sigma \sim 300\,{\rm km\,s^{-1}}$, the ratio $E/|W|$ is about $10^3\bar{\varepsilon}$, indicating that the AGN energy can easily surpass the total binding energy of the host galaxy. It is therefore very well possible that the energy feedback from AGN plays an important role in the formation and evolution of galaxies.

There are two questions to be addressed in order to quantify the impact of AGN feedback on galaxy formation. The first is the value of the feedback efficiency, ε, with which the accretion power of the SMBH is released, and the second is how effective the feedback energy is coupled to the gas in the host galaxy and its large scale environment. Clearly, the answers to these questions depend on the details of the feedback processes involved.

Roughly speaking, there are three possible processes that can channel the energy from an AGN into its surrounding gas:

- Radiative processes: As we have seen above, the radiative efficiency of an accreting SMBH is $\varepsilon_r \sim 0.1$ in bright AGN, and so the amount of energy available in this channel is enormous. This energy can, in principle, feed back into the environment of an AGN through both radiation pressure and radiative heating.
- Mechanical processes: Observations show that jets and winds associated with radio galaxies contain large amounts of kinetic energy that are comparable to the radiation energy of the source.
- Energetic particles (cosmic rays): Although there is no evidence that the total energy released via this channel is significant, the production of energetic particles could in principle contribute to the overall pressure of the gas surrounding an AGN (see §8.2).

In what follows, we describe in more detail how the radiative and mechanical feedback from AGN can affect the gas in their environments. Given our poor understanding of the primary production of cosmic rays in AGN, we will not discuss this option further.

14.4.1 Radiative Feedback

Ultraviolet and X-ray photons generated by AGN can ionize neutral atoms (or partially ionized ions) through photoionization and heat the gas through photoionization heating. As we will see in §16.1, the intergalactic medium (IGM) is highly ionized at $z \lesssim 6$, although it was almost completely neutral after recombination at $z \sim 1000$. It is believed that the IGM is re-ionized by radiative sources that can output large amounts of ionizing (UV) photons. Although part of the UV photons responsible for the ionization of the IGM is produced by star-forming galaxies, the contribution from AGN is significant, and perhaps even dominant at $z \lesssim 2$ when the number density of star-forming galaxies starts to drop rapidly. In the process of photoionization, the energy surplus of an ionizing photon is converted into kinetic energy of the electron that is removed from the atom (or ion). Hence, UV photons can not only ionize gas but also heat it. As we will see in §16.3, the UV background produced by the radiation emanating from AGN can heat the IGM to a temperature of about 10^4 K. This heating can significantly suppress gas cooling and star formation in low mass halos (see §8.1.4).

If the host galaxy of an AGN contains a significant amount of dust, the radiation from the central AGN can be effectively channeled into the surrounding gas by dust absorption. The part of the radiation energy that goes into heating the dust grains does not have a net effect, because it is rapidly re-emitted in the infrared. However, for a central source, the radiation pressure on the dust grains can transfer the momentum of the radiation field into the gas through the coupling between the gas and the dust grains. If the radiation pressure can overcome the gravitational force of the host halo, the gas can flow out from the center in the form of a momentum-driven wind (e.g. Murray et al., 2005). If all photons are absorbed by the dust grains, i.e. the absorption optical depth of the gas is $\tau \gtrsim 1$, then the pressure force on the gas is $F = L/c$, and the total momentum injected into the gas is $p = E/c$, where E is the total energy radiated by the AGN. If the gas is optically thin, then both F and p should be reduced by a factor of τ. Detailed modeling shows that the specific cross-section of dust absorption is $\sim 100 \text{cm}^2 \text{g}^{-1}$ (Draine & Lee, 1984), which implies that $\tau \sim 1$ requires a column density of dust of the order of $\Sigma_d \sim 10^{-2} \text{g cm}^{-2} \sim 50 M_\odot \text{pc}^{-2}$. Such a column density can easily be reached in the central part of an AGN host galaxy. If all the momentum of the radiation is transferred into the gas, the total energy extracted is roughly a fraction of v/c of the total radiation energy of the AGN, where v is the velocity of the accelerated gas. Thus, the energy feedback generated by radiation pressure is more efficient if the radiation pressure accelerates a smaller amount of gas to a higher velocity. In broad-line QSOs, where the gas outflow velocity can be as high as $\sim 0.2c$, acceleration by radiation pressure is an efficient energy injection mechanism.

Another coupling between photons produced by an AGN and the surrounding gas is through Compton heating. If the gas is ionized and has a temperature lower than the effective (Compton) temperature of the radiation field, T, then the Compton scattering between an electron and a photon transfers energy from the photon to the electron, thereby heating the gas. As shown in §B1.3, on average a photon loses a fraction $\sim 4k_B T/m_e c^2$ of its energy in each scattering, which corresponds to an efficiency of $\sim 10^{-1}$ for $T \sim 10 \text{keV}$. Nearby AGN that have been studied in detail, such as 3C273 and 3C279, have $T \sim 10 \text{keV}$ or higher, indicating that Compton heating may be an important form of AGN feedback (e.g. Ciotti & Ostriker, 2001).

The momentum-driven feedback and Compton heating are expected to be more important for AGN in the high-accretion phase where AGN radiate efficiently. Although the number is very uncertain, it is usually assumed that a small fraction, e.g. $\sim 5\%$, of the radiative energy is deposited as thermal energy into the surrounding gas through these processes. Numerical simulations of galaxy formation incorporating such an energy source in mergers of gas-rich galaxies (which may trigger a bright quasar) show that such an energy injection may have significant impact on the gas in the host galaxy, producing gas outflows that can effectively quench the growth of the central SMBH and suppress star formation in the host galaxy (e.g. Springel et al., 2005b; Hopkins et al., 2006).

14.4.2 Mechanical Feedback

During the low-accretion mode of an AGN, when the accretion rate of the SMBH is much lower than the Eddington rate, AGN feedback is believed to proceed mainly through radiatively inefficient, mechanical forms, such as radio jets and lobes. Evidence for such energy feedback can be seen in a number of elliptical galaxies at the centers of clusters, which contain X-ray cavities filled with relativistic gas (e.g. Fabian et al., 2006; Forman et al., 2007). These X-ray cavities, often loosely referred to as 'bubbles', are believed to be inflated by the jet launched from the central SMBH, and the power involved can be substantial. The typical kinetic power estimated for the central galaxy of a cluster is about $10^{45} \text{erg s}^{-1}$, more than enough to offset the radiative cooling of the intracluster gas (e.g. Böhringer et al., 2002).

14.4 AGN and Galaxy Formation

As a crude approximation, the evolution of a jet-powered bubble is similar to that of an adiabatic stellar wind bubble. The assumption of no radiative cooling in the bubble is valid, because the gas density is low and the temperature is high. Assuming that the energy input rate, L, is a constant, and that the bubble expands as a spherically symmetric strong shock, the bubble radius evolves as

$$R \sim \left(\frac{L}{\rho}\right)^{1/5} t^{3/5}, \tag{14.80}$$

where ρ is the density of the medium into which the bubble expands (see §8.6.2). The expansion velocity is $v \sim (L/\rho)^{1/3} R^{-2/3}$. During the supersonic expansion, the feedback energy is transferred into the gas through shocks as more and more gas is swept up by the shock front. The supersonic expansion ends when the expansion velocity becomes of the order of the sound speed of the medium. For a bubble expanding into a hot medium with temperature $T = 1\,\mathrm{keV}$ and density $n = 1\,\mathrm{cm}^{-3}$, the sonic radius is $\sim 50\,\mathrm{kpc}$ assuming $L = 10^{45}\,\mathrm{erg\,s}^{-1}$. Once the expansion becomes subsonic, the energy is being transferred into the gas through PdV work. Meanwhile, if the gas pressure is stratified, the buoyancy of the gas can cause the bubble to rise in the potential well of the host, spreading the feedback energy into gas that is far away from the central source. Numerical simulations of buoyancy-driven bubbles generated in the centers of X-ray clusters indicate that they can effectively channel the feedback energy into thermal energy of the intracluster gas (e.g. Binney & Tabor, 1995; Churazov et al., 2002; Brüggen & Kaiser, 2002; Reynolds et al., 2002). It is believed that such a feedback process may be responsible for the quenching of cooling flows in clusters of galaxies (see §8.8.1).

Since every early-type galaxy seems to host a SMBH at its center, this mechanical AGN feedback may have operated in all these galaxies, affecting their formation and evolution. Suggestions have been made that such feedback may effectively heat the gas in the halos of early-type galaxies, quenching radiative cooling and star formation in these systems (e.g. Bower et al., 2006; Croton et al., 2006). As discussed in §15.7, such quenching may have played an important role in shaping the galaxy luminosity function, and in establishing the color–magnitude relation of galaxies.

15
Statistical Properties of the Galaxy Population

15.1 Preamble

Galaxies are observed to have different intrinsic properties (luminosity, morphological type, size, color, age, nuclear activity, etc.) and to reside in different environments (field, group, cluster, etc.). In the previous four chapters, we have described how the properties of individual galaxies are determined by their formation mechanisms and the various processes that play a role during their subsequent evolution. The goal there has been to constrain the initial conditions and physical processes that lead to the formation of a galaxy with some specific intrinsic properties. In this chapter, we study the statistical properties of the galaxy population, i.e. we address the question of how galaxies are distributed with respect to their intrinsic properties and in space. The goal here is to understand the conditions for the formation and evolution of the galaxy population as a whole. Clearly, in order to achieve this goal, we need not only to understand the processes that govern the formation and evolution of individual galaxies, but also to understand how this formation and evolution is linked to the cosmological initial and boundary conditions that determine the statistical properties of the cosmic density field.

The distribution of the galaxy population can be described formally by a multivariate distribution function, ϕ, defined by

$$dn = \phi(G_1, G_2, \ldots) dG_1 dG_2 \cdots, \tag{15.1}$$

where the G_i ($i = 1, 2, \ldots$) each stand for a specific property of galaxies, such as luminosity, size, color, etc, and dn is the number density of galaxies with properties $G_1, G_2, \ldots$ in the ranges $G_1 \pm dG_1/2, G_2 \pm dG_2/2, \ldots$, respectively. An important goal of astronomy is both to determine the form of ϕ observationally and to understand its origin from physical principles.

Ideally, we would like to obtain the joint distribution function of all the important properties of galaxies. In practice, however, observational data are usually sufficient only to determine the marginal distribution functions of a few quantities. These marginal distribution functions are the projections of the full distribution function onto specific axes and planes. For example, the luminosity function is the full distribution function projected onto the luminosity axis, while the color–luminosity distribution is the projection onto the color–luminosity plane. In this chapter we focus on some of the most important observed properties of the galaxy distribution, and describe how these can be understood in the current paradigm of structure formation.

As described throughout this book, galaxies are believed to form and reside in extended dark matter halos. The properties of the galaxy population are therefore related to the cosmological density field through the properties of the dark matter halo population. Since the formation and properties of the halo population depend only on gravitational processes, and can be understood relatively easily with numerical simulations and analytical models (see Chapter 7), it is convenient to describe the galaxy population in terms of its connection to dark matter halos. In the remainder of this section, we provide a formal discussion about how to link galaxies with dark matter halos. Such discussion is useful, because it allows us to clearly see how the different processes of galaxy formation and evolution described in the earlier chapters play their role in shaping the statistical properties of the galaxy population.

A dark matter halo at a given redshift can be characterized by a set of quantities that describe its salient properties, such as mass, concentration, shape, angular momentum, formation history, and environment. For convenience, we denote these halo quantities, including the redshift at which a halo is identified, collectively by $\mathscr{H}$. Similarly, we use $\mathscr{G} = \{G_1, G_2, \ldots\}$ to denote collectively the quantities that characterize the properties of a galaxy, such as luminosity, morphology, color, etc. The connection between halos and galaxies can formally be described by a conditional distribution function, $P(\mathscr{G}|\mathscr{H})$, which gives the probability of finding a galaxy with property $\mathscr{G}$ in a halo with property $\mathscr{H}$. The distribution function of the galaxy population with respect to $\mathscr{G}$ can then be written as

$$P(\mathscr{G}) = \int P(\mathscr{G}|\mathscr{H})P(\mathscr{H})\,\mathrm{d}\mathscr{H}, \qquad (15.2)$$

where $P(\mathscr{H})$ is the distribution function of dark matter halos with respect to $\mathscr{H}$. As a simple example, the luminosity function can formally be written as

$$\phi(L) = \int \Phi(L|M)n(M)\,\mathrm{d}M, \qquad (15.3)$$

where $n(M)$ is the mass function of dark matter halos, and $\Phi(L|M)$ is the conditional luminosity function that describes the probability of finding a galaxy with luminosity L in a halo of mass M (see §15.3 for details).

When describing the properties of the galaxy population, it is useful to separate galaxies into two categories: central and satellite galaxies. The central galaxy in a halo is the one residing at (or near) the bottom of the halo potential well, while all other galaxies within the halo are satellite galaxies. The central galaxy is usually the most massive one among all member galaxies, although there are exceptions, especially in massive clusters where there are often two or more dominating galaxies of comparable masses. The reason for considering the centrals and satellites separately is that they are expected to have undergone different evolutionary processes. Based on the current theory of galaxy formation, the gas that cools inside a dark matter halo accumulates at the center of the halo's potential well. Furthermore, galaxy mergers are more likely to occur near the bottom of the halo potential well, as the orbits of satellite galaxies decay due to dynamical friction. Hence, central galaxies may continue to grow due to the accretion of cooling gas and/or the accretion of satellite galaxies. A satellite galaxy, on the other hand, is subjected to various environmental effects, such as ram-pressure stripping, tidal stripping and strangulation (see §12.5), that can strip its associated gas reservoir, thereby suppressing the formation of new stars. In addition, satellite galaxies are also subjected to various interactions with the host halo and other galaxies in the halo, which can cause them to undergo a morphological transformation. Note that mergers among satellites are expected to be rare (see §12.4).

Central and satellite galaxies are also different in their relationships with dark matter halos. The centrals, which form and reside near the centers of their host halos, are expected to be closely related to the formation histories of their hosts. On the other hand, satellites are believed to be associated with subhalos, and so their properties may be more related to the formation histories of the subhalo population. As seen in §7.5.3, subhalos are themselves independent halos before merging into bigger ones. Hence, a satellite itself is a central galaxy before its halo becomes a subhalo. This suggests that we may link the properties of satellite galaxies to those of its host halo via the following formal relation,

$$P_{\mathrm{s}}(\mathscr{G}|\mathscr{H}) = \int\int P(\mathscr{G}|\mathscr{G}_a)P_{\mathrm{c}}(\mathscr{G}_a|\mathscr{H}_a)P(\mathscr{H}_a|\mathscr{H})\,\mathrm{d}\mathscr{H}_a\mathrm{d}\mathscr{G}_a, \qquad (15.4)$$

where $\mathcal{H}_a$ denotes the property of a subhalo at the time of accretion, i.e. when it first became a subhalo, and $\mathcal{G}_a$ denotes the property of a satellite galaxy at the time when it first became a satellite. In the above equation, $P(\mathcal{H}_a|\mathcal{H})$ is the probability of finding a subhalo with property $\mathcal{H}_a$ at the time of accretion in a halo with property $\mathcal{H}$, $P_c(\mathcal{G}_a|\mathcal{H}_a)$ is the probability to form a central galaxy of property $\mathcal{G}_a$ in a halo of property $\mathcal{H}_a$, and $P(\mathcal{G}|\mathcal{G}_a)$ is the probability for a central galaxy of property $\mathcal{G}_a$ to be transformed into a satellite galaxy of property $\mathcal{G}$.

The formal relationship between galaxies and dark matter halos/subhalos as represented by Eqs. (15.2) and (15.4) provides a useful way to understand how the statistical properties of the galaxy population are determined by the various aspects of galaxy formation and evolution. First, the properties of the galaxy population depend on the properties of the halo and subhalo populations, as represented by the distribution functions, $P(\mathcal{H})$ and $P(\mathcal{H}_a|\mathcal{H})$. Second, the properties of the galaxy population depend on how individual galaxies form in the centers of dark matter halos, as indicated by the conditional distribution function, $P_c(\mathcal{G}|\mathcal{H})$. Finally, the properties of the galaxy population also depend on how galaxies are transformed by environmental effects in dark matter halos, which is indicated by $P(\mathcal{G}|\mathcal{G}_a)$. Note that the description here is based on dark matter halos; any environmental effects on superhalo scales can be taken into account by including in the halo property, $\mathcal{H}$, a set of quantities that describes the large-scale environment of dark matter halos.

Clearly, in order to model the galaxy population in detail, one needs to model all the important aspects of galaxy formation and evolution. In the following sections, we will use the formalism introduced here to demonstrate how the statistical properties of the galaxy population can be understood in terms of the physical processes described in the previous chapters.

15.2 Galaxy Luminosities and Stellar Masses

One of the most basic properties of a galaxy is its luminosity. Roughly speaking, the luminosity of a galaxy is proportional to the number of stars in the galaxy and is a measure of its stellar mass. However, the luminosity of a galaxy can change with time as stars form and evolve. The study of the distribution of galaxies with respect to their luminosities can therefore provide important clues regarding the evolution of the galaxy population.

15.2.1 Galaxy Luminosity Functions

The luminosity function of galaxies, $\phi(L)$, is defined as

$$\mathrm{d}n(L) = \phi(L)\,\mathrm{d}L, \tag{15.5}$$

where $\mathrm{d}n(L)$ is the comoving number density of galaxies with luminosities in $L \pm \mathrm{d}L/2$. Since luminosities are usually measured in a specific waveband, the luminosity function is typically also determined in a specific waveband. Terms like the B band luminosity function, K band luminosity function, etc. are frequently used (see §2.1.1 for the definitions of different bands).

Before describing the observed luminosity functions, we first discuss how a luminosity function is derived from a given sample of galaxies. Suppose we have a magnitude-limited sample which includes all galaxies, in a patch of sky, with apparent magnitudes (in a given waveband) brighter than a limit m_{lim}. For a galaxy at redshift z with an apparent magnitude m, its absolute magnitude is given by

$$\mathcal{M} = m - 5\log[d_\mathrm{L}(z)/\mathrm{Mpc}] - 25 - K(z), \tag{15.6}$$

where $d_\mathrm{L}(z)$ is the luminosity distance corresponding to redshift z (see §3.2.6). The term $K(z)$ in Eq. (15.6) is the 'K correction' and is required in order to correct the observed flux into a fixed rest-frame band, so that the absolute magnitudes are the same for identical galaxies at different redshifts (see §10.3.6). By definition, the luminosity of a galaxy (in the chosen waveband) is related to the absolute magnitude by

$$2.5\log(L/L_\odot) = \mathcal{M}_\odot - \mathcal{M}, \tag{15.7}$$

where the subscript $\odot$ denotes values for the Sun in the same waveband (see §2.1.1).

Because of the apparent magnitude limit, a galaxy with a luminosity L will only be part of the observed sample if it is located within a maximum luminosity distance $d_{\max}$ (corresponding to a redshift $z_{\max}$) given by

$$5\log[d_{\max}(L)/\mathrm{Mpc}] = m_{\lim} - \mathcal{M}_\odot - 25 + 2.5\log(L/L_\odot) - K(z_{\max}). \tag{15.8}$$

Thus, for a magnitude-limited sample covering a solid angle, ω, the expected number of galaxies with luminosities in the range $L \pm dL/2$ is

$$dN = \phi(L) V_{\max}(L)\, dL, \tag{15.9}$$

where $V_{\max}$ is the comoving volume out to $z_{\max}$. For a flat universe,

$$V_{\max}(L) = \frac{\omega}{3}\left[\frac{d_{\max}(L)}{1+z}\right]^3, \tag{15.10}$$

(see §3.2). From Eq. (15.9) we have

$$\phi(L)\, dL = \frac{dN(L)}{V_{\max}(L)} = \sum_i \frac{1}{V_{\max}(L_i)}, \tag{15.11}$$

where the summation extends over all galaxies in the luminosity interval $L \pm dL/2$. This equation outlines the basic principle for deriving the luminosity function of galaxies from a magnitude-limited sample.

The observed luminosity function of galaxies is usually fitted by a functional form known as the Schechter function [see Eq. (2.34)]. Based on the principle outlined above, one can estimate the values of the three parameters, ϕ^*, α and L^* (or $\mathcal{M}^*$), which characterize the Schechter function, from a magnitude-limited galaxy sample by fitting the measurement given by Eq. (15.11) to the Schechter function. In practice, one often uses a maximum likelihood method to determine these parameters (e.g. Sandage et al., 1979; Efstathiou et al., 1988). The probability for a galaxy with luminosity L_i and redshift z_i to be included in a magnitude-limited sample is

$$p_i = \phi(L_i) \bigg/ \int_{L_{\min}(z_i)}^{\infty} \phi(L)\, dL, \tag{15.12}$$

where $L_{\min}(z_i)$ is the lowest luminosity a galaxy at redshift z_i can have in order for it to be included in the sample. From p_i one can form a likelihood function

$$\mathscr{L} = \prod_i p_i, \tag{15.13}$$

where the product extends over all galaxies in the sample. This function depends on the free parameters characterizing the luminosity function ϕ (e.g. α and L^* in the Schechter function), and the best estimates of these parameters are obtained by requiring that they maximize $\mathscr{L}$. Note that this method does not provide an estimate for the overall amplitude ϕ^*, but its value can be obtained by requiring the luminosity function to reproduce the total number of galaxies in the sample. The errors in the parameters α and L^* can be estimated by using the fact that the maximum likelihood estimate is asymptotically normal. Hence, we can write

$$\ln \mathscr{L} = \ln \mathscr{L}_{\max} - \frac{1}{2}\chi^2_\beta(k), \tag{15.14}$$

Table 15.1. Galaxy luminosity functions in different bands.

Band	α	$\mathcal{M}^* - 5\log h$	$\phi^*/(h^3\,\mathrm{Mpc}^{-3})$	Reference
u	-0.92 ± 0.07	-17.93 ± 0.03	$(3.05 \pm 0.33) \times 10^{-2}$	Blanton et al. (2003)
g	-0.89 ± 0.03	-19.39 ± 0.02	$(2.18 \pm 0.08) \times 10^{-2}$	Blanton et al. (2003)
r	-1.05 ± 0.01	-20.44 ± 0.01	$(1.49 \pm 0.04) \times 10^{-2}$	Blanton et al. (2003)
i	-1.00 ± 0.02	-20.82 ± 0.02	$(1.47 \pm 0.04) \times 10^{-2}$	Blanton et al. (2003)
z	-1.08 ± 0.02	-21.18 ± 0.02	$(1.35 \pm 0.04) \times 10^{-2}$	Blanton et al. (2003)
J	-1.10 ± 0.04	-22.85 ± 0.04	$(0.71 \pm 0.01) \times 10^{-2}$	Jones et al. (2006)
H	-1.11 ± 0.04	-23.54 ± 0.04	$(0.72 \pm 0.01) \times 10^{-2}$	Jones et al. (2006)
K	-1.16 ± 0.04	-23.83 ± 0.03	$(0.75 \pm 0.01) \times 10^{-2}$	Jones et al. (2006)
b_J	-1.21 ± 0.03	-19.66 ± 0.07	$(1.61 \pm 0.08) \times 10^{-2}$	Norberg et al. (2002b)
Near-UV	-1.12 ± 0.10	-17.77 ± 0.15	$(1.22 \pm 0.20) \times 10^{-2}$	Budavári et al. (2005)
Far-UV	-1.10 ± 0.12	-17.20 ± 0.14	$(1.30 \pm 0.20) \times 10^{-2}$	Budavári et al. (2005)

where $\chi^2_\beta(k)$ is the β point (the confidence level) of the χ^2 distribution with k degrees of freedom, and k is equal to the number of free parameters in the fit (see Eadie et al., 1971). For example, the $\beta = 96\%$ confidence level for the values of α and L^* is just the curve, $\ln\mathscr{L}(\alpha,L^*) = \ln\mathscr{L}_{\max} - 3.0$, in the α-L^* plane.

The method outlined above can also be used to estimate the luminosity function without assuming a functional form for ϕ. One can assume that the luminosity function is represented by the values of $\phi(L)$ in a number of discrete bins in L: ϕ_k ($k = 1, 2, \ldots, K$), and obtain the values of ϕ_k by maximizing the likelihood function with respect to all the ϕ_k (e.g. Efstathiou et al., 1988).

(a) The Observed Galaxy Luminosity Functions in Different Wavebands With the advent of large redshift surveys, the luminosity function of the total galaxy population can nowadays be estimated accurately down to relatively faint absolute magnitudes in optical, infrared and UV bands. Table 15.1 list the Schechter parameters obtained from fitting the observed luminosity functions in different bands. These results are for galaxies brighter than $\sim \mathcal{M}^* + 3\,\mathrm{mag}$. The behavior of the galaxy luminosity function at the very faint end is still uncertain, mainly because in a magnitude-limited sample these galaxies can only be observed within a very small, local volume so that the errors are dominated by cosmic variance. Current results based on the SDSS suggest that the luminosity function may steepen significantly with respect to the Schechter function at $\mathcal{M} \gtrsim \mathcal{M}^* + 3\,\mathrm{mag}$ (e.g. Blanton et al., 2005).

(b) The Stellar Mass Function of Galaxies With multi-band photometry and high-quality spectroscopy, the total stellar mass of individual galaxies can be estimated using the population synthesis method described in §10.3. One can therefore also estimate the stellar mass function of galaxies from observations. One problem here is that such an estimate requires a large sample complete in stellar mass, while galaxy samples are typically either magnitude limited, or selected to be complete in luminosity (in some band). Care must therefore be taken to ensure that any incompleteness is properly corrected for. As shown in Bell & de Jong (2001), the stellar mass-to-light ratios in the near-infrared bands vary only by a factor of 2 or less over a wide range of star-formation histories, in contrast with a factor of 10 in the blue bands. Thus, the luminosity of a galaxy in a near-infrared band (e.g. the K band) is a better tracer of its total stellar mass, and it is therefore advantageous to start with a complete sample selected in the near-infrared when estimating the stellar mass function. Galaxy stellar mass functions have been estimated based on near-infrared and optical data (e.g. Cole et al., 2001; Bell et al., 2003b), and the stellar mass function thus obtained can be described by a Schechter form with a slope at the low-mass end, $\alpha \sim -1.1$, and a characteristic stellar mass $\sim 5 \times 10^{10} h^{-2} \mathrm{M}_\odot$. The estimated stellar mass density of the Universe is $\rho_\star \sim 5.5 \times 10^8 h\,\mathrm{M}_\odot\,\mathrm{Mpc}^{-3}$, corresponding to a density parameter, $\Omega_{\star,0} \sim 2 \times 10^{-3} h^{-1}$. There are still substantial uncertainties in these numbers, because of cosmic variance and because of the uncertainties in the stellar population synthesis models, in particular regarding the assumed initial mass function and the adopted spectra of individual stars (e.g. Bell et al., 2003b).

(c) Dependence on Galaxy Color As shown in Fig. 2.27, the galaxy population at low redshift exhibits a clear bimodal distribution in the color–luminosity space. One can therefore study how the galaxy luminosity function depends on galaxy color by measuring the galaxy luminosity function separately for the red and blue populations. The results of such measurements show that the red population in general has a brighter characteristic magnitude, $\mathcal{M}^*$, and a shallower faint-end slope (i.e. a larger value of α), than the blue population (see Fig. 2.33 for an example), suggesting that the fraction of red galaxies decreases with decreasing luminosity. However, this trend may not hold at the very faint end ($\mathcal{M}_r - 5\log h \gtrsim -17$), where the luminosity function of the red population appears to show an upturn (Blanton et al., 2005). These faint red galaxies usually have low surface brightness and show exponential profiles, similar to the dwarf ellipticals observed in nearby groups and clusters (see below). Unfortunately, the significance of such upturn is still difficult to quantify, again because the cosmic volume within which the faint galaxies can be observed is small.

(d) Dependence on Galaxy Morphology As we have seen in §2.3.1, the morphologies of galaxies are strongly correlated with their colors. The color dependence of the galaxy luminosity function described above thus suggests that the luminosity function should also depend on galaxy morphology. Fig. 15.1 shows schematically how the luminosity function decomposes into the contributions from galaxies of different types. The bright-end of the luminosity function is clearly dominated by ellipticals, while the spiral galaxies dominate the intermediate luminosity range. At the faint end, the luminosity function is dominated either by irregulars (in the field) or by dwarf ellipticals (in clusters). The luminosity functions of spirals, S0s and ellipticals are peaked around

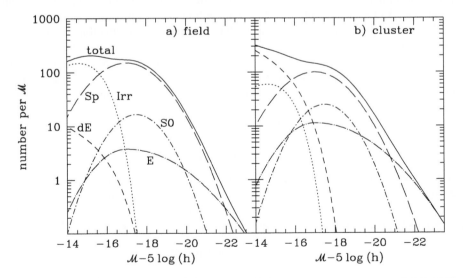

Fig. 15.1. Morphology dependence of the galaxy luminosity function. The left and right panels show the luminosity functions for galaxies in the field and in clusters, respectively. The plot for cluster galaxies is based on the results published in Jerjen & Tammann (1997). The plot for field galaxies is based on the argument by Binggeli et al. (1988) that the shapes of the luminosity functions for individual morphological classes are invariant, while their amplitudes change with environment.

some characteristic luminosities and follow roughly a Gaussian form instead of a Schechter form. On the other hand, both irregulars and dwarf ellipticals have Schechter-type luminosity functions, but the faint-end slope is steeper for dwarf ellipticals than for irregulars. As mentioned above, these dwarf ellipticals may be responsible for the upturn of the luminosity function of red galaxies at the faint end.

(e) Dependence on Environment   Is the luminosity function universal, or does it change with environment? To answer this question, one can compare the luminosity function of galaxies in clusters (high-density environment) with that in the general field. The luminosity function of cluster galaxies has been estimated for a number of nearby clusters (e.g. Jerjen & Tammann, 1997; Trentham & Hodgkin, 2002; Popesso et al., 2006). The observational results show that the luminosity function of cluster galaxies has a systematically steeper slope, with a marked upturn at the faint end ($\mathcal{M}_r - 5\log h \gtrsim -17$), and a more extended tail at the bright end than the field luminosity function. As shown by Binggeli et al. (1988), when galaxies are subdivided according to morphological type, the shapes of the luminosity functions are similar for both cluster and field galaxies within the same morphology subclass. The differences in the composite luminosity functions arise from the fact that the relative amplitudes of the luminosity functions of different subclasses change with environment, as illustrated in Fig.15.1. For instance, the steep faint-end slope for cluster galaxies is caused by the high fraction of dwarf spheroidals, while the extended tail at the bright end is caused by the increased fraction of bright ellipticals. This suggests that the environment dependence of the galaxy luminosity function may be understood in terms of the morphology dependence of the luminosity function discussed above, together with the morphology–density relation described in §2.4.5.

(f) Redshift Evolution   With the capacities of the current generation of telescopes, complete samples of galaxies to very faint magnitude limits, with spectroscopic or photometric redshifts, can be obtained in various optical and infrared bands. With such samples, the luminosity function can be determined for galaxies at different redshifts. As described in §2.6, the observational results (see Fig.2.33 for an example) indicate that the luminosity function of galaxies has undergone significant changes in the past ~ 8 Gyr, with the red and blue populations showing different evolution. In particular, the data suggests that the total stellar mass density in the red population has roughly doubled since $z \sim 1$, while that of the blue population has remained roughly constant (e.g. Borch et al., 2006; Faber et al., 2007). Apparently, while stars form in the blue population, a significant fraction of the blue galaxies must have had their star formation quenched by some physical processes so as to join the red population.

15.2.2 Galaxy Counts

Often, especially when faint galaxies are concerned, redshifts (and thus distances) of individual galaxies are not available. In these cases, one can still distill some information regarding the evolution of the galaxy population from the galaxy counts, $\mathcal{N}(m)$, defined as the number of galaxies per unit apparent magnitude (in an observational waveband) per unit solid angle:

$$\mathrm{d}^2 N(m) = \mathcal{N}(m)\,\mathrm{d}m\,\mathrm{d}\omega. \tag{15.15}$$

Although galaxy counts are easy to obtain from any galaxy catalogue with uniform photometry, we now show that a proper interpretation in terms of the evolution in the number densities or luminosities of galaxies is far from trivial.

Suppose that the luminosity function of galaxies at redshift z in a fixed passband of the rest-frame spectra is $\phi(L,z)\,\mathrm{d}L$. Suppose further that these galaxies are divided into subclasses according to their spectral energy distribution F_ν. In a given subclass, the number of galaxies per unit redshift and per unit apparent magnitude (in the same passband) in a sample complete to an apparent magnitude $m_{\lim}$ is

$$\frac{d^2 N}{dm\,dz} = \frac{\omega}{4\pi}\frac{dV(z)}{dz}\phi(L,z)\frac{dL}{dm}, \quad (m < m_{\lim}), \tag{15.16}$$

where ω is the solid angle covered by the sample, $V(z)$ is the comoving volume within redshift z, and the luminosity, L, is related to the apparent magnitude by Eq. (15.6). The total number counts is the sum over all subclasses, each with its own dL/dm due to different K-corrections. The redshift distribution of galaxies in the sample is then obtained by integrating Eq. (15.16) over m to $m_{\lim}$:

$$\frac{dN(<m_{\lim})}{dz} = \frac{\omega}{4\pi}\frac{dV(z)}{dz}\int_{L_{\lim}(z)}^{\infty}\phi(L,z)\,dL, \tag{15.17}$$

where the limiting luminosity, $L_{\lim}$, is related to $m_{\lim}$ by Eqs. (15.7) and (15.6). Similarly, the count of galaxies brighter than an apparent magnitude m can be written as

$$N(<m) = \frac{\omega}{4\pi}\int_0^{\infty}dz\,\frac{dV(z)}{dz}\int_{L_m(z)}^{\infty}\phi(L,z)\,dL, \tag{15.18}$$

with $L_m(z)$ related to m by Eqs. (15.7) and (15.6). The differential count is then

$$\mathcal{N}(m) = \frac{\ln 10}{10\pi}\int_0^{\infty}\frac{dV}{dz}\phi[L_m(z),z]L_m(z)\,dz. \tag{15.19}$$

If the luminosity function does not depend on z, so that $\phi(L,z) = \phi(L)$, and if K corrections can be neglected, we can write $L_m(z) = 4\pi f_m d_L^2(z)$, where f_m is the energy flux corresponding to apparent magnitude m ($f_m \propto 10^{-0.4m}$). In this case, Eq. (15.18) reduces to

$$N(<m) = \frac{\omega}{3}(4\pi f_m)^{-3/2}\int_0^{\infty}\frac{3V(z)}{4\pi d_L^3(z)}L^{3/2}\phi(L)\,dL, \tag{15.20}$$

where z is related to L by $L = 4\pi f_m d_L^2(z)$. If we further neglect the expansion of the Universe, this equation becomes

$$N(<m) = \frac{\omega}{3}(4\pi f_m)^{-3/2}\int_0^{\infty}L^{3/2}\phi(L)\,dL$$
$$= \frac{\omega}{3}(4\pi f_m)^{-3/2}\phi^* L^{*3/2}\Gamma\left(\alpha + \frac{5}{2}\right), \tag{15.21}$$

where the Schechter form (2.34) is used for $\phi(L)$ in the second equation. Thus, for non-evolving sources in a non-expanding Euclidean space, $N(<m) \propto f_m^{-3/2} \propto 10^{0.6m}$. In general, however, galaxy counts will be different from this relation, because of the expansion of the Universe, because of K corrections, and because of the evolution of the luminosity function.

With current telescopes and detectors we can detect galaxies to very faint magnitudes. Fig. 2.32 shows examples of the galaxy counts in the U, B, I, and K bands, along with the predictions of some theoretical models. Note that even in the no-evolution model [i.e. $\phi(L,z) = \phi(L)$], the counts $N(<m)$ increase with m slower than the $f_m^{-3/2}$ law, because the luminosity distance increases with redshift faster than the angular-diameter distance in an expanding universe. The observed galaxy counts are significantly higher than the prediction of the no-evolution model at the faint end, particularly in the bluer bands. Since the apparent magnitude of a galaxy depends both on its luminosity and its redshift, the excess can be either due to an increase in the number density of intrinsically faint galaxies at relatively low redshift, or due to an increase in the number density of intrinsically bright galaxies at relatively high redshift, or a combination of both. The nature of the faint-blue excess has been partially elucidated by the deep imaging surveys made by the HST. Because of the high image resolution, galaxies to very faint magnitudes in

the Hubble Deep Field and in the Medium Deep Survey can be assigned rough morphological types, so that counts can be estimated separately for galaxies of different morphological types (e.g. Abraham et al., 1996). The result is that objects classified as irregulars show the largest excess relative to what is expected from the no-evolution model, while the excess for ellipticals and spirals are relatively modest, suggesting that the observed excess of galaxy counts in blue bands is probably dominated by the evolution of intrinsically faint galaxies.

In broad terms, there are two different kinds of evolution that can cause an excess of galaxy counts at the faint end. The first is luminosity evolution, in which the comoving number density of galaxies remains the same but galaxies become intrinsically brighter at higher redshifts. The second is number evolution, in which galaxies are created and destroyed, so that the number density of galaxies changes with redshift. In reality, it is likely that both kinds of evolution are relevant.

In the case of pure luminosity evolution, galaxy counts can be obtained from the local luminosity function together with an assumption about how the luminosities of galaxies evolve with redshift. In practice, the galaxy counts can be calculated from Eq. (15.18) by replacing $\phi(L,z)$ with the local luminosity function and by adding an E correction term in the relation between m and $\mathcal{M}$:

$$m = \mathcal{M} + 5\log\left[\frac{d_{\rm L}(z)}{10{\rm pc}}\right] + K(z) + E(z), \qquad (15.22)$$

where $E(z)$ is the change in magnitude in the observational pass band for a galaxy at redshift z when it is evolved to the present time (see §10.3.6). Models of luminosity evolution considered in the literature generally assume that elliptical galaxies formed their stars in short bursts at an early epoch and have since evolved only passively, while disk galaxies formed their stars continuously over a substantial fraction of their lifetimes. The amount of luminosity evolution can then be calculated from the population synthesis models described in §10.3. For pure luminosity evolution, galaxies of a given stellar mass are brighter and bluer at higher redshifts. Therefore, luminosity-evolution models in general predict a significant excess in the blue counts relative to the no-evolution model, but only a moderate excess in the K-band. This is in qualitative agreement with the data, suggesting that the observed excess of galaxy counts may, at least partly, be due to luminosity evolution of the galaxy population.

In the hierarchical scenario of galaxy formation, galaxies continuously merge to build larger galaxies and one thus expects some evolution in the comoving number density of galaxies with redshift. The exact rate of evolution, however, depends on the balance between the destruction rate of existing galaxies and the formation rate of new ones. Since merging increases the stellar masses of the galaxies involved, naive expectations are that massive galaxies should be rarer at higher redshift, while low-mass galaxies were more abundant. In such a scenario, one might expect that the observed faint-blue excess is owing to an increase in the comoving number of relatively low-mass galaxies. However, luminosity evolution complicates such an interpretation. Even in the hierarchical scenario it may still be possible for the galaxy counts at the faint end to contain a significant fraction of relatively massive galaxies that were brighter in the past. It is exactly this degeneracy between luminosity evolution and number evolution that limits the power of the galaxy-count statistic for constraining the evolution of the galaxy population.

15.2.3 Extragalactic Background Light

When galaxies are too far away to be identified as discrete sources, we can only observe their contribution to the radiation background. The observed extragalactic background radiation over the entire electromagnetic spectrum is shown in Fig. 2.2. In what follows we examine how to use these data to constrain galaxy evolution.

15.2 Galaxy Luminosities and Stellar Masses

In practice, background radiation is usually measured within a unit frequency range around some frequency ν_o in the rest frame of the observer. The flux received by the observer due to a source of luminosity L at redshift z is given by

$$f(\nu_o, L, z) = \frac{\mathscr{L}(L, \nu, z)}{4\pi d_L^2(z)} \frac{d\nu}{d\nu_o} = \frac{(1+z)\mathscr{L}(L, \nu, z)}{4\pi d_L^2(z)}, \tag{15.23}$$

where $\nu = \nu_o(1+z)$ is the frequency in the rest frame of the source, $d_L(z)$ is the luminosity distance, and $\mathscr{L}(L, \nu, z) d\nu$ is the average energy emitted by a galaxy with luminosity L at redshift z in the frequency range $(\nu, \nu+d\nu)$. The specific intensity of the radiation background can be written as

$$J(\nu_o) = \frac{1}{4\pi} \int dV(z) \int dL\, f(\nu_o, L, z)\, \phi(L, z), \tag{15.24}$$

where $\phi(L, z) dL$ is the galaxy luminosity function at redshift z. Using $dV(z) = 4\pi d_L^2(z) c\, dt/(1+z)$ (see §3.2.6), we can write this as

$$J(\nu_o) = \frac{c}{4\pi} \int_0^\infty \frac{\varepsilon(\nu, z)}{(1+z)^3} \frac{dt}{dz} dz, \tag{15.25}$$

where

$$\varepsilon(\nu, z) = (1+z)^3 \int_0^\infty \mathscr{L}(L, \nu, z) \phi(L, z)\, dL \tag{15.26}$$

is the mean emissivity at redshift z. As one can see, the background radiation produced by the galaxy population is determined by the number density and mean spectrum of galaxies as functions of redshift.

As an example, let us consider how the observed background radiation can be used to constrain the global star-formation history of the Universe. Denoting the mean star-formation rate per unit *comoving* volume at time t by $\dot\rho_\star(t)$, we can write the emissivity due to the radiation of stars as

$$\varepsilon(\nu, z) = (1+z)^3 \int_0^t \dot\rho_\star(t')\, \mathscr{F}_\nu(t-t')\, dt', \tag{15.27}$$

where z is the redshift corresponding to time t, $\mathscr{F}_\nu(\Delta t)$ is the stellar population spectrum, defined as the power radiated per unit frequency per unit initial mass by a generation of stars with age Δt. For a given initial mass function, the form of $\mathscr{F}_\nu(\Delta t)$ can be obtained from the population synthesis models described in §10.3. In the optical/UV wavebands, the luminosity of a galaxy is dominated by massive stars with masses $\gtrsim 2 M_\odot$, which have lifetimes shorter than the age of the Universe. In this case, the emissivity can be written approximately as $\varepsilon(\nu, z) \sim (1+z)^3 \dot\rho_\star(z) \mathscr{E}_\nu$, where $\mathscr{E}_\nu$ is the energy radiated at rest-frame frequency ν per unit mass due to massive stars. Inserting this expression of ε into Eq. (15.25) gives

$$\int J(\nu_o) d\nu_o = \frac{c}{4\pi} \mathscr{E} \rho_{\star, 0}, \tag{15.28}$$

where $\mathscr{E} = \int \mathscr{E}_\nu d\nu$ is the total energy radiated per unit mass due to massive stars, and $\rho_{\star, 0} = \int \dot\rho_\star(z) dt$ is the comoving density of massive stars formed over the entire history of the Universe. Since the most important energy source during a star's lifetime is the conversion of hydrogen into helium in the core, during which about 0.7% of the rest mass energy of the burned hydrogen atoms is released, we can write $E \sim 0.007 M_{\rm core} c^2$, with $M_{\rm core}$ the mass of the helium core. In a massive star, the helium core is eventually converted into metals (e.g. carbon and oxygen) and ejected into the interstellar medium at the end of its lifetime. Thus, the energy released from such a star is related to the amount of metals produced by $E \sim 0.007 M_Z c^2$. This in Eq. (15.28) gives

$$\int J(\nu_o) d\nu_o = \frac{c^3}{4\pi} 0.007 \rho_{Z,0} \sim 3.0 \times 10^{-4} Z \Omega_{b,0} h^2 {\rm W\, m^{-2}\, sr^{-1}}, \tag{15.29}$$

where $\rho_{Z,0}$ is the mean mass density of metals at the present time, and $Z = \rho_{Z,0}/(\rho_{\rm crit,0}\Omega_{\rm b,0})$ is the mean metallicity of the present-day Universe. Taking $\Omega_{\rm b,0}h^2 = 0.02$ and $Z = 0.001$, we obtain $\int J(\nu_0)\mathrm{d}\nu_0 \sim 6\,[{\rm nW\,m^{-2}\,sr^{-1}}]$, which is comparable to the observed radiation background at the optical/UV wavebands (see Fig. 2.2). This suggests that the star-formation history responsible for the production of metals in the Universe may also be responsible for the production of the radiation background observed in the optical/UV wavebands.

If a significant fraction of all stars formed in dusty environments, the intensity of the background radiation in the optical/UV wavebands would be lower than that expected from the metal production, because a significant fraction of the optical/UV radiation may have been absorbed by dust grains. Since the absorbed optical/UV radiation is expected to be re-emitted in the far-infrared and submillimeter wavebands, with $\lambda \sim (100\text{--}2000)\,\mu{\rm m}$, we expect star formation in dusty environments (e.g. starbursts) to produce background radiation in these wavebands. Such background radiation has indeed been detected by the Far Infrared Absolute Spectrophotometer (FIRAS) on board of the COBE satellite (Fixsen et al., 1998). The average spectrum of the background in the wavelength range $\lambda = 125\text{--}2000\,\mu{\rm m}$ is roughly

$$J(\nu) = (1.3 \pm 0.4) \times 10^{-5}\left(\frac{\lambda}{100\,\mu{\rm m}}\right)^{-0.64 \pm 0.12} B_\nu(18.5 \pm 1.2\,{\rm K}), \qquad (15.30)$$

where $B_\nu(T)$ is the Planck function at temperature T. If this background radiation is interpreted as due to star formation, we can derive a limit on the global star-formation rate in the Universe. The stellar radiation which is absorbed by dust in a unit comoving volume can be written as

$$E_{\rm a}(z) = (1+z)^{-3} \int_0^\infty A_\nu \varepsilon(\nu,z)\,\mathrm{d}\nu, \qquad (15.31)$$

where $\varepsilon(\nu,z)$ is the emissivity of stars and A_ν is the mean fraction of photons of frequency ν absorbed by dust. This is also the energy re-radiated by the dust because of energy balance. The stellar radiation absorbed by the dust is expected to be re-radiated thermally, and so the spectrum of dust emission depends on the temperature distribution of the dust grains. If we denote the fraction of dust grains with temperatures between T and $T + \mathrm{d}T$ by $\eta(T)\,\mathrm{d}T$, the dust emissivity can be written as

$$\varepsilon_{\rm d}(\nu,z) = 4\pi \rho_{\rm d} \kappa_{\rm d}(\nu) \int_0^\infty B_\nu(T)\eta(T)\,\mathrm{d}T, \qquad (15.32)$$

where $\rho_{\rm d}$ is the mean mass density of dust grains in comoving units (usually assumed to be proportional to the mean mass density of metals, ρ_Z), and $\kappa_{\rm d}(\nu)$ is the absorption coefficient per unit mass of dust grains. Energy balance requires that $\int \varepsilon_{\rm d}(\nu,z)\,\mathrm{d}\nu = E_{\rm a}(z)$. This shows that predicting the far-infrared/submillimeter background requires models for $\kappa_{\rm d}(\nu), A_\nu$ and $\eta(T)$. These are not yet well established, so that one typically has to rely on a number of uncertain assumptions (see, e.g. Pei et al., 1999). Despite this, some interesting conclusions can already be made from such modeling. The global star-formation rate required to explain the far-infrared/submillimeter background appears to be higher than that inferred from data in the UV-optical (see §15.4.2). This suggests that there might be a large number of dust-enshrouded star-forming galaxies that are missed in UV/optical surveys due to dust extinction. Indeed, observations with the SCUBA camera (see §2.6.6) have revealed the existence of a population of ultraluminous infrared galaxies (called SCUBA sources) whose space density may be sufficient to account for nearly all observed far-infrared/submillimeter background radiation (Sanders, 1999). Clearly, detailed modeling of the background radiation can provide important constraints on the star-formation history in the Universe (see §15.4.2).

15.3 Linking Halo Mass to Galaxy Luminosity

In the current paradigm of structure formation, galaxies are believed to form and evolve within dark matter halos. It is therefore instructive to understand the galaxy luminosity function, $\phi(L)$, in terms of the mass function of dark matter halos and the occupation statistics of galaxies in individual dark matter halos.

According to the description presented in §15.1, we can formally write the galaxy luminosity function at redshift z as

$$\phi(L,z)\,dL = dL \int \Phi(L|M,z) n(M,z)\,dM, \tag{15.33}$$

where $n(M,z)$ is the mass function of dark matter halos at redshift z, and $\Phi(L|M,z)dL$ is the conditional luminosity function at z, which specifies the average number of galaxies with luminosities in the range $L \pm dL/2$ that reside in a halo of mass M at redshift z. For a given model of structure formation, $n(M,z)$ can be obtained through the models described in §7.2. As shown in §7.2.1, the number density of dark matter halos of mass M can be approximated by the Press–Schechter function,

$$n(M)\,dM = \frac{\overline{\rho}}{M} f_{\rm PS}\left[\frac{\sigma(M^*)}{\sigma(M)}\right] \frac{dM}{M}. \tag{15.34}$$

For a power-law spectrum, $P(k) \propto k^n$, we have

$$f_{\rm PS} = \sqrt{\frac{2}{\pi}} \left(1 + \frac{n}{3}\right) x^{(n+3)/6} \exp\left[-\frac{1}{2} x^{(n+3)/3}\right], \qquad (x \equiv M/M^*). \tag{15.35}$$

For a CDM spectrum, the halo mass function on galaxy scales can be approximated by the above expression with an effective power index $n \sim -2$.

Clearly, for a given model of structure formation, the galaxy luminosity function can be calculated once the conditional luminosity function, $\Phi(L|M,z)$, is known. The form of $\Phi(L|M,z)$ encapsulates all the information about galaxy formation and evolution in individual halos. In what follows, we use simplified, yet physically motivated, models for $\Phi(L|M,z)$ to demonstrate how the luminosity function of galaxies is shaped by various processes.

15.3.1 Simple Considerations

The simplest assumption about the galaxy–halo relation is that the total luminosity of a galaxy in a halo is directly proportional to the halo mass M:

$$L = \frac{\varepsilon_{\rm SF}}{\Upsilon_\star} M_{\rm baryon} = \frac{\varepsilon_{\rm SF}}{\Upsilon_\star} \frac{\Omega_{b,0}}{\Omega_{m,0}} M. \tag{15.36}$$

Here $\varepsilon_{\rm SF}$ is a constant factor describing the star-formation efficiency of the baryons associated with the halo, and $\Upsilon_\star$ is the mass-to-light ratio of the stellar population in solar units. The dashed curve in Fig. 15.2a shows the predicted galaxy luminosity function obtained assuming $\varepsilon_{\rm SF} = \Upsilon_\star = 1$ and a halo mass function given by the standard ΛCDM model. Evidently, this naive model overpredicts the observed luminosity function (indicated by the solid curve) on all mass scales, indicating that $\varepsilon_{\rm SF}/\Upsilon_\star \ll 1$. Furthermore, the shape of the predicted function is very different from that observed, suggesting that $\varepsilon_{\rm SF}$ and/or $\Upsilon_\star$ must vary as function of halo mass.

In order to obtain some insight into the relationship between L and M that is required by the observed luminosity function, let us assume that there is a monotonic relation between the luminosity of a galaxy and the mass of its host halo. In this case, the dependence of L on M can be obtained by matching the number densities of galaxies and halos, i.e. by solving L as a function of M from $\int_L^\infty \phi(L')dL' = \int_M^\infty n(M')dM'$. The L–M relation thus obtained is shown in Fig. 15.2b.

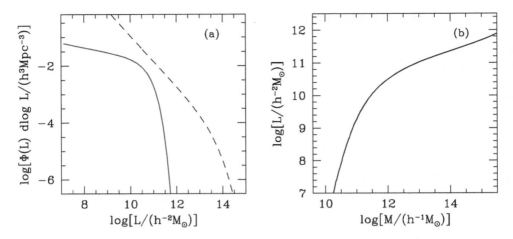

Fig. 15.2. (a) A fit to the observed luminosity function of galaxies (solid curve) compared to the model prediction (dashed curve) obtained under the assumption that each CDM halo at the present time hosts one galaxy with a luminosity, L, proportional to the total baryon mass associated with the halo. (b) The relation between the luminosity, L, and the halo mass, M, obtained by matching the number densities of the halo mass function and the galaxy luminosity function.

Since, $\varepsilon_{\rm SF}/\Upsilon_\star \propto L/M$, we immediately see that the ratio $\varepsilon_{\rm SF}/\Upsilon_\star$ is halo mass dependent, being the highest for $M \sim 10^{12} h^{-1} {\rm M}_\odot$ and becoming lower towards both the low- and high-mass ends. In what follows we will assume $\Upsilon_\star$ to be constant, and that only the star-formation efficiency, $\varepsilon_{\rm SF}$, depends on halo mass. Although this is not a realistic assumption – it is known that galaxies in more massive halos have older stellar populations hence larger $\Upsilon_\star$ – it does not have a significant impact on our qualitative discussion that follows.

The behavior of the L-M relation shown in Fig. 15.2b is consistent with what would be expected from the processes that regulate the star formation in galaxies. As seen in §8.6.3, the star-formation efficiency in low-mass halos can be affected by the energy feedback from supernova explosions and stellar winds. Consider, for instance, the simple model in which a protogalaxy can turn just enough gas into stars for the resulting supernovae to blow the rest of the gas out of the system. Since the specific binding energy of a protogalaxy is proportional to V_c^2 (where V_c is the circular velocity of the system), the fraction of gas turned into stars is also proportional to V_c^2. In this case, we have $L \propto MV_c^2 \propto M^{5/3}$. In massive halos, three important processes may act to limit the luminosities (stellar masses) of the central galaxies. First, the time scale of radiative cooling becomes longer than the age of the Universe for halos with masses $M \gtrsim 10^{12} h^{-1} {\rm M}_\odot$, and the efficiency of star formation is expected to be suppressed accordingly in these massive halos (see §8.4). Second, as discussed in §14.4, AGN feedback may be important in halos more massive than $\sim 10^{13} h^{-1} {\rm M}_\odot$, causing a further suppression of star formation. Finally, galactic cannibalism, which can increase the stellar mass of a central galaxy (see §12.5.2), is expected to become less efficient in very massive halos, simply because the time scale of dynamical friction becomes too long. Taken together, these considerations suggest a characteristic mass, $\sim 10^{12-13} h^{-1} {\rm M}_\odot$, beyond which the luminosity of the central galaxy increases only slowly with halo mass.

Evidently, in order to model the luminosity function in detail, one has to model all these processes properly. In addition, one should also take into account the fact that some halos, particularly the massive ones, may contain multiple galaxies, and that the luminosities of satellite galaxies may also be affected by environmental effects, such as ram-pressure stripping and harassment (see §12.5). Unfortunately, how exactly all these processes act together to shape the galaxy luminosity function is still not well understood and certainly too convoluted to be presented here. In what follows, we restrict ourselves to simple analytical models for the L-M relation. Although highly oversimplified, these models suffice to provide insight and understanding of the different processes by which galaxies form and evolve within dark matter halos.

15.3.2 The Luminosity Function of Central Galaxies

Let us start with the luminosity function of the central galaxies introduced in §15.1. For clarity we use a subscript 'c' to denote the quantities of central galaxies. Observational results based on galaxy groups and satellite kinematics, as well as theoretical results based on semi-analytical modeling both show that, for a given halo mass, the distribution of the luminosity (stellar mass) of the central galaxy is roughly log-normal, with a dispersion of about 0.15 dex (e.g. Yang et al., 2008; More et al., 2009). This dispersion is sufficiently small that we can ignore it (at least for the purpose of modeling the global properties of the luminosity function), which allows us to write the conditional luminosity function of central galaxies as $\Phi_c(L|M) = \delta[L - L_c(M)]$. Here $L_c(M)$ is the luminosity of the central galaxy in a halo of mass M and, for brevity, its dependence on z is included implicitly. The luminosity function of the central population can then be written as

$$\phi_c(L) = n[M(L)]\frac{\mathrm{d}M(L)}{\mathrm{d}L}, \tag{15.37}$$

where $M(L)$ is the halo mass corresponding to a central luminosity L. Based on the results presented in §15.3.1, we assume that the luminosities of central galaxies are related to their halo masses as

$$L_c = L_0 \frac{X_L^\beta}{1 + X_L^{\beta-\gamma}}, \tag{15.38}$$

where $\beta > \gamma > 0$, $X_L \equiv M/M_L$, and M_L is a characteristic halo mass such that $L_c \propto M^\beta$ for $M \ll M_L$ and $L_c \propto M^\gamma$ for $M \gg M_L$. The values of β, γ, M_L, and L_0 are determined by the combination of a number of processes that regulate the growth of stellar mass in dark matter halos (see §15.3.1). In particular, based on the various considerations given above, M_L is expected to be of the order $10^{12-13} h^{-1} \mathrm{M}_\odot$.

In general, if we write the L_c-M relation piecewisely as a power law, $L_c \propto M^\eta$, then the luminosity function given by Eq. (15.37) over the luminosity range where the power law holds can be written as

$$L\phi_c(L) = \frac{\overline{\rho}}{\eta M^*}\sqrt{\frac{2}{\pi}}\left(1 + \frac{n}{3}\right)\left(\frac{L}{L_\eta^*}\right)^{\frac{n-3}{6\eta}} \exp\left[-\frac{1}{2}\left(\frac{L}{L_\eta^*}\right)^{\frac{n+3}{3\eta}}\right], \tag{15.39}$$

where $L_\eta^* = L_0(M^*/M_L)^\eta$. For the model given by Eq. (15.38), we have $\eta = \beta$ for $M \ll M_L$, and $\eta = \gamma$ for $M \gg M_L$. Thus, for a CDM spectrum where the effective power index on the galaxy mass scale is $n \sim -2$, the predicted faint-end slope of $\phi_c(L)$ at $L \ll \min[L_\beta^*, L_0]$ is $\alpha \sim -1 - 5/(6\beta)$. The *observed* luminosity function of central galaxies has α close to -1 (Yang et al., 2009a), implying that the value of β must be large. This suggests that the luminosity of the central galaxy in a low-mass halo increases rapidly with halo mass. If L_c were directly proportional to M in low-mass halos, so that $\beta \sim 1$, the faint-end slope would be very steep, with $\alpha \sim -1.8$. The simple model of supernova feedback based on a binding-energy argument, which gives $L \propto M^{5/3}$ (see §8.6.3), leads to a faint-end slope, $\alpha \sim -1.5$, also considerably more negative than that of the observed luminosity function. This discrepancy suggests that the feedback effect that reduces the star-formation efficiency in small halos at the present time must be stronger than that based on the simple binding-energy consideration. There are a number of processes that might make the L-M relation steeper at the low-mass end. First of all, supernova feedback may be more efficient in low-mass halos than predicted by the simple energy argument. For instance, low-mass halos may contain relatively little hot halo gas (see §8.4.4), so that it is easier for the supernova-driven winds to escape. Secondly, if the intergalactic medium was preheated to a high temperature, the total amount of gas that could be accreted into low-mass halos would be reduced (e.g. Mo et al., 2005). An example of such a preheating mechanism is cosmic re-ionization (see §16.3), which heats the IGM to temperatures $\sim 10^4$ K, thereby suppressing the cooling in dark matter halos with circular velocities $V_c \lesssim 30 \mathrm{km\,s^{-1}}$ (e.g. Efstathiou, 1992). Finally, if significant amounts of star formation can occur only in disks above the surface-density threshold described in §9.5.1, a certain reduction in the cold gas mass fraction due to supernova feedback and/or preheating may imply a much larger reduction in the actual stellar mass fraction. At the present, it is still unclear which processes are the most important in shaping the faint end of the galaxy luminosity function.

At the bright end ($L \gg \max[L_\gamma^*, L_0]$), the luminosity function given by Eq. (15.39) assuming $n = -2$ goes roughly as $\exp[-\frac{1}{2}(L/L_\gamma^*)^{1/(3\gamma)}]$. This is an exponential break similar to that in the Schechter function if $\gamma \sim 1/3$. However, L_γ^*, which is the luminosity of the central galaxy in an M^* halo, *cannot* be identified as the characteristic luminosity, L^*, in the observed luminosity function. The reason is that the predicted number density at L_γ^* in the current ΛCDM model, which is $\sim n(M^*)$ in the model, is much lower than ϕ^* in the luminosity function. However, if $M_L < M^*$ (so that $L_\gamma^* > L_0$), the central luminosity function has an additional break at L_0, where the slope of the luminosity function changes from $\sim -1 - 5/(6\beta)$ to $\sim -1 - 5/(6\gamma)$. Since γ is expected to be a small, positive number, this break can be sharp. It is this sharp break that is believed to be responsible for the 'exponential' break seen in the observed luminosity function. Thus, the characteristic luminosity, L^*, is not determined by the characteristic mass in the halo mass function, M^*, but by the characteristic mass, M_L, in the L-M relation.

We can use the above model to understand how the luminosity function of central galaxies may evolve with redshift. For a CDM-like spectrum, the slope of the halo mass function at the low-mass end does not change significantly with time, while the characteristic mass, M^*, decreases with increasing redshift. Thus, any evolution in the faint-end slope of the galaxy luminosity function must be due to a change of β with redshift. The characteristic mass, M_L, may also evolve with redshift, which may cause an evolution in the characteristic luminosity at which the luminosity function breaks. For example, if M_L decreases with increasing redshift slower than M^*, the characteristic luminosity, L^*, will be determined by M^* at high redshifts where $M_L > M^*$. Since $M^* \propto [D(z)]^{6/(n+3)} \propto (1+z)^{-6}$ (assuming $z \gg 1$ and $n = -2$) decreases rapidly with increasing redshift, L^* is therefore expected to decrease with increasing z. At lower redshifts, where $M_L < M^*$, the evolution of L^* is determined by the evolution of the halo mass scale M_L.

15.3.3 The Luminosity Function of Satellite Galaxies

Next, let us consider the luminosity distribution of satellite galaxies in a halo of a given mass M. In what follows we use a subscript 's' to denote the quantities of satellite galaxies. As discussed in §15.1, satellite galaxies are associated with dark matter subhalos. According to Eq. (15.4), this allows us to formally write the conditional luminosity function of satellite galaxies as

$$\Phi_s(L|M) = \int P(L|L_a) P(L_a|m_a, z_a) n_a(m_a, z_a|M) \, dm_a \, dL_a \, dz_a. \tag{15.40}$$

Here m_a and z_a are the subhalo mass and redshift at the time of accretion (i.e. when the halo first became a subhalo), $n_a(m_a, z_a|M)$ is the distribution function of subhalos with respect to the accretion mass and redshift, $P(L_a|m_a, z_a)$ is the probability for such a subhalo to host a central galaxy of luminosity L_a at the time of accretion, and $P(L|L_a)$ is the probability for a galaxy of luminosity L_a at z_a to evolve into a galaxy of luminosity L at the present time. To proceed, we start with a simple fiducial model based on the following two assumptions: (i) the stellar mass and luminosity of a galaxy do not change after it became a satellite, so that $P(L|L_a) = \delta(L-L_a)$; (ii) the luminosity–halo mass relation is independent of redshift, so that $P(L_a|m_a, z_a) = P(L_a|m_a)$. The effects of relaxing these assumptions will be discussed in §15.3.5. With these two assumptions we can write

$$\Phi_s(L|M) = \int P(L|m_a) n_a(m_a|M) \, dm_a. \tag{15.41}$$

In order to keep the problem simple, we model $n_a(m_a|M)$ with the unevolved subhalo mass function (see §7.5.3) given by Eq. (7.151).[1] We can then write the luminosity function of satellite galaxies in a halo of mass M as:

$$L\Phi_s(L|M) = A[y(L)]^{-p} \exp\{-[y(L)]^q\} \frac{d\ln y}{d\ln L}, \tag{15.42}$$

where $A \approx 0.345$, $p \approx 0.8$, $q \approx 3$, and $y = m_a/(fM)$ (with $f \approx 0.43$) is related to L through $L = L_c(m_a)$. As an illustration, we assume that $L_c(m_a)$ has the same form as given by Eq. (15.38). Thus, at $L \ll L_0$ we have $m_a \ll M_L$, so that

$$L\Phi_s(L|M) \approx \frac{A}{\beta} \left(\frac{fM}{M_L}\right)^p \left(\frac{L}{L_0}\right)^{-p/\beta}. \tag{15.43}$$

This implies a faint-end slope of $\alpha' = -1 - p/\beta$, which is close to -1 if β is large. The behavior of $\Phi_s(L|M)$ at the bright end depends on the mass of the halo, M. For massive halos with $fM \gg M_L$ we have

$$L \sim L_M f^\gamma y^\gamma, \tag{15.44}$$

where $L_M = L_0(M/M_L)^\gamma$. In this case,

$$L\Phi_s(L|M) = \frac{A}{\gamma} \left(\frac{L}{f^\gamma L_M}\right)^{-p/\gamma} \exp\left[-\left(\frac{L}{f^\gamma L_M}\right)^{q/\gamma}\right]. \tag{15.45}$$

As mentioned above, physical considerations suggest that γ is a small positive number, so that $\Phi_s(L|M)$ is expected to have a sharp break at $L_{bk} \sim f^\gamma L_M \sim L_M$. The value of the luminosity function at the break is $L_{bk}\Phi_s(L_{bk}|M) \sim A/(e\gamma)$, which is ~ 1 assuming the observed value of $\gamma \sim 0.3$ (Yang et al., 2009a). Since the *central* galaxy in the host halo has a luminosity $L \sim L_M$, there is no significant luminosity gap between the central galaxy and the brightest satellite galaxy. For low-mass halos with $fM \ll M_L$, we have

$$L \sim L_M f^\beta y^\beta, \tag{15.46}$$

where $L_M = L_0(M/M_L)^\beta$, and

$$L\Phi_s(L|M) = \frac{A}{\beta} \left(\frac{L}{f^\beta L_M}\right)^{-p/\beta} \exp\left[-\left(\frac{L}{f^\beta L_M}\right)^{q/\beta}\right]. \tag{15.47}$$

Since β is expected to be a large positive number, the break in the satellite luminosity function occurs at $L_{bk} \sim f^\beta L_M \ll L_M$. In this case, there is a large luminosity gap between the central

---

[1] Note, however, that this function only includes the population of subhalos that are *directly* accreted onto the main progenitor of the host halo and neglects the possibility that subhalos may themselves contain 'sub-subhalos', and sub-subhalos may contain 'sub-sub-subhalos' and so on. As demonstrated in Yang et al. (2009b), neglecting these higher-order subhalos is a valid assumption to first order.

galaxy and the brightest satellite galaxy. As $L_{\text{bk}}\Phi_{\text{s}}(L_{\text{bk}}|M) \sim A/(e\beta) \ll 1$, the conditional luminosity function of all (central plus satellite) galaxies is expected to show a prominent peak at the bright end, which is dominated by the central galaxies.

These predicted trends are in qualitative agreement with observations. Using a large catalogue of galaxy groups and clusters constructed from the SDSS, Yang et al. (2008) found that the conditional luminosity function of galaxies in groups with masses lower than $\sim 10^{13} h^{-1} M_\odot$ clearly shows a prominent peak at the bright end. This peak becomes successively weaker for more massive clusters. The luminosity gap, in terms of the difference in the magnitudes of the central galaxy and the brightest satellite galaxy, was found to become increasingly larger for lower-mass systems. The simple model considered here demonstrates that such luminosity gap can be produced naturally if L depends strongly on M for low-mass halos, as is required by the observed faint-end slope of the galaxy luminosity function.

The total luminosity function of satellite galaxies can be written as

$$\phi_{\text{s}}(L) = \int \Phi_{\text{s}}(L|M) n(M)\,\mathrm{d}M. \tag{15.48}$$

Let us first consider the behavior of this function at the faint end where $L \propto M^\beta$. Since $\Phi_{\text{s}}(L|M)$ declines rapidly at $L \gtrsim f^\beta L_M$, only halos with masses $M \gtrsim M_{\min}(L) = f^{-1} M^*(L/L_\beta^*)^{1/\beta}$ make a significant contribution to $\phi_{\text{s}}(L)$ at a given L. With the expression of $\Phi_{\text{s}}(L|M)$ given in Eq. (15.43) and the halo mass function given in Eq. (15.34), one can show that

$$L\phi_{\text{s}}(L) = \frac{A}{\beta}\left(\frac{fM^*}{M_L}\right)^p \frac{\bar{\rho}}{M^*}\sqrt{\frac{2}{\pi}}\left(1+\frac{n}{3}\right)\left(\frac{L}{L_0}\right)^{-p/\beta} \mathscr{I}_\beta(L/L_\beta^*), \tag{15.49}$$

where

$$\mathscr{I}_\beta(L/L_\beta^*) = \left(\frac{3}{n+3}\right) 2^a \Gamma(b+1, y_{\min}), \tag{15.50}$$

$$a = \frac{6p+n-3}{2(n+3)},\quad b = \frac{6p-n-9}{2(n+3)},\quad y_{\min} = \frac{1}{2}\left(\frac{L}{f^\beta L_\beta^*}\right)^{\frac{(n+3)}{3\beta}}. \tag{15.51}$$

If β is large and $n = -2$, $y_{\min}$ depends only weakly on L, and so $\phi_{\text{s}}(L) \propto L^{-1-p/\beta}$. For $p = 0.8$ the implied faint-end slope is similar to that of the central galaxies described in §15.3.2.

At the bright end where $L \propto M^\gamma$, the satellite luminosity function can be written in the same form as Eq. (15.49) with β replaced by γ. Since γ is expected to be much smaller than β, there is a break at $L \sim L_0$ where the slope of the luminosity function changes from $-1-p/\beta$ to $-1-p/\gamma$. For $L \gg f^\gamma L_\gamma^*$ (i.e. $y_{\min} \gg 1$), we have $\mathscr{I}_\gamma \sim [3/(n+3)] 2^a y_{\min}^b e^{-y_{\min}}$, and so $\phi_{\text{s}}(L)$ breaks exponentially at $L \gtrsim f^\gamma L_\gamma^*$.

15.3.4 Satellite Fractions

In the previous two subsections we investigated the luminosity functions of central and satellite galaxies, respectively. In order to develop some feeling for the relative contributions of centrals and satellites to the total luminosity function, consider the satellite fraction,

$$f_{\text{sat}}(L) = \frac{\phi_{\text{s}}(L)}{\phi_{\text{c}}(L) + \phi_{\text{s}}(L)} = \frac{F_{\text{s}}(L)}{1 + F_{\text{s}}(L)} \tag{15.52}$$

where $F_s(L) = \phi_s(L)/\phi_c(L)$ is the ratio between the satellite and central luminosity functions. At the faint end, we can use the results presented above to obtain

$$F_s(L) = \mathscr{I}_\beta A \left(\frac{fM^*}{M_L}\right)^p \left(\frac{L}{L_\beta^*}\right)^{(3-n)/(6\beta)} \left(\frac{L}{L_0}\right)^{-p/\beta}. \quad (15.53)$$

If $n = -2$ then $F_s \sim Af^p \mathscr{I}_\beta$. For a large value of β, $\mathscr{I}_\beta \sim 1.6$, and so $f_{\rm sat} \sim 0.22$ ($F_s \sim 0.28$) independent of L. Thus, the luminosity function at the faint end is expected to be dominated by central galaxies. At the bright end, using the asymptotic behavior of the incomplete gamma function leads to

$$F_s(L) = Af^p \left(\frac{3}{3+n}\right) 2^{a-b} \left(\frac{L}{f^\gamma L_\gamma^*}\right)^{b(n+3)/(3\gamma)} \sim \left(\frac{L}{L_\gamma^*}\right)^{-1.2}, \quad (15.54)$$

where we have used that $\gamma \sim 0.3$. Thus, the satellite fraction becomes negligible at the bright end ($L \gg L_{\gamma*}$). These simple estimates are in good agreement with the satellite fractions inferred from galaxy–galaxy lensing, galaxy clustering and galaxy group catalogues, all of which indicate that $f_{\rm sat}$ decreases from $\sim 0.35 \pm 0.1$ at $L \sim 10^9 h^{-2} L_\odot$ to $\sim 0.05 \pm 0.05$ at $L \sim 5 \times 10^{10} h^{-2} L_\odot$ (e.g. Mandelbaum et al., 2006b; Tinker et al., 2007; van den Bosch et al., 2007, 2008a; Yang et al., 2008). These studies also find that red galaxies have significantly higher satellite fractions than blue galaxies, reaching values of $\sim 0.6 \pm 0.1$ at $L \sim 10^9 h^{-2} L_\odot$. Thus, at all luminosities (at least above $\sim 10^9 h^{-2} L_\odot$), the total luminosity function is dominated by central galaxies. Only the faint end of the luminosity function of red galaxies may actually be dominated by satellite galaxies.

15.3.5 Discussion

The results obtained above are based on two naive assumptions: (i) the relation (15.38) between luminosity and halo mass is independent of redshift, and (ii) the luminosity of a satellite galaxy remains constant. None of these assumptions is expected to hold accurately. Even in the case of passive evolution (see §10.3.3), the luminosities of galaxies are expected to evolve with redshift, and it would be highly coincidental if galaxy luminosity and halo mass evolve in exactly the same way. Indeed, using halo occupation statistics to model the observed evolution of the stellar mass function and the normalization of the observed relation between star-formation rate and stellar mass to $z \sim 1$, Conroy & Wechsler (2009) have shown that the relation between luminosity and halo mass is likely to evolve with redshift (although only mildly so between $z = 0$ and $z = 1$). The second assumption is also unlikely to hold in detail. As discussed in §12.2, satellite galaxies may lose substantial amounts of mass due to tidal stripping. In addition, dynamical friction may cause (massive) satellites to sink towards the halo center, where they can be accreted by the central galaxy (see §12.3), and strangulation and ram-pressure stripping are expected to quench the star formation of satellite galaxies, causing their luminosities to evolve passively. Clear indications that the above two assumptions cannot both be correct comes from a study by Yang et al. (2009b), who have shown that the conditional luminosity function for satellite galaxies predicted under these two assumptions [i.e. Eq. (15.42)] is in clear disagreement with the $\Phi_s(L|M)$ obtained directly from a large galaxy group catalogue. Yang et al. (2009b) argued that most likely a significant fraction of satellite galaxies is disrupted by tidal forces, giving rise to a significant stellar halo component, in quantitative agreement with the observational results obtained by Gonzalez et al. (2007).

15.4 Linking Halo Mass to Star-Formation History

Having linked galaxy luminosity to halo mass, we now proceed to link the build-up of galaxy luminosity (i.e. the star-formation history) to halo mass. After constraining the present-day specific star-formation rates of galaxies as function of halo mass using the color–magnitude relation, we describe the global star-formation history of the Universe.

15.4.1 The Color Distribution of Galaxies

In §2.4.3 we have seen that brighter galaxies in the local Universe on average have redder colors. Furthermore, for a given luminosity (or stellar mass), the color distribution is bimodal, with a relatively well defined red sequence and a broader blue sequence usually referred to as the blue cloud (see Fig. 2.27). Recent observations have shown that this color bimodality in the galaxy population persists at least out to $z \simeq 1$ (e.g. Bell et al., 2004; Willmer et al., 2006; Cooper et al., 2007), and that the total stellar mass density of galaxies on the red sequence has roughly doubled over the last 6–8 Gyr while that of blue galaxies has remained roughly constant (e.g. Bell et al., 2004; Borch et al., 2006; Faber et al., 2007). Since new stars form primarily in blue galaxies, this suggests that galaxies are being transformed from the blue to the red population. In this section, we give a very crude but insightful description of how different processes in galaxy formation and evolution may give rise to the observed color distribution of galaxies.

As seen in §10.3.2, the color of a galaxy is determined mainly by its star-formation history, assuming dust extinction is properly taken into account. Based on the spectral synthesis results shown in Fig. 10.6, the time required for a galaxy to move from the blue cloud to the red sequence after its star formation has been truncated is roughly 2 Gyr, which is short compared to the typical age of the stellar populations of present day galaxies. In addition, a galaxy with a current specific star-formation rate as low as $0.1/t_0$ (with t_0 the age of the Universe) will still have blue colors. Thus, by and large the color distribution of galaxies reflects a distribution in their current specific star-formation rates. Because of this, our discussion of galaxy colors will be based on the following parameter:

$$S \equiv \frac{\dot{M}_\star}{L} = \Upsilon_\star \frac{\dot{M}_\star}{M_\star}. \tag{15.55}$$

Here $M_\star$ is the stellar mass, $\dot{M}_\star$ is the current star-formation rate, and $\Upsilon_\star$ is the stellar mass-to-light ratio. In what follows we assume that all galaxies have the same $\Upsilon_\star$ at all times, so that S is directly proportional to the specific star-formation rate. Although clearly unrealistic, this oversimplification does not invalidate any of the quantitative results presented below.

(a) Central Galaxies Let us first consider star formation in the central galaxies of dark matter halos. As described above, the relation between the luminosity of a central galaxy and the mass of its host halo can roughly be described by Eq. (15.38). In order to model the specific star-formation rate, S, we also need a model for the star-formation rates in central galaxies. Considerations based on the radiative cooling efficiency of hot gas in dark matter halos described in §8.4, and on the energy feedback from AGNs described in §14.4, suggest that star formation is suppressed in halos with masses exceeding $\sim 10^{13} h^{-1} \rm M_\odot$. In addition, central galaxies are only expected to cannibalize gas-rich satellite galaxies, which may lead to the formation and presence of young stars in the central galaxy, in halos with $M \lesssim 10^{13} h^{-1} \rm M_\odot$; in more massive halos the dynamical friction times are simply too long (see §12.3). Based on these considerations, we consider a simple model in which the star-formation rate is related to the halo mass through

$$R \equiv \frac{\dot{M}_\star}{M} = R_0(z) \frac{X_R^{\beta'}}{1 + X_R^{\beta'-\gamma'}}, \tag{15.56}$$

15.4 Linking Halo Mass to Star-Formation History

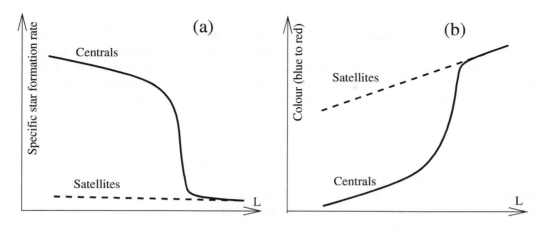

Fig. 15.3. (a) An illustration showing the specific star-formation rate as a function of luminosity. The solid curve is for central galaxies and the dashed curve for satellite galaxies. (b) The expected color–luminosity relation corresponding to the specific star-formation rates shown in panel (a). Note that the luminosity dependence of color can be enhanced by the fact that more massive galaxies are metal richer and therefore appear redder.

where $X_R \equiv M/M_R$ with $M_R \sim 10^{12-13} h^{-1} \, \mathrm{M}_\odot$ a characteristic halo mass above which star formation is heavily suppressed (i.e. γ' is expected to be a large negative number).[2] In the above expression, R_0 represents an overall normalization, which we set to be $R_0(z) \propto H(z)$.

Inserting Eqs. (15.56) and (15.38) into Eq. (15.55) we obtain, for central galaxies only,

$$S = \frac{MR_0}{L_0} \frac{X_R^{\beta'}}{1 + X_R^{\beta'-\gamma'}} \frac{1 + X_L^{\beta-\gamma}}{X_L^\beta}. \tag{15.57}$$

For $M \ll \min(M_R, M_L)$, we have

$$S = \frac{R_0}{L_0} \frac{M_L^\beta}{M_R^{\beta'}} M^{\beta'+1-\beta}. \tag{15.58}$$

Thus, in low-mass halos the specific star-formation rate of central galaxies increases (decreases) with halo mass if $\beta' > \beta + 1$ ($\beta' < \beta + 1$). For halos with $M \gg \max(M_R, M_L)$, we have

$$S = \frac{R_0}{L_0} \frac{M_L^\gamma}{M_R^{\gamma'}} M^{\gamma'-\gamma+1}. \tag{15.59}$$

Recall that $\gamma > 0$. Thus, if γ' is a large negative number, then a sharp break of S is expected at $M \sim M_R$, implying that the central galaxies in halos of $M > M_R$ have very low specific star-formation rates. As an illustration, the solid curve in Fig. 15.3a shows the expected behavior of S versus L, where we have assumed that $\beta' < \beta + 1$. The sharp break occurs at a luminosity corresponding to that of the central galaxy in a halo of mass M_R.

(b) Satellite Galaxies Next consider the present specific star-formation rate for a galaxy that became a satellite at $z = z_a$. Formally we can write

$$S(z_0) = T(z_0, z_a) S(z_a), \tag{15.60}$$

---

[2] In reality, we also expect R to be truncated at low-mass halos due to photoionization heating (see §8.1.4). Since we are not concerned with dwarf galaxies in this section, this aspect is ignored here.

where $S(z_a)$ is the specific star-formation rate of the galaxy at the time of accretion and $T(z_0, z_a)$ describes the change in the specific star-formation rate after the galaxy became a satellite. As discussed in §12.5 satellite galaxies are exposed to tidal stripping, strangulation and ram-pressure stripping, all of which cause star-formation quenching. Hence, in general we expect that $T < 1$. For simplicity, we assume that the various star-formation quenching mechanisms cause, on average, an exponential decline in the star-formation rate of satellite galaxies, i.e.

$$T(z_0, z_a) = \exp\left(-\frac{t_0 - t_a}{\tau}\right), \tag{15.61}$$

where τ is the star-formation quenching time scale, t_a is the time of accretion, and t_0 is the time corresponding to z_0. If $\tau \ll t_0 - t_a$ then the satellite galaxy is expected to have its star formation quenched (i.e. $S \sim 0$). On the other hand, if $\tau \gg t_0 - t_a$ the satellite will have roughly the same specific star-formation rate as a central galaxy of the same luminosity. If the quenching mechanisms mentioned above are efficient, we expect a satellite to be stripped of its gas (or at least a significant fraction thereof) on a time scale that is comparable to the dynamical time of its host halo, which at $z = 0$ ($z = 1$) is $\sim 2\,\mathrm{Gyr}$ ($\sim 0.7\,\mathrm{Gyr}$). Thus, except for the satellites accreted recently, we expect them to have specific star-formation rates substantially lower than for central galaxies of the same luminosities. This is illustrated by the dashed curve in Fig. 15.3(a).

The existence of a truncation halo mass for star formation in central galaxies combined with one or more star-formation quenching mechanisms for satellites is the key concept in our understanding of the color–magnitude distribution of galaxies. It is important to realize, though, that the colors of galaxies are not determined entirely by their (current) specific star-formation rates; metallicity also plays an important role. Since more massive galaxies have higher metallicities (see §2.4.4), and since metal-rich stars are redder than metal-poor stars (see §10.3.5), the dependence of color on luminosity is enhanced relative to the effect due to the specific star-formation rate alone. Hence, under the assumptions made, the color–luminosity relations for central and satellite galaxies are expected to roughly resemble the solid and dashed curves in Fig. 15.3(b), respectively.

(c) Additional Considerations Using galaxy group catalogues, van den Bosch et al. (2008a) have tested some of these ideas by comparing the colors of central and satellite galaxies of the same stellar mass. Indeed, satellite galaxies are found to be significantly redder than their central counterparts, but not by an awful lot. Furthermore, a significant fraction of satellite galaxies are still blue, especially at the low-mass end. Both these results indicate that star-formation quenching in satellite galaxies takes place on relatively long time scales (see also Baldry et al., 2006; Weinmann et al., 2006a, 2009; Kang & van den Bosch, 2008). This advocates strangulation (see §12.5.4) as the most likely quenching mechanism, as it implies a star-formation quenching time scale equal to the gas consumption time scale, $M_{\mathrm{cold}}/\dot{M}_\star$, which can be significantly longer than the dynamical time (see §9.3). Ram-pressure stripping and tidal stripping may also play a role, but are apparently unable to strip a satellite galaxy of its *entire* cold gas reservoir on a dynamical time.

In summary, the most massive central galaxies are expected to be on the red sequence because of a (currently not yet well understood) star-formation truncation mechanism, which is likely to be connected to a characteristic halo mass, $M_R \sim 10^{12-13}h^{-1}\,\mathrm{M}_\odot$. With the exception of satellite galaxies accreted recently (within the last $\sim 2\,\mathrm{Gyr}$), most satellite galaxies are also expected to be on the red sequence due to one or more star-formation quenching mechanisms (most likely dominated by strangulation). Most central galaxies in halos with $M < M_R$ are expected to be forming stars actively, and thus to have blue colors. There are, however, also central galaxies in low mass halos that are red (e.g. van den Bosch et al., 2008a). Those that are relatively close to a nearby more massive halo (roughly within three virial radii) may actually have passed through their massive neighbors, and thus have experienced the same quenching mechanisms that make the satellite galaxies red (Wang et al., 2008c). Those that are not associated with a nearby more massive halo may have experienced a major merger which has used up all gas in a starburst and/or expelled the remaining gas due to feedback processes (e.g. Springel et al., 2005a; Hopkins et al., 2006). This is supported by the fact that most red centrals are ellipticals. Thus, at least qualitatively, our current framework of galaxy formation seems able to explain the observed color–magnitude distribution of the present-day galaxy population. However, in order to make more quantitative predictions, one has to model all the relevant processes in detail. In §15.7 we will see how to achieve this goal with the use of numerical simulations and semi-analytical modeling.

15.4.2 Origin of the Cosmic Star-Formation History

As shown in §2.6.8, the global star-formation history of the Universe can be characterized by a global quantity, $\dot{\rho}_\star(z)$, which describes the average amount of mass that turns into stars per unit time per unit comoving volume at redshift z. Observations of star-forming galaxies have been used to estimate $\dot{\rho}_\star(z)$ over a wide range of redshifts, from $z \sim 0$ to $z \sim 6$ (see Fig. 2.35).

In order to understand the redshift dependence of the global star-formation history, we once again try to link (specific) star-formation rates to dark matter halos. Since star formation in satellite galaxies is expected to be suppressed, as described above (see also §15.5.1), we simplify the problem by assuming that star formation only occurs in central galaxies. We can then formally write

$$\dot{\rho}_\star(z) = \int \langle \dot{M}_\star \rangle (M,z) n(M,z) \, \mathrm{d}M, \qquad (15.62)$$

where $\langle \dot{M}_\star \rangle(M,z)$ is the mean star-formation rate in the central galaxies of halos of mass M at redshift z. As described above, physical considerations suggest that $\langle \dot{M}_\star \rangle(M,z)$ may be approximated as $\langle \dot{M}_\star \rangle(M,z) = MR(M,z)$, with R given by Eq. (15.56). Assuming a power-law spectrum, $P(k) \propto k^n$, we can then write

$$\dot{\rho}_\star(z) = \sqrt{\frac{2}{\pi}} \overline{\rho}_0 R_0(z) \int_{\nu_{\min}}^{\infty} \frac{(\nu/\nu_R)^{\beta'_n}}{1+(\nu/\nu_R)^{\beta'_n-\gamma'_n}} \exp\left(-\frac{\nu^2}{2}\right) \mathrm{d}\nu, \qquad (15.63)$$

where $\overline{\rho}_0$ is the average comoving density of the Universe, $\beta'_n = 6\beta'/(n+3)$, $\gamma'_n = 6\gamma'/(n+3)$, $\nu = [M/M^*(z)]^{(n+3)/6}$, and $\nu_R = [M_R(z)/M^*(z)]^{(n+3)/6}$. In the above expression, we have included a minimum halo mass, $M_{\min}$ (through $\nu_{\min}$), for significant star formation, to mimic the suppression of star formation by photoionization heating (see §8.1.4) in low-mass halos with $M < M_{\min}$.

In what follows we assume that $M_R(z)$ does not evolve much with redshift. Hence, at high redshifts $M_R(z) \gg M^*(z)$ (i.e. $\nu_R \gg 1$) so that Eq. (15.63) reduces to

$$\dot{\rho}_\star(z) \approx \frac{2^{\beta'_n/2}}{\sqrt{\pi}} \overline{\rho}_0 R_0(z) \nu_R^{-\beta'_n} \Gamma\left(\frac{\beta'_n+1}{2}, \frac{\nu_{\min}}{2}\right), \qquad (15.64)$$

where $\Gamma(x,y)$ is the incomplete gamma function. At sufficiently high redshifts, where $\nu_{\min} \gg 1$, we can use the properties of the incomplete gamma function to write

$$\dot{\rho}_\star(z) \approx \sqrt{\frac{2}{\pi}} \overline{\rho}_0 R_0(z) \nu_R^{-\beta'_n} \nu_{\min}^{\beta'_n-1} \exp\left(-\frac{\nu_{\min}^2}{2}\right). \qquad (15.65)$$

In the presence of photoionization, halos with virial temperatures below $\sim 10^{4.5}$ K are not able to trap gas to form stars. In this case, the minimum halo mass for star formation may be taken as $M_{\min} \sim 10^{10} h^{-1} \mathrm{M}_\odot / H(z)$, so that $\nu_{\min} \propto [H(z)]^{-1/6}[D(z)]^{-1} \propto (1+z)^{3/4}$ at $z \gg 1$ (assuming

$n = -2$). Therefore, an exponential decline of $\dot{\rho}_\star(z)$ is expected at high z if the Universe was already re-ionized by photoionization at such redshift. This decline is expected to set in roughly at $\nu_{\min} \sim \sqrt{2}$, which corresponds to a redshift of about 4, assuming $M_0^* = 10^{13} h^{-1} \, \mathrm{M}_\odot$.

At lower redshifts, $\nu_{\min} \ll 1$ and Eq. (15.63) reduces to

$$\dot{\rho}_\star(z) \approx \sqrt{\frac{2}{\pi}} \overline{\rho}_0 R_0(z) \nu_R \int_0^\infty \frac{y^{\beta_n'}}{1 + y^{\beta_n' - \gamma_n'}} \exp\left(-\frac{\nu_R^2}{2} y^2\right) dy. \qquad (15.66)$$

When ν_R is large, the contribution to the integration is dominated by $y \ll 1$, and we can write

$$\dot{\rho}_\star(z) \approx \sqrt{\frac{2}{\pi}} \overline{\rho}_0 R_0(z) \nu_R \int_0^\infty y^{\beta_n'} \exp\left(-\frac{\nu_R^2}{2} y^2\right) dy$$

$$\approx \frac{2^{\beta_n'/2}}{\sqrt{\pi}} \overline{\rho}_0 R_0(z) \nu_R^{-\beta_n'} \Gamma\left(\frac{\beta_n' + 1}{2}\right). \qquad (15.67)$$

Assuming M_R is independent of z, we have $\nu_R \propto [D(z)]^{-1}$. Thus, if $\beta' > 0$, so that $\beta_n' = 6\beta'/(n+3) \sim 6\beta'$ is a positive number, a power-law decline of $\dot{\rho}_\star(z)$ with increasing redshift is expected to occur at $\nu_R > 1$. With $M_R \sim 10^{12} h^{-1} \, \mathrm{M}_\odot$, $\nu_R = 1$ corresponds to $z \sim 2$ in the ΛCDM model, implying that the power-law decline starts at $z \sim 2$. Finally let us consider the behavior of $\dot{\rho}_\star(z)$ at low-z when $\nu_R \ll 1$. If $\gamma_n' < -1$ (i.e. $\gamma \lesssim -1/6$ assuming $n = -2$), then the exponential term in the integration in Eq. (15.66) can be replaced by 1, so that

$$\dot{\rho}_\star(z) \approx \sqrt{\frac{2}{\pi}} \overline{\rho}_0 R_0(z) \nu_R \int_0^\infty \frac{y^{\beta_n'}}{1 + y^{\beta_n' - \gamma_n'}} dy. \qquad (15.68)$$

In this case, $\dot{\rho}_\star(z) \propto R_0(z) \nu_R \propto H(z)[D(z)]^{-1}$ (for a M_R that is independent of z), which increases with increasing redshift.

To summarize, the standard ΛCDM model, together with the simple star-formation model considered here, leads to a global star-formation history with the following properties: the star-formation density, $\dot{\rho}_\star(z)$, increases with redshift at low z, reaches a maximum where $\nu_R \sim 1$ ($z \sim 2$), then declines with z as a power law before breaking exponentially at a redshift $z \sim 4$ corresponding to $\nu_{\min} \sim 1$. These properties are qualitatively consistent with the observed cosmic star-formation history shown in Fig. 2.35, suggesting that, within the CDM scenario of galaxy formation, we have a reasonable understanding of the origin of the cosmic star-formation history.

15.5 Environmental Dependence

As we have seen in §2.7, on large scales galaxies are distributed in a complex web of filaments and sheets surrounding large empty voids (see Fig. 2.36), so that galaxies can reside in a variety of environments. The properties of galaxies are correlated with their environments, as is evident from, among others, the morphology–density relation (see §2.4.5) and the color–luminosity dependence of the galaxy correlation function (see Fig. 6.4). In this section we take a closer look at the environment dependence of galaxy formation and evolution.

There are numerous parameters that can be used to characterize the environment of a galaxy. However, since galaxies are believed to form and reside in dark matter halos, it is convenient to separate these parameters into two categories: those related to properties of the host halo of a galaxy (e.g. halo mass, halo spin, halo shape), and those describing the environment on superhalo scales (e.g. mass of filament or sheet in which the host halo of a galaxy is embedded, or the mass overdensity on scales significantly larger than individual halos). As seen in the previous chapters, in our current paradigm galaxy properties are expected to depend strongly on the properties of the host halos. For example, more massive halos are expected to host more massive galaxies, while the spin of a halo is expected to control the size of the disk galaxy it hosts. In addition, the efficiency of many important processes for galaxy formation, such as cooling, star formation and feedback, are expected to scale with halo mass. Whether or not galaxy properties depend on environmental properties on superhalo scales is less clear. The dynamical times on these scales are typically longer than or comparable to the Hubble time, indicating that there has not been enough time to induce a direct environmental dependence on superhalo scales by gravitational processes alone. However, it could still be possible for non-gravitational processes to introduce some environment dependence on large scales.

Before we discuss the environmental impact on galaxy formation and evolution in more detail, it is important to realize that a correlation between galaxy properties and a particular environmental property does not necessarily imply a causal connection. Let $P(\mathscr{G}|\mathscr{H})$ be the probability distribution of galaxy property $\mathscr{G}$ given environmental property $\mathscr{H}$. In general, we can say that $\mathscr{G}$ is independent of $\mathscr{H}$ if $P(\mathscr{G}|\mathscr{H}) = P(\mathscr{G})$, i.e. if the conditional on $\mathscr{H}$ does not affect the probability distribution of $\mathscr{G}$. Note, however, that the inverse is not true: $P(\mathscr{G}|\mathscr{H}) \neq P(\mathscr{G})$ does not necessarily mean causal dependence of $\mathscr{G}$ on $\mathscr{H}$. To see this, consider two halo properties, $\mathscr{H}_1$ and $\mathscr{H}_2$. Suppose that $\mathscr{G}$ depends causally on $\mathscr{H}_1$, but not at all on $\mathscr{H}_2$. Then $P(\mathscr{G}|\mathscr{H}_1, \mathscr{H}_2) = P(\mathscr{G}|\mathscr{H}_1)$, and we have that

$$\frac{P(\mathscr{G}|\mathscr{H}_2)}{P(\mathscr{G})} = \frac{\int P(\mathscr{G}|\mathscr{H}_1) P(\mathscr{H}_1|\mathscr{H}_2) \, d\mathscr{H}_1}{\int P(\mathscr{G}|\mathscr{H}_1) P(\mathscr{H}_1) \, d\mathscr{H}_1}. \tag{15.69}$$

This is only equal to unity if $\mathscr{H}_1$ and $\mathscr{H}_2$ are independent of each other, so that

$$P(\mathscr{H}_1|\mathscr{H}_2) = P(\mathscr{H}_1, \mathscr{H}_2)/P(\mathscr{H}_2) = P(\mathscr{H}_1).$$

Otherwise $P(\mathscr{G}|\mathscr{H}_2) \neq P(\mathscr{G})$ even if there is no causal connection between $\mathscr{G}$ and $\mathscr{H}_2$; the correlation between $\mathscr{G}$ and $\mathscr{H}_2$ is entirely due to the correlation between $\mathscr{G}$ and $\mathscr{H}_1$ combined with the correlation between $\mathscr{H}_1$ and $\mathscr{H}_2$. This demonstrates that, in order to investigate whether the properties of galaxies causally depend on some environmental property $\mathscr{H}'$, one needs to use all other environmental properties that are correlated with $\mathscr{H}'$ as control variables. In general, this is not an easy task, as one is not always aware of all these environmental properties, and as not all environmental properties are readily accessible observationally. Thus, while it is in principle straightforward to investigate the correlation of some galaxy property with some environmental property, the interpretation of such a correlation in terms of a causal, physical connection can be extremely complicated.

15.5.1 Effects within Dark Matter Halos

Let us first consider the environmental dependence of the galaxy population within dark matter halos, ignoring any effects on superhalo scales. Here it is useful to once again consider central galaxies and satellite galaxies separately, simply because they are expected to be influenced by their environments in different ways. In what follows we discuss both in turn.

(a) Central Galaxies As mentioned above, the properties of central galaxies are expected to strongly depend on the mass of the halo in which they reside. First and foremost, brighter, more massive centrals reside in more massive halos. This is not only expected from the simple fact that more massive halos form from larger Lagrangian volumes, which thus contain more baryonic material, but is also supported by a large range of observations, including gravitational lensing (e.g. McKay et al., 2001; Guzik & Seljak, 2002; Hoekstra et al., 2004; Sheldon et al., 2004; Mandelbaum et al., 2006b; Cacciato et al., 2009) and satellite kinematics (Brainerd & Specian, 2003; Prada et al., 2003; van den Bosch et al., 2004; Conroy et al., 2007; More et al.,

2009). On average, central galaxies in more massive halos are also redder and more centrally concentrated. As discussed in the previous section, the colors are related to both metallicity and star-formation history. Since galactic winds are less likely to develop in more massive halos owing to their deeper potential wells (see §8.6.3), more massive halos are more likely to retain their metals, hence to have redder centrals. In addition, the progenitors of more massive halos start forming earlier (although they may assemble later, see §7.3.4), so that their central galaxies are expected to have older stellar populations. And finally, as mentioned in §15.4, cooling and AGN feedback may suppress star formation in halos with masses exceeding a characteristic mass of $M_R \sim 10^{12-13} h^{-1} \, M_\odot$. The combination of these three effects are believed to be responsible for the fact that more massive halos host redder centrals.

But why do central galaxies in more massive halos tend to be more centrally concentrated, or equivalently, why are they more likely to be ellipticals? In the standard paradigm most, if not all, galaxies are believed to form as disk galaxies, which can then be transformed into ellipticals due to major mergers (see §13.2.2). Subsequently, a new disk may form around the spheroidal merger remnant if new gas is able to cool, giving rise to a disk–bulge system (e.g. Kauffmann et al., 1993; Baugh et al., 1996). This suggests two possible explanations for the observed trend of morphology with halo mass: either more massive halos experience more major mergers, or in more massive halos no gas is able to cool after a major merger. The former is inconsistent with a CDM cosmology, in which halos of all masses are predicted to have (average) merger histories that are very similar in terms of the mass ratios of their progenitors (see §7.3).[3] The latter explanation is consistent with the existence of a characteristic mass, M_R, required to explain why almost all central galaxies in massive halos are red. As discussed above, the existence of such a characteristic mass is believed to be associated with the onset of AGN feedback which suppresses cooling in massive halos. Such feedback is invoked not only to explain both the morphologies and the colors of central galaxies in massive halos, but also to explain the cut-off in the stellar mass function at the high-mass end (see §15.3; see also Bower et al., 2006; Cattaneo et al., 2006; Croton et al., 2006; Kang et al., 2006; Somerville et al., 2008b). However, many uncertainties still remain as far as the details of this feedback mechanism are concerned (see discussion in §14.4).

In addition to halo mass, the spin and shape of dark matter halos are also expected to have some impact on the properties of central galaxies. Unfortunately, since the spin and shape of dark matter halos are not easily accessible observationally, relatively little is known about this aspect of the 'environment dependence'. An interesting exception is the observational indication that central galaxies tend to align with the major axes of their dark matter halos (Hoekstra et al., 2004; Mandelbaum et al., 2006a; Wang et al., 2008a), suggesting that the properties of central galaxies are indeed affected by other halo properties in addition to mass.

Finally, it is important to keep in mind that the central regions of dark matter halos are special environments in themselves. They are the repositories of low angular momentum material and of massive objects (satellite galaxies and/or supermassive black holes) brought to the center of the potential well by dynamical friction. Consequently, central galaxies are exposed to a number of processes that satellite galaxies are not, such as galactic cannibalism (see §12.5.2) and cooling flows (see §8.8.1), which can also have a substantial impact on their evolution.

(b) Satellite Galaxies As discussed in detail in §12.5, satellite galaxies are susceptible to a number of environmental processes, such as harassment, ram-pressure stripping, tidal stripping, dynamical friction, and strangulation, that only operate on satellite galaxies. All these processes are expected to suppress or quench star formation (see discussion in §15.4 above) and/or to transform disk components into more spheroidal morphologies.

---

[3] The main difference between massive and low-mass halos is in assembly *times*, with more massive halos assembling later.

The impact of these processes on the properties of galaxies can be investigated by comparing the properties of satellite and central galaxies, and by studying the properties of satellite galaxies located in halos of different masses and at different halo-centric radii. Such analysis has been carried out recently with the use of large samples of galaxy groups and clusters constructed from the SDSS. It is found that, on average, satellite galaxies are more massive, more centrally concentrated and redder in more massive halos and/or at smaller halo-centric radii (e.g. Weinmann et al., 2006b). Although suggestive, it is unclear whether these trends are a direct, causal result of the satellite-specific environmental effects described above. For instance, they could simply be a reflection of the fact that more massive galaxies (including central galaxies) in general are redder and more concentrated, combined with the fact that more massive halos host, on average, more massive satellite galaxies.[4] To further test whether the properties of satellite galaxies are influenced by environmental processes, van den Bosch et al. (2008a) compared the colors and concentrations of satellites galaxies to those of central galaxies of the same stellar mass, adopting the hypothesis that the latter are the progenitors of the former. They found that satellite galaxies are on average redder and more concentrated than central galaxies of the same stellar mass, indicating that satellite-specific transformation processes do operate (see also Weinmann et al., 2009). They also found that the color and concentration differences of central–satellite pairs matched in stellar mass are quite independent of the host halo mass of the satellite galaxy, indicating that satellite-specific transformation mechanisms are equally efficient in halos of different masses. This rules against mechanisms that operate only in very massive halos, such as ram-pressure stripping and harassment, suggesting that strangulation and/or tidal interactions are the dominant transformation mechanisms. As discussed in §15.4, the same conclusion can be derived from the fact that a significant fraction of satellite galaxies still have blue colors indicative of ongoing star formation.

Can strangulation also explain the fact that satellite galaxies are somewhat more concentrated than central galaxies of the same stellar mass? At first sight, the answer seems to be no, because strangulation refers only to stripping of the gaseous halo of a satellite galaxy. However, as shown by van den Bosch et al. (2008a), central–satellite pairs that are matched in both stellar mass and color show no difference in average concentration. This suggests that one and the same mechanism may well explain both the color and concentration differences between centrals and satellites. Indeed, as shown by Weinmann et al. (2009), all observed differences between satellite and central galaxies can be explained by a simple fading model, in which the star formation in the disk decreases over a time scale of 2–3 Gyr after a galaxy becomes a satellite. This not only causes an overall reddening of the galaxy, but the light distribution also becomes more centrally concentrated owing to the presence of color gradients (see §11.7.2) and the fact that blue (young) stellar populations redden faster than red (old) stellar populations.

15.5.2 Effects on Large Scales

We now turn our attention to environment dependence on superhalo scales. A convenient way to parameterize this environment is via the mass overdensity, $\delta_R = M/\overline{M} - 1$, where M is the mass within a volume $V = 4\pi R^3/3$ and $\overline{M} = \overline{\rho}V$, with $\overline{\rho}$ the average density of the Universe. Observationally, it is easier to measure the overdensity in luminosity rather than mass. In what follows we assume that these are equivalent to each other, which is a reasonable assumption as long as R is sufficiently large.

A number of studies have measured galaxy luminosity functions as function of δ_R, $\Phi(L|\delta_R)$, using $5h^{-1}\,\mathrm{Mpc} \lesssim R \lesssim 10h^{-1}\,\mathrm{Mpc}$. The results show that (i) $\Phi(L|\delta_R)$ is well fit by a Schechter

---

[4] The latter is simply a consequence of the way halos assemble, combined with the fact that central galaxies display a relatively tight relation between halo mass and stellar mass.

function, (ii) the characteristic luminosity, L^*, brightens with increasing δ_R, and (iii) the faint-end slope becomes significantly steeper with increasing δ_R for early-type galaxies, but is roughly independent of δ_R for late-type galaxies (Bromley et al., 1998; Hütsi et al., 2002; Croton et al., 2005). At first sight these results seem to suggest that some environmental processes are at work on superhalo scales. However, as shown in §7.4, regions of different mass overdensity δ_R also have a different halo mass function: at any given time halos of higher masses are biased towards higher density environments. Thus, any dependence of galaxy property, $\mathscr{G}$, on halo mass, M, will also induce a dependence on δ_R via

$$P(\mathscr{G}|\delta_R) = V \int P(\mathscr{G}|M) n(M|\delta_R) \, dM, \qquad (15.70)$$

where $n(M|\delta_R)$ is the conditional mass function of dark matter halos, which gives the number density of halos as a function of halo mass in a large-scale environment with mass overdensity δ_R. For a given model of structure formation, $n(M|\delta_R)$ can be obtained through the halo bias model described in §7.4 or from cosmological N-body simulations. As shown by Mo et al. (2004), the entire δ_R-dependence of the galaxy luminosity function can be accurately reproduced in a model where galaxy luminosity and type depend only on halo mass as described by Eq. (15.70). This indicates that the environmental dependence of the galaxy population on superhalo scales is mainly a consequence of halo bias, rather than due to causal environmental effects operating on superhalo scales. This is in quantitative agreement with Kauffmann et al. (2004), who have shown that the dependence of star formation activity on the density on scales $R > 1 h^{-1} \mathrm{Mpc}$ largely disappears once the density on smaller scales is specified (see also Blanton et al., 2006; Blanton & Berlind, 2007, for similar results).

In order to examine any environmental effects beyond the modulation of the halo mass function by large-scale structure, one has to compare the properties of galaxies in halos of the same mass but residing in different large-scale environments (i.e. one has to use halo mass as a control variable). One way to do this is to use a galaxy group catalogue and to shuffle the positions of galaxies among groups (halos) of similar masses. This changes the large-scale environment of the galaxies while leaving $P(\mathscr{G}|M)$ invariant. If this shuffling modifies the two-point correlation function of galaxies of property $\mathscr{G}$, then $\mathscr{G}$ must depend on its large-scale environment. Such a test applied to the SDSS by Blanton & Berlind (2007) has shown that groups with red centrals are more strongly clustered (and thus reside in denser large-scale environments) than groups *of the same mass* but with blue centrals. Similar results were obtained by Yang et al. (2006) and Wang et al. (2008b) using the cross-correlation function between groups and galaxies. In addition, these authors also found that the color dependence is more prominent in less massive groups and becomes insignificant in groups with $M \gtrsim 10^{14} h^{-1} \mathrm{M}_\odot$.

These observational trends seem to indicate that there is some dependence of galaxy properties on large-scale environment that is not accounted for by halo-mass dependence alone. This result is usually interpreted as a consequence of the environmental dependence of halo assembly time, i.e. the halo assembly bias described in §7.4.2. Since halos, especially low-mass ones, in higher-density environments on average have earlier assembly times, a correlation between the color of a galaxy and the assembly time of its host halo would lead to environmental dependence similar to what is observed. Note again that no environmental process operating on superhalo scales is evoked here. Rather, it is due to a dependence of color on halo assembly time (a halo property), combined with a correlation between large-scale environment (i.e. δ_R) and halo assembly time. However, this interpretation is not unique. A similar trend could result if star formation in a halo is affected by some processes operating on superhalo scales, such as the radiation and feedback from galaxies in neighboring dark matter halos (e.g. Shapiro et al., 2004; Iliev et al., 2005). Although such effects may only be effective for low-mass halos at high redshifts, it is premature to rule out that similar processes also play a role at later times for more massive halos.

15.6 Spatial Clustering and Galaxy Bias

In §2.7 we have seen that galaxies are clustered in space, and in §6.5 we have described how to quantify such clustering using various statistical measures. In this section we examine how the clustering properties of galaxies are related to the formation and evolution of the galaxy population. Since galaxies are believed to reside in dark matter halos, and since the clustering properties of dark matter halos can be modeled reliably (see §7.4), it is instructive to describe the properties of galaxy clustering in terms of the halo clustering and the distribution of galaxies in dark matter halos.

In the spirit of the halo model discussed in §7.6, we split the two-point correlation function of galaxies into a one-halo term and a two-halo term. Suppose that, for a halo of mass M, the number of galaxies contained in it (the occupation number) is N, and the (average) spatial distribution of galaxies in a halo of mass M is given by a normalized function $u(\mathbf{x}|M)$. It is then easy to show that the average number of galaxy pairs, separated by $\mathbf{r} = \mathbf{x}_2 - \mathbf{x}_1$, within such a halo is

$$Nu(\mathbf{x}_1 - \mathbf{x}_0|M)\,(N-1)u(\mathbf{x}_2 - \mathbf{x}_0|M)\,\mathrm{d}V_1\,\mathrm{d}V_2,$$

where $\mathbf{x}_0$ is the position of the halo center, $\mathrm{d}V_i = \mathrm{d}^3\mathbf{x}_i$ is the volume element at $\mathbf{x}_i$. Since pairs are specified by a vector $\mathbf{r}$, there is no double counting. Thus, the average number of pairs separated by $\mathbf{r}$ per unit volume in a halo of mass M which hosts N galaxies is

$$\begin{aligned}GG^{1h}(\mathbf{r}|M,N) &= \int Nu(\mathbf{x}_1-\mathbf{x}_0|M)\,(N-1)u(\mathbf{x}_2-\mathbf{x}_0|M)\,\mathrm{d}^3\mathbf{x}_0 \\ &= \int Nu(\mathbf{x}|M)\,(N-1)u(\mathbf{x}+\mathbf{r}|M)\,\mathrm{d}^3\mathbf{x},\end{aligned} \quad (15.71)$$

where the superscript '1h' indicates that such a pair is located within a single halo, and therefore contributes to the one-halo term. Taking account of the fact that different halos of mass M can host different N, the average number of pairs per unit volume in a halo of mass M is

$$\begin{aligned}GG^{1h}(\mathbf{r}|M) &= \sum_N P(N|M)GG^{1h}(\mathbf{x}|M,N) \\ &= \int \langle N(N-1)|M\rangle u(\mathbf{x}|M)u(\mathbf{x}+\mathbf{r}|M)\,\mathrm{d}^3\mathbf{x},\end{aligned} \quad (15.72)$$

where $P(N|M)$ is the halo occupation distribution, and

$$\langle N^k|M\rangle \equiv \sum_N N^k P(N|M). \quad (15.73)$$

Integrating over the halo mass function, we obtain the total number of galaxy pairs:

$$GG^{1h}(\mathbf{r}) = \int \mathrm{d}M\,n(M)\,\langle N(N-1)|M\rangle \int \mathrm{d}^3\mathbf{x}\,u(\mathbf{x}|M)\,u(\mathbf{x}+\mathbf{r}|M). \quad (15.74)$$

Note that GG is the number of pairs per volume-squared.

Next, consider the two-halo term. Let M_1 be the mass of a halo which contains N_1 galaxies, and M_2 the mass of a different halo hosting N_2 galaxies. The number of interhalo galaxy pairs separated by $\mathbf{r} = \mathbf{x}_2 - \mathbf{x}_1$ contributed by these halos per $\mathrm{d}V_1\,\mathrm{d}V_2$ is then given by

$$N_1\,u(\mathbf{x}_1 - \mathbf{x}|M_1)\,N_2\,u(\mathbf{x}_2 - \mathbf{x}'|M_2).$$

Since dark matter halos are correlated in space, the joint probability of finding a halo of mass M_1 at $\mathbf{x}$ and a halo of M_2 at $\mathbf{x}'$ is proportional to

$$n(M_1)\mathrm{d}M_1\,n(M_2)\mathrm{d}M_2\left[1 + \xi_{\mathrm{hh}}(\mathbf{x}-\mathbf{x}'|M_1,M_2)\right]\mathrm{d}^3\mathbf{x}\mathrm{d}^3\mathbf{x}',$$

where ξ_{hh} is the halo–halo two-point correlation function (see §7.4). Thus, the probability of having an interhalo galaxy pair, which is separated by $\mathbf{r}$ and which is hosted by an M_1-M_2 halo pair separated by $\mathbf{x} - \mathbf{x}'$, is equal to the product of the two quantities given above. The average number of interhalo galaxy pairs per $dV_1 \, dV_2$ separated by $\mathbf{r} = \mathbf{x}_2 - \mathbf{x}_1$ can then be obtained by first summing the product over N_1 and N_2, and then integrating the result over the halo masses, M_1 and M_2, and over the halo locations, $\mathbf{x}$ and $\mathbf{x}'$:

$$GG^{2h}(\mathbf{r}) = \int dM_1 \, n(M_1) \, dM_2 n(M_2) \langle N_1 | M_1 \rangle \langle N_2 | M_2 \rangle$$
$$\times \int d^3\mathbf{x} \, d^3\mathbf{x}' u(\mathbf{x}_1 - \mathbf{x} | M_1) u(\mathbf{x}_2 - \mathbf{x}' | M_2) \left[1 + \xi_{hh}(\mathbf{x} - \mathbf{x}' | M_1, M_2) \right]. \quad (15.75)$$

Finally, with the definition given in §6.1.1, the two-point correlation function of galaxies can be written as

$$\xi_{gg}(\mathbf{r}) = \frac{[GG^{1h}(\mathbf{r}) + GG^{2h}(\mathbf{r})] dV_1 \, dV_2}{RR(\mathbf{r}) dV_1 \, dV_2} - 1, \quad (15.76)$$

where $RR(\mathbf{r}) dV_1 \, dV_2 = \bar{n}_g^2 dV_1 \, dV_2$, with $\bar{n}_g = \int n(M) \langle N | M \rangle dM$ the mean number density of galaxies, is the expected number of pairs in the absence of clustering. As one can see, once a model of structure formation is adopted, so that the halo mass function, $n(M)$, and the halo–halo correlation function, ξ_{hh}, can be obtained, the two-point correlation function of galaxies is determined by the first two moments of the halo occupation distribution $P(N|M)$, i.e. $\langle N|M \rangle$ and $\langle N^2|M \rangle$, and by the galaxy distribution within individual halos, $u(\mathbf{x}|M)$. On large scales where the sizes of individual halos can be neglected, and the halo correlation function is related to the correlation function of matter, $\xi(r)$, by the linear bias relation, $\xi_{hh}(r|M_1, M_2) = b_h(M_1) b_h(M_2) \xi(r)$ (see §7.4.1), we have

$$\xi_{gg}(r) \approx b_g^2 \xi(r), \quad (15.77)$$

where

$$b_g = \int dM \, n(M) \, b_h(M) \frac{\langle N|M \rangle}{\bar{n}_g} \quad (15.78)$$

is the mean bias parameter of the galaxy population.

The Fourier transform of $\xi_{gg}(\mathbf{r})$ gives the power spectrum:

$$P_{gg}(k) = P_{gg}^{1h}(k) + P_{gg}^{2h}(k), \quad (15.79)$$

where

$$P_{gg}^{1h}(k) = \int dM \, n(M) \frac{\langle N(N-1)|M \rangle}{\bar{n}_g^2} |\tilde{u}(k|m)|^2; \quad (15.80)$$

$$P_{gg}^{2h}(k) \approx P_{\text{lin}}(k) \left[\int dM \, n(M) \, b_h(M) \frac{\langle N|M \rangle}{\bar{n}_g} \tilde{u}(k|M) \right]^2, \quad (15.81)$$

with $\tilde{u}(\mathbf{k}|M)$ the Fourier transform of $u(\mathbf{x}|M)$, and $P_{\text{lin}}(k)$ the linear power spectrum of the mass distribution (see §7.6). On large scales, where the two-halo term dominates and $\tilde{u}(k|m) \to 1$, the above expression reduces to

$$P_{gg}(k) \approx b_g^2 P_{\text{lin}}(k). \quad (15.82)$$

The model considered above assumes that every galaxy in a halo samples the profile $u(\mathbf{x}|M)$ in an unbiased way. This is unlikely to hold for all galaxies. In particular, according to the current theory of galaxy formation, a central galaxy is assumed to always sit near the center of the host halo, while the distribution of the satellite galaxies may be described by a density profile. Thus, the central galaxy in a halo does not really sample the density profile and has to be treated separately. Suppose that the profile of the satellite distribution in a halo of mass M is given by $u_s(\mathbf{x}|M)$. The number of one-halo satellite–satellite pairs is then

$$SS^{1h}(\mathbf{r}) = \int dM\, n(M) \langle N_s(N_s-1)|M\rangle \int d^3\mathbf{x}\, u_s(\mathbf{x}|M) u_s(\mathbf{x}+\mathbf{r}|M), \tag{15.83}$$

where N_s is the number of satellite galaxies. Since each halo contains either one or zero central galaxy, the central galaxies only contribute to the one-halo term through central–satellite pairs, and the number of such pairs can be written as

$$CS^{1h}(\mathbf{r}) = \int dM\, n(M) \langle N_c|M\rangle \langle N_s|M\rangle\, u_s(\mathbf{r}|M), \tag{15.84}$$

where $\langle N_c|M\rangle$ indicates the average number of central galaxies in a halo of mass M and ranges from zero to unity. Note that here we have assumed that N_c and N_s are independent random variables. It then follows that the one-halo term of the power spectrum can be written as

$$\begin{aligned}P_{gg}^{1h}(k) &= \int dM\, n(M) \frac{\langle N_s(N_s-1)|M\rangle}{\bar{n}_g^2} |\tilde{u}_s(k|m)|^2 \\ &+ \int dM\, n(M) \frac{\langle N_c|M\rangle \langle N_s|M\rangle}{\bar{n}_g^2} |\tilde{u}_s(k|M)|.\end{aligned} \tag{15.85}$$

A comparison of the above expressions for ξ_{gg} and P_{gg} with those for the dark matter in §7.4 shows that the distribution of galaxies is only unbiased relative to the mass distribution if the following three criteria are met: $\langle N|M\rangle \propto M$, $\langle N(N-1)|M\rangle \propto M^2$, and the distribution of galaxies within individual halos is identical to that of the dark matter particles. It is unlikely that all these criteria are met for any (sub)population of galaxies, so that we expect galaxies, in general, to be a biased tracer of the mass distribution. On large scales, the galaxy correlation function depends on the halo occupation distribution only through its first moment, independent of the distribution of galaxies within individual halos. Thus, if a population of galaxies forms in massive halos, the correlation function of this population on large scales should be as strong as the massive halos within which these galaxies form and reside. This is why cD galaxies, which only reside at the centers of rich clusters, reveal a correlation function similar to that of rich clusters (e.g. West & van den Bergh, 1991). This principle also allows us to understand why bright galaxies are more strongly clustered than faint galaxies (see Fig. 6.4): bright galaxies, on average, reside in more massive halos, which are more strongly clustered (see §7.4). Thus, once a model of structure formation is adopted, the correlation amplitude of galaxies on large scales can be used to constrain the masses of their host halos. This principle has been used extensively to constrain the halo occupation statistics of various subclasses of galaxies (e.g. Yang et al., 2003; van den Bosch et al., 2003, 2007; Porciani et al., 2004; Tinker et al., 2005)

For separations smaller than the size of a typical dark matter halo, most of the galaxy pairs are contained in individual halos. In this case, the two-point correlation function depends not only on the average occupation number, but also on the second moment of the halo occupation distribution, as well as on how galaxies are distributed within individual halos. Clearly, a more concentrated population is expected to have a steeper correlation function on small scales. As mentioned in §2.5.1, red satellites are more centrally concentrated than blue satellites, which explains, at least partly, why red galaxies have a steeper correlation function on small scales than blue galaxies (see Fig. 6.4).

Based on the principle outlined above, one can also develop halo occupation models for the higher-order correlation statistics of the galaxy distribution, such as the three-point correlation function and bispectrum. On large scales, these statistics depend only on the higher-order correlation functions of dark matter halos and the mean occupation $\langle N|M\rangle$. On small scales, however, they depend also on the spatial number density distribution of galaxies within individual dark matter halos, $u(\mathbf{x}|M)$, and the higher-order moments of the halo occupation distribution (e.g. Cooray & Sheth, 2002; Wang et al., 2004).

Clearly, the halo occupation distribution, $P(N|M)$, is a key element in understanding the clustering of galaxies. Physically, the form of $P(N|M)$ encapsulates all information about how galaxies form in dark matter halos, and should, in principle, be derived from the theory of galaxy formation. However, in the absence of a complete theory, one typically adopts simple models for $P(N|M)$ and uses the observed correlation function of galaxies to constrain the associated free parameters. For instance, a simple model adopted in the literature assumes that the mean occupation number for galaxies above a certain luminosity threshold changes with halo mass as

$$\langle N|M\rangle = \begin{cases} 1 + \left(\frac{M}{M_0}\right)^\alpha & \text{(for } M > M_{\min}) \\ 0 & \text{(for } M < M_{\min}). \end{cases} \quad (15.86)$$

where M_0, $M_{\min}$ and α are free parameters. In general, we expect $\alpha > 0$, because a more massive halo should, on average, host a larger number of galaxies. A minimum halo mass, $M_{\min}$, is introduced to account for the fact that only halos that are sufficiently massive can host a galaxy above a certain luminosity threshold. The form (15.86) is actually in qualitative agreement with the halo occupation statistics obtained from galaxy groups (Collister & Lahav, 2005; Yang et al., 2005a), as well as with the predictions from semi-analytical models and hydrodynamical simulations of galaxy formation (e.g. Berlind et al., 2003; van den Bosch et al., 2003). The halo occupation distribution for the satellite galaxies, $P(N_s|M)$, is usually modeled to be a Poisson distribution:

$$P(N_s|M) = \frac{e^{-\langle N_s\rangle}\langle N_s\rangle^{N_s}}{N_s!}; \quad \langle N_s(N_s-1)\rangle = \langle N_s\rangle^2, \quad (15.87)$$

where $\langle N_s\rangle = \langle N|M\rangle - 1$. This model is consistent with the observed halo occupation statistics obtained from galaxy groups (e.g. Collister & Lahav, 2005; Yang et al., 2005a, 2008). Furthermore, numerical simulations of the subhalo population (see §7.5.3) show that the number of subhalos in a halo of a given mass roughly has a Poisson distribution (Kravtsov et al., 2004). Since satellite galaxies are believed to be associated with subhalos, $P(N_s|M)$ is also expected to be approximately Poissonian. With proper choices of model parameters (M_0, $M_{\min}$, α), the above simple halo occupation model, combined with the halo population predicted by the current model of structure formation, is remarkably successful in matching the observed two-point correlation function of galaxies (e.g Jing et al., 1998; Tinker et al., 2005; Zehavi et al., 2005; Zheng et al., 2005).

More generally, we can consider the halo occupation distribution as a function of galaxy luminosity using the conditional luminosity function, $\Phi(L|M)$, introduced in §15.3. Based on the formalism outlined above, it can be shown that the mean linear bias factor for galaxies with luminosities $L_1 \leq L < L_2$ can be written as

$$b_g(L_1, L_2) = \frac{1}{n_g(L_1, L_2)} \int_{L_1}^{L_2} dL \int_0^\infty \Phi(L|M) b_h(M) n(M) dM, \quad (15.88)$$

where

$$n_g(L_1, L_2) = \int_{L_1}^{L_2} \phi(L) dL = \int_{L_1}^{L_2} dL \int_0^\infty \Phi(L|M) n(M) dM, \quad (15.89)$$

with $\phi(L)$ the galaxy luminosity function. Thus, once a cosmological model is adopted so that $n(M)$ and $b_h(M)$ can be calculated, the correlation amplitude of galaxies as a function of luminosity constrains the conditional luminosity function, $\Phi(L|M)$. As shown in Yang et al. (2003) and van den Bosch et al. (2003), in the current ΛCDM cosmogony the form of $\Phi(L|M)$ required to match the observed luminosity dependence of the two-point correlation function of galaxies is similar to that required to match the observed galaxy luminosity function, and the implied relation between the central galaxy luminosity and host halo mass is very similar to that given by Eq. (15.38) with $M_L \sim 10^{12} h^{-1} \, \rm M_\odot$, $\beta \sim 4$ and $\gamma \sim 0.3$. This suggests that the current CDM cosmogony can accommodate not only the intrinsic properties of galaxies but also their clustering in space.

15.6.1 Application to High-Redshift Galaxies

With the advent of deep imaging and spectroscopic surveys, large and homogeneous samples of high-redshift galaxies can now be constructed, allowing the statistical properties of the high-redshift galaxy population to be studied in a way similar to that in the local Universe. Since galaxies at higher redshifts are on average younger, the study of the galaxy population as a function of redshift can provide direct information about the formation and evolution of galaxies through the history of the Universe. Many of the observational aspects of high-redshift galaxies are described in §2.6. Here we concentrate on what the clustering properties of high-redshift galaxies can tell us about galaxy formation and evolution.

As described above, the large-scale clustering amplitude of a population of galaxies is determined by the mass distribution of their host halos so that the observed correlation amplitude can be used to constrain the mass distribution of the halos hosting the galaxy population in question. Consider, for example, a population of galaxies at redshift $\sim z$ with luminosity function $\phi(L,z)$. As shown in §15.3, we can write $\phi(L,z)$ in terms of the dark matter halo mass function at z, $n(M,z)$, as

$$\phi(L,z) = \int \Phi(L|M,z) n(M,z) \, \mathrm{d}M, \qquad (15.90)$$

where $\Phi(L|M,z)$ is the conditional luminosity function. According to Eq. (15.88) the linear bias factor for this population of galaxies can be written as

$$b_{\rm g}(z) = \frac{1}{n_{\rm g}(z)} \int b_{\rm g}(L,z) \phi(L,z) \, \mathrm{d}L, \qquad (15.91)$$

where $n_{\rm g}(z) = \int \phi(L,z) \, \mathrm{d}L$, and

$$b_{\rm g}(L,z) = \frac{1}{\phi(L,z)} \int \Phi(L|M,z) b_{\rm h}(M,z) n(M,z) \, \mathrm{d}M. \qquad (15.92)$$

Thus, for a given model of structure formation, one can constrain $\Phi(L|M,z)$ from the observed clustering. The simplest model for $\Phi(L|M,z)$ is the one in which L is determined entirely and deterministically by M. In this case, $\Phi(L|M,z)$ is a Dirac delta function of M, and $b_{\rm g}(L,z) = b_{\rm h}(M,z)$, with L related to M through a deterministic function, $L = L(M)$. More generally, we can define a characteristic mass, $M_{\rm c}$, so that $b_{\rm h}(M_{\rm c},z) = b_{\rm g}(L,z)$. Using Eq. (15.77), the galaxy bias $b_{\rm g}(L,z)$ can be determined from

$$b_{\rm g}(L,z) = \sqrt{\frac{\xi_{\rm gg}(r|L,z)}{\xi(r|z)}}, \qquad (15.93)$$

in the limit of large r, where $\xi_{\rm gg}(r|L,z)$ is the two-point correlation function of galaxies of luminosity L at redshift z, and $\xi(r|z)$ is the two-point correlation function of the matter at that redshift in the assumed model of structure formation. With the $b_{\rm g}(L,z)$ thus obtained, the corresponding characteristic mass, $M_{\rm c}$, follows from its definition given above.

As an example, consider the population of Lyman break galaxies (LBGs) described in §2.6.4. Bright LBGs at $z \sim 3$ are found to be strongly clustered, with a correlation length $r_0 \approx 6 h^{-1}$Mpc (in comoving units) assuming a flat universe with $\Omega_{\rm m,0} = 0.3$ and $\Omega_{\Lambda,0} = 0.7$ (e.g. Steidel et al., 1999). In the standard ΛCDM model, this correlation length corresponds to a bias factor of ~ 3.

Using the expression of the bias given in §7.4, we can infer $M_c \sim 10^{12} M_\odot$. This suggests that bright LBGs at $z \sim 3$ are typically hosted by relatively massive halos (e.g. Mo et al., 1999; Bullock et al., 2002). The correlation amplitude is found to decrease with galaxy luminosity, indicating that fainter LBGs are, on average, hosted by halos of lower masses.

According to Eq. (7.105), the bias factor, b', for the present-day descendents of a halo population of mass M at z is related to $b_h(M,z)$ by $b' = 1 + D(z)[b_h(M,z) - 1]$. The value of the bias factor quoted above for the bright LBGs gives $b' \sim 1.6$. Since normal L^* galaxies at $z = 0$ have $b \sim \sigma_8^{-1}$, the correlation amplitude for the descendents of LBGs is thus predicted to exceed that of normal bright galaxies by a factor of about $(\sigma_8 b')^2$, or ~ 1.7 assuming the standard ΛCDM model with $\sigma_8 = 0.8$. This suggests that the descendents of bright LBGs are probably among the brightest galaxies at $z \sim 0$ (e.g. Mo & Fukugita, 1996), while the fainter ones may evolve into the more typical population of galaxies at the present time.

15.7 Putting it All Together

In the previous sections of this chapter we have used statistical techniques and simple arguments, based on our current paradigm of hierarchical galaxy formation, to link galaxy properties (luminosity, stellar mass, star-formation rate, color, morphology, etc.) to halo properties. We now briefly discuss a more thorough approach based on *ab initio* modeling of the galaxy population. Starting from the initial conditions (the density perturbations discussed in Chapter 4), this method predicts the properties of galaxies by following the merger histories of dark matter halos and by modeling the various astrophysical processes within those halos that lead to the formation of galaxies (gas cooling, star formation, feedback processes, galaxy mergers, etc.). Two different approaches have been used: one treating the important processes in a semi-analytical fashion, and the other using cosmological hydrodynamical simulations. In what follows we briefly discuss these two methods in turn.

15.7.1 Semi-Analytical Models

A semi-analytical model of galaxy formation typically consists of the following steps:

(i) Choose a cosmological model, i.e. choose the cosmological parameters that specify the geometry and matter content of the universe ($\Omega_{m,0}$, $\Omega_{b,0}$, $\Omega_{\Lambda,0}$, H_0) and the amplitude and shape of the initial fluctuation spectrum (σ_8 and n, respectively).

(ii) Use N-body simulations, or the extended Press–Schechter formalism (§7.3), to trace the merger histories for a series of dark matter halos of different (present-day) masses.

(iii) Within each dark matter subunit which is present at any stage of one of these merger histories, follow three distinct baryonic components: hot gas, cold gas (presumably in a disk), and stars.

(iv) Use simple prescriptions to specify the conversion rates between these baryonic components: cooling converts hot gas into cold gas, star formation converts cold gas into stars, and feedback from massive stars and AGN either converts cold gas into hot gas, or reheats the hot gas directly. At the same time, keep track of the metallicity of each of the three different components using chemical evolution models (§10.4).

(v) Use a stellar population synthesis model (§10.3) to convert the star-formation history and metallicity of the stellar population into luminosity and color, including a model for dust extinction.

(vi) Adopt prescriptions to specify the dynamical evolution of the various components when halos merge. Typically these consist of the following: The hot gas components in the merging progenitors are heated to a new virial temperature as they merge. The least massive progenitor galaxy becomes a satellite galaxy in the new system and ceases to accrete new gas (strangulation, see §12.5.4). The most massive progenitor galaxy becomes the new central galaxy and continues to accrete new gas as long as radiative cooling in the host halo is efficient.

(vii) Assume that satellites can merge with central objects on a time scale set by dynamical friction (§12.3), and make simple assumptions about the outcome of such a merger. Presumably, if the mass of the satellite galaxy is much smaller than that of the central galaxy, the satellite is 'cannibalized' (§12.5.2) without major impact on the morphology of the central galaxy. If, on the other hand, the mass of the satellite galaxy is comparable to that of the central, the disks of both progenitors are likely to be destroyed, resulting in the formation of an elliptical remnant (§12.4.2 and §13.2.2). If new gas is able to subsequently cool, a new disk may grow around the spheroid, giving rise to a galaxy consisting of disk and bulge.

(viii) In order to build a sample of model galaxies that fairly represents the galaxy population in the present-day Universe, repeat this procedure for a large number of halos at $z=0$ that properly sample the present-day halo mass function.

Thus, in a semi-analytical model the complicated, tightly intertwined astrophysical processes associated with the formation and evolution of galaxies are modeled as a set of 'recipes' and 'prescriptions' which carry a number of free model parameters (see Baugh, 2006, for more details). In principle, these model parameters should be chosen such that the underlying physical processes are represented as adequately as possible. Unfortunately, our understanding of many of these processes is still too limited, and the prescriptions are often too oversimplified, so that the values of the model parameters cannot be derived from first principles. Rather, one typically 'normalizes' the model using some observational constraints. For example, one can set the free parameters governing the efficiency of star formation and the efficiency of stellar energy feedback to guarantee that a 'Milky Way' halo contains, on average, the same mass in stars and in cold gas as our own Galaxy, and adjust the dynamical friction time scale (which depends on the orbital properties and the efficiency of tidal stripping) so that a 'Milky Way' halo contains, on average, the right number of 'Magellanic Cloud'-sized satellites (e.g. Kauffmann et al., 1993). Another possibility is to normalize these parameters so that the model reproduces the observed luminosity function of galaxies (e.g. Cole et al., 2000) and/or the zero-point of the Tully–Fisher relation (Kauffmann et al., 1999a).

Once the free parameters are set, the semi-analytic scheme can be used to generate large samples of model galaxies, whose properties (luminosities, stellar masses, colors, sizes, morphologies, metallicities, etc.) can be compared to data. Such a comparison allows an assessment of the success of each particular model and of the importance of certain specific model ingredients (e.g. compare models with and without AGN feedback).

When the first semi-analytical models were constructed in the early 1990s, it immediately became clear that in the absence of any form of energy input an unacceptably large fraction of the baryons will cool and condense in low-mass halos at high redshifts (e.g. Cole, 1991; White & Frenk, 1991). Although this is generally interpreted as a requirement for supernova feedback and/or re-ionization, to this date fitting the faint-end slope of the galaxy luminosity function remains challenging (e.g. Benson et al., 2003). In addition to the problem with the faint-end slope of the galaxy luminosity function, early models also suffered from three other major problems: (i) they predicted a color–magnitude relation that was inverted with respect to the observations (ii) they overpredicted the number density of bright galaxies, a problem that was more severe for cosmologies with a lower value of $\Omega_{m,0}$, and (iii) the models were unable to simultaneously fit the luminosity function and the normalization of the Tully–Fisher relation (e.g. Kauffmann et al.,

1993; Cole et al., 1994; Heyl et al., 1995). It soon became clear that the first problem could be remedied by including chemical evolution and/or a treatment for dust extinction, but the problem with the bright galaxies being too blue remained (Kauffmann & Charlot, 1998a; Somerville & Primack, 1999; Cole et al., 2000). This problem, together with the overprediction of the number density of massive galaxies, seems to require a mechanism that can suppress cooling in massive halos. Several such mechanisms have been suggested, including AGN feedback (e.g. Bower et al., 2006; Cattaneo et al., 2006; Croton et al., 2006; Kang et al., 2006), thermal conduction (Benson et al., 2003; Voigt & Fabian, 2004), multi-phase cooling (Maller & Bullock, 2004), and heating by substructure or clumpy, cold accretion (Khochfar & Ostriker, 2008; Dekel & Birnboim, 2008). When one or more of these ingredients are included, the models are able to fit the observed luminosity function, as well as the color–magnitude relation. However, to this date the problem with the Tully–Fisher zero-point remains largely unsolved. Typically the models predict disk rotation velocities that are too high, unless adiabatic contraction and/or self-gravity of the disk are ignored (Cole et al., 2000; Dutton et al., 2007). The main cause of this discrepancy seems to be that CDM halos are predicted to have too concentrated mass distributions. As discussed in §11.3, it remains to be seen whether this implies a problem for the CDM framework, or it indicates that our models for disk formation are inappropriate. In addition to this persisting problem with the Tully–Fisher zero-point, it has been shown that, among others, semi-analytical models typically overpredict the fraction of red satellite galaxies (Baldry et al., 2006; Weinmann et al., 2006a), and have problems matching the evolution of the galaxy mass function with redshift (e.g. Somerville et al., 2008b; Fontanot et al., 2009).

Most semi-analytical models use the extended Press–Schechter (EPS) formalism to construct merger histories for the dark matter halos. Although this has the advantage of not being overly CPU intensive, the downside is that the EPS merger histories suffer from several systematic inaccuracies (see §7.3). Hence, several investigators have constructed semi-analytical models based on merger histories extracted directly from numerical N-body simulations (e.g. Kauffmann et al., 1999a; Kang et al., 2005; Croton et al., 2006). This has the additional advantage that the information about the spatial distribution of the dark matter halos allows one to investigate the predicted clustering properties of the model galaxies, and the model predictions are found to be in reasonable agreement with observations (e.g. Kauffmann et al., 1999b; Springel et al., 2005c).

To summarize, despite several outstanding problems, the overall agreement between the predictions of semi-analytical models and observational results is encouraging, and seems to suggest that our current comprehension of galaxy formation may already have captured the essence of the problem. Nevertheless, it is important to realize that the theory is far from complete, mainly because many of the astrophysical processes involved have not yet been treated accurately from first principles.

15.7.2 Hydrodynamical Simulations

In a cosmological hydrodynamical simulation, one again starts by choosing a cosmology. The initial density fields in both dark matter and baryonic gas are sampled either on a large spatial grid (in the Eulerian approach) or by a large number of particles (in the Lagrangian approach) (see Appendix C). The evolution of the density fields is then followed by solving the gravitational and hydrodynamical equations numerically. In principle, hydrodynamical simulations can follow the evolution of both the gas and dark matter without relying on simplified approximations of all the important processes. In practice, however, simulations are limited by numerical resolution. Consequently some of the physical processes still have to be modeled approximately on the subgrid level, using 'recipes' that are not very different from those adopted in the semi-analytical method described above. For instance, the star-formation rate within a fluid element is usually modeled with the Schmidt law based on the local cold gas density, without taking

15.7 Putting it All Together

into account any of the detailed star-formation processes described in Chapter 9. The feedback from supernova explosions and AGN is typically included as energy/momentum sources based on simple prescriptions without treating the supernova remnants and AGN-driven outflows from first principles. Despite these limitations, hydrodynamical simulations have several important advantages over semi-analytical models. First, the evolution of the dark matter halo population is followed in detail, and no assumption is needed to model the structure and formation history of individual halos. In particular, the interaction between the dark matter and baryon components is taken into account without relying on simple approximations such as the adiabatic contraction described in §11.1.3. Second, hydrodynamical simulations follow the dynamics of the diffuse cooling gas in full generality, without having to make assumptions such as spherical symmetry or quasi-static evolution. Third, once a feedback scheme is adopted on the subgrid level, hydrodynamical simulations can treat the subsequent evolution of the supernova/AGN-driven winds and their interactions with the gas component fully self-consistently without adopting simplified prescriptions to specify the impact of these winds. Finally, in a hydrodynamical simulation mergers of dark matter halos and galaxies are followed self-consistently as part of the dynamical process, rather than being modeled with simple prescriptions based on dynamical friction time scales and on the mass ratio of the merger progenitors.

With computational facilities and simulation codes currently available, hydrodynamical simulations can now be used to study galaxy formation and evolution in a full cosmological context. However, computational power is still a severe limitation at the present, and one has to compromise between simulation resolution and box size. Typically, the internal structure of galaxies can only be resolved in simulations of small volumes (containing a relatively small number of galaxies), while simulations of large volumes containing sufficient numbers of galaxies to represent the total galaxy population typically lack the spatial resolution to resolve the internal structure of the simulated galaxies. Hydrodynamical simulations of individual galaxies have already been discussed in some detail in Chapters 11 and 13. In what follows we focus on simulations of cosmologically representative volumes.

In a cosmological hydrodynamical simulation, galaxies are typically identified as massive and dense stellar clumps. Their luminosities and colors are obtained from the corresponding star-formation histories with the use of a stellar population synthesis model. With these, one can then generate galaxy 'catalogues' that contain both the intrinsic properties (stellar mass, luminosity, cold gas mass) and locations of individual galaxies, and study the luminosity (stellar mass) function, star-formation histories and the large-scale distribution of galaxies.

Simulations without supernova/AGN feedback generally overpredict the number density of galaxies at all galaxy masses, in particular at the low- and high-mass ends (e.g. Kereš et al., 2009). Massive galaxies are also predicted to be too blue in such simulations, because significant amounts of gas can cool, even in massive halos, to form stars in the central galaxies. In recent years, star formation and feedback recipes are routinely incorporated in hydrodynamical simulations of galaxy formation. It has been demonstrated that simply including the supernova energy in the thermal content of the surrounding gas is very ineffective, because supernovae typically explode in high density regions where the cooling time is short, so that the feedback energy is radiated away before it can affect the hydrodynamics (e.g. Katz, 1992; Mac Low & Ferrara, 1999b). Because of this, alternative implementations have been proposed. For example, Gerritsen & Icke (1997) proposed a scheme in which the cooling rate of the heated gas particles is reduced by turning off radiative cooling for a brief period of time after the heating; Thacker & Couchman (2000) used a scheme in which the density of heated gas is artificially reduced; Navarro & White (1993) suggested incorporating the supernova energy as outward motions imposed on surrounding gas. Although these implementations have some success in reducing the star-formation efficiency, they are fairly arbitrary and ad hoc, with results depending significantly on the details of the implementation.

More recently, Springel & Hernquist (2003) developed a smoothed-particle hydrodynamics (SPH) code (see Appendix C) which treats the ISM as a multi-phase medium with a co-spatial mixture of a cold dense phase and a hot diffuse phase. Some feedback schemes have also been implemented in this code (e.g Springel et al., 2005b; Scannapieco et al., 2006). In such a multi-phase model, supernova energy is injected mainly into the hot diffuse phase where radiative cooling is inefficient, making the feedback more effective in driving outflows and in reducing star formation in galaxies. Such a scheme seems to be able to produce disk galaxies with characteristics similar to those of real galaxies. Simulations of mergers of relatively massive gas-rich galaxies have also been carried out, incorporating feedback effects from both AGN and starbursts (e.g. Naab et al., 2007; Di Matteo et al., 2008). The results demonstrate that star formation can be effectively quenched even in a massive system after an initial burst of stars accompanied by the formation of AGN/starburst-driven outflows, producing massive red galaxies reminiscent of the bright ellipticals observed today.

In summary, it is clear that hydrodynamical simulations provide a promising avenue to study galaxy formation and evolution in a fully cosmological context. At the present, however, such simulations still suffer severely from insufficient numerical resolution, and many key processes, in particular star formation and feedback, still have to be treated approximately on subgrid level. The arbitrariness in the implementations of these processes make their validity uncertain, and so the results obtained with such simulations should be interpreted with caution.

16
The Intergalactic Medium

Galaxies are ecosystems consisting of dark matter, stars and gas. It is useful to split the gas into three broad components according to its relation to the galaxy. The first is the interstellar medium (ISM), which is the gas that is directly associated with the galaxy. The second is the halo gas, which is distributed outside the galaxy but inside the host dark matter halo of the galaxy. The third is the gas that is not associated with dark matter halos. The latter two components combined are known as the intergalactic medium (IGM). During the formation and evolution of a galaxy, the ISM and IGM interact actively with each other. Halo gas can cool and be accreted into the galaxy to become part of the ISM. The gas in the ISM can be ejected into the halo or even into the large-scale environment by galactic winds and stripping. And finally, dark matter halos can accrete gas from the IGM in their large-scale environments.

Clearly, the IGM is a crucial ingredient of any theory of galaxy formation and evolution. In fact, by definition, all baryons were part of the IGM at sufficiently early times (before the formation of stars and galaxies). At later times, more and more material of the IGM is accreted by virialized dark matter halos, where it can be converted into cold gas (ISM) or stars. However, even at the present day, a very substantial fraction of the baryons is believed to still reside in the IGM. As described in §2.10.2, the total mass in stars in the present-day Universe accounts for less than 10% of the total baryonic mass expected from cosmic nucleosynthesis and from observations of the cosmic microwave background. After adding the baryonic material detected as cold gas (atomic or molecular) associated with galaxies, and the hot gas in the ICM of clusters, still more than 50% of the baryonic material is unaccounted for. These missing baryons are believed to reside in the IGM.

The properties of the IGM (density, temperature, ionization state, chemical composition, spatial distribution, etc.) can be studied through its absorption of light from background sources or through the radiation field it produces. The study of the IGM is important for several reasons. First of all, since galaxy formation is a process of gas condensation in the cosmological density field, the IGM reflects the embryonic material out of which galaxies form. Furthermore, galaxies in turn can have a significant impact on the IGM by injecting it with energy and mass (in particular metals). Studying the properties of the IGM as function of redshift can thus provide important insight into the formation and evolution of galaxies. Secondly, since the properties of the IGM can be affected by various radiative and gas-dynamical processes, studying the IGM at different redshifts may also provide insight into cosmological events that have occurred since recombination. Finally, the gas in the IGM may interact with the CMB photons, thereby distorting the CMB spectrum. Hence, a proper interpretation of the CMB also requires a proper understanding of the IGM.

In this chapter, we first describe the global properties of the IGM and examine how various radiative and gas-dynamical processes drive the evolution of the IGM. We then describe how to probe the IGM in more detail using various QSO absorption line systems.

16.1 The Ionization State of the Intergalactic Medium

16.1.1 Physical Conditions after Recombination

We have seen in §3.5 that the gas in the Universe became predominantly neutral at $z = z_{\rm rec} \sim$ 1,300 when electrons recombined with protons to form hydrogen atoms. The Universe at such early times must also be highly uniform, as indicated by the high degree of isotropy in the CMB, so that very few, if any, bright objects could have formed. Consequently, the Universe cooled below a few thousand degree due to adiabatic expansion, and entered a 'dark age' in which the Universe was filled with CMB photons (with frequencies in the infrared), a predominantly neutral gas, and dark matter particles. Since there were no heating and cooling sources, the Universe evolved adiabatically, and the temperature of the CMB decreased with redshift as

$$T_\gamma = 2.736(1+z)\,{\rm K}. \tag{16.1}$$

Since photons have decoupled from baryons, naively one would expect the temperature of the baryonic gas to decrease as $T_{\rm gas} \propto (1+z)^2$, and so $T_{\rm gas}$ would be lower than T_γ soon after the decoupling. In reality, however, the residual ionization fraction (although small) can channel a significant amount of energy from the radiation field to the gas through Compton scattering, keeping baryons at the same temperature as the CMB until a much later time ($z \sim 100$, see §3.5.3). Since primordial nucleosynthesis produces very little heavy elements (see §3.4), the baryonic content of the Universe during the 'dark age' consists almost entirely of hydrogen and helium, with ${\rm He/H} \approx 0.07$ by number. The mean number density of hydrogen atoms in the medium at redshift z (in proper units) can then be written as

$$n_{\rm H}(z) \simeq \frac{\rho_{\rm IGM}(z)}{1.3 m_{\rm H}} \approx 8.6 \times 10^{-6} \Omega_{\rm b,0} h^2 (1+z)^3 \,{\rm cm}^{-3}, \tag{16.2}$$

where $\Omega_{\rm b,0}$ is the present-day density parameter of the baryons.

Because recombination requires the encounter of an electron with a proton, the rate of recombination is expected to decrease as the number densities of electrons and protons are reduced by recombination and by the expansion of the Universe. At the time when the recombination rate becomes much lower than the expansion rate, the number of free electrons is frozen, and the residual ionization fraction is of the order $X_e \equiv n_e/n \sim 10^{-5}$–$10^{-3}$ (see §3.5.1). As shown in §9.7, this residual of free electrons can act as a catalyst for the formation of molecular hydrogen, which may be the main source of cooling at high redshift for the formation of the first generation of luminous objects in the Universe.

The 'dark age' is expected to end when some perturbations in the cosmic density field evolve into the nonlinear regime and collapse to form objects (e.g. stars, galaxies and/or AGN) that can re-ionize the IGM. An important question is whether or not such re-ionization indeed occurred and, if the answer is yes, when and how the IGM was re-ionized. In the remainder of this section we will focus on observational constraints on the state of the IGM, leaving the discussion of the other aspects of the problem to the following two sections.

16.1.2 The Mean Optical Depth of the IGM

A powerful way to probe the ionization state of the IGM is by examining how it absorbs the radiation coming from background sources. Suppose that a source is located at some high redshift z_e (corresponding to a time t_e) and emits with an intrinsic spectral energy distribution $\mathscr{L}(\nu)\,d\nu$ (defined as the the luminosity of the source at frequency ν). If the emitted light travels in a vacuum, an observer at $z = 0$ should observe a flux given by

16.1 The Ionization State of the Intergalactic Medium

$$F(\nu)\,d\nu = \frac{\mathscr{L}[\nu(1+z_e)]}{4\pi d_L^2}(1+z_e)\,d\nu, \tag{16.3}$$

where d_L is the luminosity distance (see §3.2.6). In this case, the observed spectrum has a shape similar to the intrinsic one, except that it is stretched by the expansion of the Universe. If, however, the emitted photons travel in a medium of atoms and/or ions, they may be absorbed and re-emitted, and so the spectrum of the source is reprocessed before it is observed. If the emission from the medium is negligible, the number density of photons in the light ray from the source decreases on its way to the observer, because photons can be absorbed by the medium and scattered away from the line-of-sight. For photons with initial frequencies in the range ν_e to $\nu_e + d\nu_e$, the rate of change in the number density is given by

$$\dot{n}_\gamma(\nu_e,t) = -\Lambda\left[\nu_e \frac{a(t_e)}{a(t)},t\right] n_\gamma(\nu_e,t), \tag{16.4}$$

where $\Lambda(\nu,t)$ is the absorption coefficient at time t for photons with frequency ν. The factor $a(t_e)/a(t)$ describes the redshift of the photon frequency as they travel from t_e to t. The solution of the above equation is

$$n_\gamma(\nu_e,t) = e^{-\tau} n_\gamma(\nu_e,t_e), \tag{16.5}$$

where

$$\tau = \int_{t_e}^{t} \Lambda\left[\nu_e \frac{a(t_e)}{a(t')},t'\right] dt' \tag{16.6}$$

is the optical depth due to absorption.

In many of our following applications, we are interested in resonance absorption which involves the transition of an atom between two bound states 1 and 2. The energies of these two states will be denoted by E_1 and E_2 (we assume $E_2 > E_1$ as a convention), and the photon frequency involved in the transition is therefore $\nu_{12} \equiv (E_2 - E_1)/h_P$. The probability per unit time, P_{12}, that the atom undergoes a transition from the low state 1 to the high state 2 due to photon excitation is $P_{12} = B_{12}B_\nu$, where B_{12} is the Einstein coefficient of excitation and $B_\nu\,d\nu$ is the brightness of the radiation field (see §B1.3). Note that P_{12} is just the probability per unit time with which a photon with frequency ν is absorbed by the atom, and we can express it in terms of the absorption cross-section σ_{12}:

$$P_{12} = B_{12}B_\nu = \int \sigma_{12}(\nu)\,c\,n_\gamma(\nu)\,d\nu. \tag{16.7}$$

Conventionally, B_{12} is expressed in terms of the oscillator strength f_{12}:

$$B_{12} = \frac{4\pi a_{12}}{h_P \nu_{12}}, \quad a_{12} = \frac{\pi q_e^2}{m_e c} f_{12}. \tag{16.8}$$

For resonance absorption, $\sigma_{12}(\nu)$ is peaked sharply at the resonance frequency ν_{12}, and so the cross-section is related to a_{12} by

$$a_{12} = \int \sigma_{12}(\nu)\,d\nu. \tag{16.9}$$

The absorption rate by atoms in state 1 is proportional to the number density of atoms in that state and the absorption cross-section. Thus, the absorption rate Λ in Eq. (16.4) can be written as

$$\Lambda(\nu,t) = c\,n_1(t)\,\sigma_{12}(\nu). \tag{16.10}$$

As discussed in §B1.3, we can write the absorption cross-section as

$$\sigma_{12}(\nu) = a_{12}\phi_{12}(\nu) \quad \text{with} \quad \int \phi_{12}(\nu)\,d\nu = 1, \tag{16.11}$$

where $\phi_{12}(\nu)$ describes the profile of the absorption line in ν space.

An atom at the excited state 2 tends to decay back spontaneously to the low energy state 1, emitting a photon with energy $h_P\nu_{12}$ in a random direction. Because photons emitted in this process have random propagation direction, they will not be observed as part of the emission from the source, although they contribute to the background radiation. Thus, the net effect of photon excitation and spontaneous emission is to scatter a photon off the line-of-sight to the source, causing the source to dim at the absorption frequency ν_{12}. In addition to the spontaneous decay, an atom at state 2 can also make a transition back to state 1 due to the stimulus of a photon in the light beam. Unlike the spontaneous emission, the photon emitted from a stimulated transition propagates in the same direction as the incident photon. Such photons will be observed as part of the emission from the source, and so stimulated transitions do not lead to net absorption. The rate of stimulated emission, $\Omega(\nu,t)$, is related to the absorption rate Λ by the Einstein relations, $\Omega(\nu,t) \propto n_2 B_{21} \propto (n_2 g_1/n_1 g_2)\Lambda(\nu,t)$, where g_i is the number of internal states of level i, so that

$$\Omega(\nu,t) = \exp\left(-\frac{h_P\nu}{k_B T}\right)\Lambda(\nu,t). \tag{16.12}$$

This rate should be added to the right-hand side of Eq. (16.4). With this term, the optical depth becomes

$$\tau = \int_{t_e}^{t}\left\{1 - \exp\left[-\frac{h_P\nu_e a(t_e)}{k_B T(t')a(t')}\right]\right\}\Lambda\left[\nu_e \frac{a(t_e)}{a(t')},t'\right]dt'. \tag{16.13}$$

Now suppose that a photon from a source at redshift z_e is absorbed in resonance at redshift z_a (corresponding to a time t_a). The frequencies at emission (ν_e), at absorption ($\nu_a = \nu_{12}$), and at observation (ν_o) are related by

$$\nu_e = \frac{1+z_e}{1+z_a}\nu_{12} = (1+z_e)\nu_o. \tag{16.14}$$

Since the absorption cross-section is peaked sharply at ν_{12}, the optical depth at the observed frequency ν_o is

$$\tau(\nu_o) \approx \frac{cn_1(t_a)}{a(t_a)H_0 E(z_a)}\left\{1 - \exp\left[-\frac{h_P\nu_{12}}{k_B T(t_a)}\right]\right\}I_{12}, \tag{16.15}$$

where $n_1(t_a)$ is the proper number density of the absorbing atoms at the position of the absorption, $H_0 E(z_a) \equiv \dot{a}(t_a)/a(t_a)$ is the Hubble constant at z_a [see §3.2.3 for the expression of $E(z)$], and

$$I_{12} = \frac{1}{\nu_{12}}\int \sigma_{12}(\nu)d\nu. \tag{16.16}$$

Since the absorption can be caused only by atoms between the source and the observer, the wavelength range (in the observed spectrum) over which absorption can possibly occur is

$$\lambda_{12} \leq \lambda_o \leq (1+z_e)\lambda_{12}. \tag{16.17}$$

16.1.3 The Gunn–Peterson Test

As mentioned in the beginning of this section, intergalactic space may contain a diffuse component of neutral (mainly HI) gas. The density of this component can be estimated by measuring its optical depth in Lyα absorption. This absorption occurs when a hydrogen atom is excited from the $1s$ state to the $2p$ state by a resonant photon, and the wavelength, frequency, and temperature associated with this transition are

$$\lambda_{\text{Ly}\alpha} = 1216\,\text{Å}, \quad \nu_{\text{Ly}\alpha} = 2.47 \times 10^{15}\,\text{Hz}, \quad T_{\text{Ly}\alpha} \equiv \frac{h_P\nu_{\text{Ly}\alpha}}{k_B} = 1.18 \times 10^5\,\text{K}. \tag{16.18}$$

If the temperature of the IGM $T \ll T_{\text{Ly}\alpha}$, we can neglect the exponential term in the optical depth of Eq. (16.15) and obtain

$$\tau(\nu_0) = \frac{c n_{\text{HI}}(t_a)}{a(t_a) H_0 E(z_a)} I_{\text{Ly}\alpha}, \tag{16.19}$$

where

$$I_{\text{Ly}\alpha} = \frac{a_{\text{Ly}\alpha}}{\nu_{\text{Ly}\alpha}} = 4.5 \times 10^{-18} \, \text{cm}^2. \tag{16.20}$$

It then follows that the proper number density of HI atoms is related to the optical depth by

$$n_{\text{HI}}(t_a) = 2.4 \times 10^{-11} h E(z_a) \tau \, \text{cm}^{-3}. \tag{16.21}$$

A constraint on n_{HI} can therefore be obtained from measurements of the optical depth τ.

The optical depth of the IGM can be probed by measuring the absorption it causes in the spectra of QSOs at high redshifts. The most basic observable in a QSO spectrum is the flux decrement, D, defined as the mean value of the ratio between the observed continuum flux and the expected continuum flux in the absence of absorption. Such a decrement caused by the diffused component of the IGM is usually referred to as the Gunn–Peterson effect. In practice, one measures the mean flux decrement D_A between the Lyα and Lyβ emission lines (with rest-frame wavelengths $\lambda_{\text{Ly}\alpha} = 1216\,\text{Å}$ and $\lambda_{\text{Ly}\beta} = 1026\,\text{Å}$):

$$D_A = \left\langle 1 - \frac{F_{\text{obs}}(\lambda)}{F_{\text{cont}}(\lambda)} \right\rangle = \langle 1 - e^{-\tau} \rangle = 1 - e^{-\tau_{\text{eff}}}, \tag{16.22}$$

where $F_{\text{obs}}(\lambda)$ is the observed flux at wavelength λ, $F_{\text{cont}}(\lambda)$ is the estimated flux of the continuum in the absence of absorption, the average is over the wavelength range covering the observed Lyα and Lyβ emission lines, and the last equation defines the effective optical depth, τ_{eff}. Measurements of D_A imply an effective optical depth of

$$\tau_{\text{eff}}(z) = (0.85 \pm 0.06) \times \left(\frac{1+z}{5} \right)^{4.3 \pm 0.3} \quad \text{for} \quad z \lesssim 5.5 \tag{16.23}$$

(Fan et al., 2006). This optical depth in Eq. (16.21) implies that

$$n_{\text{HI}}(z) \sim 2.0 \times 10^{-11} h E(z) \left(\frac{1+z}{5} \right)^{4.3} \, \text{cm}^{-3}, \tag{16.24}$$

or in terms of the critical density,

$$\Omega_{\text{HI}}(z) \equiv \frac{n_{\text{HI}}(z) m_p}{\rho_{\text{crit},0}(1+z)^3} \sim 1.4 \times 10^{-8} h^{-1} E(z) \left(\frac{1+z}{5} \right)^{1.3}. \tag{16.25}$$

Comparing this with the density parameter of baryons, $\Omega_{b,0} \sim 0.02 h^{-2}$, given by cosmic nucleosynthesis (see §3.4), we see that the fraction of hydrogen in the neutral diffuse component must be smaller than 10^{-4} at $z \lesssim 5.5$! Thus, given that only a small fraction of the baryons are locked up in stars, the IGM must be highly ionized at $z \lesssim 5.5$.

Results based on the spectra of QSOs at $z \gtrsim 6$ show that the effective optical depth increases very rapidly with redshift (Fan et al., 2006). This indicates a rapid increase in the neutral content of the IGM at $z \gtrsim 6$. However, because the Lyα absorption cross-section is so large, even a small fraction of neutral hydrogen can result in an almost complete absorption of Lyα flux, making it difficult to estimate $n_{\text{HI}}(z)$ accurately. A more stringent constraint on the neutral hydrogen fraction can be obtained from the optical depth in higher order lines, such as Lyβ and Lyγ, which have lower absorption efficiencies. The effective Lyα optical depth obtained from such analyses is $\gtrsim 10$ for $z \sim 6.5$. Unfortunately, even such an optical depth corresponds to a neutral fraction of only about 10^{-3}, still far from that of a predominantly neutral gas. Thus, it is still unclear whether the IGM becomes neutral quickly at a redshift slightly above 6 or slowly so that the re-ionization of the IGM occurred at a much higher redshift.

Constraints on $n_{\rm HI}$ can also be obtained by considering the Lyman limit absorption, which involves the ionization of a neutral hydrogen from its ground state. The wavelength at the ionization threshold is $\lambda_{\rm L} = c/\nu_{\rm L} = 912\,{\rm Å}$. The photoionization (bound-free) cross-sections for hydrogenic atoms are given in §B1.3. For the Lyman-limit absorption of hydrogen atoms, we have

$$\sigma_{\rm L}(\nu) \approx \mathcal{K}_0 (\nu_{\rm L}/\nu)^3 \quad (16.26)$$

for $\nu \geq \nu_{\rm L}$, while $\sigma_{\rm L}(\nu) = 0$ for $\nu < \nu_{\rm L}$, and $\mathcal{K}_0 \approx 7.91 \times 10^{-18}\,{\rm cm}^2$. For a uniform medium, $n_{\rm HI}(t) \propto 1/a^3(t)$. Under the assumption that $T \ll T_{\rm L} \equiv h_{\rm P}\nu_{\rm L}/k_{\rm B} \approx 1.6 \times 10^5\,{\rm K}$, the optical depth can be written as

$$\tau(\nu_{\rm o}) = \frac{c}{H_0} n_{\rm HI}(t_{\rm a}) \mathcal{K}_0 \int_{z_{\rm a}}^{\infty} \frac{{\rm d}z}{(1+z)E(z)} \quad (16.27)$$
$$= 5 \times 10^{10} h^{-1} (1+z_{\rm a})^{-3/2} \left[\frac{n_{\rm HI}(t_{\rm a})}{{\rm cm}^{-3}}\right] \quad \text{(for EdS)},$$

where $z_{\rm a} = (\nu_{\rm L}/\nu_{\rm o}) - 1$. The problem with this test is again that the absorption cross-section is large for UV photons. The optical depth of the IGM is expected to be of order unity only in the extreme UV and soft X-ray, a regime difficult to observe because of Galactic absorption.

The Gunn–Peterson test can also be based on absorption by other atoms. For example, in the presence of neutral helium, its $(1s)^2 \to 1s2p$ absorption can cause a flux decrement at $584\,{\rm Å} < \lambda < 584\,{\rm Å}(1+z_{\rm e})$ in the observed spectra of QSOs. Similarly, in the presence of singly ionized helium, its $1s \to 2p$ absorption can cause a flux decrement at $304\,{\rm Å} < \lambda < 304\,{\rm Å}(1+z_{\rm e})$. Such observations require spectroscopy in the UV, and have been carried out for a number of quasars at $z \sim 3$ (e.g. Jakobsen et al., 1994; Heap et al., 2000).

16.1.4 Constraints from the Cosmic Microwave Background

An independent test of the epoch of re-ionization is provided by the anisotropy of the cosmic microwave background (CMB). As described in §6.7, once the Universe is ionized, CMB photons can interact with the free electrons through Thomson scattering as they propagate to us from the original last-scattering surface at $z \sim 1000$. This re-scattering suppresses the temperature fluctuations in the CMB and polarizes it. Both effects are proportional to the Thomson optical depth of the IGM, which is determined by the ionization state of the IGM.

The Thomson optical depth up to a redshift z along a line-of-sight can be written as

$$\tau(z) = \sigma_{\rm T} \int_0^z n_{\rm e}(z') \frac{{\rm d}l(z')}{{\rm d}z'}\,{\rm d}z', \quad (16.28)$$

where $\sigma_{\rm T}$ is the Thomson cross-section, $n_{\rm e}(z)$ is the number density (in proper units) of free electrons at redshift z along the line-of-sight in question, and ${\rm d}l$ is the proper depth corresponding to ${\rm d}z$ (see §3.2.6). We write $n_{\rm e}(z) = n_{\rm p}(z)X_{\rm e}(z)$, where $n_{\rm p}$ is the number density of protons (including those in helium), and $X_{\rm e}$ is the ionization fraction. Assuming the IGM gas to be primordial with 24% of the mass in helium, we have $n_{\rm p}(z) = 0.88\rho_{\rm b}(z)/m_{\rm p}$, where $\rho_{\rm b}$ is the baryonic mass density and $m_{\rm p}$ is the mass of a proton. With all these, the optical depth can be written as

$$\tau(z) \approx 0.063 h\Omega_{\rm b,0} \int_0^z \frac{(1+z')^2}{E(z')} \left[1 + \delta_{\rm b}(z')\right] X_{\rm e}(z')\,{\rm d}z', \quad (16.29)$$

where $\delta_b(z') \equiv [\rho_b(z')/\overline{\rho}_b(z')] - 1$ is the density fluctuation at z' along the line-of-sight in question. As one can see, the optical depth along a line-of-sight depends on both the gas density distribution and ionization fraction along the line-of-sight. Therefore, a map of τ across the sky can in principle be used to probe the ionization structure of the IGM. Consider a simple case in which the IGM is uniform so that $\delta_b = 0$, and re-ionization occurs instantaneously at $z = z_{ri}$ so that $X_e(z) = 0$ for $z > z_{ri}$ and $X_e(z) = 1$ for $z < z_{ri}$. Assume further that $z_{ri} \gg 1$. The optical depth towards the CMB generated by re-ionization is then

$$\tau(z_{ri}) \approx 0.07 \left(\frac{h}{0.7}\right)\left(\frac{\Omega_{b,0}}{0.04}\right)\left(\frac{\Omega_{m,0}}{0.3}\right)^{-1/2}\left(\frac{1+z_{ri}}{10}\right)^{3/2}. \quad (16.30)$$

Thus, if re-ionization occurs at $z_{ri} \sim 10$, the free electrons can scatter about 10% of the CMB photons. As shown in §6.7, the effect of such scattering can be observed in the CMB temperature and polarization spectra on scales comparable to the horizon size at the re-ionization epoch. The result obtained from the WMAP 3-Year data release gives $\tau = 0.1 \pm 0.03$ (Page et al., 2007), which corresponds to a re-ionization redshift $9 \lesssim z_{ri} \lesssim 14$ under the assumption of a homogeneous IGM and instantaneous re-ionization. With better data anticipated from future missions, such as PLANCK, it will be possible to probe the re-ionization process in more detail.

16.2 Ionizing Sources

The discussion in the last section shows that only a very small fraction of the baryons in the Universe reside in a neutral, diffuse component at redshift $z \lesssim 6$. Since the IGM is expected to be predominantly neutral immediately after recombination (see §16.1.1), the IGM must have been re-ionized during the epoch associated with the redshift range $6 \lesssim z \lesssim 1,000$. In this section, we examine possible sources that may have contributed to the re-ionization. Details regarding the re-ionization process itself will be described in the next section.

16.2.1 Photoionization versus Collisional Ionization

There are two main processes by which the IGM can be re-ionized: photoionization and collisional ionization. In what follows we describe both processes in turn.

(a) Photoionization Photoionization is the process in which an atom is ionized by the absorption of an ionizing photon (see §B1.3). As we will see below, the observed flux of ionizing photons can roughly be approximated by

$$J(\nu) = \left(\frac{\nu_L}{\nu}\right)^{\beta} J_{-21} \times 10^{-21} \mathrm{erg\,cm^{-2}\,s^{-1}\,Hz^{-1}\,sr^{-1}}, \quad (16.31)$$

where ν_L is the frequency of a Lyman-limit photon. It then follows that the photoionization rate for hydrogen is

$$\Gamma_{\gamma,H} = \int_{\nu_t}^{\infty} \frac{4\pi J(\nu)}{ch_P \nu} c\sigma_L(\nu)\,d\nu = \Gamma_{-12} \times 10^{-12}\mathrm{s^{-1}}, \quad (16.32)$$

where σ_L is the photoionization cross-section of a hydrogen atom, and

$$\Gamma_{-12} \approx \frac{12}{\beta+3} J_{-21}, \quad (16.33)$$

assuming σ_L given by Eq. (16.26). This rate corresponds to a time scale $t_{pi} \sim \Gamma_{\gamma,H}^{-1} \sim 10^5 \Gamma_{-12}^{-1}$ yr, which is much shorter than the expansion time scale $a/\dot{a}$ at the redshifts we are concerned with here, unless Γ_{-12} is very small. Thus, if the IGM is photoionized, the neutral fraction is determined by the condition that photoionization balances recombination between electrons and photons. The equilibrium equation in terms of the mean number density of HI is

$$\Gamma_{\gamma,\mathrm{H}}\langle n_{\mathrm{HI}}\rangle = \alpha_{\mathrm{H}}\langle n_{\mathrm{p}}n_{\mathrm{e}}\rangle = \alpha_{\mathrm{H}}\mathscr{D}\langle n_{\mathrm{p}}\rangle^2, \quad (16.34)$$

where $\alpha_{\mathrm{H}} \approx 5\times 10^{-13}(T/10^4\,\mathrm{K})^{-0.7}\,\mathrm{cm}^3\,\mathrm{s}^{-1}$ is the recombination coefficient of H^+ (see §B1.3), and we have assumed that $n_{\mathrm{e}} = n_{\mathrm{p}}$, which is a good approximation for a highly ionized IGM dominated by hydrogen. The factor $\mathscr{D} \equiv \langle n_{\mathrm{p}}^2\rangle/\langle n_{\mathrm{p}}\rangle^2 > 1$ takes account of possible clumpiness of the medium. Thus, the mean density of HI gas is

$$\langle n_{\mathrm{HI}}\rangle = \frac{\alpha_{\mathrm{H}}\mathscr{D}\langle n_{\mathrm{p}}\rangle^2}{\Gamma_{\gamma,\mathrm{H}}}. \quad (16.35)$$

The Gunn–Peterson limit on n_{HI} given in Eq. (16.24) then requires

$$\Gamma_{-12} \sim 20\mathscr{D}h^{-1}\left(\frac{\Omega_{\mathrm{p}}}{0.02h^{-2}}\right)^2\left(\frac{1+z}{5}\right)^{1.7}\frac{1}{E(z)}T_4^{-0.7}, \quad (16.36)$$

where

$$\Omega_{\mathrm{p}} \equiv \frac{n_{\mathrm{p}}m_{\mathrm{p}}}{\rho_{\mathrm{crit},0}(1+z)^3}. \quad (16.37)$$

Since the IGM is highly ionized, the density parameter of the IGM is $\Omega_{\mathrm{IGM}} \approx 1.3\Omega_{\mathrm{p}}$, assuming a mass fraction in helium of 0.24. Thus, if a non-negligible fraction of baryons is in the diffuse IGM at $z \sim 4$, then $\Gamma_{-12} \sim 1$, assuming $T_4 = (T/10^4\,\mathrm{K}) \sim 2$. An obvious question is: what are the sources that can generate such an UV flux?

(b) Collisional Ionization In addition to photoionization, the IGM can also be ionized by collisional processes. The most important one is the ionization of a hydrogen atom by collision with an electron. Neglecting photoionization, the ionization equilibrium gives

$$\frac{\Gamma_{\mathrm{e,H}}}{n_{\mathrm{e}}}\langle n_{\mathrm{HI}}\rangle = \alpha_{\mathrm{H}}\langle n_{\mathrm{p}}\rangle, \quad (16.38)$$

where $\Gamma_{\mathrm{e,H}}$ is the collisional rate (see §B1.3). Note that $\Gamma_{\mathrm{e,H}} \propto n_{\mathrm{e}}$ and so clumpiness of the medium does not enter here. Inserting the expressions of α_{H} and $\Gamma_{\mathrm{e,H}}$ into this equation, we obtain

$$\langle n_{\mathrm{HI}}\rangle = 7.5\times 10^{-13}\,\mathrm{cm}^{-3}\left(\frac{\Omega_{\mathrm{p}}}{0.02h^{-2}}\right)(1+z)^3\left(\frac{T_{\mathrm{t}}}{T}\right)^{1.2}\exp\left(\frac{T_{\mathrm{t}}}{T}\right)\frac{1+T_5^{0.5}}{1+T_6^{0.7}}, \quad (16.39)$$

where T_5 and T_6 are the temperature in units of 10^5 K and 10^6 K, respectively, and $T_{\mathrm{t}} \approx 157{,}809$ K. Thus, if a substantial fraction of the baryons at $z \sim 3$ are part of the IGM, and if collisional ionization dominates, then the observed Gunn–Peterson limit in Eq. (16.24) requires $T \gtrsim 10^6$ K. Such a high temperature is neither consistent with that inferred from the linewidths of Lyα lines in the Lyα forest (see §16.5) nor with the lack of spectral distortions in the CMB which are expected to be produced by such a hot medium (see §3.5.4). Therefore, it is nowadays generally believed that the IGM is predominately ionized by photoionization.

(c) Possible Ionizing Sources There are several obvious candidates for the sources of UV radiation that may contribute to the re-ionization of the IGM. Quasars are known to emit copious amounts of UV photons, and since they are observed out to redshifts $z \sim 6.5$, their role in re-ionizing the IGM has been examined extensively. We will describe in detail how to calculate the ionizing flux expected from quasars in §16.2.2. As we will see there, quasars alone may not be sufficient to provide all the UV flux required to re-ionize the Universe to the observed level, especially at redshifts beyond ~ 2 where the space density of bright quasars is observed to decline rapidly with increasing redshift (§14.3.2).

Young galaxies are another obvious source of Lyman continuum photons. As described in §2.6, optical surveys have revealed large numbers of Lyman-break galaxies at $z \sim 3$. The light of these galaxies observed in the visible bands, corresponding to the rest-frame UV continuum at $\sim 1500\,\text{Å}$, indicates that these galaxies are strong UV emitters. The contribution of young galaxies to the UV flux can be estimated from their UV luminosity function, as we will see in §16.2.2. The main uncertainty is that star-forming galaxies are often dusty, making it difficult to determine the exact fraction of UV photons that can escape from the sources to contribute to the UV background. Current estimates indicate that the combination of young galaxies and quasars provide sufficient UV flux to keep the Universe at the observed ionization level at $z \lesssim 5$. An interesting possibility is that the Universe has been re-ionized at $z \gtrsim 10$ by Population III stars forming from the primordial gas that manages to cool inside small collapsed dark matter halos (see §9.7). This population of objects has yet to be discovered, and its exact role in re-ionizing the IGM is still unclear. Nevertheless, as we will see in §16.3, numerous theoretical studies have investigated their potential impact on the re-ionization history of the Universe.

16.2.2 Emissivity from Quasars and Young Galaxies

The starting point of calculating the flux of ionizing photons is the radiative transfer equation (B1.18). Applying this equation to an expanding universe and integrating along the path of a light ray gives the specific intensity at a space-time position $(\mathbf{x}_o, t_o)$ in a direction $\hat{\mathbf{k}}_o$:

$$J(\nu_o, \hat{\mathbf{k}}_o, \mathbf{x}_o, t_o) = \frac{c}{4\pi} \int_0^{t_o} dt\, \frac{a^3(t)}{a^3(t_o)} \varepsilon\left[\frac{a(t_o)}{a(t)}\nu_o, \mathbf{x}(t), t\right] e^{-\tau[\nu_o, \mathbf{x}_o, t_o, \mathbf{x}(t), t]}, \qquad (16.40)$$

where

$$\tau[\nu_o, t_o, \mathbf{x}_o, \mathbf{x}(t), t] = c \int_t^{t_o} \rho[\mathbf{x}(t'), t'] \kappa\left[\frac{a(t_o)}{a(t')}\nu_o, \mathbf{x}(t'), t'\right] dt' \qquad (16.41)$$

is the optical depth at frequency ν_o [as observed by an observer at $(\mathbf{x}_o, t_o)$] for light emitted from $(\mathbf{x}, t)$. Hence, in order to obtain J in a given direction, we need to know the emissivity ε and opacity $\rho\kappa$ along the line-of-sight.

Consider a class of discrete sources with luminosity function $\phi(L, z)\,dL$, which is the comoving number density at redshift z for sources with luminosity in the range L to $L + dL$. For simplicity, we assume that the spectral energy distribution depends only on L and is $\mathscr{L}(L, \nu)\,d\nu$. The mean emissivity at z can then be written as

$$\varepsilon(\nu, z) = (1+z)^3 \int_0^\infty \mathscr{L}(L, \nu)\phi(L, z)\,dL. \qquad (16.42)$$

If the shapes of spectra are similar for all sources, we can write $\mathscr{L}(L, \nu) = Lf(\nu)$ with $\int f(\nu)\,d\nu = 1$, and

$$\varepsilon(\nu, z) = (1+z)^3 f(\nu) \int_0^\infty L\phi(L, z)\,dL. \qquad (16.43)$$

Thus, once $\phi(L, z)$ and $f(\nu)$ are known for a given class of sources, it is straightforward to calculate the mean emissivity. As one can see, if $f(\nu)$ is a power law, $f(\nu) \propto \nu^{-\alpha}$, and if there is no absorption by the IGM, the ionizing flux $J(\nu)$ is a power law with the same power index as $f(\nu)$, and with an amplitude that is determined by the luminosity density of the sources.

(a) Quasars The continuum emission spectrum of a quasar in the ultraviolet can be approximated by a power law,

$$f(\nu) = f_* \left(\frac{\nu}{\nu_*}\right)^{-\alpha}, \qquad (16.44)$$

where α is the power index, and ν_* is a reference frequency. The observed quasar luminosity function is usually parameterized as

$$\phi(L,z)\,dL = \phi^* \left\{ \left[\frac{L}{L^*(z)}\right]^{\beta_1} + \left[\frac{L}{L^*(z)}\right]^{\beta_2} \right\}^{-1} \frac{dL}{L^*(z)}, \quad (16.45)$$

where β_1 and β_2 are constant power indices, and $L^*(z)$ is the characteristic luminosity (see §14.3.2). The redshift dependence of $L^*(z)$ over the entire redshift range may be parameterized as

$$L^*(z) = L^*(0)(1+z)^{\alpha-1} \exp\left[-\frac{z(z-2z_*)}{2\sigma_*^2}\right]. \quad (16.46)$$

It then follows that the mean emissivity from quasars is

$$\varepsilon_q(\nu,z) = \varepsilon_q(\nu_*,0) \left(\frac{\nu}{\nu_*}\right)^{-\alpha} (1+z)^{\alpha+2} \exp\left[-\frac{z(z-2z_*)}{2\sigma_*^2}\right], \quad (16.47)$$

where $\varepsilon_q(\nu_*,0)$ is the emissivity at the reference frequency ν_* at the present time.

The constants in Eq. (16.47) can all be determined from the observed quasar luminosity function. Assuming $\alpha = 1$, $h = 0.5$ and $q = 0.1$, Pei (1995) obtained that $\beta_1 = 1.83$, $\beta_2 = 3.7$, $z_* = 2.77$, $\sigma_* = 0.91$, $\log(\phi^*/\mathrm{Gpc}^{-3}) = 2.37$, and $\log[L^*(0)/L_\odot] = 13.4$ (in the B-band). If we choose the reference frequency at $\nu_* = c/(4400\,\text{Å})$, then $\varepsilon_q(\nu_*,0) \approx 6.4 \times 10^{32}\,\mathrm{erg\,Gpc^{-3}\,s^{-1}\,Hz^{-1}}$. The dotted line in Fig. 16.1 shows the UV flux obtained from the quasar emissivity in a perfectly transparent universe (where $\tau = 0$). This flux is roughly a power law, $J(\nu) \propto \nu^{-1}$.

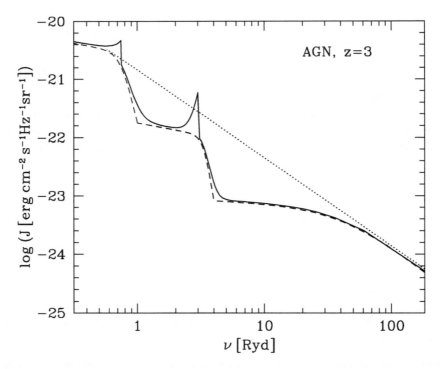

Fig. 16.1. The ionizing background between 0.1 and 200 Ryd at $z = 3$. The dotted line shows the unattenuated ($\tau = 0$) spectrum due to quasars. The dashed line shows the result after absorption due to an intervening medium is taken into account, while the solid line shows the results when also accounting for the recombination emissivity. [Based on data published in Haardt & Madau (1996)]

(b) Young Galaxies A massive (OB) star can emit a significant amount of UV photons during its main-sequence lifetime. Thus, star-forming galaxies can also contribute to the UV background. Since massive stars are short-lived, their number density should be directly proportional to the current star-formation rate density. Suppose that the mean star-formation rate per unit proper volume at time t is $\dot{\rho}_\star(t)$. The mean emissivity from star-forming galaxies can then be written as

$$\varepsilon_g(\nu,t) = \int_0^t dt' \int_0^\infty dm \frac{\phi(m)}{M_\odot} f_\nu(m, t-t') \dot{\rho}_\star(t'), \qquad (16.48)$$

where $\phi(m)$ is the IMF, and $f_\nu(m,\tau)$ is the energy spectrum of a star with mass m at age τ.

The star-formation history of the Universe, $\dot{\rho}_\star(z)$, has been determined observationally out to $z \sim 6$ (see §2.6.8). With the use of spectra of stars of different masses and ages (see §10.3.1), the total background emissivity can be calculated. It turns out that the contribution of young galaxies to the UV background radiation is comparable to that of quasars at $z \sim 2$, is dominating at $z > 3$ and is unimportant at $z < 1$. Since stellar spectra are generally soft with little emission at $\nu > 4\,\mathrm{Ryd}$, star-forming galaxies contribute mainly to the ionizing flux at $\nu < 4\,\mathrm{Ryd}$.[1]

16.2.3 Attenuation by Intervening Absorbers

As shown by Eq. (16.40), in order to obtain the ionizing flux from the emissivity, one also needs to know the effective optical depth of the medium. The Gunn–Peterson test shows that the optical depth due to the diffuse part of the IGM is very small. However, the optical depth due to discrete absorption systems may still be important. Here we show how to calculate the effective optical depth from discrete systems (e.g. clouds).

As one can inspect from Eq. (16.40), in the absence of local emission the fluxes at two adjacent points $t - dt$ (corresponding to redshift $z + dz$) and t (corresponding to z) along a particular line-of-sight are related by

$$j(\nu_o, \hat{\mathbf{k}}_o, z) = j(\nu_o, \hat{\mathbf{k}}_o, z + dz) \exp\left[-\tau(\nu_o, \hat{\mathbf{k}}_o, z, z + dz)\right], \qquad (16.49)$$

where $j(\nu_o, \hat{\mathbf{k}}_o, z) = J(\nu_o, \hat{\mathbf{k}}_o, z)/(1+z)$. Averaged over different lines-of-sight, this gives

$$j(\nu_o, z) = j(\nu_o, z+dz) \left\langle \exp\left[-\tau(\nu_o, \hat{\mathbf{k}}_o, z, z+dz)\right] \right\rangle_{\hat{\mathbf{k}}_o}. \qquad (16.50)$$

In a cloudy medium, if a line-of-sight 'hits' a cloud in the redshift range $z \to z + dz$, the flux will be reduced by a factor $\exp[-\tau_c(\nu_z)]$, where $\tau_c(\nu_z)$ is the optical depth of a cloud to photons with frequency $\nu_z = (1+z)\nu_o$. Otherwise, if the line-of-sight does not hit a cloud there is no absorption. If we denote the probability for a line-of-sight to hit a cloud in the redshift range $z \to z + dz$ to be $\eta_z dz$, we have that

$$\left\langle \exp\left[-\tau(\nu_o, \hat{\mathbf{k}}_o, z, z+dz)\right] \right\rangle_{\hat{\mathbf{k}}_o} = (1 - \eta_z dz) + \eta_z dz\, e^{-\tau_c(\nu_z)}. \qquad (16.51)$$

Inserting this into Eq. (16.50) and integrating over redshift we get

$$j(\nu_o, z_o) = j(\nu_o, z_{em}) \exp\left[-\tau_{eff}(\nu_o, z_o, z_{em})\right], \qquad (16.52)$$

where the effective optical depth is

$$\tau_{eff}(\nu_o, z_o, z_{em}) = \int_{z_o}^{z_{em}} \eta_z \left[1 - e^{-\tau_c(\nu_z)}\right] dz. \qquad (16.53)$$

---

[1] The Rydberg constant (Ryd) represents the limiting value of the highest wavenumber (the inverse wavelength) of any photon that can be emitted from the hydrogen atom and corresponds to $1.0974 \times 10^5\,\mathrm{cm}^{-1}$.

The optical depth of a single cloud depends on its HI column density, $N_{\rm HI}$. Thus, for a medium consisting of clouds with various column densities, the effective optical depth can be written as

$$\tau_{\rm eff}(\nu_0,z_0,z_{\rm em}) = \int_{z_0}^{z_{\rm em}} dz \int_0^\infty dN_{\rm HI}\,\mathscr{F}(N_{\rm HI},z)\left[1-e^{-\tau(\nu,N_{\rm HI})}\right], \qquad (16.54)$$

where $\tau(\nu,N_{\rm HI})$ [with $\nu \equiv \nu_0(1+z)/(1+z_0)$] is the optical depth of an absorber with HI column density $N_{\rm HI}$ for photons with frequency ν, and $\mathscr{F}(N_{\rm HI},z)$ is the average number of absorption systems per line-of-sight, per unit redshift interval around z, and per unit interval of HI column density around $N_{\rm HI}$ (see §16.4.1). As we will see in §16.5.2, the form of $\mathscr{F}(N_{\rm HI},z)$ can be determined from observations of quasar absorption line systems. For a hydrogen–helium gas, the optical depth can be written as

$$\begin{aligned}\tau(\nu,N_{\rm HI}) &= N_{\rm HI}\sigma_{\rm HI}(\nu) + N_{\rm HeI}\sigma_{\rm HeI}(\nu) + N_{\rm HeII}\sigma_{\rm HeII}(\nu) \\ &= N_{\rm H}Y_{\rm HI}\left[\sigma_{\rm HI} + \frac{N_{\rm He}}{N_{\rm H}}\left(\frac{Y_{\rm HeI}}{Y_{\rm HI}}\sigma_{\rm HeI} + \frac{Y_{\rm HeII}}{Y_{\rm HI}}\sigma_{\rm HeII}\right)\right],\end{aligned} \qquad (16.55)$$

where σ_i is the photoionization cross-section of species 'i', $Y_{\rm HI} \equiv N_{\rm HI}/N_{\rm H}$, $Y_{\rm HeI} \equiv N_{\rm HeI}/N_{\rm He}$, and $Y_{\rm HeII} \equiv N_{\rm HeII}/N_{\rm He}$. As expected, the optical depth of a cloud depends not only on its HI column density, but also on the column density ratios such as $N_{\rm HeI}/N_{\rm HI}$ and $N_{\rm HeII}/N_{\rm HI}$. Thus a self-consistent calculation of the effective optical depth requires a detailed understanding of the ionization state of the absorbing gas. This in turn requires a self-consistent treatment of the evolution of the thermal and ionization states of the IGM. To proceed, we assume that the absorbing gas is optically thin so that the ionization intensity does not change significantly across an absorber. We also assume that the absorbing gas is in thermal and ionization equilibrium. Under these assumptions, the ionization fractions for a hydrogen–helium gas are given by

$$1 - Y_{\rm HI} = Y_{\rm HI}I_{\rm HI}, \qquad (16.56)$$

$$Y_{\rm HeII} = Y_{\rm HeI}I_{\rm HeI}, \quad 1 - Y_{\rm HeI} - Y_{\rm HeII} = Y_{\rm HeII}I_{\rm HeII}, \qquad (16.57)$$

where

$$I_i = \frac{\Gamma_{\gamma,i}}{n_e\alpha_i(T)}, \quad \Gamma_{\gamma,i} = \int_{\nu_i}^\infty d\nu\,\frac{4\pi J(\nu)}{h_{\rm P}\nu}\sigma_i(\nu), \quad (i={\rm HI, HeI, HeII}). \qquad (16.58)$$

In the above equations, T is the temperature of the gas, α_i is the radiative recombination rate to all levels of species 'i' (the Case A recombination rate), ν_i is the ionization threshold, and σ_i is the photoionization cross-section. These rates and cross-sections for hydrogen and helium are given in §B1.3. Inserting Eqs. (16.56) and (16.57) into Eq. (16.55) we have

$$\tau(\nu,N_{\rm HI}) = N_{\rm HI}\left[\sigma_{\rm HI} + \frac{N_{\rm He}}{N_{\rm H}}\frac{(\sigma_{\rm HeI}+I_{\rm HeI}\sigma_{\rm HeII})(1+I_{\rm HI})}{1+I_{\rm HeI}(1+I_{\rm HeII})}\right]. \qquad (16.59)$$

For highly ionized gas, $Y_{\rm HI} \ll 1$, $Y_{\rm HeI} \ll Y_{\rm HeII} \ll 1$, and so $I_{\rm HI} \gg 1$, $I_{\rm HeI} \gg 1$, and $I_{\rm HeII} \gg 1$. It then follows that

$$\tau(\nu,N_{\rm HI}) = N_{\rm HI}\left[\sigma_{\rm HI}(\nu) + \frac{N_{\rm He}}{N_{\rm H}}\frac{I_{\rm HI}}{I_{\rm HeII}}\sigma_{\rm HeII}(\nu)\right]. \qquad (16.60)$$

Under these assumptions, τ depends on $N_{\rm HI}$ but does not depend on the density of the absorbing gas. Since $\sigma_{\rm HI}$ and $\sigma_{\rm HeII}$ are strongly peaked at the HI and HeII ionization edges, respectively, the absorption by hydrogen and helium in intervening clouds produces two breaks near 1 and 4 Ryd in the ionizing flux, even though the emissivity $\varepsilon(\nu)$ is a pure power law in ν (see Fig. 16.1).

So far we have only considered the *absorbing* effect of intervening clouds. Ionized and excited atoms in the clouds can also emit photons through recombination and de-excitation. Although such photons do not contribute to the emission we observe from a source (because they have random propagation direction), they contribute to the radiation background. The emissivity from such processes should therefore be included in the calculation of the ionizing flux. As an example, let us consider the recombination emissivity from hydrogen atoms. As described in §B1.3, when a hydrogen ion captures an electron, the recombination is either directly to the ground level or first to an excited level which then decays radiatively to the ground state. In general, the emissivity associated with a given transition (free–bound or bound–bound) can be written as

$$\varepsilon_{\rm rec}(\nu,z)\,{\rm d}\nu = h_{\rm P}\nu f_\gamma(\nu)\,{\rm d}\nu\,[n_{\rm HI}\Gamma_{\gamma,{\rm H}}]\,\frac{\alpha_{nn'}}{\alpha_{\rm H}}, \qquad (16.61)$$

where $f_\gamma(\nu)\,{\rm d}\nu$ is the probability for a recombination to produce a photon with frequency between ν and $\nu+{\rm d}\nu$, and the ratio $\alpha_{nn'}/\alpha_{\rm H}$ is the fraction of recombinations leading to the transition in question (see Chapter 4 of Osterbrock, 1989). The quantity in brackets is the photo-ionization rate per unit volume, and is equal to the volume recombination rate under the condition of ionization equilibrium. Given the HI column density distribution, the mean value of $n_{\rm HI}$ can be obtained as

$$n_{\rm HI}(z) = \frac{1}{c}\frac{{\rm d}z}{{\rm d}t}\int \mathscr{F}(N_{\rm HI},z)N_{\rm HI}\,{\rm d}N_{\rm HI}. \qquad (16.62)$$

For bound–bound transitions, $f_\gamma(\nu) = \delta(\nu-\nu_{nn'})$, while for free–bound transitions, $f_\gamma(\nu)$ is given by the velocity distribution of free electrons $f(v)$ (e.g. a Maxwellian distribution) as $f_\gamma(\nu) = f(v)({\rm d}v/{\rm d}\nu)$, where $\frac{1}{2}m_e v^2 = h_{\rm P}(\nu-\nu_{\rm t})$ with $\nu_{\rm t}$ the threshold frequency of the energy level in consideration. With all these, the recombination emissivity from hydrogen atoms can be calculated. Similar procedures can be used for other atoms or ions. The solid line in Fig. 16.1 shows the predicted UV background when these recombination processes are accounted for; the bumps, which sometimes look like emission lines, are mainly due to Lyα line emission from the absorbing clouds.

16.2.4 Observational Constraints on the UV Background

The UV flux in the local Universe can be measured directly from space. However, since it is difficult to isolate the contribution due to the Galactic component, it is still somewhat uncertain what this implies for the UV background radiation (see Henry, 1991, for a review).

At $z > 0$, the existence of an ionizing background can be probed by the proximity effect. As noted by Weymann et al. (1981), the number density of Lyα forest lines (produced by HI Lyα absorption) in a quasar absorption spectrum decreases as the absorption redshift approaches that of the quasar. The standard interpretation of this proximity effect is that the absorbing clouds near the quasar are more ionized than average because of the local UV radiation produced by the quasar. Since the strength of a Lyα line measures the amount of HI gas in the absorber (see §16.4), a reduction in the neutral fraction causes a decrease in the number of lines above a given threshold strength. The extent of this proximity effect can therefore be used to measure the intensity, $J(\nu)$, of the UV background. For example, if we define a 'zone of influence' for a particular quasar as the region within which the ionizing radiation from the quasar is larger than or equal to that of the ambient background, then the larger the ambient flux $J(\nu)$ is, the smaller this 'zone of influence'. The ionizing flux from a quasar can be inferred from its luminosity and spectrum. Together with a measurement of the extent of the proximity effect, we can in principle estimate the background $J(\nu)$ at the HI ionization edge. Since the mean flux, $J(\nu)$, is by definition a constant at a given redshift, in practice many quasars with similar redshifts are used to give a statistical measurement. The basic formalism for such analyses is described by Bajtlik et al. (1988). Observational results for $J(\nu)$ can be found in Lu et al. (1991), Bechtold

(1994), Giallongo et al. (1996) and Scott et al. (2000), among others. These results can be summarized as

$$J(1\,\mathrm{Ry}) = (0.1 \to 1) \times 10^{-21}\,\mathrm{erg\,s^{-1}\,cm^{-2}\,Hz^{-1}\,sr^{-1}} \quad (2 \lesssim z \lesssim 4). \tag{16.63}$$

At lower redshifts ($z \lesssim 2$), the observational results are more uncertain. The rather limited observational data show that $J(1\,\mathrm{Ry})$ decreases by a factor of 10–100 from $z \sim 2$ to $z \sim 0$ (Kulkarni & Fall, 1993; Donahue et al., 1995; Scott et al., 2000), consistent with the fact that the comoving number density of bright quasars decreases rapidly with decreasing z at $z \lesssim 2$ (see §14.3.2).

The observed optical depth can also be used to constrain the UV background, if assumptions are made about cosmology, in particular the value of $\Omega_{\mathrm{b},0}$, and about the temperature of the IGM. This is a model-dependent approach, but it can provide a useful check on whether the UV flux required in current models of structure formation to match the observed optical depth is consistent with the UV background expected from quasars and young galaxies. The standard approach adopted here is to use gas simulations or semi-analytical models to trace the structure, temperature and ionization state of the IGM, assuming a UV background, which is tuned to match the observed optical depth. The results from such analyses (e.g. McDonald & Miralda-Escudé, 2001; Fan et al., 2002; Bolton et al., 2005) show that $\Gamma_{-12} \sim 1$ at $z \sim 2$–3, dropping by a factor of ~ 10 at $z \sim 6$, and by a similar factor at $z \sim 0$.

16.3 The Evolution of the Intergalactic Medium

For regions much smaller than the Hubble radius, the IGM is adequately described by a Newtonian fluid. The basic equations governing the evolution of the IGM are therefore those described in Appendix B. To summarize briefly, the baryonic component obeys the fluid equations, and is coupled to the dark matter component through gravity and to the radiation field through radiative processes. Once the ionizing and heating sources are known, these equations can in principle be solved to determine the thermal and ionization states of the IGM, as well as the properties of the radiation field. Obviously, the solution to such a dynamical system is in general extremely complicated, and in many cases one has to resort to numerical simulations. In what follows, we describe the basic properties of IGM evolution, often based on simplifying assumptions that allow us to gain insight.

16.3.1 Thermal Evolution

Assuming that the IGM is uniform and in thermal equilibrium, the entropy equation (see §B1.2) can be written as

$$\frac{\mathrm{d}\ln T}{\mathrm{d}\ln(1+z)} = (\gamma-1)\left[3 + \frac{1}{(\gamma-1)}\frac{\mathrm{d}\ln\mu}{\mathrm{d}\ln(1+z)} - \frac{\mathscr{H}-\mathscr{C}}{H(z)nk_{\mathrm{B}}T}\right], \tag{16.64}$$

where n is the particle number density, γ is the adiabatic index, $\mu m_{\mathrm{p}} \equiv \rho/n$ is the mean molecular weight, and the gas density evolves with z as $\rho(z) = \rho(0)(1+z)^3$. For a primordial gas consisting of hydrogen and helium, the particles involved are H^0, H^+, He^0, He^+, He^{++}, and e. The electron number density and the total number density of particles are

$$n_{\mathrm{e}} = n_{\mathrm{H}^+} + n_{\mathrm{He}^+} + 2n_{\mathrm{He}^{++}}, \tag{16.65}$$

$$n = n_{\mathrm{H}^0} + 2n_{\mathrm{H}^+} + n_{\mathrm{He}^0} + 2n_{\mathrm{He}^+} + 3n_{\mathrm{He}^{++}}, \tag{16.66}$$

respectively. The number densities of ions obey the following set of ionization equations:

$$\frac{\mathrm{d}n_i}{\mathrm{d}t} = -\left(\Gamma_{\gamma,i} + \Gamma_{\mathrm{e},i}n_{\mathrm{e}}\right)n_i + \alpha_{i+1}n_{i+1}n_{\mathrm{e}}, \tag{16.67}$$

where $(i, i+1)$ are (H^0, H^+), (He^0, He^+), (He^+, He^{++}), and $\Gamma_{\gamma,i}$, $\Gamma_{e,i}$, α_i are the rates of photoionization, collisional ionization, and recombination, respectively. These rates for hydrogen and helium are given in §B1.3. The three ionization equations, together with the total mass density, $\rho = m_H n_H + m_{He} n_{He}$, and the relative number of H and He atoms, $Y \equiv n_{He}/n_H$, completely determine the number densities of ions, once the temperature, T, and the ionizing intensity, $J(\nu)$, are given. With this, the cooling rate $\mathscr{C}$ can also be calculated using the formulae given in Table B1.2.

The ionization time scale for species i and the recombination time scale for species $i+1$ are

$$t_{\text{ion}}(i) = (\Gamma_{\gamma,i} + \Gamma_{e,i} n_e)^{-1} \quad \text{and} \quad t_{\text{rec}}(i) = (\alpha_{i+1} n_e)^{-1}, \tag{16.68}$$

respectively. If these time scales are shorter than the Hubble time, then we can set $dn_i/dt = 0$ in Eq. (16.67) and the assumption of ionization equilibrium is justified. As is evident from Eq. (16.68), at low redshifts, where n_e is small, the recombination time scale may become longer than the Hubble time for the species in consideration. In this case, the assumption of ionization equilibrium is invalid, and one has to solve the time-dependent ionization equation to determine the densities of ions.

(a) Photoionization Heating In order to solve Eq. (16.64) we also need to specify the heating rate $\mathscr{H}$. If there is no heating source other than photoionization, the heating rate is

$$\mathscr{H} = n_{H^0} \varepsilon_{H^0} + n_{He^0} \varepsilon_{He^0} + n_{He^+} \varepsilon_{He^+}, \tag{16.69}$$

with

$$\varepsilon_i = \int_{\nu_i}^{\infty} \frac{4\pi J(\nu)}{h_P \nu} \sigma_i(\nu) h_P(\nu - \nu_i) \, d\nu, \tag{16.70}$$

where ν_i is the frequency at the ionizing threshold for species i and σ_i is its photoionization cross-section (see §B1.3). In this simple case, the evolution of the IGM is completely specified by ρ, Y and $J(\nu)$.

Eq. (16.64) has been integrated for various choices of ρ, $J(\nu)$ and Y. For $\rho \sim 10^{-6}(1+z)^3 m_H \text{cm}^{-3}$ and $Y \sim 0.1$, the typical values for the IGM, and for a UV flux similar to that given by Eq. (16.31) with $\Gamma_{-12} \sim 1$, the results can be summarized as follows. If the initial temperature is less than 10^4 K, the IGM is heated up quickly by photoionization to a level of a few $\times 10^4$ K and then decreases slowly due to adiabatic cooling as the Universe expands. If some (unknown) sources [other than those responsible for $J(\nu)$] once heated the gas up to $T \gg 10^4$ K at a time when Compton cooling is still efficient (i.e. at $z > z_{\text{Comp}}$, see §8.1.2), Compton cooling against the CMB photons rapidly reduces the IGM temperature to a level of about 3×10^4 K, after which the temperature continues to decrease slowly due to the expansion of the Universe. In this case, the heating does not have an important effect on the IGM. If, however, the heating epoch is at $z < z_{\text{Comp}}$, then Compton cooling is insufficient, and the temperature of the IGM can remain significantly higher than 10^4 K even up to the present time.

(b) Shock Heating Apart from the heating by photoionization, the IGM may also have been shock heated. There are basically two sources for shock heating: gravitational collapse and explosive events such as supernovae, and we will discuss both of these in turn.

As discussed in §8.3, when density perturbations collapse, the associated gas is shock heated to the virial temperature of the corresponding structure. As discussed in §5.3, the collapse of density perturbations is generically aspherical, first forming sheet-like pancakes (first axis collapse), followed by filamentary structures (second axis collapse), and eventually virialized dark matter halos (third axis collapse). Hence, we expect the space between the halos to be permeated with sheets and filaments as in numerical simulations of structure formation in hierarchical cosmogonies (see §5.6). In the process of forming these sheets and filaments, the associated gas is shock heated to temperatures in the range $10^5 \text{ K} \lesssim T \lesssim 10^6 \text{ K}$ (Mo et al., 2005). At redshifts $z \gtrsim 2$

the density of this shock heated gas is sufficiently high that it manages to cool back to its original temperature within a Hubble time. However, at $z \lesssim 2$ the cooling time exceeds the Hubble time, and since a significant fraction of the IGM is embedded within these sheets and filaments, the gravitational collapse associated with their formation is expected to cause a net heating of the IGM. This is indeed confirmed by hydrodynamical simulations, which show that at $z \sim 2$, between 30% and 40% of the IGM is heated to temperatures in the range $10^5\,\mathrm{K} \lesssim T \lesssim 10^7\,\mathrm{K}$, making up what is called the warm-hot intergalactic medium (WHIM) which is distributed in sheets and filaments (e.g. Cen & Ostriker, 1999; Davé et al., 2001). Although this WHIM is extremely difficult to detect, there is observational support for its existence from OVI absorption line systems (see §16.9.2).

In addition to gravitational collapse, shocks can also be produced by stars. In particular, as discussed in §10.5, the total kinetic energy associated with supernova explosions in young galaxies is comparable to the binding energy of the interstellar gas. If this kinetic energy is effectively thermalized, the interstellar gas can be heated up to temperatures high enough for the gas to escape from a galaxy, producing galactic winds that may heat the IGM. This possibility has been discussed in some detail by Ikeuchi & Ostriker (1986). Neglecting the details of shock propagation and thermalization, one can assume that the average heating rate due to young galaxies is proportional to the star-formation rate, $\mathscr{H}(z) = A\dot{\rho}_\star(z)$. In this case Eq. (16.64) can be solved if the form of $\dot{\rho}_\star(z)$ and the value of A are known. As mentioned above, an instantaneous heating at $z > z_\mathrm{Comp}$ has little effect on the thermal evolution of the IGM at lower redshift, but heating at a later epoch can raise the gas temperature to $T \gg 10^4\,\mathrm{K}$. In the extreme cases considered by Ikeuchi & Ostriker, the temperature of the IGM was initially heated to a value higher than $10^6\,\mathrm{K}$. Although such high temperatures is what is required to collisionally ionize the IGM to the Gunn–Peterson limit, the existence of a hot IGM is stringently constrained by the lack of spectral distortions in the CMB (see §6.7) and by the velocity widths of absorption lines in the Lyα forest (see §16.5).

16.3.2 Ionization Evolution

In the photoionization model, the IGM is ionized by the UV photons from ionizing sources. Here we describe how such photoionization proceeds. To start with, consider an isolated ionizing source embedded in an ambient medium which is initially neutral and has a mean density $\langle \rho \rangle = m_\mathrm{p} n_\mathrm{I}$ and temperature T_I. The photons from the source ionize the medium, producing an ionization front which propagates into the neutral medium. The production of electrons by ionization at a radius r from the source is governed by the equation

$$\frac{\partial n_\mathrm{e}}{\partial t} + \nabla \cdot (n_\mathrm{e} \mathbf{v}) = -\nabla \cdot \left[\frac{\dot{N}_\gamma e^{-\tau(r)}}{4\pi r^2} \hat{\mathbf{r}} \right] - \alpha_\mathrm{H} n_\mathrm{e}^2, \tag{16.71}$$

where $\dot{N}_\gamma$ is the number of ionizing photons released by the source per unit time, and $\tau(r) \sim 0$ is the optical depth from the source to radius r due to the small fraction of neutral hydrogen in the ionized region. The motion of an ionization front in a uniform medium is characterized by two distinct stages. During the first stage, the ionized volume grows rapidly in size to an equilibrium radius equal to that of a Strömgren sphere,

$$r_\mathrm{i} = \left(\frac{3\dot{N}_\gamma}{4\pi n_\mathrm{I}^2 \alpha_\mathrm{H}} \right)^{1/3}, \tag{16.72}$$

within which the gas is highly ionized and heated by photoionization to a temperature $T_\mathrm{II} \sim 10^4\,\mathrm{K}$. The second stage begins when the overpressure (due to the high temperature) in the ionized region causes a significant outward motion of the ionized gas. This motion pushes the ambient medium, shocking and compressing the swept-up gas into a thin mass shell. The expansion of the ionized gas reduces the overpressure, and the final number density of electrons, n_f, within the ionized volume is given by pressure balance with the ambient gas: $2n_f T_{\rm II} = n_{\rm I} T_{\rm I}$. The final radius of the ionized volume is

$$r_{\rm f} = \left(\frac{3\dot{N}_\gamma}{4\pi n_f^2 \alpha_{\rm H}}\right)^{1/3} = r_{\rm i}\left(\frac{2T_{\rm II}}{T_{\rm I}}\right)^{2/3}. \tag{16.73}$$

This final radius can be reached if the lifetime of the ionizing source is long enough.

The motion of the ionization front during the first stage can be easily obtained. Since little mass motion is involved, we can neglect the term containing $\mathbf{v}$ in Eq. (16.71). Integrating this equation over the volume V enclosed by the instantaneous ionization front, we have

$$\frac{\rm d}{{\rm d}t}[\langle n_{\rm e}\rangle V] = \dot{N}_\gamma - \alpha_{\rm H}\langle n_{\rm e}^2\rangle V, \tag{16.74}$$

where we have allowed possible inhomogeneity in the ionized volume and denoted the average over V by $\langle\cdots\rangle$. If the medium is expanding with the Universe, we have $\langle n_{\rm e}\rangle \approx n_{\rm I} \propto a^{-3}$, where the first relation follows from the fact that the gas within V is highly ionized. We can then write Eq. (16.74) as

$$\frac{{\rm d}V}{{\rm d}t} - \frac{3\dot{a}}{a}V = \frac{\dot{N}_\gamma}{n_{\rm I}} - \frac{V}{\bar{t}_{\rm rec}}, \tag{16.75}$$

where

$$\bar{t}_{\rm rec} \equiv \frac{\langle n_{\rm e}\rangle}{\alpha_{\rm H}\langle n_{\rm e} n_{\rm p}\rangle} \approx \frac{1}{\mathscr{Q}\alpha_{\rm H} n_{\rm I}} \tag{16.76}$$

(with $\mathscr{Q}$ the clumpiness factor; see §16.2.1) is the mean recombination time scale within V. If $\bar{t}_{\rm rec} \ll t$, so that the expansion of the Universe can be neglected, the solution to the above equation is

$$V(t) = \frac{\dot{N}_\gamma \bar{t}_{\rm rec}}{n_{\rm I}}\left[1 - \exp\left(-\frac{t}{\bar{t}_{\rm rec}}\right)\right]. \tag{16.77}$$

In this case, the ionized volume is simply a time-dependent Strömgren sphere.

If the mean number density of the ionizing sources at redshift z is $n_{\rm s}$, the volume filling factor of Strömgren spheres is $n_{\rm s}V$, and the rate per unit comoving volume at which ionizing photons are emitted is

$$\mathscr{\dot{N}}_\gamma(z) \equiv n_{\rm s}\dot{N}_\gamma. \tag{16.78}$$

In order for the Universe to be completely ionized ($n_{\rm s}V > 1$), it is required that

$$\mathscr{\dot{N}}_\gamma > \frac{n_{\rm I}(0)}{\bar{t}_{\rm rec}(z)} \approx 5.3 \times 10^{50}\,{\rm s}^{-1}\,{\rm Mpc}^{-3}\left(\frac{\mathscr{Q}}{10}\right)\left(\frac{1+z}{6}\right)^3\left(\frac{\Omega_{\rm b,0}h^2}{0.02}\right)^2. \tag{16.79}$$

16.3.3 The Epoch of Re-ionization

In reality, re-ionization is a process rather than an event. It is expected to proceed in the following phases: (i) an initial phase in which the Universe is largely neutral except for isolated HII regions produced by individual ionizing sources; (ii) a second phase in which individual HII regions start to overlap, but about half of the IGM is still neutral; (iii) a third phase in which the Universe is largely ionized, except for some pockets of neutral gas associated with high density regions; and (iv) a final phase in which the Universe is completely ionized, with a very small neutral fraction, similar to what is inferred from the Gunn–Peterson test at $z \sim 6$.

The details of this re-ionization process depend on a number of factors. First of all, it depends on the physical properties of the individual ionizing sources, such as luminosity, spectrum, and the fraction of ionizing photons that can escape from the source. Secondly, it depends on the properties of the source population, such as number density, time evolution, and spatial distribution (clustering) in the cosmic density field. Thirdly, it depends on the properties of the gas density field. In particular, since recombination is a two-body process, accurate predictions for the ionization fraction requires accurate modeling of the high-density regions, even if these only cover a small fraction of the total volume. And finally, the first generation of sources can affect the IGM not only through ionization, but also through photoheating, shock-heating, and chemical enrichment, which can affect the subsequent formation of ionizing sources. Clearly, modeling the re-ionization history of the Universe is a daunting challenge, even with modern computer simulations. On the one hand, one needs very high resolution to properly model star formation, feedback effects and small-scale clumpiness in the IGM. On the other hand, one needs a large volume to properly represent the cosmic density field. In addition, one needs to keep track of the radiation field through radiative transfer and to calculate the ionization states of different species by solving the ionization equations. In recent years, the re-ionization history of the Universe has been investigated both numerically and analytically (Loeb & Barkana, 2001; Ciardi & Ferrara, 2005). Although many of the details are still uncertain, there are a number of results that are worth mentioning.

During the early phase, re-ionization proceeds slowly because of the small number density of ionizing sources and because of the small sizes of HII regions due to the high density of the IGM. The re-ionization process accelerates as more ionizing sources form and the average density of the IGM decreases. As soon as the HI regions start to percolate, re-ionization proceeds very fast as the ionizing photons can propagate to large distances to ionize the neutral medium. Since the medium is inhomogeneous and the sources are clustered in space, re-ionization can be extremely patchy, with complicated morphologies. Naively, one might expect percolation of the HII regions to occur first in regions of high source density. However, if ionizing sources form preferentially in high density regions, the answer is not obvious. The ionization fraction of the gas depends not only on the local flux of ionizing photons, but also on the recombination rate, which is higher in regions of higher gas densities. As shown by Gnedin (2000) and Miralda-Escudé et al. (2000), the enhancement of the recombination rate in high density regions may overcome the high UV flux associated with the high source density, so that re-ionization proceeds in an outside-in fashion, i.e. from low density to high density. Unfortunately, the results depend on the details of the assumptions about source distribution and the radiative transfer around sources. For instance, Ciardi & Madau (2003) and Iliev et al. (2006) find that re-ionization proceeds basically in an inside-out fashion, i.e. from high density regions near the sources to lower density regions that are farther away.

The recombination time scale is comparable to the Hubble time at $z \sim 8$. Since this redshift is within the range in which re-ionization may occur, it allows more complexities in the re-ionization history. For instance, if the IGM was first re-ionized at $z \gtrsim 8$ by a generation of sources, it could have recombined again if feedback from the ionizing sources can effectively suppress further source formation. The IGM would then have to be re-ionized once again at a later stage when new ionizing sources manage to form. Indeed, it has been suggested that Population III stars may re-ionize the IGM a first time at $z \gtrsim 10$ and, in doing so, quench the subsequent formation of such objects. The Universe may then recombine as the ionizing sources die off, and be re-ionized a second time at $z \sim 6$–7 by young galaxies that form in halos where atomic cooling becomes important (e.g. Cen, 2003).

16.3.4 Probing Re-ionization with 21-cm Emission and Absorption

As discussed in the previous subsection, re-ionization is believed to be a complicated process. Hence, detailed observations are required in order to gain insight into how re-ionization has proceeded. The Gunn–Peterson test described in §16.1.3 provides a probe along sightlines towards high-redshift quasars. Unfortunately, the number of quasars available for such test is limited, and it is hard to use it to probe the structure of re-ionization in a statistical way. Furthermore, the large optical depth of neutral hydrogen makes it difficult to use this method to determine the ionization fraction accurately. The test based on the polarization of the cosmic microwave background (CMB), described in §6.7.4, is sensitive to the total optical depth of Thomson scattering, but the lack of detailed redshift information in the test makes it insensitive to the exact re-ionization history. Recently, a potentially powerful method has been proposed, which may probe the era of cosmological re-ionization directly. This method is based on the fact that neutral hydrogen may be directly detectable in emission or absorption against the CMB at the frequency corresponding to the redshifted HI 21-cm lines associated with the spin-flip transition in the ground ($n = 1$) state (e.g. Field, 1959). Since the CMB provides an all-sky background, and since the emission and absorption are produced by resonant transitions with a specific rest-frame wavelength (21cm), it is possible to obtain 21-cm tomography that can be used to probe the structure of gas density and ionization fraction in three-dimensional space (e.g. Madau et al., 1997). In what follows we give a brief description of the basic principles.

Due to the spin–spin coupling of proton and electron, the ground ($n = 1$) state of a neutral hydrogen atom is split into two hyperfine states; a singlet state corresponding to anti-alignment of the two spins, and a degenerate triplet state corresponding to alignment of the two spins. In what follows, these two states are referred to as the singlet state (or state 0) and the triplet state (or state 1), respectively. The energy difference, is $E_{10} \equiv E_1 - E_0 \approx 5.9 \times 10^{-6}$ eV, corresponding to a wavelength of $\lambda_{10} \approx 21$ cm, a frequency $\nu_{10} \approx 1{,}420$ MHz, and a temperature $T_{10} \equiv E_{10}/k_B \approx 0.068$ K. In equilibrium, the population ratio in these two states is determined by the spin temperature, T_s, as

$$\frac{n_1}{n_0} = \frac{g_1}{g_0} \exp\left[-\frac{T_{10}}{T_s}\right]. \tag{16.80}$$

where $g_1/g_0 = 3$. Note that T_s characterizes the thermal equilibrium between the two spin states, and should not be confused with the kinetic temperature of the gas.

As CMB photons propagate through a medium containing neutral hydrogen, some of them can be absorbed as they excite hydrogen atoms from the singlet level to the triplet level. The corresponding optical depth is

$$\tau(z) = \frac{3c^2 h_P A_{10} n_{HI}(z)}{32\pi \nu_{10}^2 k_B T_s(z) H(z)}, \tag{16.81}$$

where $A_{10} \approx 2.9 \times 10^{-15}$ s^{-1} is the spontaneous decay rate from state 1 to state 0 (Wild, 1952; Field, 1959). The brightness temperature of the CMB then becomes

$$T_b(z) = T_{CMB}(z) e^{-\tau(z)} + \left(1 - e^{-\tau(z)}\right) T_s(z), \tag{16.82}$$

where $T_{CMB}(z)$ is the brightness temperature of the CMB in the absence of absorption. Thus, the change in the brightness temperature, seen by an observer at the present time, is

$$\delta T = T_b - T_{CMB} = \frac{(1 - e^{-\tau(z)})[T_s(z) - T_{CMB}(z)]}{(1+z)}. \tag{16.83}$$

Assuming $\tau \ll 1$ and $z \gg 1$, we have

$$\delta T \approx 28\,\text{mK} \left(\frac{\Omega_{b,0} h}{0.03}\right) \left(\frac{\Omega_{m,0}}{0.3}\right)^{-1/2} \left(\frac{1+z}{10}\right)^{1/2} \frac{(T_s - T_{\text{CMB}})}{T_s} x_{\text{HI}}, \tag{16.84}$$

where x_{HI} is the neutral fraction. As one can see, if the spin temperature is the same as the CMB temperature, no net effect is produced by the gas on the CMB. In this case, the spin state is in thermal equilibrium with the CMB, so that absorption is exactly compensated by emission. On the other hand, $T_s > T_{\text{CMB}}$ leads to net emission ($\delta T > 0$), while $T_s < T_{\text{CMB}}$ leads to net absorption ($\delta T < 0$).

The spin temperature can be made different from T_{CMB} in the presence of radiation sources or when the kinetic temperature of the gas is different from T_{CMB}. Re-ionization can create such a condition, so that it can be probed by observing the 21-cm emission and absorption it produces.

There are two processes that can change the spin temperature in a neutral gas. One is the collisional excitation/de-excitation of the spin states, and the other is the mixing of the two spin states via the Wouthuysen–Field process (Wouthuysen, 1952; Field, 1958) whereby a hydrogen atom initially in the $n = 1$ electronic level with a given spin state absorbs a Lyα photon to jump to the $n = 2$ level and then spontaneously decays back to the $n = 1$ level with a different spin state. These processes can couple the spin temperature to the kinetic temperature of the gas, T_K, and to the brightness temperature of the radiation field, T_α. In general, we can write

$$T_s = \frac{T_{\text{CMB}} + y_\alpha T_\alpha + y_c T_K}{1 + y_\alpha + y_c}, \tag{16.85}$$

where y_α and y_c are the coupling factors of the Wouthuysen–Field and collisional processes, respectively (Field, 1958).

In a predominantly neutral gas, the main collisional process is that among the neutral hydrogen atoms themselves. In this case, y_c can be written as

$$y_c = \frac{C_{10}}{A_{10}} \frac{T_{10}}{T_K}, \tag{16.86}$$

where C_{10} is the rate of collisional de-excitation of the triplet state. The value of C_{10}/n_{HI} as a function of T_K can be found in Allison & Dalgarno (1969) and Zygelman (2005). For the Wouthuysen–Field process, the coupling factor can be written as

$$y_\alpha = \frac{P_{10}}{A_{10}} \frac{T_{10}}{T_\alpha}, \tag{16.87}$$

where P_{10} is the de-excitation rate of the triplet state due to the absorption of Lyα photons. This rate is related to the total scattering rate of Lyα photons by $P_{10} = 4P_\alpha/27$. In the presence of a radiation field with specific intensity J_ν, the total scattering rate is

$$P_\alpha = \int d\Omega \int \frac{J_\nu}{h_P \nu} \sigma_\nu d\nu, \tag{16.88}$$

where σ_ν is the Lyα absorption cross-section. The brightness temperature of the radiation field, T_α, is defined through the occupation number of photons, $\mathcal{N}_\nu$, near the Lyα frequency:

$$\frac{1}{T_\alpha} = -\frac{k_B}{h_P} \left(\frac{\partial \ln \mathcal{N}_\nu}{\partial \nu}\right)_{\nu = \nu_\alpha}, \tag{16.89}$$

where $\mathcal{N}_\nu = c^2 J_\nu/(2 h_P \nu^3)$, assuming an isotropic radiation field. Because of the large cross-section of neutral hydrogen near the Lyα frequency, and because photons can exchange energy with the gas through the recoil of the atoms while emitting a Lyα photon, the radiation spectrum near Lyα can be thermalized effectively with the gas (Field, 1959). Thus, the brightness temperature T_α defined above is expected to relax quickly to the kinetic temperature of the gas, T_K.

During the re-ionization process, re-ionization sources may produce non-ionizing UV photons (with energies between Lyα and the Lyman limit) which can propagate into the neutral medium and be redshifted to the Lyα frequency, providing Lyα photons required for the Wouthuysen–Field process. In addition, the collisional process may also play a role in changing the spin temperature in regions where the gas density is high. The source information, together with the gas density and ionization fraction produced by the re-ionization process can in principle be used to make detailed predictions for the 21-cm tomography of the re-ionization epoch (see Ciardi & Ferrara, 2005, and Barkana & Loeb, 2007, for reviews on this rapidly expanding field).

16.4 General Properties of Absorption Lines

An extremely powerful probe of the IGM is provided by quasar absorption lines, which are produced by gas clouds along the quasar's line-of-sight. Since quasars can be observed out to very high redshifts, these absorption line systems can be used to study the evolution of the IGM over a large part of cosmic history. In this section, we start with a general discussion about quasar absorption lines, to show what kind of information we can obtain from studying the properties of these lines. In the subsequent sections we then discuss the specific properties of the different classes of absorption line systems mentioned in §2.8.2.

An absorption line in a quasar spectrum is produced by an intervening gas cloud through the absorption of continuum light from the quasar near the resonance frequency, $\nu_{12} = |E_2 - E_1|/h_P$, between two energy levels E_1 and E_2 of the absorbing atoms in question (see §16.1.2 for a more detailed description of resonance absorption). Since photons at the resonance frequency are absorbed much more efficiently than other photons, the line profile, $\phi(\nu)$, can roughly be approximated by a delta function in photon frequency ν at ν_{12}. In reality, however, absorption lines are broadened by various effects, and the shape (profile) of an absorption line contains much information about the physical properties of the absorbing gas, such as density, temperature and ionization state. In addition, the abundance patterns of quasar absorption systems provide information regarding the nucleosynthetic origins and dust content of the gas and regarding the build-up of metals with cosmic time.

A great advantage of absorption line spectroscopy is its tremendous sensitivity. Quasar absorption lines can be used to detect low-density gas that is orders of magnitude below the detection threshold of most other techniques. While current techniques require considerable effort to detect the emission of neutral hydrogen with a column density of $N_{HI} \sim 10^{18}\,\text{cm}^{-2}$, it has become fairly straightforward to detect hydrogen with column densities as low as $N_{HI} \sim 10^{12.5}\,\text{cm}^{-2}$ in absorption. In addition, absorption lines can be measured comparably well from $z = 0$ out to $z > 4$, while current facilities only allow the detection of 21-cm emission from the nearby Universe.

16.4.1 Distribution Function

Let $n(N,z)\mathrm{d}N$ denote the comoving number density of absorption line systems with column densities in the interval $(N, N+\mathrm{d}N)$ at redshift z, and $\sigma(N,z)$ the (proper) absorption cross-section of an absorber with column density N at z. Then the expected number of absorption lines per unit column density per unit redshift along a random line-of-sight is

$$\mathscr{F}(N,z) \equiv \frac{\mathrm{d}^2 \mathscr{N}}{\mathrm{d}N\,\mathrm{d}z} = n(N,z)(1+z)^3 \sigma(N,z)\frac{\mathrm{d}l}{\mathrm{d}z} = \frac{c}{H(z)}(1+z)^2 n(N,z)\sigma(N,z), \qquad (16.90)$$

where $dl = c\,dt$ is the interval of proper distance corresponding to dz, and $H(z)$ is the Hubble constant at z (see §3.2.6). If there is no evolution in the absorbers [i.e. both $n(N,z)$ and $\sigma(N,z)$ are independent of z], then for a matter dominated universe,

$$\frac{d^2\mathcal{N}}{dN\,dz} \propto \begin{cases} 1+z & \text{for } q_0 = 0 \\ (1+z)^{1/2} & \text{for } q_0 = 1/2. \end{cases} \quad (16.91)$$

It is sometimes useful to replace z by a coordinate X defined by

$$dX = \frac{H_0}{c}(1+z)^3\,dl = \frac{(1+z)^2}{E(z)}\,dz, \quad (16.92)$$

where $E(z) = H(z)/H_0$. In this coordinate, the line density is

$$\mathcal{F}(N,X) \equiv \frac{d^2\mathcal{N}}{dN\,dX} = \frac{c}{H_0}n(N,z)\,\sigma(N,z), \quad (16.93)$$

which is independent of cosmological redshift for a non-evolving population of absorbers.

Once the distribution function $\mathcal{F}(N,z)$ is known, it can be used to compute the total mass density of absorbers. For example, consider Lyα absorbers, so that N is now the column density of neutral hydrogen, $N_{\rm HI}$. The neutral hydrogen mass of an absorber with column density $N_{\rm HI}$ at redshift z is $m_{\rm H} N_{\rm HI} \sigma(N_{\rm HI},z)$, with $m_{\rm H}$ the mass of a hydrogen atom. The comoving density of neutral hydrogen at redshift z is therefore

$$\rho_{\rm HI,com}(z) = m_{\rm H}\int N_{\rm HI}\,\sigma(N_{\rm HI},z)\,n(N_{\rm HI},z)\,dN_{\rm HI} = \frac{H_0}{c}m_{\rm H}\int \mathcal{F}[N_{\rm HI},X(z)]\,N_{\rm HI}\,dN_{\rm HI}. \quad (16.94)$$

This is often expressed in terms of the critical density at $z=0$:

$$\Omega_{\rm HI}(z) \equiv \frac{\rho_{\rm HI,com}(z)}{\rho_{\rm crit,0}}. \quad (16.95)$$

Note that this should not be confused with the cosmological density parameter of neutral hydrogen at redshift z, which is equal to $\rho_{\rm HI}(z)/\rho_{\rm crit}(z)$, with $\rho_{\rm HI} = \rho_{\rm HI,com}(1+z)^3$ the proper density of neutral hydrogen at z. Rather, it is the cosmological density parameter of neutral hydrogen the Universe would have at $z=0$ if the comoving number density and absorption cross-sections of the absorbers do not evolve from redshift z.

16.4.2 Thermal Broadening

If the absorbing gas is not at rest with the expanding background, then in addition to the redshift caused by the general expansion of the Universe, the observed frequency will be shifted by the Doppler effect by an amount

$$\nu_{\rm o} - \nu_{\rm a} = -\frac{v}{c}\nu_{\rm a}, \quad (16.96)$$

where $\nu_{\rm a}$ is the absorption frequency in the rest frame of the absorber, $\nu_{\rm o}$ is the frequency observed by a fundamental observer (who moves with the general expansion), and v is the peculiar velocity of the absorbing atom projected along the line-of-sight (v being positive when the atom is moving away from the observer). If the velocity distribution of the absorbing atoms is Maxwellian, then the distribution of v is

$$\mathcal{P}(v)\,dv = \frac{1}{\pi^{1/2}b}\exp\left(-\frac{v^2}{b^2}\right)dv, \quad (16.97)$$

where the Doppler parameter, b, is related to the velocity dispersion of the gas, σ, by $b = \sqrt{2}\sigma$. If v is caused solely by thermal motion, then $b^2 = 2k_{\rm B}T/m$, where T is the gas temperature and

m is the mass of an absorbing atom. In the more general case where turbulent motion (usually assumed to be independent of the thermal motion) with a Gaussian distribution is included, we can write

$$b^2 = \frac{2k_\mathrm{B}T}{m} + b_\mathrm{turb}^2, \tag{16.98}$$

where b_turb is the b parameter due to turbulent motion. The line profile defined in Eq. (16.11) can then be written as

$$\phi_{12}(\nu)\,d\nu = \frac{1}{\pi^{1/2}b_\nu}\exp\left[-\frac{(\nu-\nu_{12})^2}{b_\nu^2}\right]d\nu \quad \text{with} \quad b_\nu \equiv \frac{\nu_{12}}{c}b. \tag{16.99}$$

In this case, the line profile is a Gaussian centered at the resonance frequency ν_{12}, with a width given by b_ν.

16.4.3 Natural Broadening and Voigt Profiles

In the absence of any other broadening, there is a minimal (natural) amount of line broadening due to the fact that the lifetime of an excited state is finite so that the energy of the photon associated with the transition cannot be exact due to Heisenberg's uncertainty principle. This natural broadening can be calculated rigorously from quantum electrodynamics by considering an atom with two energy levels in a radiation field (see Shu, 1991b, for details). If we write $\gamma \equiv A_{21}/(4\pi)$, where A_{21} is the spontaneous transition coefficient, then the natural profile can be written as

$$\phi_{12}(\nu) = \mathscr{L}(\nu), \tag{16.100}$$

where

$$\mathscr{L}(\nu) \equiv \frac{1}{\pi}\left[\frac{\gamma}{(\nu-\nu_{12})^2 + \gamma^2}\right] \tag{16.101}$$

is the Lorentzian (or Cauchy) profile whose FWHM is equal to 2γ.

In the presence of thermal broadening, the line profile is a convolution of this Lorentzian profile with the thermal Gaussian profile:

$$\phi(\nu) = \int_{-\infty}^{\infty} \mathscr{L}\left[\nu\left(1-\frac{v}{c}\right)\right]\mathscr{P}(v)\,dv. \tag{16.102}$$

We can write

$$\phi(\nu) = \frac{1}{\sqrt{\pi}}\frac{c}{b}\frac{\mathscr{V}(A,B)}{\nu}, \tag{16.103}$$

where

$$\mathscr{V}(A,B) = \frac{A}{\pi}\int_{-\infty}^{\infty}\frac{\exp\left(-y^2\right)dy}{(B-y)^2 + A^2}, \tag{16.104}$$

with

$$A \equiv \frac{c}{b}\frac{\gamma}{\nu}; \quad B \equiv \frac{c}{b}\frac{\nu-\nu_{12}}{\nu}. \tag{16.105}$$

This profile is called the Voigt profile and $\mathscr{V}(A,B)$ is the Voigt function. A useful approximation to the Voigt function is

$$\mathscr{V}(A,B) \approx \exp\left(-B^2\right) + \frac{1}{\sqrt{\pi}}\frac{A}{A^2+B^2}, \tag{16.106}$$

which shows that near the center of an absorption line the profile is dominated by thermal broadening while the wings (where $|\nu-\nu_{12}|/\nu_{12} \gg b/c$ so that $B \gg 1$) are dominated by the

Lorentzian profile. These frequency intervals, in which natural broadening dominates Doppler broadening, are called the damping wings of the line profile.

16.4.4 Equivalent Width and Column Density

The strength of an absorption line is usually measured by its equivalent width, defined as

$$W = \frac{1}{f_0} \int [f_0 - f(\lambda)] \, d\lambda, \tag{16.107}$$

where $f(\lambda)$ is the observed spectrum as a function of wavelength, and f_0 is what would be expected if the line were absent. Since $f(\lambda) = f_0 \exp[-\tau(\lambda)]$, where $\tau(\lambda)$ is the optical depth at wavelength λ, we can write

$$W = \int_0^\infty \left[1 - e^{-\tau(\lambda)}\right] d\lambda. \tag{16.108}$$

For a Voigt profile, τ is related to the column density N by

$$\tau(\lambda) = N\sigma_{12}(\nu) = \eta_0 \mathscr{V}(A,B) \quad \text{where} \quad \eta_0 \equiv \frac{1}{\sqrt{\pi}} \frac{c}{b} \frac{a_{12}}{\nu_{12}} N. \tag{16.109}$$

The plot of W as a function of η_0 is called the theoretical curve of growth. For a given species of absorbing atoms, this curve relates W with N and b. As an example, Fig. 16.2 plots the equivalent width W of the HI Lyα line as a function of $N_{\rm HI}$ and b.

When $\eta_0 \lesssim 1$, the optical depth is small and the equivalent width reduces to

$$W \simeq \int_0^\infty \tau(\lambda) \, d\lambda = \frac{a_{12}}{\nu_{12}} N \lambda_{12}. \tag{16.110}$$

In this case, the equivalent width increases linearly with the column density, independent of the b parameter. For the Lyα absorption of HI gas, we obtain

$$W = 0.055 \left(\frac{N_{\rm HI}}{10^{13} \, {\rm cm}^{-2}}\right) \text{Å}, \tag{16.111}$$

in the rest frame.

When $\eta_0 > 10^4$, the main contribution to the integration in Eq. (16.108) comes from $B \gg 1$. As we have seen, the Voigt profile in this case is dominated by the Lorentzian profile. The equivalent width can then be written as

$$W \simeq \lambda_{12} x_0 \int_{x_0^{-1}}^\infty \left[1 - \exp\left(-y^{-2}\right)\right] dy \quad \text{with} \quad x_0^2 = \frac{2}{3\pi} \left(\frac{a_{12}}{\nu_{12}}\right)^2 \frac{N}{\lambda_{12}^2}. \tag{16.112}$$

Since x_0 is generally smaller than one and since the integrand in the above expression goes to zero rapidly for $|y| > 1$, we can replace the lower limit of the integration by $-\infty$. It then follows that $W \propto N^{1/2}$, independent of b. For Lyα absorption we have

$$W = 7.3 \left(\frac{N_{\rm HI}}{10^{20} \, {\rm cm}^{-2}}\right)^{1/2} \text{Å}, \tag{16.113}$$

in the rest frame. Because the observed equivalent width of a line formed at redshift z is $W_{\rm obs} = (1+z)W$, systems with $N_{\rm HI} > 10^{20} \, {\rm cm}^{-2}$ at $z = 2$ have $W_{\rm obs} > 22$Å. Lines this strong are easily identified, even when they are embedded in a Lyα forest. The damping wing due to Lorentzian broadening, $\phi(\nu) \propto 1/[(\nu - \nu_{12})^2 + \gamma^2]$, becomes detectable when $N_{\rm HI} \gtrsim 5 \times 10^{18} \, {\rm cm}^{-2}$ for HI Lyα lines.

In the intermediate range of η_0, the line profile is still dominated by thermal broadening, but becomes optically thick near the line center [i.e. $\tau(\lambda) > 1$ for $\lambda \sim \lambda_{12}$]. In this case, the

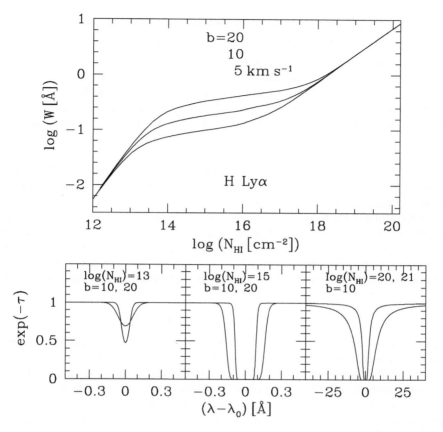

Fig. 16.2. The upper panel shows the equivalent width of the Lyα line as as a function of the HI column density N_{HI} for three different values of the b parameter, as indicated. The three lower panels show the absorption line profiles (in terms of the fractional transmitted flux) for different values of N_{HI} and b, as indicated.

integration in Eq. (16.108) is dominated by the constant part and W increases very slowly with η_0 (or N). The line is then saturated (see the bottom panels of Fig. 16.2). For $10 \lesssim \eta \lesssim 1,000$, we can approximate the relation by

$$W \approx \frac{2b}{\nu_{12}} (\ln \eta_0)^{1/2}. \tag{16.114}$$

Because of the degeneracy between a small change in the apparent line width and a large change in the column density, it is relatively difficult to infer the column density from an observed equivalent width in this case. Note that in this regime W increases almost linearly with b. For Lyα absorption, the value of η_0 defined in Eq. (16.109) is related to the HI column density by

$$N_{HI} \approx 2.6 \times 10^{13} \eta_0 \left(\frac{b}{20 \,\text{km s}^{-1}}\right) \,\text{cm}^{-2}. \tag{16.115}$$

Thus, if $b \sim 20 \,\text{km s}^{-1}$ (the characteristic value for absorbers with $N_{HI} \sim 10^{13}$–$10^{15} \,\text{cm}^{-2}$; see §16.5), an HI Lyα line becomes saturated when $N_{HI} \gtrsim 10^{14} \,\text{cm}^{-2}$.

16.4.5 Common QSO Absorption Line Systems

Some absorption line systems, such as HI, CII, CIV, OVI, MgII, SiIV, are more frequently detected in QSO spectra than other systems, because the elements of these species have relatively high cosmic abundances and because the transitions involved have large oscillator strengths (see §B1.3). In addition, real spectrographs can cover only a limited wavelength range, say from $\lambda_{\min}$ to $\lambda_{\max}$. Some absorption lines are not detected simply because they are out of the observational window. Since the observed wavelength λ_o is related to the wavelength at absorption λ_a by $\lambda_o = \lambda_a(1+z_a)$, for a fixed λ_a, only systems in the redshift range from $z_{\min} \equiv \lambda_{\min}/\lambda_a - 1$ to $z_{\max} \equiv \lambda_{\max}/\lambda_a - 1$ can be observed.

The number density of atoms (for a given element) at a particular ionization state depends not only on the element abundances, but also on the balance between ionization and recombination. This balance depends on the density and temperature of the absorbing gas, as well as on the intensity of the ionizing photons. For example, if the temperature of the gas is about 10^4 K, large number densities are only expected for low ionization species, such as HI, CII, MgII. On the other hand, if $T > 10^5$ K, only highly ionized species, such as CIV and OVI, are expected to be abundant. Therefore, the relative strengths of different absorption lines not only convey information about abundance ratios, but also about the thermal and ionization state of the absorbing gas.

16.4.6 Photoionization Models

Although individual absorption line systems may have diverse properties, in many cases we can abstract an absorption system as a gas cloud heated and ionized by a radiation field. This allows one to predict the physical conditions (temperature, density, ionization state, etc.) by solving equations which describe heating/cooling and ionization/recombination processes. Such an approach is powerful, because a large number of observables can be predicted from relatively few input parameters. The most important input parameters are (i) the shape and intensity of the UV background, (ii) the chemical composition of the gas, and (iii) the geometry (i.e. the spatial density distribution) of the gas. From these one can obtain the temperature of the gas, the ionization fractions of various elements, and the relative populations of different ionization stages, which can be used to compute the absorption and emission properties of the gas to compare with observation. Although a well-defined problem, the treatment of all relevant radiative processes for a large number of elements can be extremely tedious. Fortunately, computer programs for this kind of calculations are publicly available, a well-known example of which is CLOUDY (Ferland et al., 1998).

16.5 The Lyman α Forest

The Lyman α (Lyα) forest is the set of absorption lines arising from the absorption of Lyα resonance photons by neutral hydrogen along the line-of-sight towards a QSO. Since neutral hydrogen clouds at different positions along this line-of-sight see the photons at different wavelengths (due to the redshift), each individual cloud leaves its fingerprint as an absorption line at a different position in the observed spectrum blueward to the Lyα emission line of the QSO. Fig. 16.3 shows the spectrum of a QSO whose redshift is $z = 3.5$. The numerous lines to the left of the Lyα emission line are the 'trees' that make up the Lyα forest.

With a spectrum of sufficient spectral resolution (i.e. with FWHM $\lesssim 25\,\mathrm{km\,s^{-1}}$) the typical Ly$\alpha$ lines are resolved, and one can model the forest by fitting each individual line with a Voigt profile. The standard approach relies on χ^2 minimization to achieve a complete decomposition of the spectrum into as many independent Voigt profiles as necessary to make

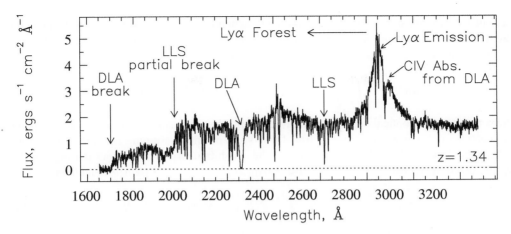

Fig. 16.3. Typical spectrum of a quasar, showing the quasar continuum and emission lines, and the absorption lines produced by galaxies and intergalactic material that lie between the quasar and the observer. This spectrum of the $z = 1.34$ quasar PKS0454+039 was obtained with the Faint Object Spectrograph on the HST. The emission lines at ~ 2400Å and ~ 2850Å are Lyβ and Lyα. The Lyα forest, absorption produced by various intergalactic clouds, is apparent at wavelengths blueward of the Lyα emission line. The two strongest absorbers are a damped Lyα absorber (DLA) at $z = 0.86$ and a Lyman limit system (LLS) at $z = 1.15$. The former produces a Lyman limit break at ~ 1700Å and the latter a partial Lyman limit break at ~ 1950Å since the neutral hydrogen column density is not large enough for it to absorb all ionizing photons. Many absorption lines are produced by the DLA at $z = 0.86$ (C IV 1548, for example, is redshifted onto the red wing of the quasar's Lyα emission line). [Adapted from Charlton & Churchill (2000)]

the χ^2 probability consistent with that expected from random fluctuations. As a result, one obtains a list of absorption lines, each with a redshift z, a column density $N_{\rm HI}$ and a Doppler parameter b.

16.5.1 Redshift Evolution

For a QSO at $z \sim 3$, Lyα forest lines can be observed over a redshift range $\Delta z \sim 1$. This range covers a significant fraction of the Hubble time at $z \sim 3$, so that we may expect to see some evolution in the properties of the absorbers over the observed redshift range.

The observed line density of Lyα forest systems above a certain column density (or rest-frame equivalent width) threshold is usually expressed in the form

$$\frac{d\mathcal{N}}{dz} = A(1+z)^{\gamma}. \tag{16.116}$$

For $2 \lesssim z \lesssim 4$ and $N_{\rm HI} \geq 10^{14}\,{\rm cm}^{-2}$, observational determinations give

$$A \sim 3.5 \quad \text{and} \quad \gamma \sim 2.7 \tag{16.117}$$

(e.g. Lu et al., 1991; Kim et al., 1997). However, the variance in both A and γ from different studies is quite large. This is mainly due to uncertainties in the method used for counting the lines. The number of lines given by the Voigt profile fitting can be affected by line blending, which depends on both spectral resolution and redshift. Furthermore, the finite redshift range available in each individual study and possible dependence of γ on column density also contribute to the uncertainty.

Some of the uncertainties in the line counting approach can be reduced by considering the effective optical depth of the forest. Using the distribution function $\mathscr{F}(N_{\rm HI}, b, z)$, which is a straightforward extension of the distribution function defined in §16.4.1 that includes the Doppler parameter b, we can write the effective optical depth caused by absorbers with $N_{\rm HI} \in [N_1, N_2]$, $b \in [b_1, b_2]$ and $z \in [z_1, z_2]$ as

$$\tau_{\rm eff} = \int_{z_1}^{z_2} \int_{b_1}^{b_2} \int_{N_1}^{N_2} \left[1 - e^{-\tau(N_{\rm HI},b)}\right] \mathscr{F}(N_{\rm HI}, b, z) \, dN_{\rm HI} \, db \, dz, \quad (16.118)$$

where $\tau(N, b)$ is the optical depth of an individual absorber with column density N and Doppler parameter b, and is given by Eq. (16.109) for a Voigt profile. Assuming that the distributions of $N_{\rm HI}$ and b are independent of z and that the redshift distribution of line density can be approximated by a power law, we can write $\mathscr{F}(N_{\rm HI}, b, z) = (1+z)^\gamma \mathscr{F}(N_{\rm HI}, b)$, and thus

$$\tau_{\rm eff} = \frac{(1+z)^{\gamma+1}}{\lambda_a} \int_{b_1}^{b_2} \int_{N_1}^{N_2} \mathscr{F}(N_{\rm HI}, b) \, W(N_{\rm HI}, b) \, dN_{\rm HI} \, db, \quad (16.119)$$

where W is the rest-frame equivalent width. This relation can be used to measure γ even if individual lines are not resolved. For $z \gtrsim 2$, this approach gives $\gamma = 2.5$–3 (e.g. Press et al., 1993; Zuo & Lu, 1993). Despite the large uncertainties, the derived values of γ imply significant evolution in the absorbing clouds that make up the Lyα forest at $z > 2$. As we will see in §16.5.7, this is in stark contrast to the Lyα forest at lower redshifts.

16.5.2 Column Density Distribution

The HI column densities of Lyα forest lines span the range from $\sim 10^{12} \, {\rm cm}^{-2}$ to $\sim 10^{17} \, {\rm cm}^{-2}$. The lower limit reflects the resolution limit of the state-of-the-art high-resolution spectroscopy, while the upper limit comes from the fact that clouds with HI column densities above $10^{17} \, {\rm cm}^{-2}$ are optically thick to Lyman continuum photons and are detected as Lyman-limit systems (to be discussed in §16.7).

The HI column density distribution is described by a distribution function $\mathscr{F}(N_{\rm HI})$ which is defined as the number of absorption lines per unit $N_{\rm HI}$ per unit X [defined in Eq. (16.92)]:

$$d\mathscr{N} = \mathscr{F}(N_{\rm HI}) \, dN_{\rm HI} \, dX. \quad (16.120)$$

In practice, $\mathscr{F}(N_{\rm HI})$ is estimated by counting the number of absorption lines in finite bins of $N_{\rm HI}$ and X:

$$\mathscr{F}(N_{\rm HI}) = \frac{\mathscr{N}}{\Delta N_{\rm HI} \sum_i (\Delta X)_i}, \quad (16.121)$$

where $\sum_i (\Delta X)_i$ is the total surveyed distance interval.

Observations show that the column density distribution is well described by a simple power law:

$$\mathscr{F}(N_{\rm HI}) \propto N_{\rm HI}^{-\beta}. \quad (16.122)$$

For the Lyα forest at $z \gtrsim 2$, E. M. Hu et al. (1995) obtained

$$\mathscr{F}(N_{\rm HI}) = 4.9 \times 10^7 \, {\rm cm}^2 \left(\frac{N_{\rm HI}}{{\rm cm}^{-2}}\right)^{-1.46} \quad \text{for } 12.3 < \log N_{\rm HI} < 14.5. \quad (16.123)$$

These results are shown in Fig. 2.40, along with the results for higher column density systems compiled by Petitjean et al. (1993) that will be discussed in the following sections. It is remarkable that the HI column density distribution over the entire observable range (nearly ten orders

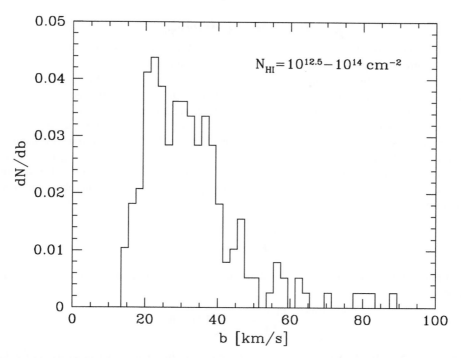

Fig. 16.4. Distribution of Doppler b parameters for Lyα forest lines with HI column densities in the range $10^{12.5} \leq N_{\rm HI} \leq 10^{14}\,{\rm cm}^{-2}$. [Based on data published in E. M. Hu et al. (1995), Lu et al. (1996b), and Kim et al. (1997)]

of magnitude in column density) is well represented by a single power law with power index $\beta \simeq 1.5$ (but see discussion in §16.8.1).

16.5.3 Doppler Parameter

From the Voigt profile fitting, we can also obtain the Doppler b parameter for each Lyα forest line. As shown in §16.4, the values of b may be used to infer the temperature and kinematics (turbulence) of the absorbing clouds.

High-resolution spectra show that many Lyα systems with $N_{\rm HI} \lesssim 10^{15}\,{\rm cm}^{-2}$ have $b = (10-50)\,{\rm km\,s}^{-1}$, consistent with the assumption that the absorbing clouds are heated by photo-ionization ($T \sim 10^{4.5}\,{\rm K}$), although some lines have b values as large as $100\,{\rm km\,s}^{-1}$. Fig. 16.4 shows the Doppler parameter distribution of Lyα forest lines obtained with KECK spectroscopy (E. M. Hu et al., 1995; Lu et al., 1996b; Kim et al., 1997). Hu et al. found that the b parameter distribution at $z \sim 3$ can be well represented by a Gaussian with a median $b_{\rm m} = 28\,{\rm km\,s}^{-1}$, a width $\sigma_b \approx 10\,{\rm km\,s}^{-1}$, and with a cutoff below $b_{\rm c} \approx 20\,{\rm km\,s}^{-1}$. There seems to be a trend of the b distribution with redshift, with both $b_{\rm m}$ and $b_{\rm c}$ decreasing with increasing redshift. For example, the median Doppler parameter for relatively strong lines ($13.8 < \log N_{\rm HI} < 16.0$) is found to change from $41\,{\rm km\,s}^{-1}$ at $\langle z \rangle \sim 2.3$ (Kim et al., 1997) to $23\,{\rm km\,s}^{-1}$ at $\langle z \rangle \sim 3.7$ (Lu et al., 1996b), with the lower cutoff $b_{\rm c}$ dropping from 24 to $15\,{\rm km\,s}^{-1}$ over the same redshift range. Kirkman & Tytler (1997) find $b_{\rm c} \approx 19\,{\rm km\,s}^{-1}$ for systems with $\log N_{\rm HI} > 13.8$ and mean redshift $\langle z \rangle \sim 2.7$, consistent with the results of Kim et al. (1997). However, for systems with $\log N_{\rm HI} > 12.5$ at the same redshift, the median and lower cutoff drop to $b_{\rm m} \sim 23\,{\rm km\,s}^{-1}$ and $b_{\rm c} \sim 14\,{\rm km\,s}^{-1}$, respectively, suggesting that the Doppler parameter may be correlated with column density (but see Rauch et al., 1992).

16.5.4 Sizes of Absorbers

One basic property of absorbing clouds is their sizes. With the knowledge of cloud sizes, we can hope to infer their gas densities from the observed column densities, thereby inferring other physical properties of the absorbing clouds. As one can see from Eq. (16.90), the line density of Lyα forest systems is proportional to the number density of absorbing clouds and their absorption cross-sections (properly averaged). Thus, in order to infer cloud sizes from the observed line density, one must know the number density of the absorbing clouds. To proceed, first consider a simple model in which the comoving number density of clouds is assumed to be the same as the number density of local galaxies. In this case, the number density is given by the galaxy luminosity function $\phi(L)\,dL$. As an illustration, let us also assume that the cross-section for absorption associated with a galaxy depends only on its luminosity, i.e. $\sigma = \sigma(L)$. The number of absorption lines per unit redshift is then

$$\frac{d\mathcal{N}}{dz} = \frac{(1+z)^2}{E(z)} \int \sigma(L)\phi(L)\,dL. \tag{16.124}$$

Using a luminosity function of the Schechter form (2.34) and assuming $\sigma(L) \equiv \pi R^{*2}(L/L^*)^\beta$, we have

$$\frac{d\mathcal{N}}{dz} = \frac{(1+z)^2}{E(z)} \phi^* R^{*2} \Gamma(1+\beta-\alpha). \tag{16.125}$$

Taking $\phi^* = 1.5 \times 10^{-2} h^3\,\mathrm{Mpc}^{-3}$ and $\alpha = -1$, and assuming that the gaseous sizes of galaxies obey the Holmberg (1975) relation $R \propto L^{0.4}$, so that $\beta = 0.8$, one finds that the observed line density for $z = 3$ requires that $R^* = 1\,h^{-1}\mathrm{Mpc}$. This radius is much larger than the optical radius of an L^* galaxies, indicating that the majority of the Lyα forest clouds at high redshift cannot be closely associated with individual galaxies.

The typical sizes of absorption clouds can be estimated directly by studying the coincidence of lines in the spectra of close quasar pairs or the images of a lensed quasar. In the case of quasar pairs, the separation (in proper units) of the two lines-of-sight at the redshift of absorption z_a is related to the angular separation of the two quasars, ϑ, as

$$l_\perp = a(z_a)r(z_a)\vartheta = 145h^{-1}\,\mathrm{kpc}\left(\frac{\vartheta}{10''}\right)\frac{H_0 a_0 r(z_a)}{c(1+z_a)}, \tag{16.126}$$

where $r(z_a)$ is the radial coordinate at redshift z_a (see §3.2.6). For an Einstein–de Sitter universe, $\vartheta = 10''$ corresponds to a separation $l_\perp \approx 40 h^{-1}\mathrm{kpc}$ at $z_a = 2$. If the absorbing clouds at redshift z_a have a typical size $D > l_\perp$, then they can intersect the two lines-of-sight simultaneously, and the spectrum of each quasar will contain an absorption line produced by the same cloud. It is therefore possible to estimate the typical cloud size by analyzing the frequency of line coincidences in the spectra of quasar pairs with various angular separations. Applications of this method to real data can be found in, Crotts & Fang (1998) and D'Odorico et al. (1998) for example. The sizes of Lyα forest clouds appear to be large, with typical sizes of the order of 200–500 h^{-1}kpc.

For a lensed quasar the geometry of the system is shown in Fig. 16.5. At redshift $z_a < z_L$ (where z_L is the redshift of the lens), the separation of the two lines-of-sight is also given by Eq. (16.126), with ϑ being the angular separation of the two images. At $z_a > z_L$, we can write $l_\perp = d_{Sa}(z_a)\vartheta'$, where d_{Sa} is the angular-diameter distance from the source to the absorber at z_a, and ϑ' is the angle between the two lines-of-sight at the source position. Since $\vartheta d_{OL} = \vartheta' d_{SL}$, we have

$$l_\perp = \frac{d_{OL} d_{Sa}}{d_{SL}}\vartheta, \quad (z_a > z_L), \tag{16.127}$$

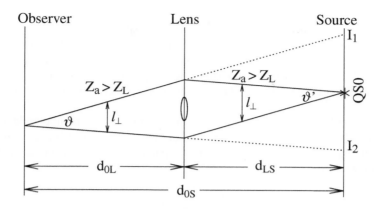

Fig. 16.5. The geometry of a lensed quasar. See text for details.

where d_{ij} is given in §3.2.6. The observed spectra of multiple images of a lensed quasar usually appear very similar, which yields a lower limit of $\sim 50\,\mathrm{kpc}$ for the sizes of the absorbing clouds (e.g. Smette et al., 1992; Impey et al., 1996a).

16.5.5 Metallicity

The discussion in §16.5.4 shows that the majority of the absorbing clouds responsible for the Lyα forest at high redshift ($z \gtrsim 2$) must be intergalactic. Therefore, in early studies it was generally assumed that the absorbing clouds of the high-redshift Lyα forest had mainly a primordial composition with little or no metal enrichment. However, the situation changed when better spectra became available which showed that some absorbing clouds, especially those with relatively large equivalent widths, have weak CIV absorption (Meyer & York, 1987). From high-resolution spectra taken with modern 10-m class telescopes we now know that most Lyα systems with $N_{\mathrm{HI}} \gtrsim 10^{15}\,\mathrm{cm}^{-2}$ and more than half of all systems with $N_{\mathrm{HI}} \gtrsim 3 \times 10^{14}\,\mathrm{cm}^{-2}$ have detectable CIV lines associated with them (Cowie et al., 1995; Tytler et al., 1995; Songaila & Cowie, 1996). The observed CIV column densities are in the range $10^{12} < N_{\mathrm{CIV}} < 10^{14}\,\mathrm{cm}^{-2}$, implying a metallicity of about $Z \sim (10^{-4}\text{--}10^{-2})Z_\odot$ for the absorbing clouds (e.g. Simcoe et al., 2004). Lu et al. (1998) found that the mean metallicity for systems with $10^{13.5} < N_{\mathrm{HI}} < 10^{14}\,\mathrm{cm}^{-2}$ is a factor of 10 lower than that inferred for Lyα clouds with $N_{\mathrm{HI}} > 3 \times 10^{14}\,\mathrm{cm}^{-2}$. However, low HI column density systems with high metallicity have also been observed (Schaye et al., 2007), suggesting that the scatter in metallicity is large.

The presence of CIV absorption in the Lyα clouds indicate that the IGM may have been enriched in heavy elements at high redshift. The level of the enrichment and its relation to the column density therefore contain important information about how the IGM may have been enriched. As we show in §16.6, hydrodynamical cosmological simulations indicate that Lyα systems with $N_{\mathrm{HI}} \gtrsim 10^{14.5}\,\mathrm{cm}^{-2}$ are mostly associated with filaments connecting collapsed objects in which star formation may have occurred, while systems with lower HI column densities are preferentially found in lower density regions farther away from collapsed objects. It is then conceivable that the high column density systems are enriched locally by star formation in the collapsed objects, while the low column density clouds remain more or less primordial. On the other hand, if the IGM was enriched by Population III stars at an early epoch, the chemical enrichment might be more uniform, and there should not be a strong cutoff in the metallicity at low N_{HI}.

16.5.6 Clustering

An important question is how the absorbers that make up the Lyα forest are distributed in space. Are they randomly distributed or clustered? Do they trace the same large-scale structure as galaxies? The answers to these questions not only help to constrain the nature of Lyα forest systems, but also tell us whether they can be used to study the mass distribution in the Universe.

For a forest of Lyα lines, a simple measure of clustering is the two-point correlation function of lines in redshift (velocity) space along the line-of-sight. For a given line at redshift z_1, the probability of finding another line whose redshift separation to the first one is in the interval $\Delta z \pm \delta z/2$ can be written as

$$\delta \mathscr{P}(z_1, \Delta z) = \left(\frac{d\mathscr{N}}{dz}\right)_{z_2} [1+\xi(\Delta z)]\,\delta z, \qquad (16.128)$$

where $\Delta z = z_2 - z_1$, and $\xi(\Delta z)$ is the two-point correlation function. If $\xi = 0$, then $\delta \mathscr{P}$ is just the expected number of lines from the mean line density $d\mathscr{N}/dz$. In real applications $\Delta z \ll \bar{z} \equiv (z_1 + z_2)/2$, and we can define the two-point correlation function in velocity space as

$$\delta \mathscr{P}(\bar{z}, \Delta v) = \left(\frac{d\mathscr{N}}{dz}\right)_{\bar{z}} (1+\bar{z})[1+\xi(\Delta v)]\frac{\delta v}{c}, \qquad (16.129)$$

where $\Delta v = c\Delta z/(1+\bar{z})$ is the velocity difference in the rest frame at the mean redshift $\bar{z}$. Thus the two-point correlation function can be estimated by counting pairs (of lines) of different velocity separations. This has been done in many studies. For Lyα forest systems at high redshift ($z \sim 3$), significant clustering is detected only at small separations, $\Delta v \lesssim 300\,\mathrm{km\,s}^{-1}$ (e.g. Rauch et al., 1992; Cristiani et al., 1997). The clustering amplitude, $\xi(100\,\mathrm{km\,s}^{-1}) \sim 0.5$, is much lower than that of local galaxies at similar separations (assuming that the velocity separation is related to the spatial separation Δr by $\Delta v \sim H_0 \Delta r$ for local galaxies). The amplitude of ξ appears to increase with the HI column density of Lyα forest systems (Cristiani et al., 1995, 1997).

There is, however, a severe uncertainty in the estimated correlation function based on individual absorption lines. Because of line blending the clustering amplitude on small scales may be dramatically underestimated (Rauch et al., 1992). There are indications that line blending is indeed playing an important role. For example, the metal lines associated with Lyα systems, which are less sensitive to blending than Lyα lines owing to the higher mass of the absorbing atoms (hence smaller b parameters), often reveal much stronger clustering on small scales, i.e. for small Δv (Cowie et al., 1995). Because of this uncertainty, two other definitions of $\xi(\Delta v)$ have been proposed. The first defines the correlation function in a similar way as given by Eq. (16.129), but each line is assigned a weight equal to its rest-frame equivalent width (in units of the average equivalent width) in the pair counting (Webb & Barcons, 1991; Zuo, 1992). The other method considers the Lyα forest as a one-dimensional intensity field sampled on pixels, rather than individual lines, and correlates the pixel intensities (Press et al., 1993; Zuo & Bond, 1994).

Nowadays the statistic of choice to study the clustering of Lyα forest is based on the power spectrum, $P_F(k,z)$, of the normalized transmitted flux, $F(\lambda) = \exp[-\tau(\lambda)]$. After removing metal lines and highly saturated Lyα lines, and after correcting for possible variations in the continuum, the normalized flux $F(\lambda)$ in a wavelength range, usually between the Lyα and Lyβ emission lines of the QSO, is Fourier transformed to obtain the power spectrum as a function of k, the wavenumber corresponding to the separation in redshift (velocity) space (e.g. Croft et al., 1998; Kim et al., 2004; McDonald et al., 2005; Jena et al., 2005). The power spectrum, $P_F(k,z)$, can now be determined quite accurately over a large range of wavenumbers, $0.2 \lesssim k/(h\,\mathrm{Mpc}^{-1}) \lesssim 5$. As we will see in §16.6, $P_F(k,z)$ is directly related to the linear power spectrum of the matter distribution, and thus gives important constraints on the model of structure formation.

In addition to the two-point correlation function and power spectrum, several other statistical methods have been used to study the clustering of Lyα forest systems. A particularly interesting statistic is the void probability function, which is designed to detect voids (gaps) in the Lyα forest. If Lyα lines have a Poisson distribution along a line-of-sight, the probability of finding a gap of Δz in the redshift distribution is

$$\mathscr{P}(z,\Delta z) = \exp\left[-\left(\frac{\mathrm{d}\mathscr{N}}{\mathrm{d}z}\right)\Delta z\right]. \qquad (16.130)$$

Any deviation from this distribution will be an indication of clustering. Based on such analysis, Carswell & Rees (1987) concluded that voids with sizes like those in the galaxy distribution ($\sim 50\,h^{-1}\mathrm{Mpc}$ comoving) are very rare in the Lyα forest. This conclusion is confirmed by subsequent studies based on larger datasets, indicating that Lyα forest systems are indeed not tightly associated with galaxies. Nevertheless, large gaps with low line densities are occasionally found along individual sightlines (Crotts, 1987; Dobrzycki & Bechtold, 1991; Cristiani et al., 1995). It is not clear, though, whether these large-scale gaps are due to the large-scale structure in the mass distribution. It is also possible, for instance, that they are produced by the ionizing radiation from quasars and star-forming galaxies close to the lines-of-sight, an analogy of the proximity effect discussed earlier. Gaps in the Lyα forest can also be produced in a sightline intersecting with the winds and outflows from star-forming galaxies, which can heat their environments (e.g. Adelberger et al., 2003).

16.5.7 Lyman α Forests at Low Redshift

Since the Lyα resonance transition has a wavelength of 1,216 Å, the Lyα forest at $z \lesssim 1.5$ is unaccessible with optical spectra. With the capabilities of UV spectroscopy onboard the Hubble Space Telescope, it became possible to also probe this low-redshift Lyα forest. This has resulted in the detection of a large number of Lyα absorption lines in the redshift range $0 < z \lesssim 1.5$ (e.g. Bahcall et al., 1991; Weymann et al., 1998; Penton et al., 2004). The number of lines per unit redshift derived from this sample can be summarized as follows:

$$\frac{\mathrm{d}\mathscr{N}}{\mathrm{d}z} \sim A(1+z)^{\gamma} \quad \text{with } \gamma \sim 0.15 \text{ and } A \sim 35 \qquad (16.131)$$

for Lyα systems with rest-frame equivalent widths $W_r > 0.24$Å. Remarkably, the value of γ for the low-z systems is much smaller than that for the Lyα forest systems at $z \gtrsim 2$ ($\gamma \sim 2.5$) and is consistent with the assumption of no evolution in the absorbers [see Eq. (16.91)]. In the range $0.1\text{Å} < W_r < 1\text{Å}$, the distribution of the rest-frame equivalent widths can be approximated by an exponential function $\exp(-W_r/W_*)$ with $W_* \approx 0.27$Å (Weymann et al., 1998). This equivalent-width distribution corresponds to a column-density distribution similar to that for high-z Lyα systems (§16.5.2).

The marked difference between the $\mathrm{d}\mathscr{N}/\mathrm{d}z$ at low and high redshifts is believed to be due to changes in the UV background intensity. At high redshifts $\mathrm{d}\mathscr{N}/\mathrm{d}z$ falls steeply with declining redshift due to the decreasing gas density caused by the expansion of the Universe. However, at $z \lesssim 2$ the UV background intensity drops due to a decline in the universal star-formation rate and in the space density of quasars (see §16.2.4). This fading of the UV background counteracts the decreasing density so that the forest does not thin out as rapidly as expected from the expansion of the Universe alone (e.g. Theuns et al., 1998; Davé et al., 1999).

If we assume that the comoving number density of the absorbers for the low-z Lyα systems is the same as the number density of local galaxies, the typical absorption radius given by Eqs. (16.131) and (16.125) is $R^* \sim 400 h^{-1}$ kpc for systems with $W_r > 0.24$Å. This value of R^* is much larger than the optical radii of normal galaxies, but is comparable to the virial radius of galactic-sized halos at $z \sim 0$. Thus it is possible that strong Lyα systems at low redshifts are associated with extended galactic halos. For the low-z systems, redshift surveys of galaxies around QSO sightlines can be carried out to directly investigate possible links between Lyα absorbers and galaxies. This has been done for several sightlines towards nearby QSOs (Salzer, 1992; Spinrad et al., 1993; Morris et al., 1993; Salpeter & Hoffman, 1995; Penton et al., 2002). Redshift coincidences between absorbers and galaxies (which are usually hundreds of kpc away from the absorbers) are found for some of the lines-of-sight, but not for all. This suggests that at least some of the low-z Lyα clouds are intergalactic. This is also supported by the work of Morris et al. (1993), who found that the cross-correlation between absorbers and galaxies is weaker than the galaxy–galaxy correlation in the same field. Furthermore, modeling Lyα absorbers as a mixture of randomly distributed objects and clouds associated with galactic halos, Mo & Morris (1994) found that only about a quarter of all Lyα systems detected along the 3C273 sightline are associated with galactic halos. In contrast, by using a larger galaxy survey and selecting stronger lines, Lanzetta et al. (1995) concluded that many Lyα systems in their sample are related to the gaseous envelopes of galaxies, which have radii up to $\sim 200 h^{-1}$ kpc. Furthermore, Lanzetta et al. also found an anticorrelation between the rest frame line equivalent width and the impact parameter of the sightlines to the centers of galaxies, suggesting that the absorbers are indeed physically connected with the galaxies. The first result of Lanzetta et al. was confirmed by subsequent observations (Le Brun et al., 1996; Bowen et al., 1996) and is consistent with the strong clustering found for strong lines at low redshift (Bahcall et al., 1996; Ulmer, 1996). The existence of an anticorrelation between W_r and impact parameter for strong lines is also seen in the data of Penton et al. (2002). Taken together, these results suggest that many strong Lyα lines at low redshift may be produced by clouds associated with galaxies, while weak lines are probably associated with intergalactic gas.

16.5.8 The Helium Lyman α Forest

Since helium is the second most abundant element in the Universe, one also expects absorption due to HeI and HeII in the IGM. For a realistic ionizing background, HeI is undetectable because most helium atoms are in HeIII and HeII. The important absorption lines are then HeII Lyα with a wavelength of 304Å, one fourth that of HI Lyα. Since the ionization threshold of HeII is at $\lambda = 228$Å, the abundance of HeII depends on the intensity of the UV background in the far-UV (at much shorter wavelengths than the HI edge). Hence, observations of HeII Lyα lines can yield independent constraints on the state of the IGM and the UV background.

The basic observable of an HeII Lyα forest is the ratio between the HeII and HI Gunn–Peterson optical depths:

$$\frac{\tau_{\text{HeII}}}{\tau_{\text{HI}}} = \frac{1}{4}\eta \quad \text{where} \quad \eta \equiv \frac{n_{\text{HeII}}}{n_{\text{HI}}}. \tag{16.132}$$

This follows from Eq. (16.15) assuming that $h_P \nu_{12} \gg k_B T$ and using the fact that the value of I_{12} for HI Lyα is 4 times that for HeII Lyα (since $\nu_{\text{HeII}} = 4\nu_{\text{HI}}$ and both transitions have the same oscillator strengths, see Table B1.1). For an optically thin gas highly ionized by photoionization we have

$$\eta = \frac{\alpha_{\text{HeIII}} n_{\text{HeIII}}}{\alpha_{\text{HII}} n_{\text{HII}}} \frac{\Gamma_{\gamma,\text{HI}}}{\Gamma_{\gamma,\text{HeII}}} \approx \frac{4 n_{\text{He}}}{n_{\text{H}}} \frac{\Gamma_{\gamma,\text{HI}}}{\Gamma_{\gamma,\text{HeII}}}, \tag{16.133}$$

where $\Gamma_{\gamma,\mathrm{HI}}$ and $\Gamma_{\gamma,\mathrm{HeII}}$ are the photoionization rates defined in Eq. (16.32). Assuming $J(\nu) \propto \nu^{-\beta_{\mathrm{HI}}}$ just above ν_{HI} and $J(\nu) \propto \nu^{-\beta_{\mathrm{HeII}}}$ just above ν_{HeII} and using the photoionization cross-sections given in §B1.3, we obtain

$$\eta \approx 1.7 \frac{(3+\beta_{\mathrm{HI}})J(\nu_{\mathrm{HI}})}{(3+\beta_{\mathrm{HeII}})J(\nu_{\mathrm{HeII}})} \approx 1.7 \frac{J(\nu_{\mathrm{HI}})}{J(\nu_{\mathrm{HeII}})}. \tag{16.134}$$

Thus the HeII/HI ratio, η, is determined by the ratio of the UV background intensities at the two absorption edges. Depending on the shape of $J(\nu)$, the value of η is generally much larger than 4, and so the optical depth in HeII is much larger than that in HI. As a result, the HeII Lyα Gunn–Peterson effect is a more sensitive measure of diffuse gas than that of HI Lyα (Miralda-Escude, 1993).

Because of its short wavelength, HeII Lyα absorption can only be observed at $z \gtrsim 2.5$ even with the far UV bands accessible to the Hubble Space Telescope. Consequently, only a small number of QSOs are suitable for the search. A non-zero HeII Lyα Gunn–Peterson optical depth ($\tau_{\mathrm{HeII}} \sim$ few) has been detected by a number of observations (Jakobsen et al., 1994; Davidsen et al., 1996; Hogan et al., 1997; Reimers et al., 1997). However, whether the observed optical depth is due to discrete Lyα forest systems or due to a diffuse component is still unclear. In addition, the data are not accurate enough to constrain the shape of $J(\nu)$.

16.6 Models of the Lyman α Forest

A number of models have been proposed to explain the origin of the Lyα forest. Since Lyα forest systems are observed over a large range of redshift, they are not just a transient phenomenon. Therefore, one important question any model should address is how these systems are sustained or replenished. In this section, we describe some of the most influential models that have been proposed.

16.6.1 Early Models

(a) Pressure Confinement As we have seen, the majority of the Lyα forest systems at $z \gtrsim 2$ are intergalactic instead of being associated closely with galaxies, and they are produced by gas that is relatively cold ($T \sim 10^4$ K), as indicated by their Doppler b parameters. One postulation for the origin of the Lyα forest systems is that there exists an ambient hot, tenuous component in the IGM which acts to 'confine' the cooler and denser Lyα clouds by its pressure. The standard version of this pressure confinement model (Sargent et al., 1980; Ostriker & Ikeuchi, 1983; Ikeuchi & Ostriker, 1986) assumes spherical and homogeneous clouds. Because of the general expansion of the Universe, the pressure of the ambient gas scales with redshift as $P \propto \rho T \propto (1+z)^5$ for adiabatic expansion. Since the temperature of Lyα clouds is roughly constant ($T_{\mathrm{cloud}} \sim 10^4$ K), their densities drop as $\rho_{\mathrm{cloud}} \propto (1+z)^5$ as the Universe expands until heating by photoionization can no longer compensate the work of expansion. Thereafter the clouds cool and expand less rapidly than the background. The sound speed drops even faster, and so eventually pressure equilibrium with the hot IGM breaks and the clouds expand freely. The range of cloud masses is constrained by the fact that massive clouds are Jeans-unstable and too small clouds are evaporated by the hot medium due to heat conduction, which gives $10^5 \lesssim M_{\mathrm{cloud}} \lesssim 10^{10} \, \mathrm{M}_\odot$ (Sargent et al., 1980; Ostriker & Ikeuchi, 1983).

One motivation for the pressure confinement model at the time when it was proposed was that the hot ambient IGM needed to confine Lyα clouds might also be responsible for the observed X-ray background. In the explosive scenario proposed by Ostriker & Cowie (1981), large-scale explosions (probably from starburst galaxies or QSOs) in the early Universe might have driven shock waves into the IGM, re-ionizing the IGM and heating it to a high temperature. Meanwhile the dense shells enclosing the hot cavities might cool and fragment to form the Lyα clouds. However, the pressure confinement model has several problems. There is now much evidence that the observed X-ray background is produced by discrete sources rather than by a diffuse medium, and observations of the Compton-y parameter in the CMB also put stringent limits on the existence of a hot diffuse component in the IGM. These leave little room for the existence of the hot gas required for the pressure confinement. Furthermore, the observed large range of HI column densities ($10^{13} \lesssim N_{\rm HI} \lesssim 10^{16} \, {\rm cm}^{-2}$) is difficult to accommodate in the pressure confinement model. It requires some nine orders of magnitude in cloud mass (since $N_{\rm HI} \sim n_{\rm HI} R_{\rm cloud} \sim M_{\rm cloud}^{1/3} J^{-1} P^{-5/3}$, with J the ionizing flux and P the confining pressure), or a factor of about 100 in pressure; both are difficult to realize in the standard pressure confinement model.

Although implausible for explaining Lyα forest systems at high redshift, pressure confinement of cold clouds may happen in the halos of galaxies, where dense clouds may have formed via local instabilities and be in pressure equilibrium with a hot halo gas at the virial temperature. Some of the observed Lyα systems at low redshift may be produced in this manner (e.g. Mo, 1994; Mo & Miralda-Escudé, 1996).

(b) Gravitational Confinement Since the large-scale structure in the Universe is generally believed to be produced by gravitational instability, a natural hypothesis is that Lyα clouds are confined by gravity. Self-gravitating gas clouds were proposed by Melott (1980) and Black (1981) as an alternative to the pressure confinement model. However, self-gravitating gas clouds are unstable in the presence of effective cooling, and it is unclear how the absorbing clouds are sustained or replenished.

In the CDM-type of cosmogonies, dark matter halos form constantly as density perturbations turn around and collapse. In 'minihalos' where the potential is too shallow for the trapped gas (photoionized by the UV background) to cool and to collapse, but deep enough to keep the photoionized gas from escaping, gas may be stably confined. Since the temperature of a photoionized gas is of the order of a few $\times 10^4$ K, the circular velocities of these minihalos are $\sim 30 \, {\rm km \, s}^{-1}$ (Rees, 1986; Ikeuchi, 1986). For an isothermal halo, the density profile is $n(r) \propto r^{-2}$ and, for a photoionized optically thin gas, the HI density profile is $n_{\rm HI} \propto r^{-4}$, assuming balance between recombination and photoionization. The resulting HI column density distribution, sampled by random sightlines through different parts of such halos, is $d\mathcal{N}/dN_{\rm HI} \propto N_{\rm HI}^{-1.5}$, similar to the observed distribution. In addition, because of the large density gradient, a large range in column densities can be produced, and since minihalos are constantly created from the collapse of density peaks and destroyed by merging, one naturally expects some evolution in the number density of absorbers. From the Press–Schechter halo mass function (see §7.2.1), we see that the comoving number density of minihalos [which has $\delta_c(z)/\sigma \ll 1$ at $z \lesssim 3$ for a realistic power spectrum] at a fixed circular velocity changes with redshift as $n(V_c, z) \propto (1+z)^{5/2}$ in an EdS universe. If the absorption cross-section does not change rapidly with z for fixed V_c, then $d\mathcal{N}/dz \sim (1+z)^3$, not very different from the observed value [see Eq. (16.117)]. Since minihalos are generally sub-M^* halos, they are also expected to be weakly clustered (e.g. Mo et al., 1993).

16.6.2 Lyman α Forest in Hierarchical Models

The simple minihalo model discussed above is obviously an oversimplification of what really happens in hierarchical clustering. As discussed in §5.6, the collapse in hierarchical models is highly non-spherical, proceeding from sheets (first axis collapsed) to filaments (first two axes collapsed) to halos (all three axes collapsed). Thus, at any given time, the Universe consists of dark matter halos, filaments and sheets, as well as density perturbations that are still in the linear regime. In principle, all these structures can produce QSO absorption systems, as long as their column densities of neutral gas are sufficiently high. Therefore, it really does not make much sense to stick to the notion of distinct objects for the Lyα absorbers. Rather, one should consider the entire cosmic density field.

This change of notion started when H. Bi and collaborators (Bi et al., 1992; Bi, 1993; Bi & Davidsen, 1997) realized that the absorption optical depth fluctuations corresponding to the linear density fluctuations in the IGM can already give a realistic representation of the Lyα forest (ignoring high column density systems which may be produced by collapsed objects, such as minihalos). Because the involved physics (quasi-linear collapse and optically thin gas) is relatively simple, model predictions can be worked out in some detail and their connections to cosmogonic parameters are easy to make.

(a) Optical Depth Fluctuations Consider a sightline through the cosmic density field. We denote the comoving position along the sightline by x – which is related to the redshift z as described in §3.2.6. The optical depth at an observed frequency ν_o due to the Lyα absorption of the HI gas in the density field can be written as

$$\tau(\nu_o) = \int_0^\infty n_{\rm HI}(x) \sigma_{\rm Ly\alpha}(\nu) \frac{dx}{1+z}; \quad \nu = \nu_o(1+z)\left[1 + \frac{v_{\rm r}(x)}{c}\right], \tag{16.135}$$

where $n_{\rm HI}(x)$ and $v_{\rm r}(x)$ are the proper number density and radial peculiar velocity of the HI gas at x, and $\sigma_{\rm Ly\alpha}(\nu)$ is the Lyα absorption cross-section at frequency ν. Assuming Voigt profiles, $\sigma_{\rm Ly\alpha}$ is given by Eqs. (16.11) and (16.103). Thus, in order to calculate $\tau(\nu_o)$ we need to know $n_{\rm HI}(x)$, $v_{\rm r}(x)$, and $b(x)$ (the velocity dispersion of the gas). In principle, these quantities can be obtained by tracing in detail the evolution of the hydrodynamical, thermal and ionizational states of the absorbing gas. In what follows we use a simple analysis, which, although oversimplified, allows one to gain useful insight.

If we assume that the absorbing gas is optically thin and highly ionized by photoionization, then $n_{\rm HI}$ can be obtained by the equation of ionization balance (assuming ionization equilibrium), $n_{\rm HI}(x) = n_{\rm H}^2(x)\alpha_{\rm H^+}(T)/\Gamma_{\gamma,\rm HI}$, where $\alpha_{\rm H^+}(T)$ is the recombination rate coefficient, and $\Gamma_{\gamma,\rm HI}$ is the photoionization rate. It then follows that

$$n_{\rm HI}(x) \approx 5.9 \times 10^{-11} \left[\frac{n_{\rm H}(x)}{\bar{n}_{\rm H}(z)}\right]^2 \mathscr{A}(T) \, {\rm cm}^{-3}, \tag{16.136}$$

where

$$\mathscr{A}(T) \equiv \left(\frac{\Omega_{\rm b,0} h^2}{0.02}\right)^2 \left(\frac{1+z}{4}\right)^6 T_4^{-0.7} \left(\frac{\Gamma_{\gamma,\rm HI}}{10^{-12}\,{\rm s}^{-1}}\right)^{-1}. \tag{16.137}$$

In the above expressions, $T_4 = T/10^4$ K, $\bar{n}_{\rm H}$ is the mean number density of hydrogen atoms in the Universe, and for simplicity we have approximated the recombination rate coefficient as $\alpha_{\rm H^+} \approx 4.3 \times 10^{-13} T_4^{-0.7}\,{\rm cm}^3\,{\rm s}^{-1}$. If $n_{\rm H}(x)$ is proportional to the baryonic gas density $\rho_{\rm b}(x)$, we can write

$$n_{\rm H}(x) = \bar{n}_{\rm H}(z)[1 + \delta_{\rm b}(x)], \tag{16.138}$$

where $\delta_{\rm b}(x)$ is the density fluctuation in the baryonic gas. For an ideal gas with a polytropic equation of state, we have that

$$T = T_0(1 + \delta_{\rm b})^{\Gamma-1}, \tag{16.139}$$

where T_0 is independent of position and Γ is the polytropic index, which should not be confused with the photoionization rate which always carries an index. If the gas evolves adiabatically, then $\Gamma = 5/3$ (see §B1.1). In reality, both T_0 and Γ are expected to depend on the thermal and ionization history of the gas, but numerical simulations show that Eq. (16.139) holds approximately for

$\delta_b \lesssim 5$, with $T_0 \sim 10^4$ K and $\Gamma \sim 1.5$ at $z \sim 3$ (e.g. Hui & Gnedin, 1997). Under these assumptions we can write

$$n_{\rm HI}(x) \approx 5.9 \times 10^{-11} \mathscr{A}(T_0) [1+\delta_b(x)]^q \text{ cm}^{-3}, \tag{16.140}$$

with

$$q \equiv 2 + 0.7(1-\Gamma). \tag{16.141}$$

Note that at a given redshift, the mean optical depth $\tau \propto (\Omega_b h^2)^2/\Gamma_{\gamma,\rm HI}$. Since $\Gamma_{\gamma,\rm HI}$ can be estimated directly from the observed ionizing background, observational constraints on the mean optical depth can in principle be used to constrain $\Omega_b h^2$ (Rauch et al., 1997).

If the Universe is dominated by CDM and if the mass density field is in the linear regime, then

$$\delta_b(\mathbf{k}) = \frac{\delta_{\rm DM}(\mathbf{k})}{1+k^2/k_{\rm J}^2}, \quad \text{where} \quad k_{\rm J} = \sqrt{\frac{2}{3}}\frac{a}{c_s t} \tag{16.142}$$

(see §4.1.6) and, from the continuity equation,

$$\mathbf{v}(\mathbf{k}) = \frac{i a \mathbf{k}}{k^2} \frac{{\rm d}\delta_b(\mathbf{k})}{{\rm d}t} \approx \frac{i a \mathbf{k}}{k^2} H f(\Omega_m) \delta_b(\mathbf{k}), \tag{16.143}$$

where $f(\Omega_m) \approx \Omega_m^{0.6}$ is given by Eq. (4.78). Thus if the initial density field is Gaussian, both $\delta_b(x)$ and $v_r(x)$ are correlated Gaussian random fields determined by the following power spectrum:

$$P_b(k) = \frac{P_{\rm DM}(k)}{(1+k^2/k_{\rm J}^2)^2}. \tag{16.144}$$

We can write these two fields as linear combinations of two independent Gaussian fields $u(k)$ and $w(k)$:

$$\delta_b(k) = D(a)[u(k) + w(k)], \quad v_r(k) = H(a) f(\Omega_m) i k \alpha(k) w(k), \tag{16.145}$$

where $D(a)$ is the linear growth factor, and

$$\alpha(k) = \frac{\int_k^\infty P_b(k') k'^{-3} {\rm d}k'}{\int_k^\infty P_b(k') k'^{-1} {\rm d}k'}. \tag{16.146}$$

The power spectra for w and u are

$$P_w(k) = \frac{2\pi}{\alpha(k)} \int_k^\infty P_b(k') \frac{{\rm d}k'}{k'}; \quad P_u(k) = 2\pi \int_k^\infty P_b(k') k' {\rm d}k' - P_w(k), \tag{16.147}$$

respectively (Bi & Davidsen, 1997). To take into account some of the nonlinear evolution in the density field, Bi and Davidsen used a log-normal distribution for the density field. In this case, Eq. (16.138) is replaced by

$$n_{\rm H}(x) = \overline{n}_{\rm H}(z) \exp\left[\delta_b(x) - \frac{\langle \delta_b^2 \rangle}{2}\right]. \tag{16.148}$$

Thus, for a given $P_{\rm DM}(k)$, one can easily construct Monte Carlo realizations of $\delta_b(x)$ and $v_r(x)$, and obtain $\tau(\nu_o)$, i.e. a model Lyα absorption spectrum. For CDM-type models, these model spectra are remarkably similar to the observed Lyα forest. When the simulated spectra are fitted by Voigt line profiles, the observed HI column density distribution and b parameter distribution can all be reproduced reasonably well. More sophisticated approximations, such as the Zel'dovich approximation and particle-mesh simulations of dark matter particles, have also been used to trace the cosmic density field (Hui et al., 1997; Petitjean et al., 1995), with basic results similar to those obtained from the linear theory.

16.6 Models of the Lyman α Forest

(b) Line Profiles Consider the gas density field in the neighborhood of a point at redshift z_1. Assuming that Lyα absorption lines are broadened by thermal motion, we can write the optical depth as

$$\tau(s_o) = \int \tilde{n}_{HI}(s) I_{Ly\alpha} \frac{c}{\sqrt{\pi}b} \exp\left[-\frac{(s-s_o)^2}{b^2}\right] ds, \tag{16.149}$$

where $I_{Ly\alpha}$ is given by Eq. (16.20),

$$\tilde{n}_{HI}(s) \equiv \frac{n_{HI}(x)}{1+z} \left|\frac{ds}{dx}\right|^{-1}, \tag{16.150}$$

$$s_o \equiv c\left[1 - \frac{\nu_o}{\nu_{Ly\alpha}}(1+z_1)\right], \tag{16.151}$$

and we have used a velocity coordinate s which is related to x by

$$s = c\frac{\nu_o}{\nu_{Ly\alpha}}\left[(z-z_1) + (1+z)\frac{v_r(x)}{c}\right] \approx \frac{H(z_1)(x-x_1)}{1+z_1} + v_r(x). \tag{16.152}$$

From Eq. (16.149) it is clear that τ is just the convolution of the HI density in velocity space, $\tilde{n}_{HI}$, with a Gaussian due to thermal broadening. Note that an absorption system can thus be produced by a velocity perturbation even without a perturbation in the HI density. For a local maximum in $\tilde{n}_{HI}$ at some $s = s_{max}$ with width much smaller than b, the exponential factor in Eq. (16.149) can be considered as a constant and taken out of the integral;

$$\tau(s_o) = \frac{c}{\sqrt{\pi}b} \exp\left[-\frac{(s_{max}-s_o)^2}{b^2}\right] I_{Ly\alpha} \int \tilde{n}_{HI}(s)\, ds. \tag{16.153}$$

In this case, the absorption line has a Gaussian profile with a strength given by the density in velocity space. If, on the other hand, the distribution of $\tilde{n}_{HI}(s)$ in velocity space has a width much larger than b, then

$$\tau(s_o) = cI_{Ly\alpha}\tilde{n}_{HI}(s_o). \tag{16.154}$$

In this case, the absorption profile is completely determined by the velocity-space density $\tilde{n}_{HI}$, independent of the thermal broadening. An absorption system is then merely a fluctuation in the optical depth, not a line in the conventional sense. A Voigt profile may still be used to model such systems, but the b parameter obtained is no longer equal to $(2k_B T/m)^{1/2}$. To see this more clearly, we can expand the optical depth, τ, as a function of s around a local maximum s_{max} to the first non-zero order, which gives

$$\tau(s) = \exp[\ln \tau(s)] \approx \tau(s_{max}) \exp\left[\frac{1}{2}(\ln \tau)''(s-s_{max})^2\right], \tag{16.155}$$

where the prime denotes differentiation with respect to s. This is just a Voigt profile (for low-N_{HI} systems) with an effective b parameter

$$b_{eff} = \sqrt{-2/\eta'} \quad \text{with} \quad \eta = (\ln \tau)'. \tag{16.156}$$

In the first case discussed above, we have $b_{eff} = b$ from Eq. (16.153), while in the second case $b_{eff} = \sqrt{-2/(\ln \tilde{n}_{HI})''}$. In what follows we adopt the shorthand notation

$$\zeta \equiv (\ln \tau)'' = -\frac{2}{b_{eff}^2}. \tag{16.157}$$

Using the properties of a Gaussian random density field, we can also gain some insight into the distribution of b (Hui & Rutledge, 1999). Suppose that the number density of peaks in $\ln[\tau(s)]$ per unit s with $\eta' < \zeta < 0$ is $d\mathcal{N}$. From Eq. (16.156) it then follows that

$$\frac{d\mathcal{N}}{db} = \frac{4}{b^3}\frac{d\mathcal{N}}{d\zeta}. \tag{16.158}$$

In general we can write the number density at a given point s as $\mathcal{N}(s) = \sum_i \delta^{(D)}(s-s_i)$ where the summation is over points where $\eta = 0$ and $\eta' < \zeta < 0$ (because the peak points are local maxima; see §7.1). Around a given peak we can expand $\eta(s)$ to get $\eta(s) \approx (s-s_i)\eta'(s_i)$, and so

$$\mathcal{N}(s) = \sum_i \delta^{(D)}(s-s_i) = |\eta'(s)|\,\delta^{(D)}[\eta(s)]\Theta[-\eta'(s)+\zeta], \tag{16.159}$$

where $\Theta(x)$ is the Heaviside step function which enforces that $\eta' < \zeta$. Performing an ensemble average over this equation and inserting it into Eq. (16.158) yields

$$\frac{d\mathcal{N}}{db} = \frac{8}{b^5}\left\langle \delta^{(D)}(\eta)\delta^{(D)}(\eta'-\zeta)\right\rangle. \tag{16.160}$$

Note that the quantity in $\langle \cdot \rangle$ is just the joint probability of η and η', and that no assumption about the Gaussianity has been made in deriving Eq. (16.160). If the underlying density field is Gaussian, then η and η' are independent Gaussian random fields, and the joint probability of η and η' is the product of two Gaussian functions with variances $\langle \eta^2 \rangle$ and $\langle \eta'^2 \rangle$. It then follows from Eq. (16.160) that

$$\frac{d\mathcal{N}}{db} \propto \frac{4b_\sigma^4}{b^5}\exp\left(-\frac{b_\sigma^4}{b^4}\right), \quad \text{with} \quad b_\sigma^4 \equiv \frac{2}{\langle \eta'^2 \rangle}. \tag{16.161}$$

This distribution exhibits a power-law form in the high-b tail ($\propto b^{-5}$) and an exponential cutoff at low b. Both these characteristics are in good agreement with observations (see Fig. 16.4).

(c) Column Density Similar consideration can also be used to calculate the column density distribution (Hui et al., 1997). The HI column density around a density peak at x_{max} can be written as

$$N_{HI} = \int_{pk} n_{HI}(x)\frac{dx}{1+z}, \tag{16.162}$$

where the integration is over a sufficient interval around x_{max}. Expanding $n_{HI}(x)$ around x_{max} to second order, we have

$$\ln[n_{HI}(x)] = \ln[n_{HI}(x_{max})] + \frac{1}{2}\left[\frac{d^2\ln n_{HI}}{dx^2}\right]_{x_{max}}(x-x_{max})^2, \tag{16.163}$$

which in Eq. (16.162) gives

$$N_{HI} = \sqrt{2\pi}\frac{n_{HI}(x_{max})}{1+z_{max}}\left[\left(-\frac{d^2\ln n_{HI}}{dx^2}\right)_{x_{max}}\right]^{-1/2}. \tag{16.164}$$

It then follows from Eq. (16.140) that

$$N_{HI} \approx 4.4 \times 10^{14}\frac{\mathscr{A}(T_0)}{1-z_{max}}\frac{(\omega/\text{Mpc})}{\sqrt{q}}\,\text{cm}^{-2}, \tag{16.165}$$

where

$$\omega \equiv [1+\delta_b(x_{max})]^q \left\{\left[-\frac{d^2\ln(1+\delta_b)}{dx^2}\right]_{x_{max}}\right\}^{-1/2}. \tag{16.166}$$

Thus, if we denote by $\mathcal{N}_{\rm pk}$ the number of peaks as a function of ω and x, the HI column density distribution defined in Eq. (16.120) can be written as

$$\mathcal{F}(N_{\rm HI},z) = \frac{{\rm d}^2\mathcal{N}_{\rm pk}}{{\rm d}\omega{\rm d}x}\left(\frac{{\rm d}\omega}{{\rm d}N_{\rm HI}}\right)\left(\frac{{\rm d}x}{{\rm d}X}\right)$$

$$= 2.5\times 10^{-15}\,{\rm cm}^2\,{\rm Mpc}\frac{c}{H_0}\frac{\sqrt{q}}{(1+z)\mathscr{A}(T_0)}\frac{{\rm d}^2\mathcal{N}_{\rm pk}}{{\rm d}\omega{\rm d}x}. \qquad (16.167)$$

Defining $\xi(x)\equiv\ln[1+\delta_{\rm b}(x)]$, and using the same argument leading to Eq. (16.160) we can write

$$\frac{{\rm d}^2\mathcal{N}_{\rm pk}}{{\rm d}\omega{\rm d}x} = \left\langle |\xi''|\,\delta^{(\rm D)}(\omega_\xi-\omega)\,\delta^{(\rm D)}(\xi')\,\Theta(-\xi'')\right\rangle$$

$$= \frac{1}{q\omega}\left\langle |\xi''|\,\delta^{(\rm D)}(\xi-\xi_\omega)\,\delta^{(\rm D)}(\xi')\,\Theta(-\xi'')\right\rangle, \qquad (16.168)$$

where

$$\omega_\xi = \frac{\exp(q\xi)}{\sqrt{-\xi''}}, \quad\text{and}\quad \xi_\omega = \frac{1}{q}\ln\left(\omega\sqrt{-\xi''}\right), \qquad (16.169)$$

with the prime denoting differentiation with respect to x. Thus, if the joint distribution function for ξ, ξ' and ξ'' is given, it is straightforward to calculate the column density distribution. In the special case where $\xi(x)$ is a Gaussian random field (so that $\delta_{\rm b}$ has a log-normal distribution), this joint distribution is a multivariate Gaussian, with covariance matrix given by the power spectrum of ξ. The final result can be written in the form (Gnedin & Hui, 1998):

$$\frac{{\rm d}^2\mathcal{N}_{\rm pk}}{{\rm d}\omega{\rm d}x} = \frac{1}{q\omega}\frac{1}{(2\pi)^{3/2}R_*\sigma_0^3\gamma^2\sqrt{9/5-\gamma^2}}\int_0^\infty y e^{-Q(\omega,y)}{\rm d}y, \qquad (16.170)$$

where

$$Q(\omega,y) = \frac{1}{2\sigma_0(9/5-\gamma)}\left(\frac{9}{5}\Delta^2-2y\Delta+\frac{y^2}{\gamma^2}\right), \qquad (16.171)$$

and

$$\Delta(\omega,y) = \frac{1}{q}\ln\left(\frac{\omega y^{1/2}}{R_*}\right)+\frac{\sigma_0^2}{2}, \qquad (16.172)$$

with σ_0, R_*, γ defined as in §7.1, and ω related to $N_{\rm HI}$ by Eq. (16.165). For a given power spectrum, $P_{\rm b}(k)$, these three quantities can be calculated, and the integral in Eq. (16.170) can be computed numerically.

As one can see from Eqs. (16.167) and (16.170), the shape of the HI column density distribution depends on σ_0, R_*, and γ. Indeed, it can be shown by direct integration of Eq. (16.170) that, for CDM-type models, the shape of $\mathcal{F}(N_{\rm HI})$ depends mainly on σ_0, with a lower value of σ_0 resulting in a steeper distribution. The overall amplitude of $\mathcal{F}(N_{\rm HI})$ depends on $\mathscr{A}\propto(\Omega_{\rm b}h^2)^2/\Gamma_{\gamma,{\rm HI}}$; an increase in $\mathscr{A}$ shifts $\mathcal{F}(N_{\rm HI})$ horizontally to larger $N_{\rm HI}$.

(d) Clustering The theoretical model of the Lyα forest described above predicts a tight relation between the optical depth of Lyα absorption and the underlying mass density field. This suggests that it is possible to infer the clustering properties of the cosmic density field from the clustering in the Lyα forest.

Consider a three-dimensional field $\delta(x,y,z)$. When this field is observed along a particular line (without losing generality, we assume it to be along the z axis), we obtain a one-dimensional field, $\delta_{\rm 1d}(z)\equiv\delta(0,0,z)$. Denoting the Fourier transform of $\delta(x,y,z)$ by $\tilde{\delta}({\bf k})$, we can write

$$\delta_{\rm 1d}(z) = \sum_{k_z}\tilde{\delta}_{\rm 1d}(k_z)e^{-ik_z z} \quad\text{with}\quad \tilde{\delta}_{\rm 1d}(k_z) = \sum_{k_x,k_y}\tilde{\delta}(k_x,k_y,k_z), \qquad (16.173)$$

where we assume that $\delta(x,y,z)$ is periodic on a cubic box with volume V_u. It then follows that the power spectrum of $\delta_{1d}(z)$ is

$$P_{1d}(k_z) \equiv V_u^{1/3} \left\langle |\tilde{\delta}_{1d}(k_z)|^2 \right\rangle = V_u^{-2/3} \sum_{k_x,k_y} P(k_x,k_y,k_z), \qquad (16.174)$$

where $P(k_x,k_y,k_z) \equiv V_u \langle |\tilde{\delta}(\mathbf{k})|^2 \rangle$ is the power spectrum of the three-dimensional field, $\delta(x,y,z)$. If the field is isotropic in the (x,y) plane, and changing the summation into integration, we can write

$$P_{1d}(k_z) = \frac{1}{2\pi} \int P(\mathbf{k}_\perp, k_z) k_\perp \, dk_\perp. \qquad (16.175)$$

If δ is also isotropic in space, we have

$$P_{1d}(k_z) = \frac{1}{2\pi} \int_{k_z}^\infty P(k) k \, dk, \quad \text{and} \quad P(k) = -\frac{2\pi}{k} \frac{d}{dk} P_{1d}(k). \qquad (16.176)$$

In this special case, one can derive the three-dimensional power spectrum from its one-dimensional counterpart.

For the Lyα forest, what we observe is the optical depth $\tau(s_o)$ as a function of velocity coordinate s_o [related to the observed frequency ν_o by Eq. (16.151)], or the flux transmission $F = e^{-\tau}$. In the limit of small perturbations, the fluctuation in flux transmission, $\delta_F \equiv (F - \overline{F})/\overline{F}$, is proportional to the fluctuation in optical depth, $\delta_\tau \equiv (\tau - \overline{\tau})/\overline{\tau}$. Keeping terms to first order in δ_b and v_r, we obtain from Eqs. (16.149) and (16.140) that

$$\delta_F(s_o) \propto \delta_\tau(s_o) = \int \left[q\delta_b - \frac{\partial v_r}{\partial s} + (\Gamma-1)\frac{b_0^2}{4}\frac{\partial^2 \delta_b}{\partial s^2} \right] W(s-s_o) \, ds, \qquad (16.177)$$

where Γ is the polytropic index, $b_0 = (2k_B T_0/m)^{1/2}$, and

$$W(s-s_o) = \frac{1}{\sqrt{\pi} b_0} \exp\left[-\frac{(s-s_o)^2}{b_0^2} \right]. \qquad (16.178)$$

Using the convolution theorem, it then follows that the one-dimensional power spectrum for the flux transmission is

$$P_F(k_z) = \frac{1}{2\pi} \int_{k_z}^\infty W(k_z,k) P_b(k) k \, dk, \qquad (16.179)$$

with

$$W(k_z,k) = A \exp\left(-\frac{k_z^2 b_0^2}{2H^2} \right) \left[1 + \frac{f(\Omega_m)}{q}\frac{k_z^2}{k^2} - \frac{\Gamma-1}{4q}\frac{k_z^2 b_0^2}{H^2} \right]^2, \qquad (16.180)$$

where A is a constant, and H is the Hubble constant at the redshift in consideration (Hui, 1999). It is clear from this expression that, in the linear regime, the power spectrum for the flux transmission is just the power spectrum of the gas distribution convolved with a kernel, W, which depends both on the thermal broadening and on the peculiar velocity distortions [the term with $f(\Omega_m)$]. Thermal broadening reduces the transmission power on small scale (large k) and is negligible on scales much larger than b_0/H. If peculiar velocity distortions are neglected, the filter W is independent of k and Eq. (16.179) can be easily inverted to obtain $P_b(k)$ – up to the constant factor A – from $P_F(k)$ by direct differentiation (Croft et al., 1998). In general, one can use an iterative procedure to obtain $P_b(k)$ from P_F for a given set of model parameters: Ω_m, Γ (or q), b_0/H. The constant A can be determined by requiring that the predicted mean optical depth of HI Lyα absorption matches the observed one. Tests with numerical simulations show that this method can recover the true mass power spectrum on comoving scales in the range $\sim 1 \to 10 h^{-1}$Mpc (Croft et al., 1998). This method has also been applied to real data (e.g. Croft et al., 1998; Kim

16.6 Models of the Lyman α Forest

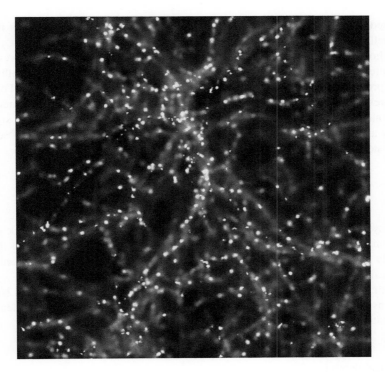

Fig. 16.6. Spatial distribution of HI column density in a simulation of the IGM at redshift $z \sim 2$. The comoving size of the box is 22.22 Mpc. The gray scale is such that the white blobs correspond to $N_{\rm HI} \geq 10^{16.5}\,{\rm cm}^{-2}$, the faint filaments correspond to $10^{14.5}\,{\rm cm}^{-2} \leq N_{\rm HI} \leq 10^{15.5}\,{\rm cm}^{-2}$, and the dark voids correspond to $N_{\rm HI} \leq 10^{14.5}\,{\rm cm}^{-2}$. [Adapted from Katz et al. (1996) by permission of AAS]

et al., 2004; McDonald et al., 2005; Jena et al., 2005), and the results put stringent constraints on the matter power spectrum on scales $\lesssim 10\,h^{-1}{\rm Mpc}$ (e.g. Spergel et al., 2007).

16.6.3 Lyman α Forest in Hydrodynamical Simulations

If Lyα forest systems are indeed produced by gravitationally induced density fluctuations in the general matter density field, the whole process can be studied in more detail by hydrodynamical cosmological simulations. Starting from cosmological initial conditions, such simulations follow the motions of gas and dark matter under their mutual gravitational field, as well as the heating, cooling, ionization, and chemical-enrichment processes of the gas component (e.g. Cen et al., 1994; Zhang et al., 1995; Miralda-Escudé et al., 1996; Hernquist et al., 1996). Despite some quantitative differences in details, the general picture emerging from these simulations are similar. Fig. 16.6 shows the spatial distribution of HI gas in one of such simulations. As one can see, depending on their HI column densities, Lyα absorbers show a variety of structures. Systems with $N_{\rm HI} \lesssim 10^{14}\,{\rm cm}^{-2}$ are generally associated with sheet-like structures while higher-$N_{\rm HI}$ clouds arise from more filamentary structures. These structures generally extend hundreds to thousands of kiloparsecs. At the lowest column density, the absorbing clouds remains unshocked and are ripples on a low-density background, and the gas is partly confined by gravity and partly by the ram pressure of the ambient gas. At the high column density end ($N_{\rm HI} \gtrsim 10^{16}\,{\rm cm}^{-2}$), the absorbing clouds become more or less spherical, reminiscent of virialized minihalos. Detailed analyses of such simulations show that the observed column density distribution, Doppler b parameter distribution, redshift evolution of line density, absorber sizes, and clustering can all be reproduced reasonably well in the current CDM cosmogonies.

There are, however, many important unresolved issues in simulating the Lyα forest in a cosmological context. Some of the problems arise from limitations of the simulations, others are due to the lack of proper understanding of the details of the physical processes involved. For example, many Lyα forest systems are found to be enriched in heavy elements. If the enrichment is due to mass loss of massive stars, the gas distribution in the IGM may have been affected significantly by the energy and momentum injection from these stars. In this case, the origin of some of the Lyα forest systems may be gas-dynamical instead of gravitational.

16.7 Lyman-Limit Systems

Lyman-limit systems are narrow line absorption systems which are optically thick to photons capable of ionizing hydrogen atoms. Although Lyman-limit systems can also be detected as strong, heavily saturated Lyα lines, their high column densities make them physically distinct from the optically thin Lyα forest lines.

To start with, consider a ray of photons with wavelengths shorter than $\lambda_{LL} = 912$Å and with an initial intensity I_0. Upon passing through an absorbing cloud with HI column density N_{HI} it will emerge with an attenuated intensity

$$I(\lambda) = I_0(\lambda)\exp[-\tau(\lambda)], \qquad (16.181)$$

where the optical depth is

$$\tau(\lambda) = N_{HI}\sigma_{pi}(\lambda) = \left(\frac{N_{HI}}{1.6\times 10^{17}\,\text{cm}^{-2}}\right)\left(\frac{\lambda}{912\text{Å}}\right)^3 \quad \text{for } \lambda \leq \lambda_{LL} \qquad (16.182)$$

and $\tau(\lambda) = 0$ for $\lambda > \lambda_{LL}$. Thus, if we define Lyman-limit systems as the ones with $\tau > 1$ for Lyman continuum radiation, then their HI column density $N_{HI} > 1.6 \times 10^{17}\,\text{cm}^{-2}$. Such systems cause a break (from $I = I_0$ to $I < 0.368I_0$) at the rest wavelength 912 Å, which is very easy to recognize in a QSO absorption line spectrum (see Fig. 16.3 for an example). For a break at observed wavelength λ_0, a strong Lyα line is expected at a wavelength $(1215/912)\lambda_0$. Because of their large optical depths to Lyman continuum photons, Lyman-limit systems may be important in attenuating the ionizing background (see §16.2.3).

The number of Lyman-limit systems per unit redshift has been found to roughly follow a power law

$$\frac{d\mathcal{N}}{dz} = 0.27(1+z)^{1.55} \quad \text{for } 0.01 < z < 5 \qquad (16.183)$$

(Storrie-Lombardi et al., 1994; Stengler-Larrea et al., 1995). Under the assumption that the comoving number density of Lyman-limit absorbers is the same as the number density of local galaxies, the typical absorption radius obtained from Eq. (16.125) is $R_* \sim 70h^{-1}$kpc at $z \sim 1$ and $R_* \sim 80h^{-1}$kpc at $z \sim 3$ (assuming a flat universe with $\Omega_{m,0} = 0.3$). These values of R_* are roughly consistent with the absorbers being clouds in galactic halos.

Although all Lyman-limit systems have HI column densities $N_{HI} > 1.6 \times 10^{17}\,\text{cm}^{-2}$, an accurate determination of their column densities is difficult because the associated Lyα lines are heavily saturated (see §16.4.4). For systems with $\tau \sim 1$, the HI column densities can be estimated from the strength of the break at the Lyman limit, but the number of such systems is small. For other systems, weak lines in the Lyman series may be used, but such lines are difficult to identify because they are usually embedded deeply in the Lyα forest. Photoionization models (see §16.4.6) may help to constrain the HI column density of a Lyman-limit system with the metal lines associated with it, but this is model dependent. Because of all these difficulties, the distribution of HI column densities for QSO absorption line systems is still poorly determined over the range $10^{17}\,\text{cm}^{-2} \lesssim N_{HI} \lesssim 10^{20}\,\text{cm}^{-2}$ (see Fig. 2.40).

Observationally, Lyman-limit systems are always associated with metal absorption line systems. This is not surprising, given the fact that virtually all Lyα forest systems with $N_{HI} \gtrsim 10^{15}\,\text{cm}^{-2}$ are known to have associated metal lines. Unfortunately, since a direct measurement of the HI column density of a Lyman-limit system is difficult, one typically requires photoionization models (§16.4.6) in order to infer their metallicities. Steidel (1990) derived a value $Z \sim 10^{-2} Z_\odot$ from modeling individual Lyman-limit systems at $z \sim 3$, consistent with the metallicities of Lyα forest systems with high column densities.

16.8 Damped Lyman α Systems

Damped Lyα systems (hereafter DLAs) are defined as systems with an HI column density $N_{HI} \geq 2 \times 10^{20}\,\text{cm}^{-2}$, and are characterized by absorption line profiles dominated by the damping wings due to natural broadening (see Fig. 16.7 for an example). This defining limit for N_{HI}, first introduced by Wolfe et al. (1986), is largely historical and was chosen to match the sensitivity limit of early 21-cm observations of local spirals by Bosma (1981). As we have seen in §16.4.4, though, technically speaking any absorption system with $N_{HI} \gtrsim 10^{19}\,\text{cm}^{-2}$ will exhibit damping wings. It has become customary to refer to absorption line systems with $10^{19}\,\text{cm}^{-2} \leq N_{HI} \leq 2 \times 10^{20}\,\text{cm}^{-2}$ as sub-DLAs (Péroux et al., 2002).[2] Although the distinction

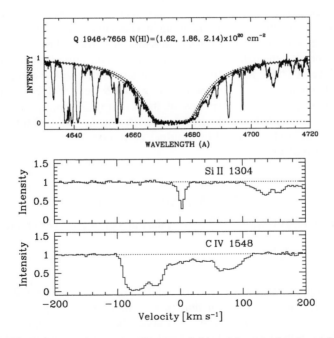

Fig. 16.7. A damped Lyα system (upper panel) at $z = 2.84$ and its associated metal lines of the highly ionized CIV (lower panel) and lowly ionized SiII (middle panel). [After Lu et al. (1996a)]

---

[2] Note that both DLAs and sub-DLAs are also Lyman-limit systems.

between DLAs and sub-DLAs is somewhat arbitrary, there is a subtle physical difference: the hydrogen in DLAs at high redshifts is largely neutral, while that in sub-DLAs may still be significantly ionized (Prochaska & Wolfe, 1996).

16.8.1 Column Density Distribution

The column density of a DLA is relatively easy to determine from its observed equivalent width [see Eq. (16.113)]. The HI column density distribution of DLAs is shown in Fig. 2.40. At first sight it seems to follow the overall power law $\mathscr{F}(N_{\rm HI}) \propto N_{\rm HI}^{-1.5}$, which also fits the Ly$\alpha$ forest and Lyman-limit systems. However, more detailed studies, based on larger data sets, have shown that the single power law is not a good fit to the data at the high column density end of the distribution. Since the data reveal a pronounced steepening of $\mathscr{F}(N_{\rm HI})$ at large $N_{\rm HI}$, it has become customary to fit the column density distribution of DLAs (and sub-DLAs) with a Schechter function

$$\mathscr{F}(N_{\rm HI}) {\rm d}N_{\rm HI} = \mathscr{F}^* \left(\frac{N_{\rm HI}}{N_{\rm HI}^*}\right)^\beta \exp\left(-\frac{N_{\rm HI}}{N_{\rm HI}^*}\right) \frac{{\rm d}N_{\rm HI}}{N_{\rm HI}^*}, \qquad (16.184)$$

which is often referred to as the 'Γ function' in the DLA community. The best-fit values of β and $N_{\rm HI}^*$ depend on both redshift and on the minimum column density considered, but typically fall in the range $-2 < \beta < -1$ and $10^{21} {\rm cm}^{-2} < N_{\rm HI}^* < 10^{21.5}$ (Storrie-Lombardi & Wolfe, 2000; Péroux et al., 2003; Prochaska et al., 2005).

Because of their high HI column densities, the gas in DLAs responsible for the absorption somehow must have cooled and collapsed, and it is thus expected that DLAs reside in dark matter halos. In order to predict the column density distribution for DLAs, we need to model how the cold gas is distributed within individual dark matter halos. Let $\Sigma(R|M_{\rm h}, \lambda)$ be the surface density distribution of a (neutral) gas disk residing in a halo of mass $M_{\rm h}$ and spin parameter λ (as discussed in §11.2, the size and surface density of a disk are expected to depend on both $M_{\rm h}$ and λ). For a given cosmology, the halo mass function $n(M_{\rm h}, z)$ and spin parameter distribution $p(\lambda)$ are known (see §7.2 and §7.5.4, respectively), and it is straightforward to compute the $\mathscr{F}(N_{\rm HI}, z)$ expected from such a population of disk galaxies (e.g. Fall & Pei, 1993; Mo et al., 1998). Under the assumption that the gas disks are exponential, the resulting $\mathscr{F}(N_{\rm HI}, z)$ are in good agreement with observations, provided halos with circular velocities as low as $\sim 50 {\rm km s}^{-1}$ are included. Similar conclusions were also reached by Gardner et al. (1997) and Nagamine et al. (2004a) based on numerical simulations.

16.8.2 Redshift Evolution

DLAs are relatively rare. For $N_{\rm HI} \geq 10^{20.3} {\rm cm}^{-2}$, the number of lines per unit redshift interval is

$$\frac{{\rm d}\mathscr{N}}{{\rm d}z} = 0.04(1+z)^{1.3 \pm 0.5} \qquad (0.1 < z < 4.7) \qquad (16.185)$$

(Storrie-Lombardi et al., 1996a), which has a redshift dependence similar to that for the Lyman-limit systems but with a normalization that is ~ 7 times lower [see Eq. (16.183)]. Using Eq. (16.125), we obtain $R^* \sim 30 h^{-1}$kpc at $z \sim 3$ under the assumption that the comoving number density of DLAs is the same as that of local galaxies. Although this radius is somewhat larger than that of HI disks for typical spiral galaxies at $z = 0$, it suggests a close connection between DLAs and (proto)-galaxies. Hence, any successful model of galaxy formation should accommodate the observational results for DLAs.

16.8 Damped Lyman α Systems

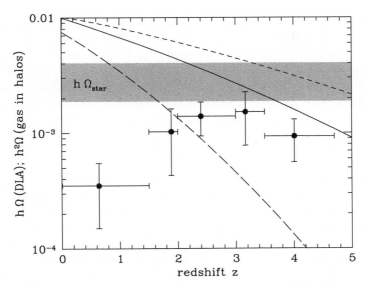

Fig. 16.8. The cosmic density parameter of baryonic material associated with DLAs as a function of redshift. The data points with error bars are taken from Storrie-Lombardi et al. (1996b). Note that $\Omega_{\rm DLA} \propto h^{-1}$. The shaded area indicates the present-day cosmic density parameter of stars and its uncertainties, and the three lines are CDM model predictions for the evolution of the cosmic density parameter of cold gas in dark matter halos with circular velocities $\geq 50\,{\rm km\,s^{-1}}$ ($\propto h^{-2}$) for three different cosmologies. Solid line: $\Omega_{\rm m,0}=1$, $h=0.5$, $\Gamma=0.5$. Long-dashed line: $\Omega_{\rm m,0}=1$, $h=0.5$, $\Gamma=0.2$. Short-dashed line: $\Omega_{\rm m,0}=0.3$, $\Omega_{\Lambda,0}=0.7$, $h=0.7$, $\Gamma=0.2$. All three models assume $\Omega_{\rm b,0}=0.015h^{-2}$ and are normalized so that $\sigma_8=0.5\Omega_{\rm m,0}^{-0.5}$.

Since the power-law slope, β, of $\mathscr{F}(N_{\rm HI},z)$ at the low column density end is larger than -2, the total mass in neutral hydrogen is dominated by systems with high column densities. To a good approximation we can therefore write

$$\Omega_{\rm HI}(z) \simeq \frac{H_0}{c}\frac{m_{\rm H}}{\rho_{\rm crit,0}} \int_{N_{\rm min}}^{\infty} \mathscr{F}(N_{\rm HI},z)\,N_{\rm HI}\,{\rm d}N_{\rm HI}, \quad (16.186)$$

with $N_{\rm min} \simeq 2\times 10^{20}\,{\rm cm^{-2}}$ (see §16.4.1). Since DLAs are largely neutral, we have that the cosmic density parameter of baryonic material associated with DLAs is equal to $\Omega_{\rm DLA}(z) = \mu\Omega_{\rm HI}(z)$ where $\mu \sim 1.3$ is the mean molecular weight taking account of the mass in helium.

Fig. 16.8 shows $\Omega_{\rm DLA}(z)$ obtained by Storrie-Lombardi et al. (1996b). For $z \sim 3$, the observed column density distribution of DLAs implies

$$\Omega_{\rm DLA} \sim 0.0015h^{-1} \qquad (z\sim 3). \quad (16.187)$$

For comparison, the mass density of stars in present-day galaxies is $\Omega_{\rm star} \sim 0.003h^{-1}$ (see §2.10.2). Thus, DLAs at $z\sim 3$ contain roughly half the amount of baryons that at $z=0$ is locked up in stars.

At $z>3$ the data of Storrie-Lombardi et al. (1996b) suggest that $\Omega_{\rm DLA}(z)$ decreases with redshift. However, as shown by Péroux et al. (2003) this is due to the fact that at $z>3.5$ up to 45% of the neutral hydrogen gas resides in sub-DLAs (at $z<3.5$ this fraction is less than 10%). After correcting for this by changing $N_{\rm min}$ in Eq. (16.186) to $10^{19}\,{\rm cm^{-2}}$, there is no evidence for any evolution in the cosmic density parameter of neutral gas at high redshifts.

In general, we can write the mass density of gas in DLAs as

$$\Omega_{\rm DLA}(z) = \Omega_{\rm b,0}F(z), \quad (16.188)$$

where $\Omega_{b,0}$ is the cosmic density parameter of baryons at the present time, and $F(z)$ is the fraction of baryons in gas clouds which are capable of producing DLAs. As argued above, it is expected that DLAs reside in dark matter halos. As discussed in §8.1.4, photoionization heating can prevent the gas from cooling in halos with virial temperatures $T_{\rm vir} \lesssim 10^5$ K, which corresponds to halos with a circular velocity $V_c \lesssim 50 \, {\rm km \, s}^{-1}$ [see Eq. (8.45)]. If we make the naive assumption that in halos with $V_c \leq V_{\rm lim} = 50 \, {\rm km \, s}^{-1}$ cooling is completely inhibited, while in halos with $V_c > V_{\rm lim}$ all gas has cooled, then the Press–Schechter model (see §7.2) yields

$$F(z) = {\rm erfc}\left[\frac{\delta_c(z)}{\sqrt{2}\sigma(M_{\rm lim})}\right], \qquad (16.189)$$

where $\delta_c(z) \simeq 1.69/D(z)$, with $D(z)$ the linear growth rate normalized to unity at the present, $\sigma(M)$ is the mass variance of the smoothed density field (see §7.2.1), and $M_{\rm lim}$ is the mass of a dark matter halo with circular velocity $V_{\rm lim}$. The three curves in Fig. 16.8 show $\Omega_{\rm DLA}(z)$ for three different cosmologies, obtained by inserting Eq. (16.189) in Eq. (16.188) with $V_{\rm lim} = 50 \, {\rm km \, s}^{-1}$.

Obviously, since not all gas in dark matter halos with $V_c > V_{\rm lim}$ can cool and since star formation consumes gas that has cooled, these curves have to be considered upper limits. A comparison with the data shows that this rules against an Einstein–de Sitter cosmology with a shape parameter $\Gamma = 0.2$ (see §4.3.2 for the definition of Γ). On the other hand, a ΛCDM cosmology with parameters that are consistent with the CMB constraints obtained from the WMAP mission (Spergel et al., 2007) yields a $\Omega_{\rm DLA}(z)$ that lies well above the data, as required. Thus, at least in the currently favored cosmologies, the total amount of gas locked up in DLAs is consistent with a picture in which they are protogalaxies residing in dark matter halos with circular velocities $V_c \gtrsim 50 \, {\rm km \, s}^{-1}$.

Since star formation in the Universe occurs mainly in high-density gas clouds, we expect that DLAs are a major source of cold gas for star formation in the Universe. As shown in Fig. 16.8, the value of $\Omega_{\rm DLA}$ drops significantly with decreasing z from its maximum value at $z \gtrsim 3$ to a value that is 5–10 times smaller at $z \sim 0$. This decrease of the HI content is likely due to the consumption by star formation, and may thus be used to infer the star-formation history in the Universe. Numerous studies have used analytical techniques or numerical simulations to predict $\Omega_{\rm DLA}(z)$ taking account of star formation, inflow (gas accretion) and outflow due to feedback processes (see Wolfe et al., 2005, for a detailed review). Typically, these models predict metallicities for the DLAs that are significantly higher than the observed values discussed below, unless significant corrections for dust obscuration are invoked (e.g. Pei & Fall, 1995; Pei et al., 1999; Somerville et al., 2001; Cen et al., 2003; Nagamine et al., 2004b).

16.8.3 Metallicities

As we have seen, the total amount of gas contained in DLAs at $z \sim 3$ is comparable to the luminous matter in galaxies. Hence, it may well be that high-redshift DLAs are the progenitors or building blocks of present day galaxies, comprising the neutral gas reservoir for star formation. Therefore, the evolution of the chemical composition of DLAs with redshift may directly probe the history of chemical enrichment and star formation in protogalaxies.

Damped Lyα systems are excellent targets for abundance studies. Their HI column density can be accurately determined from their damping wings, and since the absorbing hydrogen gas is predominantly neutral, no ionization corrections are needed to obtain their total column density of hydrogen. Furthermore, because of the high HI column densities involved, metal lines associated with DLAs are relatively easy to detect, even if the gas metallicity is low. The task is then to identify suitable metal line systems produced in the same region as the damped system and to measure their column densities. In order to have accurate estimates of the column densities, weak lines (which are not saturated) with accurate oscillator strengths are preferred (e.g. Verner et al.,

16.8 Damped Lyman α Systems

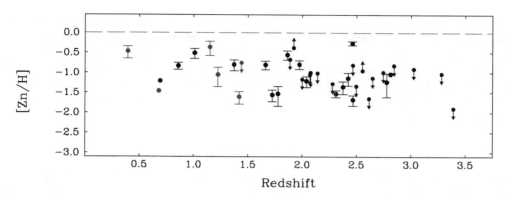

Fig. 16.9. Metallicity, expressed via the Zn abundance, of DLAs as a function of z. Abundances are measured on a log scale relative to the solar value, indicated by the dashed line at [Zn/H] = 0.0. Upper limits, corresponding to non-detections of the ZnII lines, are indicated by downward pointing arrows. [Adapted from Pettini et al. (1999) by permission of AAS]

1994; Savage & Sembach, 1996). The most important species are low (mostly singly ionized) ions, such as SiII, MnII, FeII, NiII, ZnII, CrII (Pettini et al., 1997; Lu et al., 1996a). DLAs are optically thick at photon energies $13.6\,\text{eV} \lesssim h_P\nu \lesssim 400\,\text{eV}$. Since the ionization potentials of the singly ionized elements are lower than that of hydrogen, photons with $h_P\nu < 13.6\,\text{eV}$, which can penetrate deep into the neutral gas, are capable of photoionizing the neutral state of each element to the singly ionized state. Higher energy photons, which would photoionize the elements to higher states, cannot penetrate the neutral gas. An exception are photons with $h_P\nu > 400\,\text{eV}$, but because the photoionization cross-sections are low at such photon energies, the ionization rates are low. Consequently, the total element abundances can be inferred directly from the observed column densities without significant ionization corrections. One potential problem is that some elements can be depleted onto dust grains, causing an underestimate of the inferred abundance. Consequently, elements that are less affected by dust grains (such as Zn) are preferred over elements (such as Fe) that are easily depleted by dust (Jenkins, 1987).

Fig 16.9 shows the observed Zn abundance in 40 DLAs with $0.4 \lesssim z \lesssim 3.4$. The majority of the DLAs have metallicities in the range $-2 \lesssim [\text{Zn/H}] \lesssim -0.5$, i.e. about 1/100 to 1/3 of the solar metallicity (e.g. Pettini et al., 1999; Kulkarni et al., 2005). These low metallicities are consistent with the assumption that DLAs are gas clouds that have not experienced much star formation. In addition, the metallicity does not seem to increase rapidly with decreasing redshift, at least not over the redshift range probed. To quantify this, one can estimate the total mass density of metals in DLAs as a function of redshift z:

$$\Omega_Z(z) = \Omega_{\text{DLA}}(z) Z(z), \qquad (16.190)$$

where Z is the column-density weighted mean abundance of heavy elements in the gas:

$$Z = \frac{\sum_i (N_{\text{HI}})_i Z_i}{\sum_i (N_{\text{HI}})_i}. \qquad (16.191)$$

The summation in the above equation is over all DLAs within a redshift bin centered at z. The data in Fig. 16.9 show that $Z(z)$ is roughly constant with redshift. This result, together with the observed $\Omega_{\text{DLA}}(z)$ discussed in the previous subsection, implies that the total metal content in DLAs does not increase significantly with decreasing redshift. This is not expected if DLAs trace the galaxy population in an unbiased way, because the typical metallicity of present day galaxies is roughly solar. One possibility is that DLAs arise preferentially in gas clouds with low star-formation rates, for instance if in systems with high star-formation rates the resulting energy output reduces the HI column density to a level below the detection threshold. Another possibility is that the lines-of-sight through metal-rich gas are systematically under-represented because the background QSOs are obscured by the dust associated with the metal-rich gas. In order to discriminate between these two effects, a large survey of radio selected QSOs (in which dust obscuration is not important) is required. Thus far, only one modest-size radio selected sample has been published (Ellison et al., 2001), which indicates that at most half of all DLAs are missing from optically selected samples.

16.8.4 Kinematics

One can study the internal velocity structure of DLAs from the line-of-sight velocity profiles of the associated narrow absorption lines produced by electronic transitions in low ions, such as SiII and NiII (Prochaska & Wolfe, 1997, see Fig. 16.7 for an example). These low ions are accurate tracers of neutral hydrogen at $N_{\rm HI} > 2 \times 10^{20}\,{\rm cm}^{-2}$, because the gas is effectively shielded from ionizing photons (see above). The observed line-of-sight velocity profiles often comprise multiple narrow components, suggesting that the absorbing gas is distributed among multiple discrete clouds with small internal velocity dispersions. The velocity spread among the components has a quite uniform distribution between ~ 20 and $\sim 200\,{\rm km\,s}^{-1}$. Some profiles show a pronounced skewness, reminiscent of the line profiles of rotating disks.

Detailed modeling by Prochaska & Wolfe (1998) shows that, in order to explain the observed velocity structures with isolated rotating disks, the disks must be thick (with scale heights that are at least one third of the scale lengths) and rotating fast (with rotation velocity $\gtrsim 200\,{\rm km\,s}^{-1}$). This is a problem, since as we have seen above, the observed abundances require many DLAs to reside in dark matter halos with small circular velocities. However, if DLAs indeed arise in disk galaxies, this would severely underpredict the observed line widths. Hence, it appears difficult to reconcile the observational results with a model in which DLAs are associated with disk galaxies in centrifugal equilibrium. Using hydrodynamical simulations, Haehnelt et al. (1998) demonstrated that this inconsistency within hierarchical models for galaxy formation dissapear if the gas is modeled with a more realistic spatial distribution and kinematic structure. In their simulations, DLAs arise from multiple protogalactic clumps bound to a virialized dark matter halo, and the kinematic structure is due to a mixture of rotation, random motions, infall and mergers. In addition, it has been shown that galactic winds and outflows associated with star formation in DLAs may spread out cold gas, thereby generating broad velocity structures even in halos with relatively low masses (Nulsen et al., 1998; Schaye, 2001). Although no simulation to date has been able to simultaneously and self-consistently match the observed abundances, metallicities and kinematics of DLAs from $z \sim 4$ to the present day, it is unlikely that DLAs arise solely from well-developed disks.

16.9 Metal Absorption Line Systems

In addition to absorption lines due to hydrogen and helium, QSO spectra also frequently show absorption lines due to metals. The best known examples are MgII, CIV and OVI systems, which are caused by the strong resonance-line doublets, MgII$\lambda\lambda 2796, 2800$, CIV$\lambda\lambda 1548, 1550$, and OVI$\lambda\lambda 1031, 1037$, respectively. The MgII and CIV lines have rest-frame wavelengths longer than $\lambda_{\rm Ly\alpha} = 1{,}216\,{\rm \AA}$, and can therefore appear on the red side of the Lyα emission line in the QSO spectrum. These lines are thus relatively easy to identify, as they are not embedded in the thick Lyα forest (see Fig. 16.3 for an example). On the other hand, the OVI lines are usually embedded in the Lyα forest, making them more difficult to identify, especially at high redshifts where the Lyα forest is thick.

The ionization thresholds of these species are quite different (15 eV for MgII, 65 eV for CIV, and 138 eV for OVI), which suggests that they are probably associated with gas in different environments. While MgII systems, like HI Lyα, may be produced by relatively cold gas with $T \sim 10^4$ K, CIV and OVI systems are likely associated with hotter gas with $T \sim (10^5 - 10^6)$ K. Thus, by studying different metal absorption line systems one can probe the different phases of the IGM.

16.9.1 MgII Systems

Because of their relatively long wavelengths, MgII$\lambda\lambda$2976, 2803 doublets can only be observed at relatively low redshift ($z \lesssim 2.2$ for a spectrum that can cover wavelengths up to about 10^4 Å).

(a) Line Density and Column Density In the redshift range from 0.2 to 2.2, the observed line density can be approximated by

$$\frac{d\mathcal{N}}{dz} = 0.55(1+z)^{0.8} \tag{16.192}$$

for systems with rest-frame equivalent widths W_0(MgII) ≥ 0.3Å (Steidel & Sargent, 1992). According to Eq. (16.91), this redshift dependence is consistent with no evolution in the absorbers, a conclusion that is supported by more recent observations (e.g. Nestor et al., 2005). Using Eq. (16.125), we obtain $R^* \sim 50 h^{-1}$kpc under the assumption that the comoving number density of MgII absorbers is the same as that of local galaxies. This radius is larger than that of HI disks, but smaller than that of the corresponding dark matter halos. Hence, it is possible that MgII systems are produced by gas clouds that are somehow associated with galactic halos.

It is in general quite difficult to determine the column density of a MgII system from the observed line profiles, because the observed lines are often heavily saturated. Assuming a gas temperature of a few times 10^4 K, the curve of growth (see §16.4.4) gives $N_{\rm MgII} \sim 10^{13}$ cm^{-2} for lines with $W_r = 0.3$Å. The ionization threshold for MgII (~ 15.0 eV) is not very different from that of HI. Thus, if the UV flux above 80.1 eV (the ionization threshold of MgIII) is weak, so that most Mg is in the form of MgII, the ratio $N_{\rm MgII}/N_{\rm HI}$ for a photoionized cloud should not depend strongly on the UV flux. This ratio is proportional to the metallicity of the gas, and for $Z = 0.1 Z_\odot$ gives $N_{\rm MgII}/N_{\rm HI} \sim 10^{-4}$. Thus, MgII systems with $N_{\rm MgII} \gtrsim 10^{13}$ cm^{-2} are expected to be similar to Lyman-limit systems in terms of their HI column densities.

(b) Connection to Galaxies Since MgII absorption lines can be observed at relatively low redshifts, where normal galaxies are easily detected, they are ideally suited to probe the connection between (MgII) absorbers and galaxies. Observations with deep imaging and follow-up spectroscopy around MgII absorption lines have shown that one usually finds a nearby galaxy with a redshift similar to that of the absorption line (Bergeron & Boissé, 1991; Bergeron et al., 1992; Bechtold & Ellingson, 1992; Steidel, 1995). Note, though, that such an association by itself does not necessarily imply a direct connection between the absorber and the galaxy proper. After all, since galaxies are clustered, it is possible that the association is merely a result of absorbers preferentially residing in denser regions (i.e. regions with a larger number density of galaxies), without a direct connection with the galaxies. It is therefore crucial to check whether or not the properties of absorption line systems are correlated with the properties of their identified host galaxies. Unfortunately, the observational results are mixed. While some observations indicate that the strength and kinematics of MgII absorption lines are correlated with the luminosity, morphology and/or kinematics of the host galaxies (e.g. Steidel, 1995; Steidel et al., 2002; Kacprzak et al., 2007), others do not find any compelling evidence for such correlations (e.g. Churchill et al., 1996). In addition, there is only a very weak correlation, with a large scatter, between the strength of the absorption line (i.e. its equivalent width) and the impact parameter to its host galaxy (e.g. Steidel, 1995; Tripp & Bowen, 2005). Furthermore, while early studies seemed to indicate that every bright galaxy produces a MgII absorption line system in a QSO sightline as long as its impact parameter $\lesssim 40h^{-1}$ kpc (e.g. Steidel, 1995), more recent studies (Bowen et al., 1995; Tripp & Bowen, 2005) find much lower rates of incidence. This indicates that the gas clouds responsible for the absorption must have a covering factor that is significantly smaller than unity.

16.9.2 CIV and OVI Systems

(a) CIV Systems CIV is another species commonly found in QSO absorption spectra as absorption line doublets CIV$\lambda\lambda$1548,1550. Because of their short wavelengths, CIV absorption lines can be observed at redshift $z \gtrsim 1.2$ in optical spectroscopy, and complete samples can be selected according to a rest-frame equivalent width threshold. For $W_r(\text{CIV}1548) > 0.15$ Å, the observed line density is $d\mathcal{N}/dz \sim 3$ at $z \sim 2$ (Steidel 1990), which is about two times as high as that of Lyman-limit systems. This suggests that a large fraction of the CIV systems are probably associated with gas clouds which are optically thin. The value of $d\mathcal{N}/dz$ appears to decrease with increasing redshift at $z \gtrsim 2$ for strong lines with $W_r \gtrsim 0.4$ (Sargent et al., 1988). This is contrary to the redshift evolution of Lyα forest lines and Lyman-limit systems, suggesting a build-up of the metal content in the absorbing gas with the passage of time.

The ionization threshold of CIV is 64.5 eV, and so it is a highly ionized species compared to HI and MgII. If the ionization is produced by a UV flux, $J(\nu)$, with a fixed shape, then we expect $N_{\text{CIV}}/N_{\text{HI}} \propto (J_0/n_e)^2$, where J_0 is a measure of the amplitude of J. This follows from the fact that $N_{\text{HI}}/N_{\text{H}} \propto n_e/J_0$ (because most hydrogen atoms are in HII) and $N_{\text{CIV}}/N_{\text{C}} \propto J_0/n_e$ (because most carbon atoms are in lower ions than CIV unless the UV flux is extremely hard). Thus, for a given HI column density, the CIV absorption is expected to be stronger if the gas density is lower. Because of this, strong CIV lines are probably associated with relatively low-density but highly ionized gas in the IGM. Some of the CIV absorption systems may also be produced by collisionally ionized gas with a temperature of $\sim 10^5$ K. Such gas may exist as the conduction front between a hot (10^6 K) medium and cold (10^4 K) clouds, or produced by shocks associated with the formation of sheets and filaments around galaxy-sized halos.

(b) OVI Systems and Warm-Hot IGM As mentioned in §2.10.2, the total amount of baryons in stars, cold gas and hot X-ray gas at $z \sim 0$ is only about half of all the baryons in the Universe. It is likely that the 'missing' part makes up the warm-hot intergalactic medium (WHIM), a diffuse medium with $T \sim (10^5 - 10^7)$ K that is difficult to detect in emission. This conjecture is supported by cosmological hydrodynamical simulations of structure formation in the CDM scenario, which show that a large fraction of baryons can be in this form at the present time (e.g. Cen & Ostriker, 1999; Davé et al., 2001).

For gas at such a high temperature, only highly ionized species can exist in significant amounts. Here OVI absorption lines can play an important role for two reasons. First, in collisional ionization equilibrium, the OVI ion fraction peaks at about $\sim 10^{5.5}$ K, making it a sensitive probe of the WHIM. Secondly, the doublet OVI$\lambda\lambda$1031,1037 is usually strong and relatively easy to identify in a UV spectrum at low redshifts where the Lyα forest is relatively thin. Although current data on OVI absorption line systems at low redshifts is still fairly limited, it suggests that it may contain a significant fraction of the baryons at the present epoch (Tripp et al. 2000, 2007; Savage et al. 2005; Sembach et al. 2001; Richter et al. 2004; Danforth & Shull 2005). More importantly, these observations demonstrate the potential of using OVI absorption lines to map out the distribution of the 'missing' baryons in the local Universe.

Appendix A
Basics of General Relativity

General relativity (hereafter GR) is the subject dealing with the structure of space-time and with how to describe physical laws in any given space-time. The perspective of space-time in GR is very different from that in Newtonian physics. In Newtonian physics, space is considered to be flat, infinite and eternal, time is considered to flow uniformly, and physical processes are considered to act in this external space-time frame. In the framework of GR, however, space-time is a four-dimensional manifold which may be curved and the properties of space-time itself are determined by dynamical processes.

This appendix provides a brief summary of the aspects of GR that are used in this book. More details can be found in the excellent textbooks by Weinberg (1972), Misner et al. (1973), Rindler (1977), and Carroll (2004).

A1.1 Space-time Geometry

In order to gain some insight in how to describe space-time as a four-dimensional manifold (hypersurface), consider a two-dimensional analog. To describe a two-dimensional surface, we can construct a coordinate system and label each point on the surface by its coordinates. The geometrical properties of the surface can be obtained by considering the distance between each pair of infinitesimally close points on the surface in terms of the differences in coordinates. In general, the square of this distance can be written as

$$dl^2 = \sum_{i,j=1}^{2} g_{ij}(\mathbf{x}) dx^i dx^j, \tag{A1.1}$$

where $\mathbf{x} = (x^1, x^2)$ are the coordinates and $g_{ij}(\mathbf{x})$ is the metric which gives the distance in terms of the difference in coordinates. As an example, if we use Cartesian coordinates (x,y), then $g_{ij} = \delta_{ij}$ is the metric for a plane, because $dl^2 = dx^2 + dy^2$. Similarly,

$$dl^2 = \frac{R^2 - y^2}{R^2 - x^2 - y^2} dx^2 + \frac{R^2 - x^2}{R^2 - x^2 - y^2} dy^2 + \frac{2xy}{R^2 - x^2 - y^2} dx\,dy \tag{A1.2}$$

gives the metric of a sphere with radius R. This is evident by using the spherical coordinates:

$$dl^2 = R^2(d\vartheta^2 + \sin^2\vartheta\, d\varphi^2), \tag{A1.3}$$

where (ϑ, φ) is related to (x,y) by $x = R\sin\vartheta\cos\varphi$, $y = R\sin\vartheta\sin\varphi$. This shows that the metric not only depends on the properties of the surface, but also on the choice of the coordinate system. In general, one chooses a coordinate system which simplifies the problem at hand.

The geometrical properties of the space-time can be described in a similar manner. Each point on the four-dimensional space-time hypersurface is an event, represented by a time coordinate and three spatial coordinates. The 'distance' between any two points (events) on this hypersurface is the interval ds. For a flat space-time this interval has the same form as in special relativity:

$$ds^2 = c^2 dt^2 - dx^2 - dy^2 - dz^2 = \eta_{\mu\nu} dx^\mu dx^\nu = c^2 dt^2 - \delta_{ij} dx^i dx^j, \tag{A1.4}$$

where x, y, z are the Cartesian coordinates, $(x^0, x^1, x^2, x^3) = (ct, x, y, z)$, δ_{ij} is the Kronecker delta function, and

$$\eta_{\mu\nu} = \text{diag}(1, -1, -1, -1) \tag{A1.5}$$

is the Minkowski metric. In Eq. (A1.4) and in the following, a pair of repeated upper and lower indices implies summation over their range, Greek indices run from 0 to 3 while Latin indices run from 1 to 3. For a general space-time, the interval can be written as

$$ds^2 = g_{\mu\nu} dx^\mu dx^\nu, \tag{A1.6}$$

where x^μ ($\mu = 0, 1, 2, 3$) are general space-time coordinates and the metric $g_{\mu\nu}$ gives the interval in terms of the difference in space-time coordinates. Note that $g_{\mu\nu} = g_{\nu\mu}$.

Since ds is invariant under coordinate transformation, the metric must transform as

$$g_{\mu\nu}(x) \to g'_{\mu\nu}(x') = \frac{\partial x^\alpha}{\partial x'^\mu} \frac{\partial x^\beta}{\partial x'^\nu} g_{\alpha\beta}(x), \tag{A1.7}$$

under a general coordinate transformation $x \to x'$. The inverse four-metric $g^{\mu\nu}$ is the inverse of $g_{\mu\nu}$:

$$g_{\mu\alpha} g^{\alpha\nu} = \delta_\mu^{\ \nu}, \tag{A1.8}$$

and so transforms as

$$g^{\mu\nu}(x) \to g'^{\mu\nu}(x') = \frac{\partial x'^\mu}{\partial x^\alpha} \frac{\partial x'^\nu}{\partial x^\beta} g^{\alpha\beta}(x). \tag{A1.9}$$

From the space-time metric, one can derive other useful geometric quantities. The affine connection, $\Gamma^\mu_{\ \alpha\beta}$, which connects vectors in nearby tangent spaces, is defined as

$$\Gamma^\mu_{\ \alpha\beta} = \frac{1}{2} g^{\mu\sigma} \left(\partial_\beta g_{\sigma\alpha} + \partial_\alpha g_{\sigma\beta} - \partial_\sigma g_{\alpha\beta} \right), \tag{A1.10}$$

where $\partial_\mu \equiv \partial / \partial x^\mu$. The Riemann–Christoffel curvature tensor, $R^\mu_{\ \nu\alpha\beta}$, which describes the curvature of the space-time manifold, is defined as

$$R^\mu_{\ \nu\alpha\beta} = \partial_\alpha \Gamma^\mu_{\ \nu\beta} - \partial_\beta \Gamma^\mu_{\ \nu\alpha} + \Gamma^\mu_{\ \sigma\alpha} \Gamma^\sigma_{\ \nu\beta} - \Gamma^\mu_{\ \sigma\beta} \Gamma^\sigma_{\ \nu\alpha}. \tag{A1.11}$$

The Ricci tensor and the curvature scalar are defined as

$$R_{\mu\nu} \equiv R^\sigma_{\ \mu\sigma\nu} \quad \text{and} \quad R \equiv g^{\mu\nu} R_{\mu\nu}, \tag{A1.12}$$

respectively.

For the Robertson–Walker metric,

$$ds^2 = c^2 dt^2 - a^2(t) \left[\frac{dr^2}{1 - Kr^2} + r^2 (d\vartheta^2 + \sin^2 \vartheta \, d\varphi^2) \right], \tag{A1.13}$$

the non-zero components of the affine connection are

$$\begin{aligned}
&\Gamma^0_{11} = c^{-1} a\dot{a}/(1 - Kr^2); & &\Gamma^0_{22} = c^{-1} a\dot{a} r^2; & &\Gamma^0_{33} = c^{-1} a\dot{a} r^2 \sin^2 \vartheta; \\
&\Gamma^1_{01} = \Gamma^2_{02} = \Gamma^3_{03} = \dot{a}/ca; & &\Gamma^2_{12} = \Gamma^3_{13} = 1/r; & & \\
&\Gamma^1_{11} = Kr/(1 - Kr^2); & &\Gamma^1_{22} = -r(1 - Kr^2); & &\Gamma^1_{33} = -r(1 - Kr^2) \sin^2 \vartheta; \\
&\Gamma^2_{33} = -\sin\vartheta \cos\vartheta; & &\Gamma^3_{23} = \cot\vartheta,
\end{aligned} \tag{A1.14}$$

where $(x^0, x^1, x^2, x^3) = (ct, r, \vartheta, \varphi)$ and $\dot{a} = da/dt$. The non-zero components of the Ricci tensor are

$$R_{00} = -\frac{3}{c^2}\frac{\ddot{a}}{a}, \quad R_{ij} = -\frac{1}{c^2}\left[\frac{\ddot{a}}{a} + 2\frac{\dot{a}^2}{a^2} + \frac{2c^2 K}{a^2}\right] g_{ij}, \tag{A1.15}$$

and the curvature scalar is

$$R = -\frac{6}{c^2}\left[\frac{\ddot{a}}{a} + \frac{\dot{a}^2}{a^2} + \frac{Kc^2}{a^2}\right]. \tag{A1.16}$$

For small perturbations of Minkowski space-time, the perturbed metric can in general be written as

$$ds^2 = c^2(1 + 2\Psi/c^2)\, dt^2 - 2cw_i\, dt\, dx^i - \left[(1 - 2\Phi/c^2)\delta_{ij} + H_{ij}\right] dx^i\, dx^j, \tag{A1.17}$$

where the perturbation quantities $|\Psi|/c^2$, $|\Phi|/c^2$, $|w_i|$, and $|H_{ij}|$ are all $\ll 1$. To the first order of the perturbation quantities, the non-zero components of the affine connection are:

$$\begin{aligned}
&\Gamma^0_{00} = \partial_0 \Psi; &&\Gamma^0_{i0} = \partial_i \Psi; \\
&\Gamma^i_{00} = \partial_i \Psi + \partial_0 w_i; &&\Gamma^i_{j0} = \tfrac{1}{2}(\partial_j w_i - \partial_i w_j) + \tfrac{1}{2}\partial_0 h_{ij}; \\
&\Gamma^0_{jk} = -\tfrac{1}{2}(\partial_j w_k + \partial_k w_j) + \tfrac{1}{2}\partial_0 h_{jk}; &&\Gamma^i_{jk} = \tfrac{1}{2}(\partial_j h_{ki} + \partial_k h_{ji}) - \tfrac{1}{2}\partial_i h_{jk},
\end{aligned} \tag{A1.18}$$

where $h_{ij} = H_{ij} - 2\Phi\delta_{ij}$. The components of the Ricci tensor are

$$\begin{aligned}
R_{00} &= \delta^{ij}\partial_i\partial_j\Psi + 3\partial_0^2\Phi + \partial_0\partial_j w^j; \\
R_{0j} &= -\tfrac{1}{2}\delta^{kl}\partial_k\partial_l w_j + \tfrac{1}{2}\partial_j\partial_k w^k + 2\partial_0\partial_j\Phi + \tfrac{1}{2}\partial_0\partial_k H_j{}^k; \\
R_{ij} &= \partial_i\partial_j(\Phi - \Psi) - \tfrac{1}{2}\partial_0(\partial_i w_j + \partial_j w_i) + \delta_{ij}\eta^{\mu\nu}\partial_\mu\partial_\nu\Phi - \tfrac{1}{2}\eta^{\mu\nu}\partial_\mu\partial_\nu H_{ij} + \tfrac{1}{2}\partial_k(\partial_i H_j{}^k + \partial_j H_i{}^k).
\end{aligned} \tag{A1.19}$$

These results can also be used in dealing with small metric perturbations of a flat, expanding universe. Here the perturbed metric can be written as

$$ds^2 = c^2(1 + 2\Psi/c^2)\, dt^2 - 2caw_i\, dt\, dx^i - \left[(1 - 2\Phi/c^2)\delta_{ij} + H_{ij}\right] a^2\, dx^i\, dx^j, \tag{A1.20}$$

where $a(t)$ is the scale factor. It is evident that if we use a new set of space-time coordinates, (ct, x'^1, x'^2, x'^3) where $dx'^i = a\, dx^i$, the affine connection and Ricci tensor corresponding to metric (A1.20) will have the same forms as given by Eqs. (A1.18) and (A1.19), except that all spatial derivatives are with respect to x'^i. The quantities in terms of the comoving coordinates, x^i, can then be obtained by using general coordinate transformations (see below).

A1.2 The Equivalence Principle

According to the principle of general relativity, all reference frames are equivalent, and a physical law should have the same form under general coordinate transformation (the general covariance). Because of this, physical fields defined on a space-time must transform according to a set of rules under the general coordinate transformation. Depending on whether it is a scalar, S, a vector, $\mathbf{V}$, or a tensor, $\mathbf{T}$, a physical field transforms as

$$S'(x') = S(x); \quad V'^\mu(x') = \frac{\partial x'^\mu}{\partial x^\alpha} V^\alpha(x); \quad T'^{\mu\nu}(x') = \frac{\partial x'^\mu}{\partial x^\alpha}\frac{\partial x'^\nu}{\partial x^\beta} T^{\alpha\beta}(x), \tag{A1.21}$$

under the general coordinate transformation $x^\mu \to x'^\mu$. From a given vector or a given tensor, we can define another vector or another tensor as

$$V_\mu = g_{\mu\nu} V^\nu; \quad T_{\mu\nu} = g_{\mu\alpha} g_{\nu\beta} T^{\alpha\beta}. \tag{A1.22}$$

In general, new tensors can be obtained by using $g_{\mu\nu}$ to lower indices and by using $g^{\mu\nu}$ to raise indices. As in special relativity, V_α and V^α are called the covariant and contravariant components of **V**. Generally, a lower index is called the covariant index while an upper index is called a contravariant index. It is easy to prove that under general coordinate transformation, V_μ and $T_{\mu\nu}$ transform as

$$V'_\mu(x') = \frac{\partial x^\alpha}{\partial x'^\mu} V_\alpha(x); \quad T'_{\mu\nu}(x') = \frac{\partial x^\alpha}{\partial x'^\mu} \frac{\partial x^\beta}{\partial x'^\nu} T_{\alpha\beta}(x). \tag{A1.23}$$

Thus, both $V^\mu V_\mu$ and $T^{\mu\nu} T_{\mu\nu}$ are invariant under general coordinate transformations.

From the perspective of GR, gravitation is manifested as curved space, and so the space-time must be (locally) Minkowskian in a frame which is in free fall in a gravitational field. An important aspect of GR is embodied in the equivalence principle that can be stated as follows: In a reference frame which is in free fall in a gravitational field, all physical laws have their special relativistic form, except the gravitational force which disappears. Together with the principle of general relativity, the equivalence principle enables us to find physical equations valid for any general reference frame: what we need to do is just to write the usual special relativistic equations in covariant forms.

Since physical equations generally involve derivatives with respect to the space-time coordinates, we need to find the covariant forms of the derivatives of physical fields. The covariant derivative with respect to a space-time coordinate x^μ is usually denoted by a subscript ';μ'. For a scalar field it is defined as

$$S_{;\mu} \equiv \partial_\mu S, \tag{A1.24}$$

while for vector fields it is defined as

$$V^\alpha{}_{;\beta} = \partial_\beta V^\alpha + \Gamma^\alpha{}_{\mu\beta} V^\mu, \quad V_{\alpha;\beta} = \partial_\beta V_\alpha - \Gamma^\mu{}_{\alpha\beta} V_\mu. \tag{A1.25}$$

It is easy to show that, under general coordinate transformation, $S_{;\mu}$ transforms as a vector, while $V^\alpha{}_{;\beta}$ and $V_{\alpha;\beta}$ transform as tensors. In general, to obtain the covariant derivative of the tensor $T^{...}_{...}$ with respective to x^α, we add to the ordinary derivative $\partial_\alpha T^{...}_{...}$ a term $-\Gamma^\mu{}_{\beta\alpha} T^{...}_{...\mu}$ for each covariant index β ($T^{...}_{...\beta}$), and a term $\Gamma^\beta{}_{\mu\alpha} T^{...\mu}_{...}$ for each contravariant index β ($T^{...\beta}_{...}$). Note that the affine connection itself is *not* a tensor.

Using the definition of the metric, one can show that

$$\sqrt{-g}\, d^4 x = \sqrt{-g'}\, d^4 x', \tag{A1.26}$$

where g is the determinant of $g_{\mu\nu}$. This means that $\sqrt{-g}\, d^4 x$ is an invariant volume element. Thus, if $S(x)$ is a scalar field, then $\int S(x)\sqrt{-g}\, d^4 x$ is independent of the choice of coordinates.

A1.3 Geodesic Equations

As an application of the equivalence principle, consider the motion of a free particle with non-zero mass in a gravitational field. In the reference frame comoving with the particle (where, according to the principle of equivalence, the space-time must be locally Minkowskian with metric $\eta_{\mu\nu}$), the motion of the particle is given by

$$\frac{d^2 \xi^\mu}{ds^2} = 0, \tag{A1.27}$$

where ds/c is the proper time interval measured in the free-fall frame, and ξ^μ is the space-time coordinates of the particle. For a general reference frame with coordinates x^μ related to ξ^ν by $x^\mu(\xi)$, the metric is related to $\eta_{\mu\nu}$ by

$$g_{\mu\nu} = \eta_{\alpha\beta} \frac{\partial \xi^\alpha}{\partial x^\mu} \frac{\partial \xi^\beta}{\partial x^\nu}. \tag{A1.28}$$

In the x-frame, the equation of motion (A1.27) becomes

$$\frac{d^2 x^\mu}{ds^2} = -\Gamma^\mu{}_{\alpha\beta} \frac{dx^\alpha}{ds} \frac{dx^\beta}{ds}, \tag{A1.29}$$

where

$$\Gamma^\mu{}_{\alpha\beta} = \frac{\partial x^\mu}{\partial \xi^\nu} \frac{\partial^2 \xi^\nu}{\partial x^\alpha \partial x^\beta}. \tag{A1.30}$$

One can prove that $\Gamma^\mu{}_{\alpha\beta}$ is just the affine connection defined by Eq. (A1.10). This can be done by using the relation

$$\partial_\lambda g_{\mu\nu} = \Gamma^\alpha{}_{\lambda\mu} g_{\alpha\nu} + \Gamma^\alpha{}_{\lambda\nu} g_{\alpha\mu}, \tag{A1.31}$$

and the results of cyclically permuting the three indices. Thus, in the x-frame there is a force exerting on the free-fall particle. This is gravity. But in the perspective of GR it is because the particle is moving in a curved space (non-zero affine connection). Free particles move along geodesics, and so Eq. (A1.29) is also the geodesic equation. If we define the four-momentum as

$$p^\mu = mU^\mu, \qquad U^\mu = c \frac{dx^\mu}{ds}, \tag{A1.32}$$

where m is the rest mass of the particle, Eq. (A1.29) can be written in the form

$$\frac{p^0}{c} \frac{dp^\mu}{dt} = -\Gamma^\mu{}_{\alpha\beta} p^\alpha p^\beta, \tag{A1.33}$$

where $p^0 = mc\, dx^0/ds = mc^2 dt/ds$. Another useful form of Eq. (A1.29) is

$$\frac{p^0}{c} \frac{dp_\mu}{dt} = \frac{1}{2}(\partial_\mu g_{\alpha\beta}) p^\alpha p^\beta, \tag{A1.34}$$

where $p_\mu = g_{\mu\nu} p^\nu$. Note that

$$g_{\mu\nu} p^\mu p^\nu = m^2 c^2. \tag{A1.35}$$

In time-orthogonal coordinates, where $g_{00} = 1$ and $g_{0i} = 0$, we have

$$(p^0)^2 + g_{ij} p^i p^j = m^2 c^2. \tag{A1.36}$$

If we define the magnitude of the three-momentum as $p^2 = -g_{ij} p^i p^j$, then

$$(p^0)^2 = p^2 + m^2 c^2, \tag{A1.37}$$

and cp^0 can be considered the total energy of the particle in the time-orthogonal frame.

For massless particles $ds \to 0$ and so Eq. (A1.29) is invalid. However, both Eqs. (A1.33) and (A1.34) are well defined for massless particles, provided that p^μ is properly defined. One possibility is to define the four-momentum as

$$p^\mu = \frac{p^0}{c} \frac{dx^\mu}{dt}, \tag{A1.38}$$

which is the same as Eq. (A1.32) for massive particles if we choose $p^0 = mc^2 dt/ds$. For massless particles, $g_{\mu\nu} p^\mu p^\nu = 0$. It can then be shown that in time-orthogonal coordinates, cp^0 is the energy of the particle. To cast the equation of motion for massless particles in the form of the geodesic equation (A1.29), we introduce an affine parameter λ by the equation

$$p^0 \equiv \frac{dx^0}{d\lambda}. \tag{A1.39}$$

Eq. (A1.33) can then be written in the form

$$\frac{d^2 x^\mu}{d\lambda^2} = -\Gamma^\mu{}_{\alpha\beta} \frac{dx^\alpha}{d\lambda} \frac{dx^\beta}{d\lambda}. \tag{A1.40}$$

A1.4 Energy–Momentum Tensor

If a charge Q is invariant under Lorentz transformation, the equation of charge conservation can be written in the form

$$\frac{\partial(nQ)}{\partial t} + \nabla \cdot \mathbf{j} = 0, \tag{A1.41}$$

where $\mathbf{j} = nQ\mathbf{v}$ is the current density. In covariant form this is

$$J^\mu{}_{;\mu} = 0, \tag{A1.42}$$

where J^μ is the four-current density vector. One might consider applying this to the mass to obtain a covariant form for the continuity equation. However, mass is not invariant under Lorentz transformation; it depends on momentum because of its connection to energy. Thus, a covariant continuity equation must involve both energy and momentum. The conserved quantity we are seeking is expected to have 16 components: the energy and energy current in three directions, plus momenta in three directions and their currents (each momentum has three components). Thus the quantity must be a 4×4 tensor which we call the energy–momentum tensor and denote by $T^{\mu\nu}$. The conservation of energy–momentum can then be written in the covariant form

$$T^{\mu\nu}{}_{;\mu} = 0. \tag{A1.43}$$

In many cosmological applications, the material content can be approximated by a fluid. To obtain the corresponding energy–momentum tensor, we again use the equivalence principle. A fluid is characterized by the density, $\rho(\mathbf{x})$, and pressure, $P(\mathbf{x})$, both measured by an observer *comoving* with the fluid at the point $\mathbf{x}$, and the velocity of the fluid element relative to some reference frame. Note that ρ and P defined in this way are invariant under general coordinate transformation. In the rest frame of a fluid element, the energy–momentum tensor is

$$T^{\mu\nu} = \text{diag}(\rho c^2, P, P, P) = (\rho + P/c^2)U^\mu U^\nu - P\eta^{\mu\nu}, \tag{A1.44}$$

where $U^\mu = (c, 0, 0, 0)$ is the four-velocity of the fluid element in the comoving frame. We can make a Lorentz transformation to get the energy–momentum tensor in a reference frame which is in free fall in the gravitational field at the point of the fluid element:

$$T^{\mu\nu} = (\rho + P/c^2)U^\mu U^\nu - P\eta^{\mu\nu}, \tag{A1.45}$$

where U^μ is the four-velocity of the fluid element in the free-fall reference frame. Thus, using the principle of equivalence, the energy–momentum tensor in a general coordinate system is

$$T^{\mu\nu} = (\rho + P/c^2)U^\mu U^\nu - Pg^{\mu\nu}, \tag{A1.46}$$

where $U^\mu = c\,dx^\mu/ds$.

A1.5 Newtonian Limit

One interesting question is what form the space-time metric takes in the Newtonian limit of gravity. Such a metric tells us how Newtonian gravity (the gravitational potential) is interpreted in terms of geometric quantities, thereby providing a hint how to construct the field equation in GR by generalizing the Newtonian field equation (Poisson's equation). To start with, consider a reference frame O' which is in free fall in a Newtonian gravitational potential Φ which is zero at some large distance. In this frame, the metric has the Minkowski form:

$$ds^2 = c^2 dt'^2 - dx'^2. \tag{A1.47}$$

Now consider another reference frame O relative to which O' has the free-fall velocity given by $v^2 = -2\Phi$ (assumed to be in the x-direction). According to Lorentz transformation, we have

$$dt' = (1 + 2\Phi/c^2)^{1/2} dt; \quad dx' = (1 - 2\Phi/c^2)^{1/2} dx. \tag{A1.48}$$

Thus, the metric in terms of the coordinates in the O system can be written as

$$ds^2 = c^2 \left(1 + 2\Phi/c^2\right) dt^2 - \left(1 - 2\Phi/c^2\right) \left(dx^2 + dy^2 + dz^2\right). \tag{A1.49}$$

This is the metric in the Newtonian limit.

A1.6 Einstein's Field Equation

In the Newtonian limit, the 0–0 component of the energy–momentum tensor has the form $T_{00} = \rho c^2$, and $g_{00} = (1 + 2\Phi/c^2)$. The Poisson equation for gravity therefore takes the form

$$\nabla^2 g_{00} = 8\pi G T_{00}/c^4. \tag{A1.50}$$

This is a relation between the energy–momentum tensor and the derivatives of the metric. In general, the field equation must be a covariant extension of the above relation. We therefore expect the right-hand side of the above equation to be replaced by $8\pi G T_{\mu\nu}/c^4$, and the left-hand side to be replaced by a 4×4 tensor constructed from the metric and its derivatives. Einstein proposed a tensor (the Einstein tensor) of the form

$$G_{\mu\nu} = R_{\mu\nu} - \frac{1}{2} g_{\mu\nu} R, \tag{A1.51}$$

and so the Einstein field equation takes the form

$$G_{\mu\nu} = \frac{8\pi G}{c^4} T_{\mu\nu}. \tag{A1.52}$$

Using that $g_{\mu\nu} g^{\mu\nu} = 4$, we see that the trace of the field equation is $R = -8\pi G T/c^4$, where $T = T_\mu{}^\mu$. The field equation can then be written in the form

$$R_{\mu\nu} = \frac{8\pi G}{c^4} \left(T_{\mu\nu} - \frac{1}{2} g_{\mu\nu} T \right). \tag{A1.53}$$

It can be shown that, in the Newtonian limit (A1.49), this equation reduces to the Poisson equation.

Einstein also realized that he could add to $G_{\mu\nu}$ a term $-\Lambda g_{\mu\nu}$ and write the field equation as

$$R_{\mu\nu} - \frac{1}{2} g_{\mu\nu} R - \Lambda g_{\mu\nu} = \frac{8\pi G}{c^4} T_{\mu\nu}, \tag{A1.54}$$

where Λ is a constant called the cosmological constant. Using the expression of $T_{\mu\nu}$ for an ideal fluid [see Eq. (A1.46)], we see that the Λ term can be included in the energy–momentum tensor as an ideal fluid with $\rho = -P/c^2 = c^2 \Lambda / 8\pi G$.

Appendix B
Gas and Radiative Processes

Galaxy formation and evolution involves many gaseous and radiative processes. In this appendix we review some of the basic concepts that are related to our descriptions in the main text. More detailed descriptions can be found in Rybicki & Lightman (1979), Osterbrock (1989), and Shu (1991a,b).

B1.1 Ideal Gas

An ideal gas is a hypothetical gas that consists of identical particles of zero volume that undergo perfectly elastic collisions and for which the intermolecular forces can be neglected. These assumptions are valid, to good approximation, for low and moderate density gases, and therefore have many applications in astrophysics. An ideal gas obeys the ideal gas law

$$PV = Nk_B T, \qquad (B1.1)$$

with P the absolute pressure, V the volume, N the number of particles, k_B the Boltzmann constant, and T the absolute temperature. This can also be written as

$$P = \frac{k_B T}{\mu m_p} \rho, \qquad (B1.2)$$

where $\rho = N\mu m_p/V$ is the matter density, and μ is the mean molecular mass of a gas particle in units of the proton mass, m_p. For a gas of fully ionized hydrogen, one has that $\mu = 1/2$, while $\mu \simeq 0.59$ for a fully ionized gas of primordial composition (assuming a mixture of 75% hydrogen and 25% helium, in terms of mass).

The specific internal energy (i.e. the internal energy per unit mass) of an ideal gas depends only on temperature, and is given by

$$\mathscr{E} = \frac{1}{\gamma - 1} \frac{k_B T}{\mu m_p}. \qquad (B1.3)$$

Here $\gamma = c_P/c_V$ is the adiabatic index, which is defined as the ratio of the specific heats

$$c_P = T \left(\frac{\partial S}{\partial T} \right)_P \quad \text{and} \quad c_V = T \left(\frac{\partial S}{\partial T} \right)_V, \qquad (B1.4)$$

with S the specific entropy. For an ideal gas, we have that $\gamma = (q+5)/(q+3)$, with q the *internal degrees of freedom* of the particles. Thus, for a monatomic gas of point particles, $q = 0$ and $\gamma = 5/3$.

Although the general equation of state of an ideal gas is of the form $P = P(\rho, T)$, one often considers special cases in which the equation of state is barotropic, meaning $P = P(\rho)$. The two most important cases are the ideal isothermal process, for which the temperature stays constant so that $P \propto \rho$, and the ideal isentropic process,[1] for which the entropy of the gas remains constant, so that $P \propto \rho^\gamma$. More generally, a barotropic equation of state of the form $P \propto \rho^\Gamma$ is known as a polytropic equation of state, with Γ the polytropic index. Note that $\Gamma = 1$ for an isothermal equation of state, while Γ is equal to the adiabatic index in the case of an isentropic equation of state.

B1.2 Basic Equations

(a) Fluid Equations   In many problems discussed in this book the gas component can be approximated as a fluid in terms of its macroscopic properties. The evolution of a fluid is governed by a set of hydrodynamical equations which is derived from the Boltzmann equation under the assumption that the mean free path of gas particles, l, is much smaller than the smallest macroscopic scale of the structure, L, we are interested in. In this case, the gas can be considered in local thermal equilibrium (LTE), and can be described as a continuous medium characterized by a set of thermal dynamical quantities, such as density ρ, pressure P, and temperature T, all varying with time, t, and spatial location, $\mathbf{x} = (x_1, x_2, x_3)$.

The first moment of the Boltzmann equation leads to mass conservation:

$$\frac{D\rho}{Dt} + \rho \frac{\partial u_j}{\partial x_j} = 0, \tag{B1.5}$$

where $\mathbf{u} = (u_1, u_2, u_3)$ is the velocity of the fluid, repeated indices imply summation, and

$$\frac{D}{Dt} \equiv \frac{\partial}{\partial t} + u_i \frac{\partial}{\partial x_i} \tag{B1.6}$$

is the convective time derivative. The second moment of the Boltzmann equation leads to momentum conservation:

$$\rho \frac{Du_j}{Dt} = -\frac{\partial P}{\partial x_j} + \frac{\partial \pi_{jk}}{\partial x_k} - \rho \frac{\partial \Phi}{\partial x_j}, \tag{B1.7}$$

where Φ is the gravitational potential, and π_{jk} is the viscous stress tensor defined as follows. Suppose that a particle at a given location in the fluid has velocity $\mathbf{U}$. In general we can write $\mathbf{U} = \mathbf{u} + \mathbf{w}$. By definition $\langle \mathbf{U} \rangle = \mathbf{u}$, so that $\langle \mathbf{w} \rangle = 0$ and $\langle U_i U_j \rangle = u_i u_j + \langle w_i w_j \rangle$. The pressure and the viscous stress tensor are then given by

$$P = \frac{1}{3}\rho \langle w^2 \rangle; \quad \pi_{ij} = P\delta_{ij} - \rho \langle w_i w_j \rangle. \tag{B1.8}$$

The energy equation given by the Boltzmann equation can be written as

$$\frac{\partial}{\partial t}\left[\rho\left(\frac{u^2}{2} + \mathscr{E}\right)\right] + \frac{\partial}{\partial x_k}\left[\rho\left(\frac{u^2}{2} + \mathscr{E}\right)u_k + (P\delta_{jk} - \pi_{jk})u_j + F_{\text{cond},k}\right] = -\rho u_k \frac{\partial \Phi}{\partial x_k}, \tag{B1.9}$$

where $\mathscr{E}$ is the specific internal energy, and

$$\mathbf{F}_{\text{cond}} = \frac{1}{2}\rho \langle w^2 \mathbf{w} \rangle \tag{B1.10}$$

is the conduction flux. Under the assumption of LTE, one can define a specific entropy, $S = c_V \ln(P\rho^{-\gamma})$. The energy equation can then be combined with the mass and momentum equations to give the entropy equation,

$$\rho T \frac{DS}{Dt} = -\frac{\partial F_{\text{cond},k}}{\partial x_k} + \mathscr{V}, \tag{B1.11}$$

---

[1] An isentropic process is a process that is adiabatic (i.e. no heat transfer) and reversible.

where

$$\mathscr{V} \equiv \pi_{ij}\frac{\partial u_i}{\partial x_j} \tag{B1.12}$$

is the viscous dissipation.

If the mean free path l is zero, then $\mathbf{F}_{\text{cond}} = 0$, $\pi_{ij} = 0$, and the above set of equations reduces to the normal Euler equations. In this case, the specific entropy is conserved according to Eq. (B1.11), as is expected from the fact that there is no heat flow across the border of a fluid element.

To the first order in l/L, we have

$$F_{\text{cond},k} = -\mathscr{K}\frac{\partial T}{\partial x_k}, \tag{B1.13}$$

where $\mathscr{K} \sim (3/2)k_B n v_{\text{th}} l$ is the coefficient of thermal conductivity, with $v_{\text{th}} = (k_B T/\mu m_p)^{1/2}$ the thermal velocity, and

$$\pi_{ik} = \nu\left[\frac{\partial u_i}{\partial x_k} + \frac{\partial u_k}{\partial x_i} - \frac{2}{3}\frac{\partial u_j}{\partial x_j}\delta_{ik}\right], \tag{B1.14}$$

where $\nu \sim \mu m_p v_{\text{th}}/\sigma$ is the coefficient of shear viscosity, with σ the typical collision cross-section of gas particles. Note that in this approximation both the conduction flux and the viscous stress are written in terms of other fluid quantities.

In the presence of energy sources (heating) and sinks (cooling), the following energy loss function should be added at the right-hand sides of Eqs. (B1.9) and (B1.11):

$$\mathscr{L} \equiv \mathscr{C} - \mathscr{H}, \tag{B1.15}$$

where $\mathscr{H}$ and $\mathscr{C}$ are the heating and cooling rates per unit volume, respectively. We are particularly interested in radiative heating and cooling, for which both $\mathscr{C}$ and $\mathscr{H}$ can be calculated by modeling the relevant microscopic processes (see §B1.3 below). One can also include mass and momentum source/sink terms in the mass and momentum equations.

(b) Ionization Equations The main processes for the production and destruction of ions of a specific species are photoionization, by which an atom or ion is ionized by a photon, and two-body processes, by which an atom or ion is ionized due to a collision with another particle (e.g. an electron or another ion). The number density for any given species a (including free electrons) is therefore governed by

$$\frac{\partial n_a}{\partial t} + \nabla\cdot(n_a\mathbf{v}) = \sum_{b,c}k_{bc}n_b n_c - \sum_b k_{ab}n_b n_a + \sum_c \Gamma_{\gamma,c}n_c - \Gamma_{\gamma,a}n_a, \tag{B1.16}$$

where k_{ab} is the rate coefficient for a two-body interaction between species a and b, and $\Gamma_{\gamma,a}$ is the photoionization rate for species a. The sum $\sum_{b,c}$ in the above equation is over all reactions of the kind $b+c \to a$, and $\sum_c$ is over those of the kind $c+\gamma \to a$, both representing the production of species a. The sum $\sum_b$ is over all two-body reactions involving the annihilation of an a particle. This term, together with the term $\Gamma_{\gamma,a}n_a$, represents the destruction rate of species a.

(c) Radiative Transfer Equations Since baryonic gas interacts with photons, a consistent treatment of the gas component must also include the evolution of the radiation field. The radiation field at a space-time position $(\mathbf{r},t)$ is described by the specific intensity, $J(\nu,\hat{\mathbf{k}},\mathbf{x},t)$, defined so that

$$dE \equiv J(\nu,\hat{\mathbf{k}},\mathbf{x},t)\,\hat{\mathbf{k}}\cdot\hat{\mathbf{n}}\,dA\,d\Omega\,d\nu\,dt \tag{B1.17}$$

is the amount of radiant energy carried by photons with frequencies between ν and $\nu+d\nu$ and propagating in a direction within a solid angle $d\Omega$ centered around $\hat{\mathbf{k}}$, through an area dA with unit normal $\hat{\mathbf{n}}$, in a time interval dt. The unit of J is therefore $[J] = \mathrm{erg\,s^{-1}cm^{-2}Hz^{-1}sr^{-1}}$. The time evolution of J along the path of a light ray is given by the radiative transfer equation, which can be derived as follows. We can label the position along the path of a light ray by a time t so defined that $(t_o - t)$ is the time needed for a photon package at that position to reach an observer located at a position $\mathbf{r}_o$ along the path at time t_o. The trajectory of the photon package is $\mathbf{x} = \mathbf{x}(t)$ with $\mathbf{x}(t_o) = \mathbf{x}_o$. The change of J per unit time *along the path* can be written as

$$\frac{dJ}{dt} \equiv \frac{\partial J}{\partial t} + c\hat{\mathbf{k}} \cdot \nabla_x J$$
$$= \frac{c}{4\pi}\varepsilon[\nu,\mathbf{x}(t),t] - c\rho\kappa[\nu,\mathbf{x}(t),t]J. \tag{B1.18}$$

Here $\varepsilon(\nu,\mathbf{x},t)$ is the emissivity (assumed to be isotropic) per unit proper volume at $(\mathbf{x},t)$; $\rho\kappa$ is the opacity, i.e. the fractional decrease of the specific intensity due to absorption per unit proper length along the path. So defined, κ is the absorption cross-section per unit mass.

Both the opacity and emissivity depend on the composition and ionization state, as well as thermodynamic properties (temperature, pressure, etc.) of the gas. In addition, the emissivity also depends on the properties and distribution of radiating sources.

B1.3 Radiative Processes

Radiative processes involve the interactions of photons with atoms, ions and electrons. The cross-sections of all such interactions can, in principle, be calculated from quantum electrodynamics, but such calculations are very tedious for complex atomic systems. In practice, many of the cross-sections for complex systems have to be determined through laboratory experiments. Atomic data are published and updated regularly in *Atomic Data and Nuclear Data Tables* (Academic Press); there are also projects like the Opacity Project (Seaton, 1995) that are specifically devoted to the determinations of radiative cross-sections.

Radiative processes can be classified into several broad categories:

- Bound–bound processes: These are the processes by which an electron makes a transition from one bound level to another bound level in an atom (or ion). Such transitions can be made either by collisions with electrons (collisional excitation and de-excitation) or by interactions with photons (photon excitation, spontaneous and stimulated decay).
- Bound–free processes: These processes involve the removal of an electron from a bound orbit, which can happen when an atom (or ion) collides with an electron (collisional ionization) or when it absorbs a photon (photoionization). The reverse process is recombination, by which a free electron recombines with an ion.
- Free–free processes: These processes involve electrons only in unbound (free) states. When a free electron is accelerated or decelerated, it emits photons through bremsstrahlung. A free electron can also absorb a photon through free–free absorption.
- Compton scattering: This is the process in which photons are scattered by free electrons (or ions). If the photon energy is high, electrons gain energy (the Compton process). On the other hand, if the kinetic energy of the electron is much larger than the photon energy, energy is transferred from the electron to the photon (the inverse Compton process).

In the following we describe some of these processes in more detail.

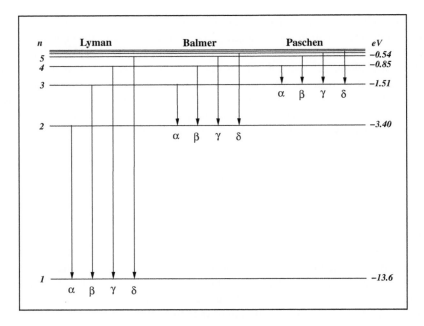

Fig. B1.1. Illustration of the energy levels of the hydrogen atom and some of the most important transitions.

B1.3.1 Einstein Coefficients and Milne Relation

The bound–bound and bound–free processes are related to the energy levels of the atom in question and the transition probabilities associated with them. As an example, Fig. B1.1 shows the basic energy levels of the hydrogen atom and some of the most important transitions. To start with, consider the transition between two bound energy states of a given atom, which we call state 'i' and state 'j'. The energies of these two states will be denoted by E_i and E_j, and as a convention we assume $E_j > E_i$. The photon frequency involved with the transition is therefore $\nu_{ij} \equiv (E_j - E_i)/h_\mathrm{P}$. The probability per unit time, P_{ij}, that the atom will undergo a transition from the low state 'i' to the high state 'j' due to photon excitation is

$$P_{ij} = B_{ij} B_\nu, \tag{B1.19}$$

where $B_\nu \, d\nu$ is the brightness of the radiation in the frequency range $\nu \pm d\nu/2$, and B_{ij} is the Einstein coefficient of the excitation due to the absorptions of photons of frequency ν. Note that $[B_{ij}] = \mathrm{g}^{-1}\,\mathrm{s}$, $[B_\nu] = \mathrm{g}\,\mathrm{s}^{-2}$ so that $[P_{ij}] = \mathrm{s}^{-1}$. An atom at the excited state 'j' tends to decay back spontaneously to the low-energy state 'i', emitting a photon with energy $h_\mathrm{P} \nu_{ij}$ but in a random direction. The probability for this to happen is given by the Einstein coefficient for spontaneous transition, A_{ji}. In addition to the spontaneous decay, an atom at state 'j' can also make a transition back to state 'i' due to the stimulus of a photon. The rate of stimulated transition is $B_{ji} B_\nu$, where B_{ji} is the Einstein stimulated emission coefficient. The total probability of emission is therefore

$$P_{ji} = A_{ji} + B_{ji} B_\nu. \tag{B1.20}$$

Under the assumption of thermodynamical equilibrium, the number of transitions from 'i' to 'j' should be equal to that from 'j' to 'i', and so

$$n_i P_{ij} = n_j P_{ji}, \tag{B1.21}$$

where n_i and n_j are the number densities of atoms in state 'i' and state 'j', respectively. Because of thermodynamical equilibrium, the number of atoms in each of the two states is given by the Boltzmann distribution, so that

$$\frac{n_j}{n_i} = \frac{g_j}{g_i} \exp\left(-\frac{h_\mathrm{P} \nu_{ij}}{k_\mathrm{B} T}\right), \tag{B1.22}$$

where g_k is the statistical weight of state 'k', and T is the equilibrium temperature. This in Eq. (B1.21) gives

$$g_i B_{ij} B_\nu = \exp\left(-\frac{h_\mathrm{P} \nu_{ij}}{k_\mathrm{B} T}\right) (g_j A_{ji} + g_j B_{ji} B_\nu). \tag{B1.23}$$

The brightness of blackbody radiation at temperature T is given by the Planck function

$$B_\nu(T) = \frac{2 h_\mathrm{P} \nu^3}{c^2} \left[\exp\left(\frac{h_\mathrm{P} \nu}{k_\mathrm{B} T}\right) - 1\right]^{-1} = \frac{h_\mathrm{P} \nu}{4\pi} c \mathcal{N}_\gamma(\nu), \tag{B1.24}$$

where $\mathcal{N}_\gamma(\nu) \, d\nu$ is the number density of photons with frequencies in the range $\nu \pm d\nu/2$. Inserting this with $\nu = \nu_{ij}$ into Eq. (B1.23) gives

$$\frac{2 h_\mathrm{P} \nu_{ij}^3}{c^2} g_i B_{ij} = \exp\left(-\frac{h_\mathrm{P} \nu_{ij}}{k_\mathrm{B} T}\right) \left[\frac{2 h_\mathrm{P} \nu_{ij}^3}{c^2} g_j B_{ji} - g_j A_{ji}\right] + g_j A_{ji}. \tag{B1.25}$$

Since this relation is expected to hold for all T, we must have the following relations between the Einstein coefficients:

$$g_j A_{ji} = \frac{2 h_\mathrm{P} \nu_{ij}^3}{c^2} g_i B_{ij} \quad \text{and} \quad g_j B_{ji} = g_i B_{ij}. \tag{B1.26}$$

These relations involve only the properties of the atoms, and so are expected to hold in general cases, although they are derived under the assumption of thermodynamical equilibrium. With these relations, the photon excitation process is completely determined by one of the Einstein coefficients, e.g. $g_i B_{ij}$. Conventionally, we write

$$B_{ij} = \frac{4\pi a_{ij}}{h_\mathrm{P} \nu_{ij}}, \tag{B1.27}$$

where a_{ij} is related to the oscillator strength of the transition, f_{ij}, by

$$a_{ij} = \frac{\pi q_\mathrm{e}^2}{m_\mathrm{e} c} f_{ij}, \tag{B1.28}$$

with q_e and m_e the charge and mass of an electron, respectively. The oscillator strength f_{ij} (or a_{ij}) of a transition can be calculated from quantum electrodynamics or obtained from laboratory experiments. Table B1.1 lists the atomic data for several important transitions; more complete lists can be found in Cox (2000) and Verner et al. (1994), for example.

Inserting the relation between B_{ij} and B_{ji} given by Eq. (B1.26) into Eq. (B1.23) we obtain

$$n_j A_{ji} = n_i B_{ij} B_\nu \left[1 - \exp\left(-\frac{h_\mathrm{P} \nu_{ij}}{k_\mathrm{B} T}\right)\right]. \tag{B1.29}$$

This is a relation between the rate of spontaneous transition and that of photo-excitation. A similar relation also holds for the case in which level 'j' is a free state. This gives a relation between the spontaneous recombination cross-section, σ_r, and the photoionization cross-section, σ_pi:

$$n_{a+1} v \sigma_\mathrm{r}(v) n_\mathrm{e} f(v) \, dv = \left[1 - \exp\left(-\frac{h_\mathrm{P} \nu}{k_\mathrm{B} T}\right)\right] n_a c \sigma_\mathrm{pi}(\nu) \mathcal{N}_\gamma(\nu) \, d\nu. \tag{B1.30}$$

Table B1.1. Data for some atomic transitions.

Species	Transition	λ (Å)	g_i	f_{ik}
H I – Lyα	$1s-2p$	1215.67	2	0.416
H I – Lyβ	$1s-3p$	1025.72	2	0.079
H I – Hα	$2s-3p$	6562.74	2	0.435
H I – Hα	$2p-3s$	6562.86	6	0.014
H I – Hα	$2p-3d$	6562.81	6	0.696
H I – Hβ	$2s-4p$	4861.29	2	0.103
H I – Hβ	$2p-4s$	4861.35	6	0.003
H I – Hβ	$2p-4d$	4861.33	6	0.122
He I – Lyα	$1s^2-1s2p$	584.33	1	0.285
He II – Lyα	$1s-2p$	303.78	2	0.416
C IV	$1s^22s-1s^22p$	1550.77	2	0.095
C IV	$1s^22s-1s^22p$	1548.20	2	0.190
O VI	$1s^22s-1s^22p$	1037.62	2	0.066
O VI	$1s^22s-1s^22p$	1031.93	2	0.133
Mg II	$2p^63s-2p^63p$	2803.53	2	0.314
Mg II	$2p^63s-2p^63p$	2796.35	2	0.629
Si IV	$2p^63s-2p^63p$	1402.77	2	0.260
Si IV	$2p^63s-2p^63p$	1393.75	2	0.524

The left-hand side is the electron-capturing (recombination) rate for an ion X at the $(a+1)^{\text{th}}$ ionization stage. It is proportional to the density of X_{a+1}, n_{a+1}, and to $v\sigma_r n_e f(v)\,dv$, the rate at which an X_{a+1} ion can capture an electron of velocity between v and $v+dv$. The right-hand side is the photoionization rate for the ath ionization stage (with an extra exponential term that is due to induced recombination). It is proportional to the number density of X_a and to $c\sigma_{\text{pi}}\mathcal{N}_\gamma(\nu)\,d\nu$, the rate at which an X_a ion absorbs a photon of frequency between ν and $\nu+d\nu$. Because of energy conservation, we have

$$\frac{1}{2}m_e v^2 = h_{\text{P}}(\nu-\nu_{\text{t}}), \tag{B1.31}$$

where ν_{t} is the threshold energy required to ionize an X_a ion, and so $m_e v\,dv = h_{\text{P}}\,d\nu$. Under the conditions of local thermodynamical equilibrium, the number densities n_{a+1} and n_a are related by the Saha equation:

$$\frac{n_{a+1}}{n_a}n_e = \frac{2U_{a+1}(T)}{U_a(T)}\left(\frac{2\pi m_e k_B T}{h_{\text{P}}^2}\right)^{3/2}\exp\left(-\frac{h\nu_{\text{t}}}{k_B T}\right). \tag{B1.32}$$

Here U_a is the partition function of the ath ionization stage, equal to the sum $\sum_k g_k e^{-E_k/k_B T}$ over all possible energy levels of the ionization stage, with g_k the statistical weight of level k, and E_k the difference between the energy of level k and that of the ground state of the ionization stage. As described by Rybicki & Lightman (1979), in many applications, the partition function U_a can be approximated by the statistical weight, g_a, of the ground state of the corresponding ionization stage. Assuming that $\mathcal{N}_\gamma(\nu)$ is given by Eq. (B1.24) and that electrons have a Maxwellian velocity distribution

$$f(v) = \left(\frac{2}{\pi}\right)^{1/2}\left(\frac{m_e}{k_B T}\right)^{3/2}v^2\exp\left(-\frac{m_e v^2}{2k_B T}\right), \tag{B1.33}$$

we obtain the Milne relation:

$$\sigma_{\text{r}}(v) = \frac{g_a}{g_{a+1}}\left(\frac{h_{\text{P}}\nu}{m_e c v}\right)^2 \sigma_{\text{pi}}(\nu), \tag{B1.34}$$

where we have used g_a to replace U_a. This is a relation between the recombination cross-section at a given v and the photoionization cross-section at the corresponding ν, and so it is valid independent of the assumption of thermodynamical equilibrium, although it is derived based on thermodynamical equilibrium.

B1.3.2 Photoionization and Photo-excitation

Photoionization is the process in which an atom is ionized by the absorption of a photon. For hydrogen, this is

$$H^0 + \gamma \to p + e, \tag{B1.35}$$

where H^0 denotes a neutral hydrogen atom. The photoionization rate is proportional to the number density of ionizing photons and to the photoionization cross-section. Thus, we can write

$$\Gamma_{\gamma,H} = \int_{\nu_t}^{\infty} c\sigma_{pi}(\nu)\mathcal{N}_{\gamma}(\nu)\,d\nu, \tag{B1.36}$$

where ν_t is the threshold frequency for ionization. $\mathcal{N}_{\gamma}(\nu)d\nu$ in the above equation is the number density of photons with frequencies in the range ν to $\nu + d\nu$, and is related to the energy flux of the radiation field, $J(\nu)$, by

$$\mathcal{N}_{\gamma}(\nu) = \frac{4\pi J(\nu)}{ch_P\nu}. \tag{B1.37}$$

The photoionization cross-sections can be obtained from quantum electrodynamics by calculating the bound–free transition probability of an atom in a radiation field (see, e.g. Rybicki & Lightman, 1979; Shu, 1991b). For hydrogenic atoms with nuclear charge Z at the nth excited state, the result is

$$\sigma_{pi}(\nu,n) = \frac{n\mathcal{K}_0}{Z^2}\left[\frac{\nu_t(Z,n)}{\nu}\right]^3 g_{bf}(n,\nu) \quad \text{for } \nu \geq \nu_t(Z,n), \tag{B1.38}$$

while $\sigma_{pi}(\nu,n) = 0$ for $\nu < \nu_t(Z,n)$. Here $\nu_t(Z,n) = Z^2\nu_t(1,n)$ is the threshold frequency of ionization, $g_{bf}(n,\nu)$ is the bound–free Gaunt factor for the nth level (Karzas & Latter, 1961), and

$$\mathcal{K}_0 = \frac{64}{3\sqrt{3}}\left(\frac{2\pi e^2}{h_P c}\right)\pi a_0^2 = 7.91 \times 10^{-18}\,\text{cm}^2, \tag{B1.39}$$

is the Krammers absorption cross-section at the Lyman edge of H^0, with $a_0 \equiv \hbar_p^2/4\pi^2 m_e q_e^2$ the Bohr radius. Detailed calculations show that the typical lifetime of a hydrogenic atom against photoionization is much longer than the lifetime of an exited state. Thus, to very good approximation, photoionization can be assumed to be from the ground state. With this assumption, the photoionization cross-section can be written as

$$\sigma_{pi}(\nu) = \frac{\mathcal{K}_0}{Z^2}\left(\frac{\nu_Z}{\nu}\right)^3 g_1(\nu) \quad (\text{for } \nu \geq \nu_Z), \tag{B1.40}$$

where $\nu_Z \equiv \nu_t(Z,1)$. The bound–free g factors are close to unity at optical frequencies (e.g. Karzas & Latter, 1961). Taking into account the ν dependence of g_1, the photoionization cross-section for hydrogenic atoms can be written as

$$\sigma_{pi}(\nu) = \frac{A_0}{Z^2}\left(\frac{\nu_Z}{\nu}\right)^4 \frac{\exp\left[4 - 4(\tan^{-1}\tau)/\tau\right]}{1 - \exp[-2\pi/\tau]}, \tag{B1.41}$$

where $A_0 = 6.30 \times 10^{-18}\,\text{cm}^2$ and $\tau \equiv [(\nu/\nu_Z) - 1]^{1/2}$ (see Osterbrock, 1989). For many-electron atoms, the photoionization cross-sections usually show large variations in very small frequency intervals at some threshold frequencies, but in many applications a smoothed-out representation is adequate. A useful approximation to the smoothed contribution of each threshold to the photoionization cross-section is

$$\sigma_{\mathrm{pi}}(\nu) = a_\mathrm{t} \left[\beta \left(\frac{\nu_\mathrm{t}}{\nu}\right)^s + (1-\beta) \left(\frac{\nu_\mathrm{t}}{\nu}\right)^{s+1} \right] \quad (\nu > \nu_\mathrm{t}), \tag{B1.42}$$

and the total cross-section is the sum over individual thresholds. A list of numerical values of ν_t, a_t, β and s for some common atoms and ions can be found in Osterbrock (1989). For He^0, the values are $\nu_\mathrm{t}/c = 1.983 \times 10^5 \,\mathrm{cm}^{-1}$, $a_\mathrm{t} = 7.83 \times 10^{-18}\,\mathrm{cm}^2$, $\beta = 1.66$, and $s = 2.05$.

In addition to photoionization, the absorption of a photon can also excite an atom (photoexcitation). This is a bound–bound process, with a rate given by

$$\Gamma_{ij,\gamma} = B_{ij} B_\nu = \int \sigma_{ij}(\nu) c \mathcal{N}_\gamma(\nu) \, d\nu, \tag{B1.43}$$

where σ_{ij} is the absorption cross-section. For resonance absorption, $\sigma_{ij}(\nu)$ is peaked sharply at the resonance frequency ν_{ij}, and so the cross-section is related to a_{ij} by

$$a_{ij} = \int \sigma_{ij}(\nu) \, d\nu. \tag{B1.44}$$

We can therefore write the absorption cross-section as

$$\sigma_{ij}(\nu) = a_{ij} \phi_{ij}(\nu) \quad \text{with} \quad \int \phi_{ij}(\nu) \, d\nu = 1, \tag{B1.45}$$

where $\phi_{ij}(\nu)$ describes the profile of the absorption line in ν-space. A more detailed discussion about absorption line profiles is presented in Chapter 16.

B1.3.3 Recombination

Recombination is the process by which an ion recombines with an electron. For hydrogen ions (i.e. protons), the process is

$$\mathrm{p} + \mathrm{e} \to \mathrm{H} + \gamma. \tag{B1.46}$$

This is clearly a cooling process, as a photon is emitted. For a hydrogenic ion, the recombination cross-section to form an atom (or ion) at level n, $\sigma_\mathrm{r}(v,n)$, is related to the corresponding photoionization cross-section by the Milne relation (B1.34):

$$\sigma_\mathrm{r}(v,n) = \frac{g_n}{g_{n+1}} \left(\frac{h_\mathrm{P} \nu}{m_\mathrm{e} c v}\right)^2 \sigma_{\mathrm{pi}}(\nu, n), \tag{B1.47}$$

where ν and v are related by $m_\mathrm{e} v^2/2 = h_\mathrm{P}(\nu - \nu_n)$, and $g_n \approx 2n^2$ for hydrogenic atoms. The recombination coefficient for a given level n is the product of the capture cross-section and velocity, $\sigma_\mathrm{r}(v,n)v$, averaged over the velocity distribution $f(v)$. For an optically thin gas where all photons produced by recombination can escape without being absorbed, the total recombination coefficient is the sum over n:

$$\alpha_\mathrm{A} = \sum_{n=1}^\infty \alpha_n = \sum_{n=1}^\infty \int \sigma_\mathrm{r}(v,n) v f(v) \, dv. \tag{B1.48}$$

This is the Case A recombination coefficient, to distinguish it from the Case B recombination in an optically thick gas. In Case B, recombinations to the ground level generate ionizing photons that are absorbed by the gas, so that they do not contribute to the overall ionization state of the gas. It is easy to see that the Case B recombination coefficient is $\alpha_\mathrm{B} \approx \alpha_\mathrm{A} - \alpha_1$.

For electrons in local thermal equilibrium, $f(v)$ is the Maxwellian distribution function and both α_A and α_B depend only on the electron temperature, T. Using the definition of α_n and the relation between σ_r and σ_{pi}, we can write

$$\alpha_n(T) = \frac{g_n}{g_{n+1}} \left(\frac{2\pi m_e k_B T}{h_P^2}\right)^{-3/2} \frac{4\pi}{c^2} \int_{\nu_n}^{\infty} \nu^2 \sigma_{pi}(\nu, n) e^{-h_P(\nu - \nu_n)/k_B T} d\nu. \quad (B1.49)$$

The values of α_A and α_B can then be calculated using Eq. (B1.38). These values for H$^+$ ($Z = 1$) and other simple ions can be found in Spitzer (1978) and Osterbrock (1989), for example. In the temperature range 10^2–10^8 K, the value of α_A for H$^+$ can be approximated by

$$\alpha_A(T) = 4.0 \times 10^{-13} T_4^{-0.7} \left(1 + T_6^{0.7}\right)^{-1} \text{cm}^3 \text{s}^{-1}, \quad (B1.50)$$

where $T_4 \equiv T/10^4$ K, and so on. This approximation is the same as that given by Black (1981) for $T \lesssim 10^5$ K, except that the last factor is included to enforce proper behavior at $T > 10^5$ K (Cen, 1992). For hydrogenic ions with $Z > 1$, $\alpha(Z, T) = Z\alpha(1, T/Z^2)$. For He$^+$,

$$\alpha_A(T) = 4.3 \times 10^{-13} T_4^{-0.6353} \text{cm}^3 \text{s}^{-1} \quad (B1.51)$$

(see Black, 1981).

For complex ions, dielectronic recombination is sometimes important. This is a recombination process in which a free electron is captured by an ion and the excess energy of the recombination is taken up by a second (ionic) electron which then also occupies an excited state. The doubly excited ion subsequently relaxes either by auto-ionizing (to give back the original ion and a free electron) or via radiative cascades. The coefficients for this kind of recombination are listed in e.g. Nussbaumer & Storey (1983) and Osterbrock (1989). For He$^+$, it can be approximated by

$$\alpha_d(T) = 6.0 \times 10^{-10} T_5^{-1.5} e^{-4.7/T_5} \left(1 + 0.3 e^{-0.94/T_5}\right) \text{cm}^3 \text{s}^{-1}. \quad (B1.52)$$

B1.3.4 Collisional Ionization and Collisional Excitation

Collisional ionization is the process by which an atom (or ion) is ionized by the collision with another particle. This is a cooling process because part of the kinetic energy is used for the ionization. The most important process is the collision with an electron. For a hydrogen atom, this process is

$$H^0 + e \to p + e + e. \quad (B1.53)$$

The collision-induced ionization rate is proportional to the number density of colliding electrons, n_e, and the collisional cross-section, $\sigma_{ci}(v)$, which depends on the velocity v of the colliding electron. Hence we can write the collisional rate as

$$\Gamma_{e,H} = n_e \langle \sigma_{ci}(v) v \rangle = n_e \int_0^{\infty} \sigma_{ci}(v) v f(v) dv, \quad (B1.54)$$

where $f(v)$ is the velocity distribution of electrons. For a given v the collisional cross-section, $\sigma_{ci}(v)$, can be calculated from quantum mechanics. If $f(v)$ is Maxwellian then $\Gamma_{e,H}$ depends only on the electron temperature and, for 10^4 K $< T < 10^8$ K, can be approximated by

$$\Gamma_{e,H} = \Gamma_0 \left(\frac{T}{T_t}\right)^{1/2} \exp\left(-\frac{T_t}{T}\right) \left(1 + T_5^{1/2}\right)^{-1} n_e \text{ cm}^3 \text{s}^{-1}, \quad (B1.55)$$

where

$$\Gamma_0 = 2.32 \times 10^{-8}, \quad T_t \equiv h_P \nu_t(H)/k_B = 157809.1 \text{ K}. \quad (B1.56)$$

Collisional ionization rates for He^0 and He^+ can be approximated by the same form, with

$$\Gamma_0 = 1.27 \times 10^{-8}, \quad T_t = 285335.4\,\text{K} \quad (\text{for } He^0), \tag{B1.57}$$

$$\Gamma_0 = 4.51 \times 10^{-9}, \quad T_t = 631515.0\,\text{K} \quad (\text{for } He^+). \tag{B1.58}$$

In addition to ionization, collisional processes can also excite atoms or ions (collisional excitation). Cooling then occurs when the excited atoms make transitions back to their ground states. The rate of excitation between two levels i and j by collisions with a given kind of particles (e.g. electrons) can be written as

$$\Gamma_{ij,e} = n_e \gamma_{ij,e}, \quad \gamma_{ij,e} = \int_0^\infty \sigma_{ij,e}(v) v f(v)\,dv, \tag{B1.59}$$

where $\gamma_{ij,e}$ is the collisional rate coefficient. The quantity $\sigma_{ij,e}(v)$ is the collisional cross-section for collisions with velocity v, which is often expressed in terms of the collision strength $\Omega_{ij,e}$:

$$\sigma_{ij,e}(v) = \frac{\pi}{g_i}\left(\frac{h_P}{2\pi m_e v}\right)^2 \Omega_{ij,e}. \tag{B1.60}$$

The values of Ω_{ij} are listed in, e.g. Mendoza (1983).

B1.3.5 Bremsstrahlung

In an ionized gas, electrons can be accelerated and emit photons when colliding with ions. This kind of emission is called bremsstrahlung. For a cosmic gas, the main process is the collision between electrons and protons:

$$p + e \to p' + e' + \gamma. \tag{B1.61}$$

The energy radiated from a plasma per unit time, per unit volume and per unit frequency interval (i.e. the volume emissivity) can be written as

$$\varepsilon_{ff}(\nu) = n_i n_e \int P(v,\nu) f(v)\,dv, \tag{B1.62}$$

where $f(v)$ is the electron velocity distribution, and $P(v,\nu)$ is the power radiated per unit frequency due to the collision of an electron of velocity v with an ion. The form of $P(v,\nu)$ can be calculated directly from electrodynamics (e.g. Shu, 1991b; Rybicki & Lightman, 1979). For thermal bremsstrahlung, where $f(v)$ is Maxwellian, the emissivity due to collisions between electrons and ions with charge number Z is

$$\varepsilon_{ff}(\nu) = \frac{32\pi}{3} \frac{Z^2 q_e^6}{m_e^2 c^4} \left(\frac{2\pi m_e c^2}{3 k_B T}\right)^{1/2} n_i n_e g(\nu, T) \exp\left(-\frac{h_P \nu}{k_B T}\right)$$

$$\approx 6.8 \times 10^{-42} Z^2 \frac{n_i n_e}{T_8^{1/2}} g(\nu, T) \exp\left(-\frac{h_P \nu}{k_B T}\right)\,\text{erg s}^{-1}\text{cm}^3\text{Hz}^{-1}, \tag{B1.63}$$

where $T_8 \equiv T/10^8\,\text{K}$, and $g(\nu, T) \sim 1$ is the Gaunt g factor. The inverse process is free–free absorption, in which a photon is absorbed by an electron when it is accelerated by an ion. The absorption opacity can be obtained by using Kirchhoff's law, $\varepsilon_{ff} = 4\pi(\rho \kappa_{ff}) B_\nu$:

$$\rho \kappa_{ff}(\nu) = \frac{4}{3} \frac{Z^2 q_e^6}{m_e^2 c^2 h_P \nu^3} \left(\frac{2\pi m_e c^2}{3 k_B T}\right)^{1/2} n_i n_e g(\nu, T)\left[1 - \exp\left(-\frac{h_P \nu}{k_B T}\right)\right]. \tag{B1.64}$$

B1.3.6 Compton Scattering

Compton scattering is the process in which a free electron collides with a photon:

$$e + \gamma \to e' + \gamma'. \tag{B1.65}$$

If the photon energy, $h_P\nu$, is much smaller than $m_e c^2$, the cross-section for this process is the classical Thomson cross-section:

$$\sigma_T = \frac{8\pi}{3}\left(\frac{q_e^2}{m_e c^2}\right)^2 \approx 6.65 \times 10^{-25}\,\text{cm}^2. \tag{B1.66}$$

During Compton scattering, energy exchange can occur between electrons and photons. To see this, consider a medium of non-relativistic electrons at temperature T_e. When photons propagate through such a medium, they will be scattered and Doppler-shifted by the electrons, and the photon occupation number $\mathcal{N}(\nu, t)$ changes with time according to the Kompaneets equation (Kompaneets, 1957; Rybicki & Lightman, 1979):

$$\frac{\partial \mathcal{N}}{\partial t} = \frac{\sigma_T n_e h_P}{m_e c}\frac{1}{\nu^2}\frac{\partial}{\partial \nu}\left\{\nu^4\left[\mathcal{N}(\mathcal{N}+1) + \frac{k_B T_e}{h_P}\frac{\partial \mathcal{N}}{\partial \nu}\right]\right\}. \tag{B1.67}$$

Using the following dimensionless variables,

$$x = \frac{h_P \nu}{k_B T_e}, \quad dy = \frac{k_B T_e}{m_e c^2}\sigma_T n_e c\,dt, \tag{B1.68}$$

the Kompaneets equation can be written in a more compact form,

$$\frac{\partial \mathcal{N}}{\partial y} = \frac{1}{x^2}\frac{\partial}{\partial x}\left\{x^4\left[\mathcal{N}(\mathcal{N}+1) + \frac{\partial \mathcal{N}}{\partial x}\right]\right\}. \tag{B1.69}$$

The Compton y-parameter introduced here can be written as

$$y = \int_{t_1}^{t} \frac{k_B T_e(t')}{m_e c^2}\sigma_T n_e(t') c\,dt', \tag{B1.70}$$

which is proportional to the pressure of the electron gas, $n_e T_e$. It can also be considered as the product of the scattering optical depth, $\int \sigma n_e c\,dt$, and a weighted average of the electron temperature.

The number of photons per unit volume, n_γ, is the integral of the occupation number, $\mathcal{N}$, over phase space:

$$n_\gamma = \int \mathcal{N}(x,y) x^2\,dx. \tag{B1.71}$$

It can be easily verified from Eq. (B1.69) that $dn_\gamma/dt = 0$, consistent with the fact that the photon number is conserved during Compton scattering. A static solution of Eq. (B1.69) is

$$\mathcal{N}(x) = \left(e^{x+x_0} - 1\right)^{-1}, \tag{B1.72}$$

where x_0 is a constant. This solution is expected, because when the photon number is conserved, the energy spectrum relaxes to a Bose–Einstein distribution at a finite chemical potential (proportional to x_0).

The energy density in the radiation is

$$u_\gamma = \int_0^\infty \mathcal{N}(x,y) x^3\,dx. \tag{B1.73}$$

Multiplying Eq. (B1.69) by x^3 and integrating over x we obtain

$$\frac{du_\gamma}{dy} = 4u_\gamma - \int x^4 \mathcal{N}(\mathcal{N}+1)\,dx. \tag{B1.74}$$

If the radiation field is close to that of a blackbody with temperature T, then

$$\mathcal{N} \approx (e^{h_\mathrm{P}\nu/k_\mathrm{B}T} - 1)^{-1}, \quad \text{and} \quad \mathcal{N}(\mathcal{N}+1) \approx -\frac{T}{T_\mathrm{e}}\frac{\partial \mathcal{N}}{\partial x}. \tag{B1.75}$$

In this case, the equation for u_γ reduces to

$$\frac{1}{u_\gamma}\frac{\mathrm{d}u_\gamma}{\mathrm{d}y} = 4\frac{T_\mathrm{e} - T}{T_\mathrm{e}}. \tag{B1.76}$$

Thus, the radiation field gains energy from electrons if $T_\mathrm{e} > T$. For $T_\mathrm{e} \gg T$, the energy gain per volume is

$$\frac{\mathrm{d}u_\gamma}{\mathrm{d}t} = 4u_\gamma\frac{\mathrm{d}y}{\mathrm{d}t} = \frac{4k_\mathrm{B}T_\mathrm{e}}{m_\mathrm{e}c^2}c\sigma_\mathrm{T}n_\mathrm{e}a_\mathrm{r}T_\gamma^4, \tag{B1.77}$$

where a_r is the radiation constant.

B1.4 Radiative Cooling

(a) Atomic Cooling For an optically thin gas, the volume cooling rate due to bremsstrahlung is obtained from the bremsstrahlung (free–free) emissivity given by Eq. (B1.63):

$$\mathscr{C}_\mathrm{ff} = \int \varepsilon_\mathrm{ff}(\nu)\,\mathrm{d}\nu$$

$$\approx 1.4 \times 10^{-23}T_8^{1/2}\left(\frac{n_\mathrm{e}}{\mathrm{cm}^{-3}}\right)^2 \mathrm{erg\,s}^{-1}\mathrm{cm}^{-3}, \tag{B1.78}$$

where in the second expression we have assumed $Z = 1$ and $n_\mathrm{i} \sim n_\mathrm{e}$, valid for a completely ionized hydrogen gas. The total cooling rate is the sum over all species of ions.

The cooling rate due to recombination of an ionization level $a+1$ to form an ion of level a is

$$\mathscr{C}_a(T) = \frac{g_a}{g_{a+1}}n_\mathrm{e}n_{a+1}\left(\frac{2\pi m_\mathrm{e}k_\mathrm{B}T}{h_\mathrm{P}^2}\right)^{-3/2}\frac{4\pi}{c^2}$$

$$\times \int_{\nu_a}^\infty \nu^2 \sigma_\mathrm{pi}(\nu,a)h_\mathrm{P}(\nu - \nu_a)\exp\left[-\frac{h_\mathrm{P}(\nu - \nu_a)}{k_\mathrm{B}T}\right]\mathrm{d}\nu, \tag{B1.79}$$

which can be derived in the same way as Eq. (B1.49). The cooling rate arising from the collisional excitation and de-excitation of a given kind of element X by a given type of particles Y (e.g. electrons) can be written as

$$\mathscr{C}_{X,Y} = n_Y n_X \sum_{b<a}(E_a - E_b)[x_b\gamma_{ba}(X,Y) - x_a\gamma_{ab}(X,Y)], \tag{B1.80}$$

where $x_a = n_a/n_X$ ($x_b = n_b/n_X$) is the fractional population of X atoms at level a (b), E_a (E_b) is the corresponding energy level, and γ_{ba} and γ_{ab} are the collisional excitation and de-excitation rate coefficients, respectively. The number densities of atoms at different energy levels can be obtained by solving the population equations,

$$\frac{\partial n_a}{\partial t} + \nabla \cdot (n_a \mathbf{v}) = \sum_Y\sum_b n_b\Gamma_{ba,Y} - n_a\sum_Y\sum_b \Gamma_{ab,Y}, \tag{B1.81}$$

where $\Gamma_{ab,Y}$ denotes the rate for ions (of a given element) on level a to make a transition to level b as a result of process Y (both collisional and photoelectric). Given the rates of transitions, the population densities can be solved, and so the cooling rate can be calculated. Table B1.2 lists the rates of the dominant cooling processes for hydrogen and helium. These formulae are adopted from Black (1981), with some modifications given by Cen (1992).

Table B1.2. H and He cooling rates.

Process	Species	Cooling rate/(erg s^{-1} cm^{-3})
collisional–excitation	H^0	$7.50 \times 10^{-19} e^{-118348/T} (1+T_5^{1/2})^{-1} n_e n_{H^0}$
	He$^+$	$5.54 \times 10^{-17} T^{-0.397} e^{-473638/T} (1+T_5^{1/2})^{-1} n_e n_{He}^+$
collisional–ionization	H^0	$1.27 \times 10^{-21} T^{1/2} e^{-157809/T} (1+T_5^{1/2})^{-1} n_e n_{H^0}$
	He^0	$9.38 \times 10^{-22} T^{1/2} e^{-285335/T} (1+T_5^{1/2})^{-1} n_e n_{He^0}$
	He$^+$	$4.95 \times 10^{-22} T^{1/2} e^{-631515/T} (1+T_5^{1/2})^{-1} n_e n_{He^+}$
recombination	H$^+$	$2.82 \times 10^{-26} T^{0.3} (1+3.54 T_6)^{-1} n_e n_{H^+}$
	He$^+$	$1.55 \times 10^{-26} T^{0.3647} n_e n_{He^+}$
	He^{++}	$1.49 \times 10^{-25} T^{0.3} (1+0.885 T_6)^{-1} n_e n_{He^{++}}$
dielectronic–recombination	He$^+$	$1.24 \times 10^{-13} T^{-1.5} e^{-470000/T} (1+0.3 e^{-94000/T}) n_e n_{He^+}$
free–free	all ions	$1.42 \times 10^{-27} g_{\rm ff} T^{1/2} n_e (n_{H^+} + n_{He^+} + 4 n_{He^{++}})$

The cooling rate of a cosmic gas (which is rich in hydrogen) is usually represented by a cooling function defined as

$$\Lambda(T) \equiv \frac{\mathscr{C}}{n_H^2}, \qquad (B1.82)$$

where n_H is the number density of hydrogen atoms (both neutral and ionized). So defined, Λ is independent of gas density for an optically thin gas. Note that other normalizations, such as $\mathscr{C}/n_e^2$, are also in use in the literature to define Λ. In order to calculate the cooling function we need to know the abundances of different ions. This can be done by solving the ionization equations, which are similar to Eq. (B1.81) except that the population densities are summed over all energy levels of a given ion. For example, for gas containing only H and He and in ionization equilibrium, the balance between collisional ionization and recombination gives:

$$\Gamma_{e,H^0} n_{H^0} n_e = \alpha_{H^+} n_{H^+} n_e, \qquad (B1.83)$$

$$\Gamma_{e,He^0} n_{He^0} n_e = \left(\alpha_{He^+} + \alpha_{He^+,d}\right) n_{He^+} n_e, \qquad (B1.84)$$

$$\Gamma_{e,He^+} n_{He^+} n_e + \left(\alpha_{He^+} + \alpha_{He^+,d}\right) n_{He^+} n_e = \alpha_{He^{++}} n_{He^{++}} n_e + \Gamma_{e,He^0} n_e, \qquad (B1.85)$$

$$n_e = n_{H^+} + n_{He^+} + 2 n_{He^{++}}, \qquad (B1.86)$$

where $\Gamma_{e,a}$ is the ionization rate of ion a due to a collision with an electron, and α_a is the recombination rate. $\alpha_{He^+,d}$ is the dielectronic recombination rate of He$^+$. The equations of these rates are given in the previous section. Thus, for given T, n_H and n_{He}, the equilibrium population densities can all be solved. The equilibrium cooling function depends on both T and n_{He}/n_H. The $Z=0$ curve in Fig. 8.1 shows the cooling function versus temperature for $n_{He}/n_H = 1/12$.

The same calculation can also be performed for gas containing heavier elements. Sutherland & Dopita (1993) have calculated the cooling functions for various metallicities. Their results for $Z = Z_\odot$, $Z = Z_\odot/10$ and $Z = Z_\odot/100$ are also shown in Fig. 8.1.

The cooling function obtained above assumes ionization equilibrium. For gas that is not in ionization equilibrium, ion densities may differ from the equilibrium values, and the cooling function has to be calculated using these non-equilibrium densities that are obtained from solving the time-dependent ionization equation (B1.81). As one can see from this equation, ionization equilibrium can be achieved if the time scales for the radiative processes in consideration are much shorter than the Hubble expansion time scale and the hydrodynamic time scale.

(b) Molecular Cooling Molecules usually possess vibrational and rotational quantum states, in addition to electronic levels. Vibrational states are due to the vibrations of the constituent atoms relative to each other, whereas rotational states come from the quantization of the total angular momentum. Normally, the energy involved in a vibrational/rotational transition is much lower than that in an atomic transition, and so molecules can be vibrationally or rotationally excited through a collision with an atom or with another molecule even in gas with relatively low temperature. The gas can then cool as a result of the energy loss due to the de-excitation of the excited molecules.

The simplest and most important molecule is H_2, molecular hydrogen, which has a binding energy of 4.48 eV. The vibrations of the two hydrogen atoms relative to each other can be considered as harmonic oscillators, and have 15 energy levels ranging from about 0.25 eV all the way to the binding energy. For each vibrational level, there are different rotational states that correspond to the quantization of the total angular momentum, J, with the alternate rotational states having nuclear spins aligned (ortho) or antiparallel (para). Because H_2 lacks a dipole moment, the vibrational/rotational transitions are quadrupole in nature (i.e. the change in the angular momentum quantum number is $\Delta J = 2$).

An H_2 molecule can be rotationally or vibrationally excited by collisions with an H^0 atom or with another H_2 molecule. The de-excitation can be either radiative, which amounts to energy loss from the gas to photons, or collisional, in which case the thermal energy lost to the excitation is returned to the gas and there is no net cooling. If the number density of colliding particles (either H or H_2) is very low, the cooling process dominates. In this case, since collisional excitation is not frequent, hydrogen molecules spend most of their time in the ground state or in the $J=1$ rotational state (as its radiative decay to the ground state is forbidden), and collisional excitation is almost always followed by a radiative decay before another collision can occur. The volume cooling rate is therefore proportional to the square of the number density of H_2, $n(H_2)$. In the high-density limit, on the other hand, collisions can occur before an excited molecule decays radiatively. In this case, the distribution of molecules in various states is expected to follow the Boltzmann distribution for the local thermal equilibrium (LTE), and the volume cooling rate – due to the occasional radiative de-excitation – is now roughly proportional to $n(H_2)$.

The calculation of H_2 cooling is, in principle, straightforward. The number densities of H_2 at different excitation levels are determined by the rates of excitation and de-excitation due to collisions with H^0 and H_2, and the cooling rate is given by Eq. (B1.80) by summing over all vibrational/rotational levels. The cooling function Λ, normalized by the total number density $n = n(H) + n(H_2)$, is then given by

$$n^2 \Lambda = n(H_2)n(H) \sum_{b<a} (E_a - E_b) [x_b \gamma_{ba}(H) - x_a \gamma_{ab}(H)]$$
$$+ n(H_2)n(H_2) \sum_{b<a} (E_a - E_b) [x_b \gamma_{ba}(H_2) - x_a \gamma_{ab}(H_2)], \quad (B1.87)$$

where $\gamma_{ab}(H)$ and $\gamma_{ba}(H)$ are the collisional excitation and de-excitation coefficients for incident H, while $\gamma_{ab}(H_2)$ and $\gamma_{ba}(H_2)$ are those for incident H_2, and the sum $\sum_{b<a}$ is over all vibrational/rotational levels. At $T < 10^4$ K, only rotational transitions within the lowest vibrational levels contribute significantly, because the temperature is not high enough to excite H_2 to high vibrational levels. The H_2 cooling function has been calculated by many authors (e.g. Lepp & Shull, 1983; Hollenbach & McKee, 1979, 1989; Forrey et al., 1997; Galli & Palla, 1998).

In order to calculate the cooling rate per unit volume (or per unit gas mass), we need to know the number density of H_2. In the absence of dust grains, the formation of H_2 proceeds mainly by the following reactions:

$$H^0 + e \rightarrow H^- + \gamma, \quad H^- + H^0 \rightarrow H_2 + e; \tag{B1.88}$$

$$H^+ + H^0 \rightarrow H_2^+ + \gamma, \quad H_2^+ + H^0 \rightarrow H_2 + H^+. \tag{B1.89}$$

These formation processes are counterbalanced by the following reactions that cause destructions of H^-, H_2^+ and H_2:

$$H^- + \gamma \rightarrow H^0 + e, \quad H^- + H^+ \rightarrow 2H^0, \quad H_2^+ + \gamma \rightarrow H^0 + H^+, \tag{B1.90}$$

$$H_2 + H^+ \rightarrow H_2^+ + H^0, \quad H_2 + e \rightarrow 2H^0 + e, \quad H_2 + \gamma \rightarrow H_2^+ + e. \tag{B1.91}$$

The rates of these reactions can be found in Abel et al. (1997) and Galli & Palla (1998), for example. The number density of H_2 in a gas can then be obtained by solving the ionization equations (B1.16). The chemistry of other common molecules and their implications for gas cooling can be found in Hollenbach & McKee (1979, 1989), for example.

Appendix C
Numerical Simulations

In the main text we have seen that galaxy formation involves many physical processes. In broad terms these processes can be divided into three main categories: gravitational, gas-dynamical and radiative. Although the physical principles governing most of these processes are well established, the dynamical systems are often so complicated that it is generally difficult to obtain analytical solutions. Thanks to the revolutionary development of powerful computers, it has become possible to tackle some of these problems using numerical simulations. In a numerical simulation, the mass distribution is usually represented by particles or sampled on a grid, and the motion of each mass element is traced numerically by taking into account its interactions with other mass elements. The solutions would be exact if we were able to simulate the motions of all individual atoms or elementary particles. Unfortunately this can never be achieved since the systems of interest (galaxies) contain of the order of 10^{68} protons. In practice, therefore, the pseudo-particles or mass elements used to represent the mass distribution each have a mass that is typically orders of magnitude larger than that of an actual atom. Such a representation is clearly an approximation, which may impose serious limitations on the reliability of the simulations.

In general, numerical simulations of galaxy formation can be divided in two broad categories: N-body simulations and hydrodynamical simulations. Given the importance of numerical simulations in modern astrophysics, this appendix briefly describes some of the basic numerical methods used.

C1.1 N-Body Simulations

Suppose that we have a system of N collisionless particles. For simplicity, we assume that each particle has the same mass $m = 1$. The state of this system at any time t is described by the positions and velocities of all particles, $(\mathbf{r}_i, \mathbf{u}_i)$ ($i = 1, 2, \ldots, N$). Since the particles interact with each other only gravitationally, the motion of the ith particle in the system can be written as

$$\frac{d\mathbf{r}_i}{dt} = \mathbf{u}_i, \tag{C1.1}$$

$$\frac{d\mathbf{u}_i}{dt} = \mathbf{F}_i = -\nabla\phi|_i, \tag{C1.2}$$

where $\mathbf{F}_i$ is the force on a unit mass (i.e. the acceleration) at the position of the particle, and ϕ is the gravitational potential which is determined by the mass density, $\rho(\mathbf{r}, t)$, through the Poisson equation,

$$\nabla^2 \phi = 4\pi G \rho(\mathbf{r}, t). \tag{C1.3}$$

In cosmological applications, where the background space is uniformly expanding with time, it is convenient to use comoving units for the coordinates, $\mathbf{r} \to \mathbf{x} = \mathbf{r}/a$ (where a is the scale factor). In terms of $\mathbf{x}$ and $\mathbf{v} \equiv d\mathbf{x}/dt$, the above equations become

$$\frac{d\mathbf{x}_i}{dt} = \mathbf{v}_i, \tag{C1.4}$$

$$\frac{d\mathbf{v}_i}{dt} + 2H(t)\mathbf{v}_i = -\frac{1}{a^2}\nabla_\mathbf{x}\Phi|_i, \tag{C1.5}$$

where $H(t) = \dot{a}/a$, and Φ is the gravitational potential due to the density perturbations:

$$\nabla_\mathbf{x}^2 \Phi = 4\pi G a^2 \left[\rho(\mathbf{x},t) - \bar{\rho}(t)\right]. \tag{C1.6}$$

As the descriptions with and without the expansion terms are very similar, we neglect the expansion in our following description.

Since gravity is a long-range force, the force on a particle depends on the positions of all other particles. The problem is then to solve a set of $2N$ coupled first-order ordinary differential equations. Numerically, this can be done straightforwardly by integrating the equations forward in finite time steps.

Suppose that at the nth time step, the time is t_n and the distribution of particles is given by $\mathbf{r}_i(t_n)$, $\mathbf{u}_i(t_n)$ where $i = 1, 2, \ldots, N$. Suppose that we can use the current position, velocity, acceleration and time to determine a time step Δt_n for the evolution so that the error introduced is within some preset limit (we will discuss how to set such a limit in §C1.1.2). Using the equations of motion (C1.1) and (C1.2), we can predict the velocities and positions at the next time

$$\mathbf{u}'_i(t_{n+1/2}) = \mathbf{u}_i(t_n) + \mathbf{F}_i(t_n)\Delta t_n/2, \tag{C1.7}$$

$$\mathbf{r}'_i(t_{n+1/2}) = \mathbf{r}_i(t_n) + \mathbf{u}(t_n)\Delta t_n/2, \tag{C1.8}$$

where $\mathbf{F}_i(t_n)$ is the force on a unit mass at the position of the ith particle at time t_n. With these new positions and velocities, we can calculate the new force on a unit mass at the position of the ith particle:

$$\mathbf{F}_i(t_{n+1/2}) = \mathbf{F}[\mathbf{u}'_i(t_{n+1/2}), \mathbf{r}'_i(t_{n+1/2}), t_n + \Delta t/2], \tag{C1.9}$$

where we have included the possible dependence of the force on velocity and time, although only the position is relevant if only gravitational interaction is involved. Finally, we can update the velocities and positions as

$$\mathbf{u}_i(t_{n+1}) = \mathbf{u}_i(t_n) + \mathbf{F}_i(t_{n+1/2})\Delta t_n, \tag{C1.10}$$

$$\mathbf{r}_i(t_{n+1}) = \mathbf{r}_i(t_n) + \frac{\Delta t_n}{2}\left[\mathbf{u}_i(t_n) + \mathbf{u}_i(t_{n+1})\right], \tag{C1.11}$$

$$t_{n+1} = t_n + \Delta t_n. \tag{C1.12}$$

This kind of time integrator is called the 'leap-frog scheme'.

Thus, if we have an efficient way to calculate the forces on all particles at each time step, the system can be integrated forward in time numerically.

C1.1.1 Force Calculations

(a) PP (Particle–Particle) Algorithm For a collisionless N-body system, the most straightforward way to calculate the gravitational force on a particle is by direct summation of particle–particle (PP) interactions:

$$\mathbf{F}_i = -\sum_{j\neq i} Gm \frac{\mathbf{r}_i - \mathbf{r}_j}{|\mathbf{r}_i - \mathbf{r}_j|^3}. \tag{C1.13}$$

Since this calculation is a pairwise operation, the total time required to calculate all the forces is proportional to N^2, which makes it difficult to integrate systems with a particle number much larger than about 10^4, even with present-day supercomputers.

(b) Tree Algorithm In order to run numerical simulations with large particle numbers ($N \gg 10^4$), one has to resort to a different method for calculating the forces. One such method is the tree method, in which particles are grouped together systematically according to their distances to the particle for which the force is to be calculated. The force from each group is then replaced by its multipole expansion. Since the higher order terms in the multipole expansion die off faster with distance, we need to keep only a smaller number of multipole components for more distant groups while maintaining the force accuracy. Equivalently, if the number of multipole components is kept fixed, more distant particles can be combined into larger groups without compromising the accuracy.

For example, in the tree method proposed by Barnes & Hut (1986), the computational domain is hierarchically partitioned into a sequence of cubes, where each cube contains eight siblings, each with half the side-length of the parent cube. These cubes form the nodes of an oct-tree structure. The tree is constructed such that each node (cube) contains either exactly one particle or is the progenitor for further nodes. In the latter case the node carries the monopole and quadrupole moments of all the particles that lie inside its cube. A force computation then proceeds by walking the tree, and summing up appropriate force contributions from tree nodes. In the standard Barnes and Hut tree walk, the multipole expansion of a node of size l is used only if $r > l/\vartheta$, where r is the distance of the point of reference to the center-of-mass of the cell and ϑ is a prescribed accuracy parameter. If a node fulfills this criterion, the tree walk along this branch can be terminated, otherwise it is 'opened', and the walk is continued with all its siblings. It can be shown that in this kind of tree algorithm, the computational time for a complete force evaluation is of the order of $N \log N$.

(c) PM (Particle–Mesh) Algorithm Another method to compute the force is to solve the Poisson equation on a grid of meshes (e.g. Hockney & Eastwood, 1988). In this scheme, the computational box (assumed to have size L) is divided up into a grid of M^3 meshes with constant size $l = L/M$. We denote the grid points by $\mathbf{X}_q = \mathbf{q}l$, where $\mathbf{q} = (q_1, q_1, q_2)$ $(q_1, q_2, q_3 = 1, 2, \ldots, M)$. Suppose that at some time step the particle positions are $\mathbf{r}_i$ ($i = 1, 2, \ldots, N$). The PM algorithm then computes the forces as follows:

(i) A mass is assigned to each grid point $\mathbf{X}_q$ according to the particle distribution as

$$\rho(\mathbf{q}) = \frac{m}{L^3} \sum_{i=1}^{N} W(\mathbf{r}_i - \mathbf{X}_q), \tag{C1.14}$$

where W is a kernel function, assumed to be normalized: $\int W(\mathbf{r}) d^3\mathbf{r} = 1$. Note that $\rho(\mathbf{q})$ is a convolution of the particle distribution with the window function, and so its Fourier transform is equal to the product of $W_\mathbf{k}$ (the Fourier transform of the window function) and $\rho_\mathbf{k}$ (the Fourier transform of the mass distribution represented by the particles).

(ii) The Poisson equation is solved on the grid. In Fourier space, the Poisson equation becomes $-k^2\phi_{\mathbf{k}} = 4\pi G\rho_{\mathbf{k}}$, where $\rho_{\mathbf{k}}$ can be obtained by the Fourier transformation of $\rho(\mathbf{q})$ deconvolved with the window function. If the computational box has a size much larger than the scales of interest, the mass distribution can be assumed to be periodic on the opposite surfaces of the box, and the Fourier transform of $\rho(\mathbf{q})$ can be obtained speedily using the fast Fourier transform (FFT) (e.g. Press et al., 1992). The gravitational force on a unit mass at each grid point can then be obtained by the Fourier transform of $\mathbf{F}_{\mathbf{k}} = -i\phi_{\mathbf{k}}\mathbf{k}$.

(iii) The forces on the grid points are interpolated at the positions of the particles:

$$\mathbf{F}(\mathbf{r}_i) = \sum_{\mathbf{q}} W(\mathbf{r}_i - \mathbf{X}_q)\mathbf{F}(\mathbf{q}), \tag{C1.15}$$

where the summation is over a designated set of grid points close to $\mathbf{r}_i$ (because W is localized, as discussed below).

The choice of the kernel is non-trivial and can affect the numerical accuracy. The kernel should be chosen such that the total momentum is conserved. It is easy to show that this is fulfilled if the same kernel is used for both the density and the force, as is done in Eqs. (C1.14) and (C1.15). The kernel should be sufficiently peaked so that the smoothing is not too heavy, but it should not change too rapidly over a mesh because otherwise a slight error in the position of a particle in a mesh would lead to a large error in the interpolation. One commonly used kernel has the form

$$W(\mathbf{r}) = \frac{1}{\pi l^3} \begin{cases} 3/4 - R^2 & (0 \leq R \leq 1/2) \\ (3/2 - R)^2/2 & (1/2 < R \leq 3/2) \\ 0 & \text{(otherwise)}, \end{cases} \tag{C1.16}$$

where $R \equiv |\mathbf{r}|/l$. More examples can be found in Hockney & Eastwood (1988), for example.

It is evident that in the PM algorithm the total computational cost for force evaluations scales linearly with N, and so a large number of particles can be used. However, the force resolution is limited by the size of the mesh; a large number of meshes has to be used to achieve a high accuracy. Consequently, the computational constraint comes mainly from memory instead of CPU time.

(d) P³M (Particle–Particle–Particle–Mesh) Algorithm   A widely used compromise between the PP and PM algorithms is the P³M scheme. In this scheme, long-range force is computed by PM while the force between particles with distance less then about $2l$ is computed directly from PP. If the particles are not strongly clustered so that only a small number of particles have distances smaller than $2l$, the scheme reduces to PM. If, on the other hand, particles are strongly clustered so that many particles have distances smaller than $2l$, the scheme is effectively PP. Thus, the simplest implementation of P³M with fixed mesh size becomes very time consuming when particles become highly clustered. This drawback can be avoided with the use of adaptive submeshes (see Couchman, 1991). In the adaptive P³M (AP³M) scheme, a finer grid is used whenever many particles are found within a parent grid. The PP algorithm is applied only to particles with distances smaller than a fixed number (e.g. two) times the finest grid, while the PM algorithm is used otherwise.

C1.1.2 Issues Related to Numerical Accuracy

From the above description we see that the mass resolution, the force resolution and the choice of time steps are critical factors controlling the overall accuracy and computational efficiency of a simulation. Here we discuss these issues in more detail.

(a) Mass and Force Resolutions An obvious limitation of using pseudo-particles with a finite mass to sample the density field is that objects with masses below the mass resolution are not represented in the simulation. Finite mass resolution also limits the ability to study the internal structures (density profiles, shapes, substructures, etc.) of objects with masses not much larger than the mass resolution. Typically one requires more than ~ 1000 particles in order to have a sufficiently reliable representation of the overall shape and density profile of a collapsed object; reliably resolving substructures requires particle numbers that are several orders of magnitude larger.

A more subtle effect of finite mass resolution concerns the computation of the gravitational forces. In a simulation, dimensionless particles of infinite mass density are used to represent three-dimensional mass elements of finite volume and finite density. The gravitational force between two such mass elements is only well approximated by that between the two particles representing them if their distance is large. Indeed, the gravitational force between two particles diverges as their distance approaches zero, while the gravitational force between two extended objects is always finite. Consequently, close encounters between simulation particles could lead to large-angle scattering events that are unrealistic. To prevent this from happening, a force softening has to be introduced. In a mesh-based algorithm, the mass of a sampling particle is partitioned to grid points with a given smoothing kernel, and so the force is automatically softened on the scale of the mesh. In the PP algorithm, artificial force softening is usually used by modifying Newton's gravity between two particles, Gm^2/r^2, by something like $Gm^2/(r^2+\varepsilon^2)$, where ε is the gravitational softening length. The value of ε is usually specified in terms of the mean interparticle separation $(\bar{l}_p)$ in the simulation. The highest density contrast that can be resolved is then roughly $(\bar{l}_p/\varepsilon)^3$. In practice $\varepsilon/\bar{l}_p$ is chosen to be in the range $1/50$–$1/20$. A choice of a much smaller ε is impractical, as it requires very small time steps to satisfy the time step criterion described below.

(b) Time Step Criterion When integrating a function $y = f(x)$ from $x_{\min}$ to $x_{\max}$ numerically, it is common practice to represent the curve by the values of $f(x)$ and its first derivatives at a set of discrete points $x_n = x_{\min} + n\Delta x$ $(n = 1, 2, \ldots, N)$, where $\Delta x = (x_{\max} - x_{\min})/N$. In order to obtain accurate results, Δx must be sufficiently small (i.e. N must be sufficiently large) so that all high-order terms in the Taylor expansion of $f(x)$ are much smaller than the first two terms. A similar criterion is used to choose the time steps in numerical simulations.

The change of the position of a particle in a time step in a simulation is given by

$$\Delta \mathbf{r} = \mathbf{r}(t_{n+1}) - \mathbf{r}(t_n) = \mathbf{u}(t_n)\Delta t + \frac{1}{2}\mathbf{F}(t_{n+1/2})\Delta t^2 + \cdots. \tag{C1.17}$$

Similarly, the change in the specific kinetic energy is

$$\Delta \varepsilon = \mathbf{u}(t_n) \cdot \mathbf{F}(t_{n+1/2})\Delta t + \frac{1}{2}\mathbf{F}(t_{n+1/2}) \cdot \mathbf{F}(t_{n+1/2})\Delta t^2 + \cdots. \tag{C1.18}$$

In order to have a reliable integration, the terms of second order in Δt should be smaller than the first order terms. This suggests a time step criterion of the form

$$\Delta t = \alpha_{\text{tol}} \frac{|\mathbf{u}|}{|\mathbf{F}|}, \tag{C1.19}$$

where α_{tol} is a dimensionless tolerance parameter. However, this criterion strongly breaks the reversibility of the integration. A better choice may be

$$\Delta t = \alpha_{\text{tol}} \frac{\sigma}{|\mathbf{F}|}, \tag{C1.20}$$

where σ is the typical velocity dispersion of particles at the position in question (see Springel et al., 2001b).

C1.1.3 Boundary Conditions

In cosmological applications one usually wishes to simulate either a 'representative' region of the Universe or a particular system, e.g. a galaxy or a cluster of galaxies, that is embedded in a dynamically active environment. In both cases appropriate modeling of the boundary conditions is important, and the limitations imposed by the fact that the simulation can only be carried out for a finite volume has to be taken into account.

When studying a typical region of the Universe, the usual choice is to apply periodic boundary conditions on opposite faces of a rectangular (most often cubic) box. This avoids any artificial boundaries and forces the mean density of the simulation to remain at the desired value. The Fourier spectrum of a periodic universe is discrete and only wavenumbers, $\mathbf{k} = (2\pi/L)(i_1, i_2, i_3)$, where i_1, i_2 and i_3 are integers, are allowed in a periodic cube. Often we are interested in effects for which the influence of long wavelength modes is important (for example, the number density of rich clusters and the large-separation behavior of the correlation function). The difference between the discrete and continuous Fourier representations can then be quite significant. It is therefore important to check the simulation results by calculating all statistics for an ensemble of equivalent models and by comparing them against those of lower resolution simulations of larger regions.

When simulating the formation of individual objects, such as galaxy halos or clusters, in a cosmological density field, it is important to represent properly the tidal effects of surrounding matter. This is usually done by sampling the large-scale density field outside the object being studied in a periodic cubic box with particles which are much more massive than the particles used to sample the object of interest and its neighborhood. Thus, the object being studied can be resolved with many more particles than it is in a simulation using equal-mass particles in a box large enough to properly represent the large-scale tidal effects. Unfortunately, this approach is quite inefficient; even a very crude representation of the tidal effects requires most of the particles to be outside the object. Tree algorithms discussed above allow a straightforward and efficient solution to this problem. The matter which always remains outside the object of interest can then be represented by relatively few 'nodes' of the tree whose internal structures need not be computed.

C1.1.4 Initial Conditions

For most galaxy formation and large-scale structure problems, the initial conditions can be split into two parts. The first is to set up a 'uniform' distribution of particles to represent the unperturbed Universe. The second is to impose density perturbations with the desired characteristics.

It is not trivial to set down a finite number of particles to represent a desired uniform mass distribution. For example if N particles are distributed randomly in a box of side L, then the fluctuation in density contrast for randomly placed spheres of radius R is given by the formula for a Poisson process,

$$\left\langle \left(\frac{\delta M}{M}\right)^2 \right\rangle^{1/2} = \left(\frac{3L^3}{4\pi R^3 N}\right)^{1/2} = \frac{1}{\sqrt{\bar{N}(R)}}, \tag{C1.21}$$

where $\bar{N}(R)$ is the mean number of particles in a sphere. The power spectrum of this 'unperturbed' universe is $P(k) \propto k^n$ with $n = 0$, i.e. a 'white noise' spectrum. If a simulation is run from such initial conditions, these fluctuations may grow rapidly into nonlinear objects even if no other fluctuations are imposed.

The most widely used solution to this problem is to represent the unperturbed Universe by a regular cubic grid of particles. This procedure works quite well. However, it introduces a strong characteristic length scale on small scales (the grid spacing) and it leads to strongly preferred directions on all scales. These effects are particularly noticeable in the simulations of hot dark matter cosmogonies where it is very important to suppress artificial small-scale noise since the theory predicts that real small-scale fluctuations should have negligible amplitude. The regularity of the grid may also affect the statistical properties of the nonlinear point distribution, particularly those that emphasize low-density regions, since the remnant of the initial grid pattern is almost always visible in such regions. While it is healthy to have a strong visual reminder of the resolution limitations imposed by the finite number of particles, alternatives to the regular mesh are valuable in allowing an evaluation of the significance of these limitations.

An extremely uniform initial particle load which has no preferred direction can be created by the following trick (see White, 1996). Particles are placed at random within the computational volume. The cosmological N-body integrator is then used to follow their motion, except that the sign of the acceleration is reversed in the equations of motion. Peculiar gravitational forces then become repulsive. If the simulation is evolved for many expansion factors, the particles settle down to a glass-like configuration in which the force on each particle is very close to zero. This state shows no discernible order or anisotropy on scales beyond a few interparticle separations. If it is used as initial condition for the standard integrator without further perturbation, no small-scale structure grows even for expansion factors as large as about 30.

Given a suitable 'unperturbed' particle distribution, any desired *linear* fluctuation distribution can be generated quite easily using the Zel'dovich approximation. First the linear density field is realized either in real space (as is simplest for simulating objects with well-defined symmetry) or in Fourier space (as is simplest for simulating a Gaussian random field). Fourier techniques can then be used to generate the perturbations in the gravitational potential, $\Phi(\mathbf{x})$, and so the displacement field $-b(t)\nabla\Phi$ which appears in the Zel'dovich approximation (see §4.1.8). This can be used to move particles from their unperturbed positions and to create a discrete realization of the desired density field. Particle velocities can be set by applying linear theory either to the displacements or to the accelerations implied by the Poisson solver used in the numerical integrator. The latter scheme works better in marginally nonlinear regions (see Efstathiou et al., 1985).

Another trick that has often been used in studies of the formation of individual objects (e.g. galaxy halos or rich galaxy clusters) is to set up initial realizations of Gaussian random fields that satisfy certain constraints. For example, in the case of a rich cluster one might require that the center of the simulation be a high (e.g. 3σ) peak of the initial density field when smoothed with a top hat window corresponding to a mass of $10^{15}\,M_\odot$. A very efficient technique for constructing such constrained realizations has been developed by Hoffman & Ribak (1991).

C1.2 Hydrodynamical Simulations

C1.2.1 Smoothed-Particle Hydrodynamics (SPH)

Smoothed-particle hydrodynamics (SPH), which was independently developed by Lucy (1977) and Gingold & Monaghan (1977), is one of the most popular techniques for solving the hydrodynamical equations for studies of galaxy formation. This technique is particle based, and it follows the motion of individual mass elements (see Monaghan, 1992, for a review). Because of this, it is convenient to write the hydrodynamical equations in the Lagrangian formulation (where coordinates are comoving with the fluid element):

$$\frac{d\rho}{dt} + \rho \nabla \cdot \mathbf{u} = 0, \tag{C1.22}$$

$$\frac{d\mathbf{u}}{dt} = -\frac{\nabla P}{\rho} - \nabla \phi, \tag{C1.23}$$

C1.2 Hydrodynamical Simulations

$$\frac{d\mathscr{E}}{dt} = -\frac{P}{\rho}\nabla \cdot \mathbf{u} - \frac{\mathscr{L}(\mathscr{E},\rho)}{\rho}, \tag{C1.24}$$

where $\mathscr{L} = \mathscr{C} - \mathscr{H}$ with $\mathscr{H}$ and $\mathscr{C}$ the heating and cooling rates per unit volume, respectively. Note that the forms of these equations are different from those in the Eulerian formulation where coordinates are fixed in space. The system is closed by the Poisson equation and an equation of state which, for an ideal gas, is

$$P = (\gamma - 1)\rho \mathscr{E}, \tag{C1.25}$$

where γ is the adiabatic index.

The fundamental idea of SPH is to represent a fluid by a Monte Carlo sampling of its mass elements using a set of N particles. Each particle is assigned with the mass of the fluid element it represents, and in the Lagrangian formulation this mass is conserved. These particles can be used to sample any field of the fluid, A, so that the field value at an arbitrary position $\mathbf{r}$ can be approximated by

$$A(\mathbf{r}) = \sum_{j=1}^{N} m_j \frac{A_j}{\rho_j} W(\mathbf{r} - \mathbf{r}_j; h), \tag{C1.26}$$

where $W(r,h)$ is a smoothing kernel of radius h. In practice $W(r,h)$ is chosen so that it equals 0 when r/h is larger than a constant η (typically of the order unity). Therefore, the summation is over all neighbors that have $|\mathbf{r} - \mathbf{r}_j| \leq \eta h$. A geometric search tree is usually employed for the neighbor search (see e.g. Springel et al., 2001b). The spatial derivative of the smoothed field is

$$\nabla A(\mathbf{r}) = \sum_{j=1}^{N} m_j \frac{A_j}{\rho_j} \nabla W(\mathbf{r} - \mathbf{r}_j; h). \tag{C1.27}$$

In this discrete representation, the equations for the changes of density, velocity and internal energy of a particle (mass element) can be written in the following forms:

$$\frac{d\rho_i}{dt} = \sum_{j=1}^{N} m_j (\mathbf{u}_i - \mathbf{u}_j) \nabla_i W(\mathbf{r}_i - \mathbf{r}_j; h), \tag{C1.28}$$

$$\frac{d\mathbf{u}_i}{dt} = -\sum_{j=1}^{N} m_j \left(\frac{P_i}{\rho_i^2} + \frac{P_j}{\rho_j^2}\right) \nabla_i W(\mathbf{r}_i - \mathbf{r}_j; h) - \nabla \phi, \tag{C1.29}$$

$$\frac{d\mathscr{E}_i}{dt} = \frac{1}{2} \sum_{j=1}^{N} m_j \left(\frac{P_i}{\rho_i^2} + \frac{P_j}{\rho_j^2}\right) (\mathbf{u}_i - \mathbf{u}_j) \cdot \nabla_i W(\mathbf{r}_i - \mathbf{r}_j; h) - \frac{\mathscr{L}}{\rho}. \tag{C1.30}$$

In these forms, the position, velocity, density and internal energy of SPH particles can be integrated forward in time as in the N-body simulations (for example, using the leap-frog scheme).

In SPH simulations the smoothing kernel W is usually taken to have the form

$$W(r,h) = \frac{1}{\pi h^3} \begin{cases} 1 - 3R/2 + 3R^3/4 & (0 \leq R \leq 1) \\ (2-R)^3/4 & (1 < R \leq 2) \\ 0 & (\text{otherwise}), \end{cases} \tag{C1.31}$$

where $R \equiv r/h$, although other forms may also be used (see Monaghan, 1992). The smoothing radius h is chosen adaptively according to the local number density of SPH particles. For example, one may choose the radius of the smoothing kernel such that it contains a fixed number of particles. Another proposal is to calculate the value of h at the position of a given particle from

$$\frac{dh_i}{dt} = -\frac{h_i}{3\rho_i} \frac{d\rho_i}{dt} \tag{C1.32}$$

at each time step (Benz, 1990). The field value at the position of a particle i can then be approximated by

$$A_i = \sum_j (m_j/\rho_j) A_j W(r_{ij}, h), \qquad (C1.33)$$

where one may either choose $h = h_i$ (the gather formulation) or $h = h_j$ (the scatter formulation). In practice, two other choices are widely used: one symmetrizes the kernel, $W = [W(r_{ij}, h_i) + W(r_{ij}, h_j)]/2$, and the other symmetrizes the smoothing radius, $h = (h_i + h_j)/2$.

As in N-body simulations, the time steps should be chosen small enough so that the $(\Delta t)^2$ term in the Taylor expansion of the position of a particle as a function of time is much smaller than the Δt term. This criterion suggests a time step of the form given by Eq. (C1.20). Furthermore, because the hydrodynamical equations also involve spatial derivatives, the integration scheme also has to fulfill a condition of the form $|v|\Delta t/\Delta r \leq 1$, which is called the Courant condition. In SPH simulations, the Courant condition implies timesteps given by

$$\Delta t_i = \frac{\alpha_C h_i}{h_i |(\nabla \cdot \mathbf{u})_i| + \max(c_{s,i}, |\mathbf{u}_i|)}, \qquad (C1.34)$$

where $c_{s,i}$ is the sound speed at the position of the ith particle and α_C is a dimensionless tolerance parameter.

As a Lagrangian formulation of hydrodynamics, SPH can achieve high resolution in dense regions, but it works poorly in low-density regions due to the small number of SPH particles available to sample the density field. Furthermore, SPH particles may stream through each other unphysically, causing unphysical oscillations around regions of converging gas flows, such as shock fronts. In order to prevent this, artificial viscosity is usually introduced. However, this may significantly broaden the structure of the interface. Another problem is that existing SPH techniques are unable to resolve several important hydrodynamical instabilities such as the Kelvin–Helmholtz and Rayleigh–Taylor instabilities (see Agertz et al., 2007, and references therein). These problems can be better treated using an Eulerian formulation of hydrodynamics as in the grid-based techniques discussed below.

C1.2.2 Grid-Based Algorithms

In Eulerian computational fluid dynamics, the standard approach is to divide space into a grid of cells and store the cell-averaged values of conserved hydrodynamical quantities at all the grid points. The hydrodynamical equations are then solved by computing the flux of mass, momentum and energy across grid cell boundaries. Since differentiation is unstable in dealing with discontinuities, such as shock fronts, one usually uses the integral form of the hydrodynamic equations applied to a cell. Neglecting the source term, the Eulerian equation for a conserved quantity, q [density ρ, momentum density ρu_i ($i = 1, 2, 3$), and the total energy density, $\rho(\mathscr{E} + u^2/2)$], can formally be written as

$$\frac{\partial q}{\partial t} + \nabla \cdot \mathbf{F} = 0, \qquad (C1.35)$$

where $\mathbf{F}$ represents the corresponding flux density. The integral form of this equation applied to a cubic cell labeled by $\mathbf{l} = (\ell_1, \ell_2, \ell_3)$ is

$$\frac{\partial q(\mathbf{l}, t)}{\partial t} + \sum_{k=1}^{3} \frac{[F_k(\ell_k + 1/2, t) - F_k(\ell_k - 1/2, t)]}{\Delta x} = 0, \qquad (C1.36)$$

where Δx is the cell size, '$\ell_k \pm 1/2$' indicates that the flux is evaluated at the corresponding cell interfaces, and

$$q(\mathbf{l},t) = \frac{1}{(\Delta x)^3}\int_{\text{cell l}} q(\mathbf{x},t)\,d^3\mathbf{x} \tag{C1.37}$$

is the cell average of q. Once the fluxes at the cell interfaces are estimated from the stored hydrodynamical quantities at a time t, Eq. (C1.36) can be used to predict q at the next time step:

$$q(\mathbf{l},t+\Delta t) = q(\mathbf{l},t) - \frac{\Delta t}{\Delta x}\sum_{k=1}^{3}\left[F_k(\ell_k+1/2,t) - F_k(\ell_k-1/2,t)\right]. \tag{C1.38}$$

The procedure can then be repeated to integrate q forward in time.

Different schemes have been proposed to obtain the fluxes at the cell interfaces. In what follows we use simple examples to demonstrate the basic ideas behind such schemes. For brevity, we consider a one-dimensional problem. We denote the cell average of a fluid quantity, Q, in cell ℓ at time step n by Q_ℓ^n:

$$Q_\ell^n = \frac{1}{\Delta x}\int_{x_{\ell-1/2}}^{x_{\ell+1/2}} Q(x)\,dx. \tag{C1.39}$$

The simplest way to estimate the values at the cell interfaces, $Q_{\ell\pm 1/2}^n$, is to use $Q_{\ell\pm 1/2}^n = (Q_{\ell\pm 1}^n + Q_\ell^n)/2$. Unfortunately this scheme is numerically unstable (see e.g. Trac & Pen, 2003). In practice, one first constructs a piecewise polynomial interpolation function, $\mathscr{Q}(x)$, to approximate $Q(x)$. Once the functional form of $\mathscr{Q}$ is chosen, the coefficients of the interpolation for a given cell ℓ are determined by Q_ℓ^n and the cell-averaged values in neighboring cells. The interpolation is usually required to reproduce the mean value in the cell, i.e. $\mathscr{Q}$ must satisfy $Q_\ell^n = \frac{1}{\Delta x}\int_{x_{\ell-1/2}}^{x_{\ell+1/2}}\mathscr{Q}(x)\,dx$. For example, in the widely used piecewise parabolic method (PPM) proposed by Colella & Woodward (1984), the interpolation uses a parabolic function:

$$\mathscr{Q}(\alpha) = Q_{\rm L} + \alpha\left[\Delta Q_\ell + (1-\alpha)Q_{6,\ell}\right], \tag{C1.40}$$

where

$$\alpha = \frac{x - x_{\ell-1/2}}{x_{\ell+1/2} - x_{\ell-1/2}} \quad (0 \le \alpha \le 1), \tag{C1.41}$$

and $Q_{\rm L}$, ΔQ_ℓ and $Q_{6,\ell}$ are interpolation coefficients. Denote the values of $\mathscr{Q}(x)$ at the two interfaces of cell ℓ by $Q_{{\rm L},\ell} = \mathscr{Q}(x\to x_{\ell-1/2})$ and $Q_{{\rm R},\ell} = \mathscr{Q}(x\to x_{\ell+1/2})$, where $x_{\ell-1/2} < x < x_{\ell+1/2}$. It is then easy to show that

$$Q_{\rm L} = Q_{{\rm L},\ell}, \quad \Delta Q_\ell = Q_{{\rm R},\ell} - Q_{{\rm L},\ell}, \quad Q_{6,\ell} = 6\left[Q_\ell^n - (Q_{{\rm L},\ell} + Q_{{\rm R},\ell})/2\right]. \tag{C1.42}$$

Thus, in the PPM the problem of interpolation is reduced to specifying $Q_{{\rm L},\ell}$ and $Q_{{\rm R},\ell}$. One can use interpolations of Q_ℓ^n together with the cell-averaged values in neighboring cells to obtain $Q_{\ell+1/2}$, the interpolated values at the two interfaces, and assume $Q_{{\rm L},\ell} = Q_{\ell-1/2}$ and $Q_{{\rm R},\ell} = Q_{\ell+1/2}$. For example, in the method proposed by Colella & Woodward (1984), one defines a sum, $S_{j+1/2} = \sum_{i\le j}Q_i^n\Delta x$, and considers it as a discrete function of $x_{j+1/2}$. A quartic polynomial interpolation of the points, $(S_{\ell+k+1/2}, x_{\ell+k+1/2})$ $(k=0,\pm 1,\pm 2)$, is then used to obtain $S(x)$, and to calculate $Q_{\ell+1/2} = dS/dx|_{x_{\ell+1/2}}$. The formula so obtained is

$$Q_{\ell+1/2} = \frac{1}{2}\left(Q_\ell^n + Q_{\ell+1}^n\right) + \frac{1}{6}(\delta Q_\ell - \delta Q_{\ell+1}), \tag{C1.43}$$

where $\delta Q_\ell = (Q_{\ell+1}^n - Q_{\ell-1}^n)/2$ is the average slope in the $\ell^{\rm th}$ cell.

Unfortunately, high-order interpolation schemes may not work properly for sharp changes in the solution domain, because it may produce spurious oscillations, making the solution unstable.

In order to avoid this, one typically adopts a flux limiter (also referred to as slope limiter) to limit the spatial derivatives to realistic values. For instance, in the PPM of Colella & Woodward (1984), this is achieved by replacing the average slope, δQ_ℓ, in the above equation with

$$\delta_m Q_\ell = \begin{cases} \min(|\delta Q_\ell|, 2|Q_\ell^n - Q_{\ell-1}^n|, 2|Q_\ell^n - Q_{\ell+1}^n|) \times \text{sgn}(\delta Q_\ell) \\ \quad \text{if } (Q_{\ell+1}^n - Q_\ell^n)(Q_\ell^n - Q_{\ell-1}^n) > 0; \\ 0 \quad \text{otherwise.} \end{cases} \qquad (C1.44)$$

In general, the aim of using a flux limiter is to make the solution satisfy the total variation diminishing (TVD) condition proposed by Harten (1983).

Once the interpolation coefficients are obtained, one can compute the average of the interpolation function in a characteristic domain for each cell edge:

$$f_{\ell+1/2,L}^Q = \frac{1}{y}\int_{x_{\ell+1/2}-y}^{x_{\ell+1/2}} \mathcal{Q}(x)\,dx; \quad f_{\ell+1/2,R}^Q = \frac{1}{y}\int_{x_{\ell+1/2}}^{x_{\ell+1/2}+y} \mathcal{Q}(x)\,dx, \qquad (C1.45)$$

where y is the size of the characteristic domain. For the PPM, it is easy to show that

$$f_{\ell+1/2,L}^Q = Q_{R,\ell} - \frac{y}{2\Delta x}\left[\Delta Q_\ell - \left(1 - \frac{2y}{3\Delta x}\right)Q_{6,\ell}\right];$$

$$f_{\ell+1/2,R}^Q = Q_{R,\ell+1} + \frac{y}{2\Delta x}\left[\Delta Q_{\ell+1} + \left(1 - \frac{2y}{3\Delta x}\right)Q_{6,\ell+1}\right]. \qquad (C1.46)$$

The (one-dimensional) difference equation corresponding to Eq. (C1.38) is then replaced by

$$q_\ell^{n+1} = q_\ell^n + \frac{\Delta t}{\Delta x}\left(\bar{F}_{\ell-1/2} - \bar{F}_{\ell+1/2}\right), \qquad (C1.47)$$

where

$$\bar{F}_{\ell-1/2} = f_{\ell-1/2,R}^F(y); \quad \bar{F}_{\ell+1/2} = f_{\ell+1/2,L}^F(y). \qquad (C1.48)$$

The value of y is usually chosen to be the characteristic scale swept out by sound waves within one time step, i.e. $y = c_{s,\ell}\Delta t$, with $c_{s,\ell}$ the sound speed in the cell in question.

Appendix D
Frequently Used Abbreviations

Symbol	Definition
AGN	active galactic nucleus/nuclei
AGB	asymptotic giant branch
CDM	cold dark matter
CMB	cosmic microwave background
DLA	damped Lyα absorber
EdS	Einstein–de Sitter
EPS	extended Press–Schechter
FIR	far-infrared
FRW	Friedmann–Robertson–Walker
FWHM	full width at half maximum
GR	general relativity
GMC	giant molecular cloud
HSB	high surface brightness
HST	Hubble Space Telescope
ICM	intracluster medium
IGM	intergalactic medium
IMF	initial mass function
ISM	interstellar medium
LBG	Lyman break galaxy
LMC	Large Magellanic Cloud
LSB	low surface brightness
MS	main sequence
NFW	Navarro–Frenk–White
RGB	red giant branch
SDSS	Sloan Digital Sky Survey
SED	spectral energy distribution
SFH	star-formation history
SFR	star-formation rate
SFE	star-formation efficiency
SMBH	supermassive black hole
SMC	Small Magellanic Cloud
SN	supernova
SSFR	specific star-formation rate
SSP	single-age stellar population
TF	Tully–Fisher
UV	ultraviolet
WHIM	warm-hot intergalactic medium
WIMP	weakly interacting massive particle
WMAP	Wilkinson Microwave Anisotropy Probe

Appendix E
Useful Numbers

Constants	
Gravitational constant	$G = 6.674 \times 10^{-8} \, \text{cm}^3 \, \text{g}^{-1} \, \text{s}^{-2}$
	$= 4.299 \times 10^{-9} \, \text{Mpc} \, \text{M}_\odot^{-1} (\text{km/s})^2$
Planck constant	$h_\text{P} = 6.626 \times 10^{-27} \, \text{cm}^2 \, \text{g} \, \text{s}^{-1}$
Speed of light	$c = 2.998 \times 10^{10} \, \text{cm} \, \text{s}^{-1}$
Boltzmann constant	$k_\text{B} = 1.381 \times 10^{-16} \, \text{erg} \, \text{K}^{-1}$
Proton mass	$m_\text{p} = 1.673 \times 10^{-24} \, \text{g} = 938.3 \, \text{MeV}/c^2$
Neutron mass	$m_\text{n} = 1.675 \times 10^{-24} \, \text{g} = 939.6 \, \text{MeV}/c^2$
Electron mass	$m_\text{e} = 9.109 \times 10^{-28} \, \text{g} = 0.511 \, \text{MeV}/c^2$
Electron charge	$e = -4.803 \times 10^{-10} \, \text{esu}$
	$= -1.602 \times 10^{-19} \, \text{C}$
Thomson cross-section	$\sigma_\text{T} = 6.652 \times 10^{-25} \, \text{cm}^2$
Stefan–Boltzmann constant	$\sigma_\text{SB} = 5.67 \times 10^{-5} \, \text{erg} \, \text{cm}^{-2} \, \text{K}^{-4} \, \text{s}^{-1}$
Radiation constant	$a_\text{r} = 4\sigma_\text{SB}/c = 7.566 \times 10^{-15} \, \text{erg} \, \text{cm}^{-3} \, \text{K}^{-4}$

Units:	
Solar mass	$1 \, \text{M}_\odot = 1.99 \times 10^{33} \, \text{g}$
Solar radius	$1 \, \text{R}_\odot = 6.960 \times 10^{10} \, \text{cm}$
Solar luminosity (bolometric)	$1 \, \text{L}_\odot = 3.827 \times 10^{33} \, \text{erg} \, \text{s}^{-1}$
Astronomical unit	$1 \, \text{AU} = 1.496 \times 10^{13} \, \text{cm}$
Parsec	$1 \, \text{pc} = 3.086 \times 10^{18} \, \text{cm}$
Electron volt	$1 \, \text{eV} = 1.602 \times 10^{-12} \, \text{erg}$
Angstrom	$1 \, \text{Å} = 1 \times 10^{-8} \, \text{cm}$

Cosmological parameters	
Hubble constant	$H_0 = 100 h \, \text{km} \, \text{s}^{-1} \, \text{Mpc}^{-1}$
Present Hubble time	$H_0^{-1} = 9.78 h^{-1} \, \text{Gyr}$
Present Hubble radius	$cH_0^{-1} = 2997.9 h^{-1} \, \text{Mpc}$
Present critical density	$\rho_\text{crit} = 1.879 \times 10^{-29} h^2 \, \text{g} \, \text{cm}^{-3}$
	$= 2.775 \times 10^{11} h^{-1} \, \text{M}_\odot / (h^{-1} \, \text{Mpc})^3$
Present photon density	$\Omega_{\gamma,0} = 2.488 \times 10^{-5} h^{-2}$

References

Aarseth S. J., Binney J., 1978, MNRAS, 185, 227
Abadi M. G., Moore B., Bower R. G., 1999, MNRAS, 308, 947
Abadi M. G., Navarro J. F., Fardal M., et al. 2009, arXiv:0902.2477
Abadi M. G., Navarro J. F., Steinmetz M., Eke V. R., 2003, ApJ, 597, 21
Abel T., Anninos P., Norman M. L., Zhang Y., 1998, ApJ, 508, 518
Abel T., Anninos P., Zhang Y., Norman M. L., 1997, New Astron., 2, 181
Abell G. O., 1958, ApJS, 3, 211
Abraham R. G., 1998, astro-ph/9809131
Abraham R. G., van den Bergh S., Glazebrook K., et al. 1996, ApJS, 107, 1
Adami C., Biviano A., Mazure A., 1998, A&A, 331, 439
Adams F. C., Fatuzzo M., 1996, ApJ, 464, 256
Adelberger K. L., Steidel C. C., 2000, ApJ, 544, 218
Adelberger K. L., Steidel C. C., Shapley A. E., Pettini M., 2003, ApJ, 584, 45
Agertz O., Moore B., Stadel J., et al. 2007, MNRAS, 380, 963
Aguerri J. A. L., Balcells M., Peletier R. F., 2001, A&A, 367, 428
Aguerri J. A. L., Debattista V. P., Corsini E. M., 2003, MNRAS, 338, 465
Aguilar L. A., White S. D. M., 1985, ApJ, 295, 374
Aguilar L. A., White S. D. M., 1986, ApJ, 307, 97
Albrecht A., Steinhardt P. J., 1982, Phys. Rev. Lett., 48, 1220
Alcock C., Allsman R. A., Alves D. R., et al. 2000, ApJ, 542, 281
Allard F., Hauschildt P. H., 1995, ApJ, 445, 433
Allen S. W., Dunn R. J. H., Fabian A. C., et al. 2006, MNRAS, 372, 21
Allen S. W., Schmidt R. W., Fabian A. C., 2002, MNRAS, 334, L11
Allgood B., Flores R. A., Primack J. R., et al. 2006, MNRAS, 367, 1781
Allison A. C., Dalgarno A., 1969, ApJ, 158, 423
Antonucci R., 1993, ARA&A, 31, 473
Antonucci R. R. J., Miller J. S., 1985, ApJ, 297, 621
Aragon-Salamanca A., Baugh C. M., Kauffmann G., 1998, MNRAS, 297, 427
Aragon-Salamanca A., Ellis R. S., Couch W. J., Carter D., 1993, MNRAS, 262, 764
Araki S., 1985, PhD thesis, MIT
Arimoto N., Yoshii Y., 1987, A&A, 173
Arnett D., 1996, Supernovae and Nucleosynthesis. An Investigation of the History of Matter, from the Big Bang to the Present. Princeton University Press, Princeton, NJ
Arnold R., de Zeeuw P. T., Hunter C., 1994, MNRAS, 271, 924
Ashman K. M., Zepf S. E., 1992, ApJ, 384, 50
Asplund M., Lambert D. L., Nissen P. E., et al. 2006, ApJ, 644, 229
Bahcall J. N., Bergeron J., Boksenberg A., et al. 1996, ApJ, 457, 19
Bahcall J. N., Jannuzi B. T., Schneider D. P., et al. 1991, ApJ, 377, L5
Bahcall J. N., Kirhakos S., Saxe D. H., Schneider D. P., 1997, ApJ, 479, 642
Bahcall N. A., Lubin L. M., 1994, ApJ, 426, 513
Bahcall N. A., McKay T. A., Annis J., et al. 2003, ApJS, 148, 243
Bahcall N. A., Soneira R. M., 1983, ApJ, 270, 20
Bailin J., Steinmetz M., 2005, ApJ, 627, 647
Bajtlik S., Duncan R. C., Ostriker J. P., 1988, ApJ, 327, 570
Balbus S. A., 1986, ApJ, 303, L79

Balbus S. A., Soker N., 1989, ApJ, 341, 611
Balcells M., Peletier R. F., 1994, AJ, 107, 135
Baldry I. K., Balogh M. L., Bower R. G., et al. 2006, MNRAS, 373, 469
Baldwin J. A., Phillips M. M., Terlevich R., 1981, PASP, 93, 5
Ballesteros-Paredes J., Hartmann L., Vázquez-Semadeni E., 1999, ApJ, 527, 285
Ballesteros-Paredes J., Klessen R. S., Mac Low M.-M., Vazquez-Semadeni E., 2007, in Reipurth B., Jewitt D., Keil K., eds., Protostars and Planets V: Molecular Cloud Turbulence and Star Formation, University of Arizona Press, Tucson, pp 63–80
Balogh M. L., Morris S. L., 2000, MNRAS, 318, 703
Balogh M. L., Morris S. L., Yee H. K. C., et al. 1997, ApJ, 488, L75
Balogh M. L., Morris S. L., Yee H. K. C., et al. 1999, ApJ, 527, 54
Balogh M. L., Navarro J. F., Morris S. L., 2000, ApJ, 540, 113
Barbanis B., Woltjer L., 1967, ApJ, 150, 461
Bardeen J. M., 1980, Phys. Rev. D., 22, 1882
Bardeen J. M., Bond J. R., Kaiser N., Szalay A. S., 1986, ApJ, 304, 15
Bardeen J. M., Steinhardt P. J., Turner M. S., 1983, Phys. Rev. D., 28, 679
Barden M., Rix H.-W., Somerville R. S., et al. 2005, ApJ, 635, 959
Barkana R., Loeb A., 2007, Rep. Prog. Phys., 70, 627
Barnes J., Efstathiou G., 1987, ApJ, 319, 575
Barnes J., Hut P., 1986, Nature, 324, 446
Barnes J., White S. D. M., 1984, MNRAS, 211, 753
Barnes J. E., 1988, ApJ, 331, 699
Barnes J. E., 1992, ApJ, 393, 484
Barnes J. E., 1998, in Kennicutt Jr. R. C., et al. eds., Saas-Fee Advanced Course 26: Galaxies: Interactions and Induced Star Formation, Springer-Verlag, Berlin, p 275
Barnes J. E., Hernquist L. E., 1991, ApJ, 370, L65
Barnes J. E., Hernquist L., 1996, ApJ, 471, 115
Bartelmann M., Steinmetz M., 1996, MNRAS, 283, 431
Baugh C. M., 2006, Rep. Prog. Phys., 69, 3101
Baugh C. M., Cole S., Frenk C. S., 1996, MNRAS, 283, 1361
Baum S. A., Zirbel E. L., O'Dea C. P., 1995, ApJ, 451, 88
Beasley M. A., Baugh C. M., Forbes D. A., et al. 2002, MNRAS, 333, 383
Bechtold J., 1994, ApJS, 91, 1
Bechtold J., Ellingson E., 1992, ApJ, 396, 20
Begelman M. C., Blandford R. D., Rees M. J., 1980, Nature, 287, 307
Begelman M. C., Blandford R. D., Rees M. J., 1984, RvMP, 56, 255
Begeman K. G., 1989, A&A, 223, 47
Begelman M. C., Blandford R. D., Rees, M. J., 1984, RvMP, 56, 255
Bekki K., Couch W. J., Drinkwater M. J., 2001, ApJ, 552, L105
Bell E. F., Baugh C. M., Cole S., et al. 2003a, MNRAS, 343, 367
Bell E. F., de Jong R. S., 2000, MNRAS, 312, 497
Bell E. F., de Jong R. S., 2001, ApJ, 550, 212
Bell E. F., McIntosh D. H., Katz N., Weinberg M. D., 2003b, ApJS, 149, 289
Bell E. F., Naab T., McIntosh D. H., et al. 2006a, ApJ, 640, 241
Bell E. F., Phleps S., Somerville R. S., et al. 2006b, ApJ, 652, 270
Bell E. F., Wolf C., Meisenheimer K., et al. 2004, ApJ, 608, 752
Bell E. F., Zucker D. B., Belokurov V., et al. 2008, ApJ, 680, 295
Bender R., 1988, A&A, 193, L7
Bender R., Burstein D., Faber S. M., 1992, ApJ, 399, 462
Bender R., Burstein D., Faber S. M., 1993, ApJ, 411, 153
Bender R., Nieto J.-L., 1990, A&A, 239, 97
Bender R., Paquet A., Nieto J.-L., 1991, A&A, 246, 349
Bender R., Surma P., Doebereiner S., et al. 1989, A&A, 217, 35
Bendo G. J., Barnes J. E., 2000, MNRAS, 316, 315
Bennert N., Falcke H., Schulz H., et al. 2002, ApJ, 574, L105
Bennett C. L., Halpern M., Hinshaw G., et al. 2003, ApJS, 148, 1
Benson A. J., Bower R. G., Frenk C. S., et al. 2003, ApJ, 599, 38
Benson A. J., Bower R. G., Frenk C. S., White S. D. M., 2000, MNRAS, 314, 557
Bentz M. C., Peterson B. M., Netzer H., et al. 2009, ApJ, 697, 160
Benz W., 1990, in Buchler J. R., ed., Numerical Modelling of Nonlinear Stellar Pulsations: Problems and Prospects, Kluwer Academic Publisher, Dordrecht, p 269

Bergeron J., Boissé P., 1991, A&A, 243, 344
Bergeron J., Cristiani S., Shaver P. A., 1992, A&A, 257, 417
Berlind A. A., Frieman J., Weinberg D. H., et al. 2006, ApJS, 167, 1
Berlind A. A., Weinberg D. H., Benson A. J., et al. 2003, ApJ, 593, 1
Bernardeau F., 1994, ApJ, 433, 1
Bernardi M., Sheth R. K., Annis J., et al. 2003a, AJ, 125, 1849
Bernardi M., Sheth R. K., Annis J., et al. 2003b, AJ, 125, 1866
Bernardi M., Sheth R. K., Annis J., et al. 2003c, AJ, 125, 1882
Bernstein J., Dodelson S., 1990, Phys. Rev. D., 41, 354
Berrier J. C., Bullock J. S., Barton E. J., et al. 2006, ApJ, 652, 56
Bertola F., Buson L. M., Zeilinger W. W., 1992, ApJ, 401, L79
Bertola F., Capaccioli M., 1975, ApJ, 200, 439
Bertschinger E., 1985, ApJS, 58, 39
Bertschinger E., 1989, ApJ, 340, 666
Bertschinger E., 1996, in Schaeffer R., et al. eds., Cosmology and Large Scale Structure, Elsevier, Amsterdam, p 273
Bertschinger E., Dekel A., Faber S. M., et al. 1990, ApJ, 364, 370
Bessell M. S., 1990, PASP, 102, 1181
Bessell M. S., Castelli F., Plez B., 1998, A&A, 333, 231
Best P. N., von der Linden A., Kauffmann G., et al. 2007, MNRAS, 379, 894
Bi H., 1993, ApJ, 405, 479
Bi H., Davidsen A. F., 1997, ApJ, 479, 523
Bi H. G., Börner G., Chu Y., 1992, A&A, 266, 1
Bigiel F., Leroy A., Walter F., et al. 2008, AJ, 136, 2846
Binggeli B., Sandage A., Tammann G. A., 1988, ARA&A, 26, 509
Binggeli B., Sandage A., Tarenghi M., 1984, AJ, 89, 64
Binggeli B., Tammann G. A., Sandage A., 1987, AJ, 94, 251
Binney J., 1976, MNRAS, 177, 19
Binney J., 1977, ApJ, 215, 483
Binney J., 2005, MNRAS, 363, 937
Binney J., Gerhard O. E., Stark A. A., et al. 1991, MNRAS, 252, 210
Binney J., Mamon G. A., 1982, MNRAS, 200, 361
Binney J., Tabor G., 1995, MNRAS, 276, 663
Binney J., Tremaine S., 1987, Galactic Dynamics. Princeton University Press, Princeton, NJ
Binney J., Tremaine S., 2008, Galactic Dynamics 2nd edn. Princeton University Press, Princeton, NJ
Binney J. J., Davies R. L., Illingworth G. D., 1990, ApJ, 361, 78
Birkinshaw M., 1999, Phys. Rep., 310, 97
Birnboim Y., Dekel A., 2003, MNRAS, 345, 349
Birnboim Y., Dekel A., Neistein E., 2007, MNRAS, 380, 339
Birzan L., Rafferty D. A., McNamara B. R., et al. 2004, ApJ, 607, 800
Black J. H., 1981, MNRAS, 197, 553
Blain A. W., Smail I., Ivison R. J., Kneib J.-P., 1999, MNRAS, 302, 632
Blakeslee J. P., Tonry J. L., 1992, AJ, 103, 1457
Blandford R., Eichler D., 1987, Phys. Rep., 154, 1
Blandford R. D., Begelman M. C., 1999, MNRAS, 303, L1
Blandford R. D., Kochanek C. S., 1987, ApJ, 321, 658
Blandford R. D., McKee C. F., 1982, ApJ, 255, 419
Blandford R. D., Payne D. G., 1982, MNRAS, 199, 883
Blandford R. D., Pyane D. G., 1982, MNRAS, 199, 883
Blandford R. D., Rees M. J., 1974, MNRAS, 169, 395
Blandford R. D., Rees M. J., 1978, in Wolfe A. M., ed., Proceedings of Pittsburgh Conference on BL Lac Objects, University of Pittsburgh, pp 328–341
Blanton M. R., Berlind A. A., 2007, ApJ, 664, 791
Blanton M. R., Eisenstein D., Hogg D. W., Zehavi I., 2006, ApJ, 645, 977
Blanton M. R., Hogg D. W., Bahcall N. A., et al. 2003, ApJ, 592, 819
Blanton M. R., Lupton R. H., Schlegel D. J., et al. 2005, ApJ, 631, 208
Blitz L., 1993, in Levy E. H., Lunine J. I., eds., Protostars and Planets: III Giant Molecular Clouds. University of Arizona Press, Tucson, pp 125–161
Blitz L., Rosolowsky E., 2006, ApJ, 650, 933
Blumenthal G. R., Faber S. M., Primack J. R., Rees M. J., 1984, Nature, 311, 517
Blumenthal G. R., Pagels H., Primack J. R., 1982, Nature, 299, 37

Boccaletti D., Pucacco G., 1996, Theory of Orbits, vol. 1: Integrable Systems and Non-perturbative Methods. Springer-Verlag, Berlin
Bodenheimer P., 1995, ARA&A, 33, 199
Böhringer H., Matsushita K., Churazov E., et al. 2002, A&A, 382, 804
Böhringer H., Voges W., Fabian A. C., et al. 1993, MNRAS, 264, L25
Boissier S., Prantzos N., 1999, MNRAS, 307, 857
Boissier S., Prantzos N., Boselli A., Gavazzi G., 2003, MNRAS, 346, 1215
Böker T., Laine S., van der Marel R. P., et al. 2002, AJ, 123, 1389
Bolton A. S., Burles S., Koopmans L. V. E., et al. 2008, ApJ, 682, 964
Bolton J. S., Haehnelt M. G., Viel M., Springel V., 2005, MNRAS, 357, 1178
Bombelli L., Couch W. E., Torrence R. J., 1991, Phys. Rev. D., 44, 2589
Bonamente M., Joy M. K., LaRoque S. J., et al. 2006, ApJ, 647, 25
Bond J. R., Cole S., Efstathiou G., Kaiser N., 1991, ApJ, 379, 440
Bond J. R., Efstathiou G., 1984, ApJ, 285, L45
Bond J. R., Efstathiou G., Silk J., 1980, Phys. Rev. Lett., 45, 1980
Bond J. R., Myers S. T., 1996, ApJS, 103, 1
Bond J. R., Szalay A. S., 1983, ApJ, 274, 443
Bond J. R., Szalay A. S., Turner M. S., 1982, Phys. Rev. Lett., 48, 1636
Bondi H., 1952, MNRAS, 112, 195
Bondi H., Gold T., 1948, MNRAS, 108, 252
Bonnor W. B., 1956, MNRAS, 116, 351
Borch A., Meisenheimer K., Bell E. F., et al. 2006, A&A, 453, 869
Borne K. D., Richstone D. O., 1991, ApJ, 369, 111
Börner G., 2003, The Early Universe: Facts and Fiction. Springer, Berlin
Bosma A., 1981, AJ, 86, 1825
Bothun G. D., Mould J. R., 1988, ApJ, 324, 123
Bottema R., 1993, A&A, 275, 16
Bottema R., 1997, A&A, 328, 517
Bouchet P., Lequeux J., Maurice E., et al. 1985, A&A, 149, 330
Bourke T. L., Myers P. C., Robinson G., Hyland A. R., 2001, ApJ, 554, 916
Bouwens R. J., Illingworth G. D., Franx M., Ford H., 2007, ApJ, 670, 928
Bowen D. V., Blades J. C., Pettini M., 1995, ApJ, 448, 662
Bowen D. V., Blades J. C., Pettini M., 1996, ApJ, 464, 141
Bower R. G., 1991, MNRAS, 248, 332
Bower R. G., Benson A. J., Malbon R., et al. 2006, MNRAS, 370, 645
Bower R. G., Lucey J. R., Ellis R. S., 1992, MNRAS, 254, 601
Boylan-Kolchin M., Ma C.-P., Quataert E., 2005, MNRAS, 362, 184
Boylan-Kolchin M., Ma C.-P., Quataert E., 2006, MNRAS, 369, 1081
Boylan-Kolchin M., Ma C.-P., Quataert E., 2008, MNRAS, 383, 93
Boylan-Kolchin M., Springel V., White S. D. M., et al. 2009, MNRAS, 398, 1150
Brainerd T. G., Specian M. A., 2003, ApJ, 593, L7
Branch D., Tammann G. A., 1992, ARA&A, 30, 359
Brandenberger R. H., 1995, astro-ph/9508159
Bressan A., Fagotto F., Bertelli G., Chiosi C., 1993, A&AS, 100, 647
Bridle A. H., Hough D. H., Lonsdale C. J., et al. 1994, AJ, 108, 766
Brighenti F., Mathews W. G., 1996, ApJ, 470, 747
Brinchmann J., Abraham R., Schade D., et al. 1998, ApJ, 499, 112
Brinchmann J., Charlot S., White S. D. M., et al. 2004, MNRAS, 351, 1151
Bromley B. C., Press W. H., Lin H., Kirshner R. P., 1998, ApJ, 505, 25
Bromm V., Larson R. B., 2004, ARA&A, 42, 79
Brook C. B., Kawata D., Gibson B. K., Flynn C., 2004a, MNRAS, 349, 52
Brook C. B., Kawata D., Gibson B. K., Freeman K. C., 2004b, ApJ, 612, 894
Brown M. J. I., Dey A., Jannuzi B. T., et al. 2007, ApJ, 654, 858
Brüggen M., Kaiser C. R., 2002, Nature, 418, 301
Bruzual G., Charlot S., 2003, MNRAS, 344, 1000
Bruzual A. G., 1983, ApJ, 273, 105
Bryan G. L., Cen R., Norman M. L., et al. 1994, ApJ, 428, 405
Bryan G. L., Norman M. L., 1998, ApJ, 495, 80
Budavári T., Szalay A. S., Charlot S., et al. 2005, ApJ, 619, L31

Bullock J. S., Dekel A., Kolatt T. S., et al. 2001a, ApJ, 555, 240
Bullock J. S., Kolatt T. S., Sigad Y., et al. 2001b, MNRAS, 321, 559
Bullock J. S., Wechsler R. H., Somerville R. S., 2002, MNRAS, 329, 246
Buote D. A., Canizares C. R., 1992, ApJ, 400, 385
Bureau M., Freeman K. C., 1999, AJ, 118, 126
Burkert A., Naab T., Johansson P. H., Jesseit R., 2008, ApJ, 685, 897
Burstein D., 1979, ApJ, 234, 829
Buson L. M., Sadler E. M., Zeilinger W. W., et al. 1993, A&A, 280, 409
Buta R., 1986, ApJS, 61, 609
Butcher H., Oemler Jr. A., 1978, ApJ, 219, 18
Byun Y. I., Freeman K. C., Kylafis N. D., 1994, ApJ, 432, 114
Cacciato M., van den Bosch F. C., More S., et al. 2009, MNRAS, 394, 929
Calzetti D., Kinney A. L., Storchi-Bergmann T., 1994, ApJ, 429, 582
Camenzind M., 1990, in Klare G., ed., Reviews in Modern Astronomy Vol. 3, Springer-Verlag, Berlin, pp 234–265
Capetti A., Axon D. J., Macchetto F., et al. 1996, ApJ, 469, 554
Cappellari M., Bacon R., Bureau M., et al. 2006, MNRAS, 366, 1126
Cappellari M., Emsellem E., Bacon R., et al. 2007, MNRAS, 379, 418
Cappellaro E., Turatto M., Tsvetkov D. Y., et al. 1997, A&A, 322, 431
Cardelli J. A., Clayton G. C., Mathis J. S., 1989, ApJ, 345, 245
Cardiel N., Gorgas J., Sánchez-Blázquez P., et al. 2003, A&A, 409, 511
Carlberg R. G., 1984, ApJ, 286, 403
Carlberg R. G., 1986, ApJ, 310, 593
Carlberg R. G., Sellwood J. A., 1985, ApJ, 292, 79
Carlberg R. G., Yee H. K. C., Ellingson E., et al. 1997, ApJ, 476, L7
Carollo C. M., Franx M., Illingworth G. D., Forbes D. A., 1997, ApJ, 481, 710
Carpintero D. D., Muzzio J. C., 1995, ApJ, 440, 5
Carr B. J., Bond J. R., Arnett W. D., 1984, ApJ, 277, 445
Carretta E., Gratton R. G., Clementini G., Fusi Pecci F., 2000, ApJ, 533, 215
Carroll S. M., 2004, Spacetime and Geometry. An Introduction to General Relativity. Addison Wesley, San Francisco
Carroll S. M., Press W. H., Turner E. L., 1992, ARA&A, 30, 499
Carswell R. F., Rees M. J., 1987, MNRAS, 224, 13P
Casertano S., 1983, MNRAS, 203, 735
Cattaneo A., Dekel A., Devriendt J., et al. 2006, MNRAS, 370, 1651
Cen R., 1992, ApJS, 78, 341
Cen R., 2003, ApJ, 591, 12
Cen R., Miralda-Escudé J., Ostriker J. P., Rauch M., 1994, ApJ, 437, L9
Cen R., Ostriker J. P., 1999, ApJ, 514, 1
Cen R., Ostriker J. P., Prochaska J. X., Wolfe A. M., 2003, ApJ, 598, 741
Chabrier G., 2003, PASP, 115, 763
Chabrier G., 2005, in Corbelli E., et al. eds., The Initial Mass Function 50 Years Later, vol. 327 of ASSL, Springer, Dordrecht, p 41
Chandrasekhar S., 1943, Rev. Mod. Phys., 15, 1
Chandrasekhar S., 1951, Roy. Soc. Lond. Proc. Ser. A, 210, 26
Chandrasekhar S., 1969, Ellipsoidal Figures of Equilibrium. Yale University Press, New Haven, CT
Charlot S., Fall S. M., 2000, ApJ, 539, 718
Charlton J., Churchill C., 2000, Quasistellar Objects: Intervening Absorption Lines. Encyclopedia of Astronomy and Astrophysics, Institute of Physics, Bristol.
Chen D. N., Jing Y. P., Yoshikaw K., 2003, ApJ, 597, 35
Chevalier R. A., 1974, ApJ, 188, 501
Chiosi C., Maeder A., 1986, ARA&A, 24, 329
Christodoulou D. M., Shlosman I., Tohline J. E., 1995, ApJ, 443, 551
Churazov E., Sunyaev R., Forman W., Böhringer H., 2002, MNRAS, 332, 729
Churchill C. W., Steidel C. C., Vogt S. S., 1996, ApJ, 471, 164
Ciardi B., Ferrara A., 2005, Space Sci. Rev., 116, 625
Ciardi B., Madau P., 2003, ApJ, 596, 1
Ciardullo R., Jacoby G. H., Dejonghe H. B., 1993, ApJ, 414, 454
Cimatti A., Daddi E., Mignoli M., et al. 2002a, A&A, 381, L68

Cimatti A., Pozzetti L., Mignoli M., et al. 2002b, A&A, 391, L1
Cinzano P., van der Marel R. P., 1994, MNRAS, 270, 325
Ciotti L., Lanzoni B., Renzini A., 1996, MNRAS, 282, 1
Ciotti L., Ostriker J. P., 2001, ApJ, 551, 131
Clayton D. D., 1983, Principles of Stellar Evolution and Nucleosynthesis. University of Chicago Press, Chicago
Cole S., 1991, ApJ, 367, 45
Cole S., Aragon-Salamanca A., Frenk C. S., et al. 1994, MNRAS, 271, 781
Cole S., Helly J., Frenk C. S., Parkinson H., 2008, MNRAS, 383, 546
Cole S., Lacey C. G., Baugh C. M., Frenk C. S., 2000, MNRAS, 319, 168
Cole S., Norberg P., Baugh C. M., et al. 2001, MNRAS, 326, 255
Colella P., Woodward P. R., 1984, J. Comp. Phys., 54, 174
Coleman G. D., Wu C.-C., Weedman D. W., 1980, ApJS, 43, 393
Coles P., Jones B., 1991, MNRAS, 248, 1
Coles P., Lucchin F., 2002, Cosmology: The Origin and Evolution of Cosmic Structure, 2nd edn. Wiley-VCH
Colless M., Dalton G., Maddox S., et al. 2001a, MNRAS, 328, 1039
Colless M., Saglia R. P., Burstein D., et al. 2001b, MNRAS, 321, 277
Collins C. A., Stott J. P., Hilton M., et al. 2009, Nature, 458, 603
Collister A. A., Lahav O., 2005, MNRAS, 361, 415
Colpi M., 1998, ApJ, 502, 167
Colpi M., Mayer L., Governato F., 1999, ApJ, 525, 720
Combes F., Debbasch F., Friedli D., Pfenniger D., 1990a, A&A, 233, 82
Combes F., Dupraz C., Gerin M., 1990b, in Wielen R., ed., Dynamics and Interactions of Galaxies, Springer-Verlag, Heidelberg, pp 205–209
Conroy C., Prada F., Newman J. A., et al. 2007, ApJ, 654, 153
Conroy C., Wechsler R. H., 2009, ApJ, 696, 620
Conselice C. J., Bershady M. A., Dickinson M., Papovich C., 2003, AJ, 126, 1183
Conselice C. J., Gallagher III J. S., Wyse R. F. G., 2001, ApJ, 559, 791
Contopoulos G., Grosbol P., 1986, A&A, 155, 11
Cooper M. C., Newman J. A., Coil A. L., et al. 2007, MNRAS, 376, 1445
Cooray A., Sheth R., 2002, Phys. Rep., 372, 1
Cora S. A., Muzzio J. C., Vergne M. M., 1997, MNRAS, 289, 253
Côté P., Ferrarese L., Jordán A., et al. 2007, ApJ, 671, 1456
Côté P., Piatek S., Ferrarese L., et al. 2006, ApJS, 165, 57
Couch W. J., Ellis R. S., Sharples R. M., Smail I., 1994, ApJ, 430, 121
Couchman H. M. P., 1991, ApJ, 368, L23
Courteau S., Dutton A. A., van den Bosch F. C., et al. 2007, ApJ, 671, 203
Cowie L. L., McKee C. F., 1977, ApJ, 211, 135
Cowie L. L., Songaila A., Kim T.-S., Hu E. M., 1995, AJ, 109, 1522
Cowsik R., McClelland J., 1972, Phys. Rev. Lett., 29, 669
Cox A. N., 2000, Allen's Astrophysical Quantities. Springer, AIP Press, New York
Cox D. P., 1972, ApJ, 178, 159
Cox T. J., Dutta S. N., Di Matteo T., et al. 2006, ApJ, 650, 791
Crawford C. S., Allen S. W., Ebeling H., et al. 1999, MNRAS, 306, 857
Cretton N., de Zeeuw P. T., van der Marel R. P., Rix H.-W., 1999, ApJS, 124, 383
Cretton N., van den Bosch F. C., 1999, ApJ, 514, 704
Cristiani S., D'Odorico S., D'Odorico V., et al. 1997, MNRAS, 285, 209
Cristiani S., D'Odorico S., Fontana A., et al. 1995, MNRAS, 273, 1016
Croft R. A. C., Weinberg D. H., Katz N., Hernquist L., 1998, ApJ, 495, 44
Croom S. M., Boyle B. J., Shanks T., et al. 2005, MNRAS, 356, 415
Croom S. M., Smith R. J., Boyle B. J., et al. 2004, MNRAS, 349, 1397
Croton D. J., Farrar G. R., Norberg P., et al. 2005, MNRAS, 356, 1155
Croton D. J., Gaztañaga E., Baugh C. M., et al. 2004, MNRAS, 352, 1232
Croton D. J., Springel V., White S. D. M., et al. 2006, MNRAS, 365, 11
Crotts A. P. S., 1987, MNRAS, 228, 41P
Crotts A. P. S., Fang Y., 1998, ApJ, 502, 16
Crutcher R. M., 1999, ApJ, 520, 706
Cyburt R. H., Fields B. D., Olive K. A., 2008, J. Cosmol. Astro-Part. Phys., 11, 12

Daddi E., Cimatti A., Renzini A., 2000, A&A, 362, L45
Daddi E., Cimatti A., Renzini A., et al. 2004, ApJ, 617, 746
Daddi E., Renzini A., Pirzkal N., et al. 2005, ApJ, 626, 680
Dalal N., White M., Bond J. R., Shirokov A., 2008, ApJ, 687, 12
Dalcanton J. J., 2007, ApJ, 658, 941
Dalcanton J. J., Spergel D. N., Summers F. J., 1997, ApJ, 482, 659
Dalgarno A., McCray R. A., 1972, ARA&A, 10, 375
Dalton G. B., Maddox S. J., Sutherland W. J., Efstathiou G., 1997, MNRAS, 289, 263
Danese L., de Zotti G., 1982, A&A, 107, 39
D'Antona F., Mazzitelli I., 1994, ApJS, 90, 467
Davé R., Cen R., Ostriker J. P., et al. 2001, ApJ, 552, 473
Davé R., Hernquist L., Katz N., Weinberg D. H., 1999, ApJ, 511, 521
David L. P., Nulsen P. E. J., McNamara B. R., et al. 2001, ApJ, 557, 546
Davidsen A. F., Kriss G. A., Zheng W., 1996, Nature, 380, 47
Davies R. L., Efstathiou G., Fall S. M., et al. 1983, ApJ, 266, 41
Davies R. L., Sadler E. M., Peletier R. F., 1993, MNRAS, 262, 650
Davis D. S., White III R. E., 1996, ApJ, 470, L35
Davis M., Efstathiou G., Frenk C. S., White S. D. M., 1985, ApJ, 292, 371
Davis M., Faber S. M., Newman J., et al. 2003, in Guhathakurta P., ed., Society of Photo-Optical Instrumentation Engineers (SPIE) Conference Series vol. 4834, pp 161–172
Davis M., Peebles P. J. E., 1983, ApJ, 267, 465
Davis R. L., Hodges H. M., Smoot G. F., et al. 1992, Phys. Rev. Lett., 69, 1856
de Bernardis P., Ade P. A. R., Bock J. J., et al. 2000, Nature, 404, 955
de Grijs R., Kregel M., Wesson K. H., 2001, MNRAS, 324, 1074
de Jong R. S., 1996a, A&A, 313, 45
de Jong R. S., 1996b, A&A, 313, 377
de Jong R. S., Davies R. L., 1997, MNRAS, 285, L1
De Lucia G., Blaizot J., 2007, MNRAS, 375, 2
De Lucia G., Kauffmann G., Springel V., et al. 2004, MNRAS, 348, 333
De Lucia G., Springel V., White S. D. M., et al. 2006, MNRAS, 366, 499
De Rijcke S., Dejonghe H., Zeilinger W. W., Hau G. K. T., 2001, ApJ, 559, L21
De Simone R., Wu X., Tremaine S., 2004, MNRAS, 350, 627
de Vaucouleurs G., 1974, in Shakeshaft J. R., ed., The Formation and Dynamics of Galaxies, vol. 58 of IAU Symposium, Reidel, Dordrecht, pp 1–52
de Zeeuw P. T., Bureau M., Emsellem E., et al. 2002, MNRAS, 329, 513
de Zeeuw T., 1985, MNRAS, 216, 273
Debattista V. P., Mayer L., Carollo C. M., et al. 2006, ApJ, 645, 209
Debattista V. P., Sellwood J. A., 2000, ApJ, 543, 704
Dehnen W., 1993, MNRAS, 265, 250
Dejonghe H., 1986, Phys. Rep., 133, 217
Dekel A., 1994, ARA&A, 32, 371
Dekel A., Birnboim Y., 2008, MNRAS, 383, 119
Dekel A., Devor J., Hetzroni G., 2003, MNRAS, 341, 326
Dekel A., Lahav O., 1999, ApJ, 520, 24
Dekel A., Silk J., 1986, ApJ, 303, 39
Dekel A., Stoehr F., Mamon G. A., et al. 2005, Nature, 437, 707
Dekel A., Woo J., 2003, MNRAS, 344, 1131
Desert F.-X., Boulanger F., Puget J. L., 1990, A&A, 237, 215
Desjacques V., 2008, MNRAS, 388, 638
Di Matteo T., Colberg J., Springel V., et al. 2008, ApJ, 676, 33
Diaferio A., Geller M. J., 1996, ApJ, 467, 19
Dicke R. H., Peebles P. J. E., 1979, in Hawking S. W., Israel W., eds., General Relativity: An Einstein Centenary Survey, Cambridge University Press, Cambridge, pp 504–517
Dicke R. H., Peebles P. J. E., Roll P. G., Wilkinson D. T., 1965, ApJ, 142, 414
Dickinson M., 1998, in Livio M., et al. eds, The Hubble Deep Field, p 219
Dickinson M., Papovich C., Ferguson H. C., Budavári T., 2003, ApJ, 587, 25
Diemand J., Kuhlen M., Madau P., 2007, ApJ, 667, 859
Djorgovski S., Davis M., 1987, ApJ, 313, 59
Dobrzycki A., Bechtold J., 1991, ApJ, 377, L69

Dodelson S., 2003, Modern Cosmology. Academic Press, Amsterdam
D'Odorico V., Cristiani S., D'Odorico S., et al. 1998, A&A, 339, 678
Donahue M., Aldering G., Stocke J. T., 1995, ApJ, 450, L45
Dopita M. A., Ryder S. D., 1994, ApJ, 430, 163
Doroshkevich A. G., 1970, Astrophysics, 6, 320
Doroshkevich A. G., 1973, Astrophys. Lett., 14, 11
Draine B. T., Lee H. M., 1984, ApJ, 285, 89
Drazin P. G., Reid W. H., 1981, NASA STI/Recon Technical Report A, 82, 17950
Dressler A., 1980a, ApJS, 42, 565
Dressler A., 1980b, ApJ, 236, 351
Dressler A., 1984, ARA&A, 22, 185
Dressler A., Gunn J. E., 1983, ApJ, 270, 7
Dressler A., Lynden-Bell D., Burstein D., et al. 1987, ApJ, 313, 42
Dressler A., Smail I., Poggianti B. M., et al. 1999, ApJS, 122, 51
Dressler A., Thompson I. B., Shectman S. A., 1985, ApJ, 288, 481
Drinkwater M. J., Gregg M. D., Hilker M., et al. 2003, Nature, 423, 519
Driver S. P., Windhorst R. A., Griffiths R. E., 1995, ApJ, 453, 48
Dubinski J., Mihos J. C., Hernquist L., 1996, ApJ, 462, 576
Dubinski J., Mihos J. C., Hernquist L., 1999, ApJ, 526, 607
Dunlop J. S., Peacock J. A., 1990, MNRAS, 247, 19
Durrell P. R., Harris W. E., Geisler D., Pudritz R. E., 1996, AJ, 112, 972
Dutton A. A., Courteau S., de Jong R., Carignan C., 2005, ApJ, 619, 218
Dutton A. A., van den Bosch F. C., 2009, MNRAS, 396, 141
Dutton A. A., van den Bosch F. C., Dekel A., Courteau S., 2007, ApJ, 654, 27
Eadie W. T., Drijard D., James F. E., 1971, Statistical Methods in Experimental Physics. Amsterdam: North-Holland
Ebert R., 1957, Z. Astrophys., 42, 263
Ebisuzaki T., Makino J., Okumura S. K., 1991, Nature, 354, 212
Edge A. C., Wilman R. J., Johnstone R. M., et al. 2002, MNRAS, 337, 49
Edge D. O., Shakeshaft J. R., McAdam W. B., et al. 1959, Mem. Roy. Astron. Soc., 68, 37
Edmunds M. G., 1990, MNRAS, 246, 678
Edvardsson B., Andersen J., Gustafsson B., et al. 1993, A&A, 275, 101
Efstathiou G., 1992, MNRAS, 256, 43P
Efstathiou G., 1995, MNRAS, 276, 1425
Efstathiou G., 2000, MNRAS, 317, 697
Efstathiou G., Bond J. R., 1986, MNRAS, 218, 103
Efstathiou G., Davis M., White S. D. M., Frenk C. S., 1985, ApJS, 57, 241
Efstathiou G., Ellis R. S., Peterson B. A., 1988, MNRAS, 232, 431
Efstathiou G., Jones B. J. T., 1979, MNRAS, 186, 133
Efstathiou G., Lake G., Negroponte J., 1982, MNRAS, 199, 1069
Efstathiou G., Silk J., 1983, Fund. Cosmic Phys., 9, 1
Efstathiou G., Sutherland W. J., Maddox S. J., 1990, Nature, 348, 705
Einasto J., 1965, Trudy Inst. Astroz. Alma-Ata, 57, 87
Einasto J., Kaasik A., Saar E., 1974, Nature, 250, 309
Einstein A., 1917, Sitzungsberichte der Königlich Preußischen Akademie der Wissenschaften (Berlin), pp 142–152
Eisenstein D. J., Hu W., 1998, ApJ, 496, 605
Eke V. R., Baugh C. M., Cole S., et al. 2004, MNRAS, 348, 866
Eke V. R., Navarro J. F., Steinmetz M., 2001, ApJ, 554, 114
Ellis R. S., Colless M., Broadhurst T., et al. 1996, MNRAS, 280, 235
Ellis R. S., Smail I., Dressler A., et al. 1997, ApJ, 483, 582
Ellison S. L., Yan L., Hook I. M., et al. 2001, A&A, 379, 393
Elmegreen B. G., 1979, ApJ, 232, 729
Elmegreen B. G., 1989, ApJ, 338, 178
Elmegreen B. G., 1993, ApJ, 411, 170
Elmegreen B. G., 1997, ApJ, 486, 944
Elmegreen B. G., 2002, ApJ, 577, 206
Elmegreen B. G., Parravano A., 1994, ApJ, 435, L121
Elmegreen B. G., Scalo J., 2004, ARA&A, 42, 211

Emsellem E., Cappellari M., Krajnović D., et al. 2007, MNRAS, 379, 401
Emsellem E., Cappellari M., Peletier R. F., et al. 2004, MNRAS, 352, 721
Eskridge P. B., Fabbiano G., Kim D.-W., 1995, ApJS, 97, 141
Ettori S., Fabian A. C., 1999, MNRAS, 305, 834
Evans N. W., 1994, MNRAS, 267, 333
Evans N. W., de Zeeuw P. T., 1994, MNRAS, 271, 202
Evrard A. E., 1990, ApJ, 363, 349
Evrard A. E., 1997, MNRAS, 292, 289
Evrard A. E., Metzler C. A., Navarro J. F., 1996, ApJ, 469, 494
Fabbiano G., 1989, ARA&A, 27, 87
Faber S. M., Friel E. D., Burstein D., Gaskell C. M., 1985, ApJS, 57, 711
Faber S. M., Jackson R. E., 1976, ApJ, 204, 668
Faber S. M., Lin D. N. C., 1983, ApJ, 266, L17
Faber S. M., Tremaine S., Ajhar E. A., et al. 1997, AJ, 114, 1771
Faber S. M., Willmer C. N. A., Wolf C., et al. 2007, ApJ, 665, 265
Fabian A. C., 1994, ARA&A, 32, 277
Fabian A. C., Celotti A., Blundell K. M., et al. 2002, MNRAS, 331, 369
Fabian A. C., Sanders J. S., Taylor G. B., et al. 2006, MNRAS, 366, 417
Fall S. M., Efstathiou G., 1980, MNRAS, 193, 189
Fall S. M., Frenk C. S., 1983, AJ, 88, 1626
Fall S. M., Pei Y. C., 1993, ApJ, 402, 479
Fan X., Hennawi J. F., Richards G. T., et al. 2004, AJ, 128, 515
Fan X., Narayanan V. K., Strauss M. A., et al. 2002, AJ, 123, 1247
Fan X., Strauss M. A., Becker R. H., et al. 2006, AJ, 132, 117
Fanaroff B. L., Riley J. M., 1974, MNRAS, 167, 31P
Fardal M. A., Katz N., Gardner J. P., et al. 2001, ApJ, 562, 605
Fardal M. A., Katz N., Weinberg D. H., Davé R., 2007, MNRAS, 379, 985
Farouki R., Shapiro S. L., 1981, ApJ, 243, 32
Farouki R. T., Shapiro S. L., 1982, ApJ, 259, 103
Fasano G., Vio R., 1991, MNRAS, 249, 629
Feast M. W., Catchpole R. M., 1997, MNRAS, 286, L1
Federman S. R., Glassgold A. E., Kwan J., 1979, ApJ, 227, 466
Ferguson A., Irwin M., Chapman S., et al. 2007, in de Jong R. S., ed., Island Universes – Structure and Evolution of Disk Galaxies, Springer, Dordrecht, p 239
Ferguson A. M. N., Irwin M. J., Ibata R. A., et al. 2002, AJ, 124, 1452
Ferguson H. C., Dickinson M., Williams R., 2000, ARA&A, 38, 667
Ferguson H. C., Sandage A., 1989, ApJ, 346, L53
Ferland G. J., Korista K. T., Verner D. A., et al. 1998, PASP, 110, 761
Fermi E., 1949, Phys. Rev., 75, 1169
Ferrarese L., Côté P., Dalla Bontà E., et al. 2006a, ApJ, 644, L21
Ferrarese L., Côté P., Jordán A., et al. 2006b, ApJS, 164, 334
Ferrarese L., Ford H., 2005, Space Sci. Rev., 116, 523
Ferrarese L., Merritt D., 2000, ApJ, 539, L9
Ferrarese L., van den Bosch F. C., Ford H. C., et al. 1994, AJ, 108, 1598
Field G. B., 1958, Proc. IRE, 46, 240
Field G. B., 1959, ApJ, 129, 536
Field G. B., 1965, ApJ, 142, 531
Field G. B., Goldsmith D. W., Habing H. J., 1969, ApJ, 155, L149
Fields B. D., Olive K. A., 1998, ApJ, 506, 177
Fillmore J. A., Goldreich P., 1984, ApJ, 281, 1
Firmani C., Avila-Reese V., 2000, MNRAS, 315, 457
Fisher K. B., Davis M., Strauss M. A., et al. 1994a, MNRAS, 267, 927
Fisher K. B., Nusser A., 1996, MNRAS, 279, L1
Fisher K. B., Scharf C. A., Lahav O., 1994b, MNRAS, 266, 219
Fixsen D. J., Cheng E. S., Gales J. M., et al. 1996, ApJ, 473, 576
Fixsen D. J., Dwek E., Mather J. C., et al. 1998, ApJ, 508, 123
Font A. S., Bower R. G., McCarthy I. G., et al. 2008, MNRAS, 389, 1619
Fontanot F., De Lucia G., Monaco P., et al. 2009, MNRAS, 397, 1776
Forbes D. A., Brodie J. P., Huchra J., 1997, AJ, 113, 887

Forbes D. A., Franx M., Illingworth G. D., 1995, AJ, 109, 1988
Forman W., Jones C., Churazov E., et al. 2007, ApJ, 665, 1057
Forman W., Jones C., Tucker W., 1985, ApJ, 293, 102
Forrey R. C., Balakrishnan N., Dalgarno A., Lepp S., 1997, ApJ, 489, 1000
Francis P. J., Hewett P. C., Foltz C. B., et al. 1991, ApJ, 373, 465
Franx M., Illingworth G., de Zeeuw T., 1991, ApJ, 383, 112
Franx M., Illingworth G., Heckman T., 1989a, ApJ, 344, 613
Franx M., Illingworth G., Heckman T., 1989b, AJ, 98, 538
Franx M., Labbé I., Rudnick G., et al. 2003, ApJ, 587, L79
Freedman W. L., Madore B. F., Gibson B. K., et al. 2001, ApJ, 553, 47
Freeman K. C., 1970, ApJ, 160, 811
Frenk C. S., White S. D. M., Bode P., et al. 1999, ApJ, 525, 554
Frenk C. S., White S. D. M., Davis M., Efstathiou G., 1988, ApJ, 327, 507
Friedli D., 1994, in Shlosman I., ed., Mass-Transfer Induced Activity in Galaxies, Cambridge University Press, Cambridge, p 268
Fry J. N., Gaztanaga E., 1993, ApJ, 413, 447
Fujii M., Funato Y., Makino J., 2006, PASJ, 58, 743
Fukugita M., Hogan C. J., Peebles P. J. E., 1998, ApJ, 503, 518
Fukugita M., Ichikawa T., Gunn J. E., et al. 1996, AJ, 111, 1748
Fukui Y., Mizuno N., Yamaguchi R., et al. 1999, PASJ, 51, 745
Gallagher III J. S., Ostriker J. P., 1972, AJ, 77, 288
Gallazzi A., Charlot S., Brinchmann J., et al. 2005, MNRAS, 362, 41
Galli D., Palla F., 1998, A&A, 335, 403
Gamow G., Teller E., 1939, Phys. Rev., 55, 654
Gao L., Navarro J. F., Cole S., et al. 2008, MNRAS, 387, 536
Gao L., Springel V., White S. D. M., 2005, MNRAS, 363, L66
Gao L., White S. D. M., 2007, MNRAS, 377, L5
Gao L., White S. D. M., Jenkins A., et al. 2004, MNRAS, 355, 819
Garcia Gomez C., Athanassoula E., 1993, A&AS, 100, 431
Gardner J. P., Katz N., Hernquist L., Weinberg D. H., 1997, ApJ, 484, 31
Garnavich P. M., Kirshner R. P., Challis P., et al. 1998, ApJ, 493, L53
Gavazzi G., Contursi A., Carrasco L., et al. 1995, A&A, 304, 325
Gebhardt K., Bender R., Bower G., et al. 2000, ApJ, 539, L13
Gebhardt K., Richstone D., Tremaine S., et al. 2003, ApJ, 583, 92
Geha M., Guhathakurta P., van der Marel R. P., 2002, AJ, 124, 3073
Geller M. J., Huchra J. P., 1983, ApJS, 52, 61
Genzel R., Pichon C., Eckart A., et al. 2000, MNRAS, 317, 348
Gerhard O., Kronawitter A., Saglia R. P., Bender R., 2001, AJ, 121, 1936
Gerhard O. E., 1981, MNRAS, 197, 179
Gerhard O. E., Binney J., 1985, MNRAS, 216, 467
Gerhard O. E., Binney J. J., 1996, MNRAS, 279, 993
Gerritsen J. P. E., Icke V., 1997, A&A, 325, 972
Gershtein S. S., Zel'dovich Y. B., 1966, ZhETF Pis ma Redaktsiiu, 4, 174
Ghez A. M., Duchêne G., Matthews K., et al. 2003, ApJ, 586, L127
Ghez A. M., Salim S., Hornstein S. D., et al. 2005, ApJ, 620, 744
Ghigna S., Moore B., Governato F., et al. 1998, MNRAS, 300, 146
Giallongo E., Cristiani S., D'Odorico S., et al. 1996, ApJ, 466, 46
Gibson B. K., 1997, MNRAS, 290, 471
Gilmore G., Reid N., 1983, MNRAS, 202, 1025
Gilmore G., Wilkinson M. I., Wyse R. F. G., et al. 2007, ApJ, 663, 948
Gingold R. A., Monaghan J. J., 1977, MNRAS, 181, 375
Giocoli C., Moreno J., Sheth R. K., Tormen G., 2007, MNRAS, 376, 977
Giocoli C., Tormen G., van den Bosch F. C., 2008, MNRAS, p. 468
Giovanelli R., Haynes M. P., Herter T., et al. 1997, AJ, 113, 53
Girardi L., Bressan A., Bertelli G., Chiosi C., 2000, A&AS, 141, 371
Glazebrook K., Ellis R., Santiago B., Griffiths R., 1995, MNRAS, 275, L19
Gloeckler G., Geiss J., 1996, Nature, 381, 210
Glover S. C. O., Mac Low M.-M., 2007, ApJ, 659, 1317
Gnedin N. Y., 2000, ApJ, 535, 530

Gnedin N. Y., Hui L., 1998, MNRAS, 296, 44
Gnedin O. Y., Hernquist L., Ostriker J. P., 1999, ApJ, 514, 109
Gnedin O. Y., Kravtsov A. V., Klypin A. A., Nagai D., 2004, ApJ, 616, 16
Gödel K., 1949, Rev. Mod. Phys., 21, 447
Goldreich P., Lynden-Bell D., 1965, MNRAS, 130, 125
Goldsmith P. F., Langer W. D., 1978, ApJ, 222, 881
Goldstein H., 1980, Classical Mechanics. 2nd edn; Addison-Wesley, Reading, MA
Gonzalez A. H., Zaritsky D., Zabludoff A. I., 2007, ApJ, 666, 147
Gorski K., 1988, ApJ, 332, L7
Gorski K. M., Davis M., Strauss M. A., et al. 1989, ApJ, 344, 1
Gott III J. R., Dickinson M., Melott A. L., 1986, ApJ, 306, 341
Gott III J. R., Thuan T. X., 1976, ApJ, 204, 649
Gould R. J., Salpeter E. E., 1963, ApJ, 138, 393
Governato F., Babul A., Quinn T., et al. 1999, MNRAS, 307, 949
Governato F., Mayer L., Wadsley J., et al. 2004, ApJ, 607, 688
Graham A., Colless M., 1997, MNRAS, 287, 221
Graham A. W., 2001, AJ, 121, 820
Graham A. W., Erwin P., Caon N., Trujillo I., 2001, ApJ, 563, L11
Graham A. W., Guzmán R., 2003, AJ, 125, 2936
Grant N. I., Kuipers J. A., Phillipps S., 2005, MNRAS, 363, 1019
Graves G. J., Faber S. M., Schiavon R. P., 2009, ApJ, 693, 486
Grebel E. K., 1999, in Whitelock P., Cannon R., eds., The Stellar Content of Local Group Galaxies, vol. 192 of IAU Symposium, ASP, San Francisco, p 17
Groth E. J., Juszkiewicz R., Ostriker J. P., 1989, ApJ, 346, 558
Groth E. J., Peebles P. J. E., 1977, ApJ, 217, 385
Gualandris A., Merritt D., 2007, arXiv:0708.3083
Guhathakurta P., Draine B. T., 1989, ApJ, 345, 230
Gunn J. E., 1977, ApJ, 218, 592
Gunn J. E., Gott J. R. I., 1972, ApJ, 176, 1
Gunn J. E., Peterson B. A., 1965, ApJ, 142, 1633
Gunn J. E., Stryker L. L., 1983, ApJS, 52, 121
Guth A. H., 1981, Phys. Rev. D., 23, 347
Guth A. H., Pi S.-Y., 1982, Phys. Rev. Lett., 49, 1110
Guth A. H., Weinberg E. J., 1981, Phys. Rev. D., 23, 876
Guzik J., Seljak U., 2002, MNRAS, 335, 311
Haardt F., Madau P., 1996, ApJ, 461, 20
Haehnelt M. G., Steinmetz M., Rauch M., 1998, ApJ, 495, 647
Hahn O., Porciani C., Carollo C. M., Dekel A., 2007, MNRAS, 375, 489
Hahn O., Porciani C., Dekel A., Carollo C. M., 2008, arXiv:0803.4211
Haiman Z., Rees M. J., Loeb A., 1996, ApJ, 467, 522
Halverson N. W., Leitch E. M., Pryke C., et al. 2002, ApJ, 568, 38
Hamilton A. J. S., 1992, ApJ, 385, L5
Hamilton A. J. S., 1993, ApJ, 417, 19
Hanany S., Ade P., Balbi A., et al. 2000, ApJ, 545, L5
Hänninen J., Flynn C., 2002, MNRAS, 337, 731
Hardcastle M. J., Evans D. A., Croston J. H., 2007, MNRAS, 376, 1849
Häring N., Rix H.-W., 2004, ApJ, 604, L89
Harms R. J., Ford H. C., Tsvetanov Z. I., et al. 1994, ApJ, 435, L35
Harris W. E., 1991, ARA&A, 29, 543
Harrison E. R., 1970, Phys. Rev. D., 1, 2726
Harten A., 1983, J. Comp. Phys., 49, 357
Hartmann L., Ballesteros-Paredes J., Bergin E. A., 2001, ApJ, 562, 852
Hatton S., Cole S., 1998, MNRAS, 296, 10
Hattori M., Habe A., 1990, MNRAS, 242, 399
Hausman M. A., Ostriker J. P., 1978, ApJ, 224, 320
Hawking S. W., 1982, Phys. Lett. B, 115, 295
Hawkins E., Maddox S., Cole S., et al. 2003, MNRAS, 346, 78
Hayashi E., Navarro J. F., 2006, MNRAS, 373, 1117
Hayashi E., White S. D. M., 2008, MNRAS, 388, 2

Heap S. R., Williger G. M., Smette A., et al. 2000, ApJ, 534, 69
Heckman T. M., Smith E. P., Baum S. A., et al. 1986, ApJ, 311, 526
Heithausen A., Bensch F., Stutzki J., et al. 1998, A&A, 331, L65
Heitsch F., Burkert A., Hartmann L. W., et al. 2005, ApJ, 633, L113
Helmi A., 2004, MNRAS, 351, 643
Helmi A., 2008, A&AR, 15, 145
Helmi A., White S. D. M., 1999, MNRAS, 307, 495
Helmi A., White S. D. M., de Zeeuw P. T., Zhao H., 1999, Nature, 402, 53
Hennebelle P., Pérault M., 1999, A&A, 351, 309
Henry J. P., Briel U. G., Nulsen P. E. J., 1993, A&A, 271, 413
Henry R. C., 1991, ARA&A, 29, 89
Herbig G. H., 1962, Adv. Astron. Astrophys., 1, 47
Hernquist L., 1989, Nature, 340, 687
Hernquist L., 1990, ApJ, 356, 359
Hernquist L., 1992, ApJ, 400, 460
Hernquist L., 1993, ApJS, 86, 389
Hernquist L., Katz N., Weinberg D. H., Miralda-Escudé J., 1996, ApJ, 457, L51
Hernquist L., Spergel D. N., 1992, ApJ, 399, L117
Hernquist L., Spergel D. N., Heyl J. S., 1993, ApJ, 416, 415
Heyer M. H., Brunt C. M., 2004, ApJ, 615, L45
Heyl J. S., Cole S., Frenk C. S., Navarro J. F., 1995, MNRAS, 274, 755
Hibbard J. E., Mihos J. C., 1995, AJ, 110, 140
Hickson P., 1982, ApJ, 255, 382
Hillenbrand L. A., Hartmann L. W., 1998, ApJ, 492, 540
Hinshaw G., Nolta M. R., Bennett C. L., et al. 2007, ApJS, 170, 288
Hirata C. M., Seljak U., 2004, Phys. Rev. D., 70, 063526
Hockney R. W., Eastwood J. W., 1988, Computer Simulation using Particles. Hilger, Bristol
Hodapp K.-W., Deane J., 1993, ApJS, 88, 119
Hoekstra H., Yee H. K. C., Gladders M. D., 2004, ApJ, 606, 67
Hoffman Y., Ribak E., 1991, ApJ, 380, L5
Hogan C. J., Anderson S. F., Rugers M. H., 1997, AJ, 113, 1495
Hohl F., 1971, ApJ, 168, 343
Holland W. S., Robson E. I., Gear W. K., et al. 1999, MNRAS, 303, 659
Hollenbach D., McKee C. F., 1979, ApJS, 41, 555
Hollenbach D., McKee C. F., 1989, ApJ, 342, 306
Hollenbach D., Salpeter E. E., 1971, ApJ, 163, 155
Hollenbach D. J., Werner M. W., Salpeter E. E., 1971, ApJ, 163, 165
Holmberg E., 1975, in Sandage A., et al. eds., Galaxies and the Universe, University of Chicago Press, Chicago, p 123
Holtzman J. A., Faber S. M., Shaya E. J., et al. 1992, AJ, 103, 691
Hopkins A. M., 2004, ApJ, 615, 209
Hopkins P. F., Hernquist L., Cox T. J., et al. 2006, ApJS, 163, 1
Hopkins P. F., Richards G. T., Hernquist L., 2007, ApJ, 654, 731
Howarth I. D., 1983, MNRAS, 203, 301
Hoyle F., 1948, MNRAS, 108, 372
Hoyle F., 1949, in Problems of Cosmical Aerodynamics. Central Air Documents Office, Dayton, OH
Hoyle F., 1954, ApJS, 1, 121
Hu E. M., Kim T.-S., Cowie L. L., et al. 1995, AJ, 110, 1526
Hu W., 1995, PhD thesis, University of California, Berkeley.
Hu W., Dodelson S., 2002, ARA&A, 40, 171
Hu W., White M., 1997, New Astron., 2, 323
Huang S., Carlberg R. G., 1997, ApJ, 480, 503
Hubble E., 1929, Proc. Natl Acad. Sci., 15, 168
Hubble E., Humason M. L., 1931, ApJ, 74, 43
Hui L., 1999, ApJ, 516, 519
Hui L., Gnedin N. Y., 1997, MNRAS, 292, 27
Hui L., Gnedin N. Y., Zhang Y., 1997, ApJ, 486, 599
Hui L., Rutledge R. E., 1999, ApJ, 517, 541
Hunter C., 1977a, AJ, 82, 271

Hunter C., 1977b, ApJ, 218, 834
Hunter C., Qian E., 1993, MNRAS, 262, 401
Hunter D., 1992, in Tenorio-Tagle G., et al. eds., Star Formation in Stellar Systems, Cambridge University Press, Cambridge, p 67
Hunter D. A., Elmegreen B. G., Baker A. L., 1998, ApJ, 493, 595
Huss A., Jain B., Steinmetz M., 1999, ApJ, 517, 64
Hütsi G., Einasto J., Tucker D. L., et al. 2002, astro-ph/0212327
Ibata R., Lewis G. F., Irwin M., et al. 2001, ApJ, 551, 294
Ibata R. A., Gilmore G., Irwin M. J., 1994, Nature, 370, 194
Iben I. J., 1965a, ApJ, 141, 993
Iben I. J., 1965b, ApJ, 142, 1447
Ikeuchi S., 1986, Astrophys. Space Sci., 118, 509
Ikeuchi S., Ostriker J. P., 1986, ApJ, 301, 522
Iliev I. T., Mellema G., Pen U.-L., et al. 2006, MNRAS, 369, 1625
Iliev I. T., Shapiro P. R., Raga A. C., 2005, MNRAS, 361, 405
Illingworth G., 1977, ApJ, 218, L43
Impey C. D., Foltz C. B., Petry C. E., et al. 1996a, ApJ, 462, L53
Impey C. D., Sprayberry D., Irwin M. J., Bothun G. D., 1996b, ApJS, 105, 209
Israel F. P., 1997, A&A, 328, 471
Jablonka P., Martin P., Arimoto N., 1996, AJ, 112, 1415
Jaffe W., 1983, MNRAS, 202, 995
Jaffe W., Bremer M. N., 1997, MNRAS, 284, L1
Jaffe W., Ford H. C., O'Connell R. W., et al. 1994, AJ, 108, 1567
Jakobsen P., Boksenberg A., Deharveng J. M., et al. 1994, Nature, 370, 35
Jeans J. H., 1902, Phil. Trans. Roy. Soc. Lond., 199, 1
Jena T., Norman M. L., Tytler D., et al. 2005, MNRAS, 361, 70
Jenkins A., Binney J., 1990, MNRAS, 245, 305
Jenkins A., Frenk C. S., White S. D. M., et al. 2001, MNRAS, 321, 372
Jenkins E. B., 1987, in Hollenbach D. J., Thronson Jr. H. A., eds., Interstellar Processes, vol. 134 of ASSL, Reidel, Dordrecht, pp 533–559
Jerjen H., Binggeli B., 1997, in Arnaboldi M., et al. eds., The Nature of Elliptical Galaxies, 2nd Stromlo Symposium vol. 116 of ASP, p 239
Jerjen H., Tammann G. A., 1997, A&A, 321, 713
Jiang C. Y., Jing Y. P., Faltenbacher A., et al. 2008, ApJ, 675, 1095
Jing Y. P., 1998, ApJ, 503, L9
Jing Y. P., 2000, ApJ, 535, 30
Jing Y. P., Börner G., 1998, ApJ, 503, 37
Jing Y. P., Mo H. J., Börner G., 1998, ApJ, 494, 1
Jing Y. P., Suto Y., 2002, ApJ, 574, 538
Jing Y. P., Suto Y., Mo H. J., 2007, ApJ, 657, 664
Jog C. J., Solomon P. M., 1984, ApJ, 276, 114
Johnston K. V., 1998, ApJ, 495, 297
Johnston K. V., Zhao H., Spergel D. N., Hernquist L., 1999, ApJ, 512, L109
Jones B. J. T., Wyse R. F. G., 1985, A&A, 149, 144
Jones D. H., Peterson B. A., Colless M., Saunders W., 2006, MNRAS, 369, 25
Jones F. C., Ellison D. C., 1991, Space Sci. Rev., 58, 259
Jørgensen I., Franx M., Kjaergaard P., 1996, MNRAS, 280, 167
Joseph R. D., Wright G. S., 1985, MNRAS, 214, 87
Jurić M., Ivezić Ž., Brooks A., et al. 2008, ApJ, 673, 864
Just A., Fuchs B., Wielen R., 1996, A&A, 309, 715
Just A., Peñarrubia J., 2005, A&A, 431, 861
Kacprzak G. G., Churchill C. W., Steidel C. C., et al. 2007, ApJ, 662, 909
Kaiser N., 1984, ApJ, 284, L9
Kaiser N., 1986, MNRAS, 222, 323
Kaiser N., 1987, MNRAS, 227, 1
Kalnajs A. J., 1965, PhD thesis, Harvard University, Cambridge, MA
Kalnajs A. J., 1972a, in Gravitational N-Body Problem, vol. 31 of ASSL, Reidel, Dordrecht, p. 13
Kalnajs A. J., 1972b, ApJ, 175, 63
Kalnajs A. J., 1973, Proc. Astron. Soc. Austral., 2, 174

Kang X., Jing Y. P., Mo H. J., Börner G., 2005, ApJ, 631, 21
Kang X., Jing Y. P., Silk J., 2006, ApJ, 648, 820
Kang X., van den Bosch F. C., 2008, ApJ, 676, L101
Kang X., van den Bosch F. C., Pasquali A., 2007, MNRAS, 381, 389
Kaplan S. A., Pikelner S. B., 1970, The Interstellar Medium. Harvard University Press, Cambridge, MA
Karzas W. J., Latter R., 1961, ApJS, 6, 167
Kasun S. F., Evrard A. E., 2005, ApJ, 629, 781
Katz N., 1992, ApJ, 391, 502
Katz N., Gunn J. E., 1991, ApJ, 377, 365
Katz N., Weinberg D. H., Hernquist L., Miralda-Escude J., 1996, ApJ, 457, L57
Kauffmann G., 1996a, MNRAS, 281, 475
Kauffmann G., 1996b, MNRAS, 281, 487
Kauffmann G., Charlot S., 1998a, MNRAS, 294, 705
Kauffmann G., Charlot S., 1998b, MNRAS, 297, L23
Kauffmann G., Charlot S., White S. D. M., 1996, MNRAS, 283, L117
Kauffmann G., Colberg J. M., Diaferio A., White S. D. M., 1999a, MNRAS, 303, 188
Kauffmann G., Colberg J. M., Diaferio A., White S. D. M., 1999b, MNRAS, 307, 529
Kauffmann G., Guiderdoni B., White S. D. M., 1994, MNRAS, 267, 981
Kauffmann G., Heckman T. M., Best P. N., 2008, MNRAS, 384, 953
Kauffmann G., Heckman T. M., Tremonti C., et al. 2003a, MNRAS, 346, 1055
Kauffmann G., Heckman T. M., White S. D. M., et al. 2003b, MNRAS, 341, 33
Kauffmann G., White S. D. M., 1993, MNRAS, 261, 921
Kauffmann G., White S. D. M., Guiderdoni B., 1993, MNRAS, 264, 201
Kauffmann G., White S. D. M., Heckman T. M., et al. 2004, MNRAS, 353, 713
Kazantzidis S., Bullock J. S., Zentner A. R., et al. 2008, ApJ, 688, 254
Kazantzidis S., Kravtsov A. V., Zentner A. R., et al. 2004, ApJ, 611, L73
Kelson D. D., van Dokkum P. G., Franx M., et al. 1997, ApJ, 478, L13
Kembhavi A. K., Narlikar J. V., 1999, Quasars and Active Galactic Nuclei: An Introduction. Cambridge University Press, Cambridge
Kennicutt Jr. R. C., 1989, ApJ, 344, 685
Kennicutt Jr. R. C., 1998a, ARA&A, 36, 189
Kennicutt Jr. R. C., 1998b, ApJ, 498, 541
Kent S. M., Gunn J. E., 1982, AJ, 87, 945
Kereš D., Katz N., Dave R., et al. 2009, MNRAS, 396, 2332
Kereš D., Katz N., Weinberg D. H., Davé R., 2005, MNRAS, 363, 2
Kewley L. J., Dopita M. A., Sutherland R. S., et al. 2001, ApJ, 556, 121
Khochfar S., Burkert A., 2003, ApJ, 597, L117
Khochfar S., Burkert A., 2005, MNRAS, 359, 1379
Khochfar S., Ostriker J. P., 2008, ApJ, 680, 54
Kim C.-G., Kim W.-T., Ostriker E. C., 2006, ApJ, 649, L13
Kim T.-S., Hu E. M., Cowie L. L., Songaila A., 1997, AJ, 114, 1
Kim T.-S., Viel M., Haehnelt M. G., et al. 2004, MNRAS, 347, 355
Kim W.-T., Ostriker E. C., 2002, ApJ, 570, 132
Kim W.-T., Ostriker E. C., Stone J. M., 2002, ApJ, 581, 1080
Kimm T., Somerville R. S., Yi S. K., et al. 2009, MNRAS, 394, 1131
King I., 1962, AJ, 67, 471
Kippenhahn R., Weigert A., 1990, Stellar Structure and Evolution. Springer-Verlag, Berlin
Kirkman D., Tytler D., 1997, ApJ, 484, 672
Kitzbichler M. G., White S. D. M., 2007, MNRAS, 376, 2
Kitzbichler M. G., White S. D. M., 2008, MNRAS, 391, 1489
Klinkhamer F. R., Norman C. A., 1981, ApJ, 243, L1
Klypin A., Gottlöber S., Kravtsov A. V., Khokhlov A. M., 1999, ApJ, 516, 530
Kochanek C. S., 1995, ApJ, 445, 559
Kodama H., Sasaki M., 1984, Prog. Theoret. Phys. Suppl., 78, 1
Kodama T., Arimoto N., 1997, A&A, 320, 41
Koester B. P., McKay T. A., Annis J., et al. 2007, ApJ, 660, 239
Kolb E. W., Turner M. S., 1990, The Early Universe. Frontiers in Physics. Addison-Wesley, Reading, MA
Komatsu E., Dunkley J., Nolta M. R., et al. 2009, ApJS, 180, 330
Komatsu E., Seljak U., 2001, MNRAS, 327, 1353

Kompaneets A. S., 1957, Soviet Phys., JETP Lett., 4, 730
Koornneef J., Code A. D., 1981, ApJ, 247, 860
Körding E. G., Jester S., Fender R., 2006, MNRAS, 372, 1366
Kormendy J., 1977, ApJ, 218, 333
Kormendy J., 1985, ApJ, 295, 73
Kormendy J., 2001, in Funes J. G., Corsini E. M., eds., Galaxy Disks and Disk Galaxies, vol. 230 of ASP, pp 247–256
Kormendy J., Bender R., 1996, ApJ, 464, L119
Kormendy J., Kennicutt Jr. R. C., 2004, ARA&A, 42, 603
Kormendy J., Richstone D., 1995, ARA&A, 33, 581
Kouchi N., Ukai M., Hatano Y., 1997, J. Phys. B Atom. Mol. Phys., 30, 2319
Koushiappas S. M., Bullock J. S., Dekel A., 2004, MNRAS, 354, 292
Koyama H., Inutsuka S.-I., 2000, ApJ, 532, 980
Kravtsov A. V., 2003, ApJ, 590, L1
Kravtsov A. V., Berlind A. A., Wechsler R. H., et al. 2004, ApJ, 609, 35
Krolik J. H., 1998, Active Galactic Nuclei: From the Central Black Hole to the Galactic Environment. Princeton University Press, Princeton, NJ
Krolik J. H., Begelman M. C., 1988, ApJ, 329, 702
Kronawitter A., Saglia R. P., Gerhard O., Bender R., 2000, A&AS, 144, 53
Kroupa P., 2002, Science, 295, 82
Krumholz M. R., Matzner C. D., McKee C. F., 2006, ApJ, 653, 361
Krumholz M. R., McKee C. F., 2005, ApJ, 630, 250
Kuchinski L. E., Terndrup D. M., Gordon K. D., Witt A. N., 1998, AJ, 115, 1438
Kuijken K., Gilmore G., 1989a, MNRAS, 239, 605
Kuijken K., Gilmore G., 1989b, MNRAS, 239, 571
Kuijken K., Merrifield M. R., 1995, ApJ, 443, L13
Kulkarni V. P., Fall S. M., 1993, ApJ, 413, L63
Kulkarni V. P., Fall S. M., Lauroesch J. T., et al. 2005, ApJ, 618, 68
Kuntschner H., Emsellem E., Bacon R., et al. 2006, MNRAS, 369, 497
Kuntschner H., Lucey J. R., Smith R. J., et al. 2001, MNRAS, 323, 615
Kurucz R. L., 1992, in Barbuy B., Renzini A., eds, The Stellar Populations of Galaxies, vol. 149 of IAU Symposium, Kluwer, Dordrecht, p 225
Labbé I., Franx M., Rudnick G., et al. 2003, AJ, 125, 1107
Lacey C., Cole S., 1993, MNRAS, 262, 627
Lacey C. G., 1984, MNRAS, 208, 687
Lacey C. G., Fall S. M., 1985, ApJ, 290, 154
Lacey C. G., Ostriker J. P., 1985, ApJ, 299, 633
Lacy M., Gates E. L., Ridgway S. E., et al. 2002, AJ, 124, 3023
Laing R. A., Riley J. M., Longair M. S., 1983, MNRAS, 204, 151
Lake G., Dressler A., 1986, ApJ, 310, 605
Lambas D. G., Maddox S. J., Loveday J., 1992, MNRAS, 258, 404
Landau L. D., Lifshitz E. M., 1959, Fluid Mechanics. Pergamon Press, Oxford
Landau L. D., Lifshitz E. M., 1975, The Classical Theory of Fields. Pergamon Press, Oxford
Lanzetta K. M., Bowen D. V., Tytler D., Webb J. K., 1995, ApJ, 442, 538
Lanzoni B., Ciotti L., 2003, A&A, 404, 819
Larson R. B., 1969, MNRAS, 145, 271
Larson R. B., 1974a, MNRAS, 166, 585
Larson R. B., 1974b, MNRAS, 169, 229
Larson R. B., 1975, MNRAS, 173, 671
Larson R. B., 1976, MNRAS, 176, 31
Larson R. B., 1981, MNRAS, 194, 809
Larson R. B., 1992, MNRAS, 256, 641
Larson R. B., 2005, MNRAS, 359, 211
Larson R. B., Tinsley B. M., 1978, ApJ, 219, 46
Larson R. B., Tinsley B. M., Caldwell C. N., 1980, ApJ, 237, 692
Lauer T. R., 1988, ApJ, 325, 49
Lauer T. R., Ajhar E. A., Byun Y.-I., et al. 1995, AJ, 110, 2622
Lauer T. R., Gebhardt K., Faber S. M., et al. 2007, ApJ, 664, 226
Le Borgne J.-F., Bruzual G., Pelló R., et al. 2003, A&A, 402, 433

Le Brun V., Bergeron J., Boisse P., 1996, A&A, 306, 691
Le Fèvre O., Abraham R., Lilly S. J., et al. 2000, MNRAS, 311, 565
Le Fèvre O., Vettolani G., Garilli B., et al. 2005, A&A, 439, 845
Ledlow M. J., Owen F. N., 1995, AJ, 110, 1959
Leisawitz D., Bash F. N., Thaddeus P., 1989, ApJS, 70, 731
Lemson G., Kauffmann G., 1999, MNRAS, 302, 111
Lepp S., Shull J. M., 1983, ApJ, 270, 578
Leroy A. K., Walter F., Brinks E., et al. 2008, AJ, 136, 2782
Levison H. F., 1987, ApJ, 320, L93
Li C., Kauffmann G., Wang L., et al. 2006a, MNRAS, 373, 457
Li P. S., Norman M. L., Mac Low M.-M., Heitsch F., 2004, ApJ, 605, 800
Li Y., Mac Low M.-M., Klessen R. S., 2006b, ApJ, 639, 879
Li Y., Mo H. J., Gao L., 2008, MNRAS, 389, 1419
Li Y., Mo H. J., van den Bosch F. C., Lin W. P., 2007, MNRAS, 379, 689
Lichtenberg A. J., Lieberman M. A., 1992, Regular and Chaotic Dynamics. Springer, Berlin
Lichti G. G., Balonek T., Courvoisier T. J.-L., et al. 1995, A&A, 298, 711
Liddle A. R., Lyth D. H., 2000, Cosmological Inflation and Large-Scale Structure. Cambridge University Press, Cambridge
Lifshitz E. M., 1946, J. Phys. (Moscow), 10, 116
Lilly S. J., Le Fevre O., Crampton D., et al. 1995, ApJ, 455, 50
Lin C. C., 1970, in Becker W., Kontopoulos G. I., eds, The Spiral Structure of our Galaxy, vol. 38 of IAU Symposium, Reidel, Dordrecht, p 377
Lin C. C., Shu F. H., 1964, ApJ, 140, 646
Lin C. C., Shu F. H., 1966, Proc. Natl Acad. Sci., 55, 229
Lin D. N. C., Faber S. M., 1983, ApJ, 266, L21
Lin D. N. C., Pringle J. E., 1987a, MNRAS, 225, 607
Lin D. N. C., Pringle J. E., 1987b, ApJ, 320, L87
Lin D. N. C., Tremaine S., 1983, ApJ, 264, 364
Lin L., Koo D. C., Willmer C. N. A., et al. 2004a, ApJ, 617, L9
Lin Y.-T., Mohr J. J., Stanford S. A., 2004b, ApJ, 610, 745
Linde A. D., 1982, Phys. Lett. B, 108, 389
Linde A. D., 1986, Phys. Lett. B, 175, 395
Lineweaver C. H., Tenorio L., Smoot G. F., et al. 1996, ApJ, 470, 38
Linsky J. L., Diplas A., Wood B. E., et al. 1995, ApJ, 451, 335
Lisker T., Grebel E. K., Binggeli B., 2006, AJ, 132, 497
Lisker T., Grebel E. K., Binggeli B., Glatt K., 2007, ApJ, 660, 1186
Loeb A., Barkana R., 2001, ARA&A, 39, 19
Loh E. D., Spillar E. J., 1986, ApJ, 307, L1
Longair, M. S., 2006, The Cosmic Century: A History of Astrophysics and Cosmology. Cambridge University Press, Cambridge
Lotz J. M., Davis M., Faber S. M., et al. 2008, ApJ, 672, 177
Lu L., Sargent W. L. W., Barlow T. A., et al. 1996a, ApJS, 107, 475
Lu L., Sargent W. L. W., Barlow T. A., Rauch M., 1998, astro-ph/9802189
Lu L., Sargent W. L. W., Womble D. S., Takada-Hidai M., 1996b, ApJ, 472, 509
Lu L., Wolfe A. M., Turnshek D. A., 1991, ApJ, 367, 19
Lu Y., Mo H. J., 2007, MNRAS, 377, 617
Lu Y., Mo H. J., Katz N., Weinberg M. D., 2006, MNRAS, 368, 1931
Lucy L. B., 1977, AJ, 82, 1013
Ludlow A. D., Navarro J. F., Springel V., et al. 2009, ApJ, 692, 931
Lumsden S. L., Nichol R. C., Collins C. A., Guzzo L., 1992, MNRAS, 258, 1
Lynden-Bell D., 1962, MNRAS, 124, 1
Lynden-Bell D., 1967, MNRAS, 136, 101
Lynden-Bell D., 1969, Nature, 223, 690
Lyth D. H., Woszczyna A., 1995, Phys. Rev. D., 52, 3338
Lyubimov V. A., Novikov E. G., Nozik V. Z., et al. 1980, Phys. Lett. B, 94, 266
Ma C.-P., 1996, ApJ, 471, 13
Ma C.-P., Bertschinger E., 1995, ApJ, 455, 7
Ma C.-P., Fry J. N., 2000a, ApJ, 543, 503
Ma C.-P., Fry J. N., 2000b, ApJ, 531, L87

Mac Low M.-M., Ferrara A., 1999a, ApJ, 513, 142
Mac Low M.-M., Ferrara A., 1999b, ApJ, 513, 142
Mac Low M.-M., Klessen R. S., 2004, Rev. Mod. Phys., 76, 125
Mac Low M.-M., Klessen R. S., Burkert A., Smith M. D., 1998, Phys. Rev. Lett., 80, 2754
MacArthur L. A., Courteau S., Bell E., Holtzman J. A., 2004, ApJS, 152, 175
MacArthur L. A., Courteau S., Holtzman J. A., 2003, ApJ, 582, 689
Macciò A. V., Dutton A. A., van den Bosch F. C., 2008, MNRAS, 391, 1940
Macciò A. V., Dutton A. A., van den Bosch F. C., et al. 2007, MNRAS, 378, 55
MacMillan J. D., Widrow L. M., Henriksen R. N., 2006, ApJ, 653, 43
Madau P., Meiksin A., Rees M. J., 1997, ApJ, 475, 429
Maddox S. J., Efstathiou G., Sutherland W. J., Loveday J., 1990a, MNRAS, 242, 43P
Maddox S. J., Efstathiou G., Sutherland W. J., Loveday J., 1990b, MNRAS, 243, 692
Madore B. F., Freedman W. L., 1991, PASP, 103, 933
Maeder A., 1992, A&A, 264, 105
Magorrian J., Tremaine S., Richstone D., et al. 1998, AJ, 115, 2285
Malagoli A., Bodo G., Rosner R., 1996, ApJ, 456, 708
Malin D. F., Carter D., 1983, ApJ, 274, 534
Maller A. H., Bullock J. S., 2004, MNRAS, 355, 694
Maller A. H., Dekel A., Somerville R., 2002, MNRAS, 329, 423
Malumuth E. M., Richstone D. O., 1984, ApJ, 276, 413
Mandelbaum R., Hirata C. M., Broderick T., et al. 2006a, MNRAS, 370, 1008
Mandelbaum R., Seljak U., Kauffmann G., et al. 2006b, MNRAS, 368, 715
Mao S., Mo H. J., 1998, MNRAS, 296, 847
Maoz D., Rix H.-W., 1993, ApJ, 416, 425
Maoz E., 1998, ApJ, 494, L181
Marchesini D., van Dokkum P. G., Forster Schreiber N. M., et al. 2008, ArXiv:0811.1773
Marconi A., Risaliti G., Gilli R., et al. 2004, MNRAS, 351, 169
Marigo P., Bressan A., Chiosi C., 1996, A&A, 313, 545
Mark J. W.-K., 1974, ApJ, 193, 539
Mark J. W. K., 1976, ApJ, 205, 363
Markevitch M., 1998, ApJ, 504, 27
Marochnik L. S., 1968, Sov. Astron., 11, 873
Martin C. L., Kennicutt Jr. R. C., 2001, ApJ, 555, 301
Martin N. F., Ibata R. A., Chapman S. C., et al. 2007, MNRAS, 380, 281
Martinelli A., Matteucci F., Colafrancesco S., 1998, MNRAS, 298, 42
Mason B. S., Cartwright J. K., Padin S., et al. 2002, in Gurzadyan V. G., et al. eds., The Ninth Marcel Grossmann Meeting, World Scientific, Singapore, pp 2171–2172
Mateo M. L., 1998, ARA&A, 36, 435
Mathews W. G., Baker J. C., 1971, ApJ, 170, 241
Mathews W. G., Brighenti F., 2003, ARA&A, 41, 191
Mathewson D. S., Cleary M. N., Murray J. D., 1974, ApJ, 190, 291
Mathis J. S., 1990, ARA&A, 28, 37
Mathis J. S., Rumpl W., Nordsieck K. H., 1977, ApJ, 217, 425
Matteucci F., 1992, ApJ, 397, 32
Mattig W., 1958, Astron. Nachr., 284, 109
Matzner C. D., McKee C. F., 2000, ApJ, 545, 364
Maulbetsch C., Avila-Reese V., Colín P., et al. 2007, ApJ, 654, 53
May A., van Albada T. S., 1984, MNRAS, 209, 15
McCarthy I. G., Frenk C. S., Font A. S., et al. 2008, MNRAS, 383, 593
McCray R., 1993, ARA&A, 31, 175
McDonald P., Miralda-Escudé J., 2001, ApJ, 549, L11
McDonald P., Seljak U., Cen R., et al. 2005, ApJ, 635, 761
McGaugh S. S., de Blok W. J. G., 1997, ApJ, 481, 689
McIntosh D. H., Guo Y., Hertzberg J., et al. 2008, MNRAS, 388, 1537
McKay T. A., Sheldon E. S., Racusin J., et al. 2001, astro-ph/0108013
McKee C. F., Cowie L. L., 1977, ApJ, 215, 213
McKee C. F., Ostriker E. C., 2007, ARA&A, 45, 565
McKee C. F., Ostriker J. P., 1977, ApJ, 218, 148
McKee C. F., Zweibel E. G., Goodman A. A., Heiles C., 1993, in Levy E. H., Lunine J. I., eds., Protostars and Planets III, University of Arizona Press, Tucson, p 327

McNamara B. R., Wise M., Nulsen P. E. J., et al. 2000, ApJ, 534, L135
Melott A. L., 1980, ApJ, 241, 889
Mendoza C., 1983, in Flower D. R., ed., Planetary Nebulae, vol. 103 of IAU Symposium, Reidel, Dordrecht, pp 143–172
Merrifield M. R., Kent S. M., 1991, AJ, 101, 783
Merritt D., 1984, ApJ, 276, 26
Merritt D., 1985, ApJ, 289, 18
Merritt D., Fridman T., 1996, ApJ, 460, 136
Mestel L., 1963, MNRAS, 126, 553
Mestel L., Spitzer Jr. L., 1956, MNRAS, 116, 503
Mészáros P., 1974, A&A, 37, 225
Meyer D. M., York D. G., 1987, ApJ, 315, L5
Mihalas D., 1978, Stellar Atmospheres. W. H. Freeman and Co., San Francisco
Mihos J. C., Hernquist L., 1996, ApJ, 464, 641
Miller B. W., Lotz J. M., Ferguson H. C., et al. 1998, ApJ, 508, L133
Miller C. J., Nichol R. C., Reichart D., et al. 2005, AJ, 130, 968
Miller G. E., Scalo J. M., 1979, ApJS, 41, 513
Milne G., 1935, Relativity, Gravitation and World Structure. Clarendon Press, Oxford
Milosavljević M., Merritt D., Rest A., van den Bosch F. C., 2002, MNRAS, 331, L51
Miralda-Escude J., 1993, MNRAS, 262, 273
Miralda-Escudé J., Cen R., Ostriker J. P., Rauch M., 1996, ApJ, 471, 582
Miralda-Escudé J., Haehnelt M., Rees M. J., 2000, ApJ, 530, 1
Misner C. W., 1968, ApJ, 151, 431
Misner C. W., Thorne K. S., Wheeler J. A., 1973, Gravitation. W. H. Freeman and Co., San Francisco
Miyoshi M., Moran J., Herrnstein J., et al. 1995, Nature, 373, 127
Mo H. J., 1994, MNRAS, 269, L49
Mo H. J., Fukugita M., 1996, ApJ, 467, L9
Mo H. J., Jing Y. P., Börner G., 1992, ApJ, 392, 452
Mo H. J., Jing Y. P., Börner G., 1997a, MNRAS, 286, 979
Mo H. J., Jing Y. P., White S. D. M., 1996, MNRAS, 282, 1096
Mo H. J., Jing Y. P., White S. D. M., 1997b, MNRAS, 284, 189
Mo H. J., Mao S., 2000, MNRAS, 318, 163
Mo H. J., Mao S., White S. D. M., 1998, MNRAS, 295, 319
Mo H. J., Mao S., White S. D. M., 1999, MNRAS, 304, 175
Mo H. J., Miralda-Escudé J., 1996, ApJ, 469, 589
Mo H. J., Miralda-Escudé J., Rees M. J., 1993, MNRAS, 264, 705
Mo H. J., Morris S. L., 1994, MNRAS, 269, 52
Mo H. J., White S. D. M., 1996, MNRAS, 282, 347
Mo H. J., Yang X., van den Bosch F. C., Jing Y. P., 2004, MNRAS, 349, 205
Mo H. J., Yang X., van den Bosch F. C., Katz N., 2005, MNRAS, 363, 1155
Monaco P., 1998, Fund. Cosmic Phys., 19, 157
Monaco P., Fontanot F., 2005, MNRAS, 359, 283
Monaghan J. J., 1992, ARA&A, 30, 543
Moore B., 2001, in Wheeler J. C., Martel H., eds., 20th Texas Symposium on Relativistic Astrophysics, vol. 586 of AIP, p 73
Moore B., Ghigna S., Governato F., et al. 1999a, ApJ, 524, L19
Moore B., Governato F., Quinn T., et al. 1998a, ApJ, 499, L5
Moore B., Katz N., Lake G., et al. 1996, Nature, 379, 613
Moore B., Lake G., Katz N., 1998b, ApJ, 495, 139
Moore B., Lake G., Quinn T., Stadel J., 1999b, MNRAS, 304, 465
More S., van den Bosch F. C., Cacciato M., et al. 2009, MNRAS, 392, 801
Morris S. L., Weymann R. J., Dressler A., et al. 1993, ApJ, 419, 524
Moster B. P., Maccio' A. V., Somerville R. S., et al. 2009, ArXiv:0906.0764
Motte F., Andre P., Neri R., 1998, A&A, 336, 150
Mukhanov V. F., Feldman H. A., Brandenberger R. H., 1992, Phys. Rep., 215, 203
Murai T., Fujimoto M., 1980, PASJ, 32, 581
Murphy D. W., Browne I. W. A., Perley R. A., 1993, MNRAS, 264, 298
Murray N., Quataert E., Thompson T. A., 2005, ApJ, 618, 569
Murray S. D., White S. D. M., Blondin J. M., Lin D. N. C., 1993, ApJ, 407, 588

Mushotzky R., 1993, in Shull J. M., Thronson H. A., eds, The Environment and Evolution of Galaxies, vol. 188 of ASSL, Kluwer, Dordrecht, p 383
Mushotzky R. F., Loewenstein M., Awaki H., et al. 1994, ApJ, 436, L79
Myers A. D., Brunner R. J., Nichol R. C., et al. 2007, ApJ, 658, 85
Myers P. C., Fuller G. A., 1992, ApJ, 396, 631
Naab T., Burkert A., 2003, ApJ, 597, 893
Naab T., Burkert A., Hernquist L., 1999, ApJ, 523, L133
Naab T., Johansson P. H., Ostriker J. P., Efstathiou G., 2007, ApJ, 658, 710
Naab T., Khochfar S., Burkert A., 2006, ApJ, 636, L81
Nagai D., Kravtsov A. V., Vikhlinin A., 2007, ApJ, 668, 1
Nagamine K., Springel V., Hernquist L., 2004a, MNRAS, 348, 421
Nagamine K., Springel V., Hernquist L., 2004b, MNRAS, 348, 435
Narayan R., Medvedev M. V., 2001, ApJ, 562, L129
Narayan R., Yi I., 1994, ApJ, 428, L13
Navarro J. F., Benz W., 1991, ApJ, 380, 320
Navarro J. F., Frenk C. S., White S. D. M., 1996, ApJ, 462, 563
Navarro J. F., Frenk C. S., White S. D. M., 1997, ApJ, 490, 493
Navarro J. F., Hayashi E., Power C., et al. 2004, MNRAS, 349, 1039
Navarro J. F., White S. D. M., 1993, MNRAS, 265, 271
Navarro J. F., White S. D. M., 1994, MNRAS, 267, 401
Negroponte J., White S. D. M., 1983, MNRAS, 205, 1009
Neistein E., Dekel A., 2008, MNRAS, 383, 615
Neistein E., van den Bosch F. C., Dekel A., 2006, MNRAS, 372, 933
Nelan J. E., Smith R. J., Hudson M. J., et al. 2005, ApJ, 632, 137
Nestor D. B., Turnshek D. A., Rao S. M., 2005, ApJ, 628, 637
Neto A. F., Gao L., Bett P., et al. 2007, MNRAS, 381, 1450
Newberg H. J., Yanny B., Rockosi C., et al. 2002, ApJ, 569, 245
Neyman J., Scott E. L., Shane C. D., 1953, ApJ, 117, 92
Nipoti C., Londrillo P., Ciotti L., 2003, MNRAS, 342, 501
Noeske K. G., Faber S. M., Weiner B. J., et al. 2007a, ApJ, 660, L47
Noeske K. G., Weiner B. J., Faber S. M., et al. 2007b, ApJ, 660, L43
Noguchi M., 1987, MNRAS, 228, 635
Noguchi M., 1988, A&A, 203, 259
Nolta M. R., Dunkley J., Hill R. S., et al. 2009, ApJS, 180, 296
Nolthenius R., White S. D. M., 1987, MNRAS, 225, 505
Nomoto K., Thielemann F.-K., Yokoi K., 1984, ApJ, 286, 644
Norberg P., Baugh C. M., Gaztanaga E., Croton D. J., 2009, MNRAS, 396, 19
Norberg P., Baugh C. M., Hawkins E., et al. 2001, MNRAS, 328, 64
Norberg P., Baugh C. M., Hawkins E., et al. 2002a, MNRAS, 332, 827
Norberg P., Cole S., Baugh C. M., et al. 2002b, MNRAS, 336, 907
Nordström B., Mayor M., Andersen J., et al. 2004, A&A, 418, 989
Norman C. A., Sellwood J. A., Hasan H., 1996, ApJ, 462, 114
Nulsen P. E. J., 1986, MNRAS, 221, 377
Nulsen P. E. J., Barcons X., Fabian A. C., 1998, MNRAS, 301, 168
Nussbaumer H., Storey P. J., 1983, A&A, 126, 75
Nusser A., 2001, MNRAS, 325, 1397
Odenkirchen M., Grebel E. K., Dehnen W., et al. 2002, AJ, 124, 1497
Oemler Jr. A., 1976, ApJ, 209, 693
Ogorodnikov K. F., 1965, Dynamics of Stellar Systems. Pergamon Press, Oxford
Okamoto T., Eke V. R., Frenk C. S., Jenkins A., 2005, MNRAS, 363, 1299
Olivier S. S., Primack J. R., Blumenthal G. R., 1991, MNRAS, 252, 102
O'Meara J. M., Tytler D., Kirkman D., et al. 2001, ApJ, 552, 718
Oppenheimer B. D., Davé R., 2006, MNRAS, 373, 1265
Orr M. J. L., Browne I. W. A., 1982, MNRAS, 200, 1067
Osterbrock D. E., 1989, Astrophysics of Gaseous Nebulae and Active Galactic Nuclei. University Science Books, Mill Valley, CA
Ostriker J. P., 1980, Comments on Astrophysics, 8, 177
Ostriker J. P., Cowie L. L., 1981, ApJ, 243, L127
Ostriker J. P., Ikeuchi S., 1983, ApJ, 268, L63

Ostriker J. P., McKee C. F., 1988, Rev. Mod. Phys., 60, 1
Ostriker J. P., Peebles P. J. E., 1973, ApJ, 186, 467
Ostriker J. P., Peebles P. J. E., Yahil A., 1974, ApJ, 193, L1
Ostriker J. P., Tremaine S. D., 1975, ApJ, 202, L113
O'Sullivan E., Forbes D. A., Ponman T. J., 2001, MNRAS, 328, 461
Padmanabhan T., 2002, Theoretical Astrophysics, vol. 3: Galaxies and Cosmology. Cambridge University Press, Cambridge
Padoan P., Nordlund Å., 1999, ApJ, 526, 279
Padoan P., Nordlund Å., 2002, ApJ, 576, 870
Padoan P., Nordlund Å., Kritsuk A. G., et al. 2007, ApJ, 661, 972
Page L., Hinshaw G., Komatsu E., et al. 2007, ApJS, 170, 335
Pahre M. A., Djorgovski S. G., de Carvalho R. R., 1998, AJ, 116, 1591
Palla F., Stahler S. W., 1992, ApJ, 392, 667
Panter B., Jimenez R., Heavens A. F., Charlot S., 2007, MNRAS, 378, 1550
Parker E. N., 1966, ApJ, 145, 811
Partridge R. B., 1995, 3K: The Cosmic Microwave Background Radiation. Cambridge University Press, Cambridge
Partridge R. B., Peebles P. J. E., 1967, ApJ, 147, 868
Pasquali A., van den Bosch F. C., Mo H. J., et al. 2009, MNRAS, 394, 38
Pasquali A., van den Bosch F. C., Rix H.-W., 2007, ApJ, 664, 738
Patton D. R., Pritchet C. J., Carlberg R. G., et al. 2002, ApJ, 565, 208
Peñarrubia J., McConnachie A., Babul A., 2006, ApJ, 650, L33
Peacock J. A., 1997, MNRAS, 284, 885
Peacock J. A., 1999, Cosmological Physics. Cambridge University Press, Cambridge
Peacock J. A., 2002, in Metcalfe N., Shanks T., eds., A New Era in Cosmology, vol. 283 of ASP, p 19
Peacock J. A., 2003, astro-ph/0309240
Peacock J. A., West M. J., 1992, MNRAS, 259, 494
Peebles P. J. E., 1965, ApJ, 142, 1317
Peebles P. J. E., 1968, ApJ, 153, 1
Peebles P. J. E., 1969, ApJ, 155, 393
Peebles P. J. E., 1971, A&A, 11, 377
Peebles P. J. E., 1980, The Large-Scale Structure of the Universe. Princeton University Press, Princeton, NJ
Peebles P. J. E., 1981, ApJ, 248, 885
Peebles P. J. E., 1982, ApJ, 263, L1
Peebles P. J. E., 1993, Principles of Physical Cosmology. Princeton University Press, Princeton, NJ
Peebles P. J. E., Yu J. T., 1970, ApJ, 162, 815
Pei Y. C., 1995, ApJ, 438, 623
Pei Y. C., Fall S. M., 1995, ApJ, 454, 69
Pei Y. C., Fall S. M., Hauser M. G., 1999, ApJ, 522, 604
Peletier R. F., Balcells M., 1996, AJ, 111, 2238
Peletier R. F., Davies R. L., Illingworth G. D., et al. 1990, AJ, 100, 1091
Peletier R. F., de Grijs R., 1998, MNRAS, 300, L3
Pen U.-L., Seljak U., Turok N., 1997, Phys. Rev. Lett., 79, 1611
Penton S. V., Stocke J. T., Shull J. M., 2002, ApJ, 565, 720
Penton S. V., Stocke J. T., Shull J. M., 2004, ApJS, 152, 29
Penzias A. A., Wilson R. W., 1965, ApJ, 142, 419
Percival W. J., Cole S., Eisenstein D. J., et al. 2007, MNRAS, 381, 1053
Percival W. J., Verde L., Peacock J. A., 2004, MNRAS, 347, 645
Perlmutter S., Aldering G., Goldhaber G., et al. 1999, ApJ, 517, 565
Péroux C., Dessauges-Zavadsky M., Kim T., et al. 2002, Astrophys. Space Sci., 281, 543
Péroux C., McMahon R. G., Storrie-Lombardi L. J., Irwin M. J., 2003, MNRAS, 346, 1103
Peterson B. M., 1993, PASP, 105, 247
Peterson J. R., Kahn S. M., Paerels F. B. S., et al. 2003, ApJ, 590, 207
Petitjean P., Mueket J. P., Kates R. E., 1995, A&A, 295, L9
Petitjean P., Webb J. K., Rauch M., et al. 1993, MNRAS, 262, 499
Pettini M., Boksenberg A., Hunstead R. W., 1990, ApJ, 348, 48
Pettini M., Ellison S. L., Steidel C. C., Bowen D. V., 1999, ApJ, 510, 576
Pettini M., Smith L. J., King D. L., Hunstead R. W., 1997, ApJ, 486, 665
Pettini M., Zych B. J., Steidel C. C., Chaffee F. H., 2008, MNRAS, 385, 2011

Pfenniger D., Friedli D., 1991, A&A, 252, 75
Phillipps S., Edmunds M. G., 1991, MNRAS, 251, 84
Phillips A. C., Illingworth G. D., MacKenty J. W., Franx M., 1996, AJ, 111, 1566
Phillips M. M., Lira P., Suntzeff N. B., et al. 1999, AJ, 118, 1766
Phinney E. S., 1994, in Shlosman I., ed., Mass-Transfer Induced Activity in Galaxies, Cambridge University Press, Cambridge, p 1
Pickles A. J., 1998, PASP, 110, 863
Pierce M. J., Tully R. B., 1992, ApJ, 387, 47
Pilyugin L. S., Vílchez J. M., Contini T., 2004, A&A, 425, 849
Plummer H. C., 1911, MNRAS, 71, 460
Pohlen M., Dettmar R.-J., Lütticke R., 2000, A&A, 357, L1
Popesso P., Biviano A., Böhringer H., Romaniello M., 2006, A&A, 445, 29
Porciani C., Dekel A., Hoffman Y., 2002a, MNRAS, 332, 325
Porciani C., Dekel A., Hoffman Y., 2002b, MNRAS, 332, 339
Porciani C., Magliocchetti M., Norberg P., 2004, MNRAS, 355, 1010
Portinari L., Chiosi C., Bressan A., 1998, A&A, 334, 505
Prada F., Vitvitska M., Klypin A., et al. 2003, ApJ, 598, 260
Prantzos N., 2008, in Charbonnel C., Zahn J.-P., eds., EAS Publications Series, vol. 32, pp 311–356
Prantzos N., Boissier S., 2000, MNRAS, 313, 338
Press W. H., Rybicki G. B., Schneider D. P., 1993, ApJ, 414, 64
Press W. H., Schechter P., 1974, ApJ, 187, 425
Press W. H., Teukolsky S. A., Vetterling W. T., Flannery B. P., 1992, Numerical Recipes in FORTRAN. The Art of Scientific Computing. Cambridge University Press, Cambridge
Pringle J. E., 1981, ARA&A, 19, 137
Prochaska J. X., Herbert-Fort S., Wolfe A. M., 2005, ApJ, 635, 123
Prochaska J. X., Wolfe A. M., 1996, ApJ, 470, 403
Prochaska J. X., Wolfe A. M., 1997, ApJ, 487, 73
Prochaska J. X., Wolfe A. M., 1998, ApJ, 507, 113
Proctor R. N., Sansom A. E., 2002, MNRAS, 333, 517
Prugniel P., Simien F., 1996, A&A, 309, 749
Pryke C., Ade P., Bock J., et al. 2009, ApJ, 692, 1247
Pudritz R. E., Ouyed R., Fendt C., Brandenburg A., 2007, in Reipurth B., et al. eds., Protostars and Planets V, University of Arizona Press, Tucson, pp 277–294
Purcell C. W., Kazantzidis S., Bullock J. S., 2009, ApJ, 694, L98
Putman M. E., Gibson B. K., Staveley-Smith L., et al. 1998, Nature, 394, 752
Quilis V., Moore B., Bower R., 2000, Science, 288, 1617
Quinn P. J., 1984, ApJ, 279, 596
Quinn P. J., Hernquist L., Fullagar D. P., 1993, ApJ, 403, 74
Quintana H., 1979, AJ, 84, 15
Quirk W. J., 1972, ApJ, 176, L9
Raban D., Jaffe W., Röttgering H., et al. 2009, MNRAS, 394, 1325
Raha N., Sellwood J. A., James R. A., Kahn F. D., 1991, Nature, 352, 411
Rakos K. D., Schombert J. M., 1995, ApJ, 439, 47
Rasia E., Ettori S., Moscardini L., et al. 2006, MNRAS, 369, 2013
Rauch M., Carswell R. F., Chaffee F. H., et al. 1992, ApJ, 390, 387
Rauch M., Miralda-Escude J., Sargent W. L. W., et al. 1997, ApJ, 489, 7
Ravindranath S., Ho L. C., Peng C. Y., et al. 2001, AJ, 122, 653
Rees M. J., 1984, ARA&A, 22, 471
Rees M. J., 1986, MNRAS, 218, 25P
Rees M. J., Ostriker J. P., 1977, MNRAS, 179, 541
Refregier A., 2003, ARA&A, 41, 645
Refregier A., Rhodes J., Groth E. J., 2002, ApJ, 572, L131
Refsdal S., 1966, MNRAS, 132, 101
Reid M. A., Wilson C. D., 2006, ApJ, 650, 970
Reimers D., Kohler S., Wisotzki L., et al. 1997, A&A, 327, 890
Reines F., Sobel H. W., Pasierb E., 1980, Phys. Rev. Lett., 45, 1307
Renzini A., Voli M., 1981, A&A, 94, 175
Rest A., van den Bosch F. C., Jaffe W., et al. 2001, AJ, 121, 2431
Reynolds C. S., Heinz S., Begelman M. C., 2002, MNRAS, 332, 271

Richards G. T., Strauss M. A., Fan X., et al. 2006, AJ, 131, 2766
Richstone D. O., 1976, ApJ, 204, 642
Richstone D. O., 1980, ApJ, 238, 103
Riess A. G., Filippenko A. V., Challis P., et al. 1998, AJ, 116, 1009
Rindler W., 1977, Essential Relativity. Special, General and Cosmological. Springer, New York
Rix H.-W., de Zeeuw P. T., Cretton N., et al. 1997, ApJ, 488, 702
Rix H.-W., White S. D. M., 1990, ApJ, 362, 52
Roberts M. S., Haynes M. P., 1994, ARA&A, 32, 115
Roberts M. S., Hogg D. E., Bregman J. N., et al. 1991, ApJS, 75, 751
Roberts M. S., Rots A. H., 1973, A&A, 26, 483
Roberts P. H., 1962, ApJ, 136, 1108
Robertson B., Bullock J. S., Cox T. J., et al. 2006a, ApJ, 645, 986
Robertson B., Cox T. J., Hernquist L., et al. 2006b, ApJ, 641, 21
Robertson B. E., Kravtsov A. V., 2008, ApJ, 680, 1083
Robson I., 1996, Active Galactic Nuclei. Wiley, New York, Praxis Publishing, Chichester
Rocha-Pinto H. J., Maciel W. J., 1996, MNRAS, 279, 447
Romanowsky A. J., Douglas N. G., Arnaboldi M., et al. 2003, Science, 301, 1696
Romanowsky A. J., Kochanek C. S., 1997, MNRAS, 287, 35
Rosenbluth M. N., MacDonald W. M., Judd D. L., 1957, Phys. Rev., 107, 1
Rosolowsky E., 2005, PASP, 117, 1403
Rubin V. C., Thonnard N., Ford Jr. W. K., 1978, ApJ, 225, L107
Rubin V. C., Thonnard N., Ford Jr. W. K., 1980, ApJ, 238, 471
Rusin D., Ma C.-P., 2001, ApJ, 549, L33
Rybicki G. B., 1987, in de Zeeuw P. T., ed., Structure and Dynamics of Elliptical Galaxies, vol. 127 of IAU Symposium, p 397
Rybicki G. B., Lightman A. P., 1979, Radiative Processes in Astrophysics. Wiley-Interscience, New York
Ryder S. D., Dopita M. A., 1994, ApJ, 430, 142
Sackett P. D., Morrison H. L., Harding P., Boroson T. A., 1994, Nature, 370, 441
Sackett P. D., Sparke L. S., 1990, ApJ, 361, 408
Saglia R. P., Burstein D., Baggley G., et al. 1997, MNRAS, 292, 499
Salim S., Rich R. M., Charlot S., et al. 2007, ApJS, 173, 267
Salpeter E. E., 1955, ApJ, 121, 161
Salpeter E. E., 1964, ApJ, 140, 796
Salpeter E. E., Hoffman G. L., 1995, ApJ, 441, 51
Salzer J. J., 1992, AJ, 103, 385
Sandage A., Binggeli B., 1984, AJ, 89, 919
Sandage A., Tammann G. A., Yahil A., 1979, ApJ, 232, 352
Sandage A., Visvanathan N., 1978, ApJ, 223, 707
Sanders D. B., 1999, Astrophys. Space Sci., 269, 381
Sanders D. B., Soifer B. T., Elias J. H., et al. 1988, ApJ, 325, 74
Sanders R. H., Huntley J. M., 1976, ApJ, 209, 53
Sandvik H. B., Möller O., Lee J., White S. D. M., 2007, MNRAS, 377, 234
Santoro F., Shull J. M., 2006, ApJ, 643, 26
Sargent W. L. W., Boksenberg A., Steidel C. C., 1988, ApJS, 68, 539
Sargent W. L. W., Young P. J., Boksenberg A., Tytler D., 1980, ApJS, 42, 41
Sato H., 1971, Prog. Theoret. Phys., 45, 370
Sato H., Takahara F., 1980, Prog. Theoret. Phys., 64, 2029
Satoh C., 1980, PASJ, 32, 41
Savage B. D., Sembach K. R., 1996, ARA&A, 34, 279
Scalo J., 1990, in Capuzzo-Dolcetta R., et al. eds., Physical Processes in Fragmentation and Star Formation, vol. 162 of ASSL, Kluwer, Dordrecht, pp 151–176
Scalo J. M., 1986, Fund. of Cosmic Phys., 11, 1
Scannapieco C., Tissera P. B., White S. D. M., Springel V., 2006, MNRAS, 371, 1125
Scannapieco C., Tissera P. B., White S. D. M., Springel V., 2008, MNRAS, 389, 1137
Schade D., Barrientos L. F., Lopez-Cruz O., 1997, ApJ, 477, L17
Schaller G., Schaerer D., Meynet G., Maeder A., 1992, A&AS, 96, 269
Schaye J., 2001, ApJ, 559, L1
Schaye J., 2004, ApJ, 609, 667
Schaye J., Carswell R. F., Kim T.-S., 2007, MNRAS, 379, 1169

Schechter P., 1976, ApJ, 203, 297
Schiminovich D., Wyder T. K., Martin D. C., et al. 2007, ApJS, 173, 315
Schmidt M., 1959, ApJ, 129, 243
Schmoldt I., Branchini E., Teodoro L., et al. 1999, MNRAS, 304, 893
Schneider D. P., Gunn J. E., Hoessel J. G., 1983, ApJ, 268, 476
Schneider P., 2006, in Meylan G., et al. eds, Saas-Fee Advanced Course 33: Gravitational Lensing: Strong, Weak and Micro, Springer Verlag, Berlin, pp 269–451
Schneider R., Omukai K., Inoue A. K., Ferrara A., 2006, MNRAS, 369, 1437
Schödel R., Ott T., Genzel R., Eckart A., et al. 2003, ApJ, 596, 1015
Schombert J. M., 1986, ApJS, 60, 603
Schramm D. N., Steigman G., 1981, ApJ, 243, 1
Schwarzschild M., 1965, Structure and Evolution of the Stars. Dover Publications, New York
Schwarzschild M., 1979, ApJ, 232, 236
Schweizer F., 1982, ApJ, 252, 455
Schweizer F., Miller B. W., Whitmore B. C., Fall S. M., 1996, AJ, 112, 1839
Schweizer F., Seitzer P., 1992, AJ, 104, 1039
Schweizer F., Seitzer P., Faber S. M., et al. 1990, ApJ, 364, L33
Scorza C., Bender R., 1995, A&A, 293, 20
Scott D., 2000, in Courteau S., Willick J., eds., Cosmic Flows Workshop, vol. 201 of ASP, p 403
Scott J., Bechtold J., Dobrzycki A., Kulkarni V. P., 2000, ApJS, 130, 67
Seaton M. J., 1979, MNRAS, 187, 73P
Seaton M. J., 1995, The Opacity Project. Institute of Physics, Bristol
Seaton M. J., Yan Y., Mihalas D., Pradhan A. K., 1994, MNRAS, 266, 805
Seitzer P., Schweizer F., 1990, in Wielen R., ed., Dynamics and Interactions of Galaxies, Springer-Verlag, Heidelberg, pp 270–271
Seljak U., Zaldarriaga M., 1996, ApJ, 469, 437
Sellwood J. A., Nelson R. W., Tremaine S., 1998, ApJ, 506, 590
Sellwood J. A., Sparke L. S., 1988, MNRAS, 231, 25P
Sellwood J. A., Wilkinson A., 1993, Rep. Prog. Phys., 56, 173
Sérsic J. L., 1968, Atlas de Galaxias Australes. Observatorio Astronomico, Cordoba, Argentina
Seth A. C., Dalcanton J. J., de Jong R. S., 2005, AJ, 130, 1574
Sevenster M. N., 1996, in Buta R., et al. eds., IAU Colloq. 157: Barred Galaxies, vol. 91 of ASP, p 536
Shakura N. I., Sunyaev R. A., 1973, A&A, 24, 337
Shandarin S. F., Zel'dovich Y. B., 1989, Rev. Mod. Phys., 61, 185
Shapiro P. R., Iliev I. T., Raga A. C., 2004, MNRAS, 348, 753
Sharma S., Steinmetz M., 2005, ApJ, 628, 21
Shaw L. D., Weller J., Ostriker J. P., Bode P., 2006, ApJ, 646, 815
Sheldon E. S., Johnston D. E., Frieman J. A., et al. 2004, AJ, 127, 2544
Shen J., Abel T., Mo H. J., Sheth R. K., 2006, ApJ, 645, 783
Shen J., Sellwood J. A., 2004, ApJ, 604, 614
Shen S., Mo H. J., White S. D. M., et al. 2003, MNRAS, 343, 978
Sheth R. K., 1996, MNRAS, 279, 1310
Sheth R. K., Lemson G., 1999, MNRAS, 305, 946
Sheth R. K., Mo H. J., Tormen G., 2001, MNRAS, 323, 1
Sheth R. K., Tormen G., 1999, MNRAS, 308, 119
Sheth R. K., Tormen G., 2004, MNRAS, 350, 1385
Shlosman I., Begelman M. C., Frank J., 1990, Nature, 345, 679
Shu F. H., 1974, A&A, 33, 55
Shu F. H., 1991a, Physics of Astrophysics, vol. II: Gas Dynamics. University Science Books, New York
Shu F. H., 1991b, Physics of Astrophysics: vol. I: Radiation. University Science Books, New York
Shu F. H., Adams F. C., Lizano S., 1987, ARA&A, 25, 23
Sievers J. L., Bond J. R., Cartwright J. K., et al. 2003, ApJ, 591, 599
Sigad Y., Eldar A., Dekel A., et al. 1998, ApJ, 495, 516
Sikora M., Wilson D. B., 1981, MNRAS, 197, 529
Silk J., 1968, ApJ, 151, 459
Silk J., 1977, ApJ, 211, 638
Silk J., 1995, ApJ, 438, L41
Silk J., 1997, ApJ, 481, 703
Silva L., Granato G. L., Bressan A., Danese L., 1998, ApJ, 509, 103

Simcoe R. A., Sargent W. L. W., Rauch M., 2004, ApJ, 606, 92
Simien F., de Vaucouleurs G., 1986, ApJ, 302, 564
Skillman E. D., 1996, in Skillman E. D., ed., The Minnesota Lectures on Extragalactic Neutral Hydrogen, vol. 106 of ASP Conference Series, Neutral Hydrogen in Dwarf Galaxies, p 208
Smail I., Ivison R. J., Blain A. W., 1997, ApJ, 490, L5
Smartt S. J., Rolleston W. R. J., 1997, ApJ, 481, L47
Smette A., Surdej J., Shaver P. A., et al. 1992, ApJ, 389, 39
Smith B. D., Sigurdsson S., 2007, ApJ, 661, L5
Smith R. E., Peacock J. A., Jenkins A., et al. 2003, MNRAS, 341, 1311
Smoot G. F., Bennett C. L., Kogut A., et al. 1992, ApJ, 396, L1
Soifer B. T., Rowan-Robinson M., Houck J. R., et al. 1984, ApJ, 278, L71
Solomon P. M., Sanders D. B., Rivolo A. R., 1985, ApJ, 292, L19
Somerville R. S., Barden M., Rix H.-W., et al. 2008a, ApJ, 672, 776
Somerville R. S., Hopkins P. F., Cox T. J., et al. 2008b, MNRAS, 391, 481
Somerville R. S., Kolatt T. S., 1999, MNRAS, 305, 1
Somerville R. S., Primack J. R., 1999, MNRAS, 310, 1087
Somerville R. S., Primack J. R., Faber S. M., 2001, MNRAS, 320, 504
Sommer-Larsen J., 1991, MNRAS, 250, 356
Sommer-Larsen J., Dolgov A., 2001, ApJ, 551, 608
Songaila A., 1998, AJ, 115, 2184
Songaila A., Cowie L. L., 1996, AJ, 112, 335
Spergel D. N., Bean R., Doré O., et al. 2007, ApJS, 170, 377
Spergel D. N., Verde L., Peiris H. V., et al. 2003, ApJS, 148, 175
Spinrad H., Filippenko A. V., Yee H. K., et al. 1993, AJ, 106, 1
Spite F., Spite M., 1982, A&A, 115, 357
Spitzer L. J., 1942, ApJ, 95, 329
Spitzer L. J., 1958, ApJ, 127, 17
Spitzer L. J., 1962, Physics of Fully Ionized Gases. Interscience, New York
Spitzer L. J., 1968, Diffuse Matter in Space. Interscience, New York
Spitzer L. J., 1978, Physical Processes in the Interstellar Medium. Wiley-Interscience, New York
Spitzer L. J., Schwarzschild M., 1953, ApJ, 118, 106
Springel V., Di Matteo T., Hernquist L., 2005a, ApJ, 620, L79
Springel V., Di Matteo T., Hernquist L., 2005b, MNRAS, 361, 776
Springel V., Hernquist L., 2003, MNRAS, 339, 289
Springel V., Wang J., Vogelsberger M., et al. 2008, MNRAS, 391, 1685
Springel V., White S. D. M., 1999, MNRAS, 307, 162
Springel V., White S. D. M., Jenkins A., et al. 2005c, Nature, 435, 629
Springel V., White S. D. M., Tormen G., Kauffmann G., 2001a, MNRAS, 328, 726
Springel V., Yoshida N., White S. D. M., 2001b, New Astronomy, 6, 79
Sridhar S., 1989, MNRAS, 238, 1159
Stahler S. W., Shu F. H., Taam R. E., 1980, ApJ, 241, 637
Stanek K. Z., Mateo M., Udalski A., et al. 1994, ApJ, 429, L73
Stanford S. A., Eisenhardt P. R., Dickinson M., 1998, ApJ, 492, 461
Stark A. A., 1977, ApJ, 213, 368
Starobinsky A. A., 1982, Phys. Lett. B, 117, 175
Starobinsky A. A., 1985, Sov. Astron. Lett., 11, 133
Statler T. S., 1987, ApJ, 321, 113
Statler T. S., 1991, AJ, 102, 882
Stecher T. P., Williams D. A., 1967, ApJ, 149, L29
Steidel C., Adelberger K., Giavalisco M., et al. 1999, Roy. Soc. Lond. Phil. Trans. Ser. A, 357, 153
Steidel C. C., 1990, ApJS, 74, 37
Steidel C. C., 1995, in Meylan G., ed., QSO Absorption Lines, Springer-Verlag, Berlin, p 139
Steidel C. C., Giavalisco M., Pettini M., et al. 1996, ApJ, 462, L17
Steidel C. C., Kollmeier J. A., Shapley A. E., et al. 2002, ApJ, 570, 526
Steidel C. C., Sargent W. L. W., 1992, ApJS, 80, 1
Steinmetz M., Mueller E., 1994, A&A, 281, L97
Steinmetz M., Navarro J., 2000, in Combes F., et al. eds., Dynamics of Galaxies: From the Early Universe to the Present, vol. 197 of ASP, p 165
Stengler-Larrea E. A., Boksenberg A., Steidel C. C., et al. 1995, ApJ, 444, 64

Stewart K. R., Bullock J. S., Wechsler R. H., et al. 2008, ApJ, 683, 597
Stone J. M., Ostriker E. C., Gammie C. F., 1998, ApJ, 508, L99
Storrie-Lombardi L. J., Irwin M. J., McMahon R. G., 1996a, MNRAS, 282, 1330
Storrie-Lombardi L. J., McMahon R. G., Irwin M. J., 1996b, MNRAS, 283, L79
Storrie-Lombardi L. J., McMahon R. G., Irwin M. J., Hazard C., 1994, ApJ, 427, L13
Storrie-Lombardi L. J., Wolfe A. M., 2000, ApJ, 543, 552
Strateva I., Ivezić Ž., Knapp G. R., et al. 2001, AJ, 122, 1861
Stringer M. J., Benson A. J., 2007, MNRAS, 382, 641
Sugerman B., Summers F. J., Kamionkowski M., 2000, MNRAS, 311, 762
Sugiyama N., 1995, ApJS, 100, 281
Sunyaev R. A., Zel'dovich Y. B., 1972, Comments Astrophys. Space Phys., 4, 173
Sutherland R. S., Dopita M. A., 1993, ApJS, 88, 253
Suto Y., Sasaki S., Makino N., 1998, ApJ, 509, 544
Swaters R. A., Madore B. F., Trewhella M., 2000, ApJ, 531, L107
Syer D., White S. D. M., 1998, MNRAS, 293, 337
Takahara F., 1976, Prog. Theoret. Phys., 56, 1665
Talbot Jr. R. J., Arnett D. W., 1974, ApJ, 190, 605
Tan J. C., 2000, ApJ, 536, 173
Tanaka Y., Nandra K., Fabian A. C., et al. 1995, Nature, 375, 659
Taylor E. N., Franx M., G van Dokkum P., et al. 2009, ApJ, 694, 1171
Taylor J. E., Babul A., 2001, ApJ, 559, 716
Tegmark M., Blanton M. R., Strauss M. A., et al. 2004, ApJ, 606, 702
Tegmark M., Silk J., Rees M. J., et al. 1997, ApJ, 474, 1
Terlevich R., Davies R. L., Faber S. M., Burstein D., 1981, MNRAS, 196, 381
Testi L., Sargent A. I., 1998, ApJ, 508, L91
Thacker R. J., Couchman H. M. P., 2000, ApJ, 545, 728
Theuns T., Leonard A., Efstathiou G., 1998, MNRAS, 297, L49
Thielemann F.-K., Nomoto K., Hashimoto M., 1993, in Prantzos, N., Vangioni-Flam, E., Cassé, M., eds., Origin and Evolution of the Elements, Cambridge University Press, Cambridge, pp 297–309
Thielemann F.-K., Nomoto K., Hashimoto M.-A., 1996, ApJ, 460, 408
Thomas D., Maraston C., Bender R., 2003, MNRAS, 339, 897
Thomas D., Maraston C., Bender R., Mendes de Oliveira C., 2005, ApJ, 621, 673
Tinker J. L., Norberg P., Weinberg D. H., Warren M. S., 2007, ApJ, 659, 877
Tinker J. L., Weinberg D. H., Zheng Z., Zehavi I., 2005, ApJ, 631, 41
Tisserand P., Le Guillou L., Afonso C., et al. 2007, A&A, 469, 387
Tolman R. C., 1938, The Principles of Statistical Mechanics. Clarendon Press, Oxford
Toomre A., 1963, ApJ, 138, 385
Toomre A., 1964, ApJ, 139, 1217
Toomre A., 1966, Geophysical Fluid Dynamics, Notes on 1966 Summer Study Prog. Woods Hole Oceanographic Institution, pp 66–46, 111
Toomre A., 1977, in Tinsley B. M., Larson R. B., eds., Evolution of Galaxies and Stellar Populations, Yale University Observatory, New Haven, p 401
Toomre A., 1981, in Fall S. M., Lynden-Bell D., eds., Structure and Evolution of Normal Galaxies, Cambridge University Press, Cambridge, pp 111–136
Toomre A., Toomre J., 1972, ApJ, 178, 623
Torres D. F., Capozziello S., Lambiase G., 2000, Phys. Rev. D., 62, 104012
Tóth G., Ostriker J. P., 1992, ApJ, 389, 5
Tozzi P., Norman C., 2001, ApJ, 546, 63
Trac H., Pen U.-L., 2003, PASP, 115, 303
Trager S. C., Faber S. M., Worthey G., González J. J., 2000a, AJ, 120, 165
Trager S. C., Faber S. M., Worthey G., González J. J., 2000b, AJ, 119, 1645
Tran H. D., 1995a, ApJ, 440, 565
Tran H. D., 1995b, ApJ, 440, 597
Tran H. D., Tsvetanov Z., Ford H. C., et al. 2001, AJ, 121, 2928
Tremaine S., 1999, MNRAS, 307, 877
Tremaine S., Gebhardt K., Bender R., et al. 2002, ApJ, 574, 740
Tremaine S., Gunn J. E., 1979, Phys. Rev. Lett., 42, 407
Tremaine S., Henon M., Lynden-Bell D., 1986, MNRAS, 219, 285
Tremaine S., Richstone D. O., Byun Y.-I., et al. 1994, AJ, 107, 634

Tremaine S., Weinberg M. D., 1984, MNRAS, 209, 729
Tremaine S. D., Ostriker J. P., Spitzer Jr. L., 1975, ApJ, 196, 407
Tremblay B., Merritt D., 1995, AJ, 110, 1039
Tremblay B., Merritt D., 1996, AJ, 111, 2243
Tremonti C. A., Heckman T. M., Kauffmann G., et al. 2004, ApJ, 613, 898
Trentham N., Hodgkin S., 2002, MNRAS, 333, 423
Treu T., Koopmans L. V. E., 2004, ApJ, 611, 739
Tripp T. M., Bowen D. V., 2005, in Williams P., et al. eds., IAU Colloq. 199: Probing Galaxies through Quasar Absorption Lines, Cambridge University Press, Cambridge, pp 5–23
Tristram K. R. W., Meisenheimer K., Jaffe W., et al. 2007, A&A, 474, 837
Trujillo I., Burkert A., Bell E. F., 2004, ApJ, 600, L39
Trujillo I., Feulner G., Goranova Y., et al. 2006, MNRAS, 373, L36
Tsai J. C., Mathews W. G., 1995, ApJ, 448, 84
Tsiklauri D., Viollier R. D., 1998, ApJ, 500, 591
Tytler D., Fan X.-M., Burles S., 1996, Nature, 381, 207
Tytler D., Fan X.-M., Burles S., et al. 1995, in Meylan G., ed., QSO Absorption Lines, Springer-Verlag, Berlin, p 289
Ulmer A., 1996, ApJ, 473, 110
Urry C. M., Padovani P., 1995, PASP, 107, 803
Vader J. P., 1986, ApJ, 305, 669
Valluri M., Merritt D., 1998, ApJ, 506, 686
van Albada T. S., 1982, MNRAS, 201, 939
van den Bergh S., 1976, ApJ, 206, 883
van den Bosch F. C., 1997, MNRAS, 287, 543
van den Bosch F. C., 1998, ApJ, 507, 601
van den Bosch F. C., 2000, ApJ, 530, 177
van den Bosch F. C., 2001, MNRAS, 327, 1334
van den Bosch F. C., 2002a, MNRAS, 332, 456
van den Bosch F. C., 2002b, MNRAS, 331, 98
van den Bosch F. C., Abel T., Croft R. A. C., et al. 2002, ApJ, 576, 21
van den Bosch F. C., Aquino D., Yang X., et al. 2008a, MNRAS, 387, 79
van den Bosch F. C., Burkert A., Swaters R. A., 2001, MNRAS, 326, 1205
van den Bosch F. C., de Zeeuw P. T., 1996, MNRAS, 283, 381
van den Bosch F. C., Jaffe W., van der Marel R. P., 1998, MNRAS, 293, 343
van den Bosch F. C., Lewis G. F., Lake G., Stadel J., 1999, ApJ, 515, 50
van den Bosch F. C., Norberg P., Mo H. J., Yang X., 2004, MNRAS, 352, 1302
van den Bosch F. C., Pasquali A., Yang X., et al. 2008b, arXiv:0805.0002
van den Bosch F. C., Robertson B. E., Dalcanton J. J., de Blok W. J. G., 2000, AJ, 119, 1579
van den Bosch F. C., Swaters R. A., 2001, MNRAS, 325, 1017
van den Bosch F. C., Tormen G., Giocoli C., 2005, MNRAS, 359, 1029
van den Bosch F. C., Yang X., Mo H. J., 2003, MNRAS, 340, 771
van den Bosch F. C., Yang X., Mo H. J., et al. 2007, MNRAS, 376, 841
van der Kruit P. C., Searle L., 1981, A&A, 95, 105
van der Marel R. P., 1991, MNRAS, 253, 710
van der Marel R. P., 1994, MNRAS, 270, 271
van der Marel R. P., Franx M., 1993, ApJ, 407, 525
van der Marel R. P., Magorrian J., Carlberg R. G., et al. 2000, AJ, 119, 2038
van der Marel R. P., van den Bosch F. C., 1998, AJ, 116, 2220
van der Wel A., Holden B. P., Zirm A. W., et al. 2008, ApJ, 688, 48
van Dokkum P. G., 2005, AJ, 130, 2647
van Dokkum P. G., Franx M., 1995, AJ, 110, 2027
van Dokkum P. G., Franx M., 1996, MNRAS, 281, 985
van Dokkum P. G., Franx M., Kelson D. D., Illingworth G. D., 1998, ApJ, 504, L17
van Dokkum P. G., Franx M., Kriek M., et al. 2008, ApJ, 677, L5
van Dokkum P. G., Quadri R., Marchesini D., et al. 2006, ApJ, 638, L59
van Gorkom J. H., 2004, in Mulchaey J. S., et al. eds., Clusters of Galaxies: Probes of Cosmological Structure and Galaxy Evolution, Cambridge University Press, Cambridge, p 305
Van Waerbeke L., Mellier Y., Radovich M., et al. 2001, A&A, 374, 757
Vazdekis A., Casuso E., Peletier R. F., Beckman J. E., 1996, ApJS, 106, 307

Vázquez-Semadeni E., Ballesteros-Paredes J., Klessen R. S., 2003, ApJ, 585, L131
Vázquez-Semadeni E., Gómez G. C., Jappsen A. K., et al. 2007, ApJ, 657, 870
Vázquez-Semadeni E., Ryu D., Passot T., et al. 2006, ApJ, 643, 245
Velazquez H., White S. D. M., 1999, MNRAS, 304, 254
Verner D. A., Barthel P. D., Tytler D., 1994, A&AS, 108, 287
Verolme E. K., Cappellari M., Copin Y., et al. 2002, MNRAS, 335, 517
Viana P. T. P., Liddle A. R., 1996, MNRAS, 281, 323
Vietri M., Ferrara A., Miniati F., 1997, ApJ, 483, 262
Vila-Costas M. B., Edmunds M. G., 1992, MNRAS, 259, 121
Vilenkin A., Shellard E. P. S., 1994, Cosmic Strings and other Topological Defects. Cambridge University Press, Cambridge
Villalobos Á., Helmi A., 2008, MNRAS, 391, 1806
Villumsen J. V., 1985, ApJ, 290, 75
Viola M., Monaco P., Borgani S., et al. 2008, MNRAS, 383, 777
Vitvitska M., Klypin A. A., Kravtsov A. V., et al. 2002, ApJ, 581, 799
Voigt L. M., Fabian A. C., 2004, MNRAS, 347, 1130
Vollmer B., Cayatte V., Balkowski C., Duschl W. J., 2001, ApJ, 561, 708
Wagoner R. V., 1973, ApJ, 179, 343
Wagoner R. V., Fowler W. A., Hoyle F., 1967, ApJ, 148, 3
Wake D. A., Miller C. J., Di Matteo T., et al. 2004, ApJ, 610, L85
Walcher C. J., van der Marel R. P., McLaughlin D., et al. 2005, ApJ, 618, 237
Waldron W. L., 1984, ApJ, 282, 256
Walker I. R., Mihos J. C., Hernquist L., 1996, ApJ, 460, 121
Walker T. P., Steigman G., Kang H.-S., et al. 1991, ApJ, 376, 51
Wang B., 1995, ApJ, 444, 590
Wang B., Silk J., 1994, ApJ, 427, 759
Wang H., Mo H. J., Jing Y. P., 2009, MNRAS, 396, 2249
Wang H. Y., Mo H. J., Jing Y. P., 2007a, MNRAS, 375, 633
Wang J., White S. D. M., 2008, arXiv:0809.1322
Wang L., Li C., Kauffmann G., De Lucia G., 2007b, MNRAS, 377, 1419
Wang Y., Yang X., Mo H. J., et al. 2004, MNRAS, 353, 287
Wang Y., Yang X., Mo H. J., et al. 2008a, MNRAS, 385, 1511
Wang Y., Yang X., Mo H. J., et al. 2008b, ApJ, 687, 919
Wang Y., Yang X., Mo H. J., et al. 2008c, ApJ, 697, 247
Wang Y., Yang X., Mo H. J., van den Bosch F. C., 2007c, ApJ, 664, 608
Wannier P., Wrixon G. T., 1972, ApJ, 173, L119
Warren M. S., Abazajian K., Holz D. E., Teodoro L., 2006, ApJ, 646, 881
Warren M. S., Quinn P. J., Salmon J. K., Zurek W. H., 1992, ApJ, 399, 405
Weaver R., McCray R., Castor J., et al. 1977, ApJ, 218, 377
Webb J. K., Barcons X., 1991, MNRAS, 250, 270
Webb J. K., Carswell R. F., Lanzetta K. M., et al. 1997, Nature, 388, 250
Wechsler R. H., Bullock J. S., Primack J. R., et al. 2002, ApJ, 568, 52
Wechsler R. H., Zentner A. R., Bullock J. S., et al. 2006, ApJ, 652, 71
Wegner G., Colless M., Saglia R. P., et al. 1999, MNRAS, 305, 259
Weil M. L., Hernquist L., 1996, ApJ, 460, 101
Weinacht J., 1924, Math. Ann., 91, 279
Weinberg D. H., Hernquist L., Katz N., 1997a, ApJ, 477, 8
Weinberg D. H., Miralda-Escude J., Hernquist L., Katz N., 1997b, ApJ, 490, 564
Weinberg M. D., 1985, MNRAS, 213, 451
Weinberg M. D., 1986, ApJ, 300, 93
Weinberg M. D., 1989, MNRAS, 239, 549
Weinberg S., 1971, ApJ, 168, 175
Weinberg S., 1972, Gravitation and Cosmology: Principles and Applications of the General Theory of Relativity. Wiley-VCH
Weinberg S., 2008, Cosmology. Oxford University Press, Oxford
Weinmann S. M., Kauffmann G., van den Bosch F. C., et al. 2009, MNRAS, 394, 1213
Weinmann S. M., van den Bosch F. C., Yang X., et al. 2006a, MNRAS, 372, 1161
Weinmann S. M., van den Bosch F. C., Yang X., Mo H. J., 2006b, MNRAS, 366, 2
West M. J., van den Bergh S., 1991, ApJ, 373, 1

Weymann R. J., Carswell R. F., Smith M. G., 1981, ARA&A, 19, 41
Weymann R. J., Jannuzi B. T., Lu L., et al. 1998, ApJ, 506, 1
Wheeler J. C., Sneden C., Truran Jr. J. W., 1989, ARA&A, 27, 279
White D. A., Jones C., Forman W., 1997, MNRAS, 292, 419
White M., 2002, ApJS, 143, 241
White S. D. M., 1976a, MNRAS, 174, 467
White S. D. M., 1976b, MNRAS, 174, 19
White S. D. M., 1978, MNRAS, 184, 185
White S. D. M., 1979, MNRAS, 189, 831
White S. D. M., 1980, MNRAS, 191, 1P
White S. D. M., 1982, in Martinet L., Mayor M., eds., Saas-Fee Advanced Course 12: Morphology and Dynamics of Galaxies, Observatiore de Genere, Sauverny, pp 289–420
White S. D. M., 1984, ApJ, 286, 38
White S. D. M., 1996, in Schaeffer R., et al. eds, Cosmology and Large Scale Structure, Elsevier, Amsterdam, p 349
White S. D. M., Davis M., Frenk C. S., 1984, MNRAS, 209, 27P
White S. D. M., Efstathiou G., Frenk C. S., 1993a, MNRAS, 262, 1023
White S. D. M., Frenk C. S., 1991, ApJ, 379, 52
White S. D. M., Frenk C. S., Davis M., 1983, ApJ, 274, L1
White S. D. M., Navarro J. F., Evrard A. E., Frenk C. S., 1993b, Nature, 366, 429
White S. D. M., Rees M. J., 1978, MNRAS, 183, 341
White S. D. M., Silk J., 1979, ApJ, 231, 1
White S. D. M., Zaritsky D., 1992, ApJ, 394, 1
Whitelock P., Catchpole R., 1992, in Blitz L., ed., The Center, Bulge, and Disk of the Milky Way, vol. 180 of ASSL, Kluwer, Dordrecht, p 103
Whitmore B. C., Zhang Q., Leitherer C., et al. 1999, AJ, 118, 1551
Wielen R., 1977, A&A, 60, 263
Wild J. P., 1952, ApJ, 115, 206
Williams J. P., Blitz L., McKee C. F., 2000, Protostars and Planets IV, University of Arizona Press, Tucson, p. 97
Williams J. P., McKee C. F., 1997, ApJ, 476, 166
Willman B., Blanton M. R., West A. A., et al. 2005, AJ, 129, 2692
Willmer C. N. A., Faber S. M., Koo D. C., et al. 2006, ApJ, 647, 853
Wirth G. D., Koo D. C., Kron R. G., 1994, ApJ, 435, L105
Wolf C., Meisenheimer K., Rix H.-W., et al. 2003, A&A, 401, 73
Wolfe A. M., Gawiser E., Prochaska J. X., 2005, ARA&A, 43, 861
Wolfe A. M., Turnshek D. A., Smith H. E., Cohen R. D., 1986, ApJS, 61, 249
Wong T., Blitz L., 2002, ApJ, 569, 157
Woosley S. E., 1990, in Supernovae Petsohek, A., ed., Springer-Verlag, Berlin, pp 182–212
Woosley S. E., Weaver T. A., 1995, ApJS, 101, 181
Worthey G., 1994, ApJS, 95, 107
Worthey G., Faber S. M., Gonzalez J. J., 1992, ApJ, 398, 69
Worthey G., Faber S. M., Gonzalez J. J., Burstein D., 1994, ApJS, 94, 687
Worthey G., Ottaviani D. L., 1997, ApJS, 111, 377
Wouthuysen S. A., 1952, AJ, 57, 31
Wyse R. F. G., Gilmore G., Franx M., 1997, ARA&A, 35, 637
Wyse R. F. G., Silk J., 1989, ApJ, 339, 700
Xu H., Kahn S. M., Peterson J. R., et al. 2002, ApJ, 579, 600
Yang X., Mo H. J., Jing Y. P., van den Bosch F. C., 2005a, MNRAS, 358, 217
Yang X., Mo H. J., van den Bosch F. C., 2003, MNRAS, 339, 1057
Yang X., Mo H. J., van den Bosch F. C., 2006, ApJ, 638, L55
Yang X., Mo H. J., van den Bosch F. C., 2008, ApJ, 676, 248
Yang X., Mo H. J., van den Bosch F. C., 2009a, ApJ, 695, 900
Yang X., Mo H. J., van den Bosch F. C., 2009b, ApJ, 693, 830
Yang X., Mo H. J., van den Bosch F. C., et al. 2005b, MNRAS, 362, 711
Yang X., Mo H. J., van den Bosch F. C., et al. 2007, ApJ, 671, 153
Yang X., Mo H. J., van den Bosch F. C., Jing Y. P., 2005c, MNRAS, 357, 608
Yanny B., Newberg H. J., Grebel E. K., et al. 2003, ApJ, 588, 824
Yi S. K., Kim Y.-C., Demarque P., 2003, ApJS, 144, 259

Yoachim P., Dalcanton J. J., 2005, ApJ, 624, 701
Yoachim P., Dalcanton J. J., 2006, AJ, 131, 226
Yoo J., Miralda-Escudé J., Weinberg D. H., et al. 2007, ApJ, 667, 813
York D. G., Adelman J., Anderson Jr. J. E., et al. 2000, AJ, 120, 1579
Yoshii Y., Arimoto N., 1987, A&A, 188, 13
Young J. S., Scoville N. Z., 1991, ARA&A, 29, 581
Zaritsky D., Kennicutt Jr. R. C., Huchra J. P., 1994, ApJ, 420, 87
Zehavi I., Blanton M. R., Frieman J. A., et al. 2002, ApJ, 571, 172
Zehavi I., Zheng Z., Weinberg D. H., et al. 2005, ApJ, 630, 1
Zel'dovich Y. B., 1970, A&A, 5, 84
Zel'dovich Y. B., 1972, MNRAS, 160, 1P
Zel'dovich Y. B., Kurt V. G., Sunyaev R. A., 1968, ZhETF, 55, 278
Zel'dovich Y. B., Novikov I. D., 1964, Sov. Phys. Dokl., 158, 811
Zentner A. R., Bullock J. S., 2003, ApJ, 598, 49
Zepf S. E., 1997, Nature, 390, 377
Zepf S. E., Ashman K. M., 1993, MNRAS, 264, 611
Zhang J., Fakhouri O., Ma C.-P., 2008, MNRAS, 389, 1521
Zhang Y., Anninos P., Norman M. L., 1995, ApJ, 453, L57
Zhao D. H., Jing Y. P., Mo H. J., Börner G., 2003a, ApJ, 597, L9
Zhao D. H., Jing Y. P., Mo H. J., Börner G., 2008, arXiv:0811.0828
Zhao D. H., Mo H. J., Jing Y. P., Börner G., 2003b, MNRAS, 339, 12
Zhao H., Spergel D. N., Rich R. M., 1995, ApJ, 440, L13
Zheng Z., Berlind A. A., Weinberg D. H., et al. 2005, ApJ, 633, 791
Zheng Z., Shang Z., Su H., et al. 1999, AJ, 117, 2757
Zibetti S., White S. D. M., Brinkmann J., 2004, MNRAS, 347, 556
Ziegler B. L., Bender R., 1997, MNRAS, 291, 527
Zinnecker H., McCaughrean M. J., Wilking B. A., 1993, in Levy E. H., Lunine J. I., eds., Protostars and Planets III, University of Arizona Press, Tucson, pp 429–495
Zuo L., 1992, MNRAS, 258, 36
Zuo L., Bond J. R., 1994, ApJ, 423, 73
Zuo L., Lu L., 1993, ApJ, 418, 601
Zygelman B., 2005, ApJ, 622, 1356

Index

4000 Å break, 474
ΛCDM, 21, 201, 259, 682
β discrepancy, 412
κ mechanism, 468
σ_8, 207
τCDM, 21
2dFGRS galaxy catalogue, 61

AB magnitude, 28
Abel integral, 277, 497
Abel transformation, 575
Abell clusters of galaxies, 67
Abell radius, 68, 410
absolute magnitude, 28
absorption
  cross-section, 691
  line profile, 691, 756
  resonance, 691
absorption line systems, 88, 709, 714
  CIV systems, 87, 740
  MgII systems, 87, 739
  OVI systems, 740
  photoionization models of, 714
absorption lines
  equivalent width of, 712
  natural broadening of, 711
  thermal broadening of, 710
abundance, 141
abundance ratios, 491
accretion disks, 624
accretion shock, 378, 381
acoustic peaks, 307, 310
  heights of, 309
  positions of, 309
acoustic waves, 167, 174, 307
action-angle variables, 238, 550
active galactic nuclei, 4, 60, 618, 619
  and galaxy interactions, 564, 642
  and star-forming galaxies, 632
  broad-line region, 637
  central engine of, 623
  clustering of, 645
  dark matter halos of, 645
  demographics, 644
  duty cycle, 646
  emission lines of, 631
  emission mechanisms of, 626
  evolution of, 640, 644

  feedback, 9, 612, 648, 664, 670, 676, 685, 687
    mechanical, 650
    radiative, 649
  formation of, 640, 641, 646
  fueling of, 641
  luminosity function of, 644
  narrow-line region, 637
  obscuring torus, 637
  supermassive black hole paradigm of, 61, 623
  unified model of, 638
active galaxies, 4, 618
adiabatic baryon model, 198
adiabatic CDM model, 200
adiabatic contraction, 501, 508
adiabatic HDM model, 201
adiabatic index, 366, 748
adiabatic invariants, 502
adiabatic perturbations, 18, 166
affine connection, 742, 743, 745
affine parameter, 746
AGB, *see* asymptotic giant branch
age–metallicity degeneracy, 475, 536, 606
AGN, *see* active galactic nuclei
alpha elements, 460
ambipolar diffusion, 425, 431
Andromeda Galaxy, 72
angular correlation functions, 290
angular momentum
  alignment of, 361
  in disk galaxies, 502
  of dark matter halos, 358
angular power spectrum, 292
angular-diameter distance, 111, 121
  in gravitational lensing, 295
angular-momentum catastrophe, 409, 511
APM galaxy catalogue, 21, 61, 68
apparent magnitude, 28
arcs
  caused by gravitational lensing, 297, 299
Arnold diffusion, 239
Arnold web, 239
assembly bias, 348, 678
asymptotic giant branch, 466

BzK galaxies, 80
BAO, *see* baryon acoustic oscillations
bar instability, 526
bars
  driving secular evolution, 528

in disk galaxies, 39, 52, 527
    pattern speed of, 528
baryogenesis, 125
baryon acoustic oscillations, 317
baryon–radiation coupling, 189
baryon-to-photon ratio, 141
    and abundances of light elements, 143
BCG, *see* brightest cluster galaxies
bending instability, 530, 615
bias
    in sampling, 266
    linear, 345, 346
    nonlinear, 346, 349
    of dark matter halos, 346
    of density peaks, 325
    of galaxies, 679
    of progenitors, 612
    scale dependence of, 346
    stochastic, 346, 350
bias parameter, 207, 325, 680
binding energy
    of nucleon, 457
Birkhoff's theorem, 92
birthline of stars, 432
BL Lac objects, 622
black dwarfs, 459, 464
black holes, 467
    demographics of, 584
    radius of influence, 583, 639
    Schwarzschild radius, 624, 639
    supermassive, 9, 47, 61, 239, 582, 623
blast waves, 399
    and supernova explosion, 401
    self-similar model of, 399
blazars, 623, 638
blue bump, 626, 627
Bohr radius, 755
bolometric luminosity, 26
Boltzmann equation, 133
    perturbations of, 186
Bondi accretion, 644, 648
Bonnor–Ebert mass, 420
bootstrap method
    of error estimate, 286
Bose–Einstein species, 127
boson stars, 639
bottom-up scenario, 11, 257
bound–bound absorption
    rate of, 756
bound–bound processes, 751
bound–free processes, 751
boxy ellipticals, 46, 576
bremsstrahlung, 8, 151, 758
    emissivity of, 758
    thermal, 758
brightest cluster galaxies, 68
brightness temperature, 627, 631
broad-line radio galaxies, 621
broad-line regions, 620, 637
brown dwarfs, 459, 464
buckling instability, *see* bending instability
bulge–disk decomposition, 51, 614
bulges, 3, 50, 509, 613
    colors of, 51
    formation of, 530, 614

pseudo-bulges, 12, 530, 615
with boxy-peanut shape, 51, 530
bulk motion, 270
    of galaxies, 272
Butcher–Oemler effect, 69, 569, 609

cannibalism
    galactic, 69, 570, 664, 676, 685
canonical momentum, 168
canonical transformation, 238
canonical variables, 237
carbon-deflagration model, 469, 488
Cauchy profile, *see* Lorentzian profile
caustics
    in gravitational lensing, 295
cD galaxies, 68, 570
CDM, *see* cold dark matter
central galaxies, 653, 675
    colors of, 670
    luminosity function of, 665
central limit theorem, 205
Cepheids, 468
    as distance indicator, 33
    as standard candle, 121
    luminosity–period relation of, 33
Chabrier initial mass function, 442
Chandrasekhar formula of dynamical friction, 556
Chandrasekhar mass, 467, 469
chaotic inflation model, 160
chaotic mixing, 250
chemical evolution, 12, 486
    closed-box model of, 488, 539
    models with inflow and outflow, 490
    of disk galaxies, 538
chemical potential, 128
chemical yield, 486
    of low-mass stars, 486
    of massive stars, 487
    of Type Ia supernovae, 488
circular frequency, 504, 517, 522
circular velocity, 499
CIV systems, 86, 740
close encounters, 231
closed orbits, 239
closed-box model, 488, 539
cluster abundance normalization, 335
clusters of galaxies, 67
    β discrepancy, 412
    β model, 410
    Abell catalogue of, 67
    biased distribution of, 347
    cooling flows, 413
    galaxy population of, 68
    mass estimates of, 69
    mass-to-light ratio of, 69
    Sunyaev–Zel'dovich effect, 315
    transformations in, 568
    X-ray emission of, 69, 410
    X-ray scaling relations, 412
CMB, *see* cosmic microwave background
CNO cycle, 459
CO-to-H_2 conversion factor, 434, 438
COBE, 19, 90, 316
coeval population, 471
cold dark matter, 6, 20, 98, 200
    bottom-up scenario, 257

cold relic particles, 133
collision strength, 758
collisional cross-section, 757
collisional damping, *see* Silk damping
collisional excitation, 757
collisional ionization, 696, 757
collisionless Boltzmann equation, 232
collisionless dynamics, 230
    direct and indirect collisions, 230
collisionless gas, 168
color excess, 479
color gradients
    in disk galaxies, 51, 537
    in elliptical galaxies, 610
color index, 29
color–magnitude relation, 64
    of disk galaxies, 536
    of elliptical galaxies, 606
comoving coordinates, 104
comoving distance, 105
comoving radius, 103
compact ellipticals, 59
compact groups, 71
Compton y-parameter, 315, 759
Compton cooling, 367
    by CMB photons, 367
Compton heating, 650
Compton scattering, 150, 151, 751, 759
    inverse, 630
Comptonization, 151
concentration index, 535
conditional luminosity function, 653, 663, 682
conduction flux, 749
conformal Newtonian gauge, 181, 184
conformal time, 105
continuity equation, 158, 163, 233, 366
contravariant index, 744
convective instability, 373
    Schwarzschild criterion for, 374
convergent point method, 33
cooling, 8
    by bremsstrahlung, 368, 760
    due to collisional excitation, 368
    due to collisional ionization, 368
    due to Compton scattering, 367
    due to recombination, 368
    effect of photoionization, 369
    molecular, 762
    overcooling problem, 387, 409
    radiative, 23, 367, 760
cooling flows, 413, 676
cooling function, 368, 386, 761
    and photoionization heating, 370
    of molecular hydrogen, 762
cooling radius, 387
cooling rate, 388, 760
cooling time, 14, 386
cooling wave, 385, 388
core radius, 26
corotation resonance, 528
correlation functions
    connected part of, 264
    in real space, 276, 277
    in redshift space, 276
    of velocity field, 270, 271
    projected, 277

cosmic density field, 262
    cluster-abundance normalization of, 335
    Euler characteristic of, 323
    genus of, 323
    mass moments of, 266
    reconstruction of, 271
    sampling of, 264
cosmic density parameter, 115, 116
cosmic energy equation, 279
cosmic microwave background, 17, 21, 25, 89, 95, 213, 302
    acoustic peaks, 307
    and cosmological parameters, 316
    anisotropy in, 19, 304
    damping on small scales, 310
    dipole, 102
    Doppler effect, 306
    energy density in, 115
    gravitational waves, 213, 318
    integrated Sachs–Wolfe effect, 307
    interaction with intergalactic medium, 314
    intrinsic perturbations, 305
    origin of, 149
    polarization of, 311, 316
    potential perturbations, 305
    re-ionization, 316, 694
    Sachs–Wolfe effect, 304, 310
    spectral distortions by hot gas, 314
    temperature autocorrelation function, 303
    temperature fluctuations in, 302
    temperature power spectrum of, 91, 303
    Thomson scattering, 311
    velocity perturbations, 305
cosmic rays, 56, 649
cosmic shear, 84, 300, 302, 361
cosmic star-formation history, 80
    origin of, 673
cosmic time, 103
cosmic variance, 259
cosmic virial theorem, 280
cosmic web, 82, 258
cosmological background radiation, 27
cosmological constant, 16, 113, 747
    energy density in, 115
cosmological density field
    realizations of, 258
    specification of, 202
cosmological parameters, 94
    constraints on, 316
cosmological phase transition, 156
cosmological principle, 6, 102
cosmology, 100
    distances and volumes, 121
    historical overview of, 16
    horizons, 119
    Hot Big Bang model of, 125
    initial condition problem, 154
    initial conditions of, 6
    particle production, 124
    perturbations, 162
    problems of standard model, 152
    relativistic, 112
    standard model of, 6
COSMOS galaxy catalogue, 68
Coulomb logarithm, 231, 555
counter-rotating cores, 579

counts-in-cells
    analysis of, 288
    distribution function of, 269, 288
    moments of, 269, 288
    shot noise, 270
Courant condition, 772
covariant derivative, 744
covariant index, 744
critical density, 94, 114, 116
critical lines
    in gravitational lensing, 295
critical surface density
    in gravitational lensing, 298
crossing time, 230
curvature perturbations, 166
curvature scalar, 742, 743

D_n-σ relation, 47, 271, 605
Damped Lyα systems, 86, 733
    Γ function, 734
    column density distribution of, 734
    evolution of, 734
    kinematics of, 738
    metallicities of, 736
damping wings, 712
dark matter halos
    subhalos, 568
dark age, 690
dark energy, 6, 318
dark matter, 6, 96
dark matter halos, 7, 23, 260, 319
    and isothermal spheres, 352
    angular momentum of, 358
    assembly bias, 348
    assembly time of, 340
    biased distribution of, 345
    circular velocity of, 352
    clustering of, 319, 345
    concentration parameter of, 353
    density profiles of, 351
    Einasto profile, 353
    formation time of, 341
    intrinsic properties of, 319
    main progenitor histories of, 339
    mass function of, 319, 326
    merger rate of, 319, 342
    merger tree of, 336
    multiplicity function of, 328
    NFW profile, 352, 373
    progenitors of, 319, 336
    shapes of, 354
    spin parameter of, 358
    subhalos, 319
    substructure of, 355
    survival time of, 343
    virial radius of, 352
    virial velocity of, 352
de Sitter universe, 118
de Vaucouleurs profile, see $R^{1/4}$-law profile
decaying particles, 137, 139
deceleration parameter, 112, 116
decoupling
    of matter and radiation, 148, 150
deformation tensor, 178
degenerate gas, 454
density cusp, 248

density peaks, 321
    bias, 325
    correlations of, 324
    modulation by background, 323
    number density of, 321
    shapes of, 325
density perturbations
    damping of, 190
    evolution of, 191
    from inflation, 209
    from topological defects, 213
    gauge-invariant approach, 181
    moments of, 203, 268
    Newtonian theory of, 162
    origin of, 209
    probability distribution function of, 203
    reconstruction from velocity field, 271
    relativistic theory of, 178
    variance of, 267
direct collisions, 230
disk galaxies, 49, 495
    and viscosity, 517
    angular momentum of, 502, 515
    bars in, 39, 52, 527
    chemical evolution of, 538
    circular velocity of, 498
    color gradients in, 51, 537
    color–magnitude relation of, 536
    colors of, 51
    formation of, 505
        numerical simulations of, 511
    gas content of, 53
    kinematics of, 53
    lop-sidedness, 51
    massive halos of, 54
    metallicity gradients in, 542
    metallicity–luminosity relation of, 540
    orbits in, 503
    potential-density pairs of, 496
    rings in, 529
    scaling relations of, 512
    spirals, see spiral arms
    stellar halos of, 52
    stellar populations of, 534
    surface brightness profiles of, 49
    vertical structure of, 51, 518
    warps, 51
    X-ray halos of, 416
disk heating
    and thick disks, 520
    due to gas clouds, 519
    due to minor mergers, 519, 565
    due to spiral arms, 519
    rate of, 518, 568
disk instabilities, 521
    bar mode, 526
    bending instability, 530
    dispersion relation for gaseous disk, 523
    global stability, 521, 525
    local stability, 521, 523
    tight-winding approximation, 523
    Toomre stability criterion, 525
disk models, 496
    double exponential disk, 496
    exponential spheroid, 497
    Kalnajs disk, 527

Maclaurin disk, 525
maximal disks, 500
thin exponential disk, 496
disk scale length, 50
disky ellipticals, 3, 46, 576
distance indicators, 33
   Cepheids, 33
   moving-cluster method, 33
   redshifts, 34, 273
   standard candles, 33
   standard rulers, 33
   trigonometric parallax, 32
   Type Ia supernovae, 33, 122
distance modulus, 29
distant red galaxies, 79
distribution function
   coarse grained, 249
   fine grained, 249
   of absorption line systems, 709
   of decoupled particle species, 132
   of Gaussian random field, 204
   of particles, 127
   of $R^{1/4}$-law profile, 244
DLA, *see* damped Lyα systems
domain walls, 153
Doppler boosting factor, 636
Doppler effect, 30, 107
   in cosmic microwave background, 306
Doppler parameter, 710, 717
double-Compton process, 151
dredge-up, 486
DRGs, *see* distant red galaxies
dry mergers, 564, 592, 596, 597
dust emission, 481
dust extinction, 478
dwarf ellipticals, 40, 57, 569, 613
   formation of, 616
   nucleated, 59
dwarf galaxies, 3, 40, 57
dwarf irregulars, 3, 57
dwarf spheroidals, 3, 40, 57, 71, 613
dynamical friction, 11, 357, 553, 565, 570, 676, 685
   and mass loss, 558
   Chandrasekhar formula, 556
      validity of, 559
   Coulomb logarithm, 555
   time scale of, 557
dynamics of statistics, 278

$E+A$ galaxies, *see* $k+a$ galaxies
E correction, 476
early integrated Sachs–Wolfe effect, 307
Eddington limit, 625, 641
Eddington luminosity, 433, 624, 635, 640
Eddington time, 640
EdS universe, *see* Einstein–de Sitter universe
effective optical depth
   from intervening clouds, 699
effective potential, 504
effective radius, 26, 42
effective yield, 490, 541
Einasto profile, 353
Einstein coefficients, 752, 753
Einstein field equation, 113, 747
   perturbation of, 185
Einstein radius, 296

Einstein ring, 296
Einstein tensor, 747
Einstein universe, 118
Einstein–de Sitter universe, 117, 119
ellipsoidal collapse, 226
   and Zel'dovich approximation, 227
elliptical galaxies, 2, 39, 41, 574
   and feedback, 612
   anti-hierarchical formation of, 612
   boxy, 46, 576
   color gradients in, 610
   color–magnitude relation of, 606
   colors of, 46
   cusps and cores, 45, 576
   D_n-σ relation, 605
   disky, 3, 46, 576
   dynamical modeling of, 579
   dynamics of, 574
   evidence for dark halos, 581
   Faber–Jackson relation, 47, 605
   fine structure of, 600
   formation of, 587
   frosting model, 608
   fundamental plane of, 48, 602
   gas content of, 49
   isophote twisting, 42, 576
   kinematic misalignment, 579
   kinematic properties of, 46, 577
   kinematically distinct cores, 579
   Kormendy relation, 605
   merger scenario, 590, 593, 604, 611
   metallicity gradients in, 610
   monolithic collapse, 588, 610
   number density of, 594
   observables of, 575
   phase-space density constraints, 598
   photometric properties of, 41, 576
   shapes of, 45, 584
   sizes of, 595
   stellar populations of, 606
   structure of, 574
   surface brightness profiles of, 41
   X-ray halos of, 414
ellipticals, *see* elliptical galaxies
emissivity, 751
   from quasars, 697
   from recombination, 701
   from young galaxies, 699
encounters
   close, 231
   high speed, 545
   prograde vs. retrograde, 551
energy equation, 366
energy–momentum tensor, 113, 746
entropy, 108, 129, 164, 456, 748, 749
   and accretion shocks, 384
   conservation law of, 130
   in gravitational systems, 254
entropy density, 130
entropy equation, 389, 702, 749
entropy perturbations, 166, 192
environment dependence
   of galaxy population, 674
epicycle approximation, 504
epicycle frequency, 423, 504, 522
epoch of matter–radiation equality, 115

EPS formalism, *see* extended Press–Schechter formalism
equation of state, 163, 164, 234, 748
    barotropic, 748
    isentropic, 749
    isothermal, 749
    of stars, 453
    of the Universe, 108
    polytropic, 749
equations of motion
    in Hamiltonian formalism, 237
equilibrium models
    axisymmetric, 244
    spherical, 240
    triaxial, 247
equipartition, 553
equivalence principle, 101, 743, 744, 746
equivalent width, 712
ergodicity, 265
EROs, *see* extremely red objects
error estimate
    bootstrap method, 286
    jackknife method, 286
Euler equations, 163, 233, 366, 521, 750
evaporation
    rate of, 397
event horizon, 119
evolutionary tracks of stars, 36, 463
excursion set
    and the Press–Schechter formalism, 328
    interpretation of, 331
exponential disks, 50
    circular velocity of, 499
    origin of, 515
extended Press–Schechter formalism, 321, 328, 684
    interpretation of, 331
extinction law, 478, 479
    for LMC, 480
    for Milky Way, 480
    for SMC, 480
extragalactic background light, 660
extremely red objects, 79

Faber–Jackson relation, 47, 605
faint blue galaxies
    excess of, 75, 659
false vacuum, 158
feedback, 9, 492
    as mass and energy sources, 493
    from active galactic nuclei, 9, 612, 648, 664, 670, 676, 685, 687
    from stellar winds, 492, 664
    from supernova explosions, 9, 493, 664, 685, 687
Fermi acceleration, 627
Fermi–Dirac species, 127
fermion balls, 639
Field criterion for thermal instability, 395
filaments, 260
finger of God, 274
flatness problem, 152, 156
fluid equations, 366, 749
    Lagrangian vs. Eulerian formulation, 770
    perturbations of, 185
flux, 26, 27
flux decrement, 693
flux limiter, 774
forbidden lines, 620, 631

force softening, 768
four-momentum, 745
four-point correlation function
    of density field, 264
Fourier transformation
    convention of, 165
free–free absorption, 758
free–free emission, *see* bremsstrahlung
free–free processes, 751
free-fall time, 14, 252, 407, 420
free-streaming damping, 171
freeze-out
    of stable particles, 133
Friedmann equations, 93, 113
Friedmann model, 16
Friedmann–Robertson–Walker cosmology, 113
friends-of-friends algorithm, 333
fundamental frequencies, 238
fundamental observer, 102, 163
fundamental plane
    in κ space, 604, 613
    of elliptical galaxies, 48, 602
    tilt of, 603

G dwarf problem, 539
galactic winds, 9, 404, 406, 611
galaxies, 2, 37
    active, 4, 618
    at high redshift, 72, 683
    centrals, 653, 675
    chemical evolution of, 486
    classification of, 38
    color–magnitude relation of, 64
    colors of, 4, 64, 670
    early-types, 2, 39
    environment dependence, 4, 65, 674
    gas mass fractions of, 4
    interactions, 544
    late-types, 2, 39
    luminosity function of, *see* luminosity function
    mass–metallicity relation of, 65
    morphological types, 2, 38
    passive evolution of, 476
    satellites, 653, 676
    sizes of, 3, 63
    statistical properties of, 61, 652
    stellar mass function of, 656
    transformations, 544
galaxy bias, 679
galaxy clustering, 283
    angular correlation functions, 290
    angular power spectrum, 292
    counts-in-cells analysis, 288
    evolution of, 278
    in real and redshift space, 273
    power spectrum analysis, 288
    redshift space distortions, 273
    shot noise, 289
    three-point correlation function, 287
    two-point correlation function, 287
galaxy counts, 73, 658
galaxy distribution
    as a point process, 265
galaxy formation
    and active galactic nuclei, 648
    and dark matter halos, 23

and supernova feedback, 406
historical overview of, 15
role of cooling, 23
time scales of, 14
galaxy harassment, 569, 617, 676
galaxy luminosity function, *see* luminosity function
galaxy stellar mass function, 656
galaxy–galaxy lensing, 85
GALEX, 536
gas accretion
  cold mode, 391
  hot mode, 391
gas consumption time, 425, 433
gaseous collapse
  effect of dark matter, 383
  self-similar model of, 379, 388
gaseous halos, 366
  application of virial theorem, 374
  around disk galaxies, 416
  around elliptical galaxies, 414
  collapse of, 379
  convective instability of, 373
  density profiles of, 371
  evolution of, 398
  formation of, 376
  hydrodynamical instabilities of, 396
  in numerical simulations, 408
  observational tests, 410
  radiative cooling in, 385
  spherical collapse of, 384
  thermal instability of, 393
  within dark matter halos, 383
gauge conditions, 180
gauge freedom, 179
gauge transformation, 179, 182
  versus general coordinate transformation, 179
gauge-invariant variables, 182, 184
Gaussian distribution, 266
Gaussian random field, 204, 266
  distribution function of, 204
  random phase of, 205
general coordinate transformation, 179, 742, 744
general relativity, 100, 741
  Einstein field equation of, 747
  Newtonian limit, 747
geodesic equation, 744
giant molecular clouds, 418
  disruption of, 428
  dynamical state of, 419
  formation of, 422
  magnetic fields in, 420, 425
  mass distribution of, 419
  turbulence in, 419, 423, 426
globular clusters, 121
  bimodal color distribution of, 600
  specific frequency of, 599, 616
GMC, *see* giant molecular clouds
graceful exit problem
  of inflation, 159
grand unified theories, 153
gravitational capture, 545
gravitational clustering
  in N-body simulations, 258
  self-similar model, 280
gravitational collapse, 215
  ellipsoidal, 226

of cosmic density field, 257
  spherical, 215
gravitational instability, 7, 18, 167
  and star formation, 422
  and star-formation threshold, 438
  in disks, 521
gravitational lensing, 70, 292
  arcs, 297, 299
  basic equations of, 292
  by a point mass, 295
  by elliptical objects, 299
  by isothermal spheres, 298
  caustics, 295
  convergence, 294, 300, 301
  critical lines, 295
  critical surface density, 298
  Einstein radius, 296
  Einstein ring, 296
  galaxy–galaxy lensing, 85
  image distortions, 292
  lens equation, 296
  magnification, 295
  microlensing, 97
  multiple images, 296
  shear, 294, 300
  strong, 70
  time delay, 295, 296
  weak, 70, 84, 300
gravitational waves, 213, 318
groups of galaxies, 71
  compact groups, 71
Gunn–Peterson effect, 85, 692
GUT, *see* grand unified theories

H-R diagram, *see* Hertzsprung–Russell diagram
half-light radius, *see* effective radius
halo assembly bias, *see* assembly bias
halo mass function, 319, 326
halo model, 362
halo occupation statistics, 663, 679, 682
Hamiltonian, 237
  integrability, 238
Harrison–Zel'dovich spectrum, 19, 206
Hayashi track, 463
HCG, *see* Hickson compact groups
HDF, *see* Hubble Deep Field
HDM, *see* hot dark matter
heat capacity of gravitational system, 548
heat conduction, 397
heating
  by accretion shocks, 378
  by photoionization, 703
  by shocks, 703
  by supernova explosions, 403
  due to photoionization, 369
Heaviside step function, 221
helium flash, 467
helium Lyman α forest, 722
Hernquist profile, 242
Hertzsprung–Russell diagram, 36, 463
  asymptotic giant branch, 466
  birthline, 432, 463
  horizontal branch, 465
  main sequence, 461
Hickson compact groups, 71
hierarchical clustering, 257, 260

hierarchical formation, 11
Higgs mechanism, 157
high-order correlation functions
  of a point process, 265
  of density field, 264
high-order perturbation theory, 176, 282
high-redshift galaxies, 72, 683
high-speed encounters, 545
HII regions, 56, 477, 483
  line luminosity, 478
  recombination lines of, 477
Holmberg radius, 26
horizon, 119
horizon problem, 152, 155
horizontal branch, 465
Hot Big Bang, 117
  chronology of, 125
  cosmology, 17
hot dark matter, 20, 201
  top-down scenario, 257
hot relic particles, 133
Hubble constant, 34, 93, 95, 106, 317
  reduced, 34
Hubble Deep Field, 74
Hubble drag, 133, 167
Hubble law, 34
Hubble parameter, 105
Hubble sequence, 3, 39
Hubble time, 14
hydrodynamical simulations, 24, 408, 686
  grid-based methods, 772
  particle-based methods, 770
  with radiative cooling, 409
hydrostatic equilibrium, 371, 582

ICM, *see* intracluster medium
ideal gas, 748
ideal gas law, 748
IGM, *see* intergalactic medium
IMF, *see* initial mass function
impact parameter, 230
impulse approximation, 546
indirect collisions, 230
inflation, 7, 17, 152
  concept of, 154
  density perturbations, 209
  Gaussian perturbations, 209
  graceful exit problem of, 159
  gravitational waves, 212
  isentropic perturbations, 209
  models of, 158
  perturbation amplitudes, 213
  realization of, 156
  scale-invariant perturbations, 209
  slow-roll approximation, 157
  tensor perturbations, 213
  tilted power spectrum, 212
inflaton, 19, 157
infrared-submillimeter background, 662
initial condition problem
  in cosmology, 154
initial mass function, 9, 417, 440
  derivation from observation, 441
  of Chabrier, 442
  of Kroupa, 442
  of Miller and Scalo, 441

  of Salpeter, 441
  of Scalo, 441
  theoretical models of, 443
  top heavy, 445, 448, 491
  universality of, 440, 444
instability strip, 468
instantaneous recycling approximation, 489
integrals of motion, 237
  classical, 238
  isolating and non-isolating, 237
integrated Sachs–Wolfe effect, 307
intergalactic medium, 13, 85, 689
  after recombination, 690
  collisional ionization, 696
  evolution of, 702
  interaction with cosmic microwave background, 314
  ionization evolution, 704
  Lyα absorption, 692
  Lyman-limit absorption, 694
  mean optical depth of, 690, 693
  photoionization, 695
  photoionization heating, 703
  residual ionization, 690
  shock heating, 703
interstellar medium, 12
  emission and absorption, 476
  multi-phase state of, 393, 422, 688
  thermal instability of, 393, 422
intracluster light, 68
intracluster medium, 69, 410
inverse Compton scattering, 630
invisible matter, *see* dark matter
involution, 237
ionization equations, 750
ionization fraction, 146, 150
ionizing background
  attenuation by intervening clouds, 699
  flux of, 697
IRAS galaxy survey, 273
iron-peak elements, 460
irregular galaxies, 39
irregular orbits, 239
isentropic initial conditions, 192
isentropic perturbations, 166
  in two non-relativistic components, 173
ISM, *see* interstellar medium
isocurvature CDM model, 202
isocurvature initial conditions, 166, 192
isocurvature models, 310
isophotal radius, 26
isophote twisting, 42, 45, 576
isophotes, 26
isothermal perturbations, 18
isothermal sheet, 498, 518, 598
isothermal sphere, 242
  modified, 242
  singular, 242, 372, 499
ISW effect, *see* integrated Sachs–Wolfe effect

J_3 integral, 263
jackknife method
  of error estimate, 286
Jaffe profile, 242
JCMT, 79
Jeans criterion, 18, 253, 523
Jeans equations, 233

Jeans instability, 167
Jeans length, 167, 253
Jeans mass, 167, 420
Jeans models, 241, 244
　anisotropy parameter, 241, 580
　of elliptical galaxies, 580
Jeans theorem, 240
　strong, 240
jets, 621, 633
　collimation of, 633
　one-sided, 637
　relativistic beaming, 636
　superluminal motion, 635

$k+a$ galaxies, 69
K correction, 475
Kalnajs disk, 527
KAM theorem, 239, 240
Kelvin–Helmholtz instability, 396, 772
Kelvin–Helmholtz time, 429
kinematic misalignment, 579
kinematically distinct cores, 579
kinetic energy tensor, 234
King model, 242
King profile, 372
King radius, 242, 598
Kirchhoff's law, 758
Kompaneets equation, 314, 759
konus densities, 576
Kormendy relation, 605
Krammers absorption cross-section, 755
Kroupa initial mass function, 442
kurtosis, 269

LAE, see Lyα emitters
Lagrangian function, 237
Landau damping, 171, 253, 563, 566
Lane–Emden equation, 372
Large Magellanic Cloud, 71
large-scale structure, 81, 262
large-scale velocity field, 83, 270
Larmor's formula, 627
last scattering surface, 89, 149, 303
late integrated Sachs–Wolfe effect, 307
latent heat, 159
Layzer–Irvine equation, 279
LBG, see Lyman-break galaxies
leap-frog integration scheme, 765
lens equation, 296
lenticular galaxies, 39
Lick indices, 474, 607
light elements
　primordial abundances, 143, 144
Lindblad resonances
　outer Lindblad resonance, 528
Lindblad resonances, 528
　inner Lindblad resonance, 528
line-of-sight velocity distribution, 575
linear growth factor, 172
linear growth rate, 172
linear power spectrum
　cluster-abundance normalization, 207
　COBE normalization, 207
　tilt of, 212
linear response theory, 560
LINERs, 620

Liouville's theorem, 168
LIRG, see luminous infrared galaxies
LMC, see Large Magellanic Cloud
Local Group, 71
longitudinal gauge, see conformal Newtonian gauge
lookback time, 119
Lorentz factor, 627
Lorentzian profile, 711
LOSVD, see line-of-sight velocity distribution
low surface brightness galaxies, 500
luminosity distance, 111, 121
luminosity function, 3, 654
　dependence on color, 657
　dependence on environment, 658
　dependence on morphology, 657
　in clusters, 658
　in the field, 658
　redshift evolution of, 658
　Schechter function, 62, 655
luminosity gap
　in groups and clusters, 668
luminous infrared galaxies, 60
Lyα emitters, 78
Lyapunov exponent, 250
Lyman α forest, 86, 714
　at low redshift, 721
　clustering of, 720, 729
　column density distribution of, 716, 728
　Doppler parameter distribution of, 717
　evolution of, 715
　for helium, 722
　in hierarchical models, 724
　in hydrodynamical simulations, 731
　line profiles, 727
　metallicity of, 719
　models of, 723
　sizes of absorbers, 718
Lyman break, 474
Lyman-break galaxies, 77, 683
Lyman-limit systems, 86, 732

Mészáros effect, 176, 195
MACHO, see massive compact halo objects
Maclaurin disk, 525
magnetic buoyancy instability, see Parker instability
magnetic fields
　and star formation, 425
　in giant molecular clouds, 420
　in molecular cores, 430
　Parker instability, 424
magnetic monopoles, 153
magnification
　by gravitational lensing, 295
magnitude-limited sample, 284
　characteristic depth of, 291
main-sequence stars, 461
　lifetimes of, 462
Malmquist bias, 61
mass moments, 266
mass segregation, 252
mass shell, 216
mass–anisotropy degeneracy, 580, 581
mass–metallicity relation, 65
massive compact halo objects, 97
massive stars
　formation of, 432

Mattig formula, 121
maximal disk model, 500
Maxwellian distribution, 754
MDM, *see* mixed dark matter
mechanical feedback, 650
merger rates, 601, 647
    of dark matter halos, 342
merger trees, 336
    binary tree, 337
    N-branch method with accretion, 338
    trees built from progenitor libraries, 338
mergers, 10, 22, 590
    and active galactic nuclei, 564
    and disk heating, 565
    and starbursts, 564
    criterion for, 561
    demographics of, 563
    dry vs. wet, 564, 592, 596, 597
    major vs. minor, 564
    of galaxies, 561
    prograde vs. retrograde, 564
    signatures of, 600
meridional plane, 244, 504
metal absorption line systems, 86, 738
metallicity, 12, 35, 489
metallicity gradients
    in disk galaxies, 542
    in elliptical galaxies, 610
metallicity–luminosity relation
    of disk galaxies, 540
metals, 12
    yield of, 489
metric, 741
    in Newtonian limit, 747
metric perturbations, 192
MgII systems, 86, 739
    connection to galaxies, 739
microlensing, 97
Milky Way, 55
    bar of, 55
    bulge of, 55
    dark halo of, 57, 551
    stellar halo of, 55
    thick disk of, 55
    thin disk of, 55
Miller–Scalo initial mass function, 441
Milne relation, 752, 754
Minkowski metric, 627, 742
missing baryons, 97
mixed dark matter, 21
mixed dark matter model, 202
mixing length theory, 456
modified Hubble profile, 242
modified isothermal sphere, 242
molecular clumps, 419
molecular cores, 419
    collapse of, 429
    role of magnetic fields, 430
molecular hydrogen, 762
    destruction and formation of, 421, 763
    self-shielding, 421, 436, 440, 447
moment of inertia tensor, 235
moments
    of density field, 268
    relation to correlation function, 269
momentum equations, 233

momentum-driven wind, 650
monolithic collapse, 22, 256, 588, 610
monopole problem, 153
morphology–density relation, 4, 12, 66, 658, 674
moving-cluster method, 33
multiplicity function
    of dark matter halos, 328

N-body simulations, 24, 764
    boundary conditions of, 769
    initial conditions of, 769
    mass and force resolution, 768
    time step criterion, 768
    tree algorithm, 766
narrow-line radio galaxies, 621
narrow-line regions, 620, 637
natural line broadening, 711
natural unit system, 125
neutrinos, 115, 124
    as dark matter, 20, 136
    decoupling of, 131, 140, 146
    Dirac type vs. Majorana type, 136
    energy density in, 115
    free-streaming of, 171, 202
    light, 171
    massive, 136, 137, 201
neutron stars, 467
neutron-to-proton ratio, 142
neutrons
    and abundances of light elements, 143
    mean lifetime, 140
new inflation model, 160
Newton's theorems, 232
NFW profile, 242, 352, 373
    circular velocity of, 499
    concentration parameter of, 353
Noether theorem, 237
non-Gaussian field, 205
    χ^2 model, 205
    log-normal model, 205
non-Gaussianity
    development of, 282
nuclear star clusters, 59
nucleosynthesis
    bottleneck of, 142
    primordial, 12, 139, 142
null geodesic, 104

oblate isotropic rotator, 245
obscuring torus, 637
observations
    distance measurements, *see* distance indicators
    photometry, 26
    spectroscopy, 29
old inflation, 158
Oort constant, 439
opacity, 751
    from bound–bound absorption, 455
    from free–free absorption, 455
    from Thomson scattering, 454
    in stellar interiors, 451
    Rosseland mean, 455
optical depth, 691, 692
optically violently variable quasars, 622
orbit-superposition method, *see* Schwarzschild models

orbits, 236
  box orbits, 248
  circularity of, 558
  classification of, 238
  closed, 239
  decay of, 556
  eccentricity of, 557
  elliptic, 561, 596
  hyperbolic, 561, 596
  in disk galaxies, 503
  inner long-axis tubes, 248
  irregular, 239
  outer long-axis tubes, 248
  parabolic, 561, 596
  regular, 238
  resonance, 238
  short-axis tubes, 244, 248
  stochastic, 239
  theory of, 236
oscillator strength, 753
overcooling problem, 387, 409
OVI systems, 740
OVV, see optically violently variable quasars

PAHs, see polycyclic aromatic hydrocarbons
pair conservation equation, 279
pairwise peculiar velocity dispersion, 278, 280
  of galaxies, 287
pancakes, 178, 259
Parker instability, 424
parsec
  definition of, 32
particle decay, 137
particle horizon, 119
particle production
  freeze-out, 133
  in early Universe, 124
  in thermal equilibrium, 127
particle–particle (PP) algorithm, 766
particle–particle–particle–mesh (P^3M)
  algorithm, 767
particle-mesh (PM) algorithm, 766
passive evolution, 472
peak formalism, 321
  cloud-in-cloud problem, 326
  modulation by background, 323
peak-background split, 324
peculiar galaxies, 3, 40
peculiar velocity, 34, 107, 163, 273
perfect sphere, 242
perfect spheroid, 247
period–luminosity relation
  of Cepheids, 468
permitted lines, 631
perturbation amplitudes
  and inflation, 213
perturbation power spectrum
  amplitude of, 206
  initial, 206
  linear, 206
  normalization of, 206
  spectral index of, 206
perturbation theory, 521
  high-order, 176, 282
  high-order moments, 283
  relativistic, 178

perturbations
  classification of, 181
  evolution of, 191
  in entropy, 192
  in metric, 192
  modes of, 165
  on a relativistic background, 175
  origin of, 209
  scalar modes, 182
  stagnation of, 176
  tensor modes, 182
  vector modes, 182
phantom energy, 98
phase mixing, 249, 250, 563
phase transitions, 153
phase-space density
  constraints from, 598
  of collisionless particles, 168
phase-space distribution function, 232
photodissociation, 421
photo-excitation, 755
photoionization, 695, 714, 750, 755
  cross-section of, 753, 755
photoionization heating, 649
  of the IGM, 703
  rate of, 369
photometric redshifts, 75
photometric systems, 28
piecewise parabolic method
  (PPM), 773
Planck function, 753
planetary nebulae, 466
Plummer sphere, 242
Poisson bracket, 237
Poisson equation, 163, 232, 747
Poisson gauge, 181, 184
Poisson sampling
  of density field, 264
polarization, 311
  B mode, 313
  E mode, 313
  power spectrum of, 316
polycyclic aromatic hydrocarbons, 481
polytropic gas, 371
polytropic index, 371, 749
Population III stars, 446
post-main-sequence evolution, 464
post-starburst galaxies, see k+a galaxies
potential energy tensor, 234
power spectral index, 206
power spectrum, 208, 260
  analysis, 288
  of density field, 263
  of density perturbations, 204
  shape parameter of, 200
pp chain, 459
pre-main-sequence
  evolution, 463
  tracks, 432
Press–Schechter formalism, 19, 327
  and excursion set, 328
  applied to clusters, 334
  extended, 321, 328
  spherical versus ellipsoidal collapse, 331
  tests of, 333
primary elements, 460, 491

primordial abundance
    of deuterium, 145
    of helium-3, 145
    of helium-4, 144
    of light elements, 143, 144
    of lithium-7, 145
primordial nucleosynthesis, 139, 142
    model predictions for, 142
    nuclear reactions in, 140
progenitor bias, 612
proper distance, 104
proper motion, 32
proper time, 104
proper volume, 123
protostellar cores, 419
protostellar outflows, 431
proximity effect, 701
pseudo-bulges, 12, 530, 615

QSO, *see* quasi-stellar objects
quantum fluctuations, 7
quark soup, 126
quasar absorption line systems, *see* absorption line systems
quasars, 621
    emissivity from, 697
    host galaxies of, 642
quasi-stellar objects, 621
quasi-stellar radio sources, *see* quasars
quintessence, 98

$R^{1/4}$ law profile, 43, 256
    distribution function of, 244
radiation
    specific intensity of, 750
radiation constant, 760
radiative cooling, 367, 760
    in gaseous halos, 385
    time scale of, 385
radiative feedback, 649
radiative processes, 751
radiative transfer, 750
radio galaxies, 620
    double lobed structure of, 621
    Fanaroff–Riley classes, 621, 648
    flat spectrum sources, 621
    hot spots, 621
    steep spectrum sources, 621
radio lobes, 633
ram-pressure stripping, 12, 571, 616, 676
Rankine–Hugoniot jump conditions, 377, 378, 381
Rayleigh–Taylor instability, 396, 433, 772
re-ionization, 86, 318, 690
    21cm tomography of, 707
    effect on cosmic microwave background, 316, 694
    epoch of, 705
    sources, 696
real distance, 273
recombination, 756
    Case A, 756
    Case B, 756
    coefficient of, 756
    cross-section of, 756
    dielectronic, 757
    of early Universe, 126, 146, 167, 190, 690

reconstruction of density field
    from velocity field, 271
red giant branch, 465
red supergiants, 466
reddening, 30
redshift, 4, 30, 93, 106
    as distance indicator, 34, 273
    photometric, 75
redshift space distortions, 83, 273
    in correlation functions, 276
redshift surveys
    $1/V_{max}$ weighting, 61
    flux limited, 61
    volume limited, 61
regular orbits, 238
reheating, 156
relativistic beaming, 636
relativistic cosmology, 112
relaxation
    chaotic mixing, 250
    definition of, 249
    end state of, 254
    in collisionless systems, 248
    in N-body simulations, 256
    Landau damping, 253
    phase mixing, 249
    violent, *see* violent relaxation
residual gauge modes, 183
resonance coupling, 528
resonant orbit, 238
reverberation mapping, 639
RGB, *see* red giant branch
Ricci tensor, 742, 743
Riemann–Christoffel curvature tensor, 742
Robertson–Walker metric, 104, 742
rotation curves, 53, 498
    cusp–core degeneracy, 501
    disk–halo degeneracy, 500
    of disk galaxies, 32
rotation-curve decomposition, 500
Rydberg constant, 699

Sérsic index, 42, 584
Sérsic profile, 42, 51, 58, 257, 603
Sachs–Wolfe effect, 305, 310
    integrated, 307
Saha equation, 146, 147, 754
Salpeter initial mass function, 441
satellite galaxies, 653, 676
    colors of, 671
    fraction of, 668
    luminosity function of, 666
scale factor, 103, 104
scale-invariant perturbations
    from inflation, 209
scale-invariant spectrum, 206
Scalo initial mass function, 441
Schechter function, 62, 655, 734
Schmidt law of star formation, 434
Schwarzschild criterion for convective instability, 374
Schwarzschild models, 580
Schwarzschild radius, 624, 639
SCUBA, 79
    sources, 662
SDSS galaxy catalogue, 61
secondary elements, 460, 491

secular evolution, 12, 528, 615
SED, *see* spectral energy distribution
selection function
　of a galaxy sample, 283
self-absorption, 629
self-similar model of gaseous collapse, 379
　with gas cooling, 388
semi-analytical models, 24, 684
semi-forbidden lines, 631
Seyfert galaxies, 619, 637
shear
　in gravitational lensing, 294, 300
shear instability, 439
shear tensor, 294
shear viscosity
　coefficient of, 750
shell crossing, 219
shock heating, 703
shocks
　accretion shock, 378, 381
　isothermal, 378
　jump conditions for, 377
　non-radiative, 376, 378
shot noise, 289
　in counts-in-cells, 270
Silk damping, 18, 175, 190
similarity model
　for spherical collapse, 220
　scale-free solutions, 222
singular isothermal sphere, 242, 372, 499
skewness, 269
Small Magellanic Cloud, 71
SMBH, *see* supermassive black holes
SMC, *see* Small Magellanic Cloud
smoothed-particle hydrodynamics, 687, 770
　smoothing kernel, 771
SN, *see* supernovae
solar abundance, 35
sonic radius, 651
sound horizon, 192
sound waves, *see* acoustic waves
space-like separation, 104
space-time metric, 742
specific heats, 748
　ratio of, 748
specific intensity of radiation, 750
specific internal energy, 748
specific star-formation rate, 494, 535
spectral energy distribution, 25, 27, 29
spectral features, 474
spectral indices, 474
spectral synthesis modeling, 36, 471
　including emission lines, 476
spectroscopic observations, 29
SPH, *see* smoothed-particle hydrodynamics
spherical collapse, 19, 215
　collapse time, 217
　critical overdensity for, 218, 219
　effect of cosmological constant, 218
　maximum expansion, 216
　of gaseous halos, 384
　similarity solutions of, 220
　with cooling, 390
　with shell crossing, 219
spin parameter, 358, 502
spin temperature, 707

spiral arms, 53, 531
　and star formation, 424
　as density waves, 532
　driven by bars, 534
　formation of, 531
　material arms, 532
　morphology of, 531
　pitch angle of, 531
　shape function of, 531
　winding problem of, 532
spiral galaxies, 2, 39, 49, 495, 531
spirals, *see* spiral galaxies
spontaneous symmetry breaking, 153, 157
spontaneous transition, 752
stable clustering, 281
standard candles, 33
standard rulers, 33
star formation, 8
　associated with flocculant spirals, 424
　birthrate, 535
　driven by galaxy interactions, 424
　driven by spiral density waves, 424
　empirical laws of, 433
　quiescent, 9
　regulation by shear, 439
　self-regulation, 428
　specific star-formation rate, 494, 535
　spontaneous versus induced, 428
　the Kennicutt–Schmidt law, 434
　the role of turbulence, 426
　the Schmidt law, 434
　time scale of, 425
star-formation diagnostics, 482
　far-infrared continuum, 484
　forbidden lines, 484
　nebular emission lines, 483
　UV continuum, 482
star-formation efficiency, 425
star-formation threshold, 438
star-formation truncation, *see* star-formation threshold
starburst, 9, 60, 564
　and mergers, 564
stars, 34
　birth point, 432
　birthline, 432, 463
　effective temperature of, 451
　evolutionary tracks of, 36, 463
　formation of, 417, 429
　luminosity classes, 35
　luminosity of, 451
　main sequence of, 36
　Population I, 56
　Population II, 56
　Population III, 446
　spectra of, 35
　spectral classes, 35
　variable, 468
steady-state cosmology, 102
stellar atmospheres, 451, 471
stellar evolution, 453
　effect of convection, 456
　gravitational time scale, 449
　Hayashi track, 463
　homologous model, 460
　hydrodynamic time scale, 449
　nuclear time scale, 450

passive, 472
post-main-sequence, 464
pre-main-sequence, 463
scaling relations, 460
thermal time scale, 450
stellar halos, 52
stellar nuclei, *see* nuclear star clusters
stellar nucleosynthesis, 457
stellar population synthesis, 13, 470
stellar populations, 449
of disk galaxies, 534
of elliptical galaxies, 606
stellar spectra
empirical, 470
theoretical, 471
stellar structure, 449
assumption of hydrostatic equilibrium, 449
basic equations of, 450, 452
boundary conditions, 451, 452
energy generation rate, 451
equation of state, 453
evolution of, 453
opacity, 451, 454
stellar winds, 404
mechanical luminosity of, 404
stimulated transition, 752
stochastic orbits, 239
Strömgren sphere, 704
strangulation, 572, 672, 676
stress tensor, 169, 233
strings, 153
strong lensing, *see* gravitational lensing
structure formation problem, 153
sub-DLAs, 86, 733
subgiants, 465
subhalo mass function, 356
unevolved, 357, 568
submillimeter sources, 78
Sunyaev–Zel'dovich effect, 152
kinematic, 315
thermal, 315
super-Eddington accretion, 641
supercooling, 159
superluminal motion, 622, 635
supermassive black holes, 9, 47, 61, 239, 582, 623
demographics of, 47, 584
evidence for, 47, 582, 639
formation of, 640
supernova feedback, 9, 406, 664, 685, 687
supernova rate, 469
supernova remnants, 401
free-expansion phase, 402
Sedov phase, 402
snowplow phase, 403
supernovae
heating by, 403
progenitors of, 468
Type Ia, 16, 33, 122, 317, 469
Type II, 470
surface brightness, 4, 26, 112
dimming, 112
surface pressure, 234
synchronous gauge, 181, 183
synchrotron emission, 627
SZ effect, *see* Sunyaev–Zel'dovich effect

textures, 153
thermal conductivity
coefficient of, 750
thermal instability, 393, 422
Field criterion of, 395
thermalization, 151
thermodynamic equilibrium
in N-body systems, 553
thermodynamics
the first law of, 108
the second law of, 108
thermonuclear reactions, 458
CNO cycle, 459
pp chain, 459
triple-α process, 460
thick disks, 52, 496, 518, 520
Thomson cross-section, 148, 759
Thomson scattering, 148, 311, 694, 707
three-point correlation function
of density field, 264
of galaxies, 287
tidal force, 544
tidal approximation, 546
tidal interactions, 12, 545
tidal radius, 548
tidal shock, 546
tidal streams, 550
tidal stripping, 12, 548, 617, 676
tidal tails, 550, 590
formation in mergers, 551
tidal tensor, 228
tidal torques, 19
theory of, 359
tight-coupling limit, 189
tight-winding approximation, 523
time delay
in gravitational lensing, 295, 296
time-like separation, 104
Toomre stability criterion, 423, 525
top hat window function, 207
top-down scenario, 257
topological defects, 153, 205, 213
and density perturbations, 213
total variation diminishing (TVD) condition, 774
transfer functions
adiabatic baryon model, 198
adiabatic CDM model, 200
adiabatic HDM model, 201
isentropic, 198
isocurvature, 198
isocurvature CDM model, 202
linear, 196
transmission function, 479
transmission of Earth's atmosphere, 25
tree algorithm, 766
triaxiality parameter, 355
trigonometric parallax, 32
triple-α process, 460
Tully–Fisher relation, 54, 271, 512, 685
tuning-fork diagram, *see* Hubble sequence
turbulence, 423
driving mechanisms, 427
in giant molecular clouds, 419, 423, 426
supersonic, 419, 426
two-body relaxation time, 231
two-photon process, 147

two-point correlation function, 82, 260
  as measure of excess of neighboring particles, 265
  error estimates of, 286
  estimators of, 284
  integral constraint on, 263
  J_3 weighting, 285
  of dark matter particles, 260
  of density field, 262
  of density perturbations, 203
  of galaxies, 82, 284, 287
  one-halo term, 363, 679
  redshift evolution, 281
  two-halo term, 363, 679

UBVRI photometric system, 28
UHURU, 69
ULIRG, see ultraluminous infrared galaxies
ultra compact dwarfs, 59
ultraluminous infrared galaxies, 60, 565
universe
  age of, 119
  flat, open, closed, 116
UV background radiation, 370
  observations of, 701
UV drop-outs, 77

vacuum energy, 7
vacuum state, 153
variable stars, 468
variance
  of density fluctuations, 267
  of mass in windows, 268
velocity ellipsoid, 234
velocity tensor, 233
velocity–distance relation, 16
vertical frequency, 504
violent relaxation, 7, 11, 251, 351, 353, 563
  time scale of, 252
virial density, 236
virial radius, 236
  of dark matter halos, 352

virial temperature, 8, 374, 375
virial theorem, 234, 374, 526, 547
  application to spherical collapse, 235
  scalar form of, 235
  tensor form of, 235
viscous disks, 517
viscous dissipation, 750
viscous stress tensor, 749
Vlasov equation, 168
  moments of, 168
voids, 258, 260
Voigt profile, 87, 711
volume-limited sample, 284

warm dark matter, 20
warm-hot intergalactic medium, 97, 704, 740
  and OVI systems, 740
WDM, see warm dark matter
weak lensing, see gravitational lensing
weakly interacting massive particles, 136
WHIM, see warm-hot intergalactic medium
white dwarfs, 467
WIMP, see weakly interacting massive particles
window function, 267
  k space top-hat, 267
  Gaussian, 267
  top-hat, 267
WMAP, 19, 317
Wouthuysen–Field process, 708

X-ray cavities, 650
X-ray scaling relations
  of clusters of galaxies, 412

yield, 489
  effective, 490, 541

Zel'dovich approximation, 19, 177, 272, 359, 770
zero-age main sequence, 463
zero-point energy, 157